Contents in Brief

D0186850

Essentials of
Anatomy &
Physiology

Kevin T. Patton, PhD

Professor of Life Sciences
St. Charles Community College
Cottleville, Missouri

Assistant Professor of Physiology (retired)
St. Louis University Medical School
St. Louis, Missouri

Gary A. Thibodeau, PhD

Chancellor Emeritus and Professor Emeritus of Biology
University of Wisconsin–River Falls
River Falls, Wisconsin

Matthew M. Douglas, PhD

Professor of Zoology and Anatomy and Physiology
Grand Rapids Community College
Grand Rapids, Michigan

Adjunct Professor of Zoology
Michigan State University
East Lansing, Michigan

ELSEVIER
MOSBY

ELSEVIER
MOSBY

3251 Riverport Lane
St. Louis, Missouri 63043

Essentials of Anatomy and Physiology   ISBN: 978-0-323-08511-3

Some material was previously published.

Library of Congress Cataloging-in-Publication Data   ISBN: 978-0-323-08511-3

Senior Editor: Jeff Downing
Senior Developmental Editor: Karen Turner
Managing Editor: Rebecca Swisher
Editorial Assistant: Chelsea Newton
Publishing Services Manager: Jeff Patterson
Senior Project Manager: Mary G. Stueck
Cover Designer: Paula Catalano

Printed in the United States of America

Last digit is the print number: 9 8 7 6

About the Authors

KEVIN T. PATTON has taught anatomy and physiology to high school, community college, and university students from various backgrounds for three decades. This experience has helped him produce a text that will be easier to understand for all students. He has earned several citations for teaching anatomy and physiology. "One thing I've learned," says Kevin, "is that most of us learn scientific concepts more easily when we can *see* what's going on." His talent for using imagery to teach is evident throughout this new text, with its extensive visual resources. Kevin found that the work that led him to a PhD in vertebrate anatomy and physiology instilled in him an appreciation for the "Big Picture" of human structure and function. Kevin's interest in promoting excellence in teaching anatomy and physiology has led him to take an active role in the Human Anatomy and Physiology Society (HAPS). He is a HAPS President Emeritus and was the founding director of HAPS Institute (HAPS-I). Kevin is an active blogger, producing **The A&P Student** blog (*theAPstudent.org*) with study tips and shortcuts and **The A&P Professor** (*theAPprofessor.org*) with content updates and teaching resources.

GARY A. THIBODEAU taught anatomy and physiology for more than three decades. Gary's teaching style encouraged active interaction with students—a style that has been incorporated into every aspect of this new text. He is a recognized pioneer in the introduction of collaborative learning strategies to the teaching of anatomy and physiology, and his focus continues to be on successful student-centered learning. Gary is active in numerous professional organizations, including HAPS and the American Association of Anatomists (AAA). While earning master's degrees in both zoology and pharmacology, as well as a PhD in physiology, Gary says that he became "fascinated by the connectedness of the life sciences." As with Kevin, that fascination has contributed to this book's focus on how each concept fits into the "big picture" of the human body.

MATTHEW M. DOUGLAS ("DR. MATT") has more than 30 years of teaching experience in zoology and anatomy and physiology. His former academic posts include undergraduate and graduate positions at California State University (Fresno), Boston University, the Boston University (Woods Hole) Marine Biology Program, as well a summer undergraduate term at Harvard University. Matt currently teaches at Grand Rapids Community College and is an adjunct research scientist at Michigan State University and the University of Kansas, where he earned his PhD. Matt's dual interest in English and the biological sciences at the University of Michigan have led him to embrace the teaching concepts of Marie Montessori, who stressed the power of language, academic independence, personal investigation, and "hands on" collaborative learning. In the past, Matt authored a biweekly newspaper science column for *The Boston Herald* and served as a consultant for BBC video productions. He also received a National Science Foundation grant to conduct his research and has published articles in journals such as *Science, Evolution, Natural History,* and *Audubon.* Matt is also a professional screenwriter and novelist, and he serves as the series book editor for the University of Michigan Press's *Great Lakes* series. Matt has a great enthusiasm for teaching, which he finds the most powerful and positive profession for examining the intricacies and interconnectedness of life from all points of view.

Reviewers

We gratefully acknowledge the following individuals for their reviews of this text at various stages of development:

Curtis J. Baird
Riverside Community College
Riverside, California

Verona A. Barr
Heartland Community College
Normal, Illinois

Rachel Venn Beecham
Mississippi Valley State University
Itta Bena, Mississippi

Mary K. Beals
Southern University A&M College
Baton Rouge, Louisiana

Patty Bostwick Taylor
Florence Darlington Technical
    College
Florence, South Carolina

Sheri L. Boyce
Messiah College
Grantham, Pennsylvania

J. Alison Brown
Wingate University
Wingate, North Carolina

Lu Anne Clark
Lansing Community College
Lansing, Michigan

Karla L. Clifton
Paris Junior College
Paris, Texas

Leslie Day
Northeastern University
Boston, Massachusetts

Nicholas A. Delamere
University of Arizona
Tucson, Arizona

Sally Flesch
Black Hawk College
Moline, Illinois

Eric D. Forman
Sauk Valley Community College
Dixon, Illinois

Louis Giacinti
Milwaukee Area Technical
    College
Milwaukee, Wisconsin

Amy D. Goode
Illinois Central College
East Peoria, Illinois

Hugh W. Harling
Methodist University
Fayetteville, North Carolina

Sharon Harris-Pelliccia
Mildred Elley College
Latham, New York

Jon-Philippe K. Hyatt
Georgetown University
Washington, DC

Rebecca Ingraham
St. Charles Community College
Cottleville, Missouri

Brian H. Kipp
Grand Valley State University
Allendale, Michigan

Kristin Moline
Lourdes College
Sylvania, Ohio

Greg Mullen
South Louisiana Community College
Lafayette, Louisiana

Karen M. Payne
Chattanooga State Community
    College
Chattanooga, Tennessee

Scott D. Schaeffer
Harford Community College
Bel Air, Maryland

Brian P. Smith
Washington Bible College
Lanham, Maryland

William G. Sproat, Jr
Walters State Community College
Morristown, Tennessee

Claudia I. Stanescu
University of Arizona
Tucson, Arizona

W. Craig Stevens
West Chester University
West Chester, Pennsylvania

Jennifer Swann
Lehigh University
Bethlehem, Pennsylvania

Preface

We are excited and humbled to offer a new textbook in the combined field of human anatomy and physiology. The true quality of a textbook is best measured by how well it supports, promotes, and encourages both good teaching and effective learning. *Essentials of Anatomy & Physiology* is a new text based on profound respect for both the teacher and the student. That respect is coupled with an excitement for the subject matter honed by the authors during decades of teaching and writing about anatomy and physiology.

We have listened carefully to input and creative suggestions from professors all over North America and beyond. We know that teachers use different techniques to convey ideas, present difficult concepts, or explain how applications of principles of anatomy and physiology can affect, for example, health, diverse personal interests of students in the class, or other areas of biology. Students, of course, learn in different ways, at different paces, and for different reasons. Success for some is largely predicated on readability of the text; others are more visual in their learning and rely heavily on excellent illustrations; many use the textbook as a resource for case-based learning; still others learn best in groups and by verbal review of concepts. A good text must be flexible enough to help accommodate, not hinder, these differing needs of both teacher and student.

In this information age, success in both teaching and learning is, in many ways, determined by how effective we are at transforming information into knowledge and application. This is especially true in anatomy and physiology, where both student and teacher continue to be confronted with an enormous accumulation of factual information. *Essentials of Anatomy & Physiology* is intended to help transform that information into a coherent knowledge base. It was written at an appropriate level to help students with divergent needs and learning styles to unify information, stimulate critical thinking, and hopefully acquire a taste for knowledge about the wonders of the human body. This new text is designed for ease of use and will encourage students to explore, question, and look for relationships not only between related facts in a single discipline but also among fields of academic inquiry and personal experience.

Essentials of Anatomy & Physiology uses many features that have proved successful in our other textbooks. However, as a new text it presents carefully chosen content and pedagogical enhancements organized in a way to effectively introduce the essential basics of human structure and function. The writing style and depth of coverage are intended to challenge, reward, and reinforce introductory students as they grasp and assimilate important concepts.

During the development of this text, each element of content and organization was evaluated by anatomy and physiology teachers working in the field—teachers currently assisting students to learn about human structure and function for the first time. The result is a text that students will read, one designed to help the teacher teach and the student learn. Emphasis is on material required for entry into more advanced courses and successful application of information in a practical, work-related environment.

UNIFYING THEMES

Essentials of Anatomy & Physiology is dominated by two major unifying themes. First, structure and function complement one another in the normal, healthy human body. Second, nearly all structure and function in the body can be explained in terms of keeping conditions in the internal environment relatively constant—in homeostasis. Repeated emphasis of these principles encourages students to integrate otherwise isolated factual information into a cohesive and understandable whole. As a result, anatomy and physiology emerge as living and dynamic topics of personal interest and importance to the student.

ORGANIZATION AND CONTENT

The 27 chapters of *Essentials of Anatomy & Physiology* present the core material of anatomy and physiology most important for introductory students. The selection of appropriate information eliminates the confusing mix of nonessential and overly specialized material that unfortunately accompanies basic information in many introductory textbooks. Information is presented in a way that makes it easy for students to know and understand what is important. Further, pedagogical aids in each chapter identify learning objectives and then reinforce successful mastery of this clearly identified core material. The sequencing of chapters in the book follows a course organization most commonly used in teaching at the undergraduate level. However, because each chapter is self-contained, instructors have the flexibility to alter the sequence of material to fit personal teaching preferences or the special content or time constraints of their courses or students.

At every level of organization, both within and among chapters, care has been taken to begin with basic structural concepts and then integrate related functional principles. The reader can then easily grasp how form follows function.

Many introductory anatomy and physiology texts organize course concepts into 20 or fewer chapters. Our experience tells us that breaking things apart into smaller chunks helps to make them easier to absorb. This textbook covers the essential course concepts in 27 relatively short chapters, each of which is easily read as a whole in one brief session. This "many short chapters" approach also makes it easier for instructors to move things around a bit to suit course structure.

We consulted with reading and learning experts to design a text that any reader can use most effectively, whether they are strong readers or not. For example, we found that most students use textbooks not only for *reading* but also for *raiding*. That is, although some students really do read through every chapter of their textbooks, many "raid" only those parts they feel they need at the moment. All readers are also occasional raiders. Therefore, each chapter has clear visual organizers (such as numerous, logical subheadings, summary tables, and chapter outlines) to help readers more easily comprehend what they are reading and to help readers quickly find what they need within a chapter.

Acquiring and using the terms necessary for the study of anatomy and physiology can be difficult, even for the best students. Anatomy and physiology students acquire new vocabulary at about the same rate as students in a foreign language class. To assist students in this process, new terms are introduced in a word list that starts at the beginning of every chapter. Each list is accompanied by the advice to read each term aloud before reading the chapter, which neuroscientists tell us helps better integrate the term into the reading centers of the brain. Pronunciation guides ensure that students are comfortable with each term as they practice saying it aloud. An online audio glossary is also available to assist in correct pronunciation. Meanings of word parts that constitute each term in the list help develop a deep and lasting understanding of scientific terminology. The listed terms are identified in boldface type when encountered for the first time in the text—yet another visual tool to help readers and raiders alike.

Every chapter also uses many illustrations that have been skillfully designed with a consistent style and color palette to complement the written information with visual reinforcement.

The presentation style in this text and its readability, accuracy, and level of coverage have been carefully developed to meet the needs of undergraduate students taking an in-

ILLUSTRATIONS AND DESIGN

A major strength of this text is the exceptional quality, accuracy, and beauty of the illustration program. We worked very closely with professional scientific illustrators to provide attractive and colorful images that clearly and accurately portray the major concepts of anatomy and physiology.

The truest test of any illustration is how effectively it can complement and strengthen the written information in the text and how successfully it can be used by the student as a learning tool. Each illustration is explicitly referred to in the

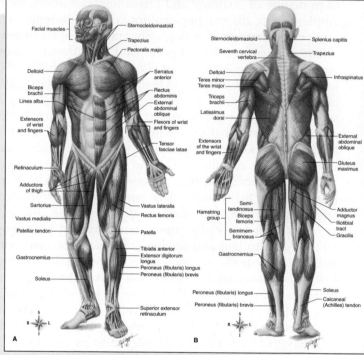

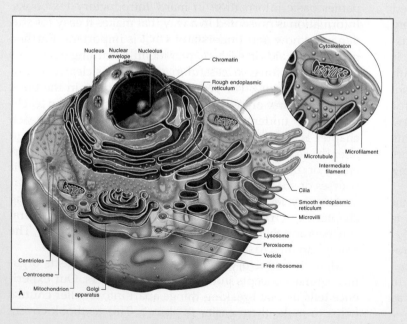

text and is designed to support the text discussion. We paid careful attention to placement and sizing of the illustrations to maximize usefulness and clarity. Each figure and all labels are relevant to and consistent with the text discussion. Each illustration has a boldface title for easy identification. Most illustrations also include a brief explanation—or even numbered steps—that guide the student through the image as a complement to the nearby text narrative.

troductory course in anatomy and physiology. *Essentials of Anatomy & Physiology* is offered as an introductory textbook—a teaching book rather than a reference text. No textbook can replace the direction and stimulation provided by an enthusiastic teacher to a curious and involved student. A good textbook, however, can and should be enjoyable to read and helpful to both.

FEATURES

Essentials of Anatomy & Physiology is a student-oriented text. Written in a very readable style, it has many features that make learning easier and more effective.

Learning Aids

Unit Introductions

Each of the six major units of the text begins with a brief overview statement. The general content of the unit is discussed, and the chapters and their topics are listed. Before

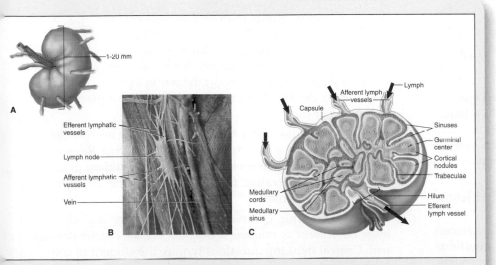

A

B

C

1-20 mm

Efferent lymphatic vessels

Lymph node

Afferent lymphatic vessels

Vein

Capsule

Afferent lymph vessels

Lymph

Sinuses

Germinal center

Cortical nodules

Trabeculae

Medullary cords

Medullary sinus

Hilum

Efferent lymph vessel

The beauty of artistically drawn, full-color artwork is both aesthetically pleasing and functional. A thoughtfully designed color palette—used consistently throughout the book—highlights specific structures within drawings to help organize or emphasize complex material in illustrated tables or conceptual flow charts.

All illustrations used in the text are an integral part of the learning process and should be carefully studied by the student.

beginning the study of material in a new unit, the student is encouraged to scan the introduction and each of the chapter outlines in the unit to understand the relationship and "connectedness" of the material to be studied. Each unit has a color-coded tab at the outside edge of every page to help the student quickly find the information he or she needs.

Chapter Learning Aids

- **Student Learning Objectives**—*provide measurable objectives that the student can work toward.* Critical to student success in any course, especially one as rich in content as anatomy and physiology, these learning objectives come directly and sequentially from major topics of the text material. They are simplified into clear recall exercises designed to integrate facts and principles of anatomy and physiology into a coherent whole. Using these student learning objectives is central and vital to understanding the Big Picture developed for students.

- **Chapter Outline**—*summarizes the contents of a chapter at a glance.* An overview outline introduces each chapter and enables the student to preview the content and direction of the chapter at the major concept level before beginning the detailed reading. Page references enable students to quickly locate topics in the chapter.

- **Language of Science and Medicine**—*introduces the student to new scientific terms in the chapter.* A comprehensive list of new terms is presented starting at the beginning of the chapter. Each term in the list has an easy-to-use pronunciation guide to help the learner easily "own" the word by being able to say it. Translated word parts help the student learn how to deduce the meaning of new terms themselves. The listed terms are defined in the text body, where they appear in **boldface** type, and may also be found in the glossary at the back of the book. The **boldface** type feature enables the student to scan the text for new words before beginning the first detailed reading of the material so it may be read without having to disrupt the flow to grapple with new, multisyllabic words or phrases.

- **Introductory Stories**—*brief case studies that challenge the student with "real-life" clinical or other practical situations so the reader can creatively apply what has been learned.* Every chapter begins with a brief Introductory Story. The story is a case study that describes a

LANGUAGE OF SCIENCE AND MEDICINE

HINT *Before reading the chapter, say each of these terms out loud. This will help you avoid stumbling over them as you read.*

adipocyte (AD-i-poh-syte)
  [*adipo-* fat, *-cyte* cell]

adipose tissue (AD-i-pohs TISH-yoo)
  [*adipo-* fat, *-ose* full of, *tissu-* fabric]

apocrine gland (AP-oh-krin gland)
  [*apo-* from, *-crin* secrete, *gland* acorn]

avascular (ah-VAS-kyoo-lar)
  [*a-* without, *-vas-* vessel, *-ula-* little, *-ar* relating to]

axon (AK-son)
  [*axon* axle]

basement membrane (BM)
  [*base-* base, *-ment* thing, *membran-* thin skin]

body composition (BOD-ee kahm-poh-ZIH-shun)

body membrane (BOD-ee MEM-brayne)

bone (bohn)
  [*bon-* bone]

canaliculus (kan-ah-LIK-yoo-lus)
  [*canal-* channel, *-iculi* little] *pl.,* canaliculi (kan-ah-LIK-yool-eye)

cancellous bone tissue (KAN-seh-lus bohn TISH-yoo)
  [*cancel-* lattice, *-ous* characterized by, *bon-* bone, *tissu-* fabric]

cardiac muscle tissue (KAR-dee-ak MUSS-el TISH-yoo)
  [*cardia-* heart, *-ac* relating to, *mus-* mouse, *-cle* small, *tissu-* fabric]

chondrocyte (KON-droh-syte)
  [*chondro-* cartilage, *-cyte* cell]

collagen (KAHL-ah-jen)
  [*colla-* glue, *-gen* produce]

collagenous fiber (kah-LAJ-eh-nus FYE-bor)
  [*colla-* glue, *-gen-* produce, *-ous* relating to, *fibr-* thread]

columnar (koh-LUM-nar)
  [*column-* column, *-ar* relating to]

compact bone tissue (kom-PAKT bohn TISH-yoo)

"real life" situation related to one or more concepts discussed in the chapter. Each story is continued later in the chapter with a series of questions that require the student to use critical thinking skills to solve the mystery by using concepts as they learn them.

- **Study Hints**—*give specific suggestions for using many of the learning aids found in each chapter.* Because many readers have never learned the special skills needed to make effective use of pedagogical resources found in science textbooks, helpful tips are embedded within each Language of Science and Medicine list, Review Questions set, Critical Thinking Questions section, and at the end of the Introductory Story question set.

- **Color-Coded Illustrations**—*help the beginning student appreciate the "Big Picture" of human structure and function.* A special feature of the illustrations in this text is the careful and consistent use of color to identify important structures and substances that recur throughout the book. Consistent use of a color key helps the beginning student appreciate the "Big Picture" of human structure and function each time a familiar structure is depicted in a new illustration. For an explanation of the color scheme, see the color key on pages xx to xxi, preceding the Unit 1 introductory text.

- **Directional Rosettes**—*help the student learn the orientation of anatomical structures.* Where appropriate, small orientation diagrams and directional rosettes are included as part of an illustration to help the student locate a structure with reference to the body as a whole or to orient a small structure in a larger view. Like training wheels, the orientation diagrams and rosettes provide help only when needed as the student becomes more familiar with the structure of the body.

- **Quick Check questions**—*test the student's knowledge of material just read.* Short objective-type questions are located immediately following major topic discussions throughout the body of the text. These simple questions quickly review important facts presented in the preceding section. The student unable to answer the questions should reread that section before proceeding. This feature therefore enhances reading comprehension. Answers to the Quick Check questions are found on the Evolve site (*http://evolve.elsevier.com/Patton/Essentials*).

- **Cycle of Life**—*describes major changes that occur over an individual's lifetime.* In many body systems, changes in structure and function are frequently related to a person's age or state of development. In appropriate chapters of the text, these changes are highlighted in this special section.

- **The Big Picture**—*explains the interactions of the system discussed in a particular chapter with the body as a whole.* This helps the student relate information about body structures or functions that are discussed in the chapter to the body as a whole. *The Big Picture* feature helps the student

improve critical thinking by focusing on how structures and functions relate to one another on a global basis.

- **Mechanisms of Disease**—*helps the student understand the basic principles of human structure and function by showing what happens when things go wrong.* Examples of pathology, or disease, are included in each chapter of the book to stimulate the student's interest and to help the student understand that the disease process is a disruption in homeostasis, a breakdown of normal integration of form and function. The intent of the Mechanisms of Disease section is to reinforce the normal structures and mechanisms of the body while highlighting the general causes of disorders for a particular body system. Expanded coverage of disease mechanisms for each chapter is found online at the Evolve site (*http://evolve.elsevier.com/Patton/Essentials*).

- **Chapter Summary**—*outlines essential information in a way that helps students organize their study.* Detailed end-of-chapter summaries in an outline format provide excellent guides for the student reviewing the text materials before examinations. Many students also find the summaries to be useful as a chapter preview in conjunction with the chapter outline.

- **Audio Chapter Summaries**—*allow students to listen and learn wherever they may be.* Chapter summaries are available in MP3 format for download at the Evolve site. Students can play them on their computer, import them into a portable media device, or burn them onto a CD for playback in a stereo or car.

- **Review Questions**—*help the student determine whether the important concepts of each chapter have been mastered.* Review questions at the end of each chapter give the student practice in using a narrative format to discuss the concepts presented in the chapter.

- **Critical Thinking Questions**—*actively engage and challenge the student to evaluate and synthesize the chapter content.* Critical thinking questions require the student to use higher-level reasoning skills and demonstrate understanding of, not just their repetition of, complex concepts.

Answers for all of the Introductory Story questions and also the review and critical thinking questions are in the *TEACH-Instructor's Resource Manual* and *Instructor's Electronic Resource (DVD)* that accompanies *Essentials of Anatomy & Physiology.* Teachers can then choose to use the questions as homework assignments or include them on tests.

Boxed Information

We made every effort to incorporate the most current anatomy and physiology research findings in this textbook. Although there continues to be an incredible explosion of knowledge in the life sciences, not all new information is appropriate for inclusion in an essentials level textbook. Therefore we were selective in choosing new clinical, pathological, or special-interest material to include. This text is focused on normal anatomy and physiology. The addition of boxed content alongside the primary narrative is intended to stimulate student interest and provide examples that reinforce the immediate personal relevance of anatomy and physiology as important disciplines for study. Each of the various categories of boxed information has a distinct design and color scheme to make it easily recognizable.

A&P CONNECT

A valuable skill is the ability to trace the flow of blood completely through the normal adult circulatory route. Check out **How to Trace the Flow of Blood** online at **A&P Connect** for some tips and a handy diagram of the route of blood flow through the body.

evolve learning system

- **A&P Connect**—*call the reader's attention to online articles that illustrate, clarify, and apply concepts encountered in the text.* Embedded within the text narrative, these boxes connect the student with interesting, brief online articles that stimulate thinking, satisfy curiosity, and help the student apply important concepts. They are often heavily illustrated with micrographs, medical images, and lavish medical illustrations.

- **General Interest Boxes**—*provide an expanded explanation of specific chapter content.* Many chapters contain boxed essays, occasionally clinical in nature, that expand on or relate to material covered in the text. Examples of subjects include the RNA revolution, the sarcomere, and local hormones.

- **Health Matters**—*present current information on diseases, disorders, treatments, and other health issues related to normal structure and function.* These boxes contain information related to health issues or clinical applications. In some instances, examples of structural anomalies or pathophysiology are presented. Information of this type is often useful in helping the student understand the mechanisms involved in maintaining the "normal" interaction of structure and function.

- **Diagnostic Study**—*keep the student abreast of developments in diagnosing diseases and disorders.* These boxes deal with specific diagnostic tests used in clinical medicine or research. Complete blood count, prostate cancer screening, and the electroencephalogram are examples of topics discussed.

- **FYI**—*give the student more in-depth information on interesting topics mentioned in the text.* Topics of current interest such as new advances in anatomy and physiology research are covered in these "for your information" boxes.

- **Sports and Fitness**—*highlight sports-related topics.* Exercise physiology, sports injury, and physical education applications are highlighted in these boxes.

Glossary

A comprehensive glossary of terms is located at the end of the text. Accurate, concise definitions and phonetic pronunciation guides are provided. An audio glossary is also available on the **Evolve** site—http://evolve.elsevier.comPatton/Essentials—with definitions and audio pronunciations for most of the key terms in the text.

Learning Supplements For Students
Clear View of the Human Body

We are particularly excited to present a full-color, semi-transparent model of the body called the Clear View of the Human Body. Found between pages 114 and 115, this feature permits the virtual dissection of male and female human bodies along several different planes of the body. Developed by Kevin Patton and Paul Krieger, this tool helps learners assimilate their knowledge of the complex structure of the human body. It also provides a unique learning resource that helps students visualize human anatomy in the manner of today's clinical and athletic body imaging technology.

Evolve Resources—http://evolve.elsevier.com/Patton/Essentials

Essentials of Anatomy & Physiology is supported by a comprehensive multimedia Evolve website, featuring the following:

- Answers to the questions posed in the **Introductory Stories**

- Answers to all of the **Quick Check** questions found in the textbook

- Quick access to all **A&P Connect** articles cited in the textbook

- **Mechanisms of Disease** are brief, illustrated articles that expand the coverage of disease processes in ways that help illustrate how normal functions of the body can go wrong

- **Audio Chapter Summaries** can be played on the student's computer or downloaded to a portable media player. These summaries can be used to preview or review chapter concepts while away from the book and notes.

- An interactive **Audio Glossary**, with definitions and pronunciations for over 1000 key terms from the textbook.

Online appendices, including:
- **Quick Guide to the Language of Science and Medicine** provides the basic principles of using scientific terminology. Learning how to build terms from word parts and pluralizing Latin-based terms, for example, helps the student get comfortable with the new language introduced

in the textbook. Comprehensive lists of common prefixes, roots, suffixes, acronyms, and eponyms are also included for handy reference.

- **Online Tutoring** offers the student one-on-one expert assistance from an experienced mentor.

- **Frequently Asked Questions** identify some of the common "trouble spots" in student understanding encountered in each chapter and provide clarifications, examples, and further explanations.

- The **Body Spectrum** electronic coloring book, which offers dozens of anatomy illustrations that can be colored online or printed out and colored by hand.

- More than 380 **Student Post-Test** questions that allow the student to get instant feedback on what has been learned in each chapter.

- **WebLinks** provide the student with access to hundreds of important sites simply by clicking on a subject in the book's table of contents.

- **Anatomy & Physiology Online** provides an interactive online learning experience for students working independently or as part of an online or web-enhanced class facilitated by the course instructor.

You can visit the Evolve site by pointing your browser to *http://evolve.elsevier.com/Patton/Essentials.*

Survival Guide for Anatomy & Physiology

The Survival Guide for Anatomy & Physiology, written by Kevin Patton, is an easy-to-read and easy-to-understand brief handbook to help students achieve success in their anatomy and physiology course. Read with greater comprehension using the 10 survival skills, study more effectively, prepare for tests and quizzes, and tap into all of the information resources available. It also includes a *Quick Reference* filled with illustrations, tables, and diagrams that convey all of the important facts and concepts students need to know to succeed in an anatomy and physiology course.

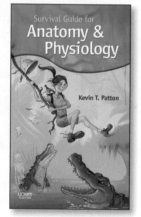

Study Guide

The *Study Guide,* written by Drew Case, is a valuable student workbook that provides the reinforcement and practice necessary for students to succeed in their study of anatomy and physiology. Important concepts from the text are reinforced through illustration labeling, fill-ins, matching, short answer, and multiple-choice

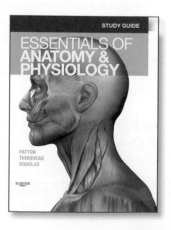

questions. Mnemonic devices, rhymes, and visual clues give students several ways to memorize topics, and special icons are included beside questions that could be made into effective flash cards.

Anatomy & Physiology Study and Review Cards

The boxed set of *Anatomy & Physiology Study and Review Cards* developed by Dan Matusiak helps students learn and retain the essentials of anatomy and physiology. Divided into 20 color-coded sections, more than 320 cards cover all of the body systems with a vivid mix of illustrations, tables, case studies, quizzes, and labeling exercises. The vibrant illustrations and supporting text will make the most of study time while improving comprehension and retention.

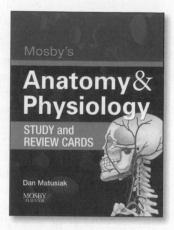

Essentials of A&P Laboratory Manual

The Essentials of A&P Laboratory Manual, authored by David Hill, is an invaluable resource for students. This fun and flexible lab complement is filled with hands-on activities and computer-based experiments that flesh out the concepts from the text while reinforcing the links between what students are learning today and what they will be performing tomorrow. Themes such as super-

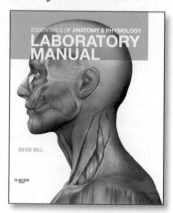

heroes and Hollywood detectives are used to provide analogies between anatomy and physiology concepts and topics familiar to students.

Gray's Atlas of Anatomy

Gray's Atlas of Anatomy presents a vivid, visual depiction of anatomical structures with beautiful, clear artwork correlated with appropriate clinical images and surface anatomy.

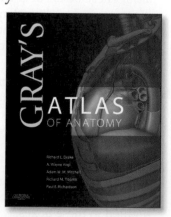

TEACHING SUPPLEMENTS FOR INSTRUCTORS

Instructor's Electronic Resource **DVD and** TEACH-Instructor's Resource Manual

The Instructor's Electronic Resource DVD includes a TEACH-Instructor's Resource Manual (also available in a print version) that links all parts of the *Essentials of Anatomy & Physiology* educational package by providing customizable lesson plans and lecture materials based on learning objectives drawn from the text. The TEACH lesson plans are keyed to the chapter-by-chapter organization of *Essentials of Anatomy & Physiology* and can be modified or combined to meet an individual course's scheduling and teaching needs. Each lesson plan features the following:

- Teaching Focus summaries that identify essential chapter concepts
- Lesson preparation checklists that make planning for class quick and easy
- Pretests and Background Assessment questions that gauge student comprehension
- Critical Thinking questions to focus and motivate students
- Teaching Resources that cross-reference several ancillaries
- Class Activities designed to engage students in the learning process

In addition to the lesson plans, you'll discover unique lecture materials including creative learning activities and suggestions for approaching difficult concepts to complete your TEACH experience. Whether you are a seasoned veteran or a new instructor, you'll certainly benefit from having these pedagogical resources at your fingertips.

The DVD also includes a **Computerized Test Bank** with almost 5500 multiple-choice, true/false, short-answer, and challenge questions (which you can also import into your **Classroom Performance System** to quickly assess student comprehension and monitor your classroom's response). An **Electronic Image Collection** to accompany *Essentials of Anatomy & Physiology* features hundreds of full-color illustrations and photographs, offered with and without labels. Customizable **PowerPoint Presentations** for each chapter include many of the images from the book directly in the slideshow. An **Animation Library** provides animated video clips that you can use in classroom presentations and other applications. A set of **Student Response Questions** helps you get the most out of your use of classroom clickers.

Laboratory Instructor's Guide

The Laboratory Instructor's Guide to accompany the student laboratory manual offers detailed information to help the instructor prepare for the laboratory exercises. Alternate activities, substitutions, student handouts, and other resources help instructors tailor the use of the Essentials of A&P Laboratory Manual to their own course. Answers for all questions in the lab reports are also provided to either check student work or to provide for students who use lab reports as self-tests.

Acknowledgments

Many people have contributed to the development of *Essentials of Anatomy & Physiology.*

We extend our thanks and deep appreciation to all the students, classroom instructors, content experts, reading teachers, and ESL specialists who have provided us with helpful suggestions. And we are grateful to our friends at Content Connections who helped us recruit and manage our very creative and capable team of reviewers.

We also thank the many contributors who have provided us with extraordinary insights and useful features that we have added to our textbook. Dan Matusiak and Izak Paul helped us produce our word lists—a huge task—and we appreciate their help. Paul Krieger helped us design the **Clear View of the Human Body,** for which we are grateful. We thank Dennis Strete for his wonderful histology photographs.

To those at Elsevier who put their best efforts into producing this textbook, we are indebted. This whole learning package would not have been possible without the nearly superhuman efforts of Karen Turner, Senior Developmental Editor, and Jeff Downing, Senior Editor. In addition, we are grateful to Tom Wilhelm, Vice President and Publisher and Sally Schrefer, Managing Director of Nursing and Health Professions for their continuing guidance and support. And where the rubber meets the road, we were fortunate to have a wonderful team of Elsevier professionals working with us to keep it all on track and moving along: Chelsea Newton, Editorial Assistant; Deborah Vogel, Publishing Services Manager; Deon Lee and Mary Stueck, Senior Project Managers; and Joy Knobbe, Marketing Manager. Editorial assistants Jenny, Andrew, and Aileen Patton at Lion Den, Inc. helped with manuscript preparation and photography. We are also grateful to our colleagues at Electronic Publishing Services, who rolled up their sleeves to work directly with the author team to develop and implement our tightly integrated design, layout, and art program.

Kevin T. Patton
Gary A. Thibodeau
Matthew M. Douglas

Contents

UNIT 5   Respiration, Nutrition, and Excretion

CHAPTER 20   Respiratory System, 454

CHAPTER 21   Digestive System, 482

Unit 6   Reproduction and Development

COLOR KEY

BIOCHEMISTRY

C Carbon

Cl Chloride

Energy Energy

ATP ATP

H Hydrogen

N Nitrogen

O Oxygen

K Potassium

Na Sodium

S Sulfur

Ca Calcium

P Organic Phosphate

Pi Inorganic Phosphate

H₂O Water

Hormone

Enzyme

Protein

Carbohydrate

Fatty acid

DNA, Nucleic Acid

RNA

C Cytosine

A Adenine

G Guanine

T Thymine / Uracil

Chromosome

CELLULAR STRUCTURES

Cytosol

Extracellular Fluid

Plasma Membrane

Nucleus

Golgi Apparatus

Mitochondrion

Endoplasmic Reticulum

Ribosome

Centrioles

Microtubule

Intermediate Filament

Microfilament

Na⁺ Channel

OTHER STRUCTURES

Artery

Vein

Capillary

Bone

Muscle

Nerve

Schwann Cell

Fat

Gland

Afferent (Sensory) Pathway

Efferent (Motor) Pathway

Sympathetic

Parasympathetic

SEEING THE BIG PICTURE

Before you begin, we hope you spend a few minutes glancing through this book. Flip through the chapters, look at the pictures and diagrams, and read a few passages that may grab your attention. We hope your curiosity about your body will be stimulated. This book, after all, is about *you*. Your studies will give you a wonderful opportunity to gain a better understanding of the underlying scientific principles of human structure and function. This will require an ability to appreciate "the parts" and "the whole" at the same time. Some of these parts are microscopic, such as cellular parts, whereas others are large, such as arms and legs. Learning to name the various body parts, to describe their detailed structures, and to explain the mechanisms that produce their functions is essential to understanding the human body. *Understanding the nature of individual body parts, however, is meaningless if you do not understand how the parts work together in a living person.*

We've written this book with *you* in mind. Our primary goal is to help you understand the *big picture* by periodically stepping back during our discussions to focus attention on the broader view of the human body. We are confident that our "whole body" approach will help you put each new fact or concept you learn into its proper place within a larger framework of understanding. You may also better appreciate why it is so important to learn some detailed facts that may at first seem to have no practical value to you. By the time you have finished, however, you will have gained a more complete understanding of the essential nature of the human body. We wish you well on your voyage!

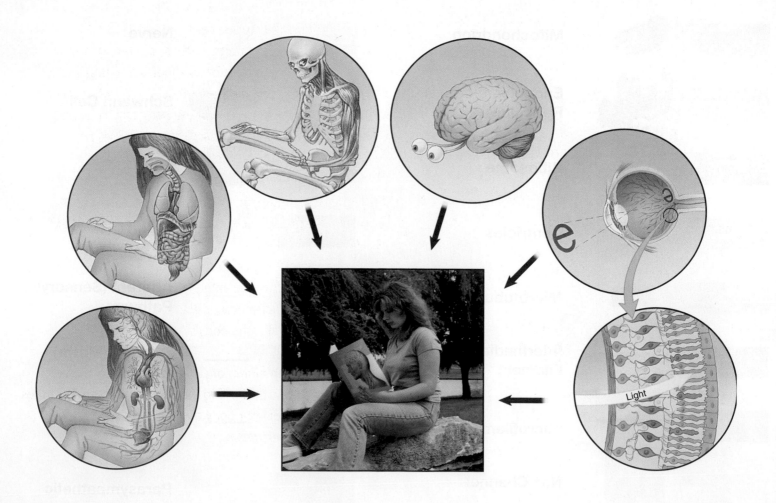

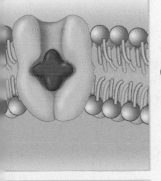

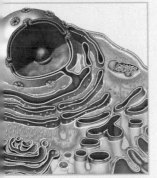
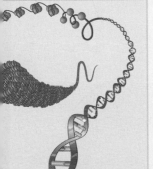
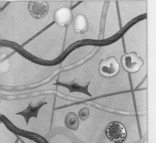
UNIT **1**

Constituents of the Human Body

We've designed these first six chapters to show you how the underlying structures and functions of the human body are intimately connected. This introductory information will help you see that our bodies are not just a jumble of isolated parts. Instead, the underlying structures and functions of our bodies are highly integrated and coordinated. In **Chapter 1** we present levels of organization within the body. You will see how the unifying theme of *homeostasis* underlies the interactions of our body's structures at every level. **Chapter 2** explores the basic chemical interactions that govern the control, integration, and regulation of all the body's processes. The three chapters that follow then build on this information to show you how all body functions are really *cellular functions.* The structure and function of cells is presented in **Chapters 3** and **4.** Types of cell and organism reproduction are discussed in **Chapter 5.** Finally, the body's tissues and their functions are discussed in **Chapter 6.** The rest of this book will focus on the remaining organ systems of our bodies, providing you with the "big picture" of how the structures and functions of your body are controlled and integrated—*to create you!*

1

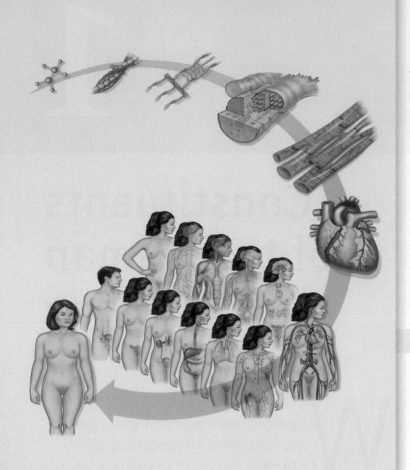

1

Organization of the Human Body

STUDENT LEARNING OBJECTIVES

At the completion of this chapter, you should be able to do the following:

1. Describe what is meant by levels of organization.
2. Describe a person standing in the correct anatomical position.
3. Describe the appearance of the body using the three different body planes.
4. Identify the major body cavities and the major organs lying within each cavity.
5. Describe the four abdominopelvic quadrants and the nine abdominopelvic regions.
6. Use directional terms to describe the relative positions of different body structures.
7. Describe the relationship between structure and function.
8. Describe the process of homeostasis and its importance.
9. Compare a negative feedback mechanism with a positive feedback mechanism.

SAM was just starting his first year in college, and he had been thinking about going into the medical field—maybe nursing. To find out what nursing was really like, Sam signed up for a "shadow a nurse" day at the local hospital. He was scheduled to follow a nurse in the emergency department to see what the job involved.

Everything seemed rather boring and quiet for the first hour. Then suddenly, an ambulance pulled into the bay and a paramedic rushed a patient in on a stretcher. The paramedic gave a quick patient report, and Sam overheard, "Stab wound to the right upper quadrant . . . additional cuts to his right brachial region . . . there's a large contusion on his right thigh, just proximal to the knee."

Sam felt like he was listening to a foreign language! You may feel likewise, but after reading this chapter, you should be able to interpret the paramedic's report and answer questions we pose at the end of this chapter.

HUMAN ANATOMY AND PHYSIOLOGY

What could be more fascinating and personally empowering than understanding our own bodies? To do so requires that we know the *structures*, or **anatomy,** of the human body as well as understand how the *functions of these structures*, or **physiology,** are integrated. We tend to separate these two fields of study, but in reality they are intimately connected: As you will see, *structure determines function*. Understanding this basic concept takes on great significance as you enter into one of the many fascinating health careers available to students today.

Sometimes anatomists use the phrase "gross anatomy" to describe the study of body parts visible to the naked eye. In contrast, we use the phrase "microscopic anatomy" to describe the study of cells. **Microscopic anatomy** includes the study of cells, *cytology*, and tissues, *histology*.

Our understanding of human anatomy and physiology is the result of thousands of years of human inquiry into the causes of human illness. Today, modern researchers use the **scientific method** to study nature in a rational, logical manner. This means scientists must conduct rigorous experiments under controlled conditions to explore a **hypothesis**—a statement used as a proposed explanation for a scientific problem. Through the scientific method, a reasonable conclusion can be obtained. If the results of experiments are repeatable, they may support the hypothesis. If the data are *not* supportive, the hypothesis is disproven. However, sometimes the original hypothesis needs to be "tweaked," sending the investigators in a more refined direction. Eventually, through rigorous and controlled testing, enough support may build for the hypothesis to elevate it to the level of a **theory.**

You should note that a hypothesis is *not* a theory and that hypotheses must always be testable in a valid, scientific manner. You also should understand that a theory in science is very rare, and most important, a theory is a well-tested statement that has never been *disproved*. This is because science depends on disproof, not proof. Also note that it is not "cheating" to return to your original hypothesis and change it when new

Before reading the chapter, say each of these terms out loud. This will help you avoid stumbling over them as you read.

HINT

abdominopelvic cavity (ab-DOM-i-no-PEL-vik KAV-ih-tee)
[*abdomin*- belly, *-pelv*- basin, *cav*- hollow, *-ity* state]

afferent (AF-fer-ent)
[*a[d]*- toward, *-fer*- carry, *-ent* relating to]

anatomical position (an-ah-TOM-i-kal po-ZISH-un)
[*ana*- apart, *-tom*- cut, *-ical*- relating to, *posit*- place, *-tion* state]

anatomy (ah-NAT-o-mee)
[*ana*- apart, *-tom*- cut, *-y* action]

anterior (an-TEER-ee-or)
[*ante*- front, *-er*- more, *-or* quality]

apical (AY-pik-al)
[*apic*- tip, *-al* relating to]

appendicular (ah-pen-DIK-yoo-lar)
[*append*- hang upon, *-ic*- relating to, *-ul*- little, *-ar* relating to]

atrophy (AT-ro-fee)
[*a*- without, *-troph*- nourishment, *-y* state]

axial (AK-see-al)
[*axi*- axis, *-al* relating to]

basal (BAY-sal)
[*bas*- base, *-al* relating to]

bilateral symmetry (bye-LAT-er-al SIM-e-tree)
[*bi*- two, *-later*- side, *-al* relating to, *sym*- together, *-metr*- measure, *-ry* condition of]

body plane (BOD-ee playn)

cell (sell)
[*cell* storeroom]

central (SEN-tral)
[*centr*- center, *-al* relating to]

coronal plane (ko-RO-nal playn)
[*corona*- crown, *-al* relating to]

cortical (KOR-tik-al)
[*cortic*- bark, *-al* relating to]

deep

dissection (dis-SEK-shun)
[*dissect*- cut apart, *-tion* process]

distal (DIS-tal)
[*dist* distance, *-al* relating to]

dorsal cavity (DOR-sal KAV-ih-tee)
[*dors*- back, *-al* relating to, *cav*- hollow, *-ity* state]

efferent (EF-fer-ent)
[*e*- away, *-fer*- carry, *-ent* relating to]

feedback control loop

homeostasis (ho-mee-o-STAY-sis)
[*homeo*- same or equal, *-stasis* standing still]

continued on page 17

evidence arises. This is the power of science—we are always improving our understanding of nature, but may never actually reach a complete understanding. Nonetheless, major scientific breakthroughs are constantly changing our lives.

Using the scientific method, anatomists learn about the structures of the human body by the process of **dissection**—literally cutting the body apart. Physiologists use many complex technologies as they attempt to discover and understand the intricate control systems that permit the body to operate and survive, often in changing, even hostile, environments.

Both anatomy and physiology are typically studied by examining specific organ systems. We will begin in this chapter with an overview of the body as a whole. Later chapters will examine the body both structurally and functionally, system by system. Our ultimate goal, of course, is the basic understanding of the *whole* body.

Such an understanding requires learning many scientific terms—often derived from Greek and Latin roots—that are standardized throughout the world. The study of anatomy and physiology, therefore, requires you to learn the definitions and uses of these terms, just as if you were learning a foreign language—and to be truthful, you are!

To help you learn the vocabulary of anatomy and physiology, we have provided several helpful tools. Within each chapter, a section entitled *Language of Science and Medicine* provides you with a list of boldface key terms. Each term in the list has a pronunciation guide, the Latin or Greek word parts that make up the term, and the meaning of the word parts. Each term also appears in boldface within the chapter text where it is defined. *Spending the time to learn these terms will help you tremendously throughout your studies.*

Scientific language evolves like any other language. This evolution of the language of science helps us understand changes that are constantly taking place in science, and it also allows us to accommodate the description of new discoveries. The International Federation of Associations of Anatomists (IFAA) now adheres to a list of "universal" anatomical terminology called the *Terminologia Anatomica (TA),* which has become a useful, standard reference.

One purpose of the *Terminologia Anatomica* is to avoid *eponyms,* or terms that are based on a person's name. For this reason, a more descriptive Latin-based term is always preferred. For example, the term *Eustachian tube* (named after Eustachius) has been replaced by the more descriptive term *auditory tube.* Likewise, the *islets of Langerhans* are now called *pancreatic islets.* However, we should note here that a few eponyms continue to be used.

In 2007 the *Terminologia Histologica (TH)* was published for microscopic anatomy, such as the study of cells and tissues.

In this textbook, we use the English terms from the published lists as our standard reference, but we do occasionally refer to the pure Latin form or an alternate term when appropriate.

Unfortunately, there are currently no standard lists of physiological terms.

Now, before we begin our tour of the human body, let's first discuss what we mean by "life."

QUICK CHECK

1. Define anatomy and physiology. How do they differ?
2. Why is the *Big Picture* of anatomy and physiology important to your studies?
3. What is the difference between a hypothesis and a theory?
4. What is the value of the *Terminologia Anatomica* and *Terminologia Histologica*?

CHARACTERISTICS OF HUMAN LIFE

Before we begin, you should understand that there is no one "perfect" definition for life. Life, as defined by modern researchers, exhibits *self-sustaining, biological processes,* which include many of the characteristics that we list below as properties of being "alive."

Human life, like all life, shares a number of basic characteristics not associated with inorganic matter. Taken together, these characteristics define life and separate life from nonlife. They all illustrate the *self-organization* needed to live. In this book, we are interested in characteristics of *human life,* and you are probably familiar with at least a few of these characteristics: *responsiveness, conductivity, growth, respiration, circulation, digestion, absorption, secretion, excretion,* and *reproduction.* These characteristics of human life are listed and described briefly for you in Table 1-1.

TABLE 1-1   Characteristics of Human Life

CHARACTERISTIC	DESCRIPTION
Responsiveness	Ability of an organism to sense, monitor, and respond to changes in both its external and internal environments
Conductivity	Capacity of living cells to transmit a wave of electrical disturbance from one point to another within the body
Growth	Organized increase in the size and number of cells, and therefore an increase in size of the individual or a particular organ or part
Respiration	Exchange of respiratory gases (oxygen and carbon dioxide) between an organism and its environment
Circulation	Movement of body fluids containing many substances from one body area to another in a continuous, complete route through hollow vessels
Digestion	Process by which complex food products are broken down into simpler substances that can be absorbed and used by individual body cells
Absorption	Movement of molecules, such as respiratory gases or digested nutrients, through a membrane and into the body fluids for transport to cells for use
Secretion	Production and release of important substances, such as digestive juices and hormones, for diverse body functions
Excretion	Removal of waste products
Reproduction	Formation of new individual offspring

Each characteristic of life is a part of all the physical and chemical processes that take place in our bodies. The sum total of all of life's chemical processes is called **metabolism.**

We mention these important characteristics of human life because we will study each characteristic of life and its integration with other life functions throughout this book.

LEVELS OF STRUCTURAL ORGANIZATION

Before you begin the study of the structures and functions of the human body, it's important to think about how the various parts fit together to function effectively. Figure 1-1 illustrates the *hierarchy* of our human organization. Refer to the illustration and note that chemical reactions form the base, followed by organelles making up cells, cells making up tissues, tissues making up organs, and organs making up organ systems. The sum of all these integrated levels working together as a coordinated whole *is* the human body. Let's look briefly at each level so you can better understand the stepwise increase of complexity from chemistry to organisms.

All of life is based on chemistry. As we will see in Chapter 2, there are more than 100 different chemical building blocks called *elements* making up all of nature. Each *element* is made of atoms consisting of varying numbers of subatomic particles: protons, electrons, and neutrons. If you took a very narrow "chemical" viewpoint of life, you could say that we are nothing more than organized protons, electrons, and neutrons that make up atoms, which then bond together to make **molecules.** In turn, these molecules combine in specific ways to make even larger, more complex molecules called **macromolecules.**

These enormous macromolecules form **organelles**—subcellular structures or packages of gel-like fluids surrounded by membranes. Organelles are the "tiny organs" that allow each cell to live. They are to cells

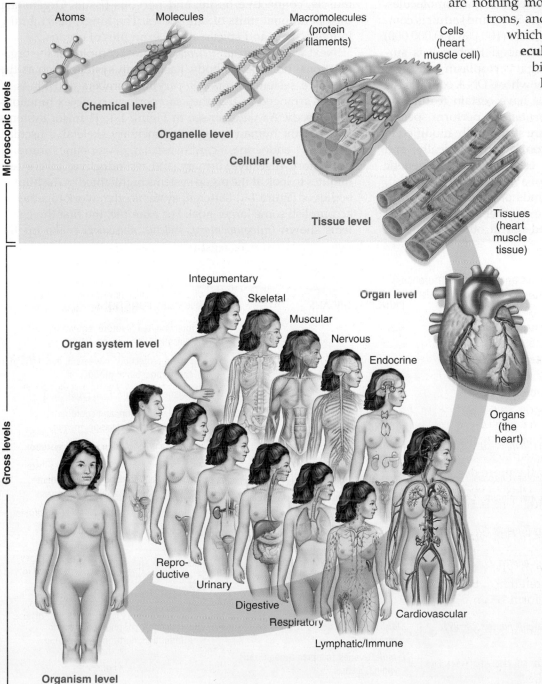

◀ **FIGURE 1-1 Levels of organization.** The smallest parts of the body are the atoms that make up our molecules. In turn, molecules make up microscopic parts called organelles that fit together to form each cell of the body. Groups of similar cells are called tissues, which combine with other tissues to form organs. Groups of organs that work together are called systems. All the systems of the body together make up an individual organism. Knowledge of the different levels of organization will help you understand the basic concepts of human anatomy and physiology.

what organs are to our bodies. Organelles cannot survive outside the cell, but without organelles a cell could not survive either. Dozens of different kinds of organelles have been identified. A few examples include *mitochondria,* the so-called powerhouses of the cell, and *cilia,* tiny whiplike extensions of specialized cells. Unique and complex relationships exist between atoms, molecules, macromolecules, and organelles to make up the *cytoplasm* of our cells. You will find a complete discussion of organelles and their functions in Chapter 3.

Cells are the foundation and basic building blocks of our bodies. They are the smallest structural units that possess and exhibit the basic characteristics of living matter. The specific characteristics of each of our many kinds of cells result from a hierarchy of structure and function that begins with the organization of atoms, molecules, macromolecules, and organelles. Incredibly, a 68-kg (150-pound) adult is composed of nearly one hundred trillion (100,000,000,000,000) cells—an almost unimaginable number! Each cell is surrounded by a membrane enclosing its cytoplasm and organelles, and has one or more nuclei whose DNA controls the cell's activities. Although all cells have certain features in common, they specialize or *differentiate* to perform specific functions. Fat cells, for example, are structurally modified to store fat; cardiac muscle cells contract rhythmically; and sperm cells are rapidly moving cells that deliver genetic material from a father to the egg of a mother.

There are over 200 different kinds of cells in the human body. (Body cells differ from one another because specific genes are active in some cells and not in others.) Cells are organized into **tissues**—groups of many related cells that develop together (from the same part of a human embryo) and perform certain functions. The cells of tissues are surrounded by a *matrix* of nonliving substances that holds them together. There are four major tissue types: *epithelial, connective, muscle,* and *nervous,* and a number of different subtypes within each major type. For example, *cardiac* muscle is a specific type of muscle found only in the heart. Chapters 3, 4, and 5 will give you a more complete understanding of the types and functions of cells and tissues making up our bodies.

As we continue up the ladder of organization, you can see that **organs** are structures made up of several different kinds of tissues arranged to perform specific functions. The heart is an example of an organ: it is made of specialized cardiac muscle, connective tissue, and nervous tissue. Organs are the operational units of our bodies. The lungs, heart, brain, kidneys, liver, and spleen are all examples of organs.

Next we have *organ systems,* the most complex of the organizational units of the body. Each **organ system,** such as the digestive system, comprises varying numbers and kinds of organs arranged so that they can perform complex functions of the body. As you can see in Figure 1-1, 11 major systems make up the human body: integumentary, skeletal, muscular, nervous, endocrine, cardiovascular, lymphatic/immune, respiratory, digestive, urinary, and reproductive. Take a few minutes to look at the organ systems highlighted in the human bodies of Figure 1-1. Different systems often work together to accomplish some larger goal. For example, the first three systems shown (*integumentary, skeletal, muscular*) make up the

TABLE 1-2   Body Systems (with Unit and Chapter References)

FUNCTIONAL CATEGORY	SYSTEM	PRINCIPAL ORGANS	PRIMARY FUNCTIONS
Support and movement (*Unit 2*)	Integumentary (*Chapter 7*)	Skin	Protection, temperature regulation, sensation
	Skeletal (*Chapters 8–9*)	Bones, ligaments	Support, protection, movement, mineral and fat storage, blood production
	Muscular (*Chapter 10*)	Skeletal muscles, tendons	Movement, posture, heat production
Communication, control, and integration (*Unit 3*)	Nervous (*Chapters 11–14*)	Brain, spinal cord, nerves, sensory organs	Control, regulation, and coordination of other systems; sensation; memory
	Endocrine (*Chapter 15*)	Pituitary gland, adrenals, pancreas, thyroid, parathyroids, and other glands	Control and regulation of other systems
Transportation and defense (*Unit 4*)	Cardiovascular (*Chapters 16–18*)	Heart, arteries, veins, capillaries	Exchange and transport of materials
	Lymphatic and immune (*Chapter 19*)	Lymph nodes, lymphatic vessels, spleen, thymus, tonsils	Immunity, fluid balance
Respiration, nutrition, and excretion (*Unit 5*)	Respiratory (*Chapter 20*)	Lungs, bronchial tree, trachea, larynx, nasal cavity	Gas exchange, acid-base balance
	Digestive (*Chapters 21–22*)	Stomach, small and large intestines, esophagus, liver, mouth, pancreas	Breakdown and absorption of nutrients, elimination of waste
	Urinary (*Chapter 23*)	Kidneys, ureters, bladder, urethra	Excretion of waste, fluid and electrolyte balance, acid-base balance
Reproduction and development (*Unit 6*)	Reproductive (*Chapters 24–27*)	*Male:* Testes, vas deferens, prostate, seminal vesicles, penis	Reproduction, continuity of genetic information, nurturing of offspring
		Female: Ovaries, fallopian tubes, uterus, vagina, breasts	

framework of the body and provide support and movement. These *body systems* correspond to the organization of this book and serve as a broad outline for how we will study the organ system components of the human body (Table 1-2).

Finally, we have the *organism level*—the incredibly unique human organism of our studies. Although composed of the 11 organ systems, you can readily see that the human body is certainly more complex than the sum of all its parts. As **organisms,** our bodies are marvelously coordinated teams of interactive and self-regulating structures capable of surviving in often very hostile environments. The human body reproduces itself (and its genetic information) and maintains ongoing repair and replacement of worn or damaged parts. It can also control—in a constant and predictable way—an incredible number of variables required to lead healthy, productive lives. We should always remember that the various systems of the human body are *interdependent*—they do not operate in isolation! For that reason alone, you should not just memorize facts. Instead, you should try to *visualize* in your mind what is being said in the text, and then *put together* the factual information presented so that you understand the human body as a complete biological system.

> ### QUICK CHECK
>
> 5. What are the characteristics of life?
> 6. Try to recall the various levels of life.
> 7. Without looking, try to list the 11 major organ systems.

LANGUAGE OF BODY ORGANIZATION

Anatomical Position

Our discussion of the human body requires that we pick a standard position as a reference point of view. This is called the **anatomical position.** This position allows us to discuss the body's structures relative to one another without confusion—as when we always hold a map in the same orientation. In this reference position the body is viewed in a *standing posture with the feet and head facing forward and the arms at the sides with the palms facing forward* (Figure 1-2). All body directions and the location of structures relative to one another require reference to this standard anatomical position. This reference frame works well because the body (for the most part) exhibits **bilateral symmetry**—the external right and left sides of the body are roughly mirror images of one another. This fact is important in describing injury to parts of the body. For example, injuries to an arm or leg require careful *comparison* of the injured side with the noninjured side.

Body Cavities

The human body may appear solid but it is not. In fact, it contains two major cavities, the *ventral* and *dorsal* cavities. In turn these major cavities are subdivided and contain com-

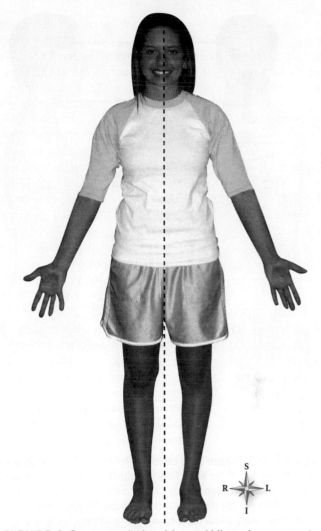

▲ FIGURE 1-2   **Anatomical position and bilateral symmetry.** In the anatomical position, the body is in an erect, or standing, posture with the arms at the sides and palms facing forward. The head and feet are also pointing forward. The dotted line shows the axis of the body's bilateral symmetry. As a result of this organizational feature, the right and left external sides of the body are roughly mirror images of each other.

pact arrangements of internal organs. As you read through the next few paragraphs, refer to Figure 1-3 and Table 1-3 to help you visualize the content.

The **ventral cavities** include the *thoracic* (chest) *cavity* and the *abdominopelvic* (abdomen and pelvis) *cavity.*

The **thoracic cavity** includes *a right and a left pleural cavity* and a middle portion called the **mediastinum.** As you might suspect, the left lung lies in the left pleural cavity and the right lung lies in the right pleural cavity. All the other organs of the thoracic cavity are located in the mediastinum. Here you will find the heart, trachea, esophagus, and thymus, as well as several major nerves, blood vessels, and lymph nodes and their ducts.

The lower **abdominopelvic cavity** is divided into an upper *abdominal cavity* and a lower *pelvic cavity.* The *abdominal cavity* contains the liver, gallbladder, stomach, pancreas, intestines, spleen, and kidneys and their ducts, whereas the *pelvic cavity* contains the reproductive organs and part of the large intestine.

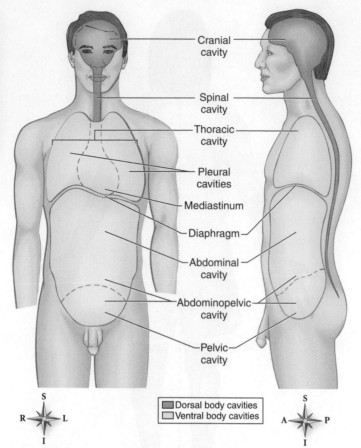

▲ **FIGURE 1-3   Major body cavities.** The dorsal body cavities are in the dorsal (back) part of the body. They include a cranial cavity above and a spinal cavity below. The ventral body cavities are on the ventral (front) side of the trunk. They include the thoracic cavity above the diaphragm and the abdominopelvic cavity below the diaphragm. The thoracic cavity is subdivided into the mediastinum in the center and pleural cavities to the sides. The abdominopelvic cavity is subdivided into the abdominal cavity above the pelvis and the pelvic cavity within the pelvis.

The **dorsal cavities** include the *cranial* and *spinal cavities.* The *cranial cavity* lies in the skull and houses the brain. The head also has smaller cavities: the *oral cavity* (teeth and tongue), the *nasal cavity* (the nose and sinuses), the *orbital cavity* (eyes and associated muscles and nerves), and the *middle ear cavities* (housing the bones of the middle ear). The *spinal cavity* lies in the spinal column and houses the spinal cord.

There are thin membranes that line the body cavities or cover the surfaces of organs within the body cavities. These membranes also have special names. The term **parietal** refers to the inside wall of a body cavity or the membrane that covers this wall. In contrast, the term **visceral** refers to the internal organs, or **viscera**), that reside within the cavity, or to the thin membrane that covers the internal organs. A thin *serous fluid* between the parietal and visceral membranes lubricates the organs so that they can move around within the body without injury.

TABLE 1-3   Organs in Ventral Body Cavities

AREAS	ORGANS
Thoracic Cavity	
Right pleural cavity	Right lung (in pleural cavity)
Mediastinum	Heart (in pericardial cavity)
	Trachea
	Right and left bronchi
	Esophagus
	Thymus gland
	Aortic arch and thoracic aorta
	Venae cavae
	Various lymph nodes and nerves
	Thoracic duct
Left pleural cavity	Left lung (in pleural cavity)
Abdominopelvic Cavity	
Abdominal cavity	Liver
	Gallbladder
	Stomach
	Pancreas
	Intestines
	Spleen
	Kidneys
	Ureters
Pelvic cavity	Urinary bladder
	Female reproductive organs
	Uterus
	Uterine tubes
	Ovaries
	Male reproductive organs
	Prostate gland
	Seminal vesicles
	Parts of vas deferens
	Part of large intestine: sigmoid colon and rectum

Body Regions

The human body can be divided into an **axial** portion that contains the head, neck, and trunk and an **appendicular** portion comprising the arms and legs. Each of these regions and appendages is further divided (Figure 1-4). Can you imagine your body described this way: solid arms and legs attached to a head, thorax, and abdomen filled with organ-containing cavities of various sizes? We should note, how-ever, that every individual (even identical twins) differs superficially from other individuals. This is because each person passes through a unique fetal development and therefore has unique identifying characteristics as an adult. The terms used to describe human body regions are stan-dardized. But even though most terms used to describe

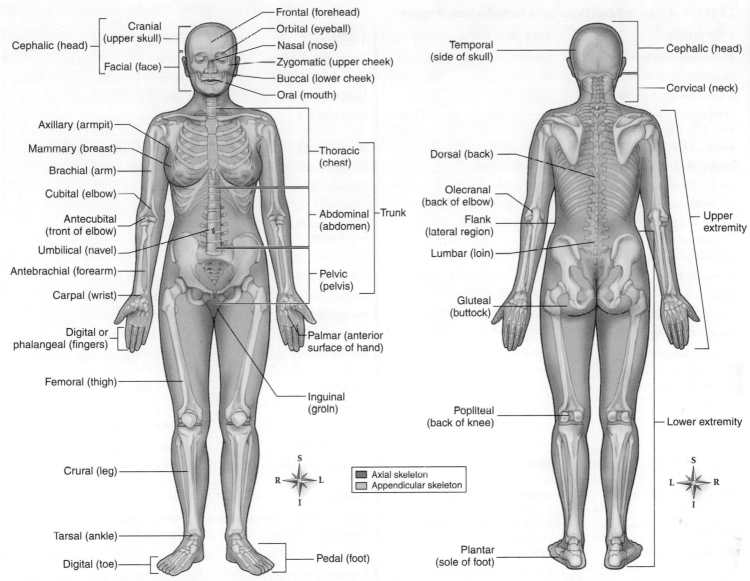

▲ FIGURE 1-4   Specific body regions. Note that the body as a whole can be subdivided into two major portions: **axial** (along the middle, or axis, of the body) and **appendicular** (the arms and legs, or appendages). Names of specific body regions follow the Latin form, with the English equivalent in parentheses.

gross body regions are familiar to us, their misuse is common. For example, to an anatomist the term *leg* refers to the area of the lower extremity between the knee and the ankle and *not* to the entire lower extremity. The complete list of descriptive terms for body regions in Table 1-4 may be useful to you as a reference.

Abdominopelvic Regions and Quadrants

For convenience in locating abdominal organs, anatomists divide the abdomen and pelvis into a grid of nine imaginary regions. These regions are shown in Figure 1-5 and help to organize the organs within them. Likewise, physicians and other health professionals often divide the abdominopelvic

area into four *quadrants* (Figure 1-6). This division helps describe the site of abdominopelvic pain and may help locate internal pathology such as a tumor or abscess. A horizontal line and a vertical line passing through the *umbilicus* (navel) divide the abdomen into *right and left upper quadrants* and *right and left lower quadrants*.

QUICK ✓ CHECK

8. Of what value is the *anatomical position*?
9. Is the human body perfectly bilateral? Explain your answer.
10. What are the major body cavities?
11. Of what value is it to divide the body into *body regions*?

TABLE 1-4   Latin-Based Descriptive Terms for Body Regions*

BODY REGION	AREA OR EXAMPLE	BODY REGION	AREA OR EXAMPLE
Abdominal (ab-DOM-in-al)	Anterior torso below diaphragm	**Mammary** (MAM-er-ee)	Breast
Acromial (ah-KRO-mee-al)	Shoulder	**Manual** (MAN-yoo-al)	Hand
Antebrachial (an-tee-BRAY-kee-al)	Forearm	**Mental** (MEN-tal)	Chin
Antecubital (an-tee-KYOO-bi-tal)	Depressed area just in front of elbow (cubital fossa)	**Nasal** (NAY-zal)	Nose
Axillary (AK-si-lair-ee)	Armpit (axilla)	**Navel** (NAY-vel)	Area around navel, or umbilicus
Brachial (BRAY-kee-al)	Upper part of arm	**Occipital** (ok-SIP-i-tal)	Back of lower part of skull
Buccal (BUK-al)	Cheek (inside)	**Olecranal** (o-LECK-ra-nal)	Back of elbow
Calcaneal (cal-CANE-ee-al)	Heel of foot	**Oral** (OR-al)	Mouth
Carpal (KAR-pal)	Wrist	**Orbital** or **ophthalmic** (OR-bi-tal or op-THAL-mik)	Eyes
Cephalic (se-FAL-ik)	Head	**Otic** (O-tick)	Ear
Cervical (SER-vi-kal)	Neck	**Palmar** (PAHL-mar)	Palm of hand
Coxal (COX-al)	Hip	**Patellar** (pa-TELL-er)	Front of knee
Cranial (KRAY-nee-al)	Skull	**Pedal** (PED-al)	Foot
Crural (KROOR-al)	Leg	**Pelvic** (PEL-vik)	Lower portion of torso
Cubital (KYOO-bi-tal)	Elbow	**Perineal** (pair-i-NEE-al)	Area (perineum) between anus and genitals
Cutaneous (kyoo-TANE-ee-us)	Skin (or body surface)	**Plantar** (PLAN-tar)	Sole of foot
Digital (DIJ-i-tal)	Fingers or toes	**Pollex** (POL-ex)	Thumb
Dorsal (DOR-sal)	Back or top	**Popliteal** (pop-li-TEE-al)	Area behind knee
Facial (FAY-shal)	Face	**Pubic** (PYOO-bik)	Pubis
Femoral (FEM-or-al)	Thigh	**Supraclavicular** (soo-pra-cla-VIK-yoo-lar)	Area above clavicle
Frontal (FRON-tal)	Forehead	**Sural** (SUR-al)	Calf
Gluteal (GLOO-tee-al)	Buttock	**Tarsal** (TAR-sal)	Ankle
Hallux (HAL-luks)	Great toe	**Temporal** (TEM-por-al)	Side of skull
Inguinal (ING-gwi-nal)	Groin	**Thoracic** (thoh-RASS-ik)	Chest
Lumbar (LUM-bar)	Lower part of back between ribs and pelvis	**Zygomatic** (zye-go-MAT-ik)	Cheek

*The left column lists English adjectives based on Latin terms that describe the body parts listed in English in the right column.

Terms Related to Anatomical Directions

We use specific terms to help us reduce confusion when discussing the relationship between body areas or the location of a particular structure. Imagine the body in the proper anatomical position and use Figure 1-7 to help you describe the location of one body part with respect to another. The following sets of opposite terms are the ones most commonly used by anatomists:

Superior/Inferior: **Superior** means "upper" or "above" and **inferior** means "lower" or "below." Thus, the head is superior to the heart and the heart is inferior to the head.

Anterior/Posterior: **Anterior** means "front" or "in front of" and **posterior** means "back" or "in back of." (The terms *ventral* and *dorsal* can be used in place of anterior and posterior for humans because we walk upright.)

For example, the nose is on the anterior (ventral) surface of the body; the shoulder blades are on the posterior (dorsal) surface of the body.

Medial/Lateral: **Medial** means "toward the midline of the body," whereas **lateral** means "toward the side of the body." Thus, the big toe is on the medial side of the foot and the little toe is on its lateral side.

Proximal/Distal: **Proximal** means "toward or closer to" the trunk or the base of an appendage. **Distal** means "farther away from" the trunk or the base of an appendage. For example, the elbow of the arm is proximal and the fingers are distal to the midline of the body.

Superficial/Deep: **Superficial** means "nearer the surface"; **deep** means "farther away from the body surface." For example, the skin is superficial to the muscles, and the muscles lie deep to the skin.

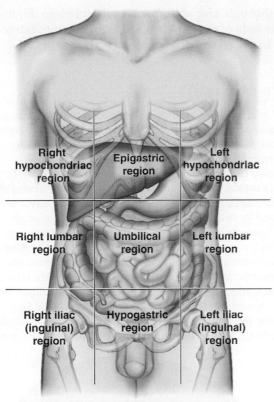

▲ **FIGURE 1-5**   Nine regions of the abdominopelvic cavity. Only the most superficial structures of the internal organs are shown here.

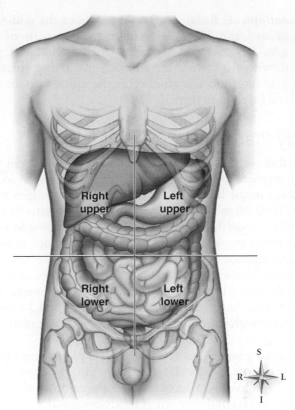

▲ **FIGURE 1-6**   Division of the abdomen into four quadrants. The diagram shows the relationship of internal organs to the four abdominopelvic quadrants.

Terms Related to Organs and Organ Systems

There are other useful terms that describe anatomical relationships among organs in a system or region, or anatomical relationships within an organ. The following are a few of these terms:

Lumen: Organs such as the stomach, small intestine, blood vessels, and the like have a hollow, often fluid-filled interior called the **lumen.** Thus, you can talk about the lumen of an artery, which is like the hollow part of a hose.

Central/peripheral: **Central** means "near the center or middle" and **peripheral** means "near the surface or outside." The central nervous system, for example, includes the brain and spinal cord, which are positioned nearer to the central line of the body. In contrast, the peripheral nervous system includes nerves that service muscles, skin, and organs on the outer boundary or periphery of the body.

Medullary/cortical: **Medullary** refers to an inner region of an organ; **cortical** refers to an outer region or layer of an organ. The inner region of the kidney is the *medulla,* and the *cortex* defines the outer layer of the kidney. As an analogy, if you were looking at a peach, the "medulla" would be the pit and the "cortex" would be the fleshy exterior surrounding the pit.

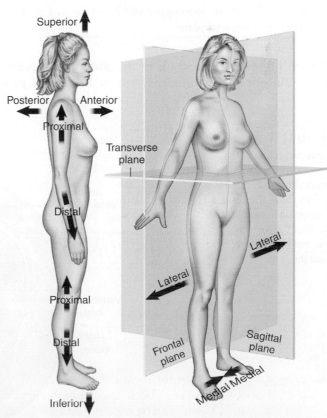

▲ **FIGURE 1-7**   Directions and planes of the body.

Basal/apical: **Basal** refers to the base or the widest part of an organ; **apical** refers to the narrower tip of an organ. For example, the apical tip of the heart points downward toward the stomach. Basal and apical may also refer to individual cells: the apical surface of the cell faces *away* from its base.

Body Planes and Sections

Now that we've finished our discussion of directional terms, take a moment to look carefully at Figure 1-7. You can see there are three major **body planes** that lie at right angles to each other. They are called the *sagittal, coronal,* and *transverse* planes. Literally hundreds of sections ("cuts") can be made in each plane. For example, a mid-transverse plane divides an individual into anterior and posterior sections near the area of the umbilicus. However, many other transverse planes are possible along the entire length of the body. Thus, a transverse section through your knee would amputate your leg and a transverse section through your neck would result in decapitation!

- *Sagittal plane:* This is a lengthwise plane running from front to back, dividing the body into *right and left sides.* If a sagittal section is made in the midline of the body (a midsagittal section), it produces equal and symmetrical right and left external halves.
- *Coronal plane:* Also called a *frontal plane,* the coronal plane divides the body into *anterior (ventral) and posterior (dorsal) portions,* which are not equal.
- *Transverse plane:* This crosswise plane divides the body or any of its parts into *upper and lower parts.* This is also called a *horizontal plane.*

Figure 1-8 shows the organs of the abdominal cavity, as they would appear in the transverse or horizontal plane,

cutting through the abdomen represented in Figure 1-7. Using the *body* of the vertebra as a bearing, test yourself and see how many other body parts you can identify.

A&P CONNECT

Why bother to learn about sections of the body? In the short term, you'll need to understand how to interpret the many illustrations like Figure 1-8 in this book. In the long term, you will use them in clinical settings—as in medical imaging.

Cadavers (preserved human bodies used for scientific study) can be cut into sagittal, frontal, or transverse sections for easy viewing of internal structures, but living bodies, of course, cannot. This fact has been troublesome for medical professionals who must determine whether internal organs are injured or diseased. In some cases the only sure way to detect a *lesion* or variation from normal is extensive *exploratory surgery.* Fortunately, recent advances in medical imaging allow physicians to visualize internal structures of the body without risking the trauma or other complications associated with extensive surgery. This figure shows a computed tomography (CT) scan similar to the perspective of Figure 1-8. CT scanning and some of the other widely used techniques are illustrated and described in **Medical Imaging of the Body** online at A&P Connect.

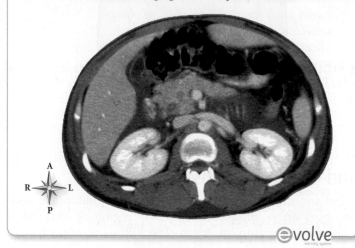

evolve

▶ **FIGURE 1-8**
Transverse section of the abdomen. A transverse, or horizontal, plane through the abdomen shows the position of various organs within the body. This is a view from below, looking in a direction toward the head.

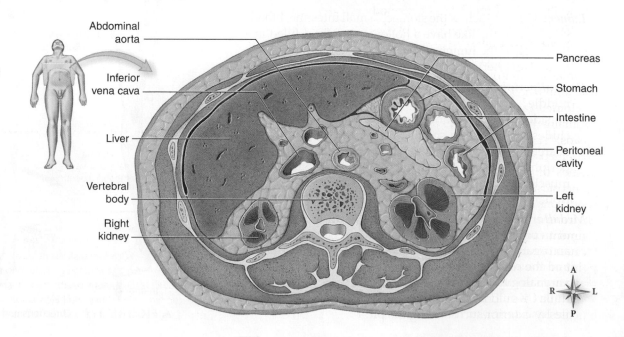

QUICK ✓ CHECK

12. Using scientific terminology, describe the *relative* position of your arm with respect to the side of your body.

13. Do you think there any other planes of the body beyond sagittal, coronal, and transverse?

14. Define and contrast each term in these pairs: superior/inferior, anterior/posterior, medial/lateral, dorsal/ventral.

15. Of what value are anatomical "directions" based on Latin and Greek root words, rather than "plain" English?

INTERACTION OF STRUCTURE AND FUNCTION

Many thousands of generations of natural selection have produced human structures honed to perform very specific functions. This is why structure is intimately connected with function. This fact will help you better understand the mechanisms of disease and structural abnormalities often associated with pathology. A great deal of current human biology research centers on the integration, interaction, development, modification, and control of functioning body structures. Understanding the relationship of structure and function in the human body will help you integrate otherwise isolated facts into a complete picture by the end of your course.

THE CONCEPT OF HOMEOSTASIS

As you know, many elements of our external environment (such as temperature and humidity) are constantly changing. Nonetheless, important aspects of our *internal environment* such as body temperature and water content usually remain remarkably stable. Conditions of our internal cellular environment, such as *pH* (whether it is acid or base) and *salinity* (the concentration of salt), also can vary. Even the contents of the extracellular fluid that bathes each body cell may experience change. Good health, indeed life itself, depends on the correct and constant amount of each substance in the blood and other body fluids. The precise and constant chemical composition of the internal environment must be maintained within very narrow limits (normal ranges) or sickness and death may result.

Homeostasis describes the *relatively* constant internal states maintained by the human body. The word *homeostasis* literally means, "staying the same." As you might suspect, the concept of homeostasis is one of the most unifying and important themes of human physiology. Looking at the big picture, homeostasis is the maintenance of relatively constant internal conditions *despite* changes in either the internal or the external environment. For example, even if the external thermal environment varies, internal *homeostatic control mechanisms* attempt to maintain the body temperature at a relatively constant **set point** of about 37° C (98.6° F). Of course, body temperature may vary slightly depending on our daily activities or state of health (Box 1-1). Likewise, the concentration of human blood glucose has a normal *set point range* between 80 and 100 mg of

BOX 1-1   FYI

Changing the Set Point

Like the set point on a furnace, the physiological set points in your body can be changed. First, not everyone's set point, or "normal," body temperature is the same. Temperatures vary widely. This explains why some people are comfortable at a temperature that is too cold for others around them—their temperature set point must be lower. However, your set point can also change under varying circumstances. During a bacterial infection, your immune system sends chemicals to signal the brain's hypothalamus to "turn up the set point temperature." Your body shivers, and you now have a fever. You may ask for a blanket as your body tries to reach this new higher set point.

The bacteria that have invaded your body did so because they could multiply quickly within the normal temperature of your body. When you are experiencing a fever, your body becomes uncomfortably warm for the bacteria, and they slow down their reproductive rate, which slows the infection. At the same time, the warmer temperature helps improve the immune system's function as it deals with the bacteria. After the infection is over, the hypothalamus returns to its usual set point and your fever goes away.

glucose per milliliter of blood depending on diet and timing of meals (Figure 1-9).

Specific regulatory mechanisms are *continually adjusting* body systems to maintain **homeostasis.** This ability of the body to "self-regulate" in order to maintain homeostasis is the most important concept in physiology and also helps us understand the mechanisms of disease. Each cell of the body, each tissue, and each organ and organ system plays an important role in homeostasis. Throughout this book you will see how the body's regulatory systems rely on homeostatic mechanisms to maintain the body's set point ranges and maintain its physiological stability (see Box 1-1).

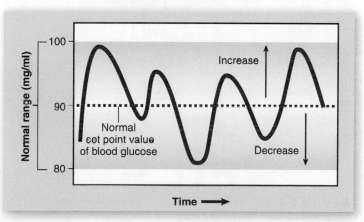

▲ FIGURE 1-9   **Homeostasis of blood glucose.** The range over which a given value, such as the blood glucose concentration, is maintained is accomplished through homeostatic mechanisms. Note that the concentration of glucose fluctuates above and below a normal set point value (90 mg/ml) within a normal set point range (80–100 mg/ml).

Look at the systems diagram of the human body outline in Figure 1-10. This is a classic model that for the sake of our studies simply depicts the body as a "bag of fluid." The fluid inside the bag is our internal environment, and it is this fluid that must be kept at a relatively constant temperature. The "bag" must also maintain a constant glucose level, carbon dioxide level, and so on, if the cells of the body are to survive.

A&P CONNECT

World events have shown us that the international transmission of disease can be used as a weapon of terror. For example, the bacterial infection anthrax usually infects grazing animals such as sheep, but has been used as a weapon against people. For more, check out **Disease as a Weapon** online at **A&P Connect**.

evolve

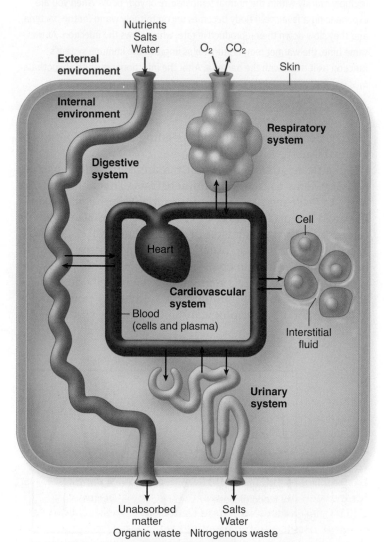

▲ **FIGURE 1-10** **Diagram of the body's internal environment.** The human body is like a bag of fluid separated from the external environment. Tubes, such as the digestive tract and respiratory tract, bring the external environment to deeper parts of the bag where substances may be absorbed into the internal fluid environment or excreted into the external environment. All the "accessories" help maintain a constant environment inside the bag, allowing the cells that live there to survive.

HOMEOSTATIC CONTROL MECHANISMS

Processes that maintain or restore homeostasis are called *homeostatic control mechanisms*. They involve virtually all of the body's organs and systems. Remember, a relatively constant internal environment is vital throughout our lives. For this reason, the body must always respond to changes in the external and internal environments. For example, exercise increases the demand for oxygen and results in accumulation of more waste carbon dioxide than when we are at rest. By increasing our breathing rate above its average of about 17 breaths per minute, we can maintain an adequate blood oxygen level and also increase the elimination of carbon dioxide. When exercise stops, the need for an increased respiratory rate no longer exists and our rate of breathing returns to normal.

To accomplish this self-regulation, a very complex and integrated communication control system called a **feedback control loop** is used. Information about changes in the internal environment is transmitted within these control loops by nervous impulses or by specific chemical messengers called *hormones*, which are secreted into the blood. Virtually all feedback control loops have the same basic components and work in similar ways.

Basic Components of Control Mechanisms

As you read the following paragraphs, refer to Figure 1-11 to help you visualize the four basic components in every feedback loop: *sensor, integrator, effector,* and *feedback.*

We use the terms *afferent* and *efferent* when discussing feedback loops to indicate the *direction* of a signal from a sensor mechanism to a particular integrator mechanism. These directional terms also indicate the *movement* of a signal from the integrator mechanism to some type of effector mechanism. **Afferent** means that a signal is traveling *toward* a particular center, whereas **efferent** means that the signal is moving *away from* a particular center. You will see that these two terms are especially important in the study of the nervous and endocrine systems in Chapters 11 and 15.

In a feedback loop the body must first be able to sense the variable that is changing. Sensory nerve cells or hormone-producing glands often act as homeostatic sensors. If deviations from a normal set point take place, the sensor generates an afferent signal (nerve impulse or hormone). This afferent signal then transmits the information to the integration center.

The diagrams in Figure 1-11 compare a furnace controlled by a thermostat to the homeostatic mechanism of temperature control in the human body. Note that changes in room temperature (the variable to be controlled by the feedback loop) are detected by a thermometer (sensor) attached to the thermostat (integrator). The thermostat contains a switch that controls the furnace (effector). When cold weather causes a decrease in room temperature, the change is detected by the thermometer and relayed to the thermostat. The thermostat compares the

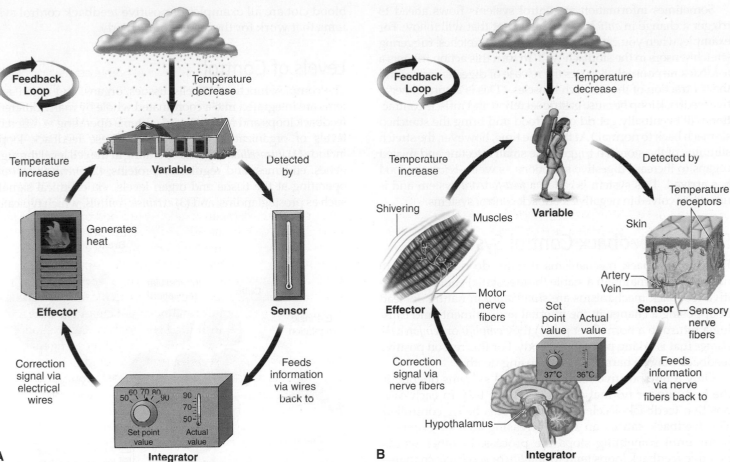

▲ **FIGURE 1-11**   **Basic components of homeostatic control mechanisms. A,** Heat regulation by a furnace controlled by a thermostat. **B,** Homeostasis of body temperature. Note that in both examples **A** and **B,** a stimulus (drop in temperature) activates a sensor mechanism (sensor in the thermostat or body temperature receptor) that sends input to an integrating, or control, center (on-off switch or hypothalamus), which then sends input to an effector mechanism (furnace or contracting muscle). The resulting heat that is produced maintains the temperature within the "normal range." Feedback of effector activity to the sensor mechanism completes the control loop. Both are examples of negative feedback loops.

actual room temperature with its set point temperature. After the integrator determines that the actual room temperature is too low, it sends out an efferent "correction" signal that switches on the furnace. The furnace produces heat and increases the room temperature back to the set point on the thermostat. If the room temperature increases above normal, feedback information from the thermometer causes the thermostat to switch off the signals to the furnace. Thus, by switching the furnace on and off, the room maintains a relatively constant temperature.

Body temperature is regulated in very much the same way as room temperature. In the human body, however, sensory receptors in the skin and superficial blood vessels act as sensors by monitoring the body temperature. When cold conditions reduce body temperature, feedback information is relayed through nerves via afferent signals to the "thermostat" in a part of the brain called the *hypothalamus.* The integrator in the hypothalamus compares the actual body temperature with the normal set point body temperature and then sends efferent nerve signal to effectors. In this example, the skeletal muscles act as effectors by shivering, thus producing metabolic heat to raise the body temperature. Shivering eventually raises the body temperature back

to normal. At this point shivering stops because of constant feedback information that causes the hypothalamus to shut off its stimulation of the skeletal muscles.

Negative Feedback Control Systems

The example of temperature regulation just described is a classic example of a *negative feedback control system.* These systems are by far the most important and common of the homeostatic control mechanisms regulating the internal conditions of the human body. **Negative feedback** control systems are always inhibitory. That is, they *oppose* a change (a drop in temperature) by creating a response (production of metabolic heat) that is *opposite* in direction to the initial fall in temperature from its normal set point. If body temperature rises too high, a negative feedback system brings it back down. All negative feedback mechanisms in the body respond in this way regardless of the variable being controlled: They produce an action that is *opposite* to the change that activated the system. Negative feedback control systems *stabilize* physiological variables and prevent them from straying too far away from their normal ranges. Thus, negative feedback control systems are responsible for maintaining a constant internal environment.

Sometimes information in control systems flows ahead to trigger a change in *anticipation* of an event that will follow. For example, when you eat a meal, the stomach stretches, triggering stretch sensors in the stomach wall. In turn this action triggers a feedback response that causes the release of digestive juices and the contraction of the stomach muscles. (This is a normal negative feedback loop because gastric secretion and muscle contraction will eventually get rid of the food and bring the stretched stomach back to normal.) At the same time, however, the stretch stimulus of the stomach triggers the small intestine and related organs to increase digestive secretions as well—*before* the food has arrived. This system is called a *feed-forward* system and is usually involved in negative feedback control systems.

Positive Feedback Control Systems

Positive feedback mechanisms usually do not function to help the body maintain a stable (homeostatic) condition. Positive feedback mechanisms are *stimulatory* in nature. Instead of opposing a change in the internal environment (and causing a return to a normal set point) they *amplify* or *reinforce the change* that is taking place in the body. For this reason positive feedback may be harmful, even causing death.

Only a few positive feedback control systems operate in the body under normal conditions (Box 1-2). In each case, positive feedback accelerates the process being controlled. The feedback causes an ever-increasing rate of events to occur until something stops the process. In other words, positive feedback loops tend to *amplify* or *accelerate* change—in contrast to negative feedback loops, which reverse a change imposed on the body. Events that lead to a sneeze or cough, the birth of a baby, milk production by a new mother, an immune response to an infection, or the formation of a

blood clot are all examples of positive feedback control systems that work for the benefit of the body.

Levels of Control

The complex functions of cells, tissues, organs, and organ systems are integrated into a coordinated whole by many different feedback loops and feed-forward systems operating at different levels of organization (Figure 1-12). These feedback loops include (1) *intracellular controls* operating at the cell level through genes, enzymes, and regulatory proteins; (2) *intrinsic controls* operating at the tissue and organ levels, via chemical signals such as prostaglandins; and (3) *extrinsic controls*, which typically

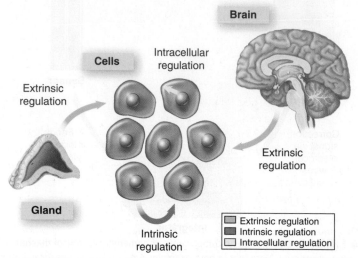

▲ **FIGURE 1-12   Levels of control.** The many complex processes of the body are coordinated at many levels: intracellular (within cells), intrinsic (within tissues/organs), and extrinsic (organ to organ).

BOX 1-2   FYI

Positive Feedback During Childbirth

One of the mechanisms that operates during delivery of a newborn illustrates the concept of positive feedback. As delivery begins, the baby is pushed from the *womb*, or *uterus*, into the birth canal, or *vagina*. Stretch receptors in the wall of the reproductive tract detect the increased stretch caused by movement of the baby. Information regarding increased stretch is fed back to the brain, which triggers the pituitary gland to secrete a hormone called *oxytocin*. Oxytocin travels through the bloodstream to the uterus, where it stimulates stronger contractions. Stronger contractions push the baby farther along the birth canal, thereby increasing stretch and stimulating the release of more oxytocin. Uterine contractions quickly get stronger and stronger until the baby is pushed out of the body and the positive feedback loop is broken. Oxytocin can also be injected therapeutically by a physician to stimulate labor contractions.

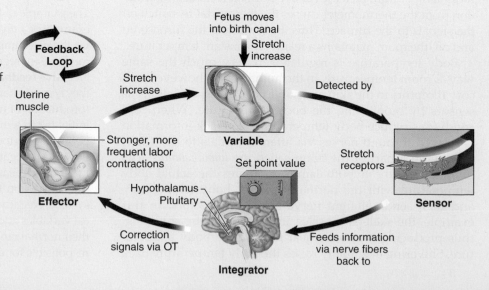

involve the nervous and endocrine (hormonal) systems. There is no need to memorize these levels of control at this point. Examples of all these control mechanisms will be thoroughly discussed at different points throughout this book.

QUICK CHECK

16. Explain the relationship of structure and function.
17. Distinguish between homeostasis and homeostatic mechanisms.
18. Briefly discuss the differences between a negative feedback mechanism and a positive feedback mechanism.
19. Are negative feedback systems more common than positive feedback systems? Why?

The BIG Picture

Your success in anatomy and physiology will require that you *understand* and *integrate* information about the structures and functions of the human body. We call this the "Big Picture." Why? Because the body really is more than just the sum of its parts, and understanding the connectedness between structure and function is both the real challenge and the greatest reward of the study of human anatomy and physiology. This means that your ability to integrate what may seem to be isolated factual information (about bones, muscles, nerves, and blood vessels, for example) will permit you to view the human body in a more cohesive and understandable way. The ability to understand and appropriately use the vocabulary of anatomy and physiology allows you to accurately describe the body's structures, their relative positions, their orientations, their functions, and how they integrate to make organ systems that then coordinate all the major functions of the human body. We also have seen that homeostasis is the glue that integrates and explains how the normal interaction of structure and function is achieved and maintained and how a breakdown of this integration results in disease.

Cycle of LIFE ⟳

As you know, every body organ, regardless of location or function, undergoes change over the years. In general, the body performs less efficiently during infancy and old age. Organs develop and grow during the years before maturity, and body functions gradually become more and more efficient and effective. In a healthy young adult all body systems are mature and fully operational. Homeostatic mechanisms tend to function most effectively during this period of life.

After maturity, effective repair and replacement of the body's structural components often decrease. The term **atrophy** is used to describe the wasting effects of advancing age. In addition to structural atrophy, the functioning of many physiological control mechanisms also decreases and becomes less effective. This is part of the *aging process*. Many specific age changes that occur during the aging process will be presented to you throughout this book. ●

MECHANISMS OF DISEASE

Disturbances to homeostasis (and the body's response to these disturbances) are the basic mechanisms that cause disease. There are a number of factors that can cause homeostatic disturbances. Briefly, these include genetic mechanisms, pathogenic organisms, prions, viruses, bacteria, fungi, protozoa, pathogenic parasitic animals, and tumors and cancers. There are a number of other disease-causing factors, including physical and chemical agents, malnutrition, autoimmunity, inflammation, and degeneration.

evolve Find out more about the causes of homeostatic disturbances online at *Mechanisms of Disease: Organization of the Human Body.*

LANGUAGE OF SCIENCE AND MEDICINE *continued from page 3*

hypothesis (hye-POTH-eh-sis)
[*hypo-* under or below, *-thesis* placing or proposition];
pl., hypotheses (hye-POTH-eh-seez)

inferior (in-FEER-ee-or)
[*infer-* lower, *or* quality]

lateral (LAT-er-al)

lumen (LOO-men)
[*lumen* light]; *pl.,* lumina (LOO-min-ah)

macromolecule (mak-roh-MOL-eh-kyool)
[*macro-* large, *molec-* mass, *-ule* small]

medial (MEE-dee-al)
[*media-* middle, *-al* relating to]

mediastinum (mee-dee-ah-STY-num)
[*mediastinus-* midway]

medullary (MED-oo-lar-ee)
[*medulla-* middle, *-ry* state]

metabolism (me-TAB-o-liz-em)
[*metabol-* change, *-ism* condition]

microscopic anatomy (my-kroh-SKOP-ik ah-NAT-o-mee)
[*micro-* small, *-scop-* see, *-ic* relating to, *ana-* apart, *-tom-* cut, *-y* action]

molecule (MOL-eh-kyool)
[*mole-* mass, *-cule* small]

negative feedback
[*negative* opposing or prohibitive]

organ (OR-gan)
[*organ* instrument]

organ system (OR-gan SIS-tem)
[*organ* instrument, *system* organized whole]

organelle (or-gah-NELL)
[*organ-* tool or instrument, *-elle* small]

organism (OR-gah-niz-im)
[*organ-* instrument, *-ism* condition]

parietal (pah-RYE-ih-tal)
[*pariet-* wall, *-al* relating to]

peripheral (pe-RIF-er-al)
[*peri-* around, *-phera-* boundary, *-al* relating to]

physiology (fiz-ee-OL-o-jee)
[*physio-* nature (function), *-o-* combining form, *-log-* words (study of), *-y* activity]

positive feedback
[*positive* place or amplify]

posterior (pos-TEER-ee-or)
[*poster-* behind, *-or* quality]

proximal (PROK-si-mal)
[*proxima-* near, *-al* relating to]

sagittal plane (SAJ-i-tal playn)
[*sagitta-* arrow, *-al* relating to]

scientific method (sye-en-TIFF-ik METH-od)
[*scienti-* knowledge, *-ic* relating to]

set point

continued

LANGUAGE OF SCIENCE AND MEDICINE *continued from page 17*

superficial (soo-per-FISH-al)
[*super-* over or above, *-fici-* face, *-al* relating to]

superior (soo-PEER-ee-or)
[*super-* over or above, *-or* quality]

theory (THEE-o-ree)
[*theor-* look at, *-y* act of]

thoracic cavity (thoh-RASS-ik KAV-ih-tee)
[*thorac-* chest (thorax), *-ic* relating to]

tissue (TISH-yoo)
[*tissue* fabric]

transverse plane (TRANS-vers playn)
[*trans-* across or through, *-vers* turn]

ventral cavity (VEN-tral KAV-ih-tee)
[*ventr-* belly, *-al* relating to, *cav-* hollow, *-ity* state]

viscera, visceral (VISS-er-ah) (VISS-er-al)
[*visc-* internal organ, *-al* relating to]; *sing.,* viscus

NOTE: Table 1-4 lists additional key terms in this chapter.

CHAPTER SUMMARY

To download an MP3 version of the chapter summary for use with your iPod or other portable media player, access the Audio Chapter Summaries online at http://evolve.elsevier.com.

HINT Scan this summary after reading the chapter to help you reinforce the key concepts. Later, use the summary as a quick review before your class or before a test.

HUMAN ANATOMY AND PHYSIOLOGY

A. Anatomy—study of the structure of the human body
  1. Gross anatomy—study of body parts visible to the naked eye
  2. Microscopic anatomy—describes the study of cells (cytology) and tissues (histology)
B. Physiology—study of the functions of these structures
C. Scientific method—logical inquiry based on experimentation
  1. Hypothesis—statement used as a proposed explanation for a scientific problem
  2. Experiments—series of controlled tests of a hypothesis
  3. Theory—a hypothesis that has gained a high level of confidence through rigorous and controlled testing
D. Scientific language
  1. *Terminologia Anatomica (TA)*—official list of gross anatomical terms
  2. *Terminologia Histologica (TH)*—official list of microscopic anatomical terms

CHARACTERISTICS OF HUMAN LIFE (TABLE 1-1)

A. Responsiveness—ability to sense, monitor, and respond to changes in our external environment
B. Conductivity—ability of specific cells to produce a chemical or electrical wave of excitation, and conduct this wave from one point to another
C. Growth—normal outcome of cell division; results in an increase in our size until adulthood is reached
D. Respiration—exchange of carbon dioxide and oxygen, as well as the physical and chemical processes that result in the absorption, transport, use, and exchange of these gases
E. Circulation—movement of body fluids and other substances from one body area to another
F. Digestion—process through which complex food products are broken down into simpler substances that can then be absorbed and utilized by individual body cells
G. Absorption—movement of nutrients from the digestive tract into the body fluids for transport to all the body's cells
H. Secretion—production and delivery of special substances, such as digestive fluids and hormones, for diverse body functions
I. Excretion—removal of waste products
J. Reproduction—formation of new offspring

LEVELS OF STRUCTURAL ORGANIZATION (FIGURE 1-1)

A. Elements—chemical building blocks
B. Molecules—atoms that have bonded together
C. Macromolecules—molecules that have combined in specific ways
D. Organelles—"tiny organs" formed from macromolecules
E. Cells—foundation and basic building blocks of our bodies
F. Tissues—groups of many related cells that develop together and perform certain functions
G. Organs—structures made up of several different kinds of tissues arranged to perform specific functions
H. Organ systems—most complex organizational units of the body
I. Organism—integrated assemblages of interactive and self-regulating structures (organ systems)

LANGUAGE OF BODY ORGANIZATION

A. Anatomical position
  1. Anatomical position—standard position as a reference point of view
    a. Standing posture with the arms at the sides, feet, head, and palms facing forward (Figure 1-2)
  2. Bilateral symmetry—external right and left sides of the body are roughly mirror images of one another
    a. Important in describing injury to external parts of the body
B. Body cavities (Figure 1-3; Table 1-3)
  1. Ventral—includes the thoracic (chest) cavity and the abdominopelvic (abdomen and pelvis) cavity
    a. Thoracic cavity—includes a right and a left pleural cavity and a middle portion called the mediastinum
    b. Abdominopelvic cavity—divided into an upper abdominal cavity and a lower pelvic cavity
  2. Dorsal—includes the cranial and spinal cavities
    a. Cranial cavity—lies in the skull and houses the brain
    b. Spinal cavity—lies in the spinal column and houses the spinal cord
C. Body regions (Figure 1-4; Table 1-4)
  1. Axial—contains the head, neck, and trunk
  2. Appendicular—comprising the arms and legs
  3. Abdominopelvic regions—a grid of nine imaginary regions
  4. Abdominopelvic quadrants—four abdominal divisions
D. Terms related to anatomical direction
  1. Superior—upper or above
  2. Inferior—lower or below
  3. Anterior—front or in front of
  4. Posterior—back or in back of
  5. Medial—toward the midline of the body
  6. Lateral—toward the side of the body
  7. Proximal—toward or closer to the trunk or the base of an appendage
  8. Distal—farther away from the trunk or the base of an appendage

9. Superficial—nearer the surface
10. Deep—farther away from the body surface
E. Terms related to organs and organ systems
1. Lumen—hollow, often fluid-filled interior of organs
2. Central—near the center or middle
3. Peripheral—near the surface or outside
4. Medullary—inner region of an organ
5. Cortical—outer region or layer of an organ
6. Basal—base or widest part of an organ
7. Apical—narrower tip of an organ
F. Body planes and sections (Figure 1-7)
1. Sagittal—lengthwise plane running from front to back: divides body into right and left sides
2. Coronal (frontal)—divides the body into anterior (ventral) and posterior (dorsal) portions, which are not equal
3. Transverse (horizontal)—divides the body or any of its parts into upper and lower parts

INTERACTION OF STRUCTURE OF FUNCTION

A. Understanding the relationship of structure and function in the human body will help you integrate otherwise isolated facts into a cohesive picture.

THE CONCEPT OF HOMEOSTASIS (FIGURE 1-9)

A. Homeostasis—relatively constant internal states maintained by the human body

HOMEOSTATIC CONTROL MECHANISMS

A. Feedback control loops (Figure 1-11)
1. Basic components—sensor, integrator, effector, and feedback
B. Negative feedback control systems
1. Produce an action that is opposite to the change that activated the system
2. Stabilize physiological variables and prevent them from straying too far away from their normal ranges
C. Positive feedback control systems (Box 1-2)
1. Do not usually function to help the body maintain a stable (homeostatic) condition
2. Instead of opposing a change in the internal environment, they amplify the change that is taking place in the body
D. Levels of control (Figure 1-12)
1. Intracellular controls—operate at the cell level through genes, enzymes, and regulatory proteins
2. Intrinsic controls—operate at the tissue and organ levels
3. Extrinsic controls—involve the nervous and endocrine (hormonal) systems

REVIEW QUESTIONS

 Write out the answers to these questions after reading the chapter and reviewing the Chapter Summary. If you simply think through the answer without writing it down, you won't retain much of your new learning.

1. Define the terms *anatomy* and *physiology*.
2. List and briefly describe the levels of organization that relate the structure of an organism to its function. Give examples characteristic of each level.
3. What is meant by the term *anatomical position*? How do the specific anatomical terms of position or direction relate to this body orientation?
4. What is *bilateral symmetry*?
5. Define briefly each of the following terms: anterior, distal, sagittal plane, medial, dorsal, coronal plane, organ, superior, tissue.

6. Locate the mediastinum.
7. What does the term *homeostasis* mean? Illustrate some generalizations about body function using homeostatic mechanisms as examples.
8. Define *homeostatic control mechanisms* and feedback control loops.
9. Identify the three basic components of a control loop.
10. List the major types of risk factors that may increase a person's chance of a specific disease developing.

CRITICAL THINKING QUESTIONS

 After finishing the Review Questions, write out the answers to these items to help you apply your new knowledge. Go back to sections of the chapter that relate to items that you find difficult.

1. An x-ray technician has been asked to take x rays of the entire large intestine (colon), including the appendix. Which of the nine abdominopelvic regions must be included in the x ray?
2. Body cavities can be subdivided into smaller and smaller sections. Identify, from largest to smallest, the cavities in which the urinary bladder can be placed.
3. When driving in traffic, it is important to stay in your own lane. If you see that you are drifting out of your lane, your brain tells your arms and hands to move in such a way that you get back in your lane. Identify the three components of a control loop in this example. Explain why this would be a negative feedback loop.

continued from page 3

Now that you have read this chapter, see if you can answer these questions about Sam's experience in the emergency department from the Introductory Story.

1. The patient was stabbed in the right upper quadrant; where is that exactly?
   a. In his thoracic cavity
   b. In his abdominopelvic cavity
   c. In his pericardial cavity
   d. In his pleural cavity
2. What organ is most likely to have been damaged by the stabbing attack?
   a. His heart
   b. His lungs
   c. His liver
   d. His spleen
3. Where is the bruise (contusion) on the patient's leg, which was said to be just proximal to the knee?
   a. His femoral region
   b. His crural region
   c. His humoral region
   d. His popliteal region
4. Where is the patient's brachial region?
   a. His armpit
   b. His lower leg
   c. His cheek
   d. His upper arm

 To solve these questions, you may have to refer to the glossary or index, other chapters in this textbook, A&P Connect, Mechanisms of Disease, and other resources.

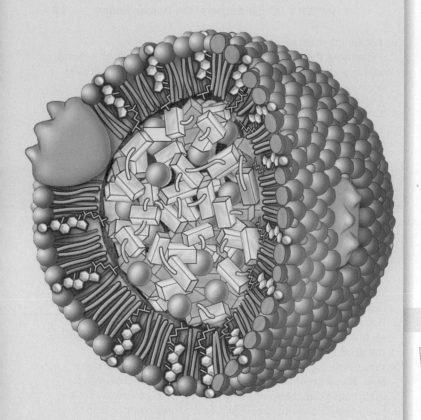

The Chemistry of Life

STUDENT LEARNING OBJECTIVES

At the completion of this chapter, you should be able to do the following:

1. Describe protons, electrons, and neutrons according to their charge, mass, and relative location in the atom.
2. Discuss the importance of the octet rule concerning the stability and reactivity of an atom.
3. Distinguish between atomic weight, atomic number, and mass number.
4. Compare and contrast these terms: elements, atoms, isotopes, molecules, and electrolytes.
5. Compare ionic bonds with covalent bonds.
6. Describe and discuss the significance of hydrogen bonds to macromolecules.
7. Identify the physical properties of water that make it physiologically important.
8. Distinguish between solvents and solutes.
9. Define pH, acid, base, and buffers, and give their physiological significance.
10. Outline the differences between inorganic and organic molecules.
11. Give examples of dehydration synthesis and hydrolysis in biochemical reactions.
12. Identify the monomers and polymers of carbohydrates, proteins, lipids, and nucleic acids.
13. Distinguish DNA from RNA.

THIS year, during her college's spring break, Calleigh visited Florida for the first time. On the first night of her vacation, she and her friends went out to dinner. Feeling rather adventurous, Calleigh ate raw oysters as an appetizer. Unfortunately, the oysters she consumed had high concentrations of bacteria, and 24 hours later, Calleigh was experiencing the "adventure" of food poisoning. Among her symptoms were nausea, vomiting, severe abdominal pain, and diarrhea.

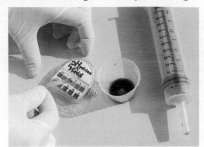

As you read through this chapter, keep Calleigh's symptoms in mind and try to answer the questions pertaining to her condition at the end of the chapter.

Think about this: Virtually all the elements of the universe are born in the furnaces of the stars, and because we are made of these elements, we are in effect, *stardust!*

Physical and chemical processes not only run the universe but also serve as the foundations of all biological processes. In fact, all cells of the human body must continually and efficiently rearrange molecules and compounds in order to function properly. It stands to reason, then, that we must first have a basic understanding of how matter interacts in our body's chemical reactions before we can begin to understand our anatomy and physiology.

The field of chemistry, like that of anatomy and physiology, is divided into many different branches. For example, biochemistry deals with living organisms and their life processes. For this reason, a basic knowledge of biochemistry is vital to your understanding of living matter and the processes that underlie life, such as cell division and muscle contraction. But modern biochemistry is in reality many subdisciplines. This chapter will help you understand the basic biochemical processes involved in life processes at every level of organization. We hope you will see how an understanding of homeostatic processes and control mechanisms (see Chapter 1) is in most cases dependent on your knowledge of basic biochemistry.

BASIC CHEMISTRY

Elements and Compounds

Matter is defined simply as anything that takes up space and has mass. On Earth, matter appears as a *solid, liquid,* or *gas.* Respective examples of these states of matter include the wood in your kitchen table, the water in your sink, and the air you are breathing right now. Each of these substances is composed of **elements**—"pure" substances that cannot be broken down or separated further into two or more different substances by ordinary chemical means. Oxygen (symbol O) in the atmosphere is an element; so is the gold (symbol Au) in Fort Knox. In most organisms, elements do not exist alone in their pure states.

LANGUAGE OF SCIENCE AND MEDICINE

HINT *Before reading the chapter, say each of these terms out loud. This will help you avoid stumbling over them as you read.*

acid (ASS-id)
[*acid* sour]

adenosine triphosphate (ATP) (ah-DEN-o-seen try-FOS-fate)
[blend of *adenine* and *ribose*, *tri-* three, *-phosph-* phosphorus, *-ate* oxygen]

amino acid (ah-MEE-no ASS-id)
[*amino* NH$_2$, *acid* sour]

anabolism (ah-NAB-o-liz-em)
[*anabol-* build up, *-ism* action]

atom (AT-om)
[*atom* indivisible]

atomic number (ah-TOM-ik)
[*atom-* indivisible, *-ic* relating to]

atomic weight (ah-TOM-ik)
[*atom-* indivisible, *-ic* relating to]

base (BAYS)
[*bas* base]

buffer (BUFF-er)
[*buffe-* cushion, *-er* actor]

carbohydrate (kar-bo-HYE-drate)
[*carbo-* carbon, *-hydr-* hydrogen, *-ate* oxygen]

catabolism (kah-TAB-o-liz-em)
[*catabol-* throw down, *-ism* action]

chemical bond (KEM-ih-kal bond)
[*chemeia-* alchemy, *bond* band]

chemical reaction (KEM-ih-kal ree-AK-shun)
[*chemeia-* alchemy, *-al* relating to, *re-* again, *-action* action]

compound (KOM-pownd)
[*compoun-* put together]

covalent bond (ko-VAYL-ent bond)
[*co-* with, *-valen-* power, *bond* band]

decomposition reaction (dee-KAHM-poh-sih-shun ree-AK-shun)
[*de-* opposite of, *-compo-* assemble, *-tion* process, *re-* again, *-action* action]

deoxyribonucleic acid (DNA) (dee-ok-see-rye-boh-noo-KLAY-ik ASS-id)
[*de-* removed, *-oxy-* oxygen, *-ribo-* ribose (sugar), *-nucle-* nucleus (kernel), *-ic* relating to, *acid* sour]

disaccharide (dye-SAK-ah-ryde)
[*di-* two, *-saccharide* sugar]

electrolyte (e-LEK-tro-lyte)
[*electro-* electricity, *-lyt-* loosening]

electron (eh-LEK-tron)
[*electro-* electricity, *-on* subatomic particle]

element (EL-em-ent)
[*element-* first principle]

continued on page 39

1																	2
H 1.008																	**He** 4.002
3 **Li** 6.941	4 **Be** 9.012											5 **B** 10.811	6 **C** 12.011	7 **N** 14.007	8 **O** 15.999	9 **F** 18.998	10 **Ne** 20.180
11 **Na** 22.990	12 **Mg** 24.305											13 **Al** 26.982	14 **Si** 28.086	15 **P** 30.974	16 **S** 32.066	17 **Cl** 35.452	18 **Ar** 39.948
19 **K** 39.098	20 **Ca** 40.078	21 **Sc** 44.956	22 **Ti** 47.867	23 **V** 50.942	24 **Cr** 51.996	25 **Mn** 54.931	26 **Fe** 55.845	27 **Co** 58.933	28 **Ni** 58.963	29 **Cu** 63.546	30 **Zn** 65.39	31 **Ga** 69.723	32 **Ge** 72.61	33 **As** 74.922	34 **Se** 78.96	35 **Br** 79.904	36 **Kr** 83.80
37 **Rb** 85.468	38 **Sr** 87.62	39 **Y** 88.906	40 **Zr** 91.224	41 **Nb** 92.906	42 **Mo** 95.94	43 **Tc** (98)	44 **Ru** 101.07	45 **Rh** 102.906	46 **Pd** 106.42	47 **Ag** 107.868	48 **Cd** 112.411	49 **In** 114.818	50 **Sn** 118.710	51 **Sb** 121.760	52 **Te** 127.60	53 **I** 126.904	54 **Xe** 131.29
55 **Cs** 132.905	56 **Ba** 137.327	57 **La** 138.905	72 **Hf** 178.49	73 **Ta** 180.948	74 **W** 183.84	75 **Re** 186.207	76 **Os** 190.23	77 **Ir** 192.217	78 **Pt** 195.08	79 **Au** 196.967	80 **Hg** 200.59	81 **Ti** 204.383	82 **Pb** 207.2	83 **Bi** 208.980	84 **Po** (209)	85 **At** (210)	86 **Rn** (222)
87 **Fr** (223)	88 **Ra** 226.025	89 **Ac** 227.028	104 **Rf** (263.113)	105 **Db** (262.114)	106 **Sg** (266.122)	107 **Bh** (264.125)	108 **Hs** (269.134)	109 **Mt** (268.139)	110 **Ds** (272.146)	111 **Rg** (272.154)	112 **Cn** (285)	113 **Uut** (284)	114 **Uuq** (289)	115 **Uup** (288)	116 **Uuh** (292)	117 **Uus** (292)	118 **Uuo** (294)

58 **Ce** 140.115	59 **Pr** 140.907	60 **Nd** 144.24	61 **Pm** (145)	62 **Sm** 150.36	63 **Eu** 151.965	64 **Eu** 157.25	65 **Gd** 158.925	66 **Tb** 162.50	67 **Ho** 164.930	68 **Er** 167.26	69 **Tm** 168.939	70 **Yb** 173.04	71 **Lu** 174.967
90 **Th** 232.038	91 **Pa** 231.036	92 **U** 238.029	93 **Np** 237.048	94 **Pu** (244)	95 **Am** (243)	96 **Cm** (247)	97 **Bk** (247)	98 **Cf** (251)	99 **Es** (252)	100 **Fm** (257)	101 **Md** (258)	102 **No** (259)	103 **Lr** (260)

Legend:
■ Major elements
□ Trace elements

6 **C** 12.011 —
- Atomic number (number of protons)
- Chemical symbol
- Atomic weight (number of protons plus average number of neutrons)

▲ **FIGURE 2-1**   **Periodic table of elements.** The major elements found in the body are highlighted in *pink*. The trace elements, found in very tiny quantities in the body, are highlighted in *orange*.

When two or more atoms of the same or different elements are joined together *chemically*, the result is called a **compound.** One common compound with which you are familiar is water—H_2O. As you might suspect, compounds can be broken down through chemical processes into the pure elements that compose them. Water, for example, can be broken down into atoms of hydrogen and oxygen in a 2:1 ratio.

Living organisms need about 26 of the 92 different elements naturally found on Earth, in the right combinations and quantities, to survive. Take a moment to review these elements in Figure 2-1. Of these, oxygen, hydrogen, carbon, and nitrogen comprise about 96% of the matter in the human body. These four elements and seven others are called major elements (Figure 2-2). The 15 remaining "trace" elements are present in amounts that total less than 0.1% of our body weight. However, all 26 elements are *essential* to life. It is the complexity, organization, and interrelationships of all elements that make life possible.

Atoms and Their Characteristics

The smallest complete unit of all elements is the **atom.** If you lined up 100 million atoms of even the densest and heaviest element, the line would measure barely 2.5 cm (1 inch). That's amazing, of course, but think of this: more than 99% of every atom is really empty space! So, what exactly *is* an atom?

Atoms are made of even smaller *subatomic* particles, of which there are three basic types: *protons, neutrons,* and *electrons.* **Protons** and **neutrons** are relatively large subatomic particles, which together make up the **nucleus**—the dense center of an atom. Protons have a positive charge, while neutrons have no charge and are *neutral.* The third type of subatomic particle, the **electron,** inhabits outer regions called electron *shells* or energy levels, and travels at up to the speed of light around the nucleus. Electrons are nearly 2,000 times smaller than protons and neutrons, but carry a negative charge as strong as the positive charge of a proton (Figure 2-3). Incredible as it may seem, these tiny electrons, which are the same in all atoms, are responsible for all chemical reactions.

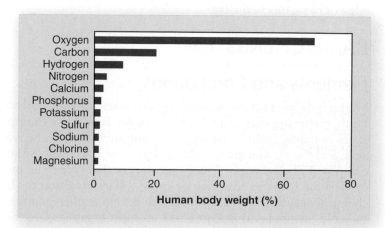

▲ **FIGURE 2-2**   **Major elements of the body.** These elements are found in great quantity in the body (see Figure 2-1). The graph shows the relative abundance of each. Notice that oxygen (O), carbon (C), hydrogen (H), and nitrogen (N) predominate.

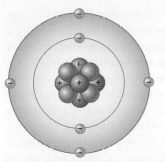

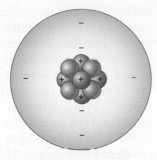

▲ **FIGURE 2-3** **Models of the atom.** The nucleus—protons (+) and neutrons—is at the core. Electrons inhabit outer regions called electron *shells* or energy levels.

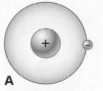
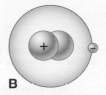

A   B   C

▲ **FIGURE 2-4** Structure of hydrogen and two of its isotopes. **A,** The most common form of hydrogen. **B,** An isotope of hydrogen called *deuterium.* **C,** The hydrogen isotope *tritium. Note that isotopes of an element differ only in the number of their neutrons.*

Because the *nucleus* of an atom contains positively charged protons and uncharged neutrons, it is positively charged as a whole. The *cloud* of space surrounding the nucleus only contains electrons, and thus, it is negatively charged. The number of protons in the nucleus determines the identity of each element and is the same as the element's **atomic number.** For example, only carbon atoms contain six protons within their nucleus, and so carbon has the atomic number of 6. Likewise, hydrogen has only one proton, and so its atomic number is 1.

Because electrons are very small, most of the *mass* of an atom comes from the protons and neutrons in the nucleus. The **atomic weight** of an atom is the sum of all the protons and the *average* number of neutrons in the nucleus. For example, the atomic weight of the element carbon is approximately 12. (The atomic weight of carbon is actually slightly greater [12.011] than this because electrons, although incredibly small, also contribute some mass.) In their neutral state, all elements have the same number of electrons surrounding the nucleus as they have protons within the nucleus. The fact that there are equal numbers of electrons and protons means that atoms are electrically *neutral* particles, because the opposite charges neutralize each other.

However, the number of neutrons within a nucleus *can* vary within the same element. So even though all atoms of the same element contain the same number of protons, they do not necessarily contain the same number of neutrons. We call these atoms *isotopes*. By definition, then, an **isotope** of an element contains the same number of protons, but a *different number of neutrons.* Isotopes of a given element have the same basic chemical properties (e.g., same boiling point, same melting point) as any other atom of that element, and they also have the same atomic number. However, because they have a different number of neutrons, they differ in atomic weight. For example, a hydrogen atom typically has only one proton and no neutrons (atomic number 1; atomic weight 1). Figure 2-4 illustrates this most common type of hydrogen and *two* of its isotopes that vary in the number of neutrons present. Note that the isotope of hydrogen called *deuterium* has one proton and one neutron for an atomic weight of 2. However, *tritium,* another isotope of hydrogen, has one proton and *two* neutrons, for an atomic weight of 3. Many isotopes are unstable and emit radiation (they are *radioactive*). Such radiation can be useful in producing medical images and treating diseases.

QUICK CHECK

1. Describe the general properties of an atom.
2. What are the characteristics of the subatomic particles of an atom?
3. Define the following: atomic weight, atomic number, and isotope.

Chemical Interactions

Elements differ in their chemical and physical properties because of the unique number of protons in their nucleus. However, it is the *electrons* of an atom that create interactions between two or more atoms. And it is the electrons that are responsible for *all chemical reactions*. For example, electrons can be gained, lost, or shared when atoms interact to form bonds.

Electrons are arranged in **energy levels** or *shells* around the nucleus, where they occur as single electrons or pairs of electrons. The first shell is considered "full" when it contains two electrons. However, the second and third shells around an atom can contain up to eight electrons each (Figure 2-5).

The electrons in an atom's outermost shell—the **valence electrons**—are those that typically participate in chemical interactions. Any atom that has its outermost shell full is generally chemically inactive and is considered to be a stable atom. In chemical interactions, atoms that are not stable attempt to gain, lose, or share electrons in order to achieve *eight electrons* in their outer shell, and thus, stability. This general principle is called the **octet rule.** There are exceptions to this rule. In the case of hydrogen atoms—because they only have one electron in the first energy level—stability occurs when there are only *two* electrons in this shell. Thus, the octet rule does not hold for hydrogen. However, having a single electron in its only shell makes lone hydrogen atoms very reactive. Helium, on the other hand, has two electrons, filling its outermost shell, and is thus *inert* or chemically inactive in its natural state.

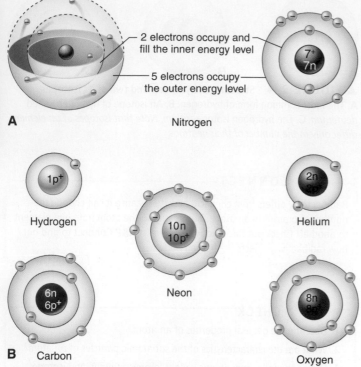

A Nitrogen

2 electrons occupy and fill the inner energy level

5 electrons occupy the outer energy level

Hydrogen 1p+

Helium 2n 2p+

Neon 10n 10p+

Carbon 6n 6p+

Oxygen 8n 8p+

B

▲ **FIGURE 2-5 Energy levels. A,** Energy levels (electron shells) surrounding the nucleus of an atom. Each concentric shell represents a different electron energy level. Note that nitrogen can accept three more electrons to satisfy the octet rule. **B,** Energy levels of five common elements. All atoms are balanced with respect to positive and negative charges. In atoms with a single energy level, two electrons are required for stability. Hydrogen with its single electron is reactive, whereas helium with its full energy level is not. In atoms with more than one energy level, eight electrons in the outermost energy level are required for stability (according to the octet rule). Neon is stable because its outer energy level has eight electrons. However, oxygen and carbon, with four and six electrons (respectively) in their outer energy shells, are chemically reactive. Oxygen can accept two more electrons and carbon can accept four more electrons to satisfy the octet rule.

Chemical Bonds

For the most part, the octet rule determines how an atom can interact with other atoms of the same element or with atoms of different elements to form stable **chemical bonds.** Bonds result when an atom *loses* an electron to another atom, or *gains* an electron from another atom, or when two or more atoms *share* electrons. According to the octet rule, the interacting atoms will all have a full outer shell by one of these processes.

When atoms interact by transferring (losing or gaining) electrons, they form an **ionic bond.** Look at Figure 2-6. You can see that sodium has an outer shell with only one electron, whereas chlorine has seven electrons in its outer shell. If sodium *loses* an electron and chlorine then *gains* that electron, both atoms will have completed their outermost shells and satisfied the octet rule. Notice that, when this happens, sodium then has *one more* proton in its nucleus than the number of electrons surrounding it. For this reason, it becomes a *positively charged* ion (there are now more protons in the nucleus than electrons surrounding it). Likewise, the chlorine atom now has

one less proton in its nucleus than the number of electrons surrounding it. As a result the chlorine atom is a *negatively charged* ion. An **ion,** then, is an atom (or group of atoms) that carries a positive or negative electric charge as a result of having lost or gained one or more electrons. The chlorine and sodium ions are *attracted* to one another because they are oppositely charged. When arranged like this, alternating Na–Cl atoms form NaCl (sodium chloride), or common table salt.

When two atoms *share* one or more pairs of electrons, a **covalent bond** results. This is the most common way atoms are joined to form molecules in the body. Consider two hydrogen atoms sharing a single pair of electrons (Figure 2-7, *A*). Each atom has filled its outer shell and in the process has formed a *single covalent bond.* When atoms share two pairs of electrons, a *double covalent bond,* or double bond, results. For example, when carbon forms covalent bonds with two oxy-

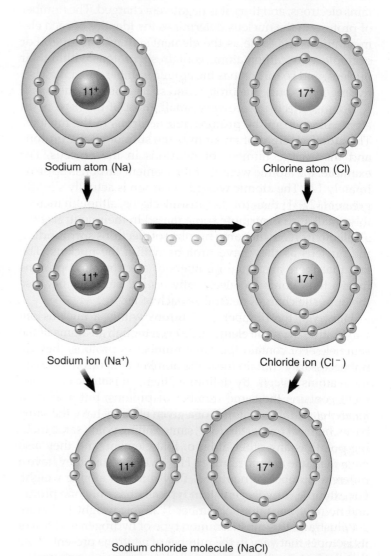

Sodium atom (Na)   Chlorine atom (Cl)

Sodium ion (Na+)   Chloride ion (Cl−)

Sodium chloride molecule (NaCl)

▲ **FIGURE 2-6 Example of an ionic bond.** Energy-level models show the steps involved in forming an ionic bond between atoms of sodium and chlorine within the internal fluid environment of the body (water). Sodium "donates" an electron to chlorine, thereby forming a positive sodium ion and a negative chloride ion. The electrical attraction between the now oppositely charged ions forms an ionic bond.

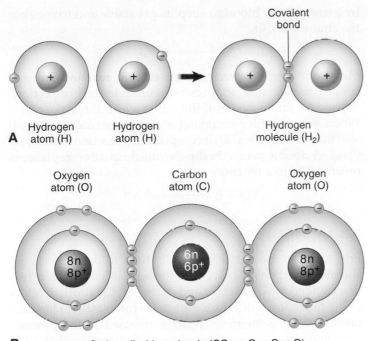

A Hydrogen atom (H)   Hydrogen atom (H)   Hydrogen molecule (H_2)

Covalent bond

Oxygen atom (O)   Carbon atom (C)   Oxygen atom (O)

8n 8p+   6n 6p+   8n 8p+

B   Carbon dioxide molecule (CO_2 or O=C=O)

▲ **FIGURE 2-7   Types of covalent bonds.** **A,** A single covalent bond formed by the sharing of one electron pair between two atoms of hydrogen results in a molecule of hydrogen gas. **B,** A double covalent bond (double bond) forms by the sharing of *two* pairs of electrons between two atoms. In this case, two double bonds form—one between carbon and *each* of the two oxygen atoms.

gen atoms, each oxygen atom forms a double bond with the carbon. This arrangement allows the atoms to fill their shells, creating a molecule of carbon dioxide (Figure 2-7, *B*).

Polar Molecules and Hydrogen Bonds

In some covalent bonds, the electrons shared between atoms are *not* shared equally. This results in an *uneven charge* around the molecule, with one side being more negative or positive than the other. A molecule in this arrangement is said to be **polar** because it has "poles" that are more positive or negative relative to each other (Figure 2-8).

Polar molecules are attracted to one another in a way similar to that in which ions are attracted to one another. Water, for example, is an electrically neutral molecule, but the oxygen atom has the shared electrons surrounding its nucleus more often than they surround either of the hydrogen atoms. As you can see from Figure 2-8, this results in a partially positive (hydrogen) side of water, and a partially negative (oxygen) side.

Water molecules will attract one another because of their polarity. These interactions, in which the negative side of one molecule is attracted to the positive side of another, create **hydrogen bonds** (Figure 2-9). However, hydrogen bonds *are not true bonds* because electrons are not being transferred or shared. Instead, they are very weak attractive interactions between two or more molecules of water. In fact, a hydrogen bond is only about $1/20$ as strong as a covalent bond.

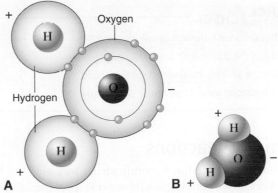

A   **B**

▲ **FIGURE 2-8   Water—example of a polar molecule.** The polar nature of water is represented in an energy-level model **(A)** and a space-filling model **(B).** The two hydrogen atoms are nearer to one end of the molecule and give that end a partial positive charge. The "oxygen end" of the molecule attracts the electrons more strongly and thus has a partial negative charge.

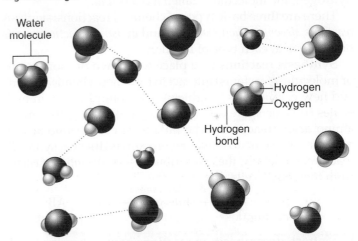

Water molecule

Hydrogen
Oxygen

Hydrogen bond

▲ **FIGURE 2-9   Hydrogen bonds between water molecules.** Hydrogen bonds serve to weakly attach the negative *(oxygen)* side of one water molecule to the positive *(hydrogen)* side of a nearby water molecule. This diagram depicts a few bonded molecules, as one would expect in liquid water. Ice would instead have more hydrogen bonds; steam would have none.

The temperature of water determines the number of hydrogen bonds created between water molecules. When water is very cold (0° C [32° F]), the molecules do not move around so much, and many hydrogen bonds can be made (see Figure 2-9). All of these bonds taken together make ice very strong. As a liquid, the molecules move around a little more, and fewer hydrogen bonds are made. In fact, hydrogen bonds of liquid water are constantly being made and broken. As steam, virtually no hydrogen bonds are formed, because the water molecules are too energetic and very far apart. As you will see, hydrogen bonds are vitally important in maintaining the three-dimensional biological structures of large macromolecules such as DNA and protein.

Together the communal strength of hydrogen bonds and similar weak attractions literally holds our bodies together! In fact, biochemists sometimes joke that hydrogen bonds are the Velcro of the biological universe.

Chemical Reactions

Now that we understand atomic structures and the formation of bonds, we can discuss the next step: *chemical reactions.* A **chemical reaction** takes place when two or more atoms bond together (either by transferring or sharing electrons) to create a **molecule.** Chemical reactions also take place when atoms of a molecule break apart. If the bonded atoms in a molecule are from different elements such as carbon and hydrogen, the molecule is called a *compound.*

There are three basic types of chemical reactions: *synthesis reactions, decomposition reactions,* and *exchange reactions.* Let's briefly look at each type of reaction.

Synthesis reactions take place when two or more atoms or molecules (reactants) interact to form new chemical bonds and new compounds (products). For example, protein molecules (products) are constructed by linking together many amino acids (reactants). The linking of each amino acid to another (and so on to make the protein) is due to a synthesis reaction. Generally, these reactions *require the input of energy from the body's cells.*

$$A + B \xrightarrow{\quad Energy \quad} AB$$
$$(reactants) \qquad\qquad (products)$$

In contrast, **decomposition reactions** occur when the bonds within a reactant molecule break and produce simpler molecules, atoms, or ions. *Energy is usually released* in these types of reactions; this is important when your body breaks down the foods you eat.

$$AB \longrightarrow A + B + Energy$$

Exchange reactions break down (or decompose) two compounds and, in exchange, synthesize two *new* compounds. As you will see in Chapter 20, exchange reactions happen frequently in the blood to keep its pH stable and to regulate the amount of CO_2.

$$AB + CD \longrightarrow AD + CB$$

Many of these reactions are **reversible reactions,** which simply means that they can proceed in *both* directions depending on the needs of the cell in which they take place. When you consider reversing a synthesis reaction, you'll see that it becomes a decomposition reaction, and vice versa. A double arrow in the chemical equation represents reversibility in a reaction.

$$A + B \longleftrightarrow AB$$

Metabolism

Physiologists use the term **metabolism** to describe *all* the chemical reactions that occur in the body. Metabolism is further divided into *catabolism* and *anabolism.* **Catabolism** refers to the breakdown of larger molecules into smaller chemical units. Energy (sometimes heat) is released in the process. *Hydrolysis* is a general type of catabolic reaction that adds water to break down large molecules. In contrast, **anabolism** refers to the construction of larger molecules from smaller molecules. This is typically accomplished by *dehydration synthesis,* the removal of water molecules during the reaction (Figure 2-10).

Catabolism reactions generally result in the production of *adenosine triphosphate (ATP),* a macromolecule that stores energy in cells. Anabolism requires the use of ATP to provide energy for the construction of larger molecules. Incredibly, nearly 6 *pounds* of ATP are produced by the cells of an average human every day—and every single molecule of ATP is used!

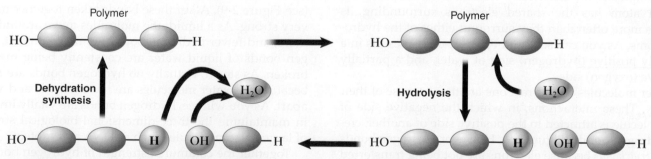

▲ **FIGURE 2-10** **Metabolic reactions.** *Hydrolysis* is a catabolic reaction that adds water to break down large molecules into smaller molecules, or subunits. *Dehydration synthesis* is an anabolic reaction that operates in the reverse fashion: small molecules are assembled into large molecules by removing water. Note that a specific example of dehydration synthesis is shown in Figure 2-15.

INORGANIC AND ORGANIC COMPOUNDS

In all living organisms, there are two kinds of compounds: **organic** and **inorganic.** Organic compounds are generally described as having carbon-carbon (C–C) or carbon-hydrogen (C–H) *covalent bonds,* or both types of bonds. Few inorganic compounds have carbon atoms in them, and none have C–C or C–H bonds. In addition, organic molecules are larger and more complex than inorganic molecules. The

properties of these molecules are often determined by **functional groups,** which are specialized arrangements of atoms attached to the carbon core of many organic molecules (Figure 2-11).

Inorganic Molecules

Water

Water has been called the molecular "cradle of life" because all living organisms require water to survive and reproduce. Not surprisingly, water is the most abundant substance in the body and the most important simply because it is involved in all cellular processes. In a very real sense, then, "water chemistry" is the chemistry of life.

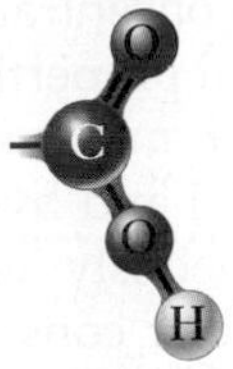

Because of its polarity and its ability to form hydrogen bonds (not only with itself but other organic molecules), water makes a very effective *solvent.* A **solvent** is something that can *dissolve* various substances *(solutes).* For example, the negative pole of water is attracted to the positive sodium ion in sodium chloride, while the positive pole of water is attracted to the negatively charged chloride ion. These ions can thus be separated and mixed in with the water, and no longer appear as crystals like the ones in your saltshaker (Figure 2-12). Many other biological compounds can *ionize* just as sodium chloride does in water. For this reason they can be transported within cells and throughout the body. By dissolving nutrients in blood, for example, water enables these materials to enter and leave the blood capillaries in the digestive organs and eventually be transported to every cell of the body.

Water also absorbs and releases heat very slowly (we say it has a high "specific heat"). This property allows the body to maintain a relatively constant temperature and resist sudden temperature changes.

Functional Group	Structural Formula	Models
Hydroxyl	–OH	
Carbonyl	$-\overset{\|}{\underset{\|}{C}}-$ (with =O)	
Carboxyl	$-C\overset{O}{\underset{OH}{}}$	
Methyl	$-\overset{H}{\underset{H}{C}}-H$	
Amino	$-N\overset{H}{\underset{H}{}}$	
Sulfhydryl	–SH	
Phosphate	$-O-\overset{OH}{\underset{\|}{P}}-OH$ (with =O)	
Acetyl	$-\overset{O}{\underset{}{C}}-\overset{H}{\underset{H}{C}}-H$	

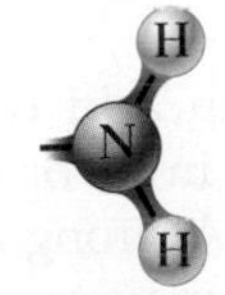

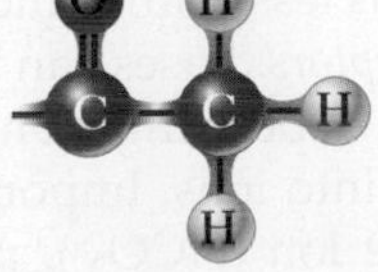

▲ FIGURE 2-11 The principal functional chemical groups. Each functional group confers specific chemical properties on the molecules that possess them.

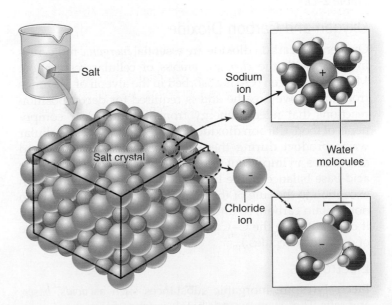

▲ FIGURE 2-12 Water as a solvent. The polar nature of water *(blue)* favors ionization of substances in solution. Sodium (Na^+) ions *(pink)* and chloride (Cl^-) ions *(green)* dissociate in the solution.

TABLE 2-1   Properties of Water

PROPERTY	DESCRIPTION	EXAMPLE OF BENEFIT TO THE BODY
Strong polarity	Polar water molecules attract other polar compounds, which causes them to dissociate	Many kinds of molecules can dissolve in cells, thereby permitting a variety of chemical reactions and allowing many substances to be transported
High specific heat	Hydrogen bonds absorb heat when they break and release heat when they form, thereby minimizing temperature changes	Body temperature stays relatively constant
High heat of vaporization	Many hydrogen bonds must be broken for water to evaporate	Evaporation of water in perspiration cools the body
Cohesion	Hydrogen bonds hold molecules of water together	Water works as lubricant or cushion to protect against damage from friction or trauma

Water not only plays a key role in virtually all cellular processes, including the transport of materials into and out of cells, but it is also involved in the formation of cellular secretions. These functions and many more are important to maintaining homeostasis, as you will see. As you progress from system to system in the chapters that follow, you will better understand the importance of water in almost every regulatory and control mechanism of the body. These and other cellular functions provided by water are essential to life and to the proper functioning of the human body (Table 2-1).

Oxygen and Carbon Dioxide

Oxygen and carbon dioxide are essential *inorganic* molecules involved with the *chemical process* of cellular respiration. Molecular oxygen (O_2) is absorbed in the alveoli of the lungs from the air we breathe and is required for decomposition reactions that release energy from the molecular components of food. Carbon dioxide (CO_2) is produced as a cellular waste product during these decomposition reactions and also serves an important role in maintaining the appropriate acid-base balance in the body. (Note: Carbon dioxide is an exception to the rule of thumb that inorganic substances do not contain carbon, so although carbon dioxide has carbon, it does not have the C–C or C–H bonds that are distinctive of organic molecules.)

Electrolytes

Electrolytes are inorganic substances such as *acids*, *bases*, and *salts*. These compounds break apart or *dissociate* when in solution to form charged ions. Ions with a positive charge

are called *cations* and those that are negatively charged are called *anions*. Figure 2-12 illustrates how water molecules dissociate the common electrolyte sodium chloride (NaCl) to create Na^+ and Cl^- anions. When an ionic compound such as NaCl dissolves in water, the positive and negative ions originally present in the crystal framework remain as ions in the solution. In this way they can carry positive or negative electrical charges from one place to another, and thus conduct an electrical current. As we will see throughout this book, this underlies many physiological processes such as nerve conduction.

Acids and Bases

Acids and bases are essentially chemical opposites, which dissociate in solution but release different complementary ions. **Acids** are substances that release a hydrogen ion (H^+) when in solution, and are frequently called *proton donors* (remember that hydrogen has only one proton in its nucleus, so a hydrogen atom that has given up its electron is really just a proton). The number of hydrogen ions a particular acid will release in solution indicates its level of "acidity." A higher concentration of hydrogen ions accounts for different chemical properties of that acid in solution.

Water molecules continually dissociate into hydrogen ions (H^+) and hydroxide ions (OH^-), which then can unite spontaneously again and again to re-form water. Because pure water consists only of water molecules (no dissolved particles), it has an equal number of these ions (that is, the concentration of each ion is the same). For this reason, pure water is neither an acid nor a base. Pure water (distilled water) is actually quite expensive to produce. Bottled water is not pure water (Box 2-1).

$$H_2O \longleftrightarrow H^+ + OH^-$$

When an acid dissociates in water, it shifts the H^+/OH^- balance in favor of excess H^+, thus increasing the level of acidity. A strong acid completely or nearly completely dissociates in water, thus releasing more H^+. In contrast, a *weak acid* dissociates very little and therefore produces only a few excess H^+ ions in solution. There are many important acids in the body and they perform many functions. Hydrochloric acid, for example, is the acid produced in the stomach that aids the digestive process (see Chapter 21).

Bases, or *alkaline* compounds, are electrolytes that, when dissociated in solution, shift the H^+/OH^- balance to produce *more hydroxide ions*. The dissociation of a common base, sodium hydroxide (NaOH), yields the cation Na^+ and the anion OH^-. Bases can accept hydrogen ions (making solutions less acidic) and are therefore often referred to as *proton acceptors*. Bases can be classified in the same way that acids are, depending on how readily and completely they dissociate into ions. Important bases in the body, such as bicarbonate ion (HCO_3^-), play vital roles in the transportation of respiratory gases and in the elimination of waste products of metabolism from our bodies.

BOX 2-1 FYI

Bottled or Tap?

As the demand for bottled water has increased over recent years, have you ever asked yourself, "Which source of water really is better for my body, bottled or tap?" It may be of interest to you that 40% to 60% of the bottled water sold worldwide is actually nothing more than packaged tap water that may have been reprocessed in some way. Overall, both sources of water in the United States are safe to drink. Some believe that bottled water is healthier than tap, but EPA regulations for tap water are stricter than FDA regulations for bottled water. Taste preference seems to be a major factor in which is chosen, and it may be this reason most of all that could impact health and homeostasis. If you don't like the way your water tastes, you won't drink as much of it, so taste and hydration are related, and this could affect your health. However, many of us can't tell the difference between bottled water and tap water in a taste test. In addition, fluoride, an important mineral that is widely accepted as being beneficial in reducing the incidence of cavities, is often missing from bottled water but generally added to tap water. So, the next time you consider whether to pay extra for bottled water or simply turn on the faucet, you may want to do a little research before you decide, "bottled or tap?"

The pH Scale

The *pH scale* is a logarithmic scale ranging from 0 to 14; **pH** is an abbreviation for a phrase meaning "the power of hydrogen." Briefly look at Figure 2-13. You can see that the pH scale measures *acidity* (how acid a solution is) or *alkalinity* (how basic a solution is). The pH measurement indicates the *relative concentration* of H^+ ions with respect to OH^- ions in a solution. As the concentration of H^+ ions increases, the pH goes down and the solution becomes more acidic. As the concentration of H^+ ions decreases, the pH goes up and the solution is said to be more basic.

A pH of 7 indicates neutrality (pure water, or equal amounts of H^+/OH^- ions), a pH greater than 7 indicates alkalinity, and a pH lower than 7 indicates acidity. Because the range of pH (0 through 14) is expressed on a logarithmic scale, a change of 1 pH unit represents a *10-fold difference* in the number of H^+ ions. For example, going from a solution of pH 6 to a solution of pH 5 means that the pH 5 solution is *10 times* as acidic as the one at pH 6. Likewise, going from pH 6 to pH 4 means a 100-fold increase in H^+ ions or acidity. You can get a good mental picture of the pH scale by remembering the pH of a few common substances with which you are familiar (such as oven cleaner, ammonia, milk, coffee, and lemon juice).

Buffers

The normal pH range of blood and other body fluids is extremely narrow. For example, the pH of blood in our veins (pH 7.36) is only slightly more acidic than the blood in our arteries (pH 7.41). The difference between venous blood and arterial blood is caused by carbon dioxide entering venous blood as a waste product of cellular metabolism. Carbon dioxide is carried as carbonic acid (H_2CO_3) and, like any acid, carbonic acid lowers the pH of blood in your veins.

But the amazing constancy of the pH homeostasis of our blood relies partly on the presence of substances called **buffers**—substances that minimize changes in concentrations of H^+ and OH^- in a solution. Buffers act as a "reservoir" for H^+ ions. They can donate *or* remove H^+ ions when necessary to maintain a relatively constant pH in our body fluids. We'll look more closely at some of the body's buffering systems in Chapter 23.

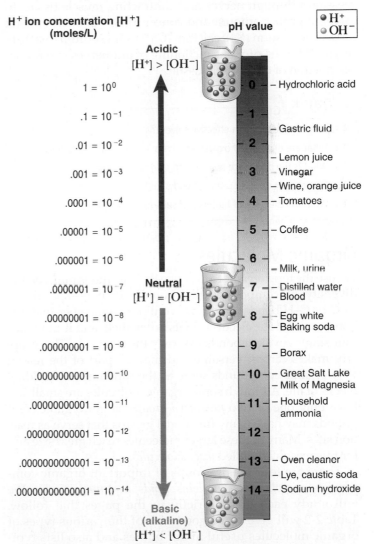

▲ **FIGURE 2-13** **The pH scale.** Note that as the concentration of H^+ increases, the solution becomes increasingly acidic and the pH value decreases. As the H^+ concentration decreases, the pH value increases, and the solution becomes more and more basic, or alkaline. (The scale on the left side of the diagram shows the actual concentrations of H^+ in moles per liter, or molar concentration, as an ordinary number, and expressed as an exponent [logarithm] of 10. You can see that the pH scale is simply the negative exponent of 10.)

Salts

Salts are compounds composed of ions with opposite charges that result from a chemical reaction between an acid and a base. Salts, like acids and bases, are electrolyte compounds and dissociate in solution to form positively and negatively charged ions. When mixed and allowed to react, the positive ion (cation) of a base and the negative ion (anion) of an acid will join to form a salt and an additional water molecule in an exchange reaction (in this case called a *neutralization reaction*). If the water in the solution is removed, the ions will crystallize and form a salt. In the case in which hydrochloric acid (HCl) is reacted with sodium hydroxide (NaOH), sodium chloride and water are formed.

Salt ions in solution are important for the transmission of messages through nerves and contracting muscle tissue. In fact, the proper amount and concentration of such mineral ions as potassium (K^+), calcium (Ca^{++}), and sodium (Na^+) are required for the proper functioning of our nerves and for the contraction of our muscle tissue.

QUICK ✓ CHECK

13. Why is water such an effective solvent?
14. What are the roles of oxygen and carbon dioxide in our bodies?
15. What is the difference between an acid and a base?
16. Give the pH of some common body fluids.
17. What are the roles of buffers and salts in our bodies?

Organic Molecules

We use the term *organic* to describe the enormous number and diversity of compounds that contain carbon—specifically C–C or C–H bonds. Before we continue, remember that carbon has only four electrons in its outer shell, and it can make four single covalent bonds to satisfy the octet rule. This property makes carbon versatile and may be part of the reason why organic compounds serve as the fundamental building blocks of life. Although some organic molecules are small and have only one or two *functional groups*, larger organic compounds may have many functional groups that serve as reaction sites. Many of these larger molecules reach great size and for this reason are called *macromolecules*.

There are four major groups of important organic compounds: *carbohydrates*, *proteins*, *lipids*, and *nucleic acids*. We will study each group briefly in the pages that follow. Table 2-2 will give you a good idea of the various types of organic molecules useful in our bodies and also lists typical functions.

Carbohydrates

All **carbohydrates** are composed of carbon, oxygen, and hydrogen, forming organic compounds in which the carbon atoms are attached to one another in chains or rings of varying lengths. Common carbohydrates such as sugars and starches serve as primary sources of energy for most living things. In addition, carbohydrates serve structural roles by forming the "backbone" of important molecules such as RNA and DNA. As you will soon see, these molecules are vital to cell reproduction and protein synthesis.

There are three important groups of carbohydrates: *monosaccharides*, *disaccharides*, and *polysaccharides*.

Monosaccharides, or simple sugars, generally have six carbon atoms in a chain (sometimes in a ring formation) and are called *hexoses* (*hexa* = six). Typically, sugars such as glucose are present as a "straight chain" while dry, but form a ring when in solution (Figure 2-14). The most common hexoses are *glucose, fructose,* and *galactose*, but there are 13 other *isomers* of hexose (a total of 16) with the same chemical formula. All hexose isomers have the same chemical formula, $C_6H_{12}O_6$, but the different forms vary in their three-dimensional structures.

There are also five-carbon sugars called *pentoses* (*penta* = five). These sugars may also form a ring and are very important parts of *nucleotides* (see below). *Ribose* and *deoxyribose* are pentose monosaccharides that make up the backbone structure of RNA and DNA, respectively. We will discuss these vital macromolecules more fully later under Nucleic Acids and Related Molecules.

When two monosaccharide sugars are bonded together (a synthesis reaction), a molecule of water is removed and the resulting carbohydrate is called a **disaccharide**—literally, "two sugars." Table sugar (sucrose), maltose, and lactose are all disaccharides. Each disaccharide has its own distinctive chemical and physical properties. Figure 2-15 shows the formation of sucrose from one molecule each of glucose and fructose. Notice how the hydrogen of the OH group of the first monosaccharide combines with the OH group of the second monosaccharide ($H^+ + OH^- = H_2O$) to form water.

Virtually all synthesis reactions in our bodies work in the same way, and of course we give the resulting compounds different names. For example, a chain of many monosaccharides bonded together in this way is called a **polysaccharide.** Such chains can be straight or have many branches. Some polysaccharides are made of many of the *same* monosaccharide subunits and are called *polymers*. *Glycogen*, an enormous macromolecule, is a common glucose polymer that our bodies use to conveniently store sugar for later use in generating energy.

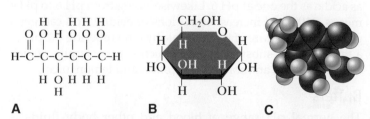

A   **B**   **C**

▲ **FIGURE 2-14** **Structure of glucose. A,** Straight chain, or linear model, of glucose. **B,** Ring model representing glucose in solution. **C,** Three-dimensional, or space-filling, model of glucose.

TABLE 2-2 Examples of Important Biomolecules

MACROMOLECULE	SUBUNIT	FUNCTION	EXAMPLE
Carbohydrates			
Glucose	Simple sugar (hexose)	Stores energy	Blood glucose
Ribose	Simple sugar (pentose)	Important in expression of hereditary information	Component of RNA
Glycogen	Glucose units	Stores energy	Liver glycogen
Lipids			
Triglycerides	Glycerol + 3 fatty acids	Store energy	Body fat
Phospholipids	Glycerol + phosphate + 2 fatty acids	Make up cell membranes	Plasma membrane of cell
Steroids	Steroid nucleus (4-carbon ring)	Make up cell membranes, hormone synthesis	Cholesterol, various steroid hormones, estrogen
Prostaglandins	20-carbon unsaturated fatty acid containing 5-carbon ring	Regulate hormone action; enhance immune system; affect inflammatory response	Prostaglandin E, prostaglandin A
Proteins			
Functional proteins	Amino acids	Regulate chemical reactions	Hemoglobin, antibodies, enzymes
Structural proteins	Amino acids	Component of body support tissues	Muscle filaments, tendons, ligaments
Nucleic Acids			
DNA	Nucleotides (sugar, phosphate, base)	Encodes hereditary information	Chromatin, chromosomes
RNA	Nucleotides (sugar, phosphate, base)	Helps decode hereditary information; acts as "RNA enzyme"; silencing of gene expression	Transfer RNA (tRNA), messenger RNA (mRNA), double-stranded RNA (dsRNA)
Nucleotides and Related Molecules			
Adenosine triphosphate (ATP)	Phosphorylated nucleotide (adenine + ribose + 3 phosphates)	Transfers energy from fuel molecules to working molecules	ATP present in every cell of the body
Creatine phosphate (CP)	Amino acid + phosphate	Transfers energy from fuel to ATP	CP present in muscle fiber as "backup" to ATP
Combined or Altered Forms			
Glycoproteins	Large proteins with small carbohydrate groups attached	Similar to functional proteins	Some hormones, antibodies, enzymes, cell membrane components
Lipoproteins	Protein complex containing lipid groups	Transport lipids in the blood	LDLs (low-density lipoproteins); HDLs (high-density lipoproteins)

Lipids

Lipids are organic biological molecules composed largely of carbon, hydrogen, and oxygen. However, the amount of oxygen in lipids is much lower than that in carbohydrates, and some lipids also contain nitrogen or phosphorus. Lipids are *hydrophobic* (literally, "water fearing") and therefore are not soluble in water. This is because lipids are nonpolar molecules, and so have no charged regions with which to cling to the polar water molecules. They generally have a greasy or oily feel and are soluble in some organic solvents such as

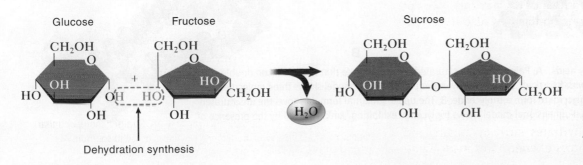

◄ FIGURE 2-15 Formation of sucrose. Glucose and fructose are joined in a synthesis reaction that involves the removal of water.

ether or alcohol. Lipids are further subdivided into four main groups: triglycerides, phospholipids, steroids, and prostaglandins. They serve as sources of energy, provide insulation, form hormones, and provide the primary structural elements of cell membranes. Examples of lipids and their functions are outlined for you in Table 2-3.

Triglycerides or **fats** are the most abundant lipids and serve primarily as energy sources. They are composed of two fundamental units: a *glycerol* molecule bonded at three sites to three *fatty acids*. These fatty acids may be the same size or different sizes. The specific types of fatty acids bonded to the glycerol molecule determine the chemical nature of any triglyceride. Fatty acids can have variable numbers of carbon in their chains as well as a variable number of hydrogen atoms attached to, or *saturating*, the available bonds holding the carbons together.

A *saturated* fatty acid is one that has all of its available carbon bonds taken up by hydrogen atoms. For this reason, we say all bonds are saturated in this type of fatty acid. An *unsaturated* fatty acid is one that has one or more *double bonds* in the chain and therefore does not have a carbon chain that is saturated with hydrogen atoms. Polyunsaturated fats are those that have more than several double bonds. For example, look at Figure 2-16, *A* and *B*. Notice how palmitic acid is fully saturated (no double bonds in the carbon chain) but linolenic acid has three double bonds (in red), where two more hydrogens could be placed at the site of each double bond.

Generally, animal fats are saturated and are solid at room temperature, whereas vegetable oils are unsaturated and usually liquid at room temperature. This is because the double bonds cause "kinks" in the carbon chain and thus prevent the molecules from fitting together as snugly. As a result, these lipids can move around more, thereby exhibiting the properties of a liquid. Figure 2-17 illustrates the for-

mation of a triglyceride. Again, notice how there are *three* reaction sites where water is removed by the combination of the H from the glycerol molecule on the left with the OH from the fatty acid on the right.

Phospholipids are fat compounds similar to triglycerides. However, one of the fatty acids in a triglyceride is *replaced by* a different chemical compound containing phosphorus (and sometimes nitrogen). This part of the phospholipid chemical structure, called a *phosphate group*, faces the opposite direction from the fatty acid chains, and forms a *head*. The phosphate head of the molecule is polar and therefore water

TABLE 2-3 Major Functions of Human Lipid Compounds

FUNCTION	EXAMPLE
Energy	Lipids can be stored and broken down later for energy; they yield more energy per unit of weight than carbohydrates or proteins do
Structure	Phospholipids and cholesterol are required components of cell membranes
Vitamins	Fat-soluble vitamins: vitamin A forms retinol (necessary for night vision); vitamin D increases calcium uptake; vitamin E promotes wound healing; and vitamin K is required for the synthesis of blood-clotting proteins
Protection	Fat surrounds and protects organs
Insulation	Fat under the skin minimizes heat loss; fatty tissue (myelin) covers nerve cells and electrically insulates them
Regulation	Steroid hormones regulate many physiological processes; for example, estrogen and testosterone are responsible for many of the differences between females and males; prostaglandins help regulate inflammation and tissue repair

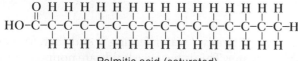

Palmitic acid (saturated)

A

B

▲ **FIGURE 2-16** **Types of fatty acids. A,** Palmitic acid, a saturated fatty acid. Note that it contains no double bonds; its hydrocarbon chain is filled with hydrogen atoms. The lower three-dimensional model shows three molecules of palmitic acid joined to a molecule of glycerol to form a triglyceride. **B,** The upper structural formula shows the unsaturated fatty acid linolenic acid. The lower three-dimensional model shows triglyceride exhibiting "kinks" caused by the presence of double bonds in the fatty acids.

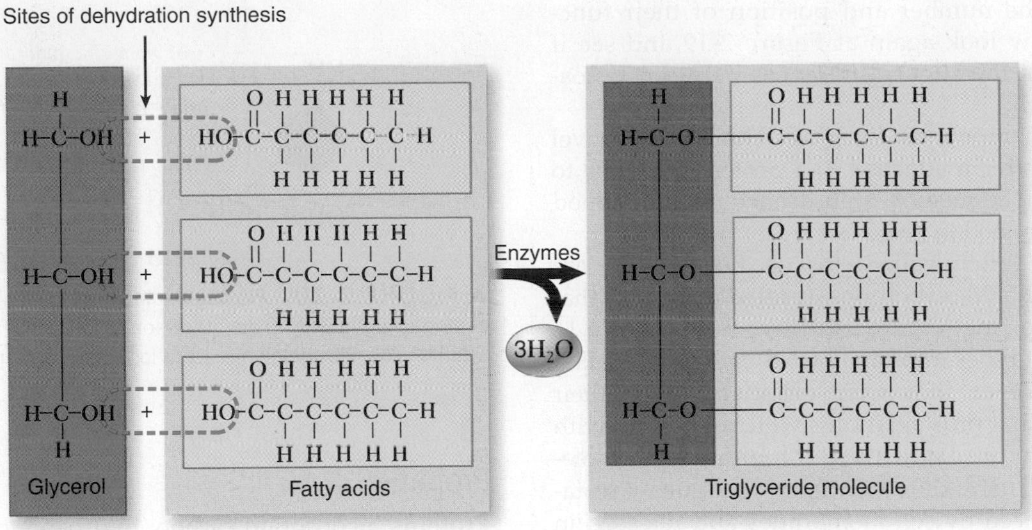

Sites of dehydration synthesis

Glycerol   +   Fatty acids   → Enzymes / 3H₂O →   Triglyceride molecule

▲ **FIGURE 2-17**   **Formation of triglyceride.** Glycerol tricaproate is a composite molecule made up of three molecules of caproic acid (a six-carbon fatty acid). The fatty acids are coupled in a dehydration synthesis reaction to a single glycerol backbone. In addition to the triglyceride, this process results in the formation of three molecules of water.

soluble (it is *hydrophilic* or "water loving"). The remainder of the molecule is nonpolar and therefore insoluble in water, but soluble in fats. This end is called *hydrophobic* or "water fearing." In water, phospholipids form a double layer or *bilayer.* In this state the hydrophilic heads face the water and the hydrophobic tails face each other, forming a double layer of phospholipids—a *membrane.* In fact, this is one of the dominant structures in all cell membranes and forms an effective barrier for a cell (Figure 2-18).

Steroids contain a *steroid nucleus,* which is composed of four attached rings of carbon. Cholesterol is an important steroid that helps stabilize the structure of cell membranes. The body can modify cholesterol to form important steroid hormones, three of which are shown in Figure 2-19.

Steroidal properties are determined by the presence of functional groups, which are attached to the four-ring steroid nucleus. The four steroids shown in Figure 2-19 all share the same steroid nucleus and vary, in some cases

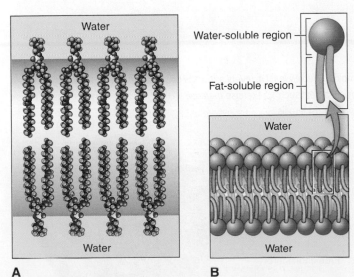

A

B

▲ **FIGURE 2-18**   **Phospholipid bilayer. A,** Orientation of phospholipid molecules when surrounded by water and forming a bilayer. **B,** Cartoon commonly used to depict a phospholipid bilayer.

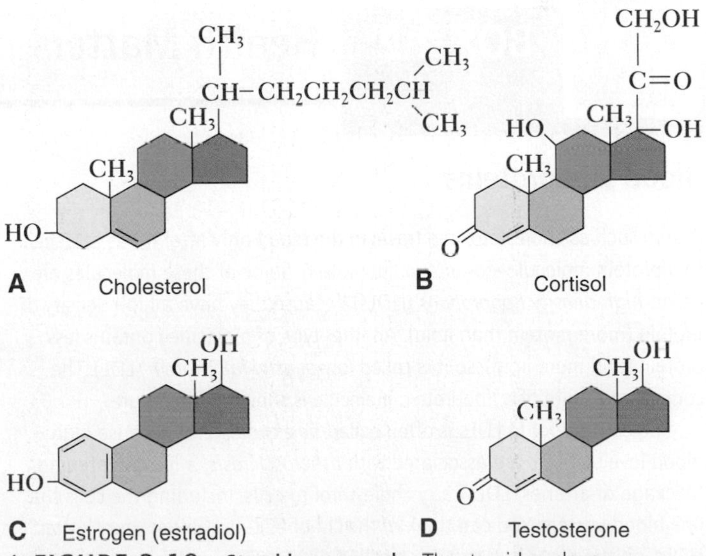

A   Cholesterol

B   Cortisol

C   Estrogen (estradiol)

D   Testosterone

▲ **FIGURE 2-19**   **Steroid compounds.** The steroid nucleus— highlighted in *yellow*—found in cholesterol **A,** forms the basis for many other important compounds such as cortisol **B,** estradiol (an estrogen) **C,** and testosterone **D.**

only slightly, in the number and position of their functional groups. Now look again at Figure 2-19 and see if you can determine the slight differences between testosterone and estrogen.

Some steroid compounds such as cholesterol travel through the bloodstream attached to a protein molecule to form a *lipoprotein.* The medical significance of these blood lipoproteins is examined in Box 2-2.

Prostaglandins (often called "tissue hormones") are lipids composed of a 20-carbon unsaturated fatty acid that contains a 5-carbon ring (Figure 2-20). We now classify 16 prostaglandin (PG) types into nine broad categories, called PGA to PGI. However, they were originally given their name because the first prostaglandins were associated with prostate tissue. Now we know these important lipid molecules are found in many different types of tissues. Prostaglandins are produced by cell membranes and released in response to particular stimuli. They cause a wide variety of effects near their release, and are then broken down. They regulate the effects of hormones, influence blood pressure, control the secretion of digestive juices, and even function in the immune system and blood clotting.

QUICK CHECK

18. How do carbohydrates function in the body?
19. What are the various types of lipids?
20. Briefly discuss the various functions of lipids.
21. How do carbohydrates differ from lipids?

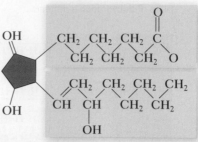

▲ **FIGURE 2-20** **Prostaglandin.** Prostaglandins such as this example of prostaglandin E (PGE) are 20-carbon unsaturated fatty acids with a 5-carbon ring. Prostaglandins act as local regulators in the body.

Proteins

Proteins incorporate carbon, hydrogen, oxygen, and nitrogen into their molecular structure, and some specialized proteins also contain sulfur, iron, and phosphorus. These huge and diverse macromolecules are *polymeric chains* whose subunits are called **amino acids.** We now have several ways of "visualizing" what proteins look like, and we present these methods for you in Box 2-3.

Proteins can be both *structural* and *functional.* **Structural proteins** form the structures of cells, tissues, and organs. They form different shapes and compositions, such as flexible rods, elastic strands, and waterproof layers, to create many different building blocks of the body.

We now understand that it is the *shape* of a protein that determines how it performs. Even at the molecular level, it

BOX 2-2 Health Matters

Blood Lipoproteins

A lipid such as cholesterol can travel in the blood only after it has attached to a protein molecule—forming a lipoprotein. Some of these molecules are called *high-density lipoproteins* (HDLs) because they have a high density of protein (more protein than lipid). Another type of molecule contains less protein (and more lipid), so it is called *low-density lipoprotein* (LDL). The composite nature of a lipoprotein molecule is shown in the figure.

The cholesterol in LDLs is often called *bad* cholesterol because high blood levels of LDL are associated with *atherosclerosis,* a life-threatening blockage of arteries. LDLs carry cholesterol to cells, including the cells that line blood vessels. You can think of the "L" of LDL as standing for "lethal." HDLs, on the other hand, carry so-called good cholesterol *away from cells* and toward the liver for elimination from the body. A high proportion of HDL in the blood is associated with a low risk for atherosclerosis. You can think of the "H" of HDL as standing for "healthful." Factors such as cigarette

smoking *decrease* HDL levels and thus contribute to the risk for atherosclerosis. Factors such as exercise *increase* HDL levels and thus decrease the risk for atherosclerosis.

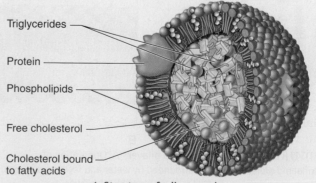

▲ Structure of a lipoprotein.

Visualizing Proteins

Only a few decades ago, the usual way for a biochemist to demonstrate the three-dimensional structure of a protein molecule was by building wooden ball-and-stick models. These models were less than ideal because they took a long time to build, often fell apart, and were not easy to handle. But now there are many sophisticated, easy-to-use, computer programs that can "build" protein molecules on the monitor screen. These "virtual protein molecules" can be rotated and viewed from nearly any angle. They can also be changed when trying to design a new protein molecule for therapeutic or other purposes.

Computer modeling has become the usual way to represent the complex shapes of protein molecules in the twenty-first century. Here, three common types of protein models are shown. The *ribbon model* shows the areas where alpha helices and folded sheets form within the molecule. The *space-filling model* shows each atom as a "cloud" filling up the space occupied by that atom. The *surface-rendering model* shows the three-dimensional boundaries of the whole protein molecule, often also color coding for charged regions on the surface of the protein.

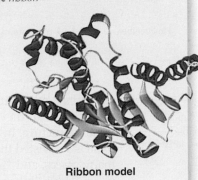

Ribbon model

- ■ Helix
- ■ Folded sheet
- □ Unorganized area

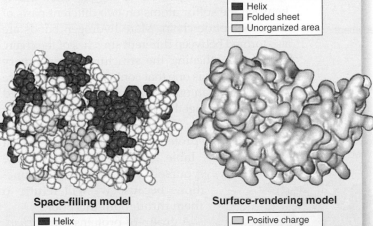

Space-filling model

- ■ Helix
- ▨ Folded sheet
- □ Unorganized area

Surface-rendering model

- □ Positive charge
- ■ Negative charge
- □ Unchanged area

▲ Three ways to visualize the same folded protein molecule.

is important for you to see that form and function go hand-in-hand—the right shape for the right job! For example, some **functional proteins** act as *accelerators* or *helpers* in chemical reactions that are necessary for cellular metabolism and other body functions. These proteins, called **enzymes**, have a unique shape that allows them to fit together with certain other molecules. Enzymes speed up

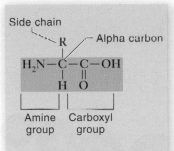

Side chain
R Alpha carbon
$H_2N - C - C - OH$
H O
Amine group   Carboxyl group

▲ **FIGURE 2-21** **Basic structural formula for an amino acid.** Note relationship of the side chain *(R)*, amine group, and carboxyl group to the alpha carbon. The amine group (NH_2) is depicted in the figure as H_2N to show that the *nitrogen* atom of the group bonds to the alpha carbon.

chemical reactions without being consumed by the reaction. They function by converting reactants (called *substrates*) into specific products. Enzymes allow chemical reaction to occur at a much lower (body) temperature than would otherwise be required. Other proteins, such as insulin, can also act as hormones and trigger chemical and activity changes in organs and cells, which affect the body in a broad sense.

Amino Acids

There are 20 commonly occurring amino acids, and many proteins have at least one of each type. Eight of these 20 are known as *essential* amino acids, because the human body cannot manufacture them on its own and so they must be obtained from food. The basic structure of all amino acids (shown in Figure 2-21) consists of the following: a central carbon, an amine group (NH_2), a carboxyl group (COOH), a hydrogen atom, and an "R" group. The R group can be thought of as the "remainder" of the molecule. The R group is unique for each of the 20 amino acids.

You can think of amino acids as the "alphabet" of proteins. As words are made up of chains of different letters that can be reused in any sequence, so proteins are made up of amino acids linked together to form long chains. Amino acids can be combined, like letters, to form an almost infinite variety of polypeptides and proteins. A **peptide bond** is necessary to link one amino acid to another. This covalent bond joins the carboxyl group of one amino acid to the amine group of another.

Look again at the general steps for *dehydration* and *hydrolysis* reactions in Figure 2-10. In this case, dehydration synthesis makes polypeptides (chains of many amino acids linked together, which are called *proteins* if they are very long). Hydrolysis, the reverse reaction, breaks peptides down into their constituent amino acids.

Levels of Protein Structure

The most complex globular proteins, such as blood hemoglobin, exhibit four levels of organization: *primary (first level)*, *secondary (second level)*, *tertiary (third level)*, and

Primary (first level)
Protein structure is a sequence of amino acids in a chain.

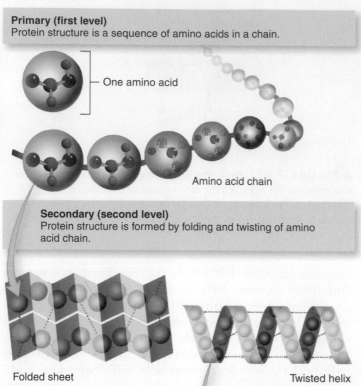

One amino acid

Amino acid chain

Secondary (second level)
Protein structure is formed by folding and twisting of amino acid chain.

Folded sheet

Twisted helix

▲ **FIGURE 2-22** **Structural levels of protein.** *Primary structure:* determined by the number, kind, and sequence of amino acids in the chain. *Secondary structure:* hydrogen bonds stabilize folds or helical spirals. *Tertiary structure:* globular shape maintained by strong (covalent) intramolecular bonding and by stabilizing hydrogen bonds. *Quaternary structure:* results from bonding between more than one polypeptide unit.

Tertiary (third level)
Protein structure is formed when the twists and folds of the secondary structure fold again to form a larger three-dimensional structure.

Folded sheet

Twisted helix

Quaternary (fourth level)
Protein structure is a protein consisting of more than one folded amino acid chain.

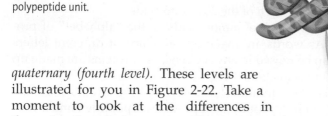

quaternary *(fourth level).* These levels are illustrated for you in Figure 2-22. Take a moment to look at the differences in these structures before you continue.

The **primary structure** of a protein refers to the number, type, and sequence of amino acids in the polypeptide chain. Parathyroid hormone (PTH), for example, found in the human parathyroid gland, is made of only 84 amino acids and exists in this primary chain structure.

However, most polypeptides do not exist in a straight chain. They have a **secondary structure** that is coiled or bent into pleated sheets. Coils usually turn clockwise and are called *alpha helices* (singular *helix*). The amino acids in these coils form hydrogen bonds with nearby amino acids, which are very important in stabilizing the structure.

TABLE 2-4 Major Functions of Human Protein Compounds

FUNCTION	EXAMPLE
Provide structure	Structural proteins include keratin of skin, hair, and nails; parts of cell membranes; tendons
Catalyze chemical reactions	Lactase (enzyme in intestinal digestive juice) catalyzes chemical reaction that changes lactose to glucose and galactose
Transport substances in blood	Proteins classified as albumins combine with fatty acids to transport them in form of lipoproteins
Communicate information to cells	Insulin, a protein hormone, serves as chemical message from islet cells of the pancreas to cells all over the body
Act as receptors	Binding sites of certain proteins on surfaces of cell membranes serve as receptors for insulin and various other hormones
Defend body against many harmful agents	Proteins called *antibodies* or *immunoglobulins* combine with various harmful agents to render those agents harmless
Provide energy	Proteins can be metabolized for energy

Secondary structures of proteins can in turn be twisted so that they are *globular proteins* with a **tertiary structure.** In this structure, the polypeptide chain is so twisted that its coils touch one another in many places, and "spot welds" or *disulfide linkages* occur. These linkages are strong, covalent bonds between the sulfur atoms on two different parts of the polypeptide chain. More hydrogen bonds are also formed between different stretches of the chain, further strengthening the structure. A **quaternary structure** protein is one that contains clusters of more than one polypeptide chain, all linked together into one massive molecule.

The major functions of human protein compounds are shown in Table 2-4. You should familiarize yourself with these basic functions because we will return to them throughout this book.

A group of proteins called *chaperones* (chaperonins), present in every cell, aid in directing each protein into its final folded shape or *conformation.* (Recall that both structural and functional proteins require precise

A&P CONNECT

To learn more about disulfide linkages and why they are important to your hair, check out **Disulfide Linkages** online at **A&P Connect.**

ⓔvolve

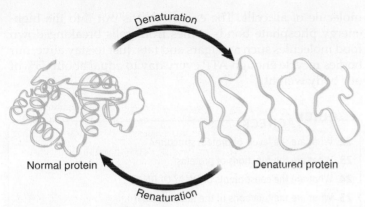

▲ FIGURE 2-23 Denatured protein. When a protein loses its normal folded organization and thus loses its functional shape, it is called a *denatured protein*. Denatured proteins are not able to function normally. However, if the protein shape is restored, the *renatured protein* may resume its normal function.

conformation to perform their functions correctly.) A protein that is incorrectly folded may not function properly, and a number of serious diseases cause this to happen on a regular basis. Additionally, if proteins are exposed to too much heat, radiation, or a radically different pH than normal, they may change conformation (Figure 2-23). Such damaged or *denatured* proteins are not able to perform their tasks. In some cases chaperone proteins may correct or reshape a denatured protein, allowing it to perform its functions normally again.

Nucleic Acids and Related Molecules

The nucleic acids literally run our lives, and in fact, the lives of all living organisms. The two fundamental nucleic acids are **deoxyribonucleic acid (DNA)** and **ribonucleic acid (RNA).** DNA and RNA are composed of subunits called **nucleotides**—deoxyribonucleotides for DNA and ribonucleotides for RNA. You can see a general comparison of DNA and RNA in Table 2-5. Each nucleotide consists of a five-carbon sugar (deoxyribose or ribose), a phosphate group, and a *nitrogenous base*. DNA is composed of the bases *thymine, guanine, adenine,* and *cytosine*, whereas RNA is composed of guanine, adenine, and cytosine, but has thymine

replaced by a closely related nitrogenous base, *uracil*. DNA encodes information that is required for all of the cells in the body to generate all proteins (whether these proteins be enzymes, hormones, or other types) that are necessary for everyday function (Figure 2-24).

Each helical chain in a DNA molecule has its phosphate-sugar backbone toward the outside and its bases pointing inward toward the bases of the other chain. Each base in one chain is joined to a base in the other chain by means of hydrogen bonds to form what is known as a *base pair*. One important principle to remember is that only two kinds of base pairs comprise DNA: adenine-thymine (A–T) and guanine-cytosine (G–C). Although a DNA molecule contains only these two kinds of base pairs, it contains millions of them—more than 100 million pairs are estimated to be in one human DNA molecule! It would be a good idea to refer to the information in Figure 2-24 as we continue with our discussion.

Notice that nucleotides are linked together by their phosphate and sugar groups to form a *polynucleotide chain*. RNA usually exists as a single chain, with some intramolecular bonding between its subunits. However, DNA typically exists as a double chain, which spirals into a springlike form called a *double helix*. The double helix of DNA is held together by thousands of hydrogen bonds between the nitrogenous bases, which face one another. During protein synthesis,

TABLE 2-5 Comparison of DNA and RNA Structure

	DNA	RNA
Polynucleotide strands	Double; very long	Single or double; short
Sugar	Deoxyribose	Ribose
Base pairing	Adenine-Thymine (A–T)	Adenine-Uracil (A–U)
	Guanine-Cytosine (G–C)	Guanine-Cytosine (G–C)

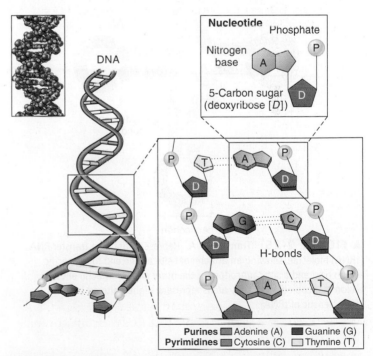

▲ FIGURE 2-24 The DNA molecule. Representation of the DNA double helix showing the general structure of a nucleotide and the two kinds of "base pairs": adenine (A) *(blue)* with thymine (T) *(yellow)*, and guanine (G) *(purple)* with cytosine (C) *(red)*. Note that the G–C base pair has three hydrogen bonds and an A–T base pair has two. Hydrogen bonds are extremely important in maintaining the structure of this molecule.

consecutive triplet bases of DNA are transcribed into complementary triplet bases of RNA *(codons)*. These codons in turn are translated triplet by triplet into proteins by the ribosomes. The structure of transfer RNA (tRNA) is shown for you in Figure 2-25, and its function is discussed briefly on page 82.

The details of how information is stored and retrieved by the cells is discussed fully in Chapter 5.

Adenosine triphosphate (ATP) is a very important nucleotide composed of the nitrogenous base adenine, a ribose sugar, and a string of three phosphate groups (Figure 2-26). ATP is really an adenine ribonucleotide with three high-energy phosphate groups attached to each other via **high-energy bonds.** When these high-energy bonds are broken, the energy released can be used to construct new compounds or to cause muscles to contract, for example. In fact, ATP is the fundamental immediate energy-storage

molecule of all cells. The energy that is put into the high-energy phosphate bonds comes from cells breaking down food molecules such as sugars and fats. Just to stay alive, our bodies recycle enough ATP every day to equal about 75% of our body weight.

QUICK ✓ CHECK

22. What are the levels of protein structure?
23. What are the functions of proteins?
24. What are the constituents of DNA? Of RNA?
25. What are the functions of the nucleic acids?

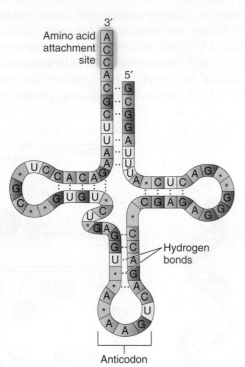

▲ **FIGURE 2-25**   **Transfer RNA.** Representation of a transfer RNA (tRNA) molecule, showing an attachment site at one end for a specific amino acid and a site (anticodon) at the other end for attachment to a codon of a copied gene. *Gray* areas represent slightly altered bases (a characteristic of tRNA).

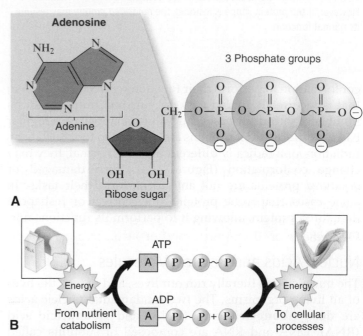

▲ **FIGURE 2-26**   **Adenosine triphosphate (ATP). A,** Structure of ATP. A single adenosine group *(A)* has three attached phosphate groups *(P)*. High-energy bonds between the phosphate groups can release chemical energy to do cellular work. **B,** General scheme of the ATP energy cycle. ATP stores energy in its last high-energy phosphate bond. When that bond is later broken, energy is transferred as important intermediate compounds are formed. The adenosine diphosphate (ADP) and phosphate groups that result can be resynthesized into ATP, thereby capturing additional energy from nutrient catabolism. Note that energy is transferred from nutrient catabolism to ADP, thus converting it to ATP. Energy is then transferred *from* ATP to provide the energy required for anabolic reactions or cellular processes as it reverts back to ADP.

The BIG Picture

We have seen the importance of molecular structure and function in this chapter. This is a critical first step, as you will see, to your understanding of the foundations of cells, tissues, and organs and their functions. How the basic chemical building blocks of the body are organized and how they relate to one another are key determinants in understanding normal structure and function. They are also important to understanding pathological anatomy and disease. As you learn about the structure and functions of the various organ systems of the body, the information contained in this chapter will take on new meaning and practical significance. It will help you fully understand and answer many questions that require you to integrate otherwise isolated factual information to make anatomy and physiology emerge as living and dynamic topics of personal interest.

our biochemical pathways. In some studies, aging has been shown to be characterized by the accumulation of potentially harmful altered proteins. The hypothesis is that these harmful proteins could lead to the weakening of cellular functions. This in turn leads to a greater probability of cell death as we grow older. As we age, the idea goes, we build up oxidized proteins that become "reactive" species that can damage and ultimately kill our cells.

The study of molecular aging is in its infancy, but you are likely to hear much more about these processes in the near future, especially as "baby boomers" draw more attention to human aging processes. ●

Cycle of LIFE ⚪

We begin life so fresh and vibrant that new parents often comment on their baby's newborn smell.

The aging process, unfortunately, begins as soon as we are born. Scientists are now studying the biomolecules known to be involved in the aging process. One avenue of pursuit involves *calorie restriction* (CR), which appears to prolong life in research animals by reducing the effects of free radicals on the proteins and other macromolecules in

MECHANISMS OF DISEASE

Chemical imbalances in the body can result in serious disease. One common example of an imbalance involves increasing carbon dioxide concentration in the blood, resulting in a condition called *hypercapnia*. Disease may also result when the body does not manufacture certain chemicals needed for body function, such as collagen in the case of *osteogenesis imperfecta*.

evolve   Find out more about diseases related to chemistry online at *Mechanisms of Disease: The Chemistry of Life.*

LANGUAGE OF SCIENCE *continued from page 21*

energy level
[*en-* in, *-erg-* work, *-y* state]

enzyme (EN-zyme)
[*en-* in, *-zyme* ferment]

exchange reaction
[*ex-* from, *-change* change, *re-* again, *-action* action]

fat

functional group (FUNK-shun-al groop)
[*function-* perform, *-al* relating to]

functional protein (FUNK-shun-al PRO-teen)
[*function-* perform, *-al* relating to, *prote-* primary, *-in* substance]

high-energy bond
[*en-* in, *-erg-* work, *-y* state, *bond* band]

hydrogen bond (HYE-droh-jen bond)
[*hydro-* water, *-gen* produce, *bond* band]

inorganic (in-or-GAN-ik)
[*in-* not, *-organic* natural]

ion (EYE-on)
[*ion* go]

ionic bond (eye-ON-ik bond)
[*ion* go, *bond* band]

isotope (EYE-so-tope)
[*iso-* equal, *-tope* place]

lipid (LIP-id)
[*lipi-* fat, *-id* form]

matter
[*matter-* something from which something is made]

metabolism (me-TAB-o-liz-em)
[*metabol-* change, *-ism* condition]

molecule (MOL-eh-kyool)
[*mole-* mass, *-cule* small]

monosaccharide (mon-oh-SAK-ah-ryde)
[*mono-* one, *-saccharide* sugar]

neutron (NOO-tron)
[*neuter-* neither]

nucleotide (NOO-klee-oh-tyde)
[*nucleo-* nut or kernel, *-ide* chemical]

nucleus (NOO-klee-us)
[*nucleus* kernel] *pl.*, nuclei (NOO-klee-eye)

octet rule (ok-TET rool)
[*octet* group of eight]

organic (or-GAN-ik)
[*organ-* tool or instrument, *-ic* relating to]

peptide bond (PEP-tyde bond)
[*pept-* digest, *-ide* chemical, *bond* band]

pH (pee AYCH)
[abbreviation for *potenz* power, *hydrogen* hydrogen]

phospholipid (fos-fo-LIP-id)
[*phospho-* phosphorus, *-lip-* fat, *-id* form]

polar (PO-lar)
[*pol-* pole, *-ar* relating to]

polysaccharide (pahl-ee-SAK-ah-ryde)
[*poly-* many, *-saccharide* sugar]

continued

LANGUAGE OF SCIENCE *continued from page 39*

primary structure (PRYE-mayr-ee STRUK-cher)
[*prim-* first, *-ary* relating to]

prostaglandin (pross-tah-GLAN-din)
[*pro-* before, *-sta-* stand, *-gland-* acorn,
-in substance]

protein (PRO-teen)
[*prote-* primary, *-in* substance]

proton (PRO-ton)
[*protos-* first, *-on* elementary atomic particle]

quaternary structure (KWAH-ter-nair-ee STRUK-cher)
[*quarti-* fourth, *-ary* relating to]

reversible reaction (ree-VER-si-bl ree-AK-shun)
[*re-* again, *-vers-* turn, *-ible* able to, *re-* again,
-action action]

ribonucleic acid (RNA) (rye-boh-noo-KLAY-ik ASS-id)
[*ribo-* ribose (sugar), *-nucle-* nucleus, *-ic* pertaining to,
acid sour]

secondary structure (SEK-on-dair-ee STRUK-cher)
[*second-* second, *-ary* relating to]

solvent (SOL-vent)
[*solver-* dissolve]

steroid (STAYR-oid)
[*ster-* sterol, *-oid* like]

structural protein (STRUK-cher-al PRO-teen)
[*structura-* arrangement, *-al* relating to,
prote- primary, *-in* substance]

synthesis reaction (SIN-the-sis ree-AK-shun)
[*synthes-* put together, *-is* process, *re-* again,
-action action]

tertiary structure (TERSH-ee-air-ee STRUK-cher)
[*terti-* third, *-ary* relating to]

triglyceride (try-GLISS-er-yde)
[*tri-* three, *-glycer-* sweet (glycerine), *-ide* chemical]

valence electron (VAY-lens eh-LEK-tron)
[*electro-* electric, *-on* subatomic particle]

CHAPTER SUMMARY

To download an MP3 version of the chapter summary for use with your iPod or other portable media player, access the Audio Chapter Summaries *online at http://evolve.elsevier.com.*

HINT   *Scan this summary after reading the chapter to help you reinforce the key concepts. Later, use the summary as a quick review before your class or before a test.*

INTRODUCTION

A. Chemical processes are the foundations of all biological processes.

BASIC CHEMISTRY

A. Elements and compounds (Figure 2-1)
   1. Matter—anything that takes up space and has mass
   2. Elements—substances that cannot be broken down or separated further into two or more different substances
   3. Compound—two or more atoms of the same or different elements chemically joined together
B. Atoms and their characteristics
   1. Atom—smallest complete unit of all elements
   2. Subatomic particles (Figure 2-3)
      a. Protons—positively charged
      b. Electrons—negatively charged
      c. Neutrons—neutrally charged
   3. Atomic number—number of protons in the nucleus
   4. Atomic weight—sum of all the protons and the average number of neutrons in the nucleus
   5. Isotope—an element containing the same number of protons, but different numbers of neutrons (Figure 2-4)
C. Chemical interactions
   1. Energy levels (shells)—electrons in an atom's outermost shell are those that participate in chemical interactions

   2. Octet rule—first shell is considered "full" when it contains two electrons; second and third shells around an atom can contain up to eight electrons each
D. Chemical bonds
   1. Ionic bond—atoms interact by transferring electrons (Figure 2-6)
   2. Covalent bond—two atoms sharing one or more pairs of electrons (Figure 2-7, *A*)
E. Polar molecules and hydrogen bonds
   1. Polar molecule—electrons shared between atoms are not shared equally; results in an uneven charge around the molecule, with one side being more negative or positive than the other (Figure 2-8)
   2. Hydrogen bond—created when the negative side of one molecule is attracted to the positive side of another (Figure 2-9)
F. Chemical reactions—take place when two or more atoms bond together to create a molecule or when atoms of a molecule break apart
   1. Synthesis reaction—occurs when two or more atoms or molecules interact to form new chemical bonds and new compounds
   2. Decomposition reaction—occurs when the bonds within a reactant molecule break and produce simpler molecules, atoms, or ions
   3. Exchange reaction—breaks down two compounds and, in exchange, synthesizes two *new* compounds
G. Metabolism—all the chemical reactions that occur in the body
   1. Catabolism—breakdown of larger molecules into smaller chemical units
      a. Hydrolysis—catabolic reaction that adds water to break down large molecules (Figure 2-10)
   2. Anabolism—construction of larger molecules from smaller molecules
      a. Dehydration synthesis—anabolic reaction that removes water molecules during the reaction (Figure 2-10)

INORGANIC AND ORGANIC COMPOUNDS

A. Organic compound—having carbon-carbon (C–C) or carbon-hydrogen (C–H) covalent bonds, or both types of bonds
B. Inorganic compound—contains no C–C or C–H bonds
C. Functional groups—specialized arrangements of atoms attached to the carbon core of many organic molecules (Figure 2-11)
D. Inorganic molecules
  1. Water—all living things need water to survive and reproduce (Table 2-1)
    a. Polarity—because of its polarity and its ability to form hydrogen bonds, water makes a very effective solvent
    b. Absorbs and releases heat very slowly
    c. Transport of materials into and out of cells
    d. Formation of cellular secretions
  2. Oxygen and carbon dioxide
    a. Essential inorganic molecules involved with the chemical process of cellular respiration
  3. Electrolytes—inorganic substances such as acids, bases, and salts
    a. Cations—ions with a positive charge
    b. Anions—ions with a negative charge
  4. Acids and bases
    a. Acids—substances that release a hydrogen ion (H^+) when in solution
    b. Bases (alkaline compounds)—electrolytes that, when dissociated in solution, shift the H^+/OH balance to produce *more* hydroxide ions
    c. The pH scale—measurement that indicates the relative concentration of H^+ ions with respect to OH^- ions in a solution
    d. pH of 7 indicates neutrality; a pH greater than 7 indicates alkalinity; a pH lower than 7 indicates acidity (Figure 2-13)
  5. Buffers
    a. Substances that minimize changes in concentrations of H^+ and OH^- in a solution
  6. Salts
    a. Compounds composed of ions with opposite charges that result from a chemical reaction between an acid and a base
    b. Reaction between an acid and a base to form a salt and water is called a *neutralization reaction*
E. Organic molecules
  1. Organic—describes the enormous number and diversity of compounds that contain carbon—specifically C–C or C–H bonds (Table 2-2)
  2. Four major groups: carbohydrates, proteins, lipids, and nucleic acids
  3. Carbohydrates
    a. Carbohydrates such as sugars and starches serve as a primary source of energy for most living things
    b. Monosaccharides—simple sugars (glucose, fructose, and galactose)
    c. Disaccharides—two monosaccharide sugars that are bonded together (table sugar [sucrose], maltose, and lactose) (Figure 2-15)
    d. Polysaccharides—a chain of many monosaccharides bonded together (glycogen)

4. Lipids
  a. Composed largely of carbon, hydrogen, and oxygen
  b. Not soluble in water (hydrophobic; nonpolar)
  c. Serve as sources of energy, provide insulation, form hormones, and provide the primary structural elements of cell membranes (Table 2-3)
  d. Triglycerides (fats)—most abundant lipids and serve primarily as energy sources
    (1) Composed of two fundamental units: a glycerol molecule bonded at three sites to three fatty acids (Figure 2-17)
    (2) Saturated fatty acid—one that has all of its available carbon bonds taken up by hydrogen atoms (Figure 2-16, *A*)
    (3) Unsaturated fatty acid—one that has one or more double bonds in the chain and therefore does not have a carbon chain that is saturated with hydrogen atoms (Figure 2-16, *B*)
    (4) Polyunsaturated fats—those that have more than several double bonds
  e. Phospholipid—fat compounds similar to triglycerides; one of the fatty acids in a triglyceride is replaced by a different chemical compound containing phosphorus (and sometimes nitrogen)
    (1) In water, phospholipids form a double or bilayer membrane (Figure 2-18)
  f. Steroids—cholesterol stabilizes cell membranes and forms the basis of many steroid hormones (Figure 2-19)
  g. Prostaglandins—lipids composed of a 20-carbon unsaturated fatty acid that contains a 5-carbon ring (Figure 2-20)
    (1) Regulate the effects of hormones, influence blood pressure, control the secretion of digestive juices, and even function in the immune system and blood clotting (Box 2-2)
5. Proteins
  a. Incorporate carbon, hydrogen, oxygen, and nitrogen into their molecular structure; some specialized proteins also contain sulfur, iron, and phosphorus
  b. Subunits are called amino acids (Figure 2-21)
  c. Proteins can be both structural and functional
    (1) Structural proteins—form the structures of cells, tissues, and organs
    (2) Functional proteins—act as accelerators or helpers in chemical reactions that are necessary for cellular metabolism and other body functions (enzymes)
  d. Peptide bond—links one amino acid to another
  e. Dehydration synthesis—OH from the carboxyl group of one amino acid and the H from the amino group of an adjacent amino acid join to form a molecule of water and a new compound called a peptide
  f. Hydrolysis—reverse reaction, breaks peptides down into their constituent amino acids
  g. Levels of protein structure (Table 2-4)
    (1) Primary structure—refers to the number, type, and sequence of amino acids in the polypeptide chain (parathyroid hormone)

    (2) Secondary structure—polypeptide that is coiled or bent into pleated sheets

    (3) Tertiary structure—polypeptide chain is so twisted that its coils touch one another in many places, and "spot welds" or disulfide linkages occur

    (4) Quaternary structure—protein that contains clusters of more than one polypeptide chain, all linked together into one massive molecule

    (5) Chaperones (chaperonins)—aid in directing each protein into its final folded shape

  6. Nucleic acids and related molecules

    a. Two fundamental nucleic acids: DNA (deoxyribonucleic acid) and RNA (ribonucleic acid)

      (1) Composed of subunits called nucleotides—five-carbon sugar, phosphate group, and nitrogenous base

    b. DNA—encodes information that is required for all of the cells in the body to generate all proteins that are necessary for everyday function (Figure 2-24)

      (1) Nitrogen bases—adenine, thymine, guanine, and cytosine

      (2) Exists as a double chain, which spirals into a spring-like form called a double helix

    c. RNA—involved in protein synthesis (transcription and translation)

      (1) Nitrogen bases—adenine, uracil, guanine, and cytosine

      (2) Exists as a single chain (Figure 2-25)

    d. Adenosine triphosphate (ATP)

      (1) Nucleotide composed of the nitrogenous base adenine, a ribose sugar, and a string of three phosphate groups (Figure 2-26)

      (2) ATP—an adenine ribonucleotide with three high-energy phosphate groups attached to each other via high-energy bonds; fundamental immediate energy-storage molecule of all cells

REVIEW QUESTIONS

 HINT *Write out the answers to these questions after reading the chapter and reviewing the Chapter Summary. If you simply think through the answer without writing it down, you won't retain much of your new learning.*

1. Define the following terms: *element, compound, atom.*
2. Name and define three kinds of subatomic particles.
3. Define and contrast meanings of the terms *atomic number* and *atomic weight.*
4. Define and give an example of an isotope.
5. Explain what the term *chemical reaction* means.
6. Identify and differentiate between the three basic types of chemical reactions.
7. Define the term *inorganic.*
8. Explain why water is said to be polar.
9. What are electrolytes and how are they formed?
10. What is a cation? an anion?
11. Define the terms *acid, base, salt,* and *buffer.*
12. Explain how pH indicates the degree of acidity or alkalinity of a solution.
13. What are the structural units, or building blocks, of proteins? of carbohydrates? of triglycerides? of DNA?
14. Describe some of the functions that proteins perform.
15. Proteins, carbohydrates, lipids—which of these are insoluble in water? contain nitrogen? include prostaglandins? include phospholipids?
16. What groups make up a nucleotide?
17. What sugar is present in a deoxyribonucleotide?
18. Describe the size, shape, and chemical structure of the DNA molecule.
19. What base is thymine always paired with in the DNA molecule? What other two bases are always paired?
20. What is the function of DNA?
21. What is catabolism? What function does it serve?
22. Compare catabolism, anabolism, and metabolism.

CRITICAL THINKING QUESTIONS

HINT *After finishing the Review Questions, write out the answers to these items to help you apply your new knowledge. Go back to sections of the chapter that relate to items that you find difficult.*

1. In modern blimps, the gas of choice used to inflate them is helium rather than hydrogen. Hydrogen would be lighter, but helium is safer. Compare and contrast the atomic structure of hydrogen and helium. What characteristics of the atomic structure of helium make it so much less reactive than hydrogen?
2. How would you contrast single covalent bonds, double covalent bonds, and ionic bonds?
3. Amino acids are the building blocks of proteins. Only 20 amino acids make up our proteins. Explain how these 20 amino acids are responsible for the billions of proteins that are used by the body.
4. How does ATP supply the cells with the energy they need to work? Outline the general scheme of the ATP energy cycle.

continued from page 21

Remember Calleigh and her "adventure" with raw oysters from the Introductory Story? See if you can answer the following questions about her now that you have read this chapter.

1. With continued vomiting, Calleigh keeps losing _____ from her stomach, which could make her entire body too _____.
   a. acid; acidic
   b. base; basic
   c. acid; basic
   d. base; acidic

2. When we talk about measuring the pH of a substance, we are measuring the concentration of what ions in that substance?
   a. Oxygen ions
   b. Carbon ions
   c. Phosphate ions
   d. Hydrogen ions

Finally, after another 24 hours, Calleigh is able to keep clear liquids down.

3. Why should she not drink just plain water?
   a. Water cannot replenish the electrolytes she has lost.
   b. Flavored liquids will more effectively stimulate her appetite.
   c. Water can irritate the stomach lining.
   d. Plain water is just fine; it will quickly replace her body's lost fluid.

When Calleigh feels well enough to try eating something, her first food items should provide energy but be easy to digest.

4. Which organic molecule best fits that description—high energy, easily digested?
   a. Protein
   b. Carbohydrates
   c. Triglycerides
   d. Nucleic acids

 HINT *To solve these questions, you may have to refer to the glossary or index, other chapters in this textbook, A&P Connect, Mechanisms of Disease and other resources.*

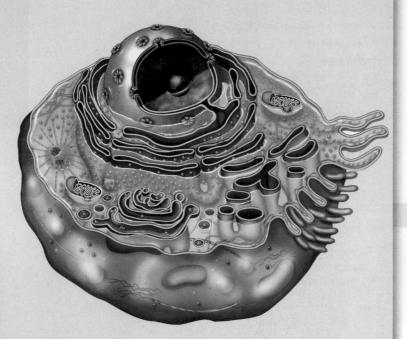

Anatomy of Cells

STUDENT LEARNING OBJECTIVES

At the completion of this chapter, you should be able to do the following:

1. Define the basic components of a typical cell.
2. Describe the fluid mosaic model of the cell membrane.
3. Describe how carbohydrates, lipids, and proteins are distributed in the cell membrane.
4. Define the term organelle.
5. List the major organelles of the cell and describe their functions.
6. Illustrate how the organelles, nucleus, cytosol, and cell membrane work together to form a functioning cell.

SUNIL had been looking forward to his trip to the Natural History Museum for weeks. His third-grade class had just finished a section on the cell, and he couldn't wait to walk through the greatly over-sized cell model that was on display. His father, a nursing student, seemed just as excited, maybe more, if that were possible. When they got to the cell model, their first challenge was figuring out how to get inside. The "membrane" of the cell had no holes in it. They walked around the cell and finally found a small doorway through which they could crawl to enter the cell.

Farther into the cell, Sunil and his father entered a branching network of tunnels. Some of the transparent walls farther away from them were studded on the outside by rocklike structures, but the branching walls near them were smooth. They followed this maze until they came to another sign telling them they had arrived at the nucleus.

After exploring the nucleus, Sunil and his father looked for a way to exit the cell's cytoskeleton. They had to work their way through a meshwork of plastic pipes, ropes, and yarn designed to represent the cell's cytoskeleton. When you are finished with this chapter, judge the accuracy of the museum's cell model.

• • •

In this chapter, you'll first learn exactly how the cell membrane regulates the passage of materials into and out of our cells. In the remainder of the chapter, we'll investigate additional components of the model and what they represent to the proper functioning of the model cell.

THE FUNCTIONAL ANATOMY OF CELLS

Cells and Cell Theory

All organisms are composed of one or more cells. In multicellular organisms such as ourselves, these cells are constantly being replaced. That is easy enough to understand, but how do these cells originate and how do they diversify? One unifying principle in biology is the **cell theory**. This theory simply states that the fundamental unit of all living things is the cell, and that all modern cells now come from previously existing cells. To develop your understanding of the cell and its internal organization, we will first investigate the basic structures of a typical cell. In Chapter 4, we will investigate the physiology of a cell and its component parts.

Typical Cells

In truth, there is no typical cell. However, most human cells are microscopic in size, ranging in diameter from 7.5 micrometers (µm) (red blood cells) to about 150 µm (egg cell or *ovum*). As a reference, the period at the end of this sentence is about 100 µm in diameter. Cells generally have a particular size or form because these characteristics relate intimately to the functions they perform in the body. For example, in order to transmit information from various parts of the body, many nerve cells have extensions that are over a meter in length! In contrast, muscle cells are adapted to contract—that is,

LANGUAGE OF SCIENCE AND MEDICINE

HINT *Before reading the chapter, say each of these terms out loud. This will help you avoid stumbling over them as you read.*

adenosine triphosphate (ATP) (ah-DEN-o-seen try-FOS-fate)
[blend of *adenine* and *ribose*, *tri-* three, *-phosph-* phosphorus, *-ate* oxygen]

cell theory (sell THEE-o-ree)
[*cell* storeroom, *theor-* look at, *-y* act of]

centriole (SEN-tree-ohl)
[*centr-* center, *-ole* small]

centrosome (SEN-troh-sohm)
[*centr-* center, *-som-* body]

chromatin (KROH-mah-tin)
[*chrom-* color, *-in* substance]

chromosome (KROH-meh-sohm)
[*chrom-* color, *-som-* body]

cilium (SIL-ee-um)
[*cili-* eyelid] *pl.*, cilia (SIL-ee-ah)

crista
[*crista* crest or fold] *pl.*, cristae

cytoplasm (SYE-toh-plaz-em)
[*cyto-* cell, *-plasm* substance]

cytoskeleton (sye-toh-SKEL-e-ton)
[*cyto-* cell, *-skeleto-* dried body]

desmosome (DES-mo-sohm)
[*desmos-* band, *-som-* body]

endoplasmic reticulum (ER) (en-doh-PLAZ-mik reh-TIK-yoo-lum)
[*endo-* inward or within, *-plasm-* substance, *-ic* relating to, *ret-* net, *-ic-* relating to, *-ul-* little, *-um* thing] *pl.*, endoplasmic reticula (en-doh-PLAZ-mik reh-TIK-yoo-lah)

flagellum (flah-JEL-um)
[*flagellum* whip] *pl.*, flagella (flah-JEL-ah)

fluid mosaic model (FLOO-id mo-ZAY-ik MAHD-el)

gap junction (gap JUNK-shen)

Golgi apparatus (GOL-jee ap-ah-RA-tus)
[*Camillo Golgi* Italian histologist]

lysosome (LYE-so-sohm)
[*lyso-* dissolution, *-som-* body]

microvillus (my-kroh VIL-us)
[*micro-* small, *-villus* shaggy hair] *pl.*, microvilli (my-kroh-VIL-eye)

mitochondrion (my-toh-KON-dree-on)
[*mito-* thread, *-chondrion* granule] *pl.*, mitochondria (my-toh-KON-dree-ah)

nuclear envelope (NOO-klee-ar)
[*nucle-* nucleus (kernel), *-ar* relating to]

nucleolus (noo-KLEE-oh-lus)
[*nucleo-* nucleus (kernel), *-olus* little] *pl.*, nucleoli (noo-KLEE-ohl-eye)

continued on page 57

TABLE 3-1 Examples of Cell Types

TYPE	STRUCTURAL FEATURES	FUNCTIONS
Nerve cells 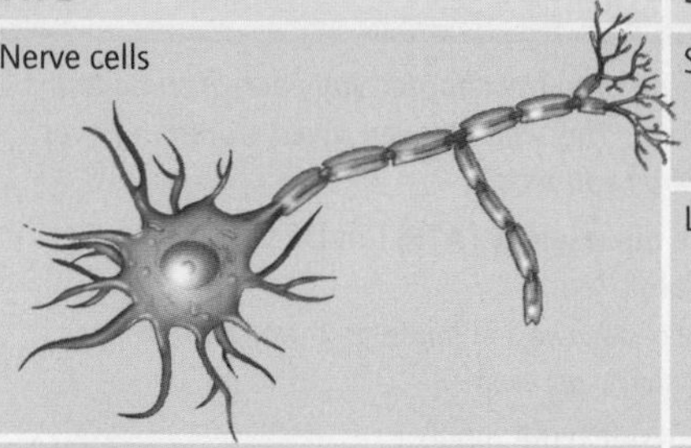	Sensitive to stimuli	Detect changes in internal or external environment
	Long extensions	Transmit nerve impulses from one part of the body to another
Muscle cells 	Elongated, threadlike Contain tiny fibers that forcefully slide together	Contract (shorten) to allow movement of body parts
Red blood cells 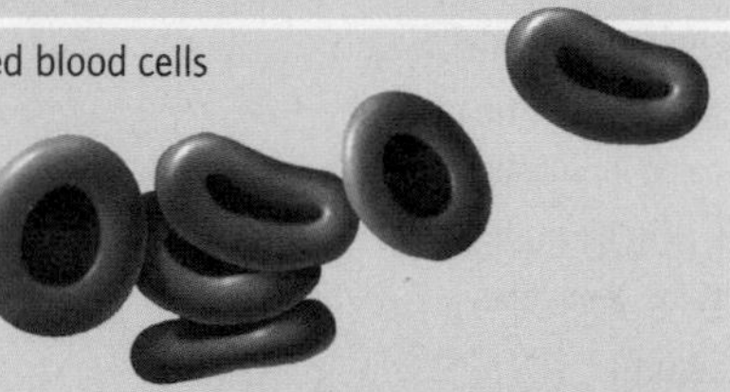	Contain hemoglobin, a red pigment that attracts, then releases, oxygen	Transport oxygen in the bloodstream (from lungs to other parts of the body)
Gland cells 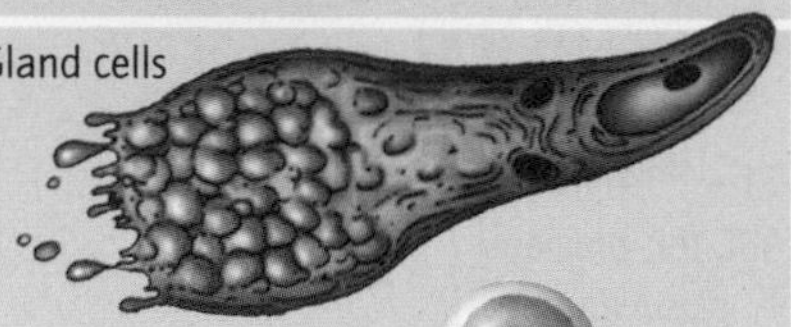	Contain sacs that release a secretion to the outside of the cell	Release substances such as hormones, enzymes, mucus, and sweat
Immune cells 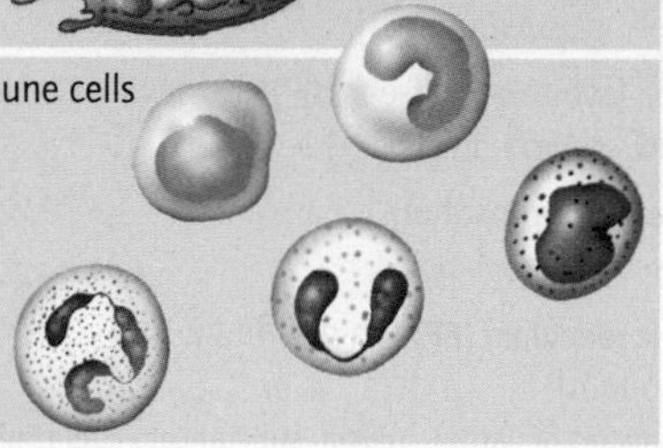	Some have outer membranes able to engulf other cells Some have systems that manufacture antibodies Some are able to destroy other cells	Recognize and destroy "nonself" cells such as cancer cells and invading bacteria

they shorten to provide pulling strength. Still other cells serve to provide protection or produce a variety of cellular secretions (Table 3-1).

Although cells can be very specialized to serve certain functions, most cells share a number of similarities. Not *all* cells have *all* of the same general components. To help you with our discussion, however, we will talk about a typical "composite cell" that *does* have all of these components. As you read through the paragraphs that follow, please refer to the diagram of the composite cell presented in Figure 3-1.

A&P CONNECT

Are you a little confused by the metric size units used in science? Check out **Metric Measurements and Their Equivalents** online at **A&P Connect**.

evolve learning system

Basic Cellular Components

Cells are surrounded by a **plasma membrane** that separates the interior of the cell (*intracellular space*) from the body's internal fluid environment (*extracellular space*). Plasma membranes are composed of phospholipids (see Chapter 2) and also contain many proteins and other molecules that allow the cell to communicate with other cells in the body. Contained within the plasma membrane is the **cytoplasm.** The cytoplasm is a gelatinous substance made up of organized subunits called **organelles,** which are embedded in a solution called *cytosol.* Many of the organelles, such as mitochondria, are enclosed by membranes. Other membrane-bound internal organelles are sometimes arranged as sacs or canals.

In most cells, a large *nucleus* lies deep within the cytoplasm. The nucleus contains DNA, which serves as the genetic "instruction booklet" for the cell. Each different cell

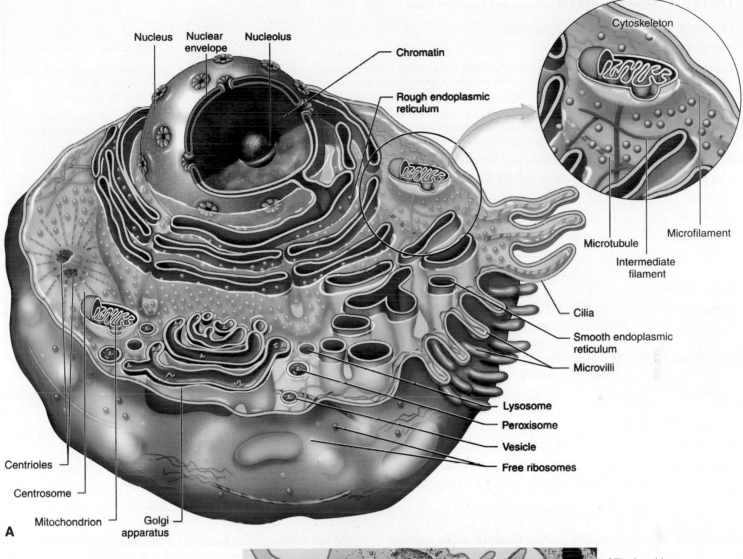

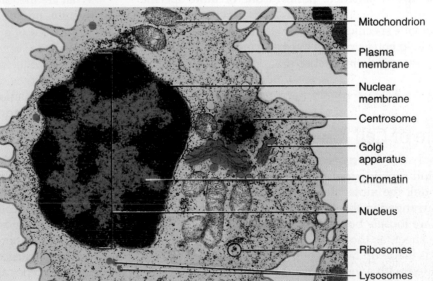

▲ FIGURE 3-1 Typical, or composite, cell. A, Artist's interpretation of cell structure. B, Color-enhanced electron micrograph of a cell. Both show the many organelles, including the mitochondria, known as the "power plants of the cell." *Note the innumerable dots bordering the endoplasmic reticulum. These are ribosomes, the cell's "protein factories."*

TABLE 3-2 Some Major Cell Structures and Their Functions

CELL STRUCTURE	FUNCTIONS
Membranous	
Plasma membrane	Serves as the boundary of the cell, maintains its integrity; protein molecules embedded in plasma membrane perform various functions—for example, they serve as markers that identify cells of each individual, as receptor molecules for certain hormones and other molecules, and as transport mechanisms
Endoplasmic reticulum (ER)	Ribosomes attached to rough ER synthesize proteins that leave cells via the Golgi apparatus; smooth ER synthesizes lipids incorporated in cell membranes, steroid hormones, and certain carbohydrates used to form glycoproteins—also removes and stores Ca^{++} from the cell's interior
Golgi apparatus	Synthesizes carbohydrate, combines it with protein, and packages the product as globules of glycoprotein
Vesicles	Membranous bags that temporarily contain molecules for transport or later use
Lysosomes	Bags of digestive enzymes break down defective cell parts and ingested particles; a cell's "digestive system"
Peroxisomes	Contain enzymes that detoxify harmful substances
Mitochondria	Catabolism; adenosine triphosphate (ATP) synthesis; a cell's "power plants"
Nucleus	Houses the genetic code, which in turn dictates protein synthesis, thereby playing an essential role in other cell activities, namely, cell transport, metabolism, and growth
Nonmembranous	
Ribosomes	Site of protein synthesis; a cell's "protein factories"
Proteasomes	Hollow protein cylinders that destroy misfolded or otherwise abnormal proteins manufactured by the cell; a "quality control" mechanism for protein synthesis
Cytoskeleton	Acts as a framework to support the cell and its organelles; functions in cell movement; forms cell extensions (microvilli, cilia, flagella)
Centrosome	The microtubule-organizing center (MTOC) of the cell; includes two centrioles that assist in forming and organizing microtubules
Microvilli	Tiny, fingerlike extensions that increase a cell's absorptive surface area
Cilia and flagella	Hairlike cell extensions that serve to move substances over the cell surface (cilia) or propel sperm cells (flagella)
Nucleolus	Part of the nucleus; plays an essential role in the formation of ribosomes

part is structurally suited to perform one or more specific functions within the cell—much as each of your organs is suited to provide specific functions within your body. Refer to Table 3-2 to help you learn the larger and more important cellular structures and their functions.

CELL MEMBRANES

Structure of Cell Membranes

All external and internal cell membranes have a basic structure and function that follow the **fluid mosaic model.** In this model, the molecules that make up the cell membrane are arranged in sheets much like an art mosaic. However, they are loosely bound so that they can move around within the sheet, as though it were a fluid. The mosaic of molecules is made up mostly of phospholipids and cholesterol, studded with a variety of proteins and hybrid molecules such as glycoproteins and glycolipids (Figure 3-2). Knowledge of the structure of the cell membrane is vital to understanding the discussion of its various functions throughout this book.

What are the forces that hold a cell membrane together? To answer this question, recall from Chapter 2 that cell membranes are composed of a double layer of phospholipids. Each phospholipid has a *hydrophilic* (water-loving) "head" that faces either the interior of the cell or the cell's external environment. Likewise, each phospholipid has a *hydrophobic* (water-fearing) "tail" that faces the interior of the membrane. Phospholipids naturally arrange themselves into these double layers, or *bilayers*, in water. When cholesterol is added to a phospholipid bilayer, it strengthens the membrane without losing flexibility, so that the cell membrane remains pliable. Membrane proteins have hydrophobic and hydrophilic regions that can easily fit among the phospholipids of the bilayer structure.

Each human cell manufactures various kinds of phospholipids that, along with cholesterol molecules, create a bilayer that forms a natural "fencing" material of varying thickness. This basic construction allows a cell membrane to be "choosy" about what types of molecules it lets through, thus forming an effective barrier. This is because the overall hydrophobic nature of the membrane makes it especially difficult for water and water-soluble molecules to pass through it. As a result, the cytosol and its molecules will not be accidentally lost.

Functions of Cell Membranes

A cell can make its membrane more stiff or flimsy in certain areas as necessary. Cells do this by altering the types of phospholipids and the concentration of cholesterol placed in their membrane. In addition, protein molecules may be anchored to one or both sides of the membrane—or all the way through the membrane. These membrane proteins have several important functions. For example, some membrane proteins facilitate the transport of molecules across the membrane. Some of these transporter proteins have "gates" that allow a cell to regulate when transport occurs.

Some proteins are modified by organelles in the cell by the addition of carbohydrates. These *glycoproteins* serve as identification markers, allowing cells to determine the identity of other cells with which they come into contact. For example, identification markers (such as glycoproteins) allow immune cells to determine the identity of bacterial or cancerous cells. They also prevent the body from accepting certain types of donated tissues or organs that bear different cellular markers. Another function of membrane proteins is their action as *receptors*. In this case, a membrane protein reacts to a *messenger molecule* (such as a hormone) and can trigger chemical reactions or other metabolic changes within

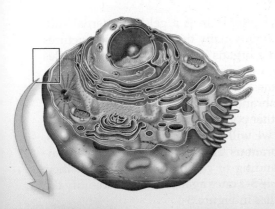

▼ **FIGURE 3-2**   **Plasma membrane.** The plasma membrane is a fluid mosaic made of a bilayer of phospholipid molecules arranged with their nonpolar "tails" pointing toward each other. Cholesterol molecules help stabilize the flexible bilayer structure. Protein molecules and hybrid molecules (e.g., glycolipids) may be found on the outer or inner surface of the bilayer. They may also extend all the way through the membrane.

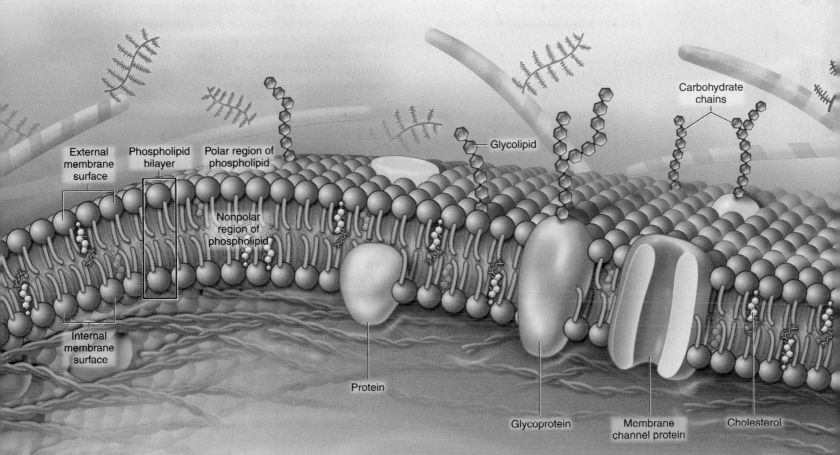

TABLE 3-3 **Functional Anatomy of Cell Membranes**

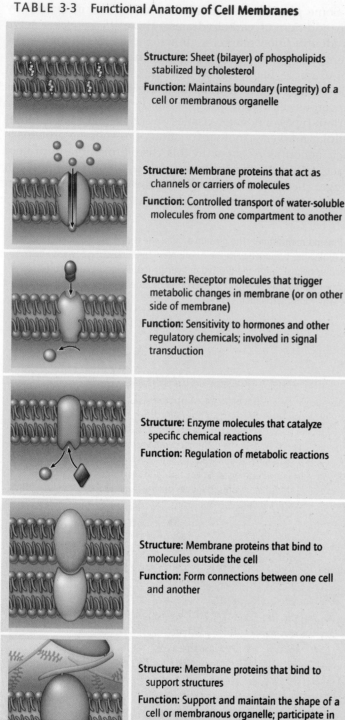

	Structure: Sheet (bilayer) of phospholipids stabilized by cholesterol **Function:** Maintains boundary (integrity) of a cell or membranous organelle
	Structure: Membrane proteins that act as channels or carriers of molecules **Function:** Controlled transport of water-soluble molecules from one compartment to another
	Structure: Receptor molecules that trigger metabolic changes in membrane (or on other side of membrane) **Function:** Sensitivity to hormones and other regulatory chemicals; involved in signal transduction
	Structure: Enzyme molecules that catalyze specific chemical reactions **Function:** Regulation of metabolic reactions
	Structure: Membrane proteins that bind to molecules outside the cell **Function:** Form connections between one cell and another
	Structure: Membrane proteins that bind to support structures **Function:** Support and maintain the shape of a cell or membranous organelle; participate in cell movement; bind to fibers of the extracellular matrix (ECM)
	Structure: Glycoproteins or proteins in the membrane that act as markers **Function:** Recognition of cells or organelles

a cell. This triggering action is called *signal transduction.* Membrane proteins can also serve to join cells together to form a larger mass of tissue. Some of the major functions associated with cell membranes are outlined for your review in Table 3-3.

QUICK ✓ CHECK

1. What is the cell theory?
2. Identify: plasma membrane, cytoplasm, and nucleus.
3. What is the basic structure of the plasma membrane?
4. What is a glycoprotein?

ORGANELLES

As we have seen, the cytoplasm is a gel-like internal substance of cells that contains many tiny suspended structures called *organelles* ("little organs"). We can group organelles into two broad categories: *membranous* and *nonmembranous.* Those with membranes have sacs or canals made of cell membrane. Nonmembranous organelles are made of microscopic filaments or other particles. Undoubtedly, as research techniques improve, we will discover even more organelles, perhaps both membranous and nonmembranous. To improve your understanding of cellular structures and their functions, review Table 3-2 once again and try to identify the membranous organelles in Figure 3-1.

A full review of all the organelles and their functions is beyond the scope of this book. However, we would like you to understand a few basic types of organelles and their most important functions. As you read about each type of organelle, refer to the figures provided to help in your understanding of these important structures.

Endoplasmic Reticulum and Ribosomes

Let's begin our discussion with one of the most obvious and extensive intracellular structures, the **endoplasmic reticulum (ER).** This extensively folded structure is really a network of canals and sacs made of cell membrane. It extends from the nucleus throughout the cytosol to the plasma membrane. There are two types of ER, the *rough* and the *smooth.*

The **rough endoplasmic reticulum (RER)** has tiny organelles called *ribosomes* attached to it. This arrangement gives it a bumpy or "rough" appearance under a microscope (Figure 3-3, *A* and *B*).

Ribosomes translate the genetic code into proteins. These proteins are then folded, altered, or joined by other proteins within the ER (Figure 3-4). The tiny ribosomes can be attached to the ER or free floating in the cytosol. Ribosomes are so small that on drawings of whole cells they look like tiny dots or balls (see Figure 3-3). However, up close you can

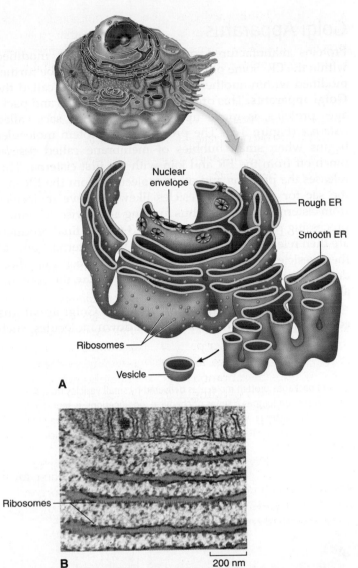

A

B

▲ **FIGURE 3-3** **Endoplasmic reticulum (ER).** In both the drawing **A,** and the transmission electron micrograph **B,** the rough ER (RER) is distinguished by the presence of tiny ribosomes dotting the boundary of flattened membrane sacs. The smooth ER (SER) is more tubular in structure and lacks ribosomes on its surface. Note also that the ER is continuous with the outer membrane of the nuclear envelope.

see that each ribosome is a nonmembranous structure made of two subunits, one larger than the other. Each subunit is composed of ribonucleic acid (RNA) bonded to protein. (NOTE: *Ribosomal RNA* is usually abbreviated as *rRNA.* Other types of RNA, which help ribosomes function, are *messenger RNA [mRNA]* and *transfer RNA [tRNA].*)

Ribosome subunits come together on an mRNA strand to "read" the mRNA's code, copied from the DNA, to form a polypeptide. As these "free" ribosomes work, they move to the surface of the ER and insert the finished polypeptide into the ER's interior. Inside the ER, the polypeptide is processed and folded into a protein. Ribosomes often work together in

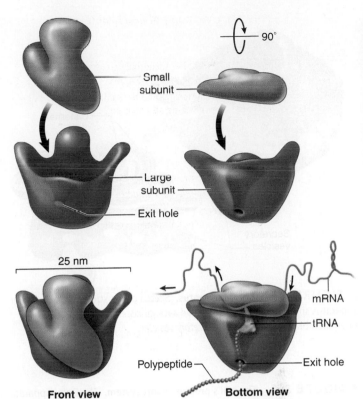

Front view   **Bottom view**

▲ **FIGURE 3-4** **Ribosome.** A ribosome is composed of a *small subunit* and a *large subunit,* shown here from two different perspectives. After the small subunit attaches to an mRNA strand containing the genetic "recipe" for a polypeptide strand, the subunits come together to form a complete ribosome. tRNA brings amino acids into the cavity between subunits, where they are assembled into a strand according to the mRNA code. As the polypeptide strand elongates, it moves out through a tunnel and a tiny exit hole in the large subunit.

chains along a single mRNA strand to generate large quantities of proteins more efficiently.

Figure 3-4 shows an overview of the process of protein synthesis, a job performed by the ribosome. We will discuss this process in detail in Chapter 5.

The **smooth endoplasmic reticulum (SER),** which lacks ribosomes, continues from the rough endoplasmic reticulum. The SER synthesizes membrane lipids, steroid hormones, and some carbohydrates that may be used to form glycoproteins. The SER also transports calcium ions (Ca^{++}) from the cytosol into ER sacs and canals, thus keeping a low concentration of Ca^{++} ions within the cytosol. This is essential for a number of vital cellular functions that we will discuss more fully in upcoming chapters.

Small membranous sacs called **peroxisomes,** which pinch off from the ER, contain enzymes that remove toxic chemicals from the cytoplasm. These toxic wastes are generally produced by the cell's own metabolic activity and are often found in kidney and liver cells.

Golgi Apparatus

Proteins manufactured on the ribosomes are modified within the ER. Some of these molecules may then be further modified within another membranous organelle called the **Golgi apparatus.** This organelle, which processes and packages proteins, is made up of small flattened sacs called *cisternae* (Figure 3-5). The processing of protein molecules begins when small bubbles of membrane called *vesicles* pinch off from the ER and join with the first cisterna. This releases the proteins and other molecules from the ER into the sac for processing. Vesicles likewise move molecules from cisterna to cisterna, thus forming a stepwise "chemical processing plant" in the cell (Figure 3-6). The final products are then released in vesicles from the final cisterna. Some of these vesicles deliver their contents inside the cell and others deliver them to the plasma membrane, where the contents are secreted outside the cell.

Some vesicles that pinch off from the Golgi apparatus contain specific enzymes that break down molecules, such

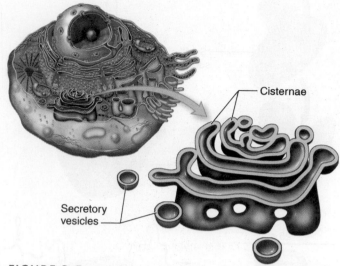

▲ **FIGURE 3-5** **Golgi apparatus.** Sketch of the structure of the Golgi apparatus showing a stack of flattened sacs, or *cisternae*, and numerous small membranous bubbles, or *secretory vesicles.*

▼ **FIGURE 3-6** **The cell's protein export system.** The Golgi apparatus processes and packages protein molecules delivered by small vesicles from the endoplasmic reticulum. After entering the first cisterna of the Golgi apparatus, a protein molecule undergoes a series of chemical modifications, is sent (by means of a vesicle) to the next cisterna for further modification, and so on, until it is ready to exit the last cisterna. When it is ready to exit, a molecule is packaged in a membranous secretory vesicle that migrates to the surface of the cell and "pops open" to release its contents into the space outside the cell. Some vesicles remain inside the cell for some time and serve as storage vessels for the substance to be secreted.

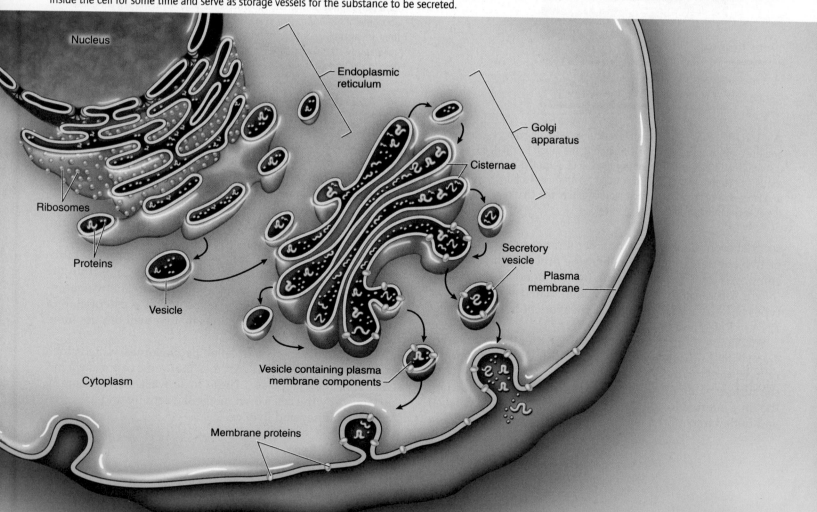

as proteins not needed by the cell. These digestive vesicles are called **lysosomes.** Lysosomes can also digest invading particles such as those from bacteria or viral infections. For these reasons, lysosomes are sometimes called "digestive bags" or "cellular garbage disposals."

A&P CONNECT

Scientists are using their knowledge of the Golgi apparatus to mimic the cell's chemical-making functions in order to manufacture therapeutic treatments more efficiently. Check out **Biomimicry** online at **A&P Connect.**

evolve

Proteasomes

Proteasomes are hollow, cylindrical, drumlike organelles found throughout the cytoplasm. Proteasomes are responsible for breaking down abnormal and misfolded proteins from the ER. They also destroy normal regulatory proteins in the cytoplasm that are no longer needed by the cell. Unlike lysosomes, which can destroy large groups of proteins all at once, proteasomes are "picky eaters" and destroy specific protein molecules one at a time.

Mitochondria

Note that the "power plants" of cells, the **mitochondria** (Figures 3-1 and 3-7), are bound by *two* membranes that form a sac-within-a-sac structure. The outer membrane is the same as our cell membranes, but the inner membrane is similar to that of certain bacteria. This fact suggests that these organelles are bacterial cells that have become incorporated into the cells of plants, fungi, and animals in a mutually beneficial relationship. In fact, each tiny sausage-shaped mitochondrion has its own set of mitochondrial DNA and replicates itself in the same manner as its bacterial relatives. Having its own DNA also enables each mitochondrion to divide and produce genetically identical daughter mitochondria. The study of mitochondrial genetics and metabolism is one of the fastest growing areas in medicine, connecting scientific disciplines ranging from the study of embryology to cancer. We now know that problems with mitochondrial metabolism likely play a role in many common diseases of aging, including diabetes, heart disease, Parkinson disease, and dementia.

In general, the more work a cell does, the more mitochondria its cytoplasm contains. Liver cells, for example, do more biochemical work and for this reason have more mitochondria

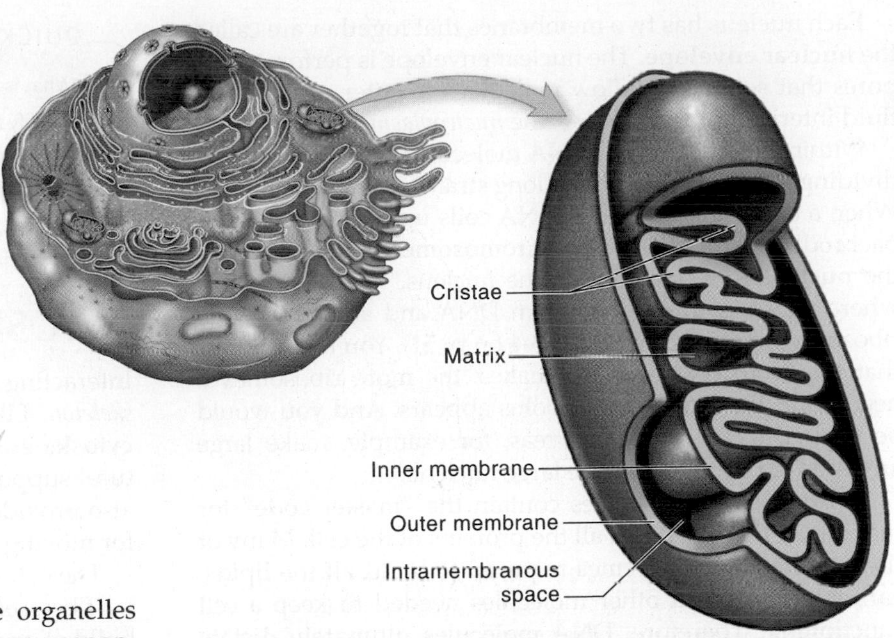

▲ **FIGURE 3-7** **Mitochondrion.** Cutaway sketch showing outer and inner membranes. Note the many folds (cristae) of the inner membrane. Although some mitochondria have the capsule shape shown here, many are round or oval.

than skin cells. The mitochondria in some cells multiply when energy consumption increases. For example, frequent aerobic exercise can increase the number of mitochondria inside skeletal muscle cells.

Intricate folds called **cristae** greatly increase the surface area of the inner membrane of mitochondria. The inner membrane is the site of many metabolic reactions that transfer energy from the breakdown of glucose to **adenosine triphosphate (ATP).** This process makes energy available to the cell. (Recall from Chapter 2 that ATP is the chemical that transfers energy from food to cellular processes.) The functional proteins (enzymes) in the membranes of the cristae are arranged precisely in the order that is required to function. This is another important example of the fact that structure and function are intimately related, even at the molecular level.

NUCLEUS

The **nucleus** (plural: *nuclei*) of the cell is commonly found near the center of the cell. It is often roughly spherical, and acts as the "control center" of the cell. Most cells have just one nucleus. However, some cells (such as skeletal muscle fibers) merge during development to form *multinucleate* cells, which have many nuclei.

Each nucleus has two membranes that together are called the **nuclear envelope.** The nuclear envelope is perforated by pores that selectively allow molecules to enter or leave the fluid interior of the nucleus—the *nucleoplasm* (Figure 3-8).

Within the nucleus are DNA molecules. When a cell isn't dividing, DNA is unwound in long strands called **chromatin.** When a cell is dividing, this DNA coils up tightly into compact rodlike structures called **chromosomes.** A smaller body, the **nucleolus,** resides within the nucleus. The nucleolus is where rRNA is transcribed from DNA and assembled into ribosome subunits (see Figure 3-4 on p. 51). You might guess that the more protein a cell makes, the more ribosomes it needs, and the larger its nucleolus appears. And you would be correct! Cells of the pancreas, for example, make large amounts of protein and have large nucleoli.

In brief, DNA molecules contain the "master code" for making all the RNA plus all the proteins of the cell. Many of these proteins are enzymes required to build all the lipids, carbohydrates, and other molecules needed to keep a cell functioning. Therefore, DNA molecules ultimately dictate both the structures and functions of all our cells. We will discuss this topic in much greater detail in Chapters 4 and 5.

CYTOSKELETON

Interacting with nearly all of the parts of the cell is the *cytoskeleton.* Like the skeletal system of a large organism, the **cytoskeleton** contains specific proteins that provide structural support for the cell and its components (Figure 3-9). It also provides a means of physically moving cells, as well as for moving the organelles and molecules within the cytosol.

The cytoskeleton is composed of many chainlike molecular fibers of different thicknesses and lengths that actively build at one end and disintegrate at the other (Figure 3-10). The smallest fibers are called *microfilaments,* composed of twisted strings of protein subunits, often lying parallel to one another. This parallel arrangement allows them to produce a shortening of the cell (as in muscle cells). *Intermediate filaments* are similar to microfilaments but are slightly thicker and perform a primarily structural role, acting as a supporting framework for the cell. The thickest fibers are *microtubules,* made of many protein subunits oriented in a spiral fashion. Microtubules help move organelles within the cell.

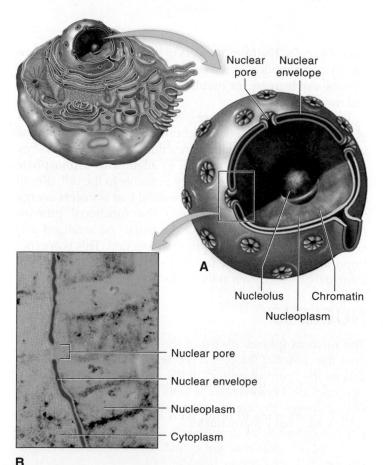

Nuclear pore   Nuclear envelope

A

Nucleolus   Chromatin
Nucleoplasm

Nuclear pore

Nuclear envelope

Nucleoplasm

Cytoplasm

B

▲ **FIGURE 3-8**   **Nucleus.** An artist's rendering **A**, and an electron micrograph **B**, show that the nuclear envelope is composed of two separate membranes and is perforated by large openings, or nuclear pores.

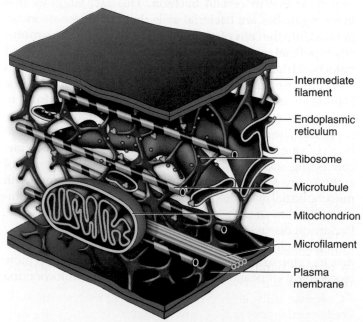

Intermediate filament

Endoplasmic reticulum

Ribosome

Microtubule

Mitochondrion

Microfilament

Plasma membrane

▲ **FIGURE 3-9**   **The cytoskeleton.** Artist's interpretation of the cell's internal framework. Notice that the "free" ribosomes and other organelles are not really free at all but are secured by the cytoskeletal elements of the cell.

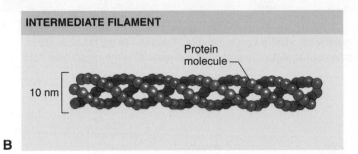

MICROFILAMENT

Protein molecule

7 nm

A

INTERMEDIATE FILAMENT

Protein molecule

10 nm

B

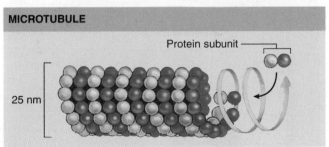

MICROTUBULE

Protein subunit

25 nm

C

▲ **FIGURE 3-10** **Cell fibers.** **A,** Microfilaments are thin, twisted strands of protein molecules. **B,** Intermediate filaments are thicker, twisted protein strands. **C,** Microtubules are hollow fibers that consist of a spiral arrangement of protein subunits.

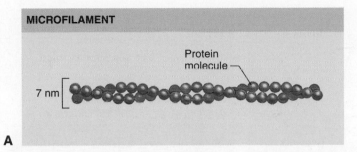

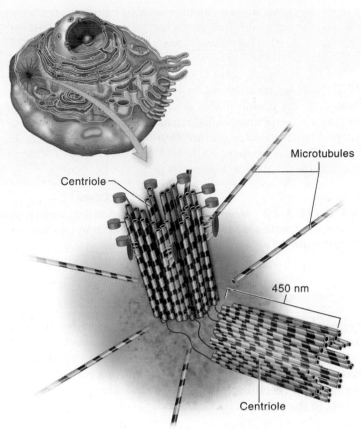

Microtubules

Centriole

450 nm

Centriole

▲ **FIGURE 3-11** **Centrosome.** Sketch showing the structure of a centrosome, which acts as a microtubule-organizing center for the cell's cytoskeleton.

Centrosome, Centrioles, and Cell Extensions

Centrosomes, centrioles, and cell extensions are components or extensions of the cytoskeleton with specific functions. Coordinating the building and breaking of microtubules near the center of the cell is the **centrosome.** The centrosome also plays an important role in cell division, when it constructs a "spindle" of microtubules to move chromosomes around the cell (Figure 3-11). In the center of the centrosome are two cylindrical structures oriented at right angles to one another called **centrioles.** The walls of these cylinders consist of nine bundles of microtubules, with three microtubules in each bundle.

Centrioles also perform an important role in the formation of *cellular extensions* of the plasma membrane called **cilia** and **flagella.** Cilia and flagella are cell extensions that have cylinders made of microtubules at their core. Each cylinder is composed of nine double microtubules arranged around

two single microtubules in the center. A type of motor protein causes the microtubule pairs to slip back and forth over each other to produce a waving motion (Figure 3-12).

Cilia are much shorter than flagella. Most cells have one cilium and some cells have many cilia. The cilia often act as sensors to detect conditions outside the cell. In groups, the cilia can brush fluid along the surfaces of cells—as they do in the airways of the lungs to remove contaminants. The only human cells that have flagella are sperm cells, each with a single flagellum that allows it to propel itself toward its target, the human egg. Refer back to Figure 3-1 for a sketch of microvilli and cilia.

Microvilli, supported by microfilaments instead of microtubules, are found in the epithelial cells of the intestines, for example. They serve to increase the surface area of the lining of the intestines, thus improving the body's absorption of nutrients.

Each year seems to bring with it the discovery of new types of cytoskeletal components. As we determine their functions, we find that the cytoskeleton is an amazingly rich network that serves literally as the "bones and muscles" of the cell. In addition, it provides a variety of many different types of cell movement, both internal and external, depending on the cell type and the circumstances. The

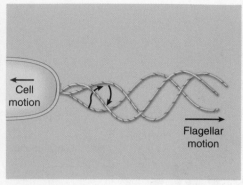
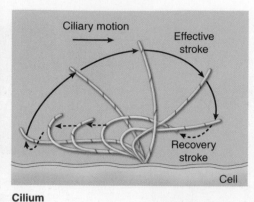
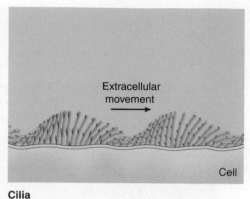

Flagellum **Cilium** **Cilia**

▲ **FIGURE 3-12** **Movement patterns.** A flagellum (*left*) produces wavelike movements, which propel a sperm cell forward—like the tail of an eel. In humans, cilia (*middle* and *right*) found in groups on stationary cells beat in a coordinated oarlike pattern to push fluid and particles in the extracellular fluid along the outer cell surface.

cytoskeleton also provides structural support for the cell and its parts; it even becomes involved in connections with other cells. Most amazing of all, perhaps, the cytoskeleton has the ability to *organize itself* by means of a complex set of signals and reactions so that it can quickly respond to the needs of the cell.

CELL CONNECTIONS

Tissues and organs of the body must be held together, and much of this is accomplished by intercellular connections. Some cells attach directly to extracellular material, called the *extracellular matrix.* Other cells are connected directly to one another. In some cases these connections allow for direct communication between cells.

The basic forms of intercellular connections are summarized for you in Figure 3-13. As you can see, **desmosomes** look like "spot welds" or "belts" that hold adjacent cells together. *Belt desmosomes* form a band linking adjacent cells together. *Spot desmosomes,* in contrast, do not form bands around cells, but are arranged in buttonlike points of contact between plasma membranes of adjacent cells. Both types of desmosomes are essential for cells to function as structural units in tissues. Desmosomes are common in adjacent skin cells, which are held together by this molecular version of Velcro.

Gap junctions form when membrane channels of adjacent plasma membranes connect to each other. These junctions form gaps or "tunnels" that join the cytoplasm of two cells by fusing the two plasma membranes into a single structure. Because of this arrangement, certain molecules can pass directly from one cell to another—thus making the cells one funtional unit. Gap junctions join heart muscle cells together so that a single impulse can travel to, and thus stimulate, many cells at the same time.

Tight junctions occur in cells that are joined near their top surfaces by "collars" of tightly fused membranes. Look again at Figure 3-13: An entire sheet of cells can be bound together

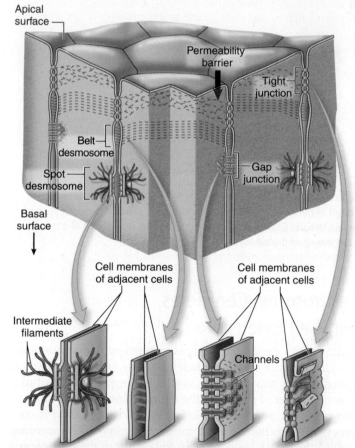

Spot desmosome **Belt desmosome** **Gap junction** **Tight junction**

▲ **FIGURE 3-13** **Cell connections.** Spot and belt desmosomes, gap junction, and tight junction.

the way soft drink cans are held in a six-pack by plastic collars—only more tightly. When tight junctions hold a sheet of cells together, molecules cannot easily *permeate* (spread through) the cracks between the cells. Tight junctions occur in the lining of the intestines and other parts of the body, where it is important to control what gets past a sheet of cells.

The BIG Picture

Probably one of the most difficult things to do when first exploring the microscopic world of the cell is to appreciate the structural significance a single cell has to the whole body. Where are these unseen cells? How do they relate to one another within my body?

One useful way to approach the structural role of cells is to compare your body to a large, complex building. For example, the building in which you take your anatomy and physiology course is made up of thousands of structural subunits. Bricks, blocks, metal or wood studs, boards, and so on are individual structures within the building. Each component has specific parts, or organelles, that somehow contribute to overall function. For example, a brick often has sides of certain dimensions that allow it to fit easily with other bricks or with other building materials. Three or four surfaces are usually textured in an aesthetically pleasing manner, and holes in two of the brick surfaces lighten the weight of the brick. These holes allow other materials such as wires or reinforcing rods to pass through and permit mortar to form a stronger joint with the brick. Furthermore, the material from which each brick is made has been formulated with certain ratios of sand, clay, pigments, or other materials. In short, each structural feature of each brick has a functional role to play. Likewise, each brick and other structural subunits of the building have important roles to play in supporting the building and making it a pleasing, functional place to study.

Like the structural features of different bricks making up a building, each organelle of each cell has a functional role to play within the cell. In fact, you can often guess a cell's function by the proportion and variety of different organelles it has. Each cell, like each brick in a building, has a role to play in providing its tiny portion of the overall support and function of the whole body. So, where are these cells? In the same place you would find bricks and other structural subunits in a building—*everywhere!* Like bricks and mortar, everything in your body is made of cells and the extracellular material surrounding them. How do cells relate to the whole body? Like bricks that are held together to form a building, cells form the body.

Just as a brick building is made of many different materials, the human body comprises many structural subunits, each with its own structural features or organelles that contribute to the function of the cell—and therefore to the whole body.

QUICK CHECK

10. What is the cytoskeleton?
11. What is the role of the centrosome?
12. Why are cytoskeletal components described as the "bones and muscles" of cells?
13. What are desmosomes?
14. What are the differences between a gap junction and a tight junction?

Cycle of LIFE

Whether we like it or not, our trillions of cells age, and as a result our bodies age along with them. Some cells replace themselves continuously, such as those of the gastrointestinal system. Others, such as the endothelial cells lining our arteries, lie dormant unless they are injured, at which time they too undergo mitosis for repair. Still others, such as red blood cells (erythrocytes) have short lives and must be replaced continuously.

The aging process of cells has to do with underlying causes within our cells. These processes affect our organelles, our DNA, and even the overall structure of our cells. Cell aging is poorly understood and there are many hypotheses that might explain this process. Some claim that aging is preprogrammed into our cells. Others claim that physical changes in the protective *telomeres* at the end of our chromosomes cause aging because they become shorter each time a cell undergoes mitosis. This process can affect the ability of the cell to divide in the future. Still others claim that environmental damage to cell structure is the root cause. However, there is no single hypothesis that can explain all the cellular and subcellular processes that contribute to our overall aging. Nonetheless, these hypotheses are helping us to better understand how we age. ●

LANGUAGE OF SCIENCE AND MEDICINE *continued from page 45*

nucleus (NOO-klee-us)
 [*nucleus* kernel] *pl.,* nuclei (NOO-klee-eye)

organelle (or-gah-NELL)
 [*organ-* tool or organ, *-elle* small]

peroxisome (per-OKS-ih-sohm)
 [*peroxi-* hydrogen peroxide, *-soma* body]

plasma membrane (PLAZ-mah MEM-brayne)
 [*plasma* substance, *membran-* thin skin]

proteasome (PROH-tee-ah-sohm)
 [*protea-* protein, *-som* body]

ribosome (RYE-boh-sohm)
 [*ribo-* ribose or RNA, *-som-* body]

rough endoplasmic reticulum (RER) (ruf en-doh-PLAZ-mik reh-TIK-yoo-lum)
 [*endo-* inward or within, *-plasm-* to mold, *-ic* relating to, *ret-* net, *-ic-* relating to, *-ul-* little, *-um* thing] *pl.,* rough endoplasmic reticula (ruf en-doh-PLAZ-mik reh-TIK-yoo-lah)

smooth endoplasmic reticulum (SER) (smooth en-doh-PLAZ-mik reh-TIK-yoo-lum)
 [*endo-* inward or within, *-plasm-* to mold, *-ic* relating to, *ret-* net, *-ic-* relating to, *-ul-* little, *-um* thing] *pl.,* smooth endoplasmic reticula (smooth en-doh-PLAZ-mik reh-TIK-yoo-lah)

tight junction (tite JUNK-shen)

CHAPTER SUMMARY

To download an MP3 version of the chapter summary for use with your iPod or other portable media player, access the Audio Chapter Summaries *online at http://evolve.elsevier.com.*

HINT *Scan this summary after reading the chapter to help you reinforce the key concepts. Later, use the summary as a quick review before your class or before a test.*

THE FUNCTIONAL ANATOMY OF CELLS

A. Cells and cell theory
  1. All organisms are composed of one or more cells
  2. Cell theory—states that the fundamental unit of all living things is the cell, and that all modern cells now come from previously existing cells
B. Typical cells
  1. Size—most human cells are microscopic in size
  2. Cell characteristics relate intimately to the functions they perform in the body
  3. Composite cell (Figure 3-1)
C. Basic cellular components (Table 3-2)
  1. Plasma membrane—separates the interior of the cell from the outside environment
  2. Cytoplasm—gel-like substance contained within the plasma membrane
     a. Cytosol—watery intracellular fluid
     b. Organelles—"little organs" suspended in the cytoplasm
  3. Nucleus—located within the cytoplasm; contains DNA

CELL MEMBRANES

A. Structure of cell membranes
  1. Each cell contains many internal organelles that are contained by or made of the same type of membrane
  2. Fluid mosaic model—the molecules that make up the cell membrane are arranged in sheets much like an art mosaic
     a. Made up mostly of phospholipids and cholesterol; studded with a variety of proteins and hybrid molecules such as glycoproteins and glycolipids (Figure 3-2)
  3. Cell membranes are composed of a double layer of phospholipids
     a. Each phospholipid has a hydrophilic "head" that faces either the interior of the cell or the cell's external environment
     b. Each phospholipid has a hydrophobic "tail" that faces the interior of the membrane
     c. Cholesterol—strengthens the membrane without losing flexibility; the cell membrane remains pliable
  4. Membrane proteins—hydrophobic and hydrophilic regions; they can easily fit among the phospholipids of the bilayer structure
B. Functions of cell membranes
  1. Specific cells may make their membrane more stiff or flimsy by altering types of phospholipids and the concentration of cholesterol
  2. Protein molecules—may be anchored to one or both sides of the membrane and facilitate the transport of important molecules from one side of the membrane to the other (Table 3-3)

  3. Specific types of proteins transport specific types of molecules
     a. Glycoproteins—serve as identification markers so that cells can determine the identity of other cells with which they come into contact
     b. Receptor proteins—protein reacts to a messenger molecule; triggers chemical reactions or other metabolic changes within a cell (signal transduction)

CYTOPLASM AND ORGANELLES

A. Cytoplasm—gel-like internal substance of cells that contains many tiny suspended structures
B. Organelles—"little organs" suspended in the cytoplasm
  1. Membranous—have sacs or canals made of cell membrane (Table 3-2)
  2. Nonmembranous—made of microscopic filaments or other particles
C. Basic organelles
  1. The endoplasmic reticulum and associated organelles
  2. Endoplasmic reticulum (ER)—network of canals and sacs made of cell membrane; extends from the nucleus throughout the cytosol to the plasma membrane
     a. Rough endoplasmic reticulum (RER)—has tiny organelles called ribosomes attached to it; site of protein synthesis (Figure 3-3)
     b. Smooth endoplasmic reticulum (SER)—lacks ribosomes; synthesizes membrane lipids, steroid hormones, and some carbohydrates that may be used to form glycoproteins
  3. Peroxisomes—harbor enzymes that remove toxic chemicals from the cytosol
  4. Golgi apparatus—processes and packages protein molecules delivered by small vesicles from the endoplasmic reticulum (Figures 3-5 and 3-6)
  5. Lysosomes—digest invading particles such as those from bacteria or viral infections; "digestive bags" or "cellular garbage disposals"
  6. Proteasomes—hollow, cylindrical, drumlike organelles responsible for breaking down abnormal and misfolded proteins from the ER
  7. Mitochondria—"power plant" of the cell; bound by two membranes that form a sac-within-a-sac structure (Figures 3-1 and 3-7)
     a. Inner membrane is the site of many metabolic reactions that transfer energy from the breakdown of glucose to ATP
     b. Cristae—intricate folds that greatly increase the surface area of the inner membrane
     c. Functional proteins in the membranes of the cristae are arranged precisely in the order that they are required to function

NUCLEUS

A. Nucleus—"control center" of the cell; commonly found near the center of the cell
B. Each nucleus has two membranes together called the nuclear envelope
  1. Nuclear envelope is perforated by pores that selectively allow molecules to enter or leave the fluid interior of the nucleus (nucleoplasm) (Figure 3-8)
C. Chromatin—long strands of DNA

D. Chromosomes—compact rodlike structures of tightly coiled DNA

E. Nucleolus—area of the nucleus that transcribes rRNA and assembles ribosome units

F. Functions of the nucleus are primarily functions of DNA molecules

CYTOSKELETON

A. Cytoskeleton—provides structural support for the cell and its components; composed of many microscopic fibers of different thicknesses
1. Microfilaments—composed of twisted strands of protein molecules, often lying parallel to one another; produce a shortening of the cell (as in muscle cells)
2. Intermediate filaments—similar to microfilaments but are slightly thicker and perform a primarily structural role
3. Microtubules—made of many protein subunits oriented in a spiral fashion; primarily move things around the cell and can even move the whole cell itself (Figure 3-10)

B. Centrosome—plays an important role in cell division; constructs a "spindle" of microtubules to move chromosomes around the cell (Figure 3-11)

C. Centrioles—cylinders consisting of nine bundles of microtubules, with three microtubules in each bundle
1. Performs important role in the formation of cellular extensions of the plasma membrane called cilia and flagella

D. Cell extensions—supported by cytoskeletal elements; present only in certain types of cells
1. Cilia—act as sensors or to brush fluid along surfaces of cells; shorter and more numerous than flagella
2. Flagella—single tail-like structure found on human sperm cell
3. Microvilli—found in the epithelial cells of the intestines; serve to increase the surface area of the lining of the intestines

CELL CONNECTIONS (FIGURE 3-13)

A. Desmosomes—Velcro-like spots or belts that hold adjacent cells together; common in adjacent skin cells

B. Gap junctions—linked channels of adjacent plasma membranes form gaps or "tunnels" that join the cytoplasm of cells into one functional unit; found in cardiac muscle

C. Tight junctions—occur in cells that are joined near their top surfaces by "collars" of tightly fused membranes; found in the lining of the intestines

REVIEW QUESTIONS

 Write out the answers to these questions after reading the chapter and reviewing the Chapter Summary. If you simply think through the answer without writing it down, you won't retain much of your new learning.

1. List the three main cell structures.
2. Explain the communication function of the plasma membrane, its transportation function, and its identification function.
3. Briefly describe the structure and function of the following cellular structures/organelles: endoplasmic reticulum, ribosomes, Golgi apparatus, mitochondria, lysosomes,

proteasomes, peroxisomes, cytoskeleton, cell fibers, centrosome, centrioles, and cell extensions.
4. Describe the three types of intercellular junctions. What are the special functional advantages of each?
5. Describe briefly the functions of the nucleus and the nucleoli.

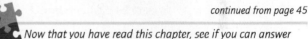

CRITICAL THINKING QUESTIONS

 After finishing the Review Questions, write out the answers to these items to help you apply your new knowledge. Go back to sections of the chapter that relate to items that you find difficult.

1. Using the complementarity principle that cell structure is related to its function, discuss how the shapes of the nerve cell and muscle cell are specific to their respective functions.
2. What is the relationship among ribosomes, endoplasmic reticulum, the Golgi apparatus, and plasma membrane? How do they work together as a system?

continued from page 45

Now that you have read this chapter, see if you can answer these questions about the experience of Sunil and his father in the oversized cell.

1. Which membrane structure did the small doorway represent, where hydrophilic molecules (people in this case) could pass through?
   a. A phospholipid
   b. A glycolipid
   c. A transmembrane cholesterol
   d. A channel protein

2. What organelle did the maze of tunnels, some bumpy, some smooth, represent?
   a. Golgi apparatus
   b. Endoplasmic reticulum
   c. Mitochondrion
   d. Microtubules

3. What were the "rocks" stuck to the outside of the walls of the maze?
   a. Lysosomes
   b. Peroxisomes
   c. Ribosomes
   d. Mitochondria

4. Which cell fiber type did the pipes (the thickest of the strands) in the cytoskeleton represent?
   a. Microtubules
   b. Microfilaments
   c. Microvilli
   d. Intermediate filaments

 To solve these questions, you may have to refer to the glossary or index, other chapters in this textbook, A&P Connect, Mechanisms of Disease, and other resources.

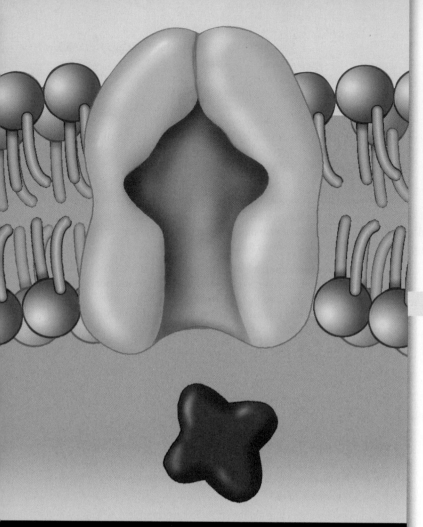

CHAPTER 4

Physiology of Cells

STUDENT LEARNING OBJECTIVES

At the completion of this chapter, you should be able to do the following:

1. Outline the basic differences between passive transport and active transport.
2. Describe simple diffusion, facilitated diffusion, and osmosis.
3. Distinguish between pinocytosis, phagocytosis, and filtration.
4. Discuss how cells use endocytosis and exocytosis.
5. Discuss the source of energy for both passive transport and active transport.
6. Demonstrate the effects of hypotonic, isotonic, and hypertonic conditions on living cells.
7. Describe the general role of enzymes.
8. Discuss how enzymes are named.
9. Define the phrase "cellular respiration."
10. Describe glycolysis, where it takes place, its energy output, and efficiency of energy production.
11. Describe the citric acid cycle, where it takes place, its energy output, and efficiency of energy production.
12. Describe the electron transport system, its location, its energy output, and efficiency of energy production.

CHAPTER OUTLINE

 Scan this outline before you begin to read the chapter, as a preview of how the concepts are organized.

TOBIE was visiting New York City for the first time. Being from a small town, he was fascinated by all the tall buildings and crowds of people. On his first day in the city, he took the subway across town. When the train finally arrived at its destination, Tobie was surprised to see a mob of people waiting on the subway platform; only a few people were already on the train. As the doors opened, the crowd surged forward through the doors and into the subway cars.

With the crush of people flooding onto the train, Tobie found himself being pushed toward the middle of the train car and he had to push his way against the crowd to get out of the train. In order for Tobie to keep contracting the muscles in his arms and legs as he pushed his way against the flow of the crowd, his muscle cells had to move sodium residing inside the cells back out of the cells after each contraction—*against the normal sodium concentration gradient.*

The transport of material across the cellular membrane is like Tobie's experience on the subway, and like Tobie, you'll see in this chapter how cell membrane transport is vital to all processes within our cells, regardless of the type of cell involved. Try to recall this analogy at the end of the chapter when we will again revisit Tobie's experiences on the subway.

• • •

The cell is the basic structural and functional unit of our bodies. So it should not be surprising that a good understanding of human physiology begins with a good overview of how cells function. You've already been introduced to the basic structures and some of the important functions of the cell, and we will build on those concepts in this and the following chapter.

We will begin by studying the transport of materials through the cell membrane. These transport mechanisms are vital to the proper function of all cells. They also provide a physical foundation for all of our physiological processes. We will then continue on to cell metabolism, the basis of human body chemistry as a whole, which will bring us to a discussion of cell growth and reproduction in Chapter 5.

MOVEMENT OF SUBSTANCES THROUGH CELL MEMBRANES

In Chapter 3 we saw how cells use the *cytoskeleton* to move organelles within the cytoplasm. But cells must also be able to communicate with one another. And they must be able to obtain nutrients from outside the cell membrane while removing waste products from inside the cell. Cells accomplish this transport and communication in one of two ways: **passive transport** or **active transport**. Simply put, *passive transport processes* do not require energy expenditure from the cell. However, *active transport processes* require the use of metabolic energy within the cell to "push" or "pull" substances across the membrane in the desired direction. Let's look at these processes more closely.

continued on page 75

BOX 4-1 Dialysis

Under certain circumstances, a type of diffusion called **dialysis** may occur. Dialysis is a form of diffusion in which the selectively permeable nature of a membrane causes the separation of smaller solute particles from larger solute particles. *Solutes* are the particles dissolved in a *solvent* such as water. Together, the solutes and solvents form a mixture called a *solution*.

Part *A* of the figure illustrates the principle of dialysis. A bag made of dialysis membrane—material with microscopic pores—is filled with a solution containing glucose, water, and albumin (protein) molecules. The bag is immersed in a container of pure water. Both water and glucose molecules are small enough to pass through the pores in the dialysis membrane. Albumin molecules, like all protein molecules, are very large and do not pass through the membrane's pores. Because of differences in concentration, glucose molecules diffuse out of the bag as small water molecules diffuse into the bag. Despite a concentration gradient, the albumin molecules do not diffuse out of the bag. Why not? Because they simply will not fit through the tiny pores in the membrane. After some time has passed, the large solutes are still trapped within the bag, but most of the smaller solutes are outside of it.

The principle of dialysis can be used in medicine to treat patients with kidney failure. In **hemodialysis** (part *B* of the figure), blood pumped from a patient is exposed to a dialysis membrane that separates the blood from a clean, osmotically balanced dialysis fluid. Small solutes such as urea and various ions can diffuse through the membrane to reach an equilibrium, thus removing them from the blood. The larger plasma proteins (including albumin) and blood cells remain in the blood, which is returned to the patient's body. In hemodialysis, the process of dialysis is used to "clean up" the patient's blood because the kidneys have failed to perform this task.

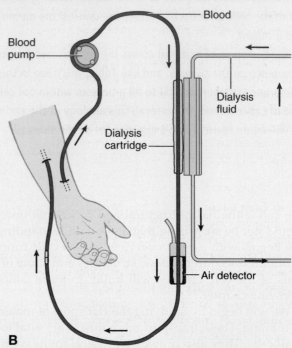

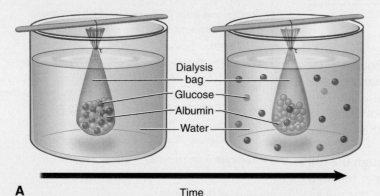

A Time

▲ **A, Dialysis.** A dialysis bag containing glucose, water, and albumin (protein) molecules is suspended in pure water. Over time, the smaller solute molecules (glucose) diffuse out of the bag. The larger solute molecules (albumin) remain trapped in the bag because the bag is impermeable to them. Thus dialysis is diffusion that results in separation of small and large solute particles.

B

▲ **B, Hemodialysis.** In hemodialysis, the patient's blood is pumped through a dialysis cartridge, which has a semipermeable membrane that separates the blood from the clean dialysis fluid. As dialysis occurs, some of the urea and other small solutes in the blood diffuse into the dialysis fluid, whereas the larger solutes (plasma proteins) and blood cells remain in the blood.

Passive Transport Processes

Diffusion

Diffusion describes the tendency of small particles to spread out and distribute themselves within any given space, from an area of higher concentration to an area of lower concentration. The energy for diffusion is provided by the *energy of the molecules* within the cell system. Even though you can't see them, molecules in a warm solution bounce around in short, chaotic paths and tend to spread out, or diffuse, while doing so. The warmer the solution, the more collisions there are, and the faster the diffusion. If you put a package of sugar in your coffee, your coffee isn't sweet right away because at first all of the sugar molecules are still packed closely together. Given enough time, however (and perhaps some stirring to speed

up the process), the sugar molecules will gradually spread out and bounce off of one another and other molecules in your coffee until they are evenly spaced throughout (Figure 4-1). In effect, the sugar molecules move away from an area of *high concentration* (where they are tightly packed crystals at the bottom of the cup) toward an area of *low concentration* (the surrounding coffee) until there is an *equilibrium* of relatively constant concentration throughout the coffee.

When molecules are dissolved together like this, the final product is called a **solution.** All solutions are made up of molecules to be dissolved—**solutes** (the sugar) that diffuse and dissolve throughout a **solvent** (the coffee, in this case).

Molecules tend to move across porous membranes in a similar fashion from places of high concentration to areas of low concentration. This action is called moving *down* a concentration gradient. A **concentration gradient** is simply the measurable difference of solute concentration from one area of a solution to another.

In Figure 4-2 we see a container with a membrane separating two solutions made up of water (the solvent) and the same type of purple particles (the solutes). In Figure 4-2, *A*, the solution on the right has a higher concentration (20%) of solute than the solution on the left (10%). Now, suppose that the membrane separating the solutions has pores that are large enough to allow both solute and water molecules to pass through them. Over time, as the solute and water molecules collide with each other and the membrane, some will inevitably go into the pores and pass through the membrane. For a while, more solute particles enter the pores from the 20% side simply because they are more numerous there than on the 10% side. The water molecules will similarly move from the 10% side (which is 90% water) to the 20% side (80% water). Both types of molecules can be said to be moving *down* their respective concentration gradients.

Eventually, diffusion of both types of molecules produces a condition in which both sides of the membrane have solutions with equal concentrations of solute. At that point in time we say that a **dynamic equilibrium** has occurred (Figure 4-2, *B*). This dynamic equilibrium is a *balanced state* in which diffusion *does* continue across the membrane, but without any change in the equilibrium.

Simple Diffusion

Recall from Chapter 2 that the phospholipid bilayer of a cell membrane is *hydrophilic* both on the outside and inside but *hydrophobic* in the middle. Because of this hydrophobic middle portion, many lipid-soluble molecules can pass through a cell membrane easily, as can smaller hydrophobic molecules such as oxygen (O_2) and carbon dioxide (CO_2). However, small charged particles such as water (H_2O) and urea can only diffuse in small amounts across membranes. This information is summarized for you in Figure 4-3. When molecules pass directly through a membrane, the process is called **simple diffusion.**

Molecules that diffuse across a membrane are said to **permeate** the membrane. For this reason they are called *permeant molecules.* As you might suspect, those molecules that cannot diffuse readily are said to be *impermeant molecules.* Similarly,

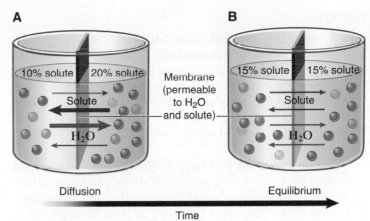

A **B**

10% solute / 20% solute Membrane (permeable to H_2O and solute) 15% solute / 15% solute

Solute Solute

H_2O H_2O

Diffusion Equilibrium

Time

▲ **FIGURE 4-2** **Diffusion through a membrane.** Note that the membrane allows solute (a dissolved particle) and water to pass and that it separates a 10% solution from a 20% solution. The container on the left (**A**) shows the two solutions separated by the membrane at the start of diffusion. The container on the right (**B**) shows the result of diffusion after time.

membranes are said to be *permeable* to molecules that readily diffuse across them, but *impermeable* to molecules that do not. A very important practical application of this concept is *dialysis*, shown for you in Box 4-1.

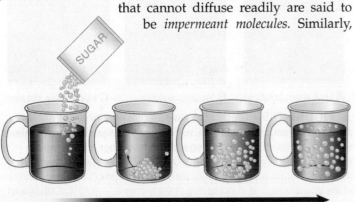

▲ **FIGURE 4-1** **Diffusion.** The molecules of sugar are very densely packed when they enter the coffee. As sugar molecules collide frequently in the area of high concentration, they gradually spread away from each other— toward the area of lower concentration. Eventually, the sugar molecules become evenly distributed. That is, they reach an equilibrium of concentration.

Time

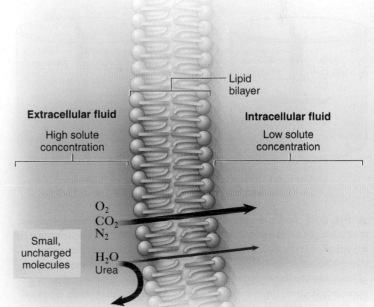

Lipid bilayer

Extracellular fluid **Intracellular fluid**

High solute concentration Low solute concentration

O_2
CO_2
N_2

Small, uncharged molecules

H_2O
Urea

▲ **FIGURE 4-3** **Simple diffusion through a phospholipid bilayer.** Some small, uncharged molecules can easily pass through the phospholipid membrane, but water and urea (a waste product of protein catabolism) rarely get through the membrane. Larger uncharged molecules and ions (charged molecules) may not pass through the phospholipid membrane at all.

Osmosis

Strictly speaking, **osmosis** is the diffusion of water through a selectively permeable membrane. (A membrane is called **selectively permeable** if it allows some molecules to diffuse through but not others.) As we've seen in Figure 4-3, water, because it is a polar molecule, does not readily diffuse through the plasma membrane. Instead, osmosis is accomplished by special water-conducting channels called *aquaporins.*

Now, imagine that you have a 20% solution of albumin separated by a selectively permeable membrane from a 10% albumin solution (Figure 4-4). Assume that the membrane has numerous aquaporins for water to pass through but is impermeable to albumin (because it is a very large protein that is too big to pass through the membrane by diffusion). In this case, the water molecules will diffuse or *osmose* down water's concentration gradient from the 10% albumin solution (90% water) to the 20% albumin solution (80% water) in a passive effort to make both sides the same concentration. As in our previous example, a dynamic equilibrium is reached. The difference here is that the membrane is *only* permeable to water; thus one side of the membrane will *gain volume* and rise and the other will *lose volume* and fall in order to generate equal concentrations of solute (15%).

Cells act a bit differently than the example in Figure 4-4 because they are closed containers (enclosed by their plasma membranes). Changes in volume of these cellular compartments will also mean a change in pressure. Think of a water balloon. If you add water to the balloon, there will be an increase in volume as well as in the pressure on the inside of the balloon walls. A similar change happens if water is added to the inside of a cell. Water pressure that develops in a solution as a result of osmosis is called **osmotic pressure.** So, we say that osmotic pressure develops in the solution that originally has the higher concentration of impermeant solute (the 20% albumin solution in Figure 4-4, for example).

Osmotic pressure becomes very important in physiology and medicine because healthy functioning of human cells requires a relatively constant volume and pressure. When the intracellular (inside the cell) and extracellular (outside the cell) fluids have the same potential osmotic pressure, they are said to be **isotonic** (*iso-* means "same," *-tonic* means "pressure") to each other (Figure 4-5, *B*). Thus, isotonic solutions have the same concentration of impermeant solutes.

In order to maintain isotonicity throughout our bodies, plasma membranes allow water to enter and exit according to whether there are more impermeant solutes in the intracellular or extracellular fluid. If the extracellular space has a lower concentration of impermeant solutes, it is said to be **hypotonic** (Figure 4-5, *A*). This will cause water to move into

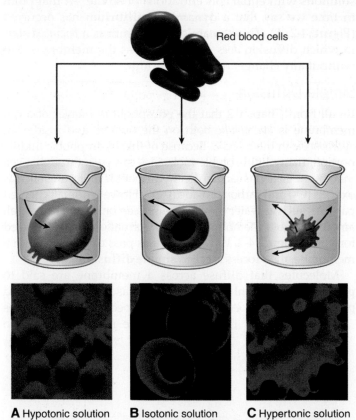

A Hypotonic solution   **B** Isotonic solution   **C** Hypertonic solution

▲ **FIGURE 4-5**   **Effects of osmosis on cells. A,** Normal red blood cells placed in a hypotonic solution may swell (as the scanning electron micrograph shows) or even burst (as the drawing shows). This change results from the inward diffusion of water (osmosis). **B,** Cells placed in an isotonic solution maintain constant volume and pressure because the potential osmotic pressure of the intracellular fluid matches that of the extracellular fluid. **C,** Cells placed in a solution that is hypertonic to the intracellular fluid lose volume and pressure as water osmoses out of the cell into the hypertonic solution. The "spikes" seen in the scanning electron micrograph are rigid microtubules of the cytoskeleton. These supports become visible as the cell "deflates."

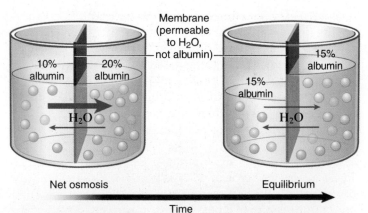

▲ **FIGURE 4-4**   **Osmosis.** Osmosis is the diffusion of water through a selectively permeable membrane. Because there are relatively more water molecules in 10% albumin than in 20% albumin, more water molecules osmose from the more dilute into the more concentrated solution (as indicated by the *larger arrow* in the left diagram) than osmose in the opposite direction. The overall direction of osmosis, in other words, is toward the more concentrated solution. Net osmosis produces the following changes in these solutions: (1) their concentrations equilibrate, (2) the volume and pressure of the originally more concentrated solution increase, and (3) the volume and pressure of the other solution decrease proportionately.

the cell, which achieves isotonicity. If the extracellular fluid has a higher concentration of impermeant solutes than the intracellular fluid, it is said to be **hypertonic** (Figure 4-5, C). This will cause water to move out of the cell, which likewise achieves isotonicity.

If the extracellular fluid is greatly hypotonic, so much water may rush into the cell that it may burst or *lyse* (see Figure 4-5, A). On the other hand, if the extracellular fluid is greatly hypertonic, it can cause a cell to shrivel and shrink. Either of these outcomes may cause cell death.

In summary, just as molecules diffuse down a concentration gradient, water will always osmose from a hypotonic (*higher* water concentration, fewer impermeant solutes) to a hypertonic (*lower* water concentration, more impermeant solutes) solution.

Facilitated Diffusion

As with water, many charged and water-soluble particles that are necessary for life cannot pass readily through a cell membrane. Many of these particles, such as sodium ions (Na^+), are brought in by a process called **facilitated diffusion.** In this passive process, special membrane channels and transport proteins in the plasma membrane carry the particles inside the cell.

Channel-Mediated Passive Transport

Membrane channels are protein pores that span (extend completely through) cell membranes. These channels allow water molecules, specific ions, or other small, water-soluble molecules to cross the membrane, moving down their concentration gradient. Most are very specific, such as *sodium channels* and *chloride channels,* which permit only sodium (Na^+) and chloride (Cl^-) ions to pass, respectively.

The relative permeability of a membrane to a molecule can also be affected by the opening and closing of membrane

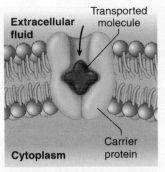
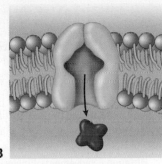

▲ **FIGURE 4-7** **Membrane carrier.** In carrier-mediated transport, a membrane-bound carrier protein attracts a solute molecule to a binding site **(A)** and changes shape in a manner that allows the solute to move to the other side of the membrane **(B)**. Passive carriers may transport molecules in either direction, depending on the concentration gradient.

channels that act like swinging gates in a fence (Figure 4-6). For this reason, they are sometimes called *gated channels*. Some gated channels are triggered to open by electrical changes (voltage). Others are stimulated to open by light, and even mechanical or chemical stimuli. Gated channels provide another way for membranes to be selectively permeable. Many membrane channels only allow certain molecules to pass in either direction (depending on the concentration gradient). Others only allow movement in one direction (when the concentration gradient permits). Because ions move down their concentration gradients as they pass through channels, this type of facilitated diffusion is passive and thus is called **channel-mediated passive transport.**

Carrier-Mediated Passive Transport

Molecules may also move down their concentration gradient in either direction across a membrane by passing through a protein called a *membrane carrier.* As you might guess, this process is termed **carrier-mediated passive transport** (Figure 4-7). These membrane *carriers* have a particular structure that attracts and binds a particular solute molecule. The carrier then changes shape to release the bound molecule inside or outside the cell.

Filtration

Filtration, another type of passive transport (but technically *not* diffusion), is discussed in Box 4-2. At this point you should understand that passive transport processes are critical in maintaining a homeostatic balance of many vital substances within the cells of our bodies. Malfunctions in these processes can result in many serious diseases and even death, as you will see later in this chapter.

Role of Passive Transport Processes

Diffusion and filtration are always *passive processes* in which the energy for transport across a membrane comes from molecular movements and collisions, and not from the cell. Because these

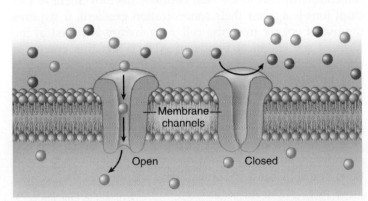

▲ **FIGURE 4-6** **Membrane channels.** Gated channel proteins form tunnels through which only specific molecules may pass—as long as the "gates" are open. Molecules that do not have a specific shape and charge are never permitted to pass through the channel. Notice that the transported molecules move from an area of high concentration to an area of low concentration. The cell membrane is said to be permeable to the type of molecule in question.

BOX 4-2 | Filtration

Another important passive process for transport in the body is **filtration.** This form of transport involves the passage of water and permeable solutes through a membrane by the *force of hydrostatic pressure.* **Hydrostatic pressure** is the force, or weight, of a fluid pushing against a surface. Strictly interpreted, this process is *not* the same as diffusion for that reason.

Filtration is movement of molecules through a membrane from an area of high hydrostatic pressure to an area of low hydrostatic pressure—that is, down a *hydrostatic pressure gradient.* Filtration most often transports substances through a sheet of cells. The force of pressure pushes the molecules through or between the cells that form the sheet. Because the filtration membrane does not allow larger particles through, filtration results in the separation of large and small particles, as you can see in the figure. This is similar to dialysis, except that dialysis is driven by a *concentration gradient.* Filtration is instead driven by a *hydrostatic pressure gradient.*

A simple model of filtration is found in many drip-type coffee makers. Ground coffee is placed in a porous paper filter cup in an upper container and boiling water is added. Gravity pulls downward on the mixture in the upper container, generating hydrostatic pressure against the bottom of the filter. The pores in the paper filter are large enough to let water molecules and other small particles pass through to a coffee pot below the filter. Most of the coffee grounds are too large to pass through the filter. The coffee in the pot below is called the *filtrate.*

How and where does filtration occur in the body? Most often, it occurs in tiny blood vessels called *capillaries,* which are found throughout the body. Hydrostatic pressure of the blood (blood pressure) generated by heart contractions, gravity, and other forces pushes water and small solutes out of the capillaries and into the interstitial spaces of a tissue. Blood cells and large blood proteins are too large to fit through pores in the capillary wall; therefore, they cannot be filtered out of the blood. Capillary filtration allows the blood vessels to supply tissues with water. Capillary filtration is also the first step used by the kidneys to form urine.

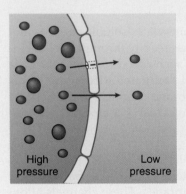

High pressure **Low pressure**

◀ **Filtration.** Particles small enough to fit through the pores in the filtration membrane move from the area of high hydrostatic pressure to the area of low hydrostatic pressure. This results in separation of small particles from larger ones.

---

substances move down their concentration or hydrostatic pressure gradients, passive transport tends to maintain equilibrium of these substances on both sides of the membrane (Table 4-1).

> **QUICK CHECK**
>
> 5. Define osmosis.
> 6. Compare a hypotonic solution with a hypertonic solution.
> 7. What is channel-mediated diffusion?
> 8. Define carrier-mediated diffusion.

Active Transport Processes

In passive transport, the energy for movement of molecules across a membrane comes from the concentration gradient and the movement of the molecules themselves. The energy for the movement of these molecules is provided by the energy of the entire solution. (The temperature of your body is a reflection of the energy contained by the molecules making up your body.) **Active transport,** however, requires the cell to expend *metabolic energy* to move particles across cell membranes *against the concentration gradient.*

Transport by Pumps

Membrane pumps use energy supplied by the cell to transport molecules *against* their concentration gradient (that is, from low concentration to high concentration). First, note that this process is exactly the opposite of facilitated diffusion. Think, for a moment, of rolling a ball uphill. Clearly, the ball will not perform

this task by itself! Rather, you must input energy to move the ball by pushing it *against* the force of gravity. This idea is similar to moving a molecule *against* its concentration gradient.

Some cells require the maintenance of a concentration gradient. Muscle cells, for instance require a very low intracellular calcium ion (Ca^{++}) concentration. They use active transport to move most of these ions into the extracellular fluid or into special compartments in the cell. Because this movement of calcium ions is against their concentration gradient, it requires the use of energy from the cell. This energy is stored in the form of the chemical bonds of ATP. When this energy is released and used, the bond is broken, the work is done (in this case, moving calcium ions), and ADP and a phosphate result.

Another type of pump that requires the use of ATP is the membrane protein called *sodium-potassium adenosine triphosphatase.* This **sodium-potassium pump,** as the name suggests, moves sodium and potassium ions across the cell membrane, but in *different* directions. Figure 4-8 shows that three Na^+ ions bind to sodium binding sites on the pump's inner face. At the same time, an ATP molecule produced by the cell's mitochondria binds to the pump. Again, when the ATP breaks apart, its stored energy is transferred to the pump and causes the pump to change shape. The pump releases the three Na^+ ions to the outside of the cell and attracts two K^+ ions to its outer face. The pump then returns to its original shape and releases the two potassium ions and the ADP and phosphate remnants of the ATP molecule to the intracellular fluid. Spend a few minutes to digest this important concept because it will be revisited throughout this book.

TABLE 4-1 Passive Transport Processes

PROCESS		DESCRIPTION	EXAMPLES
Simple diffusion		Movement of particles through the phospholipid bilayer or through channels from an area of high concentration to an area of low concentration—that is, down the concentration gradient	Movement of carbon dioxide out of all cells
Osmosis		Diffusion of water through a selectively permeable membrane in the presence of at least one solute that cannot diffuse passively through the membrane (often involves both simple and channel-mediated diffusion)	Diffusion of water molecules into and out of cells to correct imbalances in water concentration
Channel-mediated passive transport (facilitated diffusion)		Diffusion of particles through a membrane by means of channel structures in the membrane (particles move down their concentration gradient)	Diffusion of sodium ions into nerve cells during a nerve impulse
Carrier-mediated passive transport (facilitated diffusion)		Diffusion of particles through a membrane by means of carrier structures in the membrane (particles move down their concentration gradient)	Diffusion of glucose molecules into most cells

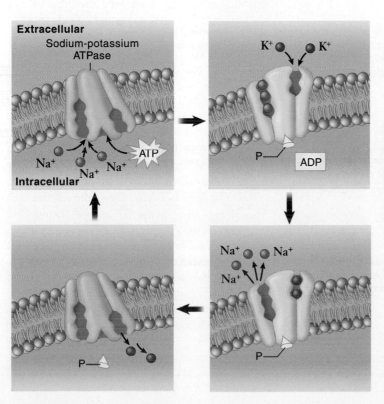

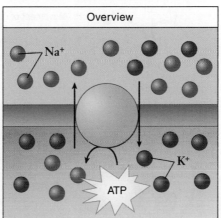

◀ **FIGURE 4-8 Sodium-potassium pump.** Three sodium ions (Na^+) bind to sodium binding sites on the pump's inner face. At the same time, an energy-containing adenosine triphosphate (ATP) molecule produced by the cell's mitochondria binds to the pump. The ATP breaks apart, and its stored energy is transferred to the pump. The pump then changes shape, releases the three Na^+ ions to the outside of the cell, and attracts two potassium ions (K^+) to its potassium binding sites. The pump then returns to its original shape, and the two K^+ ions and the remnant of the ATP molecule are released to the inside of the cell. The pump is now ready for another pumping cycle. *ATPase,* Adenosine triphosphatase. The small inset is a simplified view of Na^+-K^+-pump activity.

Transport by Vesicles

Mechanisms that carry large groups of molecules into or out of the cell by means of *vesicles* (membranous packages) also require metabolic energy. These bulk transport mechanisms differ from pump mechanisms in that they allow substances to enter or leave the interior of a cell without actually moving *through* its plasma membrane (Figure 4-9). Vesicles act like delivery trucks pulling up to the dock at a warehouse.

Major mechanisms of active transport are summarized for you in Table 4-2.

Endocytosis

In **endocytosis,** the plasma membrane "traps" extracellular material and brings it into the cell. In this process, the cytoskeleton does all the work by pulling part of the plasma membrane inward to form a bubblelike depression. The depression then creates a membrane-bound "bubble" called a vesicle that can be moved throughout the cell by the cytoskeleton.

In **receptor-mediated endocytosis,** special receptor molecules of the plasma membrane first bind to specific molecules in the extracellular fluid (Figure 4-10). This signal triggers that portion of the membrane to be pulled *inward* by the cytoskeleton, forming a vesicle containing the receptor-bound molecules. Other molecules may also be taken in during this process.

Phagocytosis means "cell eating." It is a type of endocytosis that engulfs large particles or microorganisms, forming vesicles within the cell. These vesicles then fuse with the membranous walls of lysosomes. The enzymes within the lysosomes break down these larger particles into smaller molecules.

Pinocytosis, or "cell drinking," is the endocytosis of fluids and the substances dissolved in them. Pinocytosis is also used to remove material such as membrane receptors and transporters from the external cell membrane, thus regulating the function of that cell's membrane.

Exocytosis

Exocytosis is a process by which large molecules (packaged in vesicles) can leave the cell even though they are too large to move out through the plasma membrane. Exocytosis is much like the reverse of endocytosis. Proteins and other large molecules are first enclosed in vesicles in the Golgi apparatus (see Chapter 3). Then the packed vesicles are transported through the plasma membrane by the cytoskeleton of the cell. Finally, the vesicle membranes fuse with the existing plasma membrane of the cell and release their contents outside into the intercellular space.

Through exocytosis, the cell also can add new membrane to its existing membrane. (Imagine how little soap

TABLE 4-2 Active Transport Processes

PROCESS		DESCRIPTION	EXAMPLES
Pumping		Movement of solute particles from an area of low concentration to an area of high concentration (up the concentration gradient) by means of an energy-consuming pump structure in the membrane	In muscle cells, pumping of nearly all calcium ions to special compartments—or out of the cell
Phagocytosis (endocytosis)		Movement of cells or other large particles into a cell by trapping it in a section of plasma membrane that pinches off to form an intracellular vesicle; a type of *vesicle-mediated transport*	Trapping of bacterial cells by phagocytic white blood cells
Pinocytosis (endocytosis)		Movement of fluid and dissolved molecules into a cell by trapping them in a section of plasma membrane that pinches off to form an intracellular vesicle; a type of *vesicle-mediated transport*	Trapping of large protein molecules by some body cells
Exocytosis		Movement of proteins or other cell products out of the cell by fusing a secretory vesicle with the plasma membrane; a type of *vesicle-mediated transport*	Secretion of the hormone prolactin by pituitary cells

Golgi apparatus

Lysosome

Fusion of vesicle with lysosome

Release of contents of vesicle

Membrane-bound vesicle

Digestion by enzymes

Exocytosis

Endocytosis

Particle

Membrane-bound vesicle

◄ **FIGURE 4-9   Bulk transport by vesicles.** This sketch summarizes the essential difference between endocytosis, which moves substances into the cell by means of a vesicle, and exocytosis, which moves substances out of the cell by means of a vesicle. The type of endocytosis shown here is phagocytosis, in which the endocytic vesicle fuses with a lysosome to allow digestive enzymes to break down the ingested material.

bubbles pop and join to make bigger soap bubbles with larger membranes.)

Role of Active Transport Processes

Whenever you think of active transport processes, think of the need for cellular energy. Active transport processes such as ion

pumps, endocytosis, and exocytosis all transfer substances across membranes using cellular energy (see Table 4-2). Ion pumps can be used to keep ions at high or low concentrations within organelles such as the smooth endoplasmic reticulum or mitochondria. In addition, the cytoskeleton acts on vesicles of varying sizes to engulf needed substances, or to add/remove

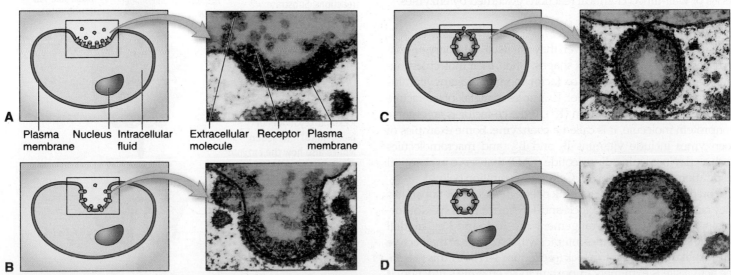

A

Plasma membrane   Nucleus   Intracellular fluid

Extracellular molecule   Receptor   Plasma membrane

B

C

D

▲ **FIGURE 4-10   Receptor-mediated endocytosis.** An artist's interpretation *(left)* and transmission electron micrographs *(right)* show the basic steps of receptor-mediated endocytosis. **A,** Membrane receptors bind to specific molecules in the extracellular fluid. **B,** A portion of the plasma membrane is pulled inward by the cytoskeleton and forms a small pocket around the material to be moved into the cell. **C,** The edges of the pocket eventually fuse and form a vesicle. **D,** The vesicle is then pulled inward—away from the plasma membrane—by the cytoskeleton. In this example, only the receptor-bound molecules enter the cell. In some cases, some free molecules or even entire cells may also be trapped within the vesicle and transported inward.

bits of membrane and receptors to or from the membrane. These activities alter membrane function. Secretion of important hormones and neurotransmitters is often accomplished by vesicles in exocytosis, as you will see throughout this book.

QUICK CHECK

9. Does active transport always require energy from the cell?
10. Give an example of transport by pumps.
11. Compare exocytosis with endocytosis.
12. What is receptor-mediated endocytosis?

CELL METABOLISM

Cell **metabolism** refers to all of the chemical reactions of a cell and is the basis for all human functions. Cell metabolism involves many different kinds of chemical reactions that often occur in a sequence called a **metabolic pathway.** These pathways can be *catabolic* if the net result is a breakdown of molecules. They also can be *anabolic*, where the chemical reactions in the pathway build larger molecules from smaller ones. Anabolic reactions usually require the input of energy while catabolic reactions usually release energy.

Role of Enzymes

As you probably know, chemical reactions generally do not happen on their own. In fact, most reactions require an input of *activation energy* to begin (even if the net result of the chemical reaction is a gain in energy). **Enzymes** are *functional proteins* that serve as **catalysts** for nearly all of our chemical reactions. They *catalyze* or lower the required energy needed to start chemical reactions (Figure 4-11). Catalysts participate in chemical reactions but are not changed by the reactions. They are so important to the essential reactions of life that you could think of life as a series of well-timed chemical reactions governed by enzymes.

Chemical Structures of Enzymes

Enzymes are proteins. However, they are usually tertiary or quaternary proteins with complex shapes. Often, their molecules contain a nonprotein part called a **co-factor.** Inorganic ions or vitamins may make up part of a co-factor. Examples of metal ion co-factors are zinc (Zn^{++}) and potassium (K^+). If the co-factor is an organic nonprotein molecule, it is called a **coenzyme.** Some examples of coenzymes include vitamins B_6 and B_{12}, and macromolecules such as flavin adenine dinucleotide (FAD) that serve in chemical respiration (see the section on Cellular Respiration below).

You may find it helpful to think of enzymes and the substrates they act upon as parts of a "lock-and-key" arrangement. Not surprisingly, we call this arrangement the **lock-and-key model!** Specific substrate molecules interact with enzymes at their **active site** just like a key fits into a lock (see Figure 4-11). For this reason, the active site is extremely important to an enzyme's function. If it is altered in some way, the substrate will not fit into the active site and the chemical reaction may not be carried out. This, in turn, may halt the entire metabolic pathway, which could be disastrous for the cell and even the entire human body. Different types of enzymes are discussed throughout this book.

Classification and Naming of Enzymes

Enzymes are named by combining the root name of the substance they act upon, or the word that describes the kind of chemical reaction that is catalyzed, with the suffix *-ase*, which is placed at the end of the word. Thus, sucrase is an enzyme that acts upon the sugar substrate *sucrose*. (This enzyme might also be called *sucrose hydrolase*, because it *hydrolyzes* sucrose.) However, you should note that some enzymes retain older names that were used before this method was adopted.

General Functions of Enzymes

Most enzymes are *specific in their action*. This means that they act only on a specific substrate or type of substrate. Look for a moment at Figure 4-12. Note that every reaction that occurs in a metabolic pathway requires one or more specific enzymes; otherwise the entire pathway will be disrupted. Metabolic pathways can be naturally turned "on" or "off." This is accomplished by the activation or inactivation of certain enzymes that control regulatory processes of the cell.

There are a variety of physical and chemical agents that can activate or inhibit enzyme action simply by changing the shape of the enzyme. These agents literally alter the "key" in the lock-and-key model. A molecule or other agent that alters enzyme function by *changing its shape* is called an **allosteric effector.** Recall that changing the shape of any protein will change its function. Changes in pH or temperature, and some antibiotic drugs, can act as allosteric effectors.

▶ **FIGURE 4-11**
Model of enzyme action. Enzymes are functional proteins whose molecular shape allows them to catalyze chemical reactions. Substrate molecule AB is acted on by a digestive enzyme to yield simpler molecules A and B as products of the reaction. Notice how the active site of the enzyme chemically fits the substrate—the lock-and-key model of biochemical interaction. Notice also how the enzyme molecule bends its shape in performing its function.

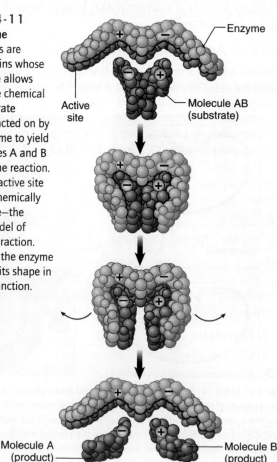

Enzyme
Active site
Molecule AB (substrate)
Molecule A (product)
Molecule B (product)

▶ FIGURE 4-12
Enzyme regulation of a metabolic pathway. In a metabolic pathway, the product of one enzyme-regulated reaction becomes the substrate for the next reaction. Thus a whole series of enzymes is required to keep the pathway functioning. Notice that these enzymes are embedded in a cell membrane, whereas other types of enzymes are mobile in the cytosol.

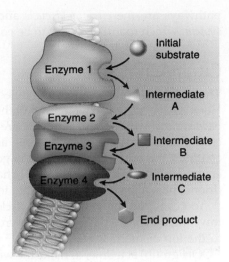

Changes in temperature, pH, and other factors may destroy the configuration of an enzyme in a process called **denaturation.** Denatured enzymes lose their shape and are unable to perform their functions.

Even though enzymes are not "used up" in the reactions they catalyze, they are continually being destroyed, and therefore they must continually be synthesized. In many cases, enzymes are synthesized as *inactive* **proenzymes,** which require substances (typically other enzymes) called **kinases** for them to become active. Kinases usually act in an allosteric fashion on proenzymes to activate them. For example, *enterokinase* changes inactive trypsinogen into active trypsin by changing the shape of the molecule (see Chapter 21).

CELLULAR RESPIRATION

Catabolism is the metabolic process that breaks down complex organic molecules into smaller ones, usually releasing energy in the process. Of all of the catabolic pathways that occur in cells, perhaps the most important is that of **cellular respiration.** This is a continuous process required of all cells by which nutrients such as carbohydrates, proteins, and fats are broken down into carbon dioxide (CO_2) and water (H_2O). As the molecule breaks down, the potential energy that had been stored in the nutrient's bonds is released. Much of this released energy is converted to heat. However, about 40% of the energy is transferred to the high-energy bonds of adenosine triphosphate (ATP), which is later used to drive many types of cellular processes.

You should always think of cellular respiration as a continuous cycle. However, for convenience, we divide cellular respiration into three smaller pathways:

1. Glycolysis
2. Citric acid cycle
3. Electron transport system

Glycolysis

Glycolysis is a catabolic process that occurs inside the cytosol of cells. Glycolysis literally means, "breaking glucose." In Figure 4-13 you can see *one* glucose molecule (six carbon

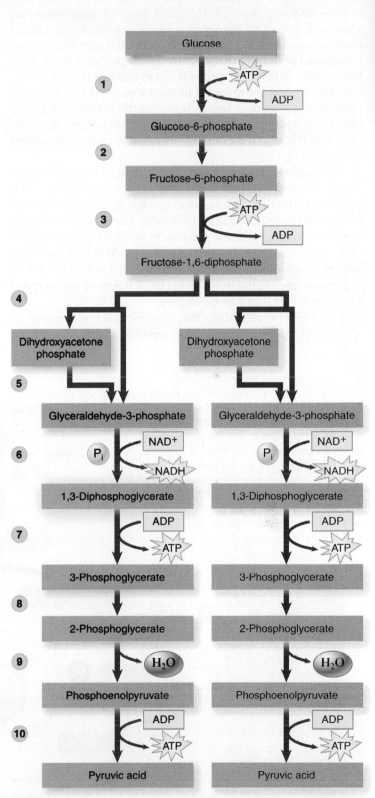

▲ FIGURE 4-13 **Glycolysis.** This diagram of the reactions involved in glycolysis represents a classic example of a catabolic pathway. Note that each of the chemical reactions in this pathway cannot proceed until the previous step has occurred. Recall from our discussion that each step also requires the presence of one or more specific enzymes. *ADP,* Adenosine diphosphate; *ATP,* adenosine triphosphate; *NAD+* and *NADH,* oxidized and reduced forms of nicotinamide adenine dinucleotide; *P$_i$,* inorganic phosphate.

atoms) broken into *two* molecules of pyruvic acid (three carbon atoms per molecule). Because glycolysis does not require oxygen, we call this first part of respiration an **anaerobic** process. Glycolysis actually *requires* energy input from cellular ATP to begin (steps 1 and 3). Later this energy is regained at the end of the pathway (steps 7 and 10). In fact, the amount of energy input early in glycolysis is greatly outweighed by the energy gain in the form of ATP in the later pathways of cellular respiration. Another energy transfer molecule, the reduced form of *nicotinamide adenine dinucleotide* (NADH), also temporarily stores some energy generated in glycolysis.

The pyruvic acid generated by glycolysis can be transferred to one of two pathways. If oxygen is not available, the pyruvic acid will continue on an anaerobic pathway to form a molecule called *lactic acid.* This molecule is later converted back to pyruvic acid or glucose in another pathway that requires oxygen and the input of energy. Much of the lactic acid diffuses out of the cells in which it formed, and it is processed later in the liver. The cells of the liver are adapted to perform this function efficiently.

If oxygen *is* available, the pyruvic acid generated in glycolysis will transfer to the **aerobic** pathways. These

pathways are the *citric acid cycle* and the *electron transport system*, which require oxygen to be completed.

Citric Acid Cycle

The **citric acid cycle** is sometimes called the *Krebs cycle* after Sir Hans Krebs, who discovered the pathway. It is called a *cycle* because some of the end products are similar to those molecules that begin the cycle and some are inevitably reused to continue the pathway. The cycle is governed by a series of enzymes that are found in the inner chamber of the mitochondrion. The major steps are illustrated for you in Figure 4-14.

You can think of the citric acid cycle as a carbon "wagon" traveling uphill to which we attach pyruvic acid. In order to get to the top of the hill, we "throw off" one carbon atom (as a COOH group) at a time, to make the wagon "lighter." In doing so, each carbon rolls back down the hill to produce a molecule of carbon dioxide (CO_2). This rolling action releases the potential energy that was holding the carbon to the wagon. The potential energy is transferred to the energy-transfer molecules: the reduced form of nicotinamide adenine dinucleotide (NADH), flavin adenine dinucleotide (FAD), and ATP. The wagon is returned to its original form and can now be used again.

▶ **FIGURE 4-14** **Citric acid cycle.** The citric acid cycle is a circular metabolic pathway that breaks down an acetyl molecule with the release of CO_2 molecules and energized electrons (which along with their protons [H^+], are shuttled away by the coenzymes nicotinamide adenine dinucleotide [NAD] and flavin adenine dinucleotide [FAD]). *ATP,* Adenosine triphosphate.

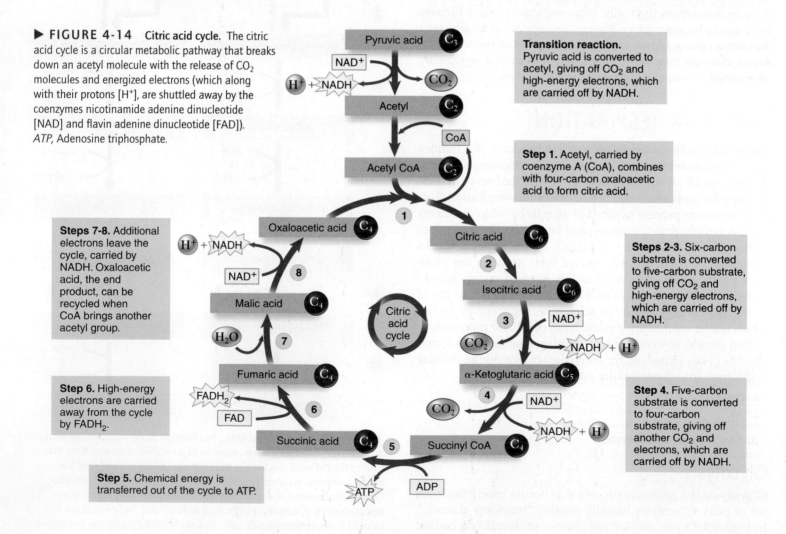

Electron Transport System

The energy transferred from the citric acid cycle is used to put protons behind a sort of "dam." Then the energy of protons flowing back through "energy generators" (ATP synthase) is stored in ATP. This is similar to how a hydroelectric dam traps water and then uses the flow of water through special "water wheels" in the dam to generate electricity. In cells, this is done in the **electron transport system** (Figure 4-15). In most cells, aerobic respiration transfers enough energy from each glucose molecule to form a net gain of 36 to 38 ATP molecules. The respiratory process also produces carbon dioxide and water as waste products. The major steps of all three aerobic respiratory pathways—glycolysis, the citric acid cycle, and the electron transport system—are summarized in Figure 4-16.

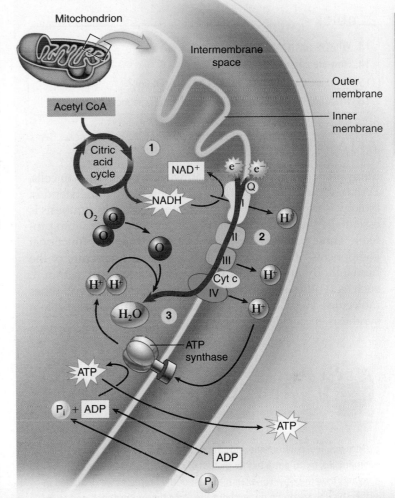

▶ **FIGURE 4-15** **Electron transport system (ETS). 1,** Pairs of high-energy electrons (e⁻) and their protons (H⁺) are shuttled from the citric acid cycle by coenzymes NAD and FAD to protein complexes *(I, II, III, IV)* embedded in the inner membrane of the mitochondrion. **2,** As the electrons are transported from molecule to molecule *(red path)*, their energy is used to pump the protons (H⁺) to the intermembrane space. **3,** As the proton gradient increases, passive movement of protons back across the membrane (through the *ATP synthase* carrier) provides the energy needed to "recharge" ADP and Pᵢ to form ATP. Notice that oxygen (from O₂) is required as the final acceptor of the electrons and protons transported through the system, thus forming H₂O as a by-product. *ADP,* Adenosine diphosphate; *ATP,* adenosine triphosphate; *FAD,* flavin adenine dinucleotide; *NAD,* nicotinamide adenine dinucleotide.

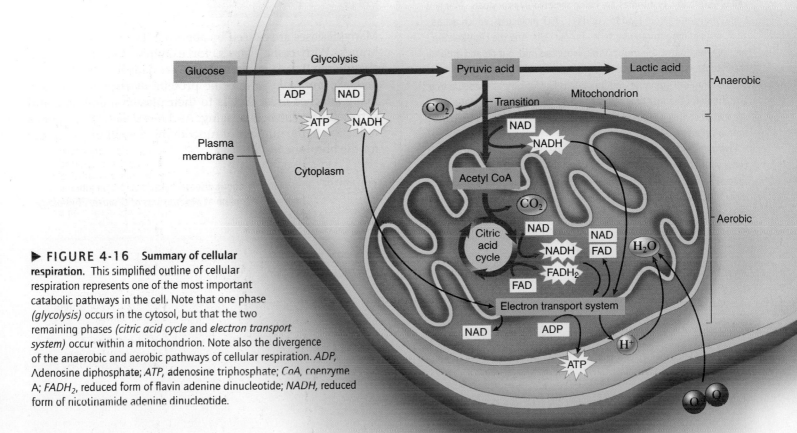

▶ **FIGURE 4-16** **Summary of cellular respiration.** This simplified outline of cellular respiration represents one of the most important catabolic pathways in the cell. Note that one phase *(glycolysis)* occurs in the cytosol, but that the two remaining phases *(citric acid cycle* and *electron transport system)* occur within a mitochondrion. Note also the divergence of the anaerobic and aerobic pathways of cellular respiration. *ADP,* Adenosine diphosphate; *ATP,* adenosine triphosphate; *CoA,* coenzyme A; *FADH₂,* reduced form of flavin adenine dinucleotide; *NADH,* reduced form of nicotinamide adenine dinucleotide.

┌───┐
│ QUICK ✓ CHECK   │
│   │
│ 13. What is an enzyme?   │
│ 14. How are enzymes named?   │
│ 15. What are the three phases of cellular respiration? │
│ 16. What phase of cellular respiration generates the most ATP? │
└───┘

ANABOLISM

Many anabolic or "building" pathways occur in human cells. Perhaps one of the most important concepts to understand in biology is the process of protein synthesis. Protein synthesis is essentially the translation of the genetic code of cells into proteins. These proteins are important structural and functional components of cells. Indeed, they dictate the structure of the human body and most of the processes that occur within it, such as synthesis of important lipids, nucleic acids, and other molecules. We will investigate one of these processes, protein synthesis, in detail in the next chapter.

The BIG Picture

Once you feel comfortable with the structure and functions of the major organelles, try to put a bigger picture of cell function together as it relates to your body. For example, most of the processes we discussed in this chapter are going on at about the same time within each and every cell of your body. Lysosomes are being secreted by exocytosis. Energy is being transferred from food molecules to ATP molecules, which act as short-term energy storage batteries for our cells. And the cytoskeleton and cell membrane are transporting materials into, out of, and around the cytoplasm.

Studying cell structure and function is like looking at the score of a symphony for the first time. All the individual parts look unrelated and somewhat confusing, but with some effort you can combine the parts to form a coherent "big picture."

Now, consider another picture of a huge "society" of trillions of cells—your body! Each cell contributes to the survival of the body by specializing in functions that help maintain the homeostasis of the internal environment. The next chapter will investigate how these huge numbers of cells work together to maintain homeostasis in our bodies.

Cycle of LIFE ⊙

Every cell in our body requires a continuous supply of energy in order to carry out numerous vital functions. The main source of this energy, of course, is cellular respiration, which requires glycolysis, the Krebs citric acid cycle, and the electron transport system. As we grow older, the ability to maintain the level of respiration we had as youths begins to decline.

The aging process may be programmed genetically or influenced by normal or abnormal cellular activities and processes.

For example, in our mitochondria, where the citric acid cycle and the electron transport system are located, free electrons react with molecular oxygen to form highly reactive chemicals called *free radicals*. One theory of aging points to an imbalance in older individuals between production and elimination of free radicals that results in mitochondrial damage. As mitochondria begin to lose their ability to combat free radicals, the free radicals accumulate and cause further damage to other cells and cellular organelles. Damage is especially serious in cellular pathways of protein synthesis and may explain in part the loss of muscle mass that occurs as we age.

Thus, aging is a part of life even at the cellular and organelle level. Investigations of the aging process currently under way are focusing on this level of organization. ●

MECHANISMS OF DISEASE

Many diseases are caused by abnormalities in our cell membranes. With *cystic fibrosis*, for example, the chloride ion channels in the plasma membrane are defective. In *Duchenne muscular dystrophy*, a specific protein, *dystrophin*, that normally attaches muscle cells to their plasma membrane and extracellular matrix is missing. And *type 2 diabetes* triggers a reduction in the number of functioning membrane receptors for the hormone *insulin*.

⊖volve   Find out more about diseases related to cell membrane abnormalities online at *Mechanisms of Disease: Physiology of Cells.*

LANGUAGE OF SCIENCE AND MEDICINE *continued from page 61*

enzyme (EN-zyme)
[*en-* in, *-zyme* ferment]

exocytosis (eks-o-sye-TOH-sis)
[*exo-* outside or outward, *-cyto-* cell, *-osis* condition]

facilitated diffusion (fah-SIL-i-tay-ted di-FYOO-zhun)
[*facili-* easy, *-ate* act of, *diffuse-* spread out, *-sion* process]

filtration (fil-TRAY-shun)
[*filtr-* strain, *-ation* process]

glycolysis (glye-KOHL-i-sis)
[*glyco-* sweet (glucose), *-o-* combining form, *-lysis* loosening]

hemodialysis (he-mo-dye-AL-i-sis)
[*hemo-* relating to blood, *-dia-* apart, *-lysis* loosening]

hydrostatic pressure (hye-droh-STAT-ik)
[*hydro-* water, *-stat-* standing, *-ic* relating to]

hypertonic (hye-per-TON-ik)
[*hyper-* excessive, *-ton-* tension, *-ic* relating to]

hypotonic (hye-poh-TON-ik)
[*hypo-* under, *-tonic* stretching]

isotonic (eye-soh-TON-ik)
[*iso-* equal, *-ton-* tension, *-ic* relating to]

kinase (KYE-nayz)
[*kin-* motion, *-ase* enzyme]

lock-and-key model (lok and kee MAHD-el)

membrane pump (MEM-brayne pump)
[*membran-* thin skin]

metabolic pathway (met-ah-BOL-ik PATH-way)
[*metabol-* change, *-ic* relating to]

metabolism (me-TAB-o-liz-em)
[*metabol-* change, *-ism* condition]

osmosis (os-MO-sis)
[*osmos-* push, *-osis* condition]

osmotic pressure (os-MOT-ik PRESH-ur)
[*osmo-* push, *-ic* relating to]

passive transport
[*passiv-* passive, *trans-* across, *-port* carry]

permeate (PERM-ee-ayt)
[*permea-* pass through, *-ate* process]

phagocytosis (fag-oh-sye-TOE-sis)
[*phago-* eat, *-cyto-* cell, *-osis* condition]

pinocytosis (pin-oh-sye-TOE-sis)
[*pino-* drink, *-cyto-* cell, *-osis* condition]

proenzyme (pro-EN-zyme)
[*pro-* first, *-en-* in, *-zyme* ferment]

receptor-mediated endocytosis (ree-SEP-tor MEE-dee-ayt-ed en-doh-sye-TOH-sis)
[*recept-* receive, *-or* agent, *medi-* middle, *-ate* process, *endo-* inward or within, *-cyto-* cell, *-osis* condition]

selectively permeable (sel-EK-tiv-lee PERM-ee-ah-bil)
[*select-* choose, *-ive* relating to, *perme-* pass through, *-able* ability]

simple diffusion (simple di-FYOO-zhun)
[*diffus-* spread out, *-sion* process]

sodium-potassium pump
[*sod-* soda, *-um* thing or substance, *potass-* potash, *-um* thing or substance]

solute (SOL-yoot)
[*solutus-* dissolved]

solution
[*solutus-* dissolved, *-ion* process]

solvent (SOL-vent)
[*solver-* dissolve]

CHAPTER SUMMARY

To download an MP3 version of the chapter summary for use with your iPod or other portable media player, access the Audio Chapter Summaries *online at http://evolve.elsevier.com.*

HINT *Scan this summary after reading the chapter to help you reinforce the key concepts. Later, use the summary as a quick review before your class or before a test.*

INTRODUCTION

A. As the basic structural and functional unit of our bodies, cells provide a physical foundation for all of our physiological processes

MOVEMENT OF SUBSTANCES THROUGH CELL MEMBRANES

A. Passive transport processes—do not require energy expenditure from the cell
  1. Diffusion—tendency of small particles to spread out and distribute themselves within any given space; from places of high concentration to areas of low concentration (Figure 4-2)
    a. Concentration gradient—measurable difference of solute concentration from one area of a solution to another
    b. Dynamic equilibrium—condition in which both sides of the membrane have solutions with equal concentrations of solute (Figure 4-2, *B*)

2. Simple diffusion—when molecules pass directly through a membrane (Figure 4-3)
3. Osmosis—diffusion of water through a selectively permeable membrane
  a. Osmosis is accomplished by special water-conducting channels called aquaporins (Figure 4-4)
  b. Osmotic pressure—water pressure that develops in a solution as a result of osmosis
  c. Isotonic—when the two fluids have the same potential osmotic pressure (Figure 4-5, *B*)
  d. Hypotonic—extracellular space has a lower concentration of impermeant solutes (Figure 4-5, *A*)
  e. Hypertonic—extracellular fluid has a higher concentration of impermeant solutes than the intracellular fluid (Figure 4-5, *C*)
4. Facilitated diffusion—passive process that occurs when special membrane channels and transport proteins in the plasma membrane carry the particles inside the cell
  a. Channel-mediated passive transport—process by which channels allow water molecules, specific ions, or other small, water-soluble molecules to cross the membrane (Figure 4-6)
    (1) Sodium channels—permit only sodium ions (Na⁺) to pass
    (2) Chloride channels—permit only chloride ions (Cl⁻) to pass
    (3) Gated channels—membrane channels that act like swinging gates in a fence (Figure 4-6)

b. Carrier-mediated passive transport—when molecules move down their concentration gradient in either direction across a membrane by passing through a protein called a membrane carrier (Figure 4-7)
   (1) Membrane carriers have a particular structure that attracts and binds a particular solute molecule; changes shape to release the bound molecule inside or outside the cell

5. Filtration (Box 4-2)
6. Role of passive transport processes (Table 4-1)
   a. Critical in maintaining a homeostatic balance of many vital substances within the cells of our bodies

B. Active transport processes—require the use of metabolic energy within the cell to "push" or "pull" substances across the membrane against the concentration gradient
   1. Transport by pumps—membrane pumps use energy supplied by the cell to transport molecules against their concentration gradient
      a. Sodium-potassium pump—moves sodium and potassium ions across the cell membrane, but in different directions (Figure 4-8)
   2. Transport by vesicles—mechanisms that carry large groups of molecules into or out of the cell by means of vesicles (membranous packages) (Figure 4-9)
      a. Endocytosis—plasma membrane "traps" extracellular material and brings it into the cell
         (1) Receptor-mediated endocytosis—special receptor molecules of the plasma membrane bind to specific molecules in the extracellular fluid (Figure 4-10)
         (2) Phagocytosis—type of endocytosis that engulfs large particles or microorganisms, forming vesicles within the cell; "cell eating"
         (3) Pinocytosis—endocytosis of fluids and the substances dissolved in it; "cell drinking"
      b. Exocytosis—process by which large molecules (packaged in vesicles) can leave the cell even though they are too large to move out through the plasma membrane
   3. Role of active transport processes (Table 4-2)
      a. Ion pumps can be used to keep ions at high or low concentrations within organelles
      b. Cytoskeleton acts on vesicles of varying sizes to engulf needed substances; add/remove bits of membrane and receptors to or from the membrane
      c. Secretion of important hormones and neurotransmitters is often accomplished by vesicles in exocytosis

CELL METABOLISM

A. Cell metabolism—all of the chemical reactions of a cell, and the basis for all human functions; involves many different kinds of chemical reactions (metabolic pathway)
   1. Catabolic—net result is a breakdown of molecules; usually releases energy
   2. Anabolic—where the chemical reactions in the pathway build larger molecules from smaller ones; requires the input of energy

B. Role of enzymes—functional proteins that serve as catalysts for nearly all of our chemical reactions
   1. Lower the required energy needed to start chemical reactions (Figure 4-11)
   2. Chemical structure of enzymes
      a. Enzymes are tertiary or quaternary proteins with complex shapes

(1) Their molecules often contain a nonprotein part called a co-factor
(2) If the co-factor is an organic nonprotein molecule, it is called a coenzyme
b. Lock-and-key model—specific substrate molecules interact with enzymes at their active site just like a key fits into a lock (Figure 4-11)

3. Classification and naming of enzymes
   a. Enzymes are named by combining the root name of the substance they act upon, or the word that describes the kind of chemical reaction that is catalyzed, with the suffix -ase placed at the end of the word (e.g., sucrase, hydrolase)

4. General functions of enzymes
   a. Every reaction that occurs in a metabolic pathway requires one or more specific enzymes (Figure 4-12)
   b. Metabolic pathways can be regulated by the activation and inactivation of certain enzymes
   c. Allosteric effector—molecule or other agent that alters enzyme function by changing its shape
   d. Denaturation—process that occurs when changes in temperature, pH, and other factors may destroy the configuration of an enzyme; denatured enzymes lose their shape and are unable to perform their functions
   e. Enzymes are continually being destroyed, and therefore they must continually be synthesized
      (1) Proenzymes—enzymes are synthesized to these molecules; require substances called kinases for them to become active

CELLULAR RESPIRATION

A. Cell respiration—process required of all cells by which nutrients such as carbohydrates, proteins, and fats are broken down into carbon dioxide and water
   1. Glycolysis—"breaking of glucose"; catabolic process that occurs inside the cytosol of cells (Figure 4-13)
      a. One glucose molecule broken into two molecules of pyruvic acid
      b. Does not require oxygen—anaerobic process
      c. If oxygen is not available, the pyruvic acid will continue on an anaerobic pathway (lactic acid)
      d. If oxygen is available, the pyruvic acid transfers to the aerobic pathways (citric acid cycle, electron transport system)
   2. Citric acid cycle (Krebs cycle)—metabolic pathway that is governed by a series of enzymes in the mitochondrion (Figure 4-14)
      a. Pyruvic acid (three carbons) is broken down into two, two-carbon molecules
      b. The broken carbon-carbon bonds release energy in the form of electrons to the energy carrier NADH, which brings them to the electron transport chain
      c. Carbon released from pyruvic acid breakdown makes two molecules of carbon dioxide
      d. Each two-carbon molecule joins a four-carbon molecule in the Krebs cycle and makes a six-carbon compound (citric acid)
      e. Enzymes work on the six-carbon compound, releasing two molecules of carbon dioxide and creating the original four-carbon compound at the top of the cycle; cycle repeats

f. Energy released by breaking the bonds of the six-carbon compound are transferred either to ATP or, in the form of electrons and their accompanying protons, to the energy carrier molecules NADH and $FADH_2$

3. Electron transport system—metabolic pathway that uses NADH and $FADH_2$ to transfer their newly gained energized electrons to a set of special enzymes (Figure 4-15), ultimately creating ATP and metabolic water

   a. Electrons "jump" from one enzyme to the next and, as they do so, they lose a small amount of their energy

   b. Energy is used to "pump" their accompanying protons from the inner chamber to the outer chamber of the mitochondria

   c. Protons travel down their concentration gradient through ATP synthase and energy is stored in the bonds of ATP

   d. Electrons that "jumped" down the electron transport system (and their associated protons) are captured by oxygen to form metabolic water (H_2O)

   e. Aerobic respiration transfers enough energy from each glucose molecule to form a net gain of 36 to 38 ATP molecules (Figure 4-16)

ANABOLISM

A. Protein synthesis—anabolic process that translates the genetic code of cells into proteins

REVIEW QUESTIONS

 HINT Write out the answers to these questions after reading the chapter and reviewing the Chapter Summary. If you simply think through the answer without writing it down, you won't retain much of your new learning.

1. Define the terms *diffusion, dialysis, facilitated diffusion,* and *osmosis.*
2. Explain how a concentration gradient relates to the process of diffusion.
3. Describe and give an example of a membrane channel.
4. Explain the terms *isotonic, hypotonic,* and *hypertonic.*
5. State the principle about the direction of active transport.
6. Name and describe the active transport pump that operates in the plasma membrane of all human cells.
7. Explain the processes of endocytosis and exocytosis.
8. Describe the classification of enzymes.
9. Discuss three general principles of enzyme function.
10. What is metabolism? Catabolism? Anabolism?
11. Describe briefly each of the three pathways that make up the process of cellular respiration.

CRITICAL THINKING QUESTIONS

 HINT After finishing the Review Questions, write out the answers to these items to help you apply your new knowledge. Go back to sections of the chapter that relate to items that you find difficult.

1. The process of diffusion and the process of a white blood cell trapping bacteria by phagocytosis are both transport processes. Compare and contrast these processes and identify them as active or passive.

2. Intravenous solutions can be isotonic to blood cells. Therefore, it is very important to know whether sugar, a nonelectrolyte, or salt (NaCl), an electrolyte, is being given in the solution so that the proper amount can be added. What would result if the tonicity of the solution is not isotonic to the blood cells?
3. White blood cells engulf bacteria and solutions that contain dissolved proteins. How would you summarize the processes that allow them to ingest both solids and liquids?
4. Explain how the shape of an enzyme determines its function. What would result if an allosteric effector changed the shape of the enzyme?
5. Compare and contrast aerobic and anaerobic respiration.
6. Why is the mitochondrion such an important organelle for survival of the cell? Explain why some cells, such as skeletal muscle cells, have more mitochondria than others do.

continued from page 61

With the knowledge you have gained from reading this chapter, see if you can answer these questions about Tobie's experience with the subway from the Introductory Story.

1. The movement of the large crowd of people through the doors into the less crowded subway could be thought of as a model of what type of cell transport?
   a. Active transport by vesicles
   b. Facilitated diffusion
   c. Simple diffusion
   d. Active transport by pumps

2. Tobie's movement against the crowd and out of the train is a model of what kind of cell transport?
   a. Active transport by vesicles
   b. Facilitated diffusion
   c. Simple diffusion
   d. Active transport by pumps

3. The movement of sodium from a low concentration to a higher concentration is an example of _____.
   a. Active transport by vesicles
   b. Facilitated diffusion
   c. Simple diffusion
   d. Active transport by pumps

After rushing around most of the day on his whirlwind tour of the city, Tobie was so thirsty, he felt that he could drink gallons of water.

4. If he did indeed drink several gallons of pure water in a short time, what effect might that have on his blood?
   a. His blood would become hypertonic.
   b. His blood would become hypotonic.
   c. His blood would become isotonic.
   d. The water would have no effect on his blood.

 HINT To solve these questions, you may have to refer to the glossary or index, other chapters in this textbook, A&P Connect, Mechanisms of Disease, and other resources.

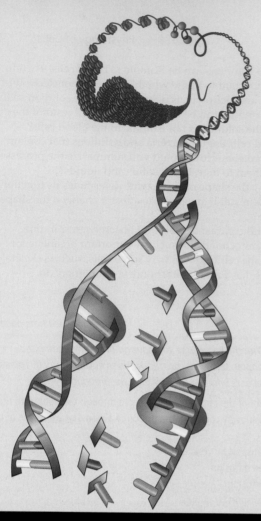

CHAPTER 5

Cell Growth and Reproduction

CHAPTER OUTLINE

 Scan this outline before you begin to read the chapter, as a preview of how the concepts are organized.

STUDENT LEARNING OBJECTIVES

At the completion of this chapter, you should be able to do the following:

1. Define what is meant by the genetic code and codons making up the code.
2. Compare and contrast the structures of DNA and RNA.
3. Describe the process of transcription.
4. Describe the process of translation.
5. Outline the roles of tRNA, mRNA, and rRNA in protein synthesis.
6. Describe the process of DNA replication
7. Identify the terms gene, chromatid, and chromosome.
8. Describe the cellular purpose of mitosis and meiosis.
9. Compare and contrast in detail the processes of mitosis and meiosis.
10. Outline the stages of mitosis.
11. Define interphase, mitosis, and cytokinesis and give the significance of each stage to the cell cycle.

FOR most of her 50 years, Wanda spent summers sunbathing along the pure white beaches of Saugatuck, Michigan. When it got too hot on the beach, she and her friends would climb the huge sand dunes and then race each other down the steep slope to dive into the cool waters of the Big Lake.

Besides enjoying the outdoors, getting tanned was a major goal of everyone in her group. Tan enhancers, including oils that literally broiled the skin, were readily available and advertised to all those who wanted to sport a "healthy" glowing tan.

Last year, Wanda attended a huge family reunion on the Big Lake. A niece, who was a medical student, noticed a number of raised, scaly patches on Wanda's forehead. Several looked like they had bled recently.

"I don't like the looks of those rough patches of skin on your face," her niece said to her with a concerned tone. "I have a dermatologist friend in town who can check those out for you."

Panicked, Wanda immediately thought of melanoma, a particularly deadly form of skin cancer. But the next day the doctor had better news: Wanda had early stage basal cell carcinoma—still cancer, but very treatable.

When you finish this chapter you should be able to make the connection between runaway cell division and the growth of cancer.

• • •

As you know, all cells come from previously existing cells, a process that requires cell division. Protein synthesis is vital to nuclear division (mitosis) and human reproduction (meiosis), as well as to all cellular processes. For these reasons, we now will examine the nucleic acids, protein synthesis, and their connection to cell division and human reproduction. In the following pages, we will help you investigate the basics of these vital processes.

DEOXYRIBONUCLEIC ACID (DNA)

In 1953, a team of American and British scientists—Francis Crick, Rosalind Franklin, James Watson, and Maurice Wilkins—won the race to solve the puzzle of DNA's molecular structure (Figure 5-1). Solving the DNA structure is perhaps the greatest biological discovery of our time. The importance of DNA's function—in a word, information—surpasses that of any other molecule.

The **deoxyribonucleic acid** molecule is a giant among molecules. It is an extremely long *polymer*—a type of large molecule made up of many smaller molecules that are similar to one another. In the case of DNA, the smaller molecules are called **nucleotides.** The nucleotides are ordered in two complementary chains that meet at their center, much like two halves of a ladder snapped together at each of its rungs. If you can picture this ladder coiling around its axis like a spiral staircase, you'll get an idea of what DNA looks like: a double spiral or **double helix** (see Figure 5-1).

Each nucleotide of a DNA molecule is a compound formed by combining phosphoric acid with a five-carbon sugar and a nitrogenous base.

continued on page 89

There are four different nitrogenous bases that make four unique nucleotides: *adenine, guanine, cytosine,* and *thymine.* These are the molecular "symbols" for the information encoded within the DNA molecule.

The sides of the DNA "ladder" comprise a chain of alternating phosphate and deoxyribose units. The "rungs" are made of pairs of bases that *always* combine with the same partner base. This is known as **obligatory base pairing.** As a result, adenine always pairs to thymine (A–T) and cytosine always pairs with guanine (C–G), and vice versa. The bases of DNA are held together by comparatively weak hydrogen bonds (which as we have seen in Chapter 2 are not true molecular bonds and only ½₀ as strong as a single covalent bond). However, the additive strength of *all* the hydrogen bonds makes DNA strong enough to create a macromolecule that is tens of thousands of base pairs long. For this reason, DNA and related molecules are amazingly resistant to breakage.

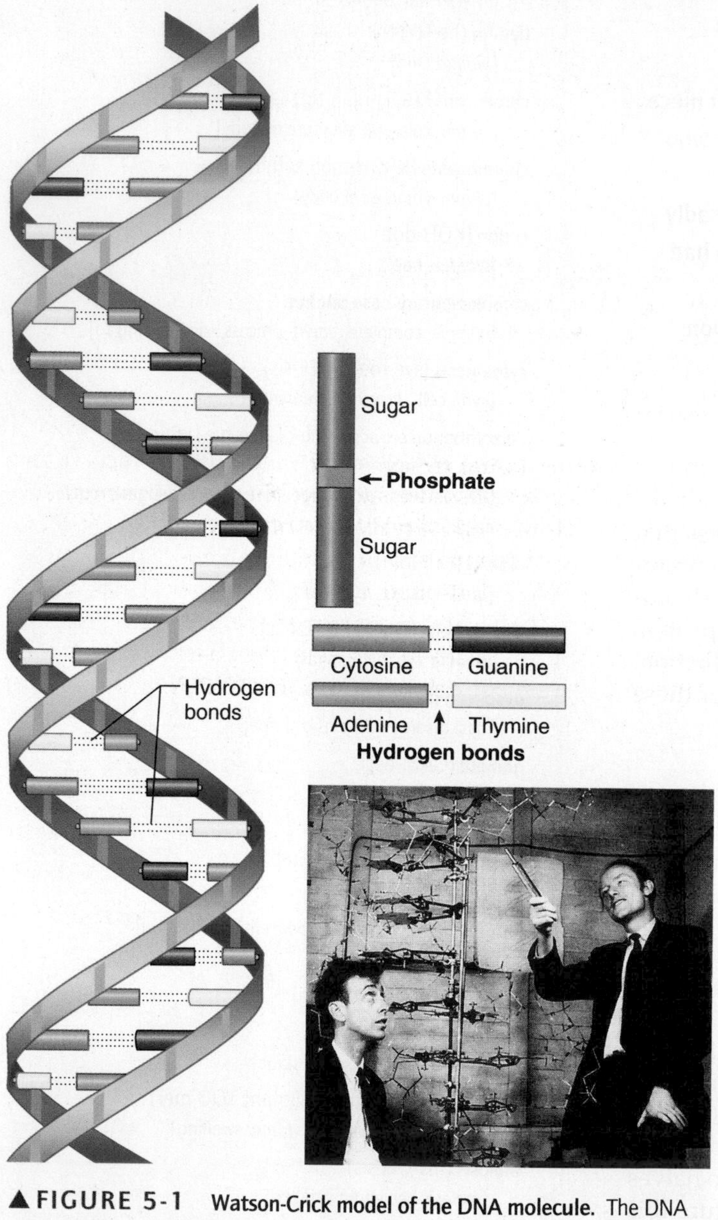

The Gene

A **gene** is a segment of DNA that codes for a specific protein. It usually consists of a chain of approximately 1,000 base pairs joined in a precise sequence. The transcript (copy) code is read in segments of three nucleotides. A series of three nucleotides is called a **codon.** This is where the complexity and diversity of genetic sequences is generated. As Figure 5-2 shows, each cell transfers a given gene's information from the nucleus to the cytoplasm by *transcribing* the genetic code onto shorter RNA molecules (*ribo*nucleic *a*cid). These RNA molecules are then transported out of the nucleus via nuclear pores.

▲ **FIGURE 5-1** **Watson-Crick model of the DNA molecule.** The DNA structure illustrated here is based on that published by James Watson *(photograph, left)* and Francis Crick *(photograph, right)* in 1953. Note that each side of the DNA molecule consists of alternating sugar and phosphate groups. Each sugar group is united to the sugar group opposite it by a pair of nitrogenous bases (adenine-thymine or cytosine-guanine). The sequence of these pairs constitutes a genetic code that determines the structure and function of a cell.

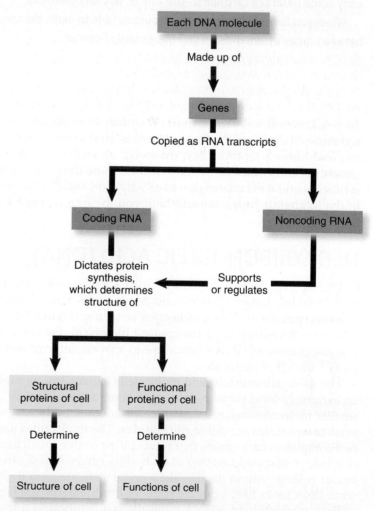

▲ **FIGURE 5-2** **Function of genes.** Genes copied from DNA are copied to RNA molecules, which use the code to determine a cell's structural and functional characteristics. There are approximately 24,000 genes in the human genome (all the DNA molecules together).

RIBONUCLEIC ACID

To make a protein, a gene encoded in nuclear DNA is first copied as a **messenger RNA (mRNA)** molecule. This mRNA molecule is a copy or *transcript* of a gene. The transcript code will later be *translated* by ribosomes in the cytoplasm into new proteins. Because it is a copy of the gene's code, we call mRNA *coding RNA*. A few RNA transcripts are also used to support or regulate protein production and are called *noncoding RNA* molecules (see Figure 5-2). Examples of non-coding RNAs are ribosomal RNA (rRNA) and transfer RNA (tRNA).

This would be a good time to take a few moments to study Table 5-1, which summarizes the major types of RNA and their functions.

Transcription from DNA to RNA

Protein synthesis, summarized for you later in Table 5-2 and Figure 5-4, begins when a single strand of mRNA forms along a segment of one strand of a DNA molecule (Figure 5-3). Note that RNA differs from DNA in that RNA molecules are smaller, they contain ribose instead of deoxyribose sugar, and use the base uracil instead of thymine. In fact, you can always tell DNA from RNA because thymine is found in DNA and uracil replaces it in RNA.

As a single strand of mRNA is formed along a strand of DNA in the nucleus, uracil (RNA) bonds to adenine (DNA), adenine (RNA) bonds to thymine (DNA), and cytosine and guanine always match up in the new strand of RNA. This process is known as **complementary base pairing.** As it forms, the mRNA strand detaches from the DNA template. The production of an mRNA molecule is called **transcription** because it copies or "transcribes" a

gene within the DNA strand. Before leaving the nucleus, the mRNA transcript is chemically edited to remove unused segments.

Translation

In the cytoplasm, the edited mRNA molecule attracts both a small and large ribosomal subunit. The two ribosomal subunits are composed largely of *ribosomal RNA* (rRNA). Together, the ribosomal subunits and the transcript of mRNA create an mRNA-ribosomal *complex* (Figure 5-4). The cell is now ready to "translate" the genetic code and form a specific sequence of amino acids in a process aptly named **translation.** From this point you should refer often to Figure 5-4 and

TABLE 5-1 Types of RNA Involved in Protein Synthesis

ACRONYM	NAME	DESCRIPTION	ROLE IN CELL FUNCTION
mRNA	Messenger RNA	Single, unfolded strand of nucleotides	Serves as working copy of one protein-coding gene
rRNA	Ribosomal RNA	Single, folded strand of nucleotides	Component of the ribosome (along with proteins); attaches to mRNA and participates in translation
tRNA	Transfer RNA	Single, folded strand of nucleotides; has an anticodon at one end and an amino acid–binding site at the other end	Carries a specific amino acid to a specific codon of mRNA at the ribosome during translation

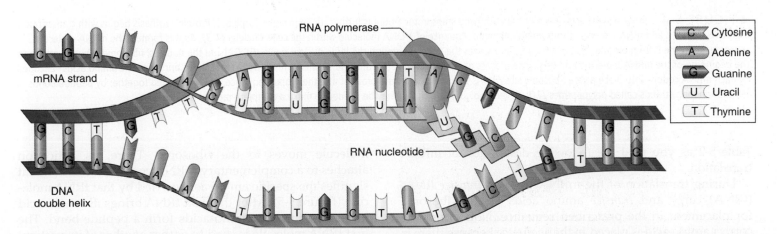

▲ **FIGURE 5-3** **Transcription of messenger RNA (mRNA).** A DNA molecule "unzips" in the region of the gene to be transcribed. RNA nucleotides already present in the nucleus temporarily attach themselves to exposed DNA bases along one strand of the unzipped DNA molecule according to the principle of complementary pairing. As the RNA nucleotides attach to the exposed DNA, they bind to each other and form a chainlike RNA strand called a messenger RNA (mRNA) molecule. Notice that the new mRNA strand is an exact copy of the base sequence on the opposite side of the DNA molecule, with the exception of U (uracil) on the mRNA replacing T (thymine) on the DNA. As in all metabolic processes, the formation of mRNA is controlled by an enzyme—in this case, the enzyme is called RNA polymerase.

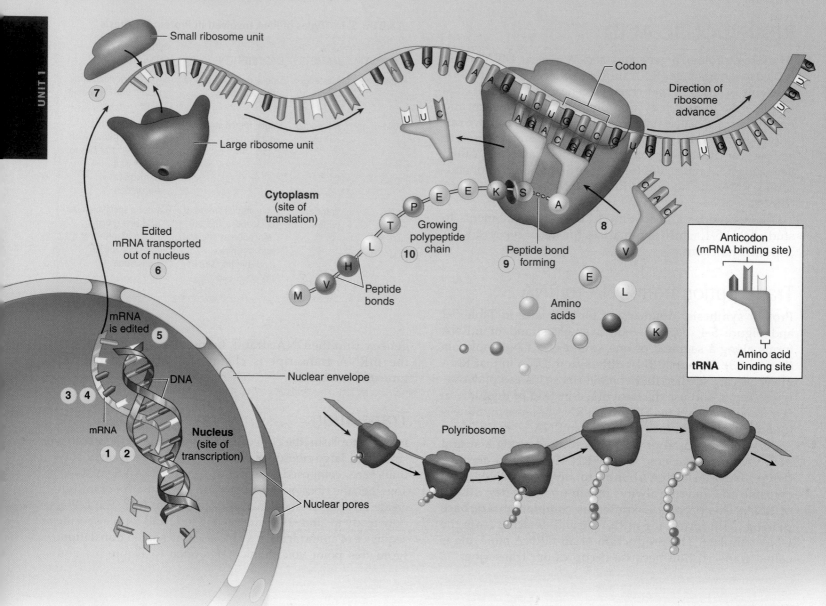

▲ **FIGURE 5-4**   **Protein synthesis.** Each of the numbered steps in the figure is further summarized in Table 5-2. Protein synthesis begins with transcription, a process in which an mRNA molecule forms along one gene sequence of a DNA molecule within the cell's nucleus *(1-3)*. As it is formed, the mRNA molecule separates from the DNA molecule *(4)*, is edited *(5)*, and leaves the nucleus through the large nuclear pores *(6)*. Outside the nucleus, ribosome subunits attach to the beginning of the mRNA molecule and begin the process of translation *(7)*. In translation, transfer RNA (tRNA) molecules bring specific amino acids—encoded by each mRNA codon—into place at the ribosome site *(8)*. As the amino acids are brought into the proper sequence, they are joined together by peptide bonds *(9)* to form long strands called polypeptides *(10)*. Several polypeptide chains may be needed to make a complete protein molecule.

Table 5-2 as you read the following description of mRNA translation.

During translation of the mRNA, specific *transfer RNAs* (tRNA) carry and *transfer* amino acids to the ribosome for placement in the prescribed sequence. Notice that the correct amino acid is placed in the sequence because there is a specific tRNA with complementary nucleotides to every mRNA codon (base triplet). Thus, each tRNA is connected to a specific amino acid and has a specific *anticodon* for any type of codon.

After picking up its amino acid from a pool of 20 different types of amino acids floating freely in the cytoplasm, a tRNA

molecule moves to the ribosome. There, its anticodon attaches to a complementary mRNA codon—the codon that signifies the specific amino acid carried by that tRNA molecule (Figure 5-5). After the next tRNA brings its amino acid into place, the two amino acids form a peptide bond. The first tRNA molecule is shed to gather another of its particular amino acid and the new tRNA stays attached to the ribosome awaiting the next tRNA. This cycle continues as the ribosome moves *down* the mRNA molecule until it reaches a codon that informs the ribosome to detach and stop. Now we have a string of amino acids hooked together by peptide bonds—a polypeptide or perhaps even a small protein.

TABLE 5-2 Summary of Protein Synthesis

STEP*	LOCATION IN THE CELL	DESCRIPTION
Transcription		
1	Nucleus	One region, or gene, of a DNA molecule "unzips" to expose its bases.
2	Nucleus	According to the principles of complementary base pairing, RNA nucleotides already present in the nucleoplasm temporarily attach themselves to the exposed bases along one side of the DNA molecule.
3	Nucleus	As RNA nucleotides align themselves along the DNA strand, they bind to each other and thus form a chainlike strand called *messenger RNA* (mRNA).
		This binding of RNA nucleotides is controlled by the enzyme RNA polymerase.
Preparation of mRNA		
4	Nucleus	As the preliminary mRNA strand is formed, it peels away from the DNA strand.
		This mRNA strand is a copy, or *transcript,* of a gene.
5	Nucleus	The mRNA transcript is edited by removing noncoding portions of the strand *(introns)* and splicing together the remaining pieces *(exons).*
6	Nuclear pores	The edited mRNA transcript is transported out of the nucleus through pores in the nuclear envelope.
Translation		
7	Cytoplasm	Two subunits sandwich the end of the mRNA molecule to form a ribosome.
8	Cytoplasm	Specific transfer RNA (tRNA) molecules bring specific amino acids into place at the ribosome, which acts as a sort of "holder" for the mRNA strand and tRNA molecules.
		The kind of tRNA (and thus the kind of amino acid) that moves into position is determined by complementary base pairing: each mRNA codon exposed at the ribosome site will only permit a tRNA with a complementary *anticodon* to attach.
9	Cytoplasm	As each amino acid is brought into place at the ribosome, an enzyme in the ribosome binds it to the amino acid that arrived just before it.
		The chemical bonds formed, called *peptide bonds,* link the amino acids together to form a long chain called a *polypeptide.*
10	Cytoplasm	As the ribosome moves along the mRNA strand, more and more amino acids are added to the growing polypeptide chain in the sequence dictated by the mRNA codons. (Each codon represents a specific amino acid to be placed in the polypeptide chain.)
		When the ribosome reaches the end of the mRNA molecule, it drops off the end and separates into large and small subunits again; often, enzymes later link two or more polypeptides together to form a whole protein molecule.

*These steps are illustrated in Figure 5-4.

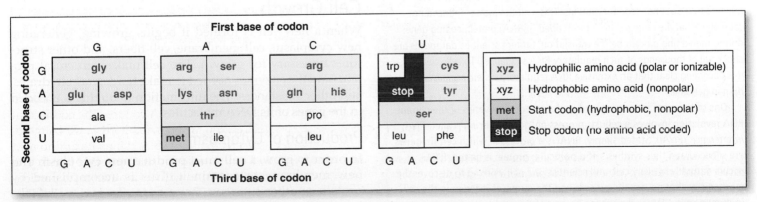

▲ **FIGURE 5-5 The genetic code.** The first graphic "phrasebook" or "decoder" of the genetic language of the cell was developed in 1966 to summarize the amino acids encoded by various codons (three-base sequences of nucleotides) in RNA. One of the clearest is this version adapted from Ben Fry's decoder, developed when he was a student at Massachusetts Institute of Technology. To read the decoder, start with any codon (for example, CGA). Find the first base along the top of the decoder to find the correct box to use (C is the third box). The second base is found in each row of that box, labeled on the left (G is the first row). Then use the third base to find the correct column, labeled at the bottom of each box (A is the second column, showing that Arg [arginine] is the amino acid encoded by CGA).

Ribosome subunits, mRNA, and tRNA can be reused again and again to form multiple copies of the same protein. In fact, as you can see in Figure 5-4, many ribosomes often position themselves in a chain along a single mRNA molecule, forming a **polyribosome.**

Translation can be inhibited or prevented by a process called **RNA interference (RNAi),** which is discussed in Box 5-1. Current research suggests that some types of RNAi are used by cells to protect against viral infections by interfering with the process of translating the viral genes.

As specific polypeptides are formed, chaperone proteins and other enzymes in the endoplasmic reticulum (ER), Golgi apparatus, or cytosol step into action. These proteins help fold and link the new polypeptides correctly and also help link them together to form secondary, tertiary, and quaternary protein molecules. Enzymes

may also promote the formation of hybrid molecules of proteins and lipids (lipoproteins) and proteins and sugars (glycoproteins).

The complete *set of proteins* synthesized by a cell is called the **proteome** of the cell. Human cells synthesize thousands of different enzymes as well as many structural and functional proteins for use within the cell or in other cells throughout the body. Liver cells, for example, synthesize blood plasma proteins such as prothrombin, fibrinogen, albumin, and globulin. The human proteome is much larger than the **genome**—the entire set of genes in a cell. This is because proteins coded by DNA can be combined with other proteins and other types of molecules. They may also undergo other modifications that make them into different proteins with different functions.

QUICK CHECK

1. Outline the overall structure of DNA.
2. What is the difference between DNA and RNA?
3. What are the phases of protein synthesis?
4. Why is the human proteome more complex than the genome?

BOX 5-1 | The RNA Revolution

For nearly a half-century after the discovery of DNA and its central role in heredity in 1953, scientific understanding of cellular physiology focused primarily on DNA—that "most golden of molecules." RNA was seen as the "go-between" molecule, acting mainly as a temporary working copy of a segment of DNA code (gene). However, we soon learned that transfer RNA (tRNA) and ribosomal RNA (rRNA) play a functional role other than direct coding for proteins. This discovery sparked some interest in looking for possible additional regulatory or supportive roles of RNA. Soon, catalytic forms of *noncoding RNA* were shown to be involved in editing mRNA (messenger RNA) transcripts before translation. The dawn of the twenty-first century saw an explosion of new discoveries regarding additional roles of RNA in regulating the function of the genome—a scientific revolution that still continues.

One aspect of the RNA revolution involves the surprising treasure trove of noncoding RNA molecules transcribed from the roughly 98% of the DNA genome that does not code for proteins. This includes *introns* (noncoding segments) that are removed from transcribed (mRNA) protein-coding genes during the editing process before translation. Current scientific databases are chock full of thousands of such regulatory RNA sequences. To cell scientists, it is becoming clear that an extensive RNA regulatory system manages the human genome.

One of the more interesting and useful recent discoveries of the current RNA revolution involves a cellular process of "gene silencing" called *RNA interference (RNAi)*. Stated simply, RNAi is a process in which certain genes are silenced and thus synthesis of a particular protein is halted. RNAi occurs naturally in every cell, and scientists are still working to discover the mechanisms that govern the process and the purposes it serves in the cell. In human cells, RNAi is thought to be part of a complex scheme of "fine-tuning" protein synthesis. The cell could also use RNAi to inhibit replication of virus components within the cell during a viral infection—thus protecting the cell.

GROWTH AND REPRODUCTION OF CELLS

Cell growth and reproduction are the most fundamental of living functions. Together, they make up the *cell life cycle* (Figure 5-6 and Table 5-3). *Cell growth* depends on using genetic information to make the structural and functional proteins needed for cell survival. *Cell reproduction* ensures that this genetic information is passed from one generation of organisms to the next. Mistakes in these processes can cause lethal genetic disorders, cancer, and other serious conditions. Developments in genetic and cellular research are providing insights into potential treatments and cures of some of these ailments.

Cell Growth

When a new cell is formed it begins growing, generating new cytoplasm, cell membrane, cell fibers, and other structures necessary for growth. The cell makes structural proteins and the enzymes needed to make lipids, carbohydrates, and other substances. The information to do this is encoded in the genes of its DNA molecules.

Production of Cytoplasm

In order to grow, a cell must produce new cytoplasm and new membrane to contain it. This is accomplished by protein synthesis, as we have seen. To review briefly, information coded in DNA is transcribed to mRNA and translated at ribosomes with rRNA and tRNA in order to generate proteins for cell growth and metabolism. Products of cell anabolism during the growth phase of the cell

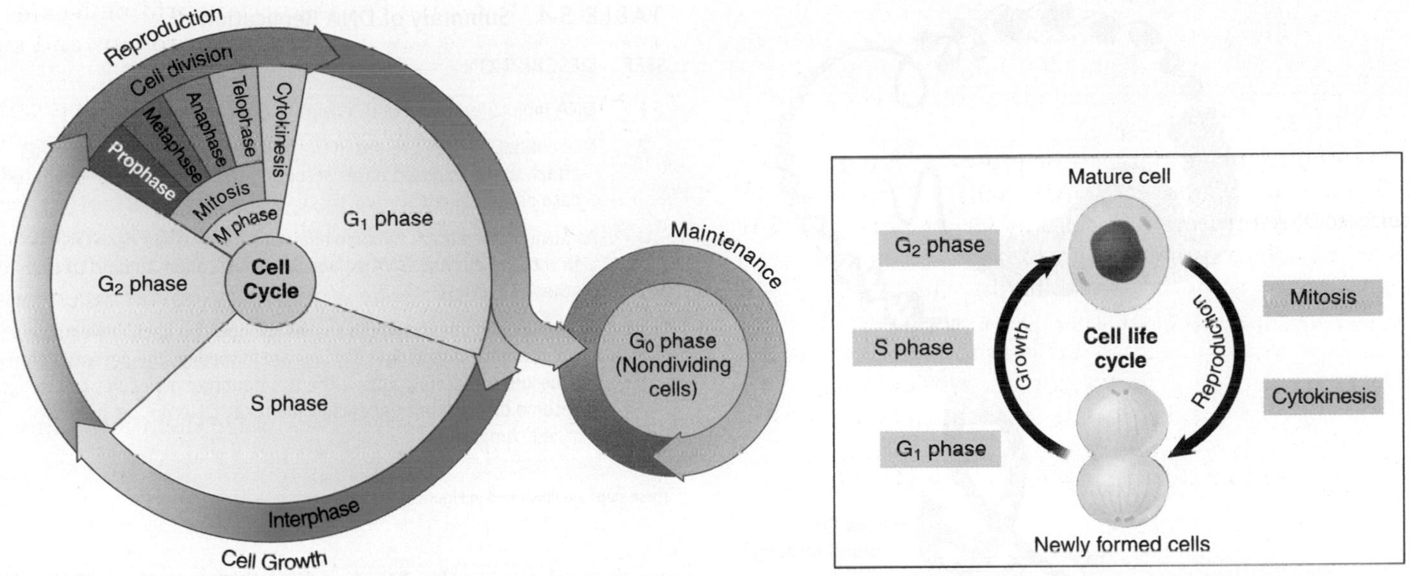

▲ **FIGURE 5-6**   **Life cycle of the cell.** The processes of growth and reproduction of successive generations of cells exhibit a cyclic pattern. Newly formed cells grow to maturity by synthesizing new molecules and organelles (G_1 and G_2 phases), including the replication of an extra set of DNA molecules (S phase) in anticipation of reproduction. Mature cells reproduce (M phase) by first distributing the two identical sets of DNA (produced during the S phase) in the orderly process of mitosis, then splitting the plasma membrane, cytoplasm, and organelles of the parent cell into two distinct daughter cells (cytokinesis). Daughter cells that do not go on to reproduce are in a maintenance phase (G_0).

TABLE 5-3   Summary of the Cell Life Cycle

PHASE OF CELL LIFE CYCLE	DESCRIPTION
Cell Growth	*Interphase*
Protein synthesis	Proteins are manufactured according to the cell's genetic code; functional proteins, the enzymes, direct the synthesis of other molecules in the cells and thus the production of more and larger organelles and plasma membrane; sometimes called the *first growth phase* or *G_1 phase* of interphase
DNA replication	Nucleotides, influenced by newly synthesized enzymes, arrange themselves along the open sides of an "unzipped" DNA molecule, thereby creating two identical daughter DNA molecules; produces two identical sets of the cell's genetic code, which enables the cell to later split into two different cells, each with its own complete set of DNA; sometimes called the *(DNA) synthesis stage* or *S phase* of interphase
Protein synthesis	After DNA is replicated, the cell continues to grow by means of protein synthesis and the resulting synthesis of other molecules and various organelles; this *second growth phase* is also called the *G_2 phase*
Cell Reproduction	*M Phase*
Mitosis or meiosis	The parent cell's replicated set of DNA is divided into two sets and separated by an orderly process into distinct cell nuclei; mitosis is subdivided into at least four phases: *prophase, metaphase, anaphase,* and *telophase*
Cytokinesis	The plasma membrane of the parent cell "pinches in" and eventually separates the cytoplasm and two daughter nuclei into two genetically identical daughter cells

life cycle may be used in the production of additional organelles such as the ER, Golgi apparatus, and plasma membrane. Other anabolic processes similar to protein synthesis produce additional enzymes that allow breakdown of molecules for energy and the construction of necessary carbohydrates, lipids, and other organic compounds needed in the cell. The mitochondrion is unique in that it can reproduce on its own by dividing in a process called *fission.* Each mitochondrion even has its own small genome consisting of about 37 genes!

DNA Replication

As a cell becomes larger, regulatory mechanisms in the cell trigger the synthesis of a complete copy of the cell's genome. Actual copying of DNA *(replication)* occurs during a period of the cell's **interphase.** Later, the cell will undergo **mitosis,** where one complete set of DNA molecules will go to one daughter cell and one complete set to the other. The mechanics of DNA replication resemble those of RNA synthesis.

In the first step of DNA replication, the tightly coiled DNA molecules uncoil except for small segments. As they uncoil, the two strands of the molecules unwind or "unzip." Then, with the help of an enzyme called *DNA polymerase,* complementary nucleotides begin to line up along the free bases of both unzipped strands. This action generates two new, identical DNA strands, called **chromatids,** from the original chromosome (Figure 5-7 and Table 5-4). However, these new, identical chromatids are still attached to each other at a point called a *centromere.* For that reason, we call both identical strands *sister chromatids.* In effect, each chromosome has doubled its genetic code. (Note that both

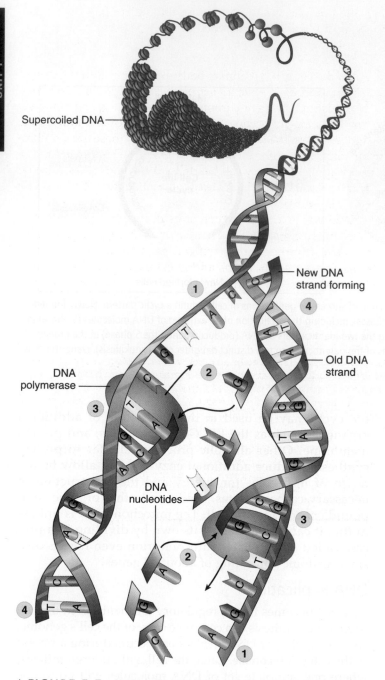

Supercoiled DNA

New DNA
strand forming

Old DNA
strand

DNA
polymerase

DNA
nucleotides

▲ **FIGURE 5-7   DNA replication.** When a DNA molecule makes a copy of itself, it "unzips" to expose its nucleotide bases. Through the mechanism of obligatory base pairing, coordinated by the enzyme DNA polymerase, new DNA nucleotides bind to the exposed bases. This forms a new "other half" to each half of the original molecule. After all the bases have new nucleotides bound to them, two identical DNA molecules will be ready for distribution to the two daughter cells. Refer to Table 5-4 for numbered steps.

chromatids are still joined by a single centromere and thus technically part of a single chromosome.)

DNA replication is really the *synthesis* of new DNA during interphase, the longest part of the cell's growth cycle. The actual period of interphase during which DNA is replicated is called the *S phase*. The other two portions of inter-

TABLE 5-4   Summary of DNA Replication

STEP	DESCRIPTION
1	DNA molecules uncoil and "unzip" to expose their bases.
2	Nucleotides already present in the intracellular fluid of the nucleus attach to the exposed bases according to the principle of obligatory base pairing.
3	As nucleotides attach to complementary bases along each DNA strand, the enzyme *DNA polymerase* causes them to bind to each other.
4	As new nucleotides fill in the spaces left open on each DNA strand, two identical *daughter molecules* are formed; as the parent DNA molecule completely unzips, the two daughter molecules coil to become distinct, but genetically identical, DNA double helices called *chromatids*.

These steps are illustrated in Figure 5-7.

phase are the growth phases called *first growth* or G_1 and G_2 or *second growth* phases. These periods occur before and after the S phase, respectively.

Cell Reproduction

Cell reproduction is the process by which a cell divides its duplicated genome and its cytoplasm to generate two new, genetically identical *daughter cells*. This portion of the cell cycle can be divided into two phases. The first is *mitosis (M phase)*, in which the nucleus disassembles and the sister chromatids break apart, leaving two new nuclei with the same genetic information. The second is **cytokinesis,** which literally translates as "cell movement." Cytokinesis takes place when the cytoskeleton pinches the cell membrane between the two new cells, dividing it roughly in half.

Mitosis

Mitosis really is the process of organizing and distributing nuclear DNA during cell division. Mitosis consists of four phases:

1. Prophase
2. Metaphase
3. Anaphase
4. Telophase

These four phases take place in a continuous sequence; there are no "rests" between them!

During **prophase** ("first phase"), the nuclear envelope falls apart as the paired chromatids generated in the S phase coil up to form dense, compact **chromosomes.** Simultaneously, the centriole pairs move toward opposite ends of the parent cell as they arrange parallel microtubules, or *spindle fibers,* across the cell.

Metaphase or "middle phase" begins when the chromosomes are aligned along a plane at the center of the centriole pairs or the "equator" of the cell. One chromatid of each chromosome faces one pole of the cell and its identical sister chromatid faces the opposite pole. Each chromatid then attaches to a spindle fiber.

During **anaphase** ("apart phase") the centromere of each chromosome splits to form two single chromosomes. Each new chromosome is pulled by a spindle fiber toward the pole it faced as a chromatid during metaphase.

As the chromosomes arrive at their respective poles, two new nuclear envelopes begin to form around the separated chromosomes. This marks the beginning of **telophase** ("end phase") Meanwhile, the DNA begins to unravel again so it can be decoded and used in the new daughter cells. At the same time, a furrow begins to form in the plasma membrane between the two new nuclei. Eventually this furrow completely separates the two new daughter cells in a process called *cytokinesis*. Cytokinesis often begins at the end of anaphase and completes during or just after telophase. The two new genetically identical cells will now enter interphase, where they will grow, produce new cytoplasm, replicate their DNA, and begin the cycle all over again.

Before you continue, review the steps of mitosis again using Table 5-5 to guide you.

Meiosis

All of the cells in the human body except for mature **gametes** (sperm and egg) have 23 *homologous pairs* of chromosomes. Homologous pairs of chromosomes contain genes that code for the same traits: one from the mother of the individual, and one from the father. (However, two of these are sex chromosomes, which are different in males.) This means that a normal **diploid** body cell has 46 total chromosomes. Diploid means "pair" and indicates that each chromosome has a homologous pair.

Meiosis is a special type of cell division that generates sex cells, or *gametes* (sperm in males; eggs in females). When a diploid parent sex cell goes through meiosis, it undergoes two divisions and effectively splits its homologous pairs of chromosomes *in half* to generate four new

TABLE 5-5 The Major Events of Mitosis

PROPHASE	METAPHASE	ANAPHASE	TELOPHASE
1. Chromosomes shorten and thicken (from coiling of the DNA molecules that compose them); each chromosome consists of two chromatids attached at the centromere	1. Chromosomes align across the equator of the spindle fiber at its centromere	1. Each centromere splits, thereby detaching two chromatids that compose each chromosome from each other, elongating in the process (DNA molecules start uncoiling)	1. Changes occurring during telophase essentially reverse those taking place during prophase; new chromosomes start elongating (DNA molecules start uncoiling)
2. Centrioles move to opposite poles of the cell; spindle fibers appear and begin to orient between opposing poles		2. Sister chromatids (now called *chromosomes*) move to opposite poles; there are now twice as many chromosomes as there were before mitosis started	2. A nuclear envelope forms again to enclose each new set of chromosomes
3. Nucleoli and the nuclear membrane disappear			3. Spindle fibers disappear

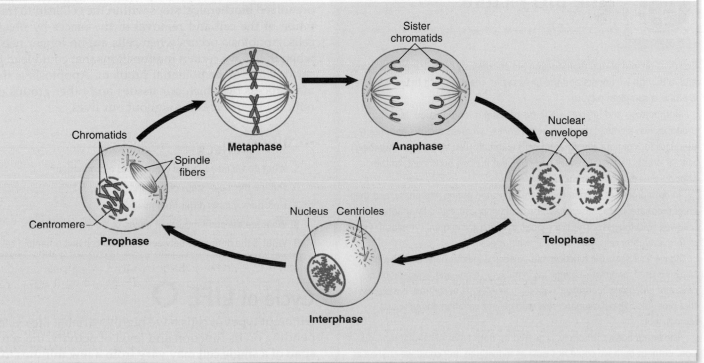

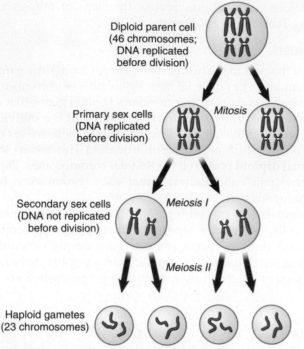

Diploid parent cell
(46 chromosomes;
DNA replicated
before division)

Mitosis

Primary sex cells
(DNA replicated
before division)

Secondary sex cells
(DNA not replicated
before division)

Meiosis I

Meiosis II

Haploid gametes
(23 chromosomes)

▲ **FIGURE 5-8   Meiosis.** Meiotic cell division takes place in two steps: *meiosis I* and *meiosis II*. Meiosis I is called *reduction division* because the number of chromosomes is reduced by half (from the diploid number to the haploid number). More detailed diagrams of meiosis are presented in Chapters 24 and 25. Only four sets of chromosomes were used in this figure for the sake of simplicity.

The BIG Picture

Cell division and sexual reproduction are enormously complicated processes and, although we understand the basic cycles, there is much to learn before we have a complete picture.

Both mitosis and meiosis are forms of cellular reproduction. Mitosis produces two identical daughter cells *(clones)*, whereas meiosis potentially produces four genetically different cells with half the chromosome number. Mitosis occurs in virtually every body cell, but meiosis occurs only in the *gonads*, the sex glands of the body.

In both processes, chromosomes are replicated during interphase, and then enter prophase, metaphase, anaphase, and telophase. Typically anaphase is followed by cytokinesis. This is a process in which the original cytoplasm is split to form daughter cells, each with its own nucleus.

As we have seen, the functions of mitosis and meiosis differ considerably. Mitosis is used for growth, change (as in puberty), repair and regeneration of cells, and potentially for asexual reproduction. Meiosis, in contrast, is a process that produces haploid daughter cells with genetic variation during sexual reproduction.

The entire body, ultimately, is geared for both types of cellular reproduction. Without both we could not exist as individuals, nor could we change as a species.

haploid cells (haploid means half). In effect, gamete-producing cells divide *twice* to generate four genetically different haploid cells.

We've outlined the general process of meiosis for you in Figure 5-8 and more fully in Chapters 24 and 25.

Changes in Cell Growth, Reproduction, and Survival

DNA and RNA are also intimately involved in cell growth, reproduction, and survival. For example, cells have the ability to use their genes to adapt to changing conditions. Cells may alter their size, reproductive rate, or other characteristics to adapt to changes in the internal environment. Such adaptations usually allow cells to work more efficiently.

If the body loses its ability to control mitosis normally, abnormal cell growth may occur. The new mass of cells thus formed is a tumor or **neoplasm.** Neoplasms may be relatively harmless growths called **benign** tumors. If tumor cells can break away and travel through the blood or lymphatic vessels to other parts of the body, the neoplasm is a **malignant tumor,** or cancer. Cells in malignant neoplasms often exhibit a characteristic called **anaplasia.** Anaplasia is a condition in which cells fail to *differentiate* into a specialized cell type. **Dysplasia** is an abnormal change in shape, size, or organization of cells in a tissue and is often associated with neoplasms.

Cells also die. In **necrosis,** cells die because of an injury or pathological condition, often causing nearby cells to die and triggering an immune response called *inflammation,* which removes the debris if possible. Nonpathological cell death, often called **apoptosis,** occurs frequently in the cells of your body. Apoptosis is a type of programmed cell death in which organized biochemical steps within the cell lead to fragmentation of the cell and removal of the pieces by phagocytic cells. Apoptosis occurs when cells are no longer needed or when they have certain malfunctions that could lead to cancer or some other potential problem. Apoptosis is the normal process by which our tissues and other groups of cells remodel themselves throughout our lives.

QUICK ✓ CHECK

5. What do we mean by cell growth and reproduction?

6. Outline the major steps of mitosis.

7. How does mitosis differ from meiosis?

8. What are the end products of meiosis?

9. What is the difference between a diploid cell and a haploid cell?

Cycle of LIFE ⟳

Different types of cells have highly variable life cycles. Depending on its function and level of activity, the active life span of a single cell may vary from a few minutes to many years. In fact, some cells remain dormant or inactive for years. For example, a woman's *ova* (eggs) are present from birth but gradually mature throughout her reproductive life span. In contrast, immune system cells are programmed to be more active when pathogenic invaders are present in the body, and to generate antibodies against these diseases. Frequently the function of cells and organ systems begins to decrease or deteriorate with age. ●

MECHANISMS OF DISEASE

Cell reproduction and genetics are involved in several diseases in humans. *Tumors*, which can lead to *cancer*, are caused by abnormalities in the mitotic division of cells. *Genetic disorders* are caused by mistakes in a cell's genetic code. When abnormal genes produce abnormal enzymes or proteins, diseases such as *sickle cell anemia* can occur.

evolve Find out more about these diseases online at *Mechanisms of Disease: Cell Growth and Reproduction.*

LANGUAGE OF SCIENCE AND MEDICINE *continued from page 79*

necrosis (ne-KROH-sis)
  [*necro-* death, *-osis* condition]

neoplasm (NEE-o-plazm)
  [*neo-* new, *-plasm* tissue or substance]

nucleotide (NOO-klee-oh-tyde)
  [*nucleo-* nut or kernel, *-ide* chemical]

obligatory base pairing (oh-BLIG-ah-tor-ee base PAIR-ing)

polyribosome (PAHL-ee-RYE-bo-sohm)
  [*poly-* many, *-ribo-* ribose (sugar), *-som* body]

prophase (PRO-fayz)
  [*pro-* first, *-phase* stage]

protein synthesis (PRO-teen SIN-the-sis)
  [*prote-* primary, *-in* substance, *synthes-* put together, *-is* process]

proteome (PRO-tee-ome)
  [*prote-* protein, *-ome* body (whole set)]

RNA interference (RNAi)
  [*RNA* ribonucleic acid]

telophase (TEL-oh-fayz)
  [*telo-* end, *-phase* stage]

transcription (trans-KRIP-shun)
  [*trans-* across, *-script-* write, *-tion* process]

translation (trans-LAY-shun)
  [*translat-* a bringing over, *-tion* process]

CHAPTER SUMMARY

To download an MP3 version of the chapter summary for use with your iPod or other portable media player, access the Audio Chapter Summaries online at http://evolve.elsevier.com.

HINT *Scan this summary after reading the chapter to help you reinforce the key concepts. Later, use the summary as a quick review before your class or before a test.*

INTRODUCTION

A. All cells come from previously existing cells; requires cell division
B. Protein synthesis is vital to nuclear division and human reproduction

DEOXYRIBONUCLEIC ACID (DNA)

A. DNA—large molecule made up of many smaller molecules that are similar to one another (nucleotides)
  1. Nucleotide—a compound formed by combining phosphoric acid with a five-carbon sugar and a nitrogenous base
  2. Nitrogenous bases:
    a. Adenine
    b. Guanine
    c. Cytosine
    d. Thymine

B. DNA looks like: a double spiral or double helix (Figure 5-1)
  1. Sides of the DNA "ladder" comprise a chain of alternating phosphate and deoxyribose units
  2. The "rungs" are made of pairs of bases that always combine with the same partner base
    a. Obligatory base pairing—adenine always pairs to thymine (A–T) and cytosine always pairs with guanine (C–G)
    b. Bases of DNA are held together by comparatively weak hydrogen bonds
C. Gene—segment of DNA that codes for a specific protein
  1. Codon—transcript (copy) code that is read in segments of three nucleotides
  2. Cell transfers a given gene's information from the nucleus to the cytoplasm by transcribing the genetic code onto shorter RNA molecules (Figure 5-2)

RIBONUCLEIC ACID

A. Major types of RNA (Table 5-1)
  1. Messenger RNA (mRNA)
  2. Transfer RNA (tRNA)
  3. Ribosomal RNA (rRNA)
  4. Transcription from DNA to RNA (Table 5-2)
    a. Protein synthesis—begins when a single strand of RNA forms along a segment of one strand of a DNA molecule (Figure 5-3)

b. A single strand of messenger RNA (mRNA) is formed along a strand of DNA

c. Complementary base pairing—uracil attaches to adenine, adenine to thymine, guanine to cytosine, and cytosine to guanine

d. Transcription—production of RNA molecule by coping or "transcribing" a gene within the DNA strand

5. Translation—"translating" the genetic code to form a specific sequence of amino acids (Figure 5-4)

a. Specific transfer RNAs (tRNAs) carry and transfer amino acids to the ribosome for placement in the prescribed sequence

b. Each tRNA is connected to a specific amino acid and has a specific anticodon for any type of codon (Figure 5-5)

c. After picking up its amino acid from a pool of 20 different types of amino acids floating freely in the cytoplasm, a tRNA molecule moves to the ribosome, where its anticodon attaches to its complementary mRNA codon

d. Ribosome moves *down* the mRNA molecule until it reaches a codon that informs the ribosome to detach and stop; forms a polyribosome

e. Translation can be inhibited or prevented by a process called RNA interference (RNAi) (Box 5-1)

f. As specific polypeptides are formed, chaperone proteins and other enzymes in the endoplasmic reticulum (ER), Golgi apparatus, or cytosol step into action; they help fold and link the new polypeptides correctly

g. The complete set of proteins synthesized by a cell is called the proteome of the cell

(1) Human proteome is much larger than the genome—the entire set of genes in a cell

(2) Proteome is larger because proteins coded by DNA can be combined with other proteins and other types of molecules

GROWTH AND REPRODUCTION OF CELLS

A. Cell growth and reproduction are the most fundamental of living functions

1. Together, they make up the cell life cycle (Figure 5-6 and Table 5-3)

B. Cell growth depends on using genetic information to make the structural and functional proteins needed for cell survival

C. Cell reproduction ensures that this genetic information is passed from one generation of organisms to the next

D. Cell growth—when a new cell is formed it begins growing, generating new cytoplasm, cell membrane, cell fibers, and other structures (structural proteins) necessary for growth

1. Production of cytoplasm

a. In order to grow, a cell must produce new cytoplasm and new membrane to contain it

b. Accomplished by protein synthesis

2. DNA replication—synthesis of new DNA during interphase (Figure 5-7 and Table 5-4)

a. As a cell becomes larger, regulatory mechanisms in the cell trigger the synthesis of a complete copy of the cell's genome

b. Cell will undergo mitosis; one complete set of DNA molecules will go to one daughter cell and one complete set to the other

c. Occurs during a period of the cell's interphase

(1) S phase—DNA is replicated

(2) G1 and G2 phases—growth phases that occur before and after S phase

E. Cell reproduction—process by which a cell divides its duplicated genome and its cytoplasm to generate two new, genetically identical daughter cells

1. Mitosis—process of organizing and distributing nuclear DNA during cell division

a. Mitosis consists of four phases (Table 5-5)

(1) Prophase—the nuclear envelope falls apart as the paired chromatids coil up to form compact chromosomes; centriole pairs move toward opposite ends of the parent cell as they arrange spindle fibers across the cell

(2) Metaphase—begins when the chromosomes are aligned along a plane at the center or the "equator" of the cell; one chromatid of each chromosome faces one pole of the cell and its identical sister chromatid faces the opposite pole

(3) Anaphase—centromere of each chromosome splits to form two single chromosomes; pulled toward the pole it faced as a chromatid during metaphase

(4) Telophase—two new nuclear envelopes begin to form around the separated chromosomes; a furrow begins to form in the plasma membrane between the two new nuclei

b. Cytokinesis—furrow completely separates the two new daughter cells; begins at the end of anaphase and completes during or just after telophase

2. Meiosis—special type of cell division that generates sex cells (Figure 5-8)

a. When a diploid parent sex cell goes through meiosis, it undergoes two divisions; splits its homologous pairs of chromosomes in half to generate four new haploid cells

F. Changes in cell growth, reproduction, and survival

1. Cells can use genes to adapt to changing conditions, increase chances of survival, or become more efficient

2. Abnormal cell growth may occur if mitosis doesn't occur normally—new mass of cells is a tumor or neoplasm

3. Cell death occurs because of injury or pathological condition (necrosis) or through programmed mechanisms if the cell is no longer needed or malfunctions (apoptosis)

REVIEW QUESTIONS

continued from page 79

Now that you have read this chapter, try to answer these questions about Wanda from the Introductory Story.

HINT *Write out the answers to these questions after reading the chapter and reviewing the Chapter Summary. If you simply think through the answer without writing it down, you won't retain much of your new learning.*

1. Describe the size and shape of a DNA molecule.
2. Where is DNA located?
3. Briefly outline the steps of protein synthesis.
4. As a cell grows, how is additional cell material added?
5. What are the steps involved in DNA replication and when does it occur?
6. Define *mitosis.*
7. Briefly describe the four distinct phases of mitosis.
8. Define *meiosis* and discuss its role in the body.
9. When does the reduction of chromosomes from the diploid to the haploid number take place?

CRITICAL THINKING QUESTIONS

HINT *After finishing the Review Questions, write out the answers to these items to help you apply your new knowledge. Go back to sections of the chapter that relate to items that you find difficult.*

1. Certain antibiotics can damage ribosomes in normal human body cells. People taking these antibiotics need to be carefully monitored. Summarize the effect of fewer ribosomes on the process of transcription and translation.
2. DNA is often called the "blueprint of life." However, DNA is composed of only four nitrogen bases as part of its nucleotide structure. Explain how these four nitrogen bases can influence the genetic makeup of an individual.
3. How would you describe the importance of the process of protein synthesis in the functioning of a cell?
4. Starting with DNA, can you identify the steps in the transcription and translation process through to the production of a protein?

1. What was the likely cause of Wanda's skin cancer?
   a. Pollution
   b. Excessive exposure to sunlight
   c. Inherited predisposition to skin cancer
   d. The cause is not known

2. Wanda's rough patches were scaly and raised. Under the microscope, it was easy to see that the skin cells were irregular and jumbled together. What word(s) in this chapter would you use to describe them? (Pick all that apply.)
   a. Benign
   b. Malignant
   c. Neoplasm
   d. Dysplasia

3. The dermatologist told Wanda that it may have taken 30 years for her to develop her skin cancer. He advised her to wear sunscreen every time she went outdoors. Would this prevent future eruptions of basal cell carcinoma? Why or why not?
   a. Yes, because she would protect herself from the ultraviolet light that caused the cancer.
   b. No, because the damage had already been done from years of exposure to her skin.
   c. Yes, because the sunscreen would prevent the cancer from reappearing by killing the cancerous skin cells.
   d. Yes, the sunscreen can cure years of skin damage caused by sunlight.

HINT *To solve these questions, you may have to refer to the glossary or index, other chapters in this textbook, A&P Connect, Mechanisms of Disease, and other resources.*

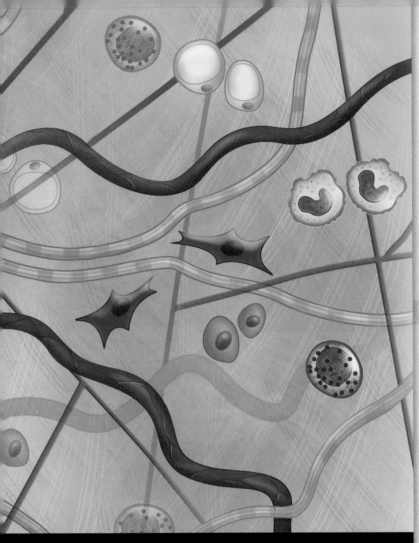

Tissues and Their Functions

CHAPTER OUTLINE

 Scan this outline before you begin to read the chapter, as a preview of how the concepts are organized.

HINT

STUDENT LEARNING OBJECTIVES

At the completion of this chapter, you should be able to do the following:

1. List the four major tissue types and give their basic functions.
2. Describe the different types of epithelial tissues and list their distinguishing features.
3. Describe the basic function of each epithelial type.
4. Compare exocrine and endocrine glands both structurally and functionally.
5. Identify the various kinds of exocrine glands.
6. Give at least one location in the body for each type of epithelium.
7. Identify the different types of connective tissues based on their cell and matrix characteristics.
8. Give locations in the body where each type of connective tissue is found.
9. Identify the different types of muscle tissues based on their structural features.
10. List the location and function of cardiac, smooth, and skeletal muscle.
11. Describe the structure and function of neurons and neuroglial cells.
12. Describe how tissues respond to injuries.
13. Describe the structure and function of the following: mucous, serous, cutaneous, and synovial membranes.

NATHAN had been mountain biking for about 2 hours, winding his way up and down the familiar trails with a devil-may-care attitude. Pedaling at a breakneck speed, he swerved around a fallen branch as he headed down the last steep incline. Suddenly, he lost control and went flying off his bike. After rolling over a couple of times, he stopped hard, jamming his right foot against a boulder. His friends helped him limp the rest of the way down the hill and back to their car; then they drove him to the urgent care center (which was almost as familiar to Nathan as the trails).

The physician's assistant ordered an x-ray film of Nathan's lower leg to confirm no break in the bones. She came back soon with the results and told Nathan that there was no break, just "soft tissue damage" (a sprained ankle). By "soft tissues," she meant the tendons and ligaments that hold all of the bones of the foot and ankle together.

After you read this chapter we'll investigate other aspects of Nathan's case and come to some conclusions as to the specifics of what happened when he fell.

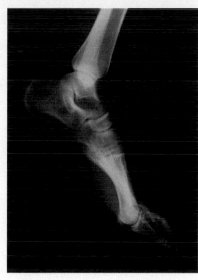

Your body is composed of dozens of interacting **tissues**—each a group of similar cells performing a common function. Together, tissues form organs, connect those organs together, and hold them in place. All of the cellular anatomy you have learned up to this point will provide you with an excellent basis for understanding tissues. After all, tissues are simply vast networks of cells, governed by the same cellular processes we have already explored.

The study of tissues is called **histology.** Every tissue performs at least one unique function that helps the body as a whole maintain homeostasis. Tissues vary in arrangement from being just one cell deep to huge masses comprising millions of cells. All tissues are surrounded by (or embedded in) a complex mix of fluids and proteins usually referred to as the **extracellular matrix (ECM).** (Sometimes we simply call it the *matrix.*)

A&P CONNECT

This brief introduction to histology involves many images made with microscopes of various types. Refer to **Tools of Microscopic Anatomy** online at **A&P Connect** for an overview of the types of microscopy used in this chapter.

evolve

LANGUAGE OF SCIENCE AND MEDICINE

HINT *Before reading the chapter, say each of these terms out loud. This will help you avoid stumbling over them as you read.*

adipocyte (AD-i-poh-syte)
[*adipo-* fat, *-cyte* cell]

adipose tissue (AD-i-pohs TISH-yoo)
[*adipo-* fat, *-ose* full of, *tissu-* fabric]

apocrine gland (AP-oh-krin gland)
[*apo-* from, *-crin* secrete, *gland* acorn]

avascular (ah-VAS-kyoo-lar)
[*a-* without, *-vas-* vessel, *-ula-* little, *-ar* relating to]

axon (AK-son)
[*axon* axle]

basement membrane (BM)
[*base-* base, *-ment* thing, *membran-* thin skin]

body composition (BOD-ee kahm-poh-ZIH-shun)

body membrane (BOD-ee MEM-brayne)

bone (bohn)
[*ban-* bone]

canaliculus (kan-ah-LIK-yoo-lus)
[*canal-* channel, *-iculi* little] *pl.,* canaliculi (kan-ah-LIK-yool-eye)

cancellous bone tissue (KAN-seh-lus bohn TISH-yoo)
[*cancel-* lattice, *-ous* characterized by, *ban-* bone, *tissu-* fabric]

cardiac muscle tissue (KAR-dee-ak MUSS-el TISH-yoo)
[*cardia-* heart, *-ac* relating to, *mus-* mouse, *-cle* small, *tissu-* fabric]

chondrocyte (KON-droh-syte)
[*chondro-* cartilage, *-cyte* cell]

collagen (KAHL-ah-jen)
[*colla-* glue, *-gen* produce]

collagenous fiber (kah-LAJ-eh-nus FYE-ber)
[*colla-* glue, *-gen-* produce, *-ous* relating to, *fibr-* thread]

columnar (koh-LUM-nar)
[*column-* column, *-ar* relating to]

compact bone tissue (kom-PAKT bohn TISH-yoo)
[*compact* put together, *ban-* bone, *tissu-* fabric]

connective tissue (koh-NEK-tiv TISH-yoo)
[*connect* bind, *-ive* relating to, *tissu-* fabric]

connective tissue membrane (koh-NEK-tiv TISH-yoo MEM-brayne)
[*connect* bind, *-ive* relating to, *tissu-* fabric, *membran-* thin skin]

cuboidal (kyoo-BOYD-al)
[*cub-* cube, *-oid-* like, *-al* relating to]

cutaneous membrane (kyoo-TAYN-ee-us MEM-brayne)
[*cut-* skin, *-aneous* relating to, *membran-* thin skin]

dendrite (DEN-dryte)
[*dendr-* tree, *-ite* part (branch) of]

continued on page 110

TABLE 6-1   Major Tissues of the Body

TISSUE TYPE	STRUCTURE	FUNCTION	EXAMPLES IN THE BODY
Epithelial tissue	One or more layers of densely arranged cells with very little extracellular matrix May form either sheets or glands	Covers and protects the body surface Lines body cavities Movement of substances (absorption, secretion, excretion) Glandular activity	Outer layer of skin Lining of the respiratory, digestive, urinary, reproductive tracts Glands of the body
Connective tissue	Sparsely arranged cells surrounded by a large proportion of extracellular matrix often containing structural fibers (and sometimes mineral crystals)	Supports body structures Transports substances throughout the body	Bones Joint cartilage Tendons and ligaments Blood Fat
Muscle tissue	Long fiberlike cells, sometimes branched, capable of pulling loads; extracellular fibers sometimes hold muscle fibers together	Produces body movements Produces movements of organs such as the stomach, heart Produces heat	Heart muscle Muscles of the head/neck, arms, legs, trunk Muscles in the walls of hollow organs such as the stomach, intestines
Nervous tissue	Mixture of many cell types, including several types of neurons (conducting cells) and neuroglia (support cells)	Communication between body parts Integration/regulation of body functions	Tissue of brain and spinal cord Nerves of the body Sensory organs of the body

Although a number of subtypes exist, all tissues can be classified by their structure and function into four general categories: *epithelium, connective, muscle,* and *nerve.* We've summarized the major tissue types for you in Table 6-1. Take a few minutes to review these tissue types and study their functions before you continue.

EXTRACELLULAR MATRIX

The *extracellular matrix* (ECM) surrounding cells within tissues is very complex and varies greatly among different tissues. Generally, the structural makeup of the ECM of

any particular tissue matches its function. For example, the ECM of bone tissue contains crystals of calcium that give strength and support to the entire body. In contrast, the ECM surrounding the cells of tough ligament tissue contains many proteins that are flexible and elastic. This arrangement allows for the controlled movement of joints, for example.

The *amount* of ECM may also vary between different tissues. The general makeup of ECM consists of water and its dissolved ions. However, depending on the tissue type, it may also contain mineral crystals (such as the calcium crystals in bone tissue), proteins, and *proteoglycans.* Proteoglycans (*proteo-*, "protein"; + *glyc-*, "sugar") are large molecules constructed of a protein backbone linked to carbohydrates such as sugars. These molecules function as shock absorbers, lubricants, and fluid thickeners (much like the oil and other fluids in your car's engine).

Two common structural proteins, *collagen* and *elastin,* provide support and elasticity in many different types of tissue. These proteins are often connected to cells by other modified proteins called **glycoproteins** that form "bridges" between cells. This complex protein system connects cells to their surrounding ECM, which then links all cells to form a unified structure—a *tissue!* By providing this interconnecting protein network, cells within a tissue can communicate and coordinate tissue development, changes in shape, and individual cell movement. As you can see, a tissue is not composed only of cells, but also of the ECM that binds these cells together into a whole unit.

Tissues with little ECM have cells that are usually connected by cell junctions, such as *desmosomes* and *tight junctions* (see Chapter 3). The tissue that forms the outer layer of skin is held together by such connections, which allows the tissue to act as a *continuous sheet*. In connective and skeletal muscle tissues, the ECM forms a fibrous network that holds cells together. In blood, the ECM does not bind the cells composing the tissue at all. Instead, it is a fluid that allows blood cells of many types to flow freely past one another in the fluid matrix.

QUICK CHECK

1. Name the four basic tissue types.
2. What is the ECM? Why is it so important to form a tissue?
3. How do cells hold themselves together?

EPITHELIAL TISSUE

Types and Locations of Epithelial Tissue

You can think of epithelium as a covering of our body as well as most of our internal body organs and tracts (*epi-* means "upon"). Epithelium is divided into two general types: (1) **membranous** epithelium, which forms coverings and linings, and (2) **glandular** epithelium, which produces secretions. Membranous epithelium covers the body and lines body cavities. It also lines blood and lymphatic vessels, as well as the respiratory, digestive, and urogenital tracts. Glandular epithelium is grouped into solid cords or hollow follicles. These form the secretory units of endocrine and exocrine glands.

Functions of Epithelial Tissue

Epithelial tissues serve several important functions throughout our bodies:

Protection. The epithelium of our organs and our body surface, including the skin, provides basic protection from potentially harmful elements in our environment and from damage due to abrasion or other mechanical insult.

Sensory functions. Modified epithelial structures allow the skin, nose, eyes, and ears to sense and interpret the environment around us.

Secretion. Glandular epithelium is adapted for secretion of hormones, mucus, digestive juices, and sweat.

Absorption. The epithelial lining of the gut allows for the absorption of nutrients from the gut. The epithelial lining of the respiratory tract allows for the exchange of important gases between air in the lungs and the fluid of the blood.

Excretion. The unique epithelial lining of kidney tubules allows for the excretion and concentration of products in the urine.

Generalizations About Epithelial Tissue

Most epithelial tissues have very little intercellular matrix material, which is part of the reason why these tissues provide efficient protection. When viewed under a microscope, epithelial tissue usually looks like sheets, because there is so little space between the cells. Much of the structural integrity of these sheets is provided by intercellular connections such as desmosomes and tight junctions.

Sheets of epithelial cells create the surface layers of our skin as well as our mucous and serous membranes. The epithelial tissue attaches to an underlying layer of connective tissue. This is accomplished by a thin, noncellular layer of adhesive, permeable material called the **basement membrane (BM)**. Epithelial tissues are said to be **avascular** (*a*, "without"; *vascular*, "vessels"). This means they must obtain nutrients and oxygen by *diffusion* from capillary vessels, through the basement membrane, and into the overlying epithelium. This is a good example of how important the process of diffusion is to your entire body.

Classification of Epithelial Tissue
Membranous Epithelium
Classification Based on Cell Shape

Membranous epithelial tissues are classified into four groups based on the shapes of their cells: *squamous, cuboidal, columnar*, and *pseudostratified columnar* (Figure 6-1). **Squamous** (Latin for "scale") cells are flat and platelike. **Cuboidal** cells are cube-shaped and tend to be a bit larger than squamous cells. **Columnar** epithelial cells are taller than they are wide and appear narrow and cylindrical. **Pseudostratified columnar** epithelium is made up of *one* layer of cells that *appears* to be at least several layers (*pseudo-*, "false"; *strata-*, "layer"). However, despite their appearance, all cells are attached to the basement membrane. Some are taller and their nuclei rest near the top of the cell. Others are shorter and have nuclei that rest closer to the basement membrane.

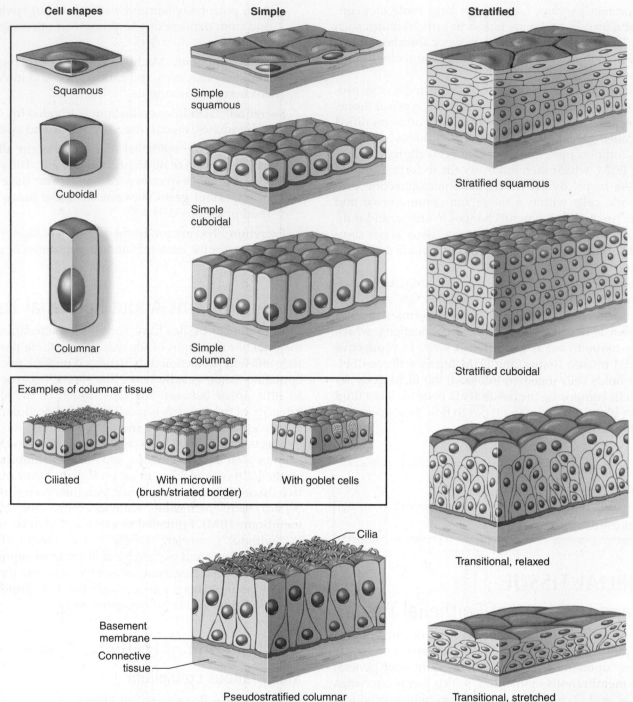

▲ **FIGURE 6-1** **Classification of epithelial tissues.** The tissues are classified according to the shape and arrangement of cells. The color scheme of these drawings is based on a common staining technique used by histologists called *hematoxylin and eosin (H&E)* staining. H&E staining usually renders the cytoplasm pink and the chromatin inside the nucleus a purplish color. The cellular membranes, including the plasma membrane and nuclear envelope, do not generally pick up any stain and are thus transparent.

Classification Based on Layers of Cells

The location and function of membranous epithelium often determines whether the cells composing a specific type will be a single layer that is one cell thick, or stacked into several or many layers. **Simple epithelium** is only one cell layer thick. **Stratified epithelium** is composed of several to many layers of cells. **Transitional epithelium** is made up of an arrangement of differing cell shapes in a stratified sheet.

Notice that when stratified tissues are classified, they are named for the shape of the cells *within the top layer.* Refer to Table 6-2 as you read the following descriptions.

Simple Epithelium

Simple squamous epithelium consists of only *one layer* of flat, scalelike cells. Because of this, substances can easily diffuse or filter through this type of tissue. The microscopic air sacs (alveoli) of the lungs and the linings of blood and lymphatic

TISSUE		LOCATION	FUNCTION
Simple Epithelium			
Simple squamous	◀ Photomicrograph of the Bowman capsule of the kidney showing thin simple squamous epithelium. — Simple squamous epithelial cells	Alveoli of lungs	Absorption by diffusion of respiratory gases between alveolar air and blood
		Lining of blood and lymphatic vessels	Absorption by diffusion, filtration, and osmosis
		Surface layer of the pleura, pericardium, and peritoneum	Absorption by diffusion and osmosis; also, secretion
Simple cuboidal	◀ Photomicrograph of kidney tubules showing the single layer of cuboidal cells touching a basement membrane. — Lumen of tubule — Cell nuclei — Cuboidal epithelial cells — Basement membrane	Ducts and tubules of many organs, including the exocrine glands and kidneys	Secretion Absorption
Simple columnar	◀ Photomicrograph of simple columnar epithelium. — Goblet cell — Columnar epithelial cells	Surface layer of the mucous lining of the stomach, intestines, and part of the respiratory tract	Protection Secretion Absorption Moving of mucus (by ciliated columnar epithelium)
Pseudostratified columnar	◀ This photomicrograph from the respiratory system shows that each irregularly shaped columnar cell touches the underlying basement membrane. Placement of cell nuclei at irregular levels in the cells gives a false (pseudo) impression of stratification. This tissue is ciliated—note the "fuzz" along the outer edge of the cells. — Cilia — Columnar cells — Basement membrane	Surface of the mucous membrane lining the trachea, large bronchi, nasal mucosa, and parts of the male reproductive tract (epididymis and vas deferens); lines the large ducts of some glands (e.g., parotid)	Absorption Secretion
Stratified Epithelium			
Stratified squamous, keratinized	◀ Photomicrograph of the thick skin showing cells becoming progressively flattened and scalelike as they approach the surface and are lost. The outer surface of this epithelial sheet contains many flattened cells that have lost their nuclei. — Basal cell — Keratinized layer — Stratified squamous epithelium — Basement membrane — Dermis	Surface of the skin (epidermis)	Protection
Stratified squamous, nonkeratinized	◀ Photomicrograph of the lining of the esophagus. Each cell in the layer is flattened near the surface and attached to the sheet. — Superficial squamous cell — Basal cell — Basement membrane	Surface of the mucous membrane lining the mouth, esophagus, and vagina	Protection

(continued)

TABLE 6-2 Membranous Epithelial Tissues (*continued*)

TISSUE		LOCATION	FUNCTION
Stratified Epithelium (continued)			
Stratified cuboidal (and columnar)	◄ Photomicrograph of the excretory duct of a gland. Note the double layer of cuboidal cells.	Ducts of the sweat glands and mammary glands; lining of the pharynx; covering part of the epiglottis; lining part of the male urethra	Protection Secretion Support
Transitional	◄ Photomicrograph from the urethra showing that its cell shape is variable, from cuboidal to squamous. Several layers of cells are present.	Surface of the mucous membrane lining the urinary bladder and ureters	Permits stretching Protection

vessels are composed of simple squamous epithelium. So are the surfaces of the sac around the heart (pericardium) as well as other membranes.

Simple cuboidal epithelium is one layer of cuboidal cells resting upon a basement membrane. Many types of glands and their ducts, as well as the tubules of organs such as the kidney, are lined with this type of epithelium. As you can see in Table 6-2, the cuboidal epithelial cells look like small bricks, which gives them strength and prevents them from being compressed.

Simple columnar epithelium forms the surface of the mucous membrane that lines the stomach, intestine, uterus, uterine tubes, and trachea. Many of the cells possess cilia or microscopic cellular projections called **microvilli.** In the intestines, for example, columnar cells produce hundreds of microvilli that increase the surface area of the cell available to absorb nutrients from digested food. Other columnar epithelial cells are modified into large *goblet cells.* (Goblet cells are shaped like a drinking goblet.) Large vesicles bursting from the surface of the goblet cells contain mucus—a solution of water, electrolytes, and proteoglycans. Mucus is thus secreted onto the surface of the epithelial membrane and acts as a lubricant and defense mechanism against bacteria and infectious particles such as viruses.

Pseudostratified columnar epithelium lines the air passages of the respiratory system and parts of the male reproductive tract, such as the urethra. Goblet cells and rhythmically moving cilia are usually present. In the respiratory tract, mucus secreted by goblet cells help trap dust and other contaminants from the air. These pollutants are then moved by cilia away from the interior of the lungs and upward toward the mouth.

Stratified Epithelium

Stratified squamous epithelium is characterized by multiple layers of cells with typical scalelike squamous cells at the *outer surface* of the epithelial sheet (farthest from the basement membrane). This type of epithelium is best exemplified by skin tissue, which has a special hard protein called *keratin* in its cells. Keratin provides a protective barrier against intruding forces from outside, as well as an effective barrier against losing water to the environment. The **keratinized** stratified squamous epithelium of your skin provides your body with its first line of defense.

Nonkeratinized *stratified squamous* epithelium lines the vagina, mouth, and esophagus. The cells farthest from the basement membrane lack keratin and are moist. Nevertheless, these tissues serve as protective barriers against abrasion in these particular parts of the body.

Stratified cuboidal epithelium consists of two or more layers of cuboidal cells arranged over a basement membrane. It serves as a protective and secretory layer in different tissues such as the epithelium of the pharynx, parts of the epiglottis, and ducts of the sweat glands and mammary glands.

Transitional epithelium is a stratified tissue typically found in organs and areas that require stretching due to stress and tension on the organ. The urinary bladder is perhaps the most obvious example. The linings of our bladders consist of layers of transitional epithelia with cells of varying shapes. These cells can flatten when tension forces the epithelium to expand. This is an important structural adaptation. Other types of epithelia are less "stretchy" and would not hold up under repeated stretching (you may be more thankful for this fact during your next lecture class!).

Glandular Epithelium

Glandular epithelium is adapted for the secretion of important substances. It may occur as unicellular glands; as clusters, solid cords, or hollow follicles; or as multicellular glands.

Glands secrete their products into ducts, into the lumen of hollow structures, onto the body, or directly into the blood. The two types of glands, endocrine and exocrine, are defined by where they deposit their products. **Endocrine glands,** such as our pituitary, thyroid, and adrenal glands, secrete their products—usually hormones—directly into the blood, which delivers the products to their ultimate destination. **Exocrine glands,** such as our salivary glands, discharge their secretions into ducts that transport the secretions to their final destination. We will only cover exocrine glands here because endocrine glands will be discussed in Chapter 15.

Structural Classification of Exocrine Glands

Multicellular exocrine glands are generally classified by their overall structure. The shape of their ducts and the branching pattern of their duct systems are used as distinguishing characteristics. Ducts can be either tubular or alveolar (saclike). They can also have a branching pattern that includes a single duct (simple) or multiple ducts (compound). For example, the stomach is lined by simple tubular gastric glands with different branching patterns.

Table 6-3 gives you an overview of the major structural classes of exocrine glands.

Functional Classification of Exocrine Glands

Exocrine glands can be further classified into three general types distinguished by the *method of secretion* from the cells within the gland. These glands are called *apocrine, holocrine,* and *merocrine* (Figure 6-2).

Apocrine glands (*apo-,* "apex" or "tip") collect their secretory products near the tip of their cells and then release them into a duct by pinching off the end of the cell. Of course, this activity results in the loss of cytoplasm and requires some cell repair, which occurs rapidly. Milk-producing mammary glands are examples of apocrine glands. **Holocrine glands** (*holo-,* "whole") collect their products inside the entire cell and rupture completely to release these secretions. Sebaceous oil glands of the skin are examples of holocrine glands. **Merocrine glands** require no damage to the cell in order to secrete their products. Merocrine glands, such as salivary glands, pass their products through the plasma membrane into their ducts. Look closely at the diagrams in Figure 6-2 and you will see how the method of secretion from these cells differs.

TABLE 6-3 Structural Classification of Multicellular Exocrine Glands

SHAPE*	COMPLEXITY†	TYPE	EXAMPLE
Tubular (single, straight)	Simple	Simple tubular	Intestinal glands
Tubular (coiled)	Simple	Simple coiled tubular	Sweat glands
Tubular (multiple)	Simple	Simple branched tubular	Gastric (stomach) glands
Alveolar (single)	Simple	Simple alveolar	Sebaceous (skin oil) glands
Alveolar (multiple)	Simple	Simple branched alveolar	Sebaceous glands
Tubular (multiple)	Compound	Compound tubular	Mammary glands
Alveolar (multiple)	Compound	Compound alveolar	Mammary glands
Some tubular; some alveolar	Compound	Compound tubuloalveolar	Salivary glands

*Shape of the distal secreting units of the gland.
†Number of ducts reaching the surface.

QUICK ✓ CHECK

4. List the basic types of epithelium.
5. What are the general functions of epithelial tissue?
6. What are membranous epithelia?
7. What are glandular epithelia?

Apocrine gland   Holocrine gland   Merocrine gland

▲ **FIGURE 6-2 Three types of exocrine glands.** Exocrine glands may be classified by the method of secretion.

CONNECTIVE TISSUE

Connective tissue is one of the most widespread and variable tissue types in the body. It literally *connects* the whole body together! Connective tissue exists in many forms, including elastic sheets, tough cords, delicate webs, rigid bones, and even as a fluid (blood).

Functions of Connective Tissue

Connective tissues as a group form a supporting framework for individual organs as well as for our entire body. Different types serve to connect, support, and defend. Perhaps the most unique connective tissue is the blood, which serves to transport important substances throughout the body and remove waste products. As part of the immune system, blood tissue also has specialized *leukocytes* (white blood cells) that defend our bodies against biological invaders.

Characteristics of Connective Tissue

Connective tissue is unique in that it consists predominantly of extracellular matrix (ECM). The ECM can contain extracellular fluid, varying numbers and types of fibers, and other material sometimes called **ground substance.** The nature of the ground substance defines the function of each type of connective tissue. The matrix of blood, for example, contains no fibers (except during clotting). As a fluid, blood matrix is ideal for transporting materials throughout the body. The fibers that make up the ECM of other types of connective tissue are *collagenous, reticular* (netlike), or *elastic.* Collagenous and elastic fibers are sometimes called "white" and "yellow" fibers, respectively, because of their color in living tissue.

Collagenous fibers are made of units of the protein **collagen,** which can exist in many different forms, depending on its function. Collagen is extremely abundant in the body and can comprise one fourth of all our body protein.

Reticular fibers occur in networks of delicate fibers. They support small structures such as capillaries and nerve fibers and are made of a special type of collagen called *reticulin.*

Elastic fibers are made of a protein called **elastin,** which, like a rubber band, returns to its original length and shape after being stretched. Elastic fibers are found in "stretchy" tissues, such as the cartilage of the external ear and in the walls of arteries. Try scrunching your ear with your hand and watch its shape return instantly as you release your grip. This is a prime example of your elastic fibers!

The matrix of connective tissues may also contain many *proteoglycans.* These protein-sugar molecules, along with *hyaluronic acid* and *chondroitin sulfate,* act like cornstarch in gravy—making the matrix thick. A thick matrix acts as a lubricant and as a barrier to bacteria and other invaders.

Classification of Connective Tissue

Connective tissues are classified by the structure of their intercellular material (note that the main types are fibrous, bone, cartilage, and blood). Look at Box 6-1 and Figure 6-3 for an overview. A complete list of connective tissues, their locations, and their functions is presented for you in Table 6-4.

BOX 6-1 | Connective Tissue Classification

1. Fibrous
   a. Loose (areolar)
   b. Adipose (fat)
   c. Reticular
   d. Dense
      (1) Irregular
      (2) Regular
         (a) Collagenous
         (b) Elastic

2. Bone
   a. Compact
   b. Cancellous (spongy)

3. Cartilage
   a. Hyaline
   b. Fibrocartilage
   c. Elastic

4. Blood

Fibrous

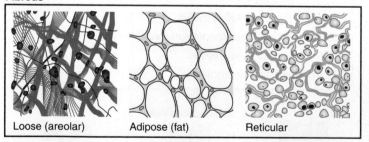

Loose (areolar)   Adipose (fat)   Reticular

Dense Fibrous

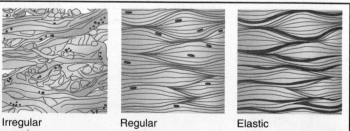

Irregular   Regular   Elastic

Bone

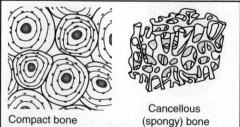

Compact bone   Cancellous (spongy) bone

Cartilage

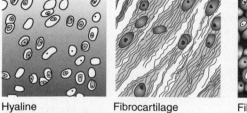

Hyaline   Fibrocartilage   Fibrocartilage

Blood

▲ FIGURE 6-3   Classification of connective tissues.

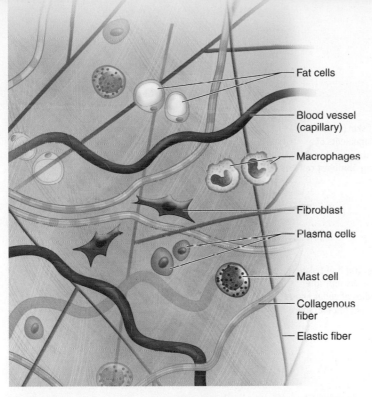

Fat cells

Blood vessel (capillary)

Macrophages

Fibroblast

Plasma cells

Mast cell

Collagenous fiber

Elastic fiber

▲ **FIGURE 6-4** **Diagram of loose (areolar) connective tissue.** This artist's sketch illustrates that loose connective tissue includes a number of different extracellular matrix (ECM) components such as collagenous fibers and elastic fibers, as well as a variety of different cell types.

Fibrous Connective Tissue
Loose (Areolar) Connective Tissue

Loose connective tissue, shown in Table 6-4, acts like an elastic glue holding adjacent body structures together. It is also called *areolar tissue* because it contains many tiny spaces within its matrix (*are-*, "open space"; *-ola*, "little"). It is one of the most widely distributed tissues within the body. It contains *elastin*, which provides its elasticity. The matrix is like a soft, thick gel. An enzyme, hyaluronidase, can change the matrix from its thick gel state to a watery consistency. Knowledge of this property is useful to physicians, who reduce the thickness of loose connective tissue when injecting drugs or other fluids. This allows the surrounding tissues to more easily absorb the drugs and reduces pain. Some bacteria, notably *pneumococci* and *streptococci*, actually spread through connective tissues by secreting hyaluronidase!

The matrix of loose connective tissue contains many interwoven collagenous and elastic fibers and about a half a dozen kinds of cells (Figure 6-4). *Fibroblasts* are usually present in the greatest numbers, with *macrophages* being less common. **Fibroblasts** synthesize the gel-like ground substance and fibers. **Macrophages** (*macro-*, "large"; *phage-*, "eater") carry on *phagocytosis*, which is an

TABLE 6-4 Connective Tissues

TISSUE			LOCATION	FUNCTION
Fibrous				
Loose (areolar)	Elastic fibers Bundles of collagenous fibers	◄ Notice that the bundles of collagen fibers are a pinkish color and the elastin fibers and cell nuclei are a darker, purplish color.	Between other tissues and organs Superficial fascia	Connection Cushions Provides flexibility
Adipose (fat)	Plasma membrane Storage area for fat Nucleus of adipose cell	◄ Note the large storage spaces for fat inside the adipose tissue cells.	Under skin Padding at various points	Protection Insulation Support Reserve food
Reticular	Blood cells Reticular fibers	◄ The supporting framework of reticular fibers are stained black in this section of lymph node.	Inner framework of spleen, lymph nodes, bone marrow	Support Filtration

(continued)

TABLE 6-4 Connective Tissues (*continued*)

TISSUE		LOCATION	FUNCTION
Dense Fibrous			
Irregular Collagenous fibers Fibroblasts	◄ Section of skin (dermis) showing arrangements of collagenous fibers (pink) and purple-staining fibroblast cell nuclei.	Deep fascia Dermis Scars Capsule of kidney, etc.	Connection Support
Regular collagenous Fibroblasts Collagenous fibers (in bundles)	◄ Photomicrograph of tissue in a tendon. Note the multiple (regular) bundles of collagenous fibers arranged in parallel rows.	Tendons Ligaments Aponeuroses	Flexible but strong connection
Elastic Elastic fibers	◄ Note the roughly parallel arrangement of short, darkly stained elastic fibers.	Walls of some arteries	Flexible, elastic support
Bone			
Compact bone Canaliculi Osteon (Haversian system) Lacunae (containing osteocytes) Central (Haversian) canal	◄ Photomicrograph of ground compact bone. Many wheel-like structural units of bone, known as *osteons* or *haversian systems*, are apparent in this section.	Skeleton (outer shell of bones)	Support Protection Calcium reservoir
Cancellous (spongy) bone Red bone marrow Trabeculae	◄ Photomicrograph of cancellous (*spongy* or *trabecular*) bone. The pink-stained mineralized bone tissue forms a lattice of irregular beams, or *trabeculae*, that support the softer reticular tissue of the bone marrow. The darkly stained nuclei of osteocytes *(OC)* are visible, as well as the dark boundaries *(arrows)* of mineralized bone layers or lamellae.	Skeleton (inside bones)	Support Provides framework for blood production
Cartilage			
Hyaline Chondrocytes (in lacunae) Matrix Perichondrium layer	◄ Photomicrograph of the trachea. Note the many spaces, or lacunae, in the gel-like matrix.	Part of nasal septum Covering articular surfaces of bones Larynx Rings in trachea and bronchi	Firm but flexible support Protection

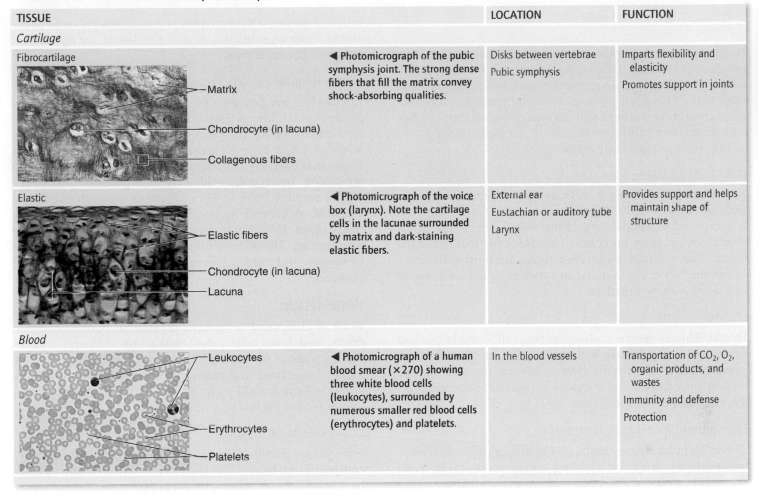

TISSUE		LOCATION	FUNCTION
Cartilage			
Fibrocartilage Matrix Chondrocyte (in lacuna) Collagenous fibers	◀ **Photomicrograph of the pubic symphysis joint. The strong dense fibers that fill the matrix convey shock-absorbing qualities.**	Disks between vertebrae Pubic symphysis	Imparts flexibility and elasticity Promotes support in joints
Elastic Elastic fibers Chondrocyte (in lacuna) Lacuna	◀ **Photomicrograph of the voice box (larynx). Note the cartilage cells in the lacunae surrounded by matrix and dark-staining elastic fibers.**	External ear Eustachian or auditory tube Larynx	Provides support and helps maintain shape of structure
Blood			
Leukocytes Erythrocytes Platelets	◀ **Photomicrograph of a human blood smear (×270) showing three white blood cells (leukocytes), surrounded by numerous smaller red blood cells (erythrocytes) and platelets.**	In the blood vessels	Transportation of CO_2, O_2, organic products, and wastes Immunity and defense Protection

important part of the body's defense against invading parasites. **Mast cells** are specialized cells that release chemicals such as *histamine, heparin, leukotrienes,* and *prostaglandins.* These substances cause inflammation in response to exposure to foreign substances entering the body (Box 6-2). Finally, small numbers of white blood cells (leukocytes) and some fat cells are also found in loose connective tissue.

BOX 6-2 | Health Matters

Hay Fever and Asthma

Mast cells in loose connective tissues are often involved in *allergy,* or *hypersensitivity,* reactions in local tissues. This is a type of inflammation in response to *allergens,* which are substances that trigger such allergic responses. For example, the inflammation that you experience after a bee sting or when your skin is exposed to poison ivy are examples of allergic or hypersensitivity reactions. Such reactions are triggered when mast cells encounter an allergen and release any of a group of chemical mediators.

For example, in hay fever (allergies to grasses and other plants), mast cells release the chemical **histamine.** Histamine increases the permeability of blood vessels in the nasal membranes, which in turn causes swelling in the lining of the nose. This gives us a "stuffy feeling" because the swelling makes it difficult to breathe easily. Histamine can also cause itchiness of the nose and eyes (see figure). The stuffiness and itchy eyes of hay fever can be relieved by antihistamines such as diphenhydramine (Benadryl) and fexofenadine (Allegra), which block histamine's action.

On the other hand, a different mast cell product is responsible for the breathing difficulties in an asthma attack. In asthma, chemicals called *leukotrienes* trigger muscles in the walls of the respiratory tract to contract—thereby constricting the airways. Thus, leukotriene-blocking drugs such as montelukast (Singulair) and zileuton (Zyflo) are often used along with other drugs that prevent or reduce contraction of airway muscles.

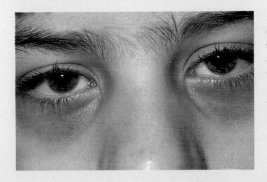

◀ Some allergies cause itchiness and watering of the eyes, which may trigger frequent rubbing of the eyes that produces these reddened features.

Adipose Tissue

Adipose tissue is similar to loose connective tissue, from which it is derived. Of course, it contains fat cells, called **adipocytes,** and far fewer fibroblasts, macrophages, and mast cells. These fatty tissues form supporting, protective pads around the kidneys and various other structures. They also store energy and serve as insulation for the body.

Box 6-3 discusses body fat composition.

Reticular Tissue

Reticular tissue forms a delicate, three-dimensional web that supports the spleen, lymph nodes, and bone marrow. It is composed of slender, branching *reticulin fibers* that originate from similarly branching reticular cells that cover the fibers. The reticular meshwork filters harmful substances out of the blood and lymph, and then reticular cells engulf and destroy these invaders.

Dense Fibrous Tissue

Dense fibrous tissue consists mainly of densely packed fibers within the matrix. Relatively few fibroblast cells are present within the matrix. Dense fibrous tissue is further broken into *regular* and *irregular* categories, depending on the arrangement of the fibers.

Dense Irregular Fibrous Tissue

Dense irregular fibrous tissue has *bundles of collagenous fibers* intertwined in irregular, swirling arrangements, as you can see in Table 6-4. This irregular pattern means that the tissue can stretch yet withstand stress applied from any direction. The capsules of tissue surrounding our kidneys and spleen,

much of the fascia that surrounds our muscles, as well as the strong inner layer of human skin, the *dermis,* are all made of irregular fibrous tissue.

Dense Regular Fibrous Tissue

Regularly arranged *parallel bundles* of fibers characterize dense regular fibrous tissue. These tough tissues form most of the connections between bones and muscles. Tendons, which connect bone to muscle, are dense regular fibrous tissue made largely of collagen bundles that can withstand stretching from both ends. By comparison, ligaments, which connect bone to bone, contain more elastic fibers. These fibers provide ligaments with some elasticity. Tissue made up of *elastic, dense, regular, fibrous* tissue forms part of our arteries. This elastic fibrous tissue allows our arteries to stretch and contract in response to changes in blood pressure.

Bone Tissue

Bone is composed of a rigid crystalline matrix, collagen fibers, and bone cells called **osteocytes.** Mineral crystals make up about two thirds of the ECM, which as you know makes bone hard and strong. These characteristics allow bones to form a rigid support system for the softer tissues and organs in the body.

Compact Bone Tissue

Bone tissue forming the hard shell of a bone is called **compact bone tissue.** The basic organizational unit of compact bone is the microscopic **osteon,** or *haversian system* (see Table 6-4). Notice that osteocytes are located in small spaces called **lacunae,** which in turn are arranged in concentric

---

BOX 6-3 Sports & Fitness

Tissues and Fitness

Achieving and maintaining ideal body weight is a health-conscious goal. However, a better indicator of health and fitness is **body composition.** Exercise physiologists assess body composition to identify the percentage of the body made of lean tissue and the percentage made of fat. Body fat percentage is often determined by using calipers to measure the thickness of skin folds at certain locations on the body (see figures). The thickness measurements, which reflect the volume of adipose tissue under the skin, are then used to estimate the percentage of fat in the entire body. A much more accurate method is to weigh a subject totally immersed in a tank of water. Fat has very low density and therefore increases the buoyancy of the body. Thus the lower a person's weight while immersed, the higher the body fat percentage.

A person with low body weight may still have a high ratio of fat to muscle, an unhealthy condition. In this case the individual is "underweight" but "overfat." In other words, fitness depends more on the percentage and ratio of specific tissue types than on the overall amount of tissue present. Therefore, one goal of a good fitness program is a desirable body fat percentage. For men, the ideal is 15% to 18%, and for women, the ideal is 20% to 22%.

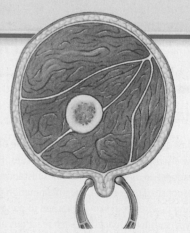

Because fat contains stored energy (measured in calories), a low fat percentage means a low energy reserve. High body fat percentages are associated with several life-threatening conditions, including diabetes and cardiovascular disease. A balanced diet and an exercise program ensure that the ratio of fat to muscle tissue stays at a level appropriate for maintaining homeostasis.

layers of bone matrix called **lamellae.** Tiny canals called **canaliculi** ("little canals") connect lacunae (and their osteocytes) with blood vessels from the central canal (the haversian canal). In this manner, all bone cells are fed directly from the circulatory system.

Bone is actually a dynamic tissue, though it may seem lifeless. For example, mature osteocytes lie immobile within the bone matrix. However, at one time they were active, bone-forming cells called **osteoblasts.** As osteoblasts surround themselves with matrix and become mature osteocytes, they become trapped and stop producing matrix. Another type of bone cell, the **osteoclast,** digests matrix from around mature osteocytes, thus releasing them to become active osteoblasts again. The interaction of bone construction by osteoblasts and bone destruction by osteoclasts allows your bones to grow and to be reshaped in response to body activity.

Cancellous (Spongy) Bone Tissue

Inside many bones is an intricate lattice of thin beams of **cancellous bone tissue** called **trabeculae.** These beams form a framework that supports the rest of the bone (much like trusses support a roof). Trabeculae may be surrounded by softer *red bone marrow* tissue, also called *myeloid tissue.* Myeloid tissue is a type of reticular tissue housing stem cells that produce the various types of blood cells.

Cartilage Tissue

Cartilage tissue only has one cell type, the **chondrocyte,** which produces the fibers and the tough, rubbery ground substance of cartilage. Chondrocytes, like osteocytes, are nestled in lacunae. No blood vessels run through cartilage, so all nutrients must reach the tissue by diffusion. This takes place through the matrix from blood vessels in a membrane that surrounds the cartilage, the **perichondrium.** See Table 6-4 for a summary of cartilage tissue.

Hyaline Cartilage Tissue

Hyaline cartilage contains little collagen in its matrix and thus appears translucent and glassy (*hyalos,* "glass"). This is the most prevalent type of cartilage. Hyaline cartilage covers the ends of long bones, creates the support rings of the trachea, and forms the rubbery tip of your nose.

Fibrocartilage Tissue

Fibrocartilage is the strongest and most durable type of cartilage. Fibrocartilage forms the intervertebral disks and part of the knee joint. It has a matrix that is rigid and densely packed with collagen fibers—which makes it an excellent shock absorber.

Elastic Cartilage Tissue

Elastic cartilage contains few collagen fibers, but large numbers of very fine elastic fibers that give the matrix material a high degree of flexibility. This type of cartilage is found in the external ear and in the larynx.

Blood Tissue

Blood is unique among connective tissues because it is a liquid that contains neither ground substance nor fibers. Its matrix, known as **plasma,** is the liquid portion of blood that surrounds blood cells. Blood cells are divided into three classes: **erythrocytes** (red blood cells), **leukocytes** (white blood cells), and **thrombocytes** (platelets) (see Table 6-4).

Blood transports respiratory gases (oxygen and carbon dioxide), nutrients, and waste products. In addition, blood plays a critical role in maintaining a constant body temperature and regulating the pH of body fluids. The leukocytes within blood tissue are an important portion of the body's immune system.

QUICK CHECK

8. What are the main types of connective tissue?
9. List common functions of connective tissue.
10. How does the ECM change in the basic types of connective tissue?
11. How does adipose tissue differ from loose connective tissue?
12. What are the unusual properties of the blood matrix?

MUSCLE TISSUE

Muscle tissue contains cells that are specialized for contractility. They exist in three forms: *skeletal, smooth,* and *cardiac,* as shown for you in Table 6-5 and Figure 6-5.

Skeletal muscle tissue, or *striated voluntary* muscle, makes up most of the muscles attached to bones. These muscles move the interlocking parts of the body. Skeletal muscles appear striated (striped) when viewed under a microscope because the organization of fibers within each cell gives it distinct light and dark bands. In Table 6-5, you can see that skeletal muscle cells tend to be long (often greater than 3.75 cm). They also have many nuclei in each cell (*multinucleate*), a condition that results from the fusion of many immature muscle cells during embryonic development.

Smooth muscle tissue has no obvious striations and is generally *involuntary.* It is found in the walls of hollow internal organs such as the stomach, intestines, and blood vessels (see Table 6-5). Individual smooth muscle cells have a single, large nucleus, and are shorter than skeletal muscle cells.

Muscle

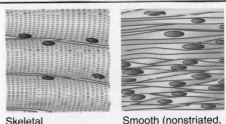
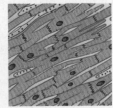

| Skeletal (striated voluntary) | Smooth (nonstriated, involuntary, or visceral) | Cardiac (striated involuntary) |

▲ **FIGURE 6-5** Types of muscle tissue.

TABLE 6-5   Muscle and Nervous Tissues

TISSUE		LOCATION	FUNCTION
Muscle			
Skeletal (striated voluntary)	◀ Note the striations of the skeletal muscle cell fibers in longitudinal section.	Muscles that attach to bones	Movement of bones
— Striations of muscle cell		Extrinsic eyeball muscles	Eye movements
— Muscle fiber — Nuclei of muscle cell		Upper third of the esophagus	First part of swallowing
Smooth (nonstriated, involuntary, or visceral)	◀ Photomicrograph, longitudinal section. Note the central placement of nuclei in the spindle-shaped smooth muscle fibers.	In the walls of tubular viscera of the digestive, respiratory, and genitourinary tracts	Movement of substances along the respective tracts
		In the walls of blood vessels and large lymphatic vessels	Change diameter of blood vessels, thereby aiding in regulation of blood pressure
— Nuclei of smooth muscle cells		In the ducts of glands	Movement of substances along ducts
		Intrinsic eye muscles (iris and ciliary body)	Change diameter of pupils and shape of the lens
		Arrector muscles of hairs	Erection of hairs (gooseflesh)
Cardiac (striated involuntary)	◀ The dark bands, called *intercalated disks*, which are characteristic of cardiac muscle, are easily identified in this tissue section.	Wall of the heart	Contraction of the heart
— Nucleus — Intercalated disks			
Nervous			
— Nerve cell body — Axon — Nuclei of neuroglia — Dendrites	◀ Photomicrograph showing multipolar neurons surrounded by smaller neuroglia in a smear of spinal cord tissue. All of the large neurons in this photomicrograph show cell bodies and multiple cell processes.	Brain Spinal cord Nerves	Excitability Conduction

However, they are still rather long when we consider the average length of most cells.

Cardiac muscle tissue is *striated* and *involuntary* muscle that makes up the walls of the heart. Some of the larger striations of cardiac muscle are different in structure and appearance than those of skeletal muscle. The darkly staining bands that join cardiac muscle cells together are called *intercalated disks*. The *gap junctions* of the intercalated disks allow direct communication between all of the cells in the heart so that the interconnected cells may contract in unison. Cardiac cells (fibers) branch in and out, abutting many neighboring cells. However, each cell is enclosed in a complete plasma membrane, with intercalated disks where one cell joins another.

NERVOUS TISSUE

Nervous tissue acts much like our Internet networks, rapidly regulating and integrating activities of different parts of the body. It is structurally adapted for electrical excitation and conductivity, which allows a nerve to literally transmit messages to and from body regions via electric pulses. **Neurons** are the individual cells of nervous tissue and are the "wires" that conduct messages (see Table 6-5 and Figure 6-6). The supporting cells, **neuroglia,** serve to connect and protect the neurons, some acting much like the insulation on an electric wire.

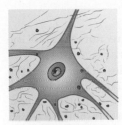

▲ FIGURE 6-6
Nervous tissue.

All neurons are made up of a cell body called the **soma** and, generally, at least two processes: one **axon,** which transmits nerve impulses *away* from the soma, and one or more **dendrites,** which carry nerve signals *toward* the soma and axon. There are many types of neuroglia that differ in structure and function. They function in several coordinating roles in the nervous system, including to protect brain tissue, destroy pathogens, and insulate axons.

QUICK ✓ CHECK

13. Name the three basic types of muscles.
14. What is the function of intercalated disks?
15. What are the basic features of a neuron?

TISSUE REPAIR

When tissues are damaged, they undergo **regeneration.** In this process, phagocytic cells remove dead or injured cells while other cells divide and begin to repair the wound. For example, epithelial and connective tissues respond to a cut in your skin by first dividing to form new daughter cells. Connective tissues then form many collagen fibers in an especially dense mass to aid in filling the gap. This mass of connective tissue may be replaced by normal tissue later. If the cut is especially large or cell damage is extensive, the dense fibrous mass may remain, becoming a **scar** (Figure 6-7).

Muscle and nervous tissues have limited abilities to regenerate. Damaged neurons, if still alive, can repair themselves. However, muscles are often left with dense collagen masses that leave them less functional. Nervous tissue in general cannot reproduce new cells, but a few specialized areas in the adult brain are now known to contain stem cells (Box 6-4). These unique precursor cells continue to divide and can provide some regeneration capabilities for the surrounding tissue.

BOX 6-4	Stem Cells

During embryonic development, new kinds of cells can be formed from a special kind of *undifferentiated cell* called a **stem cell.**

Embryonic stem cells have the potential to reproduce many different kinds of daughter cells, including more stem cells—thus populating the body with all the different cells and tissues needed for body function. Sometimes-controversial research is now underway to learn the secrets of embryonic stems cells and develop therapies in which embryonic stem cells might be used to repair or replace damaged tissue in adults.

Adult stem cells are undifferentiated cells found scattered within a differentiated, mature tissue. Many different tissues contain adult stem cells—we may soon find that all adult tissues have some stem cells. Adult stem cells can usually produce any of the specialized cell types within their particular tissue. However, recent research shows that some adult stem cells can be coaxed into producing a variety of different types of cells. For example, blood cell–producing stem cells from bone marrow have been used to help repair damaged muscle tissue in laboratory animals. Already, therapies for treating degenerative diseases of muscles, the heart, and the brain—even baldness—are being proposed by medical scientists.

BODY MEMBRANES

Body membranes are thin structures that have important supportive and protective functions in the body. **Epithelial membranes** are composed of epithelial tissue and an underlying layer of supportive connective tissue. **Connective tissue membranes** are composed exclusively of various types of connective tissue. Membranes cover and line body surfaces inside and out and also serve to anchor organs to each other or to bones. Some membranes also secrete lubricating fluids that reduce friction between bones during body movement. Other connective tissue membranes reduce friction during organ movements, such as when your lungs expand and contract.

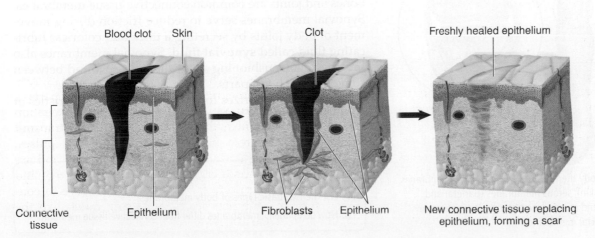

Blood clot Skin   Clot   Freshly healed epithelium

Connective tissue   Epithelium   Fibroblasts   Epithelium

New connective tissue replacing epithelium, forming a scar

◀ FIGURE 6-7 Healing of a minor wound. When a minor injury damages a layer of epithelium and the underlying connective tissue (as in a minor skin cut), the epithelial tissue and the connective tissue can repair itself.

Epithelial Membranes

There are three types of epithelial tissue membranes in the body: cutaneous, serous, and mucous membranes (Figure 6-8). Each is made up of a layer of epithelial tissue attached to a layer of fibrous connective tissue by a gluelike *basement membrane.*

Cutaneous Membrane

Cutaneous membrane is another term for the **skin**—the sole organ of the integumentary system. It contains many sweat and oil glands that produce a *surface film* over the skin. This film aids in the skin's function as a protector of the body against harmful environmental factors.

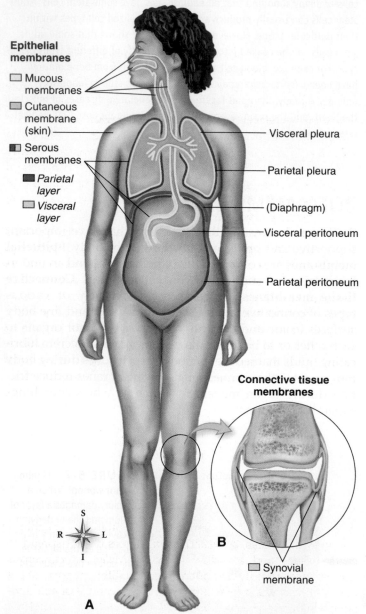

Epithelial membranes

☐ Mucous membranes
☐ Cutaneous membrane (skin)
■ Serous membranes
  ■ *Parietal layer*
  ☐ *Visceral layer*

— Visceral pleura
— Parietal pleura
— (Diaphragm)
— Visceral peritoneum
— Parietal peritoneum

Connective tissue membranes

B

☐ Synovial membrane

A

▲ **FIGURE 6-8   Types of body membranes. A,** Epithelial membranes, including cutaneous membrane (skin), serous membranes (parietal and visceral pleura and peritoneum), and mucous membranes. **B,** Connective tissue membranes, including synovial membranes.

Serous Membranes

Serous membranes line cavities that are not open to the external environment. These membranes also cover many of the organs residing inside these cavities. One layer, the *epithelial sheet,* is a thin layer of simple squamous epithelium, which is supported by a thin sheet of connective tissue. Serous membranes secrete a thin, watery fluid that lubricates organs as they rub against one another or against the walls of the cavities that contain them.

The serous membrane that lines body cavities and covers the surfaces of organs in these cavities is really a single, continuous sheet covering two different surfaces. The *parietal membrane* is the portion that lines the wall of the cavity. The *visceral membrane* covers the surfaces of the organs (the *viscera*). Two important serous membranes are shown in Figure 6-8: the *pleura,* which surrounds a lung and lines the thoracic cavity, and the *peritoneum,* which covers the abdominal viscera and lines the abdominal cavity. The *pericardium* is the serous membrane that surrounds the heart.

Mucous Membranes

Mucous membranes, or *mucosa,* are epithelial membranes that line body surfaces that open directly to the exterior, such as the respiratory, digestive, urinary, and reproductive tracts. The epithelial component of mucous membranes varies with its location. For example, the upper portion of the esophagus is lined with a tough, abrasion-resistant stratified squamous epithelium. In contrast, a thin layer of columnar epithelium covers the walls of the lower digestive tract. The fibrous connective tissue underlying the epithelium in mucous membranes is called the **lamina propria.**

All mucous membranes produce a film of **mucus** that is composed of water, electrolytes, and different types of proteoglycans called *mucins.* Some mucus serves only to protect the underlying cells, while other mucus acts as a lubricant (e.g., for food in the digestive tract) or even as a way of trapping environmental pathogens (in many of the respiratory passages).

Connective Tissue Membranes

Connective tissue membranes do not contain an epithelial layer. The **synovial membranes** lining the spaces between bones and joints are common connective tissue membranes. Synovial membranes serve to reduce friction during movement of body joints by secreting a thick and colorless lubricating fluid called **synovial fluid.** Synovial membranes also line the small, cushioning sacs called *bursae* found between some moving body parts.

Table 6-6 summarizes for you the main characteristics of the four major types of membranes in the body.

QUICK CHECK

16. What is a scar?
17. List the two basic types of body membranes.
18. How do epithelial membranes differ from connective tissue membranes?

TABLE 6-6 **Membranes of the Body**

TYPE	SUPERFICIAL LAYER	DEEP LAYER	LOCATION	FLUID SECRETION	FUNCTION
Epithelial					
Cutaneous (skin)	Keratinized stratified squamous epithelium (epidermis)	Dense irregular fibrous connective tissue (dermis)	Directly faces external environment	Sweat; sebum (skin oil)	Protection, sensation, thermoregulation
Serous	Simple squamous epithelium	Fibrous connective tissue	Lines body cavities that are not open to the external environment	Serous fluid	Lubrication
Mucous	Various types of epithelium	Fibrous connective tissue (lamina propria)	Lines tracts that open to the external environment	Mucus	Protection, lubrication
Connective					
Synovial	Dense fibrous connective tissue	Loose fibrous connective tissue	Lines joint cavities (in movable joints)	Synovial fluid	Helps hold joint together, lubricates, cushions

Cycle of LIFE ⭕

Most of our body's tissues show decreasing function with age. These dramatic changes typically lead to decreased activity, declining health, and a reduction in the overall quality of life.

The BIG Picture

Like the pieces of fabric in a garment, tissues and body membranes are portions of a larger integrated structure—your body. Each type of tissue within the body has a specific role to play in our human fabric. Together, our tissues function to maintain a state of relative constancy—*homeostasis*.

How do the major tissue types help maintain homeostasis? Epithelial tissues promote constancy of the body's internal environment. They form membranes that contain and protect the internal fluid environment. They absorb nutrients and other substances needed to maintain an optimal concentration in the body. They also serve to secrete various products that we need to regulate our body functions.

Connective tissues hold organs and systems together to form a functional body. They also form structures that support the body and permit movement. Some connective tissues, such as blood, transport nutrients, wastes, and other substances within our internal environment. Some blood cells also help protect us by functioning along with other parts of our immune system.

Muscle tissues work together with connective tissues (bones, tendons, etc.) to permit movement. They also allow us to communicate, and to locate food, shelter, and other requirements for life. Nervous tissue works with the entire body to regulate many vital processes within our bodies. All of these are important to maintaining homeostasis.

Now that you have a basic knowledge of the various types of body tissues and membranes, you are ready to study the structure and function of specific organs and systems. As you take this next step in your studies, pay close attention to the tissue types that make up each organ. If you do, you will find it much easier to understand the characteristics of a particular organ. You will also improve your understanding of the integrated nature of your body.

Understanding the biological processes involved in cellular aging will help us develop ways for treatment at the tissue level.

Accompanying tissue aging is an increase in the incidence of disease throughout our body. For example, muscles and bones often weaken as we age, which can lead to greater incidence of falling and disabling injuries from these falls. The skin also undergoes significant changes that we can readily see. In addition, these changes may impair the process of wound healing—a distinct threat to our overall health. In fact, all systems are affected, from the hormonal system to the reproductive systems.

Unfortunately, the cellular events and the mechanisms underlying the aging of tissues are not well understood. However, this is changing. A great deal of new research is now examining the underlying mechanisms concerned with tissue aging. To do this successfully, investigators are comparing the differences in aging between genders and race. They are also studying the aging process and the rate at which it affects multiple tissues. Of course, not all tissues "age" at the same rate, and this points to the possibility of different mechanisms of aging in different tissues. ●

MECHANISMS OF DISEASE

A disease that can affect tissues is *cancer*, caused by an abnormal growth of cells, a *tumor*. *Benign* tumors do not spread to other tissues and are slow growing. *Malignant* cancerous tumors grow rapidly and often spread to different parts of the body. All cancers involve a mistake in the control of cell division, which can be caused by *genetic factors*, *carcinogens*, and *age*.

ⓔvolve Find out more about cancer, including how to detect and treat it, online at *Mechanisms of Disease: Tissues and Their Functions*.

LANGUAGE OF SCIENCE AND MEDICINE *continued from page 93*

dense fibrous tissue (dense FYE-brus TISH-yoo)
[*dense* thick, *fibro-* fiber, *-ous* relating to, *tissu-* fabric]

elastic cartilage (eh-LAS-tik KAR-ti-lij)
[*elast-* drive or beat out, *-ic* relating to, *cartilage* cartilage]

elastic fiber (eh-LAS-tik FYE-ber)
[*elast-* drive or beat out, *-ic* relating to, *fibr-* thread]

elastin (e-LAS-tin)
[*elast-* drive or beat out, *-in* substance]

endocrine gland (EN-doh-krin gland)
[*endo-* inward or within, *-crin* secrete, *gland* acorn]

epithelial membrane (ep-i-THEE-lee-al MEM-brayne)
[*epi-* on or upon, *-theli-* nipple, *-al* relating to, *membran-* thin skin]

erythrocyte (eh-RITH-roh-syte)
[*erythro-* red, *-cyte* cell]

exocrine gland (EK-soh-krin gland)
[*exo-* outside or outward, *-crin* secrete, *gland* acorn]

extracellular matrix (ECM) (eks-trah-SEL-yoolar MAY-triks)
[*extra-* beyond, *-cell-* storeroom, *-ular* relating to, *matrix* womb] *pl.,* matrices (MAY-tri-sees)

fibroblast (FYE-broh-blast)
[*fibro-* fiber, *-blast* bud]

fibrocartilage (fye-broh-KAR-ti-lij)
[*fibro-* fiber, *-cartilage* cartilage]

glandular (GLAN-dyoo-lar)
[*gland-* acorn (gland), *-ula-* little, *-ar* relating to]

glycoprotein (glye-koh-PRO-teen)
[*glyco-* sweet (glucose), *-prote-* first rank, *-in* substance]

ground substance
[*substancia-* material]

histamine (HIS-tah-meen)
[*hist-* tissue, *-amine* ammonia compound]

histology (his-TOL-oh-jee)
[*hist-* tissue, *-o-* combining form, *-log-* words (study of), *-y* activity]

holocrine gland (HOH-loh-krin gland)
[*holo-* whole, *-crine* secrete, *gland* acorn]

hyaline cartilage (HYE-ah-lin KAR-ti-lij)
[*hyaline* of glass, *cartilage* cartilage]

keratinize (KEHR-ah-tin-eyes)
[*kera-* horn, *-in-* substance, *-ize* make]

lacuna (lah-KOO-nah)
[*lacuna* pit] *pl.,* lacunae (lah-KOO-nay)

lamella (lah-MEL-ah)
[*lam-* plate, *-ella* little] *pl.,* lamellae (lah-MEL-ay)

lamina propria (LAM-in-ah PROH-pree-ah)
[*lamina* thin plate, *propria* proper]

leukocyte (LOO-koh-syte)
[*leuko-* white, *-cyte* cell]

loose connective tissue (LOOS koh-NEK-tiv TISH-yoo)

macrophage (MAK-roh-fayj)
[*macro-* large, *-phag* eat]

mast cell
[*mast* fattening, *cell* storeroom]

matrix (MAY-triks)
[*matrix* womb] *pl.,* matrices (MAY-tri-sees)

membranous (MEM-brah-nus)
[*membran-* thin skin, *-ous* characterized by]

merocrine gland (MER-oh-krin gland)
[*mero-* part, *-crine* separate, *gland* acorn]

microvillus (my-kroh-VIL-us)
[*micro-* small, *-villi* shaggy hairs] *pl.,* microvilli (my-kroh-VIL-eye)

mucous membrane (MYOO-kus MEM-brayne)
[*muc-* slime, *-ous* characterized by, *membran-* thin skin]

mucus (MYOO-kus)
[*mucus* slime]

muscle tissue (MUSS-el TISH-yoo)
[*mus-* mouse, *-cle* small, *tissu-* fabric]

nervous tissue
[*nervous* relating to nerves, *tissu-* fabric]

neuroglia (noo-ROG-lee-ah; noo-roh-GLEE-ah)
[*neuro-* nerve, *-glia* glue] *sing.,* neuroglial cell

neuron (NOO-ron)
[*neuron* string or nerve]

nonkeratinize (non-KEHR-ah-tin-eyes)
[*non-* not, *-kera-* horn, *-in-* substance, *-ize* make]

osteoblast (OS-tee-oh-blast)
[*osteo-* bone, *-blast* bud]

osteoclast (OS-tee-oh-klast)
[*osteo-* bone, *-clast* break]

osteocyte (OS-tee-oh-syte)
[*osteo-* bone, *-cyte* cell]

osteon (AHS-tee-on)
[*osteo-* bone, *-on* unit]

perichondrium (pair-i-KON-dree-um)
[*peri-* around, *-chondr-* cartilage, *-um* thing] *pl.,* perichondria (pair-i-KON-dree-ah)

plasma (PLAZ-mah)
[*plasma* substance]

pseudostratified columnar (sood-oh-STRAT-i-fyed KAHL-um-nar)
[*pseudo-* false, *-strati-* layer, *-fied* be made, *column-* column, *-ar* relating to]

regeneration (ree-jen-er-AY-shun)
[*re-* again, *-generat-* produce, *-tion* process]

reticular fiber (reh-TIK-yoo-lar FYE-ber)
[*ret-* net, *-ic-* relating to, *-ul-* little, *-ar* relating to, *fibr-* thread]

reticular tissue (reh-TIK-yoo-lar TISH-yoo)
[*ret-* net, *-ic-* relating to, *-ul-* little, *-ar* relating to, *tissu-* fabric]

scar

serous membrane (SEE-rus MEM-brayne)
[*sero-* watery body fluid, *-ous* characterized by, *membran-* thin skin]

simple epithelium (ep-i-THEE-lee-um)
[*simple* not mixed, *epi-* on, *-theli-* nipple, *-um* thing] *pl.,* epithelia (ep-i-THEE-lee-ah)

skeletal muscle tissue (SKEL-eh-tal MUSS-el TISH-yoo)
[*skeleto-* dried body, *-al* relating to, *mus-* mouse, *-cle* small, *tissu-* fabric]

skin

smooth muscle tissue (smooth MUSS-el TISH-yoo)
[*mus-* mouse, *-cle* small, *tissu-* fabric]

soma (soh-mah)
[*soma* body]

squamous (SKWAY-muss)
[*squam-* scale, *-ous* characterized by]

stem cell

stratified epithelium (STRAT-i-fyde ep-i-THEE-lee-um)
[*strati-* layer, *epi-* on, *-theli-* nipple, *-um* thing] *pl.,* epithelia (ep-i-THEE-lee-ah)

synovial fluid (si-NO-vee-all FLOO-id)
[*syn-* together, *-ovi-* egg (white), *-al* relating to, *fluid* flow]

synovial membrane (si-NO-vee-all MEM-brayne)
[*syn-* together, *-ovi-* egg (white), *-al* relating to, *membran-* thin skin]

thrombocyte (THROM-boh-syte)
[*thrombo-* clot, *-cyte* cell]

tissue (TISH-yoo)
[*tissu-* fabric]

trabecula (trah-BEK-yoo-la)
[*trab-* beam, *-ula* little] *pl.,* trabeculae (trah-BEK-yoo-lay)

transitional epithelium (tranz-i-shen-al ep-i-THEE-lee-um)
[*trans-* across, *-tion* process, *epi-* on, *-theli-* nipple, *-um* thing] *pl.,* epithelia (ep-i-THEE-lee-ah)

CHAPTER SUMMARY

To download an MP3 version of the chapter summary for use with your iPod or other portable media player, access the Audio Chapter Summaries online at http://evolve.elsevier.com.

HINT *Scan this summary after reading the chapter to help you reinforce the key concepts. Later, use the summary as a quick review before your class or before a test.*

INTRODUCTION

A. Tissues—each a group of similar cells performing a common function
B. Histology—study of tissues
C. Extracellular matrix (ECM)—complex mix of fluids and proteins that surround tissue
D. All tissues can be classified by their structure and function into four general categories: epithelium, connective, muscle, and nerve (Table 6-1)

EXTRACELLULAR MATRIX

A. Extracellular matrix (ECM)—very complex and varies greatly among different tissues
   1. Structural makeup of the ECM of any particular tissue matches its function
      a. ECM of bone tissue contains crystals of calcium that give strength and support to the entire body
      b. ECM surrounding the cells of tough ligament tissue contains many proteins that are flexible and elastic
B. The amount of ECM may vary between different tissues
C. General makeup of ECM consists of water and its dissolved ions
   1. May also contain mineral crystals, proteins, and proteoglycans
      a. Proteoglycans—large molecules constructed of a protein backbone linked to carbohydrates
D. Collagen and elastin—common structural proteins; provide support and elasticity
   1. Glycoproteins—connect collagen and elastin to cells; form "bridges" between cells
E. Tissues cells, with little ECM, are connected primarily by intercellular junctions (desmosomes, tight junctions) (see Chapter 3)

EPITHELIAL TISSUE

A. Types and locations of epithelial tissue
   1. Membranous epithelium—forms coverings and linings; covers the body and lines body cavities
   2. Glandular epithelium—produces secretions; forms secretory units of endocrine and exocrine glands
B. Functions of epithelial tissue
   1. Protection
   2. Sensory
   3. Secretion
   4. Absorption
   5. Excretion
C. Generalizations about epithelial tissue
   1. Most have very little intercellular matrix material
   2. Structural integrity is provided by intercellular connections such as desmosomes and tight junctions

   3. Epithelial tissue attaches to an underlying layer of connective tissue
      a. Basement membrane—thin, noncellular layer of adhesive, permeable material (Figure 6-1)
   4. Epithelial tissues are said to be avascular
D. Classification of epithelial tissue
   1. Membranous epithelium—classification based on cell shape at surface of the membrane (Figure 6-1)
      a. Squamous—cells are flat and platelike
      b. Cuboidal—cells are cube-shaped
      c. Columnar—cells are taller than they are wide
      d. Pseudostratified columnar—epithelium is made up of one layer of cells that appears to be at least several layers
   2. Membranous epithelium—classification based on layers of cells (Figure 6-1, Table 6-2)
      a. Simple epithelium—one cell layer thick
         (1) Simple squamous—one layer of flat, scalelike cells; substances can readily diffuse or filter through this type of tissue
         (2) Simple cuboidal—one layer of cuboidal cells resting upon a basement membrane; found in glands and their ducts, and tubules of organs
         (3) Simple columnar—cylindrical cells that create the surface of the mucous membranes; lines the stomach, intestine, uterus, uterine tubes, and trachea; many of the cells possess cilia or microvilli
         (4) Pseudostratified columnar—lines the air passages of the respiratory system and parts of the male reproductive tract; goblet cells and cilia are usually present
      b. Stratified epithelium—several to many layers of cells
         (1) Stratified squamous with keratin—multiple layers of cells with typical scalelike squamous cells at the outer surface of the epithelial sheet; found in skin tissue, which provides your body with its first line of defense; contains keratin in cells
         (2) Nonkeratinized stratified squamous—lines the vagina, mouth, and esophagus; serves as a protective barrier against abrasion
         (3) Stratified cuboidal—two or more layers of cuboidal cells arranged over a basement membrane; serves as a protective and secretory layer in different tissues
         (4) Transitional—cells of this tissue alter their shape and flatten when tension is placed on the epithelium and it is forced to expand; found in organs and areas that require stretching due to stress and tension on the organ (urinary bladder)
      c. Transitional epithelium—made up of an arrangement of differing cell shapes in a stratified sheet
   3. Glandular epithelium—adapted for the secretion of important substances; may occur as unicellular or multicellular glands
      a. Endocrine—secrete their products directly into the blood
      b. Exocrine—discharge their secretions into ducts that transport the secretion to its final destination
         (1) Multicellular exocrine glands are generally classified by their structure, the shape of their ducts, or the branching pattern of the duct systems (Table 6-3)
         (2) Exocrine glands can be classified into three general types distinguished by the method of secretion: apocrine, holocrine, and merocrine (Figure 6-2)
         (3) Apocrine—collect their secretory products near the tips of their cells, then release them into a duct by pinching off the end of the cell; mammary glands

(4) Holocrine—collect their products inside the entire cell and rupture completely to release the products; sebaceous glands

(5) Merocrine—pass their products through the plasma membrane into their ducts; salivary glands

CONNECTIVE TISSUE

A. Connective tissue—one of the most widespread and variable tissue types in the body
1. Exists in many forms: elastic sheets, tough cords, delicate webs, rigid bones, and fluid
B. Functions of connective tissue
1. Connect
2. Support
3. Defend
C. Characteristics of connective tissue
1. Connective tissue—consists predominantly of extracellular matrix (ECM)
2. ECM—extracellular fluid, varying numbers and types of fibers, and other material (ground substance)
3. Ground substance—defines the function of each type of connective tissue
4. ECM fibers
   a. Collagenous fibers—made of units of the protein collagen; extremely abundant in the body
   b. Reticular fibers—networks of delicate fibers
   c. Elastin—made of a protein called elastin; found in "stretchy" tissues, such as the cartilage of the external ear and the walls of arteries
5. Proteoglycans, hyaluronic acid, and chondroitin sulfate thicken the matrix
   a. Thick matrix—acts as a lubricant and as a barrier to bacteria
D. Classification of connective tissue—classified by the structure of their intercellular material (Table 6-4)
1. Fibrous connective tissue
   a. Loose areolar connective tissue—one of the most widely distributed tissues within the body; acts like an elastic glue holding adjacent body structures together
      (1) Contains many interwoven collagenous and elastic fibers and several different kinds of cells (Figure 6-4)
      (2) Fibroblasts—synthesize the gel-like ground substance and fibers
      (3) Macrophages—carry on phagocytosis
      (4) Mast cells—specialized cells that release chemicals
   b. Adipose tissue—contains fat cells (adipocytes)
      (1) Form supporting, protective pads around various other structures
      (2) Store energy and serve as insulation for the body
   c. Reticular tissue—forms a delicate, three-dimensional web that supports the spleen, lymph nodes, and bone marrow
      (1) Composed of slender, branching reticulin fibers
      (2) Filters harmful substances out of the blood and lymph
   d. Dense fibrous tissue—consists mainly of densely packed fibers within the matrix
      (1) Dense irregular tissue—consists of bundles of collagenous fibers intertwined in irregular, swirling arrangements; forms the dermis

(2) Dense regular tissue—arranged in parallel bundles of fibers; form most of the connections between bones and muscles
2. Bone tissue—composed of a rigid crystalline matrix, collagen fibers, and bone cells (osteocytes)
   a. Compact bone tissue—bone tissue that forms the hard shell of a bone
      (1) Osteon (haversian system)—basic organizational unit of compact bone
      (2) Lacunae—small spaces that contain osteocytes
      (3) Lamellae—lacunae that are arranged in concentric layers of bone matrix
      (4) Canaliculi—tiny canals that connect lacunae to blood vessels from the central canal
   b. Bone cells
      (1) Osteocytes—mature bone cells
      (2) Osteoblasts—bone-forming cells
      (3) Osteoclasts—digest matrix from around mature osteocytes
   c. A balance of osteoblast and osteoclast activity allows bones to grow and to be reshaped
   d. Cancellous bone tissue (spongy bone)—intricate lattice of thin beams called trabeculae
      (1) Forms a framework that supports the rest of the bone
      (2) May surround red bone marrow (myeloid tissue)
3. Cartilage tissue—has one cell type (chondrocyte)
   a. Chondrocyte—produces the fibers and the tough, rubbery ground substance; each is nestled in a lacuna
   b. No blood vessels
   c. Perichondrium—membrane surrounding the cartilage
   d. Hyaline cartilage—most prevalent type of cartilage
      (1) Covers the ends of long bones; creates the support rings of the trachea; forms the rubbery tip of your nose
   e. Fibrocartilage—strongest and most durable type of cartilage
      (1) Forms the intervertebral disks and part of the knee joint
   f. Elastic cartilage—contains few collagen fibers; many very fine elastic fibers
      (1) Found in the external ear and in the larynx
4. Blood tissue—its matrix (plasma) is the liquid portion of blood that surrounds blood cells
   a. Plasma contains neither ground substance nor fibers
   b. Three types of blood cells
      (1) Erythrocytes—red blood cells
      (2) Leukocytes—white blood cells
      (3) Thrombocytes—platelets

MUSCLE TISSUE

A. Contains cells that are specialized for contractility; exist in three forms (Table 6-5, Figure 6-5):
1. Skeletal
2. Smooth
3. Cardiac
B. Skeletal muscle tissue (striated, voluntary)—makes up most of the muscles attached to bones
C. Smooth muscle tissue (involuntary)—found in the walls of hollow internal organs such as the stomach, intestines, and blood vessels

D. Cardiac muscle tissue (striated, involuntary)—makes up the walls of the heart
  1. Darkly staining bands that join cardiac muscle cells together are called intercalated disks
  2. Gap junctions of the intercalated disks allow communication among all of the cells in the heart

NERVOUS TISSUE

A. Nervous tissue—regulates and integrates activities of different parts of the body (Table 6-5; Figure 6-6)
  1. Structurally adapted for electrical excitation and conductivity
B. Neurons—cells of nervous tissue; the "wires" that conduct messages
  1. Soma—cell body
  2. Axon—transmits nerve impulses away from the soma
  3. Dendrites—carry nerve signals toward the soma and axon
C. Neuroglia—supporting cells that serve to connect and protect the neurons

TISSUE REPAIR

A. When tissues are damaged, they undergo regeneration
  1. Phagocytic cells remove dead or injured cells, while other cells divide and begin to repair the wound
B. Muscle and nervous tissues have limited abilities to regenerate

BODY MEMBRANES (TABLE 6-6)

A. Body membranes—thin structures that have important supportive and protective functions in the body
  1. Membranes cover and line body surfaces inside; serve to anchor organs to each other or to bones; secrete lubricating fluids that reduce friction between bones and organ movements
B. Epithelial membranes—composed of epithelial tissue and an underlying layer of supportive connective tissue
  1. Cutaneous membrane (skin)—the primary organ of the integumentary system
    a. Contains many sweat and oil glands that produce a surface film over the skin
  2. Serous membranes—line cavities that are not open to the external environment
    a. Epithelial sheet—a thin layer of simple squamous epithelium
    b. Secrete a thin, watery fluid that lubricates organs
    c. Serous membranes are really a single, continuous sheet covering two different surfaces (Figure 6-8)
      (1) Parietal membrane—portion that lines the wall of the cavity
      (2) Visceral membrane—covers the surfaces of the organs
  3. Mucous membranes (mucosa)—epithelial membranes that line body surfaces that open directly to the exterior
    a. Lamina propria—fibrous connective tissue underlying the epithelium in mucous membranes
    b. Mucus—composed of water, electrolytes, and different types of proteoglycans (mucins)
  4. Serves to protect the underlying cells and acts as a lubricant
C. Connective tissue membranes—composed exclusively of various types of connective tissue

1. Connective tissue membranes—do not contain an epithelial layer
  a. Synovial membranes—line the spaces between bones and joints
    (1) Synovial fluid—thick, colorless lubricating fluid
    (2) Bursae—small, cushioning sacs

REVIEW QUESTIONS

Write out the answers to these questions after reading the chapter and **HINT** *reviewing the Chapter Summary. If you simply think through the answer without writing it down, you won't retain much of your new learning.*

1. Define the term *tissue* and identify the four principal tissue types.
2. Name the primary germ layers.
3. What are the five most important functions of epithelial tissue?
4. Which of the following best describes the number of blood vessels in epithelial tissue: none, very few, very numerous?
5. Explain how the shape of epithelial cells is used for classification purposes. Identify the four types of epithelium described in this classification process.
6. Classify epithelium according to the layers of cells present.
7. List the types of simple and stratified epithelium and give examples of each.
8. What is glandular epithelium? Give examples.
9. Discuss the structural classification of exocrine glands. Give examples of each type.
10. Describe loose connective tissue.
11. How do the types of dense fibrous connective tissue differ from one another?
12. Discuss and compare the microscopic anatomy of bone and cartilage tissue.
13. Compare the structure of the three major types of cartilage tissue. Locate and give an example of each type.
14. List the three major types of muscle tissue.
15. Identify the two basic types of cells in nervous tissue.
16. Describe the regenerative capacity of muscle and nerve tissues.
17. Name the two major categories or types of body membranes. Give examples of each.

CRITICAL THINKING QUESTIONS

After finishing the Review Questions, write out the answers to these items **HINT** *to help you apply your new knowledge. Go back to sections of the chapter that relate to items that you find difficult.*

1. Summarize the structural characteristics of epithelial tissues that enable them to perform their specific functions.
2. Does the production of saliva, milk, or oil cause the most damage to the cell that produces it? Explain.
3. Describe the role of the fiber types in the classification of connective tissue. What examples can you find of these various types?
4. Many athletes work to reduce their body fat to the lowest possible percentage. What would happen if too little body fat were present?

5. If a tendon is badly damaged, it may need to be replaced surgically. Based on what you know about the structural and functional differences, explain why a tendon rather than a ligament must replace it.
6. When a joint swells, sometimes it is necessary to remove a thick, colorless liquid from the joint. What is it, where did it come from, and what is its normal function?

continued from page 93

With the knowledge you have gained from reading this chapter, see if you can answer these questions about the injuries Nathan got while mountain biking.

1. Nathan injured the "soft tissues" around his ankle; what specific tissue type was involved?
   a. Dense irregular connective tissue
   b. Skeletal muscle
   c. Dense regular connective tissue
   d. Reticular connective tissue

The bones in Nathan's ankle and foot are connected by joints (discussed in Chapter 9).

2. The ends of these bones are covered with what type of tissue?
   a. Fibrocartilage
   b. Transitional epithelium
   c. Stratified cuboidal epithelium
   d. Hyaline cartilage

In addition to spraining his ankle, Nathan cut his arm in the fall. He had a deep gash that required five stitches to close.

3. What type of membrane did the sutures (stitches) go through?
   a. Cutaneous
   b. Serous
   c. Mucous
   d. Epidermal

4. In the membrane through which the sutures passed, which layer connects the epithelial layer to the connective tissue layer?
   a. Membrane glue
   b. Basement membrane
   c. Intercellular membrane
   d. Epithelial-connective lamina

HINT *To solve these questions, you may have to refer to the glossary or index, other chapters in this textbook, A&P Connect, Mechanisms of Disease, and other resources.*

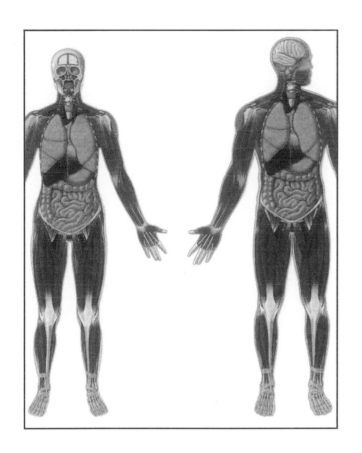

Clear View of the Human Body

Developed by: **KEVIN PATTON**
and **PAUL KRIEGER**

Illustrated by: **Dragonfly Media Group**

INTRODUCTION

A complete understanding of human anatomy and physiology requires an appreciation for how structures within the body relate to one another. Such appreciation for anatomical structure has become especially important in the 21st century with the explosion in the use of diverse methods of medical imaging that rely on the ability to interpret sectional views of the human body.

The best way to develop your understanding of overall anatomical structure is to carefully dissect a large number of male and female human cadavers—then have those dissected specimens handy while reading and learning about each system of the body. Obviously, such multiple dissections and constant access to specimens are impractical for nearly everyone. However, the experience of a simple dissection can be approximated by layering several partially transparent, two-dimensional anatomical diagrams in a way that allows a student to "virtually" dissect the human body simply by paging through the layers.

This **Clear View of the Human Body** provides a handy tool for dissecting simulated male and female bodies. It also provides views of several different parts of the human body in a variety of cross sections. The many different anterior and posterior views also give you a perspective on body structure that is not available with ordinary anatomical diagrams. This Clear View is an always-available tool to help you learn the three-dimensional structure of the body in a way that allows you to see how they relate to each other in a complete body. It will always be right here in your text-

book, so place a bookmark here and refer to the Clear View frequently as you study each of the systems of the human body.

HINTS FOR USING THE CLEAR VIEW OF THE BODY

1. Starting at the first page of the Clear View, slowly lift the page as you look at the anterior view of the male and female bodies. You will see deeper structures appear, as if you had dissected the body. As you lift each successive layer of images, you will be looking at deeper and deeper body structures. A key to the labels is found in the gray sidebar.
2. Starting with the second section of the Clear View, notice that you are looking at the posterior aspect of the male and female body. Lift each layer from the left edge to reveal body structures in successive layers from the back to the front. This very unique view will help you understand structural relationships even better.
3. On each page of the Clear View, look at the horizontal section represented in the sidebar. The section you are looking at on any one page is from the location shown in the larger diagram as a red line. In other words, if you cut the body at the red line and tilted the upper part of the body toward you, you would see what is shown in the section diagram. Notice that each section has its own labeling system that is separate from the labels used in the larger images.

KEY

1. Epicranius m.
2. Temporalis m.
3. Orbicularis oculi m.
4. Masseter m.
5. Orbicularis oris m.
6. Pectoralis major m.
7. Serratus anterior m.
8. Basilic vein
9. Brachial fascia
10. Cephalic vein
11. Rectus sheath
12. Linea alba
13. Rectus abdominis m.
14. Umbilicus
15. Abdominal oblique m., external
16. Abdominal oblique m., internal
17. Transverse abdominis m.
18. Inguinal ring, external
19. Fossa ovalis
20. Fascia of the thigh
21. Great saphenous vein
22. Parietal bone
23. Frontal bone
24. Temporal bone
25. Zygomatic bone
26. Maxilla
27. Mandible
28. Sternocleidomastoid m.
29. Sternohyoid muscle
30. Omohyoid muscle
31. Deltoid m.
32. Pectoralis minor m.
33. Sternum
34. Rib (costal) cartilage
35. Rib
36. Greater omentum
37. Frontal lobe
38. Parietal lobe
39. Temporal lobe
40. Cerebellum
41. Nasal septum
42. Brachiocephalic vein
43. Superior vena cava
44. Thymus gland
45. Right lung
46. Left lung
47. Pericardium
48. Liver
49. Gall bladder
50. Stomach
51. Transverse colon
52. Small intestines
53. Biceps brachii m.
54. Brachioradialis m.
55. Adductor longus m.
56. Sartorius m.
57. Quadriceps femoris m.
58. Patellar ligament
59. Tibialis anterior m.
60. Sup. extensor retinaculum
61. Inf. extensor retinaculum
62. Cerebrum of brain
63. Cerebellum
64. Brain stem
65. Maxillary sinus
66. Nasal cavity
67. Tongue
68. Thyroid gland
69. Heart
70. Hepatic veins
71. Esophagus
72. Spleen
73. Celiac artery
74. Portal vein
75. Duodenum
76. Pancreas
77. Mesenteric artery
78. Ascending colon
79. Transverse colon
80. Descending colon
81. Sigmoid colon
82. Mesentery
83. Appendix
84. Inguinal ligament
85. Pubic symphysis
86. Extensor carpi radialis m.
87. Pronator teres m.
88. Flexor carpi radialis m.
89. Flexor digitorum profundus m.
90. Quadraceps femoris m.
91. Extensor digitorum longus m.
92. Thyroid cartilage
93. Trachea
94. Aortic arch
95. Right lung
96. Left lung
97. Pulmonary artery
98. Right atrium
99. Right ventricle
100. Left atrium
101. Left ventricle
102. Coracobrachialis m.
103. Inferior vena cava
104. Descending aorta
105. Right kidney
106. Left kidney
107. Right ureter
108. Rectum
109. Urinary bladder
110. Prostate gland
111. Iliac artery and vein
112. Uterus
113. Parietal bone
114. Frontal sinus
115. Sphenoidal sinus
116. Occipital bone
117. Palatine process
118. Cervical vertebrae
119. Corpus callosum
120. Thalamus
121. Trapezius m.
122. Acromion process
123. Coracoid process
124. Humerus
125. Subscapularis m.
126. Deltoid m. (cut)
127. Triceps m.
128. Brachialis m.
129. Brachioradialis m.
130. Radius
131. Ulna
132. Diaphragm
133. Thoracic duct
134. Quadratus lumborum m.
135. Psoas m.
136. Lumbar vertebrae
137. Iliacus m.
138. Gluteus medius m.
139. Iliofemoral ligament
140. Sacral nerves
141. Sacrum
142. Coccyx
143. Femur
144. Vastus lateralis m.
145. Femoral artery and vein
146. Adductor magnus m.
147. Patella
148. Fibula
149. Tibia
150. Fibularis longus m.
151. Spinal cord
152. Nerve root
153. Platysma m.
154. Splenius capitis m.
155. Levator scapulae m.
156. Rhomboideus m.
157. Infraspinatis m.
158. Teres major m.
159. Lumbodorsal fascia
160. Erector spinae m.
161. Serratus post. inf. m.
162. Latissimus dorsi m.
163. Gluteus medius m.
164. Gluteus maximus m.
165. Iliotibial tract
166. Flexor carpi ulnaris m.
167. Extensor carpi ulnaris m.
168. Extensor digitorum m.
169. Carpal ligament, dorsal
170. Interosseus m.
171. Gluteus minimus m.
172. Piriformis m.
173. Gemellus sup. m.
174. Obturator internus m.
175. Gemellus inf. m.
176. Quadratus femoris m.
177. Biceps femoris m.
178. Gastrocnemius m.
179. Calcaneal (Achilles) tendon
180. Calcaneus bone
181. Subcutaneous fat
182. Corpus spongiosum
183. Corpora cavernosa
184. Umbilical ligaments
185. Epigastric artery and vein
186. Right testis
187. Transverse thoracic m.
188. Parietal pleura
189. Common bile duct
190. Lesser omentum
191. Flexor digitorum profundus
192. Epiglottis

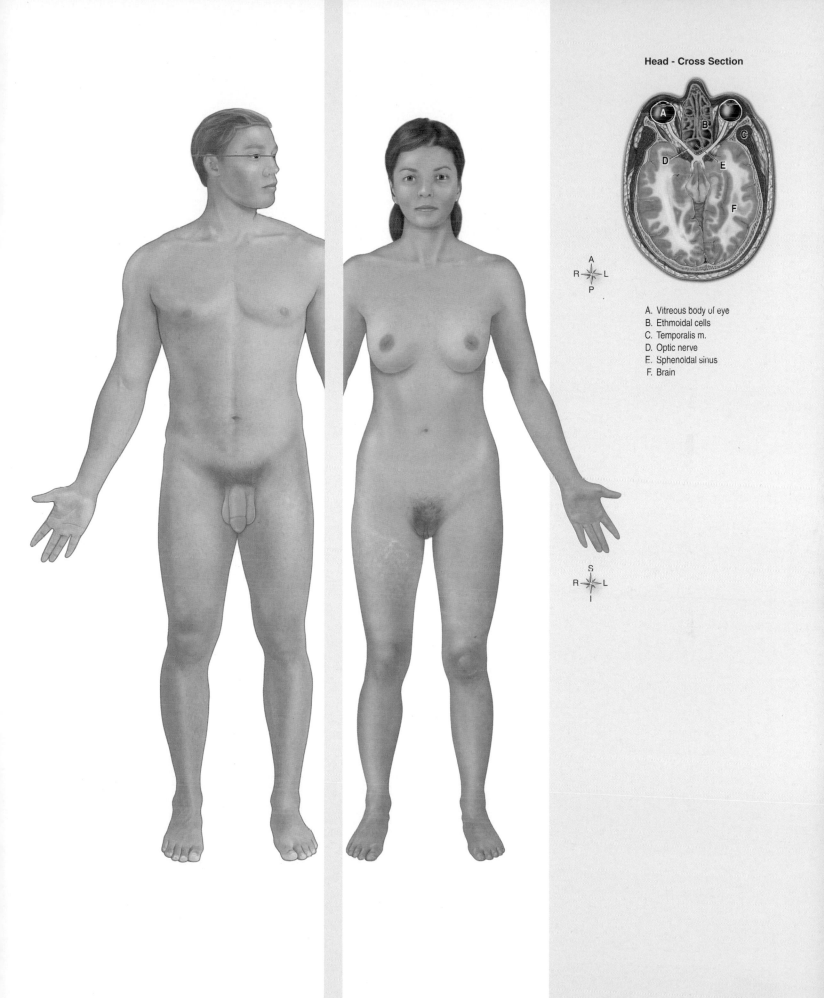

Head - Cross Section

A. Vitreous body of eye
B. Ethmoidal cells
C. Temporalis m.
D. Optic nerve
E. Sphenoidal sinus
F. Brain

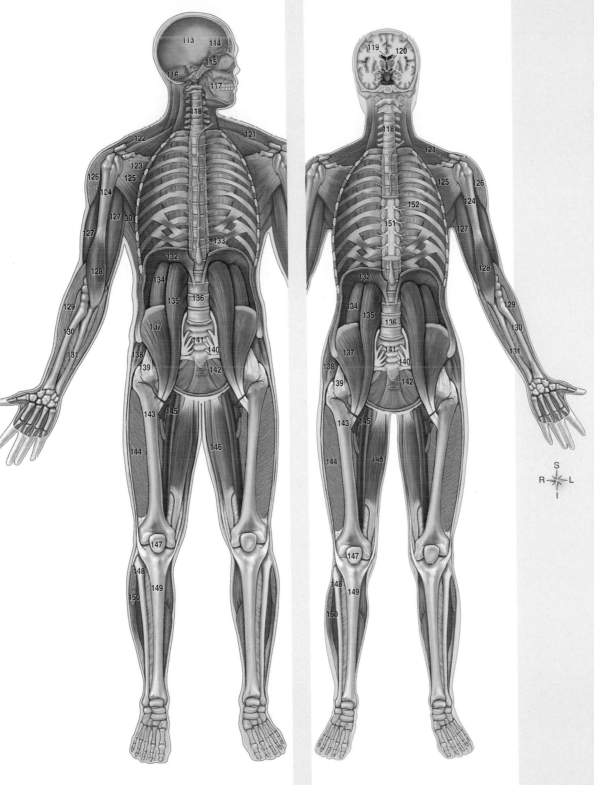

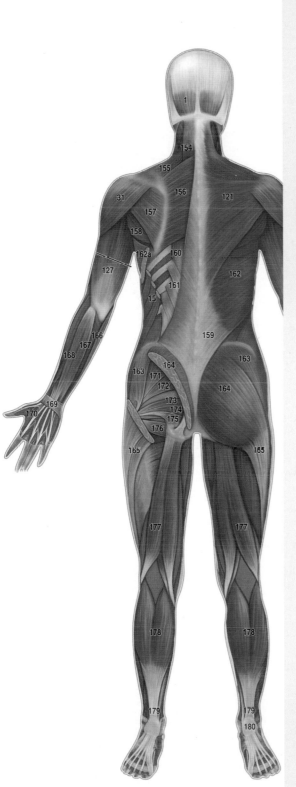

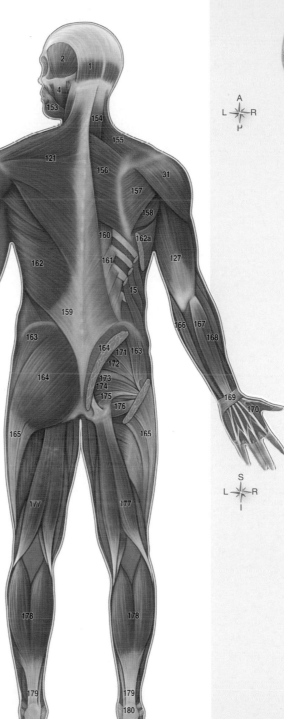

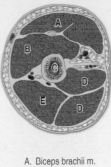

Upper Arm - Cross Section

A. Biceps brachii m.
B. Brachialis m.
C. Humerus
D. Triceps brachii m., medial
E. Triceps brachii m., lateral

Posterior View

1. Epicranius m.
2. Temporalis m.
4. Masseter m.
15. Abdominal oblique m., external
31. Deltoid m.
121. Trapezius m.
127. Triceps m.
153. Platysma m.
154. Splenius capitis m.
155. Levator scapulae m.
156. Rhomboideus m.
157. Infraspinatis m.
158. Teres major m.
159. Lumbodorsal fascia
160. Erector spinae m.
161. Serratus post. inf. m.
102. Latissimus dorsi m.
162a. Latissimus dorsi m. (cut)
163. Gluteus medius m.
164. Gluteus maximus m.
165. Iliotibial tract
166. Flexor carpi ulnaris m.
167. Extensor carpi ulnaris m.
168. Extensor digitorum m.
169. Carpal ligament, dorsal
170. Interosseus m.
171. Gluteus minimus m.
172. Piriformis m.
173. Gemellus sup. m.
174. Obturator internus m.
175. Gemellus inf. m.
176. Quadratus femoris m.
177. Biceps femoris m.
178. Gastrocnemius m.
179. Calcaneal (Achilles) tendon
100. Calcaneus bone

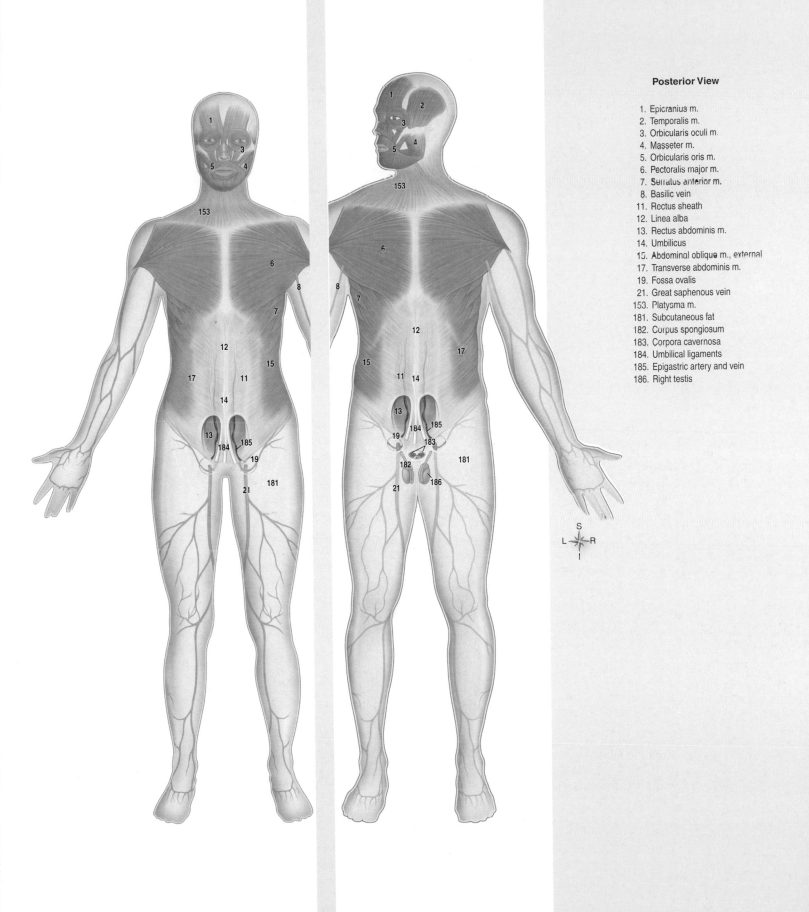

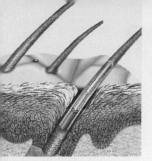

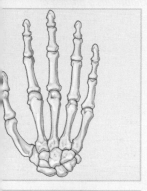

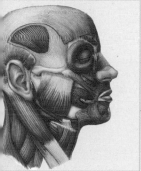

Support and Movement

The four chapters in Unit Two describe the outer covering of the body, as well as its bones, articulations (joints), and muscles.

We will begin our study with Chapter 7— Skin and Its Appendages. You will see that the skin is vital to maintaining homeostasis in our bodies. Chapter 8—Skeletal Tissues—provides information on the various types of skeletal tissues and how they are formed, grow, function, and are repaired. Skeletal tissues protect and support body structures and function as storage sites for important mineral elements vital to many body functions. Blood cell formation also occurs within the red marrow of bones.

Bones, the organs of the skeletal system, are described and organized into major subdivisions in Chapter 9. Movement between bones occurs at articulations, which are also classified in Chapter 9 according to their structure and potential for movement.

Anatomy of the major muscle groups and the basic concepts of muscle physiology are discussed in Chapter 10. The microscopic and molecular structure of muscle cells and tissues is related to function, as is the gross structure of individual muscles and muscle groups. Movement and heat production constitute the two most important functions of muscle. Discussion of muscle groups in Chapter 10 focuses on how muscles function, how they attach to bones, and how they are integrated functionally with other body organ systems.

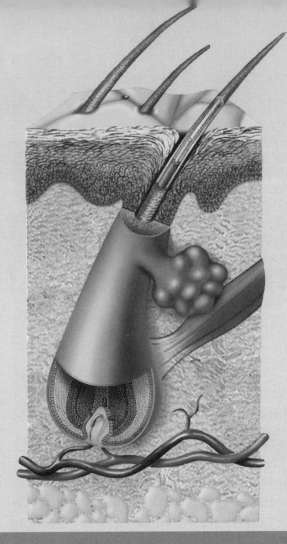

CHAPTER 7

Skin and Its Appendages

STUDENT LEARNING OBJECTIVES

At the completion of this chapter, you should be able to do the following:

1. Describe the layers and functions of the epidermis.
2. Describe the layers and functions of the dermis.
3. Discuss the most important properties of the hypodermis layer.
4. Describe the sensory receptors of the skin.
5. List the pigments producing skin color.
6. Describe the primary functions of the skin.
7. Discuss how the integumentary system functions with other body systems to maintain homeostasis.
8. Give the location of hair and nails and their structures.
9. Discuss the significance of the major eccrine, apocrine, and sebaceous glands of the skin.

CHAPTER OUTLINE

 HINT Scan this outline before you begin to read the chapter, as a preview of how the concepts are organized.

ONE afternoon, 8-year-old Aidan was playing in his tree house. While running his fingers along the railing, he suddenly yelled, snatching his hand back off the rail. When he looked down, he saw a small sliver of wood sticking out of his middle finger. The splinter hurt, but his finger wasn't bleeding.

By the end of this chapter, you should be able to outline not only what has happened, but the layers of skin that were penetrated by the sliver, and the possible implications.

* * *

Thick or thin, the skin is your largest organ—and one of the most important. It forms a self-repairing and protective boundary between the internal environment of the body and the often-hostile environment of the external world. The skin surface of an average human is nearly 2 square meters (21.5 square feet), and its thickness varies from less than 0.05 cm (1/50 inch) to slightly more than 0.3 cm (1/8 inch). As you will see, the **integumentary system**—the skin and its appendages (including hair, nails, and glands)—is one of the most vital, diverse, complex, and extensive organs of your body. Throughout this chapter, you will find many examples of how structure is related to function.

STRUCTURE OF THE SKIN

Layers of the Skin

There are two primary layers of the skin (Figure 7-1): a thin, superficial **epidermis** and a deep, thicker **dermis.** A distinct *dermoepidermal junction* forms where the dermis meets the epidermis. Beneath the underlying

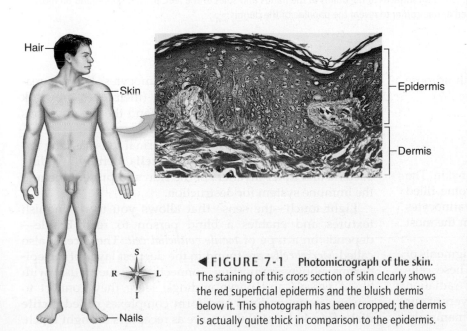

◀ FIGURE 7-1 Photomicrograph of the skin. The staining of this cross section of skin clearly shows the red superficial epidermis and the bluish dermis below it. This photograph has been cropped; the dermis is actually quite thick in comparison to the epidermis.

Labels: Hair, Skin, Nails, Epidermis, Dermis

LANGUAGE OF SCIENCE AND MEDICINE

HINT *Before reading the chapter, say each of these terms out loud. This will help you avoid stumbling over them as you read.*

apocrine sweat gland (AP-oh-krin)
  [*apo-* from, *-crin-* secrete, *gland* acorn]

basal cell carcinoma (BAY-sal sell kar-si-NOH-mah)
  [*bas-* base, *-al* relating to, *cell* storeroom, *carcin-* cancer, *-oma* tumor]

blackhead

callus (KAL-us)
  [*callus* hard skin]

cerumen (seh-ROO-men)
  [*cera* wax]

ceruminous gland (seh-ROO-mi-nus)
  [*cera-* wax, *-ous* relating to, *gland* acorn]

cyanosis (sye-ah-NO-sis)
  [*cyan-* blue, *-osis* condition]

dendritic cell (DEN-dri-tik sell)
  [*dendrit-* tree branch, *-ic* relating to, *cell* storeroom]

dermis (DER-mis)
  [*derma* skin]

dermoepidermal junction (DEJ) (DER-mo-EP-i-der-mal JUNK-shen)
  [*derm-* skin, *epi-* on or upon]

eccrine sweat gland (EK-rin)
  [*ec-* out, *-crin-* secrete, *gland* acorn]

epidermis (ep-i-DER-mis)
  [*epi-* on or upon, *-dermis* skin]

first-degree burn

fourth-degree burn

friction ridge

full-thickness burn

hair follicle (FOHL-i-kul)
  [*foll-* bag, *-icle* little]

hypodermis (hye-poh-DER-mis)
  [*hypo-* under, *-dermis* skin]

integumentary system (in-teg-yoo-MEN-tar-ee SIS-tem)
  [*integument* covering, *-ary* relating to, *system* organized whole]

Kaposi sarcoma (KAH-poh-see sar-KOH-mah)
  [Moritz K. Kaposi, Hungarian dermatologist, *sarco-* flesh, *-oma* tumor]

keratin (KER-ah-tin)
  [*kera-* horn, *-in* subtance]

keratinization (ker-ah-tin-i-ZAY-shun)
  [*kera-* horn, *-in-* substance, *-iz-* cause, *-ation* process]

keratinocyte (keh-RAT-i-no-syte)
  [*kera-* horn, *-in-* substance, *-cyte* cell]

lanugo (lah-NOO-go)
  [*lanugo* down]

continued on page 131

A Thick Skin

B Thin Skin

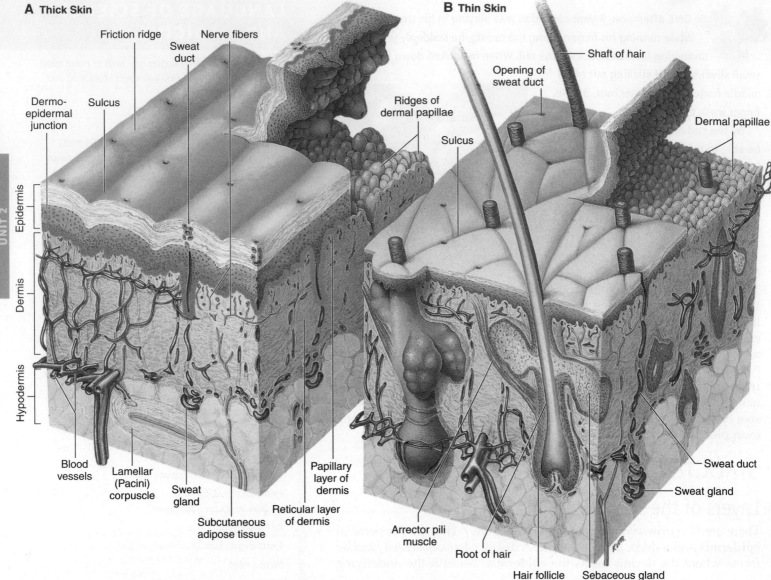

▲ **FIGURE 7-2** **Diagram of skin structure. A,** Thick skin, found on surfaces of the palms of the hands and soles of the feet. **B,** Thin skin, found on most surface areas of the body. In each diagram, the epidermis is raised at one corner to reveal the papillae of the dermis.

dermis lies a loose **hypodermis** that is filled with adipose and loose connective tissue (see Chapter 6) (Figure 7-2).

Epidermis

Epidermal Cell Types

Several types of cells make up the epidermis of our skin. The most important of these, the **keratinocytes,** become filled with a tough, fibrous protein called **keratin.** Keratinocytes make up over 90% of the epidermal cells and form the most important structural element of the outer skin.

Melanocytes give our skin its color. The dark pigments of these cells absorb ultraviolet (UV) radiation. These cells reduce the amount of potentially dangerous UV radiation that would otherwise penetrate into the deeper layers of our skin and cause cellular damage. True albino mammals,

including humans, do not have melanocytes. For this reason, they appear completely white (their eyes are pink because they lack pigment in the iris and the blood vessels are visible).

Another fascinating group of epidermal cells has long fingerlike projections. These **dendritic cells** identify invading bacteria and other invaders of the skin and present them to the immune system for destruction.

Light touch—the sense that allows you to distinguish textures and enables a blind person to read Braille—depends on a type of *tactile epithelial cell*. These cells, also called *Merkel cells,* are found in the deepest layer of the epidermis above supporting connective tissue, flush with blood vessels and nerve endings. Here they connect to sensory nerve endings and form complexes called **tactile disks (Merkel disks)** that serve as receptors for light touch.

Epidermal Cell Layers

The thickest types of epidermis have five distinct **strata** (layers): *stratum basale, stratum spinosum, stratum granulosum, stratum lucidum,* and *stratum corneum.* (Refer to these strata depicted in Table 7-1 as you read through the following paragraphs.)

The epidermal strata are named after their structural or functional characteristics. First, note that the **stratum basale**

(base layer) is a single layer of columnar cells that produces all the other cells of the epidermis. These cells migrate up from the stratum basale through the other epidermal layers. Ultimately, they are shed as dead cells from the skin surface (if you exfoliate with a luffa sponge, you speed up this process to reveal shiny "new skin" below). The **stratum spinosum** (spiny layer) typically contains about 10 layers of irregularly shaped cells that actually look "spiny" when

TABLE 7-1   Structure of the Skin*

	STRUCTURE	DESCRIPTION
A	Surface film	Thin film coating the skin; made up of a mixture of sweat, sebum, dead cells/fragments, various chemicals; protects the skin
B	Epidermis	Superficial primary layer of the skin; made up entirely of keratinized stratified squamous epithelium; structures include hairs, sweat glands, sebaceous glands
C	Stratum corneum (horny layer)	Several layers of flakelike dead cells (or *corneocytes*) mostly made up of dense networks of *keratin* fibers cemented together to form a tough, waterproof barrier; the keratinized layer: *a,* sulcus (groove); *b,* friction ridge
D	Stratum lucidum (clear layer)	A few layers of squamous cells filled with a keratin precursor that gives this layer a translucent quality (not visible in thin skin)
E	Stratum granulosum (granular layer)	2-5 layers of dying, somewhat flattened cells filled with darkly staining granules; nuclei disappear in this layer
F	Stratum spinosum (spiny layer)	8-10 layers of cells pulled by desmosomes into a spiny appearance
G	Stratum basale (base layer)	Single layer of mostly columnar cells capable of mitotic cell division; from this layer all cells of superficial layers are derived; includes keratinocytes and some melanocytes
H	Dermoepidermal junction (DEJ)	The basement membrane, a unique and complex arrangement of adhesive components that glue the epidermis and dermis together
I	Dermis	Deep primary layer of the skin; made up of fibrous tissue; also includes some blood vessels (*c*), muscles, and nerves
J	Papillary layer	Loose fibrous tissue with collagenous and elastic fibers; forms nipplelike bumps (papillae, *d*); includes tactile corpuscles (touch receptors, *e*) and other sensory receptors
K	Reticular layer	Tough network (reticulum) of collagenous dense irregular fibrous tissue (with some elastic fibers); forms most of the dermis
L	Hypodermis (subcutaneous layer; superficial fascia)	Loose (areolar) connective tissue and adipose tissue; under the skin (not part of the skin); includes fibrous bands or *skin ligaments (f)* that strongly connect the skin to underlying structures; includes lamellar corpuscles (pressure receptors, *g*) and other sensory receptors

*The capital letters in the left text column show the layers of the skin. The lower case letters on the figure show specific structures of the skin.

viewed under a microscope. These cells are dying as they are pushed further away from the blood supply below the epidermis. (Note: the term **stratum germinativum** [growth layer] is sometimes used to describe the stratum basale *and* the stratum spinosum together.)

As the cells die, they move from stratum spinosum into the two to four layers of **stratum granulosum** (granular layer). These dead cells are filled with a substance needed to produce surface keratin.

The next layer is the **stratum lucidum** (clear layer). The keratinocytes in the stratum lucidum are very flat, clear, and packed closely together. They are filled with a substance that is transformed via chemical reaction to keratin. Thick skin from the soles of the feet and the palms of the hands often has many layers of stratum lucidum. Stratum granulosum and stratum lucidum may not be well developed in thin skin.

The most superficial layer of skin, the **stratum corneum** (horny layer) is composed of very thin squamous (flat) cells. At the skin surface, the dead cells are continually shed. Much of the cytoplasm in these cells has been replaced during the process of **keratinization.** In this important process, a dense network of keratin fibers is cemented together to make a strong, waterproof barrier. The stratum corneum serves to protect the rest of the body from damage due to radiation, chemicals, abrasions, cuts, and any opportunistic bacteria and viruses that might invade. The stratum corneum also functions as a barrier to water loss and water gain. However, if you soak too long in the tub, the keratin in this layer may absorb water and make your skin appear puffy and wrinkled!

Epidermal Growth and Repair

The most important function of the integument—protection—depends on special structural features of the epidermis and its ability to create and repair itself after injury or disease. In this regard, new cells must obviously be formed at the same rate that old keratinized cells flake off from the stratum corneum in order to maintain a constant skin thickness. Cells push upward from the stratum basale into each successive layer, become keratinized, and eventually die and slough off. This is a constant process throughout life. Thick **calluses** may form in areas of constant friction or irritation when an increased rate of cell division in stratum basale dramatically increases the thickness of the stratum corneum.

Dermoepidermal Junction

Between the dermis and epidermis lies a unique **dermoepidermal junction (DEJ)** composed chiefly of an easily identified basement membrane (see Chapter 6). This junction includes fibers and unique chemical gels that, together with the basement membrane, cement or "glue" the superficial epidermis to the dermis below. If the DEJ is subjected to rubbing forces such as those experienced by the hands of gardeners working with a shovel, for example, then this junction may fill with extracellular fluid and cause a *blister* (Box 7-1).

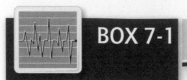

BOX 7-1 Health Matters

Blisters

Blisters (see figure) may result from injury to cells in the epidermis or from separation within the dermoepidermal junction (DEJ). A blister results in the accumulation of fluid within or beneath the epidermis. Regardless of cause, blisters represent a basic reaction of skin to injury. Any irritant that damages the physical or chemical bonds that hold adjacent skin cells or layers together, such as poison ivy, initiates blister formation.

The cell-to-cell junctions (desmosomes) that hold adjacent cells in the epidermis together are essential for the integrity of the skin. If these intercellular bridges (sometimes described as "spot welds" between adjacent cells) are weakened or destroyed, the skin literally falls apart and away from the body. Damage to the DEJ produces similar results.

Blister formation follows burns, friction injuries, exposure to primary irritants, or accumulation of toxic breakdown products after cell injury or death in the layers of the skin.

QUICK CHECK

1. What are the two main cell types of the epidermis?
2. List the five layers, or strata, of the skin.
3. Name the cement-like layer that separates the dermis from the epidermis.

Dermis and Its Components

The dermis, which is composed of a thin *papillary layer* and a thicker *reticular layer,* is sometimes called the "true skin" of the human body. As you can see from the figure in Table 7-1, the dermis is much thicker than the epidermis and may exceed 4 mm on the soles and palms. It is thinnest on the skin making up the eyelids and the penis. Thin skin coupled with a high density of touch receptors in these areas explains their great sensitivity to touch. In addition to serving a protective function against mechanical injury and compression, this layer also provides a reserve storage area for water and electrolytes important to many body functions. It is the dermis that really provides most of the great mechanical strength of our skin.

A specialized network of nerves and nerve endings called *somatic sensory receptors* reside in the dermis. These receptors process sensory information related to pain, pressure, touch, and temperature variation (see Chapter 14). A variety of muscle fibers, hair follicles, and sweat and sebaceous glands, as well as many blood vessels, extend upward at various levels of the dermis. It is the rich supply of the blood vessels in the dermis (lacking in the epidermis) that plays a critical role in the regulation of body temperature.

Note in Figure 7-2 that the superficial **papillary layer** of the dermis forms nipple-like bumps called *dermal papillae* (*papilla* = nipple) that project *into* the overlying epidermis. The dermal papillae are responsible for forming the parallel friction ridges that make up fingerprints and footprints.

The thick **reticular layer** of the dermis consists of a dense *reticulum* (network) of collagenous fibers. It is this layer from animal hides that is tanned to form leather. The dermis serves as an area of attachment for numerous skeletal (voluntary) and smooth (involuntary) muscle fibers. Because of this, skeletal muscles in your face and scalp permit a huge range of facial expressions. Hair also has a small bundle of involuntary muscles (*arrector pili muscles*) attached to it (Figure 7-3). Contraction of these muscles from cold or fright, for example, gives us "goose bumps." In the dermis of the skin of the scrotum and in the pigmented skin (*areolae*) surrounding the nipples, smooth muscle cells form a loose network that, when contracted, wrinkles the skin in these areas. This action causes the elevation of the testes and the erection of the nipples.

Millions of specialized sensory receptors are located in the dermis of all skin areas (see Figure 7-2). They permit the skin to serve as an enormous sense organ transmitting sensations of touch, pressure, pain, and temperature to the brain. Hair follicles and various skin glands (actually made of epithelial tissues that extend from the surface of the epidermis) have most of their structures embedded *within* the reticular layer of the dermis.

Unlike the epidermis, the dermis does not continually shed and regenerate. Rapid regeneration only takes place after a wound occurs. An unusually dense mass of new connective tissue forms, and if not replaced by normal tissue, becomes a *scar*.

Hypodermis

The hypodermis is sometimes called the **subcutaneous layer** or **superficial fascia.** The hypodermis is not part of the skin proper, but lies deep to the dermis (see Table 7-1). Here, it forms a connection between the skin and the underlying structures of the body. The hypodermis is composed mostly of loose fibrous and adipose tissues. The fat content of the adipose tissue varies tremendously and in obese individuals may exceed 10 cm (4 in) or more in thickness. The mobility of the skin is affected by the arrangement of fat cells and collagen fibers in this area. Bands of fibers running through the hypodermis help hold the skin to underlying structures such as deep fascia and muscles. Knowledge of the hypodermis is very important to medical professionals administering shots and some vaccinations via a *hypodermic* needle (Box 7-2).

BOX 7-2 Health Matters

Subcutaneous and Intradermal Injections

Although the hypodermis is not part of the skin itself, it carries the major blood vessels and nerves to the skin above. The rich blood supply and loose spongy texture of this area make it an ideal site for the rapid and relatively pain-free absorption of injected material. Liquid medicines, such as insulin, and pelleted implant materials are often administered by *subcutaneous (SQ) injection* with a *hypodermic needle* into this spongy and porous layer beneath the skin. *Intradermal (ID) injections*, in contrast, place the medication in the skin proper. Intradermal injections, which are ideal for some vaccines, are hard to administer with a typical needle because it is so easy to punch through to the hypodermis. Therefore, this type of injection is sometimes given with special needle-free injectors.

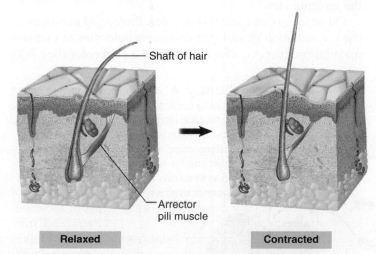

▲ **FIGURE 7-3** **Arrector pili muscle.** When the arrector pili muscle contracts, it pulls the follicle and hair into a more perpendicular position, thus "fluffing up" the hair. Notice how a "goose bump" is raised around the hair shaft.

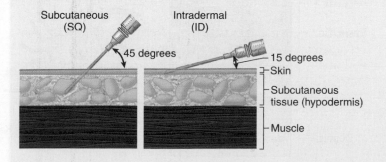

Thin and Thick Skin

Most of our body's surface is covered by *thin skin*. In contrast, the hairless skin covering the palms of the hands, fingertips, soles of the feet, and other body areas subject to friction is *thick skin*. These terms refer to the thickness of the epidermal layer *only*, not to the depth of the dermis and epidermis taken together. Note that the dermis accounts for most of the variation in the depth of the skin. For example, the skin of your back may be over 5 mm (⅕ inch) thick, whereas the skin of your eyelid is only one tenth as thick.

Thick skin does not have hair. However, our hands and feet have curved parallel **friction ridges** created by fingerlike projections of *dermal papillae* pushing upward from the underlying dermis. These dermal projections create our unique fingerprints and footprints. On our fingers, these friction ridges act like the rubber studs of garden gloves, helping us to pick up and manipulate small objects with our hands. They also help us feel textured surfaces with our fingertips. On the soles of our feet, these ridges provide slip resistance, like the treads on our sneakers. As we get older, these pads wear down and do not function as well as they do when we are young. This explains, in part, why smooth objects slip out of our hands as we age. Thin skin may lack one or more of the strata found in thick skin, and each stratum present may have fewer layers of cells. And of course, there are no fingerprint ridges in thin skin.

> **QUICK ✓ CHECK**
>
> 4. Which layer of the dermis forms the bumps that produce the ridges on the palms of the hands and the soles of the feet?
> 5. Which layer of the skin is supplied with blood vessels and nerves?
> 6. What is the function of the hypodermis?
> 7. What are the two primary layers of skin?
> 8. Briefly compare the properties of thin and thick skin.

PIGMENTS OF THE SKIN

Humans come in a virtual rainbow of colors. The main determinant of our skin color is the type and quantity of the pigment **melanin** deposited in the cells of the epidermis.

BOX 7-3   Health Matters

Sunburn and Skin

Burns caused by exposure to harmful UV radiation in sunlight are commonly called *sunburns*. As with any burn, serious sunburns can cause tissue damage and lead to secondary infections and fluid loss. Cancer researchers have recently theorized that blistering (second-degree) sunburns during childhood may trigger the development of malignant melanoma later in life. Some epidemiological studies show that adults who had more than two blistering sunburns before the age of 20 years have a greater risk for melanoma than someone who experienced no such burns. If this theory is true, it could explain the dramatic increase in skin cancer rates in the United States in recent years. Those who grew up as sunbathing and "suntans" became popular in the 1950s and 1960s are now, as adults, exhibiting melanoma and other less-deadly skin cancers at a much higher rate than in the previous generation.

In addition to obvious burns caused by excessive sun exposure, frequent and especially long-term use of "tanning beds" is now also considered a risk factor for the development of skin cancer.

The *number* of pigment-producing melanocytes (Figure 7-4) scattered throughout the stratum basale of the epidermis varies somewhat, according to a person's genetic code. However, skin pigmentation is regulated by a number of different factors. These include (1) the density of rootlike projections from melanocytes, (2) the transport of different types of **melanosomes** (tiny pigmented granules) to other cells in the skin such as *keratinocytes*, and (3) the distribution of these cells in the skin. A great deal of variation in skin color is due to the differences in pigment production and to the type and distribution of melanosomes within the melanocytes.

Other factors can affect skin color. Prolonged exposure to the UV radiation in sunlight causes melanocytes to increase melanin production, which darkens the skin color (Box 7-3).

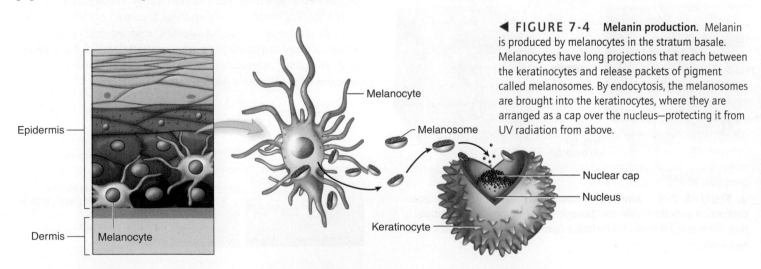

◀ **FIGURE 7-4   Melanin production.** Melanin is produced by melanocytes in the stratum basale. Melanocytes have long projections that reach between the keratinocytes and release packets of pigment called melanosomes. By endocytosis, the melanosomes are brought into the keratinocytes, where they are arranged as a cap over the nucleus—protecting it from UV radiation from above.

Epidermis

Dermis — Melanocyte

Melanocyte

Melanosome

Nuclear cap

Nucleus

Keratinocyte

Serious skin conditions and cancers can result from repeated, prolonged exposure to UV radiation, especially if such exposure has resulted in the burning, blistering, and peeling of the skin (Box 7-4).

Pigments such as yellowish *beta-carotene* (from carrots, for example) can also change skin color. This pigment, essential for skin growth, is stored in skin tissue and then converted into vitamin A. This is why extremely high consumption of carrot juice may turn an infant's skin yellow or orange. Yellowish discoloration of the skin can also be caused by bile pigments (Box 7-5). You're probably familiar with the rainbow of colors seen as a bruise heals—caused by blood leaking, clotting, then breaking down (Figure 7-5). And of course, changes in skin color can also result from blushing and excessive exercise when an increased flow of blood to the skin causes it to "flush."

In some abnormal conditions, skin color changes because of an excess amount of hemoglobin that is low in oxygen and high in carbon dioxide. For this reason, skin with relatively little melanin will appear bluish *(cyanotic)* when its blood has a high proportion of unoxygenated hemoglobin. The blue coloration results when hemoglobin changes from

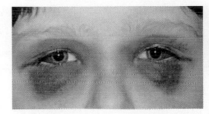

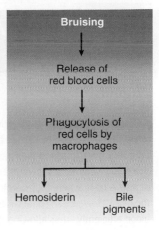

▲ **FIGURE 7-5   Color changes in a bruise.** In this light-skinned individual, different colors appear in the skin as hemoglobin becomes unoxygenated and turns bluish (see Figure 7-6) and perhaps even black as the blood clots. As macrophages consume the hemoglobin, it is broken down into brownish, greenish, and yellowish pigments—sometimes producing a rainbow of skin colors.

A&P CONNECT

How can you tell the difference between a normal mole and skin cancer? Find out at **Skin Cancer** online at **A&P Connect**.

*e*volve

BOX 7-4 | Health Matters

Skin Cancers

Each year more than 1.2 million new cases of skin cancer (of all types) are diagnosed in the United States. These *neoplasms,* or abnormal "new" growths, result from cell changes in the epidermis. They comprise the most common forms of cancer (25% in men and 14% in women).

Two forms of skin cancer, **basal cell carcinoma** (BCC) and **squamous cell carcinoma** (SCC), account for more than 95% of all reported cases of skin cancer (parts *A* and *B* of the figure). Squamous cell carcinoma tends to spread or *metastasize* less frequently than basal cell carcinoma. Both types respond well to early treatment. However, if left untreated, these cancers can cause significant damage to adjacent tissues. They can also cause disfigurement and loss of function.

A third, far more serious type of skin cancer called **malignant melanoma** has a much greater tendency to metastasize to other body areas, such as the bones and brain (part *C* of the figure). Over 50,000 new cases of malignant melanoma are reported each year, and of these, nearly 7,000 patients die.

Damage caused by ultraviolet (UV) light (either from excessive exposure to the sun or by UV lights in tanning salons) is strongly associated with BCC, SCC, and an increased incidence of malignant melanoma. However, the most common precancerous skin condition in middle-aged adults is called *actinic keratosis* (AK), which is far and away the most common UV-related skin lesion. Actinic keratoses often lead to squamous cell carcinoma. Lesions are commonly frozen off with liquid nitrogen.

Kaposi sarcoma is a rare cancer generally, but appears commonly in patients afflicted with immune deficiencies such as AIDS. It produces purple papules on the skin (part *D* of the figure) and quickly spreads to the lymph nodes and internal organs.

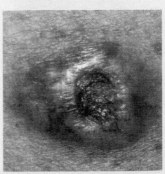
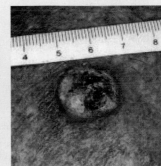

A   Basal cell carcinoma   **B**   Squamous cell carcinoma

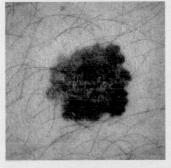
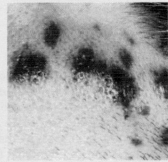

C   Melanoma   **D**   Kaposi sarcoma

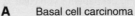
▲ Skin cancers.

BOX 7-5 | Health Matters

Jaundice

Yellowish discoloration of the skin and other tissues, such as the "white" or sclera of the eye, can be caused by bile pigments. Bile pigments are produced from a natural breakdown product when old red blood cells are destroyed. Ordinarily, bile pigments are excreted from the liver into the digestive tract, where they become part of the feces and are eliminated by the body. However, in some cases the liver is unable to remove bile from the blood efficiently—thus allowing the bile to literally "stain" the tissues of the body. An example of a condition that can cause jaundice is a liver infection, as illustrated in the photograph.

Jaundice also often occurs just after birth. In about half of all full-term newborns, jaundice occurs when old red blood cells containing an immature fetal form of hemoglobin are rapidly replaced with red blood cells containing the mature form of hemoglobin. Often, a baby's liver is simply too immature to handle the removal of such a large amount of bile pigment all at once. This form of jaundice is temporary and usually disappears without treatment. Ultraviolet light is used to break down bile pigments in the skin and can help speed recovery in infants with moderate to severe cases of jaundice.

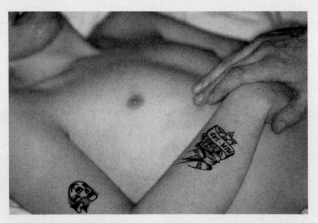

▲ **Jaundice.** Yellowish discoloration of the skin and other tissues by bile pigments was, in the case seen here, caused by a liver infection. The patient was infected with hepatitis B by a contaminated tattoo needle. The yellow tinge of the skin can be best seen by comparing the patient's skin color with that of the physician's hand.

a bright red to a dark, maroon-red color when it loses oxygen and gains carbon dioxide. Light reflected from the dark, maroon-red hemoglobin and diffused by fibers in the skin appears blue (Figure 7-6). In general, the greater amount of maroon-red hemoglobin, the greater the amount of blueness, or **cyanosis.** Pigments from cosmetics or from tattoos can also change the skin's color.

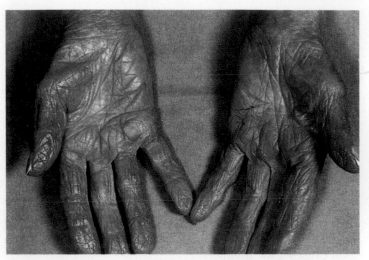

▲ **FIGURE 7-6** **Cyanosis.** The blue discoloration of the fingers of this light-skinned individual is caused by the diffusion of light reflected off dark, unoxygenated hemoglobin in the blood vessels of the skin. The fingertips have an especially high volume of blood and thus look bluer than other regions of the skin.

FUNCTIONS OF THE SKIN

Skin functions are diverse. They include processes involved in protection, sensation, growth, synthesis of important chemicals and hormones (such as vitamin D), excretion, temperature regulation, and immunity. Because it is flexible, the skin permits body growth and movement to occur without injury. (You can stretch in any direction and your skin will not tear!) Certain substances such as fat-soluble vitamins (A, D, E, and K), sex hormones, corticoid hormones, and certain drugs such as nicotine and nitroglycerin can be absorbed by the skin. Melanin production, as we have seen, screens the skin from potentially harmful ultraviolet light. And keratin forms a wonderful protective protein network.

Table 7-2 outlines the discussion of the functions of skin that follows.

Protection

There are many layers of keratinized squamous epithelial cells that cover the epidermis. These layers protect underlying tissues against invasion by microorganisms. In addition, these epithelial cells bar the entry of many harmful chemicals, and reduce the possibility of mechanical injury to underlying structures. Because it is essentially waterproof, the skin protects us from dehydration caused by the loss of internal body fluids.

The skin produces a thin, oily **surface film** from the residues of sweat and the secretions of sebaceous glands. This surface film provides lubrication and hydration. It also buffers the skin from caustic environmental irritants, blocks many toxic agents, and produces antibacterial and antifungal compounds.

TABLE 7-2   Functions of the Skin

FUNCTION	EXAMPLE	MECHANISM
Protection	From microorganisms From dehydration From ultraviolet radiation From mechanical trauma	Surface film/ mechanical barrier Keratin Melanin Tissue strength
Sensation	Pain Heat and cold Pressure Touch	Somatic sensory receptors
Permits movement and growth without injury	Body growth and change in body contours during movement	Elastic and recoil properties of skin and subcutaneous tissue
Endocrine	Vitamin D production	Activation of precursor compound in skin cells by ultraviolet light
Excretion	Water Urea Ammonia Uric acid	Regulation of sweat volume and content
Immunity	Destruction of microorganisms and interaction with immune system cells (helper T cells)	Phagocytic cells and epidermal dendritic cells
Temperature regulation	Heat loss or retention	Regulation of blood flow to the skin and evaporation of sweat

While serving as the outermost layer of protection on the body, sometimes the skin gets damaged, such as through burns (Box 7-6).

Sensation

There are millions of different sensory receptors in the skin that respond to touch, pressure, vibration, pain, and temperature. When these receptors are activated, they make it possible for the body to respond quickly to changes in both the internal and external environments.

Flexibility

For movement of the body to occur without injury, the skin must be supple and elastic. It grows as we grow, and exhibits stretch and recoil characteristics that permit easy changes in contour, without tearing.

Excretion

The skin regulates the volume and chemical content of sweat, which in turn influences the body's total fluid volume. In doing so, the skin's sweat glands excrete waste products (such as uric acid, ammonia, and urea).

Hormone Production

When the skin is exposed to ultraviolet light, chemical precursors to vitamin D are produced. These precursors then are transported in the blood to the liver and kidneys, where they are converted into an active form of vitamin D. Vitamin D acts as a hormone by regulating functions throughout the body.

Recent research suggests that vitamin D deficiency is a significant problem that occurs more frequently than expected in the general population. As a result, vitamin D supplements are now being recommended for more people than in years past. Recent research indicates that vitamin D has a significant impact on bone health and the cardiovascular system, and may reduce the prevalence of certain cancers.

Immunity

The skin has many defensive cells that destroy pathogenic microorganisms. These cells play an important role in our immunity. Other defensive cells can trigger important immune responses in certain diseases.

Homeostasis of Body Temperature

Humans maintain a remarkably constant body temperature (with a normal set point of 37° C). We owe this homeostasis largely to the temperature-regulating mechanisms of the skin. This is vital to our physiological processes because the chemical reactions within the body's cells operate most efficiently near this same set point. To maintain this precise body temperature, heat gain and heat production must balance heat loss. Heat generated by metabolism from the *interior* of the body can only be lost from the *surface* of the body. The greater the amount of body surface area relative to the body's volume, the greater the potential for heat lost from the surface. If the body temperature rises dangerously, the skin plays a critical role of reducing the body temperature. This is because its enormous surface area (relative to the body's volume) allows significant heat loss. The vast majority of this heat loss takes place by the evaporation of water and by the loss of heat by convection (air or water currents passing over the body). Box 7-7 discusses these heat loss mechanisms.

> **QUICK CHECK**
> 9. What is the function of the pigment melanin?
> 10. Name some reasons for variation in skin pigment.
> 11. List the basic functions of the skin.
> 12. How does surface film protect the skin?
> 13. How does the skin regulate body temperature?

UNIT 2

BOX 7-6 Health Matters

Burns

Burns are thermal injuries or lesions caused by contact of the skin with fire or a hot object, even water. However, overexposure to UV light can cause serious sunburns. Corrosive chemicals and electric currents can also cause burns.

When burns involve large areas of the skin, treatment and the prognosis for recovery depend largely on the total area involved and the severity of the burn. There are several different

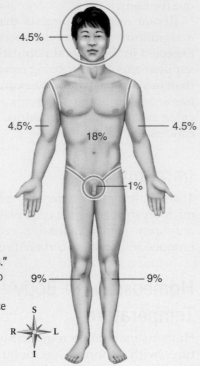

▶ **FIGURE A** "Rule of nines." Dividing the body surface area into multiples of 9% is one way to estimate the amount of skin surface burned in an adult.

methods of estimating the area of the body involved in a burn. The **"rule of nines"** is one such way of estimating the extent of burn injury and is represented in Figure A. The severity of the burn is determined by the depth and extent (percentage of body surface burned) of the lesion. The depth of a burn injury depends on the tissue layers of the skin that are involved (see Figure B).

A **first-degree burn** (such as simple sunburn) causes minor discomfort and some reddening of the skin. Although the surface layers of the burned area may peel in 1 to 2 days, no blistering occurs and actual tissue destruction is minimal. **Second-degree burns** involve the deep epidermal layers and always cause injury to the upper layers of the dermis. In deep second-degree burns, damage to sweat glands, hair follicles, and sebaceous glands may occur. However, tissue death is *not* complete. Blistering, severe pain, generalized swelling, and edema characterize this type of burn. Scarring, unfortunately, is frequent. Both first- and second-degree burns are called **partial-thickness burns.**

Third-degree (full-thickness) burns are characterized by destruction of both the epidermis and dermis. Tissue death extends below the hair follicles and sweat glands. If burning involves underlying muscles, fasciae, or bone, it may be called a **fourth-degree burn.** Third- or fourth-degree lesions are insensitive to pain immediately after injury because of the destruction of nerve endings. Scarring, in these cases, is a serious and severe problem.

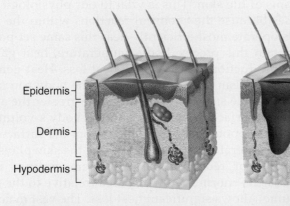

Epidermis
Dermis
Hypodermis

| Partial-thickness burns | First-degree burn: Damaged epidermis and edema | Second-degree burn: Damaged epidermis and dermis | Full-thickness burns | Third-degree burn: Deep tissue damage |

▲ **FIGURE B** **Classification of burns.** Partial-thickness burns include first- and second-degree burns. Full-thickness burns include third-degree burns. Fourth-degree burns involve tissues under the skin, such as muscle or bone.

BOX 7-7 FYI

Role of Skin in Temperature Regulation

Body heat is produced almost entirely by metabolism, especially in the muscles and glands such as the liver. However, any *ambient* (environmental) temperature above the normal temperature of the body (37° C) can elevate skin temperature as well. Body temperature also increases during exercise and serious illness. The skin must reduce this heat load via the physical mechanisms of evaporation of water and convection of colder air across the skin. In fact, nearly 80% of our heat loss takes place through the skin; the remainder is via the mucous membranes. If heat must be conserved to maintain a constant body temperature, dermal blood vessels constrict *(vasoconstriction)* to keep most of the warm blood circulating deeper in the body, as shown in part *A* of the figure. If heat loss must be increased to maintain a constant temperature, dermal blood vessels widen *(vasodilation)*

to increase the skin's supply of warm blood from deeper tissues, as shown in part *B* of the figure. Heat transferred from the warm blood to the epidermis can then be lost to the external environment.

The operation of the skin's blood vessels and sweat glands must be carefully coordinated and must take into account moment-by-moment fluctuations in body temperature. Temperature receptors in a part of the brain called the *hypothalamus* detect changes in the body's internal temperature. It acts as an integrator and sends a nervous signal to the sweat glands and blood vessels of the skin. Because internal and external conditions change constantly, the hypothalamus is always attempting to maintain the homeostatic set point of 37° C. Like most homeostatic mechanisms, heat loss by the skin is controlled by a negative feedback loop, as we have seen in Chapter 1.

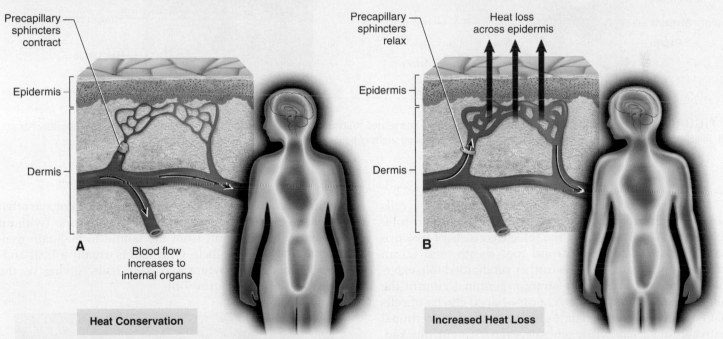

▲ **The skin as a thermoregulatory organ.** When homeostasis requires that the body conserve heat, blood flow in the warm organs of the body's core increases. **A,** When heat must be lost to maintain stability of the internal environment, flow of warm blood to the skin increases. **B,** Heat can be lost from the surface of the skin mostly by means of convection and evaporation. The head, hands, and feet are major areas of heat loss.

APPENDAGES OF THE SKIN
Hair
Development of Hair

We are mammals, and as mammals we are covered with over five million hair follicles, many of which do not produce hair. However, only a few areas of the skin are truly hairless: the palms of the hands, the soles of the feet, the lips, nipples, and some areas of the genitalia. Months before birth, tiny tubular

pockets called **hair follicles** begin to develop in the epidermis and produce very fine hair. By about the sixth month of pregnancy, the developing fetus is nearly covered by this extremely fine, soft hair called **lanugo.** Most of the lanugo is lost *before* birth. Soon after birth, any lanugo hair that remains is lost and then replaced by fine **vellus** hair, which is stronger and usually less pigmented than the lanugo. The replacement hair growth *after* birth first appears on the scalp, eyelids, and eyebrows; the coarse pubic and axillary hair that develops at puberty is called **terminal hair.**

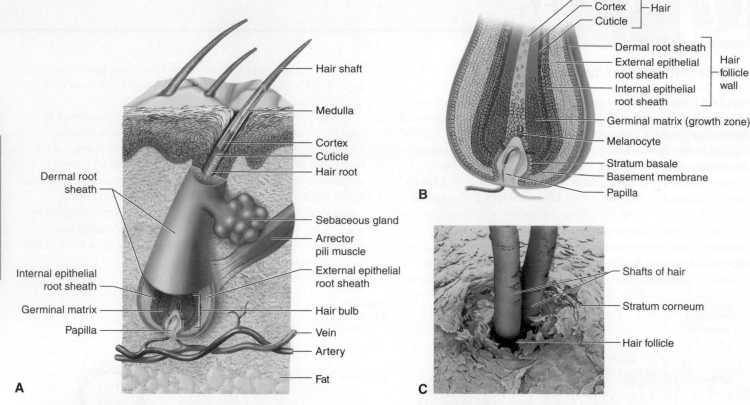

▲ **FIGURE 7-7**   **Hair follicle. A,** Relationship of a hair follicle and related structures to the epidermal and dermal layers of the skin. **B,** Enlargement of a hair follicle wall and hair bulb. **C,** Scanning electron micrograph showing shafts of hair extending from their follicles.

Hair growth begins when a pocket of epidermal cells moves into the dermis and forms a small, tubular hair follicle (Figure 7-7). The wall of the follicle consists of two primary layers: (1) an outer *dermal* root sheath, and (2) an *epithelial* root sheath, which is further subdivided into external and internal layers. The stratum germinativum of the follicle forms a *germinal matrix,* a cap-shaped cluster of cells that will actually form a hair. As long as cells of the germinal matrix are alive, hair will regenerate, even if it is cut, plucked, or otherwise removed.

The *root* of the hair lies hidden in the follicle; the visible part of a hair is called the *shaft.* The inner core of a hair is called the *medulla* and the more superficial portion around the hair is called the *cortex,* which is made of layers of keratinized cells. Finally, there is the *cuticle* that makes up the outer surface of the hair.

Appearance of Hair

Although light reflection can affect hair color somewhat, it is largely the amount, type, and distribution of melanin that determines our hair color. Varying amounts of dark *eumelanin* in the cortex and/or medulla can produce many shades of blonde and brunette hair. The lighter colored *pheomelanin* gives hair a reddish tint. Besides adding color, melanin also imparts strength to the hair shaft. Advancing age often produces increasing numbers of white hairs (without pigments), which result from the failure to maintain melanocytes in the hair follicle. Gray hair is usually a "salt-and-pepper" mixture of white and dark hairs, giving us the impression of gray (Figure 7-8).

▲ **FIGURE 7-8**   **Gray hair.** Gray hair results from a scattered arrangement of both white (pigmentless) hairs and darker (pigmented) hairs.

Whether hair is straight or wavy depends mainly on the shape of the shaft. Straight hair has a round, cylindrical shaft. Wavy hair, in contrast, has a flatter shaft that is not as strong. As a result, it is more easily broken or damaged than straight hair. Two or more small sebaceous glands secrete *sebum*, an oily substance, into each hair follicle (see Figure 7-7). Sebum lubricates and conditions the hair and surrounding skin, keeping it from becoming dry, brittle, and easily damaged. Many of us use hair and skin conditioners that also contain lipids that reduce damage to our hair and skin. Unfortunately, these products do not "repair" hair or skin. Instead, they oil our hair and reduce the dryness of our scalp.

Our hair alternates between periods of growth and rest. On average, hair on our heads grows a little less than 12 mm (½ inch) per month, or about 11 cm (5 inches) a year. Body hair grows more slowly. Head hairs live between 2 and 6 years, then die and are shed. Normally, new hairs replace those that are lost, but baldness can develop for a variety of reasons. The most common type of baldness occurs only when two requirements are met: genes for baldness must be inherited and the male sex hormone, testosterone, must be present. When the right combination of these factors exists, common baldness or **male pattern baldness** results in men. Women likewise can experience baldness, or *alopecia*.

Contrary to what many people believe, hair growth is *not* stimulated by frequent cutting or shaving. In addition, stories about hair or beard growth continuing after death are also false. After death, the skin surface dehydrates through environmental conditions or the embalming process. As a result, the beard simply *appears* more visible as the skin pulls downward and away from the shaft.

Nails

Fingernails and toenails (Figure 7-9) are composed of heavily keratinized epidermal cells. The visible part of each nail is called the *nail body*. The rest of the nail, the *root*, lies in a flat sinus hidden by a fold of skin bordered by the *cuticle*. The nail body nearest the root has a crescent-shaped white area known as the *lunula* ("little moon"). Under the nail lies a layer of epithelium called the *nail bed*, which contains many blood vessels and normally appears pink. This is useful to know. The pinkness of the nail beds is a quick way of determining the oxygenation level of blood. For this reason their color is often monitored closely during surgery or other procedures in which blood oxygen levels

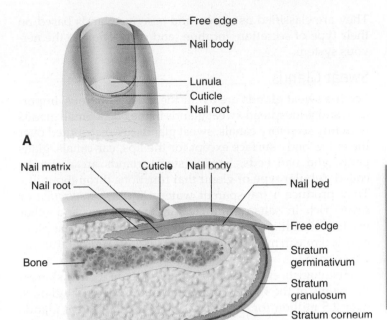

A

B
▲ **FIGURE 7-9**   **Structure of nails. A,** Fingernail viewed from above. **B,** Sagittal section of a fingernail and associated structures.

may suddenly drop. Nails naturally range in color; sometimes, lightly pigmented streaks appear in the nail beds of dark-skinned individuals.

Skin Glands

Skin glands are classified as follows: *sweat, sebaceous,* and *ceruminous* (Figure 7-10). **Sweat** or **sudoriferous** (*sudor* = sweat) **glands** are the most numerous of the skin glands.

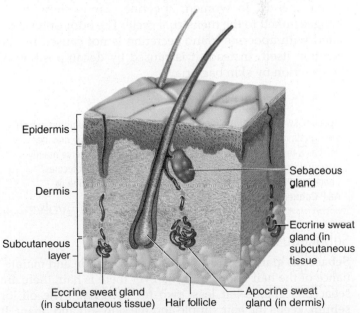

▲ **FIGURE 7-10**   **Skin glands.** Several types of exocrine glands occur in the skin.

> QUICK CHECK
> 14. What are the appendages of the skin?
> 15. Briefly describe how a hair develops.
> 16. What are the basic parts of a hair?
> 17. What are the pigments that determine hair color?

They are classified as *eccrine* and *apocrine* glands based on their type of secretion, location, and connection to the nervous system.

Sweat Glands

Eccrine sweat glands are by far the most numerous, important, and widespread sweat glands in the body. Small glands with tiny secretory canals, sweat glands are distributed over the entire body surface except for the lips, ear canals, glans penis, and nail beds. Eccrine sweat glands are a simple, coiled, tubular type of gland that functions throughout life. They produce a transparent watery liquid (*perspiration* or *sweat*) rich in salts, ammonia, uric acid, urea, and other wastes. In addition to the elimination of wastes, sweat plays a critical role in maintaining a constant body temperature, as we've seen.

Histologists estimate that a single square inch of skin on the palms of the hands contains about 3,000 sweat glands, over 2.5 million for the entire body! Eccrine sweat glands are also abundant on the soles of the feet, forehead, and upper part of the torso. Using a magnifying glass, you can locate the openings of these sweat gland ducts on the skin ridges of the palms and on the skin of the palmar surfaces of the fingers.

Apocrine sweat glands are located deep in the subcutaneous layer of the skin in the armpit (axilla), the areola of the breast, and the pigmented skin areas around the anus. These specialized sweat glands are much larger than eccrine glands and often have secretory canals that reach 5 mm or more in diameter. They are connected with hair follicles and are classified as simple, *branched*, tubular glands. Apocrine sweat glands enlarge and begin to function at the onset of puberty. During this period they produce a more viscous and colored secretion than that of eccrine glands. In women, apocrine glands show cyclic changes linked to the menstrual cycle. The odor often associated with apocrine gland secretion is not caused by the secretion itself. Instead, it is caused by decomposition of the secretion by skin bacteria.

A&P CONNECT

Apocrine sweat contains a class of molecules called pheromones, which are signaling molecules detected by other individuals. They may be used as a territorial marker, but in humans they seem to serve mainly as sexual signals. Want to know more about these "sex molecules?" Check out **Pheromones and the Vomeronasal Organ** online at **A&P Connect.**

ᴇvolve

Sebaceous Glands

Sebaceous glands secrete oil for the lubrication and maintenance of the hair and skin. Wherever there is hair, there are sebaceous glands—at least two for each hair. The oil, or **sebum,** keeps the hair supple and the skin soft and pliant. It also prevents excessive water loss and dehydration of the skin. The sebum contains triglycerides, waxes, fatty acids, and cholesterol. All of these compounds have antifungal effects and protect our skin from potential pathogens. Sebaceous glands are simple, branched glands of varying sizes. They are found in the dermis, except in the skin of the palms and soles. Although almost always associated with hair follicles, some specialized sebaceous glands do open directly onto the skin surface in such areas as the glans penis, lips, and eyelids. Sebum secretion is stimulated by increased blood levels of the sex hormones, and thus increases during adolescence.

Frequently, sebum accumulates in and enlarges some of the ducts of the sebaceous glands, forming white pimples. With oxidation, this accumulated sebum darkens and forms a **blackhead.** Common *acne*, familiar to most of us, occurs most frequently during the adolescent years, but can persist even late into adulthood. This condition is caused by overactive secretion of the sebaceous glands, with blockage and inflammation of their ducts. (The rate of sebum secretion increases by five times between 10 and 19 years of age!) As a result, sebaceous gland ducts may become plugged with sloughed skin cells and sebum contaminated with bacteria.

Ceruminous Glands

Ceruminous glands are a special type of apocrine sweat gland that appear in the external ear canal. The mixed secretions of sebaceous and ceruminous glands form a brown waxy substance called **cerumen,** commonly known as earwax. This substance normally serves a useful purpose in protecting the skin of the ear canal from pathogens, small insects, and dehydration.

QUICK **CHECK**

18. What are the two types of sweat glands?
19. How do our sweat glands differ?
20. List the basic functions of sebum.

Cycle of LIFE ○

Dramatic changes take place in the skin from birth through the mature years. As you know, infants and young children have relatively smooth and unwrinkled skin characterized by great elasticity and flexibility. Because skin tissues of the young are constantly in a stage of growth, the healing of skin injuries is typically rapid. In addition, young children have fewer sweat glands than adults do, so their bodies rely more on increased blood flow to the skin, rather than the evaporation of sweat, to maintain normal body temperature. This explains why lighter skinned preschoolers often become "red-faced" while playing outdoors on a warm day. At puberty, hormones stimulate the development and activation of sebaceous and sweat glands. "Overactive" sebaceous glands often overproduce sebum and thus give the skin an oily appearance. If the ducts to the sebaceous glands become clogged or infected, acne pimples and other sorts of blemishes may result. Activation of apocrine sweat glands during

The BIG Picture

The skin is a miracle of complexity. It is one of the major components of our body's framework. Try to imagine how the skin is continuous with the connective tissues that hold the body together. This would include the fascia, bones, tendons, and ligaments. You can easily see how the integumentary, skeletal, and muscular systems work together to protect and support the entire body.

We've seen how the skin provides a barrier that separates the internal environment from the often-harsh external environment. In fact, the skin is the boundary that separates the two. Without the protective structures and mechanisms of the skin, we would be exposed to the harshness of our environment with all its potential for physical damage and invasion by pathogens.

The dermis and epidermis work together to form a tough, waterproof envelope that protects us from drying out. It also protects us from the dangers of chemical or microbial contamination. The surface film of the skin, along with our sebaceous secretions, enhance our skin's ability to protect us. Hair, calluses, and the depth of the skin itself provide our bodies with additional protection from mechanical injury. Pigmentation in the skin and its ability to regulate this pigmentation, in part, help protect us from harmful ultraviolet radiation.

Although its primary functions are support and protection, the skin also has a vital role in many other activities, such as maintaining homeostasis. For example, the skin regulates our body temperature, serving as a radiator to the environment. The skin also helps maintain a constant level of calcium in the body by producing vitamin D. In addition, ridges on the palms and fingers literally allow us to "grip" our environment. The skin's flexibility and elasticity permit the movement we require to produce all our daily tasks. Sensory receptors in the dermis allow the skin to be our sensory "window on the world" around us. This information is relayed to our brain integration centers, where it is used to coordinate the function of our organs.

As you continue on with the skeletal and muscular systems of the next chapters, try to keep in mind that these systems could not operate properly without the structure and proper functioning of the integumentary system.

puberty causes increased sweat production. In turn, this increases the potential for body odor as organic waste products in the sweat are broken down into more offensive organic compounds by bacteria.

As one grows older, the sebaceous and sweat glands become less active. This can provide a welcome relief to those suffering from acne or other skin problems associated with overactivity of these glands. However, this can also affect normal functions of the body. For example, the reduction of sebum production can cause the skin to become less resilient and therefore more likely to wrinkle or crack. Likewise, hair becomes more brittle and more likely to split or break. Wrinkling can also be caused by an overall degeneration in the skin's ability to maintain itself as efficiently as it did during youth. Loss of function in sweat glands as adulthood advances adversely affects the body's ability to cool itself during exercise or when the external temperature is high. For this reason, the elderly are more likely to suffer severe problems during hot weather than are young adults. ●

MECHANISMS OF DISEASE

Any disorder of the skin can be called a *dermatosis*, which simply means "skin condition." There are several possible causes for dermatosis. Viruses, bacteria, and fungi can cause *impetigo, dermatophytoses, warts,* and *boils.* Vascular or inflammatory disorders can cause *pressure ulcers, hives,* and *eczema.* The skin is involved in the regulation of body temperature, but conditions such as *hypothermia, heat exhaustion,* and *frostbite* can occur when body temperature is too far from normal.

evolve Find out more about skin conditions and conditions related to body temperature online at *Mechanisms of Disease: Skin and Its Appendages.*

LANGUAGE OF SCIENCE AND MEDICINE *continued from page 117*

male pattern baldness

malignant melanoma (mah-LIG-nant mel-ah-NO-mah)
[*malign-* bad, *-ant* state, *melan-* black, *-oma* tumor]

melanin (MEL-ah-nin)
[*melan-* black, *-in* substance]

melanocyte (MEL-ah-no-syte)
[*melan-* black, *-cyte* cell]

melanosome (MEL-ah-no-sohm)
[*melan-* black, *-som-* body]

Merkel disk (MER-kuhl)
[Friedrich Sigmund Merkel, German anatomist]

papillary layer (PAP i-lair-ee)
[*papilla-* nipple, *-ary* relating to]

partial-thickness burn

reticular layer (reh-TIK-yoo-lar)
[*ret-* net, *-ic-* relating to, *-ul-* little, *-ar* relating to]

"rule of nines"

sebaceous gland (seh-BAY-shus)
[*seb-* tallow (hard animal fat), *-ous* relating to, *gland* acorn]

sebum (SEE-bum)
[*sebum* tallow (hard animal fat)]

second-degree burn

squamous cell carcinoma (SKWAY-muss sell kar-si-NO-mah)
[*squam-* scale, *-ous* characterized by, *cell* storeroom, *carcin-* cancer, *-oma* tumor]

stratum (STRAH-tum)
[*stratum* layer] *pl.,* strata (STRAH-tah)

stratum basale (STRAH-tum bay-SAH-lee)
[*stratum* layer, *bas-* base, *-ale* relating to] *pl.,* strata (STRAH-tah)

stratum corneum (STRAH-tum KOR-nee-um)
[*stratum* layer, *corneum* horn] *pl.,* strata (STRAH-tah)

continued

LANGUAGE OF SCIENCE AND MEDICINE *continued from page 131*

stratum germinativum (STRAH-tum jer-mi-nah-TIV-um)
  [*stratum* layer, *germinativum* something that sprouts]
  pl., strata (STRAH-tah)

stratum granulosum (STRAH-tum gran-yoo-LOH-sum)
  [*stratum* layer, *gran-* grain, *-ul-* little, *-osum* thing] *pl.,*
  strata (STRAH-tah)

stratum lucidum (STRAH-tum LOO-see-dum)
  [*stratum* layer, *lucid-* clear, *-um* thing] *pl.,* strata
  (STRAH-tah)

stratum spinosum (STRAH-tum spi-NO-sum)
  [*stratum* layer, *spino-* spine, *-um* thing] *pl.,* strata
  (STRAH-tah)

subcutaneous layer (sub-kyoo-TAY-nee-us)
  [*sub-* beneath, *-cut-* skin, *-ous* relating to]

sudoriferous gland (soo-doh-RIF-er-us)
  [*sudo-* sweat, *-fer-* bear or carry, *-ous* relating to, *gland*
  acorn]

superficial fascia (soo-per-FISH-all FAH-shah)
  [*super-* over or above, *-fici-* face, *-al* relating to, *fascia*
  band]

surface film

sweat gland
  [*gland* acorn]

tactile disk (TAK-tyle)
  [*tact-* touch, *-ile* relating to]

terminal hair (TUR-mih-nal)
  [*termin-* boundary, *-al* relating to]

third-degree burn

vellus (VEL-us)
  [*vellus* wool]

CHAPTER SUMMARY

To download an MP3 version of the chapter summary for use with your iPod or other portable media player, access the Audio Chapter Summaries *online at http://evolve.elsevier.com.*

HINT *Scan this summary after reading the chapter to help you reinforce the key concepts. Later, use the summary as a quick review before your class or before a test.*

INTRODUCTION

A. Skin is one of your largest and most important organs
B. Skin surface of an average human is nearly 2 square meters (20 square feet); thickness varies from less than 0.05 cm (1/50 inch) to slightly more than 0.3 cm (1/8 inch)
C. Integumentary system—the skin and its appendages (including hair, nails, glands)

STRUCTURE OF THE SKIN

A. The layers of the skin—two primary layers, epidermis and dermis (Figure 7-1), plus the hypodermis
  1. Epidermis
    a. Epidermal cell types—several types of cells make up the epidermis:
      (1) Keratinocytes—make up over 90% of the epidermal cells; filled with a tough, fibrous protein called keratin
      (2) Melanocytes—give our skin its color; dark pigments of these cells reduce the amount of potentially dangerous UV radiation
      (3) Dendritic cells—identify invading bacteria and other invaders of the skin and present them to the immune system

      (4) Merkel cells—found in the deepest layer of the epidermis; form complexes called Merkel disks that serve as receptors for light touch
    b. Epidermal cell layers—thickest types of epidermis have five distinct *strata* (layers):
      (1) Stratum basale (base layer)—single layer of columnar cells that produces all the other cells of the epidermis
      (2) Stratum spinosum (spiny layer)—cells are dying as they move away from the blood supply (stratum germinativum [growth layer] is sometimes used to describe stratum basale and the stratum spinosum together)
      (3) Stratum granulosum (granular layer)—two to four layers of dead cells that are filled with a substance needed to produce surface keratin
      (4) Stratum lucidum (clear layer)—filled with a substance needed to produce surface keratin
      (5) Stratum corneum (horny layer)—most superficial layer of skin; composed of very thin squamous (flat) cells
    c. Epidermal growth and repair—new cells must be formed at the same rate that old keratinized cells flake off from the stratum corneum
      (1) Calluses—form in areas of constant friction
    d. Dermoepidermal junction—includes fibers and unique chemical gels; together with the basement membrane, cements or "glues" the superficial epidermis to the dermis below
      (1) Blister—junction fills with extracellular fluid (Box 7-1)
  2. The dermis and its components
    a. Dermis—composed of a thin papillary layer and a thicker reticular layer; "true skin" (Table 7-1)

b. Somatic sensory receptors—specialized network of nerves and nerve endings in the dermis
   (1) Processes sensory information related to pain, pressure, touch, and temperature variation
c. Papillary layer—superficial layer of the dermis; forms nipplelike bumps (dermal papillae) (Figure 7-2)
d. Reticular layer—consists of a dense reticulum (network) of collagenous fibers
e. Dermis serves as an area of attachment for numerous skeletal and smooth muscle fibers
f. Hair has a small bundle of involuntary muscles (arrector pili muscles) attached to the dermis (Figure 7-3)
g. Millions of specialized sensory receptors are located in the dermis of all skin areas (Figure 7-2)
h. Dermis does not continually shed and regenerate; regeneration only takes place after a wound occurs
   (1) A dense mass of new connective tissue forms; if not replaced by normal tissue, a scar is formed
3. Hypodermis
   a. Region below the dermis; filled with adipose and loose fibrous connective tissue
   b. Called subcutaneous layer or superficial fascia
   c. Not part of the skin proper; forms a connection between the skin and the underlying structures of the body
B. Thin and thick skin
   1. *Thick* and *thin skin* refers to the thickness of the epidermal layer, not to the depth of the dermis and epidermis taken together
      a. Thin skin—most of our body's surface is covered by thin skin
      b. Thick skin—hairless skin covering the palms of the hands, fingertips, soles of the feet, and other body areas subject to friction
   2. Friction ridges—fingerlike projections of dermal papillae pushing upward from the underlying dermis
      a. Dermal projections create our unique fingerprints and footprints

PIGMENTS OF THE SKIN

A. Melanin—pigment that determines skin color
   1. Melanocytes—pigment-producing cells
B. Skin pigmentation is regulated by different factors:
   1. Density of rootlike projections from melanocytes
   2. Transport of different types of melanosomes (tiny pigmented granules) to other cells in the skin; keratinocytes
   3. Distribution of pigmented cells in the skin
C. Variation in skin color is due to:
   1. Differences in pigment production
   2. Type and distribution of melanosomes within the melanocytes
D. Prolonged exposure to ultraviolet radiation causes melanocytes to increase melanin production
E. Other pigments of the skin
   1. Beta-carotene—pigment essential for skin growth; stored in skin tissue and then converted into vitamin A (Figure 7-5)
   2. Bile pigments—these cause jaundice by staining the tissues of the body when the liver is unable to remove bile pigments from the blood efficiently (Box 7-5)
   3. Excess amount of hemoglobin that is low in oxygen and high in carbon dioxide

a. Blood with a high proportion of unoxygenated hemoglobin will result in skin appearing bluish (cyanotic) (Figure 7-6)

FUNCTIONS OF THE SKIN (TABLE 7-2)

A. Protection—protect underlying tissues against invasion of microorganisms; bar the entry of many harmful chemicals; reduce the possibility of mechanical injury to underlying structures
   1. Surface film—produced from the residues and secretions from sweat and sebaceous glands mixed with the dead epithelial cells
      a. Provides lubrication and hydration for the skin
      b. Buffers the skin from caustic environmental irritants
      c. Blocks many toxic agents
B. Sensation—sensory receptors that respond to touch, pressure, vibration, pain, and temperature are embedded in the skin
C. Flexibility—skin exhibits stretch and recoil characteristics that permit easy changes in contour, without tearing
D. Excretion—skin regulates the volume and chemical content of sweat; influences body's total fluid volume; regulates the amounts of waste products that are excreted
E. Endocrine/hormone (vitamin D) production—when the skin is exposed to ultraviolet light, chemical precursors to vitamin D are produced
F. Immunity—skin has many defensive cells that destroy pathogenic microorganisms
G. Homeostasis of body temperature—temperature-regulating mechanisms of the skin maintain homeostasis of body temperature (Box 7-7)

APPENDAGES OF THE SKIN

A. Hair and its development
   1. Before birth, hair follicles begin to develop in the epidermis and produce very fine hair
   2. Lanugo—extremely fine, soft hair; covers developing fetus
   3. Soon after birth, any lanugo hair that remains is lost and then replaced by fine vellus hair
      a. Stronger and usually less pigmented than the lanugo
   4. Terminal hair—coarse pubic and axillary hair that develops at puberty
   5. Hair growth begins when epidermal cells move into the dermis and form a small, tubular hair follicle (Figure 7-7)
   6. Wall of the follicle consists of two primary layers:
      a. Dermal root sheath
      b. Epithelial root sheath
   7. Root—lies hidden in the follicle
   8. Shaft—visible part of hair
   9. Medulla—inner core of a hair
   10. Cortex—superficial portion around the hair
B. Appearance of hair
   1. Appearance of hair—amount, type, and distribution of melanin determines our hair color
      a. Eumelanin—produces many shades of blonde and brunette hair
      b. Pheomelanin—gives hair a reddish tint
      c. White hairs (without pigments)—result from the failure to maintain melanocytes in the hair follicle
      d. Gray hairs—a "salt-and-pepper" mixture of white and dark hairs; gives the impression of gray (Figure 7-8)

2. Whether hair is straight or wavy depends mainly on the shape of the shaft
3. Sebaceous glands—secrete sebum, an oily substance, into each hair follicle (Figure 7-7)
   a. Sebum lubricates and conditions the hair and surrounding skin
4. Hair alternates between periods of growth and rest
   a. On average, hair on our heads grows a little less than 12 mm (½ inch) per month; 11 cm (5 inches) a year
   b. Head hairs live between 2 and 6 years
5. Male pattern baldness—most common type of baldness; occurs when genes for baldness are inherited and the male sex hormone, testosterone, is present

C. Nails
1. Fingernails and toenails are composed of heavily keratinized epidermal cells (Figure 7-9)
   a. Nail body—visible part of nail
   b. Root—lies in a flat sinus
   c. Lunula—crescent-shaped white area
   d. Nail bed—contains many blood vessels and normally appears pink

D. Skin glands
1. Classification of skin glands (Figure 7-10):
   a. Sweat
   b. Sebaceous
   c. Ceruminous
2. Sweat or sudoriferous glands—most numerous of the skin glands
   a. Eccrine
   b. Apocrine
3. Eccrine—most numerous, important, and widespread sweat glands in the body
   a. Simple, coiled, tubular type of gland
   b. Produce a transparent watery liquid (perspiration or sweat) rich in salts, ammonia, uric acid, urea, and other wastes
4. Apocrine—located deep in the subcutaneous layer of the skin; armpit (axilla), areola of the breast, and pigmented skin areas around the anus
   a. Much larger than eccrine glands and often have secretory canals
   b. Simple, branched, tubular glands
   c. Enlarge and begin to function at the onset of puberty
5. Sebaceous glands—secrete oil (sebum) for lubrication and maintenance of the hair and skin
   a. Keeps the hair supple and the skin soft and pliant
   b. Prevents excessive water loss and dehydration of the skin
   c. Simple, branched glands of varying sizes
   d. Sebum accumulates in and enlarges some of the ducts of the sebaceous glands; forms white pimples
   e. With oxidation, this accumulated sebum darkens and forms a blackhead
6. Ceruminous glands—special type of apocrine sweat glands that appear in the external ear canal
   a. Cerumen (earwax)—a brown waxy substance; from mixed secretions of sebaceous and ceruminous glands
   b. Protects the skin of the ear canal from pathogens, small insects, and dehydration

REVIEW QUESTIONS

Write out the answers to these questions after reading the chapter and reviewing the Chapter Summary. If you simply think through the answer without writing it down, you won't retain much of your new learning.

1. List and briefly discuss several of the different functions of the skin.
2. List three cell types found in the epidermis.
3. List and describe the cell layers of the epidermis from superficial to deep.
4. Discuss the process of epidermal growth and repair.
5. What is keratin? Where is it found?
6. What part of the skin contains blood vessels?
7. Why is the process of blister formation a good example of the relationship between the skin structure and function?
8. List the two layers of the dermis.
9. What are arrector pili muscles?
10. Discuss the importance of the surface film of the skin.
11. List the appendages of the skin.
12. Identify each of the following: hair papilla, germinal matrix, hair root, hair shaft, follicle.
13. List the three primary types of skin glands.
14. What is the difference between eccrine and apocrine sweat glands?
15. How is the "rule of nines" used in determining the extent of a burn injury?

CRITICAL THINKING QUESTIONS

After finishing the Review Questions, write out the answers to these items to help you apply your new knowledge. Go back to sections of the chapter that relate to items that you find difficult.

1. The stratum corneum, hair follicle, nails, stratum basale, sweat glands, and oil glands are all discussed in this chapter. Can you make a distinction regarding whether these structures are part of the integumentary system only or both the integumentary system and the integument?
2. Identify what affects the thickness of the skin (dermis and epidermis). What is the relationship between what affects the thickness of the skin and the thickness of the hypodermis?
3. What is the dermoepidermal junction (DEJ)? By analyzing its anatomical components, what inference can you make regarding how it functions?
4. Concern about skin cancer is reducing the amount of time people spend in the sun. If this caution is carried to the extreme, how would you explain the impact on skin function?
5. How would you explain why a light-skinned individual would be more susceptible to malignant melanoma?
6. An individual running a marathon expends a great deal of energy. Much of this energy generates heat, which increases core body temperature. What is the role of sweat glands in balancing body temperature during this strenuous exercise? With this loss of fluid, what suggestions do you have to avoid dehydration?

continued from page 117

Now that you have read this chapter, try to answer these questions about Aidan's splinter from the Introductory Story.

1. What layer or layers of the skin has the splinter gone through in Aidan's finger? (HINT: There's no bleeding.)
   a. Epidermis only
   b. Dermis only
   c. Both epidermis and dermis
   d. Stratum basale only

2. If the splinter went into Aidan's arm instead of his finger, which of these layers would the splinter NOT have pierced?
   a. Stratum corneum
   b. Stratum spinosum
   c. Stratum lucidum

This puncture has broken Aidan's intact skin barrier and now may allow bacteria or other pathogens to enter.

3. What cells will help identify these pathogens for destruction?
   a. Keratinocytes
   b. Melanocytes
   c. Dendritic cells
   d. Lamellar (pacinian) corpuscles

HINT   *To solve these questions, you may have to refer to the glossary or index, other chapters in this textbook, A&P Connect, Mechanisms of Disease, and other resources.*

UNIT 2

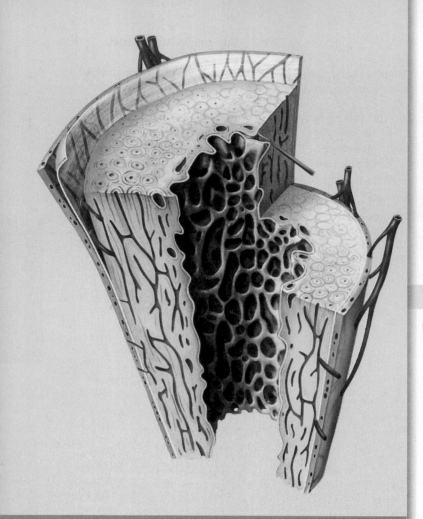

CHAPTER 8

Skeletal Tissues

STUDENT LEARNING OBJECTIVES

At the completion of this chapter, you should be able to do the following:

1. Explain the function of connective tissue in the skeleton.
2. Identify the components of a long bone.
3. Identify the components of flat bones.
4. List and describe the cellular and extracellular components of bone tissue.
5. Compare and contrast compact bone with spongy bone.
6. Discuss how blood calcium level is controlled in the body.
7. Outline the process of bone growth in a long bone.
8. Outline the process of membrane bone growth.
9. Compare the bone remodeling process in a child with that of an adult.
10. Describe the functions of osteoblasts and osteoclasts during bone growth, repair, and remodeling.
11. Discuss methods for the repair for several different types of bone fractures.
12. Describe the role of cartilage in the skeletal system.

CHAPTER OUTLINE

 Scan this outline before you begin to read the chapter, as a preview of how the concepts are organized.

HINT

ELEANOR stepped off the curb by her mailbox, just as she had done thousands of times before. But this time, her hip gave way, and she suddenly found herself on the ground, unable to get back up. A radiograph (x-ray) film at the hospital showed a fracture in the neck of her femur. Eleanor wondered how she could have broken a bone by simply stepping off a curb. The doctor was talking about surgery and a 6-month recovery time. She would definitely have to reschedule the cruise she had booked to celebrate retirement!

After you finish with this chapter you will have a much better idea of what happened to Eleanor, and exactly why it will take her so long to mend.

• • •

Bones seem to be such hard, lifeless structures. Only when you break one do you quickly realize that they are very much alive! Bones give your body form, provide you with support and protection, and allow you to move. They also serve as reservoirs for calcium and phosphorus, and they produce the major cellular components of the blood. We have devoted two chapters to the study of bones. This chapter will introduce you to the basic structures and functions of bones. Chapter 9 will then examine the divisions of the skeleton.

TYPES OF BONES

The names of the five types of bones also describe their general shapes: *long bones, short bones, flat bones, irregular bones,* and *sesamoid bones* (Figure 8-1). Together they comprise the 206 or more bones of our skeleton.

Bones differ not only in size and shape but also in the amount and proportion of the two different types of bone tissue that compose them. **Compact bone** is dense and solid in appearance. In contrast, **spongy** (*cancellous*) **bone** has open spaces partially filled by a network of fine, branched crossbeams. All five types of bones have varying amounts of compact and spongy bone in their structure. Let's look briefly at each type of bone.

Long bones (e.g., the femur of the thigh) have long shafts and often uniquely shaped ends that form complex joints such as your knee. **Short bones** are often cube or box shaped. They typically appear in coordinated groups, such as the carpal bones of our wrists and the tarsal bones of our ankles. **Flat bones** (e.g., the sternum), as the name suggests, are relatively flat. **Irregular bones** (e.g., vertebral bones) are odd shaped and may appear in groups such as in the vertebral column. Finally, **sesamoid bones** are rounded bones that resemble the shape of a sesame seed. The number and size of sesamoid bones actually may vary from person to person. They develop in the tendons close to joints (e.g., the patella of the knee).

We can now begin our tour of the skeletal system with the study of the basic composition of long bones and flat bones.

LANGUAGE OF SCIENCE AND MEDICINE

HINT *Before reading the chapter, say each of these terms out loud. This will help you avoid stumbling over them as you read.*

articular cartilage (ar-TIK-yoo-lar KAR-ti-lij)
[*artic-* joint, *-ul-* little, *ar* relating to, *cartilag-* cartilage]

avascular (ah-VAS-kyoo-lar)
[*a-* without, *-vas-* vessel, *-ula-* little, *-ar* relating to]

bone marrow (bohn MAIR-oh)

calcification (kal-sih-fih-KAY-shun)
[*calci-* lime, *-fication* make]

calcitonin (kal-sih-TOH-nin)
[*calci-* lime (calcium), *-ton-* tone, *-in* substance]

canaliculus (kan-ah-LIK-yoo-lus)
[*canal-* channel, *-icul-* little] *pl.,* canaliculi
(kan-ah-LIK-yool-eye)

cancellous bone (KAN-seh-lus bohn)
[*cancel-* lattice, *-ous* characterized by]

central canal

chondrocyte (KON-droh-syte)
[*chondro-* cartilage, *-cyte* cell]

compact bone

diaphysis (dye-AF-i-sis)
[*dia-* through or apart, *-physis* growth] *pl.,* diaphyses
(dye-AF-i-sees)

diploë (DIP-lo-EE)
[*diploë* folded over (doubled)]

elastic cartilage (eh-LAS-tik KAR-ti-lij)
[*elast-* drive or beat out, *-ic* relating to, *cartilage* cartilage]

endochondral ossification (en-doh-KON-dral os-i-fi-KAY-shun)
[*endo-* within, *-chondr-* cartilage, *-al* relating to, *os-* bone, *-fication* make]

endosteum (en-DOS-tee-um)
[*end-* within, *-osteum* bone]

epiphyseal line (ep-i-FEEZ-ee-al)
[*epi-* on, *-phys-* growth, *-al* relating to]

epiphyseal plate (ep-i-FEEZ-ee-al)
[*epi-* on, *-phys-* growth, *-al* relating to, *plate* flat]

epiphysis (eh-PIF-i-sis)
[*epi-* on, *-physis* growth] *pl.,* epiphyses (eh-PIF-i-sees)

extracellular bone matrix (eks-trah-SELL-yoo-lar bohn MAY-triks)
[*extra-* outside, *-cell-* storeroom, *-ular* relating to, *matrix* womb]

fibrocartilage (fye-broh-KAR-ti-lij)
[*fibro-* fiber, *-cartilage* cartilage]

flat bone

fracture (FRAK-chur)
[*fracture* break]

hyaline cartilage (HYE-ah-lin KAR-ti-lij)
[*hyaline* of glass, *cartilage* cartilage]

continued on page 148

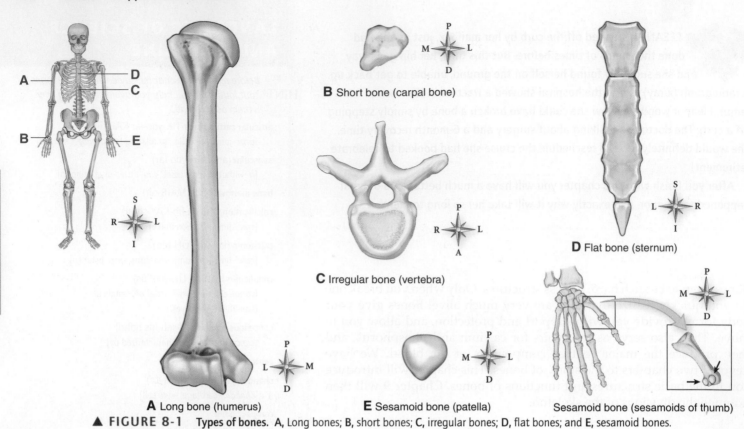

▲ **FIGURE 8-1   Types of bones.   A,** Long bones; **B,** short bones; **C,** irregular bones; **D,** flat bones; and **E,** sesamoid bones.

B Short bone (carpal bone)

C Irregular bone (vertebra)

D Flat bone (sternum)

A Long bone (humerus)

E Sesamoid bone (patella)

Sesamoid bone (sesamoids of thumb)

Major Parts of a Long Bone

The femur in Figure 8-2 highlights the major parts of a long bone: *diaphysis, epiphyses, articular cartilage, periosteum, medullary (marrow) cavity,* and *endosteum*. As you read the following passages, refer often to this figure to help you familiarize yourself with the various parts.

The **diaphysis** is a hollowed shaft. Note that the walls of the shaft are made of thick *compact bone* that provides great weight-bearing strength. The **epiphyses** (singular, *epiphysis*) are the enlarged ends of a long bone. These enlargements provide areas near joints for muscle attachments as well as areas for bone-to-bone articulation. The *spongy bone* of the epiphyses provides further support without much additional weight. In some bones, *red marrow,* a type of soft connective tissue, fills the tiny spaces within the spongy bone. Early in our development, epiphyses are separated from the shaftlike diaphysis by a layer of cartilage called the *epiphyseal plate.* As we will see, this is where bone growth takes place.

Articular cartilage—a thin layer of hyaline cartilage—covers and protects the articular (joint) surfaces of the epiphyses. It cushions jolts and blows to the joints.

Now look at the **periosteum.** This dense, white fibrous membrane covers the bone's exterior (except at the joint surfaces, where the articular cartilage forms the covering). Many of the periosteum's fibers interlace with tendon fibers from muscles to anchor them firmly to the bone. The periosteum contains bone-forming and bone-destroying cells as well as blood vessels.

The **medullary** *(marrow)* **cavity** creates a hollow space throughout the diaphysis of a long bone. It joins the smaller spaces in the spongy bone of the epiphyses. In adults, the medullary cavity is filled with a connective tissue rich in fat—a substance called *yellow marrow.* The thin, fibrous membrane lining the medullary cavity and the spaces of spongy bone is the **endosteum.** Like the periosteum, the endosteum contains both bone-forming and bone-destroying cells.

Major Parts of a Flat Bone

The structure of a *flat bone* (Figure 8-3) is somewhat similar to that of a long bone, but simpler. In flat bone, a layer of spongy bone called **diploë** is sandwiched between an *internal table* and an *external table* made of compact bone. Like long bones, flat bones are covered in a periosteum and the inner spaces are lined with endosteum. Red marrow fills the spaces of many flat bones, such as the sternum. Short bones, irregular bones, and sesamoid bones all have features similar to those of flat bones.

QUICK✓CHECK

1. Discuss the different types of bones.
2. Name the two major types of connective tissue found in the skeletal system.
3. Name the major parts of a long bone.
4. Compare the structure of a long bone with that of a flat bone.

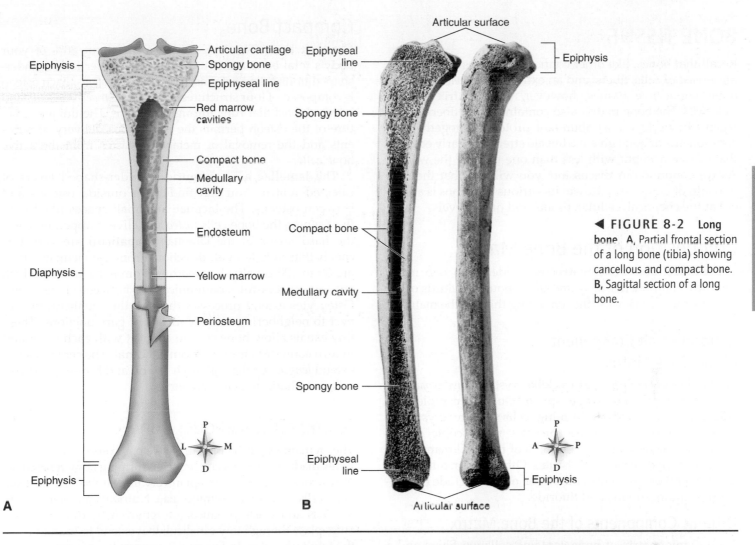

A

Epiphysis

Diaphysis

Epiphysis

Articular cartilage
Spongy bone
Epiphyseal line
Red marrow cavities
Compact bone
Medullary cavity
Endosteum
Yellow marrow
Periosteum

B

Epiphyseal line
Spongy bone
Compact bone
Medullary cavity
Spongy bone
Epiphyseal line
Epiphysis
Articular surface
Articular surface
Epiphysis

◀ **FIGURE 8-2** **Long bone.** **A,** Partial frontal section of a long bone (tibia) showing cancellous and compact bone. **B,** Sagittal section of a long bone.

UNIT 2

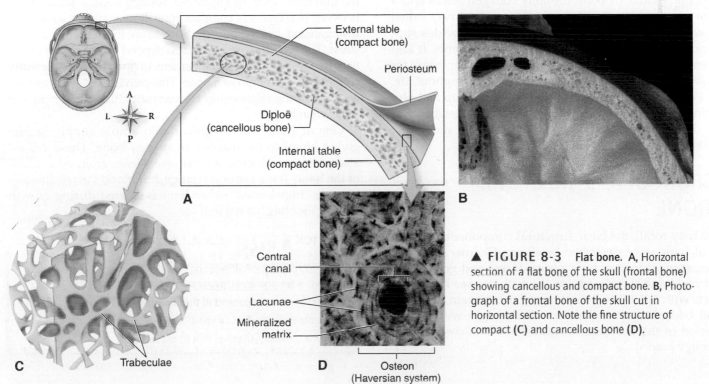

A

External table (compact bone)
Periosteum
Diploë (cancellous bone)
Internal table (compact bone)

B

C

Trabeculae

D

Central canal
Lacunae
Mineralized matrix
Osteon (Haversian system)

▲ **FIGURE 8-3** **Flat bone.** **A,** Horizontal section of a flat bone of the skull (frontal bone) showing cancellous and compact bone. **B,** Photograph of a frontal bone of the skull cut in horizontal section. Note the fine structure of compact **(C)** and cancellous bone **(D).**

BONE TISSUE

Recall that bones, like other connective tissues (see Chapter 6), consist of cells, fibers, and an extracellular *matrix*. Unlike other connective tissues, however, bone matrix is hard (*calcified*). The bone matrix also contains many fibers of collagen (the body's most abundant protein). Altogether, the components of bone give it a tensile strength nearly equal to that of cast iron but with less than one third of the weight! As we continue our discussion, you will see that the relationship of bone's structure to its various functions is apparent at the chemical, cellular, tissue, and organ levels.

Composition of the Bone Matrix

The **extracellular bone matrix** is divided into two major chemical components: *inorganic salts* (about two thirds of the matrix) and *organic salts* (the remaining third of the matrix).

Inorganic Salt Components of the Bone Matrix

The hardness of bone is due to rocklike crystals of *hydroxyapatite*, formed from calcium and phosphate. During a complex **calcification** process, bone-forming cells secrete crystalline deposits of hydroxyapatite into the microscopic spaces between collagen fibers. About 85% of the total matrix is in the form of hydroxyapatite crystals. Other minor, but important mineral components of the bone matrix include magnesium, sodium, sulfate, and fluoride.

Organic Components of the Bone Matrix

The organic matrix of bone contains collagen fibers and a mixture of protein and polysaccharides collectively called the *ground substance*. The ground substance provides support and adhesion for cellular and fibrous elements. It also plays an active role in many cellular functions important for the growth, repair, and remodeling of bone throughout life. Several compounds, including chondroitin sulfate and glucosamine, help cartilage remain smooth, compressible, and elastic. They add to the overall strength of bone and give it additional plastic-like resilience.

MICROSCOPIC STRUCTURE OF BONE

As you may recall, the basic structural components and cell types of bone were described briefly in Chapter 6. In this chapter, we will look at how a bone forms and grows, and how it repairs itself after injury. We will also see how bone interacts with other tissues and organs to maintain a number of vital homeostatic mechanisms. Let's begin with a brief discussion of the different categories of bone: *compact bone* and *spongy bone.*

Compact Bone

Compact bone (Figure 8-4) makes up about 80% of your body's total bone mass. It contains many vertical, cylinder-shaped units called **osteons,** or *haversian systems.* Each osteon is composed of four structures: *lamellae, lacunae, canaliculi,* and a *central canal* that runs through the bone. The unique structure of the osteon permits the continuous delivery of nutrients and the removal of metabolic wastes from the active bone cells.

The **lamellae** are concentric, cylinder-shaped layers of calcified matrix. You can find them outside osteons and between osteons. The **lacunae** are small spaces filled with fluid where the bone cells (*osteocytes*) live, trapped between the hard layers of the lamellae. **Canaliculi** are very fine canals that radiate in all directions from the lacunae. There are 50 to 100 canaliculi extending from each lacuna, which ensures successful communication between bone cells. Osteocytes extend processes through the canaliculi to connect to neighboring cells by means of gap junctions. These tiny canals allow bone cells to connect with each other and in turn connect to the larger **central canal.** The central canals extend *lengthwise* through each osteon and house blood vessels, lymphatic vessels, and nerves.

Spongy (Cancellous) Bone

The remaining 20% of our skeleton consists of **spongy bone,** made of crisscrossing branches forming bony **trabeculae.** No osteons are present, except in very thick trabeculae. Instead, bone cells lie within the trabeculae. Nutrients are delivered to the cells and waste products are removed by diffusion. This takes place through tiny canaliculi that extend to the surface of the thin trabeculae. In Figure 8-3, spongy bone is sandwiched between two layers of compact bone. It may appear somewhat disorganized, but the bony trabeculae are thicker along lines of stress. The type and extent of stress depend on each bone's location in the body. The arrangement of the trabeculae greatly enhances the strength of a bone. This provides us with another example of the relationship between structure and function in the human body.

One or more arteries provide an ample supply of life-giving blood to the marrow of spongy bone. These vessels also provide nutrients and remove wastes from other areas of the bone. It may sound strange, but blood vessels literally become "imprisoned" when bone is created during growth and remodeling, as we will see.

QUICK ✓ CHECK

5. Briefly describe the composition of bone matrix.

6. What is the primary inorganic salt of bones?

7. What are the components of the organic matrix?

8. List the major parts of compact bone.

9. Name the main structures that form the osteon.

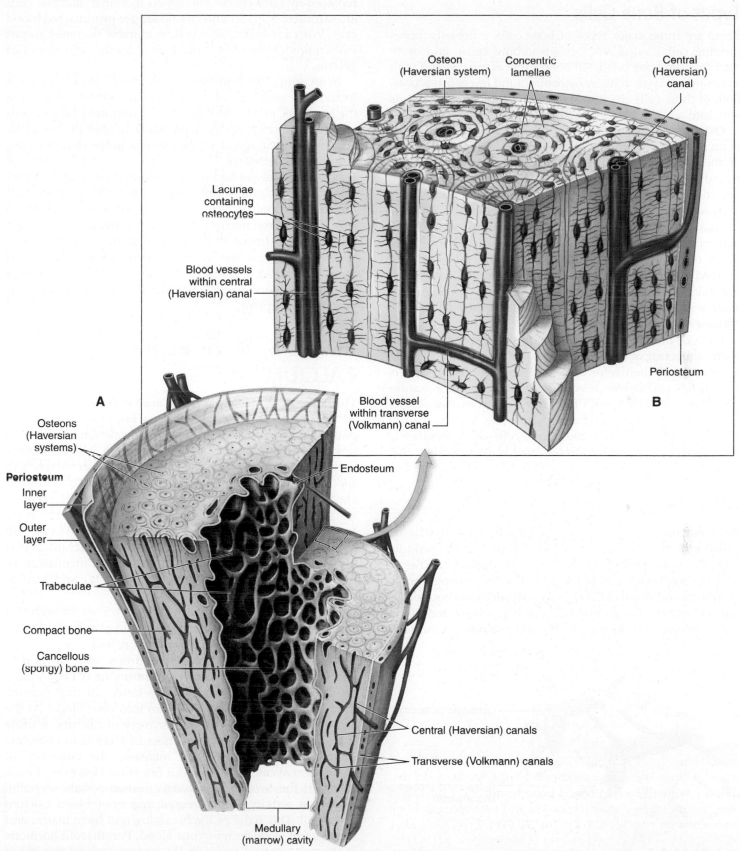

▲ FIGURE 8-4 Compact and cancellous bone in a long bone. A, Longitudinal section of a long bone showing both cancellous and compact bone.
B, Magnified view of compact bone.

Types of Bone Cells

There are three major types of bone cells: *osteoblasts* (bone-forming cells), *osteoclasts* (bone-resorbing cells), and *osteocytes* (mature bone cells). All bone surfaces are covered with a continuous layer of many osteoblasts and fewer osteoclasts. Both of these cells are critical to the survival of bone and its structure.

Osteoblasts are small cells in the inner layer of the periosteum and endosteum that secrete an important organic part of the ground substance of bone called *osteoid*. Collagen strands in the osteoid matrix serve as a framework for the deposition of calcium and phosphate compounds—the hydroxyapatite crystals of mineralized bone.

Stem cells in the endosteum and central canals form **osteoclasts.** These giant cells have many nuclei and break down old bone by dissolving the bone minerals. They are derived not from osteoblasts but from *monocytes*, cells of the immune system responsible for phagocytosis of foreign substances in the body. This function explains their phagocytic "bone-eating" activities.

Each 24-hour period sees osteoblast activity alternating with osteoclast activity so that bone is constantly being resorbed and then reformed. This continuous process is sometimes called *bone remodeling*. **Osteocytes** are mature (nondividing) bone cells that develop from osteoblasts within the lacunae, surrounded by bone matrix (Figure 8-5). The ways in which these bone cells work together is described for you under *Development of Bone*, later in this chapter.

BONE MARROW

Bone marrow is a specialized type of soft connective tissue called **myeloid tissue.** It serves as the site for the production of blood cells and is found in the medullary cavities of certain long bones and in the spaces within spongy bone. In infants and small children, virtually all of the bones contain *red marrow* (so named because it produces red blood cells, *erythrocytes*). As we age, the red marrow is gradually

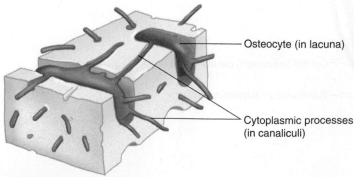

▲ **FIGURE 8-5   Osteocyte.** Sketch showing osteocytes trapped inside hollow lacunae within hard bone matrix.

Osteocyte (in lacuna)

Cytoplasmic processes (in canaliculi)

replaced by fatty *yellow marrow*. At this time, marrow cells are saturated with fat and can no longer produce red blood cells. With advancing age, yellow marrow becomes almost rust-colored, is less fatty, and resembles the consistency of gelatin.

In an adult, the main bones still containing red marrow include the sternum, ribs, bodies of the vertebrae, ends of the humerus, pelvis, and femur. The sternum plays a major role in blood cell formation, particularly after the age of 18. However, by the age of 40, the marrow in the sternum, ribs, pelvis, and vertebrae is composed of equal amounts of *hematopoietic* (blood-forming) tissue and fat. During times of decreased blood supply (e.g., chronic blood loss, anemia, and certain diseases), yellow marrow in an adult can revert back to red marrow. *Bone marrow transplants* can be a lifesaving treatment if the bone marrow of a patient is severely damaged and unable to perform its functions. If there is no adverse immune reaction, the donor cells may establish a colony of new, healthy tissue in the bone marrow of the recipient.

REGULATION OF BLOOD CALCIUM LEVELS

The bones of the skeletal system are a storehouse for nearly 98% of our body's calcium reserves. For this reason, bones play a key role in maintaining a blood calcium level within a very narrow (homeostatic) range. It is the balance between the deposition of new bone by osteoblasts and the breakdown and resorption of bone matrix by osteoclasts that helps regulate blood calcium levels. Homeostasis of the calcium concentration is essential for bone formation and reformation (described below). As you will see later in this book, homeostasis of blood calcium level is also required for normal blood clotting, transmission of nerve impulses, and contraction of skeletal and cardiac muscle.

Regulation of blood calcium levels involves the secretion of *parathyroid hormone* by the parathyroid glands and *calcitonin hormone* by the thyroid gland (Figure 8-6).

When your blood calcium level drops *below* the body's homeostatic set point, **parathyroid hormone (PTH)** induces osteoclasts to break down bone at a faster rate than normal. This activity releases more calcium into your blood. At the same time, PTH increases the recovery of calcium in urine from your kidneys. Another effect of PTH is to stimulate vitamin D synthesis, which increases the efficiency of absorption of calcium from your intestine. However, if your blood calcium level increases *above* the homeostatic set point, osteoclast activity is suppressed and your blood calcium levels fall. This reduces the breakdown of bone matrix and the amount of calcium in your blood. Parathyroid hormone is the most critical factor in the homeostasis of our blood calcium levels.

Calcitonin, a protein hormone produced by the thyroid gland, is produced in response to high blood calcium levels. It functions to stimulate bone deposition by osteoblasts, while also inhibiting the bone-dismantling activity of osteoclasts. Calcitonin thus plays a part in the homeostasis of our blood calcium levels, but its importance is much less than that of parathyroid hormone.

DEVELOPMENT OF BONE

When an infant's skeleton begins to develop—well before birth—it consists not of bones but of cartilage and fibrous structures *shaped* like bones (Figure 8-7). Gradually, these cartilage "models" are transformed into real bone through the process of **ossification**—the laying down of calcium salts by the osteoblasts. However, bone growth and development is a give-and-take process. The growing bone is constantly remodeled from birth through adulthood by the bone-forming osteoblasts and the bone-resorbing osteoclasts. The combined action of osteoblasts and osteoclasts sculpts bones into their final shape in a process called **osteogenesis.** Thus, bones can respond to stress or injury by changing their size, shape, and density. People who stress their skeletons, such as athletes, ballet dancers, and physical laborers, often have denser, stronger bones than those of less active people (Box 8-1).

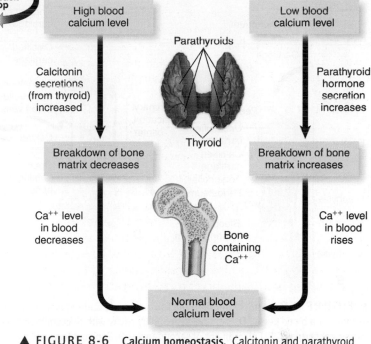

▲ **FIGURE 8-6** **Calcium homeostasis.** Calcitonin and parathyroid hormone have antagonistic (opposing) effects that help to maintain a homeostatic balance of calcium in the blood.

BOX 8-1   Sports & Fitness

Exercise and Bone Density

Walking, jogging, and other forms of exercise subject bones to stress. They respond by laying down more collagen fibers and mineral salts in the bone matrix. This, in turn, makes bones stronger. But inactivity and lack of exercise tend to weaken bones because of decreased collagen formation and excessive calcium withdrawal. To prevent these changes, as well as many others, astronauts regularly perform special exercises in space because of the lack of gravity. Regular weight-bearing exercise is also an important part of the prevention and treatment of osteoporosis.

Recent research shows conclusively that weight-bearing exercise *at any age* not only increases bone density but also reduces the risk of osteoporosis. However, not all exercise is equal when it comes to building strong, healthy bones. To maintain strong bones, the magnitude, rate, and frequency of strain during exercise all play important roles in developing greater bone density.

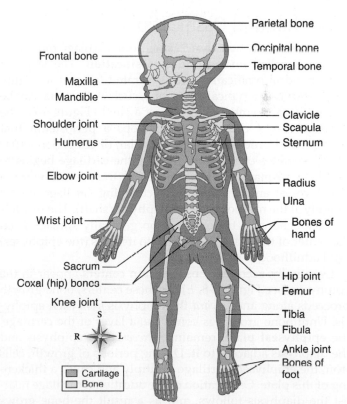

▲ **FIGURE 8-7** **Bone development.** Diagram showing osseous development at birth.

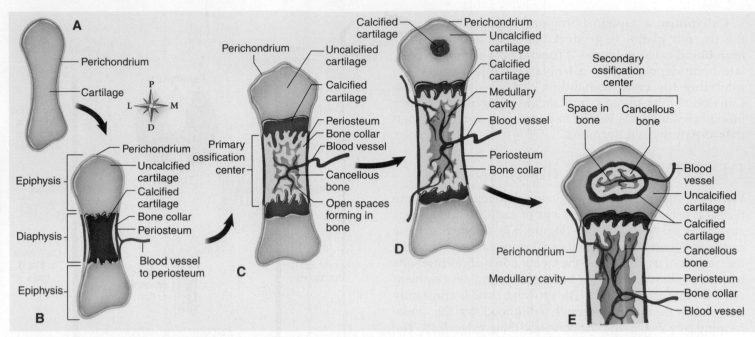

▲ **FIGURE 8-8** **Endochondral bone formation. A,** Cartilage model. **B,** Bone collar formation. **C,** Development of the primary ossification center and entrance of a blood vessel. **D,** Prominent medullary cavity with thickening and lengthening of the collar. **E,** Development of secondary ossification centers in epiphyseal cartilage. **F,** Enlargement of secondary ossification centers, with bone growth proceeding toward the diaphysis from each end. **G,** With cessation of bone growth, the lower and then the upper epiphyseal plates disappear (only the epiphyseal lines remain).

Endochondral Ossification

Most bones of the body are formed from cartilage models in a process called **endochondral ossification.** The steps of endochondral ossification are illustrated for you in Figure 8-8. Notice that a typical long bone, such as the tibia, can be seen early in an embryo's life (refer again to Figure 8-7). The cartilage model of the tibia develops a periosteum that enlarges and produces a ring, or collar, of bone. Soon after the appearance of the ring of bone, the cartilage begins to calcify. A **primary ossification center** then forms when a blood vessel enters the rapidly changing cartilage model roughly at the midpoint of the diaphysis (shaft). The result is that endochondral bone formation generally spreads from the center of the shaft (diaphysis) to its ends (the epiphyses) until adulthood is reached.

Eventually, **secondary ossification centers** appear in the epiphyses (see Figure 8-8, *E*). At these centers, bone growth proceeds along and *toward* the diaphysis *from* each epiphysis. Until bone growth is complete, a layer of the cartilage, the **epiphyseal plate,** remains *between* each epiphysis and the diaphysis adjacent to it. During periods of growth, cells from the epiphyseal cartilage multiply and cause a thickening of this plate. Ossification of the additional cartilage nearest the diaphysis follows, and as a result the bone grows longer. When the cartilage cells of epiphyseal cartilage stop multiplying and the cartilage has become completely ossified, bone growth ends. Note, however, that it is the epiphyseal plate that allows the diaphysis of a long bone to increase

in length. As long as the epiphyseal plate of the long bones remains active, bone growth will take place and your height will increase.

The number of ossification centers increases with age. Bone growth stops when the epiphyseal cartilage cells stop multiplying and the cartilage has become completely ossified. At this point, the epiphyseal cartilage disappears. The epiphyseal plate, replaced by bone, thus forms an **epiphyseal line.** Bone is now continuous between the epiphyses and the diaphysis. The epiphyseal plate area, however, is susceptible to fracture in young and preadolescent adults (Figure 8-9).

Intramembraneous Ossification

Unlike most bones, most flat bones are formed within *fibrous membranes*, rather than by replacing a cartilage framework. For this reason, they are called **membrane bones.** This process of their formation is called **intramembranous ossification** (Figure 8-10). It is responsible for the formation of parts of the lower jaw, the clavicle, and all the bones of the skull. The flat bones of the skull, for example, begin to take shape when groups of stem cells within the membrane differentiate into osteoblasts. These clusters form *ossification centers* that secrete matrix that includes collagenous fibers. In time, the ground substance and collagenous fibers make up the organic bone matrix in which calcification takes place. As calcification of bone matrix continues, the trabeculae appear and join in a network to form spongy bone. Only later will the core layer of spongy bone (the diploë) be

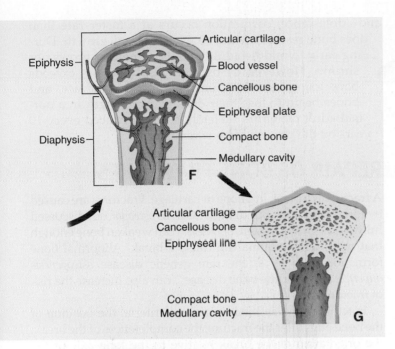

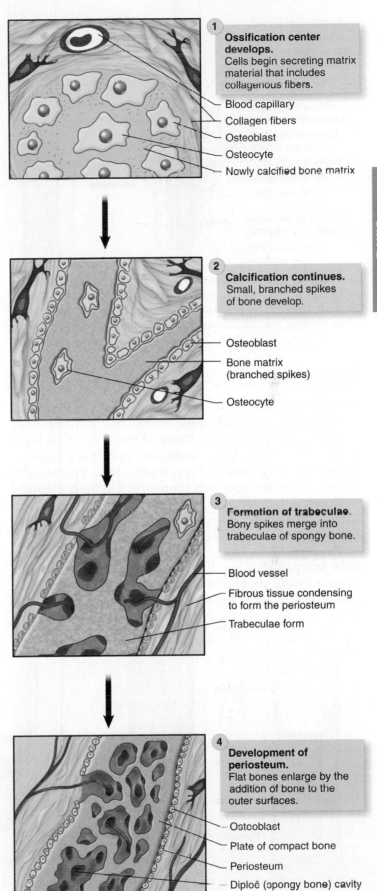

1 **Ossification center develops.** Cells begin secreting matrix material that includes collagenous fibers.

- Blood capillary
- Collagen fibers
- Osteoblast
- Osteocyte
- Newly calcified bone matrix

2 **Calcification continues.** Small, branched spikes of bone develop.

- Osteoblast
- Bone matrix (branched spikes)
- Osteocyte

3 **Formation of trabeculae.** Bony spikes merge into trabeculae of spongy bone.

- Blood vessel
- Fibrous tissue condensing to form the periosteum
- Trabeculae form

4 **Development of periosteum.** Flat bones enlarge by the addition of bone to the outer surfaces.

- Osteoblast
- Plate of compact bone
- Periosteum
- Diploë (spongy bone) cavity

▲ FIGURE 8-10 **Intramembranous ossification.** A simplified overview of the steps in forming flat bones.

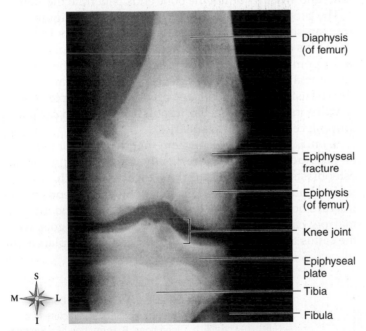

▲ FIGURE 8-9 **Epiphyseal fracture.** Radiograph showing an epiphyseal fracture of the distal end of the femur in a young athlete. Note the separation of the diaphysis and epiphysis at the level of the growth plate.

covered *on each side* by internal and external plates of compact bone. Thus a flat bone grows by adding bone to its outer boundary.

QUICK CHECK

10. Name the three major types of bone cells and give their functions.
11. Name the two types of bone marrow. Why is there a difference?
12. What are the two types of ossification?
13. What is an epiphyseal plate? What is its significance?

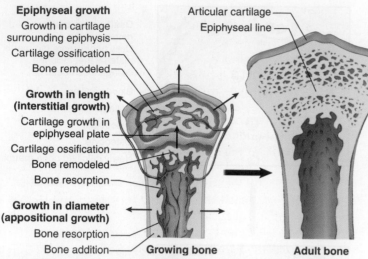

Epiphyseal growth
Growth in cartilage surrounding epiphysis
Cartilage ossification
Bone remodeled

Growth in length (interstitial growth)
Cartilage growth in epiphyseal plate
Cartilage ossification
Bone remodeled
Bone resorption

Growth in diameter (appositional growth)
Bone resorption
Bone addition

Articular cartilage
Epiphyseal line

Growing bone

Adult bone

▲ **FIGURE 8-11** **Bone remodeling.** Bone formation on the outside of the shaft coupled with bone resorption on the inside increases the bone's diameter. Endochondral growth during bone remodeling increases the length of the diaphysis and causes the epiphyses to enlarge.

BONE REMODELING

Both osteoblasts and osteoclasts are involved in the growth of bone diameter. Osteoclasts enlarge the diameter of the medullary cavity by eating away at the bone of its inside walls. At the same time, osteoblasts from the periosteum build new bone around the outside of the bone. This remodeling activity, as we have seen, is important in the homeostasis of blood calcium levels. It also permits our bones to grow in length and diameter, change their overall shape, and alter the size of the marrow cavity (Figure 8-11).

The formation of bone tissue continues long after bones have stopped growing. It is a lifelong process of bone formation (ossification) and bone destruction (reabsorption), both taking place at the same time. During our childhood and adolescence, ossification occurs at a faster rate than does bone resorption and so we have bone growth. During early to middle adulthood, bone neither grows nor shrinks. However, by the late 30s, bone loss exceeds bone gains, especially on the endosteal surface, and bones begin to lose their density. Amazingly, in a normal adult the entire skeleton is fully replaced every 10 years by the process of bone remodeling.

REPAIR OF BONE FRACTURES

A **fracture** is a break in a bone or cartilage. Fractures are caused by forceful trauma. But disease (such as *osteoporosis*, discussed fully online in *Mechanisms of Disease*) can weaken bone enough that only minor trauma can cause a break. Abnormal bone formation such as in the rare genetic disease *osteogenesis imperfecta* (the "brittle bone disease") can also increase the risk of fractures.

Fractures are classified by several criteria: the position of the bone ends after the fracture, the completeness of the break, the orientation of the break relative to the long axis of the bone, and whether or not the bone ends penetrate the skin.

The process of healing a fracture is shown in Figure 8-12. You can see that damage to blood vessels occurs immediately. This damage creates a hemorrhage and a pool of blood at the point of injury called a *fracture hematoma*. The fracture hematoma quickly develops a mesh of fibrin filaments and then transforms into *granulation tissue*. This complex tissue contains inflammatory cells, fibroblasts, bone- and cartilage-forming cells, and new capillaries.

Soon, islands of cartilage form that help anchor the ends of the fractured bone more firmly. Growing numbers of osteoblasts continue the healing process with the formation of bony *callus* tissue. This binds the broken ends of the fracture both on the outside surface and along the internal marrow cavity. Now healing can proceed. If all goes well, the callus tissue serves as a model for the bone replacement

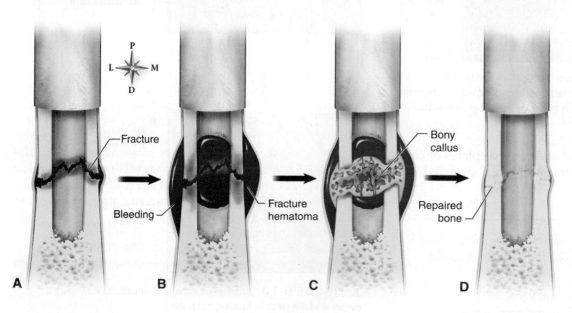

Fracture

Bleeding

Fracture hematoma

Bony callus

Repaired bone

A   **B**   **C**   **D**

◀ **FIGURE 8-12** **Bone fracture healing. A,** Fracture of a long bone. **B,** Formation of a fracture hematoma. **C,** Formation of internal and external bony callus. **D,** Bone remodeling complete.

that follows. Several synthetic bone repair materials (made of calcium in a spongy, porous matrix) are now available that help to stabilize the fracture site. These "patching" materials also enhance the movement of bone-repairing cells and nutrients to the site of repair. The patches of artificial bone degrade naturally in the body and do not require surgical removal.

When uncomplicated fractures occur in healthy children and young adults, the healing process takes about 8 weeks and often results in a repair that is all but impossible to detect just 6 months after the injury. However, repair of a bone fracture may be complicated by older age, especially in those people with underlying disease or other health problems such as osteoporosis, diabetes, infection, or a diminished blood supply to the injured area.

A&P CONNECT

What are the types of bone fractures? How are they treated? Learn more about bone fractures in **Bone Fractures** online at **A&P Connect**.

evolve

CARTILAGE

Cartilage is a connective tissue with three specialized subtypes: *hyaline cartilage, elastic cartilage,* and *fibrocartilage* (see Table 6-4 for photomicrographs of the different types of cartilage). The three cartilage subtypes differ from one another largely by the amount of matrix present. They also vary by the relative amounts of elastic and collagenous fibers embedded in the matrix. Hyaline cartilage is the most abundant type; both elastic cartilage and fibrocartilage are probably modifications of this type. Collagen fibers are most numerous in fibrocartilage, providing it with the greatest tensile strength. Elastic cartilage matrix has both elastic fibers and collagen fibers, providing it with both elasticity and firmness.

Hyaline cartilage is somewhat transparent, with a bluish opal-like cast. It is sometimes called "gristle." In the embryo, hyaline cartilage forms *centers of chondrification.* In this process, production of the collagen matrix separates and isolates the cells into depressions called *lacunae.* Like bone, the organic matrix is a mixture of ground substance and collagen fibers. Hyaline cartilage covers the articulating surfaces of long bones and forms the cartilage that connects the anterior ends of the ribs with the sternum. It also forms the cartilage rings of the trachea, the bronchi of the lungs, and the tip of the nose. **Elastic cartilage** gives flexible but definite form to the external ear, the epiglottis (covering the opening of the trachea), and the auditory (eustachian) tubes that connect the middle ear and the nasal cavity. **Fibrocartilage** has less matrix and more fibrous elements than the other two types of cartilage. It is strong, rigid, and most often found within regions of dense connective tissue. Fibrocartilage is found in the symphysis pubis, in the intervertebral disks, and near the points of attachment of large tendons to bone.

Like bone, cartilage has many collagen fibers reinforcing its matrix and also has more extracellular substance than it has cells. However, unlike bone, the fibers of cartilage are embedded in a firm gel instead of a calcified cement base. This gel imparts a plastic-like flexibility to the cartilage, which is not possible with solid, inflexible bone.

Another difference between bone and cartilage is that bone has an abundant supply of blood vessels and nerves, whereas cartilage is largely **avascular** (without blood vessels). In effect, there are no canals and blood vessels weaving their way through the cartilage matrix. As a result, nutrients and oxygen can reach the scattered, isolated **chondrocytes** (cartilage cells) only through the relatively slow process of diffusion. Nutrients and gasses thus diffuse slowly into the matrix gel from capillaries in the fibrous covering **(perichondrium)** of the cartilage, or from synovial fluid, in the case of articular cartilage of joints.

Cartilage is an excellent skeletal support tissue in the developing embryo. It forms rapidly and is rigid but flexible. A majority of the bones that eventually form the axial and appendicular skeleton (see Figure 8-7) first appear as cartilage "models." After birth, there is a decrease in the total amount of cartilage tissue present in the body. Recall, of course, that cartilage permits growth in the length of long bones and is largely responsible for their adult shape and size.

The gristlelike (cartilaginous) nature of cartilage permits it to bear great weight where it covers the articulating surfaces of bones, or to serve as a shock-absorbing pad between the vertebrae of the spine. In other areas, such as the external ear, nose, or respiratory passages, cartilage provides a strong but pliable support structure that resists deformation or collapse of tubular passageways.

QUICK CHECK

14. What is meant by "bone remodeling?" What is its significance?
15. Describe briefly how bones are repaired.
16. List the different types of cartilage and their functions.

Cycle of LIFE ⚪

Throughout this chapter, we've focused on the changes that occur in cartilage and bone tissues from before birth and throughout the remainder of your life. For example, we have outlined the basic processes by which the soft cartilage and membranous skeleton become ossified over a period of years. In fact, by the time a person has become a young adult in the mid-20s, the skeleton has become fully ossified. A few areas of softer tissues—the cartilaginous areas of the nose and ears, for example—may continue to grow and ossify slowly even through advancing age. This explains why noses and ears get longer and larger as we age!

Changes in skeletal tissue that occur during adulthood usually result from specific conditions. For example, the

The BIG Picture

Our skeletal tissues influence many important body functions crucial to our overall health and survival. The protective and support functions of our skeletons are immediately obvious. Bone and cartilage tissues working together are organized to protect the body and its internal organs from injury. In addition to providing a supporting and protective framework, these tissues also contribute to the shape, alignment, and positioning of the body and all its parts. As we will see in the next chapter, it is the organization of skeletal tissues into bones and joints that allow us to conduct coordinated and meaningful movement.

Skeletal tissues provide many homeostatic functions. For example, bone remodeling permits changes in both their structure and related functions throughout our lives. And skeletal tissues play a vital role in mineral storage and release. Regulation of blood calcium levels is important in such diverse areas as nerve transmission, muscle contraction, and normal clotting of blood. Because bones contain red marrow, they also are vital in hematopoiesis, or blood cell formation. This function ties skeletal tissues to such diverse homeostatic functions as the regulation of body pH and the transport of respiratory gases and vital nutrients. When viewed in this broad and interrelated functional context, you can easily see the "connectedness" that unites what seem to be isolated body structures into a "big picture" required for our health and survival.

mechanical stress of weight-bearing exercise can cause dramatic increases in the density and strength of bone tissue. Pregnancy, nutritional deficiencies, and illness can all cause loss of bone density. This is sometimes accompanied by loss of structural strength. As we age, degeneration of bone and cartilage tissue becomes apparent. The replacement of hard bone matrix by softer connective tissue results in a loss of strength that makes us more susceptible to injury. This is especially true in older women who suffer from osteoporosis—an abnormal degeneration of bone often seen in women when estrogen level decreases. Although osteoporosis is not reversible, even light exercise by elderly persons can counteract some of the skeletal tissue degeneration we experience normally as we age. ●

MECHANISMS OF DISEASE

Probably the most common disorder of skeletal tissue is the bone fracture. Bone fractures, which can be either *compound* (the broken bone penetrates through the skin) or *simple* (the broken bone does not penetrate through the skin) are often caused by physical injury. However, they also can result from certain bone or body disorders. Tumors of the bone (for example, *osteosarcoma*) or of cartilage (for example, *chondrosarcoma*) also may occur. Metabolic bone diseases, such as *osteoporosis*, typically occur because of degenerative processes.

evolve Find out more about these bone tissue diseases and others online at *Mechanisms of Disease: Skeletal Tissues.*

LANGUAGE OF SCIENCE AND MEDICINE *continued from page 137*

intramembranous ossification (in-trah-MEM-brah-nus os-i-fi-KAY-shun)
[*intra-* within, *-membran-* thin skin, *-ous* characterized by, *os-* bone, *-fication* make]

irregular bone
[*ir-* not, *-regula* rule, *-ar* relating to]

lacuna (lah-KOO-nah)
[*lacuna* pit] *pl.,* lacunae (lah-KOO-nay)

lamella (lah-MEL-ah)
[*lam-* plate, *-ella* little] *pl.,* lamellae (lah-MEL-eye)

long bone

medullary cavity (MED-oo-lair-ee KAV-ih-tee)
[*medulla* middle, *-ary* relating to, *cav-* hollow, *-ity* state]

membrane bone
[*membran-* thin skin]

myeloid tissue (MY-eh-loyd TISH-yoo)
[*myel-* marrow, *-oid* of or like, *tissu-* fabric]

ossification (os-i-fi-KAY-shun)
[*os-* bone, *-fication* make]

osteoblast (OS-tee-oh-blast)
[*osteo-* bone, *-blast* bud]

osteoclast (OS-tee-oh-klast)
[*osteo-* bone, *-clast* break]

osteocyte (OS-tee-oh-syte)
[*osteo-* bone, *-cyte* cell]

osteogenesis (os-tee-oh-JEN-eh-sis)
[*osteo-* bone, *-genesis* origin]

osteon (AHS-tee-on)
[*osteo-* bone, *-on* unit]

parathyroid hormone (PTH) (pair-ah-THYE-royd HOR-mohn)
[*para-* besides, *-thyr-* shield, *-oid* like, *hormon-* excite]

perichondrium (pair-i-KON-dree-um)
[*peri-* around, *-chondr-* cartilage, *-um* thing] *pl.,* perichondria (pair-i-KON-dree-ah)

periosteum (pair-ee-OS-tee-um)
[*peri-* around, *-osteum* bone]

primary ossification center (PRYE-mayr-ee os-i-fi-KAY-shun SEN-tur)
[*primary* first order, *os-* bone, *-fication* make]

secondary ossification center (SEK-on-dair-ee os-i-fi-KAY-shun SEN-tur)
[*secondary* second order, *os-* bone, *-fication* make]

sesamoid bone (SES-ah-moyd)
[*sesam-* sesame seed, *-oid* like]

short bone
[*short-* stunted]

spongy bone
[*spongia-* sponge]

trabecula (trah-BEK-yoo-la)
[*trab-* beam, *-ula* little] *pl.,* trabeculae (trah-BEK-yoo-lay)

CHAPTER SUMMARY

To download an MP3 version of the chapter summary for use with your iPod or other portable media player, access the Audio Chapter Summaries *online at http://evolve.elsevier.com.*

HINT *Scan this summary after reading the chapter to help you reinforce the key concepts. Later, use the summary as a quick review before your class or before a test.*

INTRODUCTION

A. Bones give your body form, provide you with support and protection, and allow you to move; also serve as reservoirs for calcium and phosphorus and produce the major cellular components of the blood

TYPES OF BONES

A. Types of bones (Figure 8-1)
  1. Long—long shafts and often uniquely shaped ends that form complex joints; femur
  2. Short—cube or box shaped; carpals, tarsals
  3. Flat—relatively flat; sternum
  4. Irregular—odd shaped and may appear in groups; vertebral bones
  5. Sesamoid—develop in the tendons close to joints; patella
B. Bone tissue
  1. Compact bone—dense and solid in appearance
  2. Spongy (cancellous) bone—has open spaces partially filled by a network of fine, needlelike struts
C. Major parts of a long bone (Figure 8-2)
  1. Diaphysis—hollowed shaft
  2. Epiphyses—enlarged ends of a long bone
  3. Articular cartilage—a thin layer of hyaline cartilage; covers and protects the articular (joint) surfaces
  4. Periosteum—dense, white fibrous membrane covers the bone's exterior (except at the joint surfaces, where the articular cartilage forms the covering)
  5. Medullary (marrow) cavity—hollow space throughout the diaphysis of a long bone
  6. Endosteum—thin, fibrous membrane lining the medullary cavity and the spaces of spongy bone
D. Major parts of flat bone (Figure 8-3)
  1. A layer of spongy bone is sandwiched between an internal table and an external table made of compact bone
  2. Covered in a periosteum and the inner spaces are lined with endosteum
  3. Red marrow fills the spaces of many flat bones

BONE TISSUE

A. Bones, like other connective tissues, consists of cells, fibers, and an extracellular matrix
  1. Bone matrix is hard; also contains many fibers of collagen
B. Composition of bone matrix
  1. Extracellular bone matrix is divided into two major chemical components:
    a. Inorganic salts
    b. Organic salts

  2. Inorganic salts of the bone matrix—hardness of bone is due to rocklike crystals of hydroxyapatite
    a. Other mineral components of the bone matrix include magnesium, sodium, sulfate, and fluoride
  3. Organic components of bone matrix—collagen fibers and a mixture of protein and polysaccharides (ground substance)
    a. Ground substance provides support and adhesion for cellular and fibrous elements

MICROSCOPIC STRUCTURE OF BONE

A. Compact bone—makes up about 80% of your body's total bone mass; it contains many vertical, cylinder-shaped units called osteons or haversian systems (Figure 8-4)
B. Osteon is composed of four structures:
  1. Lamellae—concentric, cylinder-shaped layers of calcified matrix
  2. Lacunae—small spaces filled with fluid where the bone cells live
  3. Canaliculi—very fine canals that radiate in all directions from the lacunae
  4. Central canal—extends lengthwise through each osteon and houses blood vessels, lymphatic vessels, and nerves
C. Spongy (cancellous) bone—needlelike branches forming bony trabeculae; sandwiched between two layers of compact bone (Figure 8-3)
  1. Arrangement of the trabeculae greatly enhances the strength of a bone
D. Types of bone cells—three major types
  1. Osteoblasts—small cells in the inner layer of the periosteum that secrete an important organic part of the ground substance
  2. Osteoclasts—break down old bone by dissolving the bone minerals
  3. Osteocytes—mature bone cells (Figure 8-5)

BONE MARROW

A. Bone marrow—specialized type of soft connective tissue called myeloid tissue
  1. Red marrow—produces red blood cells
  2. Yellow marrow—saturated with fat and can no longer produce red blood cells
B. Bone marrow transplant—lifesaving treatment if the bone marrow of a patient is severely damaged and unable to perform its functions

REGULATION OF BLOOD CALCIUM LEVELS

A. Bones of the skeletal system are a storehouse for nearly 98% of our body's calcium reserves
B. Homeostasis of the calcium concentration is essential for bone formation and reformation
C. Regulation of blood calcium levels involves the secretion of parathyroid hormone and calcitonin (Figure 8-6)
  1. Parathyroid hormone—induces osteoclasts to break down bone at a faster rate than normal; releases calcium into the blood
  2. Calcitonin—produced in response to high blood calcium levels
    a. Functions to stimulate bone deposition by osteoblasts, while also inhibiting the bone-dismantling activity of osteoclasts

DEVELOPMENT OF BONE

A. Cartilage "models" are transformed into real bone through the process of calcification (Figure 8-7)
B. Osteogenesis—combined action of osteoblasts and osteoclasts sculpts bones into their final shape
C. Most bones of the body are formed from cartilage models in a process called endochondral ossification (Figure 8-8)
   1. Primary ossification center—forms when a blood vessel enters the rapidly changing cartilage model roughly at the midpoint of the diaphysis
   2. Secondary ossification center—bone growth proceeds along and down toward the diaphysis from each epiphysis
   3. Epiphyseal plate—a layer of the cartilage between each epiphysis and the diaphysis
D. Intramembranous ossification—bones are formed within fibrous membranes, rather than by replacing a cartilage framework (Figure 8-10)
   1. Responsible for the formation of parts of the lower jaw, the clavicle, and all the bones of the skull
   2. Flat bone grows by adding bone to its outer boundary

BONE REMODELING

A. Both osteoblasts and osteoclasts are involved in the growth of bone diameter
B. The formation of bone tissue continues long after bones have stopped growing
C. During childhood and adolescence, ossification occurs at a faster rate than does bone resorption (Figure 8-11)

REPAIR OF BONE FRACTURES

A. Fracture—break in a bone or cartilage
B. Fractures are classified by several criteria:
   1. Position of the bone ends after the fracture
   2. Completeness of the break
   3. Orientation of the break relative to the long axis of the bone
   4. Whether or not the bone ends penetrate the skin
C. Following a fracture, a complex repair process begins
   1. Healing process takes about 8 weeks (Figure 8-12)
      a. Fracture hematoma—a hemorrhage and a pool of blood at the point of injury
      b. Granulation tissue—contains inflammatory cells, fibroblasts, bone- and cartilage-forming cells, and new capillaries
      c. Bony callus tissue—binds the broken ends of the fracture both on the outside surface and along the internal marrow cavity

CARTILAGE

A. Cartilage—connective tissue with three specialized subtypes:
   1. Hyaline
   2. Elastic
   3. Fibrocartilage
B. Hyaline cartilage—"glassy" (like gristle) and somewhat transparent, with a bluish opal-like cast
   1. Covers the articulating surfaces of long bones; forms the cartilage that connects the anterior ends of the ribs with the sternum, cartilage rings of the trachea, the bronchi of the lungs, and the tip of the nose

C. Elastic cartilage—gives flexible but definite form to the external ear, the epiglottis, and the auditory tubes
D. Fibrocartilage—less matrix and more fibrous elements; strong, rigid, and most often found within regions of dense connective tissue
   1. Found in the symphysis pubis and intervertebral disks
E. Features of cartilage
   1. Many collagen fibers reinforcing its matrix
   2. Fibers are embedded in a firm gel instead of a calcified cement base
   3. Cartilage is avascular
   4. Chondrocytes—cartilage cells
   5. Perichondrium—fibrous covering of cartilage
   6. Cartilage is an excellent skeletal support tissue in the developing embryo
      a. Gristlelike nature of cartilage permits it to bear great weight

REVIEW QUESTIONS

 HINT *Write out the answers to these questions after reading the chapter and reviewing the Chapter Summary. If you simply think through the answer without writing it down, you won't retain much of your new learning.*

1. Describe the microscopic structure of bone and cartilage.
2. Describe the structure of a long bone.
3. Explain the functions of the periosteum.
4. Describe the two principal chemical components of extracellular bone.
5. List and discuss each of the major anatomical components that together constitute an osteon.
6. Compare and contrast the three major types of cells found in bone.
7. Discuss and discriminate between the sequence of steps characteristic of fracture healing.
8. Compare and contrast the basic structural elements of bone and cartilage.
9. Compare the structure and function of the three types of cartilage.

CRITICAL THINKING QUESTIONS

 HINT *After finishing the Review Questions, write out the answers to these items to help you apply your new knowledge. Go back to sections of the chapter that relate to items that you find difficult.*

1. Compare and contrast bone formation in intramembranous and endochondral ossification.
2. Explain why a bone fracture along the epiphyseal plate may have serious implications in children and young adults.
3. During the aging process, adults face the issue of a changing skeletal framework. Describe these changes and explain how these skeletal framework changes affect the health of older adults.

continued from page 137

With the knowledge you have gained from reading this chapter, try to answer these questions about Eleanor's broken femur from the Introductory Story.

1. Which of the following conditions likely caused the extreme bone fragility that contributed to Eleanor's fracture?
   a. Osteosarcoma
   b. Rickets
   c. Osteoporosis
   d. Osteomyelitis

2. What type of bone did Eleanor fracture?
   a. Long bone
   b. Short bone
   c. Flat bone
   d. Irregular bone

3. What type of cartilage would you find covering the ends of that bone?
   a. Elastic cartilage
   b. Hylophil cartilage
   c. Fibrocartilage
   d. Hyaline cartilage

After her surgery, Eleanor gets a prescription from her doctor for calcitonin nasal spray.

4. What is the primary effect of that medication?
   a. Stimulates osteoclasts
   b. Stimulates osteoblasts
   c. Stimulates osteocytes
   d. All of the above

HINT *To solve these questions, you may have to refer to the glossary or index, other chapters in this textbook, A&P Connect, Mechanisms of Disease, and other resources.*

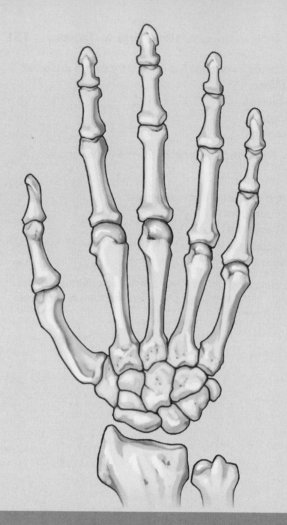

CHAPTER 9

Bones and Joints

CHAPTER OUTLINE

 Scan this outline before you begin to read the chapter, as a preview of how the concepts are organized.

STUDENT LEARNING OBJECTIVES

At the completion of this chapter, you should be able to do the following:

1. Distinguish between the axial skeleton and the appendicular skeleton.
2. Identify by position the major bones of the cranium and face.
3. Name the three types of vertebrae and the number of each.
4. List the bones that make up the thoracic cage.
5. List the bones of the pectoral girdle, the arm, and the hand.
6. List the bones of the pelvic girdle, the thigh, the leg, and the foot.
7. Describe the distinguishing features of a male and female skeleton.
8. Classify joints and specify basic groups according to function and structure.
9. Define the terms fibrous joints, cartilaginous joints, and synovial joints, and give examples of each.
10. Name and define four kinds of angular movements permitted by some synovial joints.
11. Define and give an example of the following: rotation, circumduction, pronation, supination.
12. Describe and differentiate between the following joints: shoulder, elbow, hip, knee, ankle, vertebral.

NO matter how much he turned the steering wheel, Robert's car kept sliding on the black ice and veered directly toward the guardrail. The impact from the car slamming against an immovable object was horrendous. Even though he was wearing his seat belt, Robert was thrown violently forward and sideways as the car and railing collided. After the car came to a jolting stop, Robert's first thought was to unbuckle his seat belt so he could get out of the car, but a severe pain in his chest kept him from moving. Thankfully, the paramedics were on the scene in a matter of minutes. They gently moved Robert from his car to a backboard and strapped his head in place.

In the emergency department, a nurse gave Robert the news: his spine was fractured at "T6"; his fourth and fifth ribs were cracked; and he had a broken mandible and a shattered tibia.

Before reading through this chapter, think about what you know about the bones that Robert has broken. After you finish reading, compare what you have written to the answers of the questions at the end of the chapter.

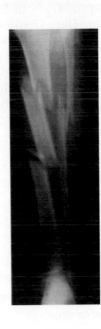

...

As we have seen, skeletal tissues are organized to form bones. Now we will see how bones are grouped to form the major subdivisions and joints of the skeletal system. Our rigid bones are surrounded by muscles and other soft tissues, which provide support and shape to our bodies. An understanding of the relationship of bones to each other and to other body structures will provide you with a foundation for understanding the functions of many other organ systems. Coordinated movement, for example, is possible only because of the way bones form joints and the way muscles are attached to provide motion in those joints. In addition, knowledge of the placement of bones within the soft tissues assists in locating and identifying other body structures.

The adult skeleton typically is composed of 206 separate bones. However, variation in the total number of bones is common. This may result from extra ribs, the formation of additional sesamoid bones, or the failure of certain small bones to fuse in the course of development. We will begin with the study of the divisions of the skeleton and end with the study of joints.

LANGUAGE OF SCIENCE AND MEDICINE

HINT *Before reading the chapter, say each of these terms out loud. This will help you avoid stumbling over them as you read.*

amphiarthrosis (am-fee-ar-THROH-sis)
[*amphi-* both sides, *-arthr-* joint, *-osis* condition] *pl.*, amphiarthroses (am-fee-ar-THROH-seez)

angular movement

appendicular skeleton (ah-pen-DIK-yoo-lar SKEL-eh-ton)
[*append-* hang upon, *-ic-* relating to, *-ul-* little, *-ar* relating to, *skeleto-* dried body]

arthroplasty (AR-throh-plas-tee)
[*arthr-* joint, *-plasty* surgical repair]

articular cartilage (ar-TIK-yoo-lar KAR-ti-lij)
[*artic-* joint, *-ul-* little, *-ar* relating to, *cartilag* cartilage]

articulation (ar-tik-yoo-LAY-shun)
[*artic-* joint, *-ul-* little, *-ation* state]

axial skeleton (AK-see-al SKEL-eh-ton)
[*axi-* axis, *-al* relating to, *skeleto-* dried body]

ball-and-socket joint

biaxial joint (bye-AK-see-al)
[*bi-* two, *-axi-* axle, *-al* relating to]

body

bursa (BER-sah)
[*bursa* purse] *pl.*, bursae (BER-see or BER-say)

capitulum (kah-PITCH-uh-lum)
[*capit-* head, *-ulum* little]

carpal bone (KAR-pul bohn)
[*carp-* wrist, *-al* relating to]

cartilaginous joint (kar-tih-LAJ-in-us joynt)
[*cartilag-* cartilage, *-in-* substance, *-ous* characterized by]

cervical vertebra (SER-vi-kal VER-teh-bra)
[*cervi-* neck, *-al* relating to, *vertebra* that which turns] *pl.*, vertebrae

circular movement

clavicle (KLAV-i-kul)
[*clavi-* key, *-cle* little]

coccyx (KOK-sis)
[*coccyx* cuckoo (beak)]

condyloid joints (KON-di-loyd)
[*condylo-* knuckle, *-oid* resembling]

coronoid fossa (KOR-uh-noyd FOSS-ah)
[*coron-* crown, *-oid* like, *fossa* ditch] *pl.*, fossae

coxal bone (KOK-sal bohn)
[*coxa-* hip, *-al* relating to]

cranium (KRAY-nee-um)
[*cranium* skull]

cribriform plate (KRIB-ri-form)
[*cribri-* sieve, *-form* shape]

dens (denz)
[*dens* tooth]

continued on page 181

UNIT 2

DIVISIONS OF THE SKELETON

Bones are organized into two major subdivisions: the *axial skeleton* and the *appendicular skeleton* (Figure 9-1). Eighty bones make up the **axial skeleton:** 74 bones that form the upright axis of the body and 6 tiny bones of the middle ear. The **appendicular skeleton** consists of 126 bones, which includes those of the shoulder girdles, arms, wrists, and hands, as well as those of the hip girdles, thighs, legs, ankles, and feet.

Descriptive information on bones and their basic markings is provided in Tables 9-1 and 9-2.

Axial Skeleton
Skull

Your skull comprises 28 irregularly shaped bones shown in Figures 9-2 through 9-6. Please refer to these figures often as you study the information below. (Note that the bones illustrated in these figures are color-coded so you can easily see how they look from different viewing angles.) Two major divisions, the **cranium** (brain case) and the **face,** make up the skull. The cranium comprises eight bones: the *frontal,* two *parietal,* two *temporal,* the

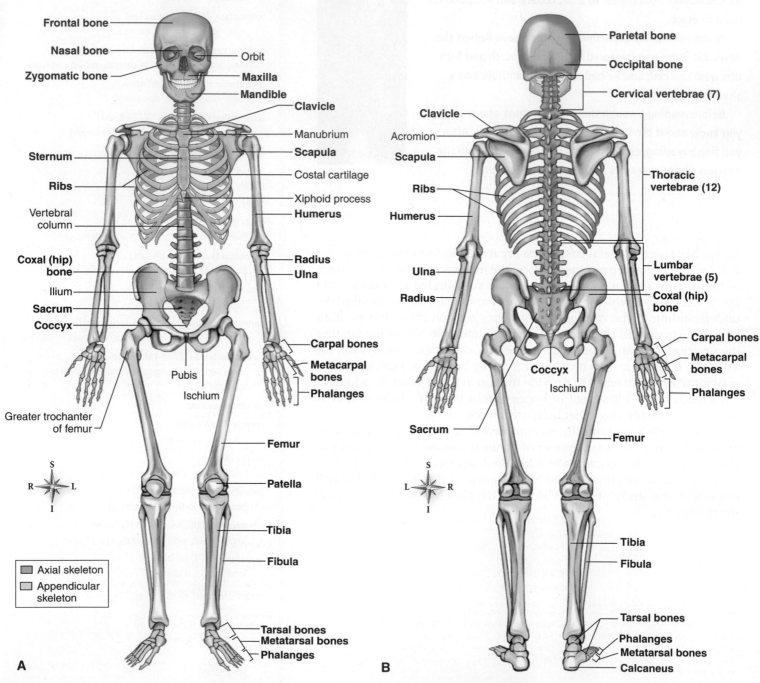

▲ **FIGURE 9-1** Skeleton. **A,** Anterior view. **B,** Posterior view.

occipital, the *sphenoid*, and the *ethmoid*. The 14 bones that form your face include two *maxillae*, two *zygomatic*, two *nasal*, the *mandible*, two *lacrimal*, two *palatine*, two inferior *nasal conchae* (turbinates), and the *vomer*. All the face

bones are paired except the mandible and vomer. However, most of the cranial bones are singled and unpaired (except for the parietal and temporal bones, which are paired). The frontal and ethmoid bones of the skull help shape the face and are considered among the facial bones. Each of these bones is described fully below.

Cranial Bones

The **frontal bone** forms our forehead and the anterior part of the top of our cranium (see Figure 9-2). It contains mucosa-lined, air-filled spaces, or **sinuses**—the *frontal sinuses*. These sinuses, like other sinuses in the sphenoid, ethmoid, and maxillae, are often called *paranasal sinuses*. This is because they have narrow channels that open into

TABLE 9-1 Bones of the Skeleton (206 Total)*

PART OF BODY	NAME OF BONE
Axial Skeleton (80 Bones Total)	
Skull (28 bones total)	
Cranium (8 bones)	Frontal (1) Parietal (2) Temporal (2) Occipital (1) Sphenoid (1) Ethmoid (1)
Face (14 bones)	Nasal (2) Maxillary (2) Zygomatic (2) Mandible (1) Lacrimal (2) Palatine (2) Inferior nasal conchae (turbinates) (2) Vomer (1)
Ear bones (6 bones)	Malleus (hammer) (2) Incus (anvil) (2) Stapes (stirrup) (2)
Neck	Hyoid bone (1)
Spinal column (26 bones total)	Cervical vertebrae (7) Thoracic vertebrae (12) Lumbar vertebrae (5) Sacrum (1) Coccyx (1)
Sternum and ribs (25 bones total)	Sternum (1) True ribs (14) False ribs (10)
Appendicular Skeleton (126 Bones Total)	
Upper extremities (including shoulder girdle) (64 bones total)	Clavicle (2) Scapula (2) Humerus (2) Radius (2) Ulna (2) Carpal bones (16) Metacarpal bones (10) Phalanges (28)
Lower extremities (including hip girdle) (62 bones total)	Coxal bones (2) Femur (2) Patella (2) Tibia (2) Fibula (2) Tarsal bones (14) Metatarsal bones (10) Phalanges (28)

*A variable number of small, flat, round bones known as *sesamoid bones* (because of their resemblance to sesame seeds) are found in tendons in which considerable pressure develops. Because the number of these bones varies greatly among individuals, only two of them, the patellae, have been counted among the 206 bones of the body. Generally, two of them can be found in each thumb (in the flexor tendon near the metacarpophalangeal and interphalangeal joints) and great toe, plus several others in the upper and lower extremities.

Sutural bones, the small islets of bone frequently found in some of the cranial sutures, also have not been counted in this list of 206 bones because of their variable occurrence.

TABLE 9-2 Terms Used to Describe Bone Markings

TERM	MEANING
Angle	Corner
Body	Main portion of a bone
Condyle	Rounded bump; usually fits into a fossa on another bone to form a joint
Crest	Moderately raised ridge; generally a site for muscle attachment
Epicondyle	Bump on or above a condyle; often gives the appearance of a "bump on a bump"; usually for ligament or tendon attachment
Facet	Small flat surface that forms a joint with another facet or flat bone
Fissure	Long, cracklike hole allowing for the passage of blood vessels and nerves
Foramen	Round hole for the passage of vessels and nerves (*pl.*, foramina)
Fossa	Depression; often receives an articulating bone (*pl.*, fossae)
Head	Distinct epiphysis on a long bone, separated from the shaft by a narrowed portion (or neck)
Line	Similar to a crest but not raised as much (is often rather faint)
Margin	Edge of a flat bone or flat portion of the edge of a flat area
Meatus	Tubelike opening or channel (*pl.*, meatus or meatuses)
Neck	Narrowed portion, usually at the base of a head
Notch	V-shaped depression in the margin or edge of a flat area
Process	Raised area or projection
Ramus	Curved portion of a bone, like a ram's horn (*pl.*, rami)
Sinus	Cavity within a bone
Spine	Similar to a crest but raised more; a sharp, pointed process; for muscle attachment
Sulcus	Groove or elongated depression (*pl.*, sulci)
Trochanter	Large bump for muscle attachment (larger than a tubercle or tuberosity)
Tuberosity	Oblong, raised bump, usually for muscle attachment; also called a *tuber*; a small tuberosity is called a *tubercle*

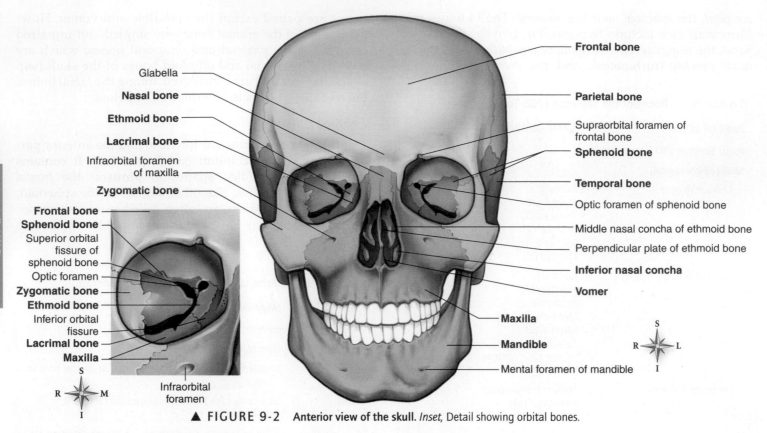

Glabella

Nasal bone

Ethmoid bone

Lacrimal bone

Infraorbital foramen
of maxilla

Zygomatic bone

Frontal bone
Sphenoid bone
Superior orbital
fissure of
sphenoid bone
Optic foramen
Zygomatic bone
Ethmoid bone
Inferior orbital
fissure
Lacrimal bone
Maxilla

Infraorbital
foramen

Frontal bone

Parietal bone

Supraorbital foramen of
frontal bone

Sphenoid bone

Temporal bone

Optic foramen of sphenoid bone

Middle nasal concha of ethmoid bone

Perpendicular plate of ethmoid bone

Inferior nasal concha

Vomer

Maxilla

Mandible

Mental foramen of mandible

▲ FIGURE 9-2   Anterior view of the skull. *Inset,* Detail showing orbital bones.

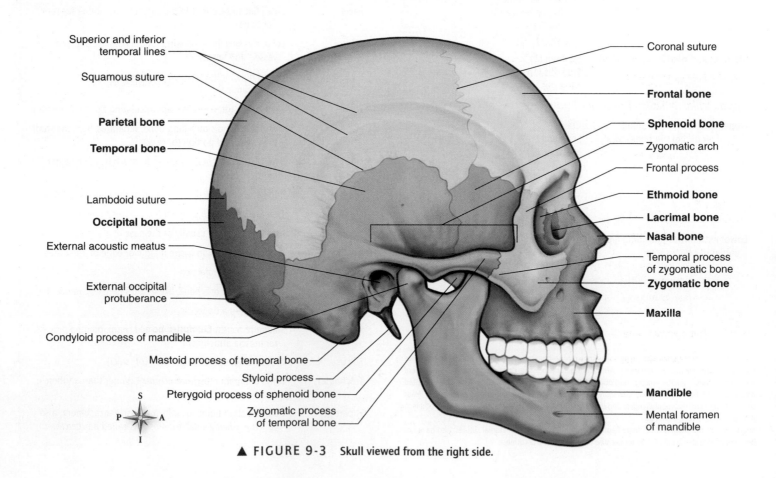

Superior and inferior
temporal lines

Squamous suture

Parietal bone

Temporal bone

Lambdoid suture

Occipital bone

External acoustic meatus

External occipital
protuberance

Condyloid process of mandible

Mastoid process of temporal bone

Styloid process

Pterygoid process of sphenoid bone

Zygomatic process
of temporal bone

Coronal suture

Frontal bone

Sphenoid bone

Zygomatic arch

Frontal process

Ethmoid bone

Lacrimal bone

Nasal bone

Temporal process
of zygomatic bone

Zygomatic bone

Maxilla

Mandible

Mental foramen
of mandible

▲ FIGURE 9-3   Skull viewed from the right side.

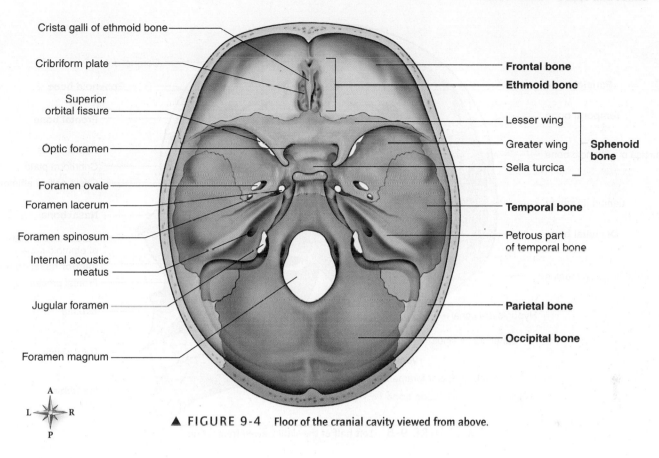

Crista galli of ethmoid bone

Cribriform plate

Superior orbital fissure

Optic foramen

Foramen ovale

Foramen lacerum

Foramen spinosum

Internal acoustic meatus

Jugular foramen

Foramen magnum

Frontal bone

Ethmoid bone

Lesser wing

Greater wing — **Sphenoid bone**

Sella turcica

Temporal bone

Petrous part of temporal bone

Parietal bone

Occipital bone

A
L · R
P

▲ **FIGURE 9-4**   Floor of the cranial cavity viewed from above.

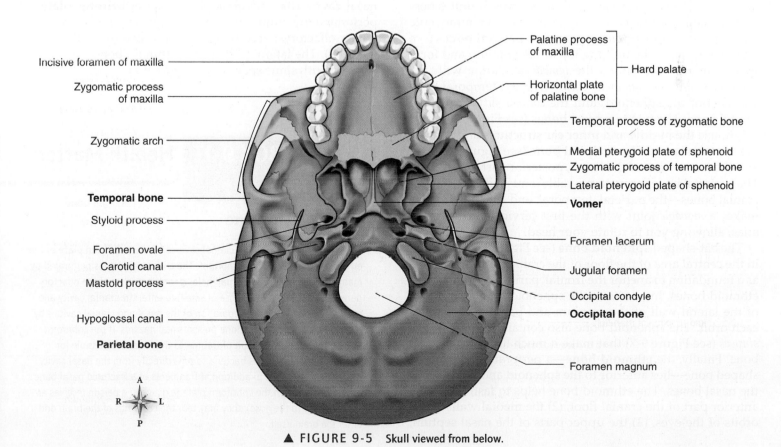

Incisive foramen of maxilla

Zygomatic process of maxilla

Zygomatic arch

Temporal bone

Styloid process

Foramen ovale

Carotid canal

Mastoid process

Hypoglossal canal

Parietal bone

Palatine process of maxilla

Horizontal plate of palatine bone — Hard palate

Temporal process of zygomatic bone

Medial pterygoid plate of sphenoid

Zygomatic process of temporal bone

Lateral pterygoid plate of sphenoid

Vomer

Foramen lacerum

Jugular foramen

Occipital condyle

Occipital bone

Foramen magnum

A
R · L
P

▲ **FIGURE 9-5**   Skull viewed from below.

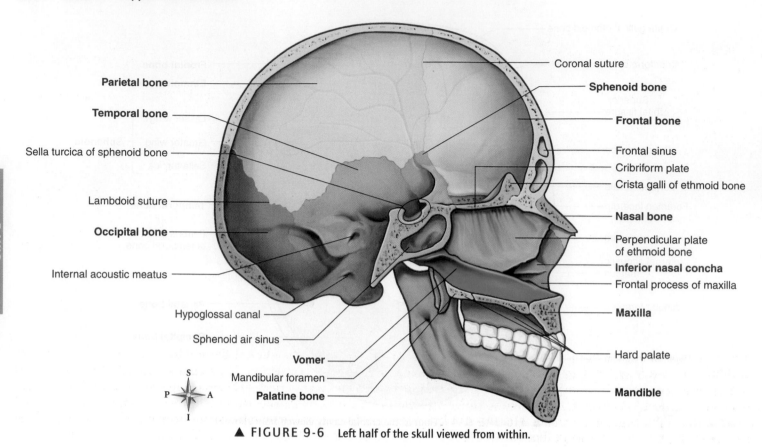

▲ FIGURE 9-6   Left half of the skull viewed from within.

the nasal cavity (Figure 9-7). A portion of the frontal bone forms the upper part of the orbits of our eyes. It unites posteriorly with the two parietal bones to create an immovable joint, or *suture*—the *coronal suture*. Two **parietal bones** (see Figure 9-3) create the bulging top of the cranium and form several immovable joints: the *lambdoidal suture* with the occipital bone, the *squamous suture* with the temporal bone and part of the sphenoid, and the *coronal suture* with the frontal bone. The two **temporal bones** (see Figures 9-3 and 9-5) house the middle and inner ear structures and contain the *mastoid sinuses*. The **occipital bone** (see Figure 9-3) creates the framework of the lower, posterior part of the skull. The occipital bone forms immovable joints with three other cranial bones—the parietal, temporal, and sphenoid. It also makes a *movable* joint with the first cervical vertebra, the atlas, allowing you to rotate your head.

The bat-shaped **sphenoid bone** (see Figure 9-4) is located in the central area of the floor of the cranium. Here it serves as a foundation to anchor the frontal, parietal, occipital, and ethmoid bones. In addition, the sphenoid bone forms part of the lateral wall of the cranium and part of the floor of each orbit. The sphenoid bone also contains large *sphenoid sinuses* (see Figure 9-7) that make it much lighter than solid bone. Finally, the **ethmoid bone**—a perforated, irregularly shaped bone—lies anterior to the sphenoid and posterior to the nasal bones. The ethmoid bone helps to fashion (1) the anterior part of the cranial floor, (2) the medial walls of the orbits of the eyes, (3) the upper parts of the nasal septum,

(4) the sides of the nasal cavity, and (5) part of the roof of the nasal cavity (the *cribriform plate*). The **cribriform plate** is perforated by numerous small *foramina* (holes) through which olfactory nerve branches pass directly to the brain (Box 9-1). The lateral parts of the ethmoid bone are honeycombed with sinus spaces (see Figure 9-7).

The Cribriform Plate

Separation of the nasal and cranial cavities by the **cribriform plate** of the ethmoid bone is clinically important. The cribriform plate is perforated by many tiny openings. These small holes permit branches of the olfactory nerve (responsible for our sense of smell) to enter the cranial cavity and reach the brain. However, separation of the nasal and cranial cavities by a thin, perforated plate of bone presents real hazards. If the cribriform plate is damaged as a result of trauma to the nose, it is possible for infectious agents such as bacteria to pass directly from the nasal cavity into the cranial fossa. In addition, if fragments of a fractured nasal bone are pushed *through* the cribriform plate (e.g., when a person receives an upward blow to the nose), they may tear the coverings of the brain and enter the brain itself.

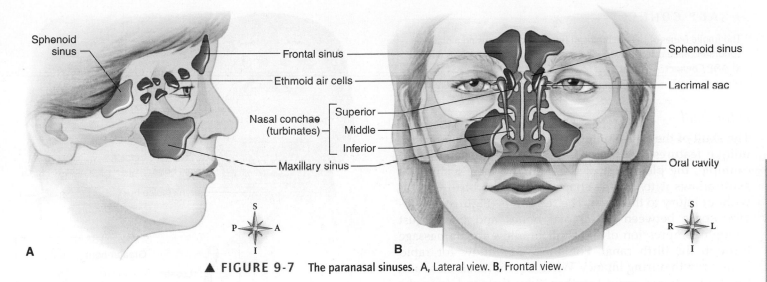

▲ FIGURE 9-7 The paranasal sinuses. **A,** Lateral view. **B,** Frontal view.

Facial Bones

The two **maxillae** serve as the foundation for our face. In addition to articulating with each other, each maxilla joins with a nasal, a zygomatic, an inferior concha, and a palatine bone (see Figures 9-2 and 9-6). You should note, however, that the mandible does *not* articulate with the maxillae. The maxillae form part of the floor of the orbits of the eyes, part of the roof of the mouth, and part of the floor and sides of the nose. Each maxilla contains a mucosa-lined *maxillary sinus* (see Figure 9-7), which is the largest of the paranasal sinuses. The lower jaw—because of the fusion of its halves during our infancy—consists of a single bone, the **mandible** (see Figure 9-3). The mandible is the largest and strongest bone of the face. It articulates with the temporal bone to form the *only* movable joint of our skull.

Our cheek is shaped by the **zygomatic** bone (see Figure 9-2), which also forms the outer margin of the orbit. With the zygomatic process of the temporal bone, it creates the *zygomatic arch.* It articulates with four other facial bones: the maxillary, temporal, frontal, and sphenoid bones.

The two **nasal bones** and the *septal cartilage* respectively form the bridge and the lower part of our nose (see Figure 9-3). The nasal bones are small, but they have several articulations. They articulate with the perpendicular plate of the ethmoid bone, the cartilaginous part of the nasal septum, the frontal bone, the maxillae, and of course, each other. The paper-thin **lacrimal bones,** shaped and sized like a fingernail, lie just posterior and lateral to each nasal bone. These tiny bones bear a groove for the *nasolacrimal* (tear) *duct.* They also form the sides of the nasal cavity and the medial wall of the orbit.

A&P CONNECT

The sinuses of the skull are often visible in x-ray images and other radiographs, as you can see in **Skeletal Radiography** online at **A&P Connect.**

Eye Orbits

The orbital cavities of the skull contain our eyes and their associated muscles. They also contain the *lacrimal* (tear) *glands* and important blood vessels and nerves. These structures are separated from the cranial cavity, nose, paranasal sinuses, and mouth by the thin and fragile orbital walls (see Figures 9-2 and 9-3). Note: Blunt trauma to your eyes may produce so-called *blowout fractures,* which result in the "raccoon eye" look. A much less serious blow generally causes the common "black eye."

The orbit of the eye consists of all or part of seven bones. The two **palatine bones,** which articulate with the maxillae and the sphenoid bone, meet like two Ls facing each other. Their horizontal portions unite to form the posterior part of the hard palate (see Figure 9-5). The vertical portion of each palatine bone rises to form the lateral wall of the posterior part of each nasal cavity. The orbital process extends from the vertical portion to create a small part of the orbit.

There are two **inferior nasal conchae** *(turbinates)* shaped like little scrolls of paper that form *conchal* ledges within the nasal cavity. There are three *conchae* (ledges) for each nasal cavity. The superior and middle conchae (projections of the ethmoid bone) form the upper and middle ledges. The inferior concha (a separate bone) forms the lower ledge. The conchae are covered by mucosa and divide each nasal cavity into three narrow, irregular channels. The inferior nasal conchae form immovable joints with the ethmoid, lacrimal, maxilla, and palatine bones.

The **vomer bone** (see Figures 9-2 and 9-5), along with the perpendicular plate of the ethmoid bone and the septal cartilage, completes the posterior portion of the septum. The vomer forms immovable joints with four bones: the sphenoid, ethmoid, palatine, and maxillae.

This would be a good time to stop and take a few moments to review the cranial and facial bones just described. As you describe them to yourself and commit them to memory, try to position them three-dimensionally in your mind.

Fetal Skull

The skull of the fetus and newborn infant has a number of unique features that are not seen in the adult skull. For example, the placement of the cranial bones in the fetal skull allows it to change shape during the birth process without injury to the brain. The **fontanels** (Figure 9-8) provide space between the cranial bones and permit just enough compression of the skull to allow for its passage through the birth canal. Fontanels also allow for rapid brain growth during infancy. When the fontanels close and the cranial bones grow together, they fuse and form the suture lines of adulthood that remain visible throughout life. Other changes also take place after birth as the newborn skull develops until maturity. For example, the paranasal sinuses undergo dramatic changes in sizes and placement. In addition, the small mounds that cover the developing *deciduous* ("baby") teeth can also be seen in the mandible. Over time these teeth erupt and eventually become the baby's first set of teeth. Finally, an infant's face grows quickly. Initially it forms an area equivalent to only 12% of the cranium, whereas the adult face is equal to an area of about 50% of the cranium!

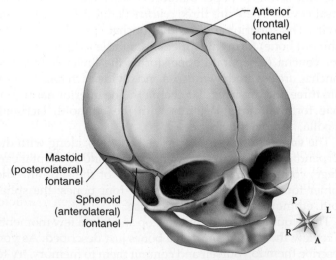

▲ **FIGURE 9-8** Skull at birth.

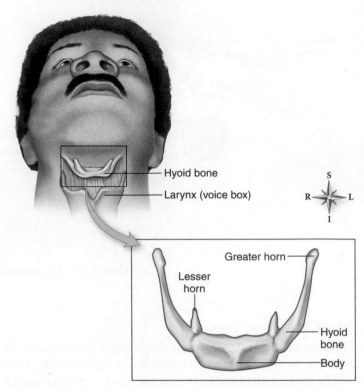

▲ **FIGURE 9-9** **Hyoid bone.** The upper part of the figure shows the relationship of the hyoid bone to the base of the skull and larynx. The inset shows details of an isolated hyoid bone. Note that the hyoid does not articulate with any other bony structure.

Hyoid Bone

The **hyoid bone** is a single, U-shaped bone below the mandible and just above the *larynx* (voice box). This unique bone, suspended from the styloid processes of the temporal bones, has no articulations with any other bones (Figure 9-9). Instead, ligaments and muscles suspend the hyoid. In addition, several muscles attach to the hyoid bone. These include an extrinsic tongue muscle that pulls your tongue back and muscles that are attached to the floor of your mouth. The hyoid bone is present in many mammals, but its lower position in humans (just above the larynx) permits a wide range of movements by the tongue, larynx, and pharynx. This in turn allows for many of the unique sound productions of human speech. Although it is hard to fracture the hyoid bone in children, in cases of suspicious death of an adult, a fractured hyoid is a strong sign of strangulation.

QUICK CHECK

1. What are the major divisions of the skeletal system?
2. List the major bones of the cranium.
3. List the major facial bones.
4. How does the fetal skull differ from that of a mature person?

Right lateral view

Anterior view

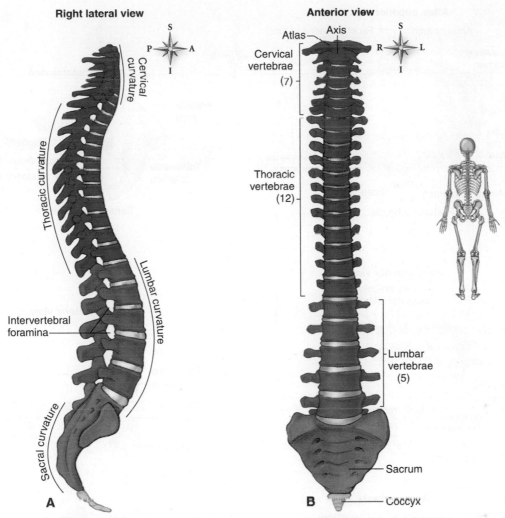

- Atlas
- Axis
- Cervical vertebrae (7)
- Thoracic vertebrae (12)
- Lumbar vertebrae (5)
- Sacrum
- Coccyx

Cervical curvature

Thoracic curvature

Lumbar curvature

Sacral curvature

Intervertebral foramina

A

B

▲ **FIGURE 9-10**   **The vertebral column. A,** Right lateral view. **B,** Anterior view.

Vertebral Column

Your vertebral (or spinal) "column" is actually a *flexible, curved chain* of 24 single **vertebrae** (Figure 9-10). This vertebral chain forms the longitudinal axis of the skeleton and includes the *inflexible sacrum* and the tail-like *coccyx*, which together consist of an additional 9 (fused) vertebrae for a total of 33. Joints between the vertebrae permit forward, backward, and sideways movements of the chainlike column. Now, consider these facts about the vertebral column: The head is balanced on top, the ribs are suspended in front, the lower extremities are attached below, and the spinal cord is enclosed within. Truly, it is the "backbone" of the body!

Single and Fused Vertebrae

Seven **cervical vertebrae** form the neck, followed by 12 **thoracic vertebrae** of the chest region and 5 **lumbar vertebrae** that support the small of the back (see Figure 9-10). Below the lumbar vertebrae lies the **sacrum**—a single broad and massive bone in the adult created from the fusion of five separate vertebrae. Last in line is the fragile **coccyx**—the "tailbone"—created from the fusion of the last four or five vertebrae of the

spine. The curvature of the vertebral column allows for compression and makes it possible to support and carry the body.

Special Features of Vertebrae

All vertebrae resemble each other in some features and differ considerably in others. For example, all vertebrae except the first cervical vertebra have a flat and rounded central **body**. These vertebrae also have a sharp or blunt **spinous process** that projects along the posterior midline, and two **transverse processes** that project from the sides of each vertebral body (Figure 9-11). Likewise, all but the sacrum and coccyx have a central opening, the **vertebral foramen.** In contrast to their obvious similarities, some vertebrae also have unique features. For example, an upward projection, the **dens** (or *odontoid process*), from the body of the second cervical vertebra, furnishes an *axis* for rotating the head. In addition, each thoracic vertebra has special **facets** (small faces), which provide for their articulation with the ribs. The vertebrae nest together on top of one another along their superior and inferior articular surfaces.

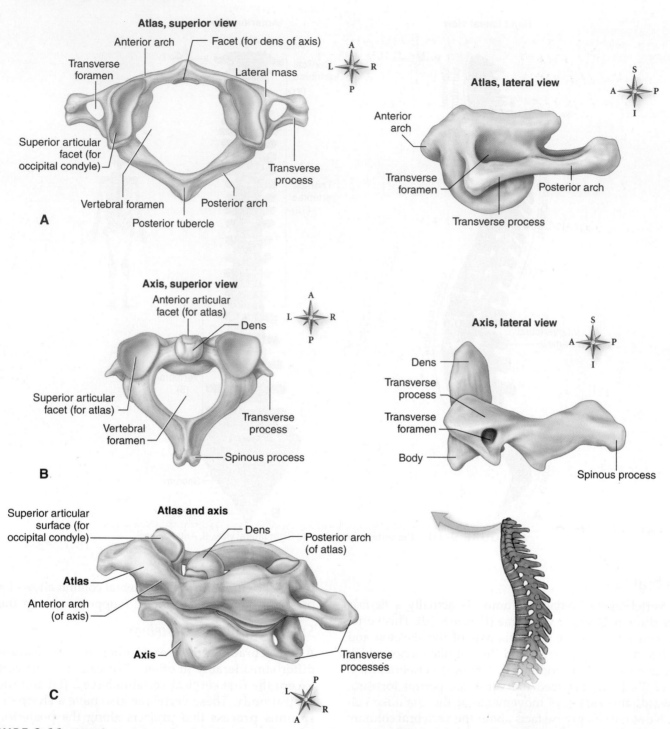

Atlas, superior view

Anterior arch — Facet (for dens of axis)

Transverse foramen

Lateral mass

Superior articular facet (for occipital condyle)

Transverse process

Vertebral foramen

Posterior arch

A

Posterior tubercle

Atlas, lateral view

Anterior arch

Transverse foramen

Posterior arch

Transverse process

Axis, superior view

Anterior articular facet (for atlas)

Dens

Superior articular facet (for atlas)

Vertebral foramen

Transverse process

B

Spinous process

Axis, lateral view

Dens

Transverse process

Transverse foramen

Body

Spinous process

Atlas and axis

Superior articular surface (for occipital condyle)

Dens

Posterior arch (of atlas)

Atlas

Anterior arch (of axis)

Axis

Transverse processes

C

▲ **FIGURE 9-11** **Vertebrae. A,** Lateral and superior views of C1, the *atlas.* **B,** Lateral and superior views of C2, the *axis.* **C,** C1 (atlas) and C2 (axis) together.

Comparison of Vertebrae

As you've just seen, the vertebral column is composed of several different types of vertebrae: cervical, thoracic, lumbar, sacral, and coccygeal. You can see representatives of these vertebral types as well as the unique atlas and axis vertebrae in Figure 9-11.

The cervical vertebrae (C1-C7) are generally smaller and more delicate than other vertebrae and have short spinous processes. Their transverse processes are also unusual in

that they have *transverse foramina,* which allow for the passage of arteries leading to the brain. Several cervical vertebrae are quite distinct. For example, the atlas (C1) does not have a body and is further modified to fit into the occipital condyles of the skull to support the entire head. The atlas allows you to nod your head in an up-and-down manner (see Figure 9-11). In contrast, the axis (C2) serves as a pivot, swiveling on its dens, thus allowing you to rotate your head from side to side. The remaining cervical vertebrae (C3-C7)

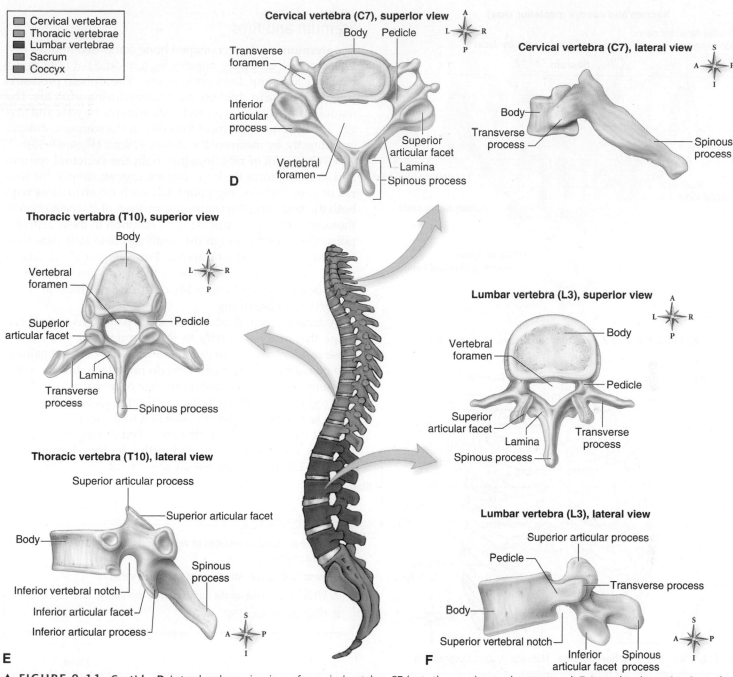

Cervical vertebrae
Thoracic vertebrae
Lumbar vertebrae
Sacrum
Coccyx

Cervical vertebra (C7), superior view
Body Pedicle
Transverse foramen
Inferior articular process
Superior articular facet
Vertebral foramen
Lamina
Spinous process
D

Cervical vertebra (C7), lateral view
Body
Transverse process
Spinous process

Thoracic vertabra (T10), superior view
Body
Vertebral foramen
Superior articular facet
Pedicle
Lamina
Transverse process
Spinous process

Lumbar vertebra (L3), superior view
Vertebral foramen
Body
Pedicle
Superior articular facet
Lamina
Transverse process
Spinous process

Thoracic vertebra (T10), lateral view
Superior articular process
Superior articular facet
Body
Spinous process
Inferior vertebral notch
Inferior articular facet
Inferior articular process
E

Lumbar vertebra (L3), lateral view
Superior articular process
Pedicle
Transverse process
Body
Superior vertebral notch
Inferior articular facet
Spinous process
F

▲ **FIGURE 9-11, Cont'd** **D,** Lateral and superior views of a cervical vertebra, C7 (note the prominent spinous process). **E,** Lateral and superior views of a thoracic vertebra, T10. **F,** Lateral and superior views of a lumbar vertebra, L3.

have short spinous processes that often have a slight fork or split, except for C7, whose fused spinous process is large and can be seen protruding at the back of the neck.

The thoracic vertebrae are larger than the cervical vertebrae and have long, downward-pointing spinous processes. Their transverse processes articulate with a rib on each side. The bodies of these vertebrae become more massive as they approach the lumbar region.

The lumbar vertebrae have the most massive bodies of the single vertebrae because they are adapted to supporting the weight of the body above them.

The sacrum is a triangle-shaped bone that results from the fusion of five vertebrae. The vertebral foramina of the five segments align to form an interior *sacral canal.* At the junctions of the five segments, lateral to the vertebral bodies, four pairs of spaces called the *anterior sacral foramina* and the *posterior sacral foramina* allow for the passage of blood vessels and spinal nerves (Figure 9-12).

The coccyx, or tailbone, is the end of the vertebral column and results from the fusion of four, sometimes five, vertebrae.

UNIT 2

Sacrum and coccyx (posterior view)

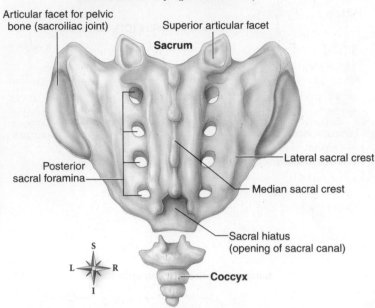

▲ **FIGURE 9-12** Sacrum and coccyx. Posterior view.

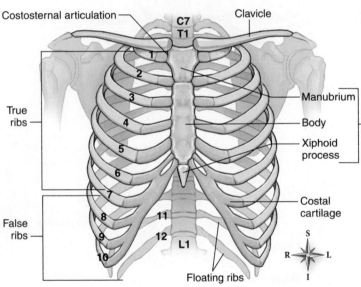

▲ **FIGURE 9-13** Thoracic cage. Note the costal cartilages and their articulations with the body of the sternum.

▶ **FIGURE 9-14**
Articulation of a rib and vertebra. A, Note the head of the rib articulating with the vertebral body and the tubercle of the rib articulating with the transverse process of the vertebra. **B,** Anatomical components of a typical rib (fifth rib) viewed from behind.

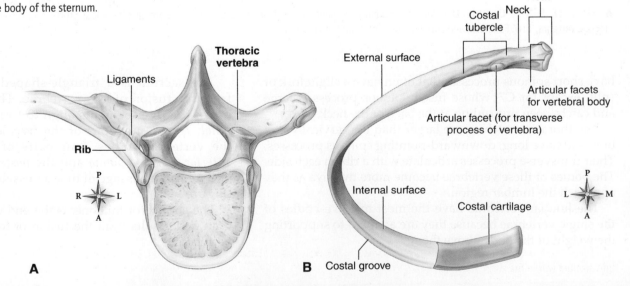

Sternum and Ribs

The **sternum** is a dagger-shaped bone composed of three distinct sections: an upper *manubrium,* a middle flat *body,* and a bladelike lower tip, the *xiphoid process.* The xiphoid process is soft and flexible in children, but ossifies during adult life. The manubrium articulates on each side with the clavicle and first rib. The next nine ribs meet the body of the sternum, directly or indirectly, by means of the *costal cartilages* (Figure 9-13).

Twelve pairs of ribs, together with the vertebral column and sternum, form the bony *thoracic cage* or, simply, the **thorax** of your body. Along your back, each rib articulates with both the body and the transverse process of its *corresponding* thoracic vertebra (Figure 9-14). In addition to these articulations, the second through the ninth ribs also articulate with the *body* of the vertebra above. From its vertebral attachment, each rib curves outward, then forward and downward (see Figures 9-1 and 9-14), creating a mechanical "box" important for breathing.

Anteriorly, each rib of the first seven pairs joins a costal cartilage that attaches *directly* to the sternum. For this reason, these seven ribs are often called *true ribs.* Ribs of the remaining five pairs (the so-called *false ribs*) do not attach directly to the sternum. Instead, each costal cartilage of rib pairs 8, 9, and 10 fuse together and attach to the costal cartilage of rib 7—which *is* the last "true rib" directly attached to the sternum. Ribs of the last two pairs of false ribs are called *floating ribs* because they do not attach (even indirectly) to the sternum (see Figure 9-13). This is why we say they "float."

> **QUICK ✓ CHECK**
>
> 5. Name the basic three types of vertebrae. How many are there of each type?
> 6. Describe the basic structures of a thoracic vertebra.
> 7. What bones make up the bony cagelike thorax?
> 8. What are floating ribs?

Appendicular Skeleton

Pectoral Girdle

The **pectoral girdle** or *shoulder girdle* consists of two *clavicles* and two *scapulae*, which together form an incomplete ring, not quite a continuous girdle (belt).

The **clavicles** (collarbones) are rodlike and shaped somewhat like a flattened "S." Each clavicle extends horizontally from its articulation with the manubrium of the sternum to its articulation with the scapula. The clavicles serve to support the scapulae and also provide areas for the attachment of muscles from the upper arms, upper chest, and back. On each side of the body, between the sternum and each clavicle, the pectoral girdle forms one bony joint (the *sternoclavicular joint*).

The **scapulae,** or shoulder blades, are broad, somewhat triangular bones that help form the upper part of the back. A large ridgelike *spine* runs at an upward angle across the upper third of each scapula. This spine forms two anterior processes. The first process, the *acromion*, forms the top of the shoulder and sometimes appears as a prominent shoulder bump in slender people. The second process, the *coracoid process*, curves anteriorly in front of the body of the clavicle. The acromion process articulates with the clavicle at its distal end and serves as a site for the attachment of muscles from the upper limb. The coracoid process also provides sites for the attachment of limb muscles from the upper chest and limb (Figure 9-15). Lying between the acromial and coracoid processes is a depression called the *glenoid cavity*. As you can see, this

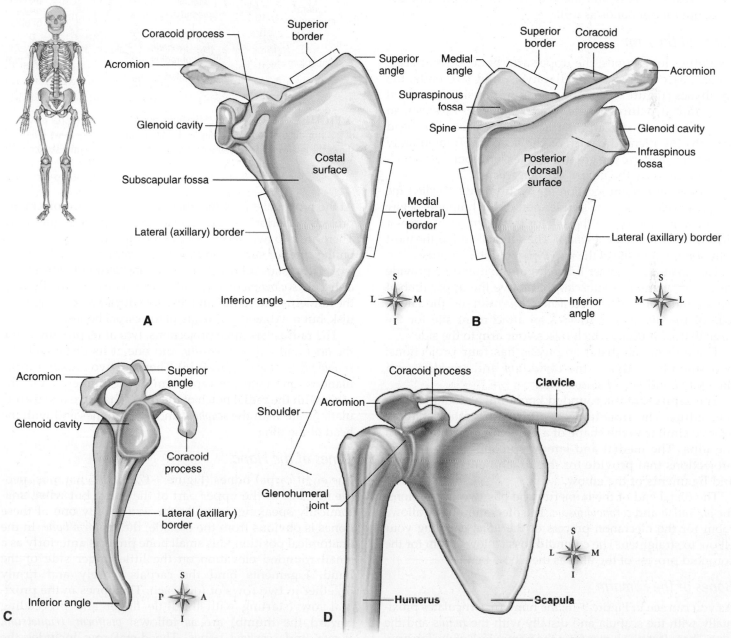

▲ **FIGURE 9-15** **Right scapula. A,** Anterior view. **B,** Posterior view. **C,** Lateral view. **D,** Right shoulder girdle. (The inset shows the relative position of the right scapula within the entire skeleton.)

depression, formed within the lateral end of the scapula, articulates with the head of the humerus.

The scapula is suspended on the thoracic wall by muscles that act to stabilize as well as actively move the scapula. These muscles provide a wide range of movements to the scapula, including *elevation* (raising), *depression* (lowering), and *rotation* (moving in a circular manner) in several planes, and the ability to tip forward and backward. Along with similarly wide-ranging movements of the clavicle, this range of motion gives the shoulder girdle considerable flexibility.

Upper Extremities

Each of our *upper extremities* consists of a scapula, clavicle, upper arm, lower arm (forearm), wrist, and hand. As you will see, the "open" arrangement of the pectoral girdle provides great flexibility and maneuverability not only for our arms, but for our hands as well.

Bones of the Arm

Like other long bones, the upper arm bone, or **humerus,** is composed of a shaft, or diaphysis, and two ends, the epiphyses (Figure 9-16). The upper epiphysis has several important structures: a *head, greater and lesser tubercles,* an *anatomical neck,* and an *intertubercular groove.* The head fits into the glenoid cavity of the scapula. Immediately below the head are the greater and lesser tubercles, which are separated by the *intertubercular groove.*

Another important location on the humerus is called the **surgical neck.** This is the slender part of the humerus below the tubercles and below the site for the insertion of a major chest muscle, the *pectoralis major.* The surgical neck is the most common fracture site of the upper part of the humerus.

The greater and lesser tubercles of the humerus provide for the attachment of muscles that move the upper limb at the shoulder. The V-shaped *deltoid tuberosity* on the lateral side of the diaphysis provides an attachment site for the large deltoid muscle, which raises your arm to the side.

The end of the distal epiphysis has four projections: two smooth condyles (the *capitulum* and *trochlea*), and the *medial* and *lateral epicondyles.*

The **capitulum** is a rounded knob that articulates with the radius. The **trochlea** is a projection with a shallow groove similar to the shape of a pulley. It articulates with the ulna. The medial and lateral epicondyles are rough projections that provide for the attachments for muscles and ligaments of the elbow.

The distal end of the humerus also has two depressions, the *olecranon* and *coronoid fossae.* The **olecranon fossa** allows room for the olecranon process of the *ulna,* enabling your elbow to straighten. The **coronoid fossa** allows room for the coronoid process of the ulna as the elbow bends.

Bones of the Forearm

As you can see in Figure 9-16, the humerus articulates proximally with the scapula and distally with the *radius* and the *ulna.* These last two bones form the framework of the forearm.

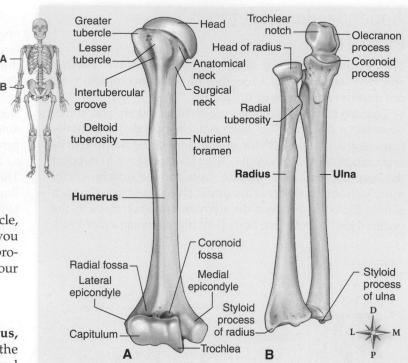

▲ **FIGURE 9-16** **Bones of the right arm. A,** Humerus (upper part of the arm, anterior view). **B,** Radius and ulna (forearm, anterior view). (The inset shows the relative position of the right arm bones within the entire skeleton.)

At the proximal end of the **ulna,** the scoop-shaped **olecranon** projects posteriorly and the *coronoid process* projects anteriorly. There are also two depressions in the ulna: the *semilunar notch* on the anterior surface and the *radial notch* on the lateral surface. The distal end has two projections: a rounded head and a sharper *styloid process.* The ulna articulates proximally with the humerus and radius and distally with a fibrocartilaginous disk, but not directly with any of the carpal bones.

The **radius** has three projections: two at its proximal end, the *head* and *radial tuberosity,* and one at its distal end, the *styloid process* (see Figure 9-16). There are two proximal articulations: one with the capitulum of the humerus and the other with the radial notch of the ulna. The three distal articulations are with the *scaphoid* and *lunate* bones and with the head of the ulna.

Bones of the Hand

The eight **carpal bones** (Figure 9-17) form what most people think of as the upper part of the hand, but what, anatomically speaking, is really the wrist. Only one of these bones is obvious from the outside, the *pisiform bone.* In the anatomical position, this small bone projects anteriorly as a small rounded elevation on the little finger side of the hand. Ligaments bind the carpals closely and firmly together in two rows of four each. The bones in the proximal row (starting with the little finger and proceeding toward the thumb) are as follows: *pisiform, triquetrum, lunate,* and *scaphoid* bones. The distal row includes the

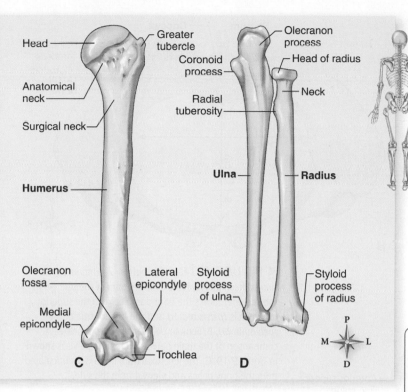

▲ **FIGURE 9-16, Cont'd** **C,** Humerus (posterior view). **D,** Radius and ulna (posterior view). (The inset shows the relative position of the right arm bones within the entire skeleton.)

hamate, capitate, trapezoid, and *trapezium* bones. The joints between the carpals and radius permit a variety of wrist and hand movements, including many different ways of flexing and extending our hands.

There are five **metacarpal bones** that form the palm of your hand. Each finger has its own metacarpal bone—

cylindrical bones rounded at each end. The number one metacarpal forms the base of the thumb; the number two metacarpal forms the base of the index finger, and so on. When you clench your fist, the distal ends of these bones form the knuckles, potentially exposing them to damage. The thumb metacarpal has the most freely movable joint with the carpals. This allows a wide range of possible movement between the thumb metacarpal and the trapezium. As a result, the thumb is *opposable* to any of the fingers. For this reason, your hand has much greater dexterity than that of any other mammal and allows you to manipulate even small objects easily and quickly.

Each finger comprises three **phalanges** (proximal, middle, and distal), except the thumb, which does not have a middle *phalanx* (see Figure 9-17).

QUICK CHECK

9. What bones make up the shoulder girdle?
10. Where does the shoulder girdle form a joint with the axial skeleton?
11. Name the major bones of the upper and lower arm.
12. Name the major types of bones found in the hand and wrist.

Pelvic Girdle

The **pelvic girdle** comprises two **coxal bones** (hipbones or *os coxae*) bound to each other anteriorly (by strong ligaments) and to the *sacrum* (Figure 9-18, *A* and *B*). The pelvic girdle provides a stable, roughly circular base that supports the trunk of the body and the anchor for the lower extremities. Before puberty, each coxal bone is made of three separate

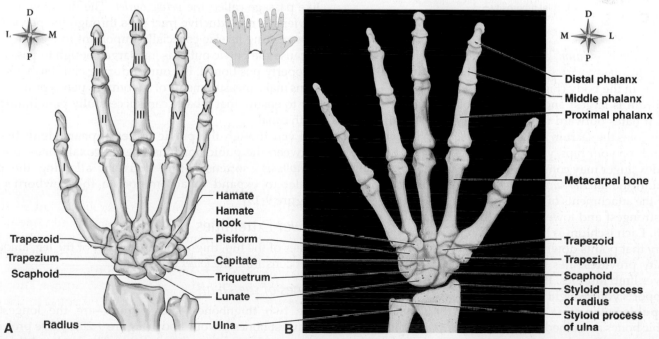

▲ **FIGURE 9-17** Bones of the hand and wrist. **A,** Illustration of dorsal view of the right hand and wrist. **B,** Palmar view of the right hand and wrist.

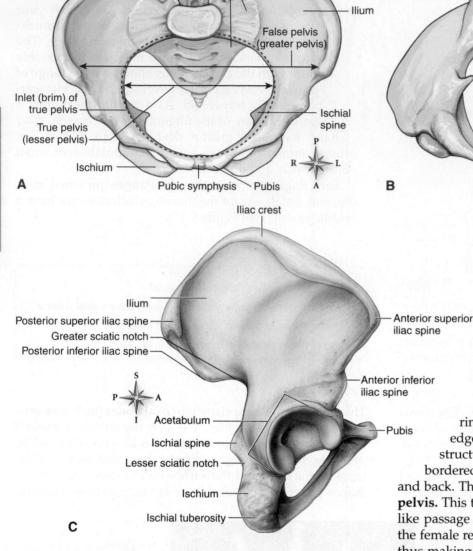

A

Sacrum

Ilium

False pelvis (greater pelvis)

Inlet (brim) of true pelvis

True pelvis (lesser pelvis)

Ischium

Ischial spine

Pubic symphysis Pubis

C

Iliac crest

Ilium

Posterior superior iliac spine

Greater sciatic notch

Posterior inferior iliac spine

Anterior superior iliac spine

Anterior inferior iliac spine

Acetabulum

Ischial spine Pubis

Lesser sciatic notch

Ischium

Ischial tuberosity

B

Sacrum Coccyx

Coxal bone

11 cm

9-11.5 cm

◀ FIGURE 9-18 **The pelvis. A,** Female pelvis viewed from above. Note that the brim of the true pelvis *(dotted line)* marks the boundary between the superior false pelvis *(pelvis major)* and the inferior true pelvis *(pelvis minor)*. **B,** Female pelvis viewed from below. A comparison of the male pelvis and female pelvis is shown in Figure 9-19. **C,** The left coxal (hip) bone is disarticulated from the bony pelvis and viewed from the side.

ischium and form the largest opening in the skeleton, the *obturator foramen.*

The **pelvis** is often divided into two parts by a ringlike passageway called the *pelvic inlet,* the edges of which form the *brim of the true pelvis.* The structure above the pelvic inlet, the **false pelvis,** is bordered by muscle in the front and bone on the sides and back. The space below the pelvic inlet is called the **true pelvis.** This true pelvis forms the boundary of another ring-like passage called the *pelvic outlet.* The digestive tract and the female reproductive tract pass through the pelvic outlet, thus making its size especially important to women. This is because the pelvic outlet is just large enough to pass a baby (properly positioned, of course) during childbirth. Obstetricians make measurements of a woman's pelvis prior to delivery to ensure that a baby can successfully pass through the birth canal.

Even though the pelvic outlet appears rigid, the joint between the pubic portions of each coxal bone—the *pubic symphysis*—softens before delivery, allowing the pelvic outlet to expand to accommodate the newborn's head (Figure 9-19).

Lower Extremities

Bones of the hip, thigh, and lower part of the leg, ankle, and foot constitute the *lower extremity.*

Bones of the Thigh

The two thighbones—the **femurs**—are the longest and heaviest bones in your body (Figure 9-20). Three projections are conspicuous at each epiphysis. The *head* and the *greater*

bones: the *ilium, ischium,* and *pubis.* Later, these bones fuse into a single, massive irregular bone that is broader than any other bone in the body (Figure 9-18, *C*).

The **ilium** is largest and uppermost of those three major regions. It flares outward to form the widest part of the hip. The ilium joins the sacrum at the *sacroiliac joint.* When you put your hands on your hips, you can feel the *iliac crest.* The tip of your index finger may come to rest on the anterior projection of the ilium, the *anterior superior iliac spine.* This spine provides sites for the attachments of ligaments and muscles.

The strongest and lowermost of the pelvic regions is the **ischium.** Each ischium is an L-shaped region with an *ischial tuberosity* that points downward and posteriorly. The ischial tuberosity provides sites for the attachments of ligaments and lower limb muscles. It is also the part of the coxal bone that supports your weight while sitting.

The **pubis** is the most anterior part of the pelvic girdle. The pubic bones are joined anteriorly by the *pubic symphysis.* The lower part of each pubis passes posteriorly to join the

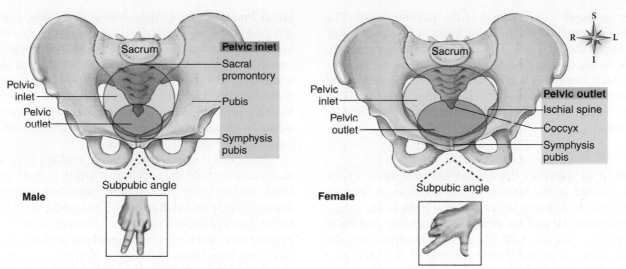

▲ **FIGURE 9-19** **Comparison of the bony pelvis of the male and female skeletons.** Notice the narrower width of the male pelvis, giving it a more funnel-like shape than the female pelvis. The finger positions shown give you an easy way to represent the approximate pubic angles in the male and female pelvis.

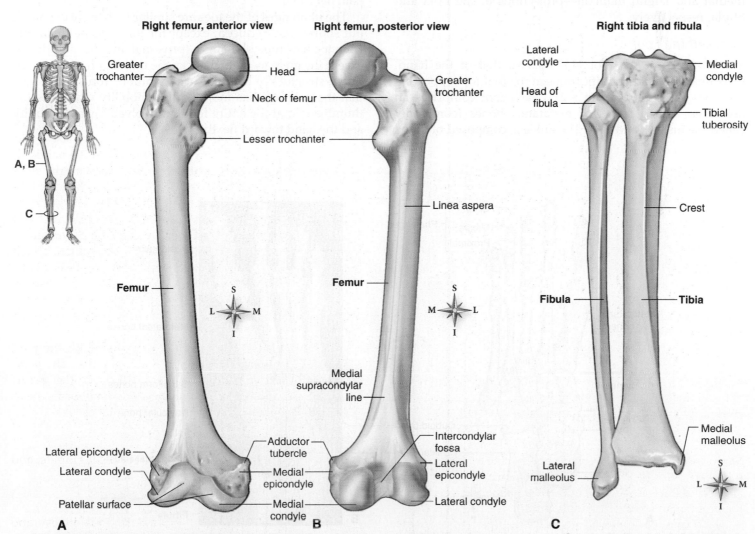

▲ **FIGURE 9-20** **Bones of the thigh and leg.** **A,** Right femur, anterior surface. **B,** Right femur, posterior view. **C,** Right tibia and fibula, anterior surface. (The inset shows the relative position of the bones of the thigh and leg within the entire skeleton.)

and *lesser trochanters* are located at the proximal end. The head of the femur fits into the *acetabulum* of the coxal bone, where it forms the hip joint. The greater and lesser trochanters serve as important sites for the attachment of muscles from the coxal bone that form the buttocks. The *medial* and *lateral condyles* and *adductor tubercle* are found at the distal end of the femur. The lateral and medial condyles articulate with the tibia.

Bones of the Leg

The kneecap or **patella**—the largest *sesamoid* bone in the body—is located in the tendon of the quadriceps femoris muscle. Of the two lower leg bones, the **tibia** is the larger and more superficial and the **fibula** is the smaller and more laterally placed (Figure 9-20, *C*). At its proximal end, the fibula articulates with the lateral condyle of the tibia, and the proximal end of the tibia articulates with the femur to form the knee joint—the largest and one of the most complex and easily injured joints in the human body. At its distal end, the tibia articulates with the fibula and with the talus, forming a boxlike socket (the ankle joint) with the medial and lateral malleoli—projections of the tibia and fibula, respectively.

Bones of the Foot

The structure of the foot is similar to that of the hand. Of course, there are modifications in the foot for supporting the body's weight (Figure 9-21). Each foot comprises an ankle, an arched series of metatarsal bones (commonly called the instep), and toes. The ankle is composed of seven **tarsal bones.** One of these bones, the **talus,** moves freely where it articulates with the tibia and fibula. The other tarsal bones are bound firmly together. One major difference between the foot and the hand is that the great toe is much wider and less mobile than the thumb. The various arches (both longitudinal and transverse) formed by the bones of the foot provide support and flexibility (Figure 9-22). This allows us to stand on our tiptoes in ballet, pivot to catch a football, and spring high in a jump.

Sometimes the strong ligaments and leg muscle tendons that normally hold the foot bones firmly in their arched positions weaken to the point where the arches flatten. This condition is aptly called **flatfoot** (Figure 9-22, *B*). Eager young ballet dancers should never be allowed to be *en pointe* (on tiptoes) until the bones and ligaments of the ankles and feet have developed their full weight-bearing capacity. Likewise, wearing high heels can also cause damage to the foot and even the lower back over time (Figure 9-22, *D*). This is because very high heels thrust the body forward. This forces an undue amount of weight on the heads of the metatarsals, and can cause permanent injury, bone deformity, and chronic pain overtime.

The **phalanges** of the toes are similar to those of the hand, but shorter and reduced, except for those of the great toe. Each toe has three phalanges (proximal, middle, and distal) except the great toe, which lacks the middle phalanx, a condition structurally like that of the thumb. You should note that the great toe has more limited mobility than does the thumb and that the entire foot has evolved toward stability and the hand toward flexibility.

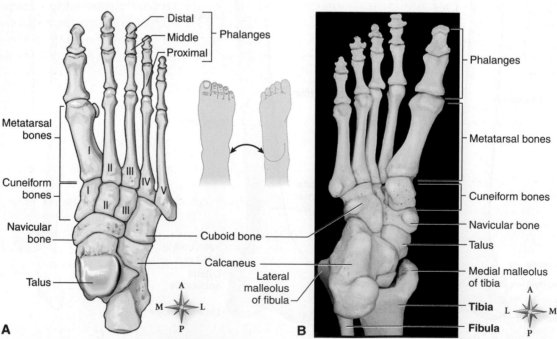

▲ **FIGURE 9-21   The foot. A,** Illustration of bones of the *right* foot viewed from above. The tarsal bones consist of the cuneiforms, navicular, talus, cuboid, and calcaneus. **B,** Bones of right foot viewed from below.

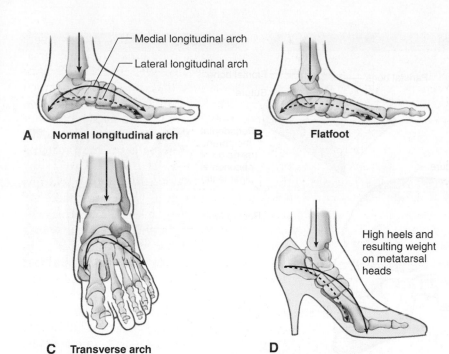

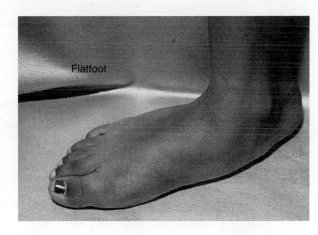

▲ **FIGURE 9-22 Arches of the foot. A,** Longitudinal arch. The medial portion is formed by the calcaneus, talus, navicular, cuneiforms, and three metatarsal bones; the lateral portion is formed by the calcaneus, cuboid, and two lateral metatarsal bones. **B,** "Flatfoot" results when the tendons and ligaments attached to the tarsal bones are weakened. Downward pressure by the weight of the body gradually flattens out the normal arch of the bones. The photo shows the clinical appearance of a flatfoot. **C,** Transverse arch in the metatarsal region of the left foot. **D,** High heels throw the weight forward and cause the heads of the metatarsal bones to bear most of the body's weight. (*Arrows* show direction of force.)

DIFFERENCES BETWEEN MALE AND FEMALE SKELETONS

As you undoubtedly know, general and specific differences typically exist between male and female skeletons. The general difference is one of size and weight. The male skeleton is larger and heavier, on average. The specific differences concern the shape of the pelvic bones and cavity. The male pelvis is deep and funnel-shaped with a narrow subpubic angle (usually less than 90 degrees). In contrast, the female pelvis is shallow, broad, and flaring. It also has a wider subpubic angle (usually greater than 90 degrees). The childbearing function obviously explains the need for these and certain other modifications of the female pelvis. These are shown for you in Figure 9-19.

> **A & P CONNECT**
>
> Check out the pictures of other skeletal differences between men and women in **Skeletal Variations** online at **A&P Connect**.
>
> *evolve*

> **QUICK ✔ CHECK**
>
> 13. Which three bones fuse together during skeletal development to form the coxal (hip) bone?
> 14. List the bones of the lower extremity.
> 15. What is the functional advantage of having an arch in your foot?
> 16. Name two major differences between the typical male and female skeletons.

ARTICULATIONS (JOINTS)

As we have seen, an **articulation** (joint) is a point of contact between two or more bones. Although most joints in the body allow considerable movement, others are completely immovable or permit only limited motion. In the case of immovable joints, such as the sutures of the skull, adjacent bones are fused together into a strong and rigid protective plate. In other joints— between the bodies of the spinal vertebrae, for example—movement is possible but highly restricted. However, most joints allow a considerable range of movement, depending on the associated musculature. These joints allow us to execute complex, highly coordinated, and deliberate movements.

CLASSIFICATION OF JOINTS

Joints can be classified *structurally* according to the type of connective tissue that joins the bones together (*fibrous joints* and *cartilaginous joints*) *or* by the presence of a fluid-filled joint capsule (*synovial joints*). Joints can also be classified according to the *degree of movement* they permit. In this system, s*ynarthroses* are immovable, *amphiarthroses* are slightly movable, and *diarthroses* are freely movable.

Fibrous Joints (Synarthroses)

Fibrous joints (synarthroses) comprise bones that fit closely together, fixing the position of the bones making up the joints. There are three subtypes of fibrous joints

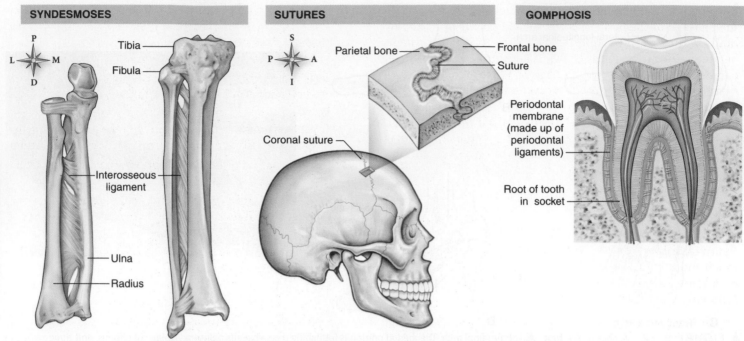

SYNDESMOSES

Tibia

Fibula

Interosseous ligament

Ulna

Radius

SUTURES

Parietal bone

Frontal bone

Suture

Coronal suture

GOMPHOSIS

Periodontal membrane (made up of periodontal ligaments)

Root of tooth in socket

▲ FIGURE 9-23   **Fibrous joints.** Examples of the types of fibrous joints.

(Figure 9-23). **Syndesmoses** are joints in which fibrous bands (ligaments) connect two bones, such as between the distal ends of the radius and ulna. **Sutures** are closely knit joints found only in the skull, becoming completely fused together by adulthood. **Gomphoses** are unique joints that occur between the root of a tooth and the mandible (or maxilla) of the jaw.

Cartilaginous Joints (Amphiarthroses)

Cartilaginous joints (amphiarthroses) comprise bones that are joined together by either hyaline cartilage or fibro-cartilage (Figure 9-24). Cartilaginous joints permit limited movement between articulating bones under specific circumstances. For example, during childbirth, the cartilagi-

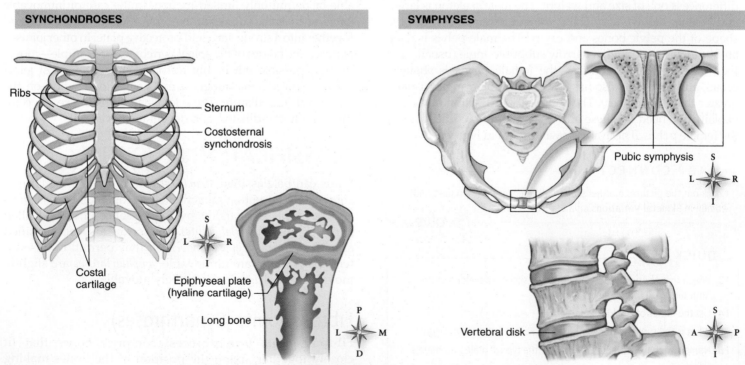

SYNCHONDROSES

Ribs

Sternum

Costosternal synchondrosis

Costal cartilage

Epiphyseal plate (hyaline cartilage)

Long bone

SYMPHYSES

Pubic symphysis

Vertebral disk

▲ FIGURE 9-24   **Cartilaginous joints.** Examples of the types of cartilaginous joints.

nous joint of the pubic symphysis moves slightly to permit the passage of the baby through the pelvis. Some cartilaginous joints, the **synchondroses,** have hyaline cartilage between the articulating bones. *Temporary synchondroses* can be found between the epiphyses of a long bone and its diaphysis during the growth years. A **symphysis** is a joint in which a pad or disk of fibrocartilage connects two bones. Most symphyses are located in the midline of the body and include the symphysis pubis and the articulation between the *bodies* of adjacent vertebrae.

Table 9-3 summarizes the different kinds of fibrous and cartilaginous joints.

Synovial Joints (Diarthroses)

Synovial joints (diarthroses) (Figure 9-25), as a group, are much more complex than either fibrous joints or cartilaginous joints. Synovial joints are typically characterized by the following:

1. **Joint capsule:** Forms a complete casing around the ends of the bones (thereby binding them to each other)

2. **Synovial membrane:** A moist, slippery membrane that lines the inner surface of the joint capsule (this secretes *synovial fluid,* which lubricates and nourishes the inner joint surfaces)

3. **Articular cartilage:** A thin layer of hyaline cartilage that covers and cushions the articulating surfaces of the bones

4. **Joint cavity:** A small space between the articulating surfaces of the two bones that permits extensive movement

5. **Meniscus:** Pad of fibrocartilage located between articulating ends of bones

6. **Ligament:** Strong cord of dense, white fibrous tissue that grows between the bones and lashes them more firmly together

7. **Bursa:** Found near some synovial joints—pillowlike structure that consists of a synovial membrane filled with synovial fluid. This fluid cushions the joint and facilitates the movement of tendons.

As you can readily see, synovial joints are complicated articulations. Let's look at several distinct types of synovial joints (Table 9-4).

TABLE 9-3 Classification of Fibrous and Cartilaginous Joints

TYPES	EXAMPLES	STRUCTURAL FEATURES	MOVEMENT
Fibrous Joints			
Syndesmoses	Joints between the distal ends of the radius and ulna	Fibrous bands (ligaments) connect articulating bones	Slight
Sutures	Joints between the skull bones	Teethlike projections of articulating bones interlock with a thin layer of fibrous tissue connecting them	None
Gomphoses	Joints between the roots of the teeth and the jaw bones	Fibrous tissue connects the roots of the teeth to the alveolar processes	None
Cartilaginous Joints			
Synchondroses	Costal cartilage attachments of the first rib to the sternum; epiphyseal plate between the diaphysis and epiphysis of a growing long bone	Hyaline cartilage connects articulating bones	Slight
Symphyses	Pubic symphysis; joints between *bodies* of vertebrae	Fibrocartilage between articulating bones	Slight

QUICK CHECK

17. How are joints classified?
18. Name three types of fibrous joints.
19. What is a cartilaginous joint?
20. Compare a synchondrosis with a symphysis.
21. What are the major parts of a synovial joint?

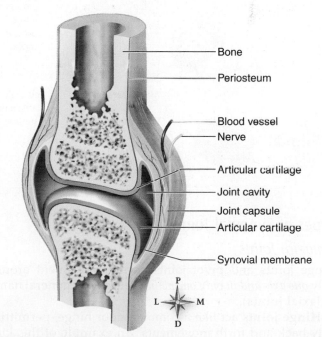

▲ **FIGURE 9-25 Structure of synovial joints.** Artist's interpretation (composite drawing) of a typical synovial joint.

Bone
Periosteum
Blood vessel
Nerve
Articular cartilage
Joint cavity
Joint capsule
Articular cartilage
Synovial membrane

TABLE 9-4 Classification of Synovial Joints

TYPES	EXAMPLES	STRUCTURE	MOVEMENT
Uniaxial Around one axis; in one place			
Hinge	Elbow joint	Spool-shaped process fits into a concave socket	Flexion and extension only
Pivot	Joint between the first and second cervical vertebrae	Arch-shaped process fits around a peglike process	Rotation
Biaxial Around two axes, perpendicular to each other; in two planes			
Saddle	Thumb joint between the first metacarpal and carpal bone	Saddle-shaped bone fits into a socket that is concave-convex-concave	Flexion, extension in one plane; abduction, adduction in the other plane; opposing the thumb to the fingers
Condyloid (ellipsoidal)	Joint between the radius and carpal bones	Oval condyle fits into an elliptical socket	Flexion, extension in one plane; abduction, adduction in the other plane
Multiaxial Around many axes			
Ball and socket	Shoulder joint and hip	Ball-shaped process fits into a concave socket	Widest range of movement: flexion, extension, abduction, adduction, rotation, circumduction
Gliding	Joints between the articular facets of adjacent vertebrae; joints between the carpal and tarsal bones	Relatively flat articulating surfaces	Gliding movements without any angular or circular movements

Types of Synovial Joints

Uniaxial Joints

Hinge joints and pivot joints permit movement around only *one axis and in only one plane* (hence their general name, **uniaxial joints**).

Hinge joints act like a common door hinge, permitting only back-and-forth movements. An example of this kind of joint can be readily observed in your elbow joint as you flex and extend your forearm: The shape of the trochlea and the semilunar notch (review Figure 9-16) permit

motion *only* in the horizontal plane. Other hinge joints include the knee and the joints between the phalanges of the fingers and toes.

Pivot joints include those in which a projection from one bone articulates with a ring or notch in another bone. An example of a pivot joint is the projection (dens) of the second cervical vertebra, which articulates with a ring-shaped portion of the first cervical vertebra.

Biaxial Joints

Other synovial joints called **biaxial joints** permit movement around *two perpendicular axes and in two perpendicular planes*. These include saddle and condyloid joints (see Table 9-4). Only two **saddle joints,** one in each thumb, are present in our bodies. Here the thumb's metacarpal bone articulates in the wrist with the trapezium (carpal) bone, allowing us to oppose the thumb with the rest of our fingers. The opposable thumb (and three extra muscles within the thumb) gives us much more dexterity with our fingers than even great apes, allowing us to draw, write, and perform intricate surgery.

Condyloid joints are biaxial joints in which a projection or *condyle* fits into an elliptical socket (a cavity). One example of this kind of joint is found where the occipital condyles at the base of the skull fit into the elliptical depressions of the atlas.

Multiaxial Joints

Finally, there are synovial joints that permit movement around *three or more axes and in three or more planes* (see Table 9-4). These are called **multiaxial joints.** The **ball-and-socket joints** are formed when the ball-shaped head of one bone fits into a concave depression of another, such as at the shoulder and hip joints. **Gliding joints** are characterized by relatively flat articulating surfaces that allow limited gliding movements. Examples of gliding joints occur between the articular surfaces of successive vertebrae of the spine. As a group, gliding joints are the least movable of all synovial joints.

Some Representative Synovial Joints

Shoulder Joint

The shoulder joint (Figure 9-26) is our most mobile ball-and-socket synovial joint. It is created by the joining of the head of the humerus and the glenoid cavity of the scapula. The *shallowness* of the glenoid cavity is primarily responsible for

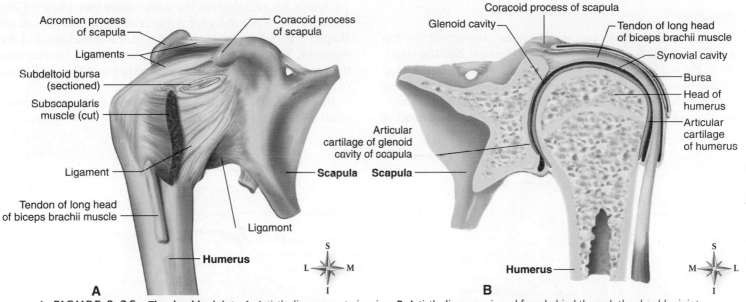

▲ **FIGURE 9-26** **The shoulder joint.** **A,** Artist's diagram, anterior view. **B,** Artist's diagram, viewed from behind through the shoulder joint.

this joint's flexibility. However, a number of structures strengthen the shoulder joint and give it a degree of stability. These include ligaments, muscles, tendons, and bursae.

Along with shoulder muscles and tendons, these components together form a cufflike arrangement around the joint, called the **rotator cuff.** Generally, an injury to the rotator cuff involves injury to the muscles and tendons. This is largely

because the structure of this joint makes it more mobile than stable. As a result, dislocations of the head of the humerus from the glenoid cavity frequently occur.

Elbow Joint

The elbow joint is a classic hinge joint formed by two articulations, between the distal end of the humerus and the proximal ends of the radius and ulna (Figure 9-27). The elbow joint provides several types of movements. Most important, it permits the hingelike bending and straightening of the joint between the humerus and the ulna, as is necessary to lift something. It also allows for the turning of the forearm, such as when a boxer straightens his arm to deliver a punch.

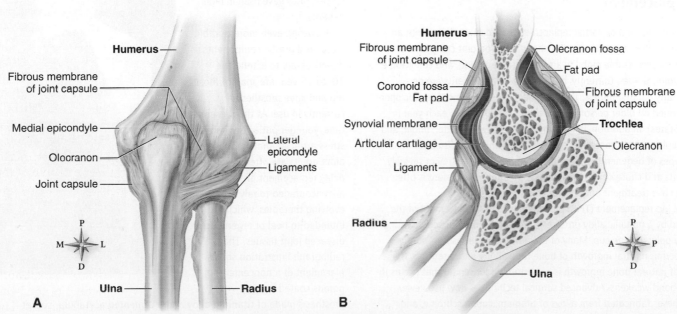

▲ **FIGURE 9-27** **The elbow joint.** **A,** Artist's diagram, posterior view. **B,** Artist's diagram, sagittal section.

UNIT 2

Hip Joint

The first characteristic you should remember about the hip joint is *stability;* the second is *mobility* (Figure 9-28). The stability of the hip joint derives largely from the shapes of the head of the femur and the *acetabulum* (the socket of the hipbone into which the femur head fits). Note the deep, cuplike shape of the acetabulum, and then notice that the ball-like head of the femur fits snugly into the acetabulum (see Figure 9-28). Compare this synovial joint with the shallow, almost saucer-shaped glenoid cavity (see Figure 9-15) and the head of the humerus. You should be able to see from these figures why the hip joint structurally has a somewhat more limited range of movement than the shoulder joint.

Box 9-2 discusses joint replacement, a procedure most often done on the hip and knee joints.

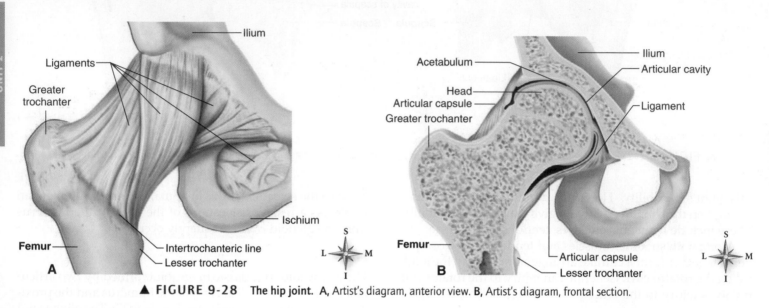

▲ **FIGURE 9-28** **The hip joint.** **A,** Artist's diagram, anterior view. **B,** Artist's diagram, frontal section.

 BOX 9-2 **Sports & Fitness**

Joint Replacement

Arthroplasty is the total or partial replacement of a diseased joint with an artificial device called a **prosthesis.** Prosthetic joints or joint components must allow for considerable mobility and be durable, strong, and compatible with surrounding tissues. Currently, hip and knee replacements are performed in nearly equal numbers. They are the most common orthopedic operations performed on older persons (more than 300,000 times each year in the United States). In addition, large numbers of elbow, shoulder, finger, and other joint replacement procedures are performed annually. *Osteoarthritis* and other types of degenerative bone disease frequently destroy or severely damage joints and cause continuous pain. Total joint replacement is often the only effective treatment option available.

The total hip replacement (THR) procedure involves replacement of the femoral head by a metallic alloy prosthesis and the acetabular socket with a high-density polyethylene cup. Many of the newer prostheses are "porous coated" to permit natural ingrowth of bone for stability and retention. The advantage of natural bone ingrowth is prevention of loosening that occurs if the cement bond weakens. Advanced surgical techniques now use newer metal prostheses fabricated from alloys of titanium, cobalt/chrome, and surgical steel. As a result, 10- to 15-year THR and total knee replacement success rates have risen in older patients.

However, even more durable metal and plastic components are necessary to extend the 10- to 15-year life span of most hip and knee prostheses currently in use. At the present time, younger patients who stress artificial joints more than older adults are often asked to delay replacement. Instead, they are encouraged to rely on evolving therapies, which are intended to heal or regenerate diseased joint tissues. The radiograph illustration shows placement of a noncemented, porous-coated femoral prosthesis made of titanium alloy and a cemented acetabular socket made of high-density plastic.

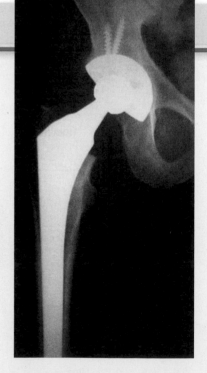

Knee Joint

The knee joint is the largest and one of the most complex and frequently injured joints in the body (Figure 9-29). Most obvious is that the condyles of the femur articulate with the flat upper surface of the tibia. Of the many ligaments that hold the femur bound to the tibia, five major ligaments are shown in Figure 9-29. Many of these (especially the anterior cruciate ligament, or ACL) are often torn by players in contact sports (Box 9-3). Thirteen bursae serve as pads around the knee joint: four anterior, four lateral, and five medial.

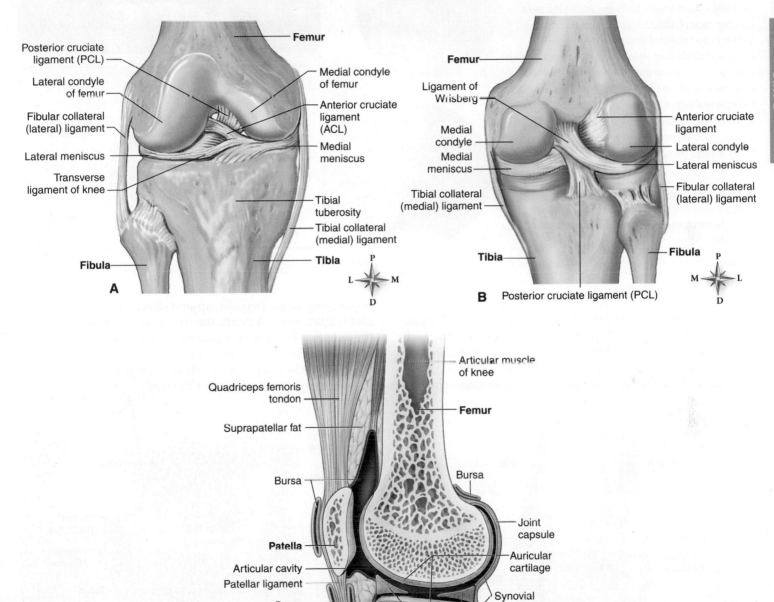

▲ **FIGURE 9-29** **The right knee joint.** **A,** Artist's diagram, dissection of the right knee viewed from in front. **B,** Viewed from behind. **C,** Artist's diagram, sagittal section through the knee joint.

BOX 9-3 Sports & Fitness

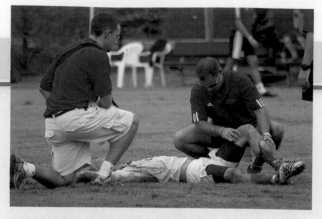

The Knee Joint

The knee is the largest and most vulnerable joint in our body. In fact, knee injuries are among the most common type of athletic injury. This is because the knee is often subjected to sudden, strong, and unusual forces during athletic activity. Sometimes the articular cartilages on the tibia become torn when the knee twists while bearing weight. The ligaments holding the tibia and femur together and those holding the medial and lateral menisci can be injured in this way. Knee injuries also may occur when a weight-bearing knee is hit by another person, especially from the side.

▶ **Knee arthrogram. A,** Normal medial meniscus. Spot film showing the normal triangular shape of the meniscus *(arrows)*. **B,** Linear tear of the medial meniscus.

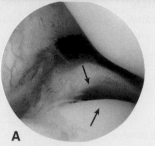

A

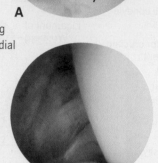

B

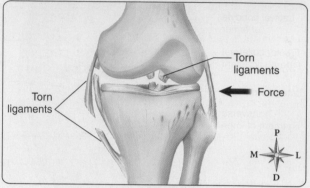

Torn ligaments

Torn ligaments

Force

Ankle Joints

The ankle is really a complex of two joints: the *true ankle joint* and the *subtalar joint*. These joints work together, but each provides the ankle with different movements. The **true ankle joint,** comprising the distal ends of the medial malleolus of the tibia and the lateral malleolus of the fibula embracing the underlying talus, permits up-and-down movement of the foot (Figure 9-30). Beneath the true ankle is the second joint, the **subtalar joint,** which consists of the talus overlying the calcaneus. This second joint permits side-to-side motion of the foot. Working together, the true ankle joint and the subtalar joint provide for rotation of the foot.

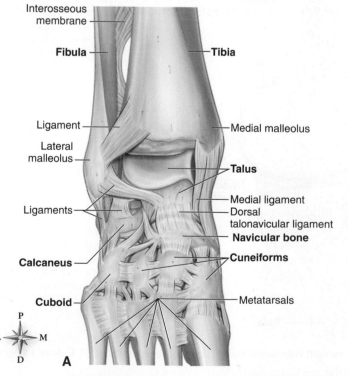

Interosseous membrane — Fibula — Tibia
Ligament — Medial malleolus
Lateral malleolus — Talus
Ligaments — Medial ligament — Dorsal talonavicular ligament — Navicular bone
Calcaneus — Cuneiforms
Cuboid — Metatarsals

A

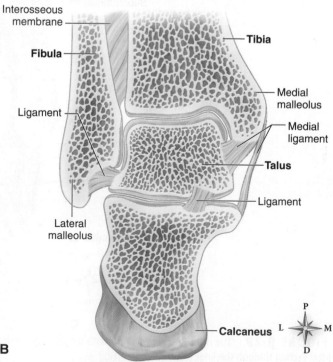

Interosseous membrane — Tibia
Fibula — Medial malleolus
Ligament — Medial ligament — Talus
Lateral malleolus — Ligament
Calcaneus

B

▲ **FIGURE 9-30** **The ankle joint. A,** Artist's diagram, anterior view. **B,** Artist's diagram, frontal section, anterior view.

Vertebral Joints

One vertebra connects to another by several joints: between their *bodies*, as well as between their articular, transverse, and spinous *processes*. Recall that the cartilaginous joints between the *bodies* of adjacent vertebrae permit only very slight movement and are classified as symphyses. However, the synovial joints between the articulating surfaces of the vertebral *processes* are more movable and are classified as gliding joints. These joints hold the vertebrae firmly together so that they are not easily dislocated. The result is the formation of a flexible column—our spine.

The *bodies* of adjacent vertebrae are structurally linked and connected by tough intervertebral disks (Figure 9-31, *A*) and strong ligaments. However, sudden compression of these disks by exertion or trauma may cause a part of a disk to become damaged and protrude into the spinal canal. This protrusion in turn presses on spinal nerves or the spinal cord itself (Figure 9-31, *B*). Severe and debilitating back pain may result. This condition, called a *herniated disk* ("slipped disk"), is one of the most common and chronically disabling conditions afflicting humanity. More workdays are lost to lower back injury, specifically herniated disks, than to any other malady.

Range of Movement at Synovial Joints

All synovial joints permit one or more of the following *types of movements:* (1) angular, (2) circular, (3) gliding, and (4) special. The actual types of movement possible at a given synovial joint depend on the shapes of the articulating surfaces of the bones forming the joints and on the positions of the joint's ligaments, nearby muscles, and tendons.

Typically there are several different *types of motion* possible at a given synovial joint (Table 9-5). These movements

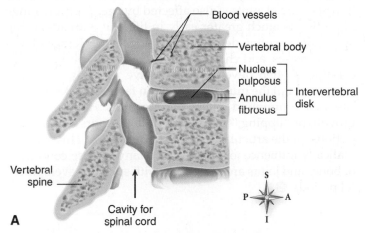

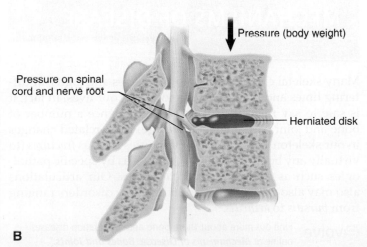

▲ **FIGURE 9-31** **Vertebrae and intervertebral disks.** Sagittal section of vertebrae showing normal **(A)** and herniated **(B)** disks.

TABLE 9-5 **Major Types of Joint Movements**

MOVEMENT	EXAMPLE	DESCRIPTION
Flexion (to flex a joint)	Flexion	Reduces the angle of the joint, as in bending the elbow
Extension (to extend a joint)	Extension	Increases the angle of a joint, as in straightening a bent elbow
Rotation (to rotate a joint)	Rotation	Spins one bone relative to another, as in rotating the head at the neck joint
Circumduction (to circumduct a joint)	Circumduction	Moves the distal end of a bone in a circle, keeping the proximal end relatively stable, as in moving the arm in a circle and thus circumducting the shoulder joint
Abduction (to abduct a joint)	Abduction	Increases the angle of a joint to move a part away from the midline, as in moving the arm up and away from the side of the body
Adduction (to adduct a joint)	Adduction	Decreases the angle of a joint to move a part toward the midline, as in moving the arm in and down toward the side of the body

TABLE 9-6   Additional Types of Joint Movement

MOVEMENT	DESCRIPTION
Hyperextension	Increase the angle of a joint beyond its usual anatomical position, as in tilting the neck backward; can also refer to abnormally extending a part, causing injury
Plantar flexion	Bending the ankle to point the toes and foot downward
Dorsiflexion	Bending the ankle to point the toes and foot upward
Supination	Twists the forearm to move the hand to a thumb-outward (lateral) position; also applies to a similar twisting movement of the leg
Pronation	Twists the forearm to move the hand to a thumb-inward (medial) position; also applies to a similar twisting movement of the leg
Inversion	Bends the ankle to move the sole of the foot inward (medially)
Eversion	Bends the ankle to move the sole of the foot outward (laterally)
Protraction	Moves a part forward, as in thrusting the mandible anteriorly
Retraction	Moves a part backward, as in pulling the mandible inward (posteriorly)
Elevation	Moves a part upward, as in raising the mandible to close the mouth
Depression	Moves a part downward, as in lowering the mandible to open the mouth

are classified according to opposing action. For example, *extending* and *flexing* the arm are opposing actions.

Angular movements include flexion, extension and *hyperextension*, *plantar flexion* and *dorsiflexion*, and abduction and adduction (Table 9-6). Other movements include **circular movements** such as rotation and circumduction, and *supination* and *pronation*. **Gliding movements** include those between the vertebral processes, creating both rigidity and flexibility. **Special movements** such as *inversion* and *eversion*, *protraction* and *retraction*, and *elevation* and *depression*, are unique or unusual movements that occur only in a few joints.

QUICK CHECK

22. List some representative synovial joints.
23. What are the major components of the hip joint?
24. What are the major components of the knee joint?
25. Briefly compare a vertebral joint with an ankle joint.
26. Which joint is the most complex and most frequently injured? Why?

Cycle of LIFE ⭘

Many changes take place in our body's skeletal framework over the course of our lives. These changes result largely from structural changes in bone, cartilage, and their accompanying muscles. For example, the bones of young children are not completely ossified and therefore can withstand the stress of falling during the process of learning how to walk. In contrast, the density and strength of bone and cartilage in young and middle-aged adults permits us to carry great loads. As we age, however, loss of bone density can make us prone to fractures from simply walking or lifting with moderate force. In addition, the loss of skeletal tissue density may result in the compression of weight-bearing bones. This inevitably causes a loss of height and may even result in the

inability to maintain an erect posture. All this may be accompanied by degeneration of muscle tissue as well.

Changes in the way bones develop and ossify from infancy to old age also affect our joints. For example, the fontanels in an infant's skull disappear and the epiphyseal growth plates ossify as our skeletons mature. Range of motion at synovial joints is also affected by age. Typically range of motion is much greater early in life, but with advancing age, joints are often described as being "stiff." Range of motion decreases, and changes in our gait are typical. Disease conditions often develop that involve not only the bones and joints, but the muscles as well. In fact, some skeletal diseases have profound effects on our joints. Abnormal bone growth, or "lipping," may result in bone spurs—sharp projections on the articular surfaces of our bones. This can dramatically influence joint function. Many disease conditions of bones and joints are associated with specific developmental periods. ●

MECHANISMS OF DISEASE

Many skeletal changes occur in our bones and joints, at differing times and rates, over the course of our lives. In fact, it is safe to say that most of us will experience a number of bone and joint pathologies as we age. Bone-related changes in our skeleton can be caused by injuries such as *fractures* (to virtually any bone in the body), as well as by specific pathologies such as *abnormal spinal curvatures*. Our articulations also may also exhibit many inflammatory disorders ranging from *bursitis* to *arthritis*.

ᐧevolve   Find out more about these bone and articulation diseases online at *Mechanisms of Disease: Bones and Joints.*

The BIG Picture

The skeletal system and its subdivisions are good examples of increasing complexity in our bodies. We've grouped skeletal tissues into discrete organs (bones) and then joined groups of individual bones together with varying numbers and kinds of other structures such as blood vessels and nerves, to form a complex operational unit—the skeletal system. As we've seen, the skeletal system is much more than an assemblage of individual bones—it is a complex and interdependent functional unit essential for life.

In this chapter, you have also seen many examples of the intimate relationship that exists between structure and function. Proper functioning of the joints is crucial for controlled and purposeful movements to occur. The fine movements of our synovial joints in particular permit movements that allow us to respond to and control our environments in ways that are impossible for other animals. For example, the great mobility of the human upper extremity is possible because of the arrangement of the bones in the shoulder girdle, arm, forearm, wrist, and hand; the location and method of attachment of muscles to these bones; and the proper range of motion of the joints involved.

All of this is required to provide the mobility and extensive range of motion to the hand. In a peculiar way, then, the hand is "the reason for the upper extremity" as many anatomists are fond of saying. After all, it is the proper functioning of the thumb that allows us to grasp and manipulate objects with great dexterity and control. By enabling us to effectively interact with our external environment, the joints of the upper extremity contribute in significant way to our maintenance of homeostasis.

LANGUAGE OF SCIENCE AND MEDICINE *continued from page 153*

diarthrosis (dye-ar-THROH-sis)
[*dia-* between, *-arthr-* joint, *-osis* condition] *pl.,* diarthroses (dye-ar-THROH-seez)

ethmoid (ETH-moyd)
[*ethmo-* sieve, *-oid* like]

face
[*fac-* appearance]

facet (fah-SET or FASS-et)
[*fac-* appearance (face), *-et* small]

false pelvis (PEL-vis)
[*pelvis* basin] *pl.,* pelves or pelvises (PEL-veez, PEL-vis-ez)

femur (FEE-mur)
[*femur* thigh]

fibrous joint (FYE-brus joynt)
[*fibr-* fiber, *-ous* relating to]

fibula (FIB-yoo-lah)
[*fibula* clasp] *pl.,* fibulae or fibulas (FIB-yoo-lee, FIB-yoo-lahz)

flatfoot

fontanel (FON-tah-nel)
[*fontan-* fountain, *-el* little]

frontal bone (FRON-tal bohn)
[*front-* forehead, *-al* relating to]

gliding joint

gliding movement

gomphosis (gom-FOH-sis)
[*gomphos-* bolt, *-osis* condition] *pl.,* gomphoses (gom-FOH-seez)

hinge joint
[*hinj* joint]

humerus (HYOO-mer-us)
[*humerus* upper arm] *pl.,* humeri (HYOO-mer-eye)

hyoid bone (HYE-oyd bohn)
[*hy-* Greek letter upsilon (Y or υ, *-oid* like]

ilium (IL-ee-um)
[*ilium* flank] *pl.,* ilia (IL-ee-ah)

inferior nasal concha (in-FEER-ee-or NAY-zal KONG-kah)
[*infer-* lower, *-or* quality, *nas-* nose, *-al* relating to, *concha* sea shell] *pl.,* conchae (KONG-kee or KONG-kay)

ischium (IS-kee-um)
[*ischium* hip joint] *pl.,* ischia (IS-kee-ah)

joint capsule

joint cavity

lacrimal bone (LAK-ri-mal)
[*lacrima-* tear, *-al* relating to]

ligament (LIG-ah-ment)
[*liga-* bind, *-ment* result of action]

lumbar vertebra (LUM-bar VER-teh-bra)
[*lumb-* loin, *-ar* relating to, *vertebra* that which turns] *pl.,* vertebrae

mandible (MAN-di-bal)
[*mandere* chew]

maxilla (mak-SIH-lah)
[*maxilla* upper jaw] *pl.,* maxillae (mak-SIH-lee)

meniscus (meh-NIS-kus)
[*meniscus* crescent] *pl.,* menisci (meh-NIS-eye or meh-NIS-kye)

metacarpal bone (met-ah-KAR-pal bohn)
[*meta-* beyond, *-carp-* wrist, *-al* relating to]

multiaxial joint (mul-tee-AK-see-al)
[*multi-* many, *-axi-* axle, *-al* relating to]

nasal bone (NAY-zal bohn)
[*nas-* nose, *-al* relating to]

occipital bone (awk-SIP-it-al bohn)
[*occipit-* back of head, *-al* relating to]

olecranon (oh-LEK-rah-non)
[*olecranon* elbow]

olecranon fossa (oh-LEK-rah-non FOSS-ah)
[*olecranon* elbow, *fossa* ditch] *pl.,* fossae (FOSS-ee)

palatine bone (PAL-ah-tyne bohn)
[*palat-* palate (roof of mouth), *-ine* relating to]

parietal bone (pah-RYE-ah-tal bohn)
[*parie-* wall, *-al* relating to]

patella (pah-TEL-ah)
[*pat-* dish, *-ella* small] *pl.,* patellae (pah-TEL-ee)

pectoral girdle (PEK-toh-ral GER-dul)
[*pector-* breast, *-al* relating to, *girdle* belt]

pelvic girdle (PEL-vic GER-dul)
[*pelvi-* basin, *-ic* relating to, *girdle* belt]

pelvis (PEL-vis)
[*pelvis* basin]

phalanx (fah-LANKS)
[*phalanx* formation of soldiers in rows] *pl.,* phalanges (fah-LAN-jeez)

pivot joint (PIV-it joynt)

prosthesis (pros-THEE-sis)
[*prosthesis* addition] *pl.,* prostheses (pros-THEE-seez)

pubis (PYOO-bis)
[*pubis* groin] *pl.,* pubes (PYOO-beez)

radius (RAY-dee-us)
[*radius* ray] *pl.,* radii (RAY-dee-eye)

rotator cuff (roh-TAY-tor)
[*rot-* turn, *-ator* agent]

sacrum (SAY-krum)
[*sacr-* sacred, *-um* thing]

continued

LANGUAGE OF SCIENCE AND MEDICINE *continued from page 181*

scapula (SKAP-yoo-lah)
[*scapula* shoulder blade] *pl.,* scapulae (SKAP-yoo-lee)

sinus (SYE-nus)
[*sinus* hollow]

special movement

sphenoid bone (SFEE-noyd bohn)
[*spheno-* wedge, *-oid* like]

spinous process (SPY-nus PRAH-ses)
[*spino-* backbone, *-ous* relating to, *process* project (from)] *pl.,* processes (PRAH-ses-eez)

sternum (STER-num)
[*sternum* breastbone] *pl.,* sterna or sternums (STER-nah, STER-numz)

subtalar joint (sub-TAY-ler joynt)
[*sub-* below, *-tal-* ankle (talus), *-ar* relating to]

surgical neck (SER-jik-el nek)
[*surg-* handwork, *-ical* relating to]

suture (SOO-chur)
[*sutur-* seam]

symphysis (SIM-fih-sis)
[*sym-* together, *-physis* growth] *pl.,* symphyses (SIM-fih-seez)

synarthrosis (sin-ar-THROH-sis)
[*syn-* together, *-arthr-* joint, *-osis* condition] *pl.,* synarthroses (sin-ar-THROH-seez)

synchondrosis (sin-kon-DROH-sis)
[*syn-* together, *-chondr-* cartilage, *-osis* condition] *pl.,* synchondroses (sin-kon-DROH-seez)

syndesmosis (sin-dez-MOH-sis)
[*syn-* together, *-desmo-* bond, *-osis* condition] *pl.,* syndesmoses (sin-dez-MO-seez)

synovial joint (si-NOH-vee-all joynt)
[*syn-* together, *-ovi-* egg (white), *-al* relating to]

synovial membrane (si-NOH-vee-all MEM-brayne)
[*syn-* together, *-ovi-* egg (white), *-al* relating to, *membran-* thin skin]

talus (TAY-lus)
[*talus* ankle] *pl.,* tali (TAY-lye)

tarsal bone (TAR-sal bohn)
[*tars-* ankle, *-al* relating to]

temporal bone (TEM-poh-ral bohn)
[*tempora-* temple (of head), *-al* relating to]

thoracic vertebra (thoh-RASS-ik VER-teh-bra)
[*thorac-* chest, *-ic* relating to, *vertebra* that which turns] *pl.,* vertebrae

thorax (THOH-raks)
[*thorax* chest] *pl.,* thoraces (THOH-rah-seez)

tibia (TIB-ee-ah)
[*tibia* shinbone] *pl.,* tibiae or tibias (TIB-ee-ee, TIB-ee-ahz)

transverse process (tranz-VERS PRAH-ses)
[*trans-* across or through, *-vers* turn, *process* project (from)] *pl.,* processes (PRAH-ses-eez)

trochlea (TROK-lee-ah)
[*trochlea* pulley] *pl.,* trochleae (TROK-lee-ee)

true ankle joint (tru ANG-kel joynt)

true pelvis (tru PEL-vis)
[*pelvis* basin] *pl.,* pelves or pelvises (PEL-veez, PEL-vis-ez)

ulna (UL-nah)
[*ulna* elbow] *pl.,* ulnae or ulnas (UL-nee, UL-nahz)

uniaxial joint (yoo-nee-AK-see-al joynt)
[*uni-* one, *-axi-* axle, *-al* relating to]

vertebra (VER-teh-bra)
[*vertebra* that which turns] *pl.,* vertebrae (VER-teh-bray or VER-teh-bree)

vertebral foramen (ver-TEE-bral for-AY-men)
[*vertebra* that which turns, *-al* relating to, *foramen* hole] *pl.,* foramina (foh-RAM-in-ah) or foramens (foh-RAY-menz)

vomer bone (VOH-mer bohn)
[*vomer* plowshare]

zygomatic bone (zye-goh-MAT-ik bohn)
[*zygo-* union or yoke, *-ic* relating to]

CHAPTER SUMMARY

To download an MP3 version of the chapter summary for use with your iPod or other portable media player, access the Audio Chapter Summaries online at http://evolve.elsevier.com.

HINT Scan this summary after reading the chapter to help you reinforce the key concepts. Later, use the summary as a quick review before your class or before a test.

INTRODUCTION

A. Skeletal tissues are organized to form bones
B. Adult skeleton is composed of 206 bones
  1. Variations in total number are common

DIVISIONS OF THE SKELETON (TABLES 9-1 AND 9-2)

A. Bones are organized into two major subdivisions (Figure 9-1)
  1. Axial—80 bones—form the upright axis of the body and 6 tiny bones of the ear
  2. Appendicular—126 bones; form the shoulder girdles, arms, wrists, and hands, hip girdles, legs, ankles, and feet

B. Axial skeleton
  1. Skull—comprises 28 irregularly shaped bones in two major divisions (Figures 9-2 through 9-6; Table 9-1)
    a. Cranial bones
      (1) Frontal bone—forms forehead, anterior part of the top of the cranium, upper part of the orbits of the eyes; contains sinuses (Figure 9-7)
      (2) Parietal bones—form the bulging top of the cranium and form several immovable joints
      (3) Temporal bones—house the middle and inner ear structures and contain the mastoid sinuses
      (4) Occipital bone—creates the framework of the lower, posterior part of the skull
      (5) Sphenoid bone—bat-shaped bone located in the central area of the floor of the cranium; anchors the frontal, parietal, occipital, and ethmoid bones (Figure 9-4)
      (6) Ethmoid bone—a complicated, irregularly shaped bone; helps to fashion the anterior part of the cranial floor, the medial walls of the orbits of the eyes, the upper parts of the nasal septum, the sides of the

nasal cavity, and part of the roof of the nasal cavity (Figures 9-2, 9-4, and 9-7)
b. Facial bones
(1) Maxillae—serve as the foundation for the face; each maxilla joins with a nasal, a zygomatic, an inferior concha, and a palatine bone
(2) Mandible—lower jaw; largest and strongest bone of the face; only movable joint of our skull
(3) Zygomatic bone—cheekbone; forms the outer margin of the orbit
(4) Nasal bones—form the bridge of the nose
(5) Lacrimal bones—paper-thin; lie just posterior and lateral to each nasal bone; each contains a groove for the nasolacrimal (tear) duct
(6) Eye orbits—orbital cavities of the skull; contain our eyes and their associated muscles
(7) Blunt trauma to eyes may produce so-called blowout fractures, which result in the "raccoon eye" look
c. Palatine bones—form the posterior part of the hard palate
d. Inferior nasal conchae (turbinates)—form ledges within the nasal cavity
e. Vomer—along with the perpendicular plate of the ethmoid bone and the septal cartilage, completes the posterior of the septum
f. Fetal skull—skull of the fetus and newborn infant has a number of unique features
(1) Fontanels—provide space between the cranial bones; permit just enough compression of the skull to allow for its passage through the birth canal (Figure 9-8)
2. Hyoid bone—single, U-shaped bone below the mandible and just above the larynx (voice box); no articulations with any other bones (Figure 9-9)
3. Vertebral column—flexible, curved chain of 24 vertebrae; forms the longitudinal axis of the skeleton (Figure 9-10)
a. Single and fused vertebrae
(1) Cervical vertebrae (7)—form the neck
(2) Thoracic vertebrae (12)—chest region
(3) Lumbar vertebrae (5)—support the small of the back
(4) Sacrum—a single broad and massive bone in the adult created from the fusion of five separate vertebrae
(5) Coccyx—"tailbone"; created from the fusion of the last four or five vertebrae of the spine
b. Vertebral characteristics
(1) Body
(2) Spinous process
(3) Transverse process
(4) Vertebral foramen
(5) Odontoid process—second cervical vertebrae
(6) Facets—thoracic vertebrae
(7) Superior and inferior articular surfaces
c. Comparison of vertebrae
(1) Cervical vertebrae are smaller and more delicate than other vertebrae, and have short spinous processes
(2) Thoracic vertebrae are larger than cervical and have downward-pointing spinous processes
(3) Lumbar vertebrae have the largest bodies
(4) Sacrum results from fusion of five vertebrae
(5) Coccyx (tailbone) results from the fusion of four or five vertebrae

4. Sternum—dagger-shaped bone composed of three distinct sections (Figure 9-13)
a. Manubrium
b. Body
c. Xiphoid process
5. Ribs—12 pairs; together with the vertebral column and sternum, form the bony thoracic cage (thorax)
a. True ribs—join a costal cartilage that attaches directly to the sternum
b. False ribs—do not attach directly to the sternum; each of the costal cartilages of rib pairs 8, 9, and 10 fuse together and attach to the costal cartilage of rib 7
c. Floating ribs—do not attach (even indirectly) to the sternum (Figure 9-13)
C. Appendicular skeleton
1. Pectoral (shoulder) girdle—consists of two clavicles and two scapulae
a. Clavicles (collarbones)—rodlike and shaped somewhat like a flattened S; support the scapulae and also provide areas for the attachment of muscles from the upper arms, upper chest, and back
b. Scapulae (shoulder blades)—broad, somewhat triangular bones that help form the upper part of the back; spine forms two anterior processes
(1) Acromion process—creates the top of the shoulder; serves as a site for the attachment of muscles from the upper limb
(2) Coracoid process—provides sites for the attachment of limb muscles from the upper chest and limb (Figure 9-15)
2. Upper extremities—each consists of a scapula, clavicle, upper arm, lower arm (forearm), wrist, and hand
a. Humerus (upper arm bone)—articulates proximally with the glenoid cavity of the scapula and distally with the ulna and radius (Figure 9-16)
b. Ulna—long bone found on little finger side of the forearm (Figure 9-16)
c. Radius—long bone found on thumb side of the forearm (Figure 9-16)
d. Carpals (wrist bones)—bound closely and firmly by ligaments (Figure 9-17)
e. Metacarpals—bones that form the palm of the hand
f. Phalanges—bones of the fingers (Figure 9-17)
3. Pelvic girdle—comprises two coxal bones (hipbones or *os coxae*) bound to each other anteriorly (by strong ligaments) and to the sacrum posteriorly; provides a stable, somewhat circular base that supports the trunk of the body and attaches the lower extremities to it (Figures 9-18 and 9-19)
a. Each coxal bone is made of three separate bones
(1) Ilium—largest and uppermost
(2) Ischium—strongest and lowermost of the fused pelvic bones
(3) Pubis—most anterior bone
4. Lower extremities
a. Lower extremity—bones of the hip, thigh, leg, ankle, and foot (Figure 9-20)
(1) Femurs (thighbones)—longest and heaviest bones in the body
(2) Patella (kneecap)—largest sesamoid bone in the body
(3) Tibia (shinbone)—larger and more superficial than fibula
(4) Fibula—smaller than tibia and more laterally placed

(5) Tarsals (ankle bones)—(Figure 9-21)
(6) Phalanges—bones of the toes
b. Arches of the foot—provide support and flexibility
(1) Longitudinal
(2) Transverse

DIFFERENCES BETWEEN MALE AND FEMALE SKELETONS

A. Difference between male and female skeletons—general difference is one of size and weight (Figure 9-19)
1. On average, male skeleton is larger and heavier
2. Male pelvis is deep and funnel-shaped with a narrow subpubic angle
3. Female pelvis is shallow, broad, and flaring

ARTICULATIONS (JOINTS)

A. Articulation—point of contact between two or more bones; allow us to execute complex, highly coordinated, and deliberate movements

CLASSIFICATION OF JOINTS

A. Joints can be classified structurally according to the type of connective tissue that joins the bones together (Table 9-3)
B. Fibrous joints (synarthroses)—bones that fit closely together, fixing the position of the bones making up the joints (Figure 9-23)
1. Syndesmoses—joints in which fibrous bands (ligaments) connect two bones
2. Sutures—closely knit joints found only in the skull
3. Gomphoses—unique joints that occur between the root of a tooth and the mandible (or maxilla) of the jaw
C. Cartilaginous joints—bones that are joined together by either hyaline cartilage or fibrocartilage
1. Synchondroses—have hyaline cartilage between the articulating bones (Figure 9-24)
2. Symphysis—joint in which a pad or disk of fibrocartilage connects two bones (Table 9-3)
D. Synovial joints (diarthroses)—much more complex than either fibrous joints or cartilaginous joints (Figure 9-25)
1. Characterized by:
a. Joint capsule
b. Synovial membrane
c. Articular cartilage
d. Joint cavity
e. Meniscus
f. Ligament
g. Bursa
2. Types of synovial joints (Table 9-4)
a. Uniaxial joints—permit movement around only one axis and in only one plane
(1) Hinge joints—permit only back-and-forth movements such as flexion and extension
(2) Pivot joints—a projection from one bone articulates with a ring or notch in another bone
b. Biaxial joints—permit movement around two perpendicular axes and in two perpendicular planes
(1) Saddle joint—located in thumb
(2) Condyloid joints—biaxial joints in which a condyle fits into an elliptical socket

c. Multiaxial joints—permit movement around three or more axes and in three or more planes
(1) Ball-and-socket joints—formed when ball-shaped head fits into concave depression
(2) Gliding joints—characterized by relatively flat articulating surfaces that allow limited gliding movements; least movable of all synovial joints
3. Some representative synovial joints
a. Shoulder joint—most mobile joint; created by the joining of the head of the humerus and the glenoid cavity of the scapula (Figure 9-26)
b. Elbow joint—formed by two articulations, between the distal end of the humerus and the proximal ends of the radius and ulna (Figure 9-27)
c. Hip joint—stable joint because of the shape of the head of the femur and the acetabulum (Figure 9-28)
d. Knee joint—largest and one of the most complex and frequently injured joints in the body (Figure 9-29; Box 9-3)
(1) Stabilizing forces are supplied by a joint capsule, cartilages, and numerous ligaments and muscle tendons (Figure 9-29)
e. Ankle joint—bony structures and ligaments enhance stability; provide an excellent platform to bear the weight of the entire body during standing and walking (Figure 9-30)
f. Vertebral joints—one vertebra connects to another by several joints; between their bodies, as well as between their articular, transverse, and spinous processes
4. Range of movement at synovial joints
a. All synovial joints permit one or more of the following types of movements:
(1) Angular—flexion, extension and hyperextension, plantar flexion and dorsiflexion, and abduction and adduction
(2) Circular—rotation and circumduction, and supination and pronation
(3) Gliding
(4) Special—inversion and eversion, protraction and retraction, and elevation and depression
b. Typically there are several different types of motion possible at a given synovial joint (Table 9-5)
(1) These movements are classified according to opposing action (e.g., extending and flexing the arm)

REVIEW QUESTIONS

Write out the answers to these questions after reading the chapter and reviewing the Chapter Summary. If you simply think through the answer without writing it down, you won't retain much of your new learning.

HINT

1. Describe the skeleton as a whole and identify its two major subdivisions.
2. Identify and differentiate the bones in the cranium and face.
3. Name the five pairs of bony sinuses in the skull.
4. Identify the bony components of the thorax.
5. Identify the bones of the shoulder and pelvic girdles.

6. Identify, compare, and organize the bones of the arm, forearm, wrist, and hand versus those of the thigh, lower part of the leg, ankle, and foot.
7. Classify joints and specify each group according to function and structure.
8. Define the terms *fibrous joints*, *cartilaginous joints*, and *synovial joints*.
9. Name and define three types of fibrous joints. Give an example of each.
10. Name and define two types of cartilaginous joints. Give an example of each.
11. List and describe the different types of synovial joints. Give an example of each.
12. What joint makes possible much of the dexterity of the human hand? Describe it and the movements it permits.
13. Describe vertebral joints.
14. Describe and differentiate between the following joints: hip, knee.

CRITICAL THINKING QUESTIONS

 HINT *After finishing the Review Questions, write out the answers to these items to help you apply your new knowledge. Go back to sections of the chapter that relate to items that you find difficult.*

1. How would you compare and contrast the differences between the skeletal structure of males and females? What are the physiological reasons for these differences?
2. What are some functional advantages to the fontanels of a baby's skull during the birth process?
3. Explain how the shoulder and pelvic girdles stabilize the appendages.
4. The elbow, the joint between the bodies of vertebrae, and the root of a tooth in the jaw are all joints given different functional names. What are these names and what characteristic distinguishes them from one another?
5. A synovial joint is the most freely moving type of joint in the body, but joints require stability and protection also. Evaluate the seven structural characteristics of a synovial joint as primarily necessary for stability, movement, and protection.
6. Compare the range of movement at different joints.
7. Describe the anatomical structure of the knee and explain why the knee joint is the most frequently injured joint in the body.

continued from page 153

With the knowledge you have gained from reading this chapter, see if you can answer these questions about Robert's injuries from the Introductory Story.

1. Which bone in the spine was fractured?
   a. Thyroid vertebra
   b. Tibia
   c. Tarsal
   d. Thoracic vertebra
   e. Lumbar vertebrae

2. What is another name for the fourth and fifth ribs that were cracked in the accident?
   a. Floating ribs
   b. False ribs
   c. True ribs
   d. Costal ribs
   e. Cartilage ribs

3. What daily activity will be affected most by his broken mandible?
   a. Walking
   b. Eating
   c. Writing
   d. Sitting
   e. Hearing

4. What daily activity will be affected most by his shattered tibia?
   a. Walking
   b. Eating
   c. Writing
   d. Sitting
   e. Watching television

 HINT *To solve a case study, you may have to refer to the glossary or index, other chapters in this textbook, A&P Connect, Mechanisms of Disease, and other resources.*

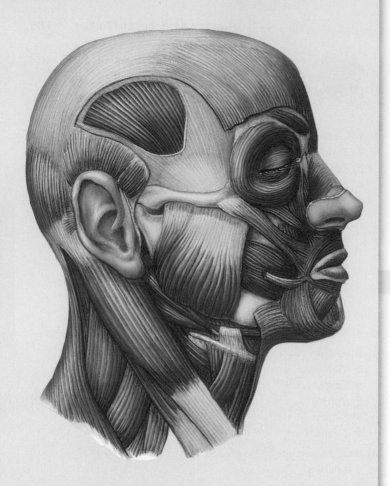

Muscular System

STUDENT LEARNING OBJECTIVES

At the completion of this chapter, you should be able to do the following:

1. Outline the major unique features of the muscle cell.
2. Outline the structures and function of the sarcomere.
3. Define: motor neuron, neuromuscular junction, motor unit.
4. Discuss the mechanism of muscle contraction.
5. List and discuss the energy sources for muscle contraction and distinguish between anaerobic and aerobic respiration.
6. Compare cardiac muscle with smooth muscle.
7. Describe the structure of skeletal muscle.
8. Define: origin, insertion, muscle action, lever system.
9. List the major muscles of mastication and facial expression.
10. List the major muscles that move the head and trunk.
11. List the major muscles that move the upper limb (from the pectoral girdle to the hand).
12. List the major muscles that move the lower limb (from the pelvic girdle to the foot).
13. Discuss muscle tone, strength, and conditioning.

FOR years now, Miguel Cervantes, age 59, has suffered sleepless nights because of chronic muscle pain. Getting up in the morning is an agonizing process, slowly moving his legs over the sides of the bed until the pain in his lower back and shoulders subsides. He could no longer do yard work because of the unremitting pain in his muscles. Finally, his daughters convinced him to get a physical.

After extensive tests to rule out the presence of more dangerous diseases, the doctor was confident in diagnosing Miguel's problem as fibromyalgia. This is a chronic and painful disease, more common in the elderly, characterized by widespread aching, stiffness, fatigue, and tender spots in the body. Pain from fibromyalgia originates in the muscles and connective tissues of the body.

The doctor noticed a loss of muscle mass, which is vital to maintaining body tone and muscle power. She went on to explain to Miguel that muscle power—the product of the force and velocity of muscle shortening—declines faster and to a greater extent than muscle strength. The doctor also noted that loss of muscle power due to normal aging has a greater functional impact than loss of strength alone. In addition, Miguel was losing muscle mass as he aged, and a number of things beyond his control were implicated.

As you read this chapter, try to understand Miguel's chronic disease. After reading this chapter, propose some simple, safe methods that might offer some relief from his fibromyalgia.

* * *

Human survival depends on *movement,* and we move by means of muscles. It is the skeletal and muscular systems acting together that actually produce most of your body movements. Now that you have a basic understanding of your skeletal system and the articulations made between your bones, you can better understand the functions of the **muscular system.** Your body has over *600 muscles* comprising 40% to 50% of your body weight! Together with the structural scaffolding provided by the skeleton, muscles also determine the form and contours of your body. In the following pages, you will learn how *contractile units* are grouped into complex functioning organs—*the muscles.* The way in which muscles are grouped, the relationship of muscles to joints, and how muscles attach to the skeleton determine their movement.

Although there are several different types of muscles (cardiac, smooth, and skeletal), we'll begin our discussion with a general description of the structure and function of skeletal muscle cells, which are responsible for movements of our body's frame.

INTRODUCTION TO SKELETAL MUSCLES

General Functions of Skeletal Muscle

Our skeletal muscles *permit movement, produce heat, stabilize our joints,* and *maintain our posture.* These functions all rely on the characteristic of **excitability**—the ability of the muscle to respond to nervous and endocrine signals. Muscles can both *contract* and *extend.* The term **contraction,** when applied to muscles, means that the ends are being pulled together, *regardless* of whether the muscle cells are actually getting shorter. Muscles also are *extensible,* which means that they can stretch or lengthen, thereby

continued on page 219

returning to their resting length after contraction. Muscles may also extend while still exerting force, such as when you lower a heavy object in your hand.

Basic Structure of a Muscle Organ

A **muscle** is an organ and, like any organ, is made up of several types of tissue—as you can see in Figure 10-1. Muscle cells—muscle fibers—are covered by a fine connective tissue called the **endomysium.** Groups of these muscle fibers, called **fascicles,** are then bound together by a thicker connective tissue called **perimysium.** Finally, a coarse sheath called the **epimysium** covers the entire muscle. All three sheaths of connective tissue are continuous with the fibrous structures that attach muscles to bones or other structures (Table 10-1).

TABLE 10-1 Fibrous Coverings of Muscle Organs

FIBROUS STRUCTURE*	PART COVERED
Fascia	External to muscles, bones, and other organs
Superficial fascia	Under the skin
Deep fascia	Surrounds deeper organs, including epimysium of muscle
Tendon sheath	Tubelike tunnel around tendon of muscle; lined with synovial membrane
Epimysium†	Surrounds entire muscle organ
Perimysium†	Surrounds a fascicle (bundle) of muscle fibers
Endomysium†	Surrounds an individual muscle fiber

*Listed here from superficial to deep; all of these fibrous structures are continuous with one another (that is, their fibers blend together).
†Continue and fuse together to form a tendon or aponeurosis.

Coursing through these layers of muscle and connective tissues are nerves that detect the tension of the muscle organ and nerves that trigger its contraction.

The epimysium, perimysium, and endomysium join together to form a dense band that extends from the muscle as a **tendon.** This tough cord joins at its other end with the fibrous periosteum covering a bone. In a few muscles, the fibrous wrapping of a muscle instead extends as a broad, flat sheet of connective tissue called an **aponeurosis.** These flattened structures usually merge with the fibrous wrappings of another muscle. So strong are tendons and aponeuroses that they rarely tear, even under stress great enough to tear muscles or break bones!

Fibrous connective tissue surrounding muscle and outside the epimysium and tendon is called **fascia.**

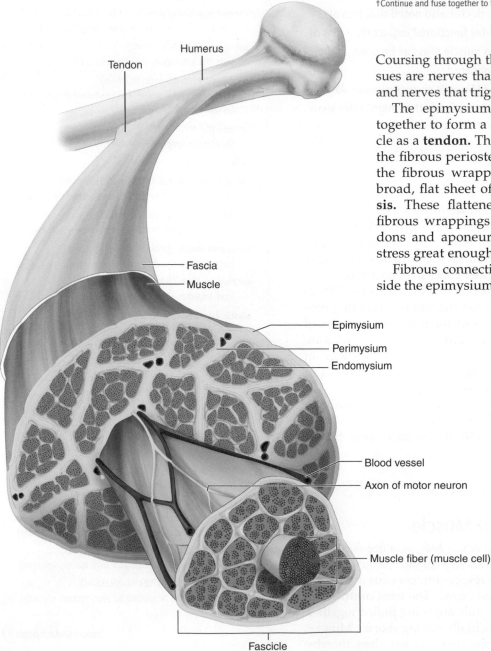

Humerus

Tendon

Fascia

Muscle

Epimysium

Perimysium

Endomysium

Blood vessel

Axon of motor neuron

Muscle fiber (muscle cell)

Fascicle

◀ **FIGURE 10-1** **Structure of a muscle organ.** Note that the connective tissue coverings—the epimysium, perimysium, and endomysium—are continuous with each other and with the tendon. The muscle fibers are held together by the perimysium in groups called fascicles.

Overview of the Muscle Cell

The ability of muscle cells to contract and extend is related to the microscopic structure of the skeletal muscle cells. This topic is somewhat complex, so follow along carefully using Figure 10-1 as your guide during our discussion. First, notice that a skeletal muscle is composed of bundles of skeletal muscle *fibers*. These **fibers** are actually individual muscle cells that have fused together during development to form a long, threadlike *multinucleated cell* (a cell with many nuclei). *Satellite (stem) cells* (located on the outer boundaries of muscle fibers) can, after muscle trauma, be called into action to produce new muscle cells during the repair process.

Skeletal muscle fibers have many of the same structural components of other cells. However, some of these structures have different names (Figure 10-2). For example, the **sarcolemma** is roughly equivalent to the plasma membrane of a typical cell. Likewise, the **sarcoplasm** is equivalent to the muscle's cytoplasm. As you might suspect (because of their intense metabolic activity), muscle fibers have many more mitochondria than typical body cells. Muscle fibers also contain networks of tubules and sacs known as the **sarcoplasmic reticulum (SR).** This is a more complex version of the smooth endoplasmic reticulum found in all other cells. The function of the SR is to temporarily store calcium ions (Ca^{++}). The membrane of the SR continually pumps Ca^{++} from the sarcoplasm and stores these ions within its sacs.

A structure unique to muscle cells is a system of transverse tubules, or **T tubules,** which extend across the sarcoplasm at *right angles* to the long axis of the cell. These T tubules are really inward extensions of the sarcolemma. Their primary function is to allow electrical signals, or *impulses* traveling along the sarcolemma, to move deeper into the cell.

The cytoskeleton of a muscle fiber is also unique. Typically it consists of thousands of very fine filaments called **myofibrils.** These filaments extend lengthwise along a skeletal muscle fiber and nearly fill the sarcoplasm. Myofibrils, in turn, are made of still finer fibers called **myofilaments** (see Figure 10-2). The myofilaments are further classified as *thick filaments* and *thin filaments.* (Note: These *colorless* filaments are colored for you in the diagram to give contrast to the illustration.)

Now refer to the figure in Box 10-1 for a more detailed look at the **sarcomere**—the basic *contractile unit* of the muscle cell. Each sarcomere is a segment of the myofibril that lies between two successive Z "lines." Rather than being a "line," it is actually a three-dimensional *disk*, more accurately called the **Z disk.** Besides serving as an anchor for myofibrils, the Z disks create useful landmarks that separate one sarcomere from the

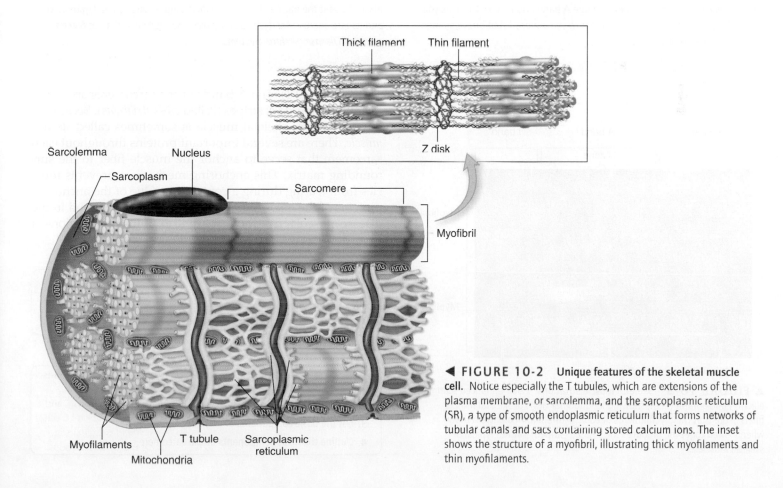

◀ **FIGURE 10-2**   **Unique features of the skeletal muscle cell.** Notice especially the T tubules, which are extensions of the plasma membrane, or sarcolemma, and the sarcoplasmic reticulum (SR), a type of smooth endoplasmic reticulum that forms networks of tubular canals and sacs containing stored calcium ions. The inset shows the structure of a myofibril, illustrating thick myofilaments and thin myofilaments.

BOX 10-1 A More Detailed Look at the Sarcomere

The **sarcomere** is the basic contractile unit of the muscle cell. To understand the function of the sarcomere, it is important that you fist visualize its three-dimensional nature. You will then realize that what used to be called the Z line (which often looks like a zigzag line in a flat diagram) is really a dense plate or disk (Z disk). The thin filaments anchor directly to the Z disk. Besides being an anchor for myofibrils, the Z disk is useful as a landmark: It separates one sarcomere from the next. (The name "Z disk" comes from the German word *zwischen*, meaning "between.")

Detailed analysis of the sarcomere also shows that the thick (myosin) filaments are held together and stabilized by protein molecules that form a middle line called the **M line**. Note that the regions of the sarcomere are identified by specific zones or bands:

A band—the segment that runs the entire length of the thick filaments

I band—the segment that includes the Z disk and the ends of the thin filaments where they do not overlap the thick filaments

H zone—the middle region of the thick filaments where they do not overlap the thin filaments

Note in Figure 10-2 that the T tubules in human muscle fibers align themselves along the borders between the A band and I band. Later, as you review the process of contraction, note how the regions listed above change during each step of the process.

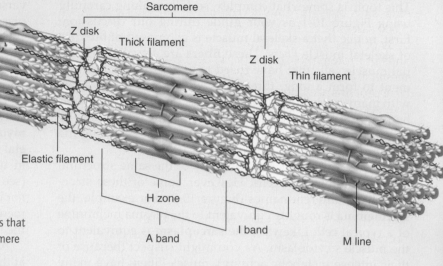

▲ **Myofibril.**

In addition to thin and thick filaments, each sarcomere has numerous **elastic filaments.** The elastic filaments are believed to give myofibrils, and thus muscle fibers, their characteristic elasticity. Dystrophin, not shown here, is a protein that holds the actin filaments to the sarcolemma. Dystrophin and a complex of connected molecules anchors the muscle fiber to surrounding matrix so that the muscle does not break during a contraction. Dystrophin and its role in muscular dystrophy are discussed further online at *Mechanisms of Disease: Skeletal System*.

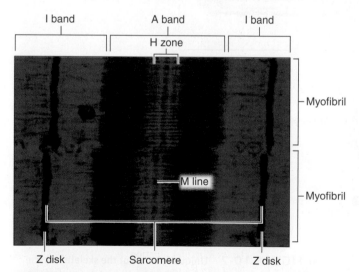

▲ **FIGURE 10-3** **Skeletal muscle striations.** Color-enhanced scanning electron micrograph (SEM) showing longitudinal views of skeletal muscle fibers. Note that the myofilaments of each myofibril form a pattern that, when viewed together, produces the striated (striped) pattern typical of skeletal muscle.

next (Figure 10-3). The **A bands** of the sarcomeres appear as relatively wide, dark stripes (called *cross striations*). Because of these striations, skeletal muscle is sometimes called *striated muscle*. There are several important proteins throughout each sarcomere that serve to anchor the muscle fiber to the surrounding matrix. This anchoring mechanism prevents muscles from tearing during a contraction. One of the anchoring proteins of the inner membrane, *dystrophin,* is affected in the disease muscular dystrophy. We will examine these regions in more detail as we learn about the process of conduction.

A&P CONNECT

The fibers of skeletal muscles are classified into three types according to their structural and functional characteristics. To learn more about these classifications, check out **Types of Muscle Fibers** on **A&P Connect**.

evolve

QUICK CHECK

1. What are the general functions of muscles?
2. What are the tissues that make up a muscle organ?
3. List the basic components of a typical muscle cell.
4. Outline the basic components of the sarcomere.

Components of Muscle Contraction

Each muscle *fiber* (remember, a muscle fiber is a multinucleate muscle cell) contains a thousand or more myofibrils lying side by side. In each myofibril are approximately 15,000 sarcomeres. In turn, each sarcomere is made of hundreds of thick and thin filaments. There are four important kinds of protein molecules that make up the myofilaments: *myosin, actin, tropomyosin,* and *troponin* (Figure 10-4). All of these proteins are necessary for muscle contraction. As you have seen in the previous section, the thick and thin filaments alternate within a myofibril (refer back a moment to review Figure 10-2). They are also very close to one another—an arrangement critical for muscle contraction, as we will see in a moment.

The contraction of a muscle fiber involves a complex process that must be coordinated in a stepwise fashion. Look at Figure 10-5 as we discuss the components of the system that controls this vital process.

First, a skeletal muscle fiber normally remains "at rest" until it is stimulated to contract by a signal from the axon terminal of a **motor neuron.** The distal ends of motor neurons connect to part of the sarcolemma of a muscle fiber called the *motor end plate.* The junction of the motor neuron and the motor end plate forms a **neuromuscular junction (NMJ).**

The NMJ is a specialized *synapse* (a junction) characterized by a tiny gap—the *synaptic cleft.* Specific "signal molecules" called neurotransmitters must *bridge this gap* in order to pass on their signal from the end of the neuron to the muscle fiber. This happens when nerve impulses reach the end of a motor neuron.

Small vesicles in the cytoplasm of the nerve ending release neurotransmitters **(acetylcholine,** or **ACh)** into the synaptic cleft. The neurotransmitter molecules diffuse quickly across this microscopic gap and come into contact with the sarcolemma of the muscle fiber. Here acetylcholine *receptors* bind with the acetylcholine. This action of acetylcholine binding with its receptor molecules produces an impulse in the sarcolemma by a process called **excitation.** After the signal is sent, the enzyme *acetylcholinesterase* rapidly breaks down—at a rate of 5,000 molecules per second!—acetylcholine into choline and an acetate group. The choline is transported back into the nerve terminals, where it is used to synthesize new acetylcholine molecules.

We will discuss neurons, impulses, and synapses more completely in the next chapter. However, before going on, read the previous paragraphs and again refer to the diagram until you understand the structure and function of the neuromuscular junction.

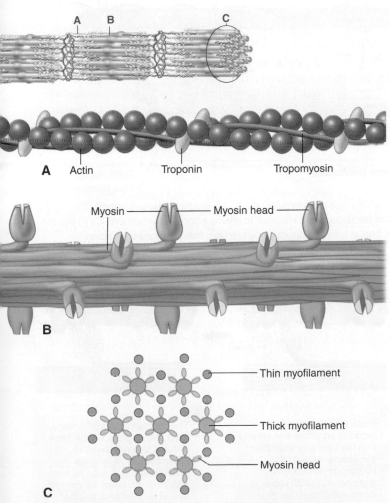

▲ FIGURE 10-4 Structure of myofilaments. **A,** Thin myofilament. **B,** Thick myofilament. **C,** Cross section of several thick and thin myofilaments showing the relative positions of myofilaments and the myosin heads that will form cross bridges between them.

A Actin   Troponin   Tropomyosin

Myosin — Myosin head

B

Thin myofilament

Thick myofilament

Myosin head

C

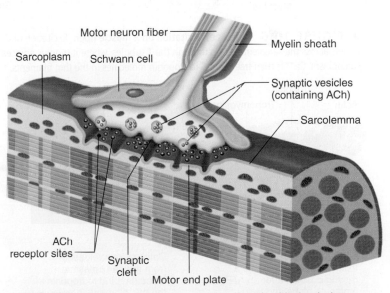

Motor neuron fiber

Myelin sheath

Sarcoplasm   Schwann cell

Synaptic vesicles (containing ACh)

Sarcolemma

ACh receptor sites   Synaptic cleft   Motor end plate

▲ FIGURE 10-5 Neuromuscular junction (NMJ). This sketch shows a side view of the NMJ. Note how the distal end of a motor neuron fiber forms a synapse, or "chemical junction," with an adjacent muscle fiber. Neurotransmitter molecules (specifically, acetylcholine, or ACh) are released from the neuron's synaptic vesicles and diffuse across the synaptic cleft. There they stimulate receptors in the motor end plate region of the sarcolemma.

Mechanism of Muscle Contraction

As we've just seen, muscle fibers contract after the impulse received by the sarcolemma creates an *electric impulse*—a temporary electrical voltage imbalance (discussed in Chapter 11). The impulse is conducted over the muscle fiber's sarcolemma and then *inward* along the T tubules (Figure 10-6). The impulse in the T tubules then triggers the release of a flood of calcium ions from the adjacent sacs of the SR. The calcium ions immediately combine with troponin in the thin filaments of the myofibrils. As you can see in Figure 10-7, this causes the tropomyosin to shift and expose the active *binding sites* on the actin molecules. Once the active sites for contraction are exposed, *myosin heads* of the thick filaments bind to actin molecules in the nearby thin filaments. These myosin heads bend with great force, literally *pulling* the thin filaments past them. Each head then releases itself, binds to the next active site, and pulls again.

Interestingly, the myosin head acts both as a binding site for actin *and* a binding site for adenosine triphosphate (ATP). This is how the myosin heads obtain the energy to "grab" onto the actin-binding sites. The myosin head breaks down ATP to release the energy.

Figure 10-8 shows how sliding of the thin filaments toward the center of each sarcomere subunit quickly shortens the entire myofibril—and thus the entire muscle fiber. This model of muscle contraction is called the **sliding-filament model.** However, it might be more appropriately named the *ratcheting-filament model* because the myosin heads actively "ratchet" the thin filaments toward the center of the sarcomere.

In essence, the contracting sarcomeres are working hard to pull the ends of the muscle fiber toward one another, as if there were a molecular tug-of-war (Figure 10-9). Because the actin-myosin bond is not permanent, the sarcomere cannot "lock up" to hold its position passively, once it has shortened.

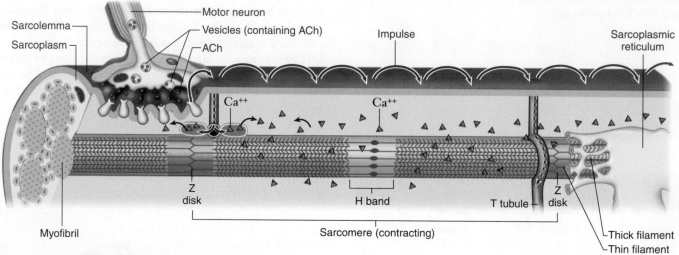

▲ **FIGURE 10-6** **Effects of excitation on a muscle fiber.** Excitation of the sarcolemma by a nerve impulse initiates an impulse in the sarcolemma. The impulse travels across the sarcolemma and through the T tubules, where it triggers adjacent sacs of the sarcoplasmic reticulum to release a flood of calcium ions (Ca^{++}) into the sarcoplasm. Ca^{++} is then free to bind to troponin molecules in the thin filaments. This binding, in turn, initiates chemical reactions that produce a contraction.

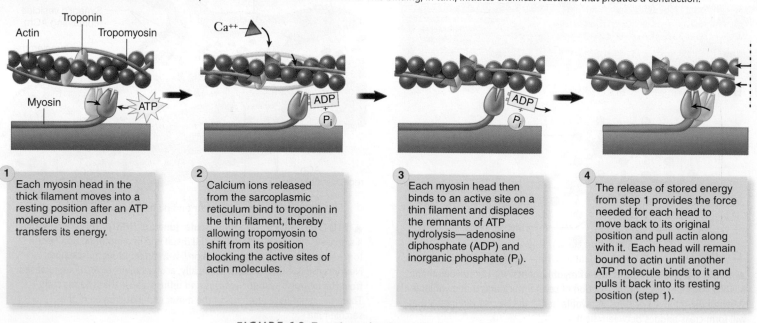

1 Each myosin head in the thick filament moves into a resting position after an ATP molecule binds and transfers its energy.

2 Calcium ions released from the sarcoplasmic reticulum bind to troponin in the thin filament, thereby allowing tropomyosin to shift from its position blocking the active sites of actin molecules.

3 Each myosin head then binds to an active site on a thin filament and displaces the remnants of ATP hydrolysis—adenosine diphosphate (ADP) and inorganic phosphate (P_i).

4 The release of stored energy from step 1 provides the force needed for each head to move back to its original position and pull actin along with it. Each head will remain bound to actin until another ATP molecule binds to it and pulls it back into its resting position (step 1).

▲ **FIGURE 10-7** The molecular basis of muscle contraction.

Thus, it takes energy in the form of ATP to actively *maintain* a shortened position. A great deal of energy is expended in intense muscular contraction, releasing an important by-product of muscle metabolism—*heat!*

Almost immediately after the SR releases its flood of calcium ions into the sarcoplasm, it begins actively pumping them back into the sacs once again. Within a few milliseconds, much of the calcium is recovered. As you might suspect, this shuts down the entire process of contraction. If no other nerve impulses follow immediately, the muscle fiber can relax. However, a muscle fiber may stay in contraction for some time. This can happen if there are *many nervous stimuli in rapid succession*, thus permitting calcium ions to remain available in the sarcoplasm for a longer period of time.

A summary of the major events of muscle contraction and relaxation is given in Box 10-2. Before continuing on, read through the different steps several times to make sure you understand the sequence of events.

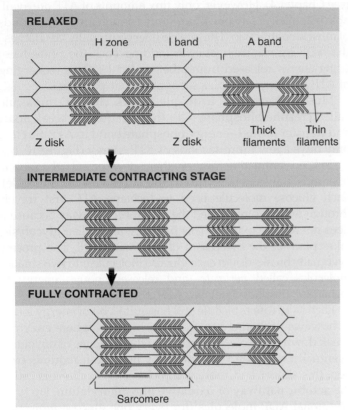

▲ **FIGURE 10-8 Sliding-filament model.** During contraction, myosin cross bridges pull the thin filaments toward the center of each sarcomere, thus shortening the myofibril and the entire muscle fiber.

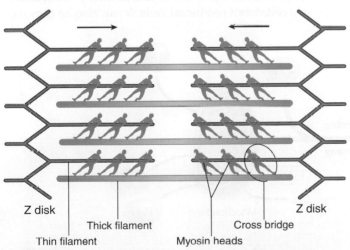

▲ **FIGURE 10-9 Simplified contracting sarcomere.** This diagram illustrates the concept of muscle contraction as a sort of tug-of-war game in which the myosin heads (shown here as little people) hold onto thin filament "ropes"—thus forming cross bridges. As the myosin heads pull on the thin filaments, the Z disks (Z lines) get closer together—thus shortening the sarcomere. Likewise, the short length of a sarcomere may be held in position by the continued effort of the myosin heads.

> **QUICK CHECK**
>
> 5. What are myofilaments?
> 6. What is the role of acetylcholine in muscle contraction?
> 7. What is a neuromuscular junction (NMJ)? How does it work?
> 8. Briefly, outline the sliding-filament model of muscle contraction.

BOX 10-2 Major Events of Muscle Contraction and Relaxation

Excitation and Contraction

1. A nerve impulse reaches the end of a motor neuron and triggers the release of the neurotransmitter acetylcholine (ACh).

2. ACh diffuses rapidly across the gap of the neuromuscular junction and binds to ACh receptors on the motor end plate of the muscle fiber.

3. Stimulation of ACh receptors creates an impulse that travels along the sarcolemma, through the T tubules, to the sacs of the sarcoplasmic reticulum (SR).

4. Ca^{++} is released from the SR into the sarcoplasm, where it binds to troponin molecules in the thin myofilaments.

5. Tropomyosin molecules in the thin myofilaments shift and thereby expose actin's active sites.

6. Energized myosin cross bridges of the thick myofilaments bind to actin and use their energy to pull the thin myofilaments toward the center of each sarcomere. This cycle repeats itself many times per second, as long as adenosine triphosphate is available.

7. As the filaments slide past the thick myofilaments, the entire muscle fiber shortens.

Relaxation

1. After the impulse is over, the SR begins actively pumping Ca^{++} back into its sacs.

2. As Ca^{++} is stripped from troponin molecules in the thin myofilaments, tropomyosin returns to its position and blocks actin's active sites.

3. Myosin cross bridges are prevented from binding to actin and thus can no longer sustain the contraction.

4. Because the thick and thin myofilaments are no longer connected, the muscle fiber may return to its longer, resting length.

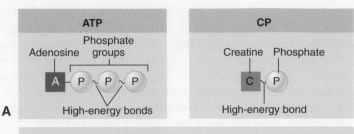

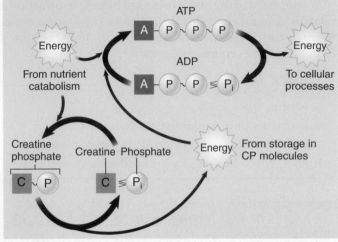

▲ **FIGURE 10-10**   **Energy sources for muscle contraction.**   **A,** The basic structure of two high-energy molecules in the sarcoplasm: adenosine triphosphate (ATP) and creatine phosphate (CP). **B,** This diagram shows how energy released during the catabolism of nutrients can be transferred to the high-energy bonds of ATP directly or, instead, stored temporarily in the high-energy bond of CP. During contraction, ATP is hydrolyzed and the energy of the broken bond is transferred to a myosin head.

ENERGY SOURCES FOR MUSCLE CONTRACTION

The energy required for muscular contraction is obtained from energy released from ATP after it binds to the myosin heads and breaks into adenosine diphosphate (ADP) and phosphate (see Chapter 4). However, only tiny amounts of ATP are available in the cell at any one time—only enough for a few seconds of muscle contraction. The muscle fiber continually "recharges" the ADP by *phosphorylating* it (adding back the phosphate) using energy released from other molecules. This recharging of ATP requires a complex set of metabolic pathways.

Muscle fibers have a unique "quick charger" molecule called creatine phosphate (CP) that can quickly break apart and transfer its high-energy phosphate bond to ATP. But this only buys a few more seconds of ATP energy (Figure 10-10).

To keep contracting for more than a few seconds, the muscle fiber must transfer energy from glucose (or a related molecule). Each glucose molecule used to recharge ATP first travels through what is often called the **anaerobic pathway** because it does not require oxygen. This pathway, also called **glycolysis,** takes place in the cytoplasm of the muscle cell. The process eventually breaks down each glucose molecule into two smaller molecules called *pyruvate*—and releases enough energy to recharge (phosphorylate) two ATP molecules.

Each pyruvate molecule still has quite a bit of energy left in it, however. It then enters the mitochondria, where enzymes break down the molecule further in the presence of oxygen—eventually releasing CO_2 and H_2O. Because it requires oxygen, this slower and more complex set of reactions is called the **aerobic pathway** or **oxidative phosphorylation.** Up to 36 additional ADP molecules (for every one glucose molecule) can be phosphorylated to make 36 molecules of ATP by the aerobic pathway (Figure 10-11).

The oxygen needed for the aerobic pathway diffuses into the muscle cells from red blood cells. It can also be removed

▲ **FIGURE 10-11**   **Summary of glucose metabolism.**   Glucose is broken down to pyruvate in the process of glycolysis. If oxygen is available, pyruvate is quickly moved into the aerobic pathway. If oxygen is not available, pyruvate is converted to lactate, creating an oxygen debt. The oxygen debt is later repaid when ATP is used to convert lactate back into pyruvate. Sometimes the pyruvate is converted all the way back to glucose. If there is an excess of glucose, the cell may convert it to glycogen. Later, individual glucose molecules can be removed from the glycogen chain when needed by the muscle fiber. You may want to review *cellular respiration* in Chapter 4.

from storage in a red muscle protein called **myoglobin (Mb).** This oxygen-storing molecule resembles the hemoglobin (Hb) in red blood cells. It is myoglobin that gives muscle tissue its distinctive red color.

The glucose needed to recharge ATP comes from blood plasma. However, like oxygen stored in myoglobin during rest for later use during muscle contraction, glucose can be stored during rest in the form of *glycogen*. **Glycogen,** a starch-like compound, consists of a branched chain of glucose molecules bonded together.

Although the anaerobic pathway in the cytoplasm recharges only a minimal amount of ATP, it has several important functions. Because it is faster than the aerobic pathway, it is useful for brief, powerful muscle contractions that are not sustained for a long time. Because it does not require oxygen, anaerobic respiration can help produce some additional ATP molecules during prolonged "endurance" activities when oxygen becomes depleted by the aerobic pathway. However, when anaerobic processes supplement the aerobic pathway in this manner, the resulting pyruvate is temporarily converted to **lactate** (the dissolved form of *lactic acid*). Lactate is eventually converted back to pyruvate—or even converted all the way back to glucose—by the muscle fiber, or by neighboring muscle fibers or other cells in the body. The buildup of lactate was once thought to cause the buildup of acid, and the resulting burning sensation, in muscles during intense exercise. However, the acid burn actually comes from different mechanisms in the contracting muscle fibers.

A person usually breathes rapidly and deeply during or after heavy exercise, when the lack of oxygen in some tissues has caused the production of excessive lactate. The body thus repays this so-called **oxygen debt** by using the extra oxygen gained by heavy breathing to process the excessive lactate that was produced during anaerobic respiration.

Because the metabolic processes of cells are never 100% efficient, much of the energy released—about 70%—is lost as heat. In fact, it is the skeletal muscles that produce most of our body's heat even when we are "at rest." Activity of the skeletal muscles can elevate the temperature of the body to dangerous levels, such as when a person works too vigorously on a hot day. In contrast, intense muscle shivering also can produce enough heat to protect the body from dangerously low body temperatures on a very cold day.

A&P CONNECT

There is more to the oxygen debt than first meets the eye! Discover more about this important process of physical exercise in **The Oxygen Debt** online at **A&P Connect.**

℮volve
learning system

QUICK ✓ CHECK

9. Briefly describe the role of ATP in muscle contraction.
10. Contrast aerobic and anaerobic pathways in muscle fibers.
11. What is the role of myoglobin in muscle fibers?
12. Why is heat produced especially during aerobic respiration?

FUNCTION OF SKELETAL MUSCLE ORGANS

The Motor Unit

Although each skeletal muscle fiber is distinct from all other fibers, it operates as part of the larger group of fibers that together form a *skeletal muscle organ* or *muscle*. Each muscle fiber receives its stimulus from a somatic motor neuron. This neuron is one of several nerve cells that enter a muscle organ together in a bundle called a **motor nerve.** One of these motor neurons, plus the muscle fibers to which it attaches, makes up a functional unit called a **motor unit.** Now, take a moment to review this information in Figure 10-12.

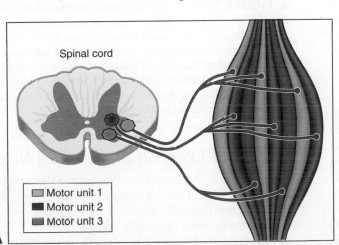

Motor unit 1
Motor unit 2
Motor unit 3

A

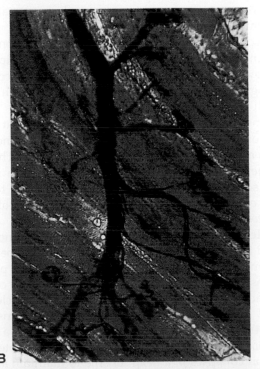

B

▲ **FIGURE 10-12   Motor unit.** A motor unit consists of one somatic motor neuron and the muscle fibers supplied by its branches. **A,** Diagram showing several motor units within the same muscle organ. **B,** Photomicrograph showing a nerve *(black)* branching to supply several dozen individual muscle fibers *(red)*.

Notice how the axon of a motor neuron divides into a number of branches where it enters the skeletal muscle. These branches may range from just a few dozen to over a thousand. As a result, the impulse conducted by one motor unit may stimulate only a small part of a muscle. However, a different motor unit may conduct an impulse to a significant part of the muscle organ (Figure 10-12, *A*). As a general rule, the *fewer* the number of muscle fibers supplied by a motor unit, the more finely coordinated the movements that muscle can produce. For example, in certain small muscles of the hand, each motor unit includes only a few muscle fibers. As a result, these muscle fibers produce very precise finger movements. In contrast, motor units in large abdominal muscles (that do not produce precise movements) may conduct an impulse to thousands of muscle fibers.

Muscle Contraction, Fatigue, and Strength

A muscle fiber contracts only after it receives a **threshold stimulus** from the fiber of a neuron. The muscle fiber does not begin to contract immediately at the instant of stimulation, but a fraction of a second later (*latent period*). During the latent period, the impulse initiated by the stimulus travels *through* the sarcolemma and T tubules to the SR. Here it triggers the release of calcium ions into the sarcoplasm, as we have seen. The muscle fiber then increases its tension and shortens until a peak of contraction (*contraction phase*) is reached. It quickly relaxes and returns to its resting state (*relaxation phase*).

Active muscles in our bodies rarely "twitch" because our nervous system subconsciously "smooths out" the movements. This prevents injury to our muscles and makes our movements more coordinated and meaningful.

A muscle contracts more forcefully after it has contracted a few times than when it first contracts—a principle well known by athletes who "warm up" before they compete. Physiologists call this stepwise increase of strength **treppe** (Figure 10-13, *B*).

It is smooth, sustained contractions, however, that control most of our body's muscles. These normal, sustained contractions are called *tetanic contractions* or **tetanus.** This happens when a series of successive stimuli come in rapidly enough so that the muscle does not have time to relax between contraction events. Muscle physiologists call this a *multiple wave summation* (Figure 10-13, *C* and *D*).

Muscle tone is the continual, partial contraction normally observed in a muscle. At any moment a small number of the total number of fibers in a muscle contract and produce slight tension in the muscle rather than a significant contraction or movement. In fact, different groups of muscle fibers scattered throughout the muscle organ "take turns" contracting, thus maintaining our body posture. When a person loses consciousness, muscles lose their tone completely, and the person collapses in a heap, unable to maintain a sitting or standing posture.

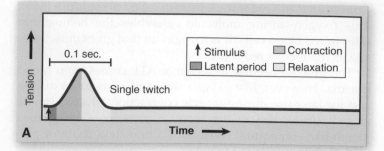

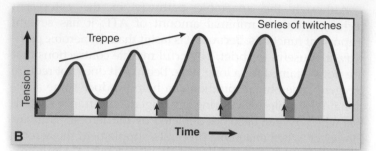

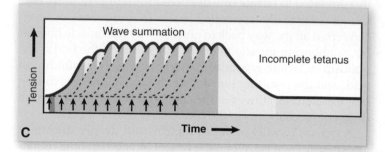

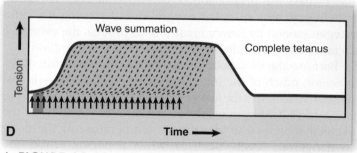

▲ **FIGURE 10-13** **Various types of muscle contractions. A,** A single-twitch contraction. **B,** The treppe phenomenon, or "staircase effect," is a steplike increase in the force of contraction over the first few in a series of twitches. **C,** Incomplete tetanus occurs when a rapid succession of stimuli produces "twitches" that seem to add together (wave summation) to produce a rather sustained contraction. **D,** Complete tetanus is a smoother sustained contraction produced by the summation of "twitches" that occur so close together that the muscle cannot relax at all.

Effects of Exercise on Skeletal Muscle

Many people have experienced **muscle fatigue** during intense exercise. This is a complete loss of strength or endurance that may result from any combination of failed steps during the process of muscle contraction.

The *strength* of a muscle is related to the *number* of fibers that contract simultaneously: the more muscle fibers that contract at

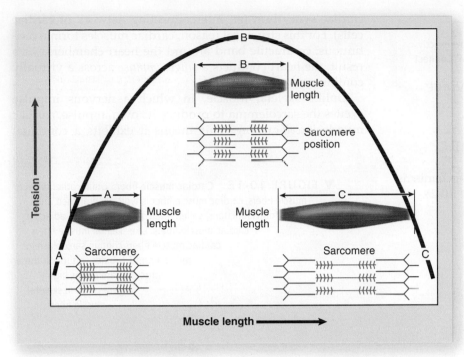

▲ **FIGURE 10-14**   **The length-tension relationship.**   As this graph of muscle tension shows, the maximum strength that a muscle can develop is directly related to the initial length of its fibers. At a short initial length, the sarcomeres are already compressed, and thus the muscle cannot develop much tension (position *A*). Conversely, the thick and thin myofilaments are too far apart in an overstretched muscle to generate much tension (position *C*). Maximum tension can be generated only when the muscle has been stretched to a moderate, optimal length (position *B*).

the same time, the greater the strength. However, the maximal strength that a muscle can develop also is directly related to the initial *length* of its fibers (Figure 10-14). The *position* in which the muscle is contracting is also important. The greatest strength of the biceps, for example, occurs when the elbow is partly flexed and the biceps only moderately stretched. Another factor that

influences the strength of a skeletal muscle contraction is the amount of load imposed on the muscle. Within certain limits, the heavier the load, the stronger the contraction. Of course, regular exercise improves muscle strength and tone. Proper exercise thus affords better posture, provides more efficient heart and lung function, reduces fatigue, helps us sleep soundly, and makes us look and feel better.

Muscle mass can increase or decrease by adding or losing myofibrils and myofilaments during exercise or the lack of it. During prolonged disuse, muscles **atrophy.** This can happen if you become a couch potato, when a cast immobilizes a body part, or when a nerve is damaged. In contrast, extensive resistance or **strength training** (e.g., weight lifting) may cause an increase in muscle size called **hypertrophy. Endurance training (aerobic training)** does not usually result in as much muscle hypertrophy as strength training does. Aerobic activities (such as running, bicycling, or swimming) increase blood flow and allow more efficient delivery of oxygen and glucose to muscle fibers during exercise. Resistance training and endurance training are excellent forms of physical therapy for persons suffering from painful muscular and connective tissue disorders such as fibromyalgia.

Both *isotonic* (*iso-* same, *tonic-* tone or force) and *isometric* (*iso-* same, *metric-* length) contractions can be used to strengthen muscle (Figure 10-15). In **isotonic contraction,** the muscle shortens and produces movement. In **isometric contraction,** the muscle pulls forcefully against a load but does *not* shorten because it cannot overcome the resistance.

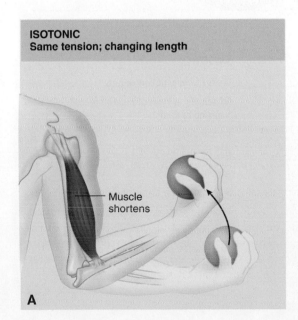

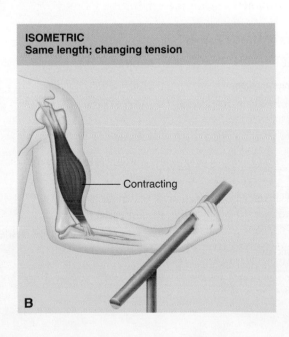

◄ **FIGURE 10-15**
Isotonic and isometric contractions. A, In isotonic contraction, the muscle shortens and produces movement. **B,** In isometric contraction, the muscle pulls forcefully against a load but does not shorten because it cannot overcome the resistance.

A&P CONNECT

You have probably experienced muscle fatigue before, but how exactly does it happen? Check it out in **Muscle Fatigue** online at **A&P Connect** to find out the details.

evolve

FUNCTION OF CARDIAC AND SMOOTH MUSCLE TISSUES

Cardiac and smooth muscle tissues operate by mechanisms similar to those of skeletal muscle tissue (Table 10-2). However, the detailed study of cardiac and smooth muscle will be set aside until we discuss specific smooth and cardiac muscle organs in later chapters. At this point we will briefly compare both muscle types.

Cardiac Muscle

Cardiac muscle is found only in the heart; it contracts rhythmically and continuously to provide the pumping action necessary to maintain a relative constancy of blood flow. Heart muscle has specialized features related to its role of pumping blood. As Figure 10-16 shows, each cardiac muscle fiber contains parallel myofibrils comprising sarcomeres. This arrangement gives the whole fiber a striated appearance. However, cardiac muscle fiber does *not taper* like skeletal muscle fiber. Instead, the fibers make strong, electrically coupled junctions at **intercalated disks.** The result is a

cardiac **syncytium** (an interconnecting network of cardiac cells). For this structural reason, cardiac muscles form a continuous, contractile band around the heart chambers. As a result, the heart conducts a *single impulse* across a virtually continuous sarcolemma.

Unlike skeletal muscle, in which a nervous impulse excites the sarcolemma to produce its own impulse, cardiac muscle is *self-exciting*. This means it exhibits a continual

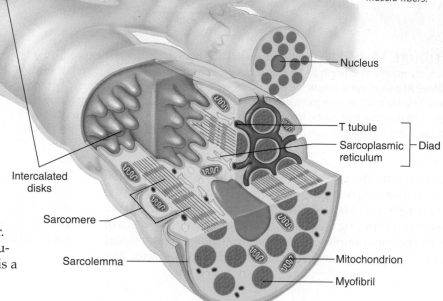

▼ **FIGURE 10-16** **Cardiac muscle fiber.** Unlike other types of muscle fibers, cardiac muscle fiber is typically branched and forms junctions, called *intercalated disks*, with adjacent cardiac muscle fibers. Like skeletal muscle fibers, cardiac muscle fibers contain sarcoplasmic reticula and T tubules—although these structures are not as highly organized as in skeletal muscle fibers.

TABLE 10-2 **Characteristics of Muscle Tissues**

	SKELETAL	**CARDIAC**	**SMOOTH**
Principal location	Skeletal muscle organs	Wall of heart	Walls of many hollow organs
Principal functions	Movement of bones, heat production, posture	Pumping of blood	Movement in walls of hollow organs (peristalsis, mixing of fluids)
Type of control	Voluntary	Involuntary	Involuntary
Structural Features			
Striations	Present	Present	Absent
Nucleus	Many; near the sarcolemma	Single (sometimes double); near the center of the cell	Single; near the center of the cell
T tubules	Narrow	Large diameter	Absent
Sarcoplasmic reticulum	Extensive	Less extensive than in skeletal muscle	Very poorly developed
Cell junctions	No gap junctions	Intercalated disks	*Single-unit**: many gap junctions *Multiunit*: few gap junctions.

*Also referred to as *visceral smooth muscle tissue.*

rhythm of excitation and contraction on its own, although the *rate* of self-induced impulses can be altered by nervous or hormonal input. Because the heart cannot sustain long tetanic contractions, it does not normally run low on ATP and thus does not experience fatigue.

Smooth Muscle

Smooth muscle is composed of small, tapered cells with large, single nuclei. Cells of smooth muscle do not have T tubules and have only a loosely organized sarcoplasmic network. Smooth muscle not only lacks the striations of skeletal and cardiac muscles, it also has thick and thin myofilaments arranged quite differently. As you can see in Figure 10-17, thin arrangements of myofilaments crisscross the cell and attach at their ends to the cell's plasma membrane. When cross bridges pull the thin filaments together, the muscle "balls up" and thus contracts the cell. Because the myofilaments are not organized into distinct sarcomeres, they have more freedom of movement. As a result, smooth muscle fiber can contract to shorter lengths than possible in skeletal and cardiac muscle.

There are two main types of smooth muscle: *single-unit* smooth muscle and *multiunit* smooth muscle. In **single-unit smooth muscle** (*visceral smooth muscle*), fibers form large, continuous sheets, much like the syncytium of heart muscle. This type of smooth muscle is the most common type, and like cardiac muscle, exhibits a rhythmic self-excitation. When these rhythmic, spreading waves of contraction become strong enough, they can push the contents of a hollow organ (such as the large intestine) progressively along its lumen. This phenomenon, called **peristalsis,** moves food along the digestive tract, assists the flow of urine to the bladder, and pushes a baby out of the uterus during labor. Such contractions can also be coordinated to produce mixing movements in the stomach and other organs.

Multiunit smooth muscle tissue does not act as a single unit (as in visceral smooth muscle). Instead, it is composed of many independent single-cell units. Each independent fiber does not usually generate its own impulse but rather responds only to nervous input. Although this type of smooth muscle can form thin sheets, as in the walls of large blood vessels, it is more often arranged in discrete bundles. Multiunit smooth muscle bundles can be found in the tiny *arrector pili* muscles of the skin, in the muscles that control the lens of the eye, or as single fibers (such as those surrounding small blood vessels).

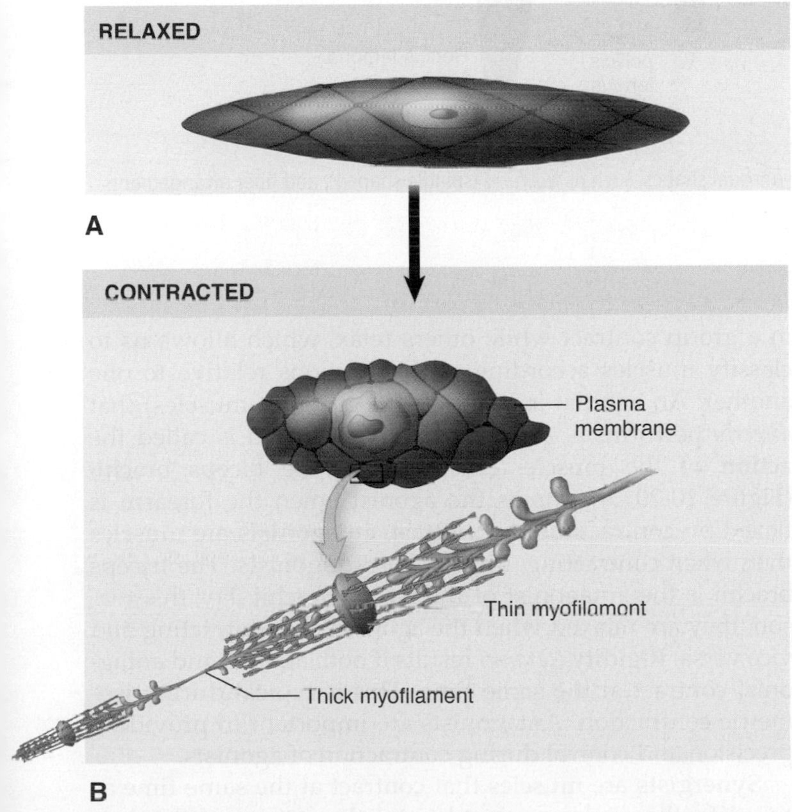

> **QUICK CHECK**
> 13. What is a motor unit?
> 14. What is meant by the term muscle tone?
> 15. What is the connection between muscle contraction and muscle strength?
> 16. Briefly outline the differences between smooth muscle and cardiac muscle.

SKELETAL MUSCLE ORGANS

Skeletal muscle organs range in size from extremely small strands such as the tiny *stapedius* muscle of the middle ear, to large masses, such as those of the thigh. Muscles also range in shape from broad and wide to long and narrow. However, many are of irregular shape and still others form sheets or bulky masses. The strength and type of movement produced by the contraction of a muscle is related to its overall shape and the orientation of its fibers. Strength and movement also depend on a muscle's attachment to bones and its arrangement with joints.

There are various descriptive terms for muscles. For example, *parallel muscles* are long and straplike; *circular muscles (sphincters)* encircle body tubes or openings (such as the *orbicularis oris* around the mouth); and *convergent muscles,* such as the *pectoralis major* (the "pecs") of the chest look like the blades of a fan. These and other shapes and fiber arrangements, such as those of the featherlike *pennate* form, are illustrated for you in Figure 10-18.

RELAXED

A

CONTRACTED

- Plasma membrane
- Thin myofilament
- Thick myofilament

B

▲ **FIGURE 10-17** **Smooth muscle fiber. A,** Thin bundles of myofilaments span the diameter of a relaxed fiber. **B,** During contraction, sliding of the myofilaments causes the fiber to shorten by "balling up."

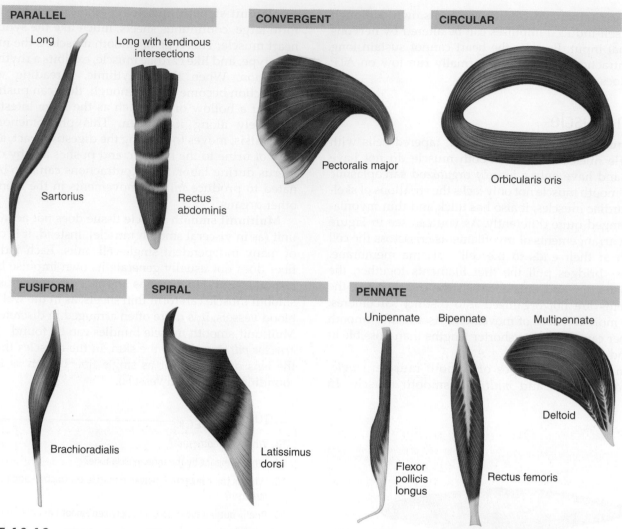

▲ FIGURE 10-18 **Muscle shape and fiber arrangement.** Muscles can have unusual shapes, such as *fusiform* (spindle shaped), and fiber arrangements, such as *pennate* (feather shaped).

Attachment of Muscles

Most of your muscles span at least one joint and attach to both articulating bones making up that joint. When contraction of a muscle takes place, one bone usually remains fixed and the other moves. The **origin** is the point of attachment that does *not move* when the muscle contracts. In contrast, the **insertion** is the point of attachment that *moves* when the muscle contracts (Figure 10-19). Typically, the insertion bone moves toward the origin bone when the muscle shortens. Many muscles have multiple points of origin or insertion, and although these are largely beyond the scope of this book, we will look at the origins and insertions of some of the major muscles.

Muscle Actions

Skeletal muscles almost always act in groups rather than singly. *In this way, coordinated movements are produced by the combined action of several muscles.* In fact, some of the muscles

in a group contract while others relax, which allows us to classify muscles according to their actions relative to one another: An **agonist** is a muscle (or group of muscles) that *directly* performs a specific movement, which is called the **action** of the muscle. For example, the biceps brachii (Figure 10-20, *A*) acts as the agonist when the forearm is flexed (in contraction). In contrast, **antagonists** are muscles that, when contracting, directly *oppose* agonists. The triceps brachii is the antagonist of the biceps brachii. For this reason, they are relaxed when the agonists are contracting and vice versa. Rigidity *(tetany)* results if both agonist and antagonist contract at the same time. This is typical during isometric contraction. Antagonists are important in providing precision and control during contraction of agonists.

Synergists are muscles that contract at the same time as agonists; they help or complement the actions of the agonists. Consider the elbow, for example. In this joint, two well-known muscles, the biceps brachii and the brachioradialis muscles, are synergistic. Contraction of both of these muscles leads to flexion.

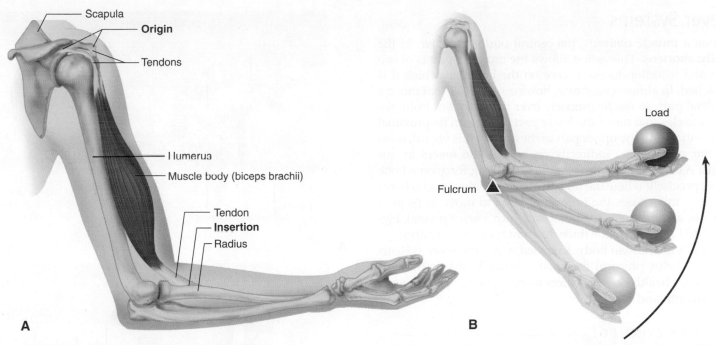

▲ FIGURE 10-19 **Attachments of a skeletal muscle.** **A,** Origin and insertion of a skeletal muscle. A muscle originates at a relatively stable part of the skeleton (origin) and inserts at the skeletal part that is moved when the muscle contracts (insertion). **B,** Movement of the forearm during weight lifting. Muscle contraction moves bones, which serve as levers, and by acting on joints, which serve as fulcrums for those levers. See the text for discussion and review Figure 10-21, which illustrates types of levers.

In contrast, **fixator muscles** serve to stabilize joints and often help maintain posture or balance when agonists are contracting, especially those of the arms and legs (Figure 10-20, *B*). Considering the arm again, the biceps brachii and the triceps brachii are antagonists: their contractions gener-ally provide opposite effects. Of course, even the simplest movement is quite complex, and most muscles function not only as agonists, but also sometimes as antagonists, synergists, or even fixators! Like life, muscle movement is very complex.

▶ FIGURE 10-20 **Muscle actions.** **A,** The flexor muscle (biceps brachii) is the prime mover in flexing the elbow. The extensor muscle (triceps brachii) is the antagonist, which in this case must relax to permit easy flexion of the elbow. The pronator teres muscle acts as a synergist by also flexing the elbow. **B,** Here the biceps brachii is again the prime mover of flexion of the elbow. The pronator teres muscle acts as a synergist by also flexing the elbow. To prevent the biceps from also moving the shoulder while straining against a heavy weight, the posterior portion of the deltoid muscle tenses to stabilize the shoulder joint—thus acting as a fixator muscle in this action.

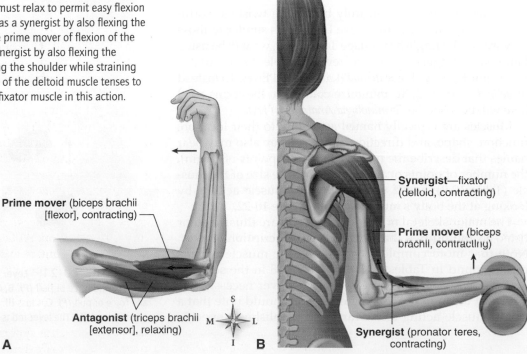

Lever Systems

When a muscle *contracts,* the central portion, known as the **belly,** shortens. This action allows the muscle's points of origin and insertion to exert force on the bones to which it is attached. In almost every case, however, muscles that move a skeletal part do not lie precisely over that part. For example, the muscles that move the lower part of the arm lie proximal to it—that is, in the upper part of the arm. In this regard, most of our muscles act indirectly on bones—the **levers** in our body. A contracting muscle applies a pulling force on a bone lever precisely where the muscle attaches to the (origin) bone. In turn, this causes the (insertion) bone to move at its joint (that is, as a *fulcrum*—the point on which a lever pivots). Figure 10-21 shows you the three different types of lever arrangements in the human body. Knowledge of how lever systems operate helps physical therapists, certified athletic trainers, and other health care providers more accurately assess muscle strength and function.

A&P CONNECT

Leverage is the change in effort needed when using a lever to do work. Want to know more about how muscles power the levers of the body? Check it out in **Leverage** online at **A&P Connect.**

evolve

QUICK ✓ CHECK

17. Describe what is meant by "origin" and "insertion."
18. What is meant by "muscle antagonists?"
19. Describe how muscles act on bones as levers.

HOW MUSCLES ARE NAMED

The names of muscles can truly be tongue twisters. Fortunately, many muscle names have Latin roots similar to those of words of the English language. In this text, we will be using Latin-based English names wherever possible (Figure 10-22). For example, we will use *deltoid* (Latin-based English) instead of *deltoideus* (Latin). To minimize confusion, the terms used here will be from the *Terminologia Anatomica (TA).*

Muscles are typically named according to their location, function, shape, and direction of fibers. They also may bear names that describe the number of heads (points of origin), the number of points of attachment, and the size of the muscle (Table 10-3). See if you can deduce muscle actions by looking at the body's musculature in Figure 10-22.

The major skeletal muscles of the body are illustrated or listed in the figures that follow their descriptions below. Note that more complete lists of specific muscle groups can be found in Tables 10-5 through 10-18 in the sections below. Please refer to these tables whenever necessary as you read through the text material. You should note that a single muscle acting alone rarely accomplishes a specific

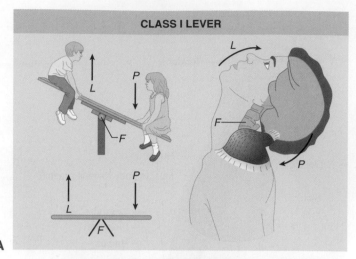

A

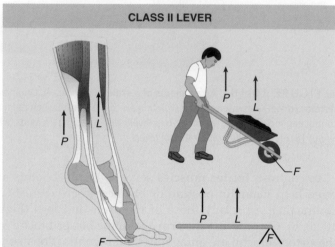

B

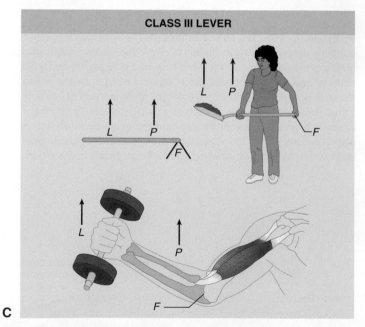

C

▲ **FIGURE 10-21**   **Lever classes. A,** Class I: fulcrum *(F)* between the load *(L)* and force or pull *(P).* **B,** Class II: load *(L)* between the fulcrum *(F)* and force or pull *(P).* **C,** Class III: force or pull *(P)* between the fulcrum *(F)* and the load *(L).* The lever rod is yellow in each.

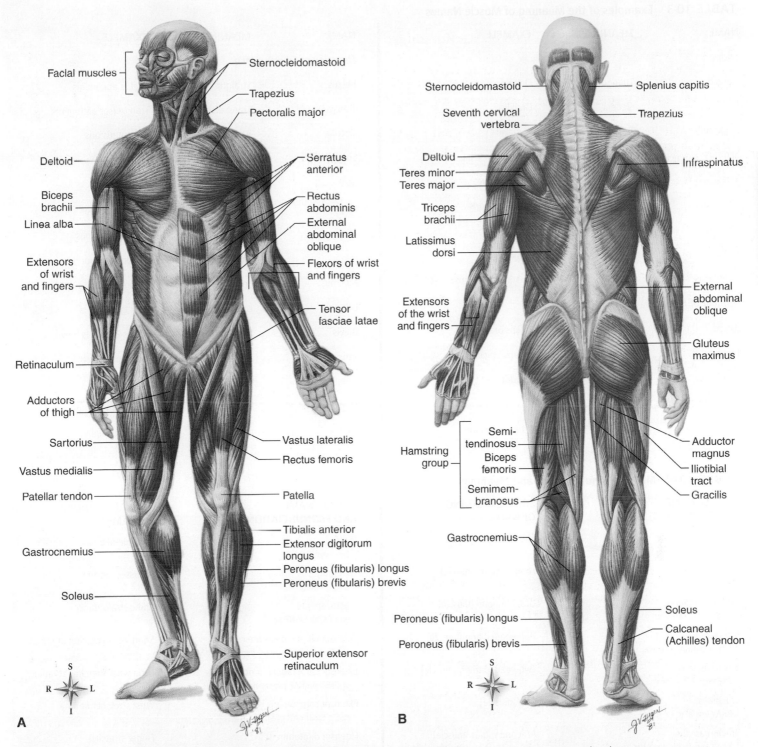

▲ **FIGURE 10-22** **General overview of the body's musculature.** **A,** Anterior view. **B,** Posterior view.

action. Instead, muscles act in groups as agonists, synergists, antagonists, and fixators, to bring about effective movements.

You'll notice that throughout the discussions presented below we have taken the opportunity to list on the figures a few additional muscles not described in the text. There are between 600 and 800 muscles in the human body (depending on how they are classified), and discussing each muscle of each group is beyond the scope of this book. We've also created Table 10-4 for you, which summarizes the pronunciations and literal meanings for many of the most important muscles of your body.

TABLE 10-3 Examples of the Meaning of Muscle Names

NAME	MEANING	EXAMPLE	NAME	MEANING	EXAMPLE
Shape			*Direction of Fibers—cont'd*		
Deltoid	Triangular	Deltoid	Rectus	Straight	Rectus abdominis
Gracilis	Slender	Gracilis	Transverse	Transverse	Transversus abdominis
Trapezius	Trapezoid	Trapezius	Circular	Around	Orbicularis oris
Serratus	Notched	Serratus anterior	Spiral	Oblique	Supinator
Teres	Round	Pronator teres	*Size*		
Orbicularis	Round or circular	Orbicularis oris	Major	Large	Pectoralis major
Pectinate	Comblike	Pectineus	Maximus	Largest	Gluteus maximus
Platys	Flat	Platysma	Minor	Small	Pectoralis minor
Quadratus	Square	Quadratus femoris	Minimus	Smallest	Gluteus minimus
Number of Heads			Longus	Long	Adductor longus
Biceps	Two heads	Biceps brachii	Brevis	Short	Extensor pollicis brevis
Triceps	Three heads	Triceps brachii	Latissimus	Very wide	Latissimus dorsi
Quadriceps	Four heads	Quadriceps	Longissimus	Very long	Longissimus
Direction of Fibers			Magnus	Very large	Adductor magnus
Oblique	Diagonal	External oblique	Vastus	Vast or huge	Vastus medialis

TABLE 10-4 Muscle Pronunciations and Meanings

MUSCLE NAME AND PRONUNCIATION	LITERAL MEANING OF NAME*	MUSCLE NAME AND PRONUNCIATION	LITERAL MEANING OF NAME*
Adductor brevis ad-DUK-ter BREV-is	Short bring-in [muscle]	Deltoid DEL-toyd	Triangle-like
Adductor longus ad-DUK-ter LONG-us	Long bring-in [muscle]	Diaphragm DYE-ah-fram	Across-enclosure
Adductor magnus ad-DUK-ter MAG-nus	Great bring-in [muscle]	Erector spinae eh-REK-tor SPINE-ee	Spine straightener
Biceps brachii BYE-seps BRAY-kee-eye	Two-headed arm [muscle]	Extensor carpi radialis brevis ek-STEN-ser KAR-pye ray-dee-AL-is BREV-is	Short wrist stretcher at radius
Biceps femoris BYE-seps FEM-uh-ris	Two-headed thigh [muscle]	Extensor carpi radialis longus ek-STEN-ser KAR-pye ray-dee-AL-is LONG-us	Long wrist stretcher at radius
Brachialis BRAY-kee-al-is	Arm [muscle]	Extensor carpi ulnaris ek-STEN-ser KAR-pye ul-NAIR-is	Wrist stretcher at ulna
Brachioradialis BRAY-kee-oh-ray-dee-AL-is	Arm and radius [muscle]	Extensor digitorum ek-STEN-ser dij-ih-TOH-rum	Finger stretcher
Buccinator BUK-si-NAY-tor	Trumpeter	Extensor digitorum longus ek-STEN-ser dij-ih-TOH-rum LONG-gus	Toe stretcher
Bulbospongiosus bul-boh-spun-jee-OH-ses	Spongy bulb (base of penis)	External intercostal eks-TER-nal in-ter-KOS-tal	Outer [muscle] between ribs
Coracobrachialis KOR-uh-koh-BRAY-kee-al-is	Coracoid process—arm [muscle]	External oblique eks-TER-nal oh-BLEEK	Outer slanted [muscle]
Corrugator supercilii KOR-uh-gay-tor soo-per-SIL-ee-eye	Wrinkler above eyelashes	Flexor carpi radialis FLEK-ser KAR-pye ray-dee-AL-is	Wrist bender at radius

*Keep in mind that, in Latin, modifiers *follow* the terms they describe. Thus, *extensor carpi radialis longus* can be translated in exact parallel as "wrist stretcher at radius, long" but is best rendered as "long wrist stretcher at radius."

TABLE 10-4   Muscle Pronunciations and Meanings—cont'd

MUSCLE NAME AND PRONUNCIATION	LITERAL MEANING OF NAME*	MUSCLE NAME AND PRONUNCIATION	LITERAL MEANING OF NAME*
Flexor carpi ulnaris FLEK-ser KAR-pye ul-NAIR-is	Wrist bender at ulna	Pectoralis major pek-toh-RAL-is MAY-jer	Greater chest [muscle]
Flexor digitorum profundus FLEK-ser dij-ih-TOH-rum pro-FUN-dis	Deep finger bender	Pectoralis minor pek-toh-RAL-is MYE-ner	Lesser chest [muscle]
Flexor digitorum superficialis FLEK-ser dij-ih-TOH-rum soo-per-fish-ee-AL-is	Superficial finger bender	Perineal pair-in-EE-al	[Muscle] around the excretion [openings]
Gastrocnemius GAS-trok-NEE-mee-us	Belly of leg	Peroneus brevis (fibularis brevis) per-oh-NEE-us BREV-is (fib-yoo-LAIR-is BREV-is)	Short boot [muscle]
Gluteus maximus GLOO-tee-us MAK-sim-us	Large butt [muscle]	Peroneus longus (fibularis longus) per-oh-NEE-us LONG-us (fib-yoo-LAIR-is LONG-us)	Long boot [muscle]
Gluteus medius GLOO-tee-us MEE-dee-us	Medium butt [muscle]	Peroneus tertius (fibularis tertius) per oh NEE-us TER-shee-us (fib-yoo-LAIR-is TER-shee-us)	Third boot [muscle]
Gluteus minimus GLOO-tee-us MIN-ih-mus	Small butt [muscle]	Plantaris plan-TAIR-is	[Muscle] of the sole
Gracilis GRASS-ih-lis	Slender [muscle]	Pronator quadratus PRO-nay-ter kwah-DRAT-is	Four-sided pronator
Iliacus ih-LYE-ah-kus	Ilium of pelvic bone [muscle]	Pronator teres PRO-nay-ter TAIR-eez	Rounded pronator
Iliocostalis ILL-ee-oh-KOS-tal-is	[Pelvic] ilium to rib [muscle]	Psoas major SO-is MAY-jer	Greater lumbar [muscle]
Iliopsoas ILL-ee-oh-SOH-is	Iliacus and psoas [muscles]	Psoas minor SO-is MYE-ner	Lesser lumbar [muscle]
Infraspinatus IN-frah-spy-nah-tus	Below the spine	Pterygoid TER-i-goyd	Winglike
Internal intercostal in-TER-nal in-ter-KOS-tal	Inner [muscle] between the ribs	Quadratus lumborum kwah-DRAT-is lum-BOR-um	Four-sided lumbar [muscle]
Internal oblique in-TER-nal oh-BLEEK	Inner slanted [muscle]	Quadriceps femoris KWAH-drih-seps FEM-uh-ris	Four-headed thigh [muscle]
Interspinales in-ter-spye-NAL-eez	Between spines	Rectus abdominis REK-tus ab-DOM-ih-nis	Straight belly [muscle]
Ischiocavernosus is-kee-oh-KAV-er-no-sus	[Pelvic] ischium to the crus (cross) of the penis's cavernosum	Rectus femoris REK-tus FEM-uh-ris	Straight thigh [muscle]
Latissimus dorsi lat-ISS-im-is DOR-sye	Very wide back [muscle]	Rhomboid major ROM-boyd MAY-jer	Greater rhombus (equilateral parallelogram)
Levator ani leh-VAY-tor A-nye	Anus lifter	Rhomboid minor ROM-boyd MYE-ner	Lesser rhombus (equilateral parallelogram)
Levator scapulae leh-VAY-tor SCAP-yoo-lee	Scapula lifter	Sartorius sar-TOR-ee-us	Tailor [muscle]
Longissimus lon-JIS-ih-mus	Very long (longest)	Semimembranosus sem-ee-mem-brah-NOH-sis	Half-membrane [muscle]
Longissimus capitis lon-JIS-ih-mus KAP-ih-tis	Longest head [muscle]	Semispinalis capitis sem-ee-spy-NAY-lis KAP-ih-tis	Half-spine (muscle) of the head
Masseter mah-SEE-ter	Chewer	Semitendinosus sem-ee-ten-din-OH-sis	Half-tendon [muscle]
Multifidus mul-TIFF-ih-dus	Many splits	Serratus anterior ser-RAY-tus an-TEER-ee-or	Front sawtooth [muscle]
Opponens pollicis oh-POH-nenz POL-i-sis (or POL-i-kiss)	Thumb opposer	Soleus SOH-lee-us	Sole [of foot]
Orbicularis oculi or-bik-yoo-LAIR-is OK-yoo-lye	Little circle [around] eye	Sphincter ani externus SFINGK-ter A-nye eks-TER-nus	Outer anus tightener
Orbicularis oris or-bik-yoo-LAIR-is OR-is	Little circle [around] mouth	Spinalis spy-NAY-lis	Spine [muscle]
Palmaris longus PAL-mar-is LONG-us	Long palm [muscle]		

(continued)

TABLE 10-4 Muscle Pronunciations and Meanings—cont'd

MUSCLE NAME AND PRONUNCIATION	LITERAL MEANING OF NAME*
Splenius capitis SPLEH-nee-us KAP-ih-tis	Head patch
Sternocleidomastoid STERN-oh-KLYE-doh-MAS-toyd	Sternum-clavicle-mastoid process
Subscapularis sub-SKAP-yoo-lar-is	Under the scapula
Supinator SOO-pin-ayt-er	Supinator [muscle]
Supraspinatus SOO-prah-spy-nah-tus	Above spine [muscle]
Temporalis tem-poh-RAL-is	Temple (of head) [muscle]
Tensor fasciae latae TEN-sor FASH-ee LAT-tee	Puller of side bundle
Teres major TER-eez MAY-jer	Greater rounded [muscle]
Teres minor TER-eez MYE-ner	Lesser rounded [muscle]
Tibialis anterior tib-ee-AL-is an-TEER-ee-or	Front shinbone [muscle]
Transversus abdominis tranz-VERS-us ab-DOM-ih-nis	Belly-crossing [muscle]
Trapezius trah-PEE-zee-us	Table-shaped [muscle]
Triceps brachii TRY-seps BRAY-kee-eye	Three-headed arm [muscle]
Urethral sphincter yoo-REE-thral SFINGK-ter	Urethra tightener
Vastus intermedius VAS-tus in-ter-MEE-dee-us	Enormous (vast) in-between [muscle]
Vastus lateralis VAS-tus lat-er-AL-is	Enormous (vast) on-the-side [muscle]
Vastus medialis VAS-tus mee-dee-AL-is	Enormous (vast) toward-the-middle [muscle]
Zygomaticus major zye-goh-MAT-ik-us MAY-jer	Greater [muscle] of zygomatic [bone]

*Keep in mind that, in Latin, modifiers *follow* the terms they describe. Thus, *extensor carpi radialis longus* can be translated in exact parallel as "wrist stretcher at radius, long" but is best rendered as "long wrist stretcher at radius."

SPECIFIC MUSCLE GROUPS
Muscles of the Head
Muscles of Facial Expression

The muscles of facial expression are unique. At least one of their points of attachment is to the *deep layers of the skin* over the face or neck rather than directly to bone. We can make an enormous number of facial expressions because of the number of facial muscles, their arrangement, and their unique points of attachment. Let's look at a few representative examples illustrated in Figure 10-23.

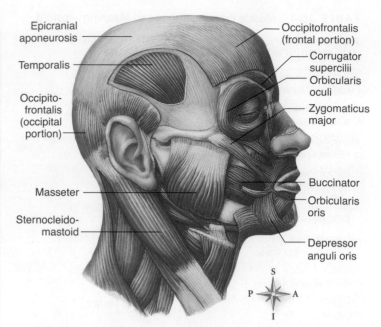

▲ **FIGURE 10-23** **Muscles of facial expression and mastication.** Lateral view.

The **occipitofrontalis** is really two muscles. One portion lies over the forehead (frontal bone); the other covers the occipital bone in the back of the head. The two muscular parts, or bellies, are joined by a connective tissue aponeurosis that covers the top of the skull. The frontal portion of the occipitofrontalis raises the eyebrows (surprise) and wrinkles the skin of the forehead horizontally (exclamation!).

The **corrugator supercilii** pulls the eyebrows together and produces the vertical wrinkles above the nose (frowning). The **orbicularis oculi** encircles and closes the eye (blinking). The **orbicularis oris** and **buccinator** pucker the mouth (kissing) and press the lips and cheeks against the teeth. The **zygomaticus major** draws the corner of the mouth upward (laughing).

You can imagine the number of facial expressions possible with just the various combinations of these few muscles (a more complete list is presented in Table 10-5).

TABLE 10-5 Muscles of Facial Expression

MUSCLE	ORIGIN	INSERTION	FUNCTION
Epicranius	Occipital bone	Skin and muscles around eye	Raises eyebrows
Corrugator supercilii	Frontal bone	Skin of eyebrow	Wrinkles forehead
Orbicularis oculi	Surrounds eyelids		Closes eye
Zygomaticus major	Zygomatic bone	Angle of mouth	Laughing (elevates angle of mouth)
Platysma	Fascia in upper chest	Lower border of mandible	Draws angle of mouth downward
Orbicularis oris	Encircles mouth		Draws lips together
Buccinator	Maxillae	Skin of sides of mouth	Facilitates smiling; blowing, as in playing a trumpet

TABLE 10-6 Muscles of Mastication

MUSCLE	ORIGIN	INSERTION	FUNCTION
Masseter	Zygomatic arch	Mandible (external surface)	Closes jaw
Temporalis	Temporal bone	Mandible	Elevates mandible and closes jaw
Pterygoids (lateral and medial)	Undersurface of skull	Mandible (medial surface)	Grates teeth

Muscles of Mastication

Again using Figure 10-23, notice that the muscles of **mastication** are responsible for our chewing movements. Powerful muscles (the **masseter** and **temporalis**) either elevate or retract the mandible. In contrast, the **pterygoids** open and protrude the jaw while making sideways chewing movements. Finally, the *buccinator* muscles hold food between the teeth as the mandible moves up and down and from side to side (Table 10-6).

Muscles that Move the Head

Paired muscles on either side of the neck are responsible for head movements (Figure 10-24). When both **sternocleidomastoid** muscles contract at the same time, the head flexes forward, hence the name "prayer muscle." Acting together, the **splenius capitis** muscles serve as strong extensors that return the head to the upright position after flexion. When either muscle contracts alone, the result is rotation and tilting of the head on that side. See if you can determine the function of the **longissimus capitis** muscle (Table 10-7).

QUICK CHECK

20. List some of the muscles responsible for facial expressions.
21. What are the major muscles for chewing?
22. What is the action of the sternocleidomastoid muscle?

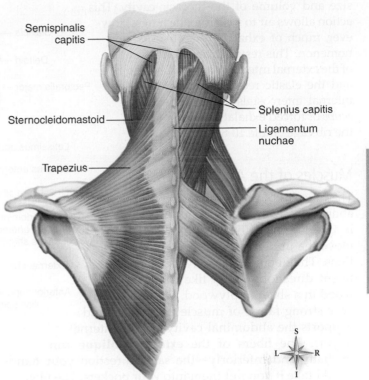

▲ **FIGURE 10-24** **Muscles that move the head.** Posterior view of muscles of the neck and the back.

Trunk Muscles
Muscles of the Thorax

Muscles of the thorax are vitally important for inhalation and exhalation. The **internal intercostal** and **external intercostal** muscles attach to the ribs at different places and their fibers are oriented in different directions. As a result, contraction of the external intercostals *elevates* the ribs during inhalation and contraction of the internal intercostals *depresses* the ribs during exhalation. During inhalation, the dome-shaped **diaphragm** flattens as it contracts, thereby increasing the

TABLE 10-7 Muscles that Move the Head

MUSCLE	ORIGIN	INSERTION	FUNCTION
Sternocleidomastoid	Sternum	Temporal bone (mastoid process)	Flexes neck, bending head forward; (so-called "prayer muscle")
	Clavicle		A single muscle; rotates neck and head toward opposite side
Semispinalis capitis	Vertebrae (transverse processes of upper six thoracic; articular processes of lower four cervical)	Occipital bone (between superior and inferior nuchal lines)	Extends neck and head, rotates head laterally to the contracting side, or brings head into an upright position
Splenius capitis	Ligamentum nuchae	Temporal bone (mastoid process)	Extends head
	Vertebrae (spinous processes of upper three or four thoracic)	Occipital bone	Bends neck and rotates head toward same side as contracting muscle
Longissimus capitis	Vertebrae (transverse processes of upper six thoracic, articular processes of lower four cervical)	Temporal bone (mastoid process)	Extends head; bends and rotates head toward contracting side

size and volume of the thoracic cavity. This action allows air to rush into the lungs. However, much of exhalation is a *passive* phenomenon. This results from the relaxation of the external intercostals and diaphragm and the elastic recoil of the lungs. The internal intercostals are responsible for *active* or forced exhalation by depressing the rib cage (Table 10-8).

Muscles of the Abdominal Wall

The muscles of the anterior and lateral abdominal wall are arranged in three layers (Figure 10-25), helping us to move the upper body in multiple directions. The fibers in each layer run in different directions much like the layers of wood in a sheet of plywood. The result is a very strong fabric of muscle that covers and supports the abdominal cavity and its internal organs. The fibers of the **external oblique** run medially and inferiorly—the same direction your hands would take if you put them into your pockets. The fibers of the **internal oblique** run almost at right angles to those of the external oblique above it. Finally, the fibers of the **transversus abdominis,** the innermost muscle layer of the abdomen, run transversely.

In addition to these three sheets of muscles, the band-shaped **rectus abdominis** runs down the midline of the abdomen from the thorax to the pubis. You can see in Figure 10-25 that the parallel fibers of this muscle seem to be interrupted by three tendinous intersections, creating the so-called six-pack look of buff abdomens. Note that the aponeuroses of the external oblique, internal oblique, and transverses abdominis muscles form the *rectus sheaths* that cover the rectus abdominis muscles. These sheaths fuse in the midline to form a tough band of connective tissue—the **linea alba** ("white line"),

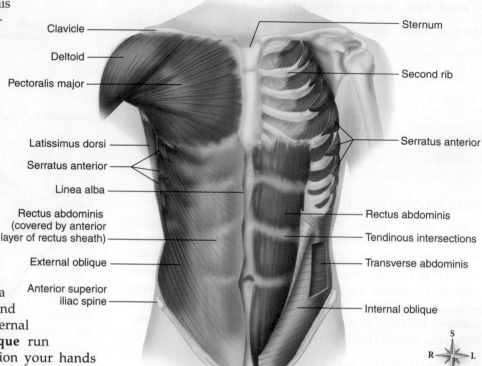

▲ **FIGURE 10-25** **Muscles of the trunk and abdominal wall.** Superficial muscles are visible on the right side of the body and deeper muscles on the left side of the body.

which stretches from the xiphoid process of the sternum to the pubis (Table 10-9).

Working as a group, the abdominal muscles not only protect and hold the abdominal viscera in place, they are also responsible for a number of vertebral column movements, including flexion, lateral bending, and some rotation. These very important muscles are involved in respiration, help to "push" a baby through the birth canal during delivery, and assist in urination, defecation, and vomiting.

A&P CONNECT

Weakness of abdominal muscles can lead to a *hernia*, or protrusion, of an abdominal organ (commonly the small intestine or stomach) through an opening in the abdominal wall. Even if you have not experienced a hernia before now, one day you may. Hernias are quite common. See some examples of the common hernia types in **Hernias** online at **A&P Connect.**

*e*volve

Muscles of the Back

The superficial back muscles play a major role in moving the head and limbs (Figure 10-26, *A*). The deep back muscles (Figure 10-26, *B*) allow us to move our vertebral column in a variety of ways. They also serve to stabilize our trunk, helping us maintain a stable posture. The **erector spinae** group consists

TABLE 10-8 **Muscles of the Thorax**

MUSCLE	ORIGIN	INSERTION	FUNCTION
External intercostals	Rib (lower border; forward fibers)	Rib (upper border of rib, below origin)	Elevate ribs; aid in forced inspiration
Internal intercostals	Rib (inner surface, lower border; backward fibers)	Rib (upper border of rib below origin)	Depress ribs; aid in forced expiration
Diaphragm	Lower circumference of thorax (of rib cage)	Central tendon of diaphragm	Enlarges thorax, serving as primary muscle of inspiration

TABLE 10-9   Muscles of the Abdominal Wall

MUSCLE	ORIGIN	INSERTION	FUNCTION
External oblique	Ribs (lower eight)	Pelvis (iliac crest and pubis by way of the inguinal ligament)	Pulls the chest downward and compresses abdomen and organs within
		Linea alba by way of an aponeurosis	Rotates trunk laterally
Internal oblique	Pelvis (iliac crest and iliopsoas fascia)	Ribs (lower three)	Compresses the abdomen, aiding expiration
	Lumbodorsal fascia	Linea alba	Important for posture: when these muscles lose their tone, common figure faults of protruding abdomen and lordosis develop
Transversus abdominis	Ribs (lower six) Pelvis (iliac crest, iliopsoas fascia)	Pubic bone Linea alba	Stabilizes the pelvis and lower back prior to movements of the body
	Lumbodorsal fascia	Ribs (costal cartilage of fifth, sixth, and seventh ribs)	Stabilizes the pelvis and lowers the back prior to movements of the body
Rectus abdominis	Pelvis (pubic bone and pubic symphysis)	Sternum (xiphoid process)	Similar to external oblique; compresses the abdominal cavity, aids in straining, defecation, forced expiration, childbirth, etc.; abdominal muscles are antagonists of the diaphragm, relaxing as it contracts and vice versa
			Flexes vertebral column of trunk
Quadratus lumborum	Iliolumbar ligament; iliac crest	Last rib; transverse process of vertebrae (L1-L4)	Flexes vertebral column laterally; depresses last rib

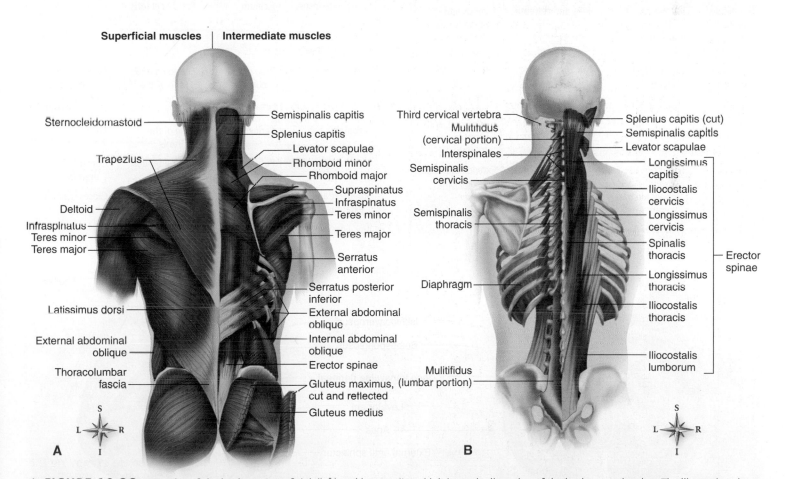

▲ **FIGURE 10-26**   **Muscles of the back.** **A,** Superficial *(left)* and intermediate *(right)* muscle dissection of the back—posterior view. The illustration shows a two-stage dissection. Superficial muscles of the neck and back are shown on the left side and an intermediate-depth dissection is shown on the right. **B,** Deep muscle dissection of the back—posterior view. The superficial and intermediate muscles have been removed. The muscles in the gluteal region have been removed to expose the pelvic insertion of the multifidus.

of a number of long, thin muscles that travel all the way down our backs (see Figure 10-26). These muscles extend (straighten or pull back) the vertebral column and also provide some rotation and flex the back laterally (Table 10-10).

Even deeper lie the **interspinales** and **multifidus** groups, which connect one vertebra to the next. These muscles also help extend the back and neck or flex them to the side. Eighty percent of the world's population experiences backache, especially involving lower back muscles, at some time in their lives. As we age, back pain from injured muscles can become extreme and debilitating, creating significant financial, social, and health care problems.

Muscles of the Pelvic Floor

The reinforced muscular floor of the pelvic cavity guards the various body outlets that project below. The muscular pelvic floor filling this diamond-shaped outlet is called the **perineum.** Passing through the pelvic floor are the anal canal and urethra in both sexes and the vagina in the female (Figure 10-27). The two **levator ani** and **coccygeus** muscles stretch across the pelvic cavity like a hammock to form most of the pelvic floor. Several unique muscles, the **ischiocavernosus** and **bulbospongiosus,** are associated with the penis in the male and the vagina in the female. Constriction of the **urethral sphincters,** which encircle the urethra in both sexes, helps control urine flow. Defecation from the terminal portion of the anal canal is controlled by another circular muscle, the **external anal sphincter** (Table 10-11).

TABLE 10-10   Muscles Groups of the Erector Spinae Group

MUSCLE	ORIGIN	INSERTION	FUNCTION
Iliocostalis group	Various regions of the pelvis and ribs	Ribs and vertebra	Extends and laterally flexes the vertebral column
Longissimus group	Cervical and thoracic vertebrae, ribs	Mastoid process, upper cervical vertebrae, or upper lumbar vertebrae	Primary muscle set of the erector group; extends neck and head, and vertebral column
Spinalis group	Lower cervical or lower thoracic/ upper lumbar vertebrae	Upper cervical or middle/upper thoracic vertebrae	Extends the neck or vertebral column

TABLE 10-11   Muscles of the Pelvic Floor

MUSCLE	ORIGIN	INSERTION	FUNCTION
Levator ani	Pubis and spine of the ischium	Coccyx	Together with the coccygeus muscles, this muscle forms the floor of the pelvic cavity and supports the pelvic organs
Ischiocavernosus	Ischium	Penis or clitoris	Compress the base of the penis or clitoris
Bulbospongiosus			
Male	Bulb of the penis	Perineum and bulb of the penis	Constricts the urethra and erects the penis
Female	Perineum	Base of the clitoris	Erects the clitoris
Urethral sphincter	Pubic ramus	Central tendon	Constricts the urethra
Sphincter ani externus	Coccyx	Central tendon	Closes the anal canal

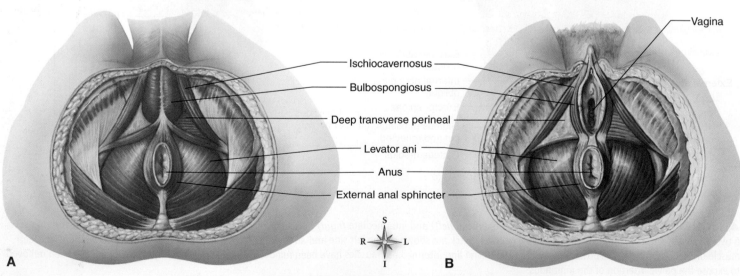

A   B

▲ FIGURE 10-27   Muscles of the pelvic floor. A, Male, inferior view. **B,** Female, inferior view.

Upper Limb Muscles

Muscles Acting on the Shoulder Girdle

The upper extremity is attached to the torso by muscles (Figure 10-28) that have an anterior location (chest) or posterior placement (back and neck). The **pectoralis minor** lies under the larger *pectoralis major* muscle on the anterior chest wall. The pectoralis minor helps to position the scapula against the thorax and also raises the ribs during forced inspiration. Another anterior chest wall muscle—the **serratus anterior**—helps hold the scapula against the thorax. This prevents "winging" by the scapula outward, and is a strong abductor that is useful in pushing or punching movements.

The posterior muscles acting on the shoulder girdle include the **levator scapulae,** which elevates the scapula; the **trapezius,** which is used to "shrug" the shoulders; and the **rhomboid major** and **rhomboid minor** muscles, which serve to adduct and elevate the scapula (Table 10-12).

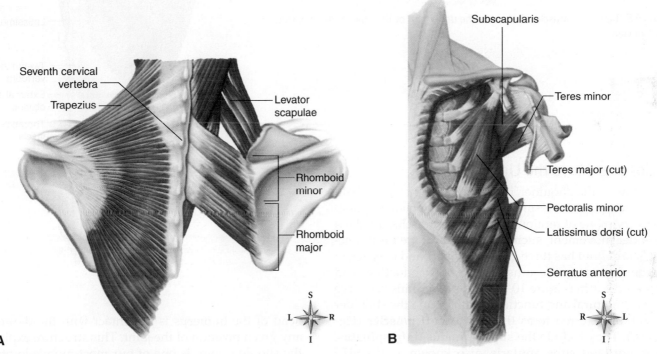

▲ **FIGURE 10-28** **Muscles acting on the shoulder girdle. A,** Posterior view. The trapezius has been removed on the right to reveal the deeper muscles. **B,** Anterior view. The pectoralis major has been removed.

TABLE 10-12 **Muscles Acting on the Shoulder Girdle**

MUSCLE	ORIGIN	INSERTION	FUNCTION
Trapezius	Occipital bone and spines of the cervical and thoracic vertebrae	Clavicle, and spine and acromion process of scapula	Rotates scapula, raises arm and scapula; pulls scapula and shoulder downward
Pectoralis minor	Ribs (second to fifth)	Scapula (coracoid)	Pulls the shoulder girdle down and forward
Serratus anterior	Ribs (upper eight or nine)	Scapula (anterior surface)	Pulls the shoulder down and forward; abducts and rotates it upward
Levator scapulae	C1-C4 (transverse processes)	Scapula (superior angle)	Elevates and retracts the scapula and abducts the neck
Rhomboid			
Major	T1-T4	Scapula (medial border)	Retracts, rotates, and fixes the scapula
Minor	C6-C7	Scapula (medial border)	Retracts, rotates, elevates, and fixes the scapula

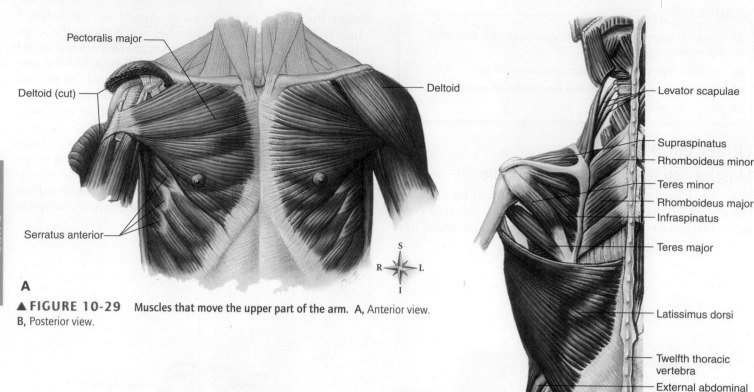

Pectoralis major

Deltoid (cut)

Serratus anterior

Deltoid

Levator scapulae

Supraspinatus

Rhomboideus minor

Teres minor

Rhomboideus major

Infraspinatus

Teres major

Latissimus dorsi

Twelfth thoracic vertebra

External abdominal oblique

Thoracolumbar fascia

A

B

▲ FIGURE 10-29   Muscles that move the upper part of the arm.   **A,** Anterior view. **B,** Posterior view.

Muscles that Move the Upper Arm

As you know, the shoulder is a synovial joint of the ball-and-socket type. As a result, extensive movement is possible in every plane of motion. Some of the muscles involved in shoulder movement, such as the **deltoid,** are multifunctional. The deltoid has three groups of fibers and may act, in effect, as three separate muscles, flexing, abducting, and extending the arm (Figure 10-29). Four other muscles serve as both a structural and functional cuff around the shoulder joint and are referred to as the **rotator cuff muscles** (Figure 10-30). They include the **supraspinatus, infraspinatus, teres minor,** and **subscapularis,** also known as the SITS muscles because their first letters spell out the acronym "SITS."

A number of other muscles are involved in the movement of the upper part of the arm, including the **pectoralis major,** which flexes the upper arm and adducts the upper arm anteriorly, and the **latissimus dorsi,** which extends the upper arm and adducts the upper arm posteriorly (Table 10-13).

Shoulder joint stability is greatly affected by the difference in size between the large (and nearly hemispheric) head of the humerus and the much smaller and shallow glenoid cavity of the scapula. Because of this difference in size, only about a quarter of the articular surface of the

head of the humerus is in contact with the depression in any given position of the joint. This structure explains why the shoulder joint is one of our most mobile joints. It also explains the inherent *instability* of the shoulder. This inherent instability is magnified because only a thin articular capsule surrounds the shoulder joint. This articular covering is extremely loose and cannot keep the articulating bones of the joint in contact. A great *range of motion* (ROM) is therefore possible, but this sometimes results in dislocation due to trauma.

There is some stability in the shoulder joint, however. The combination of muscles and tendons associated with the shoulder joint fuses into the rotator cuff, or *rotator* as it is commonly called. The rotator provides the necessary support to help prevent displacement of the head of the humerus during most types of activity.

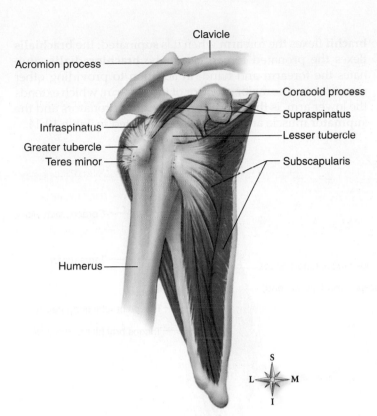

Clavicle
Acromion process
Coracoid process
Supraspinatus
Infraspinatus
Lesser tubercle
Greater tubercle
Teres minor
Subscapularis
Humerus

S
L — M
I

◄ **FIGURE 10-30**   **Rotator cuff muscles.** Note the tendons of the teres minor, infraspinatus, supraspinatus, and subscapularis muscles surrounding the head of the humerus—the so-called SITS muscles.

TABLE 10-13   **Muscles that Move the Upper Arm**

MUSCLE	ORIGIN	INSERTION	FUNCTION
*Axial**			
Pectoralis major	Clavicle (medial half), sternum costal cartilages of the true ribs	Humerus (greater tubercle)	Flexes the upper arm; adducts the upper arm anteriorly; draws it across the chest
Latissimus dorsi	Vertebrae (spines of the lower thoracic, lumbar, and sacral) Ilium (crest) Lumbodorsal fascia	Humerus (intertubercular groove)	Extends the upper arm Adducts the upper arm posteriorly
*Scapular**			
Deltoid	Clavicle Scapula (spine and acromion)	Humerus (lateral side about halfway down—deltoid tubercle)	Abducts the upper arm Assists in flexing and extending the upper arm (humerus)
Coracobrachialis	Scapula (coracoid process)	Humerus (middle third, medial surface)	Adducts arm; assists in flexing and rotating the arm medially
Supraspinatus†	Scapula (supraspinous fossa)	Humerus (greater tubercle)	Assists in abducting the arm
Teres minor†	Scapula (axillary border)	Humerus (greater tubercle)	Rotates the arm outward
Teres major	Scapula (lower part, axillary border)	Humerus (upper part, anterior surface)	Assists in extending, adducting, and rotating the arm medially
Infraspinatus†	Scapula (infraspinatus border)	Humerus (greater tubercle)	Rotates the arm laterally and outwardly
Subscapularis†	Scapula (subscapular)	Humerus (lesser tubercle)	Rotates arm medially

Axial muscles originate on the axial skeleton. *Scapular* muscles originate on the scapula.
† Muscles of the rotator cuff (SITS muscles).

Muscles that Move the Forearm

A few superficial and deep muscles of the upper extremity are shown in Figure 10-31. Remember that most muscles acting on a joint lie proximal to that joint. This is also true of the forearm muscles, which are found proximal to the elbow and attach to the ulna and radius of the forearm. The **biceps** brachii flexes the forearm when it is supinated; the **brachialis** flexes the pronated forearm; and the **brachioradialis** supinates the forearm and hand, in addition to providing other movements. The major extensor of the forearm, which extends the lower arm, is the **triceps brachii.** Other pronators and the **supinator** muscle and their actions are listed in Table 10-14.

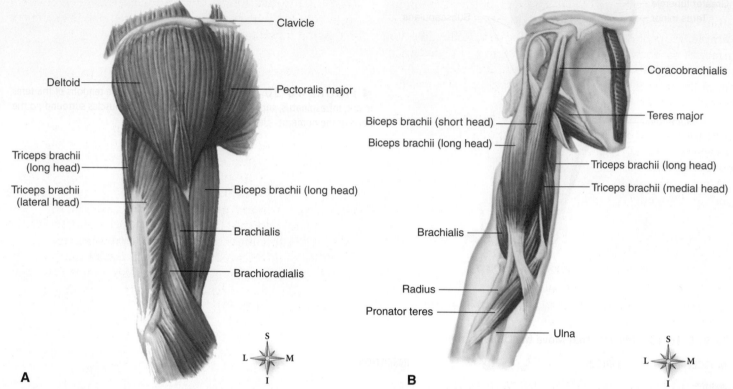

▲ **FIGURE 10-31**   **Muscles acting on the forearm. A,** Lateral view of the right shoulder and arm. **B,** Anterior view of the right shoulder and arm (deep). The deltoid and pectoralis major muscles have been removed to reveal deeper structures.

TABLE 10-14   Muscles that Move the Forearm

MUSCLE	ORIGIN	INSERTION	FUNCTION
Flexors			
Biceps brachii	Scapula (supraglenoid tuberosity) Scapula (coracoid)	Radius (tuberosity at the proximal end)	Flexes the forearm at elbow; rotates the hand laterally
Brachialis	Humerus (distal half, anterior surface)	Ulna (front of the coronoid process)	Flexes the forearm at the elbow
Brachioradialis	Humerus (above the lateral epicondyle)	Radius (styloid process)	Flexes the forearm at the elbow
Extensor			
Triceps brachii	Scapula (infraglenoid tuberosity) Humerus (posterior surface—lateral head above the radial groove; medial head, below)	Ulna (olecranon process)	Extends the lower arm at the elbow
Pronators			
Pronator teres	Humerus (medial epicondyle) Ulna (coronoid process)	Radius (middle third of the lateral surface)	Pronates the forearm
Pronator quadratus	Ulna (distal fourth, anterior surface)	Radius (distal fourth, anterior surface)	Pronates the forearm
Supinator			
Supinator	Humerus (lateral epicondyle) Ulna (proximal fifth)	Radius (proximal third)	Supinates the forearm

Muscles that Move the Wrist, Hand, and Fingers

Muscles that move the wrist, hand, and fingers can be classified as *extrinsic muscles* or *intrinsic muscles.* **Extrinsic muscles** originate *outside* the part of the skeleton moved. For example, extrinsic muscles originating in the forearm can pull on their insertions in the wrist, hand, and fingers to move them. **Intrinsic muscles** are the muscles that are actually *within* the part moved. Muscles that begin and end at different points within the hand can produce fine finger movements, for example. In most cases, the muscles located on the anterior surface of the forearm are flexors and those on the posterior surface are extensors of the wrist, hand, and fingers. As a group, the intrinsic muscles abduct and adduct the fingers and aid in flexing them.

The **opponens pollicis** of the thumb allows it to be drawn across the palm to touch the tip of any finger. This is a critical movement for many activities requiring fine motor skills and manipulation.

A summary of the major muscles that move the wrist, hand, and fingers is given in Table 10-15.

QUICK ✓ CHECK

27. What are the functions of the deltoid muscles?
28. What are the functions of the "biceps" and "triceps?"
29. Distinguish extrinsic from intrinsic muscles of the hand and wrist.

Lower Limb Muscles

The muscles, bones, and joints of the pelvic girdle and lower extremity are important for locomotion and stability of the entire body. Powerful muscles at the back of the hip, at the front of the thigh, and at the back of the leg also serve to raise the body from a sitting to a standing position. Unlike the highly mobile shoulder girdle, however, the pelvic girdle is essentially *fixed.* Our study of muscles in the lower extremity will begin with those arising from the pelvic girdle and extending down to the femur.

Muscles that Move the Thigh and Leg

Muscles acting on the thigh can be divided into three main groups: (1) muscles crossing the front of the hip, (2) the three **gluteal** muscles of the upper thigh and buttocks and the **tensor fasciae latae,** and (3) the thigh adductors. For placement of these muscles please refer to Figure 10-32.

Two major groups of muscles, the so-called *quads* and *hamstrings,* are involved in major movements of the leg. In the **quadriceps femoris** group are the **rectus femoris** (flexes the thigh in addition to extending the leg), and the **vastus lateralis, vastus medialis,** and **vastus intermedius.** All the quads are involved in extending the leg. For the most part, the **biceps femoris, semitendinosus,** and **semimembranosus** are involved in either extending the thigh or flexing the leg (Tables 10-16 and 10-17).

TABLE 10-15 Muscles that Move the Wrist, Hand, and Fingers

MUSCLE	ORIGIN	INSERTION	FUNCTION
Flexor carpi radialis	Humerus (medial epicondyle)	Second metacarpal (base of)	Flexes the hand Flexes the forearm
Palmaris longus	Humerus (medial epicondyle)	Fascia of the palm	Flexes the hand
Flexor carpi ulnaris	Humerus (medial epicondyle)	Pisiform bone	Flexes the hand
	Ulna (proximal two thirds)	Third, fourth, and fifth metacarpals	Adducts the hand
Extensor carpi radialis longus	Humerus (ridge above the lateral epicondyle)	Second metacarpal (base of)	Extends the wrist and hand Abducts the hand (moves toward the thumb side when the hand is supinated)
Extensor carpi radialis brevis	Humerus (lateral epicondyle)	Second, third metacarpals (bases of)	Extends the hand
Extensor carpi ulnaris	Humerus (lateral epicondyle) Ulna (proximal three fourths)	Fifth metacarpal (base of)	Extends and adducts the wrist and hand
Flexor digitorum profundus	Ulna (anterior surface)	Distal phalanges (fingers 2 to 5)	Flexes the distal joints of the fingers
Flexor digitorum superficialis	Humerus (medial epicondyle) Radius Ulna (coronoid process)	Tendons of the fingers	Flexes the fingers
Extensor digitorum	Humerus (lateral epicondyle)	Phalanges (fingers 2 to 5)	Extends the fingers

UNIT 2

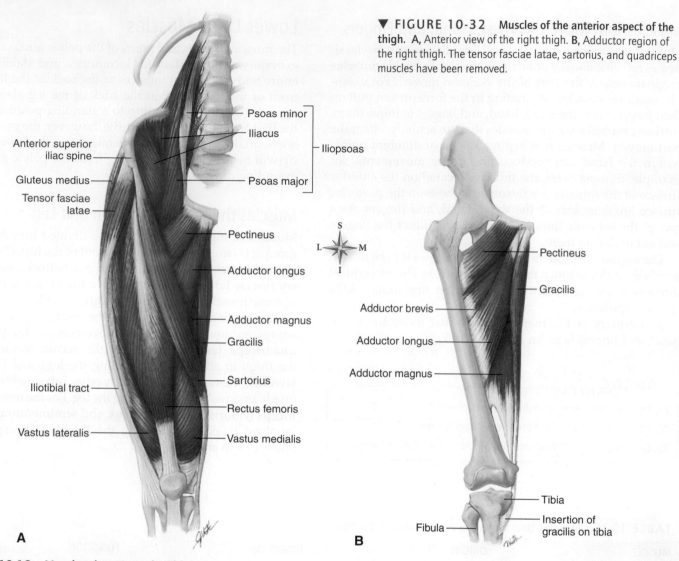

▼ **FIGURE 10-32**   Muscles of the anterior aspect of the thigh. **A,** Anterior view of the right thigh. **B,** Adductor region of the right thigh. The tensor fasciae latae, sartorius, and quadriceps muscles have been removed.

TABLE 10-16   Muscles that Move the Thigh

MUSCLE	ORIGIN	INSERTION	FUNCTION
Iliopsoas (iliacus, psoas major, and psoas minor)	Ilium (iliac fossa)	Femur (lesser trochanter)	Flexes the thigh
	Vertebrae (bodies of twelfth thoracic to fifth lumbar)		Flexes the trunk (when the femur acts as the origin)
Rectus femoris	Ilium (anterior, inferior spine)	Tibia (by way of the patellar tendon)	Flexes the thigh, extends the lower leg
Gluteal Group			
Maximus	Ilium (crest and posterior surface) Sacrum and coccyx (posterior surface) Sacrotuberous ligament	Femur (gluteal tuberosity) Iliotibial tract	Extends the thigh—rotates outward
Medius	Ilium (lateral surface)	Femur (greater trochanter)	Abducts the thigh—rotates medially
Minimus	Ilium (lateral surface)	Femur (greater trochanter)	Abducts the thigh Rotates the thigh medially
Tensor fasciae latae	Ilium (anterior part of the crest)	Tibia (by way of the iliotibial tract)	Abducts, flexes, and rotates thigh medially
Adductor Group			
Brevis	Pubic bone	Femur (linea aspera)	Adducts the thigh
Longus	Pubic bone	Femur (linea aspera)	Adducts the thigh
Magnus	Pubic bone	Femur (linea aspera)	Adducts the thigh
Gracilis	Pubic bone (just below the symphysis)	Tibia (medial surface behind the sartorius)	Adducts the thigh and flexes and adducts the leg

TABLE 10-17   Muscles that Move the Leg

MUSCLE	ORIGIN	INSERTION	FUNCTION
Sartorius	Coxal (anterior superior iliac spines)	Tibia (medial surface of the upper end of the shaft)	Adducts and flexes the leg (Permits crossing of the legs)
Quadriceps Femoris Group			
Rectus femoris	Ilium (anterior inferior spine)	Tibia (by way of patellar tendon) Extends the leg	Flexes the thigh
Vastus lateralis	Femur (linea aspera)	Tibia (by way of the patellar tendon)	Extends the leg
Vastus medialis	Femur	Tibia (by way of the patellar tendon)	Extends the leg
Vastus intermedius	Femur (anterior surface)	Tibia (by way of the patellar tendon)	Extends the leg
Hamstring Group			
Biceps femoris	Ischium (tuberosity)	Fibula (head of)	Extends the thigh; flexes the leg
	Femur (linea aspera)	Tibia (lateral condyle)	Flexes the leg
Semitendinosus	Ischium (tuberosity)	Tibia (proximal end, medial surface)	Extends the thigh; flexes the leg
Semimembranosus	Ischium (tuberosity)	Tibia (medial condyle)	Extends the thigh; flexes the leg

Muscles that Move the Ankle and Foot

As the name suggests, the **extrinsic foot muscles** (Figure 10-33) are located *outside* the foot, in the leg. However, they exert their actions by *pulling* on tendons that insert on the bones of the ankle and foot. The extrinsic muscles are responsible for a number of movements such as dorsiflexion (up) and plantar flexion (down). In contrast, **intrinsic foot muscles** are located *within* the foot and are responsible for flexing, extending, abducting, and adducting the toes.

The superficial muscles located on the posterior surface of the leg form the bulging "calf." The common tendon of the **gastrocnemius, soleus,** and **plantaris** muscles is called the **calcaneal** (Achilles) **tendon.** This tendon inserts onto the calcaneus, or heel bone. By acting together, these muscles produce plantar flexion of the foot.

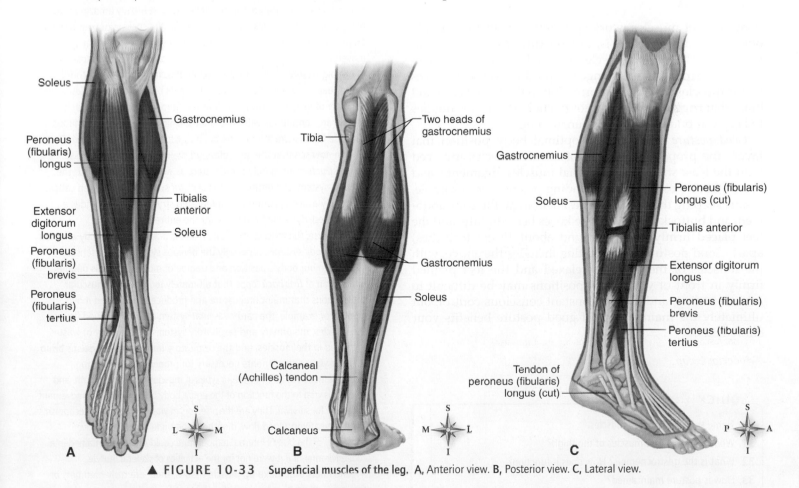

▲ **FIGURE 10-33   Superficial muscles of the leg.** A, Anterior view. B, Posterior view. C, Lateral view.

TABLE 10-18 Muscles that Move the Foot

MUSCLE	ORIGIN	INSERTION	FUNCTION
Tibialis anterior	Lateral condyle and lateral surface of tibia	Tarsal (first cuneiform)	Flexes the foot
		Base of first metatarsal bone	Inverts the foot
Gastrocnemius	Lateral and medial condyles of femur	Tarsal (calcaneus by way of the Achilles tendon)	Extends the foot Flexes lower leg
Soleus	Head and shaft of fibula and posterior surface of tibia	Tarsal (calcaneus by way of the Achilles tendon)	Extends the foot (plantar flexion)
Peroneus longus (fibularis longus)	Tibia (lateral condyle)	First cuneiform	Extends the foot (plantar flexion)
	Fibula (head and shaft)	Base of the first metatarsal	Everts the foot
Peroneus brevis (fibularis brevis)	Fibula (lower two thirds of the lateral surface of the shaft)	Fifth metatarsal (tubercle, dorsal surface)	Everts the foot Flexes the foot
Peroneus tertius (fibularis tertius)	Fibula (distal third)	Fourth and fifth metatarsals (bases of)	Flexes the foot Everts the foot
Extensor digitorum longus	Tibia (lateral condyle) Fibula (anterior surface)	Second and third phalanges (four lateral toes)	Dorsiflexion of the foot; extension of the toes

Dorsal flexors of the foot, located on the *anterior* surface of the leg, include the **tibialis anterior, peroneus tertius,** and **extensor digitorum longus**—which also serves to evert the foot and extend the toes (Table 10-18).

POSTURE

A number of muscles, working together, help to maintain our body position—that is, our **posture.** *Muscle tone* provides the basic tension muscles need to keep our skeleton in a stable position. As we change our position, sensory nerves in our muscles feed back information to the spinal cord and brain that triggers additional contraction of various muscles to keep our bodies stable and functioning.

Good posture maintains an optimal body position that favors the proper functioning of all the muscles involved with the least strain on individual muscles, ligaments, and bones. Good posture in the standing position, for example, means keeping the head and chest held high; the chin, abdomen, and buttocks pulled in; the knees bent slightly; and the feet placed firmly on the ground about 15 cm (6 inches) apart. Good posture during sitting includes the above, with the muscles of the buttocks relaxed and the feet planted firmly in front of you. These positions may be difficult to achieve at first, and require constant conscious control, but ultimately the maintenance of good posture benefits your entire body.

QUICK CHECK

30. Name the three gluteal muscles.

31. What are the major muscles of the thigh?

32. What is the gastrocnemius? How does it function?

33. How is posture maintained?

The BIG Picture

Acting together, the muscular, skeletal, and nervous systems permit us to move in a coordinated and controlled way. It is truly amazing how many individual muscles are responsible for the success of your body as an integrated whole. So many muscle names to learn, so many types of movement, so many sites of origin and insertion—it's easy to lose sight of the big picture! Remember, however, that muscles work in coordinated teams that move various components of the flexible skeleton. As a matter of fact, you now know that the fibrous wrappings of each muscle are continuous with its tendons, which in turn are continuous with the fibrous structure of the bone to which they are attached. Thus, we can now see that the muscular and skeletal systems are in essence a *single structure* intimately coordinated, as we will soon see, with the nervous system. The entire *musculoskeletal system,* as it is often called, is actually a single, continuous integrated structure that provides a coordinated, dynamic framework for the entire body.

Of course, the musculoskeletal system is directly or indirectly affected by other body systems, especially the nervous system, which senses changes in our body's position and degree of movement. This permits integration of *feedback loops* that ultimately regulate the muscular contractions that maintain posture and produce coordinated movements. For example, the cardiovascular system maintains blood flow to the muscles, the urinary and respiratory systems rid the body of wastes produced in the muscles, and the respiratory and digestive systems bring in the oxygen and nutrients necessary for proper muscle function.

The function of all three major types of muscles (skeletal, smooth, and cardiac) is vital to the function of the entire body by providing the movement necessary for survival. They are the effectors in vital homeostatic mechanisms such as breathing, blood flow, digestion, and urine flow. The relative constancy of the body's internal temperature could not be maintained in a cool environment if it were not for the activities of skeletal muscle.

The fibers that make up muscle tissues, then, are truly members of the large, interactive "society of cells" that forms the human body.

Cycle of LIFE ⟳

Our muscles develop increasing strength and power during our youth, reach their peak performance during early adulthood, but may decline as we age. In fact, we often begin to lose lean body mass because of muscle loss during middle age. This loss of muscle mass may continue at the rate of 8% to 16% per decade after the age of 50. Muscle power (equal to the product of force and velocity of muscle contraction) declines even faster than muscle strength. This decline in muscle power affects our ability to climb stairs, walk steadily, and rise from a chair.

Why do we lose muscle mass as we age? A number of factors are likely involved. But recent findings show that most of the age-related decrease in muscle strength is the result of decreased muscle *use*—a phenomenon called disuse atrophy (see p. 197). ●

MECHANISMS OF DISEASE

Muscle disorders, or *myopathies*, generally disrupt the normal movement of the body. Relatively minor muscle injuries such as *strains* are common, and even more severe muscle injuries, including *bruising* and *sprains*, are not rare. In addition, muscles can exhibit a number of common but abnormal contractions such as *cramps* and *spasms*. Most significant, muscles can also be the sites for life-threatening infection and disease.

evolve Find out more about these muscle tissue diseases online at *Mechanisms of Disease: Skeletal System.*

LANGUAGE OF SCIENCE AND MEDICINE *continued from page 187*

fascia (FAY-shah)
[*fascia* band or bundle]

fascicle (FAS-i-kul)
[*fasci-* band or bundle, *-cle* little]

fiber (FYE-ber)
[*fibr-* thread or fiber]

fixator muscle (fik-SAY-tor MUSS-el)
[*fixator* fastener, *mus-* mouse, *-cle* little]

gluteal (GLOO-tee-al)
[*glut-* buttocks, *-al* relating to]

glycogen (GLYE-koh-jen)
[*glyco-* sweet, *-gen* produce]

glycolysis (glye-KAHL-ih-sis)
[*glyco-* sweet (glucose), *-o-* combining form, *-lysis* loosening]

hypertrophy (hye-PER-troh-fee)
[*hyper-* excessive, *-troph-* nourishment, *-y* state]

H zone

I band

insertion
[*in-* in, *-ser-* join, *-tion* process]

intercalated disk (in-TER-kah-lay-ted)
[*intercalate-* insert]

intrinsic foot muscle (in-TRIN-sik MUSS-el)
[*intrins-* inward, *-ic* relating to, *mus-* mouse, *-cle* little]

intrinsic muscle (in-TRIN-sik MUSS-el)
[*intrins-* inward, *-ic* relating to, *mus-* mouse, *-cle* little]

isometric contraction (eye-soh-MET-rik)
[*iso-* equal, *-metr-* measure, *-ic* relating to, *con-* together, *-tract-* drag or draw, *-tion* process]

isotonic contraction (eye-soh-TON-ik)
[*iso-* equal, *-ton-* stretch or tension, *-ic* relating to, *con-* together, *-tract-* drag or draw, *-tion* process]

lactate (LAK-tayt)
[*lact-* milk, *-ate* chemical]

lever
[*lev-* lift, *-er* agent]

linea alba (LIN-ee-ah AL-bah)
[*linea* line, *alba* white]

mastication (mass-ti-KAY-shun)
[*mastica-* chew, *-ation* process]

M line (EM lyne)
[*M: mittel* middle]

motor nerve (MOH-ter nerv)
[*mot-* movement, *-or* agent]

motor neuron (MOH-tor NOO-ron)
[*mot-* movement, *-or* agent, *neuron* nerve]

motor unit (MOH-tor YOO-nit)
[*mot-* movement, *-or* agent]

multiunit smooth muscle
[*multi-* many, *mus-* mouse, *-cle* little]

muscle (MUSS-el)
[*mus-* mouse, *-cle* little]

muscle fatigue (MUSS-el fah-TEEG)
[*mus-* mouse, *-cle* little, *fatig-* tire]

muscle tone
[*mus-* mouse, *-cle* little, *ton-* stretch or tension]

muscular system
[*mus-* mouse, *cul-* little, *-ar* relating to]

myofibril (my-oh-FYE-bril)
[*myo-* muscle, *-fibr-* fiber, *-il* little]

myofilament (my-oh-FIL-ah-ment)
[*myo-* muscle, *-fila-* thread, *-ment* thing]

myoglobin (Mb) (my-oh-GLOH-bin)
[*myo-* muscle, *-glob-* ball, *-in* substance]

neuromuscular junction (NMJ) (noo-roh-MUSS-kyoo-lar)
[*neuro-* nerve, *-mus-* mouse, *-cul-* little, *-ar* relating to]

origin (OR-ih-jin)

oxidative phosphorylation (ahk-si-DAY-tiv fos-for-ih-LAY-shun)
[*oxi-* sharp (oxygen), *-id-* chemical (*-ide*), *-at-* action of (*-ate*), *-ive* relating to, *phos-* light, *-phor-* carry, *-yl-* chemical, *-ation* process]

oxygen debt (AHK-si-jen)
[*oxy-* sharp, *-gen* produce, debt thing owed]

perimysium (pair-ih-MEE-see-um)
[*peri-* around, *-mysium* muscle]

perineum (pair-ih-NEE-um)
[*peri-* around, *-ine-* excrete, *-um* thing] *pl.,* perinea (pair-IH-nee-ah)

peristalsis (pair-ih-STAL-sis)
[*peri-* around, *-stalsis* contraction]

posture (POS-chur)
[*postur-* position]

rotator cuff muscle (roh-TAY-tor MUSS-el)
[*rot-* turn, *-ator* agent, *mus-* mouse, *-cle* little]

sarcolemma (sar-koh-LEM-ah)
[*sarco-* flesh, *-lemma* sheath] *pl.,* sarcolemmae (sar-koh-LEM-ee)

sarcomere (SAR-koh-meer)
[*sarco-* flesh, *-mere* part]

continued

sarcoplasm (SAR-koh-plaz-em)
[*sarco-* flesh, *-plasm* substance]

sarcoplasmic reticulum (SR) (sar-koh-PLAZ-mik reh-TIK-yoo-lum)
[*sarco-* flesh, *-plasm-* substance, *-ic* relating to, *ret-* net, *-ic-* relating to, *-ul-* little, *-um* thing] *pl.,* reticula (reh-TIK-yoo-lah)

single-unit smooth muscle
[*mus-* mouse, *-cle* little]

sliding-filament model (SLY-ding FILL-ah-ment MOD-uhl)

smooth muscle (smooth MUSS-el)
[*mus-* mouse, *-cle* little]

strength training

syncytium (sin-SISH-ee-em)
[*syn-* together, *-cyt-* cell, *-um* a thing] *pl.,* syncytia (sin-SISH-ee-ah)

synergist (SIN-er-jist)
[*syn-* together, *-erg-* work, *-ist* agent]

T tubule (TEE TOOB-yool)
[*T* transverse]

tendon (TEN-dun)
[*tend-* pulled tight, *-on* unit]

tetanus (TET-ah-nus)
[*tetanus* tension]

threshold stimulus (THRESH-hold STIM-yoo-lus)
[*stimul-* excite, *-us* thing] *pl.,* stimuli (STIM-yool-eye)

treppe (TREP-ee)
[*treppe* staircase]

urethral sphincter (yoo-REE-thral SFINGK-ter)
[*ure-* urine, *-thr-* agent or channel (urethra), *-al* relating to, *sphinc-* bind tight, *-er* agent, *mus-* mouse, *-cle* little]

Z disk (also called **Z line**) (ZEE lyne)
[*Z: zwischen* between]

CHAPTER SUMMARY

To download an MP3 version of the chapter summary for use with your iPod or other portable media player, access the Audio Chapter Summaries online at http://evolve.elsevier.com.

HINT   *Scan this summary after reading the chapter to help you reinforce the key concepts. Later, use the summary as a quick review before your class or before a test.*

INTRODUCTION

A. Skeletal and muscular systems act together to actually produce most of your body movements
B. The body has over 600 muscles comprising 40% to 50% of your body weight

INTRODUCTION TO SKELETAL MUSCLES

A. General functions of skeletal muscles:
   1. Permit movement
   2. Produce heat
   3. Stabilize our joints
   4. Maintain our posture
B. These functions all rely on the characteristic of excitability—the ability of the muscle to respond to nervous and endocrine signals
C. Muscles also are extensible; they can stretch or lengthen
D. Basic structure of a muscle organ (Figure 10-1):
   1. Endomysium—fine connective tissue that covers muscle fibers
   2. Fascicles—group of muscle fibers
   3. Perimysium—thick connective tissue that covers fascicles
   4. Epimysium—coarse sheath that covers entire muscle
E. Epimysium, perimysium, and endomysium may join with fibrous tissue that extends from the muscle as a tendon
   1. Aponeurosis—broad, flat sheet of connective tissue
   2. Fascia—fibrous connective tissue surrounding muscle and outside the epimysium and tendon

F. Overview of the muscle cell (Figure 10-1)
   1. Skeletal muscle is composed of bundles of skeletal muscle fibers
      a. Muscle fibers—individual muscle cells that have fused together during development to form a long, threadlike multinucleated cell
   2. Structural components of muscle fibers
      a. Sarcolemma—equivalent to the plasma membrane of a typical cell
      b. Sarcoplasm—equivalent to the muscle's cytoplasm
      c. Sarcoplasmic reticulum (SR)—network of tubules and sacs; temporarily stores calcium ions
      d. T tubules—inward extensions of the sarcolemma; allow electrical signals to move deeper into the cell (Figure 10-2)
      e. Myofibrils—extend lengthwise along a skeletal muscle fiber
      f. Myofilaments—classified as thick filaments and thin filaments (Figure 10-2)
      g. Sarcomere—the basic contractile unit of the muscle cell (Figure 10-3; Box 10-1)
G. Components of muscle contraction
   1. The contraction of a muscle fiber involves a complex process coordinated in a stepwise fashion (Figure 10-5)
   2. Skeletal muscle fiber normally remains "at rest" until it is stimulated to contract
   3. Neuromuscular junction—motor neurons connect to the sarcolemma at the motor end plate
   4. At neuromuscular junction, small vesicles in the cytoplasm of the nerve ending release neurotransmitters (acetylcholine or ACh)
   5. Neurotransmitter molecules diffuse quickly across the synaptic cleft and come into contact with the sarcolemma; acetylcholine receptors bind with the acetylcholine
   6. Action of acetylcholine binding with its receptor molecules produces an impulse; excitation
   7. Acetylcholine fuses with receptor sites on the sarcolemma; opens sodium ion gates

8. Depolarization wave spreads across the surface of the muscle cell membrane and down into the T tubules (Figure 10-6)

H. Mechanism of muscle contraction (Box 10-2)

1. Impulse in the T tubules then triggers the release of a flood of calcium ions from the adjacent sacs of the SR
2. Calcium ions immediately combine with troponin in the thin filaments of the myofibrils; causes the tropomyosin to shift and expose the active binding sites on the actin molecules (Figure 10-7)
3. Once the active sites for contraction are exposed, myosin heads of the thick filaments bind to actin molecules in the nearby thin filaments
4. Sliding of the thin filaments toward the center of each sarcomere subunit quickly shortens the entire myofibril—and thus the entire muscle fiber (Figure 10-8)
5. Contracting sarcomeres work hard to pull the ends of the muscle fiber toward one another (Figure 10-9)

ENERGY SOURCES FOR MUSCLE CONTRACTION

A. Energy required for muscular contraction is obtained from ATP after it breaks into ADP and phosphate—only enough ATP for a few seconds of muscle contraction

1. Muscle "recharges" ADP by adding phosphate back—phosphorylating
2. Creatine phosphate is a "quick charger" that buys a few more seconds of ATP energy
3. For contraction of more than a few seconds, glucose travels through anaerobic pathway called glycolysis to create pyruvate
4. Pyruvate enters aerobic pathway called oxidative phosphorylation to create ATP

B. Myoglobin—oxygen storage molecule with a red color

C. Glucose is stored in the form of glycogen

D. Aerobic pathway—produces the maximal amount of energy available from each glucose molecule

E. Anaerobic pathway—produces minimal amount of ATP

1. Faster than aerobic pathway, used for powerful, short-lived muscle contractions
2. Can produce ATP during endurance activity when oxygen is depleted
3. When anaerobic processes supplement aerobic processes, pyruvate is converted to lactate—dissolved form of lactic acid
4. Oxygen debt—lack of oxygen in some tissues causing the production of excessive lactic acid

FUNCTION OF SKELETAL MUSCLE ORGANS

A. Motor unit—motor neuron, plus the muscle fibers to which it attaches (Figure 10-12)

1. The fewer the number of muscle fibers supplied by a motor unit, the more finely coordinated the movements that muscle can produce

B. Muscle contraction, fatigue, and strength

1. Muscle fiber does not begin to contract immediately at the instant of stimulation, but a fraction of a second later (latent period)
2. Contraction phase—muscle fiber increases its tension and shortens
3. Relaxation phase—resting state
4. Treppe—stepwise increase in strength at beginning of series of contractions

5. Tetanic contraction (tetanus)—smooth, sustained contractions
6. Muscle tone—continual, partial contraction in a muscle organ
7. Muscle fatigue—complete loss of strength or endurance that results during the process of muscle contraction

C. Strength of a muscle is related to:

1. Number of fibers
2. Initial length of muscle fibers (Figure 10-14)
3. Position in which muscle is contracting

D. Atrophy—decrease in muscle mass

E. Hypertrophy—increase in muscle mass

F. Isotonic and isometric contractions can be used to strengthen muscle (Figure 10-15)

1. Isotonic—muscle shortens and produces movement
2. Isometric—muscle pulls forcefully against a load but does not shorten

FUNCTION OF CARDIAC AND SMOOTH MUSCLE

A. Cardiac muscle—found only in the heart; contracts rhythmically and continuously (Figure 10-16)

1. Does not taper like skeletal muscle fiber
2. Contains intercalated disks—strong, electrically coupled junctions
3. Cardiac muscle is self-exciting; exhibits a continual rhythm of excitation and contraction on its own

B. Smooth muscle—composed of small, tapered cells with large, single nuclei

1. Lacks the striations of skeletal and cardiac muscles
2. Several types of smooth muscle:
   a. Single-unit or visceral smooth muscle—most common type; rhythmic self-exciting (peristalsis)
   b. Multiunit smooth muscle tissue—composed of many independent single-cell units; found in tiny arrector pili muscles, in muscles that control the lens of the eye, or as single fibers

SKELETAL MUSCLE ORGANS

A. Muscle size, shape, and fiber arrangement

1. Muscles range in size from extremely small strands to large masses; also range in shape from broad and wide to long and narrow
2. Strength and type of movement is related to muscle's overall shape and the orientation of its fibers
3. Descriptive terms for muscles: (Figure 10-18):
   a. Parallel—long and straplike
   b. Circular—encircle body tubes or openings
   c. Convergent—look like the blades of a fan

B. Attachment of muscles (Figure 10-19)

A. Origin—point of attachment that does not move when the muscle contracts

B. Insertion—point of attachment that moves when the muscle contracts

C. Muscle actions—skeletal muscles almost always act in groups rather than singly

1. Agonists—muscle (or group of muscles) that directly performs a specific movement (Figure 10-20)
2. Antagonists—muscles that, when contracting, directly oppose agonists
3. Synergists—muscles that contract at the same time as agonists; complement the actions of the agonists
4. Fixator muscles—serve to stabilize joints; often help maintain posture or balance when agonists are contracting

D. Lever systems
  1. A contracting muscle applies a pulling force on a bone lever precisely where the muscle attaches to the (origin) bone
    a. This action causes the (insertion) bone to move at its joint (that is, as a fulcrum—the point on which a lever pivots) (Figure 10-21)

HOW MUSCLES ARE NAMED

A. Many muscle names have Latin roots similar to those of words of the English language (Figure 10-22)
B. Muscles are typically named according to their location, function, shape, and direction of fibers
C. Muscles may be named according to the number of points of heads (points of origin), the number of points of attachment, and the size of the muscle (Table 10-3)

MUSCLES OF THE HEAD

A. Muscles of facial expression (Table 10-5):
  1. Occipitofrontalis—two muscles; one portion lies over the frontal bone; the other covers the occipital bone in the back of the head
  2. Corrugator supercilii—pulls the eyebrows together and produces the vertical wrinkles above the nose (frowning)
  3. Orbicularis oculi—encircles and closes the eye (blinking)
  4. Orbicularis oris and buccinator—pucker the mouth (kissing) and press the lips and cheeks against the teeth
  5. Zygomaticus major—draws the corner of the mouth upward (laughing)
B. Muscles of mastication—responsible for our chewing movements
  1. Masseter and temporalis—either elevate or retract the mandible
  2. Pterygoids—open and protrude the jaw while making sideways chewing movements
  3. Buccinator—holds food between the teeth as the mandible moves up and down and from side to side
C. Muscles that move the head—paired muscles on either side of the neck are responsible for head movements (Figure 10-24; Table 10-7)
  1. Sternocleidomastoid—"prayer muscle"; muscles contract at the same time the head flexes forward
  2. Splenius capitis—serve as strong extensors that return the head to the upright position after flexion

TRUNK MUSCLES

A. Muscles of the thorax—vitally important for inhalation and exhalation (Table 10-8)
  1. Internal and external intercostals—elevate and depress the ribs during inhalation and exhalation
  2. Diaphragm—flattens as it contracts, thereby increasing the size and volume of the thoracic cavity
B. Muscles of the abdominal wall—very strong fabric of muscles that cover and support the abdominal cavity and its internal organs (Table 10-9)
  1. External oblique—run medially and inferiorly
  2. Internal oblique—run almost at right angles to those of the external oblique
  3. Transversus abdominis—innermost muscle layer of the abdomen

  4. Rectus abdominis—runs down the midline of the abdomen from the thorax to the pubis (Figure 10-25)
    a. Linea alba—tough band of connective tissue formed from sheaths that fuse in the midline of the rectus abdominis

MUSCLES OF THE BACK

A. Muscles of the back—play a major role in moving the head and limbs (Figure 10-26, *A*)
B. Deep back muscles—allow us to move vertebral column in a variety of ways; stabilize trunk and help maintain a stable posture (Figure 10-26, *B*)
  1. Erector spinae—provide some rotation and flex the back laterally (Figure 10-26; Table 10-10)
  2. Interspinales and multifidus groups—connect one vertebra to the next; help extend the back and neck or flex them to the side

MUSCLES OF THE PELVIC FLOOR

A. Muscles of the pelvic floor—guards the various body outlets that project below (Table 10-11)
  1. Levator ani and coccygeus—stretch across the pelvic cavity like a hammock to form most of the pelvic floor
  2. Ischiocavernosus and bulbospongiosus—associated with the penis in the male and the vagina in the female
  3. Sphincter urethrae—encircle the urethra in both sexes; help control urine flow
  4. External anal sphincter—controls defecation

UPPER LIMB MUSCLES

A. Muscles acting on the shoulder girdle—(Figure 10-28; Table 10-12)
  1. Pectoralis minor—helps to position the scapula against the thorax and also raises the ribs during forced inspiration
  2. Pectoralis major—muscle on the anterior chest wall
  3. Serratus anterior—helps hold the scapula against the thorax
  4. Levator scapulae—elevates the scapula
  5. Trapezius—used to "shrug" the shoulders
  6. Rhomboid major and minor—serve to adduct and elevate the scapula
B. Muscles that move the upper arm—(Table 10-13)
  1. Deltoid—flexes, abducts, and extends the arm (Figure 10-29)
  2. Rotator cuff muscles—serve as both a structural and functional cap or cuff around the shoulder joint
    a. Infraspinatus
    b. Supraspinatus
    c. Subscapularis
    d. Teres minor
C. Muscles that move the forearm—found proximal to the elbow and attach to the ulna and radius of the forearm (Table 10-14; Figure 10-31)
D. Muscles that move the wrist, hand, and fingers—classified as extrinsic muscles or intrinsic muscles (Table 10-15)
  1. Extrinsic muscles—originate outside the part of the skeleton moved

2. Intrinsic muscles—refer to muscles that are actually within the part moved
   a. Abduct and adduct the fingers and aid in flexing them

E. In most cases, the muscles located on the anterior surface of the forearm are flexors; those on the posterior surface are extensors of the wrist, hand, and fingers

F. Opponens pollicis—allows thumb to be drawn across the palm to touch the tip of any finger

LOWER LIMB MUSCLES

A. Muscles that move the thigh and leg (Figure 10-32; Tables 10-16 and 10-17)

B. Muscles that move the ankle and foot (Table 10-18)

1. Extrinsic foot muscles—exert their actions by pulling on tendons that insert on the bones of the ankle and foot (Figure 10-33)

2. Intrinsic foot muscles—responsible for flexing, extending, abducting, and adducting the toes

3. Gastrocnemius and soleus—form the calf
   a. Calcaneal (Achilles) tendon—common tendon of gastrocnemius and soleus

4. Tibialis anterior and peroneus tertius—dorsal flexors of the foot

5. Extensor digitorum longus—serves to evert the foot and extend the toes

POSTURE

A. Muscles working together to maintain body position
   1. Muscle tone keeps skeleton in stable position

B. Good posture—maintains an optimal body position that favors the proper functioning of all the muscles involved with the least strain on individual muscles, ligaments, and bones

REVIEW QUESTIONS

Write out the answers to these questions after reading the chapter and reviewing the Chapter Summary. If you simply think through the answer without writing it down, you won't retain much of your new learning.

1. Define the terms *sarcolemma, sarcoplasm,* and *sarcoplasmic reticulum.*
2. Describe the function of the sarcoplasmic reticulum.
3. How are acetylcholine, Ca^{++}, and adenosine triphosphate (ATP) involved in the excitation and contraction of skeletal muscle?
4. Describe the general structure of ATP and tell how it relates to its function.
5. How does ATP provide energy for muscle contraction?
6. Describe the anatomical arrangement of a motor unit.
7. Explain the difference between an isotonic contraction and an isometric contraction.
8. Define the terms *endomysium, perimysium,* and *epimysium.*
9. Give an example of a muscle named by location, function, shape, fiber direction, and number of heads.

10. Give examples of the muscles of the back, chest, abdomen, neck, shoulder, upper part of the arm, lower part of the arm, thigh, buttocks, leg, and pelvic floor.
11. Give examples of the muscles that flex, extend, abduct, and adduct the upper arm; that raise and lower the shoulder.
12. Give examples of the muscles that flex and extend the lower part of the arm; that flex and extend the wrist and hand.
13. Give examples of the muscles that flex, extend, abduct, and adduct the thigh; that flex and extend the lower part of the leg and thigh; that flex and extend the foot.
14. Give examples of the muscles that flex, extend, abduct, and adduct the head.
15. Give examples of the muscles that move the abdominal wall; that move the chest wall.

CRITICAL THINKING QUESTIONS

After finishing the Review Questions, write out the answers to these items to help you apply your new knowledge. Go back to sections of the chapter that relate to items that you find difficult.

1. Explain how skeletal muscles provide movement, heat, and posture. Are all of these functions unique to muscles? Explain your answer.
2. What structures are unique to skeletal muscle fibers? Which of the structures are involved primarily in contractility and which are involved in excitability?
3. Explain how the structure of myofilaments is related to their function.
4. Explain how the sliding-filament model allows for the shortening of a muscle fiber.
5. Compare and contrast the role of Ca^{++} in excitation, contraction, and relaxation of skeletal muscle.
6. People who exercise seriously are sometimes told to work a muscle until they "feel the burn." In terms of how the muscle is able to release energy, explain what is going on in the muscle early in the exercise and when the muscle is "burning."
7. Using fiber types, design a muscle for a marathon runner and a different muscle for a 100-yard-dash sprinter. Explain your choice.
8. Explain the meaning of a "cardiac syncytium" as it relates to cardiac muscle. How does this structural arrangement affect its function?
9. Identify the muscles of facial expression. What muscles facilitate smiling and frowning?
10. How do the origin and insertion of a muscle relate to each other in regard to actual movement?
11. When the biceps brachii contracts, the elbow flexes. When the triceps brachii contracts, the elbow extends. Explain the role of both muscles in terms of agonist and antagonist in both of these movements.
12. Describe how the body maintains posture.
13. Baseball players, particularly pitchers, often incur rotator cuff injuries. List the muscles that make up the rotator cuff and explain the importance of these muscles and their role in joint stability.

continued from page 187

Now that you have read this chapter, try to answer these questions about Mr. Cervantes' fibromyalgia pain from the Introductory Story.

1. Based on information in this chapter, what type of exercise would you recommend for people such as Mr. Cervantes who are afflicted with fibromyalgia?
   a. Strength training with heavy barbells once a week
   b. Strength training with increased amounts of weight done at regular intervals throughout the week.
   c. Aerobic exercise pushing Mr. Cervantes to his physiological limit
   d. No exercise at all: Mr. Cervantes should avoid all muscle training and learn to relax

2. Which of the following might be improved if Mr. Cervantes engaged in a regular exercise program?
   a. Stiffness and muscle pain should be reduced
   b. Fatigue and depression should be reduced
   c. Sleep habits should improve
   d. All of the above could happen

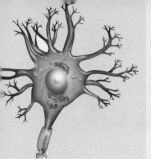

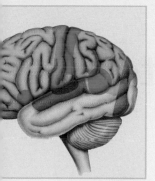

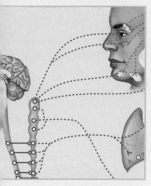

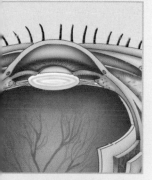

UNIT **3**

Communication, Control, and Integration

The structures and mechanisms that permit communication, control, and integration of our bodily functions are discussed in the chapters of Unit Three. To maintain homeostasis, the body must continuously monitor and respond appropriately to changes that may occur in either the internal or external environment. The nervous and endocrine systems provide this capability. Information originating in sensory nerve endings found in complex special sense organs such as the eye and in simple receptors located in skin or other body tissues provides the body with the necessary environmental input (Chapters 11 and 14). Nervous signals traveling rapidly from the brain and spinal cord over nerves to muscles and glands then initiate immediate coordinating and regulating responses (Chapters 12 and 13). Slower-acting chemical messengers—hormones produced by endocrine glands—serve to effect more long-term changes in physiological activities to maintain homeostasis (Chapter 15).

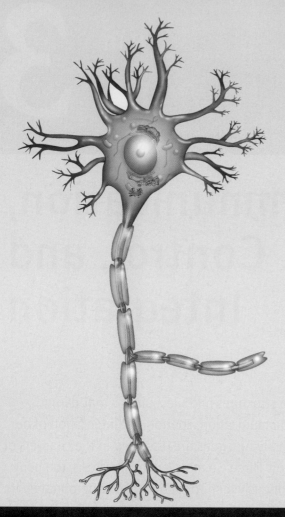

Cells of the Nervous System

STUDENT LEARNING OBJECTIVES

At the completion of this chapter, you should be able to do the following:

1. Outline the general organization of the nervous system.
2. Explain the difference between the central and peripheral nervous systems.
3. Explain the roles of the afferent and efferent divisions.
4. Compare the somatic and autonomic nervous systems.
5. Discuss the significance of the blood-brain barrier (BBB).
6. Discuss the significance of myelin sheaths and their importance in demyelinating diseases.
7. Discuss the major structures of a neuron.
8. Describe the structures and function of a reflex arc.
9. Explain how nerve impulses are conducted.
10. Identify and explain the significance of the following: resting membrane potential, local potential, action potential, refractory period.
11. Outline the components of the synapse and explain how it functions.
12. List several neurotransmitters and explain their roles in the nervous system.

KELLY had been having difficulty focusing her eyes for a couple of weeks. She assumed her recent lack of sleep was the cause. But when she started dropping her keys and feeling unsteady when walking down the stairs, she became worried enough to go to the doctor. After the results of several tests were known, the doctor told her she has myasthenia gravis. This is a disease in which the body's own white blood cells start attacking acetylcholine receptors.

These receptors may seem foreign to you, but by the end of this chapter you'll be able to put together "the big picture" and understand the foundations of this disease.

* * *

ORGANIZATION OF THE NERVOUS SYSTEM

The nervous system and the endocrine system together integrate one of the most vital functions of the human body: communication. In fact, homeostasis and our very survival depend on communication that integrates information and regulates body functions every second of our lives. Perhaps this is why so many students find the **nervous system**—comprising the brain, spinal cord, and nerves—as the most interesting body system (Figure 11-1). As you will see, the nervous system is subdivided in several ways according to its structure, the direction of information flow, and the control of effectors. In this chapter, we will investigate how the body integrates communication and control. To help you understand the various subdivisions of the nervous system we've prepared an organizational diagram in Figure 11-2. Please refer to this diagram often to review its content as you continue through our discussion.

Central and Peripheral Nervous Systems

First, our broad organizational plan simply categorizes all nervous system tissues according to their *relative position* in the body: *central* or *peripheral*. The **central nervous system (CNS),** as its name implies, is the structural and functional center of the entire nervous system. Consisting of the *brain* and *spinal cord*, the CNS integrates incoming information from the senses, evaluates the information, and initiates an outgoing response, if necessary. The **peripheral nervous system (PNS)** consists of the nerve tissues that lie on the periphery or regions outside the CNS. Nerves that originate from the brain (or pass through the skull) are called *cranial nerves*. Nerves that originate from the spinal cord are called *spinal nerves*.

LANGUAGE OF SCIENCE AND MEDICINE

HINT · Before reading the chapter, say each of these terms out loud. This will help you avoid stumbling over them as you read.

acetylcholine (ass-ee-til-KOH-leen)
[*acetyl-* vinegar, *-chole-* bile, *-ine* made of]

action potential (AK-shun poh-TEN-shal)
[*potent-* power, *-ial* relating to]

afferent division (AF-fer-ent)
[*a-* toward, *-fer-* carry, *-ent* relating to]

afferent (sensory) neuron (AF-fer-ent NOO-ron)
[*a-* toward, *-fer-* carry, *-ent* relating to, *neuron* nerve]

anesthesia (an-es-THEE-zhah)
[*an-* absence, *-esthesia* feeling]

antidepressant (an-tee-deh-PRESS-ant)
[*anti-* against, *-de-* down, *-press-* press, *-ant* agent]

astrocyte (ASS-troh-syte)
[*astro-* star shaped, *-cyte* cell]

autonomic nervous system (ANS) (aw-toh-NOM-ik SIS-tem)
[*auto-* self, *-nom-* rule, *-ic* relating to, *nerv-* nerves, *-ous* relating to]

axon (AK-son)
[*axon* axle]

axon hillock (AK-son HILL-ok)
[*axon* axle, *hill-* hill, *-ock* little]

bipolar neuron (bye-POH-lar NOO-ron)
[*bi-* two, *-pol-* pole, *-ar* relating to, *neuron* nerve]

blood-brain barrier (BBB)

cell body (sell BOD-ee)
[*cell* storeroom]

central nervous system (CNS) (SEN-tral SIS-tem)
[*centr-* center, *-al* relating to, *nerv-* nerves, *-ous* relating to]

chemical synapse (KEM-ih-kal SIN-aps)
[*chemic-* alchemy, *-al* relating to, *syn-* together, *-aps* join]

dendrite (DEN-dryte)
[*dendr-* tree, *-ite* part (branch) of]

depolarization (dee-poh-lah-ri-ZAY-shun)
[*de-* opposite, *-pol-* pole, *-ar-* relating to, *-ization* process]

efferent division (EF-fer-ent)
[*e-* away, *-fer-* carry, *-ent* relating to]

efferent (motor) neuron (EF-fer-ent NOO-ron)
[*e-* away, *-fer-* carry, *-ent* relating to, *neuron* nerve]

electrical synapse (el-EK-trih-kal SIN-aps)
[*electr-* amber, *-ical* relating to, *syn-* together, *-aps* join]

endoneurium (en-doh-NOO-ree-um)
[*endo-* inward, *-neuri-* nerve, *-um* thing] *pl.*, endoneuria (en-doh-NOO-ree-ah)

endorphin (en-DOR-fin)
[*endo-* within, *-morphin* shape]

continued on page 242

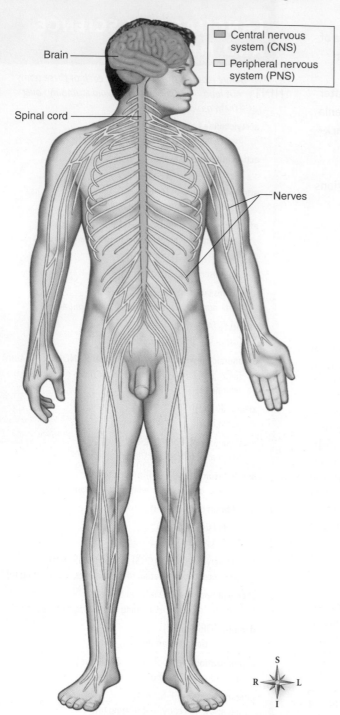

Brain

Spinal cord

Central nervous
system (CNS)

Peripheral nervous
system (PNS)

Nerves

▲ **FIGURE 11-1** **The nervous system.** Major anatomical features of the human nervous system include the brain, the spinal cord, and the individual nerves. The brain and spinal cord make up the central nervous system (CNS). All the nerves and their branches make up the peripheral nervous system (PNS). Nerves originating from the brain or through the skull are classified as *cranial nerves.* Nerves originating from the spinal cord are called *spinal nerves.*

Afferent and Efferent Divisions

The central and peripheral nervous systems include nerve cells that form *both* incoming information pathways and outgoing information pathways. For this reason, we also classify the nervous system into divisions that reflect the *direction* in which information is carried. The **afferent division** consists of all *incoming sensory* or *afferent* pathways. The **efferent division** consists of all *outgoing motor* or *efferent* pathways. (Afferent means "carry toward" and efferent means "carry away.") In Figure 11-2 and throughout the rest of this book, afferent pathways are represented in blue and efferent pathways in red.

Somatic and Autonomic Nervous Systems

Still another way of organizing the components of the nervous system is to categorize them according to the type of *effectors* they regulate. Some pathways of the **somatic nervous system (SNS)** carry information to the *somatic effectors,* which are the skeletal muscles. These motor pathways make up the **somatic motor division.** As Figure 11-2 shows, the somatic nervous system also includes the afferent pathways, making up the **somatic sensory division.** These afferent pathways provide feedback from the somatic effectors.

Pathways of the **autonomic nervous system (ANS)** carry information to the *autonomic (visceral)* effectors. These are the smooth muscles, cardiac muscle, glands, and other tissue under involuntary control. We now know that the ANS is largely "autonomous" but also influenced by the conscious mind.

The efferent pathways of the ANS are divided into the *sympathetic division* and the *parasympathetic division.* The **sympathetic division** is involved in preparing the body to deal with immediate threats. It produces the so-called "fight-or-flight" response. The **parasympathetic division** coordinates the body's normal resting activities. The parasympathetic division is sometimes called the "rest-and-repair" division.

The afferent pathways of the ANS belong to the **visceral sensory division,** which carries feedback information to the autonomic integrating centers in the central nervous system. All this may at first be confusing, so to get your bearings, take the time to review the previous paragraphs and relate the written description to the visual outline in Figure 11-2.

QUICK CHECK

1. What are the major subdivisions of the human nervous system?
2. What two organs make up the central nervous system?
3. Compare the afferent and efferent divisions.
4. Contrast the functions of the somatic and autonomic nervous systems.

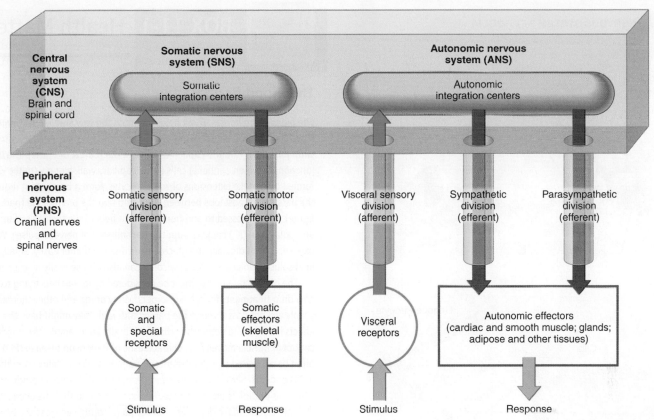

▲ FIGURE 11-2 Organizational plan of the nervous system. This diagram summarizes the scheme used by most neurobiologists studying the nervous system. Both the somatic nervous system (SNS) and the autonomic nervous system (ANS) include components in the CNS and PNS. In the SNS, somatic sensory pathways conduct information toward integrators in the CNS. Somatic motor pathways conduct information toward somatic effectors. In the ANS, visceral sensory pathways conduct information toward CNS integrators, whereas the sympathetic and parasympathetic pathways conduct information toward autonomic effectors.

CELLS OF THE NERVOUS SYSTEM

There are two main types of cells comprising the nervous system: *neurons* and *glia*. **Neurons** are *excitable* cells that conduct the impulses making all nervous system functions possible. They are the "wires" in the nervous system's informational circuits. **Glia** (glial cells), in contrast, do not usually conduct information by themselves. Instead, their job is to support the functions of neurons in various ways. Let's look in some detail at the different kinds of neurons and glia.

Neuroglia (Glia)

At first scientists thought **neuroglia** or glia (*glia* literally means "glue") served as a sort of packing material for the real cells of the nervous system—the neurons. However, the sheer number of these cells—over 900 billion—was enough for scientists to reconsider that they must serve *some* function! We now know that they do. Unlike most neurons, glial cells retain the ability to divide, which is a double-edged sword, making them also susceptible to cancer. In fact, most benign and malignant tumors of nervous tissue originate in glial cells.

There are five basic types of glia: *astrocytes, microglia, ependymal cells, oligodendrocytes,* and *Schwann cells* (Figure 11-3).

The star-shaped **astrocytes** (Figure 11-3, *A*) are found only in the CNS. Here they are the largest and most common type of glia. Their numerous, long and delicate projections extend throughout brain tissue, attaching to both neurons and tiny blood capillaries of the brain. The astrocytes actually "feed" the neurons by extracting glucose from the blood, converting it to lactate, and passing it along to the neurons to which they are connected. Because webs of astrocytes form tight sheaths around the brain's blood capillaries, they help form the **blood-brain barrier (BBB)** (Box 11-1). Recent research suggests that astrocytes influence the growth of neurons and how they connect to form circuits, and may even form pathways that transmit information in the CNS.

Microglia (Figure 11-3, *B*) are small cells found in the CNS. Sometimes stationary, microglia can enlarge and move around to serve a protective (immune surveillance) function when the brain is under attack by microorganisms, destroying them and their cellular debris.

Ependymal cells (Figure 11-3, *C*) resemble epithelial cells. They form thin sheets that line fluid-filled cavities in the brain and spinal cord. Some ependymal cells produce the

CENTRAL NERVOUS SYSTEM NEUROGLIA

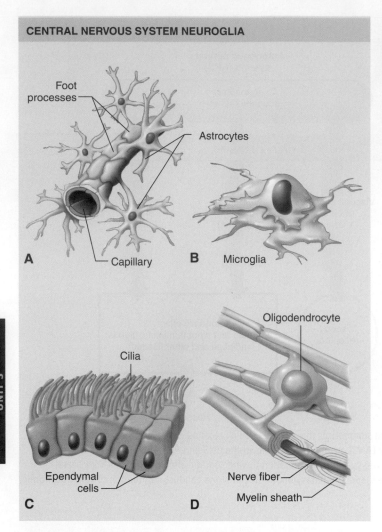

A

B Microglia

C

D

Foot processes
Astrocytes
Capillary
Oligodendrocyte
Cilia
Ependymal cells
Nerve fiber
Myelin sheath

PERIPHERAL NERVOUS SYSTEM NEUROGLIA

Node of Ranvier
Schwann cell
Nucleus of Schwann cell
Nucleus of Schwann cell
Myelin sheath
Neurilemma
Unmyelinated nerve fibers
Myelinated nerve fiber

E

F

▲ **FIGURE 11-3** **Types of neuroglia.** Neuroglia of the CNS: **A,** Astrocytes attached to the outside of a capillary blood vessel in the brain. **B,** A phagocytic microglial cell. **C,** Ciliated ependymal cells forming a sheet that usually lines fluid cavities in the brain. **D,** An oligodendrocyte with processes that wrap around nerve fibers in the CNS to form myelin sheaths. Neuroglia of the peripheral nervous system (PNS): **E,** A Schwann cell supporting a bundle of nerve fibers in the PNS. **F,** Another type of Schwann cell wrapping around a peripheral nerve fiber to form a thick myelin sheath.

BOX 11-1 | Health Matters

The Blood-Brain Barrier

The *blood-brain barrier (BBB)* helps maintain the very stable environment required for normal functioning of the brain. The BBB is formed as astrocytes wrap their "feet" around capillaries in the brain (part *A* of figure). The tight junctions between epithelial cells in the capillary wall, along with the covering formed by footlike extensions of the astrocytes, form a barrier that regulates the passage of most ions between the blood and the brain tissue (part *B* of figure). If they crossed to and from the brain freely, ions such as sodium (Na^+) and potassium (K^+) could disrupt the transmission of nerve impulses. Water, oxygen, carbon dioxide, and glucose can cross the barrier easily. Small, lipid-soluble molecules such as alcohol can also diffuse easily across the barrier.

The blood-brain barrier must be considered by researchers trying to develop new drug treatments for brain disorders. Many drugs and other chemicals simply will not pass through the barrier, although they might have therapeutic effects if they could get to the cells of the brain. For example, the abnormal control of muscle movements characteristic of **Parkinson disease (PD)** can often be alleviated by the substance dopamine, which is deficient in the brains of PD patients. Since dopamine cannot cross the blood-brain barrier, dopamine injections or tablets are ineffective. Researchers found that the chemical used by brain cells to make dopamine, **levodopa (L-dopa)**, can cross the barrier. Levodopa administered to patients with Parkinson disease crosses the barrier and converts to dopamine, and the effects of the condition are reduced.

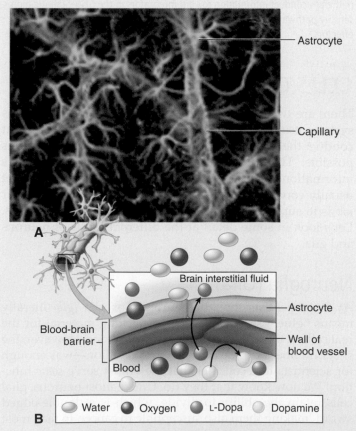

Astrocyte
Capillary
Brain interstitial fluid
Astrocyte
Blood-brain barrier
Wall of blood vessel
Blood

A

○ Water ● Oxygen ● L-Dopa ○ Dopamine

B

▲ **Blood-brain barrier.** **A,** Micrograph of fluorescent-stained astrocytes shows how they attach to capillaries, forming a coating around these tiny blood vessels. **B,** Diagram showing a cross section of the blood-brain barrier, made up of the foot processes of astrocytes and the wall of the blood capillary.

fluid that fills these spaces and others have cilia that keep the fluid circulating within the cavities.

Oligodendrocytes (Figure 11-3, *D*) help hold nerve fibers together and also produce the **myelin sheath,** which is critical to "insulating" nerve impulse conduction. (**Myelin** is a sheet of lipid with very high cholesterol content.) Some oligodendrocytes are clustered around nerve cell bodies. Others are arranged in rows between nerve fibers in the brain.

Schwann cells (Figure 11-3, *E* and *F*) are found only in the peripheral nervous system. Here they function like the oligodendrocytes of the brain, supporting nerve fibers and sometimes forming a myelin sheath around them. Microscopic gaps in the sheath, between adjacent glial cells, are called **nodes of Ranvier.** The myelin sheath and its tiny gaps are vital to the proper conduction of impulses along nerve fibers in the peripheral nervous system. Notice that the Schwann cell's nucleus and cytoplasm (see Figure 11-3, *F*) are squeezed to the perimeter to form the **neurilemma,** which is essential to the regeneration of injured nerve fibers. **Myelination** in humans is one of the most important and most vulnerable processes of brain development as we age.

Nerve fibers with many Schwann cells forming a thick myelin sheath are called **myelinated fibers** (white fibers). The fibers are called **unmyelinated fibers** (gray fibers) when several nerve fibers are held by a single Schwann cell that does *not* wrap around them to form a thick myelin sheath. The most common primary disease of the CNS is a *myelin disorder* called **multiple sclerosis (MS).** This is a serious disorder of glial cells involved in myelin formation. Recent research also shows that the age-related breakdown of myelin correlates strongly with the presence of a key genetic risk factor for Alzheimer disease.

QUICK CHECK

5. Describe neurons and glial cells.
6. What is meant by the blood-brain barrier?
7. Compare myelinated and unmyelinated nerve fibers.
8. List several different types of neuroglia cells and their functions.

Neurons

Structure of Neurons

The human brain has about 100 billion neurons, which is equal to only about 10% of the number of glial cells. All neurons consist of a **cell body,** and at least two kinds of nerve cell fibers: *one axon* and *one or more dendrites* (Figure 11-4). These fibers branch out like tiny trees or roots from the neuron body. The **axon** is usually a single long extension of the nerve fiber that conducts impulses *away* from the nerve cell body. The **dendrite** is typically a branching extension of a neuron that conducts impulses from adjacent neurons *toward* the cell body. A single neuron may have many dendrites.

The distal ends of dendrites of sensory neurons are called *receptors* because they *receive* the stimuli that initiate nerve signals. The axon usually extends from a tapered portion of the cell body called the **axon hillock.** Axons terminate in **synaptic knobs,** which contain many vesicles and their neurotransmitters, as well as many mitochondria to supply ATP for transmitting nerve signals.

Axons vary tremendously in both length and diameter. In general, axons with larger diameters conduct nervous impulses faster than those with smaller diameters. Whether

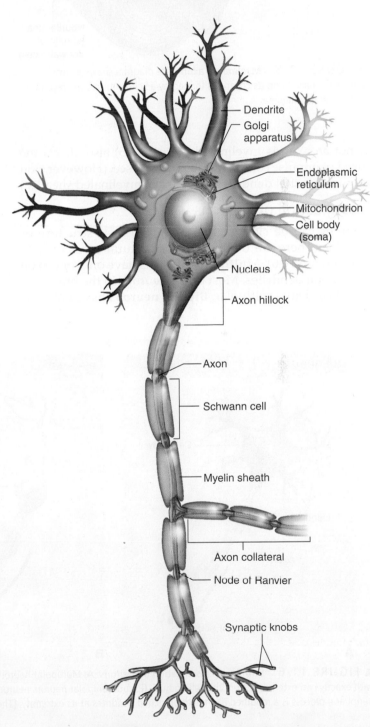

▲ **FIGURE 11-4**   Structure of a typical neuron.

Labels: Dendrite · Golgi apparatus · Endoplasmic reticulum · Mitochondrion · Cell body (soma) · Nucleus · Axon hillock · Axon · Schwann cell · Myelin sheath · Axon collateral · Node of Ranvier · Synaptic knobs

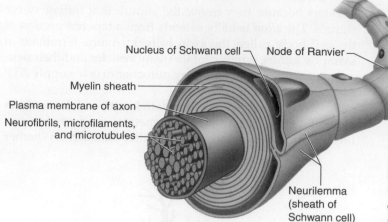

▲ **FIGURE 11-5** **Myelinated axon.** The diagram shows a cross section of an axon and its coverings formed by a Schwann cell: the myelin sheath and neurilemma.

or not an axon is myelinated (Figure 11-5) also affects the speed of impulse conduction, as we shall see. (However, you should note that dendrites do *not* have myelinated sheaths.)

Classification of Neurons

Neurons can be classified according to their structure or function. There are three distinct structural types of neurons (Figure 11-6). **Multipolar neurons** have only one axon but several dendrites. Most of the neurons in the brain and spinal cord are multipolar. **Bipolar neurons** have only one

axon and also only one, highly branched dendrite. They are the least common type of neuron, found in our specialized sense organs such as the retina of the eye, the inner ear, and the olfactory pathway. **Unipolar neurons** are sensory neurons with a single process extending from the cell body. This single process branches to form a *central* process (toward the CNS) and a side or *peripheral* process (away from the CNS). However, these two processes are actually derived from a single axon.

Neurons can also be classified according to the *direction* in which they conduct impulses (Figure 11-7). **Afferent (sensory) neurons** transmit nerve impulses *to* the spinal cord or brain. **Efferent (motor) neurons** transmit nerve impulses *away* from the brain or spinal cord to or toward muscles or glands. **Interneurons** form networks that conduct impulses from afferent neurons to or toward motor neurons. Interneurons lie entirely within the CNS (brain and spinal cord).

Reflex Arc

Notice in Figure 11-7 that neurons are often arranged in a distinct pattern called a **reflex arc.** Basically, such arcs are *automatic signal conduction routes to and from the CNS* (brain and spinal cord). The most common form of reflex arc is the three-neuron arc (Figure 11-8). It consists of an afferent neuron, an interneuron, and an efferent neuron. The afferent (sensory) neuron conducts a signal from the peripheral nervous system to the CNS. An interneuron then relays

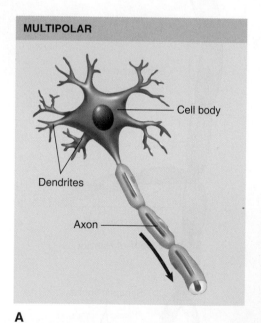

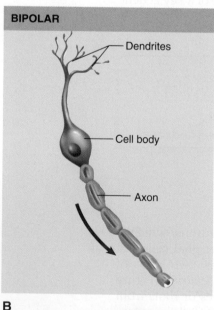

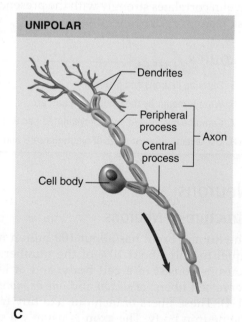

▲ **FIGURE 11-6** **Structural classification of neurons** **A,** Multipolar neuron: neuron with multiple extensions from the cell body. **B,** Bipolar neuron: neuron with exactly two extensions from the cell body. **C,** (Pseudo)unipolar neuron: neuron with only one extension from the cell body. The central process is an axon; the peripheral process is a modified axon with branched dendrites at its extremity. (The *red arrows* show the direction of impulse travel.)

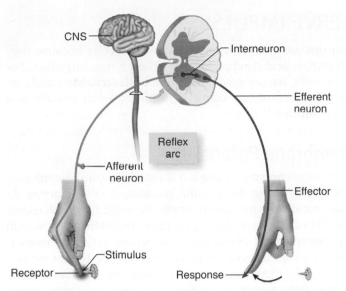

▲ **FIGURE 11-7** **Functional classification of neurons in a reflex arc.** Neurons can be classified according to the direction in which they conduct impulses. Notice that the most basic route of signal conduction follows a pattern called the reflex arc.

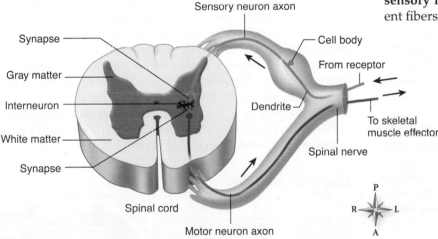

▲ **FIGURE 11-8** **Example of a reflex arc.** Three-neuron reflex arc. Sensory information enters on the same side of the central nervous system (CNS) as the motor information leaves the CNS.

this signal (from the incoming afferent neuron) to a motor neuron. Finally, an efferent (motor) neuron then conducts a signal from the CNS to an effector such as a muscle or gland.

As you look at Figure 11-8, notice the location of the *synapse*, where information is transmitted from one neuron to another. A **synapse** is a junction between the synaptic knobs of one neuron and the dendrites (or cell body) of another neuron. In this example, the first synapse lies between the synaptic knobs of the sensory neuron and the dendrites of the interneuron.

NERVES AND TRACTS

Nerves are bundles of peripheral nerve fibers (axons) held together by several layers of connective tissues (Figure 11-9). Beginning from the interior of a nerve, each nerve fiber (axon) is surrounded by a delicate layer of fibrous connective tissue called the **endoneurium.** Bundles of fibers (each with its own endoneurium) called *fascicles,* are then held together by a connective tissue layer called the **perineurium.** In turn, numerous fascicles, along with the blood vessels that nourish them, are held together to form a *complete nerve* by another fibrous coat called the **epineurium.** Within the CNS, bundles of nerve fibers without connective tissue coverings are called **tracts** rather than nerves. A nerve operates in the same manner as a tract.

Bundles of myelinated fibers make up the so-called **white matter** (myelinated *nerves* in the PNS; myelinated *tracts* in the CNS). Cell bodies and unmyelinated fibers make up the darker **gray matter** of the nervous system. Small, distinct regions of gray matter within the CNS are usually called **nuclei,** but in the PNS similar regions are called **ganglia.** Most nerves in the human nervous system are **mixed nerves** and carry both sensory (afferent) and motor (efferent) fibers. Nerves that contain mostly afferent fibers are often called **sensory nerves.** Likewise, nerves that contain mostly efferent fibers are called **motor nerves.**

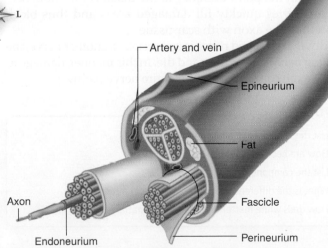

▲ **FIGURE 11-9** **The nerve.** Each nerve contains axons bundled into fascicles. A connective tissue epineurium wraps the entire nerve. Perineurium surrounds each fascicle.

BOX 11-2 Health Matters

Reducing Damage to Nerve Fibers

Crushing and bruising cause most injuries to the spinal cord—often damaging nerve fibers irreparably. This usually results in **paralysis** or loss of function in the muscles normally supplied by the damaged fibers. Unfortunately, the inflammation of the injury site usually damages even more fibers and thus increases the extent of the paralysis. However, early treatment of the injury with the anti-inflammatory drug *methylprednisolone* can reduce the inflammatory response in the damaged tissue and thus limit the severity of a spinal cord injury. Although early studies failed to confirm the effectiveness of standard doses of this steroid drug, later studies showed that very large doses administered within 8 hours of the injury reduced the extent of nerve cell damage dramatically. Since about 95% of the 10,000 Americans suffering spinal cord injuries each year are admitted for treatment well before the 8-hour limit, this drug may prove to be the first effective therapy for spinal cord injuries.

Until recently, neuroscientists thought that mature neurons could not be replaced. New evidence now clearly shows that this is not true. In fact, new neurons are often added to the existing network. However, because many mature neurons cannot divide, the only option is to repair injured or diseased nervous tissue already present. Nerve fibers can sometimes be repaired if the damage is not extensive (Box 11-2).

Researchers continue to look for new avenues to repair neurons in the CNS when they are damaged, but unfortunately, most injuries to the brain and spinal cord remain permanent. There are at least two reasons for this: First, neurons in the CNS lack the neurilemma needed to form the guiding tunnel from the point of injury to the distal connection. Second, astrocytes quickly fill damaged areas and thus block regrowth of the axon with scar tissue.

Sadly, if a damaged nerve is connected to another nerve, the receiving nerve may wither and die. In this manner, damage to a single axon can shut down an entire nerve pathway.

QUICK CHECK

9. List the components of a typical neuron.

10. How are neurons classified?

11. List the components of a reflex arc.

12. What is the difference between a nerve and a tract?

13. How does white matter differ from gray matter?

NERVE IMPULSES

Neurons are unique among the body's cells because they can initiate and conduct signals called **nerve impulses.** For this reason, we say neurons exhibit both *excitability* and *conductivity.* Definitions being definitions, what exactly is a nerve impulse?

Membrane Potentials

Physically, a **nerve impulse** is a wave of electrical disturbance that travels along the plasma membrane of the nerve. To understand how this can happen, we must first understand that all living cells, including neurons, maintain a *difference* in the concentration of certain ions across their membranes. There are slightly more positive ions outside the membrane and slightly more negative ions inside the membrane. This normal state results in a *difference in electrical charge* across their plasma membranes called the cell **membrane potential.** The membrane potential refers to the electrical *potential energy* stored. When a membrane potential is maintained, charged ions are held on opposite sides of the membrane like water behind a dam—ready to rush through toward opposite charges with force when membrane channels open.

A cell membrane with a membrane potential is, by definition, *polarized* (it has an overall negative charge on the inside and overall positive charge on the outside). The magnitude of the potential difference between the two sides of a polarized membrane is measured in volts (V) or millivolts (mV). A voltage difference of about −70 mV exists across the plasma membrane of the nerve. This number means two things: (1) there is a potential difference with a magnitude of 70 mV and (2) the inside of the membrane is *negative* with respect to the outside surface (Figure 11-10).

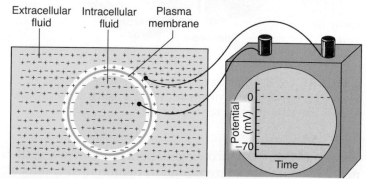

▲ **FIGURE 11-10** **Membrane potential.** The diagram on the left represents a cell maintaining a very slight difference in the concentration of oppositely charged ions across its plasma membrane. The voltmeter records the magnitude of electrical difference over time, which, in this case, does not fluctuate from −70 mV (voltage recorded over time as a red line).

Because this potential voltage persists when the neuron is *not* in an excited state, we say that the neuron is "at rest." The mechanisms that maintain this **resting membrane potential (RMP)** involve the characteristics of the plasma membrane itself.

Recall from Chapter 4 that there is a characteristic *permeability* of the plasma membrane of each cell type. This permeability is determined in part by the presence of specific channels that transport ions across the membrane. Many of these channels are *gated channels* that allow specific charged molecules to diffuse across the membrane only when the "gate" of each channel is open. As you read the following paragraphs, please refer to Figure 11-11 for clarity.

In the neuron's plasma membrane, channels for the transport of the major *anions* (negatively charged ions) are either nonexistent or closed. Therefore, the only ions that can move efficiently across a neuron's membrane are the positive ions sodium (Na^+) and potassium (K^+).

In a resting neuron, there are some channels for potassium that are open but most of the sodium channels are closed. This means that some of the potassium ions continually pushed into the neuron by the sodium-potassium pump (see pp. 67–68) can diffuse back out of the cell in an attempt to equalize its concentration gradient. However, very little of the sodium pumped out of the cell can diffuse back into the neuron (because their gated channels are closed). Thus, the membrane's selective permeability characteristics create and maintain a slight excess of positive ions on the *outer surface* of the membrane. (Note in Figure 11-11 that there are more positive ions on the outside of the membrane.)

Local Potentials

Membrane potentials of neurons can fluctuate above or below the resting membrane potential in response to certain stimuli. A slight shift away from the resting membrane potential in a specific region of the plasma membrane is often called a **local potential.**

Excitation of a neuron occurs when a stimulus triggers the opening of *stimulus-gated* Na^+ channels, located in the dendrites and body of the neuron. This could be due to either a sensory stimulus or chemical stimulus from another neuron. The stimulus permits more Na^+ to enter the cell, making the charge inside the cell more positive. This movement of the membrane potential toward zero is called **depolarization.**

In another process called **inhibition,** a stimulus triggers the opening of stimulus-gated K^+ channels. As more K^+ diffuses out of the cell, the excess of positive ions

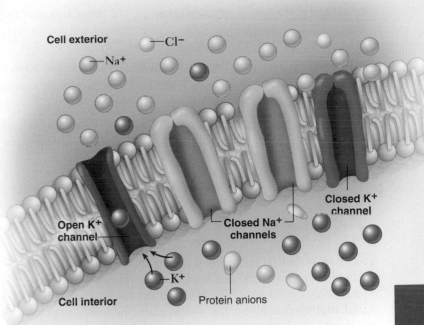

▲ FIGURE 11-11   Role of ion channels in maintaining the resting membrane potential (RMP). Some K^+ channels are open in a "resting" membrane. This allows K^+ to diffuse down its concentration gradient (out of the cell) and thus add to the excess of positive ions on the outer surface of the plasma membrane. Diffusion of Na^+ in the opposite direction would counteract this effect but is prevented from doing so by closed Na^+ channels.

increases outside the cell. As a result, this increases the magnitude of the membrane potential and results in *hyperpolarization.*

Local potentials are called *graded potentials* because the amount of deviation from the RMP varies directly with the magnitude of the stimulus. Local potentials are more or less found only in a particular region of the plasma membrane. They usually do *not* spread all the way to the end of a neuron's axon.

QUICK CHECK

14. Describe the mechanisms that produce the resting membrane potential.

15. In a resting neuron, what positive ion is most abundant outside the plasma membrane? What positive ion is most abundant inside the plasma membrane?

16. What is a local potential?

TABLE 11-1   Types of Membrane Potentials

MEMBRANE POTENTIAL	POLARIZATION	TYPICAL VOLTAGE*	DESCRIPTION
Resting membrane potential (RMP)	Polarized	−70 mV	Membrane voltage when the neuron is not excited and not conducting an impulse
Local potential	Depolarized (excitatory)	Graded; varies higher than −70 mV	Temporary fluctuation in a local region of the membrane in response to a sensory or nerve stimulus; may be an upward or downward fluctuation in voltage; loses amplitude as it spreads along membrane
	Hyperpolarized (inhibitory)	Graded; varies lower than −70 mV	
Threshold potential	Depolarized	−59 mV (but can vary between −50 and −60 mV)	Minimum local depolarization needed to trigger voltage-gated channels that produce the action potential
Action potential	Depolarized	+30 mV	Temporary maximum depolarization of membrane voltage that travels to end of axon without losing amplitude

*Example used in this chapter; actual values in body vary depending on many diverse factors.

ACTION POTENTIAL

An **action potential** *(nerve impulse),* as the term suggests, is the membrane potential of an *active* neuron (that is, one that is *conducting an impulse*). The action potential is an electrical signal that travels along the surface of a neuron's plasma membrane. A step-by-step description follows (please refer to Table 11-1 and Figure 11-12 as you read the following paragraphs).

1. When an adequate stimulus is applied to a neuron, the stimulus-gated Na⁺ channels open at the point of stimulation. Na⁺ diffuses rapidly into the cell at the site of this local depolarization (Figure 11-12, *B*).

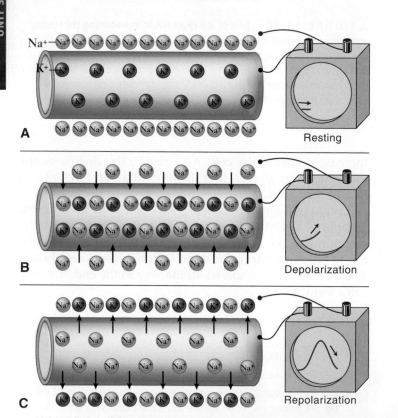

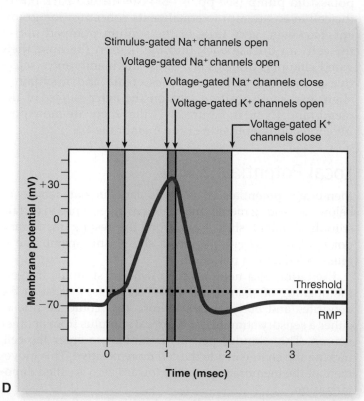

▲ **FIGURE 11-12**   **Depolarization and repolarization. A,** RMP results from an excess of positive ions on the outer surface of the plasma membrane. More Na⁺ ions are on the outside of the membrane than there are K⁺ ions on the inside of the membrane. **B,** Depolarization of a membrane occurs when Na⁺ channels open, allowing Na⁺ to move to an area of lower concentration (and more negative charge) *inside* the cell—reversing the polarity to an inside-positive state. **C,** Repolarization of a membrane occurs when K⁺ channels then open, allowing K⁺ to move to an area of lower concentration (and more negative charge) *outside* the cell. This reverses the polarity back to a negative state inside. Each voltmeter records the changing membrane potential as a *red line.* **D,** The graph shows the specific actions that occur during an action potential.

2. If the magnitude of the local depolarization exceeds a limit called the **threshold potential** (–59 mV in this example), which then triggers additional *voltage-gated* Na^+ channels to open.

3. As more Na^+ rushes into the cell, the membrane moves rapidly toward 0 mV and then continues in a positive direction to a peak of +30 mV. (Remember, the positive value at the peak of the action potential indicates that there is an excess of positive ions *inside* the membrane.) If the local potential fails to cross the threshold of –59 mV, the voltage-gated Na^+ channels do not open and the membrane simply recovers back to the resting potential of –70 mV *without* producing an action potential.

4. The action potential is an *all-or-none* response. If the threshold potential is surpassed, *all* the voltage-gated Na^+ channels open and the full peak of the action potential is *always* reached.

5. Once the peak of the action potential is reached, the membrane potential begins to move back toward the resting potential of –70 mV in a process called **repolarization** (Figure 11-12, C). A delayed opening of voltage-gated K^+ channels permits an outward rush of K^+ that restores the original excess of positive ions on the outside surface of the membrane.

6. Because the K^+ channels often remain open when the membrane reaches its resting potential, too much K^+ may rush out of the cell. This may cause a brief period of *hyperpolarization* before the resting potential is restored by the action of the **sodium-potassium pump.**

Refractory Period

The **refractory period** is a very brief period when a local area of an axon's membrane *resists* re-stimulation. For about 0.5 milliseconds the membrane will not respond to any other stimulus, no matter how strong. This is called the *absolute refractory period.* Only very strong stimuli can cause depolarization for a short time *after* this absolute refractory period—a stage called the *relative refractory period.* The stronger the stimulus, the sooner another action potential can occur. Thus the nervous system uses the frequency of nerve impulses to code for the strength of a stimulus—not changes in the magnitude of the action potential (which can only be an all-or-none response).

Conduction of the Action Potential

At the peak of the action potential, the inside of the axon's plasma membrane is positive relative to the outside. That is, its polarity is now the *reverse* of that of the resting membrane potential. This reversal in polarity causes electrical current

to flow between the site of the action potential and the adjacent regions of membrane. Now the local current flow triggers voltage-gated Na^+ channels in the *next segment* of membrane to open. As Na^+ rushes inward, this next segment develops an action potential. The action potential thus moves from one point to the next along the axon's membrane (Figure 11-13). This cycle repeats itself down the length of the axon. Because each local depolarization is an all-or-none response, the membrane potential moves along the membrane without any decrease in magnitude.

The action potential never moves backward, re-stimulating the region from which it just came. This is because the previous segment of membrane remains in the refractory stage, and cannot be re-stimulated. In myelinated fibers, the insulating properties of the thick myelin sheath resist ion movement and the resulting flow of current. Electrical changes in the membrane can only occur at gaps in the myelin sheath, that is, at the

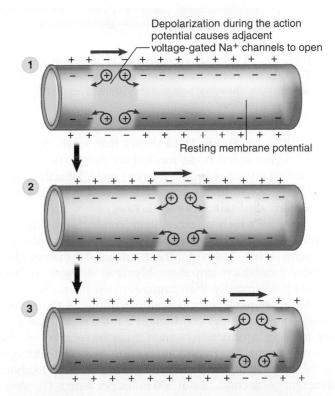

▲ **FIGURE 11-13 Conduction of the action potential.** The reverse polarity characteristic of the peak of the action potential causes local current flow to adjacent regions of the membrane *(small arrows)*. This stimulates voltage-gated Na^+ channels to open and thus create a new action potential. This cycle continues, producing wavelike conduction of the action potential from point to point along a nerve fiber. Adjacent regions of membrane behind the action potential do not depolarize again because they are still in their refractory period. This figure shows an unmyelinated fiber.

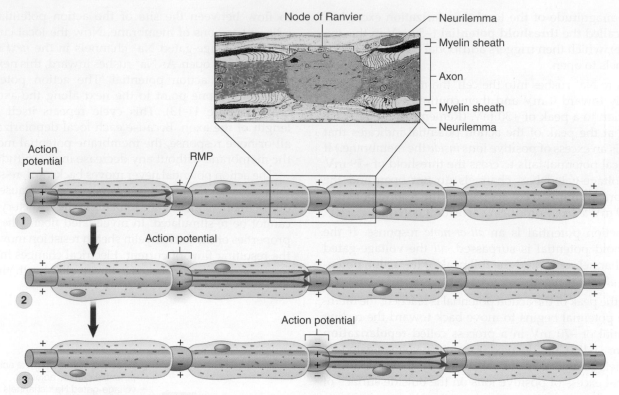

▲ **FIGURE 11-14** **Saltatory conduction.** This series of diagrams shows that the insulating nature of the myelin sheath prevents ion movement everywhere but at the nodes of Ranvier. The action potential at one node triggers current flow *(arrows)* across the myelin sheath to the next node—producing an action potential there. The action potential thus seems to "leap" rapidly from node to node. The inset is a transmission electron micrograph showing a node of Ranvier in a myelinated fiber.

nodes of Ranvier. Figure 11-14 shows that, when an action potential occurs at one node, most of the current flows *under* the insulating myelin sheath to the next node. Thus the action potential seems to "jump" from node to node in an impulse regeneration called **saltatory conduction.**

The rate at which a nerve fiber conducts an impulse depends on its diameter and also on the presence or absence of a myelin sheath. The larger the diameter of the nerve fiber, the faster it conducts impulses. Myelinated fibers conduct impulses more rapidly than unmyelinated fibers. So large-diameter, myelinated nerves conduct impulses the fastest. This is because saltatory conduction is more rapid than local point-to-point conduction, which must travel through the entire membrane. The fastest fibers, such as those that innervate the skeletal muscles, can conduct impulses up to about 130 meters per second (about 300 miles per hour)! The slowest fibers, such as those from sensory receptors in the skin, can conduct impulses at only about 0.5 meter per second (about 1 mile per hour).

QUICK ✓ CHECK

17. Describe the events that lead to the initiation of an action potential.

18. What is meant by the term *threshold potential?*

19. How does impulse conduction in an unmyelinated fiber differ from impulse conduction in a myelinated fiber?

20. What is meant by *saltatory conduction?*

SYNAPTIC TRANSMISSION

Structure of the Synapse

A *synapse* is the place where signals are transmitted from one neuron, called the *presynaptic neuron* to another neuron, called the *postsynaptic neuron*. The postsynaptic cell could also be an effector, such as a muscle.

Types of Synapses

There are two types of synapses: *electrical synapses* and *chemical synapses*.

Electrical synapses occur where two cells are joined end-to-end by gap junctions (Figure 11-15, *A*). Recall that, in this type of junction, the plasma membranes and cytoplasm of adjacent cells are functionally continuous. As a result, an action potential simply continues along the postsynaptic plasma membrane as if it belonged to the same cell.

Chemical synapses, by contrast, use a chemical called a **neurotransmitter** to send the message to the postsynaptic cell (Figure 11-15, *B*). A chemical synapse requires a *synaptic knob*, a *synaptic cleft*, and the *plasma membrane of the postsynaptic neuron*.

Each *synaptic knob,* a tiny bulge at the end of the axon, contains many small sacs (vesicles) filled with neurotransmitter molecules. When an action potential reaches the end of the axon, some of the neurotransmitters are released from the presynaptic neuron into the **synaptic cleft.** This is

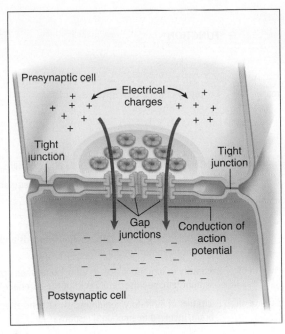

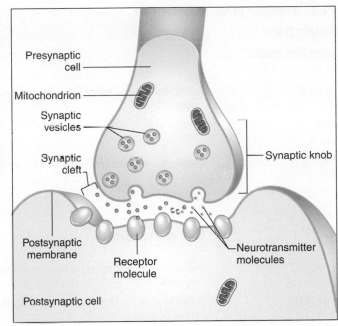

A   **B**

▲ **FIGURE 11-15** **Electrical and chemical synapses. A,** Electrical synapses involve gap junctions that allow action potentials to move from cell to cell directly by allowing electrical current to flow between cells. **B,** Chemical synapses involve transmitter chemicals (neurotransmitters) that signal postsynaptic cells, possibly inducing an action potential.

a minute, fluid-filled space (about one millionth of an inch in width) between a synaptic knob and the plasma membrane of a postsynaptic neuron. An axon may form a synapse at several different points along a postsynaptic neuron. The plasma membrane of a **postsynaptic neuron** has protein molecules embedded in it, each facing toward the synaptic knob and its vesicles. These membrane molecules serve as *receptors* to which the neurotransmitters bind. At least several, often thousands, of presynaptic knobs (and in some cases more than 100,000) synapse with a single postsynaptic neuron.

The amount of neurotransmitter released by a single knob is *not* enough to trigger an action potential in the postsynaptic neuron, but it may facilitate the initiation of an action potential by producing a local depolarization of the synaptic membrane.

After the synaptic transmission is complete, any further signaling must be stopped by removing neurotransmitters from the synaptic cleft. Many neurotransmitters are immediately transported back into the presynaptic neuron in a process called *reuptake*. As you might suspect, enzymes in the synaptic cleft break down some neurotransmitters. The resulting molecules are transported back into the presynaptic neuron for recycling.

Role of Calcium in Synapses

Calcium plays a major role in triggering the release of neurotransmitters into the synaptic cleft. Calcium channels located on the plasma membrane of the presynaptic terminal open as the membrane is depolarized by the arrival of an action potential. As a result, calcium quickly enters the cell and builds up its concentration intracellularly in the presynaptic terminal. Calcium then binds to a special intracellular protein, which through a series of events ultimately results in the fusion of the neurotransmitter vesicles with the presynaptic membrane. The neurotransmitters are then released into the synaptic cleft. For this reason, you can think of calcium as a "trigger" for the conversion of the electrical signal into the chemical information transmitted across the synaptic cleft.

Neurotransmitters

Neurotransmitters are the means by which neurons "talk" to one another. At trillions of synapses throughout our bodies, presynaptic neurons release neurotransmitters that serve to facilitate, stimulate, or inhibit postsynaptic neurons and effector cells. More than 50 compounds are now known to be neurotransmitters, and this number is expected to double as research continues. Generally, neurotransmitters are not distributed randomly throughout the nervous system. Instead, specific neurotransmitters are localized in discrete groups of neurons. Thus, they are released in specific nerve pathways.

Neurotransmitters are commonly classified by their function or by their chemical structure. For example, there are *excitatory neurotransmitters* and *inhibitory neurotransmitters*. In fact, some neurotransmitters can have an inhibitory effect at some synapses and an excitatory effect at other synapses. Such is the case with the neurotransmitter **acetylcholine**, which excites skeletal muscle cells but inhibits cardiac muscle cells. The precise function of neurotransmitters is

TABLE 11-2 **Examples of Neurotransmitters**

NEUROTRANSMITTER	LOCATION*	FUNCTION*
Small-Molecule Transmitters		
Class I		
Acetylcholine (ACh)	Junctions with motor effectors (muscles, glands); many parts of brain	Excitatory or inhibitory; involved in memory
Class II: Amines		
Monoamines		
Serotonin (5-HT†)	Several regions of the CNS	Mostly inhibitory; involved in moods and emotions, sleep
Histamine	Brain	Mostly excitatory; involved in emotions and regulation of body temperature and water balance
Catecholamines		
Dopamine (DA)	Brain; autonomic system	Mostly inhibitory; involved in emotions/moods and in regulating motor control
Epinephrine	Several areas of the CNS and in the sympathetic division of the ANS	Excitatory or inhibitory; acts as a hormone when secreted by sympathetic neurosecretory cells of the adrenal gland
Norepinephrine (NE)	Several areas of the CNS and in the sympathetic division of the ANS	Excitatory or inhibitory; regulates sympathetic effectors; in brain, involved in emotional responses
Class III: Amino Acids		
Glutamate (glutamic acid, Glu)	CNS	Excitatory; most common excitatory neurotransmitter in CNS
Gamma-aminobutyric acid (GABA)	Brain	Inhibitory; most common inhibitory neurotransmitter in brain
Glycine (Gly)	Spinal cord	Inhibitory; most common inhibitory neurotransmitter in spinal cord
Class IV: Other Small Molecules		
Nitric oxide (NO)	Several regions of the nervous system	May be a signal from postsynaptic to presynaptic neuron
Large-Molecule Transmitters		
Neuropeptides		
Substance P	Brain, spinal cord, sensory pain pathways; gastrointestinal tract	Mostly excitatory; transmits pain information
Enkephalins	Several regions of CNS; retina; intestinal tract	Mostly inhibitory; act like opiates to block pain
Endorphins	Several regions of CNS; retina; intestinal tract	Mostly inhibitory; act like opiates to block pain

*These are examples only; most of these neurotransmitters are also found in other locations, and many have additional functions.
†5-Hydroxytryptamine (synonym for serotonin).
ANS, Autonomic nervous system; *CNS*, central nervous system.

actually determined by the receptors of the postsynaptic neuron, *not* by the neurotransmitter itself. A few other important neurotransmitters include amines (dopamine, epinephrine, and norepinephrine), amino acids (glutamate, gamma-aminobutyric acid [GABA], glycine), nitric oxide (NO), and neuropeptides (short strands of amino acids), summarized for you in Table 11-2.

Severe psychological depression occurs when a deficit of norepinephrine, dopamine, serotonin, and other amines exists in certain brain synapses. This fact led to the development of some **antidepressant** drugs. One class of antidepressant drugs called *monoamine oxidase inhibitors (MAOIs)* increases the concentration of these amines in the brain synapses by inhibiting the action of an enzyme that would otherwise be involved in their breakdown. The resulting increase in concentration of these amines is the basis for

MAOI drug antidepressant activity. Other antidepressant drugs called *selective serotonin reuptake inhibitors (SSRIs)* produce antidepressant effects by inhibiting the uptake of serotonin. Cocaine, which is sometimes used in medical practice as a local anesthetic, produces a temporary feeling of well-being in cocaine abusers by inhibiting the reuptake of dopamine, resulting in an overabundance of dopamine in the synapse. Unfortunately, cocaine and similar drugs can also adversely affect blood flow and heart activity. This is especially true when cocaine is taken in amounts far beyond what is required in clinical practice. Such overdoses can rapidly cause death in susceptible individuals. See Box 11-3 for more information on anesthetics.

When we are exposed to extreme stress or pain, exercise strenuously for long periods, or engage in sexual activity, polypeptides called *endogenous opioids* are produced that

BOX 11-3 | Health Matters

General and Local Anesthetics

Anesthetics are substances that are administered to reduce or eliminate the sensation of pain, thus producing a state called **anesthesia**. Many anesthetics produce their effects by inhibiting the opening of sodium channels and thus blocking the initiation and conduction of nerve impulses. Anesthetics such as bupivacaine (Marcaine) are often used in dentistry to minimize pain involved in tooth extractions and other dental procedures. Procaine has likewise been used to block the transmission of electrical signals in sensory pathways of the spinal cord. Benzocaine and phenol, local anesthetics found in several over-the-counter products that relieve pain associated with teething in infants, sore throat pain, and other ailments, also produce their effect by blocking initiation and conduction of nerve impulses.

believe that the sensory stimulation that serves as the essence of early learning in infants and children has a critical role in directing the formation of synapses in our nervous system. The formation of new synapses, selective strengthening of existing synapses, and the selective removal of synapses are all thought to be important in the physiological mechanisms of learning and memory.

As we advance into old age, we do increase in wisdom—but we may also experience some degeneration of neurons, glia, and the blood vessels that supply them. This, coupled with age-related syndromes such as *Alzheimer disease (AD)*, may eventually produce some loss of memory, coordination, and other neural functions. ●

The BIG Picture

Neurons, the conducting cells of our nervous system, act as the "wiring" that connects the structures required to maintain homeostasis in our bodies. Neurons also form processing "circuits" that make decisions regarding appropriate responses to stimuli that threaten our internal constancy. In particular, we've seen how sensory neurons act as sensors or receptors that detect changes in our external and internal environments that may be potentially threatening. Sensory neurons then relay this information to integrator mechanisms in the CNS. There, the information is processed, often by one or more interneurons, and an outgoing response signal is relayed to effectors by way of motor neurons. At the effector, a chemical messenger or neurotransmitter triggers a response that tends to restore homeostatic balance. Neurotransmitters released into the bloodstream, where they are called hormones, can enhance and prolong such homeostatic responses.

Of course, neurons do much more than simply respond to stimuli in a preprogrammed way. Circuits of interneurons are capable of remembering, learning new responses, generating rational and creative thought, and implementing other complex processes. The exact mechanisms of many of these complex integrative functions are presently unknown, but you will learn some of what we already know in our next chapter.

give us a wonderful feeling of well-being and exhilaration. These endogenous opioids generally are called **endorphins.** They resemble opiates that are derived from the opium poppy, which include the narcotic alkaloids morphine, heroin, and codeine. These drugs serve to depress the central nervous system and in medical practice are used with great care to reduce pain. Abused, they produce a number of undesirable results, including addiction.

QUICK / CHECK

21. What are the three structural components of a synapse?
22. List the steps of synaptic transmission.

Cycle of LIFE ○

Our nervous tissue develops from embryonic *ectoderm* during the first weeks after conception. However, the most rapid and obvious development of nervous tissue occurs during fetal development in the womb and for several years shortly after birth. Remarkably, nerve cells become organized to form a coordinated and integrated network spread throughout the body. Although we know very little about how these processes take place, we do know that neurons require the coordinated action of several agents to create the proper wiring of our nervous system. For example, we now know that *nerve growth factors* released by effector cells during development stimulate the growth of neuron processes and help direct the growth of their axons to their proper target destinations.

During the first years of neural development, synapses are made, broken, and re-formed until the basic organization of the nervous system is established. Neurobiologists

MECHANISMS OF DISEASE

There are many disorders of the nervous system, most of which involve glial cells rather than neurons. From *multiple sclerosis (MS)* to *brain tumors*, our nervous systems are hardly immune to a number of debilitating, disfiguring, and deadly diseases.

evolve   Find out more about these nervous tissue diseases online at *Mechanisms of Disease: Cells of the Nervous System.*

LANGUAGE OF SCIENCE AND MEDICINE *continued from page 227*

ependymal cell (eh-PEN-di-mal sell)
[*ep-* over, *-en-* on, *-dyma-* put, *-al* relating to, *cell* storeroom]

epineurium (ep-i-NOO-ree-um)
[*epi-* upon, *-neuri-* nerve, *-um* thing] *pl.,* epineuria (ep-i-NOO-ree-ah)

excitation (ek-sye-TAY-shun)
[*excit-* arouse, *-ation* process]

ganglion (GANG-glee-on)
[*gangli-* knot, *-on* unit] *pl.,* ganglia (GANG-glee-ah)

glia (GLEE-ah)
[*glia* glue] *sing.,* glial cell (GLEE-al sell)

gray matter

inhibition (in-hib-ISH-un)

interneuron (in-ter-NOO-ron)
[*inter-* between, *-neuron* nerve]

levodopa (ʟ-dopa) (LEEV-oh-doh-pah)
[*levo-* left (form of molecule), *-dopa* acronym denoting 3,4- dihydroxyphenylalanine]

local potential
[*potent-* power, *-ial* relating to]

membrane potential (MEM-brayne)
[*membran-* thin skin, *potent-* power, *-ial* relating to]

microglia (my-KROG-lee-ah)
[*micro-* small, *-glia* glue] *sing.,* microglial cell (my-KROG-lee-al sell)

mixed nerve

motor nerve (MOH-tor)
[*mot-* movement, *-or* agent]

multiple sclerosis (MS) (MUL-ti-pul skleh-ROH-sis)
[*multi-* many, *-pl-* fold, *sclera-* hard, *-osis* condition]

multipolar neuron (mul-ti-POL-ar NOO-ron)
[*multi-* many, *-pol-* pole, *-ar* relating to, *neuron* nerve]

myelin (MY-eh-lin)
[*myel-* marrow, *-in* substance]

myelin sheath (MY-eh-lin sheeth)
[*myel-* marrow, *-in* substance]

myelinated fiber (MY-eh-lin-ay-ted FYE-ber)
[*myel-* marrow, *-in-* substance, *-ate* act of, *fiber* thread]

myelination (my-eh-li-NAY-shun)
[*myel-* marrow, *-in-* substance, *-ation* process]

nerve

nerve impulse
[*impulse-* drive]

nervous system
[*nerv-* nerves, *-ous* relating to]

neurilemma (noo-ri-LEM-mah)
[*neuri-* neuron, *-lemma* sheath] *pl.,* neurilemmae (noo-ri-LEM-mee)

neuroglia (noo-ROG-lee-ah)
[*neuro-* nerve, *-glia* glue] *sing.,* neuroglial cell (noo-ROG-lee-al sell)

neuron (NOO-ron)
[*neuron* nerve]

neurotransmitter (noo-roh-trans-MIT-ter)
[*neuro-* nerves, *-trans-* across, *-mitt-* send, *-er* agent]

node of Ranvier (rahn-vee-AY)
[*nod-* knot, *Louis A. Ranvier* French pathologist]

nucleus (NOO-klee-us)
[*nucleus* kernel] *pl.,* nuclei

oligodendrocyte (ohl-i-go-DEN-droh-syte)
[*oligo-* few, *-dendr-* part (branch) of, *-cyte* cell]

paralysis (pah-RAL-ih-sis)
[*para-* beside, *-lysis* loosening]

parasympathetic division (pair-ah-sim-pah-THET-ik)
[*para-* beside, *-sym-* together, *-pathe-* feel, *-ic* relating to]

Parkinson disease (PD) (PARK-in-son)
[*James Parkinson* English physician]

perineurium (pair-ih-NOO-ree-um)
[*peri-* around, *-neuri-* nerve, *-um* thing] *pl.,* perineuria (pair-i-NOO-ree-ah)

peripheral nervous system (PNS) (peh-RIF-er-al SIS-tem)
[*peri-* around, *-phera-* boundary, *-al* relating to, *nerv-* nerves, *-ous* relating to]

postsynaptic neuron (post-sih-NAP-tik NOO-ron)
[*post-* after, *-syn-* together, *-apt-* join, *-ic* relating to, *neuron* nerve]

reflex arc
[*re-* back or again, *-flex* bend, *arc* curve]

refractory period (ree-FRAK-toh-ree)
[*refract-* break apart, *-ory* relating to, *period* circuit]

repolarization (ree-poh-lah-ri-ZAY-shun)
[*re-* back or again, *-pol-* pole, *-ar* relating to, *-zation* process]

resting membrane potential (RMP)
[*membran-* thin skin, *potent-* power, *-ial* relating to]

saltatory conduction (SAL-tah-tor-ee)
[*salta-* leap, *-ory* relating to]

Schwann cell (shwon or shvon sell)
[*Theodor Schwann* German anatomist, *cell* storeroom]

sensory nerve (SEN-soh-ree nerv)
[*sens-* feel, *-ory* relating to]

sodium-potassium pump
[*sod-* soda, *-um* thing or substance, *potass-* potash, *-um* thing or substance]

somatic motor division (so-MAH-tik MOH-tor di-VIH-zhun)
[*soma-* body, *-ic* relating to, *mot-* movement, *-or* agent, *divis-* divide, *-ion* process of]

somatic nervous system (SNS) (so-MAH-tik SIS-tem)
[*soma-* body, *-ic* relating to, *nerv-* nerves, *-ous* relating to]

somatic sensory division (so-MAH-tik)
[*soma-* body, *-ic* relating to, *sens-* feel, *-ory* relating to]

sympathetic division (sim-pah-THET-ik)
[*sym-* together, *-pathe-* feel, *-ic* relating to]

synapse (SIN-aps)
[*syn-* together, *-aps* join]

synaptic cleft (sih-NAP-tik kleft)
[*syn-* together, *-apt-* join, *-ic* relating to]

synaptic knob (sih-NAP-tik nob)
[*syn-* together, *-apt-* join, *-ic* relating to]

threshold potential (THRESH-hold poh-TEN-shal)
[*potent-* power, *-ial* relating to]

tract (trakt)

unipolar neuron (yoo-nee-POH-lar NOO-ron)
[*uni-* single, *-pol-* pole, *-ar* relating to, *neuron* nerve]

unmyelinated fiber (un-MY-eh-lin-ay-ted FYE-ber)
[*un-* not, *-myel-* marrow, *-in-* substance, *-ate* act of, *fiber-* thread]

visceral sensory division (VISS-er-al)
[*viscer-* internal organs, *-al* relating to, *sens-* feel, *-ory* relating to]

white matter

CHAPTER SUMMARY

To download an MP3 version of the chapter summary for use with your iPod or other portable media player, access the Audio Chapter Summaries online at http://evolve.elsevier.com.

HINT Scan this summary after reading the chapter to help you reinforce the key concepts. Later, use the summary as a quick review before your class or before a test.

INTRODUCTION: ORGANIZATION OF THE NERVOUS SYSTEM

A. The nervous system and the endocrine system together integrate the communication functions of the human body (Figure 11-2)
B. Nervous system is subdivided in a variety of ways according to its structure, the direction of information flow, and the control of effectors (Figure 11-2)
C. Central nervous system (CNS)
   1. Structural and functional center of the entire nervous system
   2. Consists of the brain and spinal cord
   3. Integrates incoming information from the senses, evaluates the information, and initiates an outgoing response
D. Peripheral nervous system (PNS)
   1. Consists of the nerve tissues that lie on the periphery or regions outside the CNS
      a. Cranial nerves—nerves that originate from the brain (or through the skull)
      b. Spinal nerves—nerves that originate from the spinal cord
E. Afferent division—consists of all incoming sensory or afferent pathways
F. Efferent division—consists of all outgoing motor or efferent pathways
G. Somatic nervous system—carries information to the somatic effectors (skeletal muscles)
   1. Somatic motor division—efferent pathways
   2. Somatic sensory division—afferent pathways
      a. Integrating centers—receive the sensory information and generate the efferent response signal
H. Autonomic nervous system—carries information to the autonomic (visceral) effectors
   1. Sympathetic division—pathways that exit from the middle portions of the spinal cord; is involved in preparing the body to deal with immediate threats; "fight-or-flight" response
   2. Parasympathetic division—pathways exit from the brain or lower portions of the spinal cord and coordinate the body's normal resting activities; "rest-and-repair" division

CELLS OF THE NERVOUS SYSTEM

A. Neuroglia—provide structural and functional support to neurons
   1. Five basic types (Figure 11-3):
      a. Astrocytes—largest and most numerous type of glia; help form the blood-brain barrier (BBB) (Box 11-1; Figure 11-3, A)
      b. Microglia—serve a protective function when the brain is under attack by microorganisms (Figure 11-3, B)
      c. Ependymal cells—produce the fluid that fills the cavities in the brain and spinal cord (Figure 11-3, C)
      d. Oligodendrocytes—help hold nerve fibers together and also produce the vitally important myelin sheath in CNS (Figure 11-3, D)
      e. Schwann cells—supporting nerve fibers in the PNS and sometimes form a myelin sheath around them (Figure 11-3, F)

NEURONS

A. Neurons consist of a cell body, one or more dendrites, and one axon (Figure 11-4)
   1. Distal ends of dendrites of sensory neurons are called receptors because they receive the stimuli that initiate nerve signals
   2. Axon hillock—tapered portion of the cell body; "decides" whether to send the impulse any farther in the neuron
   3. Axons with larger diameters conduct nervous impulses faster than those with smaller diameters (Figure 11-5)
   4. Synaptic knobs—release neurotransmitter molecules to stimulate adjacent neurons or effectors such as muscles or glands
B. Classification of neurons—three distinct structural types of neurons (Figure 11-6)
   1. Multipolar neurons—have only one axon but several dendrites
   2. Bipolar neurons—have only one axon and also only one, highly branched dendrite
   3. Unipolar neurons—sensory neurons with a single process extending from the cell body
C. Classification of neurons—according to the direction in which they conduct impulses (Figure 11-7)
   1. Afferent (sensory) neurons—transmit nerve impulses to the spinal cord or brain
   2. Efferent (motor) neurons—transmit nerve impulses away from the brain or spinal cord to or toward muscles or glands
   3. Interneurons—conduct impulses from afferent neurons to or toward motor neurons
D. Reflex arc—automatic signal conduction route to and from the CNS
   1. Most common form of reflex arc is the three-neuron arc; consists of an afferent neuron, an interneuron, and an efferent neuron (Figure 11-8)
E. Synapse—junction between the synaptic knobs of one neuron and the dendrites (or cell body) of another neuron

NERVES AND TRACTS

A. Nerves—bundles of peripheral nerve fibers (axons) held together by several layers of connective tissues (Figure 11-9)
   1. Endoneurium—delicate layer of fibrous connective tissue surrounding each nerve fiber
   2. Perineurium—connective tissue layer surrounding each bundle of nerve fibers (fascicles)
   3. Epineurium—fibrous coat surrounding numerous fascicles
B. Tracts—bundles of nerve fibers within the CNS
C. White matter—bundles of myelinated fibers
D. Gray matter—cell bodies and unmyelinated fibers
   1. Nuclei—distinct regions of gray matter within the CNS
   2. Ganglia—distinct regions of gray matter within the PNS
E. Mixed nerves—carry both sensory (afferent) and motor (efferent) fibers
F. Sensory nerves—contain mostly afferent fibers

G. Motor nerves—contain mostly efferent fibers

H. Nerve fibers can sometimes be repaired if the damage is not extensive

NERVE IMPULSE

A. Nerve impulse—wave of electrical energy that travels along the plasma membrane of the nerve

B. Membrane potential—difference in electrical charge (voltage) across their plasma membrane
   1. The slight excess of positive ions on the outer surface is produced by ion transport mechanisms and the membrane's permeability characteristics
   2. The membrane's selective permeability characteristics help maintain a slight excess of positive ions on the outer surface of the membrane (Figure 11-11)
   3. Resting membrane potential (RMP)—membrane potential of an excitable cell when it is "at rest" (not excited)

C. Local potentials—slight shift away from the resting membrane potential in a specific region of the plasma membrane
   1. Excitation—occurs when a stimulus triggers the opening of stimulus-gated Na$^+$ channels
   2. Inhibition—stimulus triggers the opening of stimulus-gated K$^+$ channels
   3. Local potentials are called graded potentials because the amount of deviation from the RMP varies directly with the magnitude of the stimulus

ACTION POTENTIAL

A. Action potential—an electrical signal that travels along the surface of a neuron's plasma membrane (Figure 11-12)
   1. When an adequate stimulus is applied to a neuron, the Na$^+$ channels open at the point of stimulation. Na$^+$ diffuses rapidly into the cell at the site of this local depolarization
   2. If the magnitude of the local depolarization exceeds a limit called the threshold potential (about –59 mV), then additional voltage-gated Na$^+$ channels are opened
   3. As more Na$^+$ rushes into the cell, the membrane moves rapidly toward 0 mV and then continues in a positive direction to a peak of +30 mV
   4. The action potential is an all-or-none response
   5. Once the peak of the action potential is reached, membrane potential moves back toward RMP of –70 mV; this repolarization is caused by delayed opening of voltage-gated K$^+$ channels
   6. Because the voltage-gated K$^+$ channels often remain open when the membrane reaches its resting potential, too much K$^+$ may rush out of the cell; temporary hyperpolarization

B. Absolute refractory period—very brief period when a local area of an axon's membrane resists re-stimulation absolutely

C. Relative refractory period—additional brief phase during which only strong stimuli may trigger an action potential; allows higher frequency of signals when stimulus is strong

D. Conduction of the action potential
   1. At the peak of the action potential, the plasma membrane's polarity is now the reverse of the RMP
   2. Reversal in polarity causes electrical current to flow between the site of the action potential and the adjacent regions of membrane
   3. This local current flow triggers voltage-gated Na$^+$ channels in the *next segment* of membrane to open
   4. The action potential thus moves from one point to the next along the axon's membrane (Figure 11-13)
   5. The action potential never moves backward, re-stimulating the region from which it just came because of the refractory period
   6. In myelinated fibers, the insulating properties of the thick myelin sheath resist ion movement and the resulting flow of current
   7. Electrical changes in the membrane can only occur at gaps in the myelin sheath, that is, at the nodes of Ranvier; saltatory conduction (Figure 11-14)
   8. The rate at which a nerve fiber conducts an impulse depends on its diameter and also on the presence or absence of a myelin sheath

SYNAPTIC TRANSMISSION

A. Synapse—place where signals are transmitted from one neuron, called the presynaptic neuron, to another neuron, called the postsynaptic neuron

B. Types of synapses—electrical synapses and chemical synapses
   1. Electrical synapse—where two cells are joined end-to-end by gap junctions; as a result, an action potential simply continues along the postsynaptic plasma membrane as if it belonged to the same cell (Figure 11-15, *A*)
   2. Chemical synapses—use a chemical neurotransmitter to send the message to the postsynaptic cell (Figure 11-15, *B*)
      a. Synaptic knob—contains many small sacs (vesicles) filled with neurotransmitter molecules
      b. Synaptic cleft—fluid-filled space (about one millionth of an inch in width) between a synaptic knob and the plasma membrane of a postsynaptic neuron
      c. Postsynaptic neuron—has protein molecules embedded in it, each facing toward the synaptic knob and its vesicles

C. Role of calcium—calcium ions enter the synaptic knob in response to an action potential, triggering the release of neurotransmitters from presynaptic vesicles

NEUROTRANSMITTERS

A. Neurotransmitters—means by which neurons "talk" to one another; more than 50 compounds are now known to be neurotransmitters

B. Neurotransmitters are commonly classified by their function or by their chemical structure
   1. Excitatory neurotransmitters
   2. Inhibitory neurotransmitters

C. Severe psychological depression occurs when a deficit of norepinephrine, dopamine, serotonin, and other amines exists in certain brain synapses
   1. Antidepressant drugs—some of these inactivate dopamine and serotonin; others called selective serotonin reuptake inhibitors (SSRIs), produce antidepressant effects by inhibiting the uptake of serotonin

D. Cocaine—produces a temporary feeling of well-being in cocaine abusers by blocking the uptake of dopamine

E. Anesthetics—produce their effects by inhibiting the opening of sodium channels in the nerve cell membrane, thus blocking the initiation and conduction of nerve impulses (Box 11-3)

REVIEW QUESTIONS

 HINT *Write out the answers to these questions after reading the chapter and reviewing the Chapter Summary. If you simply think through the answer without writing it down, you won't retain much of your new learning.*

1. Briefly explain the general function the nervous system performs for the body.
2. Identify the other body system that performs the same general function.
3. Describe the function of sodium and potassium in the generation of an action potential.
4. Why is an action potential an all-or-none response?
5. How does the myelin sheath affect the speed of an action potential? What about the diameter of the nerve fiber?
6. Describe the structure of a synapse.
7. Describe the series of events that mediate conduction across synapses.
8. List and describe the four main chemical classes of transmitters.
9. Describe the actions of acetylcholine.
10. List and describe disorders of the nervous system cells.

CRITICAL THINKING QUESTIONS

 HINT *After finishing the Review Questions, write out the answers to these items to help you apply your new knowledge. Go back to sections of the chapter that relate to items that you find difficult.*

1. Compare and contrast the characteristics of the central nervous system and the peripheral nervous system. Also, explain how the somatic and autonomic nervous systems could be included in the afferent and efferent divisions.
2. There are relatively few neurons in the brain. How would you describe the other cells in the brain? What do they do?
3. All neurons have axons and dendrites. Explain where these parts of the neuron are located in multipolar, bipolar, and unipolar neurons.
4. Compare and contrast white matter and gray matter. What would result if there were a loss of myelination?
5. Nerve fibers in the peripheral nervous system are much more successful than nerve fibers in the central nervous system in repairing themselves. How would you explain this difference?
6. Define *potential*. The following is a list of times and charges taken during a nerve impulse. Identify them as being either an action potential or not and as occurring either during the

absolute or the relative refractory period. (Assume the stimulus starts at 0 msec.)
   0 mV at 1 msec
   −50 mV at 0.5 msec
   +25 mV at 1.5 msec
   +25 mV at 3 msec
7. Many antibiotics that should kill the causative agents of meningitis are ineffective if given orally. Can you explain the anatomical mechanism that could cause this ineffectiveness?
8. Explain one of the ways an anesthetic may relieve the sensation of pain.

continued from page 227

Now that you have read this chapter, see if you can answer these questions about Kelly and her diagnosis of myasthenia gravis.

1. What's the connection between acetylcholine and Kelly's symptoms?
   a. Acetylcholine is a hormone released during stress
   b. Acetylcholine is found primarily in sensory neuron connections
   c. Acetylcholine is found primarily in motor neuron connections
   d. Acetylcholine affects communication primarily with smooth muscles

2. If the binding of acetylcholine opened sodium channels on the postsynaptic membrane, this would cause a temporary _____, called an _____.
   a. depolarization; excitatory postsynaptic potential
   b. hyperpolarization; excitatory postsynaptic potential
   c. depolarization; inhibitory postsynaptic potential
   d. hyperpolarization; inhibitory postsynaptic potential

Some muscle relaxants used during surgery bind to acetylcholine receptors, effectively blocking the action of acetylcholine—just as in myasthenia gravis. The drugs (anticholinesterases) used after surgery to reverse the relaxants are the same as those used to treat myasthenia gravis.

3. What do you think the action of these drugs would be?
   a. Blocking the release of acetylcholine
   b. Blocking the action of acetylcholinesterase
   c. Increasing the production of choline
   d. Increasing the action of acetylcholinesterase

HINT *To solve a case study, you may have to refer to the glossary or index, other chapters in this textbook, A&P Connect, Mechanisms of Disease, and other resources.*

CHAPTER 12

Central Nervous System

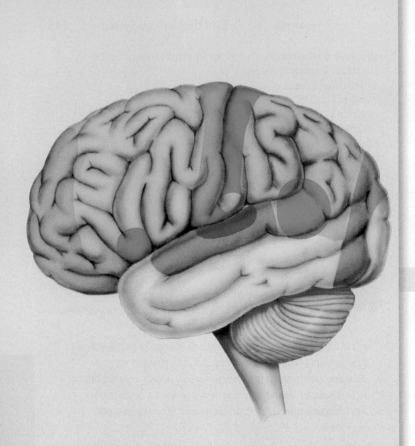

STUDENT LEARNING OBJECTIVES

At the completion of this chapter, you should be able to do the following:

1. Describe and list the functions of the coverings of the brain and spinal cord.
2. Discuss the formation and role of cerebrospinal fluid.
3. Outline the basic structures and functions of the spinal cord.
4. Discuss the significance of ascending and descending tracts in the spinal cord.
5. Discuss the *general* organization of the brain.
6. Outline the basic functions of the following: medulla oblongata, pons, midbrain, cerebellum, cerebrum.
7. List *detailed* structures and functions of the cerebellum.
8. Discuss several important roles of the hypothalamus and the pineal gland.
9. List *detailed* structures and functions of the cerebrum.
10. Describe the structures and functions of the cerebral tracts and basal nuclei.
11. Outline the functions of the cerebral cortex.
12. Discuss the significance of the electroencephalogram (EEG).
13. Discuss the structures of the central nervous system (CNS) providing for language, emotions, and memory.
14. Compare a generalized somatic sensory pathway with a somatic motor pathway in the CNS.

CHAPTER OUTLINE

Scan this outline before you begin to read the chapter, as a preview of how the concepts are organized.

AFTER several days of a pounding headache, Polai went to the university clinic, hoping for a prescription that would provide a cure. But after she explained her symptoms, instead of being sent home with a prescription, she was sent to the hospital for a lumbar puncture (also called a spinal tap) to determine whether she had contracted meningitis. In this procedure, a needle is inserted into the lumbar region to obtain a sample of cerebrospinal fluid.

Before going to the end of this chapter to answer questions about this story, try to use what you learn as you read this chapter to imagine how Polai's cerebrospinal fluid might differ from normal if she indeed had meningitis.

• • •

We've seen that the nervous system is composed of two major divisions: the central nervous system (CNS) and the peripheral nervous system (PNS). This division makes our study of the nervous system easier, although you should understand that there really is no absolute division in reality. This chapter discusses the CNS.

The CNS comprises both the brain and the spinal cord and is the principal integrator of sensory input and motor output (Figure 12-1). Thus the CNS evaluates incoming information and formulates appropriate responses to changes that may threaten our body's homeostatic balance. In Chapter 13 we will explore the nerve pathways that lead to and from the CNS, at which point we will also address structures and functions of the PNS. This study of the peripheral nerve pathways will clarify and expand your understanding of the nervous system.

COVERINGS OF THE BRAIN AND SPINAL CORD

There are two protective coverings of the brain and spinal cord. The outer covering consists of bone. Cranial bones encase the brain and vertebrae encase the spinal cord. The inner covering consists of membranes known as **meninges.** There are three distinct meningeal layers: (1) *dura mater,* (2) *arachnoid mater,* and (3) *pia mater* (Figures 12-2 and 12-3).

The tough, fibrous **dura mater** is the outer layer of the meninges and also the inner layer of the cranium's periosteum. The dura mater is composed of dense connective tissue and contains many blood vessels and nerves. The **arachnoid mater** is a delicate, spiderweb-like layer, lacking blood vessels and nerves. It lies between the dura mater and the innermost layer, the **pia mater.** The very fine pia mater adheres tightly to the outer surface of the brain and spinal cord and provides the brain and spinal cord with many blood vessels and nerves.

The dura mater has three important inward extensions that help to subdivide the brain: the *falx cerebri,* the *falx cerebelli,* and the *tentorium cerebelli.* The **falx cerebri** projects downward into the longitudinal

LANGUAGE OF SCIENCE AND MEDICINE

HINT Before reading the chapter, say each of these terms out loud. This will help you avoid stumbling over them as you read.

anesthesia (an-es-THEE-zhah)
[*an-* absence, *-esthesia* feeling]

aphasia (ah-FAY-zhah)
[*a* without, *phasia* speech]

arachnoid mater (ah-RAK-noyd MAH-ter)
[*arachn-* spider(web), *-oid* like, *mater* mother]

arbor vitae (AR-bor VI-tay)
[*arbor* tree, *vitae* of life] *pl.,* arbores vitae (AR-bor-eez VI-tay)

ascending tract (ah-SEND-ing)
[*ascend* climb, *tract* trail]

basal nuclei (BAY-sal NOO-klee-eye)
[*bas-* foundation, *-al* relating to, *nucle-* nut or kernel] *sing.,* nucleus (NOO-klee-us)

biological clock (bye-oh-LOJ-ih-kal)
[*bio-* life, *-o-* combining form, *-log-* words (study of), *-ical* relating to]

brainstem

cauda equina (KAW-da eh-KWINE-ah)
[*caud-* tail, *equina* of a horse] *pl.,* caudae equinae (KAW-dee eh-KWINE-ee)

cerebellum (sair-eh-BELL-um)
[*cereb-* brain, *-ellum* small thing] *pl.,* cerebella or cerebellums (sair-eh-BELL-ah, sair-eh-BELL-umz)

cerebral cortex (seh-REE-bral KOR-teks)
[*cerebr-* brain (cerebrum), *-al* relating to, *cortex* bark] *pl.,* cortices (KOR-tih sees)

cerebral hemisphere (seh-REE-bral HEM-ih-sfeer)
[*cerebr-* brain, *-al* relating to, *hemi-* half, *-sphere* globe]

cerebral localization (seh-REE-bral)
[*cerebr-* brain, *-al* relating to, *loc-* point, *-al-* relating to, *-ization* process]

cerebral peduncle (seh-REE-bral peh-DUNG-kul)
[*cerebr-* brain, *-al* relating to, *ped-* foot, *-uncl* little]

cerebral plasticity (seh-REE-bral plas-TIS-ih-tee)
[*cerebr-* brain, *-al* relating to, *plastic-* moldable, *-ity* state]

cerebrospinal fluid (CSF) (seh-ree-broh-SPY-nal FLOO-id)
[*cerebr-* brain, *-spin-* backbone, *-al* relating to, *fluid* flow]

cerebrovascular accident (CVA) (seh-ree-broh-VAS-kyoo-lar)
[*cerebr-* brain, *-vas-* vessel, *cul* little, *ar* relating to]

cerebrum (seh-REE-brum or SAIR-eh-brum)
[*cerebrum* brain] *pl.,* cerebra or cerebrums (seh-REE-bra [SAIR-eh-brah], seh-REE-brumz [SAIR-eh-brumz])

choroid plexus (KOH-royd PLEK-sus)
[*chorio-* skin, *-oid* like, *plexus* network] *pl.,* plexi or plexuses (PLEKS-eye, PLEK-suh-sez)

coma (KOH-mah)
[*coma* deep sleep]

continued on page 270

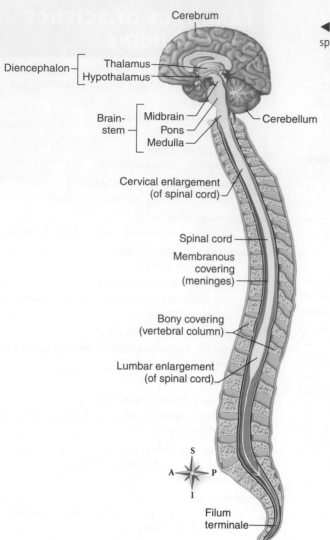

Cerebrum

Diencephalon ⎧ Thalamus
⎩ Hypothalamus

Brain-
stem ⎧ Midbrain
⎨ Pons
⎩ Medulla

Cerebellum

Cervical enlargement
(of spinal cord)

Spinal cord

Membranous
covering
(meninges)

Bony covering
(vertebral column)

Lumbar enlargement
(of spinal cord)

Filum
terminale

◀ **FIGURE 12-1** **The central nervous system.** Details of both the brain and the spinal cord are easily seen in this figure.

fissure (a "crack") to form a tough partition between the two cerebral hemispheres. The **falx cerebelli** is a sickle-shaped extension that separates the two halves, or hemispheres, of the cerebellum. The **tentorium cerebelli** separates the cerebellum from the cerebrum (a *tentorium* is a tentlike covering). Within the dura mater is a large space where the falx cerebri begins to descend between the right and left cerebral hemispheres.

Now, let's look briefly at several important spaces between and around the meninges. The **epidural space** is above the dura mater and just below the bony coverings of the spinal cord. It contains a supporting cushion of fat and other connective tissues and also contains nerves and blood vessels. (Normally there is no epidural space present around the brain, because the dura mater is continuous with the periosteum on the inside of the cranial bones.) The **subdural space** contains a small amount of lubricating *(serous)* fluid and lies between the dura mater and the arachnoid mater. The **subarachnoid space,** as its name suggests, lies beneath the arachnoid mater and outside the pia mater. This space contains a large amount of cerebrospinal fluid, which is discussed in more detail below.

The meninges of the spinal cord (see Figure 12-3) continue down inside the spinal cavity for some distance *below*

▶ **FIGURE 12-2** **Coverings of the brain.** Frontal section of the superior portion of the head, as viewed from the front. Both the bony and the membranous coverings of the brain can be seen.

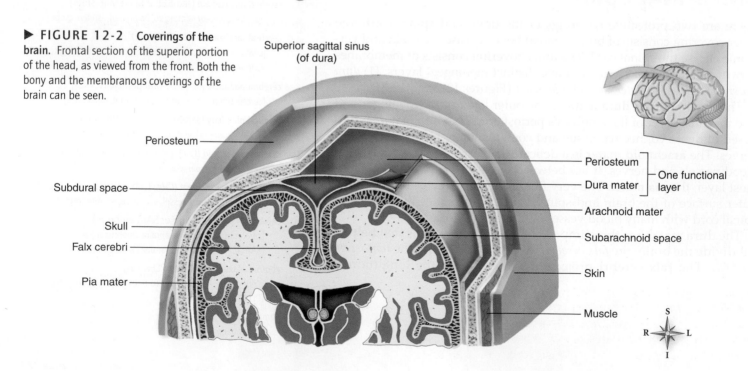

Superior sagittal sinus
(of dura)

Periosteum

Subdural space

Skull

Falx cerebri

Pia mater

Periosteum
Dura mater ⎤ One functional
⎦ layer

Arachnoid mater

Subarachnoid space

Skin

Muscle

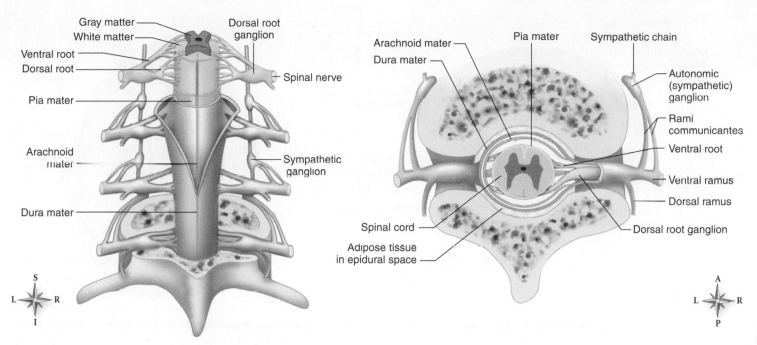

▲ **FIGURE 12-3** **Coverings of the spinal cord.** The dura mater is shown in purple. Note how it extends to cover the spinal nerve roots and nerves. The arachnoid is highlighted in orange and the pia mater in pink.

the end of the spinal cord. The pia mater forms a slender filament known as the **filum terminale** (see Figure 12-1).

Infection or inflammation of the meninges, called **meningitis,** can be life threatening (Box 12-1).

CEREBROSPINAL FLUID

Beyond bony and membranous coverings, the brain and spinal cord are also protected with a *cushion* of circulating serous fluid called the **cerebrospinal fluid (CSF).**

The CSF also provides a *reservoir* of circulating fluid. Along with other fluid body tissues such as the blood, the CSF has *homeostatic functions.* For example, it is monitored by the brain for changes in the internal environment, such as the concentration of carbon dioxide.

Cerebrospinal fluid circulates within the *subarachnoid space* around the brain and spinal cord. It also circulates *within* the fluid-filled canals and cavities **(ventricles)** of the brain and spinal cord (Figure 12-4). There are two large ventricles located in each hemisphere of the

BOX 12-1 | Health Matters

Meningitis

Infection or inflammation of the meninges is termed *meningitis.* It most often involves the arachnoid and pia mater, or the *leptomeninges* ("thin meninges"). Meningitis is most commonly caused by bacteria such as *Neisseria meningitidis* (meningococcus), *Streptococcus pneumoniae,* or *Haemophilus influenzae.* However, viral infections, mycoses (fungal infections), and tumors also may cause inflammation of the meninges. Individuals with meningitis usually complain of fever and severe headaches, as well as neck stiffness and pain. Signs also include avoidance of bright lights and loud sounds, lethargy, and confusion. Persons experiencing these symptoms, some of which mimic the flu, should seek medical help immediately. In infants and very young children, so-called projectile vomiting is common. College campuses are often the sites for localized outbreaks of meningitis, and several bacterial forms are highly contagious.

Depending on the primary cause, meningitis may be mild and self-limiting or may progress to a severe, even fatal, condition. If only the spinal meninges are involved, the condition is called *spinal meningitis.* A lumbar puncture ("spinal tap") can be diagnostic, especially for bacterial meningitis such as the type caused by *H. influenzae.* Spinal meningitis must be treated immediately to avoid an increase in intracranial pressure and damage to the nervous system. If not treated quickly, meningitis can lead to deafness, deficits in cognitive ability, and permanent brain disorders such as epilepsy. Meningitis caused by *H. influenzae* type B, pneumococci bacteria, or the mumps virus can be prevented by immunization.

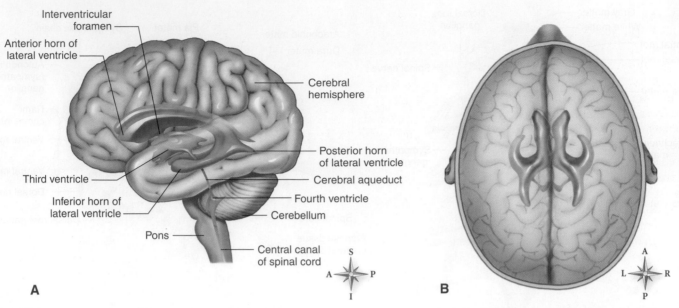

▲ **FIGURE 12-4**   **Fluid spaces of the brain.** **A,** Ventricles highlighted in blue within a translucent brain in a left lateral view. **B,** Ventricles as seen from above.

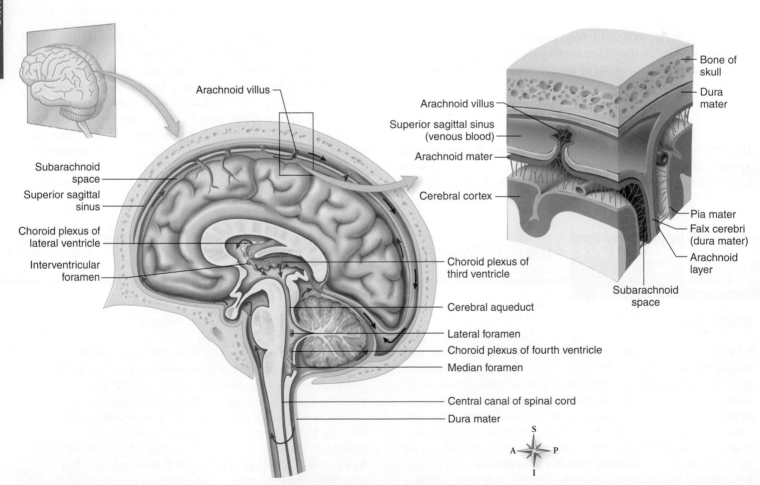

▲ **FIGURE 12-5**   **Flow of cerebrospinal fluid.** The fluid produced by filtration of blood by the choroid plexus of each ventricle flows inferiorly through the lateral ventricles, interventricular foramen, third ventricle, cerebral aqueduct, fourth ventricle, and subarachnoid space and to the blood.

cerebrum and a smaller third ventricle lying vertically between them. There is also a fourth diamond-shaped ventricle attached to the brainstem. This ventricle is actually an expansion of the central canal of the ascending spinal cord.

Cerebrospinal fluid forms primarily by separation of fluid from blood. This separation takes place in the **choroid plexuses**—networks of capillaries that project from the pia mater into the ventricles. Each choroid plexus is covered with a sheet of a special type of **ependymal** (glial) **cell** that releases the cerebrospinal fluid into the fluid spaces. The flow of CSF is continuous through a variety of openings connecting the ventricles. Ultimately, the CSF flows into the subarachnoid space, and then is absorbed by blood in the brain's large venous sinuses from the arachnoid villi (Figure 12-5).

Your body makes about 500 ml of cerebrospinal fluid per day, but only about 140 ml are circulating at any given moment.

The extension of the meninges beyond the spinal cord provides an excellent way of sampling the CSF without injury to the spinal cord. A **lumbar puncture** *(spinal tap)* is performed, which withdraws some of the CSF from the subarachnoid space in the lumbar region of the vertebral column. After the sample is taken, the CSF is tested for the presence of blood cells, bacteria, or other abnormal characteristics that may indicate an injury or infection. Infection is often indicated by increased pressure within the fluid and by changes in color of the CSF from nearly clear to a milky, cloudy, or colored appearance.

QUICK CHECK

1. Name the three fibrous coverings of the brain and spinal cord, beginning with the outermost layer.
2. Trace the path of cerebrospinal fluid throughout the brain and the spinal cord.
3. How can you sample the CSF? If a patient had an infection, what would this serous fluid look like?

SPINAL CORD

Structure of the Spinal Cord

The spinal cord lies within the spinal cavity but does not completely fill it. Accompanying the spinal cord are the meninges, CSF, a cushion of adipose tissue, and blood vessels. Averaging a length of about 18 inches (45 cm), the spinal cord extends from the foramen magnum in the cranium to the lower border of the first lumbar vertebra (Figure 12-6). The spinal cord is an oval-shaped cylinder that tapers slightly as it descends. However, there are two bulges, one in the cervical region of the neck and the other within the lumbar region of the lower spine. There are also two deep grooves: a deep, wide *anterior median fissure* and a *posterior median sulcus* that together nearly divide the cord into separate, symmetrical halves.

As you can see in Figure 12-6, two bundles of nerve fibers called *nerve roots* project from each side of the spinal cord. Fibers comprising the **dorsal (posterior) nerve root** carry

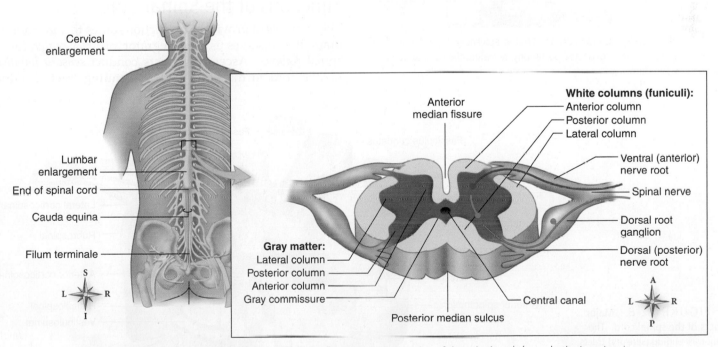

▲ FIGURE 12-6 **Spinal cord.** The inset illustrates a transverse section of the spinal cord shown in the broader view.

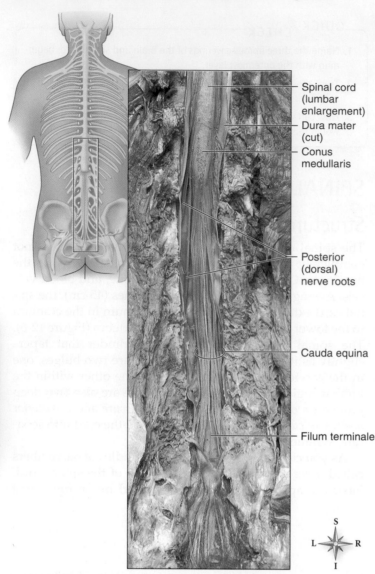

Spinal cord
(lumbar
enlargement)

Dura mater
(cut)

Conus
medullaris

Posterior
(dorsal)
nerve roots

Cauda equina

Filum terminale

▲ **FIGURE 12-7** **Cauda equina.** Photograph shows the inferior portion of the dura mater dissected posteriorly, revealing the cauda equina and nearby structures.

information *into* the spinal cord. Cell bodies of these unipolar, sensory neurons make up a small region of gray matter in the dorsal nerve root. Fibers of the **ventral (anterior) nerve root** carry motor information *out of* the spinal cord. Cell bodies of these multipolar, motor neurons are in the gray matter that composes the inner core of the spinal cord. Numerous interneurons are also located in the spinal cord's gray matter core. On each side of the spinal cord, the dorsal and ventral nerve roots join together to form a single *mixed* nerve simply called a **spinal nerve.** The spinal cord ends at vertebra L1 in a tapered cone that then branches out into a "horse tail" of spinal nerve roots called the **cauda equina.** A long cordlike *filum terminale,* formed from the trailing spinal meninges, flows from the cauda equina (Figure 12-7).

Although the spinal cord's core of gray matter looks like a flat letter "H" in a cross section of the cord, it actually has three dimensions. This is because the gray matter *extends the length of the cord.* The H-shaped rod of gray matter is made up of anterior, lateral, and posterior **gray columns.** When viewed in a cross section (see Figure 12-6), the columns forming the "H" appear to spread out like an animal horn. For this reason they are called the *anterior, posterior,* and *lateral gray* **horns.** The left and right gray columns are joined in the middle by a band called the *gray commissure.* The central canal running through the gray commissure carries the CSF throughout the spinal cord.

White matter surrounding the gray matter is also subdivided into anterior, posterior, and lateral columns. Each white column, or **funiculus,** consists of a large bundle of nerve fibers (axons) divided into smaller bundles called **spinal tracts** (Figure 12-8).

Functions of the Spinal Cord

The spinal cord provides conduction routes to and from the brain. It also serves as the integrator, or *reflex center,* for all spinal reflexes. **Ascending tracts** conduct *sensory impulses* up the cord to the brain and **descending tracts** conduct

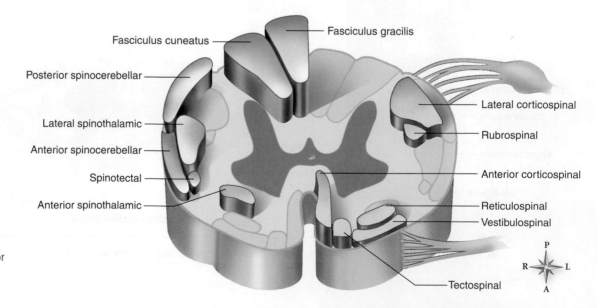

Fasciculus cuneatus

Fasciculus gracilis

Posterior spinocerebellar

Lateral corticospinal

Lateral spinothalamic

Rubrospinal

Anterior spinocerebellar

Spinotectal

Anterior corticospinal

Anterior spinothalamic

Reticulospinal

Vestibulospinal

Tectospinal

▶ **FIGURE 12-8** **Major tracts of the spinal cord.** The major ascending (sensory) tracts are highlighted in blue. The major descending (motor) tracts are highlighted in red.

motor impulses down the cord from the brain. (For clarity, we've color-coded ascending tracts in blue and descending tracts in red in Figure 12-8.) All tracts are composed of bundles of axons. Tracts are both structural and functional organizations for these nerve fibers. For example, all the axons of one tract *structurally* originate from neuron cell bodies located in the same area of the CNS. Tracts are *functional* in that all the axons that compose one tract serve one general function. There are a number of tracts with different functions that terminate in specific parts of the brain, such as the thalamus, medulla, or cerebellum (see Figure 12-8).

The spinal cord also serves as the reflex center for all spinal reflexes, the place where incoming sensory impulses become outgoing motor impulses. These arcs switch impulses from afferent to efferent neurons, as we have seen earlier (see page 233). The spinal cord can also serve as a gateway to the receiving and sending of pain information to the brain.

A&P CONNECT

Reflex centers can act as pain control areas. Pain control areas can inhibit the pain information heading toward the conscious processing centers of the brain. Identifying pain control areas and how they work has led to the development of transcutaneous electrical nerve stimulation (TENS) units and other therapies to reduce pain. To learn more about this, check out **Pain Control Areas** online at **A&P Connect**.

evolve

QUICK CHECK

4. What are spinal nerve roots? How does the dorsal root differ from the ventral root?

5. Name the regions of the white and gray matter seen in a cross section of the spinal cord (hint: think of the letter "H").

6. Distinguish between ascending tracts and descending tracts of the spinal cord.

BRAIN

The brain is one of the largest organs in adult humans, consisting of roughly 100 *billion* neurons and 900 *billion* glia cells. In most adults, it weighs about 1.4 kg (3 pounds). Most of the new neurons in the brain are produced during prenatal development and during the first few months of postnatal life. After that, the neurons grow mainly in size rather than in number. However, some regions of the brain retain neural stem cells that can continue to add small numbers of new neurons to the brain throughout adulthood. The brain attains its full size by about the eighteenth year but grows rapidly only during the first 10 years.

There are six major divisions of the brain, beginning with the base and working upward: *medulla oblongata, pons, midbrain, cerebellum, diencephalons,* and *cerebrum* (Figure 12-9).

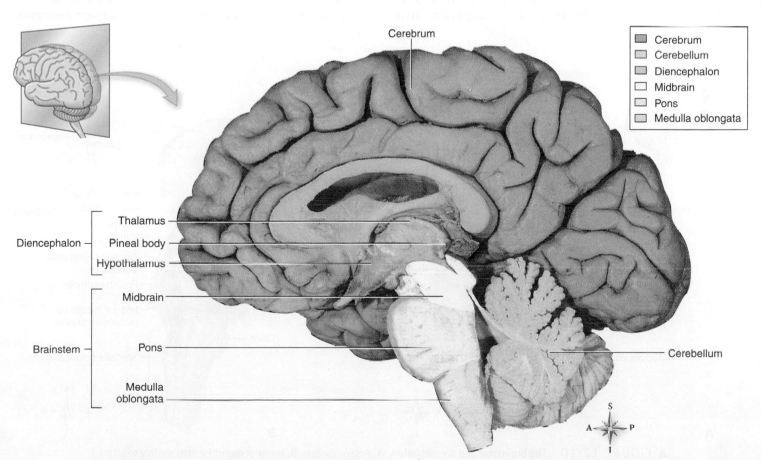

▲ **FIGURE 12-9** **Divisions of the brain.** A midsagittal section of the brain reveals features of its major divisions.

Often the medulla oblongata, pons, and midbrain are collectively called the **brainstem.**

As you read through the following passages, refer often to the diagrams to get your visual bearings. This will greatly speed your understanding of the CNS. Let's begin with Figure 12-10, which illustrates the structures of the brainstem and diencephalon.

Structure of the Brainstem

The **medulla oblongata** is really an enlargement of the upper spinal cord located just above the foramen magnum. It is composed of white matter and a network of gray and white matter (the *reticular formation*). There are two bulges or **pyramids** of white matter located on the ventral surface of the medulla. Located within the medulla's reticular formation are various *nuclei,* or clusters of neuron cell bodies. Some of these nuclei serve as cardiac, respiratory, and vasomotor control centers. All of these are obviously essential to our survival. For this reason they are called the "vital centers," and disease or injury to the medulla often proves fatal. The medulla is separated from the *pons* by a horizontal groove.

The **pons** lies just above the medulla. It also is composed largely of white matter with some gray matter nuclei. Fibers that run transversely across the pons and into the cerebellum, give it an arching, bridgelike appearance.

The **midbrain,** or *mesencephalon,* lies above the pons and below the cerebrum. Both white matter (tracts) and reticular formations make up the bulk of the midbrain. Two ropelike masses of white matter named **cerebral peduncles** extend through the midbrain. Tracts in the peduncles conduct impulses between the midbrain and the cerebrum. In addition to the cerebral peduncles, another landmark of the midbrain is the **corpora quadrigemina.** This forms the posterior, upper part of the midbrain, lying just above the cerebellum. Within specific regions of the corpora quadrigemina lie *auditory and visual centers.* Two other important midbrain structures are the *red nucleus* and the *substantia nigra,* each of which consists of clusters of cell bodies of neurons involved in muscular control. The substantia nigra (literally, "black matter") gets its name from the dark pigment in some of its cells.

Functions of the Brainstem

The brainstem, like the spinal cord, performs sensory, motor, and reflex functions. Some *spinothalamic tracts* (because they run from the spinal cord to the thalamus) are important sensory tracts in the brainstem. Nuclei in the medulla contain a number of reflex centers, the most important of which are the cardiac, vasomotor, and respiratory centers. Other centers present in the medulla function in various nonvital reflexes such as vomiting, coughing, sneezing, hiccupping, and swallowing. The pons contains centers for reflexes mediated by some of the cranial nerves, as well as important centers that help regulate respiration. The midbrain, like the pons, contains reflex centers for certain cranial nerve reflexes, for example, pupillary reflexes and eye movements.

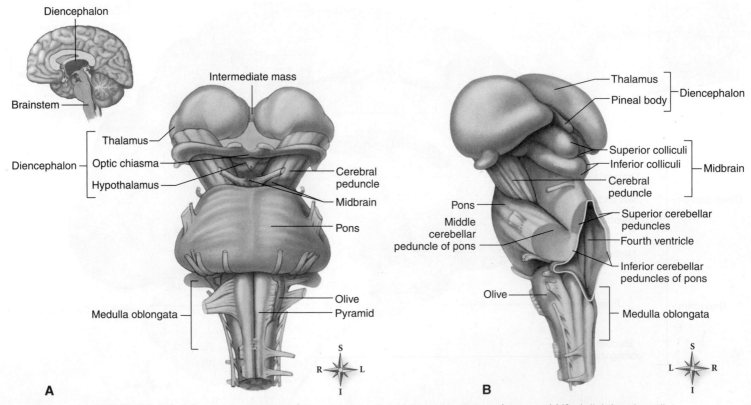

▲ FIGURE 12-10   The brainstem and diencephalon. A, Anterior aspect. B, Posterior aspect (shifted slightly to lateral).

Structure of the Cerebellum

The second largest part of the brain, the **cerebellum** (little brain), is located just below the posterior portion of the cerebrum and is partially covered by it (Figure 12-11). A transverse *fissure* separates the cerebellum from the cerebrum. Amazingly enough, the cerebellum has more neurons than all the other parts of the nervous system combined! You can imagine the "computing power" it must have compared with other parts of the brain.

The cerebrum and cerebellum have several structural characteristics in common. In each, gray matter makes up the outer portion or *cortex*, and white matter predominates in the interior regions. In Figure 12-11, locate the internal white matter of the cerebellum, called the **arbor vitae** (because it looks like the "tree of life"). Note that the surfaces of both the cerebellum and the cerebrum have numerous grooves *(sulci)* and raised areas *(gyri)*—also called **folia.** Like the cerebrum, the cerebellum consists of two large lateral masses, the left and right *cerebellar hemispheres,* and a central section called the **vermis** ("worm").

An important pair of cerebellar nuclei is the **dentate nuclei,** one of which lies in each hemisphere. Tracts connect these nuclei with the thalamus and with motor areas of the cerebral cortex. By means of these tracts, cerebellar impulses influence the motor cortex.

Functions of the Cerebellum

The cerebellum shares structural and functional similarities with the cerebrum. Although much needs to be clarified, the current consensus is that the cerebellum performs

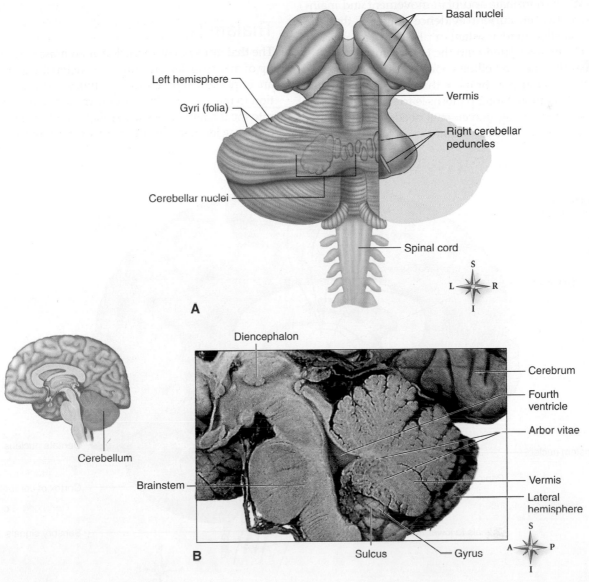

A

Basal nuclei
Left hemisphere
Gyri (folia)
Vermis
Right cerebellar peduncles
Cerebellar nuclei
Spinal cord

B

Diencephalon
Cerebrum
Fourth ventricle
Arbor vitae
Vermis
Lateral hemisphere
Brainstem
Cerebellum
Sulcus
Gyrus

▲ **FIGURE 12-11** **Cerebellum. A,** Posterior view of the surface of the cerebellum. **B,** Photograph of midsagittal brain section shows internal features of the cerebellum and surrounding structures of the brain.

a variety of different functions that complement or assist the cerebrum. Many of these functions involve planning and coordinating skeletal muscle activity and maintaining balance in the body. However, the achievement of coordinated movements results from the *combined efforts* of the cerebrum and cerebellum (Figure 12-12). Impulses from the cerebrum may trigger the action, but those from the cerebellum plan and coordinate the contractions and relaxations of the various muscles once they have begun. Interestingly, the cerebellum becomes involved when a person is just "thinking" about doing some activity, thus "getting us ready" for possible later movement. Some physiologists consider the planning and coordination of movement to be the main functions of the cerebellum. The cerebellum is also thought to be concerned with both exciting and inhibiting the postural reflexes that help us maintain stable body positions.

The cerebellum works with the cerebrum as a sort of "executive assistant" to coordinate and plan movement and maintain balance. However, current evidence suggests that the cerebellum is an all-around assistant or planner of a variety of functions normally associated with the cerebrum. In fact, it is becoming clear that the cerebellum coordinates incoming sensory information as much as or more than it coordinates outgoing motor information. Motor patterns learned in childhood, such as the sequence of leg movements needed for walking, are learned and stored in the cerebellum.

DIENCEPHALON

The **diencephalon** (literally "between brain") is located between the cerebrum and the midbrain (mesencephalon). Although the diencephalon consists of several structures located around the third ventricle, the main ones are the *thalamus* and *hypothalamus*. The diencephalon also includes the *optic chiasma,* the *pineal gland,* and several small, but important structures that we will review in the following discussion (Figures 12-9 and 12-13).

Thalamus

The **thalamus** is a dumbbell-shaped mass of gray matter made up of many nuclei. Two of these nuclei (the **geniculate bodies**) are important in processing auditory and visual input. In addition to these nuclei, large numbers of axons conduct impulses into the thalamus from the spinal cord, brainstem, cerebellum, basal nuclei, and various parts of the cerebrum. These axons

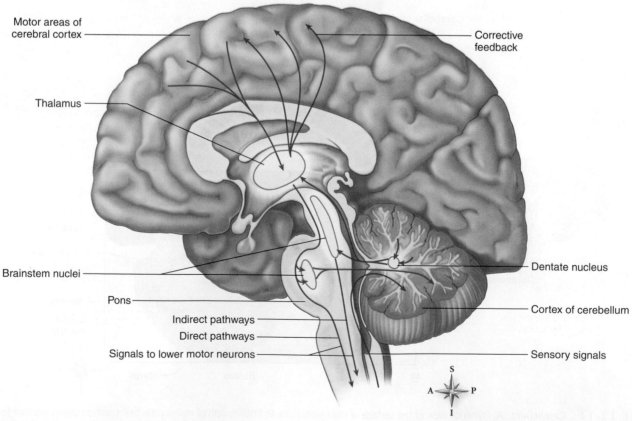

▲ **FIGURE 12-12** **Coordinating function of the cerebellum.** Impulses from the motor control areas of the cerebrum travel down to skeletal muscle tissue and to the cerebellum at the same time. The cerebellum, which also receives and evaluates sensory information, compares the intended movement with the actual movement. It then sends impulses to both the cerebrum and the muscles, thus coordinating and "smoothing" muscle activity.

terminate in *thalamic nuclei,* where they synapse with neurons whose axons conduct impulses out of the thalamus to virtually all areas of the cerebral cortex. In this regard, the thalamus serves as the major relay station for sensory impulses (such as sensations of pain, temperature, and touch) on their way to the cerebral cortex. The thalamus also plays important roles in complex reflex movements, body alerting mechanisms, and even emotions by associating sensory impulses with feelings of pleasantness or unpleasantness.

Hypothalamus

As its name suggests, the **hypothalamus** (see Figure 12-13) lies *beneath* the thalamus. The hypothalamus is composed of several structures, including several nuclei and the *mammillary bodies,* which look like miniature breasts and are involved in smell. There are also supraoptic nuclei that consist of gray matter located just above and on either side of the *optic chiasma.* The *optic chiasma* is the X-shaped junction of the optic tracts and optic nerves. The midportion of the hypothalamus gives rise to the **infundibulum,** a stalk-like attachment leading to the posterior lobe of the pituitary gland.

Although it is relatively small, the hypothalamus performs many functions of great importance not only for survival but also for the enjoyment of life. Certain areas of the hypothalamus function as "pleasure centers" or "reward centers" for the primary drives such as eating, drinking, and sex. For this reason, you might say that the hypothalamus functions as a

link between the *psyche* (mind) and *soma* (body). It also links the nervous system to the endocrine system.

The hypothalamus serves a variety of other vitally important functions. It serves as a regulator and coordinator of *autonomic* activities. In this regard, it controls and integrates the responses made by autonomic (visceral) effectors all over the body. The hypothalamus also functions as the major *relay station* between the cerebral cortex and the lower autonomic centers (the psyche-soma connection). Yet another function has to do with homeostasis. By controlling the amount of urine output, the hypothalamus plays an indirect but essential role in maintaining the water balance of our bodies.

Perhaps most important, the axons of some hypothalamic neurons produce a variety of hormones. Some are *releasing hormones* that are sent into the blood; these releasing hormones then control the release of specific hormones from the anterior pituitary gland. Others are hormones secreted by axons that end in the posterior pituitary gland. The hormone-producing hypothalamic neurons thus control the release of hormones for growth, reproductive function, and regulation of various glands that ultimately affect the functions of every cell in our bodies.

Because the hypothalamus functions as part of our *arousal system,* it plays an essential role in maintaining our waking state. The hypothalamus also helps *regulate our appetite* and therefore the amount of food intake (it has an "appetite center"). Finally, the hypothalamus functions as a crucial part of the mechanism for maintaining normal *body temperature.*

Clearly, the hypothalamus is a small part of the brain that packs a punch far out of proportion to its size.

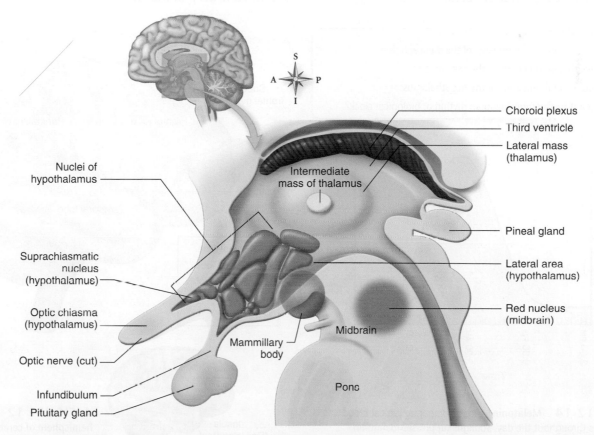

▲ **FIGURE 12-13 Diencephalon.** This midsagittal section highlights the largest regions of the diencephalon, the thalamus, and hypothalamus, but also shows the smaller optic chiasma and pineal gland. Note the position of the diencephalon between the midbrain and the cerebrum.

Pineal Gland

Although the thalamus and hypothalamus account for most of the tissue that makes up the diencephalon, there are several smaller structures of importance. For example, recall that the **optic chiasma** is a region where the right and left *optic nerves* cross each other before entering the brain. After entering the brain, they are called *optic tracts.*

In addition, the important **pineal gland** ("pine nut") controls our **biological clock,** by varying its secretion of the hormone **melatonin.** As Figure 12-14 shows, changing light levels throughout the day and night (creating a circadian [daily] rhythm) trigger changes in the amount of melatonin secretion. (Interestingly, melatonin is really an altered form of the neurotransmitter serotonin). When sunlight levels are high, melatonin secretion decreases; when light levels are low, melatonin levels increase proportionally.

The changing levels of blood melatonin during the day synchronize various body functions with each other and with external stimuli. For this reason, melatonin is often called the "timekeeping hormone." However, melatonin is also called the "sleep hormone" because high blood levels of melatonin signal the body that it is time to sleep! When a person travels quickly to another time zone, or when the seasons change, the altered sunlight patterns cause corresponding time shifts to the melatonin cycle. Likewise, subtle changes may occur in each melatonin cycle depending on how much moonlight is present each evening. Thus, the body can sometimes tell what time of the (lunar) month it is—a mechanism that may help regulate the female reproductive cycle.

QUICK CHECK

11. Describe the two main components of the diencephalon.
12. Name three general functions of the thalamus.
13. Name three general functions of the hypothalamus.
14. What is meant by the term circadian rhythm or biological clock?

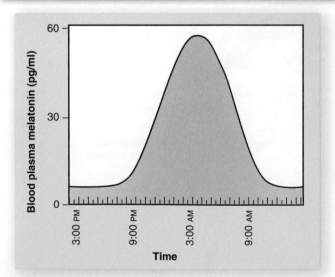

▲ **FIGURE 12-14** **Melatonin.** Graph comparing typical blood melatonin levels throughout the day. Sunlight suppresses melatonin secretion during the day. As the sun goes down, however, melatonin levels begin to rise—dropping again sharply when the sun comes up.

STRUCTURE OF THE CEREBRUM

Cerebral Cortex

The **cerebrum,** the largest and uppermost division of the brain, consists of two halves, the right and left **cerebral hemispheres.** The surface of the cerebrum—called the **cerebral cortex**—is made up of gray matter only 2 to 4 mm (¹⁄₁₂ to ⅙ inch) thick. But despite its thinness, the cortex actually has six distinct layers. Each layer is composed of millions of axon terminals synapsing with millions of dendrites and cell bodies of other neurons.

The surface of the cerebral cortex is highly convoluted, looking like a bowl of thin sausages winding back and forth (Figure 12-15). Each "sausage" is actually a **gyrus.** Because each gyrus is convoluted, it may also be called a **convolution,** and each has its own name. The convolutions greatly increase the *surface area* of the cerebrum and hence the "computing power" available. This enables the cerebrum to perform quite astounding functions far beyond the capability of even the most powerful supercomputer.

Lying between the gyri are either shallow grooves called *sulci* or deeper grooves called *fissures* (Box 12-2). Fissures and a sulcus, as well as a few largely imaginary boundaries, divide each cerebral hemisphere into *five lobes.* Four of these lobes are named for the bones that lie over them: **frontal lobe, parietal lobe, temporal lobe,** and **occipital lobe** (see Figure 12-15). The fifth lobe, the **insula** (island), lies hidden from view deep in the lateral fissure.

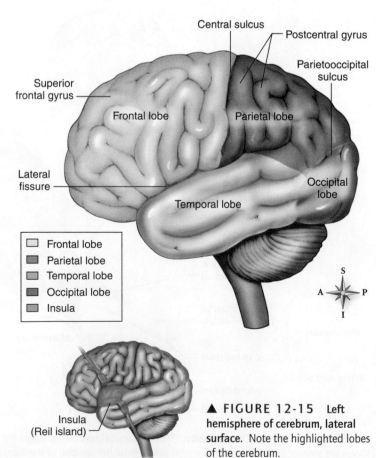

▲ **FIGURE 12-15** **Left hemisphere of cerebrum, lateral surface.** Note the highlighted lobes of the cerebrum.

Cerebral Tracts and Basal Nuclei

Beneath the cerebral cortex lies the large interior of the cerebrum, made up largely of white matter and numerous tracts (remember, tracts of the brain are like nerves). The **corpus callosum,** the prominent white curved structure above the lateral ventricle and along the longitudinal fissure seen in

Figure 12-16, *B*, is made of tracts that connect one hemisphere to the other.

A few islands of gray matter, however, lie deep inside the white matter of each hemisphere. Collectively these are called **basal nuclei** (these are like ganglia). The tracts of the corpus callosum provide a vital pathway of communication between the two hemispheres (Figure 12-16). Researchers

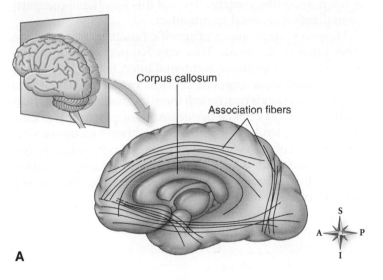

A

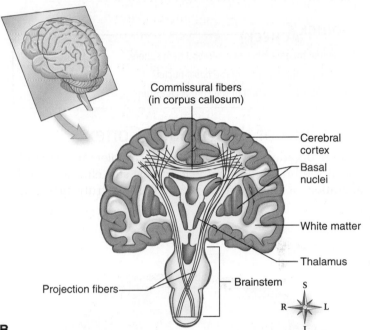

B

▲ **FIGURE 12-16 Cerebral tracts. A,** Lateral perspective, showing various association fibers. **B,** Frontal (coronal) perspective. This shows commissural fibers that make up the corpus callosum and the projection fibers that communicate with lower regions of the nervous system.

 BOX 12-2 FYI

Brain Wrinkles

One of the first things that people notice about the brain is the very wrinkled appearance of the two largest regions of the brain: the cerebellum and cerebrum (Figure A). Wrinkling of the skin is associated with aging and degeneration, but the wrinkles in the surfaces of these two brain regions provide great advantages. If you could flatten out the gray matter gyri (folia) of the cerebellar cortex and the gray matter gyri (convolutions) of both cerebral hemispheres, the advantages become clear. As Figure B shows, the surface area of these brain

structures is much larger than it first appears. The chaotic wrinkling of these brain surfaces allows more gray matter to be packed into the small space of the cranial cavity than would be the case if the brain were smooth surfaced. Thus many times more "computing power" is available to the cerebellum and cerebrum—enabling them to perform their very complex functions.

◀ **A, Wrinkles and grooves of the brain.** Gray matter on the surface of the cerebrum is folded to form bumps or gyri. The valleys between the bumps are called sulci. Larger sulci are often called fissures.

▶ **B, Surface area of the cerebellum and cerebrum.** The gray matter on the surface of the cerebellum (blue) and the cerebrum (orange) is larger than it first appears because it is highly folded.

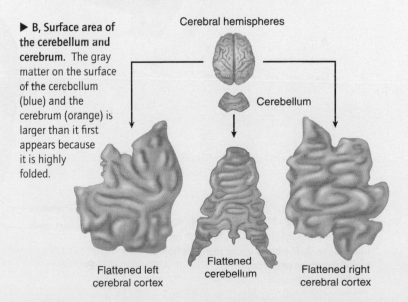

are actively investigating the exact functions of the *basal nuclei* (Figure 12-17). We do know that the nuclei play important roles in regulating voluntary motor functions. For example, the basal nuclei control and monitor the muscle contractions involved in maintaining posture, walking, and performing other gross or repetitive movements. Serious disorders, such as *Parkinson disease (PD)*, may result from the destruction of cells in the basal nuclei (Box 12-3).

> **QUICK ✓ CHECK**
>
> 15. Name the five lobes of a cerebral hemisphere.
> 16. What is the function of the basal nuclei?
> 17. What is the function of the corpus callosum?

Functions of the Cerebral Cortex

Despite decades of research, a clear, complete understanding of the brain and how it functions still eludes us. The operation of the brain is one of the last truly unknown frontiers of science. Perhaps the capacity of our human brain will always fall a little short of a complete understanding of its own complexity! However, we are making progress.

For example, we do know that certain areas of the cortex in each hemisphere of the cerebrum engage largely in one function, at least on average (Figure 12-18). Nonetheless, differences between genders and among individuals of both genders are quite common. We call this functional compartmentalization **cerebral localization.**

However, the locations of specific functions may move to other areas in the brain. This may happen when a person recovers from a stroke or accidental brain injury. This ability to move functional locations is called **cerebral plasticity.** So as far as the brain is concerned, there is always a backup Plan B, although it may not be perfect. And, of course, it is important to understand that no area of the brain functions alone. The function of each region of the cerebral cortex depends on the structures with which it communicates. Many structures of the central nervous system must function together for any one part of the brain to function normally.

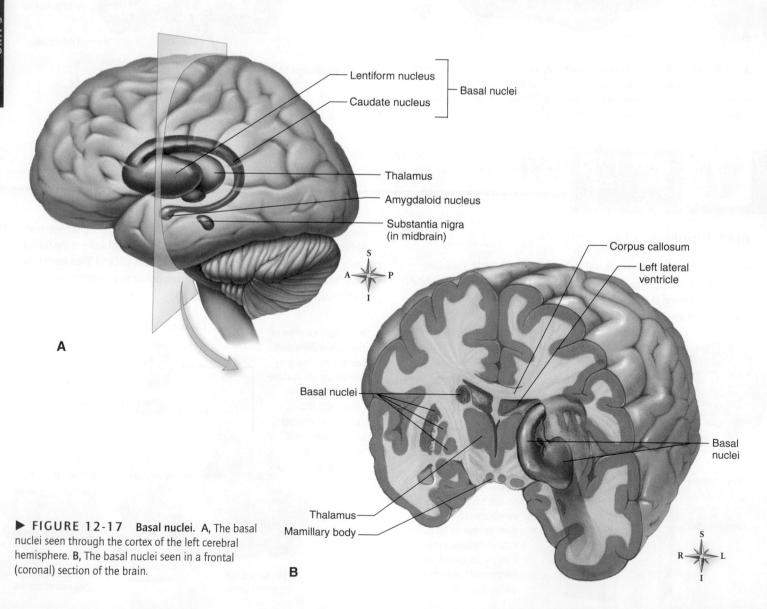

▶ **FIGURE 12-17 Basal nuclei. A,** The basal nuclei seen through the cortex of the left cerebral hemisphere. **B,** The basal nuclei seen in a frontal (coronal) section of the brain.

BOX 12-3 Health Matters

Parkinson Disease

The importance of the basal nuclei in regulating voluntary motor functions is made clear in cases of **Parkinson disease (PD)**. Normally, neurons that lead from the substantia nigra (Figure 12-17, *A*) to the basal nuclei secrete dopamine. Dopamine inhibits the excitatory effects of acetylcholine produced by other neurons in the basal nuclei. Such inhibition by *dopaminergic* (dopamine-producing) neurons produces a balanced, restrained output of muscle-regulating signals from the basal nuclei. In PD, however, neurons leading from the substantia nigra degenerate and thus do not release normal amounts of dopamine. Without dopamine, the excitatory effects of acetylcholine are not restrained, and the basal nuclei produce an excess of signals that affect voluntary muscles in several areas of the body. Overstimulation of postural muscles in the neck, trunk, and upper limbs produces the syndrome of effects that typify this disease: rigidity and tremors of the head and limbs; an abnormal, shuffling gait; absence of relaxed arm-swinging while walking; and a forward tilting of the trunk.

Frontal lobe

Basal nuclei

◀ A, Dopaminergic pathways of the brain.

S

A ✦ P

I

A

Dopamine pathways

Forward tilt of trunk

Rigidity and trembling of head

Reduced arm-swinging

Rigidity and trembling of extremities

Shuffling gait with short steps

▶ B, Signs of Parkinson disease (PD).   B

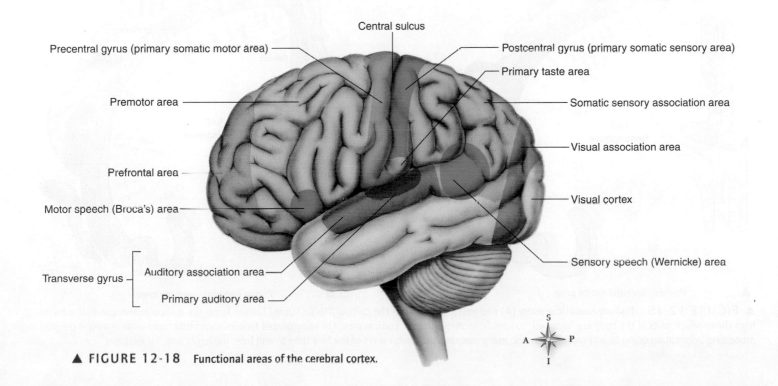

Central sulcus

Precentral gyrus (primary somatic motor area)

Postcentral gyrus (primary somatic sensory area)

Primary taste area

Premotor area

Somatic sensory association area

Visual association area

Prefrontal area

Visual cortex

Motor speech (Broca's) area

Sensory speech (Wernicke) area

Transverse gyrus

Auditory association area

Primary auditory area

S

A ✦ P

I

▲ FIGURE 12-18   Functional areas of the cerebral cortex.

Sensory and Motor Functions of the Cortex

Various areas of the cerebral cortex are essential for normal functioning of the somatic, or general, senses as well as for the so-called special senses. The somatic senses include sensations of touch, pressure, temperature, body position (proprioception), and similar perceptions that do not require complex sensory organs. The special senses include vision, hearing, and other types of perception that require complex sensory organs—for example, the eye and the ear.

The *postcentral gyrus* serves as a primary area for the general somatic senses (see Figures 12-18 and 12-19, *A*). In a way, the cortex contains a "somatic sensory map" of the body. Areas such as the face and hand have a *proportionally larger number* of sensory receptors, so their part of the somatic sensory map is larger. Likewise, information regarding vision is mapped in the visual cortex, and auditory information is mapped in the primary auditory area (see Figure 12-19).

The cortex does more than just register separate and simple sensations, however. Information sent to the primary sensory areas is in turn relayed to the various sensory association areas, as well as to other parts of the brain. There the sensory information is compared and evaluated. Eventually, the cortex *integrates* separate bits of information to produce whole perceptions.

Mechanisms that control voluntary movements are extremely complex and imperfectly understood. It is known, however, that for normal movements to take place, many parts of the nervous system—including certain areas of the cerebral cortex—must function. The *precentral gyrus*, the most posterior gyrus of the frontal lobe, constitutes the primary somatic motor area (see Figures 12-18 and 12-19, *B*).

The cerebral cortex is also involved in integrative functions. This consists of all events that take place in the cerebrum between its receiving of sensory impulses and its sending out of motor impulses. These integrative functions include consciousness and mental activities of all kinds. Consciousness, use of language, emotions, and memory are all "integrative" functions.

Our cerebral activities go on for as long as we are alive. You can easily see this from recordings of brain electrical potentials known as *electroencephalograms* or *EEGs* (Box 12-4). These and other brain-imaging techniques can be used to

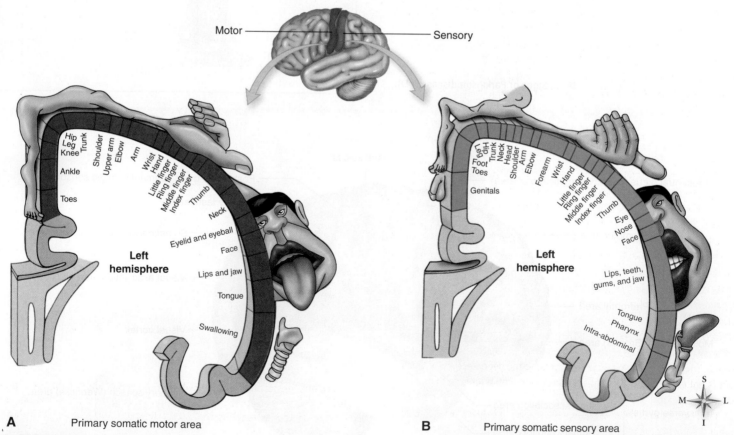

A   Primary somatic motor area

B   Primary somatic sensory area

▲ **FIGURE 12-19**   **Primary somatic sensory (A) and motor (B) areas of the cortex.** The disfigured human form—the *cortical homunculus*—illustrated here shows which parts of the body are "mapped" to specific locations in each cortical area. The exaggerated face indicates that more cortical area is devoted to processing information going to and coming from the many receptors and motor units of the face than to and from the leg or arm, for example.

discover new concepts of brain function. They can also assess structural and functional problems in the brains of patients.

Consciousness

There is no precise definition of **consciousness.** We may attempt to define it as a state of awareness of one's self, one's environment, and other humans around us. However, we know virtually nothing of the neural mechanisms that produce consciousness. We do know that consciousness depends on the continuous excitation of cortical neurons by impulses conducted to them by a network of neurons known as the **reticular activating system (RAS).** The RAS consists of centers in the brainstem's *reticular formation* that receive impulses from the spinal cord and relay them to the

---

BOX 12-4 Diagnostic Study

The Electroencephalogram (EEG)

Cerebral activity goes on as long as life itself. Only when life ceases (or moments before) does the cerebrum cease its functioning. Only then do all its neurons stop conducting impulses. Proof of this has come from records of brain electrical potentials known as **electroencephalograms,** or **EEGs.** These records are usually made from data detected by a number of electrodes placed on different regions of the scalp; they are records of wave activity—*brain waves* (parts *A* and *B* of the figure).

Four types of brain waves are recognized based on frequency and amplitude of the waves. Frequency, or the number of wave cycles per second, is usually referred to as *hertz* (Hz, named after Heinrich Hertz, a German physicist). Amplitude means voltage. Listed in order of frequency from fastest to slowest, brain waves are designated as *beta, alpha, theta,* and *delta. Beta waves* have a frequency of more than 13 Hz and a relatively low voltage. *Alpha waves* have a frequency of 8 to 13 Hz and a relatively high voltage. *Theta waves* have both a relatively low frequency—4 to 7 Hz—and a low voltage. *Delta waves* have the slowest frequency—less than 4 Hz—but a high voltage. Brain waves vary in different regions of the brain, in different states of awareness, and in abnormal conditions of the cerebrum.

Fast, low voltage beta waves characterize EEGs recorded from the frontal and central regions of the cerebrum when an individual is awake, alert, and attentive, with eyes open. Beta waves predominate when the cerebrum is busiest, that is, when it is engaged with sensory stimulation or mental activities. In short, beta waves are "busy waves." Alpha waves, in contrast, are "relaxed waves." They are moderately fast, relatively high-voltage waves that dominate EEGs recorded from the parietal lobe, occipital lobe, and posterior parts of the temporal lobes when the cerebrum is idling, so to speak. The individual is awake but has eyes closed and is in a relaxed, nonattentive state. This state is sometimes called the "alpha state." When drowsiness descends, moderately slow, low-voltage theta waves appear. Theta waves are "drowsy waves." "Deep sleep waves," on the other hand, are known as delta waves. These slowest brain waves characterize the deep sleep from which one is not easily aroused. For this reason, deep sleep is referred to as slow-wave sleep.

Physicians use electroencephalograms (EEGs) to help localize areas of brain dysfunction, to identify altered states of consciousness, and often to establish death. Two flat EEG recordings (no brain waves) taken 24 hours apart in conjunction with no spontaneous respiration and total absence of somatic reflexes are criteria accepted as evidence of brain death.

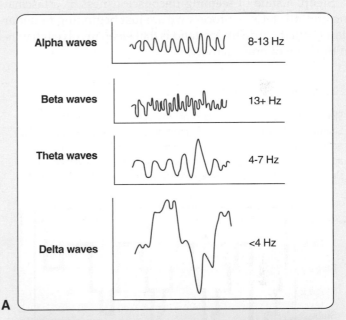

A

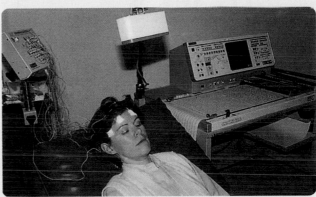

B

▲ The electroencephalogram (EEG). **A,** Examples of alpha, beta, theta, and delta waves seen on an EEG. **B,** Photograph showing a person undergoing an EEG test. Notice the scalp electrodes that detect voltage fluctuations within the cranium.

thalamus and then from the thalamus to all parts of the cerebral cortex (Figure 12-20). Without continual excitation of cortical neurons by reticular activating impulses, an individual is *unconscious* and cannot be aroused.

Drugs known to depress the RAS, such as *barbiturates,* decrease alertness and induce sleep. In contrast, *amphetamines,* which stimulate the cerebrum and enhance alertness, probably act by stimulating the RAS. Anesthetic drugs produce an altered state of consciousness, namely **anesthesia,** and disease or injury of the brain may produce an altered state called **coma.** People of various cultures who practice *meditation* are said to enter an altered state of a higher level of consciousness. This higher consciousness is accompanied, ironically, by a high degree of both relaxation and alertness. With training in meditation techniques and practice, an individual can enter the meditative state at will and remain in it for an extended period of time.

Sleep, a state of seeming unconsciousness, is a fascinating and mysterious process we are just beginning to understand. What we do understand is that sleep involves changes in the level of brain activity.

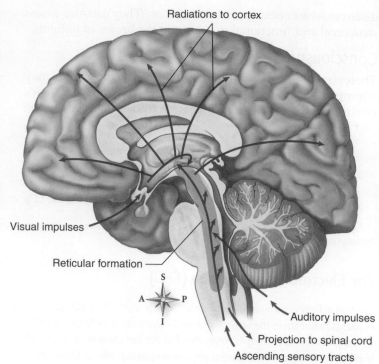

▲ **FIGURE 12-20** **Reticular activating system (RAS).** Consists of centers in the brainstem's *reticular formation* plus fibers that conduct to the centers from below and fibers that conduct from the centers to widespread areas of the cerebral cortex. Functioning of the RAS is essential for regulating levels of consciousness.

A&P CONNECT

Sleep involves changes in the level of brain activity, as seen in the figure. Shortly after falling asleep, we fall into a deep slow-wave sleep (SWS) characterized by slow EEG waves. Approximately every 90 minutes or so, brain activity increases during a dream stage called rapid-eye-movement (REM) sleep. Learn more about the sleep cycle in **Sleep** online at **A&P Connect.**

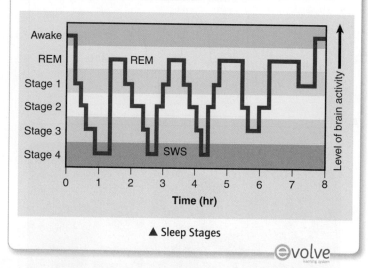

▲ Sleep Stages

evolve

Language

Language functions consist of the ability to speak and write words and the ability to understand spoken and written words. Certain areas in the frontal, parietal, and temporal lobes serve as speech centers—as crucial areas, that is, for language functions. The left cerebral hemisphere contains these areas in about 90% of the population. In the remaining 10%, either the right hemisphere or both hemispheres contain the language areas.

Lesions in speech centers give rise to language defects called **aphasias.** For example, with damage to an area in the inferior gyrus of the frontal lobe (Broca's area; see Figure 12-18), a person becomes unable to articulate words but can still make vocal sounds and understand words both heard and read.

Box 12-5 discusses a class of neurons that may help us learn and use language.

Emotions

Emotions—both the subjective experiencing and objective expression of them—involve functions of the cerebrum's **limbic system.** This unique network forms a curving border around the corpus callosum—the structure that connects the two cerebral hemispheres. The limbic system (the so-called *emotional brain*) functions in some way to make us experience many kinds of emotions—anger, fear, sexual feelings, pleasure, and sorrow, for example. To bring about the normal expression of emotions, however, parts of the cerebral cortex other than the limbic system must also function. It appears that limbic activity without the modulating influence of the other cortical areas may bring on attacks of abnormal, uncontrollable rage suffered periodically by some unfortunate individuals.

BOX 12-5 FYI

Mirror Neurons

Recent findings show that a special functional class of neurons called *mirror neurons* exists in the cortex. These neurons exhibit action potentials both when we experience something ourselves and also when someone else does. In other words, certain circuits in our cortex become active whether we perform an action or we observe someone else doing so—thus enabling the brain to "mirror" the activity of another person's brain. Although much more is yet to be learned about the so-called *mirror-neuron system*, it seems clear that the action of these neurons may explain how humans are able to learn spoken language; interpret complex body language; empathize with the feelings of others; and learn to walk, write, or ride a bicycle.

Memory

Memory is one of our major mental activities. The cerebral cortex is capable of storing and retrieving both *short-term memory* and *long-term memory*. **Short-term memory** involves the storage of information over a few seconds or minutes. Somehow, short-term memories can be consolidated by the brain and stored as long-term memories that can be retrieved days, or even years, later. Both short-term memory and long-term memory are functions of many parts of the cerebral cortex, especially of the temporal, parietal, and occipital lobes.

Long-term memory is believed to consist of some kind of structural changes—called *engrams*—in the cerebral cortex. Widely accepted today is the hypothesis that an engram consists of some kind of *permanent* change in the synapses in a specific circuit of neurons. More recent data suggest that different kinds of memories may be stored in different ways. Some memories are perhaps stored as changes at synapses and some by changes in the neurons themselves. In addition, current research suggests that the cerebrum's limbic system—the emotional brain—plays a key role in memory. For example, when the hippocampus (part of the limbic system) is removed, you may lose the ability to recall new information. Thus, your own personal experience substantiates a relationship between emotion and memory.

Table 12-1 briefly summarizes the major structures and functions of the central nervous system. Take a moment now to review it. This will help you reinforce what you have just learned about the CNS before moving on to the last topics of the chapter.

A&P CONNECT

Have you ever heard someone being called "left-brained" or "right-brained" in their approach to work or learning? What does that mean? Find out in **Specialization of Cerebral Hemispheres** online at **A&P Connect**.

evolve

QUICK CHECK

18. Give a basic outline of how sensory functions and motor functions are mapped on the cerebral cortex.
19. What does the RAS have to do with alertness?
20. What is the function of the limbic system?

TABLE 12-1 Summary of CNS Structures and Functions

REGION	STRUCTURE*	FUNCTION*
Spinal cord	Elongated cylinder extending from the brainstem through the foramen magnum of the skull; gray matter interior surrounded by white matter; 31 pairs of spinal nerves attached by dorsal and ventral nerve roots	Integration of simple, subconscious spinal reflexes; conduction of nerve impulses
Gray matter	Numerous synapses and interneurons arranged into anterior, lateral, and posterior gray columns linked by a gray commissure	Integration of spinal reflexes and filtering of information going to higher centers (as in gated pain control)
White matter	Myelinated nerve tracts arranged into anterior, lateral, and posterior white columns (funiculi)	Ascending tracts conduct sensory information to higher CNS centers; descending tracts conduct motor information from higher CNS centers

(In the Spinal cord row, the structure cell includes an illustration labeled "Spinal cord")

*Summary only; see chapter text and figures for detailed descriptions of structure and function.
CNS, Central nervous system.

(continued)

UNIT 3

TABLE 12-1 Summary of CNS Structures and Functions—cont'd

REGION	STRUCTURE*	FUNCTION*
Brainstem	Extends inferiorly from diencephalon to foramen magnum of skull, where it meets the spinal cord; central gray matter nuclei surrounded and connected by white matter tracts; 10 of the 12 pairs of cranial nerves attached here	Subconscious integration of basic vital functions
Medulla oblongata Medulla oblongata	Inferior region of brainstem between the spinal cord and the pons	Integration of cardiac, vasomotor (vessel muscle), respiratory, digestive, and other reflexes
Pons Pons	Intermediate region of brainstem between the medulla and the midbrain	Integration of numerous autonomic reflexes mediated by cranial nerves V, VI, VII, and VIII (see Chapter 13) and respiration
Midbrain Midbrain	Superior region of the brainstem between the pons and the diencephalon	Integration of numerous cranial nerve reflexes, such as eye movements, pupillary reflex, ear (sound muffling) reflexes
Reticular formation Reticular formation	Roughly cylindrical network of nerve pathways and centers extending through the brainstem and into the diencephalon	Operates the reticular activating system (RAS) that regulates state of consciousness
Cerebellum Cerebellum	Roughly spherical structure attached at the posterior of the brainstem; wrinkled gray matter cortex, branched network of white fibers inside (arbor vitae), and several small gray nuclei	Coordinates many functions of cerebrum, including planning and control of skilled movements, posture, balance, coordination of sensory information relating to body position and movement

TABLE 12-1   Summary of CNS Structures and Functions—cont'd

REGION	STRUCTURE*	FUNCTION*
Diencephalon Thalamus Hypothalamus—Pineal gland	Brain region in the central part of the brain, between the cerebrum and brainstem (midbrain); made up of various gray matter nuclei	Numerous coordinating and integrating functions
Thalamus	Large dumbbell-shaped mass of gray matter, divided into two large lateral masses connected by an intermediate mass	Crude sensations, coordination of sensory information relayed to cerebrum; involved in emotional response to sensory information; involved in arousal; general processing of information to/from cerebrum
Hypothalamus	Numerous gray matter nuclei clustered below the thalamus	Integration/coordination of many autonomic reflexes, hormonal functions; involved in arousal, appetite, thermoregulation
Pineal gland	Single nucleus of neuroendocrine tissue posterior to the thalamus	Produces melatonin, a timekeeping hormone, as part of the body's biological clock
Cerebrum—Cerebrum	Largest, most superior region of brain; divided into right and left hemispheres, connected by the corpus callosum	Complex processing of sensory and motor information; complex integrative functions
Cerebral cortex Cerebral cortex—Basal nuclei—White matter tracts	Highly wrinkled gray matter surface of the cerebrum; divided into five major lobes per hemisphere; functionally mapped based on concept of localization	Higher level processing of sensory and motor information, including conscious sensation and motor control; complex integrative functions such as consciousness, language, memory, emotions
Cerebral tracts	White matter tracts connect various regions of the cortex with each other and with inferior CNS structures	Conduction information between CNS areas to facilitate complex processing and integration
Basal nuclei	Gray matter nuclei deep in the cerebrum	Integration and regulation of conscious motor control, especially posture, walking, other repetitive movements; possible roles in thinking and learning

*Summary only; see chapter text and figures for detailed descriptions of structure and function.
CNS, Central nervous system.

SOMATIC SENSORY PATHWAYS OF THE CNS

For the cerebral cortex to perform its *sensory* functions, impulses must first be conducted to its sensory areas by *relays* of neurons called **sensory pathways**. One of these pathways is outlined for you in Figure 12-21. Referring to this diagram as you read each sentence will greatly aid in your understanding of these pathways.

First you should understand that most impulses reaching the sensory areas of the cerebral cortex have traveled over at least *three pools* of sensory neurons. These are highlighted for you on the diagrams in Figure 12-21. *Primary sensory neurons* of the relay conduct impulses from the periphery (e.g.,

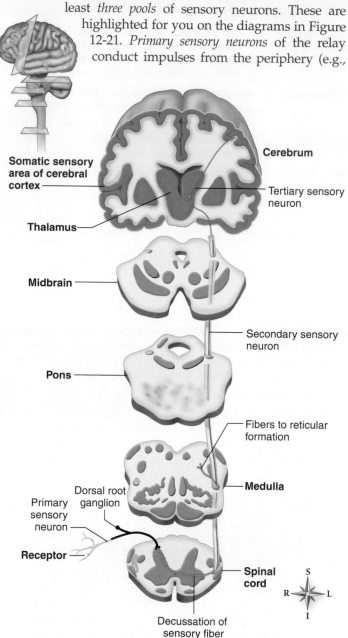

FIGURE 12-21 Example of somatic sensory pathway.
A pathway that conducts information about pain and temperature. Note that this pathway decussates in the spinal cord. Other pathways instead decussate in the brainstem.

skin) of our bodies to the CNS. Next, *secondary sensory neurons* conduct these impulses from the spinal cord or brainstem up to the thalamus. (The dendrites and cell bodies of the secondary sensory neurons are located in the gray matter of the spinal cord or brainstem. Their axons form *ascending tracts* that travel up the spinal cord, through the brainstem, and terminate in the thalamus.) Tertiary sensory neurons then conduct impulses from the thalamus to the sensory area of the parietal lobe.

There's just one twist, and this is particularly interesting. Usually the axons of secondary sensory neurons *cross over* or **decussate** at some level in their ascent to the thalamus. This means that the *left* somatic sensory area now predominantly experiences general sensations from the *right* side of the body. In a similar manner, the *right* somatic sensory area of the brain predominantly experiences general sensations of the *left* side of the body. In short, what you experience on the left side of your body you interpret on the right side of your brain, and vice versa.

SOMATIC MOTOR PATHWAYS OF THE CNS

For the cerebral cortex to perform its *motor* functions, impulses must be conducted from its motor areas to skeletal muscles by relays of neurons referred to as *somatic motor pathways*. Somatic motor pathways consist of motor neurons that conduct impulses from the CNS to somatic effectors—the skeletal muscles. Spinal cord reflex arcs are simple and well understood for the most part, but some motor pathways are extremely complex and not at all clearly defined. The motor system of our nervous system is responsible for our facial expressions (created by facial muscles), for maintaining our posture, and for moving the muscles of the trunk, head, limbs, tongue, and eyes. Reflexes, as we've seen, can create involuntary movement, but *voluntary movements* are those specific movements we control. These movements result from the stimulation of the upper and lower motor neurons.

Look at Figure 12-22 and note that *only one final common pathway* (that is, each single motor neuron from the anterior gray horn of the spinal cord) conducts *impulses* to a specific motor unit within a skeletal muscle. This is because axons from the anterior gray horn are the *only ones* that terminate in skeletal muscles. This has important practical implications. For example, it means that any condition that makes motor neurons from the anterior horn unable to conduct impulses also makes skeletal muscles supplied by these neurons unable to contract. (Obviously, you cannot *will* your muscles to contract; the proper neural network must stimulate them to contract.) As a result, muscles stay *flaccid* (flabby) and paralyzed. Historically, the most famous disease producing this condition is *poliomyelitis*, a scourge throughout the world only 60 years ago.

Numerous somatic motor pathways conduct impulses from motor areas of the cerebrum down to the anterior horn motor neurons at all levels of the spinal cord. Injury of *upper motor neurons* produces symptoms called *spastic paralysis*, often accompanied by exaggerated deep reflexes. In babies, injury to upper motor neurons can produce a positive *Babinski reflex* (where the great toe curls up instead of down and the smaller toes fan out when a probe is run along the bottom of the foot). Injury to upper motor neurons often happens after **cerebrovascular accidents (CVAs,** or strokes), producing a variety of possible disruptions of motor control.

Injury to *lower motor neurons* produces symptoms different from those of the upper motor neuron injury. If these neurons are injured, impulses can no longer reach the skeletal muscles they supply. This, in turn, results in the absence of *all* reflexive and willed movements produced by contraction of the muscles involved. Unused, these muscles soon lose their normal tone and become soft and flaccid. Thus, absence of reflexes and *flaccid paralysis* are the most important signs of lower motor neuron problems.

The set of coordinated commands that control the muscle activity mediated by the motor pathways is often called a **motor program.** Traditionally, the primary somatic motor areas of the cerebral cortex were thought to be the principal organizer of motor programs sent along each motor pathway. However, a newer concept holds that some motor programs result from the interaction of several different centers in the brain. Apparently, many voluntary motor programs are organized in the basal nuclei and cerebellum—perhaps in response to a willed command by the cerebral cortex. Impulses that constitute the motor program are then channeled through the thalamus and back to the cortex, specifically to the primary motor area. From there, the motor program is sent down to the brainstem. Signals from the brainstem then continue on down one or more spinal tracts and out to the muscles by way of the lower motor neurons. All along the way, neural connections among the various motor control centers allow refinement and adjustment of the motor program.

QUICK CHECK

21. What do we mean by the "final common pathway" as it relates to the somatic motor pathway?

22. Briefly describe how neural conduction flows in a somatic sensory pathway.

23. Briefly describe how neural conduction flows in a somatic motor pathway.

Cycle of LIFE ⟳

Perhaps the most obvious *functional* changes during our lives have to do with the development and then degeneration of our nervous system. This is due, certainly, to its incredible complexity relating to both structure and function.

Although the development of the brain and spinal cord begins in the womb, a great deal of additional development continues beyond the time of birth. A newborn lacks the more complex integrative functions such as language, complex memory, comprehension of spatial skills, and motor skills such as walking. By the time a person reaches adulthood, however, most of these complex functions are fully developed. We use them throughout our lives to help us maintain internal stability in an ever changing world.

As we enter late adulthood, the tissues of the brain and spinal cord may degenerate. This degeneration, as you probably know, varies considerably from person to person, and depends on many complex interactions between our genetic constitution and the stresses of life on our CNS. In some cases, degeneration is profound and an older person is no longer able to communicate effectively, to walk, or to perform many of the mundane tasks of everyday life. As our understanding of this process increases, we are finding ways to reduce or avoid entirely such changes associated with an aging nervous system. ●

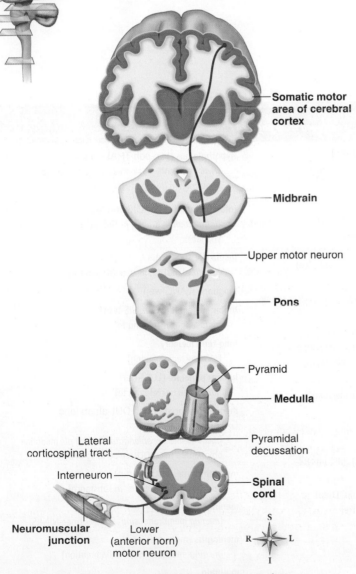

▲ **FIGURE 12-22 Example of somatic motor pathway.** A pyramidal pathway, through the lateral corticospinal tract.

Labels in figure:
- Somatic motor area of cerebral cortex
- Midbrain
- Upper motor neuron
- Pons
- Pyramid
- Medulla
- Lateral corticospinal tract
- Pyramidal decussation
- Interneuron
- Spinal cord
- Neuromuscular junction
- Lower (anterior horn) motor neuron

The BIG Picture

Organization of the Body

The central nervous system is the ultimate regulator of our entire body. It serves as the anatomical and functional center of the countless feedback loops that maintain the relative constancy of our internal environment. In fact, the CNS directly or indirectly regulates (or at least influences) nearly every organ in the body.

It is still largely a mystery how the CNS is able to integrate literally millions of bits of information every second throughout the entire body—and make sense of it all. The CNS not only makes sense of all this information, but also compares it with previously stored memories. In a fraction of a second, it can make decisions based on its own conclusions about the incoming data.

The complex integrative functions of human language, consciousness, learning, and memory enable us to adapt to situations that less complex organisms cannot.

MECHANISMS OF DISEASE

Injury or disease can destroy neurons and therefore disrupt nervous system functions. For example, damage from disrupted blood supply can cause a *stroke (cerebrovascular accident or CVA)* or *cerebral palsy*, perhaps causing a type of *paralysis*. Degeneration of tissue can cause dementia, as in *Alzheimer disease (AD)*, *Huntington disease (HD)*, *AIDS-related dementia*, or even so-called *mad cow disease*. Sometimes brain damage or other factors can trigger *seizure disorders* such as *epilepsy*.

evolve Find out more about these and other brain and spinal cord injuries and diseases online at *Mechanisms of Disease: Central Nervous System.*

LANGUAGE OF SCIENCE AND MEDICINE *continued from page 247*

consciousness
[*conscire* know wrong]

convolution (kon-voh-LOO-shun)
[*con-* together, *-volut-* roll, *-tion* process]

corpora quadrigemina (KOHR-pohr-ah kwod-ri-JEM-ih-nah)
[*corpora* bodies, *quadri-* fourfold, *-gemina* twin]

corpus callosum (KOHR-pus kah-LOH-sum)
[*corpus* body, *callosum* callous] *pl.*, corpora callosa (KOHR-poh-rah kah-LOH-sah)

decussate (de-KUS-sayt)
[*dec-* X (Roman numeral ten), *-as-* as, *-ate* process of]

dentate nucleus (DEN-tayt NOO-klee-us)
[*dent-* tooth, *-ate* of or like, *nucleus* nut or kernel] *pl.*, nuclei (NOO-klee-eye)

descending tract
[*descend-* move downward, *tract* trail]

diencephalon (dye-en-SEF-ah-lon)
[*di-* between, *-en-* within, *-cephalon* head] *pl.*, diencephala or diencephalons (dye-en-SEF-ah-lah, dye-en-SEF-ah-lonz)

dorsal (posterior) nerve root (DOR-sal)
[*dors-* back, *-al* relating to]

dura mater (DOO-rah MAH-ter)
[*durus* hard, *mater* mother]

electroencephalogram (EEG) (eh-lek-troh-en-SEF-ah-loh-gram)
[*electro-* electricity, *-en-* within, *-cephal-* head, *-gram* drawing]

ependymal cell (eh-PEN-di-mal sell)
[*ep-* over, *-en-* on, *-dyma-* put, *-al* relating to, *cell* storeroom]

epidural space (ep-ih-DOO-ral)
[*epi-* upon, *-dura-* hard, *-al* relating to]

falx cerebelli (falks ser-eh-BEL-lee)
[*falx* sickle, *cerebelli* of the cerebellum (small brain)] *pl.*, falces cerebelli (FAL-sees ser-eh-BEL-lee)

falx cerebri (falks SER-eh-bree)
[*falx* sickle, *cerebri* of the cerebrum] *pl.*, falces cerebri (FAL-sees SER-eh-bree)

filum terminale (FYE-lum ter-mih-NAL-ee)
[*filum* thread, *termin-* boundary, *-al* relating to] *pl.*, fila terminales (FYE-lah ter-mih-NAL-eez)

folia (FOH-lee-ah)
[*folia* leaves] *sing.*, folium (FOH-lee-um)

frontal lobe (FRON-tal)
[*front-* forehead, *-al* relating to]

funiculus (fuh-NIK-yoo-lus)
[*funi-* rope, *-icul-* little] *pl.*, funiculi (fuh-NIK-yool-eye)

geniculate body (jeh-NIK-ye-lit BOD-ee)
[*geni-* knee, *-icul-* small, *-ate* like]

gray column

gyrus (JYE-rus)
[*gyro-* circle]

horn

hypothalamus (hye-poh-THAL-ah-muss)
[*hypo-* beneath, *-thalamus* inner chamber] *pl.*, hypothalami (hye-poh-THAL-ah-meye)

infundibulum (in-fun-DIB-yoo-lum)
[*infundibulum* funnel] *pl.*, infundibula (in-fun-DIB-yoo-lah)

insula (IN-soo-lah)
[*insula* island] *pl.*, insulae (IN-soo-lee)

language (LANG-wej)

limbic system (LIM-bik SIS-tem)
[*limb-* edge, *-ic* relating to]

long-term memory
[*memory* remembering]

lumbar puncture (LUM-bar)
[*lumb-* loin, *-ar* relating to]

medulla oblongata (meh-DUL-ah ob-long-GAH-tah)
[*medulla* middle, *oblongata* oblong] *pl.*, medullae oblongatae (meh-DUL-ee ob-long-GAH-tee)

melatonin (mel-ah-TOH-nin)
[*mela-* black, *-ton-* tone, *-in* substance]

meninges (meh-NIN-jeez)
[*mening* membrane] *sing.*, meninx (mee-NINKS)

meningitis (men-in-JYE-tis)
[*mening-* membrane, *-itis* inflammation]

midbrain

motor program (MOH-tor)
[*mot-* movement, *-or* agent]

occipital lobe (ok-SIP-it-al)
[*occipit-* back of head, *-al* relating to]

optic chiasma (OP-tik kye-AS-mah)
[*opti-* vision, *ic* relating to, *chiasma* crossed lines (from Greek letter chi, χ)] *pl.*, chiasmata, chiasms, or chiasmas (kye-AS-mah-tah, KYE-as-zumz, kye-AS-mahz)

parietal lobe (pah-RYE-eh-tal)
[*pariet-* wall, *-al* relating to]

Parkinson disease (PD) (PAR-kin-son)
[*James Parkinson* English physician]

pia mater (PEE-ah MAH-ter)
[*pia* tender, *mater* mother]

pineal gland (PIN-ee-al)
[*pine-* pine, *-al* relating to, *gland* acorn]

pons (ponz)
[*pons* bridge] *pl.*, pontes (PON-teez)

pyramid (PEER-ah-mid)

reticular activating system (RAS) (reh-TIK-yoo-lar SIS-tem)
[*ret-* net, *-ic-* relating to, *-ul-* little, *-ar* characterized by]

sensory pathway (SEN-soh-ree PATH-way)
[*sens-* feel, *-ory* relating to]

short-term memory
[*memory* remembering]

sleep

spinal nerve (SPY-nal)
[*spin-* backbone, *-al* relating to]

spinal tract (SPY-nal)
[*spin-* backbone, *-al* relating to, *tract* trail]

subarachnoid space (sub-ah-RAK-noyd)
[*sub-* beneath, *-arachn-* spider, *-oid* like]

subdural space (sub-DOO-ral)
[*sub-* beneath, *-dura-* hard or tough, *-al* relating to]

temporal lobe (TEM-poh-ral)
[*tempor-* temple of head, *-al* relating to]

tentorium cerebelli (ten-TOR-ee-um sair-eh-BEL-lee)
[*tentorium* tent, *cerebelli* of the cerebellum (small brain)] *pl.*, tentoria cerebelli (ten-TOR-ee-ah sair-eh-BELL-eye)

thalamus (THAL-ah-muss)
[*thalamus* inner chamber] *pl.*, thalami (THAL-ah-meye)

ventral (anterior) nerve root (VEN-tral)
[*ventr-* belly, *-al* relating to]

ventricle (VEN-tri-kul)
[*ventr-* belly, *-icle* little]

vermis
[*vermis* worm] *pl.*, vermes (VER-meez)

CHAPTER SUMMARY

To download an MP3 version of the chapter summary for use with your iPod or other portable media player, access the Audio Chapter Summaries *online at http://evolve.elsevier.com.*

HINT *Scan this summary after reading the chapter to help you reinforce the key concepts. Later, use the summary as a quick review before your class or before a test.*

INTRODUCTION: THE CENTRAL NERVOUS SYSTEM

A. The central nervous system (CNS) comprises both the brain and the spinal cord and is the principal integrator of sensory input and motor output (Figure 12-1)

B. The CNS is capable of evaluating incoming information and formulating appropriate responses to changes that may threaten our body's homeostatic balance

COVERINGS OF THE BRAIN AND SPINAL CORD

A. Two protective coverings of the brain and spinal cord
1. Bone—outer covering
2. Meninges—inner covering (membranes)

B. Three distinct meningeal layers: (Figures 12-2 and 12-3)
1. Dura mater—outer layer of the meninges; inner layer of the cranium's periosteum
2. Arachnoid—delicate, spiderweb-like layer, lying between the dura mater and the innermost layer
3. Pia mater—innermost layer; adheres tightly to the outer surface of the brain and spinal cord

C. Dura mater has three important inward extensions:
1. Falx cerebri—projects downward into the longitudinal fissure (a "crack") to form a tough partition between the two cerebral hemispheres
2. Falx cerebelli—sickle-shaped extension that separates the two halves, or hemispheres, of the cerebellum
3. Tentorium cerebelli—separates the cerebellum from the cerebrum

D. Several important spaces between and around the meninges:
1. Epidural space—above the dura mater and just below the bony coverings of the spinal cord
2. Subdural space—contains a small amount of lubricating fluid and lies between the dura mater and the arachnoid mater
3. Subarachnoid space—lies underneath the arachnoid mater and outside the pia mater; contains a large amount of cerebrospinal fluid

E. The meninges of the spinal cord continue down inside the spinal cavity (Figure 12-3)
1. Filum terminale—slender filament formed by the pia mater (Figure 12-1)

F. Meningitis—infection or inflammation of the meninges

CEREBROSPINAL FLUID

A. Cerebrospinal fluid (CSF)—circulating serous fluid; protects by cushioning brain and spinal cord

B. CSF has homeostatic functions; it is monitored by the brain for changes in the internal environment, such as the concentration of carbon dioxide

C. Cerebrospinal fluid circulates within the subarachnoid space around the brain and spinal cord; also circulates within the

fluid-filled canals and cavities (ventricles) of the brain and spinal cord (Figure 12-4)
D. Ventricles—four fluid-filled spaces within the brain
1. Two large ventricles located in each hemisphere of the cerebrum
2. A smaller third ventricle lying vertically between them
3. Fourth diamond-shaped ventricle is attached to the brainstem
E. Cerebrospinal fluid forms primarily by separation of fluid from blood
1. Choroid plexuses—networks of capillaries that project from the pia mater into the ventricles; separation of CSF takes place in these networks
F. Flow of CSF is continuous through a variety of openings connecting the ventricles
G. CSF flows into the subarachnoid space, where it is absorbed by blood in the brain's venous sinuses (Figure 12-5)

SPINAL CORD

A. Structure of the spinal cord
1. Lies within the spinal cavity; oval-shaped cylinder that tapers slightly as it descends
2. Extends from the foramen magnum in the cranium to the lower border of the first lumbar vertebra (Figure 12-6)
3. Two bulges, one in the cervical region of the neck and the other within the lumbar region of the lower spine
4. Two deep grooves; a deep, wide anterior median fissure and a posterior median sulcus
5. Two bundles of nerve fibers called nerve roots project from each side of the spinal cord
   a. Dorsal (posterior) nerve root—carries information into the spinal cord
   b. Ventral (anterior) nerve root—carries motor information out of the spinal cord
6. Numerous interneurons are also located in the spinal cord's gray matter core
7. Spinal nerve—a single mixed nerve on each side of the spinal cord where the dorsal and ventral nerve roots join together
8. Cauda equina—bundle of nerve roots extending (along with the filum terminale) from the inferior end of the spinal cord (Figure 12-7)
9. Gray matter—extends the length of the cord; spinal cord's core of gray matter looks like a flat letter "H" in a cross section of the cord
   a. H-shaped rod of gray matter is made up of anterior, lateral, and posterior gray columns
   b. Viewed in a cross section, the columns forming the "H" called the anterior, posterior, and lateral gray horns
   c. The left and right gray columns are joined in the middle by a band called the gray commissure
10. White matter—surrounds the gray matter; subdivided into three funiculi: anterior, posterior, and lateral columns
    a. Funiculus—consists of a large bundle of nerve fibers (axons) divided into smaller bundles called spinal tracts (Figure 12-8)
B. Functions of the spinal cord
1. Spinal cord provides conduction routes to and from the brain
2. Also serves as the integrator, or reflex center, for all spinal reflexes
   a. Ascending tracts—conduct sensory impulses up the cord to the brain

b. Descending tracts—conduct motor impulses down the cord from the brain
   c. Tracts are composed of bundles of axons
3. Spinal cord also serves as the reflex center for all spinal reflexes where incoming sensory impulses become outgoing motor impulses
4. Spinal cord can also serve as a gateway to the receiving and sending of pain information to the brain cells

BRAIN

A. Brain—one of the largest organs in adult humans, consisting of roughly 100 billion neurons and 900 billion glia
1. Six major divisions of the brain: medulla oblongata, pons, midbrain, cerebellum, diencephalons, and cerebrum (Figure 12-9)
B. Brainstem—medulla oblongata, pons, and midbrain
1. Medulla oblongata—enlargement of the upper spinal cord located just above the foramen magnum; composed of white matter and a network of gray and white matter (reticular formation)
   a. Pyramids—two bulges of white matter located on the ventral surface of the medulla
   b. Nuclei—clusters of neuron cell bodies located within reticular formation
2. Pons—lies just above the medulla; composed largely of white matter and with some gray matter nuclei
3. Midbrain—lies above the pons and below the cerebrum; white matter (tracts) and reticular formations make up the bulk of the midbrain
   a. Cerebral peduncles—ropelike masses of white matter extending through the midbrain
   b. Corpora quadrigemina—forms the posterior, upper part of the midbrain, lying just above the cerebellum; visual and auditory centers
   c. Two other important midbrain structures are the red nucleus and the substantia nigra; each consists of clusters of cell bodies of neurons involved in muscular control
4. Functions of the brainstem—performs sensory, motor, and reflex functions
   a. Nuclei in the medulla contain a number of reflex centers; vomiting, coughing, sneezing, hiccupping, and swallowing
   b. Pons contains centers for reflexes mediated by some of the cranial nerves; help regulate respiration
   c. Midbrain, like the pons, contains reflex centers for certain cranial nerve reflexes; papillary reflexes and eye movements
C. Structure of the cerebellum—located just below the posterior portion of the cerebrum and is partially covered by it (Figure 12-11)
1. Transverse fissure separates the cerebellum from the cerebrum
2. Gray matter makes up the outer portion or cortex
3. White matter predominates in the interior regions; arbor vitae
4. Surfaces of the cerebrum have numerous grooves (sulci) and raised areas (gyri)
5. Cerebellum consists of two large lateral masses
   a. Left and right cerebellar hemispheres
   b. Vermis—central section
   c. Functions of the cerebellum—performs a variety of different functions that complement or assist the cerebrum; involve planning and coordinating skeletal muscle activity and maintaining balance in the body (Figure 12-12)

(1) Impulses from the cerebrum may trigger the action, but those from the cerebellum plan and coordinate the contractions and relaxations of the various muscles once they have begun

(2) Planning and coordination of movement may be the main functions of the cerebellum

(3) Also thought to be concerned with both exciting and inhibiting the postural reflexes; maintaining stable body positions

(4) Cerebellum coordinates incoming sensory information as much or more than it coordinates outgoing motor information

DIENCEPHALON

A. Diencephalon—located between the cerebrum and the midbrain; main structures are thalamus and hypothalamus (Figure 12-13)

1. Thalamus—dumbbell-shaped mass of gray matter made up of many nuclei
   a. Geniculate bodies—two nuclei important in processing auditory and visual input
   b. Serves as the major relay station for sensory impulses
2. Hypothalamus—lies beneath the thalamus; composed of several structures, including several nuclei and the mammillary bodies
   a. Midportion of the hypothalamus gives rise to the infundibulum
   b. Performs many functions of great importance not only for survival but also for the enjoyment of life
   c. Functions as a link between the psyche (mind) and soma (body)
   d. Links the nervous system to the endocrine system
   e. Serves as a regulator and coordinator of autonomic activities
   f. Functions as the major relay station between the cerebral cortex and the lower autonomic centers
   g. Plays an indirect but essential role in maintaining the water balance of our bodies
   h. Produces releasing hormones that are sent into the blood; controls the release of specific hormones from the anterior pituitary gland; also produces hormones secreted from axons that end in the posterior pituitary
   i. Plays an essential role in maintaining our waking state
   j. Functions as a crucial part of the mechanism for maintaining normal body temperature
3. Pineal gland—controls our biological clock; varies its secretion of the hormone melatonin (Figure 12-14)

STRUCTURES OF THE CEREBRUM

A. Cerebrum—largest and uppermost division of the brain; consists of two halves
1. Cerebral hemispheres
B. Cerebral cortex—surface of the cerebrum
1. Made up of gray matter
2. Highly convoluted (Figure 12-15)
3. Convolutions (gyri) greatly increase the surface area of the cerebrum (Figure A of Box 12-2)
4. Lying between the gyri are either shallow grooves called sulci or deeper grooves called fissures
5. Fissures divide each cerebral hemisphere into five lobes (Figure 12-15)
   a. Frontal
   b. Parietal

   c. Temporal
   d. Occipital
   e. Insula
C. Cerebral tracts and basal nuclei
1. Corpus callosum—prominent white curved structure; made of tracts that connect one hemisphere to the other (Figure 12-9)
   a. Tracts of the corpus callosum provide a vital pathway of communication between the two hemispheres (Figure 12-16)
2. Basal nuclei—islands of gray matter deep inside the white matter of each hemisphere
D. Functions of the cerebral cortex
1. Certain areas of the cortex in each hemisphere of the cerebrum engage largely in one function (cerebral localization) (Figure 12-18)
2. Locations of specific functions may move to other areas in the brain (cerebral plasticity)
3. Function of each region of the cerebral cortex depends on the structures with which it communicates
E. Sensory and motor functions of the cortex
1. Postcentral gyrus serves as a primary area for the general somatic senses (Figures 12-18 and 12-19, A)
2. Cortex contains a "somatic sensory map" of the body (Figure 12-19, A)
3. Information sent to the primary sensory areas is relayed to the various sensory association areas, as well as to other parts of the brain
4. Sensory information is compared and evaluated and the cortex integrates separate bits of information to produce whole perceptions
5. Precentral gyrus constitutes the primary somatic motor area (Figures 12-18 and 12-19, B)
6. Cerebral cortex is also involved in integrative functions; include consciousness and mental activities of all kinds
   a. Consciousness—state of awareness of one's self, one's environment, and other humans around us
      (1) Depends on the continuous excitation of cortical neurons by impulses conducted to them by a network of neurons (reticular activating system)
      (2) RAS—receives impulses from the spinal cord and relays them to the thalamus and then from the thalamus to all parts of the cerebral cortex (Figure 12-20)
      (3) Anesthetic drugs produce an altered state of consciousness
      (4) Sleep—state of seeming unconsciousness
   b. Language—Language functions consist of the ability to speak and write words and the ability to understand spoken and written words
      (1) Areas in the frontal, parietal, and temporal lobes serve as speech centers—as crucial areas, that is, for language functions
      (2) Aphasias—language defects as a result of lesions in the speech centers
   c. Emotions
      (1) Subjective experiencing and objective expression of emotions involve the functioning of the limbic system
      (2) Limbic system ("emotional brain")—functions to make us experience many kinds of emotions—anger, fear, sexual feelings, pleasure, and sorrow

(3) Limbic activity without the modulating influence of the other cortical areas may bring on attacks of abnormal, uncontrollable rage

(4) Recent research indicates that a special functional class of neurons called mirror neurons exists in the cortex

(5) Mirror neurons exhibit action potentials both when we experience something ourselves and also when we observe others having similar experiences; "mirror" the activity of another person's brain

d. Memory—one of our major mental activities

(1) Short-term memory—involves the storage of information over a few seconds or minutes

(2) Long-term memory—believed to consist of some kind of structural changes—called engrams—in the cerebral cortex; some kind of permanent change in the synapses in a specific circuit of neurons

(3) Both short-term memory and long-term memory are functions of many parts of the cerebral cortex

SOMATIC SENSORY PATHWAYS OF THE CNS

A. Somatic sensory pathways—for the cerebral cortex to perform its sensory functions, impulses must first be conducted to its sensory areas of the brain by relays of neurons (Figure 12-21)

B. Most impulses reaching the sensory areas of the cerebral cortex have traveled over at least three pools of sensory neurons

1. Primary sensory neurons—conduct impulses from the periphery (e.g., skin) of our bodies to the CNS

2. Secondary sensory neurons—conduct these impulses from the spinal cord or brainstem up to the thalamus

3. Tertiary sensory neurons—conduct impulses from the thalamus to the sensory area of the parietal lobe

a. Thalamocortical tracts—bundles of axons of tertiary sensory neurons that extend to the white matter of the cerebral cortex

C. Sensory pathways to the cerebral cortex are crossed

SOMATIC MOTOR PATHWAYS OF THE CNS

A. Somatic motor pathways—consist of motor neurons that conduct impulses from the CNS to somatic effectors—the skeletal muscles

1. Motor pathways are extremely complex and not at all clearly defined

2. Only one final common pathway conducts impulses to a specific motor unit within a skeletal muscle

3. Numerous somatic motor pathways conduct impulses from motor areas of the cerebrum down to the anterior horn motor neurons at all levels of the spinal cord

4. Motor program—set of coordinated commands that control the programmed muscle activity mediated by extrapyramidal pathways

REVIEW QUESTIONS

Write out the answers to these questions after reading the chapter and reviewing the Chapter Summary. If you simply think through the answer without writing it down, you won't retain much of your new learning.

HINT

1. What term refers to the membranous covering of the brain and cord? What three layers compose this covering?
2. What are the large fluid-filled spaces within the brain called? How many are there? What do they contain?
3. Describe the formation and reabsorption of cerebrospinal fluid.
4. Describe the spinal cord's structure and general functions.
5. List the major components of the brainstem, and identify their general functions.
6. Describe the general functions of the cerebellum.
7. Describe the general functions of the thalamus.
8. Describe the general functions of the hypothalamus.
9. Describe the general functions of the cerebrum.
10. What general functions does the cerebral cortex perform?
11. Define consciousness.
12. Name some altered states of consciousness.

CRITICAL THINKING QUESTIONS

After finishing the Review Questions, write out the answers to these items to help you apply your new knowledge. Go back to sections of the chapter that relate to items that you find difficult.

HINT

1. Explain what the term *reflex center* means. Is an interneuron necessary for a reflex center? Explain your answer.
2. Explain briefly what is meant by the arousal or alerting mechanism.
3. If a researcher discovered that a substantial reduction in neurotransmitter concentration caused difficulty in forming memory, what theory of memory formation would be refuted?
4. A person having an absence of any reflex and a person having exaggerated deep tendon reflexes are showing signs of different motor pathway injuries. Using these symptoms as the basis, explain the motor pathway that was damaged and what other symptoms each person might have.
5. A patient with a brain infection can be diagnosed by culturing cerebrospinal fluid. The greatest concentration of the disease-causing organism can be drawn from the fluid as soon as it leaves the brain at the level of the third or fourth vertebra. Why would this be an unwise place to take the sample? Where would a better location be? Explain your answer.

continued from page 247

Remember Polai and her headache from the Introductory Story? See if you can answer the following questions about her lumbar puncture now that you have read this chapter.

1. Where exactly will the cerebrospinal fluid be drawn from?
   a. Epidural space
   b. Subdural space
   c. Epiarachnoid space
   d. Subarachnoid space

2. Use Figure 12-5 (p. 250) to help you trace the flow of CSF from its origin to the site of the lumbar puncture:
   a. Subarachnoid space, median foramen of brainstem, choroid plexus of fourth ventricle
   b. Choroid plexus of lateral ventricle, subarachnoid space, cerebral aqueduct
   c. Choroid plexus of the lateral ventricle, cerebral aqueduct, subarachnoid space
   d. Cerebral aqueduct, epidural space, interventricular foramen

One possible (though rare) complication of bacterial meningitis (inflammation of the meninges) could be a spread of bacteria to the cerebellum.

3. Which of the following is NOT a function of the cerebellum?
   a. Coordinates skilled movement
   b. Interprets language
   c. Helps control posture
   d. Controls balance

Bleeding of the meninges can irritate tissues, which triggers inflammation of the meninges—meningitis.

4. Which of these CSF samples taken by lumbar puncture could be Polai's?

To solve a case study, you may have to refer to the glossary or index, other chapters in this textbook, A&P Connect, Mechanisms of Disease, and other resources.

HINT

UNIT 3

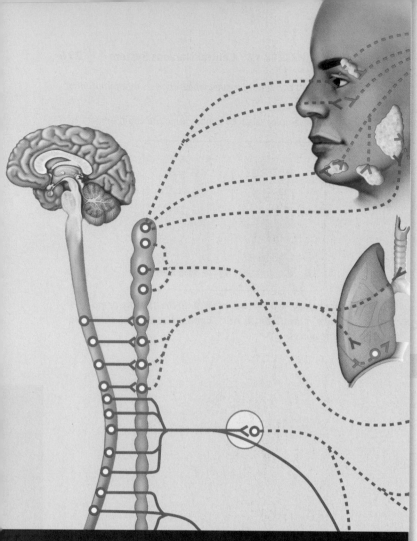

CHAPTER 13

Peripheral Nervous System

STUDENT LEARNING OBJECTIVES

At the completion of this chapter, you should be able to do the following:

1. Discuss the basic structure of a spinal nerve.
2. Outline the *basic plan* of the brachial plexus, the lumbar plexus, and the sacral plexus.
3. List the 12 cranial nerves and their basic roles in the body.
4. Outline the divisions of the peripheral nervous system.
5. Discuss the significance of the somatic motor nervous system.
6. Discuss the significance and role of reflex arcs in the body.
7. List the basic functions of the autonomic nervous system (ANS).
8. Describe the basic structures and functions of the autonomic pathways.
9. Describe the basic structures and functions of the sympathetic pathways.
10. Describe the structures and functions of the parasympathetic pathways.
11. List several of the neurotransmitters of the autonomic nervous system.
12. Describe the various functions of the sympathetic division.
13. Describe the various functions of the parasympathetic division.

CHAPTER OUTLINE

HINT *Scan this outline before you begin to read the chapter, as a preview of how the concepts are organized.*

THE store was only two blocks away. Tomasina was in such a hurry, and it was so close, she didn't think she really needed to rummage through the garage for her bike helmet. She jumped on her bike and peddled quickly toward the store. From out of nowhere, a car sped by—too close to the side of the road—and clipped Tomasina with the side mirror, knocking her to the ground. She hit her head sharply on the curb. When she regained consciousness, Tomasina was in the hospital. She was told some tests would be conducted to see whether her cranial nerves had been affected by the head injury.

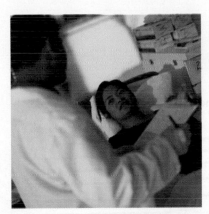

Do you think her injuries could be life threatening? After you read this chapter you should have a clearer picture.

* * *

We are now ready to explore the nerve pathways that lead to and from the CNS. Together these nerve pathways make up the **peripheral nervous system (PNS).** The PNS comprises 31 pairs of spinal nerves that emerge from the spinal cord, the 12 pairs of cranial nerves that emerge from the brain or skull, and all of the smaller nerves that branch from these primary nerves.

Before we begin, remember that *afferent fibers* carry information *into* the CNS and help us maintain homeostasis by sensing changes in our internal or external environments. Some afferent fibers (belonging to the somatic sensory or special sensory nervous system) feed back information regarding changes detected by receptors in the skin, skeletal muscles, and special sense organs. However, other afferent fibers (belonging to the autonomic nervous system) feed back information regarding the effects of autonomic control of the viscera.

Recall also that *efferent fibers* carry information *away* from the CNS. Efferent fibers belonging to the somatic nervous system (SNS) regulate skeletal muscles, allowing us to defend ourselves, gather food, and perform other essential tasks. However, efferent fibers belonging to the autonomic nervous system (ANS) allow us to control our internal environment by regulating the activities of smooth and cardiac muscles and glands.

We will first explore concepts regarding the structure and function of the spinal and cranial nerves. Following this, we will discuss the peripheral elements of the SNS and the ANS.

SPINAL NERVES

Thirty-one pairs of **spinal nerves** (designated by *number,* not names) connect directly to the spinal cord. There are eight cervical nerve pairs (C1-C8), twelve thoracic nerve pairs (T1-T12), five lumbar nerve pairs (L1-L5), five sacral nerve pairs (S1-S5), and one coccygeal pair of spinal nerves

LANGUAGE OF SCIENCE AND MEDICINE

HINT *Before reading the chapter, say each of these terms out loud. This will help you avoid stumbling over them as you read.*

abdominal reflex (ab-DOM-ih-nal REE-fleks)
[*abdomin-* belly, *-al* relating to, *re-* again, *-flex* bend]

abducens nerve (ab-DOO-sens)
[*ab-* away, *-duc-* lead, *-ens* process]

accessory nerve (ak-SES-oh-ree)

acetylcholine (ACh) (ass-ee-til-KOH-leen)
[*acetyl-* vinegar, *-chol-* bile, *-ine* made of]

ankle jerk reflex (REE-fleks)
[*re-* again, *-flex* bend]

autonomic nervous system (ANS) (aw-toh-NOM-ik SIS-tem)
[*auto-* self, *-nomo-* law, *-ic* relating to]

autonomic (visceral) reflex (aw-toh-NOM-ik REE-fleks)
[*auto-* self, *-nomo-* law, *-ic* relating to, *re-* again, *-flex* bend]

Babinski reflex (bah-BIN-skee REE-fleks)
[*Joseph F.F. Babinski* French neurologist, *re-* again, *-flex* bend]

brachial plexus (BRAY-kee-al PLEK-sus)
[*brachi-* arm, *-al* relating to, *plexus* network] *pl.,* plexi or plexuses (PLEKS-eye, PLEK-suh-sez)

cervical plexus (SER-vih-kal PLEK-sus)
[*cervic-* neck, *-al* relating to, *plexus* network] *pl.,* plexi or plexuses (PLEKS-eye, PLEK-suh-sez)

corneal reflex (KOR-nee-al REE-fleks)
[*corn-* horn, *-al* relating to, *re-* again, *-flex* bend]

cranial nerve (KRAY-nee-al)
[*crani-* skull, *-al* relating to]

dorsal ramus (DOR-sal RAY-mus)
[*dors-* the back, *-al* relating to, *ramus* branch] *pl.,* rami (RAYM-eye)

dorsal root (DOR-sal root)
[*dors-* the back, *-al* relating to]

dorsal root ganglion (DOR-sal root GANG-glee-on)
[*dors-* the back, *-al* relating to, *gangli-* knot, *-on* unit] *pl.,* ganglia (GANG-glee-ah)

facial nerve (FAY-shal)
[*faci-* face, *-al* relating to]

"fight-or-flight" reaction

glossopharyngeal nerve (glos-oh-fah-RIN-jee-al)
[*glosso-* tongue, *-pharyng-* throat, *-al* relating to]

hypoglossal nerve (hye-poh-GLOSS-al)
[*hypo-* below, *-gloss-* tongue, *-al* relating to]

knee jerk reflex
[*re-* again, *-flex* bend]

lumbar plexus (LUM-bar PLEK-sus)
[*lumb-* loin, *-ar* relating to, *plexus* network] *pl.,* plexi or plexuses (PLEKS-eye, PLEK-suh-sez)

mixed cranial nerve (KRAY-nee-al)
[*crani-* skull, *-al* relating to]

continued on page 292

(Figure 13-1). The first pair of cervical nerves emerges from the spinal cord in the space above the first cervical vertebra. The eighth pair of cervical nerves emerges between the last cervical vertebra and the first thoracic vertebra. All the thoracic nerves erupt horizontally through the *intervertebral foramina* below their respective vertebrae. In contrast, lumbar, sacral, and coccygeal nerve roots at first *descend* from their point of origin at the lower end of the spinal cord. They then reach the intervertebral foramina below their respective vertebrae, through which the nerves emerge. This arrangement gives the lower end of the spinal cord, with its attached spinal nerve roots, the appearance of a horse's tail (hence the name, *cauda equina*).

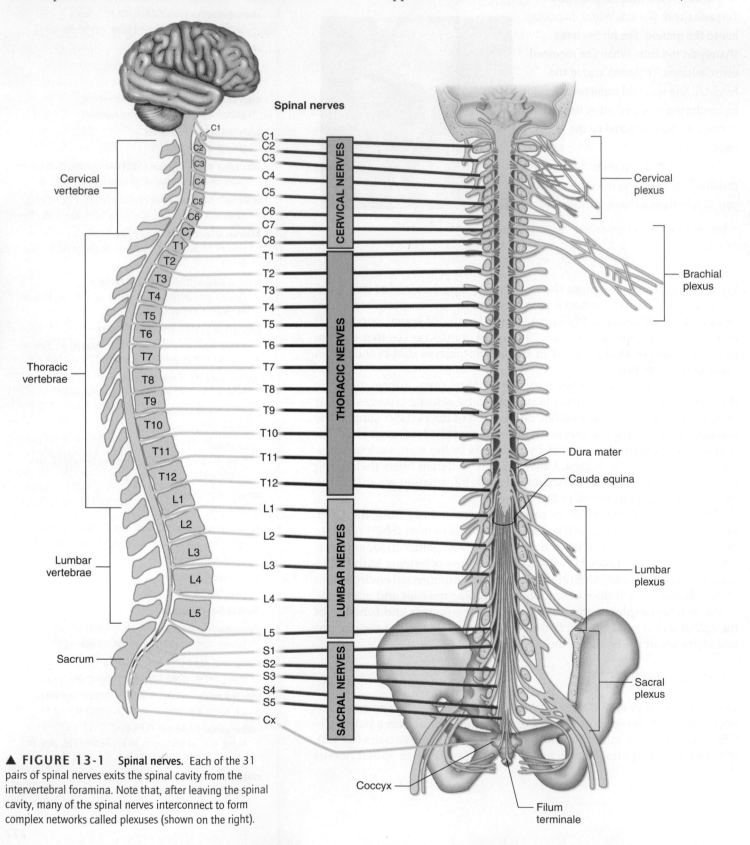

▲ FIGURE 13-1 Spinal nerves. Each of the 31 pairs of spinal nerves exits the spinal cavity from the intervertebral foramina. Note that, after leaving the spinal cavity, many of the spinal nerves interconnect to form complex networks called plexuses (shown on the right).

As we have seen in Chapter 12, each spinal nerve attaches to the spinal cord by two short roots, a **ventral** (anterior) **root** and a **dorsal** (posterior) **root.** The dorsal root of each spinal nerve is easily recognized by a swelling called the **dorsal root ganglion** (Figure 12-3 on p. 249). You can see that the roots and dorsal ganglia lie *within* the spinal cavity.

The ventral root includes motor neurons that carry information from the CNS toward effectors (e.g., muscles and glands). In each somatic motor pathway, a single motor fiber stretches from the anterior gray horn of the spinal cord, through the ventral root, and on through the spinal nerve toward a skeletal muscle. Autonomic fibers, which also carry motor information toward effectors, may also pass through the ventral root to become part of a spinal nerve.

The dorsal root of each spinal nerve includes sensory fibers that carry information from receptors in the peripheral nerves. The dorsal root ganglion contains the cell bodies of the sensory neurons. Because all spinal nerves contain *both* motor and sensory fibers, they are called **mixed nerves.**

Soon after a spinal nerve emerges from the spinal cavity, it forms two branches called **rami** (singular, *ramus*), specifically a **dorsal ramus** and a **ventral ramus** (Figure 13-2). The *dorsal ramus* supplies somatic motor and sensory fibers to several smaller nerves. These smaller nerves, in turn, innervate the muscles and skin of the *posterior surface* of the head, neck, and trunk.

The structure of the *ventral ramus* is a little more complex. Autonomic motor fibers split away from the ventral ramus, heading toward a ganglion of the *sympathetic chain.* There, some of the autonomic fibers synapse with autonomic neurons that eventually continue on to autonomic effectors by way of *splanchnic nerves.* However, some fibers synapse with autonomic neurons whose fibers *rejoin* the ventral ramus. The two thin rami formed by this splitting away,

then rejoining of autonomic fibers, are together called the *sympathetic rami.* Motor (autonomic and somatic) and sensory fibers of the ventral rami innervate muscles and glands in the extremities (arms and legs) and in the lateral and ventral portions of the neck and trunk.

Nerve Plexuses

The ventral rami of many spinal nerves (except nerves T2-T12) subdivide to form braided networks called **plexuses.** Nearly 100 plexuses exist in the human body, but there are only four major pairs: the *cervical plexus*, the *brachial plexus*, the *lumbar plexus*, and the *sacral plexus* (see Figure 13-1). Note that the name of the plexus is derived from the destination of each nerve. For example, the ventral rami of the first four cervical spinal nerves (C1-C4) exchange fibers in the *cervical plexus* found deep in the neck (see discussion below). In each of these, fibers of several different rami join together to form individual nerves. Each individual nerve that emerges from a plexus contains all the fibers that innervate a particular region of the body (Table 13-1). In a way, each braidlike plexus reduces the number of nerves needed to supply each body part.

Cervical Plexus

The **cervical plexus** (see Figure 13-1) is found deep within the neck. Ventral rami of the first four cervical spinal nerves (C1-C4), along with a branch of the ventral ramus of C5, exchange fibers in the cervical plexus. Also included in the upper part of this plexus are extensions from cranial nerves XI and XII. Individual nerves emerging from this plexus innervate the muscles and skin of the neck, upper shoulders, and part of the head.

Also exiting this plexus is the vitally important **phrenic nerve,** which innervates the diaphragm and thus directly regulates our breathing. Any disease or injury that damages the spinal cord between the third and fifth cervical segments may paralyze the phrenic nerve, paralyzing the diaphragm and causing death by asphyxiation. Two cranial nerves, the accessory nerve (XI) and the hypoglossal nerve (XII), receive small branches emerging from the cervical plexus.

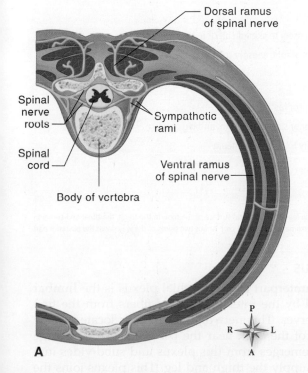

Dorsal ramus of spinal nerve

Spinal nerve roots

Sympathetic rami

Spinal cord

Body of vertebra

Ventral ramus of spinal nerve

A

P
R · L
A

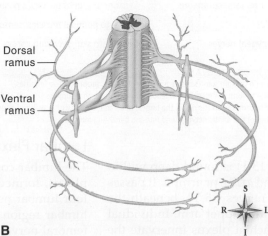

Dorsal ramus

Ventral ramus

B

S
R · L
I

◀ **FIGURE 13-2 Rami of the spinal nerves.** Note that ventral and dorsal roots join to form a spinal nerve. The spinal nerve then splits into a dorsal *ramus* (plural, *rami*) and ventral *ramus*. The ventral ramus communicates with a chain of sympathetic (autonomic) ganglia by way of a pair of thin sympathetic rami. **A,** Superior view of a pair of thoracic spinal nerves. **B,** Anterior view of several pairs of thoracic spinal nerves.

UNIT 3

TABLE 13-1 Spinal Nerves and Peripheral Branches

SPINAL NERVES	PLEXUSES FORMED FROM ANTERIOR RAMI	SPINAL NERVE BRANCHES FROM PLEXUSES		BODY PARTS SUPPLIED
Cervical 1 2 3 4	Cervical plexus	Lesser occipital Greater auricular Cutaneous nerve of neck Supraclavicular nerves Branches to muscles		Sensory to back of head, front of neck, and upper part of shoulder; motor to numerous neck muscles
Cervical 5 6 7 8	Brachial plexus	Phrenic		Diaphragm
		Suprascapular and dorsoscapular		Superficial muscles* of scapula
		Thoracic nerves, medial and lateral branches		Pectoralis major and minor
Thoracic (or Dorsal) 1 2 3 4 5 6 7 8 9 10 11 12	No plexus formed; branches run directly to intercostal muscles and skin of thorax	Long thoracic		Serratus anterior
		Thoracodorsal		Latissimus dorsi
		Subscapular		Subscapular and teres major muscles
		Axillary (circumflex)		Deltoid and teres minor muscles and skin over deltoid
		Musculocutaneous		Muscles of front of arm (biceps brachii, coracobrachialis, brachialis) and skin on outer side of forearm
		Ulnar		Flexor carpi ulnaris and part of flexor digitorum profundus; some muscles of hand; sensory to medial side of hand, little finger, and medial half of fourth finger
		Median		Rest of muscles of front of forearm and hand; sensory to skin of palmar surface of thumb, index, and middle fingers
		Radial		Triceps muscle and muscles of back of forearm; sensory to skin of back of forearm and hand
		Medial cutaneous		Sensory to inner surface of arm and forearm
Lumbar 1 2 3 4 5	Lumbosacral plexus	Iliohypogastric	Sometimes fused	Sensory to anterior abdominal wall
		Ilioinguinal		Sensory to anterior abdominal wall and external genitalia; motor to muscles of abdominal wall
		Genitofemoral		Sensory to skin of external genitalia and inguinal region
		Lateral femoral cutaneous		Sensory to outer side of thigh
Sacral 1 2 3 4 5		Femoral		Motor to quadriceps, sartorius, and iliacus muscles; sensory to front of thigh and medial side of lower leg (saphenous nerve)
		Obturator		Motor to adductor muscles of thigh
		Tibial† (medial popliteal)		Motor to muscles of calf of leg; sensory to skin of calf of leg and sole of foot
		Common peroneal (lateral popliteal)		Motor to evertors and dorsiflexors of foot; sensory to lateral surface of leg and dorsal surface of foot
		Nerves to hamstring muscles		Motor to muscles of back of thigh
		Gluteal nerves		Motor to buttock muscles and tensor fasciae latae
		Posterior femoral cutaneous		Sensory to skin of buttocks, posterior surface of thigh, and leg
		Pudendal		Motor to perineal muscles; sensory to skin of perineum
Coccygeal 1	Coccygeal plexus	Anococcygeal nerves		Sensory to skin overlying coccyx

*Although nerves to muscles are considered motor, they do contain some sensory fibers that transmit proprioceptive impulses.

†Sensory fibers from the tibial and peroneal nerves unite to form the medial cutaneous (or sural) nerve that supplies the calf of the leg and the lateral surface of the foot. In the thigh, the tibial and common peroneal nerves are usually enclosed in a single sheath to form the sciatic nerve, the largest nerve in the body with a width of approximately ¾ inch (2 cm). About two thirds of the way down the posterior part of the thigh, it divides into its component parts. Branches of the sciatic nerve extend into the hamstring muscles.

Brachial Plexus

The **brachial plexus** (see Figure 13-1) is found deep within the shoulder, between the neck and *axilla,* or armpit. It passes from the ventral rami of spinal nerves C5-T1, beneath the *clavicle* (collarbone), and toward the upper arm. Individual nerves that emerge from the brachial plexus innervate the lower part of the shoulder and the entire arm.

Lumbar Plexus

The lumbar counterpart of the brachial plexus is the **lumbar plexus,** formed by the intermingling of fibers from the first four lumbar nerves. This network of nerves is located in the lumbar region of the back near the psoas muscle. The large femoral nerve emerges from this plexus and subdivides into branches that supply the thigh and leg. This plexus joins the sacral plexus and is sometimes discussed as the *lumbosacral plexus* (see Figure 13-1).

Sacral Plexus and Coccygeal Plexus

Fibers from L4 and L5 and S1 and S4 form the **sacral plexus,** which lies in the pelvic cavity. Among other nerves that emerge from the sacral plexus are the tibial and common peroneal nerves. In the thigh, these form the largest nerve in the body, the great *sciatic nerve*, which pierces the buttock and runs down the back of the thigh. Its many branches supply nearly all the skin of the leg, the posterior thigh muscles, and the leg and foot muscles. (*Sciatica*, or neuralgia of the sciatic nerve, is a fairly common and very painful condition, in which the patient feels intermittent burning, tingling, or piercing pain running down the back of the leg to the feet.)

The last sacral spinal nerve (S5), along with a few fibers from S4, joins with the coccygeal nerve. Together they form a small *coccygeal plexus*, responsible for innervating the floor of the pelvic cavity.

> **QUICK CHECK**
> 1. How many spinal nerves are there? How are they named?
> 2. What is the basic structure of a spinal nerve?
> 3. What is a nerve plexus? List the larger plexuses.

CRANIAL NERVES

Twelve pairs of **cranial nerves** emerge mainly from the undersurface of the brainstem (Figure 13-3). These nerves pass through small *foramina* (holes) in the cranial cavity of

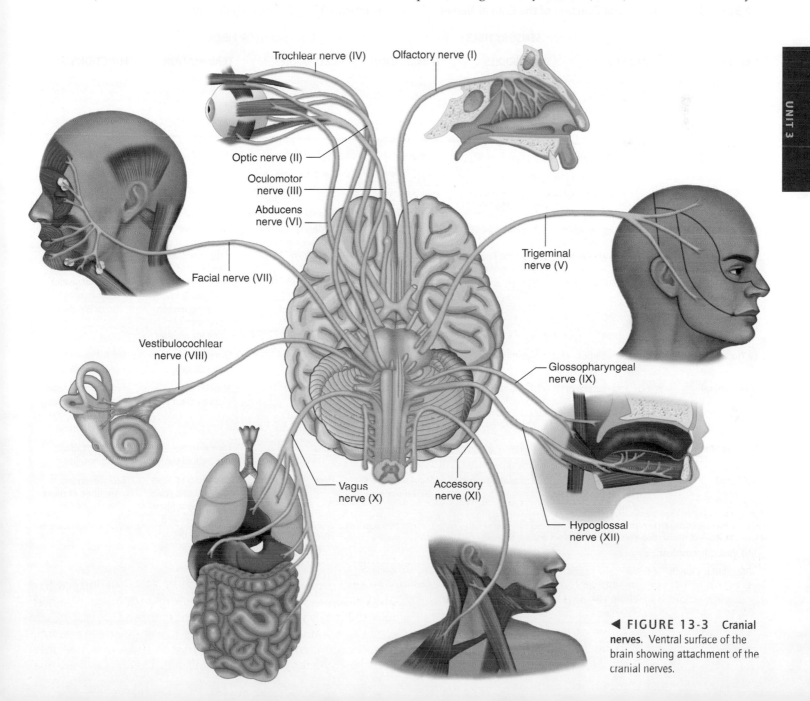

Trochlear nerve (IV)

Olfactory nerve (I)

Optic nerve (II)

Oculomotor nerve (III)

Abducens nerve (VI)

Facial nerve (VII)

Trigeminal nerve (V)

Vestibulocochlear nerve (VIII)

Glossopharyngeal nerve (IX)

Vagus nerve (X)

Accessory nerve (XI)

Hypoglossal nerve (XII)

◀ **FIGURE 13-3 Cranial nerves.** Ventral surface of the brain showing attachment of the cranial nerves.

UNIT 3

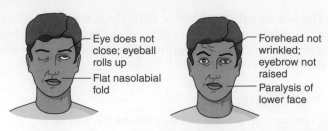

▲ FIGURE 13-4   Signs of damage to the facial nerve (VII).

Eye does not close; eyeball rolls up
Flat nasolabial fold

Forehead not wrinkled; eyebrow not raised
Paralysis of lower face

the skull. We use both names and numbers (sometimes together) to identify the cranial nerves by their functions or position. Like all nerves, cranial nerves are made from bundles of axons. **Mixed cranial nerves** have axons from both sensory and motor neurons, **sensory cranial nerves** have sensory axons only, and **motor cranial nerves** have mostly motor axons (but also have a few sensory fibers as well).

Severe head injuries often damage one or more of the cranial nerves, producing symptoms that reflect loss of the functions mediated by the affected nerve or nerves. For example, injury of the sixth cranial nerve causes the eye to turn in because of paralysis of the abducting muscle of the eye. Injury of the eighth cranial nerve produces deafness. Injury to the facial nerve results in a poker-faced expression and a drooping of the corner of the mouth from paralysis of the facial muscles (Figure 13-4).

The names, numbers, and functions of the cranial nerves are summarized in Table 13-2.

TABLE 13-2   Structure and Function of the Cranial Nerves   ☐ sensory (afferent)   ☐ motor (efferent)

| NERVE | SENSORY FIBERS | | | MOTOR FIBERS | | FUNCTIONS |
	RECEPTORS	CELL BODIES	TERMINATION	CELL BODIES	TERMINATION	
I Olfactory	Nasal mucosa	Nasal mucosa	Olfactory bulbs (new relay of neurons to olfactory cortex)			Sense of smell
II Optic	Retina (proprioceptive)	Retina	Nucleus in thalamus (lateral geniculate); some fibers terminate in superior colliculus of midbrain			Vision
III Oculomotor	External eye muscles except superior oblique and lateral rectus	Trigeminal ganglion	Midbrain (oculomotor nucleus)	Midbrain (oculomotor nucleus)	External eye muscles except superior oblique and lateral rectus; autonomic fibers terminate in ciliary ganglion and then to ciliary and iris muscles	Eye movements, regulation of size of pupil, accommodation (for near vision), proprioception (muscle sense)
IV Trochlear	Superior oblique (proprioceptive)	Trigeminal ganglion	Midbrain	Midbrain	Superior oblique muscle of eye	Eye movements, proprioception
V Trigeminal	Skin and mucosa of head, teeth	Trigeminal ganglion	Pons (sensory nucleus)	Pons (motor nucleus)	Muscles of mastication	Sensations of head and face, chewing movements, proprioception
VI Abducens	Lateral rectus (proprioceptive)	Trigeminal ganglion	Pons	Pons	Lateral rectus muscle of eye	Abduction of eye, proprioception
VII Facial	Taste buds of anterior two thirds of tongue	Geniculate ganglion	Medulla (nucleus solitarius)	Pons	Superficial muscles of face and scalp; autonomic fibers to salivary and lacrimal glands	Facial expressions, secretion of saliva and tears, taste
VIII Vestibulocochlear						
Vestibular branch	Semicircular canals and vestibule (utricle and saccule)	Vestibular ganglion	Pons and medulla (vestibular nuclei)			Balance or equilibrium sense

4. List the names and numbers of the 12 pairs of cranial nerves.
5. Identify the primary function of each pair of cranial nerves.
6. Distinguish between a motor nerve, a sensory nerve, and a mixed nerve.

DIVISIONS OF THE PERIPHERAL NERVOUS SYSTEM

Earlier we learned that the PNS includes all the nervous pathways *outside* the brain and spinal cord. Thus, the entire PNS comprises the fibers present in the cranial nerves, the spinal nerves, and all their individual branches. As we've seen, many of these nerves are *mixed nerves* containing both sensory and motor fibers. However, it is often convenient to consider the PNS as having two *functional divisions:* the sensory (afferent) division and the motor (efferent) division. We will study the details of the sensory division in the next chapter. At this point in our discussion, we will concentrate on some of the essential details of the somatic motor division of the PNS.

Somatic Motor Nervous System

The somatic motor nervous system includes all the voluntary motor pathways *outside* the CNS. This means it involves the peripheral pathways to the skeletal muscles, which are

TABLE 13-2 Structure and Function of the Cranial Nerves—cont'd

| NERVE | SENSORY FIBERS | | | MOTOR FIBERS | | |
	RECEPTORS	CELL BODIES	TERMINATION	CELL BODIES	TERMINATION	FUNCTIONS
IX Glossopharyngeal	Pharynx; taste buds and other receptors of posterior one third of tongue	Jugular and petrous ganglia	Medulla (nucleus solitarius)	Medulla (nucleus ambiguus)	Muscles of pharynx	Sensations of tongue, swallowing movements, secretion of saliva, aid in reflex control of blood pressure and respiration
	Carotid sinus and carotid body	Jugular and petrous ganglia	Medulla (respiratory and vasomotor centers)	Medulla at junction of pons (nucleus salivatorius)	Otic ganglion and then to parotid salivary gland	
X Vagus	Pharynx, larynx, carotid body, thoracic and abdominal viscera	Jugular and nodose ganglia	Medulla (nucleus solitarius), pons (nucleus of fifth cranial nerve)	Medulla (dorsal motor nucleus)	Ganglia of vagal plexus and then to muscles of pharynx, larynx, and autonomic fibers to thoracic and abdominal viscera	Sensations and movements of organs supplied; e.g., slows heart, increases peristalsis, contracts muscles for voice production
XI Accessory	Trapezius and sterno-cleidomastoid (proprioceptive)	Upper, cervical ganglia	Spinal cord	*Some people* have fibers originating in the medulla (dorsal motor nucleus of vagus and nucleus ambiguus); *most people* have fibers originating in the upper cervical spinal cord and none from the brain)	Muscles of thoracic and abdominal viscera (autonomic), pharynx, larynx, trapezius, and sterno-cleidomastoid muscle	Shoulder movements, turning movements of head, movements of viscera, voice production, proprioception
XII Hypoglossal	Tongue muscles (proprioceptive)	Trigeminal ganglion	Medulla (hypoglossal nucleus)	Medulla (hypoglossal nucleus)	Muscles of tongue and throat	Tongue movements, proprioception

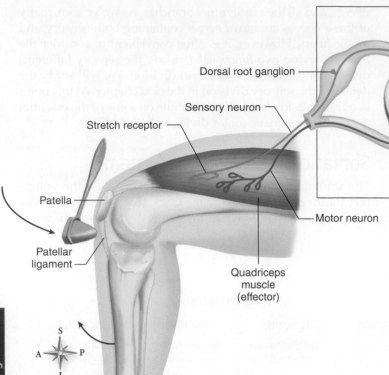

▲ **FIGURE 13-5** **Patellar reflex.** Neural pathway involved in the patellar (knee jerk) reflex.

axon extends from the anterior gray horn, through the ventral nerve root, and out to a skeletal muscle. The axon of the last somatic motor neuron in a somatic pathway always stimulates its effector cells with the neurotransmitter *acetylcholine* (see Chapter 11, p. 239).

Somatic Reflexes

A **reflex** (see Figures 11-7 and 11-8) is a *predictable response* to a stimulus. It may or may not be conscious. A *cranial reflex* has its center in the brain and a *spinal reflex* has its center in the spinal cord. There are two further classifications. **Somatic reflexes** involve contractions of skeletal muscles due to somatic motor neurons. **Autonomic (visceral) reflexes** involve contractions of smooth *or* cardiac muscle *or* secretion from glands. At this point, we will describe only somatic reflexes.

Somatic Reflexes of Clinical Importance

Understanding reflexes is important because their malfunction can indicate certain disease states. There are several common reflexes often tested by physicians: *knee jerk, ankle jerk, Babinski reflex, corneal reflex,* and *abdominal reflex* are good examples.

The well-known **knee jerk reflex,** or *patellar reflex,* results in a rapid extension of the lower leg when the patellar ligament is tapped with a rubber mallet (Figure 13-6, *A*). The administered tap stretches the ligament and stretches the quadriceps femoris muscles. This action

the *somatic effectors.* If you remember, all these pathways operate according to the principle of *final common pathway.* This means that all the somatic motor pathways involve a *single motor neuron* whose axon stretches from the cell body in the CNS to the effector (e.g., a muscle) innervated by that neuron. If the neuron originates in the spinal cord, then the

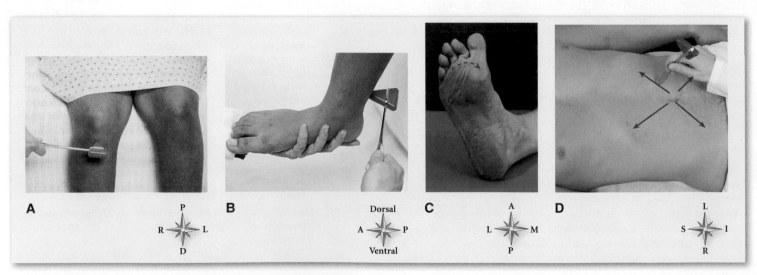

▲ **FIGURE 13-6** **Common clinical reflex tests.** **A,** Knee jerk reflex. **B,** Ankle jerk reflex. **C,** Plantar reflex showing the Babinski sign (*arrow* shows path of stimulation along plantar surface). **D,** Abdominal reflexes (one of several ways to test this type of reflex). As an object is moved away from the umbilicus and toward the side, the rectus abdominis muscles pull the umbilicus toward the stroked side.

stimulates muscle spindles (receptors) in the muscle and starts the two-neuron reflex arc (Figure 13-5). A knee reflex test is interpreted by a physician based on what should be a normal reflex.

The **ankle jerk reflex** (*Achilles reflex*) causes an extension of the foot in response to tapping of the Achilles tendon (Figure 13-6, *B*). Like the knee jerk reflex, this is a two-neuron spinal reflex arc. The **Babinski reflex** creates an extension or pointing upward of the great toe, often with a fanning of the other toes, in response to stimulation of the outer margin of the sole of the foot. Normal infants, until about 1½ years of age, exhibit this Babinski reflex, but after this age such a reflex is considered to be abnormal (Figure 13-6, *C*). The normal response after this age is the **plantar reflex,** which consists of a slight turning in and flexion of the anterior part of the foot.

Additional common reflexes include the **corneal reflex**—blinking in response to touching the cornea—and the **abdominal reflex**—moving the umbilicus (belly button) in response to stroking the side of the abdomen (Figure 13-6, *D*). The abdominal reflex often disappears during pregnancy.

Autonomic Nervous System

The **autonomic nervous system (ANS)** is a subdivision of the nervous system that *regulates involuntary effectors* by *efferent* signals. These effectors can be cardiac muscle, smooth muscle, and glandular epithelial tissue (Box 13-1 and Table 13-3). The major functions of the ANS are to regulate heartbeat, smooth muscle contraction, and glandular secretion to maintain homeostasis. Although the ANS also includes sensory (afferent) pathways that provide the feedback necessary to regulate effectors, we will emphasize the motor (efferent) pathways in this chapter.

The motor pathways of the ANS comprise two efferent divisions: the **sympathetic division** and the **parasympathetic division.** Usually the effects of the two systems are *antagonistic* to one another. This means that one system inhibits the effector and the other stimulates the effector. An antagonistic system allows for a *dually innervated* effector, such as the heart, to be controlled with great precision, just as the accelerator and brake pedals alternate to control the speed of your car.

The sympathetic division of the ANS prepares us for physical activity. It does so by increasing our heart rate and blood pressure and by dilating the passageways of our respiratory system. It also stimulates the process of perspiration to control our rising body temperature as our activity increases. This same system also induces the release of glucose from the liver and inhibits the digestive processes. For these reasons, the sympathetic system is sometimes nicknamed the *"fight-or-flight"* system. It prepares us for physical activity, which sometimes is required to defend our lives. In contrast, the parasympathetic system stimulates activities such as digestion, defecation, and urination. It also slows down the heart

BOX 13-1	Autonomic Effector Tissues and Organs

Smooth Muscle
Blood vessels
Bronchial tubes
Stomach
Gallbladder
Intestines
Urinary bladder
Spleen
Eye (iris, ciliary muscles)
Hair follicles

Cardiac Muscle
Heart

Glandular Epithelium
Sweat glands
Lacrimal glands
Digestive glands (salivary, gastric, pancreas, liver)
Adrenal medulla

Other Tissues
Adipose tissue
Kidneys

rate and respiration. For these reasons, the parasympathetic system is sometimes called the *"rest-and-repair"* system because it maintains the body's routine functions.

Structure of the Autonomic Nervous System

Each efferent autonomic pathway, whether sympathetic or parasympathetic, is made up of *autonomic* nerves, ganglia, and plexuses. These structures, in turn, are made up of efferent autonomic neurons that conduct impulses *away* from the brainstem or spinal cord and down to the autonomic effectors. A relay of *two* autonomic neurons conducts information

TABLE 13-3 Comparison of Somatic Motor and Autonomic Pathways

FEATURE	SOMATIC MOTOR PATHWAYS	AUTONOMIC EFFERENT PATHWAYS
Direction of information flow	Efferent	Efferent
Number of neurons between CNS and effector	One (somatic motor neuron)	Two (preganglionic and postganglionic)
Myelin sheath present	Yes	Preganglionic: yes Postganglionic: no
Location of peripheral fibers	Most cranial nerves and all spinal nerves	Most cranial nerves and all spinal nerves
Effector innervated	Skeletal muscle (voluntary)	Smooth and cardiac muscle, glands, and adipose and other tissues (involuntary)
Neurotransmitter	Acetylcholine	Acetylcholine or norepinephrine

CNS, Central nervous system.

from the CNS to the autonomic effectors. The first neuron is called a **preganglionic neuron** because these neurons conduct impulses from the brainstem or spinal cord to an autonomic ganglion. Preganglionic neurons are positioned *before the ganglion*, thus they are *preganglionic*. Within an autonomic ganglion, the preganglionic neuron synapses with a second efferent neuron. Because this second neuron conducts impulses *away from* the ganglion to the effector, it is called the **postganglionic neuron.**

Look at Figure 13-7 and you will see that this plan is quite different from the efferent pathways of the somatic motor nervous system. Conduction to somatic effectors requires only *one efferent neuron*—namely, the somatic motor neuron that originates in the anterior gray horn of the spinal cord. However, as we've just seen, conduction to autonomic effectors requires a sequence of *two efferent neurons* from the CNS to the effector. A comparison of the structural features of the sympathetic and parasympathetic pathways is given for you in Table 13-4. Notice that the common neurotransmitter for both preganglionic and postganglionic neurons is acetylcholine.

Structure of the Sympathetic (Thoracolumbar) Pathways

Sympathetic preganglionic neurons begin within the gray matter of the spinal cord, where their cell bodies and dendrites are located within the *thoracic and lumbar segments* (hence, "thoracolumbar" pathway) (see Figure 13-7). The axons of these neurons leave the spinal cord via the *ventral roots* of spinal nerves located in the first thoracic through the second lumbar segments. The axons then leave the spinal nerves and enter one of a *chain of paravertebral ganglia* that extend longitudinally and alongside the vertebral column. The linked ganglia are often referred to as the *sympathetic chain ganglia*, or the **sympathetic trunk.**

Preganglionic neurons that pass through chain ganglia without synapsing continue on through **splanchnic nerves** to other sympathetic ganglia. Inside these ganglia, a synapse forms between a preganglionic fiber and a second neuron. The axon of this second, postganglionic fiber, then usually returns to a spinal nerve and extends it to a visceral effector (Figure 13-8).

► **FIGURE 13-7** **Autonomic conduction pathways.** The left side of the diagram shows that one somatic motor neuron conducts impulses all the way from the spinal cord to a somatic effector. Conduction from the spinal cord to any visceral effector, however, requires a relay of at least two autonomic motor neurons—a preganglionic and a postganglionic neuron, shown on the right side of the diagram.

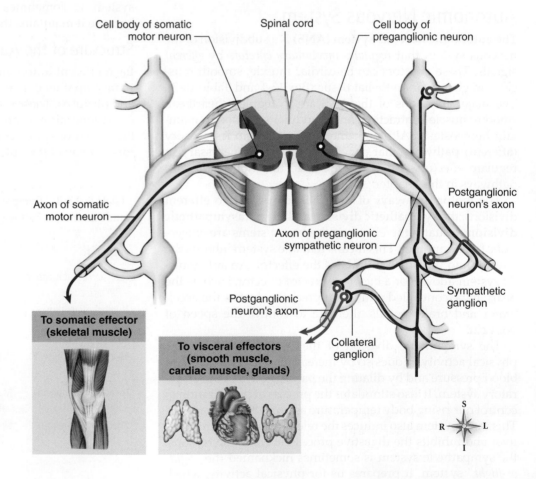

Sympathetic postganglionic neurons have their dendrites and cell bodies in the sympathetic chain ganglia. But some postganglionic axons return to a spinal nerve as unmyelinated fibers. Once in the spinal nerve, the postganglionic fibers are distributed with other nerve fibers to the various sympathetic effectors. The axon of any one sympathetic preganglionic neuron synapses with many postganglionic neurons, and these frequently terminate in widely separated organs of the body. This anatomical fact partially explains a well-known physiological principle: sympathetic responses are usually widespread, involving many organs, not just one.

Structure of the Parasympathetic (Craniosacral) Pathways

Parasympathetic preganglionic neurons have their cell bodies in *nuclei* in the *brainstem* or in the lateral gray columns of the *sacral spinal cord* (hence "craniosacral" pathway). Axons of parasympathetic preganglionic neurons are found in cranial nerves III, VII, IX, and X and in some pelvic nerves. However, at least 75% of all parasympathetic preganglionic fibers travel in the *vagus nerve (X)* before synapsing with postganglionic fibers in **terminal ganglia** near effectors in the chest and abdomen (see Figure 13-7 and Table 13-4).

Parasympathetic postganglionic neurons have their dendrites and cell bodies in parasympathetic ganglia. Unlike sympathetic ganglia that lie near the spinal column, parasympathetic ganglia lie near or are embedded in autonomic effectors. A parasympathetic preganglionic neuron usually synapses with postganglionic neurons extending to a single effector. For this reason, parasympathetic stimulation frequently involves response by only one organ. Sympathetic stimulation, on the other hand, usually evokes responses by numerous organs.

A summary and comparison of the structural features of the sympathetic and parasympathetic pathways is given for you in Table 13-4 and Figure 13-8.

TABLE 13-4 Comparison of Structural Features of the Sympathetic and Parasympathetic Pathways

NEURONS	SYMPATHETIC PATHWAYS	PARASYMPATHETIC PATHWAYS
Preganglionic Neurons		
Dendrites and cell bodies	In lateral gray columns of thoracic and first two or three lumbar segments of spinal cord	In nuclei of brainstem and in lateral gray columns of sacral segments of cord
Axons	In anterior roots of spinal nerves, to spinal nerves (thoracic and first four lumbar); to and through white rami to terminate in sympathetic ganglia at various levels or to extend through sympathetic ganglia; to and through splanchnic nerves to terminate in collateral ganglia	From brainstem nuclei through cranial nerve III to ciliary ganglion From nuclei in pons through cranial nerve VII to sphenopalatine or submaxillary ganglion From nuclei in medulla through cranial nerve IX to otic ganglion or through cranial nerves X and XI to cardiac and celiac ganglia, respectively
Distribution	Short preganglionic fibers from CNS to ganglion	Long preganglionic fibers from CNS to ganglion
Neurotransmitter	Acetylcholine	Acetylcholine
Ganglia	Sympathetic chain ganglia (22 pairs); collateral ganglia (celiac, superior, inferior mesenteric)	Terminal ganglia (in or near effector)
Postganglionic Neurons		
Dendrites and cell bodies	In sympathetic and collateral ganglia	In parasympathetic ganglia (e.g., ciliary, sphenopalatine, submaxillary, otic, cardiac, celiac) located in or near visceral effector organs
Receptors	Cholinergic (nicotinic)	Cholinergic (nicotinic)
Axons	In autonomic nerves and plexuses that innervate thoracic and abdominal viscera and blood vessels in these cavities In gray rami to spinal nerves, to smooth muscle of skin blood vessels and hair follicles, and to sweat glands	In short nerves to various visceral effector organs
Distribution	Long postganglionic fibers from ganglion to widespread effectors	Short postganglionic fibers from ganglion to single effector
Neurotransmitter	Norepinephrine (many); acetylcholine (few)	Acetylcholine

CNS, Central nervous system.

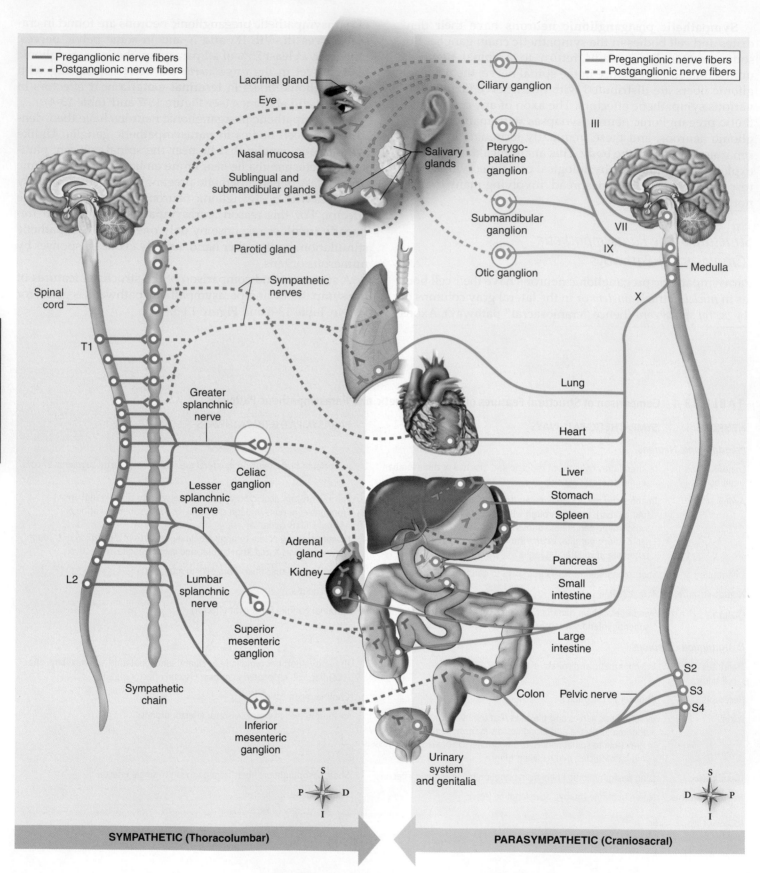

Preganglionic nerve fibers
Postganglionic nerve fibers

Lacrimal gland

Eye

Ciliary ganglion

III

Pterygo-palatine ganglion

Nasal mucosa

Salivary glands

Sublingual and submandibular glands

Submandibular ganglion

VII

IX

Parotid gland

Otic ganglion

Medulla

X

Spinal cord

Sympathetic nerves

T1

Lung

Greater splanchnic nerve

Heart

Celiac ganglion

Liver

Lesser splanchnic nerve

Stomach

Spleen

Adrenal gland

Pancreas

Kidney

L2

Small intestine

Lumbar splanchnic nerve

Superior mesenteric ganglion

Large intestine

S2

Sympathetic chain

S3

S4

Colon   Pelvic nerve

Inferior mesenteric ganglion

Urinary system and genitalia

SYMPATHETIC (Thoracolumbar)

PARASYMPATHETIC (Craniosacral)

▲ **FIGURE 13-8**   Major autonomic pathways.

QUICK CHECK

7. Do autonomic pathways follow the principle of *final common pathway*?

8. Discuss the two efferent divisions of the ANS.

9. What is the difference between a preganglionic neuron and a post-ganglionic neuron?

10. Describe the structure of the sympathetic pathway.

11. Describe the path generally taken by an impulse along a sympathetic pathway from the CNS to an autonomic effector.

Autonomic Neurotransmitters and Receptors

The axon terminals of autonomic neurons release either **norepinephrine (NE)** or **acetylcholine (ACh).** Axons that release norepinephrine are called *adrenergic fibers,* whereas axons that release acetylcholine are called *cholinergic fibers.* Autonomic cholinergic fibers are the axons of preganglionic sympathetic neurons and of both preganglionic and postganglionic parasympathetic neurons. The axons of postganglionic sympathetic neurons are typically autonomic adrenergic fibers (Figure 13-9). Sympathetic postganglionic axons to sweat glands and some blood vessels are cholinergic fibers.

As you can see in Figure 13-9, norepinephrine affects visceral effectors by first binding to different types of *adrenergic receptors* in their plasma membranes. For example, the binding of norepinephrine to *alpha* receptors in the smooth muscle of blood vessels has a stimulating effect on the muscle that causes the vessel to constrict. However, the binding of norepinephrine to *beta* receptors in smooth muscle of a different blood vessel may produce the opposite effects. In this case, it inhibits muscle contraction, causing the vessel to

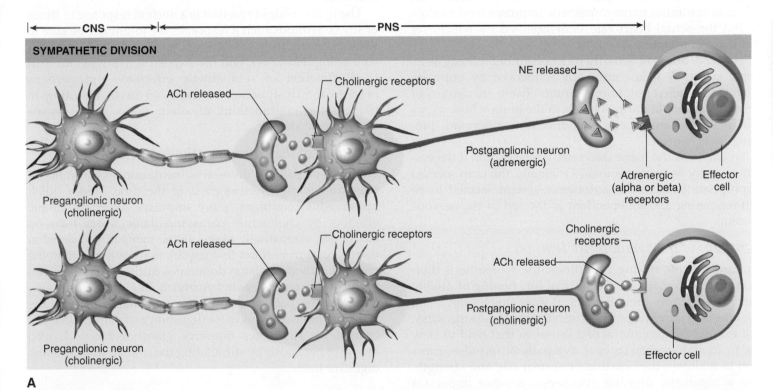

CNS	PNS

SYMPATHETIC DIVISION

ACh released — Cholinergic receptors — NE released

Preganglionic neuron (cholinergic)

Postganglionic neuron (adrenergic)

Adrenergic (alpha or beta) receptors — Effector cell

ACh released — Cholinergic receptors — Cholinergic receptors — ACh released

Preganglionic neuron (cholinergic)

Postganglionic neuron (cholinergic)

Effector cell

A

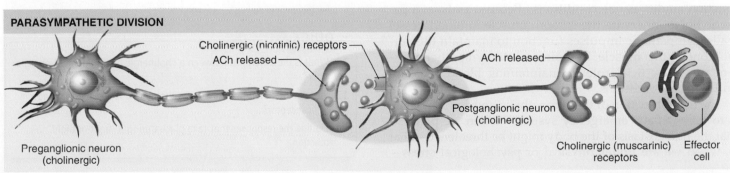

PARASYMPATHETIC DIVISION

Cholinergic (nicotinic) receptors — ACh released — ACh released

Preganglionic neuron (cholinergic)

Postganglionic neuron (cholinergic)

Cholinergic (muscarinic) receptors — Effector cell

B

▲ **FIGURE 13-9** **Locations of neurotransmitters and receptors of the autonomic nervous system.** In all pathways, preganglionic fibers are cholinergic, secreting acetylcholine *(ACh)*, which stimulates nicotinic receptors in the postganglionic neuron. Most sympathetic postganglionic fibers are adrenergic **(A)**, secreting norepinephrine *(NE)*, thus stimulating alpha- or beta-adrenergic receptors. A few sympathetic postganglionic fibers are cholinergic, stimulating cholinergic receptors in effector cells **(B)**. All parasympathetic postganglionic fibers are cholinergic **(C)**, stimulating cholinergic receptors in effector cells.

dilate. But the binding of norepinephrine to *beta* receptors in cardiac muscle has a *stimulating effect* that results in a faster and stronger heartbeat. These facts illustrate an important aspect about nervous regulation: *The effect of a neurotransmitter on any postsynaptic cell is determined by the characteristics of the receptor and not by the neurotransmitter itself.*

Functions of the Autonomic Nervous System

As we stated earlier in our discussion, the ANS as a whole functions to regulate autonomic effectors in ways that tend to maintain or quickly restore homeostasis. Doubly innervated effectors continually receive both sympathetic and parasympathetic impulses. It is the summation of the two opposing influences that determines the dominating or controlling effect. For example, continual sympathetic impulses to the heart tend to accelerate the heart rate, whereas continual parasympathetic impulses tend to slow it. But the actual heart rate is determined by whichever influence dominates.

The ANS does *not* actually function autonomously as its name suggests. It is continually influenced by impulses from the so-called autonomic centers. These are clusters of neurons located at various levels in the brain whose axons conduct impulses directly or indirectly to autonomic preganglionic neurons.

So why was the name *autonomic system* chosen if the system is really not autonomous? Originally, the term seemed appropriate because the autonomic system seemed to be self-regulating and independent of the rest of the nervous system.

Functions of the Sympathetic Division

Under ordinary, resting conditions, the sympathetic division can act to maintain the normal functioning of doubly innervated autonomic effectors. It does this by opposing the effects of parasympathetic impulses to these structures. For example, by counteracting impulses that tend to slow the heart and weaken its beat, sympathetic impulses function to maintain the heartbeat's normal rate and strength. The sympathetic division also serves another important function under usual conditions. Because only sympathetic fibers innervate the smooth muscle in blood vessel walls, sympathetic impulses function to maintain the normal tone of this muscle. By doing so, the sympathetic system plays a crucial role in maintaining blood pressure under usual conditions.

The major function of the sympathetic division, however, is to serve as an "emergency system." When we perceive that the homeostasis of the body might be threatened—that is, when we are under physical or psychological stress—

outgoing sympathetic signals increase greatly. In fact, one of the very first steps in the body's complex defense mechanism against stress is a sudden and marked increase in sympathetic activity. This brings about a group of responses that all go on at the same time. Together they make the body ready to expend maximum energy and thus to engage in the maximum muscular exertion needed to deal with the perceived threat—as, for example, in running or fighting. We now use the phrase **"fight-or-flight" reaction** for this group of sympathetic responses. Some especially important changes include faster contraction of skeletal muscles, stronger heartbeat, dilated blood vessels in the skeletal muscles, dilated bronchi, and increased blood sugar levels. Also, sympathetic impulses to the medulla of each adrenal gland stimulate its secretion of epinephrine and some norepinephrine. These hormones reinforce and prolong effects of the norepinephrine releases by sympathetic postganglionic fibers.

The fight-or-flight reaction is a normal response in times of stress. Without such a response, we might not be able to resist or retreat from something that actually threatens our well-being. However, chronic exposure to stress can lead to dysfunction of sympathetic effectors—and perhaps even to the dysfunction of the ANS itself, resulting in homeostatic malfunctions affecting many body organs and organ systems.

Functions of the Parasympathetic Division

The parasympathetic division is the dominant controller of most autonomic effectors most of the time. Under quiet, nonstressful conditions, more impulses reach autonomic effectors by cholinergic parasympathetic fibers than by adrenergic sympathetic fibers. If the sympathetic division dominates during times that require fight-or-flight, then the parasympathetic division dominates during the in-between times of **"rest-and-repair."** Acetylcholine, the neurotransmitter of the parasympathetic system, tends to slow the heartbeat but acts to promote digestion and elimination. For example, it stimulates digestive gland secretion. It also increases peristalsis by stimulating the smooth muscle of the digestive tract.

QUICK ✓ CHECK

12. What is the difference between a cholinergic fiber and an adrenergic fiber?

13. What is the difference between a cholinergic receptor and an adrenergic receptor?

14. Outline the responses that take place during a "fight-or-flight" reaction.

The BIG Picture

We can now review the entire story of the nervous system. Neurons, the conducting cells of the nervous system, act as the "wiring" that connects the structures needed to maintain the homeostasis so important to our survival. Neurons also form processing "circuits" that make decisions regarding appropriate responses to stimuli that may threaten the constancy of our internal state.

Sensory neurons act as sensors or receptors that detect changes in our external and internal environments that may be potentially threatening to homeostasis. Sensory neurons then relay this information to integrator mechanisms in the CNS. There, the information is processed—often by one or more interneurons—and an outgoing response signal is relayed to effectors such as muscles or glands by way of motor neurons. At the effector, a chemical messenger or neurotransmitter triggers a response that tends to restore homeostatic balance. Neurotransmitters released into the blood, where they are called *hormones,* can enhance and prolong such homeostatic responses.

Of course, neurons do much more than simply respond to stimuli in a preprogrammed manner. Circuits of interneurons are capable of remembering, learning new responses, generating rational and creative thought, and implementing other complex processes, such as taking tests! The exact mechanisms of many of these complex integrative functions are yet to be discovered.

The CNS is the ultimate regulator of the entire body, and it serves as the anatomical and functional center of countless feedback loops that help to maintain our body's homeostasis. The CNS directly or indirectly regulates, or at least influences, nearly every organ in the body. The CNS is able to integrate literally millions of bits of information from all over the body every second—and make sense of it all! Not only does the CNS make sense of all this information, it also compares the input with previously stored memories, and makes decisions based on its own conclusions about the data. The complex integrative functions of human language, consciousness, learning, and memory enable us to adapt to situations that less complex organisms could not.

The peripheral nervous system is made up of all the afferent nervous pathways coming into the CNS and all the efferent pathways going out of the CNS. Although we emphasized the efferent, or motor, pathways in this chapter, we will revisit the peripheral sensory pathways in the next chapter. As we have seen, the main role of the nervous system (as a whole) is to detect changes in the internal and external environment, to evaluate those changes in terms of their effect on homeostatic balance, and to regulate effectors accordingly.

The peripheral motor pathways are simply those nervous pathways that lead from the integrator (CNS) to the effectors. The somatic motor pathways lead to skeletal muscle effectors, and the autonomic efferent pathways lead to the cardiac muscle effectors, smooth muscle effectors, and glandular effectors. Thus all together the peripheral motor pathways serve as an information-carrying network that allows the CNS to communicate regulatory information to all the nervous effectors in the body. Because some motor pathways communicate with endocrine glands, endocrine effectors throughout the body can also be regulated through nervous mechanisms. Every major organ of the body is thus influenced, directly or indirectly, by nervous output. In essence, the CNS is the ultimate controller of the major homeostatic functions of the body, and the peripheral motor pathways are the means to exert that control.

Cycle of LIFE ◯

The development of nerve tissue begins from the *ectoderm* during the first weeks after conception and passes through many complicated developmental stages before birth and throughout childhood. Although we are woefully unsure of the entire process, we do know that *nerve growth factors* released by effector cells (muscles and glands) stimulate the growth of neuron processes *toward* them. During the first years of neural development, synapses are made, broken, and re-formed until the basic organization of the nervous system is intact. ●

MECHANISMS OF DISEASE

Severe head injuries can damage one or more of the cranial nerves, producing symptoms that reflect loss of the functions mediated by the affected cranial nerve or nerves, such as loss of taste, smell (olfaction), or hearing. Likewise, damage to peripheral nerves, called *peripheral neuropathy,* can be caused by diabetes and other diseases, and a number of conditions of the peripheral nervous system are caused by viruses. Compression, inflammation, and degeneration may also affect peripheral nerves.

⊖volve Find out more about these and injuries to and diseases of the peripheral nervous system online at *Mechanisms of Disease: Peripheral Nervous System.*

LANGUAGE OF SCIENCE AND MEDICINE *continued from page 277*

mixed nerve

motor cranial nerve (MOH-tor KRAY-nee-al)
  [*mot-* movement, *-or* agent, *crani-* skull, *-al* relating to]

norepinephrine (NE) (nor-ep-ih-NEF-rin)
  [*nor-* chemical prefix (unbranched C chain), *-epi-* upon, *-nephr-* kidney, *-ine* substance]

oculomotor nerve (awk-yoo-loh-MOH-tor)
  [*oculo-* eye, *mot-* movement, *-or* agent]

olfactory nerve (ol-FAK-tor-ee)
  [*olfact-* smell, *-ory* relating to]

optic nerve (OP-tik)
  [*opt-* vision, *-ic* relating to]

parasympathetic division (pair-ah-sim-pah-THET-ik)
  [*para-* beside, *-sym-* together, *-pathe-* feel, *-ic* relating to]

peripheral nervous system (PNS) (peh-RIF-er-al SIS-tem)
  [*peri-* around, *-phera-* boundary, *-al* relating to, *nerv-* nerves, *-ous* relating to]

phrenic nerve (FREN-ik)
  [*phren-* mind, *-ic* relating to]

plantar reflex (PLAN-tar REE-fleks)
  [*planta-* sole, *-ar* relating to, *re-* again, *-flex* bend]

plexus (PLEK-sus)
  [*plexus* network] *pl.,* plexi or plexuses (PLEKS-eye, PLEK-suh-sez)

postganglionic neuron (post-gang-glee-ON-ik NOO-ron)
  [*post-* after, *-ganglion-* knot, *-ic* relating to, *neuron* nerve]

preganglionic neuron (pree-gang-glee-ON-ik NOO-ron)
  [*pre-* before, *-ganglion-* knot, *-ic* relating to, *neuron* nerve]

ramus (RAY-mus)
  [*ramus* branch] *pl.,* rami (RAYM-eye)

reflex (REE-fleks)
  [*re-* again, *-flex* bend]

"rest-and-repair"

sacral plexus (SAY-kral PLEK-sus)
  [*sacr-* sacred, *-al* relating to, *plexus* network] *pl.,* plexi or plexuses (PLEKS-eye, PLEK-suh-sez)

sensory cranial nerve (SEN-sor-ee KRAY-nee-al)
  [*sens-* feel, *-ory* relating to, *crani-* skull, *-al* relating to]

somatic reflex (so-MAH-tik REE-fleks)
  [*soma-* body, *-ic* relating to, *re-* again, *-flex* bend]

spinal nerve (SPY-nal)
  [*spin-* backbone, *-al* relating to]

splanchnic nerve (SPLANK-nik)
  [*splanchn-* internal organ, *-ic* relating to]

sympathetic division (sim-pah-THET-ik)
  [*sym-* together, *-pathe-* feel, *-ic* relating to]

sympathetic postganglionic neuron (sim-pah-THET-ik post-gang-glee-ON-ik NOO-ron)
  [*sym-* together, *-pathe-* feel, *-ic* relating to, *post-* after, *-ganglion-* knot, *-ic* relating to, *neuron* nerve]

sympathetic preganglionic neuron (sim-pah-THET-ik pree-gang-glee-ON-ik NOO-ron)
  [*sym-* together, *-pathe-* feel, *-ic* relating to, *pre-* before, *-ganglion-* knot, *-ic* relating to, *neuron* nerve]

sympathetic trunk (sim-pah-THET-ik)
  [*sym-* together, *-pathe-* feel, *-ic* relating to]

terminal ganglion (TER-mih-nal GANG-glee-on)
  [*termin-* boundary, *-al* relating to, *gangli-* knot, *-on* unit] *pl.,* ganglia (GANG-glee-ah)

trigeminal nerve (try-JEM-ih-nal)
  [*tri-* three, *-gemina-* twin or pair, *-al* relating to]

trochlear nerve (TROK-lee-ar)
  [*trochlea-* pulley, *-al* relating to]

vagus nerve (VAY-gus)
  [*vagus* wanderer]

ventral ramus (VEN-tral RAY-mus)
  [*ventr-* belly, *-al* relating to, *ramus* branch] *pl.,* rami (RAYM-eye)

ventral root (VEN-tral root)
  [*ventr-* belly, *-al* relating to]

vestibulocochlear nerve (ves-TIB-yoo-loh-kok-lee-ar)
  [*vestibulo-* entrance hall, *-cochle-* sea shell, *-ar* relating to]

CHAPTER SUMMARY

To download an MP3 version of the chapter summary for use with your iPod or other portable media player, access the Audio Chapter Summaries online at http://evolve.elsevier.com.

HINT   *Scan this summary after reading the chapter to help you reinforce the key concepts. Later, use the summary as a quick review before your class or before a test.*

INTRODUCTION: THE PERIPHERAL NERVOUS SYSTEM

A. Peripheral nervous system (PNS)—nerve pathways that lead to and from the CNS
B. Comprises 31 pairs of spinal nerves, 12 pairs of cranial nerves, and all of the smaller nerves that branch from these primary nerves

SPINAL NERVES

A. Thirty-one pairs of spinal nerves (designated by number, not names) connect directly to the spinal cord (Figure 13-1)
  1. Eight cervical nerve pairs (C1-C8)
  2. Twelve thoracic nerve pairs (T1-T12)
  3. Five lumbar nerve pairs (L1-L5)
  4. Five sacral nerve pairs (S1-S5)
  5. One coccygeal pair of spinal nerves
B. Each spinal nerve attaches to the spinal cord by two short roots
  1. Ventral (anterior) root—includes motor neurons that carry information from the CNS toward effectors
  2. Dorsal (posterior) root—includes sensory fibers that carry information from receptors in the peripheral nerves
    a. Dorsal root of each spinal nerve is easily recognized by a swelling called the dorsal root ganglion; contains the cell bodies of the sensory neurons
C. Mixed nerves—contain both motor and sensory fibers
D. After a spinal nerve emerges from the spinal cavity, it forms two branches called rami

1. Dorsal ramus—supplies somatic motor and sensory fibers to several smaller nerves that innervate the muscles and skin of the posterior surface of the head, neck, and trunk
2. Ventral ramus—structure is more complex
   a. Autonomic motor fibers split away from the ventral ramus, heading toward a ganglion of the sympathetic chain
   b. Some of the autonomic fibers synapse with autonomic neurons that eventually continue on to autonomic effectors by way of splanchnic nerves
   c. Some fibers synapse with autonomic neurons whose fibers rejoin the ventral ramus; sympathetic rami
   d. Motor and sensory fibers of the ventral rami innervate muscles and glands in the extremities and in the lateral and ventral portions of the neck and trunk
E. Nerve plexuses—braided networks formed from the ventral rami of many spinal nerves (Figure 13-1)
   1. Four major pairs of plexuses: cervical plexus, brachial plexus, lumbar plexus and sacral plexus (Table 13-1)
F. Cervical plexus—found deep within the neck
   1. Nerves emerging from this plexus innervate the muscles and skin of the neck, upper shoulders, and part of the head
   2. Phrenic nerve—innervates the diaphragm and thus directly regulates our breathing
G. Brachial plexus—found deep within the shoulder
   1. Nerves that emerge from the brachial plexus innervate the lower part of the shoulder and the entire arm
H. Lumbar plexus—located in the lumbar region of the back near the psoas muscle
   1. Large femoral nerve emerges from this plexus and subdivides into branches that supply the thigh and leg
I. Sacral and coccygeal plexus—lie in the pelvic cavity
   1. Nerves that emerge from the sacral plexus are the tibial and common peroneal nerves; form the largest nerve in the body, the great sciatic nerve

CRANIAL NERVES

A. Twelve pairs of cranial nerves emerge mainly from the undersurface of the brainstem (Figure 13-3)
B. Both names and numbers (sometimes together) are used to identify the cranial nerves by their functions or position
C. Cranial nerves are made from bundles of axons (Figure 13-3 and Table 13-2)
   1. Mixed cranial nerves—axons from both sensory and motor neurons
   2. Sensory cranial nerves—sensory axons only
   3. Motor cranial nerves—motor axons only

DIVISIONS OF THE PERIPHERAL NERVOUS SYSTEM

A. Two functional divisions of the PNS:
   1. Afferent (sensory) division
   2. Efferent (motor) division
B. Somatic nervous system—includes all the voluntary motor pathways *outside* the CNS
   1. Somatic reflexes—involve contractions of skeletal muscles due to somatic motor neurons
   2. Autonomic (visceral) reflexes—involve contractions of smooth *or* cardiac muscle *or* secretion from glands

C. Somatic reflexes of clinical importance—understanding reflexes is important because their malfunction can indicate certain disease states
   1. Several common reflexes often tested by physicians:
      a. Knee jerk (patellar) reflex (Figure 13-6, *A*)—the administered tap stretches the tendon and its muscles, the quadriceps femoris; stimulates muscle spindles (receptors) in the muscle and starts the two-neuron reflex arc (Figure 13-5)
      b. Ankle jerk reflex (Achilles reflex) (Figure 13-6, *B*)—causes an extension of the foot in response to tapping of the Achilles tendon
      c. Babinski reflex (Figure 13-6, *C*)—creates an extension or upward pointing of the great toe, often with a fanning of the other toes; response to stimulation of the outer margin of the sole of the foot
      d. Corneal reflex—blinking in response to touching the cornea
      e. Abdominal reflex (Figure 13-6, *D*)—stroking skin of abdomen near the umbilicus causes umbilicus to move toward the stimulated site
D. Autonomic nervous system—subdivision of the nervous system that regulates involuntary effectors by efferent signals
   1. Effectors can be cardiac muscle, smooth muscle, and glandular epithelial tissue (Box 13-1 and Table 13-3)
   2. Major functions of the ANS are to regulate heartbeat, smooth muscle contraction, and glandular secretion to maintain homeostasis
   3. ANS includes both sensory (afferent) pathways and motor (efferent) pathways
   4. The motor pathways comprise two efferent divisions:
      a. Sympathetic division—"fight-or-flight" system; prepares us for physical activity
      b. Parasympathetic division—"rest-and-repair" system; stimulates activities such as digestion, defecation, and urination; also slows down the heart rate and respiration
E. Structure of the autonomic nervous system—each efferent autonomic pathway, whether sympathetic or parasympathetic, is made up of autonomic nerves, ganglia, and plexuses
   1. A relay of two autonomic neurons conducts information from the CNS to the autonomic effectors (Table 13-4)
      a. Preganglionic neuron—conducts impulses from the brainstem or spinal cord to an autonomic ganglion
      b. Postganglionic neuron—conducts impulses away from the ganglion to the effector
F. Structure of the sympathetic (thoracolumbar) pathways (Figure 13-8)
   1. Sympathetic preganglionic neurons begin within the gray matter of the spinal cord (thoracic and lumbar segments) (Figure 13-7)
   2. Axons of these neurons leave the spinal cord via the ventral roots of spinal nerves
   3. Axons then leave the spinal nerves and enter one of a chain of paravertebral ganglia
      a. Linked ganglia are often referred to as the sympathetic chain ganglia (sympathetic trunk)
   4. Preganglionic neurons that pass through chain ganglia without synapsing continue on through splanchnic nerves to other sympathetic ganglia
   5. Axon of this second, postganglionic fiber usually returns to a spinal nerve and extends it to a visceral effector (Figures 13-7 and 13-8)

6. Sympathetic postganglionic neurons have their dendrites and cell bodies in the sympathetic chain ganglia
7. Once in the spinal nerve, the postganglionic fibers are distributed with other nerve fibers to the various sympathetic effectors
8. The axon of any one sympathetic preganglionic neuron synapses with many postganglionic neurons, and these frequently terminate in widely separated organs of the body

G. Structure of the parasympathetic (craniosacral) pathways—parasympathetic preganglionic neurons have their cell bodies in nuclei in the brainstem or in the lateral gray columns of the sacral spinal cord (Figure 13-8 and Table 13-4)
   1. At least 75% of all parasympathetic preganglionic fibers travel in the vagus nerve (X) before synapsing with postganglionic fibers in terminal ganglia near effectors in the chest and abdomen (Figure 13-7)
   2. Parasympathetic postganglionic neurons have their dendrites and cell bodies in parasympathetic ganglia
   3. A parasympathetic preganglionic neuron usually synapses with postganglionic neurons extending to a single effector

H. Autonomic neurotransmitters and receptors
   1. Axon terminals of autonomic neurons release either norepinephrine (NE) or acetylcholine (ACh)
      a. Adrenergic fibers—axons that release norepinephrine
      b. Cholinergic fibers—axons that release acetylcholine
      c. Autonomic cholinergic—axons of preganglionic sympathetic neurons and of both preganglionic and postganglionic parasympathetic neurons
      d. Axons of postganglionic sympathetic neurons are typically autonomic adrenergic fibers (Figure 13-9)
      e. Norepinephrine affects visceral effectors by first binding to different types of adrenergic receptors in their plasma membranes
      f. The effect of a neurotransmitter on any postsynaptic cell is determined by the characteristics of the receptor and not by the neurotransmitter itself
   2. Functions of the autonomic nervous system—ANS as a whole functions to regulate autonomic effectors in ways that tend to maintain or quickly restore homeostasis
   3. Sympathetic and parasympathetic divisions are tonically active
      a. They continually conduct impulses to autonomic effectors, and exert opposite, or antagonistic, influences on them
      b. Doubly innervated effectors continually receive both sympathetic and parasympathetic impulses
   4. Functions of the sympathetic division—acts to maintain the normal functioning of doubly innervated autonomic effectors
      a. Does this by opposing the effects of parasympathetic impulses to these structures
      b. Sympathetic impulses function to maintain the normal tone of the smooth muscle lining the blood vessels
      c. The major function of the sympathetic division, however, is to serve as an "emergency system"; fight-or-flight reaction
   5. Functions of the parasympathetic division—dominant controller of most autonomic effectors most of the time
      a. Parasympathetic division dominates during the in-between times of "rest-and-repair"
      b. Acetylcholine, the neurotransmitter of the parasympathetic system, tends to slow the heartbeat but acts to promote digestion and elimination

REVIEW QUESTIONS

 HINT *Write out the answers to these questions after reading the chapter and reviewing the Chapter Summary. If you simply think through the answer without writing it down, you won't retain much of your new learning.*

1. Identify the direction of the information carried by the ventral root of a spinal nerve.
2. Define *mixed nerves.*
3. Identify the areas innervated by the individual nerves emerging from the cervical plexus.
4. What pathways are found in the somatic motor nervous system?
5. Describe an autonomic preganglionic neuron.
6. Identify the paths that a nerve signal may take once it is inside the sympathetic chain ganglion.
7. Differentiate between the sympathetic preganglionic and postganglionic neurons in terms of length.
8. How do parasympathetic ganglia differ from sympathetic ganglia in terms of location?
9. Describe the actions of norepinephrine. How are these actions terminated?
10. Describe the responses caused by acetylcholine release. How are these actions terminated?
11. Name the main types of adrenergic receptors.
12. Which efferent division of the autonomic nervous system is the dominant controller of most autonomic effectors most of the time?

CRITICAL THINKING QUESTIONS

 HINT *After finishing the Review Questions, write out the answers to these items to help you apply your new knowledge. Go back to sections of the chapter that relate to items that you find difficult.*

1. Can you distinguish between the reflexes that cause cardiac muscle to react and those that cause skeletal muscles to react?
2. An adult has an extension of the great toe and the fanning of the other toes in response to stimulation of the outer margin of the sole of the foot. Why is this a concern, and where would the problem most likely be?
3. A single reflex can be classified in several ways. What example can you find that would show this? In what ways can this reflex be classified?
4. Using the control of the speed of a car as an example, explain dual innervation in the autonomic nervous system.
5. Can you distinguish between the function of autonomic transmitters and receptors? How do they regulate dually innervated effector cells?
6. Based on what you know, explain why the autonomic nervous system is not an "automatic" nervous system with little control from the higher brain centers.
7. In times of stress, there are reports of people performing almost superhuman feats of strength. Specifically, what is happening in the body that would allow this to happen?
8. Can you predict the respiratory consequences of an injury that damaged the spinal cord between the third and fifth cervical segments? What specific mechanism would cause these consequences?

9. The drug propranolol (Inderal) blocks the effect of a portion of the sympathetic nervous system. To what class of drugs does propranolol (Inderal) belong, and what are its therapeutic effects?

continued from page 277

With the knowledge you have gained from reading this chapter, try to answer these questions about Tomasina's head injury from the Introductory Story.

1. In one of the tests, Tomasina was asked to stick out her tongue. Which cranial nerve was being tested?
   a. Glossopharyngeal
   b. Hypoglossal
   c. Vagus
   d. Trigeminal

2. In a second test, Tomasina's hearing was evaluated. What cranial nerve is involved in the sense of hearing?
   a. Accessory
   b. Vagus
   c. Vestibulocochlear
   d. Trochlear

3. Tomasina was having difficulty focusing her eyes and following the doctor's finger on command. Which of the following cranial nerves does *not* control eye movement?
   a. Optic
   b. Oculomotor
   c. Trochlear
   d. Abducens

4. The nurse touched a series of swabs with different tastes to the front portion of Tomasina's tongue—sugar, salt, lemon juice, and dissolved aspirin. Which cranial nerve was the nurse testing?
   a. Glossopharyngeal
   b. Olfactory
   c. Trigeminal
   d. Facial

HINT *To solve a case study, you may have to refer to the glossary or index, other chapters in this textbook, A&P Connect, Mechanisms of Disease, and other resources.*

UNIT 3

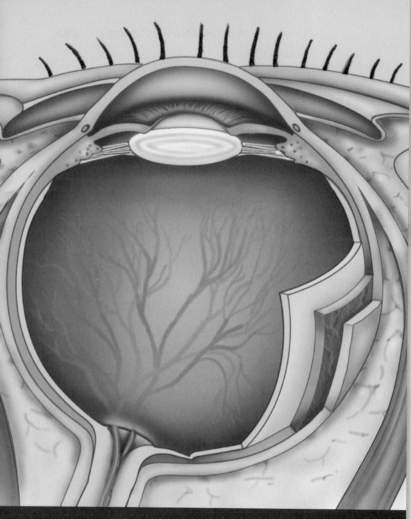

Sense Organs

STUDENT LEARNING OBJECTIVES

At the completion of this chapter, you should be able to do the following:

1. Discuss the three classifications of receptors according to location, structure, and stimulus detected.
2. Outline the differences between receptors with free nerve endings and those with encapsulated nerve endings.
3. List the general somatic sensory receptors and the types of sensations they register.
4. List the primary structures involved in the sense of smell.
5. Discuss the basic physiology of the olfactory pathway.
6. List the primary structures involved in the sense of taste.
7. Discuss the basic physiology of the gustatory pathway.
8. List the primary structures involved in the senses of hearing and balance.
9. Discuss the basic physiology of the sense of hearing.
10. Discuss the basic physiology of the sense of balance.
11. List the primary structures involved in the sense of vision.
12. Discuss the basic physiology of the sense of vision.

CHAPTER OUTLINE

HINT · *Scan this outline before you begin to read the chapter, as a preview of how the concepts are organized.*

WHY were the headlights of the oncoming cars so bright, she wondered. Rita flashed her car lights as the next car approached, but when the driver flashed his lights back in return, she saw his lights had been on the low setting all along. She noticed too that the street signs were difficult to read tonight. When she got to her daughter's house, Rita was quite relieved to be off the road! Over the years, her vision had always been excellent; not being able to see well was a new experience for her. Her daughter suggested she have her eyes checked as soon as possible.

The next week, as Rita tried to read one of the magazines in the optometrist's waiting room, she found that she could not get the letters to come into focus unless she held the magazine at arm's length—yet another sign of trouble with her eyesight.

When you have finished with this chapter, you should be able to understand what has been going wrong with Rita's eyesight, as well as describe the source of the problem.

Our bodies house millions of sense organs with which we perceive our internal and external environments. They fall into two main categories: *general sense organs* and *special sense organs*. The **general sense organs** function to produce the general, or somatic, senses, such as touch, temperature, and pain. Other general receptors initiate various reflexes necessary for maintaining homeostasis. **Special sense organs,** in contrast, function to produce our senses of smell, taste, hearing, vision, and balance. They also initiate reflexes important for homeostasis, but on a much larger scale. In this chapter, we will first describe general, somatic receptors, and then discuss our special sense organs.

SENSORY RECEPTORS

Our sensory receptors are responsive to changes in our internal and external environments. Of course, these functions are critical to our survival. Internal sensations ranging from pain and pressure to hunger and thirst help us maintain homeostasis. Likewise, being aware of our external environment (through our senses of smell, sight, and hearing, for example) provides us with direct input from our surroundings—input that can save our lives.

Receptor Response

Generally, receptors respond to stimuli by converting them to nerve impulses. As a rule, each type of receptor responds to a single type of stimulus. For example, stretch receptors respond to stretching but do not respond to heat. Likewise, heat receptors respond to heat but do not respond to stretch. When a receptor is stimulated, it develops a *local potential* in its membrane, called a **receptor potential.** When a receptor potential reaches a threshold, it triggers an action potential in the sensory neuron. These impulses then travel over sensory pathways up the spinal cord to the brain, where they are interpreted as a particular **sensation** (such as heat or cold).

LANGUAGE OF SCIENCE AND MEDICINE

HINT Before reading the chapter, say each of these terms out loud. This will help you avoid stumbling over them as you read.

accommodation (ah-kom-oh-DAY-shun)
[*accommoda-* adjust, *-ation* process]

acute pain fiber (ah-KYOOT FYE-ber)
[*acut* sharp, *fibr-* thread]

adaptation (ad-ap-TAY-shun)
[*adapt-* fit to, *-tion* process]

anterior cavity (an-TEER-ee-or KAV-i-tee)
[*ante-* front, *-er-* more, *-or* quality, *cav-* hollow, *-ity* state]

aqueous humor (AY-kwee-us HYOO-mor)
[*aqu-* water, *-ous* relating to]

auditory ossicle (AW-di-toh-ree OS-ik-ul)
[*audit-* hear, *-ory* relating to, *os-* bone, *-icle* little]

auditory tube (AW-di-toh-ree)
[*audit-* hear, *-ory* relating to]

auricle (AW-ri-kul)
[*auricula* little ear]

basilar membrane (BAYS-ih-lar MEM-brayne)
[*bas-* foundation, *-ar* relating to, *membran-* thin skin]

bipolar cell (bye-POH-lar sell)
[*bi-* two, *-pol-* pole, *-ar* relating to]

blind spot

bulbous corpuscle (BUL-bus KOR-pus-ul)
[*bulb-* swollen root, *-ous* relating to, *corpus-* body, *-cle* little]

canthus (KAN-thus)
[*canthus* corner of eye]

cerumen (seh-ROO-men)
[*cera* wax]

chemoreceptor (kee-moh-ree-SEP-tor)
[*chemo-* chemical, *-recept-* receive, *-or* agent]

chronic pain fiber (FYE-ber)
[*chron-* time, *-ic* relating to, *fibr-* thread]

ciliary body (SIL-ee-air-ee BOD-ee)
[*ciliary* eyelids or eyelashes]

cochlea (KOHK-lee-ah)
[*cochlea-* snail shell]

cochlear nerve (KOHK-lee-ar)
[*cochlea-* snail shell]

color blindness

cone

constriction (kon-STRIK-shun)
[*constrict-* draw tight, *-tion* state]

convergence (kon-VER-jens)
[*con-* together, *-verg-* incline, *ence* state]

cornea (KOR-nee-ah)
[*corneus* horny]

continued on page 318

Some sensory impulses that terminate in the brainstem may affect so-called vital sign reflexes, which help regulate heart or respiratory rates, for example. Other impulses may end in the thalamus or cerebral cortex. These inputs may then trigger "crude" sensation awareness, such as the general sensation of being touched, when they are first interpreted in the thalamus. However, they may also trigger very accurate and specific awareness of a specific *type* of sensation and the *location* and *intensity* of that sensation in the body. This complex interpretation takes place in our cerebral cortex.

Many receptors exhibit **adaptation.** In this process, the magnitude of the receptor potential *decreases over time* in response to a continuous stimulus. As a result, the rate of impulse conduction by the sensory neuron's axon also decreases. As you might suspect, the intensity of the sensation also decreases. In effect, we become less aware of the stimulus. For example, touch receptors adapt rapidly. The sensation of clothes touching your skin seems to disappear moments after you put on your clothes. This happens because touch receptors have adapted to the constant presence of your clothes. In contrast, the stretch receptors in our muscles, tendons, and joints adapt slowly. They continue sending impulses to the brain as long as stimulation of the sensors continues.

Our sensory receptors constantly receive input from our internal and external environments. However, this does not mean that we are consciously aware of this continuous input. In fact, **perception**—our ability to remain *aware* of a particular sensation over time and to *interpret* that sensation—varies considerably. For example, our bodies continuously monitor our levels of blood glucose and carbon dioxide, but we are not conscious of this continuous monitoring.

Distribution of Receptors

Receptors responsible for our **special senses** of smell, taste, vision, hearing, and balance (equilibrium) are grouped into specialized areas of our bodies. These include the surface of our tongue as well as highly complex sensory organs such as our eyes. In contrast, the *general sense organs* consist of microscopic receptors embedded within the skin, muscles, and other organs widely distributed throughout the body (Table 14-1).

The sensations generated by the general sense organs are usually called the *somatic senses*. These are not uniformly distributed. In some areas they are very dense; in others, sparse. For example, the skin covering your fingertips has many more somatic receptors than the skin on your back.

CLASSIFICATION OF RECEPTORS

We classify somatic receptors by three methods: (1) location in the body, (2) the stimulus that induces them to respond, and (3) their structure.

Classification by Location

Receptors are classified by location as *exteroceptors, interoceptors,* and *proprioceptors.* **Exteroceptors,** as you might suspect are located on or very near the surface of the body. They respond to stimuli that are *external* to the body. Examples include exteroceptors that detect pressure, touch, pain, and temperature. In contrast, **interoceptors** are located internally, within our organs. When stimulated, these receptors provide information about our internal environment. They are activated by stimuli such as pressure, stretching, and chemical changes that are experienced in many different organs.

Proprioceptors are a special type of interoceptors. You can find proprioceptors in skeletal muscle, joint capsules, and tendons. Here they inform us about body movement, orientation in space, and muscle stretching.

Classification by Stimulus Detected

We can also classify receptors based on the stimuli that activate them:

1. **Mechanoreceptors** respond to stimuli that in some manner measure changes in pressure or internal movement. These include mechanoreceptors that inform us of pressure on our skin or within our blood vessels (*baroreceptors*), as well as stretch or pressure changes in muscles, tendons, and lungs.
2. **Chemoreceptors** are activated by the amount or changing concentration of certain chemicals. Our senses of taste and smell depend on chemoreceptors.
3. **Thermoreceptors** are activated by changes in temperature.
4. **Nociceptors** perceive intense stimuli (toxic chemicals and intense light, for example) of any type that causes tissue damage. The overall sensation is one of *pain.*
5. **Photoreceptors** are found in our eyes and respond to light stimuli in the visible spectrum.
6. **Osmoreceptors** are concentrated in the hypothalamus and sense levels of osmotic pressure in our body fluids. They are important in detecting changes in the concentration of electrolytes and in stimulating the thirst-center.

Classification by Structure

Somatic sensory receptors are also classified as either having (1) *free nerve endings* or (2) *encapsulated nerve endings*. To help with your understanding, refer to summary Table 14-1 as you read through the following paragraphs.

Free Nerve Endings

Free nerve endings are the simplest, most common, and most widely distributed sensory receptors. They may be exteroceptors or interoceptors. These slender sensory fibers often terminate in small swellings called *dendritic knobs* (Figure 14-1, *A*). They serve as the primary sensory receptors for pain, heat, cold, and various forms of touch.

TABLE 14-1 General Sense Organs

BY STRUCTURE	BY LOCATION AND TYPE	BY ACTIVATION STIMULUS	BY SENSATION OR FUNCTION
Free Nerve Endings			
Nociceptor 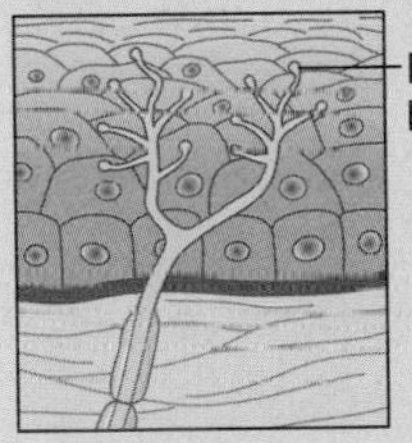 Dendrite knob	Either exteroceptor or interoceptor (e.g., visceroceptor—most body tissues)	Almost any irritating stimulus; temperature change; mechanical	Pain; temperature; itch; tickle
Tactile (Merkel) disk (meniscus) 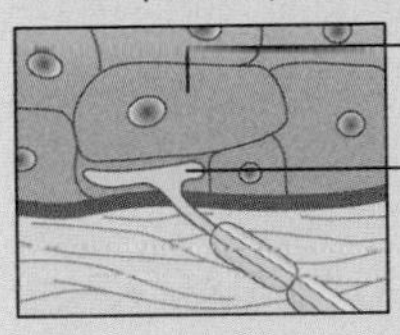 Tactile epithelial cell Tactile disk	Exteroceptor	Light pressure; mechanical	Discriminating touch
Root hair plexus 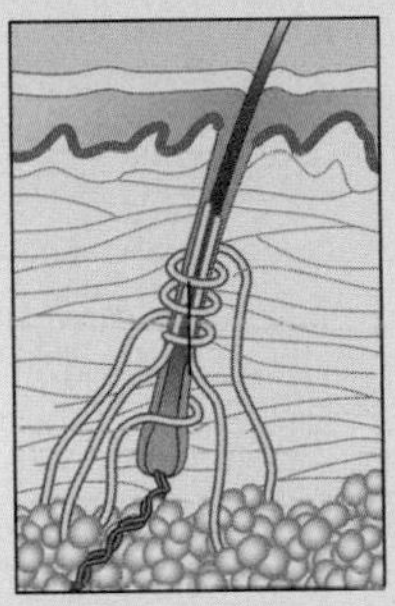	Exteroceptor	Hair movement; mechanical	Very light touch
Encapsulated Nerve Endings Touch and Pressure Receptors			
Stretch Receptors			
Tactile (Meissner) corpuscle 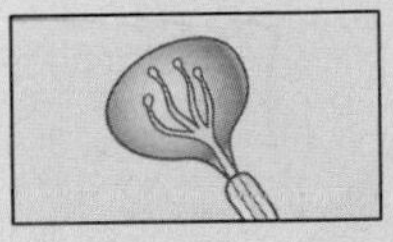	Exteroceptor; epidermis, hairless skin	Light pressure, mechanical	Touch; low-frequency vibration
Bulbous (Ruffini) corpuscle 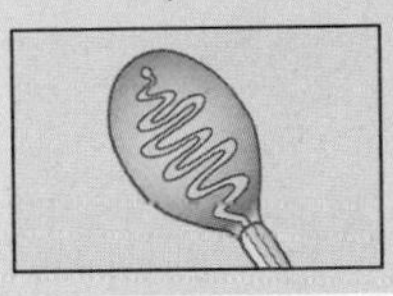	Exteroceptor; dermis of skin	Mechanical	Crude and persistent touch
Lamellar (Pacini) corpuscle 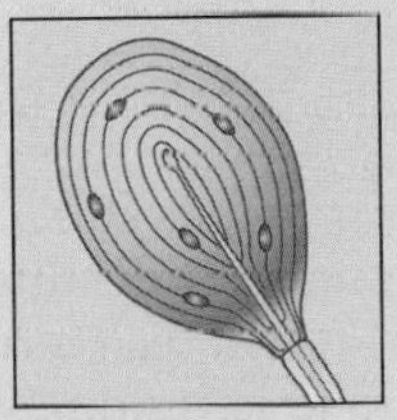	Exteroceptor; dermis of skin, joint capsules	Deep pressure, mechanical	Deep pressure; high-frequency vibration; stretch

(Continued on p. 300)

UNIT 3

TABLE 14-1 General Sense Organs—cont'd

BY STRUCTURE	BY LOCATION AND TYPE	BY ACTIVATION STIMULUS	BY SENSATION OR FUNCTION
Muscle spindles Intrafusal fibers	Interoceptor; skeletal muscle	Stretch; mechanical	Sense of muscle length
Tendon organs	Interoceptor; tendon (near muscle tissue)	Force of contraction and tendon stretch; mechanical	Sense of muscle tension

EXTEROCEPTORS

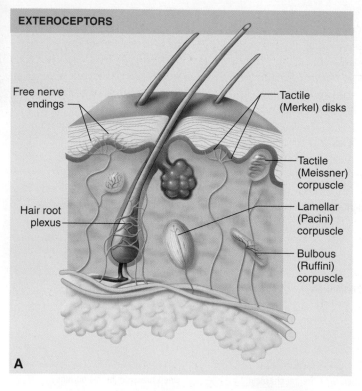

PROPRIOCEPTORS

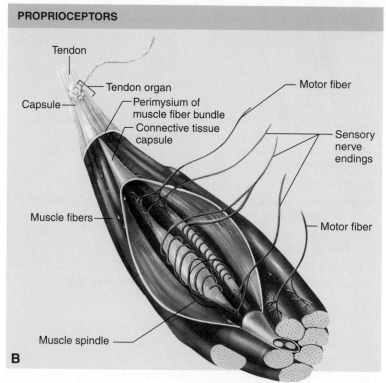

▲ FIGURE 14-1   Somatic sensory receptors.

Pain Sensations

Nerve fibers that carry pain impulses from *nociceptors*, the free nerve endings that detect pain, can be either **acute pain fibers** or **chronic pain fibers.** Acute fibers are common in the skin, mucous membranes, and other superficial areas. They respond to sharp pain sometimes described as the type that "takes your breath away." If you've ever stabbed yourself with a needle or slammed a car door on your finger, you've experienced this type of *acute somatic*

pain! In contrast, *visceral pain* is often described as dull or chronic aching pain. It could be due to cramping, intestinal obstruction, or menstrual pain. This type of pain reception travels over chronic pain fibers. Brain tissue is unique in that it lacks nociceptors that can experience pain. For this reason, it is incapable of sensing painful stimuli within the brain itself.

People vary tremendously in their ability to perceive and register pain. The brain's filtering and processing mechanisms for sensory information must play a large role, but this phenomenon is poorly understood. Pain has been called the "gift that nobody wants." This is because pain can alert us to disease and injury.

Sometimes the stimulation of pain receptors in deep viscera can be felt in areas of the skin overlying the organ, or even in areas far removed from the origin of the pain. This is called **referred pain.** It is caused by a *mixing* of sensory nerve impulses from both the diseased organs and the skin in the area of referred pain. A classic example is the referred pain often associated with a heart attack. Sensory fibers from the skin on the chest over the heart *and* from the tissue of the heart itself enter the first to the fifth thoracic spinal cord segments. Sensory fibers from the skin areas over the left shoulder and inner surface of the left arm also enter the spinal cord here. The result is that sensory impulses from all these areas travel to the brain over a *common pathway*. Thus, the brain may feel the pain of a heart attack in the shoulder or arm.

Temperature Sensations

Your skin has many *receptive fields* that are sensitive to *either* cold or warmth, but not both. This means that these *thermoreceptors* are not spread uniformly across the skin surface. Both types of thermoreceptors undergo rapid adaptation. As a result, the initial intensity of warm or cold sensations soon fades. This explains why we adapt to the heat of a sauna or a cool shower in a relatively short time. Beyond a temperature of about 48° C (118° F), a sensation of burning pain begins. In contrast, cold receptors, which are located in the deepest layer of the epidermis, have a broader temperature response than do the warm temperature receptors. Below 10° C (50° F), the firing of cold receptors decreases dramatically. This at first acts like a local anesthetic. At still lower temperatures, nociceptors that register a sensation of piercing, freezing pain are activated.

Tactile Sensations

In addition to their role in registering pain and temperature, free nerve endings are also involved in certain tactile sensations. For example, *root hair plexuses* are rapidly adapting free nerve endings activated when even very slight skin movement bends or deforms a hair. When you feel a mosquito bite, you may feel both the irritation of your hair bending as well as the pain from the piercing mouthparts of the mosquito.

An *itch* is another tactile sensation registered by free nerve endings. The cause of most itches is the release of inflammatory biochemicals (such as histamine) by the free nerve endings.

A *tickle* is a rather unique sensation. It results from tactile stimulation of the skin, not by you, but by someone else. We can't explain why people cannot tickle themselves!

Finally, we can experience *discriminative touch*, a subtle sensation that can often be quite pleasant (e.g., "soft touch") (Figure 14-2). Discriminative touch is mediated by a flattened or disk-shaped free nerve ending called a **tactile disk** (or *Merkel disk*) (see Figure 14-1, *A*).

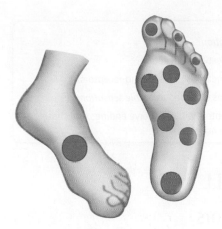

◀ **FIGURE 14-2**
Discriminative touch.
Examples of specific sites used to verify sensation of discriminative touch.

Encapsulated Nerve Endings

Encapsulated nerve endings have a *connective tissue capsule* surrounding their dendritic ends. Most of these receptors act as mechanoreceptors. They are activated by stimuli that deform structures in the skin (look again at Figure 14-1, *A*).

Tactile Receptors

Tactile corpuscles, or *Meissner corpuscles*, are egg-shaped mechanoreceptors that are larger than tactile disks. They are located in, or very close to, the dermal papillae of hairless skin areas found, for example, in the fingertips, lips, nipples, and genitals. Here they pick up sensations of light touch in addition to sensations of smoothness or roughness.

Another variant of the tactile (Meissner) corpuscle is the **bulbous corpuscle,** also called the *Ruffini corpuscle*. These corpuscles are slow-adapting receptors and permit the skin of the fingers to continuously remain sensitive to deep pressure. The ability to grasp an object, such as the steering wheel of a car, for long periods of time and still be able to "sense" its presence in our hands depends on these receptors.

Lamellar corpuscles, or *Pacini corpuscles*, are large mechanoreceptors. They are found in the deep dermis of the skin, especially in the hands and feet. They respond quickly to sensations of deep pressure, high-frequency vibration, and stretch. However, they adapt quickly, and the sensations they register rarely last for long.

Stretch Receptors

The most important stretch receptors are associated with muscles and tendons. These are classified as *proprioceptors*. They are found in tiny organs called **muscle spindles** and **tendon organs** (see Figure 14-1, *B*).

Muscle spindles are stimulated if the length of a muscle is stretched and exceeds a certain limit. This results in a *stretch reflex* that shortens a muscle or muscle group.

Tendon organs trigger a reaction opposite that of muscle spindles. They are stimulated by muscle tension so extreme that it stretches the tendon. When they are activated, tendon organs cause muscles to relax. This response, called the *tendon reflex*, protects our muscles from tearing internally or from pulling away from their tendons at points of attachment to bone. The tendon reflex is sometimes called the *Golgi tendon reflex*.

UNIT 3

QUICK CHECK

1. What is meant by a *sensation*?
2. How does adaptation affect our perception?
3. Briefly discuss the three methods of classifying receptors.
4. List a few of the sensations classified as *tactile sensations*.
5. Distinguish between receptors with *free* nerve endings and those with *encapsulated* nerve endings.

SENSE OF SMELL

Olfactory Receptors

The **olfactory epithelium** consists of *epithelial support cells, basal cells,* and bipolar-type *olfactory sensory neurons.* These neurons have unique **olfactory cilia,** which touch the surface of the olfactory epithelium lining the upper surface of the nasal cavity. These sensory neurons are chemoreceptors. They are unique because they are replaced on a regular basis by basal cells in the olfactory epithelium (Figure 14-3).

Receptor potentials are generated in olfactory sensory neurons when airborne **odorant molecules** dissolve in mucus covering the olfactory epithelium (see Figure 14-3, *B*). The cilia from these epithelial cells help to "mix" the mucus solvent and the dissolving odorant molecules in the air. The dissolved molecules then bind to protein *odor receptors* located in the cell membranes of the olfactory cells.

Note, however, that the olfactory epithelium is located high in the nasal cavity (see Figure 14-3, *A*). For this reason, much of our inspired air bypasses the olfactory receptor cells. This is why we must often sniff to discover smells. Sniffing brings short blasts of air laden with odorant molecules up to the superior portion of the nasal cavity, where they can be detected.

Our olfactory receptors are extremely sensitive to even low concentrations of odorant molecules, but they also rapidly adapt to the stimulant molecules. As a result, our sense of smell for a given odorant molecule often diminishes quickly.

Humans in general have a sense of smell far less keen than many other animals. However, most of us can distinguish at least several hundred smells, and a few can distinguish thousands of smells. Most of us can quickly identify things that smell putrid or floral, or have musky odors. In fact, as you know, many of our most pleasantly perceived smells are complex mixtures of odorant molecules.

In 2004, American researchers Richard Axel and Linda Buck received the Nobel Prize in Physiology or Medicine for a series of pioneering studies that clarified for the first time how our olfactory system works. They found that different odors are registered by the stimulation of one or a combination of more than 1,000 different odorant receptors in the cell membranes of our olfactory epithelium. The conscious identification of each odorant is then provided by the sensory cortex of the temporal lobe.

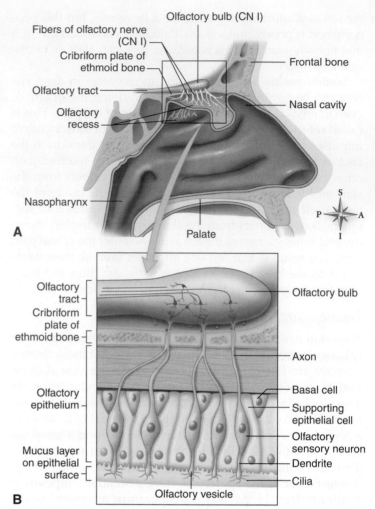

▲ **FIGURE 14-3   Olfaction.** Location of olfactory epithelium, olfactory bulb, and neural pathways involved in olfaction. **A,** Midsagittal section of the nasal area shows the locations of major olfactory sensory structures. **B,** Details of the olfactory bulb and olfactory epithelium.

Olfactory Pathway

If the level of odorant molecules dissolved in the mucus reaches a threshold level, a *receptor potential* and then an *action potential* are generated. Action potentials are passed to the olfactory nerves in the olfactory bulb, and then via the olfactory tract to the thalamic and olfactory centers. Here the odorant molecules are interpreted, and the memory stored.

Our sense of smell creates powerful and long-lasting memories. Not only can specific odors trigger long-term memory, they often allow us to recall emotions associated with our memories. For example, the advent of spring with its rich smell of damp earth and flowers can trigger intensely emotional memories of childhood. We are still not sure how such associations are maintained, even into old age.

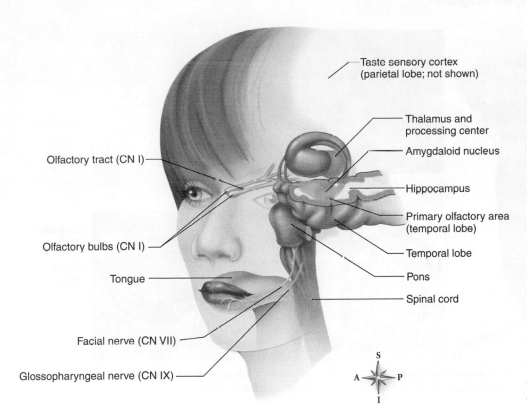

Taste sensory cortex
(parietal lobe; not shown)

Thalamus and
processing center

Olfactory tract (CN I)

Amygdaloid nucleus

Hippocampus

Primary olfactory area
(temporal lobe)

Olfactory bulbs (CN I)

Temporal lobe

Tongue

Pons

Spinal cord

Facial nerve (CN VII)

Glossopharyngeal nerve (CN IX)

◄ **FIGURE 14-4** Relationship of olfactory and gustatory pathways.

Our senses of smell and taste are closely related. Note in Figure 14-4 that the neural inputs from both olfactory and gustatory (taste) receptors travel in several common areas of the brain. Ultimately, however, olfactory sensations are interpreted in the sensory cortex of the temporal lobes. Those for taste are perceived in the sensory cortex of the parietal lobes.

A&P CONNECT

A sense related to olfaction is our ability to detect *pheromones*, or sexual signal molecules, from other people. The receptors for pheromones are just inside our nose, near the olfactory receptors. Find out more about this sexual signaling in **Pheromones and the Vomeronasal Organ** online at **A&P Connect.**

ℰvolve

SENSE OF TASTE

Taste Buds

The **taste buds** are sense organs that respond to **gustatory** (taste) stimuli. A few taste buds are located in the lining of the mouth and on the soft palate, but most are found in small, elevated projections of the tongue, called **papillae.** Refer to Figure 14-5 as you read through the following classification.

Tongue papillae are classified by their structure (see Figure 14-5, *A*). *Fungiform papillae* are large, mushroom-shaped bumps found in the anterior two thirds of the tongue surface. Each fungiform papilla contains one or a few taste buds. *Circumvallate papillae* are huge, dome-shaped bumps arranged in a row that cross the back of the tongue. Each of these papillae may contain thousands of taste buds. *Foliate papillae* are reddish, leaflike ridges of mucosa found on the lateral edges and rear of the tongue surface. Each of these bears about a hundred taste buds. Finally, *filiform papillae* are bumps with threadlike projections. They are scattered among the fungiform papillae. However, their main function is to determine food "texture."

Although the different forms of papillae are distributed differently across the tongue surface, there really is no regional specialization for specific tastes, as was formerly believed. For example, there is no "taste map" that accurately indicates the location of different tastes. All tastes can be detected in all areas of the tongue that contain taste buds.

Taste buds house the chemoreceptors responsible for taste. These receptors are stimulated by chemicals, called *tastants,* dissolved in the saliva. Each taste bud is like a banana cluster containing 50 to 125 or so of these chemoreceptors, called **gustatory cells.** Tiny threadlike structures, called *gustatory hairs,* extend from each of the gustatory

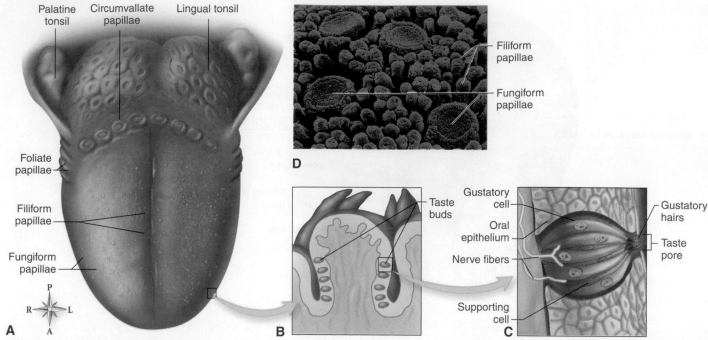

▲ **FIGURE 14-5** **The tongue.** **A,** Dorsal surface of tongue and adjacent structures. **B,** Section through a papilla with taste buds on the side. **C,** Enlarged view of a section through a taste bud. **D,** Scanning electron micrograph of the tongue surface showing the papillae in detail.

cells. These hairs project into an opening, or *taste pore,* which is bathed in saliva (see Figure 14-5, *C*). Our sense of taste depends on the creation of a receptor potential in gustatory cells. The nature and concentration of the chemicals that bind to these receptors determine how quickly a receptor potential is generated.

Taste cells are structurally similar to one another, and all of them can respond, at least in some degree, to most tastant chemicals. Each taste receptor, however, responds *most* effectively to only one of five "primary" taste sensations: *sour, sweet, bitter, salty,* and *umami (savory).* Look at the labeled line model in Figure 14-6 to see how signals from different types of receptors are conducted along different neural pathways. Our ability to experience a large variety of tastes results from a *combination of input* from these five primary sensations. Additional secondary taste sensations may also exist.

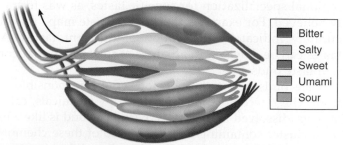

▲ **FIGURE 14-6** **Taste receptors.** The labeled line model of gustation (taste) holds that each distinct taste has a separate group of taste receptors, with each group sending its impulses along a distinct "line" or neural pathway.

As chemoreceptors, the taste buds, like olfactory receptors, tend to be quite sensitive initially. However, they fatigue easily. Adaptation often begins within a few seconds after a taste sensation is perceived. Of course, you can imagine the adaptive value of such a system. Taste allows us to chemically test our food before swallowing it. We can then selectively choose or avoid foods with specific qualities, such as variations in salt content.

Neural Pathway for Taste

Nervous impulses generated in the anterior two thirds of the tongue travel over the facial nerve (cranial nerve [CN] VII). Those generated from the posterior third of the tongue are conducted by fibers of the glossopharyngeal (CN IX) nerve. A third cranial nerve, the vagus (CN X) nerve, plays a minor role in taste.

All three cranial nerves carry impulses into the medulla oblongata. Thus, unlike any other special sense, taste has several different pathways to the brain. Once taste information is processed in the medulla, it is relayed into the thalamus and then on to the gustatory area of the cerebral cortex of the parietal lobe.

We generally think of taste in terms of the primary sensations, but touch, texture, and temperature are also important to whether or not we perceive something as pleasant, neutral, or unpleasant.

Our chemical sense of *olfaction* (smell) and *gustation* (taste) are intertwined. *Flavor,* for example, is really a combination of our senses of smell and taste. As we eat, air passing out through the pharynx also stimulates olfactory receptors. This information is integrated in our brain and we sense the combination of smell and taste as a single, complex *flavor* experience.

QUICK ✓ CHECK

6. What are odorant molecules?

7. How are receptor potentials generated in olfactory or gustatory cells?

8. What are the five "primary" taste sensations?

9. What is the perception of *flavor*?

SENSE OF HEARING

As you undoubtedly know, the ear has dual sensory functions. In addition to its role in hearing, it also functions as the sense organ of balance or equilibrium. The stimulation responsible for hearing and balance involves activation of *mechanoreceptors* called *hair cells*. Sound waves and fluid movement are the physical forces that act on these hair cells.

The ear is divided into three anatomical parts: **external ear, middle ear,** and **inner ear** (Figure 14-7). Note, however, that the structures drawn in Figure 14-7 are not represented according to scale. This rendering will help you visualize the internal structures more clearly.

External Ear

The external ear has two divisions: the *auricle* or *pinna,* and the *external acoustic meatus* (ear canal). The **auricle** is the visible part of your ear surrounding the opening of the *external acoustic*

meatus (see Figure 14-7). The ear canal, just behind the *tragus* of the ear, is about 3 cm (1¼ inch) long. It first slants upward a bit and then curves downward before passing into the temporal bone. The canal ends at the **tympanic membrane** (eardrum), which separates the canal from the middle ear.

The auditory canal has specialized glands, called *ceruminous glands,* that secret the waxlike substance **cerumen** (earwax). Normal amounts of cerumen protect the skin of the ear canal. Cerumen also assists in cleaning and lubricating the canal. In this way it offers some protection from bacteria, fungal spores, insects, and water.

Middle Ear

The middle ear is a tiny cavity hollowed out of the temporal bone. It is lined by epithelium and contains the three **auditory ossicles:** the *malleus, incus,* and *stapes* (see Figure 14-7). The names of these very small bones describe their shapes: hammer, anvil, and stirrup, respectively.

The "handle" of the malleus is attached to the inner surface of the tympanic membrane. The head of the malleus attaches to the incus, which in turn attaches to the stapes.

There are several openings into the middle ear cavity. The first is created by the external acoustic meatus, covered by the tympanic membrane, as we've seen. There are also two openings that penetrate the internal ear: the **oval window** (into which the stapes fits) and the **round window,** which is covered by a membrane. This is illustrated for you later in Figure 14-9.

Finally, there is the **auditory tube (eustachian tube),** composed partly of bone and partly of cartilage and fibrous

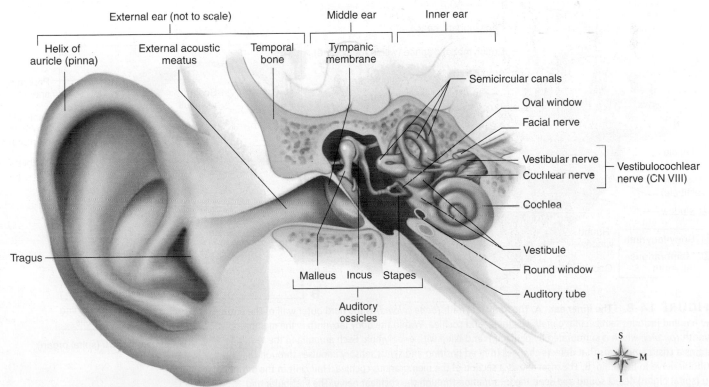

▲ **FIGURE 14-7** **The ear.** External, middle, and inner ear. (Anatomical structures are not drawn to scale.)

tissue. The auditory tube is lined with mucosa. It extends from the middle ear down to the nasopharynx.

The auditory tube makes it possible to equalize the pressure against the inner and outer surfaces of the tympanic membrane. This equalization prevents the membrane from rupturing as well as producing the discomfort that marked changes in atmospheric pressure produce.

Rapid changes in atmospheric pressure may happen when you are flying, especially during takeoff and landing. You equalize pressure on the tympanic membrane by swallowing or yawning, which allows air to spread rapidly through the open tube. Because atmospheric pressure is always exerted against its outer surface, these actions create equal pressure on both sides of the tympanic membrane.

Inner Ear

The inner ear is also called the **labyrinth** ("maze") because of its complicated shape. As you read through the following paragraphs, please refer often to Figure 14-8 to help you understand its structural components.

First, the inner ear consists of a *bony* labyrinth and, inside this, a *membranous* labyrinth. The bony labyrinth consists of three parts: the *vestibule, cochlea,* and *semicircular canals.*

The membranous labyrinth consists of the *utricle* and *saccule* (inside the vestibule), the *cochlear duct* (inside the cochlea), and the membranous *semicircular ducts* inside the bony semicircular canals.

The **vestibule** (containing the membranous utricle and saccule) and the **semicircular canals** (and their internal membranous ducts) are involved in balance. The **cochlea** (and membranous cochlear duct) is involved in hearing.

A clear, potassium-rich fluid called *endolymph* fills the membranous labyrinth. Another fluid, *perilymph,* surrounds the membranous labyrinth and fills the space between this membranous tunnel (and its contents) and the bony walls that surround it (see Figure 14-8, *A*).

Cochlea and Cochlear Duct

The word *cochlea* means "snail," which aptly describes the outer appearance of this part of the bony labyrinth. The membranous *cochlear duct*—the only part of the internal ear concerned with hearing—lies inside the cochlea. The cochlear duct is shaped somewhat like a triangular tube. It forms a continuous shelf across the inside of the bony cochlea, dividing it into upper and lower sections. The upper section (above the cochlear duct) is the *scala vestibuli* (vestibular duct). The lower section below the cochlear duct is the *scala tympani* (tympanic duct). The roof of the cochlear duct is the **vestibular membrane** *(Reissner membrane).* The **basilar membrane** is located on the floor of the cochlear duct. The basilar membrane is supported by bony and fibrous projections from the wall of the cochlea.

As we've seen, perilymph fills the scala vestibuli and scala tympani, and endolymph fills the cochlear duct. Refer again to Figure 14-8, *B,* to review the location of these important structures.

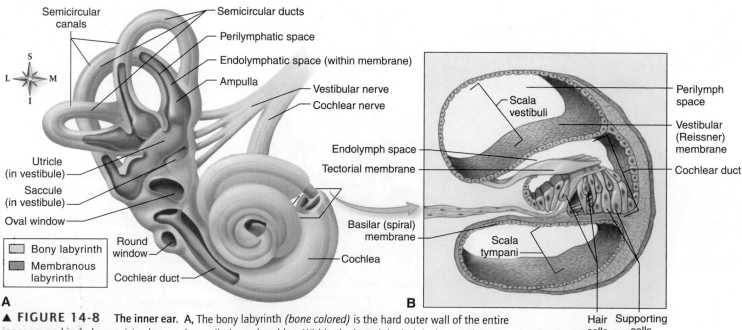

A

Semicircular canals
Semicircular ducts
Perilymphatic space
Endolymphatic space (within membrane)
Ampulla
Vestibular nerve
Cochlear nerve
Utricle (in vestibule)
Saccule (in vestibule)
Oval window
Round window
Bony labyrinth
Membranous labyrinth
Cochlear duct
Cochlea

B
Scala vestibuli
Endolymph space
Tectorial membrane
Basilar (spiral) membrane
Scala tympani
Perilymph space
Vestibular (Reissner) membrane
Cochlear duct
Hair cells
Supporting cells
Organ of Corti (spiral organ)

S L M I

▲ **FIGURE 14-8** **The inner ear. A,** The bony labyrinth *(bone colored)* is the hard outer wall of the entire inner ear and includes semicircular canals, vestibule, and cochlea. Within the bony labyrinth is the membranous labyrinth *(purple),* which is surrounded by perilymph and filled with endolymph. Each ampulla in the vestibule contains a crista ampullaris that detects changes in head position and sends sensory impulses through the vestibular nerve to the brain. **B,** The inset shows a section of the membranous cochlea. Hair cells in the organ of Corti (spiral organ) detect sound and send the information through the cochlear nerve. The vestibular and cochlear nerves join to form the eighth cranial nerve.

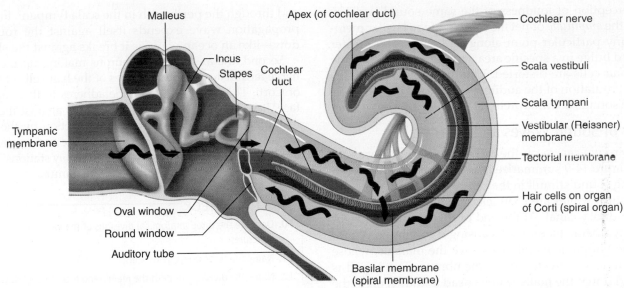

▲ **FIGURE 14-9** **Effect of sound waves on cochlear structures.** The tympanic membrane vibrates when sound waves strike it. This vibration then causes the membrane of the oval window to vibrate. In turn, vibration of the oval window produces movement in the perilymph in the bony labyrinth of the cochlea and the endolymph in the membranous labyrinth of the cochlea, or cochlear duct. Movement of endolymph causes the basilar (spiral) membrane to vibrate, which in turn stimulates hair cells on the organ of Corti (spiral organ) to transmit nerve impulses along the cranial nerve. Eventually, nerve impulses reach the auditory cortex and are interpreted as sound..

The hearing sense organ, the **organ of Corti,** rests on the basilar membrane throughout the entire length of the cochlear duct. The organ of Corti is also called the *spiral organ* because it curls in a spiral within the cochlea. The organ of Corti consists of supporting cells, as well as important hair cells that project into the endolymph. A gelatinous membrane, the *tectorial membrane,* covers these cells. Dendrites of the sensory neurons start around the bases of the hair cells of the organ of Corti. Axons of these neurons form the **cochlear nerve** (a branch of CN VIII), which extends to the brain. The cochlear nerve conducts impulses that produce our sense of hearing.

Process of Hearing

Sound is created by vibrations that may occur in air, fluid, or solid material. When we speak, for example, the vibrating vocal cords create sound waves by producing vibrations in the air passing over them. The number of sound waves that occur during a specific time unit is the *frequency.* To be heard, sound waves must have a frequency range that is capable of stimulating the hair cells in the organ of Corti at some point along the basilar membrane.

As you can see in Figure 14-9, the basilar membrane is *not the same width and thickness* throughout its coiled length. For this reason, different frequencies of sound will cause the basilar membrane to vibrate and bulge upward at different places along its length. Two "bulges" are shown for you in Figure 14-9. When a particular portion of the basilar membrane bulges upward, the cilia on the hair cells attached to that specific area are stimulated. As a result, a particular pitch is relayed to and perceived by the brain.

BOX 14-1 Health Matters

Cochlear Implants

Advances in electronic circuitry are being used to correct some forms of nerve deafness. If the hairs on the organ of Corti (spiral organ) are damaged, nerve deafness results—even if the vestibulocochlear nerve is healthy. A surgically implanted device can improve this form of hearing loss by eliminating the need for the sensory hairs.

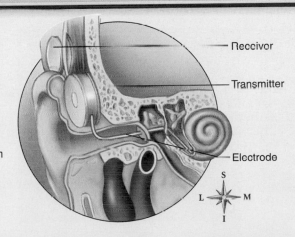

As you can see in the figure, a transmitter just outside the scalp sends external sound information to a receiver under the scalp (behind the auricle). The receiver translates the information into an electrical code that is relayed down an electrode to the cochlea. The electrode, wired to the organ of Corti, stimulates the vestibulocochlear nerve endings directly. Thus even though the cochlear hair cells are damaged, sound can be perceived.

Our perception of loudness of the same sound is determined by the *amplitude* or net movement of the basilar membrane at any particular point along its length. The greater the upward bulge in a specific area, the more the cilia on the attached hair cells are distorted. Hearing ultimately results from the stimulation of the auditory area of the cerebral cortex, where sounds are interpreted.

Pathway of Sound Waves and Hearing

At this point we can recap the rather multi-step process of hearing. Figure 14-9 summarizes for you the steps involved in detecting sound stimuli in the ear.

First, sound waves are channeled by the auricle into the external auditory canal. At the end of this canal, the waves strike the tympanic membrane, causing it to vibrate. Vibrations of the tympanic membrane move the malleus, whose "handle" attaches directly to the membrane. The sound is transmitted down the malleus to its head and then on to the incus and stapes. The stapes, in turn, vibrates the oval window, exerting pressure waves into the perilymph of the scala vestibuli of the cochlea. This action starts a "ripple" in the perilymph that is transmitted through the vestibular membrane (the roof of the cochlear duct) to the endolymph inside the duct. From here, the transmission proceeds to the organ of Corti and to the basilar membrane (which supports the organ of Corti and forms the floor of the cochlear duct). From the basilar membrane, the ripple is next transmitted to

and through the perilymph in the scala tympani. Finally, the propagation wave expends itself against the round window—like an ocean wave as it breaks against the shore.

Second, the axons of the neurons making up the cochlear nerve terminate around the bases of the hair cells of the organ of Corti. The tectorial membrane adheres to their upper surface. It is the movement of the hair cells against the tectorial membrane that initiates impulse conduction by the cochlear nerve to the brainstem. Before reaching the auditory area of the temporal lobe, impulses pass through "relay stations" in nuclei in the medulla, pons, midbrain, and thalamus.

> **QUICK ✓ CHECK**
>
> 10. Describe the three anatomical divisions of the ear and their basic structures.
> 11. Name the three auditory ossicles.
> 12. Name the divisions of both the membranous and bony labyrinths.
> 13. What is the specific sense organ responsible for the perception of hearing?

SENSE OF BALANCE

Sense of Equilibrium

The sense organs involved with balance, or **equilibrium,** are found in the vestibule and semicircular canals. The sense organs located within the vestibule (the utricle and

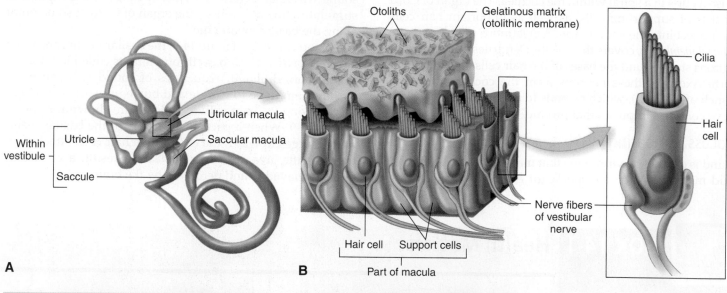

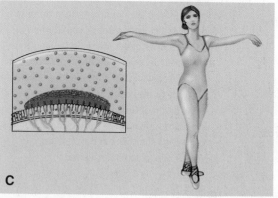

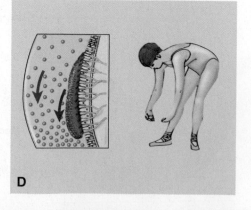

◀ **FIGURE 14-10**   The macula. **A,** Structure of vestibule showing placement of utricular and saccular maculae. **B,** Section of macula showing otoliths. **C,** Macula stationary in upright position. **D,** Macula displaced by gravity as person bends over.

saccule) maintain a state of *static equilibrium.* This state provides a reference point to sense the position of the head relative to gravity or to sense acceleration or deceleration (such as you might experience while driving a car). In contrast, the sense organs associated with the semicircular ducts function in *dynamic equilibrium.* This function is required to maintain balance when the head or body *itself* is rotated or suddenly moved. Let's look at each of these in greater detail.

Static Equilibrium

A small patch of *sensory epithelium* called the **macula** (plural, *maculae*) is found within both the utricle and the saccule of the vestibule (Figure 14-10, *A*). These *maculae,* located at nearly right angles to each other, contain receptor hair cells covered by a gelatinous matrix. The macula in the utricle is *parallel* to the base of the skull. The macula in the saccule is *perpendicular* to the base of the skull. **Otoliths,** or tiny "ear stones" composed of protein and calcium carbonate, are located within the matrix *above* each macula (Figure 14-10, *B*).

Changing the position of the head produces a change in the amount of pressure on the otolith-weighted matrix. This in turn stimulates the hair cells (Figure 14-10, *C* and *D*), which then stimulate the adjacent receptors of the vestibular nerve. It is the movement of the receptor cells within the macular patches that provides information related to head position or acceleration.

In addition to maintaining a sense of static equilibrium, stimulation of the macula creates our *righting reflexes.* These are neuromuscular responses that restore the body and its parts to their normal position after they have been displaced. Impulses from proprioreceptors and from the eyes also activate righting reflexes. Disruption of our righting reflexes can cause bouts of nausea, vomiting, and other uncomfortable symptoms to take place. These symptoms can be caused by the interruption of the vestibular, visual, or proprioceptive impulses.

Dynamic Equilibrium

Dynamic equilibrium depends on the **crista ampullaris,** located in the ampulla of each semicircular duct. This unique structure is a form of sensory epithelium similar to that of the maculae. Each ridgelike crista is dotted with many hair cells, and each has its processes embedded in a gelatinous flap called the **cupula** (Figure 14-11, *A* and *B*). Note, however, that the cupula is not weighted with otoliths and does not respond to the pull of gravity. Instead, it acts much like a

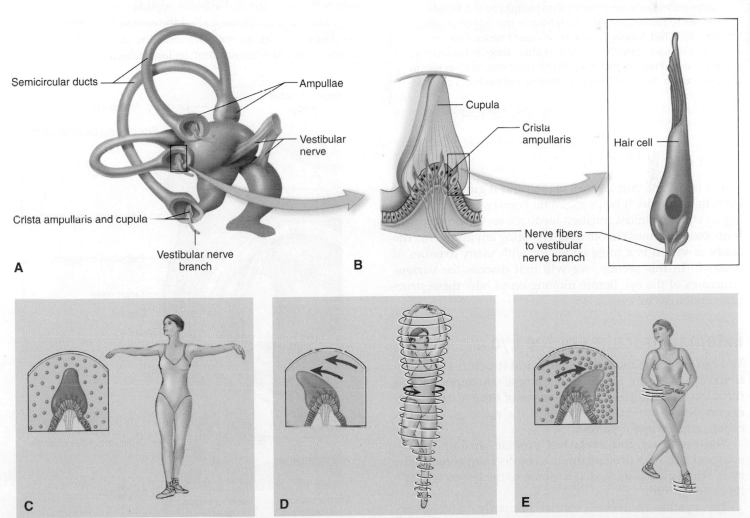

▲ **FIGURE 14-11**   **Structure and function of the crista ampullaris. A,** Semicircular ducts showing location of the crista ampullaris in ampullae. **B,** Enlargement of crista ampullaris and cupula. **C,** When a person is at rest, the crista ampullaris does not move. **D** and **E,** As a person begins to spin, the crista ampullaris is displaced by the endolymph in a direction opposite to the direction of spin.

float that moves with the flow of endolymph in the semicircular ducts.

Like the maculae, the semicircular ducts are placed nearly at right angles to each other. This arrangement enables us to detect movement in all dimensions and directions. As the cupula moves, it bends the hairs embedded in it. A receptor potential followed by an action potential is created that passes through the vestibular portion of the eighth cranial nerve. This impulse is transmitted to the medulla oblongata, and from there to other areas of the brain and spinal cord for interpretation, integration, and response.

When a person spins, such as in a pirouette (see Figure 14-11, *C* to *E*), the semicircular ducts move with the body. However, *inertia* (the tendency of a body to resist acceleration) keeps the endolymph in the ducts from moving at the same rate. The cupula therefore moves in a direction *opposite to head movement* until after the initial movement stops. Dynamic equilibrium is thus able to detect changes in both the direction and the rate at which movement occurs.

A&P CONNECT

Vertigo involves a sensation of one's own body spinning in space or of the external world spinning around the individual. Vertigo can be a frightening and recurring problem often precipitated in affected individuals by sudden changes in body position that may occur when rolling over in bed, bending to pick up an object from the floor, or simply sitting up quickly. This condition is often related to the abnormal displacement of rocks in your head. Really! If you want to know how to fix the "rocks in your head," check out **Vertigo and Ear Rocks** online at **A&P Connect.**

ⓔvolve

VISION: THE EYE

Vision is an amazing sensory ability, and it guides virtually everything we do. It helps maintain homeostasis by providing us with an almost constant feedback on various types of form and movement in our ever-changing environment. The study of vision is a huge endeavor with many avenues of research. In this section, we will first discuss the various structures of the eye before moving on to how these structures enable us to see.

External Structures of the Eye

You can locate the following anatomical structures of the external eye in Figure 14-12. Some of these structures, such as the *sclera, iris,* and *pupil,* are structures of the eye itself. Others are accessory structures such as the *eyebrows, eyelashes, eyelids,* and *lacrimal apparatus.*

The **eyebrows** and **eyelashes** provide shading for the eyes and also help prevent foreign objects from entering the eyes. Small glands located at the base of the lashes secrete a lubricating fluid.

The **eyelids** *(palpebrae)* consist primarily of voluntary muscle and skin. There is a border of thick connective tissue—the *tarsal plate*—at the free edge of each lid. You can feel this

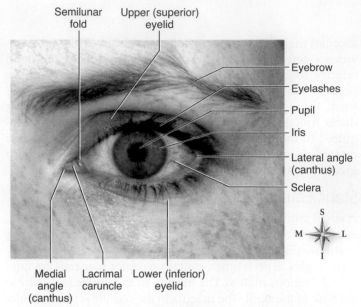

▲ **FIGURE 14-12**   **External eye structures.** Visible portion of the adult eye and surrounding structures.

ridgelike plate when you turn back the eyelid to remove a foreign object. A lateral and medial angle or **canthus** forms where the superior and inferior eyelids meet.

A mucous membrane, called the *conjunctiva,* lines each eyelid (Figure 14-13). It continues over the surface of the eyeball, where it becomes transparent. Inflammation of the conjunctiva

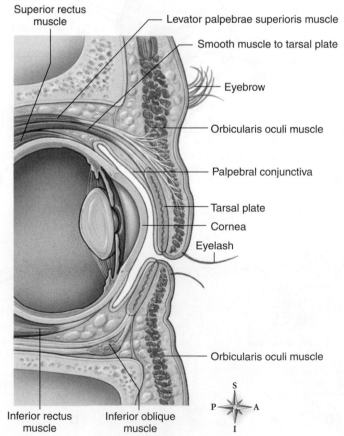

▲ **FIGURE 14-13**   **Accessory structures of the eye.** Lateral view with eyelids closed.

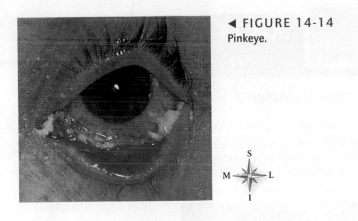

◄ **FIGURE 14-14**
Pinkeye.

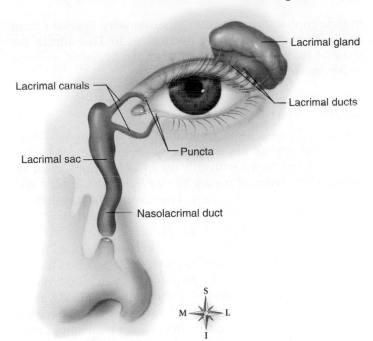

▲ **FIGURE 14-15** **Lacrimal apparatus.** Fluid produced by lacrimal glands (tears) streams across the eye surface, enters the canals, and then passes through the lacrimal sac and nasolacrimal duct to enter the nose.

(conjunctivitis) is a fairly common and irritating infection. If an infection produces a pinkish discoloration of the eye's surface, it is commonly called *pinkeye* (Figure 14-14).

The **lacrimal apparatus** comprises structures that secrete tears and drain them from the surface of the eyeball. We have illustrated lacrimal glands, lacrimal ducts, lacrimal sacs, and nasolacrimal ducts for you in Figure 14-15.

The *lacrimal glands,* each about the size of a small almond, are at the upper outer margin of each eye orbit. About a dozen small ducts lead from each lacrimal gland, draining the tears into the conjunctiva at the upper outer corner of the eye.

The *lacrimal canals* empty into the lacrimal sacs, and open into an opening or **punctum** (plural, *puncta*). You can see the puncta as a small dot at the medial angle (inner canthus) of each eye. The *lacrimal sacs* are located in a groove in the lacrimal bone. The *nasolacrimal ducts* are small tubes that extend from the lacrimal sac into the inferior meatus of the nose.

All of our tear ducts are lined with mucous membranes, extensions of the mucosa that lines the nose. When this membrane becomes inflamed and swollen, the nasolacrimal ducts become plugged, causing the tears to overflow from the eyes instead of draining into the nose as they normally do. This explains why our eyes "water" when we have a bad cold.

Muscles of the Eye

Muscles of the eye can be either *extrinsic* or *intrinsic.*

Extrinsic eye muscles are skeletal muscles that attach to the outside of the eyeball and to the bones of the orbit. They are striated voluntary muscles that collectively can move our eyeball in virtually any direction. Four of them are straight muscles and two are oblique, and their names describe their positions on the eyeball. They are the *superior, inferior, medial,* and *lateral rectus* muscles and *superior* and *inferior oblique* muscles (Figure 14-16).

Intrinsic eye muscles, in contrast, are *smooth (involuntary)* muscles located *within* the eyes. As you probably know, the *iris* regulates the size of the pupil. However it is the *ciliary muscle* that controls the shape of the lens. When the ciliary

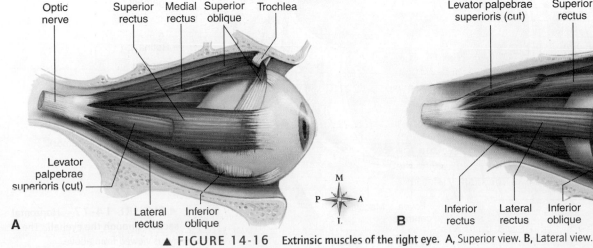

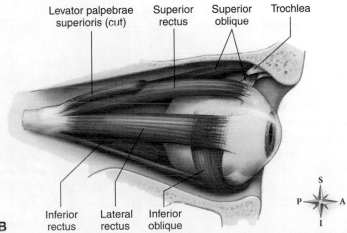

▲ **FIGURE 14-16** **Extrinsic muscles of the right eye.** **A,** Superior view. **B,** Lateral view.

muscle contracts, it releases the suspensory ligament from the backward pull usually exerted on it. This allows the somewhat elastic lens, suspended in the ligament, to bulge or become more *convex*.

Layers of the Eyeball

The eyeball forms a globe about 25 mm in diameter. About 80% of the eyeball lies in the bony orbit of the eye socket. Here, it lies protected with only the small anterior surface exposed. Three layers of tissues (*fibrous layer, vascular layer,* and *inner layer*) compose the external membrane of the eyeball. As you read about these layers, please refer to Figure 14-17.

The anterior portion of the fibrous layer, the *sclera*, is called the **cornea** and lies over the colored *iris* (see Figures 14-12 and 14-17). The cornea is transparent, whereas the rest of the sclera is white and opaque. Although blood vessels may form spiderlike patterns in the sclera, no blood vessels are found in the cornea or in the lens. Corneas that have lost their transparency enough to affect vision sometimes can be replaced by transplants from a donor (Box 14-2).

The vascular layer consists of three components—the choroid, ciliary body, and iris. This layer has many blood vessels and a large amount of the dark pigment *melanin*. Most of this layer, forming the middle and posterior coating just inside the fibrous layer, is the highly pigmented **choroid.**

The **ciliary body** is a thickening of the choroid. It fits like a collar into the area between the anterior margin of the

BOX 14-2 Health Matters

Corneal Transplants

Surgical removal of an opaque or deteriorating cornea and replacement with a donor transplant is a common medical practice. Corneal tissue is avascular; that is, the cornea is free of blood vessels. Therefore, corneal tissue is seldom rejected by the body's immune system. Antibodies carried in the blood have no way to reach the transplanted tissue, and therefore long-term success following

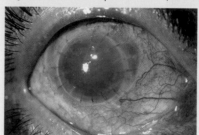 implant surgery is excellent. The figure is a photo of a patient with a recent corneal transplant. Note sutures and mild edema.

retina and the posterior margin of the iris (Figure 14-18). The small *ciliary muscle,* composed of both radial and circular *smooth* muscle fibers, lies in the anterior part of the ciliary body. The lens is held suspended by suspensory ligaments that attach between the folds of the ciliary body and the lens.

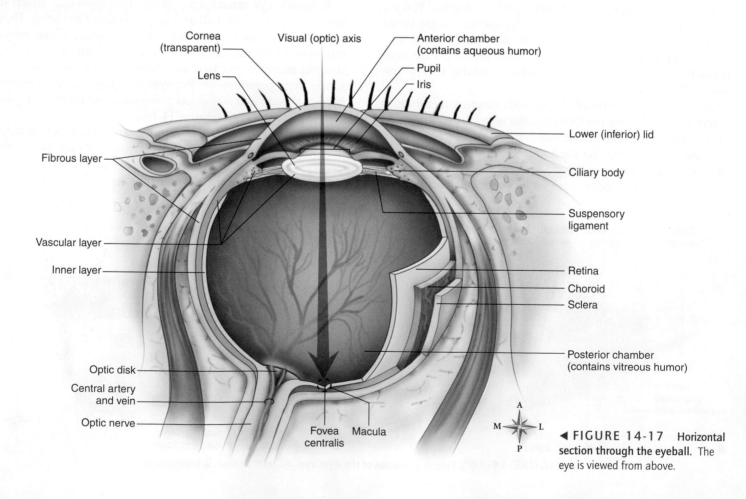

◄ FIGURE 14-17 Horizontal **section through the eyeball.** The eye is viewed from above.

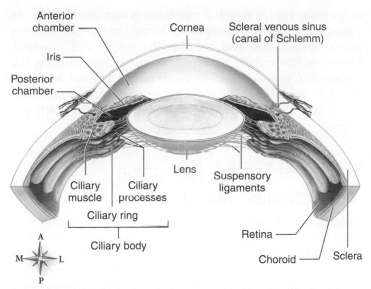

▲ **FIGURE 14-18** **Lens, cornea, iris, and ciliary body.** Note the suspensory ligaments that attach the lens to the ciliary body.

The **iris,** or the colored part of the eye, consists of circular and radial smooth muscles that form a ringlike structure that lies between the cornea and the lens. The iris is continuous with the ciliary body posteriorly. The opening in the middle of the iris is the *pupil.* The **pupil** functions like the aperture of a camera. It controls the amount of light that enters the eye by adjusting its diameter. You can watch your iris contract and expand, thereby controlling the size of the pupil, by looking in the mirror as you close your eyes, then open them quickly.

The actual eye color of the iris is determined by the amount, placement, and type of melanin in the iris.

The **retina** is the innermost lining of the eyeball and has no anterior portion. Epithelial cells containing melanin form the layer of the retina next to the choroid coat. This pigmented area is called the *pigmented retina.* However, most of the retina consists of nervous tissue and is called the *sensory retina.* Within this sensory layer, three types of neurons are found: *main photoreceptor cells, bipolar cells,* and *ganglion cells.* You can see how each of these cells is arranged in Figure 14-19, *A.* Note that the main photoreceptor cells are the deepest.

The ends of the dendrites of the main photoreceptor neurons are named after their basic shapes. As their names suggest, **rods** look like tiny rods, and the **cones** look like tiny cones (see Figure 14-19, *B* and *C*). Rods and cones are the principle visual receptors and differ in function, number, and distribution. Cones are less numerous than rods and are most densely concentrated in the **fovea centralis.** This is a small depression in the middle of the yellowish **macula lutea** found near the center of the retina (Figures 14-17 and 14-20). The cones become less and less dense from the fovea outward. Rods, in contrast, are absent entirely from the fovea and macula and increase in density toward the periphery of the retina.

Note in Figure 14-19 that a *bipolar neuron* connects either to one cone or to many rods. The **bipolar cells** are neurons that receive impulses from the rods and cones. They then pass these impulses to the **ganglion neurons.** These neurons also act like photoreceptors themselves. There are also

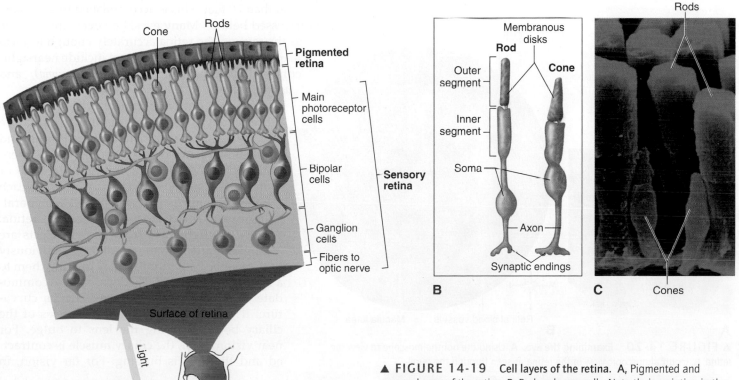

▲ **FIGURE 14-19** **Cell layers of the retina. A,** Pigmented and sensory layers of the retina. **B,** Rod and cone cells. Note their variation in the general structure of a neuron. **C,** Electron micrograph of rod and cone cells.

lateral connecting cells among the bipolar and ganglion cells. These lateral connecting cells are thought to help us detect patterns of movement. Ultimately, all the axons of ganglion neurons extend back to a small circular area in the posterior part of eyeball called the **optic disk.** This part of the sclera contains perforations through which the fibers emerge from the eyeball as the **optic nerve.** We also call the optic disk the **blind spot** because it does not have rods or cones and thus cannot respond to light rays.

Physicians use an **ophthalmoscope** to examine the retinal surface called the *fundus* at the back of the eye. With an ophthalmoscope, you can see the branching patterns of the retinal blood vessels (see Figure 14-20), which are critical for normal vision. Diseases such as diabetes mellitus and atherosclerosis can reduce blood flow to these vessels, causing loss of vision.

Cavities and Humors

Within the eyeball is a large interior space divided into **anterior** and **posterior cavities.** In turn, the *anterior cavity* is subdivided into two other chambers, both filled with **aqueous humor.** This clear, watery substance often leaks out when the eye is injured. The *posterior cavity* is much larger than the anterior cavity and occupies the entire space behind the lens and ciliary body. This cavity contains **vitreous humor,** with a consistency similar to that of soft gelatin. Both aqueous and vitreous humors help maintain enough internal pressure within the eye to prevent the eyeball from collapsing.

Aqueous humor is derived from blood in capillaries located largely in the ciliary body. It is secreted into the posterior chamber and then filters back into the anterior chamber. Usually aqueous humor drains out of the anterior chamber at the same rate at which it enters the posterior chamber. As a result, the amount of aqueous humor in the eye remains relatively constant and so does the intraocular pressure. However, if something happens to upset this balance, the intraocular pressure can increase above normal levels of about 20 mm Hg. This creates a condition called **glaucoma,** which can lead to retinal damage and blindness if left untreated.

Process of Seeing: Refraction and Accommodation

Refraction is the bending of light produced as light rays pass *at an angle* from one transparent medium into another of different optical density. The more convex the light-bending surface, the greater will be its refractive power. The refracting media of the eye are the cornea, aqueous humor, lens, and vitreous humor. Light rays are bent, or *refracted,* at the anterior surface of the cornea as they pass from the (less dense) air into the (more dense) cornea. The light rays then pass through the aqueous humor in the anterior chamber into the denser lens, and finally into the less dense vitreous humor. An optometrist or ophthalmologist can measure the refraction or light-bending power of the eye during an eye examination.

As you might suspect, tests of visual acuity also rely on a person's ability to *refract light properly.* Under normal conditions, the four refracting media together bend light rays enough to bring to focus (on the retina) the virtually parallel rays reflected from an object 20 or more feet (6 m) away. Of course, a normal eye can also focus on objects located much closer to you than 20 feet. This is accomplished by *accommodation* (discussed below). Many eyes, however, are not able to focus the rays on the retina accurately enough to form a clear image. These physical errors include nearsightedness *(myopia),* farsightedness *(hyperopia),* and *astigmatism.*

Accommodation for near vision requires three changes: (1) an increase in the curvature of the lens, (2) constriction of the pupils, and (3) convergence of the two eyes. As we mentioned above, light rays from objects 20 or more feet away are nearly *parallel* to each other. The normal eye can refract such parallel rays to focus them clearly on the retina. However, light rays from nearer objects are *divergent* rather than parallel. So obviously they must be bent more acutely to bring them to a focus on the retina. The lens must accommodate such conditions by increasing its curvature. It does so by *relaxing* the muscles of the ciliary body, allowing the lens to bulge. For near vision, then, the ciliary muscle is contracted and the lens is bulging. For far vision, in

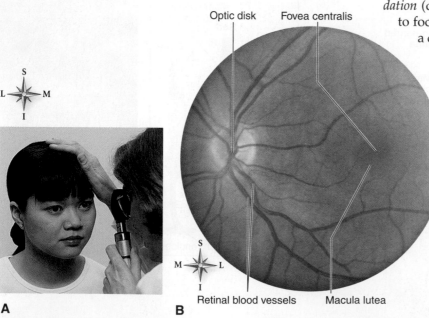

A   **B**

▲ **FIGURE 14-20**   **Examining the eye. A,** Using the ophthalmoscope to view the retina. **B,** Ophthalmoscopic view of the retina as seen through the pupil.

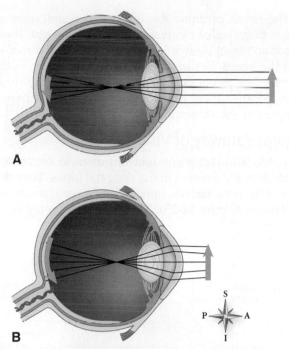

A

B

▲ **FIGURE 14-21** **Accommodation of lens. A,** Distant image: the lens is flattened (ciliary muscle relaxed), and the image is focused on the retina. **B,** Close image: the lens is rounded (ciliary muscle contracted), and the image is focused on the retina.

contrast, the ciliary muscle is relaxed and the lens is comparatively flat (Figure 14-21).

In middle adulthood, our lenses become less flexible, making it difficult to accommodate our vision to near objects. For example, reading becomes difficult unless the written material is far enough away from the eye to require little or no accommodation of the lens. This condition, which worsens with increased age, is called *presbyopia* (literally, "oldsightedness").

The muscles of the iris also play an important part in the formation of clear retinal images. Part of the accommodation process involves the contraction of the circular fibers of the iris, which constricts the pupil. This **constriction** of the pupil prevents divergent rays from the object from entering the eye through the periphery of the cornea and lens. This is important because divergent rays cannot be focused clearly on the retina. Constriction of the pupil takes place at the same time as accommodation of the lens. The pupil also constricts during sudden exposure to bright light. This *pupillary light reflex*, a rapid constriction of the iris, protects the retina from stimulation that is too intense or sudden.

Our eyes are set up for binocular vision, which gives us the ability to have depth perception and see things in a three-dimensional manner. To do this, we must converge and integrate images received at slightly different angles from each eye. **Convergence** is the movement of both eyes

inward so that their visual axes come together, or converge, at the object being viewed. The nearer the object is, the greater the degree of convergence necessary to maintain focused, single vision.

Role of Photopigments

Both rods and cones contain **photopigments,** or light-sensitive pigmented compounds. These photopigments are found in the distal end of the photoreceptor cells near the pigmented retina (see Figure 14-19). Chemically, all photopigments can be reduced to a glycoprotein called an **opsin** and a vitamin A derivative called **retinal.** It is the retinal that acts as the light-absorbing portion of all photopigments.

Rhodopsin is the only photopigment found in rods. It is so light sensitive that even dim light causes a rapid breakdown of the photopigment into its opsin and retinal components. This action triggers a change in the membrane potential of the rod cells, which then travels as a neural impulse to the brain for interpretation. Energy is required to bring back opsin to its original molecular shape. Until this takes place, the photopigment remains "bleached" and cannot respond to light.

The color of light is measured by its wavelengths (usually in nanometers or *nm*). Thus, we see different wavelengths of light energy as different colors. Figure 14-22 shows you that rods are most sensitive to light in the green range of wavelengths.

Three types of cones are found in the retina. Each cone contains its own version of the rhodopsin photopigment, and

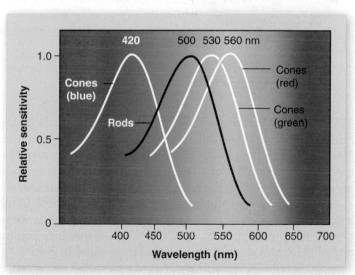

▲ **FIGURE 14-22** **Color sensitivity of rods and cones.** Rods *(black line)* are sensitive only in dim-light conditions in the range of greenish wavelengths—but their information is perceived only as light and dark, not green. Three types of cones *(white lines)* perceive colors in bright-light conditions. Blue (short-wave) cones are sensitive in the blue range, green (medium-wave) cones in the green range, and red (long-wave) cones in the red range. Information from all three cone types is combined to produce a large palette of color perception in the brain.

UNIT 3

BOX 14-3 FYI

Color Blindness

Color blindness, usually an inherited condition, is caused by mistakes in the functioning of the three *photopigments* in the cones. Each photopigment is sensitive to one of the three primary colors of light: green, blue, and red (see Figure 14-22). In many cases, the green-sensitive photopigment is missing or deficient; other times, the red-sensitive photopigment is abnormal. (Dysfunction of the blue-sensitive cone is rare.) Color-blind individuals see colors, but they cannot distinguish between them normally. Figures such as parts *A* and *B* shown here (part of the Ishihara test) are often used to screen individuals for color blindness. A person with red-green color blindness cannot see the *74* in part *A* of the figure, whereas a person with normal vision can. To determine which photopigment is deficient, a color-blind person may try a figure similar to part *B*. Persons with a deficiency of red-sensitive photopigment can distinguish only the *2;* those deficient in green-sensitive photopigment can only see the *4*.

A

B

The fovea contains the greatest concentration of cones and is the point of clearest vision in good light. If you want to see an object clearly in the daytime, look directly at it so that the image focuses on the fovea. However, in dim light you will see an object better if you look at it slightly from the side. Because there are more rods in the periphery of the retina, you should see the object more clearly.

Neural Pathway of Vision

The rods and cones conduct impulses to nerve fibers that reach the visual cortex in the occipital lobes. They do this by way of the optic nerves, optic chiasma, optic tracts, and optic radiations (Figure 14-23). There are also "relay stations" in

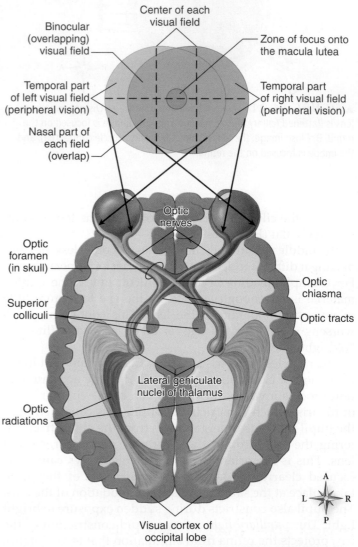

▲ **FIGURE 14-23** **Visual fields and neural pathways of the eye.** Note the structures that make up each pathway: optic nerve, optic chiasma, lateral geniculate body of thalamus, optic radiations, and visual cortex of occipital lobe. Fibers from the nasal portion of each retina cross over to the opposite side at the optic chiasma and terminate in the lateral geniculate nuclei.

each is different from the rhodopsin found in rod cells. Blue-sensitive cones are sensitive to the short wavelengths in the "blue" range of the visual spectrum, green-sensitive cones to medium wavelengths of visual light, and red-sensitive cones are sensitive to reds in the long wavelengths of the spectrum.

Our perception of the spectrum of colors available to us results from the combined neural input from varying numbers of the three different cone types. Abnormal function in any of the cones disrupts the normal perception of colors—a condition called **color blindness** (see Box 14-3 for a more complete explanation).

Because cone photopigments are less sensitive to light than the rhodopsin in rods, brighter light is necessary for their breakdown. For this reason, cones function to produce color vision only in bright light. Cones also contribute more than rods to the perception of sharp images.

the thalamus that conduct the visual impulses. Note that each optic nerve contains fibers from only one retina but that the optic chiasma contains fibers from both retinas. Each optic tract also contains fibers from both retinas.

A&P CONNECT

You have probably heard of eye surgeries such as RK, CK, PRK, or LASIK. But you may have wondered, *what is* a LASIK? Check out **Refractive Eye Surgery** online at **A&P Connect** to explore the common refractive eye surgeries and how they are done.

evolve

QUICK CHECK

14. What is the difference between static and dynamic equilibrium?
15. Outline the basic structures of the external eye.
16. Identify the layers of the retina.
17. Name the four processes that function to focus a clear image on the retina.
18. What causes color blindness?

Cycle of LIFE

The ability of our sense organs to respond to both internal and external stimuli is vital to our lives. Ultimately, all sensory information is acquired by the depolarization of our sensory nerve endings. Anything that interferes with our ability to receive stimulation or transfer information to the central nervous system can greatly reduce the acuity of our senses. Age, disease, structural defects, and lack of maturation all affect our ability to identify and respond to sensory input.

It is important to understand that the sensory structures and their development and response abilities are related to our own overall development. For example, newborn babies have more limited vision, hearing, and tactile abilities than young adults. As children mature, however, their sensory abilities become much more acute. Unfortunately, as we age, all of our senses age with us. By late adulthood, we suffer progressive hearing loss, a reduced ability to focus clearly, and a reduced sense of smell and taste. This is in large part due to structural deterioration. The lenses of our eyes become harder and less able to change shape, taste buds become less functional, and exteroreceptors of all types become less responsive to stimuli. ●

The BIG Picture

As always, seeing the "big picture" involves an understanding of how each system affects homeostasis. As such, you can see that our sense organs dramatically illustrate the relationship of structure and function. In fact, the proper functioning of our sensory systems is intimately tied to our success and livelihood.

We have seen that our somatic sense organs are widely distributed throughout the body. Their proper functioning serves to provide the body with vital input from both external and internal stimuli that ultimately affect our homeostasis. For example, pain—regardless of cause—is very often the first indicator of homeostatic imbalance. Furthermore, the ability to sense touch, pressure, vibration, stretch, or temperature changes on or in the body before these stimuli reach levels that may cause injury is vital to our survival.

How do our special senses play homeostatic roles? Classic examples include vision and hearing. These senses help us monitor our often hostile external environment so that we can avoid or respond to dangers that might otherwise be life threatening. Consider the sensation of thirst, for example. Controlling our thirst helps us regulate our water intake, allowing us to avoid dehydration while regulating important molecules such as sodium. Other examples might include a bitter taste with food or an offensive odor. These sensations are often associated with poisonous materials or toxic chemicals. Our sensations of them thus serve as important defense mechanisms.

At this point, you should take a few minutes to reflect on and review each of the somatic and special senses by relating them to the "big picture" of your body's homeostasis and survival.

MECHANISMS OF DISEASE

There are numerous diseases of the major sense organs, but perhaps the most significant and common ones affect our senses of hearing, balance, and vision. Hearing problems can involve conduction impairment or nerve impairment. In addition, hearing loss can occur for a variety of genetic, physical, and neural reasons. Likewise, there are many disorders of the eye, ranging from common refraction disorders such as nearsightedness and farsightedness to astigmatism and cataracts. Infections of the eye are also common and have the potential to impair vision significantly, sometimes even causing blindness. Finally, disorders of the visual pathways can occur because of trauma or stroke.

evolve Find out more about disorders and diseases of the major sense organs online at *Mechanisms of Disease: Sense Organs*.

LANGUAGE OF SCIENCE AND MEDICINE continued from page 297

crista ampullaris (KRIS-tah am-pyoo-LAIR-iss)
[*crista* ridge, *ampu-* flask, *-ulla-* little, *-aris* relating to]
pl., cristae ampulares (KRIS-tee am-pyoo-LAIR-ees)

cupula (KYOO-pyoo-lah)
[*cup-* tub, *-ula* little] *pl.*, cupulae (KYOO-pyoo-lee)

equilibrium (ee-kwi-LIB-ree-um)
[*equi-* equal, *-libr-* balance, *-um* thing] *pl.*, equilibria (ee-kwi-LIB-ree-ah)

eustachian tube (yoo-STAY-shun)
[*Bartolomeo Eustachio* Italian anatomist, *-an* relating to]

external ear (eks-TER-nal)
[*extern-* outside, *-al* relating to]

exteroceptor (eks-ter-oh-SEP-tor)
[*exter-* outside, *-cept-* receive, *-or* agent]

extrinsic eye muscle (eks-TRIN-sik eye MUSS-el)
[*extr-* outside, *-sic* beside, *mus-* mouse, *-cle* small]

eyebrow (EYE-brow)

eyelash (EYE-lash)

eyelid (EYE-lid)

fovea centralis (FOH-vee-ah sen-TRAL-is)
[*fovea* pit, *centralis* center] *pl.*, foveae centrales (FOH-vee-ee sen-TRAL-ees)

ganglion neuron (GANG-glee-on NOO-ron)
[*ganglion* knot, *neuron* nerve]

general sense organ (OR-gan)

glaucoma (glaw-KOH-mah)
[*glauco-* gray or silver, *-oma* tumor (growth)]

gustatory (GUS-tah-tor-ee)
[*gusta-* taste, *-ory* relating to]

gustatory cell (GUS-tah-tor-ee sell)
[*gusta-* taste, *-ory* relating to, *cell* storeroom]

inner ear

interoceptor (in-ter-oh-SEP-tor)
[*inter-* inward, *-cept-* receive, *-or* agent]

intrinsic eye muscle (in-TRIN-sik eye MUSS-el)
[*intr-* within, *-sic* beside, *mus-* mouse, *-cle* small]

iris
[*iris* rainbow]

labyrinth (LAB-ih-rinth)
[*labyrinth* maze]

lacrimal apparatus (LAK-ri-mal app-ah-RAT-us)
[*lacrima-* tear, *-al* relating to]

lamellar corpuscle (lah-MEL-ar KOR-pus-ul)
[*lam-* plate, *-ella-* little, *-ar* relating to, *corpus-* body, *-cle* little]

lateral connecting cell (LAT-er-al kon-NEKT-ing sell)
[*later-* side, *-al* relating to, *cell* storeroom]

macula (MAK-yoo-lah)
[*macula* spot] *pl.*, maculae or maculas (MAK-yoo-lee or MAK-yoo-lahz)

macula lutea (MAK-yoo-lah LOO-tee-ah)
[*macula* spot, *lutea* yellow] *pl.*, maculae luteae (MAK-yoo-lee LOO-tee-ee)

mechanoreceptor (mek-an-oh-ree-SEP-tor)
[*mechano-* machine (mechanical), *-cept-* receive, *-or* agent]

middle ear

muscle spindle (MUSS-el)
[*mus-* mouse, *-cle* small]

nociceptor (noh-see-SEP-tor)
[*noci-* harm, *-cept-* receive, *-or* agent]

odorant molecule (OH-doh-rent MOL-eh-kyool)
[*odor-* a smell, *-ant* agent, *mole-* mass, *-cul* small]

olfactory cilia (ohl-FAK-tor-ee SIL-ee-ah)
[*olfact-* smell, *-ory* relating to, *cili-* eyelid] *sing.*, cilium (SIL-ee-um)

olfactory epithelium (ohl-FAK-tor-ee ep-ih-THEE-lee-um)
[*olfact-* smell, *-ory* relating to, *epi-* upon, *-theli-* nipple, *-um* thing] *pl.*, epithelia (ep-ih-THEE-lee-ah)

ophthalmoscope (off-THAL-mah-skohp)
[*oph-* eye or vision, *-thalmo-* inner chamber, *-scop* see]

opsin (OP-sin)
[*ops-* vision, *-in* substance]

optic disk (OP-tik)
[*opti-* vision, *-ic* relating to]

optic nerve (OP-tik)
[*opti-* vision, *-ic* relating to]

organ of Corti (OR-gan of KOR-tee)
[*Alfonso Corti* Italian anatomist]

osmoreceptor (os-moh-ree-SEP-tor)
[*osmo-* push (osmosis), *-cept-* receive, *-or* agent]

otolith (OH-toh-lith)
[*oto-* ear, *-lith* stone]

oval window

papilla (pah-PIL-ah)
[*papilla* nipple] *pl.*, papillae (pah-PIL-ee)

perception

photopigment (foh-toh-PIG-ment)
[*photo-* light, *-pigment* paint]

photoreceptor (foh-toh-ree-SEP-tor)
[*photo-* light, *-cept-* receive, *-or* agent]

plexus (PLEKS-us)
[*plexus* network]

posterior cavity (pohs-TEER-ee-or KAV-ih-tee)
[*poster-* behind, *-or* quality, *cav-* hollow, *-ity* state]

proprioceptor (proh-pree-oh-SEP-tor)
[*propri-* one's own, *-cept-* receive, *-or* agent]

punctum (PUNK-tum)
[*punctum* small point] *pl.*, puncta (PUNK-tah)

pupil (PYOO-pil)
[*pup-* doll, *-il* little]

receptor potential (ree-SEP-tor poh-TEN-shal)
[*recept-* receive, *-or* agent, *potent-* power, *-ial* relating to]

referred pain

refraction (ree-FRAK-shun)
[*refract-* break apart, *-tion* process]

retina (RET-ih-nah)
[*ret-* net, *-ina* relating to]

retinal (RET-ih-nal)
[*ret-* net, *-ina-* relating to, *-al* relating to]

rhodopsin (roh-DOP-sin)
[*rhodo-* red, *-ops-* vision, *-in* substance]

rod

round window

semicircular canal (sem-ih-SIR-kyoo-lar)
[*semi-* half, *-circulare-* go around, *-lar* relating to, *canal* channel]

sensation
[*sens-* feel, *-ation* process]

special sense

special sense organ (OR-gan)

tactile corpuscle (TAK-tyle KOR-pus-ul)
[*tact-* touch, *-ile* relating to, *corpus-* body, *-cle* little]

tactile disk (TAK-tyle)
[*tact-* touch, *-ile* relating to]

taste bud

tendon organ (TEN-don OR-gan)
[*tend-* pulled tight, *-on* unit, *organ* instrument]

thermoreceptor (ther-moh-ree-SEP-tor)
[*thermo-* heat, *-cept-* receive, *-or* agent]

tympanic membrane (tim-PAN-ik MEM-brayne)
[*tympan-* drum, *-ic* relating to, *membran-* thin skin]

vertigo (VER-ti-go)
[*vertigo* turning]

vestibular membrane (ves-TIB-yoo-lar MEM-brayne)
[*vestibule-* entrance hall, *-ar* relating to, *membran-* thin skin]

vestibule (VES-tih-byool)
[*vestibule* entrance hall]

vitreous humor (VIT-ree-us HYOO-mor)
[*vitreous* glassy]

CHAPTER SUMMARY

To download an MP3 version of the chapter summary for use with your iPod or other portable media player, access the Audio Chapter Summaries *online at http://evolve.elsevier.com.*

HINT *Scan this summary after reading the chapter to help you reinforce the key concepts. Later, use the summary as a quick review before your class or before a test.*

INTRODUCTION

A. Sense organs fall into two main categories:
1. General sense organs—function to produce the general, or somatic, senses, such as touch, temperature, and pain
2. Special sense organs—function to produce our sense of vision, hearing, balance, taste, and smell

SENSORY RECEPTORS

A. Sensory receptors—responsive to changes in our internal and external environments; critical to our survival
B. Receptor response—receptors respond to stimuli by converting them to nerve impulses
1. When a receptor is stimulated, it develops a local potential in its membrane called a receptor potential
   a. When a receptor potential reaches a threshold, it triggers an action potential in the sensory neuron
   b. These impulses then travel to the brain and spinal cord, where they are interpreted as a particular sensation
2. Many receptors exhibit adaptation; the magnitude of the receptor potential decreases over time in response to a continuous stimulus
3. Perception—ability to remain aware of a particular sensation over time and to interpret that sensation
C. Distribution of receptors—receptors responsible for our special senses are grouped into specialized areas of our bodies
D. General sense organs are widely distributed throughout our bodies
1. The sensations generated by the general sense organs are usually called the somatic senses

CLASSIFICATION OF RECEPTORS

A. Classification by location—receptors are classified by location as exteroceptors, interoceptors, and proprioceptors
1. Exteroceptors—respond to stimuli that are external to the body
2. Interoceptors—provide information about the internal environment
3. Proprioceptors—inform us about body movement, orientation in space, and muscle stretching
B. Classification by stimulus detected—receptors can be classified based on the stimuli that activate them
1. Mechanoreceptors—respond to stimuli that in some manner measure changes in pressure or internal movement
2. Chemoreceptors—activated by the amount or changing concentration of certain chemicals
3. Thermoreceptors—activated by changes in temperature

4. Nociceptors—perceive intense stimuli of any type that causes tissue damage and pain
5. Photoreceptors—found in our eyes; respond to light stimuli in the visible spectrum
6. Osmoreceptors—important in detecting changes in the concentration of electrolytes and stimulating the thirst center
C. Classification by structure—somatic sensory receptors are also classified as either having (1) free nerve endings or (2) encapsulated nerve endings (Table 14-1)
1. Free nerve endings—simplest, most common, and most widely distributed sensory receptors
   a. They may be exteroceptors or interoreceptors
   b. These slender sensory fibers, or nociceptors, terminate in small swellings called dendritic knobs
   c. Pain sensations
      (1) Acute pain fibers—respond to sharp pain sometimes described as the type that "takes your breath away"
      (2) Chronic pain fibers—dull or aching pain
      (3) Referred pain—stimulation of pain receptors in deep viscera can be felt in areas of the skin overlying the organ, or even in areas far removed from the origin of the pain
   d. Temperature sensations—skin has many receptive fields that are sensitive to either cold or warmth, but not both
      (1) Thermoreceptors are not spread uniformly across the skin surface
   e. Tactile sensations—skin movement, itch, tickle, and discriminative touch
      (1) Root hair plexuses
      (2) Tactile (Merkel) disks (Figure 14-1, *A*)
2. Encapsulated nerve endings—have a connective tissue capsule surrounding their dendritic ends and act as mechanoreceptors
   a. Tactile receptors
      (1) Tactile (Meissner) corpuscles—pick up sensations of light touch in addition to sensations of smoothness or roughness
      (2) Bulbous (Ruffini) corpuscles—slow-adapting receptors; permit the skin of the fingers to continuously remain sensitive to deep pressure
      (3) Lamellar (Pacini) corpuscles—respond quickly to sensations of deep pressure, high-frequency vibration, and stretch
   b. Stretch receptors—associated with muscles and tendons; stimulated if the length of a muscle is stretched beyond a certain limit (Figure 14-1, *B*)
      (1) Classified as proprioceptors
      (2) Categorized as muscle spindles and tendon receptors

SENSE OF SMELL

A. Olfactory receptors—these sensory neurons are chemoreceptors
1. Receptor potentials are generated in olfactory sensory neurons when airborne odorant molecules dissolve in mucus covering the olfactory epithelium (Figure 14-3)
2. Cilia from these epithelial cells help to "mix" the mucus solvent and the dissolving odorant molecules in the air

3. Dissolved molecules then bind to protein odor receptors located in the cell membranes of the olfactory cells

B. Olfactory pathway
   1. If the level of odorant molecules dissolved in the mucus reaches a threshold level, a receptor potential and then an action potential are generated
   2. Senses of smell and taste are closely related (Figure 14-4)

C. Taste buds—sense organs that respond to gustatory (taste) stimuli
   1. A few taste buds are located in the lining of the mouth and on the soft palate
   2. Most are found in small, elevated projections of the tongue (papillae)
   3. Tongue papillae are classified by their structure (Figure 14-5, *A* and *B*)
      a. Fungiform papillae—large, mushroom-shaped bumps found in the anterior two thirds of the tongue surface
      b. Circumvallate papillae—huge, dome-shaped bumps arranged in a row that cross the back of the tongue
      c. Foliate papillae—reddish, leaflike ridges of mucosa found on the lateral edges and rear of the tongue surface
      d. Filiform papillae—bumps with threadlike projections; scattered among the fungiform papillae; main function is to determine food "texture"
   4. Taste buds house the chemoreceptors responsible for taste; stimulated by chemicals, called tastants, dissolved in the saliva
   5. Gustatory cells—cluster of chemoreceptors in taste buds
      a. Gustatory hairs—threadlike structures that extend from each gustatory cell
      b. Taste pore—opening in which gustatory hairs extend; bathed in saliva (Figure 14-5, *C*)
   6. Each taste receptor responds most effectively to only one of five "primary" taste sensations:
      a. Sour
      b. Sweet
      c. Bitter
      d. Salty
      e. Umami (savory)
   7. Our ability to experience a large variety of tastes results from a combination of input from these five primary sensations (Figure 14-6)

D. Neural pathway for taste
   1. Nervous impulses generated in the anterior two thirds of the tongue travel over the facial (CN VII) nerve
   2. Nervous impulses generated from the posterior third of the tongue are conducted by fibers of the glossopharyngeal (CN IX) nerve
   3. The vagus (CN X) nerve plays a minor role in taste

SENSE OF HEARING

A. The ear is divided into three anatomical parts: external ear, middle ear, and inner ear (Figure 14-7)

B. External ear has two divisions: auricle or pinna, and external acoustic meatus
   1. Auricle—visible part of your ear surrounding the opening of the external acoustic meatus
   2. Ear canal—slants upward a bit and then curves downward before passing into the temporal bone

3. Tympanic membrane—separates the canal from the middle ear

C. Middle ear—tiny cavity hollowed out of the temporal bone
   1. Lined by epithelium and contains three auditory ossicles (Figure 14-7):
      a. Malleus
      b. Incus
      c. Stapes
   2. Two openings penetrate the internal ear (Figure 14-9):
      a. Oval window
      b. Round window
   3. Auditory (eustachian) tube—extends from the middle ear down to the nasopharynx; makes it possible to equalize the pressure against the inner and outer surfaces of the tympanic membrane

D. Inner ear—also called the labyrinth (Figure 14-8)
   1. The bony labyrinth consists of three parts
      a. Vestibule—contains the utricle and saccule; involved in balance
      b. Semicircular canals—involved in balance
      c. Cochlea—involved in hearing

E. Cochlea and cochlear duct
   1. Cochlea—bony labyrinth
   2. Cochlear duct—lies inside the cochlea; only part of the internal ear concerned with hearing
      a. Shaped like a triangular tube
      b. Scala vestibuli—upper section
      c. Scala tympani—lower section
      d. Vestibular membrane (Reissner membrane)—roof of the cochlear duct
      e. Basilar membrane—located on floor of the cochlear duct
   3. Organ of Corti—rests on the basilar membrane throughout the entire length of the cochlear duct; also called spiral organ
      a. Axons of the neurons that begin around the organ of Corti, extend in the cochlear nerve to the brain to produce the sensation of hearing

F. Process of hearing
   1. Sound is created by vibrations
   2. Number of sound waves during a specific time is the frequency
   3. Sound waves must have a frequency range capable of stimulating hair cells in the spiral organ
   4. Basilar membrane is not the same width and thickness throughout its coiled length
      a. Different frequencies of sound will cause the basilar membrane to vibrate and bulge upward at different places along its length
   5. Perception of loudness of the same sound is determined by the amplitude or net movement of the basilar membrane at any particular point along its length
   6. Hearing ultimately results from the stimulation of the auditory area of the cerebral cortex

G. Pathway of sound waves and hearing (Figure 14-9)
   1. Sound waves are channeled by the auricle into the external auditory canal
   2. Sound waves strike the tympanic membrane, causing it to vibrate

3. Vibrations of the tympanic membrane move the malleus, whose "handle" attaches directly to the membrane
4. Sound is transmitted down the malleus to its head and then on to the incus and stapes
5. The stapes vibrates the oval window, exerting pressure waves into the perilymph of the scala vestibuli of the cochlea
6. This action starts a "ripple" in the perilymph that is transmitted through the vestibular membrane to the endolymph inside the duct
7. Transmission proceeds to the organ of Corti and to the basilar membrane
8. From the basilar membrane, the ripple is next transmitted to and through the perilymph in the scala tympani
9. Finally, the propagation wave expends itself against the round window
10. Neural pathway of hearing
    a. A movement of hair cells against the tectorial membrane initiates impulse conduction by the cochlear nerve to the brainstem
    b. Impulses pass through "relay stations" in nuclei in the medulla, pons, midbrain, and thalamus

SENSE OF BALANCE

A. Sense of equilibrium
  1. The sense organs involved with balance (equilibrium) are found in the vestibule and semicircular canals
B. Static equilibrium—ability to sense the position of the head relative to gravity or to sense acceleration or deceleration
  1. Macula—found within both the utricle and the saccule of the vestibule; contain receptor hair cells (Figure 14-10, A)
  2. Otoliths—"ear stones" composed of protein and calcium carbonate; located within the matrix above each macula (Figure 14-10, B)
  3. Changing the position of the head produces a change in the amount of pressure on the otolith-weighted matrix
    a. This stimulates the hair cells, which then stimulate the adjacent receptors of the vestibular nerve (Figure 14-10, C and D)
    b. Movement of the receptor cells within the macular patches provides information related to head position or acceleration
  4. Righting reflexes—muscular responses that restore the body and its parts to their normal position after they have been displaced
C. Dynamic equilibrium—needed to maintain balance when the head or body itself is rotated or suddenly moved
  1. Dynamic equilibrium depends on the crista ampullaris
  2. Cupula—gelatinous flap in which hair cells of each crista are embedded (Figure 14-11, A and B)
    a. Not weighted with otoliths and does not respond to the pull of gravity
    b. Acts much like a float that moves with the flow of endolymph in the semicircular ducts
  3. Semicircular ducts are placed nearly at right angles to each other; enables us to detect movement in all dimensions and directions
  4. Hair cells bend as cupula moves, producing a receptor potential followed by an action potential

a. Impulse is transmitted to the medulla oblongata; then to other areas of the brain and spinal cord for interpretation, integration, and response

VISION: THE EYE

A. Structure of the eye (Figure 14-12)
B. Accessory structures of the eye
  1. Eyebrows and eyelashes—provide shading for the eyes and also help prevent foreign objects from entering the eyes
  2. Eyelids—consist primarily of voluntary muscle and skin
    a. Tarsal plate—border of thick connective tissue at the free edge of each lid
    b. Canthus—where the superior and inferior eyelids meet
    c. Conjunctiva—mucous membrane (Figure 14-13)
  3. Lacrimal apparatus—structures that secrete tears and drain them from the surface of the eyeball (Figure 14-15)
    a. Lacrimal glands—located in a depression of the frontal bone at the upper outer margin of each eye orbit; about a dozen small ducts lead from each gland
    b. Lacrimal canals—empty into the lacrimal sacs and open into the punctum
    c. Lacrimal sacs—located in a groove in the lacrimal bone
    d. Nasolacrimal ducts—small tubes that extend from the lacrimal sac into the inferior meatus of the nose
C. Muscles of the eye—can be either extrinsic or intrinsic
  1. Extrinsic eye muscles—skeletal muscles that attach to the outside of the eyeball and to the bones of the orbit (Figure 14-16)
  2. Intrinsic eye muscles—smooth (involuntary) muscles located within the eyes
D. Layers of the eyeball—three layers of tissues compose the external membrane of the eyeball (Figure 14-17)
  1. Fibrous layer—sclera
    a. Cornea—anterior portion of fibrous coat; transparent; no blood vessels are found in the cornea
  2. Vascular layer—consist of three components
    a. Choroid—middle and posterior coating just inside the fibrous layer; pigmented
    b. Ciliary body—a thickening of the choroid; fits like a collar into the area between the anterior margin of the retina and the posterior margin of the iris (Figure 14-18)
    c. Iris—colored part of the eye; consists of circular and radial smooth muscles that form a ringlike structure
  3. Inner layer
    a. Retina—innermost lining of the eyeball; has no anterior portion
    b. Most of the retina consists of nervous tissue and is called the sensory retina
    c. Three types of neurons are found: main photoreceptor cells, bipolar cells, and ganglion cells (Figure 14-19, A)
    d. The ends of the dendrites of the main photoreceptor neurons are named after their basic shapes: rods and cones (Figure 14-19, B and C)
    e. Rods and cones are principle visual receptors and differ in function, number, and distribution
    f. Cones are less numerous than rods; most densely concentrated in the fovea centralis (Figures 14-17 and 14-20)

g. Bipolar cells are neurons that receive impulses from the rods and cones

h. Ganglion neurons—act like photoreceptors themselves

i. Lateral connecting cells—help us detect patterns of movement

j. Optic disk (blind spot)—circular area in the posterior part of eyeball; contains perforations through which the fibers emerge from the eyeball as the optic nerve

E. Cavities and humors

1. Cavities—eyeball has a large interior space divided into two cavities

   a. Anterior cavity—lies in front of the lens

      (1) Anterior chamber—space anterior to the iris and posterior to the cornea

      (2) Posterior chamber—space posterior to the iris and anterior to the lens

   b. Posterior cavity—occupies the entire space behind the lens and ciliary body

2. Humors—help maintain enough internal pressure within the eye to prevent the eyeball from collapsing

   a. Aqueous humor—clear, watery substance; often leaks out when the eye is injured

   b. Vitreous humor—fills the posterior cavity filling the anterior cavity; consistency similar to that of soft gelatin

F. Process of seeing: refraction and accommodation

1. Refraction of light rays—bending of light produced as light rays pass at an angle from one transparent medium into another of different optical density

2. Accommodation of lens—an increase in the curvature of the lens to achieve the greater refraction needed for near vision (Figure 14-21)

3. Constriction of the pupil—prevents divergent rays from the object from entering the eye through the periphery of the cornea and lens

4. Convergence of eyes—movement of both eyes inward so that their visual axes come together, or converge, at the object being viewed

G. Role of photopigments—light-sensitive pigmented compounds; found in the distal end of the photoreceptor cells near the pigmented retina (Figure 14-19)

1. Rods—photopigment in rods is rhodopsin

2. Cones—three types of cones are found in the retina

   a. Blue-sensitive cones are sensitive to short wavelengths

   b. Green-sensitive cones are sensitive to medium wavelengths

   c. Red-sensitive cones are sensitive to long wavelengths

H. Neural pathway of vision

1. Rods and cones conduct impulses to nerve fibers that reach the visual cortex in the occipital lobes

   a. Fibers reach the visual cortex by way of the optic nerves, optic chiasma, optic tracts, and optic radiations (Figure 14-23)

REVIEW QUESTIONS

HINT *Write out the answers to these questions after reading the chapter and reviewing the Chapter Summary. If you simply think through the answer without writing it down, you won't retain much of your new learning.*

1. Define adaptation.
2. Describe each of the following: mechanoreceptors, chemoreceptors, thermoreceptors, nociceptors, photoreceptors.
3. Identify the pathway involved for the production of the sense of smell.
4. Describe the main features of the middle ear.
5. Name the parts of the bony and membranous labyrinths and describe the relationship of those parts.
6. In what ear structure or structures is the hearing sense organ located? Location of the equilibrium sense organs?
7. What is the name of the hearing sense organ? Of the equilibrium sense organs?
8. Describe the path of sound waves as they enter the ear.
9. Define vertigo.
10. Describe the role of the basilar (spiral) membrane in hearing.
11. What is the difference between static and dynamic equilibrium? Describe the general mechanisms by which each is maintained.
12. Name two involuntary muscles in the eye. Explain their functions.
13. Define the term *refraction*. Name the refractory media of the eye.
14. Explain briefly the mechanism for accommodation for near vision.
15. Name the photopigments present in rods and in cones. Explain their role in vision.

CRITICAL THINKING QUESTIONS

HINT *After finishing the Review Questions, write out the answers to these items to help you apply your new knowledge. Go back to sections of the chapter that relate to items that you find difficult.*

1. Which free nerve ending is able to respond to light touch?
2. Describe the neural pathways for taste and smell. What would be most likely to stimulate a memory: the taste of apple pie or the smell of apple pie? How can you explain your answer?
3. Identify the three layers of the eye. Include the specific functions of each within your answer.
4. How would you compare and contrast the receptors for vision in dim light and those for vision in bright light?
5. How is sensory response related to age?

continued from page 297

Now that you have read this chapter, try to answer these questions about Rita's eyesight from the Introductory Story.

1. What is the scientific name for Rita's inability to focus clearly?
   a. Myopia
   b. Presbyopia
   c. Conjunctivitis
   d. Scotoma

After sitting in the optometrist's waiting room for a few minutes, she heard, "Rita!" The receptionist was calling her back to the examination room.

2. What type of receptors in the ear responded to the receptionist's voice?
   a. Mechanoreceptors
   b. Photoreceptors
   c. Thermoreceptors
   d. Chemoreceptors

As the optometrist leaned closer to get a better look at her eyes, Rita could smell his cologne.

3. What type of receptors are responding to the cologne smell?
   a. Mechanoreceptors
   b. Photoreceptors
   c. Thermoreceptors
   d. Chemoreceptors

When the examination was completed, the optometrist told Rita that the lenses of her eyes had some cloudy spots in them. That would account for the vision problems she had when driving at night.

4. What condition could the spots indicate?
   a. Presbyopia
   b. Glaucoma
   c. Cataracts
   d. Myopia

HINT *To solve a case study, you may have to refer to the glossary or index, other chapters in this textbook, A&P Connect, Mechanisms of Disease, and other resources.*

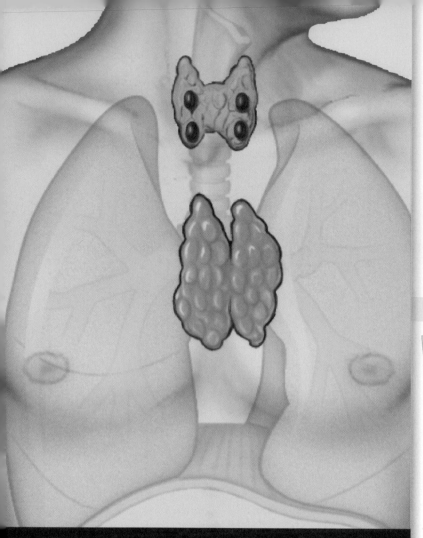

Endocrine System

STUDENT LEARNING OBJECTIVES

At the completion of this chapter, you should be able to do the following:

1. Discuss the general organization of the human endocrine system.
2. Explain how hormones are classified.
3. Distinguish between steroid and nonsteroid hormones.
4. Outline the general principles of hormone action (both steroid and nonsteroid).
5. Explain how a steroid hormone mechanism operates.
6. Explain how a second messenger mechanism operates.
7. Define the various functions of prostaglandins.
8. Outline the structure of the pituitary gland and list its hormones.
9. Explain how secretion of hormones is controlled in the anterior pituitary gland.
10. Give a basic outline of the hypothalamic hormones.
11. Give an example of negative feedback control by the hypothalamus.
12. List the hormones and their actions of the following glands: pineal, thyroid, parathyroid, adrenal, pancreatic islets.
13. List the hormones secreted by the thymus, the intestinal mucosa, and the heart.

"WHY are you so grumpy lately?" Sharon asked Cara. Sharon was the third person to ask that in the past week. When she stopped to think about it, Cara realized that she slept poorly because she was awakened every night by intense thirst. Getting up in the morning and functioning during the day was difficult too because of the chronic fatigue and headaches that plagued her.

It was December and Cara hated the gloom of winter: She lived in the perpetually cloudy snow shadow of Lake Ontario. Perhaps that was it: Just the winter blues! Her annual checkup was coming up in a few days. She would ask her doctor about it then, just to be sure before she bought a special light box to help her with what she thought was seasonal affective disorder (SAD). A lot of her friends suffered from that.

At the doctor's office, Cara filled out the usual paperwork, checking off symptoms that she had recently noted: fatigue, increased urination, irritability, headache. Now it was up to the doctor to ferret out the real problem.

As you read through this chapter, keep in mind Cara's symptoms, and at some point, perhaps you'll be asking questions that the doctor also asked.

· · ·

ORGANIZATION OF THE ENDOCRINE SYSTEM

The **endocrine system** produces and secretes compounds called **hormones** that aid in controlling the body's metabolic activities. In addition, it produces the chemical messengers that allow cells to communicate with one another. Together, the endocrine system and the nervous system both help to maintain the stability of our internal environment. Each system may also work alone or at various levels of cooperation to achieve and maintain this homeostasis. But together they provide an amazingly integrated communication and control system for the entire body. For this reason, we sometimes combine the nervous and endocrine system into a single term: the **neuroendocrine system.**

Signaling Molecules

Both the endocrine system and the nervous system perform their regulatory functions by sending chemical messengers (signals) to specific cells. As we have seen, neurons secrete neurotransmitter molecules to influence nearby cells bearing the appropriate receptor molecules on their cell membranes. In the endocrine system, secreting cells send hormone molecules by way of the bloodstream. These hormones signal specific **target cells** that may exist in a single organ or be distributed throughout the body.

Despite these overall similarities, there are *differences* between the nervous system and the endocrine system. For example, neurotransmitters are sent over minute distances across a synapse. In contrast, hormones typically *diffuse* into the bloodstream and can be carried to nearly every

continued on page 344

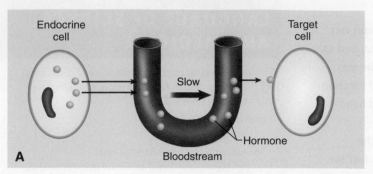
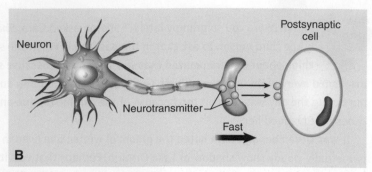

▲ **FIGURE 15-1** Mechanisms of endocrine (A) and nervous (B) signals.

point in the body (Figure 15-1). In addition, the nervous system can directly control only those muscles and glands that are innervated with efferent fibers. The endocrine system, however, can regulate most of the cells of the body. Finally,

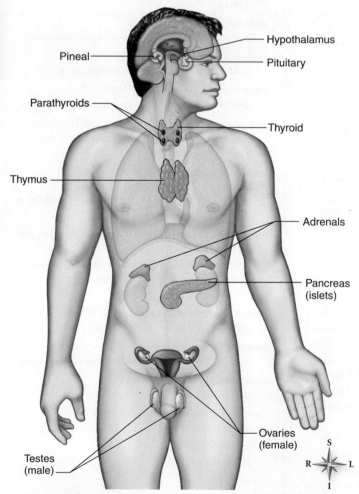

▲ **FIGURE 15-2** Locations of some major endocrine glands.

the effects of neurotransmitters are extremely rapid and short lived. The effects of hormones, in many cases, appear more slowly and have longer lasting effects on the body. Endocrine glands are widely separated throughout the body. Take a moment to look at some of the major endocrine glands illustrated for you in Figure 15-2.

Most endocrine glands are made of glandular epithelium that secretes only hormones. However, some endocrine glands are made of **neurosecretory tissue.** These are modified neurons that secrete chemical messengers that diffuse into the bloodstream rather than across a synapse. For this reason, the hormone *norepinephrine* can be called both a hormone *and* a neurotransmitter. What we call a signaling molecule depends on its source and its action in the body. When norepinephrine is released by a neuron, it is called a neurotransmitter because it quickly diffuses across a synapse and then binds to a receptor in a postsynaptic neuron. But when norepinephrine diffuses into the bloodstream from an endocrine gland, and binds to a receptor in a *distant* target cell, we call it a hormone.

New discoveries in *endocrinology* continue to add to the long list of hormone-secreting tissues. In this chapter, we will focus on how hormones function and the principal hormones of major endocrine glands. Later chapters will expose you to other hormones and their actions.

HORMONES

Hormones can be classified into a number of categories according to their general function. **Tropic hormones,** for example, are hormones produced and secreted by the anterior pituitary gland. These hormones target other endocrine glands and stimulate their growth and secretion. In contrast, **sex hormones** target reproductive tissues and **anabolic hormones** stimulate anabolism in their target cells. In addition to these three, there are many other functional categories of hormones.

We can also classify hormones according to their chemical structure as *steroid* or *nonsteroid hormones.* **Steroid hormones** are manufactured from *cholesterol,* a four-ring lipid. For this reason, steroid hormones are soluble in the phospholipid plasma membranes of their target cells.

Most **nonsteroid hormones** are synthesized primarily from proteins, peptides, and amino acids rather than from cholesterol.

How Hormones Function

Hormones "signal" a target cell by binding to specific receptors either *on the cell membrane* or *inside the cytoplasm* of the cell. In a "lock-and-key" mechanism, hormones will bind *only* to receptor molecules that "fit" them exactly. Any cell with one or more receptors for a particular hormone is a *target* of that hormone (Figure 15-3). Cells usually have many different types of receptors capable of responding to many different hormones.

In a complex process called **signal transduction,** each different hormone-receptor interaction produces different regulatory changes within the target cell. These cellular changes are usually accomplished by altering the chemical reactions taking place within the target cell. For example, some hormone-receptor interactions initiate the synthesis of new proteins. Others trigger the activation or inactivation of specific enzymes. In this manner, they can affect the target cell's metabolic processes. Still other hormone-receptor interactions regulate cells by opening or closing specific ion channels in the plasma membrane, as we shall see.

In the following pages, we will focus on the *primary actions* of just a few, selected hormones. But keep in mind that hormones often have many diverse *secondary functions* in the body. For example, prolactin (PRL) is a hormone with a primary function of regulating milk production (lactation) and reproduction. However, it also has many other secondary actions in the body that help influence the activity of other regulatory mechanisms.

The bloodstream carries hormones nearly everywhere in the body, even where there are no target cells. For this reason, endocrine glands produce more hormone molecules than the number that actually hit their target. These unused hormones are quickly excreted by the kidneys or broken down by the metabolic processes of the body.

Steroid Hormones

Steroid hormones are lipids and thus not very soluble in blood plasma (which is mostly water). For this reason, steroids must be taken to their target cells by *carrier molecules* in the blood plasma. A steroid hormone separates from its carrier before approaching the target cell. Steroid hormones are lipid soluble, so they can easily pass through the phospholipid bilayer comprising the target cell's membrane. After a steroid diffuses into its target cell, it passes into the nucleus, where it binds to receptors moving freely in the nucleoplasm. This model

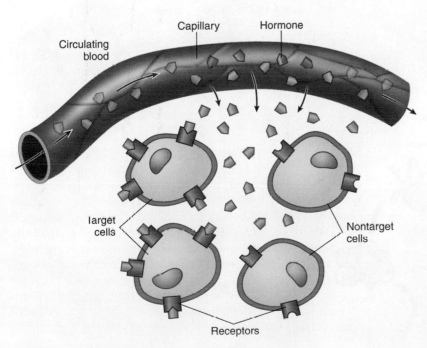

◀ **FIGURE 15-3 The target cell concept.** A hormone acts only on cells that have receptors specific to that hormone because the shape of the receptor determines which hormone can react with it. This is an example of the lock-and-key model of biochemical reactions.

Circulating blood

Capillary Hormone

Target cells

Nontarget cells

Receptors

of hormone action is called the **mobile-receptor model** (Figure 15-4).

Steroid hormones regulate the production of critical proteins. These include enzymes that control reactions inside the cell or membrane proteins that alter the permeability of the cell to specific compounds. It is the amount of steroid hormone present that determines the magnitude of a target cell's response, but the full effect of the steroid hormone may take hours to days to complete.

However, we should point out that steroid hormones may have a secondary effect on target cells: They can change the messages sent by other regulatory molecules, such as other hormones or neurotransmitters. This allows steroid hormones (under these conditions) to have rapid effects in target cells—sometimes just seconds or minutes.

Nonsteroid Hormones

Nonsteroid hormones typically operate according to the **second messenger model** (Figure 15-5). In this case, a nonsteroid hormone molecule acts as a "first messenger," delivering its chemical message to *fixed receptors* in the target cell's plasma membrane. This triggers a series of chemical reactions inside the cell. The "message" is then passed into the cell, where a "second messenger" triggers the appropriate cellular changes via a cascade of reactions influenced by enzymes called *kinases*.

▼ **FIGURE 15-4** **Steroid hormone mechanism.** According to the mobile-receptor model, lipid-soluble steroid hormone molecules detach from a carrier protein *(1)* and pass through the plasma membrane *(2)*. The hormone molecules then pass into the nucleus, where they bind with a mobile receptor to form a hormone-receptor complex *(3)*. This complex then binds to a specific site on a DNA molecule *(4)*, triggering transcription of the genetic information encoded there *(5)*. The resulting mRNA molecule moves to the cytosol, where it associates with a ribosome, initiating synthesis of a new protein *(6)*. This new protein—usually an enzyme or channel protein—produces specific effects in the target cell *(7)*. Some steroid hormones also have additional secondary effects such as influencing signal transduction pathways at the plasma membrane. *mRNA,* Messenger ribonucleic acid.

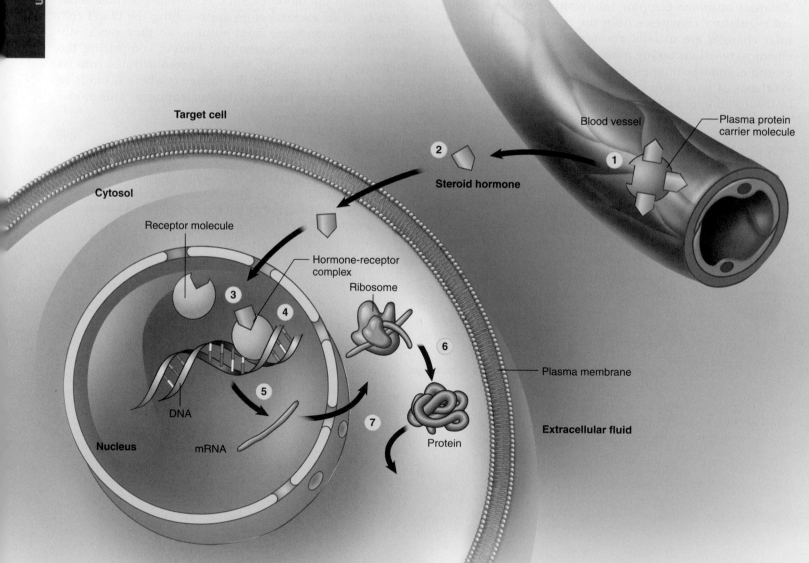

Target cell
Cytosol
Receptor molecule
Hormone-receptor complex
Ribosome
Nucleus
DNA
mRNA
Protein
Blood vessel
Plasma protein carrier molecule
Steroid hormone
Plasma membrane
Extracellular fluid

▼ **FIGURE 15-5** **Example of a second messenger mechanism.** A nonsteroid hormone (first messenger) binds to a fixed receptor in the plasma membrane of the target cell *(1)*. The hormone-receptor complex activates the G protein *(2)*. The activated G protein reacts with GTP, which in turn activates the membrane-bound enzyme adenyl cyclase *(3)*. Adenyl cyclase removes phosphates from ATP, converting it to cAMP (second messenger) *(4)*. cAMP activates or inactivates protein kinases *(5)*. Protein kinases activate specific intracellular enzymes *(6)*. These activated enzymes then influence specific cellular reactions, thus producing the target cell's response to the hormone *(7)*. *ATP,* Adenosine triphosphate; *cAMP,* cyclic adenosine monophosphate; *GTP,* guanosine triphosphate.

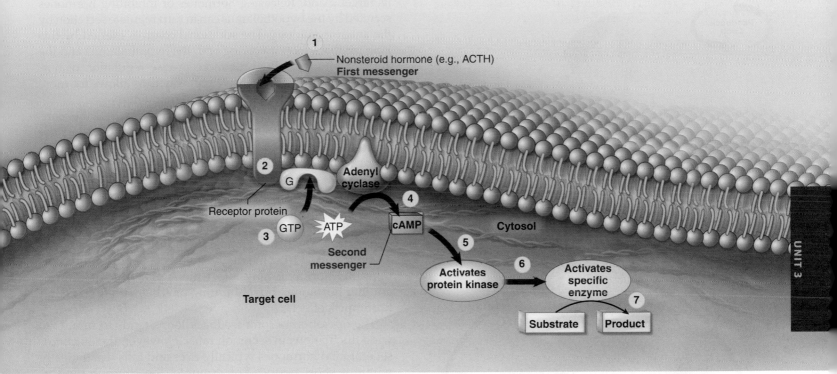

The second messenger mechanism of signal transduction produces target cell effects that differ from steroid hormone effects in several important ways (Table 15-1). First, the cascade of reactions produced in the second messenger mechanism greatly amplifies the effects of the hormone. Thus, the effects of many nonsteroid hormones are disproportionately great when compared with the amount of hormone present. (Recall that steroid hormones produce effects *in proportion to* the amount of hormone present.)

Also, the second messenger mechanism operates much more quickly than the steroid mechanism. Many nonsteroid hormones produce their full effects within seconds or minutes of initial binding to the target cell receptors, not the hours or days sometimes seen with steroid hormones.

Regulation of Hormone Secretion

The control of hormone secretion is usually part of a negative feedback loop. This tends to reverse any deviation of the internal environment away from its set point value. Responses that result from the operation of feedback loops (either negative or positive within the endocrine system) are called *endocrine reflexes*. These are similar to nervous reflex arcs. The simplest endocrine reflex is when an endocrine cell

TABLE 15-1 Comparison of Steroid and Nonsteroid Hormones

CHARACTERISTIC	STEROID HORMONES	NONSTEROID HORMONES*
Chemical structure	Lipid	One or more amino acids, sometimes with added sugar groups
Stored in secretory cell	No	Yes; stored in secretory vesicles before release
Interaction with plasma membrane	No; simple diffusion through plasma membrane and into target cell	Yes; binds to specific plasma membrane receptor
Receptor	Mobile receptor in cytoplasm or nucleus	Embedded in plasma membrane
Action	Regulates gene activity (transcription of new proteins that eventually produce effects in the cell)	Triggers signal transduction cascade, producing internal "second messengers" that trigger rapid effects in the target cell
Response time	One hour to several days	Several seconds to a few minutes

*Some nonsteroid hormones derived from amino acids (e.g., thyroid hormones T_3 and T_4) have gene-activating actions similar to steroid hormones.

is sensitive to the physiological changes produced by its target cells (Figure 15-6). In this example, parathyroid hormone (PTH) produces responses in its target cells that increase the calcium concentration in the blood. When blood calcium concentrations exceed the set point value, parathyroid cells sense it and reduce their output of PTH.

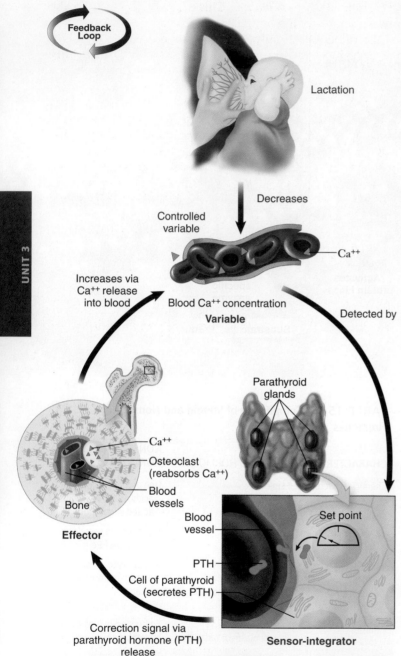

Feedback Loop

Lactation

Decreases

Controlled variable

Ca++

Increases via Ca++ release into blood

Blood Ca++ concentration
Variable

Detected by

Parathyroid glands

Ca++

Osteoclast (reabsorbs Ca++)

Blood vessels

Bone

Effector

Blood vessel

Set point

PTH

Cell of parathyroid (secretes PTH)

Correction signal via parathyroid hormone (PTH) release

Sensor-integrator

▲ **FIGURE 15-6** **Endocrine feedback loop.** In this example of a short feedback loop, each parathyroid gland is sensitive to changes in the physiological variable its hormone (parathyroid hormone) controls—blood calcium (Ca++) concentration. When lactation (milk production) in a pregnant woman consumes Ca++ and thus lowers blood Ca++ concentration, the parathyroids sense the change and respond by increasing their secretion of PTH. PTH stimulates osteoclasts in bone to release more Ca++ from storage in bone tissue (among other effects), which increases maternal blood Ca++ concentration to the set point level.

Secretions by many endocrine glands are regulated by hormones produced by other glands. For example, the anterior portion of the pituitary gland produces thyroid-stimulating hormone (TSH). This hormone stimulates the thyroid gland to release its hormones. The anterior pituitary responds to changes in a particular controlled physiological variable and to changes in the blood concentration of hormones secreted by its target gland. Releasing hormones or inhibiting hormones secreted by the hypothalamus can in turn regulate secretion by the anterior pituitary. The additional controls exerted by this long feedback loop involving both the hypothalamus and the anterior pituitary allow for more precise regulation of hormone secretion. This extensive pathway thus allows for more precise regulation of the internal environment.

Secretion of hormones by a gland can also be controlled by the nervous system (remember, it's called the *neuroendocrine* system!). For example, secretion by the posterior pituitary is *not* regulated by releasing hormones as you might suspect. Instead, it is regulated by direct nervous input from the hypothalamus. Likewise, it is sympathetic nerve impulses that reach the medulla of the adrenal gland that trigger the secretion of epinephrine and norepinephrine. Many other glands, including the pancreas, are also influenced to some degree by nervous input. In addition, one hormone may affect the signal transduction of another hormone, thus inhibiting or enhancing the second hormone's effect.

The operation of long feedback loops tends to minimize wide fluctuations in secretion rates. However, the output of several vital hormones typically rises and falls dramatically within a short period. For example, the concentration of insulin—a hormone that can correct a rise in blood glucose concentration—increases to a high level just after a meal high in carbohydrates. The level of insulin decreases only after the blood glucose concentration returns to its set point. Likewise, threatening stimuli can cause a sudden, dramatic increase in the secretion of epinephrine from the adrenal medulla (as part of the so-called fight-or-flight response).

QUICK ✓ CHECK

4. List the three basic ways in which hormones are secreted or controlled.

5. What is the difference between steroid and nonsteroid hormones?

6. What is the *mobile-receptor model?*

7. What is the *second messenger model?*

PROSTAGLANDINS AND RELATED COMPOUNDS

The **prostaglandins (PGs)** are a unique group of lipids that act as "local hormones." They serve important and widespread integrative functions throughout the body. However, they do not meet the usual definition of a true hormone. These signal molecules are rapidly metabolized

BOX 15-1 | Local Hormones

The classical view of hormones is that they are secreted from a tissue into the bloodstream and have their effects in target cells at some distance from their source. This is the standard definition of an **endocrine hormone**. That is, hormones have "global" effects in the body.

However, some hormones and related agents have primarily local effects—that is, effects *within* the source tissue. To distinguish classical endocrine hormones from local regulators such as prostaglandins and related compounds, scientists often use more precise terms for "local" hormones. Here are examples of terms used to designate local regulators:

Paracrine hormones—hormones that regulate activity in nearby cells within the same tissue as their source

Autocrine hormones—hormones that regulate activity in the secreting cell itself

To avoid confusion, scientists most often refer only to endocrine hormones simply as "hormones." For local regulators, scientists use general terms such as "tissue hormone" or "local regulator"—or more specific terms such as "paracrine" or "autocrine" factor.

---

by the body, so the level of prostaglandins circulating in the blood typically remains low. Prostaglandins and related hormonal compounds such as *thromboxanes* and *leukotrienes* are sometimes referred to as *tissue hormones.* This is because their secretions diffuse only a *short distance* to other cells within the same tissue (Box 15-1).

The term *prostaglandin* was originally given to these compounds because researchers first discovered a prostaglandin in semen and thought it was from the prostate gland. Later they found out this "prostaglandin" is actually produced by the seminal vesicles that produce much of the fluid for a man's semen.

As a group, the prostaglandins have diverse physiological effects. In fact, they are among the most varied and potent of any naturally occurring biological compounds. As signaling molecules, they play important roles in regulating blood pressure, metabolism, and bodily functions. For example, prostaglandins A (PGAs) help relax the smooth muscle fibers in the walls of certain arteries and arterioles. In contrast, prostaglandins E (PGEs) have an important role in systemic inflammation, such as fever. Common anti-inflammatory agents such as aspirin and ibuprofen produce some of their effects by inhibiting PGE synthesis. Prostaglandins F (PGFs) have an especially important role in the reproductive system. Here they cause uterine muscle contractions to induce labor and accelerate the delivery of a baby.

PITUITARY GLAND

The *pituitary gland* is a tiny but vitally important structure weighing only 0.5 gram (0.02 ounce). Sometimes it is referred to as the "master gland." It resides within the pituitary fossa of the brain's sella turcica. Covered by a portion of the dura mater called the pituitary diaphragm, the gland has a stem-like stalk, the *infundibulum*. This stalk connects it to the hypothalamus of the brain (Figure 15-7).

The pituitary actually consists of two separate glands—the **anterior pituitary gland** and the **posterior pituitary gland.** In the embryo, the anterior pituitary gland develops from an upward projection of the pharynx and is composed of *endocrine tissue.* However, the posterior pituitary gland develops from a downward projection of the brain and is composed of *neurosecretory cells.* For these developmental reasons, you will see that the hormones secreted by the anterior pituitary gland serve very different functions than those released by the posterior pituitary gland. Take a moment to review some of the major hormones from the pituitary presented in Figure 15-8.

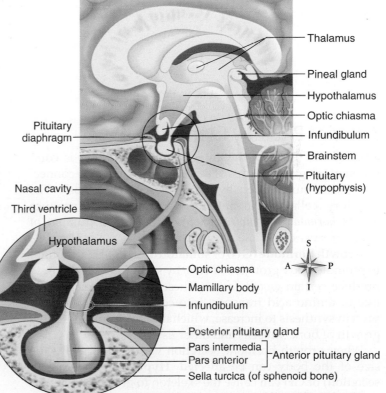

▲ **FIGURE 15-7** **Location and structure of the pituitary gland.** The pituitary gland is located within the sella turcica of the skull's sphenoid bone. It is connected to the hypothalamus by a stalk-like infundibulum. The infundibulum passes through a gap in the portion of the dura mater that covers the pituitary (the pituitary diaphragm). The inset shows that the pituitary is divided into an anterior portion, the anterior pituitary gland, and a posterior portion, the posterior pituitary gland.

▶ FIGURE 15-8 Pituitary hormones. Some of the major hormones of the anterior pituitary and posterior pituitary and their principal target organs.

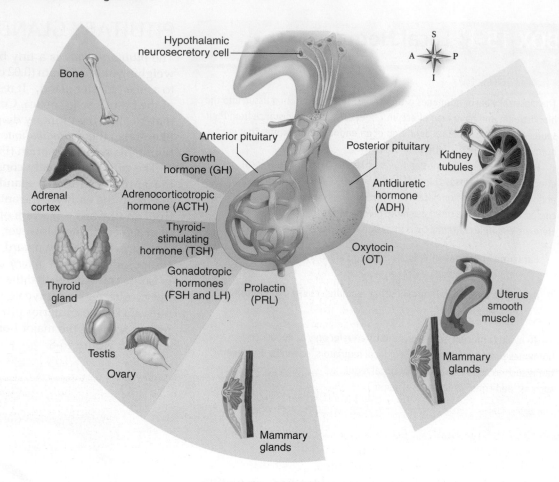

Anterior Pituitary Gland

Most of the tissue of the anterior pituitary gland is composed of irregular clumps of secretory cells and fine connective tissue surrounded by a rich vascular network. The secretory cells of the anterior pituitary gland produce *growth hormone (somatotropin), prolactin,* and a number of *tropic hormones.*

Growth hormone (GH) or **somatotropin (STH)** is thought to promote body growth *indirectly* by stimulating the liver to produce certain *growth factors.* These in turn accelerate the rate of amino acid transport into cells. This action allows protein synthesis to increase, which ultimately promotes the growth of bone, muscle, and other tissues.

When growth hormone secretion varies abnormally, the size of the body can be affected. **Hypersecretion** (excess secretion) of GH can cause the skeleton to grow at an abnormally rapid rate, resulting in *gigantism.* **Hyposecretion** (undersecretion) of GH may result in stunted body growth, or *pituitary dwarfism* (Figure 15-9).

GH also stimulates fat metabolism by mobilizing lipids from storage in adipose cells. In this way, GH tends to shift a cell's use of nutrients away from carbohydrates such as glucose toward lipids as an energy source. As a result, the levels of blood glucose tend to rise, producing a *hyperglycemic effect.* Insulin from the pancreas has an antagonistic *hypoglycemic effect,* so you can readily see the importance of the feedback loops that balance these two hormones.

Prolactin (PRL) is another hormone produced by the anterior pituitary gland. This hormone is vital to the promotion of breast development and the secretion of milk after pregnancy. At the birth of an infant, PRL in the mother stimulates the mammary glands to begin milk secretion. Hypersecretion of PRL may cause unwanted lactation in non-nursing women, disruption of the menstrual cycle, and impotence in men. Hyposecretion of PRL prevents a woman from nursing because milk production cannot be initiated or maintained without PRL.

The anterior pituitary gland also secretes a number of important **tropic hormones,** including the following:

1. **Thyroid-stimulating hormone (TSH)** promotes and maintains the growth and development of the thyroid gland. TSH causes this gland to secrete its hormones.

2. **Adrenocorticotropic hormone (ACTH)** promotes and maintains normal growth and development of the

▲ **FIGURE 15-9** **Growth hormone abnormalities.** The man on the left exhibits gigantism, and the man on the right exhibits pituitary dwarfism. The two men in the middle are of average height.

cortex of the adrenal gland. It also stimulates the adrenal cortex to synthesize and secrete some of its hormones.

3. **Follicle-stimulating hormone (FSH)** stimulates the follicles of the ovaries to grow and mature. In females, FSH also stimulates the developing follicle cells to synthesize and secrete estrogens. In males, FSH stimulates the seminiferous tubules of the testes to maintain or increase the rate of spermatogenesis (sperm production) by them.

4. **Luteinizing hormone (LH)** stimulates the formation and activity of the *corpus luteum* of the ovary. The corpus luteum (literally, "yellow body") is a glandular structure that forms once a follicle ruptures and releases the egg. The corpus luteum secretes large amounts of progesterone and some estrogens when stimulated by LH. LH also supports FSH in stimulating the maturation of follicles. In males, LH stimulates interstitial cells in the testes to develop and then synthesize and secrete testosterone.

FSH and LH are called **gonadotropins** because they stimulate the growth and maintenance of the *gonads* (ovaries and testes). During childhood, the anterior pituitary gland secretes insignificant amounts of the gonadotropins, but a few years before puberty, gonadotropin levels gradually increase. At puberty, their secretion spikes and the gonads are stimulated to develop and begin their normal functions.

Hypothalamus: Control of Secretion in the Anterior Pituitary Gland

The **hypothalamus** is a tiny region of the brain located immediately below the thalamus and just above the brainstem. It controls many of the body's basic processes associated with thirst, hunger, fatigue, body temperature, and even our daily (circadian) rhythms.

Some of the small *nuclei* (groups of neuron cell bodies) in the hypothalamus produce a number of hormones that are sometimes called *neurohormones* because they are produced by neurons. A group of neurohormones called the **hypothalamic releasing hormones** travel through a complex of small blood vessels called the *hypophyseal portal system* (Figure 15-10). (A *portal system* is an arrangement of blood vessels in which blood exiting one tissue is immediately carried to a second tissue.) The **hypophyseal portal system** carries blood from the hypothalamus *directly* to the anterior pituitary gland, where the target cells of the different releasing hormones are located. In this way, a small amount of hormone is delivered *directly* to its target tissue without first being diluted in the general circulation. As a result, the hypothalamus *directly* regulates the secretions of the anterior pituitary gland. (You might say the hypothalamus is the "master" of the so-called master gland—the pituitary!)

The hypothalamic releasing hormones may have releasing *or* inhibiting effects on various cells of the anterior pituitary gland. Through negative feedback mechanisms, the hypothalamus adjusts the secretions of the anterior pituitary gland. The anterior pituitary gland then adjusts the secretions of its

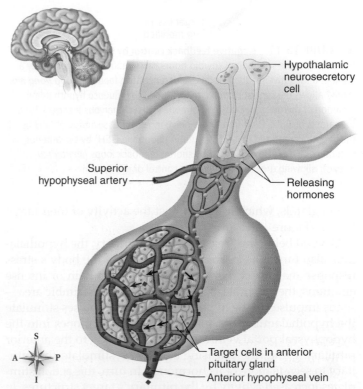

Hypothalamic neurosecretory cell

Superior hypophyseal artery

Releasing hormones

S

A —✦— P

I

Target cells in anterior pituitary gland

Anterior hypophyseal vein

▲ **FIGURE 15-10** **Hypophyseal portal system.** Neurons in the hypothalamus secrete releasing hormones into veins that carry the releasing hormones directly to the vessels of the anterior pituitary gland, thus bypassing the normal circulatory route.

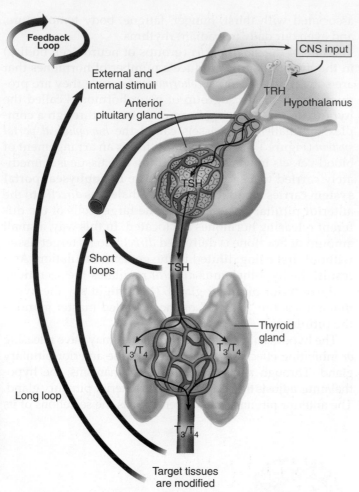

▲ **FIGURE 15-11** **Negative feedback control by the hypothalamus.** In this example, the secretion of thyroid hormone (T_3 and T_4) is regulated by a number of negative feedback loops. A long negative feedback loop *(long red arrow)* allows the central nervous system (CNS) to influence hypothalamic secretion of thyrotropin-releasing hormone (TRH) by nervous feedback from the targets of T_3/T_4 (and from other nerve inputs). The secretion of TRH by the hypothalamus and thyroid-stimulating hormone (TSH) by the anterior pituitary gland is also influenced by shorter feedback loops *(shorter red arrows)*, allowing great precision in the control of this system.

target glands, which in turn adjust the activity of their target tissues (Figure 15-11).

Beyond being the "master" of the pituitary, the hypothalamus also functions as an important part of the body's stress response machinery. For example, in severe pain or intense emotions, the cerebral cortex—especially in the limbic area—sends impulses to the hypothalamus. The impulses stimulate the hypothalamus to secrete its releasing hormones into the hypophyseal portal veins. Circulating quickly to the anterior pituitary gland, the releasing hormones stimulate the pituitary to secrete more of its hormones. In turn, this action stimulates increased activity by the pituitary's target structures. *In effect, the releasing hormones of the hypothalamus translate nervous impulse into hormone secretion by endocrine glands.* Thus, the hypothalamus links the nervous system to the endocrine system and integrates their activities. When survival is threatened, the hypothalamus can take over the anterior pituitary gland and thus gain control of literally every cell in the body.

Posterior Pituitary Gland

The posterior pituitary gland serves as a storage and release site for two important hormones: *antidiuretic hormone (ADH)* and *oxytocin (OT)*. The cells of the posterior pituitary gland do not actually make these neurohormones. Instead, neurons whose cell bodies lie in the hypothalamus synthesize them and pass the hormones down along axons into the posterior pituitary gland (Figure 15-12). In effect, the release of ADH and OT into the blood is controlled by nervous stimulation. Let's look at both hormones in some detail.

As its name implies, **antidiuretic hormone (ADH)** inhibits the production of large amounts of urine (*anti* = against; *diuresis* = urine production), thus conserving body water whenever necessary. When the body dehydrates, the increased osmotic pressure of the blood (because it has lost water and become denser with solutes) triggers the release of ADH from the posterior pituitary gland. This in turn causes water to be reabsorbed from the tubules of the kidney and returned to the blood.

ADH has many other effects in the body as well. For example, ADH also stimulates *contraction* of muscles in the walls of small arteries, thus increasing blood pressure. For this reason, ADH is also called **vasopressin** (or *arginine vasopressin*) because it causes the muscles to contract and decrease the diameter of arteries, thus increasing arterial blood pressure. Vasopressin literally means "vessel pressure stuff."

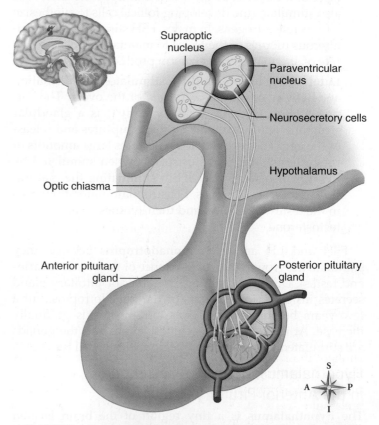

▲ **FIGURE 15-12** **Relationship of the hypothalamus and posterior pituitary gland.** Neurosecretory cells have their cell bodies in the hypothalamus and their axon terminals in the posterior pituitary gland. Thus hormones synthesized in the hypothalamus are actually released from the posterior pituitary gland.

Hyposecretion of ADH can lead to **diabetes insipidus,** an uncommon condition in which the patient produces abnormally large amounts of urine because the kidneys are unable to conserve water. As a result, severe dehydration can occur, leading to extreme thirst, fatigue, chronic headaches, and irritability. ADH administered under the name vasopressin can alleviate these symptoms by regulating urine output.

Oxytocin (OT) has two primary actions: (1) it stimulates contraction of uterine muscles during labor, and (2) it causes milk ejection into the breast ducts of lactating women.

Oxytocin's first action—the stimulation of uterine contractions—is the reason for its name. Oxytocin literally means "swift childbirth," something most women hope for. Oxytocin secretion during childbirth is a good example of a *positive feedback mechanism.* Uterine contractions push on pressure receptors in the pelvis, which triggers the release of more OT, which again pushes on the pelvic receptors, and so on. The wavelike contractions continue to some degree after childbirth, which helps the uterus expel the placenta and then return to its normal shape.

Oxytocin's second action causes alveolar cells of the breast to release their milk secretion into the ducts of the breast. This is very important because a suckling child cannot remove milk unless it has first been *ejected* into the ducts. Throughout nursing, the mechanical and psychological stimulation of the baby's suckling action triggers the release of more OT in another *positive feedback mechanism:* More suckling, more OT production, and more milk ejected.

Oxytocin also has many other effects, such as social bonding (as between a mother and infant).

QUICK ✓ CHECK

8. What is a prostaglandin?
9. How are the divisions of the pituitary distinguished by location and function?
10. Name three hormones produced by the anterior pituitary gland.
11. What is the function of the hypothalamus?
12. Name two hormones produced by the posterior pituitary gland.

PINEAL GLAND

The **pineal gland** is a tiny structure (about 1 cm or ⅜ inch) resembling a pine nut (hence its name). It is located on the dorsal area of the brain's diencephalon region (see Figure 15-7). The principal secretion of the pineal gland is the hormone *melatonin.* **Melatonin** acts as a hormone because it is released by neurosecretory cells of the pineal gland into the blood to regulate functions throughout the body. Melatonin levels rise and fall in a daily (circadian) cycle related to the changing levels of sunlight throughout the day. Melatonin levels rise when sunlight is absent (triggering sleepiness) and fall when sunlight is present (triggering wakefulness). Thus, the pineal gland and melatonin act as important parts of a person's biological clock—the timekeeping mechanism of the body.

Melatonin, whose secretion is inhibited by the presence of sunlight, may also affect a person's mood. A psychological condition called **seasonal affective disorder (SAD),** which may cause severe depression (during winter when day lengths

are shorter), has been linked to the pineal gland. Patients suffering from SAD are often advised to expose themselves to special high-intensity lights that simulate sunlight for several hours each evening during the winter months.

THYROID GLAND

The **thyroid gland** is located in the neck, on the anterior and lateral surfaces of the trachea, just below the larynx (Figure 15-13, *A*). It consists of two large *lateral lobes* and a narrow connecting isthmus that may extend anteriorly as a *pyramidal lobe.* The weight of the gland in an adult is typically about 30 grams (1 ounce). The thyroid tissue is composed of tiny structural units called **follicles,** the site of thyroid hormone synthesis (Figure 15-13, *B*).

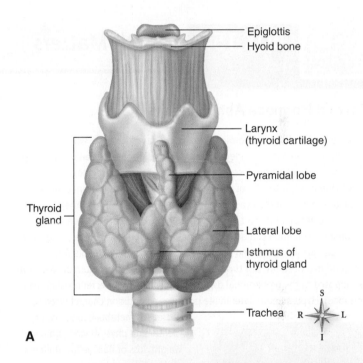

A

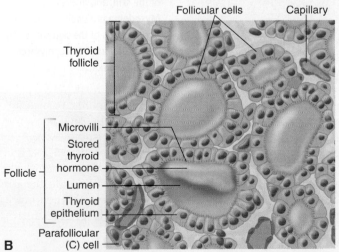

B

▲ **FIGURE 15-13** **Thyroid gland and tissue. A,** In this drawing, the relationship of the thyroid to the larynx (voice box) and to the trachea is easily seen. **B,** In the drawing of thyroid gland tissue, note that each thyroid follicle is filled with a thick fluid—the stored form of thyroid hormone.

The hormone that is called *thyroid hormone (TH)* is actually two different hormones. The most abundant type of TH is **tetraiodothyronine (T_4)** or thyroxine. The other TH is triiodothyronine (T_3). One molecule of T_4 contains four iodine atoms, and one molecule of T_3 contains three iodine atoms.

Unlike other endocrine glands, the thyroid stores a precursor to the hormones T_3 and T_4 prior to their attachment to globulin molecules to form thyroglobulin complexes. When they are to be released, T_3 and T_4 detach from the thyroglobulin complex and enter the blood. Once in the plasma, the free hormones bind to another globulin and albumin. These hormone-globulin complexes circulate throughout the bloodstream until they near their target cells, where once again they detach from their globulin carriers. A number of serious conditions are caused by the hypersecretion and the hyposecretion of thyroid hormones (Box 15-2).

The thyroid also secretes **calcitonin (CT)**, produced by *parafollicular cells* called *C cells*. Calcitonin apparently controls calcium content of the blood by increasing bone formation by osteoblasts and inhibiting bone breakdown by osteoclasts. In humans, calcitonin may slightly decrease blood calcium levels and promote conservation of hard bone matrix. *Parathyroid hormone (PTH)*, discussed below, is an antagonist to calcitonin. Together these hormones help to maintain calcium homeostasis.

BOX 15-2 | Health Matters

Thyroid Hormone Abnormalities

Hypersecretion of thyroid hormone occurs in **Graves disease,** which is thought to be an autoimmune condition. Graves disease patients may suffer from unexplained weight loss, nervousness, increased heart rate, and **exophthalmos** (protrusion of the eyeballs resulting, in part, from edema of tissue at the back of the eye socket; see parts *B* and *C* of the figure).

Hyposecretion of thyroid hormone during growth years may lead to **cretinism,** a condition characterized by a low metabolic rate, retarded growth and sexual development, and possibly mental retardation. People with profound manifestations of this condition are said to have deformed *dwarfism* (as opposed to the proportional dwarfism caused by hyposecretion of growth hormone). Hyposecretion later in life produces a condition characterized by decreased metabolic rate, loss of mental and physical vigor, gain in weight, loss of hair, yellow dullness of the skin, and myxedema. **Myxedema** is a swelling (edema) and firmness of the skin caused by accumulation of mucopolysaccharides in the skin.

In a condition called **simple goiter,** the thyroid enlarges when there is a lack of iodine in the diet (part *A* of figure). This condition is an interesting example of how the feedback control mechanisms illustrated in Figure 15-11 operate.

Because iodine is required for the synthesis of T_3 and T_4, lack of iodine in the diet results in a drop in the production of these hormones. When the reserve (stored in thyroid colloid) is exhausted, feedback informs the hypothalamus and anterior pituitary gland of the deficiency. In response, the secretion of thyrotropin-releasing hormone (TRH) and thyroid-stimulating hormone (TSH) increases in an attempt to stimulate the thyroid to produce more thyroid hormone. Because there is no iodine available to do this, the only effect is to increase the size of the thyroid gland. This information feeds back to the hypothalamus and anterior pituitary gland, and both increase their secretions in response. Thus the thyroid gets larger and larger and larger—all in a futile attempt to increase thyroid hormone secretion to normal levels.

This condition is still common in areas of the world where the soil and water contain little or no iodine. The use of iodized salt has dramatically reduced the incidence of simple goiter in much of the United States, but it still persists in populations where access to adequate nutrition is a problem.

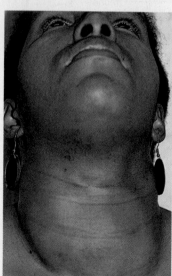

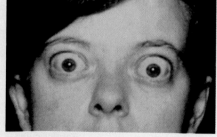

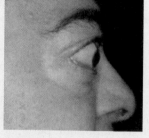

▲ **Thyroid hormone abnormalities. A,** Simple goiter. **B,** Exophthalmos goiter (anterior view). **C,** Exophthalmos goiter (lateral view).

PARATHYROID GLANDS

There are usually four small, rounded **parathyroid glands** embedded in the posterior surface of the thyroid's lateral lobes (Figure 15-14). The **parathyroid hormone (PTH)** acts on bone and kidney cells to *increase* the release of calcium into the blood. For this reason, it is the main hormone the body uses to maintain calcium homeostasis. Less *new bone* is formed and more *old bone* is dissolved. The result is more calcium and phosphate circulating in the blood. In the kidney, however, only calcium is reabsorbed from urine and returned to the blood. Under the influence of PTH, phosphate is secreted by kidney cells *out* of the blood and into the urine to be excreted. PTH also increases the body's absorption of calcium from food, which then permits calcium to be transported through intestinal cells and into the blood.

The maintenance of calcium homeostasis, achieved through the interaction of the antagonistic PTH from the parathyroid and calcitonin from the thyroid, is very important. This is because interaction of these hormones is required for normal neuromuscular excitability, blood clotting, cell membrane permeability, and normal functioning of certain enzymes (Figure 15-15). For example, hyposecretion of PTH can cause *hypocalcemia,* which increases neuromuscular irritability—sometimes so much that it produces muscle spasms and convulsions. On the other hand, high blood calcium levels decrease the irritability of muscle and nerve tissue so that constipation, lethargy, and even coma can result.

QUICK CHECK

13. How does the pineal gland adjust the body's biological clock?

14. Where are the thyroid and parathyroid glands located?

15. Name the major hormones secreted by the thyroid gland.

16. What is the function of parathyroid hormone?

ADRENAL GLANDS

The **adrenal glands** are located above the kidneys, fitting like a cap over these organs. The outer portion of each gland is the **adrenal cortex,** and the inner portion of each gland is the **adrenal medulla** (Figure 15-16). Why this distinction? Because even though the adrenal cortex and adrenal medulla are part of the same organ, they have different embryological origins and are structurally and functionally so different that they are often spoken of as if they were separate glands. The adrenal cortex is composed of regular *endocrine tissue,* but the adrenal medulla is made of *neurosecretory tissue.* Let's look at both portions of the adrenal gland in more detail.

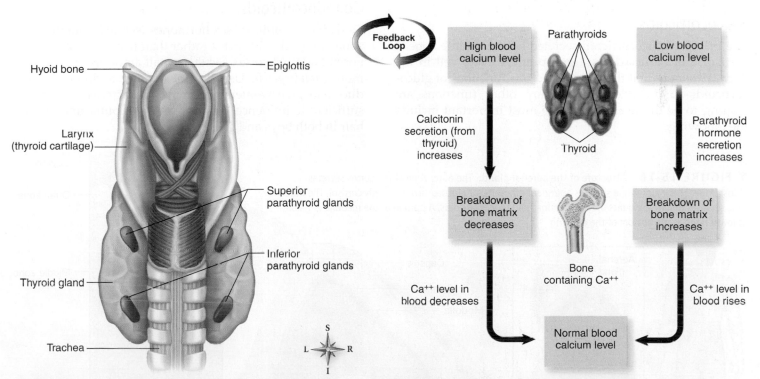

▲ **FIGURE 15-14 Parathyroid gland.** In this drawing from a posterior view, note the relationship of the parathyroid glands to each other, to the thyroid gland, to the larynx (voice box), and to the trachea.

▲ **FIGURE 15-15 Regulation of blood calcium levels.** Calcitonin and parathyroid hormones have antagonistic (opposite) effects on calcium concentration in the blood. (Also see Figure 15-6.)

Adrenal Cortex

The adrenal cortex consists of an outer zone, a middle zone, and an inner zone. Cells of the outer zone secrete a class of hormones called *mineralocorticoids.* Cells of the middle zone secrete *glucocorticoids.* The inner zone secretes small amounts of *glucocorticoids* and *gonadocorticoids* (sex hormones). All these cortical hormones are known as **corticosteroids**—a class of steroid hormones involved in many different physiological processes. These processes involve our reaction to stress, regulation of the immune system, and protein and carbohydrate metabolism.

Mineralocorticoids

The **mineralocorticoids,** as their name suggests, have an important role in regulating how mineral salts (electrolytes) are processed in the body. In humans, **aldosterone** is the only physiologically important mineralocorticoid. Its primary function is to maintain the sodium homeostasis of the blood. Aldosterone accomplishes this by increasing sodium reabsorption in the kidneys. Sodium ions are reabsorbed from the urine back into the blood in exchange for potassium or hydrogen ions. In this way, aldosterone not only adjusts blood sodium levels but also can influence potassium and pH levels of the blood. Because the reabsorption of sodium ions causes water to also be reabsorbed (partly by triggering the secretion of ADH), aldosterone promotes water retention by the body. Some aldosterone is also synthesized in the heart and blood vessels.

Glucocorticoids

The primary **glucocorticoid** secreted by the middle zone is **cortisol,** but corticoids as a group affect every cell in the body. We know that there are numerous functions of glucocorticoids—and undoubtedly many other functions are waiting to be discovered—but the most important include the following:

1. Glucocorticoids accelerate the breakdown of proteins into amino acids. The amino acids are circulated to the liver cells, where they are chemically changed to glucose in a process called *gluconeogenesis.*

2. Glucocorticoids tend to cause a shift from carbohydrate catabolism to lipid catabolism as an energy source.

3. Glucocorticoids are essential for maintaining normal blood pressure. Without adequate amounts of glucocorticoids in the blood, the hormones norepinephrine pand epinephrine cannot produce their *vasoconstricting* (vessel-constricting) effect on blood vessels. As a result, blood pressure falls.

4. A high blood concentration of glucocorticoids may quickly cause a marked decrease in the number of white blood cells. This can cause *atrophy* (shrinkage) of lymphatic tissue. As a result, antibody production decreases. However, glucocorticoids are also needed for the body to recover normally from injury.

5. Glucocorticoid secretion increases as part of the stress response. This action allows glucose to be available for skeletal muscles needed in fight-or-flight responses. We should note, however, that prolonged stress can lead to immune dysfunction.

Hypersecretion of cortisol from the adrenal cortex often produces a collection of symptoms called *Cushing syndrome* (Box 15-3).

Gonadocorticoids

Gonadocorticoids are sex hormones secreted from the inner zone of the adrenal cortex rather than from the gonads. The normal adrenal cortex secretes small amounts of male hormones *(androgens).* Usually not enough androgen is produced to give women masculine characteristics, but it is sufficient to influence the appearance of pubic and axillary hair in both boys and girls.

▼ **FIGURE 15-16** **Structure of the adrenal gland.** The outer zone of the cortex secretes aldosterone. The middle zone secretes abundant amounts of glucocorticoids, chiefly cortisol. The inner zone secretes minute amounts of sex hormones and glucocorticoids. A portion of the medulla is visible at lower right at the bottom of the drawing.

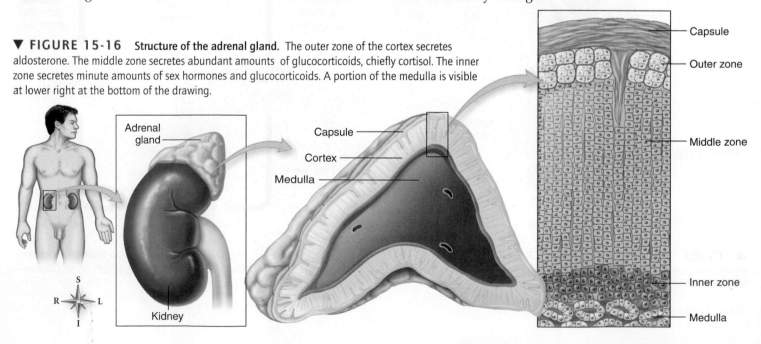

BOX 15-3 Health Matters

Adrenal Cortical Hormone Abnormalities

Hypersecretion of cortisol from the adrenal cortex often produces a collection of symptoms called **Cushing syndrome.** Hypersecretion of glucocorticoids results in a redistribution of body fat. The fatty "moon face" and thin, reddened skin characteristic of Cushing syndrome are shown in part *A* of the figure. Part *B* shows the face of the same boy 4 months after treatment. Hypersecretion of aldosterone—*aldosteronism*—leads to increased water retention and muscle weakness resulting from potassium loss. Hypersecretion of androgens can result from tumors of the adrenal cortex, called *virilizing tumors.* They are given this name because the increased blood level of male hormones in women can cause them to acquire male characteristics, such as facial hair. People with Cushing syndrome may suffer from all the symptoms described above.

Hyposecretion of mineralocorticoids and glucocorticoids, as in *Addison disease,* may lead to a drop in blood sodium and blood glucose, an increase in blood potassium levels, dehydration, and weight loss.

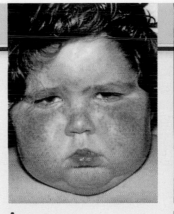 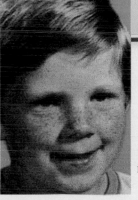

A **B**

▲ **Cushing syndrome. A,** Fatty "moon face" in a boy with Cushing syndrome. **B,** The face of the same boy 4 months after treatment.

Pharmacological preparations of glucocorticoids have been used for many years to temporarily relieve the symptoms of severe inflammatory conditions such as rheumatoid arthritis. Over-the-counter creams and ointments containing hydrocortisone are widely available for use in treating the pain, itching, swelling, and redness of skin rashes.

Adrenal Medulla

The adrenal medulla is composed of neurosecretory tissue (neurons specialized to secrete their products into the blood rather than across a synapse). The adrenal medulla secretes two important hormones, both of which are classified as nonsteroid hormones called *catecholamines* that are secreted directly into the blood. **Epinephrine** *(adrenaline)* accounts for about 80% of the medulla's secretion; **norepinephrine** accounts for about 20%. Recall that sympathetic effectors such as the heart, smooth muscle, and glands have receptors for norepinephrine. Both epinephrine and norepinephrine produced by the adrenal medulla can bind to the receptors of sympathetic effectors. This action prolongs and enhances the effects of sympathetic stimulation by the autonomic nervous system (Figure 15-17).

The hormones of the adrenal glands, their targets, and their principal actions are listed for you in Table 15-2.

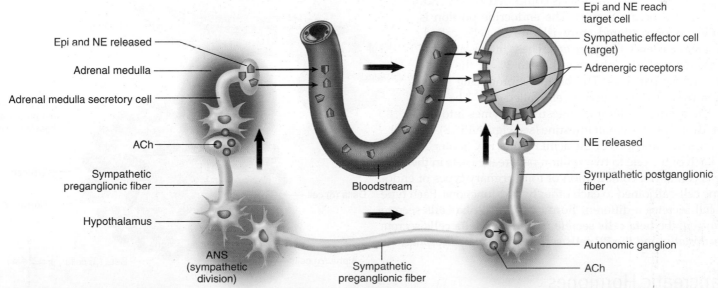

▲ **FIGURE 15-17** **Combined nervous and endocrine influence on sympathetic effectors.** A sympathetic center in the hypothalamus sends efferent impulses through preganglionic fibers. Some preganglionic fibers synapse with postganglionic fibers that deliver norepinephrine *(NE)* across a synapse with the effector cell. Other preganglionic fibers synapse with postganglionic neurosecretory cells in the adrenal medulla. These neurosecretory cells secrete epinephrine *(Epi)* and norepinephrine into the bloodstream, where they travel to the target cells (sympathetic effectors). Compare this figure with Figure 15-1. *ACh,* Acetylcholine; *ANS,* autonomic nervous system.

TABLE 15-2 Hormones of the Adrenal Glands

HORMONE	SOURCE	TARGET	PRINCIPAL ACTION
Aldosterone	Adrenal cortex (outer zone)	Kidney	Stimulates kidney tubules to conserve sodium, which in turn triggers the release of ADH and the resulting conservation of water by the kidney
Cortisol (hydrocortisone)	Adrenal cortex (middle zone)	General	Influences metabolism of food molecules; in large amounts, it has an anti-inflammatory effect
Adrenal androgens	Adrenal cortex (inner zone)	Sex organs, other effectors	Exact role uncertain but may support sexual function
Adrenal estrogens	Adrenal cortex (inner zone)	Sex organs	Thought to be physiologically insignificant
Epinephrine (Epi) (adrenaline)	Adrenal medulla	Sympathetic effectors	Enhances and prolongs the effects of the sympathetic division of the autonomic nervous system
Norepinephrine (NE)	Adrenal medulla	Sympathetic effectors	Enhances and prolongs the effects of the sympathetic division of the autonomic nervous system

ADH, Antidiuretic hormone.

QUICK ✓ CHECK

17. Where are the adrenal glands located?
18. What are the hormones produced by the various zones of the adrenal glands?
19. What is the primary glucocorticoid secreted by the middle zone of the adrenal gland?

PANCREATIC ISLETS

The pancreas (Figure 15-18) is an elongated gland (12 to 15 cm; about 5 to 6 inches in length) weighing up to 100 grams (3.5 ounces). The "head" of the gland lies in the **C**-shaped beginning of the small intestine *(duodenum)*; the body of the pancreas extends horizontally behind the stomach, and its tail touches the spleen.

The tissue of the pancreas is composed of *both* endocrine and exocrine tissues. The endocrine portion is made up of scattered, tiny islands of cells called **pancreatic islets** *(of Langerhans).* These little islands of tissue account for only about 2% to 3% of the total mass of the pancreas. The hormone-producing islets are surrounded by exocrine cells called *acini.* The **acini** secrete a fluid containing digestive enzymes into ducts that drain into the small intestine (see Figure 15-18). We will discuss the functions of the acini further in Chapter 21.

Each of the one to two million pancreatic islets in the pancreas contains a combination of four primary types of endocrine cells, all joined to each other by gap junctions. Each type of cell secretes a different hormone: The **alpha cells** secrete *glucagon,* the **beta cells** secrete *insulin,* the **delta cells** secrete *somatostatin,* and the **PP cells** secrete *pancreatic polypeptide.*

Pancreatic Hormones

Glucagon, produced by alpha cells, tends to increase blood glucose levels by stimulating the conversion of glycogen to glucose in liver cells. It also stimulates

gluconeogenesis (the transformation of fatty acids and amino acids into glucose) in liver cells. This "new" glucose, when released into the bloodstream, has a *hyperglycemic effect* and raises blood glucose levels. **Insulin,** produced by beta cells, tends to promote the movement of glucose, amino acids, and fatty acids *out* of the blood and into tissue cells. Thus, insulin tends to *lower* the blood concentration of these food molecules and to promote their metabolism by tissue cells. The antagonistic effects that glucagon and insulin have on blood glucose levels are summarized for you in Figure 15-19.

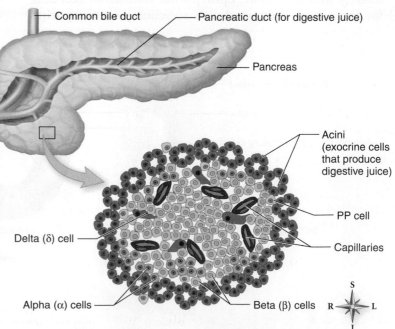

▲ **FIGURE 15-18** **Pancreas.** A pancreatic islet, or hormone-producing area, is shown among the pancreatic cells that produce the pancreatic digestive juice. The pancreatic islets are more abundant in the tail of the pancreas than in the body or head.

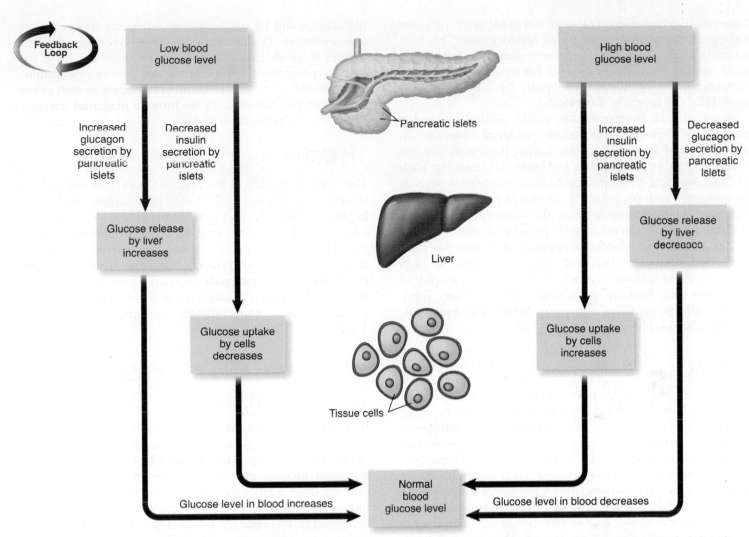

▲ **FIGURE 15-19** **Regulation of blood glucose levels.** Insulin and glucagon, two of the major pancreatic hormones, have antagonistic (opposite) effects on glucose concentration in the blood. Of course, many other hormones, such as GH and cortisol, also influence blood glucose levels.

PP cells produce **pancreatic polypeptide,** which apparently influences the digestion and distribution of food molecules. **Somatostatin,** produced by delta cells, may affect many different tissues in the body, but its primary role seems to be regulating the other endocrine cells of the pancreatic islets. Somatostatin inhibits the secretion of glucagon, insulin, and pancreatic polypeptide. It also inhibits the secretion of GH (somatotropin) from the anterior pituitary. Table 15-3 lists hormones of the pancreatic islets, their targets, and their principal actions.

GONADS

Testes

The **testes** are paired organs residing within the *scrotum,* a sac of skin that hangs from the groin area of the male (review Figure 15-2). They are composed mainly of coils of sperm-producing *seminiferous tubules.* There is also a scattering of endocrine **interstitial cells** found in areas

TABLE 15-3 Hormones of the Pancreatic Islets

HORMONE	SOURCE	TARGET	PRINCIPAL ACTION
Glucagon	Pancreatic islets (alpha [α] cells, or A cells)	General	Promotes movement of glucose from storage and into the blood
Insulin	Pancreatic islets (beta [β] cells, or B cells)	General	Promotes movement of glucose out of the blood and into cells
Somatostatin (SS)	Pancreatic islets (delta [δ] cells, or D cells)	Pancreatic cells and other effectors	Can have general effects in the body, but primary role seems to be regulation of secretion of other pancreatic hormones
Pancreatic polypeptide (PP)	Pancreatic islets (pancreatic polypeptide cells)	Intestinal cells and other effectors	Exact function uncertain but seems to influence absorption in the digestive tract

UNIT 3

between the tubules. These interstitial cells produce *androgens*, the principal one being **testosterone.** This hormone is responsible for the growth and maintenance of male sexual characteristics and for sperm production. Testosterone is regulated principally by gonadotropins, especially LH levels in the blood.

Some steroid hormones are called **anabolic steroids** because they stimulate the building of large molecules (in the process of anabolism). Specifically, they stimulate the building of proteins in muscle and bone. Athletes who wish to increase their performance may abuse steroids such as testosterone and its synthetic derivatives. The anabolic effects of the hormones increase the mass and strength of skeletal muscles. Unfortunately, prolonged steroid abuse will cause a negative feedback response of the anterior pituitary. Gonadotropin levels will drop in response to high blood levels of testosterone, which may lead to atrophy of the testes and possibly permanent sterility. Many other adverse effects, including aggressive behavioral abnormalities, are also possible.

Ovaries

The **ovaries** are a set of paired glands in the pelvis (see Figure 15-2) of the female that produce several types of hormones, including *estrogens* and *progesterone.* **Estrogens,** including *estradiol* and *estrone,* are steroid hormones secreted by the cells of the ovarian follicles. These hormones promote the development and maintenance of female sexual characteristics (see Chapter 25 for complete details of the female reproductive cycle). **Progesterone** is secreted by the corpus luteum (the tissue left behind after the follicle ruptures to release the egg during ovulation). This hormone maintains the lining of the uterus necessary for a successful pregnancy. Regulation of ovarian hormone secretion depends largely on the changing levels of FSH and LH from the anterior pituitary.

PLACENTA

The **placenta** is an important *temporary* endocrine gland that forms on the inner lining of the uterus. It serves as an interface between the circulatory system of the mother and that of the developing child. The placenta produces **human chorionic gonadotropin (hCG)** from the *chorion*—one of the extraembryonic membranes of the fetus. This hormone stimulates development and hormone secretion by maternal ovarian tissues, and remains at very high levels during the early part of pregnancy. Human chorionic gonadotropin serves as a signal to the mother's gonads to maintain the uterine lining rather than allow it to degenerate and slough off as it would during the normal menstrual cycle.

Human chorionic gonadotropin is also the hormone that serves as the pregnancy indicator (in urine) in over-the-counter pregnancy tests. As the placenta develops past the first trimester (3 months) of pregnancy, its production of

hCG drops and its production of estrogens and progesterone increases. The placenta thus takes over the job of the ovaries in producing these hormones necessary for a successful pregnancy. The placenta also produces additional estrogen and progesterone during pregnancy as well as several other hormones such as **human placental lactogen (hPL)** and **relaxin** (Table 15-4).

THYMUS

The **thymus** resides in the mediastinum, just beneath the sternum (see Figure 15-2). Large in children, the thymus begins to atrophy and continues to do so throughout adulthood. By the time an individual reaches old age, the gland is reduced to mostly fat and fibrous tissue. Although it is primarily a lymphatic organ, the thymus secretes **thymosin** and **thymopoietin,** two families of peptides that together play a critical role in the development of the immune system, especially the immune response of the T cells.

OTHER ORGANS AND TISSUES THAT PRODUCE HORMONES

In this chapter, we have discussed only a few of the more central endocrine glands and their principal hormones. However, you'll see in the paragraphs below that even the intestinal mucosa and the heart produce hormones. Not only are there many other known hormones, but many hormones have additional effects besides the principal effects described in this chapter. For example, **leptin,** a protein, is secreted by adipose tissue and plays a role in energy balance, regulation of immunity, neuroendocrine function, and development. Use Table 15-4 as a resource for examples of additional hormones produced by the body.

Gastric and Intestinal Mucosa

The mucous lining of the gastrointestinal (GI) tract, like the pancreas, contains cells that produce both endocrine and exocrine secretions (see Table 15-4). Gastrointestinal hormones such as *gastrin, secretin,* and *cholecystokinin (CCK)* have important regulatory roles in coordinating the secretory and motor activities involved in the digestive process. Endocrine cells in the gastric mucosa also secrete the hormone **ghrelin.** It acts by stimulating the hypothalamus to boost appetite. Since it also acts on other body tissues to slow metabolism and reduce fat burning, it is now believed to play an important role in contributing to obesity.

Heart

Amazing as it may seem, the heart also has hormone-producing cells secreting *atrial natriuretic peptides (ANPs),* specifically **atrial natriuretic hormone (ANH).** This hormone is secreted by the atrium, the upper chamber of the

TABLE 15-4   **Examples of Additional Hormones of the Body**

HORMONE	SOURCE	TARGET	PRINCIPAL ACTION
Prostaglandins (PGs)	Many diverse tissues of the body	Local cells within the source tissue	Diverse local (paracrine/autocrine) effects such as regulation of inflammation, muscle contraction in blood vessels
Cholecalciferol (vitamin D_3)	Skin, liver, kidney (in progressive steps)	Intestines, bones, most other tissues	Promotes calcium absorption from food, regulates mineral balance in bones, regulates growth and differentiation of many cell types
Dehydroepiandrosterone (DHEA)	Adrenal gland, testis, ovary, other tissues	Converted to other hormones	Eventually converted to estrogens, testosterone, or both
Melatonin	Pineal gland	Timekeeping tissues of the nervous system	Helps "set" the biological clock mechanisms of the body by signaling light changes during the day, month, and seasons; may help induce sleep
Testosterone	Testis (small amounts in adrenal and ovary)	Sperm-producing tissues of testis, muscles, other tissues	Stimulates sperm production, stimulates growth and maintenance of male sexual characteristics, promotes muscle growth
Estrogens, including estradiol (E_2) and estrone	Ovary and placenta (small amounts in adrenal and testis)	Uterus, breasts, other tissues	Stimulate development of female sexual characteristics, breast development, bone and nervous system maintenance
Progesterone	Ovary and placenta	Uterus, mammary glands, other tissues	Helps maintain proper conditions for pregnancy
Human chorionic gonadotropin (hCG)	Placenta	Ovary	Stimulates secretion of estrogen and progesterone during pregnancy
Human placental lactogens (hPLs)	Placenta	Mammary glands; pancreas and other tissues	Promote development of mammary glands during pregnancy; help regulate energy balance in fetus
Relaxin	Placenta	Uterus and joints	Inhibits uterine contractions during pregnancy and softens pelvic joints to facilitate childbirth
Thymosins and thymopoietins	Thymus gland	Certain lymphocytes (type of white blood cell)	Stimulate development of T lymphocytes, which are involved in immunity
Gastrin	Stomach mucosa	Exocrine glands of stomach	Triggers increased gastric juice secretion
Secretin	Intestinal mucosa	Stomach and pancreas	Increases alkaline secretions of the pancreas and slows emptying of stomach
Cholecystokinin (CCK)	Intestinal mucosa	Gallbladder and pancreas	Triggers the release of bile from gallbladder and enzymes from the pancreas
Ghrelin (GHRL)	Stomach mucosa	Hypothalamus; other diverse tissues	Stimulates hypothalamus to boost appetite; affects energy balance in various tissues
Atrial natriuretic hormone (ANH) and other atrial natriuretic peptides (ANPs)	Heart muscle	Kidney	Promote loss of sodium from body into urine, thus promoting water loss from the body and a resulting decrease in blood volume and pressure
Inhibins	Ovary and testis	Anterior pituitary	Inhibit secretion of FSH by the anterior pituitary, thus helping to regulate the female reproductive cycle
Leptin	Adipose tissue	Hypothalamus; other diverse tissues	Affects energy balance, perhaps as a signal of how much fat is stored; affects various immune, neuroendocrine, and developmental functions throughout body
Resistin	Adipose tissue and macrophages	Liver and other tissues	Reduces sensitivity to insulin (a pancreatic islet hormone), thus increasing blood glucose levels
Insulin-like growth factor 1 (IGF-1)	Liver, kidney, and other tissues	Bone, muscle, and other tissues	Secreted in response to growth hormone (GH); IGF-1 carries out many of the functions attributed to GH

heart, and functions to promote the loss of sodium from the body through the urine. When sodium is lost from the internal environment, water follows, decreasing blood volume and therefore decreasing blood pressure. ANH opposes increase in blood volume or blood pressure, and thus its actions are antagonistic to ADH and aldosterone, which tend to increase blood volume and blood pressure.

QUICK CHECK

20. Name several of the principal hormones secreted by the pancreatic islets.

21. What are the major hormones secreted by human gonads?

22. What gland produces a hormone that regulates the development of cells important to the immune system?

The BIG Picture

Together the nervous system and the endocrine system enable our body to exercise precise control over its operations. For example, the *neuroendocrine system* is able to finely adjust the availability and processing of nutrients. This is accomplished through the actions of the autonomic nervous system as well as a variety of hormones such as insulin, growth hormone, thyroid hormone, cortisol, epinephrine, and somatostatin. As another example, the absorption, storage, and transport of calcium ions are kept in balance by the antagonistic actions of calcitonin and parathyroid hormone. As a final example, our entire reproductive cycle depends on the extensive interaction of our nervous and endocrine systems. Reproduction is triggered, maintained, and timed by a diverse array of hormones: follicle-stimulating hormone, luteinizing hormone, estrogen, progesterone, testosterone, chorionic gonadotropin, prolactin, oxytocin, and melatonin. The mutual functions of these hormones and the many interactive cycles they produce in our bodies seem almost impossibly complex.

In this chapter, we have reviewed a number of different structures and regulatory mechanisms that make up the endocrine system. Our list here is by no means exclusive. There undoubtedly are many other hormones and interactions yet to be discovered. We have seen how our hormones interact with each other. We have also seen how their functions complement those of the nervous system. And yet, we are still in the infancy of our investigations. Even so, you can readily see that a basic understanding of hormonal regulatory mechanisms is required to fully appreciate the full nature of homeostasis in our bodies. In the following chapters, we will continue to explore various aspects of our endocrine system.

Cycle of LIFE ⭕

Endocrine regulation of body processes first begins during early development in the uterus. By the time a baby is born, many of the hormones are already at work influencing the activity of target cells throughout the body. In fact, new evidence suggests that it is a hormone from the fetus that instigates the onset of labor and delivery in the mother. Many of the basic hormones are active from birth, but most of the hormones related to reproductive functions are not produced or secreted until puberty. Secretion of male reproductive hormones follows the same pattern as most nonreproductive hormones: continuous secretion from puberty until there is a slight tapering off in late adulthood. The secretion of female reproductive hormones such as estrogens also declines later in life, but more suddenly and completely—often during or just at the end of middle adulthood. ●

MECHANISMS OF DISEASE

Endocrine disorders typically result from either elevated (*hypersecretion*) or depressed (*hyposecretion*) hormone levels. However, a variety of mechanisms may produce hypersecretion or hyposecretion, thus making the diagnosis of many endocrine conditions both difficult and challenging. For this reason, the diagnosis of these conditions can be very difficult.

 Find out more about these endocrine tissue diseases online at *Mechanisms of Disease: Endocrine System.*

LANGUAGE OF SCIENCE AND MEDICINE *continued from page 325*

cretinism (KREE-tin-iz-em)
   [*cretin-* idiot, *-ism* condition]

Cushing syndrome (KOOSH-ing SIN-drohm)
   [*Harvey W. Cushing* American neurosurgeon, *syn-* together, *-drome* running or (race)course]

delta cell
   [*delta (δ)* fourth letter of Greek alphabet, *cell* storeroom]

diabetes insipidus (dye-ah-BEE-teez in-SIP-ih-dus)
   [*diabetes* siphon, *insipidus* without zest]

endocrine hormone (EN-doh-krin HOR-mohn)
   [*endo-* within, *-crin* secrete *hormon-* excite]

endocrine system (EN-doh-krin SIS-tem)
   [*endo-* within, *-crin* secrete]

epinephrine (ep-ih-NEF-rin)
   [*epi-* upon, *-nephr-* kidney, *-ine* substance]

estrogen (ES-troh-jen)
   [*estro-* frenzy, *-gen* produce]

exophthalmos (ek-sof-THAL-mus)
   [*ex-* outside, *-oph-* eye, *-thalm-* inner chamber, *-os* relating to]

follicle (FOL-ih-kul)
   [*foll-* bag, *-icle* little]

follicle-stimulating hormone (FSH) (FOL-ih-kul-STIM-yoo-lay-ting HOR-mohn)
   [*foll-* bag, *-icle* little, *hormon-* excite]

ghrelin (GRAY-lin)
   [*ghrel-* grow (also acronym for growth hormone–releasing peptide), *-in* substance]

glucagon (GLOO-kah-gon)
   [*gluc-* glucose, *-agon* drive]

glucocorticoid (gloo-koh-KOR-tih-koyd)
   [*gluco-* sweet (glucose), *-cortic-* bark or cortex, *-oid* like]

gonadocorticoid (go-nad-oh-KOR-tih-koyd)
   [*gon-* offspring, *-ad-* relating to, *-cortic-* cortex or bark, *-oid* like]

gonadotropin (go-nad-oh-TROH-pin)
   [*gon-* offspring, *-ad-* relating to, *-trop-* nourish, *-in* substance]

Graves disease (gravz)
[*Robert J. Graves* Irish physician]

growth hormone (GH) (HOR-mohn)
[*hormon* excite]

hormone (HOR-mohn)
[*hormon-* excite]

human chorionic gonadotropin (hCG) (kor-ee-ON-ik go-nad-oh-TROH-pin)
[*chorion-* skin, *-ic* relating to, *gon-* offspring, *-ad-* relating to, *-trop-* nourish, *-in* substance]

human placental lactogen (hPL) (plah-SEN-tal lak-TOH-jen)
[*placenta-* flat cake, *-al* relating to, *lacto-* milk, *-gen* produce]

hypersecretion (hye-per-seh-KREE-shun)
[*hyper-* excessive, *-secret-* separate, *-tion* process]

hypophyseal portal system (hye-poh-FIZ-ee-al POR-tal)
[*hypo-* beneath, *-physis-* growth, *-al* relating to, *portal* doorway]

hyposecretion (hye-poh-seh-KREE-shun)
[*hypo-* deficient, *-secret-* separate, *-tion* process]

hypothalamic releasing hormone (hye-poh-THAL-ah-mik re-LEE-sing HOR-mohn)
[*hypo-* beneath, *-thalam-* inner chamber, *-ic* relating to, *hormon-* excite]

hypothalamus (hye-poh-THAL-ah-muss)
[*hypo-* beneath, *-thalamus* inner chamber] *pl.*, hypothalami (hye-poh-THAL-ah-meye)

insulin (IN-suh-lin)
[*insul-* island, *-in* substance]

interstitial cell (in-ter-STISH-al sell)
[*inter-* between, *-stit-* stand, *-al* relating to]

leptin (LEP-tin)
[*lept-* thin, *-in* substance]

luteinizing hormone (LH) (loo-tee-in-EYE-zing HOR-mohn)
[*lute-* yellow, *-in-* substance, *-iz-* cause, *hormon-* excite]

melatonin (mel-ah-TOH-nin)
[*mela-* black, *-ton-* tone, *-in* substance]

mineralocorticoid (min-er-al-oh-KOR-tih-koyd)
[*mineral-* mine, *-cortic-* bark or cortex, *-oid* like]

mobile-receptor model (MO-bil-ree-SEP-tor MOD-uhl)
[*recept-* receive, *-or* agent]

myxedema (mik-seh-DEE-mah)
[*myx-* mucus, *-edema* swelling]

neuroendocrine system (noo-roh-EN-doh-krin SIS-tem)
[*neuro-* nerve, *-endo-* within, *-crin-* secrete]

neurosecretory tissue (noo-roh-SEK-reh-tor-ee TISH-yoo)
[*neuro-* nerve, *-secret-* secretion, *-ory* relating to, *tissu-* fabric]

nonsteroid hormone (nahn-STAYR-oyd HOR-mohn)
[*non-* not, *-stero-* solid, *-oid* like, *hormon-* excite]

norepinephrine (nor-ep-ih-NEF-rin)
[*nor-* chemical prefix (unbranched C chain), *-epi-* upon, *-nephr-* kidney, *-ine* substance]

ovary (OH-var-ee)
[*ov-* egg, *-ar-* relating to, *-y* location of process]

oxytocin (OT) (ahk-see-TOH-sin)
[*oxy-* oxygen, *-toc-* birth, *-in* substance]

pancreatic islet (pan-kree-AT-ik EYE-let)
[*pan-* all, *-creat-* flesh, *-ic* relating to, *isl-* island, *-et* little]

pancreatic polypeptide (pan-kree-AT-ik pol-ee-PEP-tyde)
[*pan-* all, *-creat-* flesh, *-ic* relating to, *poly-* many, *-pept-* digest, *-ide* chemical]

paracrine hormone (PAIR-ah-krin HOR-mohn)
[*para-* beside, *-crin* secrete *hormon-* excite]

parathyroid gland (pair-ah-THYE-royd)
[*para-* beside, *-thyr-* shield, *-oid* like, *gland* acorn]

parathyroid hormone (PTH) (pair-ah-THYE-royd HOR-mohn)
[*para-* beside, *-thyr-* shield, *-oid* like, *hormon-* excite]

pineal gland (PIN-ee-al)
[*pine-* pine, *-al* relating to, *gland* acorn]

placenta (plah-SEN-tah)
[*placenta* flat cake] *pl.*, placentae or placentas

posterior pituitary gland (pohs-TEER-ee-or pih-TOO-ih-tair-ee)
[*poster-* behind, *-or* quality, *pituit-* phlegm, *-ary* relating to, *gland* acorn]

PP cell

progesterone (proh-JES-ter-ohn)
[*pro-* provide for, *-gester-* bearing (pregnancy), *-stero-* solid or steroid derivative, *-one* chemical]

prolactin (PRL) (proh-LAK-tin)
[*pro-* provide for, *-lact-* milk, *-in* substance]

prostaglandin (PG) (pross-tah-GLAN-din)
[*pro-* before, *-stat-* set or place (prostate), *-gland-* acorn (gland), *-in* substance]

protein hormone (PRO-teen HOR-mohn)
[*prote-* first rank, *-in* substance, *hormon-* excite]

relaxin (reh-LAK-sin)
[*relax-* relaxation, *-in* substance]

seasonal affective disorder (SAD)
[*season-* period of year, *-al* relating to, *affect-* act on, *-ive* relating to, *dis-* without, *-order* arrangement]

second messenger model

sex hormone (HOR-mohn)
[*hormon-* excite]

signal transduction (SIG-nal tranz-DUK-shen)
[*trans-* across, *-duc-* transfer, *-tion* process]

simple goiter (GOY-ter)
[from *gutter* throat]

somatostatin (soh-mah-toh-STAT-in)
[*soma-* body, *-stat-* stand, *-in* substance]

somatotroph (soh-mah toh-TROHF)
[*soma-* body, *-troph* nourish]

somatotropin (STH) (soh-mah-toh-TROH-pin)
[*soma-* body, *-trop-* nourish, *-in* substance]

steroid hormone (STAYR-oyd HOR-mohn)
[*stero-* solid, *-oid* like, *hormon-* excite]

target cell
[*cell* storeroom]

testis (TES-tis)
[*testis* witness (male gonad)] *pl.*, testes (TESS-teez)

testosterone (tes-TOS-teh-rohn)
[*testo-* witness (testis), *-stero-* solid or steroid derivative, *-one* chemical]

tetraiodothyronine (T₄) (tet-rah-eye-oh-doh-THY-roh-neen)
[*tetra-* four, *-iodo-* violet (iodine), *-thyro-* shield (thyroid gland), *-ine* chemical]

thymopoietin (thy-moh-POY-eh-tin)
[*thymo-* thymus gland, *-poiet-* make, *-in* substance]

thymosin (THY-moh-sin)
[*thymos-* thyme flower (thymus gland), *-in* substance]

thymus (THY-mus)
[*thymus* thyme flower]

thyroid gland (THY-royd)
[*thyro-* shield, *-oid* like, *gland* acorn]

thyroid-stimulating hormone (TSH) (THY-royd STIM-yoo-lay-ting HOR-mohn)
[*thyro-* shield, *-oid* like, *hormon-* excite]

thyroxine (thy-ROK-sin)
[*thyro-* shield (thyroid gland), *-ox-* oxygen, *-ine* chemical]

triiodothyronine (T₃) (try-eye-oh-doh-THY-roh-neen)
[*tri-* three, *-iodo-* violet (iodine), *-thyro-* shield (thyroid gland), *-nine* chemical]

tropic hormone (TROH-pik HOR-mohn)
[*trop-* nourish, *-ic* relating to, *hormon-* excite]

vasopressin (vas-oh-PRES-in)
[*vas-* vessel, *-press-* pressure, *-in* substance]

CHAPTER SUMMARY

To download an MP3 version of the chapter summary for use with your iPod or other portable media player, access the Audio Chapter Summaries *online at http://evolve.elsevier.com.*

HINT *Scan this summary after reading the chapter to help you reinforce the key concepts. Later, use the summary as a quick review before your class or before a test.*

ORGANIZATION OF THE ENDOCRINE SYSTEM

A. The endocrine system and the nervous system both help to maintain the stability of our internal environment
B. When the two systems work together, referred to as the neuroendocrine system, they provide an amazingly integrated communication and control system for the entire body
C. Signaling molecules
1. Both the endocrine system and the nervous system perform their regulatory functions by sending chemical messengers (signals) to specific cells
   a. Hormones—sent by secreting cells by way of the bloodstream
      (1) Hormones typically diffuse into the bloodstream and can be carried to nearly every point in the body; target cells (Figure 15-1)
      (2) Target cells—specified by hormones; may exist in a single organ or be distributed throughout the body
   b. Most endocrine glands are made of glandular epithelium that secretes only hormones; some endocrine glands are made of neurosecretory tissue (Figure 15-2)
      (1) Neurosecretory tissue—modified neurons that secrete chemical messengers that diffuse into the bloodstream rather than across a synapse

HORMONES

A. Classification of hormones according to their general function:
1. Tropic hormones—target other endocrine glands and stimulate their growth and secretion
2. Sex hormones—target reproductive tissue
3. Anabolic hormones—stimulate anabolism in their target cells
B. Classification of hormones according to their chemical structure:
1. Steroid hormones—manufactured from cholesterol; soluble in the phospholipid plasma membranes of their target cells
2. Nonsteroid hormones—made from proteins, peptides, and amino acids
C. How hormones function
1. Hormones "signal" a target cell by binding to specific receptors either on the cell membrane or inside the cytoplasm of the cell; "lock-and-key" mechanism (Figure 15-3)
2. Different hormone-receptor interactions produces different regulatory changes within the target cell; signal transduction
3. Hormones have primary functions; however, hormones often have many diverse secondary functions in the body
4. Steroid hormones
   a. Mobile-receptor model—hormone diffuses into its target cell and passes into the nucleus, where it binds to receptors moving freely in the nucleoplasm (Figure 15-4)

5. Nonsteroid hormones
   a. Second messenger model (Figure 15-5)
      (1) Nonsteroid hormone molecule acts as a "first messenger," delivering its chemical message to fixed receptors in the target cell's plasma membrane
      (2) The "message" is then passed into the cell, where a "second messenger" triggers the appropriate cellular changes via a cascade of reactions influenced by enzymes called kinases
      (3) The second messenger mechanism of signal transduction produces target cell effects that differ from steroid hormone effects in several important ways (Table 15-1)
D. Regulation of hormone secretion
1. The control of hormonal secretion is usually part of a negative feedback loop called an endocrine reflex (Figure 15-6)
2. Secretions by many endocrine glands are regulated by hormones produced by other glands
3. Secretion of hormones by a gland can also be controlled by the nervous system
4. The operation of long feedback loops tends to minimize wide fluctuations in secretion rates

PROSTAGLANDINS AND RELATED COMPOUNDS

A. Prostaglandins—unique group of lipids that act as "local hormones"
1. Serve important and widespread integrative functions throughout the body
2. Called tissue hormones because their secretions diffuse only a short distance to other cells within the same tissue (Box 15-1)
3. Prostaglandins are among the most varied and potent of any naturally occurring biological compounds

PITUITARY GLAND

A. Pituitary gland—tiny but vitally important structure; sometimes referred to as the "master gland" (Figure 15-7)
1. Consists of two separate glands:
   a. Anterior pituitary gland
   b. Posterior pituitary gland
B. Anterior pituitary gland
1. Composed of irregular clumps of secretory cells and fine connective tissue surround by a rich vascular network
2. The secretory cells produce growth hormone (somatotropin) and prolactin
   a. Growth hormone (GH) or somatotropin (STH)—thought to promote body growth indirectly by stimulating the liver to produce certain growth factors
   b. Prolactin—vital to the promotion of breast development and the secretion of milk after pregnancy
3. The anterior pituitary gland also secretes a number of important tropic hormones:
   a. Thyroid-stimulating hormone (TSH)—promotes and maintains the growth and development of the thyroid gland
   b. Adrenocorticotropic hormone (ACTH)—promotes and maintains normal growth and development of the cortex of the adrenal gland

c. Follicle-stimulating hormone (FSH)—stimulates follicles of the ovaries to grow and mature (female) or stimulates testes to produce sperm (male)

d. Luteinizing hormone (LH)—stimulates the formation and activity of the corpus luteum of the ovary (female) or stimulates testosterone secretion (male)

C. Hypothalamus: control of secretion in the anterior pituitary gland

1. Hypothalamus—almond-sized part of brain that controls processes associated with thirst, hunger, fatigue, body temperature, and circadian rhythms
2. Cell bodies of neurons in specific parts of the hypothalamus produce neurohormones called releasing hormones (Figure 15-10)
3. The hypophyseal portal system carries blood from the hypothalamus directly to the anterior pituitary gland
   a. Small amount of hormone is delivered directly to its target tissue without first being diluted in the general circulation
4. Through negative feedback mechanisms, the hypothalamus adjusts the secretions of the anterior pituitary gland
   a. Anterior pituitary gland then adjusts the secretions of its target glands, which in turn adjust the activity of their target tissues (Figure 15-11)
5. Beyond being the "master" of the pituitary, the hypothalamus also functions as an important part of the body's stress response machinery
   a. Nerve impulses stimulate the hypothalamus to secrete its releasing hormones into the hypophyseal portal veins

D. Posterior pituitary gland

1. Posterior pituitary gland—serves as a storage and release site for two important neurohormones:
   a. Antidiuretic hormone (ADH)
   b. Oxytocin (OT)
2. Antidiuretic hormone—inhibits the production of large amounts of urine; stimulates contraction of muscles in the walls of small arteries, thus increasing blood pressure; vasopressin
3. Oxytocin—has two primary actions:
   a. Stimulates contraction of uterine muscles during labor
   b. Causes milk ejection into the breast ducts of lactating women

PINEAL GLAND

A. Pineal gland—tiny structure located on the dorsal area of the brain's diencephalon region (Figure 15-7)

1. Principal secretion is melatonin
2. Pineal gland and melatonin act as important parts of a person's biological clock

THYROID GLAND

A. Thyroid gland—located in the neck, on the anterior and lateral surfaces of the trachea, just below the larynx (Figure 15-13, A)

1. Consists of two large lateral lobes and a narrow connecting isthmus
2. Composed of tiny structural units called follicles; site of thyroid hormone synthesis (Figure 15-13, B)
3. Thyroid hormone (TH)—actually two different hormones
   a. Tetraiodothyronine (T_4) or thyroxine
   b. Triiodothyronine (T_3)

4. Thyroid also secretes calcitonin (CT)—controls calcium content of the blood by increasing bone formation by osteoblasts and inhibiting bone breakdown by osteoclasts

PARATHYROID GLANDS

A. Parathyroid glands—generally four rounded glands embedded in the posterior surface of the thyroid's lateral lobes

1. Parathyroid hormone (PTH)—acts on bone and kidney cells to increase the release of calcium into the blood
   a. Main hormone the body uses to maintain calcium homeostasis
   b. Maintenance of calcium homeostasis, achieved through the interaction of the antagonistic PTH from the parathyroid and calcitonin
   c. PTH also increases the body's absorption of calcium from food

ADRENAL GLANDS

A. Adrenal glands—located above the kidneys, fitting like a cap over these organs (Figure 15-16, Table 15-2)

1. Adrenal cortex—outer portion of gland; composed of regular endocrine tissue
2. Adrenal medulla—inner portion of gland; composed of neurosecretory tissue

B. Adrenal cortex—consists of an outer zone, a middle zone, and an inner zone

1. Mineralocorticoids—secreted from outer zone; regulate how mineral salts (electrolytes) are processed in the body
   a. Aldosterone—maintains the sodium homeostasis of the blood
2. Glucocorticoids—secreted by middle zone; primary glucocorticoid is cortisol
   a. Functions of glucocorticoids:
      (1) Accelerate the breakdown of proteins into amino acids
      (2) Cause a shift from carbohydrate catabolism to lipid catabolism as an energy source
      (3) Essential for maintaining normal blood pressure
      (4) High blood concentration of glucocorticoids may quickly cause a marked decrease in the number of white blood cells
      (5) Increases as part of the stress response
3. Gonadocorticoids—sex hormones released from the inner zone of the adrenal cortex rather than from the gonads

C. Adrenal medulla—secretes two important hormones, both of which are classified as nonsteroid hormones called catecholamines (Figure 15-17)

1. Epinephrine (adrenaline)
2. Norepinephrine

PANCREATIC ISLETS

A. Pancreas—elongated gland located in the C-shaped beginning of the small intestine; body of the pancreas extends horizontally behind the stomach and then touches the spleen (Figure 15-18)

1. Tissue of the pancreas is composed of both endocrine and exocrine tissues
2. Endocrine portion is made up of scattered, tiny islands of cells called pancreatic islets (islets of Langerhans)

3. Each of the pancreatic islets in the pancreas contains a combination of four primary types of endocrine cells
   a. Alpha cells—secrete glucagon
   b. Beta cells—secrete insulin
   c. Delta cells—secrete somatostatin
   d. PP cells—secrete pancreatic polypeptide
B. Pancreatic hormones (Table 15-3)
   1. Glucagon—tends to increase blood glucose levels by stimulating the conversion of glycogen to glucose in liver cells; also stimulates gluconeogenesis
   2. Insulin—tends to promote the movement of glucose, amino acids, and fatty acids out of the blood and into tissue cells
   3. Pancreatic polypeptide—influences the digestion and distribution of food molecules
   4. Somatostatin—regulates the other endocrine cells of the pancreatic islets

GONADS

A. Testes—paired organs residing within the scrotum (Figure 15-2)
   1. Composed mainly of coils of sperm-producing seminiferous tubules and a scattering of interstitial cells
   2. Interstitial cells produce testosterone
      a. Testosterone is responsible for the growth and maintenance of male sexual characteristics and for sperm production
      b. Testosterone is regulated principally by gonadotropins, especially LH levels in the blood
   3. Anabolic steroids—stimulate the building of large molecules
      a. Anabolic effects of the hormones increase the mass and strength of skeletal muscles
B. Ovaries—set of paired glands in the pelvis (Figure 15-2)
   1. Produce several types of hormones:
      a. Estrogens—promote the development and maintenance of female sexual characteristics
      b. Progesterone—maintains the lining of the uterus necessary for a successful pregnancy
      c. Regulation of ovarian hormone secretion depends largely on the changing levels of FSH and LH from the anterior pituitary

PLACENTA

A. Placenta—important temporary endocrine gland that forms on the inner lining of the uterus
   1. Placenta produces human chorionic gonadotropin (hCG)
      a. Human chorionic gonadotropin serves as a signal to the mother's gonads to maintain the uterine lining
      b. Human chorionic gonadotropin is also the hormone that serves as the pregnancy indicator (in urine) in over-the-counter pregnancy tests
   2. Placenta also produces additional estrogen and progesterone during pregnancy; also human placental lactogen (hPL) and relaxin (Table 15-4)

THYMUS

A. Thymus—resides in the mediastinum, just beneath the sternum (Figure 15-2)
   1. Primarily a lymphatic organ; secretes thymosin and thymopoietin

OTHER ORGANS AND TISSUES THAT PRODUCE HORMONES

A. Many hormones have additional effects besides the principal effects given in this chapter (Table 15-4)
   1. Leptin—secreted by adipose tissue and plays a role in energy balance, regulation of immunity, neuroendocrine function, and development
B. Gastric and intestinal mucosa
   1. Mucous lining of the gastrointestinal (GI) tract contains cells that produce both endocrine and exocrine secretions (Table 15-4)
      a. GI hormones include gastrin, secretin, cholecystokinin (CCK), and ghrelin
         (1) Important regulatory roles in coordinating the secretory and motor activities involved in the digestive process
         (2) Ghrelin—stimulates the hypothalamus to boost appetite
C. Heart
   1. Heart also has hormone-producing cells secreting atrial natriuretic peptides (ANPs), specifically atrial natriuretic hormone (ANH)
      a. ANH—secreted by the atrium (the upper chamber of the heart); functions to promote the loss of sodium from the body through the urine

REVIEW QUESTIONS

 HINT *Write out the answers to these questions after reading the chapter and reviewing the Chapter Summary. If you simply think through the answer without writing it down, you won't retain much of your new learning.*

1. Define the terms *hormone* and *target cell.*
2. Describe the characteristic chemical group found at the core of each steroid hormone.
3. Identify the major categories of nonsteroid hormones.
4. Name the two subdivisions of the anterior pituitary gland.
5. Discuss the functions of growth hormone.
6. List the four tropic hormones secreted by the anterior pituitary gland. Which of the tropic hormones are also called *gonadotropins?*
7. How does antidiuretic hormone act to alter urine volume?
8. Describe the positive feedback associated with oxytocin.
9. Discuss the synthesis and storage of thyroxine and triiodothyronine. How are they transported in the blood?
10. Discuss the functions of parathyroid hormone.
11. List the hormones produced by each "zone" of the adrenal cortex, and describe the actions of these hormones.
12. Discuss the normal function of hormones produced by the adrenal medulla.
13. Identify the hormones produced by each of the cell types in the pancreatic islets.
14. Identify the "pregnancy-promoting" hormone.
15. Where is human chorionic gonadotropin produced? What does it do?
16. Describe the role of atrial natriuretic hormone.
17. Identify the conditions resulting from both hypersecretion and hyposecretion of growth hormone during growth years.

CRITICAL THINKING QUESTIONS

HINT   *After finishing the Review Questions, write out the answers to these items to help you apply your new knowledge. Go back to sections of the chapter that relate to items that you find difficult.*

1. Driving a car requires rapid response of selected muscles. The regulation of blood sugar level requires regulating almost every cell in the body. Based on the characteristics of each system, explain why driving would be a nervous system function and blood sugar regulation would be an endocrine function.

2. How would you explain the ways in which one hormone interacts with another and its impact on the cell?

3. What examples can you find that apply the concept of a negative feedback loop to the regulation of hormone secretion?

4. A hyposecretion of which hormone would make it difficult for a mother to nurse her child? How would you summarize the effects of this hormone?

5. Why do you think the hypothalamus can be called the "mind-body link"?

6. How would you explain the hormonal interaction that helps maintain the set point value for glucose in the blood?

7. Explain how the cell is able to become more or less sensitive to a specific hormone.

8. What is the relationship between increased blood level concentrations of FSH and menopause?

9. A lack of iodine in the diet will cause a simple goiter. Describe the feedback loop that will cause its formation.

10. A friend of yours is considering the use of anabolic steroids. Based on what you know of their effects, how would you explain why using these steroids could be harmful?

continued from page 325

Remember Cara and her bad mood from the Introductory Story? See if you can answer the following questions about her now that you have read this chapter.

1. The nurse called Cara in to take her vital signs and weight. Cara was surprised to see that she had lost over 5 pounds. The nurse remarked that she looked very dehydrated. Cara countered that she drank constantly but also urinated constantly. The nurse thought to herself that something was wrong with Cara's:
   a. Liver
   b. Thyroid gland
   c. Kidneys
   d. Ovaries

2. Increased urination and dehydration are symptoms associated with which one of the following endocrine disorders?
   a. Diabetes insipidus
   b. Acromegaly
   c. Hyperthyroidism
   d. Cretinism

3. What hormone could you use to treat Cara's condition?
   a. Oxytocin
   b. Melatonin
   c. Luteinizing hormone
   d. ADH

4. What are the major clues that Cara's condition is probably due to diabetes insipidus and not seasonal affective disorder?
   a. Frequent urination
   b. Constant thirst
   c. Chronic headaches
   d. Dehydration

HINT   *To solve a case study, you may have to refer to the glossary or index, other chapters in this textbook, A&P Connect, Mechanisms of Disease, and other resources.*

UNIT 3

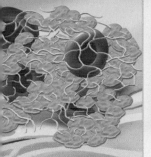
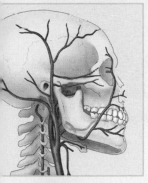

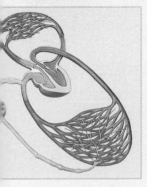

UNIT

4

Transportation and Defense

The next four chapters deal specifically with *transportation* of materials within your body. Transportation is vital to continuous homeostasis of our internal environment and must be monitored every second of our lives. **Chapter 16** discusses *blood*—the major transportation fluid. **Chapters 17** and **18** continue with the structure and functions of the cardiovascular system. Finally, in **Chapter 19** we conclude with the lymphatic system—a supplementary drainage system intimately connected with fluid homeostasis and immunity.

Blood

STUDENT LEARNING OBJECTIVES

At the completion of this chapter, you should be able to do the following:

1. Summarize the basic functions of blood.
2. Describe the components of blood and discuss their functions.
3. List the formed elements of blood and discuss their functions.
4. Discuss the origin and significance of sickle cell anemia in the world.
5. Outline the formation of erythrocytes, leukocytes, and thrombocytes from the stem cell hemocytoblast.
6. Discuss how blood doping could be dangerous.
7. List the different leukocytes and describe their functions.
8. Describe in detail the ABO blood group system and discuss its significance.
9. Discuss the physiological significance of the Rh system.
10. List the major components of blood plasma.
11. Outline the basic mechanism of blood clotting.

DUNCAN was slicing a bagel to put in the toaster. When the microwave beeped, he glanced in that direction, taking his eyes off the bagel. In that split second, the knife slipped and cut deeply into his finger. Immediately blood started spurting out of the damaged blood vessels. Duncan grabbed a towel and wrapped it tightly around the cut while holding his hand above his heart.

We've all done something similar by not paying attention, but did you ever wonder about all the complex physical and physiological processes that take place immediately after we cut ourselves? In this chapter, as you follow Duncan's story, you'll find out what really happens.

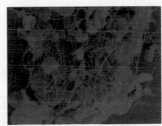

You have undoubtedly seen blood, but have you ever wondered about its properties? Blood is a wonderfully fluid transport medium that serves as a pickup and delivery system that services the entire body. For example, it picks up food and oxygen from the digestive and respiratory systems and delivers these vital elements to the cells throughout the body. At the same time it picks up wastes from cells for delivery to excretory organs. But blood does more than this. It also transports hormones, enzymes, buffers, and other important biochemicals. Finally, the flow of blood is vital to temperature regulation in our bodies. Blood exhibits a physical property called *specific heat*, which allows it to absorb heat energy while at the same time resisting significant temperature change. This property permits blood temperature to remain relatively constant and within very narrow limits even when burdened with a significant heat load. Because of its high specific heat, blood can efficiently absorb and then safely transfer large amounts of heat energy from metabolism to the body's surface where it is dissipated by evaporation, convection, and radiation to the environment (see box on p. 127 for a review of this process).

BLOOD COMPOSITION

First and foremost, blood is a *liquid connective tissue* consisting not only of fluid *plasma*, but also of cells. Plasma is the third major fluid in our bodies (the other two are the interstitial fluids and intracellular fluids). Our *blood volume* is often expressed as a *percentage of our total body weight*. However, the measurement of the plasma and formed elements is typically expressed as a percentage of the **whole blood volume.** Using this method, whole blood is equal to about 8% of total body weight. Plasma accounts for 55% and **formed elements** such as various blood cells account for 45% of the total volume (Figure 16-1).

Blood Volume

Males have about 5 to 6 liters of blood circulating in their bodies and females have about 4 to 5 liters. In addition to gender differences, blood volume varies with age and body composition. A *unit* of blood (about

HINT *Before reading the chapter, say each of these terms out loud. This will help you avoid stumbling over them as you read.*

agglutinate (ah-GLOO-tin-ayt)
[*agglutin-* glue, *-ate* process]

agranulocyte (ah-GRAN-yoo-loh-syte)
[*a-* without, *-gran-* grain, *-ul-* little, *-cyte* cell]

anemia (ah-NEE-mee-ah)
[*an-* without, *-emia* blood condition]

anticoagulant drug (an-tee-koh-AG-yoo-lant)
[*anti-* against, *-coagul-* curdle, *-ant* agent]

antigen (AN-tih-jen)
[*anti-* against, *-gen* produce]

antigen A (AN-tih-jen)
[*anti-* against, *-gen* produce]

antigen B (AN-tih-jen)
[*anti-* against, *-gen* produce]

antiplatelet drug (an-tee-PLAYT-let)
[*anti-* against, *-plate-* flat, *-let* small]

basophil (BAY-soh-fil)
[*bas-* foundation, *-phil* love]

blood boosting

blood doping

blood serum (SEER-um)
[*serum* watery fluid] *pl.,* sera (SEER-ah)

blood type
[*tupos-* impression]

B lymphocyte (B LIM-foh-syte)
[*B* bursa-equivalent tissue, *lympho-* the lymph, *-cyte* cell]

coagulation (koh-ag-yoo-LAY-shun)
[*coagul-* curdle, *-ation* process]

complete blood cell count (CBC)

coumarin (KOO-mar-in)
[*coumarou-* tonka bean tree]

diapedesis (dye-ah-peh-DEE-sis)
[*dia-* apart or through, *-pedesis* oozing]

differential white blood cell (WBC) count (dif-er-EN-shal)

electrolyte (eh-LEK-troh-lyte)
[*electro-* electricity, *-lyt-* loosening]

eosinophil (ee-oh-SIN-oh-fil)
[*eosin-* reddish color, *-phil* love]

erythroblastosis fetalis (eh-rith-roh-blas-TOH-sis feh-TAL-is)
[*erythro-* red, *-blast-* bud, *-osis* condition]

erythrocyte (eh-RITH-roh-syte)
[*erythro-* red, *-cyte* cell]

erythropoiesis (eh-rith-roh-poy-EE-sis)
[*erythro-* red, *-poiesis* making]

erythropoietin (EPO) (eh-rith-roh-POY-eh-tin)
[*erythro-* red, *-poiet-* make, *in* substance]

continued on page 370

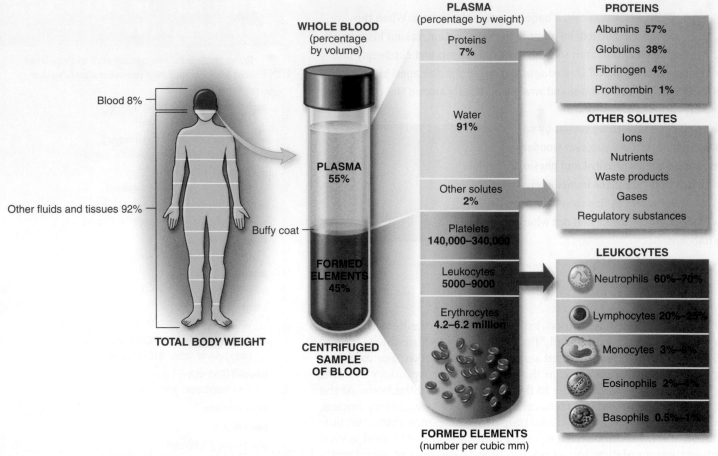

▲ FIGURE 16-1 **Composition of whole blood.** Approximate values for the components of blood in a normal adult.

0.5 liter or 1 pint) is the amount collected from blood donors for blood transfusion. One unit is equal to about 10% of the total blood volume for an average adult. There are several methods of measuring blood volume. Regardless of which method is used, it is important to have an accurate measurement in case blood volume must be replaced for a variety of conditions, including hemorrhage and shock.

One of the most important variables influencing blood volume is the amount of body fat. Blood volume per kilogram of body weight varies *inversely* with the amount of excess body fat. This means that leaner people have more blood per kilogram of body weight than obese people. Because females typically have somewhat more body fat than males (per kilogram of weight), they have slightly lower blood volumes.

FORMED ELEMENTS OF BLOOD

As you can see from Figure 16-1, blood consists of about 55% plasma and 45% of a variety of formed elements. These include *erythrocytes* (red blood cells or RBCs), *thrombocytes* (platelets), and *leukocytes* (white blood cells or WBCs). The leukocytes are further broken down into *granular leukocytes,* whose cytoplasm appears granular, and *nongranular leukocytes,* whose cytoplasm lacks granular components (Table 16-1).

In Figure 16-2, *A,* you see the results of *centrifuging* whole blood (spinning a vial at a high rate of speed). The lighter

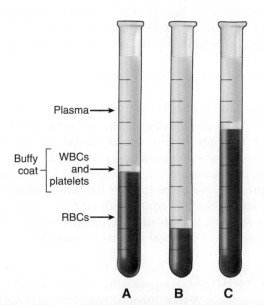

▲ FIGURE 16-2 **Hematocrit tubes showing normal blood, anemia, and polycythemia.** Note the buffy coat located between the packed RBCs and the plasma. **A,** A normal percentage of red blood cells. **B,** Anemia (a low percentage of red blood cells). **C,** Polycythemia (a high percentage of red blood cells).

TABLE 16-1 Classes of Blood Cells

CELL TYPE	DESCRIPTION	FUNCTION	LIFE SPAN
Red Blood Cells			
Erythrocyte	7 microns (μm) in diameter; concave disk shape; entire cell stains pale pink; no nucleus	Transportation of respiratory gases (O_2 and some CO_2)	105-120 days
Granular White Blood Cells			
Neutrophil	12-15 μm in diameter; spherical shape; multilobed nucleus; small, pink-purple–staining cytoplasmic granules	Cellular defense—phagocytosis of *small* pathogenic microorganisms such as bacteria	Hours to 3 days
Basophil	11-14 μm in diameter; spherical shape; generally two-lobed nucleus; large purple-staining cytoplasmic granules	Secretes heparin (anticoagulant) and histamine (important in the inflammatory response)	Hours to 3 days
Eosinophil	10-12 μm in diameter; spherical shape; generally two-lobed nucleus; large, orange-red–staining cytoplasmic granule	Cellular defense—phagocytosis of *large* pathogenic microorganisms, such as protozoa and parasites; releases anti-inflammatory substances in allergic reactions	10-12 days
Nongranular White Blood Cells			
Lymphocyte	6-9 μm in diameter; spherical shape; round (single-lobed) nucleus; small lymphocytes have scant cytoplasm	Humoral defense—secretes antibodies; involved in immune system response and regulation	Days to years
Monocyte	12-17 μm in diameter; spherical shape; nucleus generally kidney bean or horseshoe shaped with convoluted surface; ample cytoplasm often "steel blue" in color	Capable of migrating out of the blood to enter tissue spaces as a macrophage—an aggressive phagocytic cell capable of ingesting bacteria, cellular debris, and cancerous cells	Months
Platelets			
Thrombocyte	2-5 μm in diameter; irregularly shaped fragments; cytoplasm contains very small, pink-staining granules	Releases clot-activating substances and helps in formation of actual blood clot by forming platelet "plugs"	7-10 days

plasma remains at the top, and the middle-weight leukocytes and platelets form a so-called *buffy coat* in the middle. Erythrocytes are heavier and concentrate at the bottom of the test tube. The volume of packed red blood cells at the bottom of the test tube is called the **hematocrit.**

Average hematocrits vary but are normally around 45% for men and 42% for women. Conditions that result in decreased RBC numbers (Figure 16-2, *B*) are **anemias.** A reduced hematocrit number characterizes these disorders. However, healthy individuals living and working in high altitudes may have elevated RBC numbers and hematocrit values—a condition called **physiological polycythemia** (Figure 16-2, *C*).

Note that leukocytes and platelets make up less than 1% of blood volume.

QUICK ✓ CHECK

1. What is the fluid portion of whole blood?
2. What constitutes the formed elements of whole blood?
3. What factors might influence blood volume?
4. What are the average component percentages of a normal hematocrit?

Red Blood Cells (Erythrocytes)

A normal, mature erythrocyte (RBC) is only about 7.5 μm in diameter. Amazingly, more than 1,500 of them can fit side by side in a 1-cm space. Before the cell reaches maturity in the bone marrow, it loses its nucleus. Unlike other cells, it also loses its ribosomes, mitochondria, and other organelles. In their place, nearly 35% of its volume is filled with hemoglobin, the protein responsible for transporting oxygen in the blood.

As you can see in Figure 16-3, erythrocytes are shaped like tiny biconcave disks. The microscopic depression on each flat surface of the cell creates a cell with a thin center and thicker edges. This unique shape gives an erythrocyte a very large surface area relative to its volume. RBCs can passively change their shapes as they are forced through capillaries under pressure. This ability is vital to the survival of RBCs, which are under almost constant mechanical stress and strain as they rush through the capillaries of our bodies. Their shape also allows faster blood flow throughout the circulatory system.

RBCs are the most numerous of all the formed elements of blood. In men, RBC counts average about 5.5 million per

▲ FIGURE 16-3 **Erythrocytes.** Color-enhanced scanning electron micrograph shows normal erythrocytes. Note the biconcave shape.

cubic millimeter (mm³) of blood. In contrast, women have about 4.8 million/mm³.

Function of Red Blood Cells

RBCs play a critical role in the transport of oxygen and carbon dioxide in the body (this topic is discussed more fully in Chapter 18).

Altogether, the total surface area of all the RBCs in an adult is equivalent to an area larger than a football field. This is an enormous area for the efficient exchange of the respiratory gases between the RBCs (via their hemoglobin) and the interstitial fluid that bathes our body cells. (This is yet another excellent example of the relationship between form and function.)

Hemoglobin

Within each RBC are an estimated 200 to 300 *million* molecules of hemoglobin. Hemoglobin molecules are composed of four protein chains, each called a **globin.** Every globin molecule is bound to a heme group, each of which contains one atom of iron. This means that each hemoglobin molecule contains four iron atoms. Because of this arrangement, one hemoglobin molecule chemically bonds with four oxygen molecules to form *oxyhemoglobin*. This is a reversible reaction. Hemoglobin can also combine with carbon dioxide to form *carbaminohemoglobin* (also reversible). However, in this reaction, it is the globins, *not* the heme groups, that allow carbon dioxide to bond.

As we've seen, a man's blood usually contains more RBCs (and thus more hemoglobin) than a woman's blood. This is because higher levels of testosterone in men tend to stimulate erythrocyte production and cause an increase in RBC numbers. Normally, a man has 14 to 16 grams of hemoglobin for every 100 milliliters of blood in his system. An adult male who has a hemoglobin content of less than 10 g/100 ml of blood is diagnosed as having **anemia** (literally, a lack of blood). The term *anemia* is also used to describe a low RBC count. Anemias are classified according to the size and hemoglobin content of RBCs. Box 16-1 describes a specific type of anemia—*sickle cell anemia*—that is caused by the production of an abnormal type of hemoglobin due to a genetic error.

Formation of Red Blood Cells

The term **erythropoiesis** describes the entire process of RBC formation. Erythrocytes begin their maturation process in the red bone marrow from *nucleated* hematopoietic stem cells called **hemocytoblasts** (Figure 16-4). These adult stem cells have the ability to maintain a constant population of newly differentiating cells of a specific type. Note, however, that adult stem cells are *not* the same as embryonic stem cells (see Chapter 26), which are involved in embryonic and fetal development. Adult blood-forming stem cells divide by mitosis. Some of the daughter cells remain as undifferentiated adult stem cells. Others continue to develop into erythrocytes. You can follow this transformation in Figure 16-4.

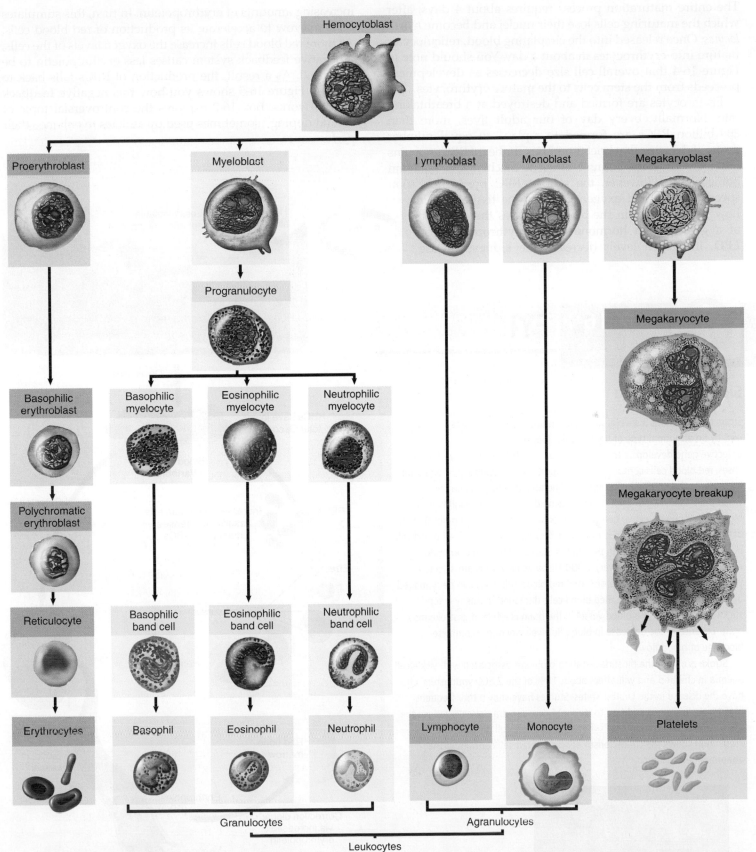

▲ FIGURE 16-4 **Formation of blood cells.** The hematopoietic stem cell, called the *hemocytoblast,* serves as the original stem cell from which all formed elements of the blood are derived. Note that all five precursor cells, which ultimately produce the different components of the formed elements, are derived from the hemocytoblast.

The entire maturation process requires about 4 days, after which the maturing cells lose their nuclei and become *reticulocytes*. Once released into the circulating blood, reticulocytes mature into erythrocytes in about a day. You should note in Figure 16-4 that overall cell size decreases as development proceeds from the stem cells to the mature erythrocytes.

Erythrocytes are formed and destroyed at a breathtaking rate. Normally, every day of our adult lives, more than 200 billion RBCs are formed to replace an equal number destroyed during that brief time. The number of RBCs remains relatively constant because efficient mechanisms maintain homeostasis. However, the rate of RBC production soon speeds up if blood oxygen levels in the tissues decline. Low oxygen level in the blood increases the secretion of a glycoprotein hormone called **erythropoietin** or **EPO.** If oxygen levels decrease, the kidneys release

increasing amounts of erythropoietin. In turn, this stimulates bone marrow to accelerate its production of red blood cells. As more red blood cells increase the oxygen levels of the cells, a negative feedback system causes less erythropoietin to be produced. As a result, the production of RBCs falls back to normal. Figure 16-5 shows you how this negative feedback system works. Box 16-2 explores the controversial topic of "blood doping" sometimes used by athletes to enhance their performance.

BOX 16-1 FYI

Sickle Cell Anemia

Sickle cell anemia is a severe, sometimes fatal, hereditary disease characterized by an abnormal type of hemoglobin. A person who inherits only one defective gene develops a form of the disease called *sickle cell trait.* In these cases, red blood cells contain a small proportion of a hemoglobin type that is less soluble than normal. This abnormal hemoglobin forms solid crystals when the blood oxygen level is low, causing distortion and fragility of the red blood cell. If two defective genes are inherited (one from each parent), more of the defective hemoglobin is produced, and the distortion of red blood cells becomes even more severe. In the United States, about 1 in every 500 African-American and 1 in every 1,000 Hispanic newborns are affected each year. In these individuals, the distorted red blood cell walls can be damaged by drastic changes in shape. Red blood cells damaged in this way tend to stick to vessel walls. If a blood vessel in the brain is affected, a stroke may occur because of the decrease in blood flow velocity or the complete blockage of blood flow.

Stroke is one of the most devastating problems associated with sickle cell anemia in children and will affect about 10% of the 2,500 youngsters who have the disease in the United States. Studies have shown that frequent blood transfusions in addition to standard care can dramatically reduce the risk of stroke in many children suffering from sickle cell anemia. The illustration shows the characteristic shape of a red cell containing the abnormal hemoglobin.

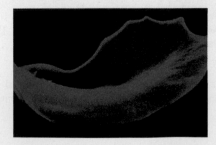

◀ **Sickle cell anemia.**

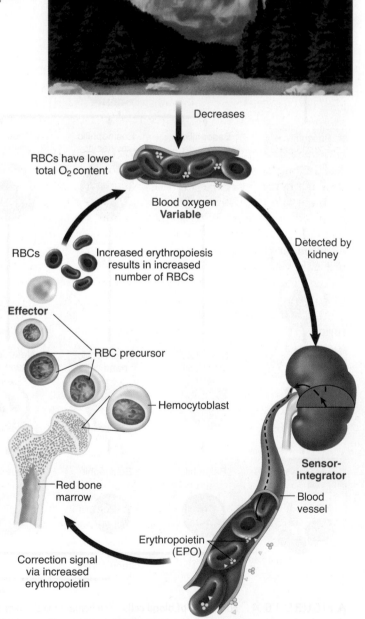

▲ **FIGURE 16-5** **Erythropoiesis.** In response to decreased blood oxygen, the kidneys release erythropoietin (EPO). This stimulates erythrocyte production in the red bone marrow.

BOX 16-2 Sports & Fitness

Blood Doping

Reports that some Olympic and other elite athletes use transfusions of their own blood to improve performance have surfaced repeatedly in the past several decades. The practice—called **blood doping** or **blood boosting**—is intended to increase oxygen delivery to muscles. A few weeks before competition, blood is drawn from the athlete and the red blood cells (RBCs) are separated and frozen. Just before competition, the RBCs are thawed and injected. Theoretically, infused RBCs and elevation of hemoglobin levels after transfusion should increase oxygen consumption and muscle performance during exercise. In practice, however, the advantage appears to be minimal. All blood transfusions carry some risk, and unnecessary or questionably indicated transfusions are medically and ethically unacceptable.

In addition to blood transfusions, injection of substances that increase RBC levels in an attempt to improve athletic performance has also been condemned by leading authorities in the area of sports medicine and by athletic organizations around the world. "Doping" with either the naturally occurring hormone erythropoietin (EPO) or with synthetic drugs that have similar biological effects—such as Epogen and Procrit—can result in devastating medical outcomes. For example, EPO abuse can produce dangerously high blood pressure that may lead to a heart attack or stroke.

Destruction of Red Blood Cells

The life span of RBCs circulating in the bloodstream averages between 105 and 120 days. They often break apart, or fragment, in the capillaries as they age. Macrophage cells in the lining of the blood vessels, especially those in the liver and spleen, *phagocytose* (ingest and destroy) the aged, abnormal, or fragmented RBCs. This process results in the breakdown of hemoglobin. As a result, amino acids, iron, and the pigment *bilirubin* are released into the bloodstream. Iron is returned to the bone marrow for use in the synthesis of new hemoglobin. Bilirubin is transported to the liver, where it is excreted as part of bile. Amino acids, released from the globin part of the hemoglobin, are reused by the body for energy or for the synthesis of new proteins.

For the RBC homeostatic mechanism to succeed in maintaining a normal number of RBCs, the bone marrow must function properly. To do this, the blood must supply it with the proper building components and catalysts with which to create new RBCs. In addition, the gastric mucosa of the stomach must provide intrinsic factor and perhaps other undiscovered factors necessary for the absorption of vitamin B_{12}. This vitamin is vital to the formation of new erythrocytes.

> ### QUICK CHECK
> 5. What are the components of hemoglobin?
> 6. How many molecules of hemoglobin are in the average RBC?
> 7. Trace the formation of a mature erythrocyte from its stem cell precursor.
> 8. Explain the negative feedback loop that controls erythropoiesis.

White Blood Cells (Leukocytes)

There are five basic types of white blood cells, or **leukocytes.** They are classified according to the presence or absence of granules as well as the staining characteristics of their cytoplasm. **Granulocytes** include the three types of WBCs that have *granules* in their cytoplasm. They are named according to their cytoplasmic staining properties: *basophils, neutrophils,* and *eosinophils.* There are two types of **agranulocytes** (WBCs without cytoplasmic granules): *lymphocytes* and *monocytes.*

As a group, the leukocytes appear brightly colored in stained preparations. In addition, they all have nuclei and are generally larger than RBCs. Before continuing with the following discussion of each type, please look at Table 16-1 and briefly familiarize yourself with each cell type, its description, and function.

Granulocytes

Neutrophils

The cytoplasmic granules of **neutrophils** (Figure 16-6) stain a light purple with neutral dyes. The granules in these cells are small and numerous. They tend to give the cytoplasm a coarse appearance. The cytoplasmic granules contain powerful lysosomes that allow them to destroy most bacterial cells.

Neutrophils make up about 65% of the WBC count in a normal blood sample. They are highly mobile, active phagocytic cells that can migrate out of blood vessels and enter into the tissue spaces. This process is called **diapedesis.** It is vital to the body's fight against invading bacteria. It works like this: Bacterial infections produce an inflammatory response. In this process, damaged cells of the body release chemicals that attract neutrophils and other phagocytic WBCs to the infection site. The swelling, pain, and heat from the infection site are indications that the battle is underway.

|←——12-15 µm——→| ◄ FIGURE 16-6 Neutrophil.

Eosinophils

├── 10-12 μm ──┤

▲ **FIGURE 16-7**
Eosinophil.

Eosinophils (Figure 16-7) contain many large cytoplasmic granules that stain orange with acid dyes such as eosin. Their nuclei generally have just two lobes. Eosinophils equal about 2% to 5% of circulating WBCs. They are abundant in the linings of the respiratory and digestive tracts. Eosinophils can ingest inflammatory chemicals and proteins associated with antigen-antibody reaction complexes. Perhaps their most important functions involve protection against infections caused by parasitic worms. They are also involved in allergic reactions, as we shall see in Chapter 19.

Basophils

Basophils (Figure 16-8) have few, but relatively large, cytoplasmic granules that stain dark purple with basic dyes. The cytoplasmic granules of basophils contain *histamine* (an inflammatory chemical) and *heparin* (an anticoagulant). Basophils have indistinct, S-shaped nuclei. They are the least numerous of the WBCs, numbering only 0.5% to 1% of the total leukocyte count. Like neutrophils, basophils are both mobile and capable of diapedesis.

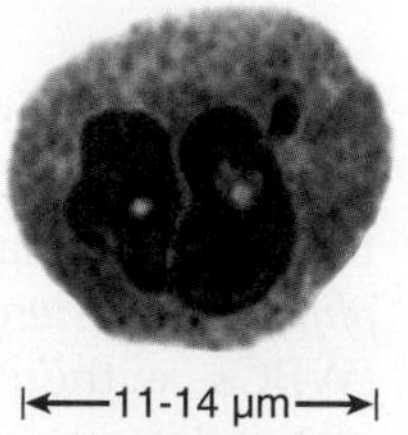

├──11-14 μm──┤

▲ **FIGURE 16-8**
Basophil.

Agranulocytes

Lymphocytes

├── 6-9 μm ──┤

▲ **FIGURE 16-9**
Lymphocyte.

Lymphocytes (Figure 16-9) are the smallest of the leukocytes, averaging only about 6 to 9 μm in diameter. They have large, spherical nuclei surrounded by a small amount of cytoplasm that stains a pale blue. After neutrophils, lymphocytes are the most numerous WBCs. They account for about 25% of all the leukocytes in our bodies.

There are two general types of lymphocytes: *T lymphocytes* and *B lymphocytes*. Both forms have important roles in our immunity. **T lymphocytes** function by directly attacking an infected or cancerous cell. **B lymphocytes,** in contrast, produce antibodies against specific antigens.

Monocytes

Monocytes (Figure 16-10) are the largest of the leukocytes. They have dark, kidney bean–shaped nuclei surrounded by large quantities of distinctive blue-gray cytoplasm. Monocytes are mobile and highly phagocytic: They can engulf large bacterial organisms and virus-infected cells.

├── 12-17 μm ──┤

▲ **FIGURE 16-10**   Monocyte.

BOX 16-3   **Diagnostic Study**

Complete Blood Cell Count

One of the most useful and frequently performed clinical blood tests is called the **complete blood cell count** or simply the **CBC**. The CBC is a collection of tests whose results, when interpreted as a whole, can yield an enormous amount of information regarding a person's health. Standard red blood cell, white blood cell, and thrombocyte counts, the differential white blood cell count, hematocrit, hemoglobin content, and other characteristics of the formed elements are usually included in this battery of tests.

White Blood Cell Numbers

Compared to erythrocytes, leukocytes are relatively rare. One cubic millimeter of normal blood usually contains only about 5,000 to 9,000 leukocytes. As we've seen, there are different percentages of each type. These numbers have clinical significance because they may change drastically under abnormal conditions such as infections or specific blood cancers. In acute appendicitis, for example, the percentage of neutrophils increases dramatically. So does the total WBC count. In fact, these characteristic changes may be deciding points for surgery to remove the infected organ.

An overall decrease in the number of WBCs is called **leukopenia.** An increase in the number of WBCs is **leukocytosis.** The number of each type of white blood cell can be determined by a **differential white blood cell (WBC) count.** In this special count (Table 16-2), the *proportion* of each type of white blood cell is reported as a percentage of the total WBC count. Because all disorders do not affect each type of WBC the same way, the differential WBC count is a valuable diagnostic tool. For example, some parasite infestations do not cause an increase in the total WBC count. However, they often do cause an increase in the proportion of eosinophils. Why? Because this type of WBC specializes in fighting large parasites such as parasitic nematode "worms." Table 16-2 presents a differential count of the major white blood cell types in the blood of an average person.

Formation of White Blood Cells

Hematopoietic stem cells serve as the precursors not only of erythrocytes, but also of leukocytes and platelets. Refer to Figure 16-4 again and follow the formation and maturation of the various leukocytes from the precursor hematopoietic stem cells (*hemocytoblasts*). Like erythrocytes, neutrophils, eosinophils, basophils, and a few lymphocytes and monocytes originate in red bone marrow **(myeloid tissue).** However, note that most lymphocytes and monocytes are derived from hematopoietic adult stem cells in *lymphatic tissue*. So although many lymphocytes are found in bone marrow, most are formed in lymphatic tissue and later carried to the bone marrow by the bloodstream.

TABLE 16-2   Differential Count of White Blood Cells

CLASS	DIFFERENTIAL COUNT*	
	NORMAL RANGE (%)	TYPICAL VALUE (%)†
Neutrophils	65–75	65
Lymphocytes (large and small)	20–25	25
Monocytes	0–3	6
Eosinophils	0–2	3
Basophils	½–1	1
TOTAL	100	100

*In any differential count the sum of the percentages of the different kinds of WBCs must, of course, total 100%.

†This mnemonic phrase may help you remember percent values in decreasing order: "Never Let Monkeys Eat Bananas."

Myeloid tissue and lymphatic tissue together constitute the *hematopoietic*, or blood cell–forming, tissues of the body. Red bone marrow is myeloid tissue that actually produces (red) blood cells. Yellow marrow is yellow because it stores a large amount of fat. Yellow marrow remains yellow except during times of disease, when it can become active and red in color because it also produces red blood cells.

Platelets (Thrombocytes)

Platelets or *thrombocytes* are really tiny fragments of cells (see Table 16-1). They are nearly colorless bodies that appear as irregular spindles or oval disks about 2 to 4 µm in diameter. Their functions are varied and have to do with *clotting*: cell aggregation, adhesiveness, and agglutination. It's difficult to see them in a slide presentation because, as soon as blood is drawn, the platelets adhere to each other and to every surface they contact. This phenomenon makes them assume many irregular forms.

A range of 150,000 to 400,000 platelets/mm^3 is considered normal for adults, but newborns often show reduced numbers. Unlike erythrocytes, there are no differences between the sexes in platelet count.

Function of Platelets

Platelets play vital roles in *hemostasis* and *coagulation*. **Hemostasis** refers to the *stoppage of blood flow* from an injured vessel. This may occur as a result of any one of several body defense mechanisms. One of these mechanisms is formation of a *platelet plug*, which temporarily reduces or stops blood flow. Formation of a platelet plug is usually followed by **coagulation,** which forms a more solid *clot*.

Within 1 to 5 seconds after injury to a blood capillary, a **platelet plug** is formed when platelets adhere to the damaged wall of the vessel. This plug helps stop the flow of blood into the tissues. The formation of the plug generally follows vascular spasms caused by the constriction of smooth muscle fibers in the wall of the damaged blood vessel, which also helps reduce blood flow.

When platelets encounter collagen in damaged vessel walls and surrounding tissue, they become sticky platelets. These sticky platelets then bind to underlying tissues and each other, forming the plug. In addition, sticky platelets secrete several biochemicals, including adenosine diphosphate (ADP), thromboxane (a local hormone), and a fatty acid (arachidonic acid). When these chemicals are released, they affect both local blood flow (by vasoconstriction) and platelet aggregation at the site of injury. If the injury is extensive, the blood-clotting mechanism (coagulation) is also activated.

Platelet plugs are also vital in controlling so-called *microhemorrhages,* which may involve a break in a single capillary. Failure to stop hemorrhage from minor but numerous and widespread capillary breaks can result in life-threatening internal blood loss. In certain types of peripheral vascular disease, platelet plugs may also be involved in creating blockage in small vessels, including arterioles.

Formation and Life Span of Platelets

Thrombopoiesis is the formation of platelets (see Figure 16-4). Mature *megakaryocytes* are huge cells that often have a bizarre shape. The abundant cytoplasm is blue to pink in color and contains a variable number of very fine granules. Between 2,000 and 3,000 platelets are created when the irregular cytoplasmic membrane surrounding the mature megakaryocyte ruptures. The resulting platelets have a plasma membrane but no nucleus. Platelets have a short life span of about 7 days.

BLOOD TYPES (BLOOD GROUPS)

The term **blood type** refers to the types of biochemical markers or **antigens** present on the plasma membranes of erythrocytes. (You can find a complete discussion of the concept of *antigens* and their associated *antibodies* in Chapter 19.) For example, there are blood **antigens A** and **B** in the ABO system. There is also a group of six **Rh antigens.** To date, researchers have isolated nearly two dozen additional blood antigens that vary from person to person. This variability is important because our immune system may "attack" donated blood cells (from a transfusion) if they have antigens different than our own. As you may know, antigens A, B, and Rh are the most important blood antigens as far as transfusions and newborn survival are concerned. The other blood antigens are less important clinically but may still cause occasional problems with transfusions.

Why do different people have different antigens on their RBCs? We don't have a complete answer to that question. However, a good working hypothesis is that their presence or absence may give some biological advantage to groups of people living under different environmental conditions. For example, an antigen called *Duffy* (after the patient in whom it was first discovered) is often missing in populations that have lived with the threat of malaria for many generations. The Duffy antigen is used by the malaria parasite to enter RBCs. So, its absence *protects* a person against developing malaria. This is because the parasite cannot "identify" its host red blood cells and, therefore, it cannot reproduce itself in the body.

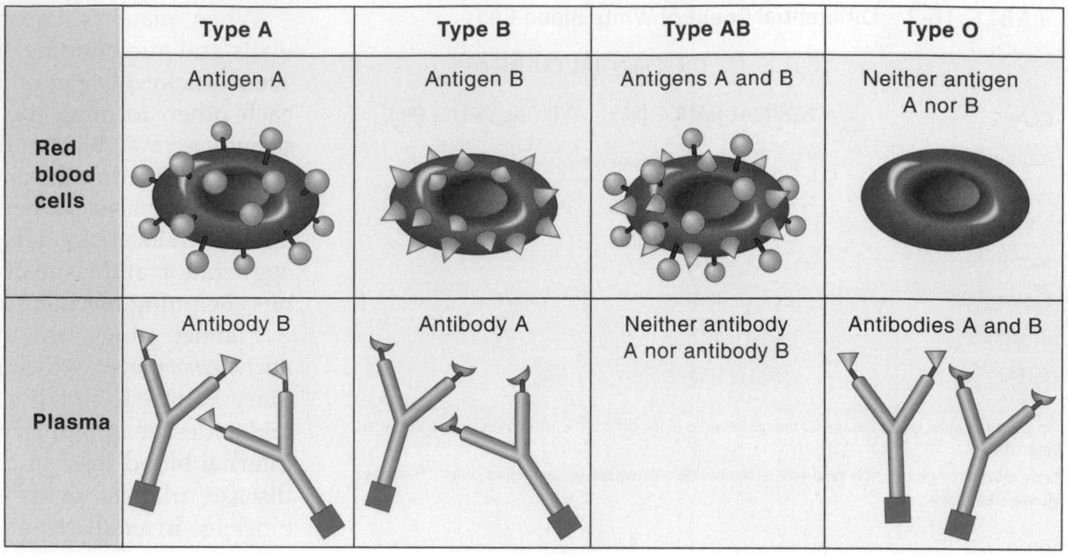

▶ **FIGURE 16-11** **ABO blood types.** Note that antigens characteristic of each blood type are bound to the surface of RBCs. The antibodies of each blood type are found in the plasma and exhibit unique structural features. This permits agglutination to occur if exposure to the appropriate antigen occurs.

	Type A	Type B	Type AB	Type O
Red blood cells	Antigen A	Antigen B	Antigens A and B	Neither antigen A nor B
Plasma	Antibody B	Antibody A	Neither antibody A nor antibody B	Antibodies A and B

The term *agglutinin* is often used to describe the antibodies dissolved in plasma that react with specific blood group antigens, or *agglutinogens.* When they combine and react, they cause RBCs to clump together or **agglutinate.** When a blood transfusion is given, great care must be taken to prevent a mixture of agglutinogens (antigens) and agglutinins (antibodies) from agglutinating. This is especially true with the ABO and Rh blood groups. If the wrong blood types are mixed together during a blood transfusion, a **transfusion reaction** may take place. As the different blood types agglutinate, blood clots form that block blood vessels and cause serious problems in the body. Clinical laboratory tests, called *blood typing* and *crossmatching,* ensure the proper identification of blood group antigens and antibodies in both donor and recipient blood.

A&P CONNECT

Blood transfusions are an important therapeutic tool. Learn more about blood transfusions, blood banking, and even *artificial* blood in **Blood Transfusions** online at **A&P Connect.**

evolve

The ABO System

Every person's blood belongs to one of the four ABO blood types (groups). These blood types are named according to the *antigen present* on the membranes of the RBCs:

1. Type A—antigen A on RBCs
2. Type B—antigen B on RBCs
3. Type AB—*both* antigen A and B on RBCs
4. Type O—*neither* antigen A nor B on RBCs

Blood plasma *may or may not* contain antibodies (agglutinins) that can react with RBC antigen A or antigen B. An important principle related to this is that plasma *never* contains antibodies against the antigens present on its own red blood cells. If it did, the antibody would react with the antigen and destroy the RBCs by agglutination. However, plasma *does* contain antibodies against antigen A or antigen B if they are *not* present on its RBCs.

With this in mind, we can deduce the following: In type A blood, antigen A is present on its RBCs. Therefore, its plasma contains no anti-A antibodies but does contain anti-B antibodies. Similarly, in type B blood, antigen B is present on its RBCs. Therefore, its plasma contains no anti-B antibodies

▶ **FIGURE 16-12**

Agglutination. A, When mixing of donor and recipient blood of the same type *(A)* occurs, there is no agglutination because only type B antibodies are present. **B,** If type A donor blood is mixed with type B recipient blood, agglutination will occur because of the presence of type A antibodies in the type B recipient blood.

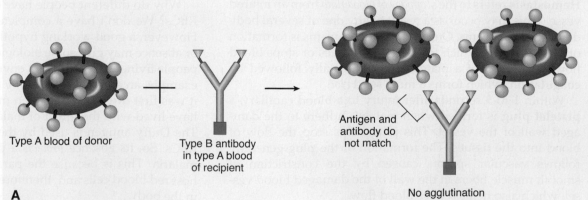

Type A blood of donor

Type B antibody in type A blood of recipient

Antigen and antibody do not match

A

No agglutination

Recipient's blood		Reactions with donor's blood			
RBC antigens	Plasma antibodies	Donor type O	Donor type A	Donor type B	Donor type AB
None (Type O)	Anti-A Anti-B				
A (Type A)	Anti-B				
B (Type B)	Anti-A				
AB (Type AB)	(None)				

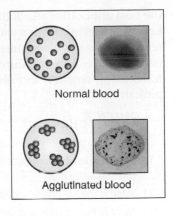

Normal blood

Agglutinated blood

▲ FIGURE 16-13   Results of (crossmatching) different combinations (types) of donor and recipient blood. The left columns show the antigen and antibody characteristics that define the recipient's blood type, and the top row shows the donor's blood type. Crossmatching identifies either a compatible combination of donor-recipient blood (no agglutination) or an incompatible combination (agglutinated blood). Photo inset shows drops of blood showing appearance of agglutinated and nonagglutinated red blood cells.

but does contain anti-A antibodies (Figure 16-11). Before going on, re-read the last two paragraphs to make sure you have an understanding of antigen and antibody.

Now look at Figure 16-12, *A.* You can see that type A blood donated to a type A recipient does *not* cause an agglutination transfusion reaction. This is because the type B antibodies in the recipient do not combine with the type A antigens in the donated blood. However, type A blood donated to a type B recipient causes an agglutination transfusion reaction. This is because the type A antibodies in the recipient combine with the type A antigens in the donated blood (Figure 16-12, *B*). Figure 16-13 shows you the results of different combinations of donor and recipient blood.

Because type O blood does not contain either antigen A or B, it has often been called the *universal donor.* This is not quite

true because the recipient's blood may contain agglutinins other than anti-A or anti-B antibodies. This is why the recipient's and the donor's blood—even if it is type O—should be crossmatched to check for agglutination. In contrast, *universal recipient* (type AB) blood contains neither anti-A nor anti-B antibodies. For this reason, it cannot agglutinate type A or type B donor red blood cells. However, other agglutinins may be present in this so-called universal recipient blood and may clump unidentified antigens in the donor's blood. Again, as with type O blood, crossmatching tests should be conducted to make sure there is no agglutination due to other agglutinins.

As you can see from the examples above, improperly typed and crossmatched blood given during a blood transfusion can cause a transfusion reaction in the recipient. As

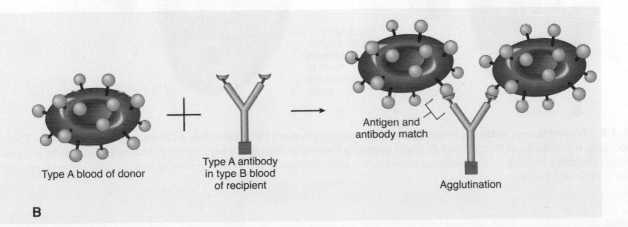

Type A blood of donor

Type A antibody in type B blood of recipient

Antigen and antibody match

Agglutination

B

the host antibodies attack the donor RBCs, the RBCs are broken apart in a process called **hemolysis.** Hemoglobin is released into the bloodstream, which may overload the kidneys and cause their failure and death. Signs of this type of transfusion reaction include fever, difficulty breathing, and pink urine.

QUICK ✓ CHECK

9. Name the granulocytic and agranulocytic leukocytes.
10. List the normal percentages of the different types of WBCs found in a differential count.
11. What is the ABO blood group system?
12. Identify the antigens and antibodies (if any) associated with the ABO blood groups.

The Rh System

The term *Rh-positive blood* means that an Rh antigen is present on the blood's RBCs. In contrast, *Rh-negative blood* does not have Rh antigens present on its red blood cells. We should note here that blood does not normally contain anti-Rh antibodies. However, anti-Rh antibodies can appear in the blood of an Rh-negative person *if* Rh-positive RBCs have *at one time in the past* entered the bloodstream. One way this can happen is by giving an Rh-negative person a transfusion of Rh-positive blood. In a short time, the person's immune system makes anti-Rh antibodies, and these remain in the blood.

The other way in which Rh-positive RBCs can enter the bloodstream of an Rh-negative individual can happen *to a*

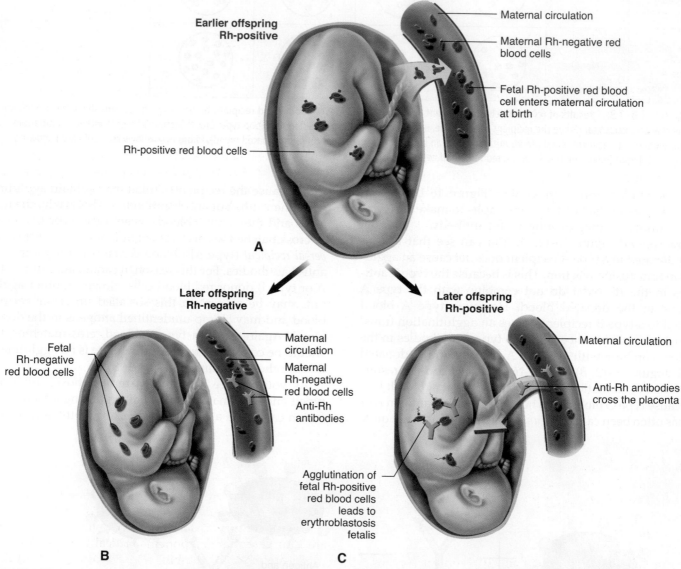

Earlier offspring Rh-positive

Maternal circulation

Maternal Rh-negative red blood cells

Fetal Rh-positive red blood cell enters maternal circulation at birth

Rh-positive red blood cells

A

Later offspring Rh-negative

Fetal Rh-negative red blood cells

Maternal circulation

Maternal Rh-negative red blood cells

Anti-Rh antibodies

B

Later offspring Rh-positive

Maternal circulation

Anti-Rh antibodies cross the placenta

Agglutination of fetal Rh-positive red blood cells leads to erythroblastosis fetalis

C

▲ **FIGURE 16-14**   **Erythroblastosis fetalis. A,** Rh-positive blood cells enter the mother's bloodstream during delivery of an Rh-positive baby. If not treated, the mother's body will produce anti-Rh antibodies. **B,** A later pregnancy involving an Rh-negative baby is normal because there are no Rh antigens in the baby's blood. **C,** A later pregnancy involving an Rh-positive baby may result in erythroblastosis fetalis. Anti-Rh antibodies enter the baby's blood supply and cause agglutination of RBCs with the Rh antigen.

TABLE 16-3 Blood Typing

BLOOD TYPE (ABO, RH)	ANTIGENS PRESENT*	ANTIBODIES PRESENT*	PERCENTAGE OF GENERAL POPULATION
O, +	Rh	A, B	35%
O, −†	None	A, B, Rh?	7%
A, +	A, Rh	B	35%
A, −	A	B, Rh?	7%
B, +	B, Rh	A	8%
B, −	B	A, Rh?	2%
AB, +‡	A, B, Rh	None	4%
AB, −	A, B	Rh?	2%

From Pagana KD, Pagana TJ: *Mosby's Manual of Diagnostic and Laboratory Tests*, ed 4. St. Louis: Mosby, 2010.

*Anti-Rh antibodies *may* be present, depending on exposure to Rh antigens.

†Universal donor.

‡Universal recipient.

woman during pregnancy. Herein lies the danger for a baby born to an Rh-negative mother and an Rh-positive father: If the offspring inherits the Rh-positive trait from the father, the Rh factor on the offspring's RBCs may stimulate the mother's body to form anti-Rh antibodies. Then, if the mother carries another Rh-positive fetus in a future pregnancy, the fetus may develop a disease called **erythroblastosis fetalis.** This is a serious hemolytic condition caused by the mother's anti-Rh antibodies reacting with the offspring's Rh-positive cells (Figure 16-14). All Rh-negative mothers who carry an Rh-positive baby should be treated with a protein marketed under the name RhoGAM. This product stops the mother's body from forming anti-Rh antibodies and thus prevents the possibility of harm to the *next Rh-positive offspring she may have.*

Table 16-3 summarizes for you the ABO and Rh blood types, including the frequency of each in the general population. Of course, the frequency of these and other blood types may be different within a family or ethnic group based on regional differences in human populations.

BLOOD PLASMA

Plasma is the liquid part of the blood. That is, plasma is whole blood without the formed elements (see again Figures 16-1 and 16-2). Plasma is prepared by spinning whole blood down in a centrifuge at a high rate of speed. The end result is a clear, straw-colored fluid—blood plasma—lying above the cell layer in the test tube.

Plasma consists of 90% water and 10% solutes. Normally, about 6% to 8% of the solutes consist of proteins. These proteins include some clotting factors, gamma globulins (important in treating weakened immune systems), and albumin (a blood volume expander). Other solutes present in much

smaller amounts include glucose, amino acids, and lipids, as well as urea, uric acid, creatinine, and lactic acid; oxygen and carbon dioxide; and hormones and enzymes.

Blood solutes are classified as **electrolytes** (molecules that ionize in solution, such as proteins and inorganic salts) or **nonelectrolytes** (molecules that do not ionize, such as glucose and lipids).

The proteins in blood plasma consist of three main kinds of compounds: *albumins, globulins,* and *clotting proteins* (principally *fibrinogen*). A total of approximately 6 to 8 grams of proteins occupy a blood plasma volume of 100 ml. Albumins constitute about 55% of this total, globulins about 38%, and fibrinogen about 7%.

Plasma proteins are critically important substances. For example, fibrinogen and a clotting protein named **prothrombin** are vital to our blood-clotting mechanism. Globulins function as essential components of the immunity mechanism. Many modified globulins, called *gamma globulins,* serve important roles as circulating antibodies (see Chapter 19). All plasma proteins contribute to the maintenance of normal (1) blood viscosity, (2) blood osmotic pressure, and (3) blood volume. As you might surmise, therefore, plasma proteins play an essential part in maintaining normal circulation.

Synthesis of most plasma proteins occurs in our liver cells. These cells form many of the plasma proteins—except some of the gamma globulin antibodies synthesized by plasma cells (recall that plasma cells are a type of lymphocyte). Cancer of plasma cells, called *multiple myeloma,* results in the production of an abnormal myeloma antibody. These gamma globulin antibodies cause a number of very serious disease symptoms.

BLOOD CLOTTING (COAGULATION)

The coagulation of blood seals ruptured vessels to stop bleeding in a process called hemostasis, as we have seen. A secondary function of coagulation is to prevent bacteria from invading our tissues. Somehow, our bodies must know *when* to coagulate. After all, coagulation of blood when it is not necessary can lead to blood clots and blockage of vessels. Such random clotting would deprive our tissues from life-sustaining oxygen. Such abnormal clotting is a frequent cause of heart attacks and strokes.

Although we have an abundance of information about the process of blood clotting, we are still shy of a complete understanding. Our best efforts to summarize what we know are presented for you in Figure 16-15.

Over a century ago, researchers determined that there are four essential components critical to coagulation: (1) prothrombin, (2) thrombin, (3) fibrinogen, and (4) fibrin. However, many coagulation factors and their functions have been discovered in recent decades. Here we divide the basic process into an *extrinsic clotting pathway* and *intrinsic clotting pathway.*

As you read through the following paragraphs, please refer to Figure 16-15. Notice that there are two pathways in the process of blood clotting. In both pathways, a series of chemical reactions called a *clotting cascade* precedes the formation of *prothrombin activator*.

In the **extrinsic pathway,** chemicals are released from damaged tissue *outside the blood* that ultimately results in the formation of prothrombin activator.

In contrast, the **intrinsic pathway** involves a series of reactions that begin with factors normally present in, or intrinsic to, the blood. For example, damage to the endothelial lining of blood vessels exposes collagen fibers. In turn, exposure of these fibers triggers the activation of a number of coagulation factors in the plasma. Sticky platelets participate in the intrinsic pathway, ultimately inducing the production of prothrombin activator.

Regardless of the pathway involved, after prothrombin activator is produced, a clot will form. Thrombin accelerates conversion of the soluble plasma protein fibrinogen to insoluble fibrin. Then formation of fibrin strands forms a fibrin clot. Fibrin appears in blood as fine tangled threads. As blood flows through the fibrin mesh, its formed elements are caught in the mesh. Because most of the cells are RBCs, clotted blood has a red color. The pale yellowish liquid left after a clot forms is **blood serum.** This serum is different from plasma because it has lost its clotting elements.

The overall reactions of clotting can be summarized as follows:

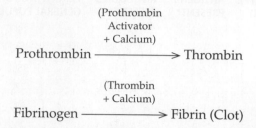

$$\text{Prothrombin} \xrightarrow{\substack{\text{(Prothrombin} \\ \text{Activator} \\ + \text{ Calcium)}}} \text{Thrombin}$$

$$\text{Fibrinogen} \xrightarrow{\substack{\text{(Thrombin} \\ + \text{ Calcium)}}} \text{Fibrin (Clot)}$$

Liver cells synthesize both prothrombin and fibrinogen, as they do almost all other plasma proteins. For the liver to synthesize prothrombin at a normal rate, blood must contain an adequate amount of vitamin K. This vitamin is absorbed into the blood from the intestine. Some foods contain vitamin K, but it is also synthesized in the intestine by certain bacteria (not present for a time in newborn infants). Because vitamin K is fat soluble, its absorption requires bile (from the liver). Therefore, if the bile ducts become obstructed and bile cannot enter the intestine, a vitamin K deficiency develops. As a result, the liver cannot produce prothrombin at its normal rate, and the blood's prothrombin concentration soon falls below its normal level. A prothrombin deficiency gives rise to a bleeding tendency. As a preoperative safeguard, therefore, surgical patients with jaundice caused by

▼ **FIGURE 16-15** **Blood-clotting mechanism. A,** The complex clotting mechanism can be summarized into three basic steps: *(1)* release of clotting factors from both injured tissue cells and sticky platelets at the injury site (which forms a temporary platelet plug); *(2)* a series of chemical reactions that eventually result in the formation of thrombin; and *(3)* formation of fibrin and trapping of blood cells to form a clot. **B,** Photo inset is a colorized electron micrograph showing RBCs and platelets *(blue)* entrapped in a fibrin *(yellow)* mesh during clot formation.

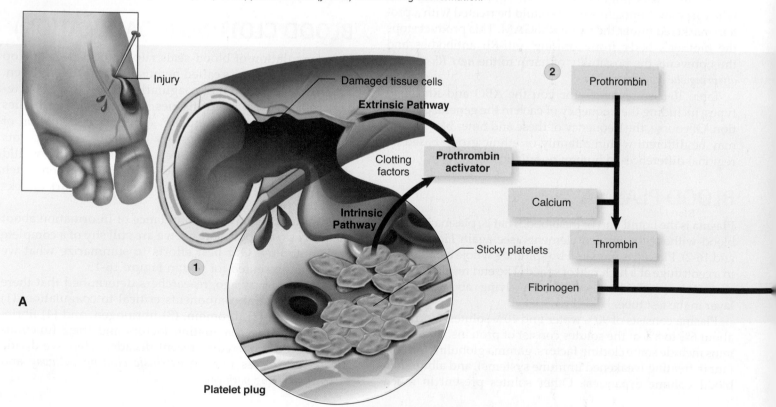

obstruction of the bile ducts are generally given some kind of vitamin K preparation.

Conditions that Oppose Clotting

There are several conditions that oppose clot formation in intact vessels to prevent abnormal, unnecessary clots from forming. The most important of these anti-clotting mechanisms is the perfectly smooth surface of the interior of the vessel created by the endothelial lining. Platelets cannot adhere to this lining and therefore do not activate and release platelet factors into the blood.

Additional deterrents to clotting are *antithrombins*, which inactivate thrombin. In this manner, antithrombins prevent thrombin from converting fibrinogen to fibrin. **Heparin,** a natural constituent of blood, acts as an antithrombin. It was first discovered in the liver (*heparin* means "liver substance"), but other organs also contain heparin. Injections of heparin are used to prevent abnormal clots from forming in vessels.

Coumarin compounds impair the liver's use of vitamin K and slow its synthesis of prothrombin and clotting factors. Indirectly, therefore, coumarin compounds retard coagulation. Citrates keep donor blood from clotting before transfusion. Aspirin and other drugs, such as clopidogrel (Plavix) or cilostazol (Pletal), that inhibit platelet aggregation, also inhibit coagulation (Box 16-4).

BOX 16-4 Health Matters

Anticoagulant and Antiplatelet Drug Treatments

If an individual is at risk for abnormal clot formation, selecting a so-called targeted or rational drug treatment plan may depend on the location in the vascular system in which the clots may form. Research has shown that abnormal clots in veins consist mainly of fibrin and red blood cells, whereas in arteries they consist mainly of platelet aggregates. This information provides a theoretical basis for selecting different types of drug treatment for conditions caused by either venous or arterial clots.

For example, **anticoagulant drugs** such as heparin and warfarin (Coumadin) should be more effective in prevention of venous thrombi. However, drugs that decrease the tendency for platelets to become sticky and form aggregates **(antiplatelet drugs)** should be more effective in preventing arterial thrombi.

A number of antiplatelet drugs, such as cilostazol (Pletal) and ticlopidine (Ticlid), exert effects by inhibiting an enzyme called *phosphodiesterase*, which is involved in platelet aggregation activity. Another popular antiplatelet drug, clopidogrel (Plavix), is used to prevent arterial clots that may cause a heart attack or stroke.

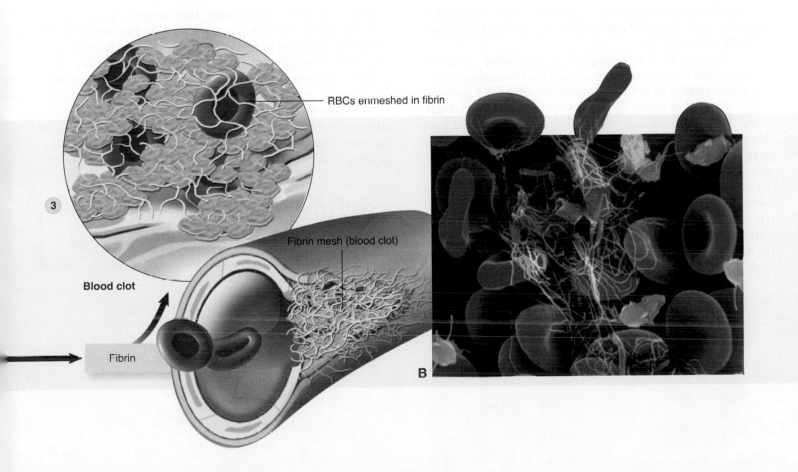

RBCs enmeshed in fibrin

Fibrin mesh (blood clot)

Blood clot

Fibrin

B

BOX 16-5 | Health Matters

Clinical Methods of Hastening Clotting

One way of treating excessive bleeding is to speed up the blood-clotting mechanism. This can be accomplished by increasing any of the substances essential for clotting, for example:

- By applying a rough surface such as gauze or by gently squeezing the tissues around a cut vessel. Each procedure causes more platelets to activate and release more platelet factors. This in turn accelerates the first of the clotting reactions.

- By applying purified thrombin (in the form of sprays or impregnated gelatin sponges that can be left in a wound). Which stage of the clotting mechanism do you think this treatment should accelerate?

- By applying fibrin foam, films, and similar applications.

Another useful strategy for speeding up the blood-clotting mechanism is the application of cold, which causes vasoconstriction and slows blood flow.

Conditions that Hasten Clotting

Two conditions especially favor clot formation: (1) a rough spot in the endothelial lining of a blood vessel and (2) abnormally slow blood flow.

Atherosclerosis, for example, is associated with an increase tendency toward **thrombosis** (clot formation). This is because *plaques* of accumulated cholesterol lipid material in the endothelial lining of arteries form rough spots. Body immobility may also lead to thrombosis because blood flow slows down as movements decrease. (This explains why physicians insist that bed patients must either move or be moved frequently, cruel as this may sound.)

Once started, a clot tends to grow. Platelets enmeshed in the fibrin threads activate, releasing more clotting factors, which in turns causes more clotting. Clot-hastening substances have proved valuable for speeding up this process. We discuss these methods more fully for you in Box 16-5.

Clot Dissolution

Blood clots are dissolved by the physiological process of **fibrinolysis.** Fibrinolysis occurs slowly, eventually dissolving the clot as the underlying vessel wall repairs itself. Blood clotting occurs continuously and simultaneously with clot dissolution (fibrinolysis). Normal blood contains an *inactive* plasma protein called **plasminogen** that can be activated by several substances released from damaged cells. These converting substances include thrombin, clotting factors, tissue plasminogen activator (t-PA), and lysosomal enzymes. Plasminogen *hydrolyzes* (breaks down) fibrin strands and dissolves the clot.

In today's clinical practice, several different kinds of proteins are used to dissolve blood clots that can cause an acute medical crisis. These are enzymes that generate plasmin when injected into patients. **Streptokinase** (SK) is a plasminogen-activating factor made by certain bacteria. This factor and recombinant t-PA can be used to dissolve clots in the large arteries of the heart. As you may have heard, blockage of these vital arteries can produce a myocardial infarction or heart attack (for further information, see Chapter 18). In addition, t-PA is a promising drug for the early treatment of strokes caused by a blood clot in a cerebrovascular accident (CVA). If given within the first 6 hours after a clot forms in a cerebral vessel, it can often improve blood flow and greatly reduce the serious aftereffects of this type of stroke.

You may find it interesting that streptococcal bacteria produce blood-dissolving factors such as SK. Recall that a secondary function of blood clotting is to trap bacteria that attempt to enter our tissues. To make their attack more effective, many bacteria such as *Streptococcus* (strep), *Staphylococcus* (staph), *Escherichia coli* (E. coli), and others, release anti-clotting agents to overcome our defenses. Such agents often activate plasminogen and thus disrupt formation of the initial blood clot. In contrast, other bacterial agents bind fibrinogen instead, which disrupts normal clotting as well.

> **QUICK CHECK**
>
> 13. What is the significance of the Rh system in pregnancy?
> 14. Briefly outline the process of clotting.

The BIG Picture

Throughout this book we've stressed the importance of *homeostasis* and the *homeostatic processes* that maintain the stability of the internal environment of our bodies. It makes sense that the fluids feeding and bathing our cells must also be kept stable. This is especially true of the blood plasma fluid. It is the blood plasma that transports substances, and even heat energy, around the entire body so that all our tissues are intimately linked together. This, of course, means that substances such as nutrients, wastes, dissolved gases, water, antibodies, and hormones can be transported between almost any two points in the body.

As we've seen, however, blood tissue is not just plasma. It contains the formed elements—blood cells and platelets. The RBCs permit the efficient transport of gaseous oxygen and carbon dioxide. WBCs are vital components of our defense mechanisms. Their presence in blood ensures that they are available to all parts of the body, all of the time, to fight cancer, resist infectious agents, and clean up injured tissues. Platelets provide mechanisms for preventing loss of the fluid that constitutes our internal environment.

All organs and organ systems of the body rely on blood to perform their functions. It's a two-way street. In fact, no organ or organ system can maintain the proper levels of nutrients, dissolved gases, or water without direct or indirect help from the blood. The list is long. For example, the respiratory system excretes carbon dioxide from the blood and picks up oxygen for the entire body via the bloodstream. Likewise, organs of the digestive system pick up nutrients, remove some toxins, and dispose of old blood cells through the flow of blood. The endocrine system regulates the production of blood cells and water content of the plasma. Besides removing toxic wastes such as urea, the urinary system has a vital role in maintaining homeostasis of plasma water concentration and pH.

Of course, blood is useless unless it continually and rapidly flows throughout the entire body. It must transport, defend, and maintain balance for us to survive. The following chapters will show you how blood circulation takes place and how the physiology of the circulatory system is maintained. We will then end our tour of transportation and defense with a thorough review of the lymphatic and immune systems.

Cycle of LIFE ○

The moment you are born, your body must quickly destroy all the erythrocytes containing fetal hemoglobin and replace them with erythrocytes containing adult hemoglobin. Fortunately, your body is capable of this tremendous feat! Your body destroys over 2.5 million erythrocytes every second. Once you reach adulthood, this is just a tiny fraction of your body's estimated 25 trillion cells. Incredibly, new erythrocytes take only 4 days to develop from stem cells in the bone marrow to mature erythrocytes.

However, because they lack a nucleus and organelles for cellular repair, erythrocytes live for only about 120 days. Damaged and aged blood cells are continually removed from the circulatory system by macrophages in our lymph nodes, spleen, and liver and replaced with newer cells from red bone marrow This process continues throughout our life span.

As we age, the amount of fat in the bone marrow increases, reducing the volume of blood-forming cells. Ordinarily this is not a problem, but under stress caused by disease, the body may require additional erythrocytes, leukocytes, and thrombocytes. When the body cannot meet the additional demand for higher production of erythrocytes, for example, anemia often results. ●

MECHANISMS OF DISEASE

There are numerous disorders of the formed elements of the blood. These include red blood cell disorders such as anemias, and white blood cell disorders that range from abnormally low or high white blood cell counts to a variety of cancers called leukemias and myelomas. There are also clotting disorders and serious inherited disorders such as sickle cell anemia. Many of these disorders are very difficult to treat.

⊖volve   Find out more about these blood tissue diseases online at *Mechanisms of Disease: Blood.*

LANGUAGE OF SCIENCE AND MEDICINE *continued from page 353*

extrinsic pathway (eks-TRIN-sik PATH-way)
  [*extr-* outside, *-sic* beside]

fibrinolysis (fye-brin-OL-ih-sis)
  [*fibr-* fiber, *-lysis* loosening]

formed element (EL-em-ent)

globin (GLOH-bin)
  [*glob-* ball, *-in* substance]

granulocyte (GRAN-yoo-loh-syte)
  [*gran-* grain, *-ul-* little, *-cyte* cell]

hematocrit (hee-MAT-oh-krit)
  [*hemato-* blood, *-crit* separate]

hemocytoblast (hee-moh-SYE-toh-blast)
  [*hemo-* blood, *-cyto-* cell, *-blast* embryonic state of development]

hemoglobin (hee-moh-GLOH-bin)
  [*hem-* blood, *-globus* ball]

hemolysis (hee-MAHL-ih-sis)
  [*hemo-* blood, *-lysis* loosening]

hemostasis (hee-moh-STAY-sis)
  [*hemo-* blood, *-stasis* standing]

heparin (HEP-ah-rin)
  [*hepar-* liver, *-in* substance]

intrinsic pathway (in-TRIN-sik)
  [*intr-* within, *-sic* beside]

leukocyte (LOO-koh-syte)
  [*leuko-* white, *-cyte* cell]

leukocytosis (loo-koh-sye-TOH-sis)
  [*leuko-* white, *-cyt-* cell, *-osis* condition]

leukopenia (loo-koh-PEE-nee-ah)
  [*leuko-* white, *-penia* lack]

lymphocyte (LIM-foh-syte)
  [*lymph-* water (lymphatic system), *-cyte* cell]

monocyte (MON-oh-syte)
  [*mono-* single, *-cyte* cell]

myeloid tissue (MY-eh-loyd TISH-yoo)
  [*myel-* marrow, *-oid* like, *tissue-* fabric]

neutrophil (NOO-troh-fil)
  [*neuter-* neither, *-phil* love]

nonelectrolyte (non-ee-LEK-troh-lyte)
  [*non-* not, *-electro-* electricity, *-lyt-* loosening]

physiological polycythemia (fiz-ee-oh-LOJ-ih-kal pol-ee-sye-THEE-mee-ah)
  [*physi-* nature, *-o-* combining form, *-log-* words (study of), *-y* activity, *poly-* many, *-cyt-* cell, *-emia* blood condition]

plasma (PLAZ-mah)
  [*plasma* substance]

plasminogen (plaz-MIN-oh-jen)
  [*plasm-* substance (plasma), *-in-* substance, *-gen* produce]

platelet (PLAYT-let)
  [*plate-* flat, *-let* small]

platelet plug (PLAYT-let)
  [*plate-* flat, *-let* small]

prothrombin (pro-THROM-bin)
  [*pro-* first, *-thromb-* clot, *-in* substance]

Rh antigen (R-H AN-tih-jen)
  [*Rh* Rhesus (monkey), *anti-* against, *-gen* produce]

streptokinase (strep-toh-KIN-ayz)
  [*strepto-* twisted, *-kin-* motion, *-ase* enzyme]

thrombopoiesis (throm-boh-poy-EE-sis)
  [*thromb-* clot, *-poiesis* making]

thrombosis (throm-BOH-sis)
  [*thromb-* clot, *-osis* condition]

T lymphocyte (LIM-foh-syte)
  [*T* thymus gland, *lymph-* water (lymphatic system), *-cyte* cell]

transfusion reaction (tranz-FYOO-zhun ree-AK-shun)
  [*trans-* across, *-fus-* pour, *-sion* process, *re-* again, *-action* action]

whole blood volume

CHAPTER SUMMARY

To download an MP3 version of the chapter summary for use with your iPod or other portable media player, access the Audio Chapter Summaries *online at http://evolve.elsevier.com.*

HINT *Scan this summary after reading the chapter to help you reinforce the key concepts. Later, use the summary as a quick review before your class or before a test.*

INTRODUCTION

A. Blood is a fluid transport medium that serves as a pickup and delivery system throughout the body
B. Blood also transports hormones, enzymes, buffers, and other important biochemicals
C. Flow of blood is vital to temperature regulation in our bodies

BLOOD COMPOSITION

A. Blood is a liquid connective tissue consisting not only of fluid plasma, but also of cells (formed elements)
  1. Plasma—third major fluid in our bodies; accounts for 55% of total volume (Figure 16-1)
  2. Formed elements—blood cells; accounts for 45% of total volume
B. Blood volume—males have about 5 to 6 liters of blood circulating in their bodies; females have about 4 to 5 liters
  1. Blood volume varies with age and body composition

FORMED ELEMENTS OF BLOOD

A. Formed elements include:
  1. Erythrocytes—red blood cells (RBCs)
  2. Thrombocytes—platelets
  3. Leukocytes—white blood cells (WBCs)
B. Red blood cells (erythrocytes)
  1. Erythrocytes—shaped like tiny biconcave disks; before the cell reaches maturity in the bone marrow, it extrudes its nucleus (Figure 16-3)
    a. Loses its ribosomes, mitochondria, and other organelles
    b. Primary component is hemoglobin
    c. RBCs are the most numerous of all the formed elements of blood
  2. Function of red blood cells—play a critical role in the transport of oxygen and carbon dioxide in the body
  3. Hemoglobin—within each RBC are an estimated 200 to 300 million molecules of hemoglobin

a. Composed of four protein (globin) chains with each attached to a heme group

b. One hemoglobin molecule chemically bonds with four oxygen molecules to form oxyhemoglobin

c. Hemoglobin can also combine with carbon dioxide to form carbaminohemoglobin

d. Males' blood usually contains more hemoglobin than that of females

e. Anemia—low RBC count (Box 16-1)

4. Formation of red blood cells (erythropoiesis)

a. Erythrocytes begin their maturation process in the red bone marrow from hematopoietic stem cells called hemocytoblasts

b. Adult blood-forming stem cells divide by mitosis

c. Some of the daughter cells remain as undifferentiated adult stem cells; others continue to develop into erythrocytes (Figure 16-4)

d. Every day of our adult lives, more than 200 billion RBCs are formed to replace an equal number destroyed

e. Homeostatic mechanisms operate to balance the number of cells formed against the number destroyed (Figure 16-5)

5. Destruction of red blood cells—life span of RBCs circulating in the bloodstream averages between 105 and 120 days

a. Macrophage cells phagocytose the aged, abnormal, or fragmented RBCs

(1) This process results in the breakdown of hemoglobin; iron, bilirubin, and amino acids are released

C. White blood cells (leukocytes)

1. Five basic types of white blood cells—classified according to the presence or absence of granules as well as the staining characteristics of their cytoplasm (Table 16-1)

2. Granulocytes—include the three WBCs that have large granules in their cytoplasm

a. Neutrophils—make up about 65% of the WBC count in a normal blood sample (Figure 16-6)

(1) Highly mobile and active phagocytic cells; can migrate out of blood vessels and enter into the tissue spaces (diapedesis)

b. Eosinophils—account for about 2% to 5% of circulating WBCs (Figure 16-7)

(1) Abundant in the linings of the respiratory and digestive tracts

(2) Can ingest inflammatory chemicals and proteins associated with antigen-antibody reaction complexes

(3) Their most important functions involve protection against infections caused by parasitic worms; also involved in allergic reactions

c. Basophils—least numerous of the WBCs, numbering only 0.5% to 1% of the total leukocyte count (Figure 16-8)

(1) Contain histamine and heparin

(2) Mobile and capable of diapedesis

3. Agranulocytes

a. Lymphocytes—account for about 25% of all the leukocytes in our bodies (Figure 16-9)

(1) Two general types of lymphocytes: T lymphocytes and B lymphocytes

b. Monocytes—largest of the leukocytes (Figure 16-10)

(1) Mobile and highly phagocytic

4. White blood cell numbers—one cubic millimeter of normal blood usually contains only about 5,000 to 9,000 leukocytes

a. These numbers have clinical significance because they may change drastically under abnormal conditions (Box 16-3)

5. Formation of white blood cells—granulocytes and agranulocytes mature from the hematopoietic stem cells (Figure 16-4)

a. Neutrophils, eosinophils, basophils, and a few lymphocytes and monocytes originate in red bone marrow

b. Most lymphocytes and monocytes are derived from hematopoietic adult stem cells in lymphatic tissue

c. Myeloid tissue and lymphatic tissue together constitute the hematopoietic tissues of the body

D. Platelets (thrombocytes)—really tiny shards of cells; nearly colorless bodies that appear as irregular spindles or oval disks about 2 to 4 μm in diameter (Table 16-1)

1. A range of 150,000 to 400,000 platelets/mm^3 is considered normal for adults; newborns often show reduced numbers

2. Functions are varied and have to do with clotting: cell aggregation, adhesiveness, and agglutination

3. Function of platelets—play vital roles in hemostasis and coagulation

a. Hemostasis—stoppage of blood flow from an injured vessel

b. Platelet plug—helps stop the flow of blood into the tissues

(1) Formed when platelets adhere to the damaged wall of the vessel

(2) Formation of the plug generally follows vascular spasms of smooth muscle fibers in the wall of the damaged blood vessel; helps reduce blood flow

c. When platelets encounter collagen in damaged vessel walls and surrounding tissue, they become sticky platelets

(1) Sticky platelets then form the plug

(2) Sticky platelets secrete several biochemicals that affect local blood flow

d. If the injury is extensive, the coagulation mechanism is also activated

4. Formation and life span of platelets—thrombopoiesis is the formation of platelets

a. Between 2,000 and 3,000 platelets are created when the cytoplasmic membrane of huge mature megakaryocytes rupture

b. Platelets have a short life span of about 7 days

BLOOD TYPES (BLOOD GROUPS)

A. Blood type—refers to the types of biochemical markers or antigens present on the plasma membranes of erythrocytes

1. A, B, and Rh are the most important blood antigens as far as transfusions and newborn survival are concerned

2. Agglutinin—the antibodies dissolved in plasma that react with specific blood group antigens or agglutinogens

a. When they combine and react, they cause RBCs to clump together or agglutinate

B. The ABO system—every person's blood belongs to one of the four ABO blood types (Table 16-3)

1. Blood types are named according to the antigen present on the membranes of the RBCs

a. Type A—antigen A on RBCs

b. Type B—antigen B on RBCs

c. Type AB—both antigen A and B on RBCs

d. Type O—neither antigen A nor B on RBCs

2. Universal donor—type O; does not contain either antigen A or B
3. Universal recipient—type AB; contains neither anti-A nor anti-B antibodies

C. The Rh system (Figure 16-14)
1. Rh positive—Rh antigen is present on its RBCs
2. Rh negative—RBCs have no Rh antigen present
3. Blood does not normally contain anti-Rh antibodies; anti-Rh antibodies can appear in the blood of an Rh-negative person if it has come in contact with Rh-positive RBCs

BLOOD PLASMA

A. Plasma—liquid part of the blood; consists of 90% water and 10% solutes (Figure 16-2)
B. Solutes—6% to 8% of the solutes consist of proteins; three main compounds:
1. Albumins
2. Globulins
3. Clotting proteins (fibrinogen)
C. Plasma proteins contribute to the maintenance of normal blood viscosity, blood osmotic pressure, and blood volume

BLOOD CLOTTING (COAGULATION)

A. Coagulation of blood plugs ruptured vessels to stop bleeding and prevents bacteria from invading our tissues
B. Four essential components critical to coagulation (Figure 16-15):
1. Prothrombin
2. Thrombin
3. Fibrinogen
4. Fibrin
C. Two basic processes of coagulation
1. Extrinsic clotting pathway—chemicals are released from damaged tissue outside the blood that ultimately results in the formation of prothrombin activator
2. Intrinsic clotting pathway—involves a series of reactions that begin with factors normally present in, or intrinsic to, the blood
   a. After prothrombin activator is produced, a clot will form
   b. Thrombin accelerates conversion of the soluble plasma protein fibrinogen to insoluble fibrin
   c. Polymerization of fibrin strands forms a fibrin clot
D. Conditions that oppose clotting
1. Perfectly smooth surface of the interior of a blood vessel
2. Antithrombins—prevent thrombin from converting fibrinogen to fibrin; example: heparin
E. Conditions that hasten clotting
1. Abnormally slow blood flow
F. Clot dissolution—clots are dissolved by the physiological process of fibrinolysis
1. Plasminogen—hydrolyzes fibrin strands and dissolves the clot
2. Streptokinase (SK)—plasminogen-activating factor made by certain *streptococci* bacteria; used to dissolve clots in the large arteries of the heart

REVIEW QUESTIONS

HINT *Write out the answers to these questions after reading the chapter and reviewing the Chapter Summary. If you simply think through the answer without writing it down, you won't retain much of your new learning.*

1. What are the formed elements of blood?
2. Describe the structure of hemoglobin.
3. How does the structure of hemoglobin allow it to combine with oxygen?
4. Discuss the steps involved in erythropoiesis.
5. What is the average life span of a circulating red blood cell?
6. How are granulocytes similar to agranulocytes? How do they differ?
7. List the important physical properties of platelets.
8. Explain what is meant by *type AB blood*. Explain *Rh-negative blood*.
9. Which organ is responsible for the synthesis of most plasma proteins?
10. What is the normal plasma protein concentration?
11. What are some functions served by plasma proteins?
12. What triggers blood clotting?
13. Identify factors that oppose blood clotting. Do the same for factors that hasten blood clotting.
14. Describe the physiological mechanism that dissolves clots.

CRITICAL THINKING QUESTIONS

HINT *After finishing the Review Questions, write out the answers to these items to help you apply your new knowledge. Go back to sections of the chapter that relate to items that you find difficult.*

1. A friend received a report on his physical examination that his hematocrit was below normal. Because he knows you are taking anatomy and physiology, he has come to you for an explanation. Based on what you know, explain to him what a hematocrit value is, how it is determined, and what value would put him below normal.
2. You are a medical examiner, and a body is brought to you to determine the cause of death. You find a very large agglutination (not a clot) in a major vein. What judgment would you make regarding the cause of death?
3. Suppose a person had a thyroid disorder that caused the production of calcitonin to be many times higher than it should be. Can you elaborate on why a possible side effect of this condition might be a very slow blood-clotting time?
4. A patient comes into the emergency department with severe bleeding. What can be done to speed the formation of a clot?
5. Some athletes, seeking a competitive edge, may resort to blood doping. How would you explain blood doping? What information would you use to discourage athletes from participating in the practice of blood doping?
6. How would you summarize the condition *erythroblastosis fetalis*?

continued from page 353

Now that you have read this chapter, try to answer these questions about Duncan's cut from the Introductory Story.

1. What is the main component of the blood coming out of Duncan's finger?
   a. Erythrocytes
   b. Leukocytes
   c. Plasma
   d. Thrombocytes

Because of the damage to his blood vessels, Duncan's body will immediately start the blood clotting process.

2. What's the first step in hemostasis (stopping bleeding)?
   a. Vascular spasm
   b. Platelet plug
   c. Coagulation
   d. Leukocytic plug

3. What is the last step in clot formation?
   a. Fibrinogen converted to fibrin
   b. Prothrombin converted to thrombin
   c. Profibrin converted to fibrin
   d. Collagen fibers trap RBCs

4. If Duncan were missing factor VIII, what condition would he have?
   a. Thrombocytopenia
   b. Pernicious anemia
   c. Polycythemia
   d. Hemophilia

HINT *To solve a case study, you may have to refer to the glossary or index, other chapters in this textbook, A&P Connect, Mechanisms of Disease, and other resources.*

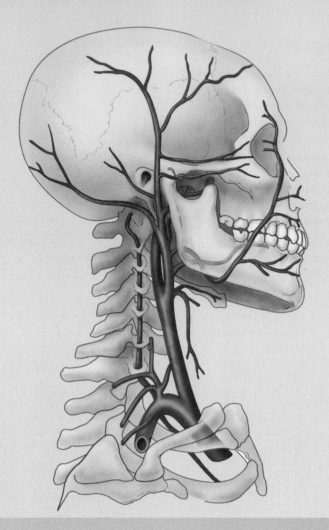

Anatomy of the Cardiovascular System

STUDENT LEARNING OBJECTIVES

At the completion of this chapter, you should be able to do the following:

1. Describe the position of the heart and its coverings.
2. Outline the major chambers and valves of the heart and give their functions.
3. Trace a drop of blood as it travels through the heart.
4. Discuss the role and operation of the coronary arteries and veins.
5. Compare the physical properties of arteries, veins, and capillaries.
6. Compare systemic circulation with pulmonary circulation.
7. List the major arteries and veins servicing the thoracic and abdominal regions.
8. List the major arteries and veins servicing the head and neck regions.
9. List the major arteries and veins servicing the arms and legs.
10. Discuss the significance of the hepatic portal system.
11. Outline the basic plan of fetal circulation.
12. Describe the changes in fetal circulation that take place at birth.

CHAPTER OUTLINE

 Scan this outline before you begin to read the chapter, as a preview of how the concepts are organized.

KYLE (45 years old) finally gave in to his wife's insistence and stopped by his local health clinic. After all, it was in the same building where he was working on a construction job. He had been having some minor chest pain for a couple of days. But, he'd been telling himself the pain was just sore muscles caused by his recent weight lifting. Kyle was expecting the receptionist to make an appointment for him. However, as he described his symptoms (chest pain, some sweating, slight nausea), she interrupted him and called over her shoulder to the nurse. As soon as the nurse was made aware of his symptoms, Kyle was rushed into an exam room, where his heart rate and blood pressure were checked and the electrical activity of his heart was measured (by performing an ECG).

"What's going on?" Kyle asked the doctor a few minutes later. The physician replied, "Based on your symptoms, we think you may have some blockage in your coronary arteries. We'd like to do an angiogram to see what's going on."

Perhaps you already have an idea what may be taking place in Kyle's body, but certainly you'll know after reading this chapter exactly why the nurse and physician acted immediately.

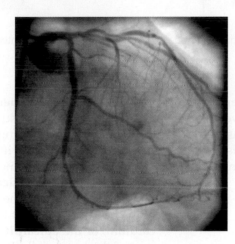

The **cardiovascular system,** or circulatory system, consists of a muscular heart and a closed system of vessels (*arteries, veins,* and *capillaries*). As the name suggests, blood within the circulatory system is pumped by the heart through a *closed* circuit of vessels.

As in the adult, survival of the developing embryo also depends on the circulation of blood to maintain homeostasis. In response to this need, the cardiovascular system develops early and reaches a functional state long before any other major organ system. Incredible as it seems, the heart begins to beat regularly early in the fourth week after fertilization.

HEART

Location, Shape, and Size of the Heart

The human **heart** is a four-chambered muscular organ, shaped and sized roughly like a person's closed fist. It lies in the *mediastinum,* or middle region of the thorax, just behind the body of the sternum.

You can see the anatomical position of the heart in the thoracic cavity in Figure 17-1, *A.* The lower border of the heart, forming the *apex,* lies on the diaphragm. The apex points to the left. To count the apical beat, a physician places a stethoscope directly over the apex, in the space between the fifth and sixth ribs.

At birth, the heart is wide and appears large in proportion to the diameter of the chest cavity. In infants, the heart is 1/130 of the total body weight

LANGUAGE OF SCIENCE AND MEDICINE

HINT *Before reading the chapter, say each of these terms out loud. This will help you avoid stumbling over them as you read.*

abdominal aorta (ab-DOM-ih-nal ay-OR-tah)
[*abdomin-* belly, *-al* relating to, *aort-* lifted, *-a* thing] *pl.,* aortae or aortas (ay-OR-tee, ay-OR-tahz)

angiography (an-jee-AH-graf-ee)
[*angi-* vessel, *-graph-* draw, *-y* process]

aorta (ay-OR-tah)
[*aort-* lifted, *-a* thing] *pl.,* aortae or aortas (ay-OR-tee, ay-OR-tahz)

aortic aneurysm (ay-OR-tik AN-yoo-riz-em)
[*aort-* lifted, *-ic* relating to, *aneurysm* widening]

aortic arch (ay-OR-tik)
[*aort-* lifted, *-ic* relating to]

arterial anastomosis (ar-TEER-ee-al ah-nas-toh-MOH-sis)
[*arteria-* vessel, *-al* relating to, *ana-* anew, *-stomo-* mouth, *-osis* condition] *pl.,* anastomoses (ah-nas-toh-MOH-seez)

arteriole (ar-TEER-ee-ohl)
[*arteri-* vessel, *-ole* little]

artery (AR-ter-ee)
[*arteri-* vessel]

ascending aorta (ah-SEND-ing ay-OR-tah)
[*ascend-* climb, *aort-* lifted, *-a* thing] *pl.,* aortae or aortas (ay-OR-tee, ay-OR-tahz)

ascites (a-SYT-eez)
[*acites* baglike]

atherosclerosis (ath-er-oh-skleh-ROH-sis)
[*athero-* gruel, *-scler-* hardening, *-osis* condition]

atrioventricular (AV) valve (ay-tree-oh-ven-TRIK-yoo-lar)
[*atrio-* entrance courtyard, *-ventr-* belly, *-icul-* little, *-ar* relating to]

atrium (AY-tree-um)
[*atrium* entrance courtyard] *pl.,* atria (AY-tree-ah)

autorhythmic (aw-toh-RITH-mic)
[*auto-* self, *-rhythm-* rhythm, *-ic* relating to]

avascular (ah-VAS-kyoo-lar)
[*a-* without, *-vas-* vessel, *-ula-* little, *-ar* relating to]

axillary vein (AK-sih-lair-ee)
[*axilla* wing, *-ary* relating to]

bicuspid valve (bye-KUS-pid)
[*bi-* double, *-cusp-* point, *-id* characterized by]

brachial vein (BRAY-kee-al)
[*brachi-* arm, *-al* relating to]

brachiocephalic artery (brayk-ee-oh-seh-FAL-ik AR-ter-ee)
[*brachi-* arm, *-cephal-* head, *-ic* relating to, *arteri-* vessel]

brachiocephalic vein (brayk-ee-oh-seh-FAL-ik)
[*brachi-* arm, *-cephal-* head, *-ic* relating to]

capacitance (kah-PASS-ih-tens)
[*capacit-* space or volume, *ance* state]

capillary (KAP-ih-lair-ee)
[*capill-* hair, *-ary* relating to]
continued on page 399

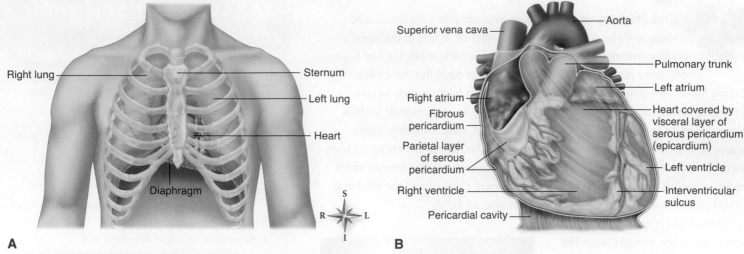

▲ **FIGURE 17-1** **Location of the heart. A,** Heart in mediastinum showing relationship to lungs and other anterior thoracic structures. **B,** Detail of heart with pericardial sac opened.

compared with about ⅓₀₀ in the adult. Between puberty and 25 years of age, the heart attains its adult shape and weight—about 310 grams in the average male and 225 grams in the average female. We've illustrated the external details of the heart and great vessels for you in Figures 17-1, *B* and 17-2. Take a moment to review those before continuing.

Coverings of the Heart

An outer sac, the **pericardium,** encloses your heart, as you can see in Figure 17-1, *B.* The loosely fitting outer layer of

this sac is the **fibrous pericardium.** This layer is made of tough, white fibrous tissue and protects the heart and also anchors it to surrounding structures. It also prevents the heart from overfilling with blood. The fibrous pericardium is lined with a smooth, moist *serous* membrane—the *parietal layer* of the **serous pericardium** (Figure 17-3). The same kind of serous membrane directly covers the entire surface of the heart, so we call it the *visceral layer* of the serous pericardium, or the **epicardium.** The epicardium is an integral part of the heart wall. (It is important to note that the two layers of the serous pericardium are *continuous:* At the

▶ **FIGURE 17-2**
The heart and great vessels. A, Anterior view. **B,** Posterior view.

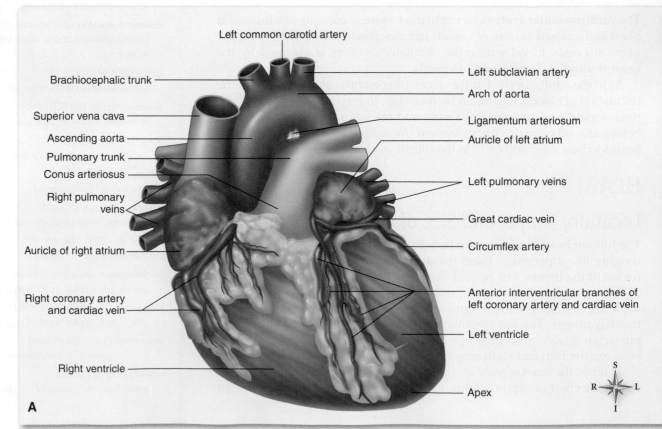

superior margin of the heart, the parietal layer attaches to the large arteries leaving the heart, and then turns inferiorly and continues over the external heart surfaces as the visceral layer.)

Look again at Figure 17-1, *B*. Note that the fibrous part of the pericardial sac attaches to the large blood vessels emerging from the top of the heart, but does *not* attach to the heart itself. Thus, the sac fits *loosely* around the heart, with a slight space between the visceral layer that adheres to the heart wall and the parietal layer that adheres to the inside of the fibrous sac. The space in between these two layers is called the **pericardial space** and contains 10 to 15 ml of **pericardial fluid** (see Figure 17-3). This fluid lubricates the space between the parietal layer of the pericardium and the visceral layer forming the (serous) epicardium.

The fibrous pericardial sac with its smooth, well-lubricated lining provides protection against friction as the heart beats.

Structures of the Heart
Wall of the Heart

Epicardium

The outer layer of the heart wall is called the epicardium, as we've just seen. The epicardium is actually the visceral layer of the serous pericardium already described. In other words, the same structure has two different names: epicardium and serous pericardium.

Myocardium

A thick, contractile, middle layer comprises the bulk of the heart wall. This **myocardium** is composed largely of cardiac muscle (take a moment to review the structure of cardiac muscle in Chapter 6, page 106). Because

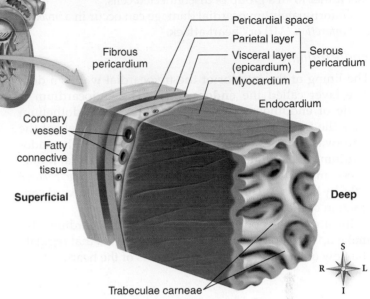

▲ **FIGURE 17-3**   **Wall of the heart.** The cutout section of the heart wall shows the outer fibrous pericardium and the parietal and visceral layers of the serous pericardium (with the pericardial space between them). A layer of fatty connective tissue is located between the visceral layer of the serous pericardium (epicardium) and the myocardium. Note that the endocardium covers beamlike projections of myocardial muscle tissue, called *trabeculae carneae.*

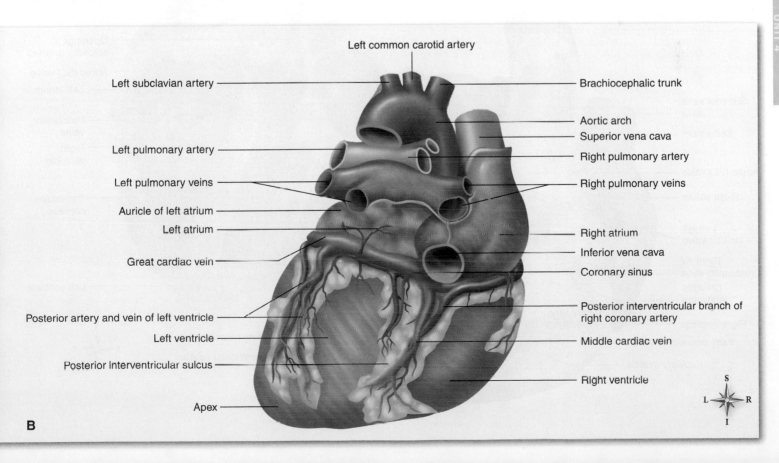

intercalated disks join adjacent cells of the heart (Table 6-5, page 106), large areas of cardiac muscle are electrically coupled into a single functioning unit. This allows your heart to conduct action potentials quickly, thereby ensuring that the chambers contract rhythmically, with great force, rather than as a flutter from a group of disconnected cells.

Unfortunately, myocardial damage can occur in a *myocardial infarction (MI)* or "heart attack."

Endocardium

The lining of the interior of the myocardial wall is a delicate layer called the **endocardium.** The endocardium is made of endothelial tissue, or *endothelium.* Endothelium lines the heart and continues to line all the vessels of the cardiovascular system. Note in Figure 17-3 that the endocardium covers branched projections of myocardial tissue. These muscular projections are called *trabeculae carneae* ("fleshy beams"). They help to add force to the inward contraction of the heart wall.

Inward folds or pockets formed by the endocardium also make up the flaps or cusps of the major valves that regulate the flow of blood through the chambers of the heart.

A&P CONNECT

How does a heart attack develop? Take a tour through an illustrated description of the process of an MI in **Heart Attack!** online at **A&P Connect.**

evolve

QUICK CHECK

1. Describe the position of the heart in anatomical terms.
2. Describe the shape of the heart.
3. What are the major coverings of the heart?
4. What is the primary function of each heart covering?

Chambers of the Heart

The interior of the heart is divided into four cavities, or heart chambers (Figure 17-4). The two upper chambers are called **atria** (singular, *atrium*). The two lower chambers are called **ventricles.** An extension of the heart wall, the **septum,** separates the left chambers from the right chambers.

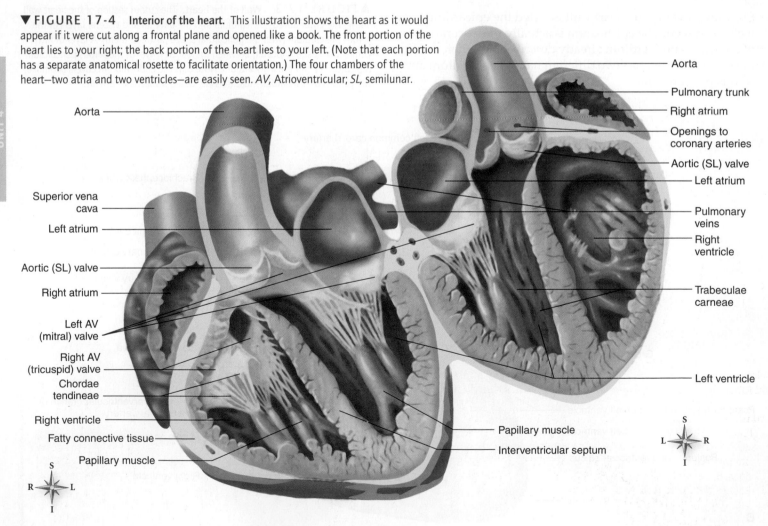

▼ **FIGURE 17-4** **Interior of the heart.** This illustration shows the heart as it would appear if it were cut along a frontal plane and opened like a book. The front portion of the heart lies to your right; the back portion of the heart lies to your left. (Note that each portion has a separate anatomical rosette to facilitate orientation.) The four chambers of the heart—two atria and two ventricles—are easily seen. *AV,* Atrioventricular; *SL,* semilunar.

The two atria are separated into left and right chambers by the *interatrial septum.* These chambers receive blood from veins—large blood vessels that return blood to the heart from the entire body. Figure 17-5 shows you how the atria alternately relax to receive blood and then contract to push the blood into the ventricles below. The atria are not very muscular because not much force is needed to deliver the blood to the chambers below. Thus, the muscular walls of

the atria are not very thick. Return to Figure 17-2 for a moment. Notice that the *auricle* (meaning "little ear") is *only* the visible earlike flap protruding from each atrium. For this reason, you should not use *auricle* and *atria* as synonyms.

Like the atria above them, the two ventricles are also separated into left and right chambers. This very muscular separation is called the *interventricular septum.* Because the ventricles receive blood from the atria and pump blood out

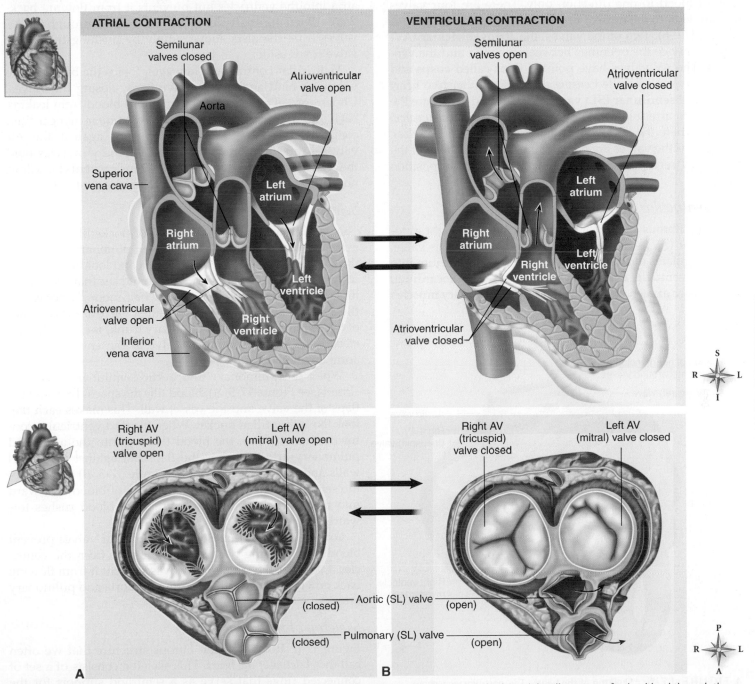

▲ FIGURE 17-5   **Chambers and valves of the heart. A,** During atrial contraction, cardiac muscle in the atrial wall contracts, forcing blood through the atrioventricular *(AV)* valves and into the ventricles. Bottom illustration shows superior view of all four valves, with semilunar *(SL)* valves closed and AV valves open. **B,** During ventricular contraction that follows, the AV valves close and the blood is forced out of the ventricles through the SL valves and into the arteries. Bottom illustration shows superior view of SL valves open and AV valves closed.

of the heart into the arteries, the ventricles are really the primary "pumping chambers" of the heart. Because more force is required to pump blood from the ventricles than from the atria, the myocardium of the ventricles is quite thick.

The pumping action of the heart chambers is summarized for you in Figure 17-5. We describe it in much greater detail in Chapter 18.

Valves of the Heart

The heart valves are tough, fibrous structures that permit the flow of blood in one direction only. There are four valves that are vital to the normal functioning of the heart (Figures 17-5 and 17-6). Two of the valves, the **atrioventricular (AV) valves,** service the openings between the atria and the ventricles. The AV valves have pointed flaps called **cusps** and for this reason are called *cuspid valves.* The other two heart valves, the **semilunar (SL) valves,** are located (1) where the pulmonary artery joins the right ventricle *(pulmonary valve)* and (2) where the aorta joins the left ventricle *(aortic valve).* Notice that the valves are named simply for the areas of the heart they service, so you can memorize them by position alone.

Atrioventricular Valves

A strong, fibrous ring encircles and anchors the right atrioventricular (AV) valve within the myocardium. This valve, which regulates the flow of blood from the right atrium into the right ventricle, consists of three cusps of endocardium. The *free edge* of each flap is anchored to the **papillary muscles**

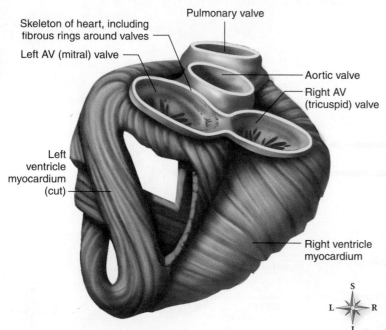

Skeleton of heart, including fibrous rings around valves

Left AV (mitral) valve

Pulmonary valve

Aortic valve

Right AV (tricuspid) valve

Left ventricle myocardium (cut)

Right ventricle myocardium

S
L — R
I

▲ **FIGURE 17-6** **Skeleton of the heart.** This posterior view shows part of the ventricular myocardium with the heart valves still attached. The rim of each heart valve is supported by a fibrous structure (the skeleton of the heart) that encircles all four valves.

of the right ventricle by several tendinous cords called **chordae tendineae.** In a way, these cords are the true "heartstrings" of our hearts.

Because the right AV valve has three cusps, it is also called the **tricuspid valve.** The valve that regulates the left AV opening is similar in structure to the right AV valve, except that it has only two flaps. For this reason we call it the **bicuspid valve.** More commonly, it is called the **mitral valve** because it resembles the hat (miter) worn by bishops. The construction of both AV valves allows blood to flow from the atria into the ventricles but prevents it from flowing backward. When the ventricles relax, blood flows through the AV valves from the atria above simply by pushing the flimsy valve cusps aside.

Ventricular contraction, however, forces the blood in the ventricles hard against the valve flaps, closing the valves. Under normal conditions, this prevents blood from leaking back into the atria. The harder the ventricular myocardium contracts, the more strongly it pushes against the AV valves—and the more strongly the papillary muscles hold the AV valves shut. This mechanism thus prevents backflow, no matter how strongly the heart ventricles contract.

Semilunar Valves

The **semilunar (SL) valves** consist of pocketlike flaps that extend inward from the lining of the pulmonary artery and the aorta. If you were facing a person and looking at a frontal section of his or her heart, you would see that each semilunar valve looks very much like a "half-moon," after which these valves are named. The semilunar valve at the entrance of the pulmonary artery (pulmonary trunk) is called the *pulmonary valve.* The semilunar valve at the entrance of the aorta is called the *aortic valve.*

When the pulmonary and aortic semilunar valves are closed (see Figure 17-5, *A*), blood fills the spaces between the flaps of the valve and the vessel wall. This makes each flap look like a tiny, filled bucket. When the next ventricular contraction takes place, the blood flowing into the aorta and pulmonary artery pushes the flaps flat against the vessel walls and opens the valves (see Figure 17-5, *B*). Closure of the semilunar valves prevents the flow of blood backward into the ventricles and ensures that the blood rushes forward.

You should note that the atrioventricular valves prevent blood from flowing back up into the atria from the ventricles. Likewise, the semilunar valves prevent it from flowing back down into the ventricles from the aorta and pulmonary arteries.

Skeleton of the Heart

Figure 17-6 shows you the fibrous structure that we often call the *skeleton of the heart.* This skeleton consists of a set of connected rings that serve as a semirigid support for the heart valves (on the inside of the rings). It also serves as sites for the attachment of cardiac muscle of the myocardium (on the outside of the rings). The skeleton of the heart also serves as an *electrical barrier* between the myocardium of the atria and the myocardium of the ventricles. This arrangement allows the ventricles to contract separately from the atria, ensuring the effective pumping of the blood.

Flow of Blood Through the Heart

Try tracing the path of blood flow with your finger, using Figure 17-5 as your guide.

Beginning with the right atrium, blood flows through the right AV (tricuspid) valve into the right ventricle. From the right ventricle, blood then flows through the pulmonary semilunar valve into the first portion of the pulmonary artery, the pulmonary trunk. The pulmonary trunk branches to form the left and right pulmonary arteries. These arteries conduct blood with carbon dioxide to the gas exchange tissues of the lungs. Here they will dispose of the carbon dioxide and pick up oxygen.

Blood flows from the lungs via the pulmonary veins back to the heart. (Note that these veins are carrying blood that is oxygenated from its journey through the lungs.) Oxygenated blood from the pulmonary veins flows into the left atrium of the heart. From the left atrium, blood flows through the left atrioventricular (mitral) valve into the left ventricle. From the left ventricle, blood then flows through the aortic semilunar valve into the aorta. Branches of the aorta then supply all the tissues of the body except the gas exchange tissues of the lungs.

Blood leaving the head and neck is deoxygenated and empties into the **superior vena cava.** Deoxygenated blood from the lower body empties into the **inferior vena cava.** Both of these large vessels then conduct blood into the right atrium, bringing us back to our beginning point.

Blood Supply of the Heart Tissue

Coronary Arteries

Myocardial cells receive blood via the right and left **coronary arteries** (Figure 17-7, *A*). The openings from the aorta into these vitally important vessels lie *behind* the flaps of the aortic semilunar valve. As a result, they are the first branches off the aorta and supply the heart muscle first. Ordinarily, arteries that branch from the aorta fill during ventricular contraction when the great force of ventricular pressure pushes blood into the arteries. However, the coronary arteries are *squeezed* during ventricular contraction and cannot fill during this time (Figure 17-7, *B*). Because the coronary artery openings are located behind the flaps of the aortic valve, blood flow is largely prevented from entering these openings during ventricular contraction. This is because, when the blood rushes out of the ventricle, the valve flaps are compressed flat against the wall of the aorta and cover the openings to the coronaries. When the ventricle relaxes, however, the coronary arteries expand somewhat, and blood flow is diverted into their openings as the aortic valve closes. This allows the coronary arteries to fill. Both right and left coronary arteries have two main branches, as shown in Figure 17-7.

More than 500,000 Americans die every year from coronary disease and another 3,500,000 suffer some degree of incapacitation. Knowledge about the distribution of coronary artery branches therefore has great practical importance. Here are some principles related to the heart's own blood supply that you should know:

1. Both ventricles receive their blood supply from branches of the right and left coronary arteries.

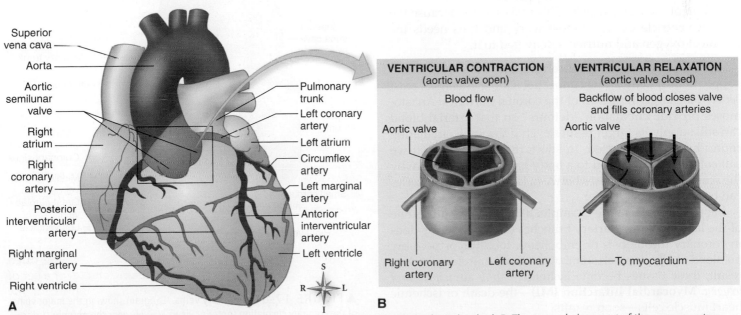

▲ FIGURE 17-7   **Coronary arteries. A,** Diagram showing the major coronary arteries (anterior view). **B,** The unusual placement of the coronary artery opening behind the leaflets of the aortic valve allows the coronary arteries to fill during ventricular relaxation.

BOX 17-1　Diagnostic Study

Angiography

A special type of radiography called *angiography* is often used to visualize arteries. A radiopaque dye—a substance that cannot be penetrated by x rays—is injected into an artery to better visualize vessels that would otherwise be invisible in a radiograph. This dye is often called *contrast medium.*

Sometimes the dye is released through a long, thin tube called a *catheter*—a procedure called *catheterization.* The catheter can be pushed through arteries until its tip is in just the right location to release the dye. As the dye begins to circulate, an *angiogram* (radiograph) will show the outline of the arteries as clearly as if they were made of bone or other dense material (see figure).

▶ **Coronary arteriogram.** This angiogram of the coronary arteries shows a narrowing *(arrow)* of the channel in the anterior ventricular (left anterior descending) artery of the heart.

small arteries of the heart. Given time, new anastomoses between the very small coronary arteries develop and provide some collateral circulation to ischemic areas. Currently, several surgical procedures are used to aid this process. In Box 17-1, we show you a special type of radiography called **angiography**—a method of visualizing the state of the coronary arteries.

Cardiac Veins

After blood has passed through the capillary networks in the myocardium, most of it enters a series of cardiac veins before draining into the right atrium. It does so through a common venous channel called the **coronary sinus.** However, several small veins from the right ventricle drain *directly* into the right atrium. As a rule, the cardiac veins (Figure 17-8) follow a pattern that closely parallels that of the coronary arteries.

Nerve Supply of the Heart

The myocardium is **autorhythmic,** which means that it creates its own internal rhythm for contraction and relaxation even without external nervous control. In fact, the heart has a system of myocardial fibers specialized for rapid electrical conduction. This pathway starts at the top of the heart and extends to the bottom of the heart. The myocardial system that generates and conducts action potentials is called the **conduction system of the heart.** In this system, both sympathetic fibers and parasympathetic fibers (from branches of

2. Each atrium, in contrast, receives blood only from a small branch of the corresponding coronary artery.

3. The most abundant blood supply goes to the myocardium of the left ventricle. This makes sense because the left ventricle does the most work and thus needs the most oxygen and nutrients delivered to it.

Another important fact concerning the heart's own blood supply is this: Only a few connections, or *anastomoses*, exist *between* the larger branches of the coronary arteries. Anastomoses provide circulation "detours" so that arterial blood can still supply its target tissues even if the main circulation routes become obstructed. In short, they provide what we call **collateral circulation** to a body part. This explains why the scarcity of anastomoses between larger coronary arteries can be such a threat to life.

Let's look at a real life example: If a blood clot plugs one of the larger coronary artery branches (as it frequently does in coronary thrombosis or embolism), too little or possibly no blood at all can reach some of the heart muscle cells. As a result, these tissues become **ischemic**—they are *deprived of oxygen.* **Myocardial infarction (MI)**—the death of ischemic heart muscle cells—soon results.

Although few anastomoses exist between the larger coronary arteries, many anastomoses exist between the very

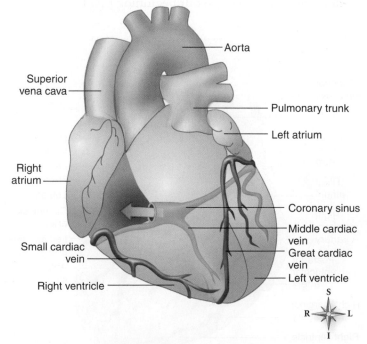

▲ **FIGURE 17-8**　**Coronary veins.** Diagram showing the major veins of the coronary circulation (anterior view). Vessels near the anterior surface are more darkly colored than vessels of the posterior surface (seen *through* the heart).

Labels: Aorta; Superior vena cava; Pulmonary trunk; Left atrium; Right atrium; Coronary sinus; Middle cardiac vein; Great cardiac vein; Small cardiac vein; Left ventricle; Right ventricle

the vagus nerve) combine to form *cardiac plexuses* located close to the arch of the aorta.

From the cardiac plexuses, fibers follow the right and left coronary arteries to enter the heart. Here most of the fibers terminate in the *sinoatrial (SA) node*. This structure is found near the junction of the superior vena cava and the right atrial wall. The SA node is part of the heart's own conduction system (see Figure 18-1, page 406). The SA node is the heart's *pacemaker*—a function we will look at more closely in Chapter 18.

QUICK ✓ CHECK

5. Briefly describe the general structure and function of coronary circulation.

6. Why is a full understanding of the coronary circulation so critical to understanding major types of heart disease?

7. Describe, in general, the conduction system of the heart.

8. What is a myocardial infarction?

BLOOD VESSELS

There are nearly 100,000 km (60,000 miles) of vessels carrying blood through your body right now! As you may know, **arteries** conduct blood *away* from the heart, **capillaries** conduct blood *through* tissues, and **veins** conduct blood *back* to the heart. At this point we'll examine these vessels in more detail (Figure 17-9).

Arteries

There are several types of arteries in the cardiovascular system.

Elastic arteries are the largest and include the aorta and some of its major branches. These huge arteries can stretch without injury. This allows them to accommodate the surge of blood forced into them when the heart ventricles contract and then recoil when the ventricles relax.

Muscular arteries, in contrast, carry blood farther away from the heart to specific organs and areas of the body. They are smaller in diameter than elastic arteries, and the muscular layer in their walls is proportionately *thicker*. Because of this, these arteries also have thicker muscular walls than similarly sized veins. Examples include *brachial* (arm), *femoral* (thigh), and *gastric* (stomach) arteries.

Arterioles are the smallest arteries. They are not named individually, as are the larger arteries, but as a *group* of arterioles. They are critically important in regulating blood flow throughout the body. They function by variable contraction of the smooth muscle in their walls. This in turn increases resistance to blood flow and helps regulate blood pressure. Their regulatory action also determines the quantity of

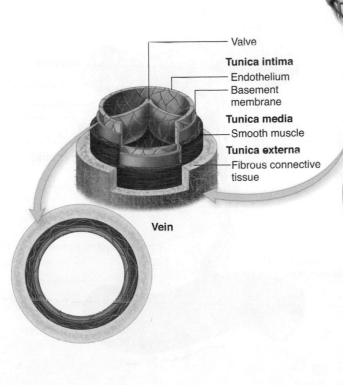

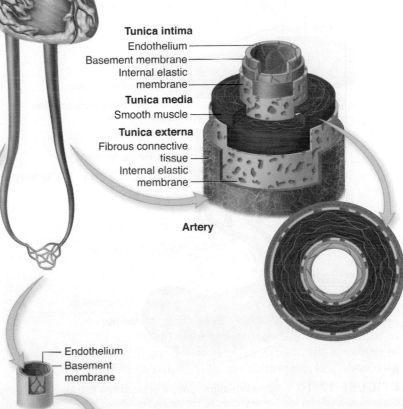

Valve

Tunica intima
Endothelium
Basement membrane

Tunica media
Smooth muscle

Tunica externa
Fibrous connective tissue

Vein

Tunica intima
Endothelium
Basement membrane
Internal elastic membrane

Tunica media
Smooth muscle

Tunica externa
Fibrous connective tissue
Internal elastic membrane

Artery

Endothelium
Basement membrane

Capillary

▲ **FIGURE 17-9** **Structure of blood vessels.** The tunica externa of the vein is color-coded blue and the artery red.

blood that enters a particular organ. For this reason they are sometimes called *resistance vessels*.

Metarterioles are short *connecting vessels* that connect true arterioles with the proximal ends of between 20 and 100 capillaries (Figure 17-10). Special "regulatory valves" encircle the proximal ends of metarterioles. These are actually smooth muscle cells called **precapillary sphincters.** These valves contract and relax, thereby regulating blood flow into specific capillaries and networks as they do so. The distal end of a metarteriole lacks precapillary sphincters and is called a **thoroughfare channel.** It is possible for blood passing directly through a metarteriole into a thoroughfare channel to bypass the intervening capillary bed.

As you will see shortly, all arteries except the pulmonary artery and its branches carry *oxygenated blood* after birth.

Capillaries

Capillaries are the microscopic vessels that carry blood from arterioles to small veins called *venules*. Blood flow through the arterioles, capillary beds, and venules is called **microcirculation** (see Figure 17-10). Transfer of nutrients and other vital substances between blood and tissue cells occurs at or very near a capillary, particularly within the capillary beds. This is why we often call capillaries the *primary exchange vessels* of the cardiovascular system. So vital is this function to our bodies that no cell is far removed from a capillary.

Although capillaries are small in size, it is estimated that our bodies carry more than 1 billion of them! However, they are not uniformly distributed. Some body tissues, such as those of the liver or cardiac muscle, or tissues with high metabolic rates, have many more of them. Therefore they require large numbers of capillary vessels to service them. Other tissues, such as cartilage and some types of epithelium, are **avascular** and lack capillary networks altogether.

True capillaries receive blood flowing out of metarterioles or other small arterioles. Precapillary sphincters regulate the volume of inflowing blood and its rate of passage through a true capillary. If the sphincter is "open," blood flows into the capillary. If the sphincter is closed, or partially so, blood flow into the capillary bed decreases significantly (see Figure 17-10).

Veins

After passing through the complex capillary networks of our bodies, blood from several capillaries flows into the distal end of the metarteriole, the thoroughfare channel, or directly enters the first of a series of venous vessels that will eventually return blood to the heart.

The first venous structures are small-diameter vessels called **venules.** Initially, these tiny veins have very narrow lumens and porous, thin walls. As in capillaries, fluid can be

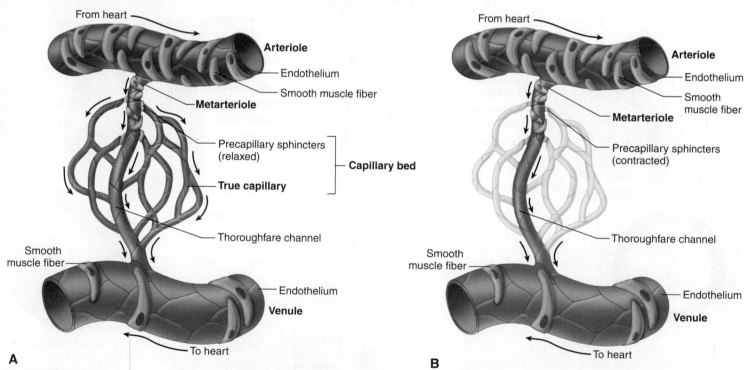

From heart — Arteriole
— Endothelium
— Smooth muscle fiber
Metarteriole
— Precapillary sphincters (relaxed) — **Capillary bed**
True capillary
— Thoroughfare channel
Smooth muscle fiber —
— Endothelium
Venule
To heart

A

From heart — Arteriole
— Endothelium
— Smooth muscle fiber
Metarteriole
— Precapillary sphincters (contracted)
— Thoroughfare channel
Smooth muscle fiber —
— Endothelium
Venule
To heart

B

▲ **FIGURE 17-10** **Microcirculation.** Control of blood flow through a capillary network is regulated by the contraction of precapillary sphincters surrounding arterioles and metarterioles. **A,** Sphincters are relaxed, permitting blood flow to enter the capillary bed. **B,** With sphincters contracted, blood flows from the metarteriole directly into the thoroughfare channel, bypassing the capillary bed.

exchanged between blood in the smallest venules and the tissue spaces. Their walls consist of little more than endothelial cells, and a few muscle cells.

The blood in the venules enters progressively larger venous channels—the veins. The names of veins correspond to their arterial counterparts. For example, there is a subclavian artery and a subclavian vein. The veins become larger as they approach the heart. As they increase in size, structural changes take place in the walls of the larger veins to accommodate and regulate the increasing blood volume. Veins do this with virtually no increase in blood pressure. This is because veins can stretch and increase their *capacity*—we say they have **capacitance.** For this reason they are called *capacitance vessels.* This feature permits veins to serve as reservoirs for blood as well as conduits for blood's passage back to the heart. Veins also have one-way valves that keep blood moving toward the heart and prevent potential backflow.

Structure of Blood Vessels

Four types of tissue "fabrics" make up a typical vessel wall: (1) a lining of endothelial cells, (2) collagen fibers, (3) elastic fibers, and (4) smooth muscle cells (see Figure 17-9).

Endothelial Cells

The endothelial cells that line the entire vascular system perform a number of different functions, depending on their location. First, they provide a smooth surface, reducing friction that creates turbulent blood flow. A smooth surface also inhibits coagulation and thus reduces clot formation. Some capillary vessels have intercellular clefts in their endothelium. These clefts vary in size and number and greatly influence the diffusion and movement of substances or cells into and out of the circulating blood.

In veins, the endothelium also forms valves that help maintain the one-way flow of blood. The capillaries are composed only of endothelium.

Fibers

Collagen fibers in the vascular wall are woven together much like the reinforcing strands found in the wall of a tire or hose. They function largely to strengthen the wall and keep the lumen of the vessel open.

Elastic fibers are secreted into the extracellular matrix and form a rubberlike network. In your largest arteries, elastic fibers are organized into nearly circular patterns, and allow for recoil after distention from blood flow. This elasticity plays a vital role in maintaining *passive tension* in the vessels of our cardiovascular system.

Smooth muscle fibers are found in the walls of the entire vascular system *except in capillaries.* They are most numerous in elastic and muscular arteries. Smooth muscle exerts *active tension* in these vessels when they contract.

Layers of Blood Vessels

The walls of arteries and veins consist of three separate layers: (1) *tunica externa,* (2) *tunica media,* and (3) *tunica intima.* As the names suggest, these layers are arranged from the outside of the vessel to the interior of the vessel. As blood vessels decrease in diameter, the relative thickness of their walls also decreases. There are also differences in the thickness of the layers, depending on the vessel's function in the body (see Figure 17-9).

Tunica Externa: The Outer Layer

The outermost layer is the **tunica externa** (external coat). It is composed of strong, flexible fibrous connective tissue. It functions to prevent tearing of the vessels walls during body movements. Collagen fibers extend outward from this layer to connect to nearby structures. The fibers anchor the vessel and help to keep the lumen open.

Tunica Media: The Middle Layer

The middle layer, the **tunica media,** is made of a layer of smooth muscle tissue sandwiched together with a layer of elastic connective tissue. The encircling smooth muscles of the tunica media permit changes in the blood vessel diameter. The tunica media is innervated by autonomic nerves and is supplied with its own blood circulation. As a rule, arteries have a thicker layer of smooth muscles than do veins.

Tunica Intima: The Inner Layer

The innermost layer of a blood vessel is the **tunica intima** (inside coat). The tunica intima is made of endothelium that is *continuous* with the endothelium that lines the heart. As such, it has a basement membrane to support it. Elastic arteries also have an *internal elastic membrane* that helps to accommodate changes in vessel diameter.

QUICK CHECK

9. Discuss the flow of blood through an artery, a capillary, and a vein.
10. How do elastic arteries differ from muscular arteries?
11. What is the difference between capacitance vessels and resistance vessels?
12. What are the major layers of arteries, veins, and capillaries?

MAJOR BLOOD VESSELS

The complexity of our cardiovascular system is amazing. It is simply not possible in an essentials text to name and describe all the blood vessels of your body. For this reason, we have concentrated our efforts on the major components.

General Circulatory Routes

Our enclosed **systemic circulation** route conducts blood from the heart through blood vessels to all parts of the body (*except* the tissues of the lungs), and then back to the heart. Refer to Figure 17-11 as we continue our discussion.

The left ventricle pumps blood into the ascending aorta. From here it flows into arteries that carry the blood into the various tissues and organs of the body. (Notice that blood moves from arteries to arterioles to capillaries as we progress.) Here the vital two-way exchange of substances occurs between the blood and cells. Blood then flows out of each organ by way of venules. These enlarge into veins that drain ultimately into the inferior or superior vena cava. The two enormous veins return the venous blood to the right atrium to complete the systemic circulation. It's almost a complete circuit, but not quite, because we don't reach our original starting point, the left ventricle.

This is where the **pulmonary circulation** route takes over. The venous blood from the right atrium is pumped to the right ventricle and then out the pulmonary trunk and its arteries to the lungs. Here, exchange of gases between blood and air takes place, converting deoxygenated blood to oxygenated blood. This oxygenated blood then flows on through lung venules into four pulmonary veins and returns to the left atrium of the heart. From the left atrium, it enters the left ventricle to be pumped again through the systemic circulation.

In a simplified circuit, blood passes through only one capillary network. However, there are several important exceptions to this general rule. In a **portal system,** blood flowing through the systemic circulation passes through *two consecutive capillary beds* rather than one. For example, notice in Figure 17-11 that blood coming from the digestive organs passes through a second capillary network in the liver before return to the heart. We will discuss this *hepatic portal system* in more detail later in this chapter.

There is a second exception to our general systemic circulation pattern. This involves **vascular anastomosis,** a direct connection or merger of blood vessels to one another. In vascular anastomoses, blood moves from veins to other veins or arteries to other arteries *without passing through a capillary network.* For example, **arterial anastomoses** involve the merger of one artery directly into another artery. **Venous anastomoses** often produce a direct connection between different veins.

Systemic Circulation

The systemic circulatory route includes the majority of the body's vessels. Please consult the figures and tables as you read through the following discussion. At this point, it is more important to understand the "big picture" rather than memorizing the content of any table.

Systemic Arteries

As you learn the names of the main arteries, keep in mind that the entire vascular tree is complex, but beyond the scope of this book. Most arteries ultimately diverge into capillaries. These *end-arteries* are vital to all of our organs. For example, permanent blindness results when the central artery of the retina, an end-artery, is blocked. Thus, blockage of end-arteries almost always has some negative effect. As we've seen, our bodies also have a number of vital arterial anastomoses. Their incidence increases as distance from the heart increases. Examples of arterial anastomoses are the *palmar* and *plantar* arches and the *cerebral arterial circle (circle of Willis)* at the base of the brain (see Figure 17-15).

Note in Figures 17-12 and 17-13 that the **aorta** is the major artery that serves as the main trunk for the entire systemic arterial system. However, the different segments of the aorta are known by different names, much like the names of streets may change as you drive into a new neighborhood. The first few centimeters of the aorta conduct blood upward out of the left ventricle and are called the **ascending aorta.** As you've seen, the coronary arteries are branches of the ascending aorta (see Figure 17-7 and Table 17-1). The aorta then turns 180 degrees to the left, forming a curved segment called the **aortic arch.** Arterial blood is then conducted downward from

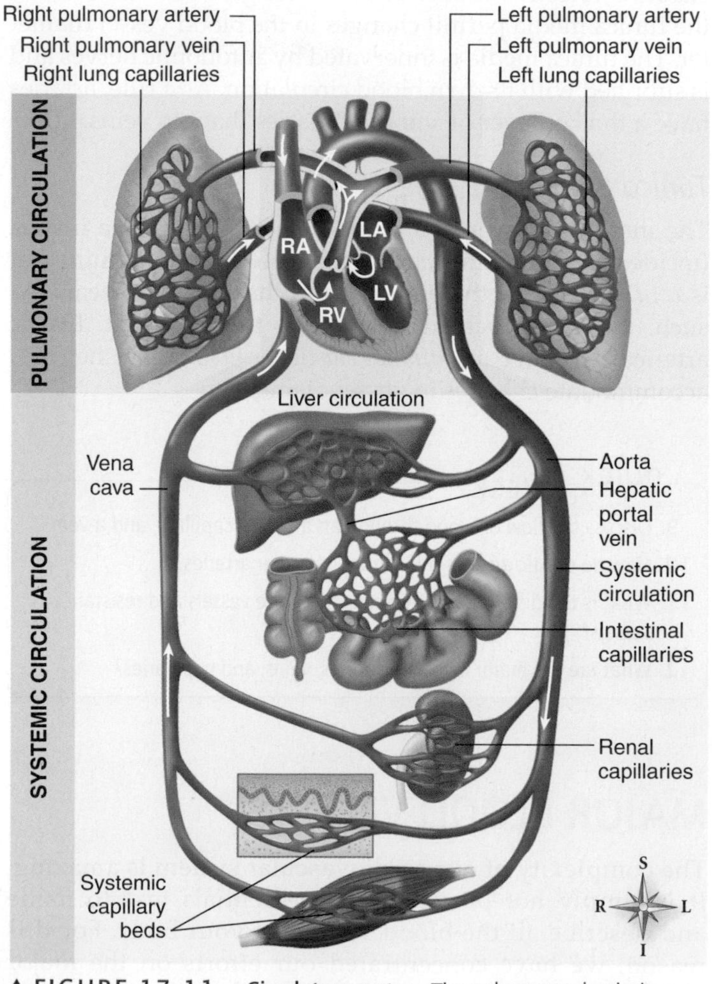

▲ FIGURE 17-11 **Circulatory routes.** The pulmonary circulation routes blood flow to and from the gas-exchange tissues of the lungs. The systemic circulation, on the other hand, routes blood flow to and from the oxygen-consuming tissues of the body.

Labels in figure:
Right pulmonary artery, Right pulmonary vein, Right lung capillaries, Left pulmonary artery, Left pulmonary vein, Left lung capillaries, PULMONARY CIRCULATION, RA, LA, LV, RV, SYSTEMIC CIRCULATION, Liver circulation, Vena cava, Aorta, Hepatic portal vein, Systemic circulation, Intestinal capillaries, Renal capillaries, Systemic capillary beds

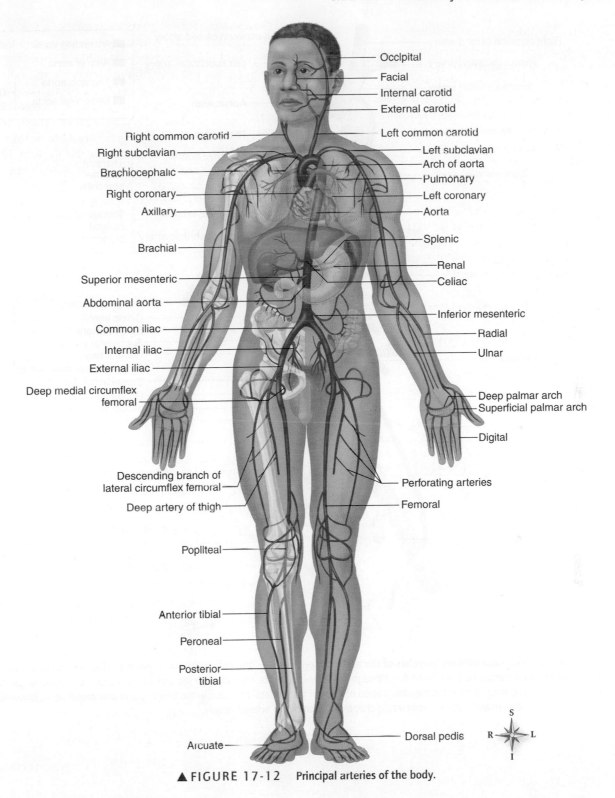

- Occipital
- Facial
- Internal carotid
- External carotid
- Right common carotid
- Left common carotid
- Right subclavian
- Left subclavian
- Brachiocephalic
- Arch of aorta
- Right coronary
- Pulmonary
- Axillary
- Left coronary
- Aorta
- Brachial
- Splenic
- Superior mesenteric
- Renal
- Abdominal aorta
- Celiac
- Common iliac
- Inferior mesenteric
- Internal iliac
- Radial
- External iliac
- Ulnar
- Deep medial circumflex femoral
- Deep palmar arch
- Superficial palmar arch
- Digital
- Descending branch of lateral circumflex femoral
- Perforating arteries
- Deep artery of thigh
- Femoral
- Popliteal
- Anterior tibial
- Peroneal
- Posterior tibial
- Arcuate
- Dorsal pedis

▲ FIGURE 17-12   Principal arteries of the body.

the aortic arch through the **descending aorta.** In the thorax, the descending aorta is called the **thoracic aorta;** in the abdomen, it is called the **abdominal aorta.** Note from Figure 17-1, *B*, and Figure 17-12 that all systemic arteries clearly branch from the aorta, or one of its branches.

Look again at the figures; notice how the main branches from the aortic arch on the right differ from those on the left. The right side of the head and neck are supplied by the **brachiocephalic artery,** which branches to become the right **subclavian artery,** and the right **common carotid artery.** On the left, however, the left subclavian artery and the left common carotid artery branch directly from the arch of the aorta—without an intervening brachiocephalic artery.

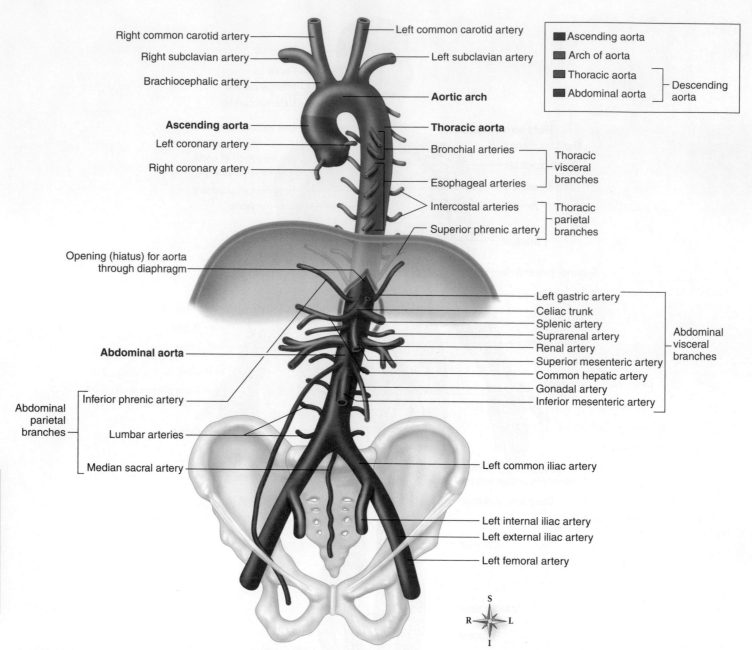

Right common carotid artery

Right subclavian artery

Brachiocephalic artery

Ascending aorta

Left coronary artery

Right coronary artery

Opening (hiatus) for aorta through diaphragm

Abdominal aorta

Abdominal parietal branches

Inferior phrenic artery

Lumbar arteries

Median sacral artery

Left common carotid artery

Left subclavian artery

Aortic arch

Thoracic aorta

Bronchial arteries ⎤
⎥ Thoracic
⎥ visceral
Esophageal arteries ⎦ branches

Intercostal arteries ⎤ Thoracic
⎥ parietal
Superior phrenic artery ⎦ branches

Left gastric artery ⎤
Celiac trunk ⎥
Splenic artery ⎥
Suprarenal artery ⎥ Abdominal
Renal artery ⎥ visceral
Superior mesenteric artery ⎥ branches
Common hepatic artery ⎥
Gonadal artery ⎥
Inferior mesenteric artery ⎦

Left common iliac artery

Left internal iliac artery

Left external iliac artery

Left femoral artery

■ Ascending aorta
■ Arch of aorta
■ Thoracic aorta ⎤ Descending
■ Abdominal aorta ⎦ aorta

▲ FIGURE 17-13 **Divisions and primary branches of the aorta (anterior view).** The aorta is the main systemic artery, serving as a trunk from which other arteries branch. Blood is conducted from the heart first through the ascending aorta, then through the arch of the aorta, and then through the thoracic and abdominal segments of the descending aorta. Note the designation of visceral and parietal branches in the thoracic and abdominal aortic divisions. Table 17-1 lists branches of the aortic divisions to assist you in interpreting chapter illustrations of arterial vessels.

TABLE 17-1 Major Systemic Arteries

ARTERY*	REGION SUPPLIED	ARTERY*	REGION SUPPLIED
Ascending Aorta		*Visceral Branches—cont'd*	
Coronary arteries	Myocardium	Celiac artery (trunk)	Abdominal viscera
Arch of Aorta		Left gastric	Stomach, esophagus
Brachiocephalic (Innominate)	Head, upper extremity	Common hepatic	Liver
Right subclavian	Head, upper extremity	Splenic	Spleen, pancreas, stomach
Right vertebral†	Spinal cord, brain	Superior mesenteric	Pancreas, small intestine, colon
Right axillary (continuation of subclavian)	Shoulder, chest, axillary region	Inferior mesenteric	Descending colon, rectum
Right brachial (continuation of axillary)	Arm, hand	Suprarenal	Adrenal (suprarenal) gland
Right radial	Lower arm, hand (lateral)	Renal	Kidney
Right ulnar	Lower arm, hand (medial)	Ovarian	Ovary, uterine tube, ureter
Superficial and deep palmar arches (formed by anastomosis of branches of radial and ulnar)	Hand, fingers	Testicular	Testis, ureter
		Parietal Branches	Walls of abdomen
Digital	Fingers	Inferior phrenic	Inferior surface of diaphragm, adrenal gland
Right common carotid	Head, neck	Lumbar	Lumbar vertebrae, muscles of back
Right internal carotid†	Brain, eye, forehead, nose	Median sacral	Lower vertebrae
Right external carotid†	Thyroid, tongue, tonsils, ear, etc.	*Common Iliac (Formed by terminal branches of aorta)*	Pelvis, lower extremity
Left Subclavian	Head, upper extremity		
Left vertebral†	Spinal cord, brain	*External Iliac*	Thigh, leg, foot
Left axillary (continuation of subclavian)	Shoulder, chest, axillary region	Femoral (continuation of external iliac)	Thigh, leg, foot
Left brachial (continuation of axillary)	Arm, hand	Popliteal (continuation of femoral)	Leg, foot
Left radial	Lower arm, hand (lateral)	Anterior tibial	Leg, foot
Left ulnar	Lower arm, hand (medial)	Posterior tibial	Leg, foot
Superficial and deep palmar arches (formed by anastomosis of branches of radial and ulnar)	Hand, fingers	Plantar arch (formed by anastomosis of branches of anterior and posterior tibial arteries)	Foot, toes
Digital	Fingers	Digital	Toes
Left Common Carotid	Head, neck	*Internal Iliac*	Pelvis
Left internal carotid†	Brain, eye, forehead, nose	*Visceral Branches*	Pelvic viscera
Left external carotid†	Thyroid, tongue, tonsils, ear, etc.	Middle rectal	Rectum
Descending Thoracic Aorta		Vaginal	Vagina, uterus
Visceral Branches	Thoracic viscera	Uterine	Uterus, vagina, uterine tube, ovary
Bronchial	Lungs, bronchi	*Parietal Branches*	Pelvic wall, external regions
Esophageal	Esophagus	Lateral sacral	Sacrum
Parietal Branches	Thoracic walls	Superior gluteal	Gluteal muscles
Intercostal	Lateral thoracic walls (rib cage)	Obturator	Pubic region, hip joint, groin
Superior phrenic	Superior surface of diaphragm	Internal pudendal	Rectum, external genitals, floor of pelvis
Descending Abdominal Aorta		Inferior gluteal	Lower gluteal region, coccyx, upper thigh
Visceral Branches	Abdominal viscera		

*Branches of each artery are indented below its name.
†See text and/or figures for branches of the artery.

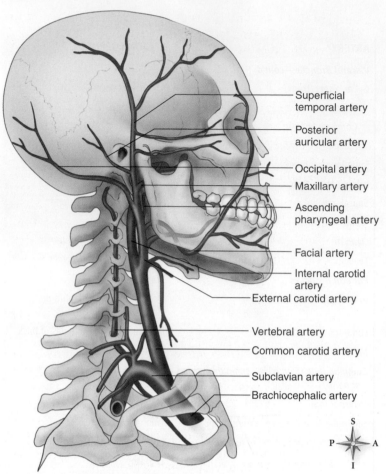

▲ **FIGURE 17-14** **Major arteries of the head and neck.** See Figure 17-15 for arteries at the base of the brain.

Major Arteries of the Head and Neck

Figure 17-14 illustrates the major arteries of the head, neck, and face. Trace the branching of the arteries in the figure with your finger as you read through Table 17-1. Notice in this figure how the right and left vertebral arteries extend from their origin as branches of the subclavian arteries up the neck, through openings *(foramina)* in the transverse processes of the cervical vertebrae, through the foramen magnum, and into the cranial cavity.

Next, look at Figure 17-15 and look at the arteries servicing the base of the brain. Note how the vertebral arteries unite on the undersurface of the brainstem to form the *basilar artery.* This artery branches quickly into the right and left *posterior cerebral arteries.* The basilar artery also branches to the pons and cerebellum. The internal carotid arteries enter the cranial cavity in the midpart of the cranial floor. Here they are known as the anterior cerebral arteries. Small vessels, the *communicating arteries,* join the *anterior and posterior cerebral arteries* to form the *cerebral arterial circle (circle of Willis)* at the base of the brain. This is a fine example of arterial anastomosis.

Major Arteries of the Thorax and Abdomen

The aortic arch continues downward as the *thoracic aorta,* which begins at the level of the fifth thoracic vertebra and ends at the diaphragm. The *abdominal aorta* is a downward continuation of the thoracic portion above it. It extends from the diaphragm to the point where it divides into the right and left *common iliac arteries* at the fourth lumbar vertebra.

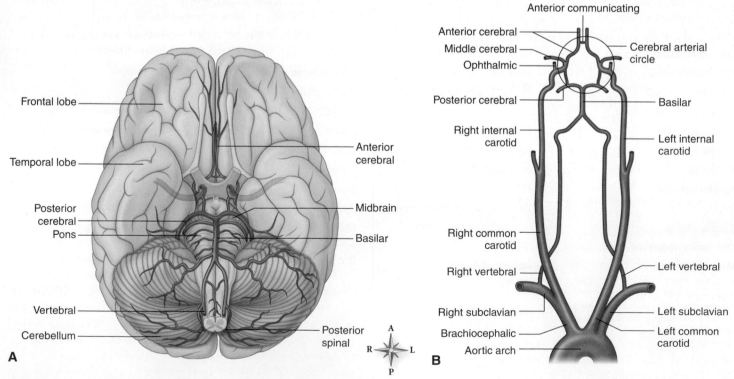

▲ **FIGURE 17-15** **Arteries at the base of the brain. A,** Diagram shows the cerebral arterial circle (circle of Willis) and related structures on the base of the brain. Note the arterial anastomoses. **B,** Origins of blood vessels that form the cerebral arterial circle.

Note in Figure 17-13 that this segment of the aorta lies just *anterior* to the vertebral bodies. As a result, a physician can feel the aortic pulse during deep palpation of the abdomen where this vessel is compressed against the underlying vertebrae. The presence of a pulsating swelling (a weak spot in the aorta)—an **aortic aneurysm**—is often diagnosed in this manner.

Major abdominal and pelvic branches of the abdominal aorta may also be described as parietal or visceral depending on the location of the organ or structures they service. You can see the branches of this segment of the aorta illustrated in Figure 17-13.

Major Arteries of the Extremities

Now, take a look back at Figure 17-12, which includes the arteries of the upper and lower extremity. Trace these arteries with your finger as you read through Table 17-1. Not every artery listed in the table appears in every illustration because of differences in view.

QUICK CHECK

13. Compare the systemic circulation route with the pulmonary circulation route.
14. What is the significance of vascular anastomoses?
15. List the major arteries coming off the aortic arch.
16. What is the significance of blockage in an end-artery?

Systemic Veins

There are a number of facts to keep in mind as you read about the names and locations of the major veins presented in the following discussion. First, veins are really larger extensions of capillaries, just as capillaries are the extensions of arteries. Thus, veins are enlargements of venules.

Second, although all vessels vary considerably in location and branches—and whether or not they are even present—the veins are especially variable. For example, the *median cubital vein* in the forearm is absent in many individuals.

Third, many of the main arteries have corresponding veins bearing the *same name* that are located alongside or near the arteries. These veins, like the arteries, lie in deep, well-protected areas, often running close to the bones. One example is the *femoral artery* and the *femoral vein,* both located along the femur bone.

You should note that some veins—for example, the large veins of the cranial cavity—are called *sinuses.* This is because they are venous enlargements that collect blood as it exits a tissue.

Finally, veins anastomose with each other in the same way as arteries. In fact, the venous portion of the systemic circulation has even more anastomoses than does the arterial portion. These anastomoses provide collateral return blood flow in cases of venous obstruction.

The major veins are listed for you Table 17-2. As you read the following discussion, follow along with Figures 17-16

through 17-19. As with the arteries, you may find it easier to learn the names of the major veins and their anatomical relation to each other from the diagrams and tables presented here.

Veins of the Head and Neck

The deep veins of the head and neck lie mostly within the cranial cavity. These are mainly *dural sinuses* and other veins that drain into the **internal jugular vein.**

Note that the superficial veins of the head and neck drain into the right and left **external jugular veins** in the neck (Figure 17-17). These veins receive blood from small superficial veins of the face, scalp, and neck, and terminate in subclavian veins. Small veins connect veins of the scalp and face with blood sinuses of the cranial cavity. This is of clinical interest because this arrangement presents a possible avenue for infections to enter the cranial cavity.

Veins of the Upper Extremity

Deep veins of the upper extremity drain into the **brachial vein,** which in turn drains into the **axillary vein.** From there blood flows into the **subclavian vein** before joining the **brachiocephalic vein**—a major tributary of the **superior vena cava.** You can review the major veins of the upper extremity in Figure 17-16. You'll see that the tributaries of the superior vena cava are more symmetrical from left to right than the nearby branches of the aorta. Compare Figures 17-12 and 17-16 to verify this for yourself.

Veins of the Thorax

Several small veins—such as the *bronchial vein, esophageal vein,* and *pericardial vein*—return blood from thoracic organs directly into the superior vena cava. The superior vena cava lies to the right of the spinal column and extends from the inferior vena cava through the diaphragm to the terminal part of the superior vena cava (Figure 17-18).

Veins of the Abdomen

The abdominal tributaries are another example of the differences between the left and right portions of the systemic venous circulation. For example, Figure 17-18 shows that the left *ovarian vein* or *testicular vein* and left *suprarenal vein* usually drain into the *left renal vein* instead of into the *inferior vena cava*—as occurs on the right side. The return of blood from the abdominal digestive organs follows below.

Hepatic Portal Circulation

Veins from the spleen, stomach, pancreas, gallbladder, and intestines do not pour their blood directly into the inferior vena cava, as do the veins from other abdominal organs. Instead they send their blood to the liver by means of the **hepatic portal vein.** Here the blood mingles with the arterial blood in the capillaries. Eventually it is drained from the liver by the *hepatic veins* that join the inferior vena cava. Any arrangement in which venous

TABLE 17-2 Major Systemic Veins

VEIN*	REGION DRAINED	VEIN*	REGION DRAINED
Superior Vena Cava	Head, neck, thorax, upper extremity	Hepatic portal system	Upper abdominal viscera
Brachiocephalic (Innominate)	Head, neck, upper extremity	Hepatic veins (continuations of liver venules and sinusoids and, ultimately, the hepatic portal vein)	Liver
Internal jugular (continuation of sigmoid sinus)	Brain		
Lingual	Tongue, mouth	Hepatic portal vein	Gastrointestinal organs, pancreas, spleen, gallbladder
Superior thyroid	Thyroid, deep face	Cystic	Gallbladder
Facial	Superficial face	Gastric	Stomach
Sigmoid sinus (continuation of transverse sinus/direct tributary of internal jugular)	Brain, meninges, skull	Splenic	Spleen
		Inferior mesenteric	Descending colon, rectum
Superior and inferior petrosal sinuses	Anterior brain, skull	Pancreatic	Pancreas
Cavernous sinus	Anterior brain, skull	Superior mesenteric	Small intestine, most of colon
Ophthalmic veins	Eye, orbit	Gastroepiploic	Stomach
Transverse sinus (direct tributary of sigmoid sinus)	Brain, meninges, skull	*Renal*	Kidneys
Occipital sinus	Inferior, central region of cranial cavity	Suprarenal	Adrenal (suprarenal) gland
		Left ovarian	Left ovary
Straight sinus	Central region of brain, meninges	Left testicular	Left testis
Inferior sagittal sinus	Central region of brain, meninges	Left ascending lumbar (anastomoses with hemiazygos)	Left lumbar region
Superior sagittal (longitudinal) sinus	Superior region of cranial cavity	Right ovarian	Right ovary
External jugular	Superficial, posterior head, neck	Right testicular	Right testis
Subclavian (continuation of axillary/direct tributary of brachiocephalic)	Axilla, lower extremity	Right ascending lumbar (anastomoses with azygos)	Right lumbar region
Axillary (continuation of basilic direct tributary of subclavian)	Axilla, lower extremity	Common iliac (continuation of external iliac; common iliacs unite to form inferior vena cava)	Lower extremity
Cephalic	Lateral and lower arm, hand		
Brachial	Deep arm	External iliac (continuation of femoral direct tributary of common iliac)	Thigh, leg, foot
Radial	Deep lateral forearm		
Ulnar	Deep medial forearm	Femoral (continuation of popliteal direct tributary of external iliac)	Thigh, leg, foot
Basilic (direct tributary of axillary)	Medial and lower arm, hand		
Median cubital (basilic) (formed by anastomosis of cephalic and basilic)	Arm, hand	Popliteal	Leg, foot
		Posterior tibial	Deep posterior leg
Deep and superficial palmar venous arches (formed by anastomosis of cephalic and basilic)	Hand	Medial and lateral plantar	Sole of foot
		Fibular (peroneal) (continuation of anterior tibial)	Lateral and anterior leg, foot
Digital	Fingers	Anterior tibial	Anterior leg, foot
Azygos (anastomoses with right ascending lumbar)	Right posterior wall of thorax and abdomen, esophagus, bronchi, pericardium, mediastinum	Dorsal veins of foot	Anterior (dorsal) foot, toes
		Small (external, short)	Superficial posterior leg, lateral foot
Hemiazygos (anastomoses with left renal)	Left inferior posterior wall of thorax and abdomen, esophagus, mediastinum	Great (internal, long) saphenous	Superficial medial and anterior thigh, leg, foot
		Dorsal veins of foot	Anterior (dorsal) foot, toes
Accessory hemiazygos	Left superior posterior wall of thorax	Dorsal venous arch	Anterior (dorsal) foot, toes
Inferior Vena Cava	Lower trunk and extremity	Digital	Toes
Phrenic	Diaphragm	*Internal Iliac*	Pelvic region

*Tributaries of each vein are identified below its name; deep veins are printed in dark blue, and superficial veins are printed in light blue.

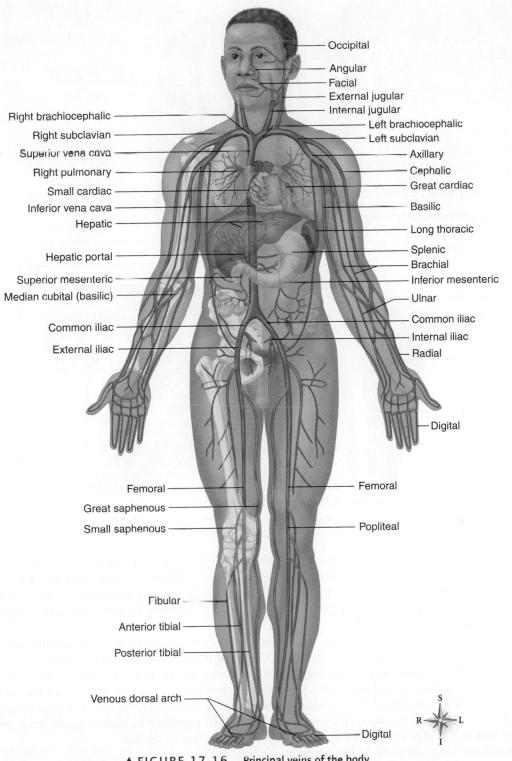

Occipital
Angular
Facial
External jugular
Internal jugular
Right brachiocephalic
Left brachiocephalic
Right subclavian
Left subclavian
Superior vena cava
Axillary
Right pulmonary
Cephalic
Small cardiac
Great cardiac
Inferior vena cava
Basilic
Hepatic
Long thoracic
Hepatic portal
Splenic
Brachial
Superior mesenteric
Inferior mesenteric
Median cubital (basilic)
Ulnar
Common iliac
Common iliac
External iliac
Internal iliac
Radial
Digital
Femoral
Femoral
Great saphenous
Small saphenous
Popliteal
Fibular
Anterior tibial
Posterior tibial
Venous dorsal arch
Digital

S
R   L
I

▲ FIGURE 17-16   Principal veins of the body.

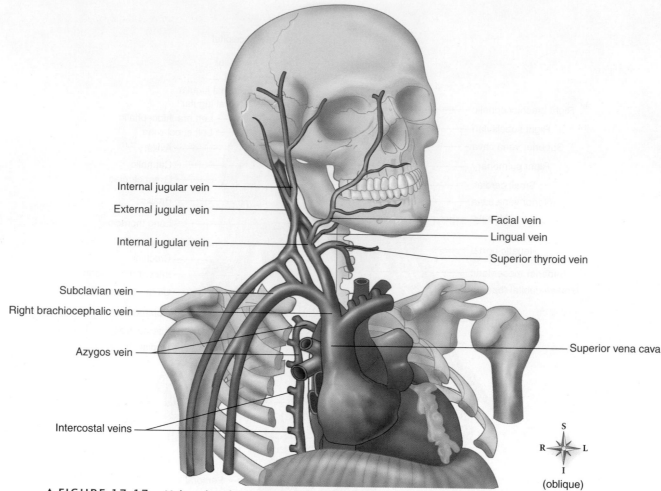

▲ FIGURE 17-17 **Major veins of the head and neck.** Anterior view showing veins on the right side of the head and neck.

Internal jugular vein

External jugular vein

Internal jugular vein

Subclavian vein

Right brachiocephalic vein

Azygos vein

Intercostal veins

Facial vein

Lingual vein

Superior thyroid vein

Superior vena cava

(oblique)

blood flows through a *second capillary network* before returning to the heart is called a *portal circulatory system.* Portal means "gateway," and the liver acts as a gateway through which blood returning from the digestive tract must pass before it returns to the heart.

There are several advantages to this "detour" from the digestive tract through the liver before returning the blood to the heart. Shortly after a meal, blood flowing through digestive organs begins absorbing glucose and other simple nutrients. The result is a tremendous increase in the blood glucose level. As the blood travels through the liver, however, excess glucose is removed and stored in liver cells as the polysaccharide **glycogen.** Thus blood returned to the heart carries only a moderate level of glucose. Many hours after food has yielded its nutrients, low-glucose blood coming from the digestive organs can pick up glucose released from the breakdown of glycogen stores. It's a simple solution to prevent the glucose from flooding the body immediately after digestion.

Yet another advantage of the hepatic portal system is that toxic molecules such as alcohol can be partially removed or detoxified before the blood is distributed to the remainder of the body. Figure 17-11 shows the basic layout of the hepatic

portal system in relation to the overall pattern of circulation. Figure 17-19 shows the details of the veins involved in the hepatic portal circulation. In most of us, the hepatic portal vein is formed by the union of the splenic and superior mesenteric veins.

Sometimes the hepatic portal system or the venous return of blood from the liver is slowed or blocked (as sometimes happens in liver disease or heart disease). In such cases, venous drainage from most of the other abdominal organs may also be obstructed. The accompanying increase in capillary pressure in these organs may account for abdominal bloating or **ascites.**

Veins of the Lower Extremity

Deep veins of the lower leg drain from the *anterior tibial vein,* which continues as the *fibular (peroneal) vein,* and the *posterior tibial vein.* The fibular and posterior tibial veins join to form the **popliteal vein.** This vein runs behind the knee joint and continues up along the femur as the deep **femoral vein.** The femoral vein then continues as the **external iliac vein,** draining into the common iliac vein and from there into the *inferior vena cava.*

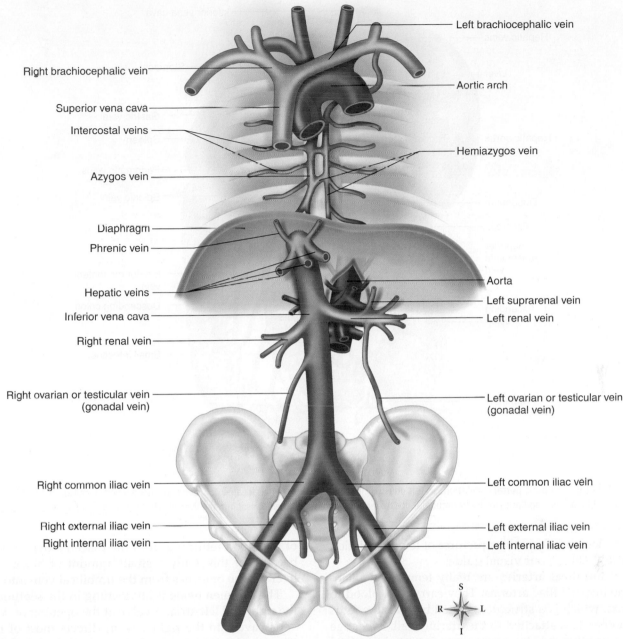

Left brachiocephalic vein

Right brachiocephalic vein

Superior vena cava

Intercostal veins

Azygos vein

Diaphragm
Phrenic vein

Hepatic veins

Inferior vena cava

Right renal vein

Right ovarian or testicular vein
(gonadal vein)

Right common iliac vein

Right external iliac vein
Right internal iliac vein

Aortic arch

Hemiazygos vein

Aorta
Left suprarenal vein
Left renal vein

Left ovarian or testicular vein
(gonadal vein)

Left common iliac vein

Left external iliac vein
Left internal iliac vein

▲ FIGURE 17-18 **Inferior vena cava and its abdominopelvic tributaries.** Anterior view of ventral body cavity with many of the viscera removed. Note the close anatomical relationship between the inferior vena cava and the descending aorta. Smaller veins of the thorax drain blood into the inferior vena cava or into the azygos vein—both are shown here.

Superficial veins of the lower extremity include the *small saphenous vein,* a tributary of the popliteal vein, and the **great saphenous vein,** which drains much of the superficial leg and foot. Saphenous means "apparent" and refers to the location of these veins near the surface of the skin.

Fetal Circulation

Basic Plan of Fetal Circulation

Of necessity, the circulation of blood before birth must be different from that after birth. This is because the fetus obtains oxygen and food from maternal blood and digestive

QUICK CHECK

17. Is the distribution of arteries and veins throughout the circulation completely symmetrical?

18. Do vein location and number vary from individual to individual?

19. What are the advantages of the hepatic portal system in the liver?

A&P CONNECT

A valuable skill is the ability to trace the flow of blood completely through the normal adult circulatory route. Check out **How to Trace the Flow of Blood** online at **A&P Connect** for some tips and a handy diagram of the route of blood flow through the body.

evolve

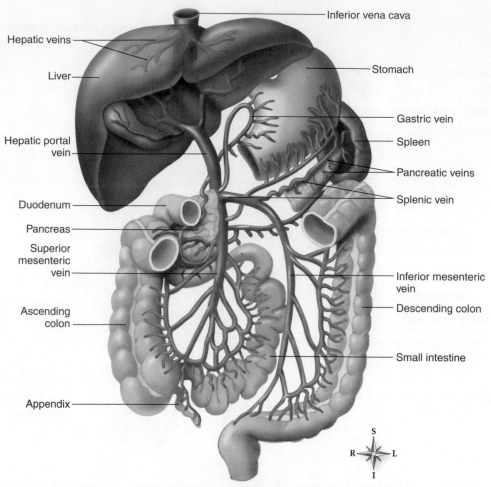

Inferior vena cava
Hepatic veins
Liver
Stomach
Gastric vein
Hepatic portal vein
Spleen
Pancreatic veins
Splenic vein
Duodenum
Pancreas
Superior mesenteric vein
Inferior mesenteric vein
Ascending colon
Descending colon
Small intestine
Appendix

▲ FIGURE 17-19 **Hepatic portal circulation.** In this unusual circulatory route, a vein is located between two capillary beds (see Figure 17-11). The hepatic portal vein collects blood from capillaries in visceral structures located in the abdomen and empties it into the liver. Hepatic veins return blood to the inferior vena cava.

organs. Let's look briefly at the structures of fetal circulation using Figure 17-20 as your visual guide.

The two **umbilical arteries** are really temporary extensions of the internal iliac arteries. They carry fetal blood to the **placenta,** which is a structure created by both the fetus and the mother. It is attached to the uterine wall. Exchange of oxygen and other substances between maternal and fetal blood takes place in the placenta. However, note that *no mixing* of maternal and fetal blood occurs under normal conditions of pregnancy. The fetus is vulnerable, however, to many substances ingested or inhaled by the mother. Box 17-2 shows you how alcohol from the maternal blood can damage a developing fetus.

The **umbilical vein** returns oxygenated blood from the placenta. It enters the fetal body through the umbilicus and extends up to the underside of the liver. Here it gives off two or three branches to the liver, and then continues on as the *ductus venosus.* Two umbilical arteries and the umbilical vein together constitute the **umbilical cord.** These are shed after birth along with the placenta when the umbilical cord is cut.

The other three structures of fetal circulation are involved as ducts or detours of circulation. The **ductus venosus** is a continuation of the umbilical vein. It runs along the underside of the liver and drains into the inferior vena cava. Most

of the blood returning from the placenta bypasses the liver because of this. Only a small amount of blood enters the liver via the branches from the umbilical vein into the liver.

The **foramen ovale** is an opening in the septum between the right and left atria. A valve at the opening of the inferior vena cava into the right atrium directs most of the blood through the foramen ovale into the left atrium so that it bypasses the fetal lungs. Thus, most of the blood does not flow into the lungs. This is because of another detour, the **ductus arteriosus,** a small vessel that connects the pulmonary artery with the descending thoracic aorta.

Changes in Circulation at Birth

Dramatic changes in the circulatory system must take place immediately upon the birth of a child. First and foremost, the six structures that serve fetal circulation are no longer needed.

As soon as the umbilical cord is cut, the two umbilical arteries, the placenta, and the umbilical vein obviously can no longer function. After the baby is born, the placenta is shed from the mother's body as the so-called *afterbirth,* along with the severed ends of the umbilical vessels still attached. The sections of these vessels that remain in the infant's body eventually become fibrous cords. These cords

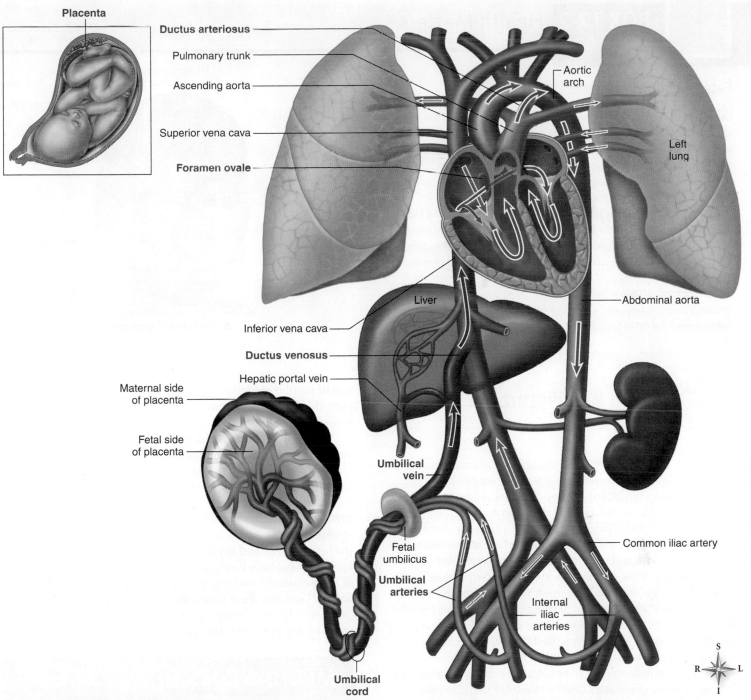

▲ **FIGURE 17-20** **Plan of fetal circulation.** Before birth, the human circulatory system has several special features that adapt the body to life in the womb. These features *(labeled in red type)* include two umbilical arteries, one umbilical vein, ductus venosus, foramen ovale, ductus arteriosus, and umbilical cord.

remain throughout life. For example, the umbilical vein becomes the *round ligament* of the liver and the ductus venosus (no longer needed to bypass blood around the liver) eventually becomes the *ligamentum venosum* of the liver.

Beyond this, the foramen ovale normally becomes at least *functionally closed* soon after birth. This takes place when the baby takes his or her first breath and full circulation is established through the lungs. Complete *structural closure,* however, usually requires 9 months or more. Eventually, the foramen ovale becomes a small depression *(fossa ovalis)* in

the wall of the right atrial septum. The ductus arteriosus contracts as soon as respiration is established. Eventually, it also turns into a fibrous cord, the *ligamentum arteriosum.*

QUICK ✓ CHECK

20. Identify the six structures unique to fetal circulation.

21. What is the function of the placenta and the umbilical vessels?

22. What major changes occur in an infant's circulation at the time of birth?

BOX 17-2 | **Health Matters**

Fetal Alcohol Syndrome

Consumption of alcohol by a woman during her pregnancy can have tragic effects on the developing fetus. Educational efforts to inform pregnant women about the dangers of alcohol continue to receive national attention. Even very limited consumption of alcohol during pregnancy poses significant hazards to the developing baby because alcohol can easily cross the placental barrier and enter the fetal bloodstream.

When alcohol enters the fetal blood, the potential result, called **fetal alcohol syndrome (FAS)**, can cause serious congenital abnormalities such as "small head," or *microcephaly*; low birth weight; cardiovascular defects; developmental disabilities such as physical and mental retardation; and even fetal death.

The photograph shows the small head, thinned upper lip, small eye openings *(palpebral fissures)*, epicanthal folds, and receded upper jaw *(retrognathia)* typical of fetal alcohol syndrome.

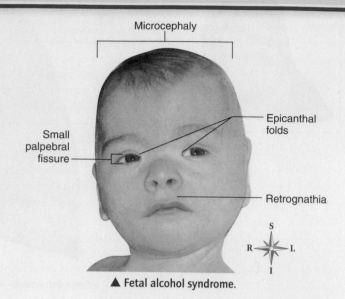

▲ Fetal alcohol syndrome.

 ## The BIG Picture

In this chapter, you've observed the basic structural and functional anatomy of the heart, major arteries, and major veins. You've also examined the structure and function of arterioles, venules, and capillaries. In essence, we have described a widely distributed, interconnected network of transport vessels for blood.

You've seen how the components of our circulatory system are integrated to form a functioning, closed circulatory loop. Now, step back from this collection of facts for a moment and think about how each component of the heart works in this important system. Use your mind's eye to envision the entire system: Keep in mind the "big picture" of the circulatory system. What are the components? How are they integrated? What is the significance of a properly functioning circulatory system for the entire body? If you continue in this vein, you'll discover how putting the big picture together really aids in your understanding of this system.

Your understanding of the components of the circulatory system are very important to your understanding of the functional anatomy of the cardiovascular system in the next chapter. Chapter 18 will deal with issues that relate to the function of the cardiovascular system and its importance in maintaining the homeostatic mechanisms of our bodies.

Cycle of LIFE ⭕

As we've seen, the heart and blood vessels undergo profound anatomical changes during early fetal development. Then at birth, the abrupt switch from a placenta-dependent circulatory system to an independent circulatory system causes yet another set of profound anatomical changes. Throughout childhood, adolescence, and adulthood, the heart and blood vessels normally maintain their basic structure and function. The only apparent changes in these structures occur as a re-

sult of regular exercise. After rigorous exercise over prolonged periods, the myocardium thickens and the supply of blood vessels in skeletal muscle tissues increases.

As we pass through adulthood, especially later adulthood, various degenerative changes can occur in the heart and the blood vessels. For example, **atherosclerosis** ("hardening of the arteries") can result in the blockage or weakening of critical arteries. Such blockage in the coronary arteries can cause a myocardial infarction. In the brain, such blockage can cause a stroke. Both can be life altering or life threatening. In addition, the heart valves and myocardial tissues often degenerate with age. They become hardened and more fibrous and lose some of their elasticity and regenerative capacity. They are thus less able to perform their functions properly. This, of course, reduces the heart's pumping efficiency and threatens homeostasis of our entire internal environment. ●

MECHANISMS OF DISEASE

There are numerous diseases of the heart and circulatory system, including *disorders involving the pericardial lining,* the heart valves, the *pericardium,* the *arteries,* and the *veins.* A number of these are congenital and apparent at birth, but many others develop as injury, disease, and age take their toll on our circulation system. The structural defects can often be remedied by surgery. However, many heart diseases require exercise, behavioral changes, and medications, which have proven to be effective tools.

🌀**volve** Find out more about these cardiovascular disorders and diseases online at *Mechanisms of Disease: Anatomy of the Cardiovascular System.*

LANGUAGE OF SCIENCE AND MEDICINE *continued from page 375*

cardiovascular system (kar-dee-oh-VAS-kyoo-lar SIS-tem)
   [*cardi-* heart, *-vas-* vessel, *-cul-* little, *-ar* relating to]

chordae tendineae (KOR-dee ten-DIN ee-ee)
   [*chorda* string or cord, *tendinea* pulled tight] *sing.,* chorda tendinea (KOR-dah ten-DIN-ee-ah)

collateral circulation (koh-LAT-er-al ser-kyoo-LAY shun)
   [*co-* together, *-later-* side, *-al* relating to, *circulat-* go around, *-tion* process]

common carotid artery (kah-ROT-id AR-ter-ee)
   [*caro-* heavy sleep, *-id* relating to, *arteri-* vessel]

conduction system of the heart (kon-DUK-shen SIS-tem)
   [*conduct-* lead, *-tion* process, *system* organized whole]

coronary artery (KOHR-oh-nair-ee AR-ter-ee)
   [*corona-* crown, *-ary* relating to, *arteri-* vessel]

coronary sinus (KOR-oh-nair-ee SYE-nus)
   [*corona-* crown, *-ary* relating to, *sinus-* hollow]

cusp (kusp)
   [*cusp* point]

descending aorta (ay-OR-tah)
   [*aort-* lifted, *-a* thing] *pl.,* aortae or aortas (ay-OR-tee, ay-OR-tahz)

ductus arteriosus (DUK-tus ar-teer-ee-OH-sus)
   [*ductus* duct, *arteri-* vessel, *-osus* relating to]

ductus venosus (DUK-tus veh-NO-sus)
   [*ductus* duct, *ven-* vein, *-osus* relating to]

elastic artery (eh-LAS-tik AR-ter-ee)
   [*elast-* drive or beat out, *-ic* relating to, *arteri-* vessel]

endocardium (en-doh-KAR-dee-um)
   [*endo-* within, *-cardi-* heart, *-um* thing]

epicardium (ep-ih-KAR-dee-um)
   [*epi-* on or upon, *-cardi-* heart, *-um* thing]

external iliac vein (eks-TER-nal IL-ee-ak)
   [*extern-* outside, *-al* relating to, *ilium* flank]

external jugular vein (eks-TER-nal JUG-yoo-lar)
   [*extern-* outside, *-al* relating to, *jugul-* neck, *-ar* relating to]

femoral vein (FEM-or-al)
   [*femor-* thigh, *-al* relating to]

fetal alcohol syndrome (FAS) (FEE-tal AL-koh-hol SIN-drohm)
   [*fet-* offspring, *-al* relating to, *syn-* together, *-drome* running or (race) course]

fibrous pericardium (FYE-brus pair-ih-KAR-dee-um)
   [*fibr-* fiber, *-ous* relating to, *peri-* around, *-cardi-* heart, *-um* thing]

foramen ovale (foh-RAY-men oh-VAL-ee)
   [*foramen* opening, *ovale* egg shaped] *pl.,* foramina ovales (foh-RAM-ih-nah oh-VAL-eez)

glycogen (GLYE-koh-jen)
   [*glyco-* sweet, *-gen* produce]

great saphenous vein (sah-FEE-nus)
   [*saphen-* manifest, *-ous* relating to]

heart

hepatic portal vein (heh-PAT-ik POR-tal)
   [*hepa-* liver, *-ic* relating to, *port-* doorway, *-al* relating to]

inferior vena cava (in-FEER-ee-or VEE-nah KAY-vah)
   [*infer-* lower, *-or* quality, *vena* vein, *cava* hollow] *pl.,* venae cavae (VEE-nee KAY-vee)

internal jugular vein (in-TER-nal JUG-yoo-lar)
   [*intern-* inside, *-al* relating to, *jugul* neck, *-ar* relating to]

ischemic (is-KEE-mik)
   [*ischem-* hold back, *-ic* relating to]

metarteriole (met-ar-TEER-ee-ohl)
   [*meta-* change or exchange, *arteri-* vessel, *-ole* little]

microcirculation (my-kroh-ser-kyoo-LAY-shun)
   [*micro-* small, *circulat-* go around, *-tion* process]

mitral valve (MY-tral)
   [*mitr-* bishop's hat, *-al* relating to]

muscular artery (MUSS-kyoo-lar AR-ter-ee)
   [*mus-* mouse, *-cul-* little, *-ar* relating to, *arteri-* vessel]

myocardial infarction (MI) (my-oh-KAR-dee-al in-FARK-shun)
   [*myo-* muscle, *-cardi-* heart, *-al* relating to, *in-* in, *-farc-* stuff *-tion-* process]

myocardium (my-oh-KAR-dee-um)
   [*myo-* muscle, *-cardi-* heart, *-um* thing] *pl.,* myocardia (my-oh-KAR-dee-ah)

papillary muscle (PAP-ih-lair-ee MUSS-el)
   [*papilla-* nipple, *-ary* relating to, *mus-* mouse, *-cle* small]

pericardial fluid (pair-ih-KAR-dee-al FLOO-id)
   [*peri-* around, *-cardi-* heart, *-al* relating to, *fluid* flow]

pericardial space (pair-ih-KAR-dee-al)
   [*peri-* around, *-cardi-* heart, *-al* relating to]

pericardium (pair-ih-KAR-dee-um)
   [*peri-* around, *-cardi-* heart, *-um* thing] *pl.,* pericardia (pair-ih-KAR-dee-ah)

placenta (plah-SEN-tah)
   [*placenta* flat cake] *pl.,* placentae or placentas (plah-SEN-tee, plah-SEN-tahz)

popliteal vein (pop-lih-TEE-al)
   [*poplit-* back of knee, *-al* relating to]

portal system (POR-tal SIS-tem)
   [*port-* doorway, *-al* relating to]

precapillary sphincter (pree-KAP-ih-lair-ee SFINGK-ter)
   [*pre-* before, *-capill-* hair, *-ary* relating to]

pulmonary circulation (PUL-moh-nair-ee ser-kyoo-LAY-shun)
   [*pulmon-* lung, *-ary* relating to, *circulat-* go around, *-tion* process]

semilunar (SL) valve (sem-ih-LOO-nar)
   [*semi-* half, *-luna* moon]

septum (SEP-tum)
   [*septum* fence] *pl.,* septa (SEP-tah)

serous pericardium (SEER-us pair-ih-KAR-dee-um)
   [*sero-* watery fluid, *-ous* relating to, *peri-* around, *-cardi-* heart, *um* thing] *pl.,* pericardia (pair-ih-KAR-dee-ah)

subclavian artery (sub-KLAY-vee-an AR-ter-ee)
   [*sub-* below, *-clavi-* key (clavicle bone), *-ula* little, *arteri-* vessel]

subclavian vein (sub-KLAY-vee-an)
   [*sub-* below, *-clavi-* key (clavicle bone), *-an* relating to]

superior vena cava (soo-PEER-ee-or VEE-nah KAY-vah)
   [*super-* over or above, *-or* quality, *vena* vein, *cava* hollow] *pl.,* venae cavae (VEE-nee KAY-vee)

systemic circulation (sis-TEM-ik ser-kyoo-LAY-shun)
   [*system-* organized whole, *-ic* relating to, *circulat-* go around, *-tion* process]

thoracic aorta (tho-RASS-ik ay-OR-tah)
   [*thorac-* chest, *-ic* relating to, *aort-* lifted, *-a* thing] *pl.,* aortae or aortas (ay-OR-tee, ay-OR-tahz)

thoroughfare channel (THUR-oh-fair CHAN-el)
   [*thoroughfare* main road, *chanel-* groove]

tricuspid valve (try-KUS-pid)
   [*tri-* three, *-cusp-* point, *-id* characterized by]

true capillary (KAP-ih-lair-ee)
   [*capill-* hair, *-ary* relating to]

tunica externa (TOO-nih-kah ex-TER-nah)
   [*tunica* tunic or coat, *extern-* outside] *pl.,* tunicae externae (TOO-nih-kee ex-TER-nee)

tunica intima (TOO-nih-kah IN-tih-mah)
   [*tunica* tunic or coat, *intima* innermost] *pl.,* tunicae intimae (TOO-nih-kee IN-tih-mee)

tunica media (TOO-nih-kah MEE-dee-ah)
   [*tunica* tunic or coat, *media* middle] *pl.,* tunicae mediae (TOO-nih-kee MEE-dee-ee)

umbilical artery (um-BIL-ih-kul AR-ter-ee)
   [*umbilic-* navel, *-al* relating to, *arteri-* vessel]

umbilical cord (um-BIL-ih-kul)
   [*umbilic-* navel, *-al* relating to]

umbilical vein (um-BIL-ih-kul)
   [*umbilic-* navel, *-al* relating to]

vascular anastomosis (VAS-kyoo-lar ah-nas-toh-MOH-sis)
   [*vas-* vessel, *-ular* relating to, *ana-* anew, *-stomo-* mouth, *-osis* condition] *pl.,* anastomoses (ah-nas-toh-MOH-seez)

vein

venous anastomosis (VEE-nus ah-nas-toh-MOH-sis)
   [*ven-* vein, *-ous* relating to, *ana-* anew, *-stomo-* mouth, *-osis* condition] *pl.,* anastomoses (ah-nas-toh-MOH-seez)

ventricle (VEN-trih-kul)
   [*ventr-* belly, *-icle* little]

venule (VEN-yool)
   [*ven-* vein, *-ule* little]

CHAPTER SUMMARY

To download an MP3 version of the chapter summary for use with your iPod or other portable media player, access the Audio Chapter Summaries online at http://evolve.elsevier.com.

HINT *Scan this summary after reading the chapter to help you reinforce the key concepts. Later, use the summary as a quick review before your class or before a test.*

INTRODUCTION

A. Cardiovascular system (circulatory system)—consists of a muscular heart and a closed system of vessels
B. Cardiovascular system develops early and reaches a functional state long before any other major organ system

HEART

A. Location, shape and size of the heart (Figure 17-1)
  1. Lies in the mediastinum just behind the body of the sternum
  2. In infants, it is $\frac{1}{130}$ of the total body weight; $\frac{1}{300}$ in the adult
  3. Between puberty and 25 years of age, the heart attains its adult shape and weight
B. Coverings of the heart
  1. Pericardium—outer sac that encloses the heart
    a. Fibrous pericardium—loosely fitting outer layer of this sac
    b. Serous pericardium—smooth, moist serous membrane (parietal layer) that lines the fibrous pericardium; visceral layer (epicardium) directly covers the surface of the heart (Figure 17-3)
    c. Pericardial space—space between the two layers; contains pericardial fluid
    d. Pericardial sac provides protection against friction as the heart beats
C. Wall of the heart
  1. Epicardium—outer layer of the heart wall; visceral layer of the serous pericardium
  2. Myocardium—thick, contractile, middle layer; comprises the bulk of the heart wall
  3. Endocardium—delicate layer lining the interior of the myocardial wall
D. Chambers of the heart—interior of the heart is divided into four cavities (chambers) (Figure 17-4)
  1. Atria—two upper chambers
    a. Separated into left and right chambers by the interatrial septum
    b. Receive blood from veins
    c. Muscular walls are not very thick
  2. Ventricles—two lower chambers
    a. Separated into left and right chambers by the interventricular septum
    b. Receive blood from the atria and pump blood out of the heart into the arteries; "pumping chambers"
    c. Myocardium is quite thick

E. Valves of the heart—four tough, fibrous structures that permit the flow of blood in one direction only and are vital to the normal functioning of the heart (Figures 17-5 and 17-6)
  1. Atrioventricular (AV) valves—service the openings between the atria and the ventricles; have pointed flaps called cusps
    a. AV valves allow blood to flow from the atria into the ventricles but prevents it from flowing backward
    b. Right AV (tricuspid) valve—consists of three cusps of endocardium
      (2) Free edge of each flap is anchored to the papillary muscles of the right ventricle by several tendinous cords (chordae tendineae)
    c. Left AV (bicuspid/mitral) valve—two cusps of endocardium
  2. Semilunar (SL) valves—located where the pulmonary artery joins the right ventricle (pulmonary valve) and where the aorta joins the left ventricle (aortic valve)
    a. Pocketlike flaps that extend inward from the lining of the pulmonary artery and the aorta; look very much like a "half-moon" (Figure 17-5)
    b. Pulmonary valve—semilunar valve at the entrance of the pulmonary artery
    c. Aortic valve—semilunar valve at the entrance of the aorta
  3. Skeleton of the heart (Figure 17-6)
    a. Consists of a set of connected rings that serve as a semirigid support for the heart valves
    b. Serves as sites for the attachment of cardiac muscle of the myocardium
    c. Also serves as an electrical barrier between the myocardium of the atria and the myocardium of the ventricles
  4. Flow of blood through the heart (Figure 17-5)
    a. Beginning with the right atrium, blood flows through the right AV (tricuspid) valve into the right ventricle
    b. From the right ventricle, blood then flows into the first portion of the pulmonary artery
    c. Pulmonary arteries carry blood to the lungs for gas exchange
    d. From the lungs, blood flows back through pulmonary veins into the left atrium of the heart
    e. From the left atrium, blood flows into the left ventricle
    f. From the left ventricle, blood then flows into the aorta
F. Blood supply of the heart tissue (Figures 17-7 and 17-8)
  1. Coronary arteries (Figure 17-7, *A*)
    a. First branches off the aorta and supplies the heart muscle first
    b. Both right and left coronary arteries have two main branches
    c. Both ventricles receive their blood supply from branches of the right and left coronary arteries
    d. Each atrium receives blood only from a small branch of the corresponding coronary artery
    e. The most abundant blood supply goes to the myocardium of the left ventricle
    f. Few connections (anastomoses) exist between the larger branches of the coronary arteries

G. Nerve supply of the heart
1. Myocardium is autorhythmic—creates its own internal rhythm for contraction and relaxation even without external nervous control
2. Sympathetic fibers and parasympathetic fibers combine to form cardiac plexuses
   a. From the cardiac plexuses, fibers follow the right and left coronary arteries to enter the heart
   b. Most of the fibers terminate in the sinoatrial (SA) node

BLOOD VESSELS

A. Arteries—conduct blood away from the heart
1. Elastic arteries—largest; include the aorta and some of its major branches
   a. Can stretch without injury
2. Muscular arteries—carry blood farther away from the heart to specific organs and areas of the body
   a. Smaller in diameter than elastic arteries
   b. Muscular layer in their walls is proportionately thicker
3. Arterioles—smallest arteries
   a. Important in regulating blood flow throughout the body
4. Metarterioles—short connecting vessels; connect true arterioles with the proximal ends of between 20 and 100 capillaries (Figure 17-10)
   a. Precapillary sphincters—regulating blood flow into specific capillaries and networks
B. Capillaries—conduct blood through tissues
1. Microscopic vessels
2. Carry blood from arterioles to small veins called venules; constitute microcirculation (Figure 17-10)
3. Primary exchange vessels of the cardiovascular system
4. Not uniformly distributed; body tissues with high metabolic rates have many more of them
5. True capillaries—receive blood flowing out of metarterioles or other small arterioles (Figure 17-10)
C. Veins—conduct blood back to the heart
1. Venules—first venous structures; very narrow lumens and porous, thin walls
2. Veins can stretch and increase their capacity; capacitance vessels
3. Venous sinus—large venous structure with very thin endothelial walls
D. Structure of blood vessels—four types of tissue "fabrics" make up a typical vessel wall
1. Endothelial cells
2. Collagen fibers
3. Elastic fibers
4. Smooth muscle cells
E. Endothelial cells—line the entire vascular system
1. Provide a smooth surface; reducing friction that creates turbulent blood flow
2. In capillaries, allows the efficient exchange of materials between the blood plasma and the interstitial fluid of the surrounding tissues

F. Fibers
1. Collagen fibers—reinforcing strands found in the vascular wall; strengthen the wall and keep the lumen of the vessel open
2. Elastic fibers—secreted into the extracellular matrix and form a rubberlike network; help maintain passive tension in the vessels of our cardiovascular system
3. Smooth muscle fibers—found in the walls of the entire vascular system except in capillaries; exert active tension in these vessels when they contract
G. Layers of blood vessels—walls of arteries and veins consist of three separate layers (Figure 17-9)
1. Tunica externa—outermost layer
   a. Composed of strong, flexible fibrous connective tissue
   b. Functions to prevent tearing of the vessel walls during body movements
2. Tunica media—middle layer
   a. Made of a layer of smooth muscle tissue sandwiched together with a layer of elastic connective tissue
   b. Arteries have a thicker layer of smooth muscles than do veins
3. Tunica intima—inner layer
   a. Made of endothelium that is continuous with the endothelium that lines the heart

MAJOR BLOOD VESSELS

A. General circulatory routes
1. Systemic circulation—blood flow from the heart through blood vessels to all parts of the body (except the tissues of the lungs), and then back to the heart (Figure 17-11)
2. Pulmonary circulation—venous blood flow from the right atrium to the right ventricle and then out the pulmonary trunk and its arteries to the lungs
3. Portal system—blood flowing through the systemic circulation passes through two consecutive capillary beds rather than one (Figure 17-11)
4. Vascular anastomosis—direct connection or merger of blood vessels to one another
   a. Arterial anastomosis—involves the merger of one artery directly into another artery
   b. Venous anastomosis—produces a direct connection between different veins
B. Systemic circulation—includes the majority of the body's vessels
C. Systemic arteries (Table 17-1 and Figures 17-12 to 17-15)
1. End-arteries—arteries that diverge into capillaries
2. Aorta—major artery that serves as the main trunk for the entire systemic arterial system
3. Major arteries of the head and neck (Figures 17-14 and 17-15)
4. Major arteries of the thorax and abdomen (Figure 17-13)
5. Major arteries of the extremities (Figure 17-12)
D. Systemic veins (Table 17-2 and Figures 17-16 to 17-19)
1. Veins are really larger extensions of capillaries
2. Veins are especially variable
3. Many of the main arteries have corresponding veins bearing the same name that are located alongside or near the arteries
4. Large veins of the cranial cavity are called sinuses
5. Veins of the head and neck (Figure 17-17)

6. Veins of the upper extremity (Figure 17-2)
7. Veins of the thorax (Figure 17-18)
8. Veins of the abdomen (Figure 17-18)
9. Hepatic portal circulation (Figure 17-19)
   a. Veins of the spleen, stomach, pancreas, gallbladder, and intestines send their blood to the liver by way of the hepatic portal vein
   b. Blood travels through the liver and excess glucose is removed and stored in liver cells as glycogen
   c. Toxic molecules such as alcohol can be partially removed or detoxified before the blood is distributed to the remainder of the body
10. Veins of the lower extremity
E. Fetal circulation—additional blood vessels in the fetus are needed to carry the fetal blood into close approximation with the maternal blood and then to return it to the fetal body
   1. Structures that allow this to happen are two umbilical arteries, the umbilical vein, and the ductus venosus
   2. Placenta—provides a place where there can be an interchange of gas, foods, and wastes between the fetal and maternal circulation
   3. Three structures located within the fetus play a vital role in fetal circulation:
      a. Ductus venous
      b. Foramen ovale
      c. Ductus arteriosus
   4. Changes in circulation at birth
      a. When umbilical cord is cut, the two umbilical arteries, the placenta, and the umbilical vein obviously can no longer function
      b. Placenta is shed from the mother's body
      c. Umbilical vein becomes the round ligament of the liver
      d. Ductus venosus—becomes the ligamentum venosum of the liver
      e. Foramen ovale normally becomes at least functionally closed soon after birth (fossa ovalis)
      f. Ductus arteriosus contracts as soon as respiration is established; turns into a fibrous cord, the ligamentum arteriosum

REVIEW QUESTIONS

 HINT Write out the answers to these questions after reading the chapter and reviewing the Chapter Summary. If you simply think through the answer without writing it down, you won't retain much of your new learning.

1. Discuss the size, position, and location of the heart in the thoracic cavity.
2. Describe the pericardium, differentiating between the fibrous and serous portions.
3. Exactly where is pericardial fluid found? Explain its function.
4. Define the following terms: *intercalated disks* and *autorhythmic*.
5. Name and locate the chambers and valves of the heart.
6. Trace the flow of blood through the heart.
7. Describe the six unique structures necessary for fetal circulation.

8. Explain how the separation of oxygenated and deoxygenated blood occurs after birth.
9. Briefly define the following terms: *aneurysm* and *atherosclerosis*.

CRITICAL THINKING QUESTIONS

 HINT *After finishing the Review Questions, write out the answers to these items to help you apply your new knowledge. Go back to sections of the chapter that relate to items that you find difficult.*

1. How is cardiopulmonary resuscitation (CPR) accomplished? What is the significance of the placement of the heart in the thoracic cavity and successful CPR?
2. What would result if there were a lack of anastomosis in the arteries of the heart?
3. The general public thinks the most important structure in the cardiovascular system is the heart. Anatomists know it is the capillary. What information would you use to support this view?
4. Compare and contrast arterial blood in the systemic circulation and arterial blood in the pulmonary circulation.
5. Can you make the distinction between an occlusion of an end-artery and an occlusion of other small arteries?
6. Describe the functional advantage of a portal system.

continued from page 375

With the knowledge you have gained from reading this chapter, see if you can answer these questions about Kyle from the Introductory Story.

1. The doctor suspects the potential blockage is in what part of Kyle's body?
   a. His brain
   b. His liver
   c. His neck
   d. His heart

"Let's take him over to the Cath Lab," the physician ordered. Kyle said, "I've heard of a cath lab, but I'm not sure what that means." "Cardiac catheterization lab," clarified the nurse. "We're going to insert a small tube . . ." She kept talking, but Kyle couldn't concentrate on her words. He was suddenly feeling a little anxious. He signed the consent form without really reading it.

In the lab, Kyle changed into a hospital gown as instructed; next he was asked to lie on the table. A nurse began cleaning a spot on his thigh in preparation for inserting a catheter. Kyle was confused—why were they cleaning his leg when it seemed like his heart was the problem?

2. Into which artery will the catheter be inserted?
   a. Brachial
   b. Popliteal
   c. Femoral
   d. Tibial

3. From this artery, the catheter will be moved toward the heart through which path?
    a. External iliac artery, abdominal aorta, descending aorta, aortic arch, ascending aorta
    b. Internal iliac artery, abdominal aorta, ascending aorta, aortic arch, descending aorta
    c. Abdominal aorta, descending aorta, aortic arch, ascending aorta
    d. Popliteal artery, external iliac artery, abdominal aorta, thoracic aorta, aortic arch

After some dye was injected, the screen monitor showed that Kyle's right coronary artery was partially blocked. The surgeon inserted a balloon through the catheter, which was then inflated to press against the sides of the artery and enlarge its diameter. Next she inserted a metal stent to keep the artery open.

4. The coronary arteries supply oxygen and nutrients for cardiac muscle contraction. The myocardium of which heart chamber receives the most abundant blood supply from the coronary arteries?
    a. Left atrium
    b. Left ventricle
    c. Right atrium
    d. Right ventricle

HINT *To solve a case study, you may have to refer to the glossary or index, other chapters in this textbook, A&P Connect, Mechanisms of Disease, and other resources.*

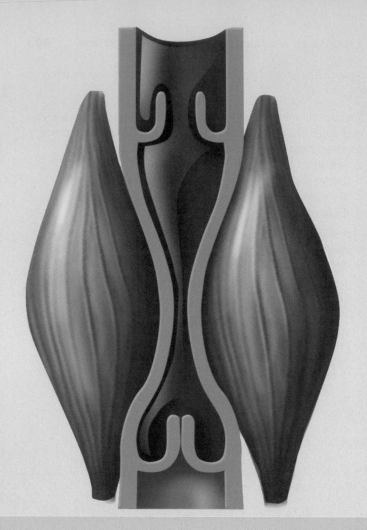

CHAPTER 18

Physiology of the Cardiovascular System

STUDENT LEARNING OBJECTIVES

At the completion of this chapter, you should be able to do the following:

1. List the basic components of the cardiovascular system.
2. Describe the structures of the heart and how they serve to pump blood.
3. Discuss the deflection waves of a normal ECG and explain their significance.
4. Outline the basic steps of the cardiac cycle.
5. Discuss the primary principle of circulation and the significance of high arterial blood pressure.
6. Describe factors that affect stroke volume and heart rate.
7. Discuss the significance of peripheral resistance in the cardiovascular system.
8. Discuss the significance of the vasomotor control mechanism.
9. Discuss how the respiratory pump and skeletal muscle pump work to assist the venous return of blood to the heart.
10. Outline the hormonal mechanisms that regulate blood volume.
11. Discuss the significance of the pulse mechanism.
12. List several major pulse points in the body.

CHAPTER OUTLINE

Scan this outline before you begin to read the chapter, as a preview of how the concepts are organized.

BOBBY was in a hurry to finish the last job of the day. He had been called in to help complete and inspect the wiring in a museum that was due to open the next week. Bobby called to his coworker, Jerry, to confirm the current was off to the electrical box on which he was working. When he heard a positive response, he climbed the ladder and reached up to tighten a few loose screws. As he was tightening the screws with his right hand, he lost his balance and reached up with his left hand to catch hold of the ladder . . . but instead caught a live wire that sent a jolt of electricity through his body.

Jerry came running and knocked the ladder out from under Bobby so he would fall to the floor, breaking the electric arc. "Are you okay?" he asked Bobby, who appeared dazed, but was conscious. "I don't feel so great," he replied, smiling weakly. Then he collapsed. Jerry yelled at the other workers in the room to call 911, checked Bobby's pulse, and started CPR. The foreman rushed in with an automated external defibrillator (AED) and attached the electrodes to Bobby's chest. The AED's mechanical voice said, "Press button when clear."

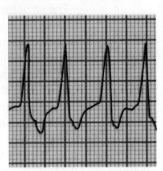

As you read through this chapter, you'll understand the electrical mechanism that operates your heart.

FUNCTION OF THE HEART AND BLOOD VESSELS

Our bodies are never at rest. In fact, a continuous supply of energy is required to maintain a relatively constant internal environment. This homeostatic maintenance of our body is due in large part to the continuous and controlled movement of blood throughout our circulatory system. Blood carries out most of its important transport functions in the thousands of miles of capillaries that comprise much of this system. As you might suspect, however, the body's total blood volume is not evenly distributed. The regulation of blood pressure and blood flow must therefore change in response to cellular activity.

There are many control mechanisms that help to regulate and integrate the diverse functions and component parts of our cardiovascular system. This system must work to supply blood to specific body areas according to their immediate needs. These mechanisms ensure a relatively constant internal environment surrounding each body cell. In this chapter we will explore the *control mechanisms* that regulate the pumping action of the heart. We will also see how these mechanisms ensure the smooth and directed flow of blood throughout our circulatory system.

HEMODYNAMICS

Hemodynamics refers to the various mechanisms that influence the movement of blood. This is vital, of course, because different organs may need vastly different amounts of blood flow, depending on their metabolic

LANGUAGE OF SCIENCE AND MEDICINE

HINT *Before reading the chapter, say each of these terms out loud. This will help you avoid stumbling over them as you read.*

ADH mechanism (A-D-H MEK-ah-nih-zem)
[*ADH* antidiuretic hormone, *mechan-* machine, *-ism* state]

ANH mechanism (A-N-H MEK-ah-nih-zem)
[*ANH* atrial natriuretic hormone, *mechan-* machine, *-ism* state]

atrioventricular (AV) bundle (ay-tree-oh-ven-TRIK-yoo-lar BUN-del)
[*atrio-* entrance courtyard, *-ventr-* belly, *-icul-* little, *-ar* relating to]

atrioventricular (AV) node (ay-tree-oh-ven-TRIK-yoo-lar)
[*atrio-* entrance courtyard, *-ventr-* belly, *-icul-* little, *-ar* relating to, *nod-* knot]

baroreceptor (bar-oh-ree-SEP-tor)
[*baro-* pressure, *-recept-* receive, *-or* agent]

cardiac cycle (KAR-dee-ak SYE-kul)
[*cardi-* heart, *-ac* relating to, *cycle* circle]

cardiac output (CO) (KAR-dee-ak)
[*cardi-* heart, *-ac* relating to]

conduction system of the heart (kon-DUK-shen SIS-tem)
[*conduct-* lead, *-tion* process, *system* organized whole]

contractility (kon-trak-TIL-ih-tee)
[*con-* together, *-tract-* drag or draw, *-il-* of or like, *-ity* quality of]

diastole (dye-ASS-toh-lee)
[*dia-* through, *-stole* contraction]

diastolic blood pressure (dye-ah-STOL-ik PRESH-ur)
[*dia-* apart or through, *-stol-* contraction, *-ic* relating to]

ectopic pacemaker (ek-TOP-ik PAYS-may-ker)
[*ec-* out of, *-top-* place, *-ic* relating to]

electrocardiogram (ECG or EKG) (eh-lek-troh-KAR-dee-oh-gram)
[*electro-* electricity, *-cardio-* heart, *-gram* drawing]

heart murmur
[*murmur* hum]

heart rate (HR)

hemodynamics (hee-moh-dye-NAM-iks)
[*hemo-* blood, *-dynami-* force, *-ic* relating to]

peripheral resistance (peh-RIF-er-al)
[*peri-* around, *-phera-* boundary, *-al* relating to]

primary principle of circulation (PRY-mair-ee PRIN-sip-al of ser kyoo-LAY-shun)
[*prim-* first, *-ary* relating to, *princip-* foundation, *circulat-* go around, *-tion* process]

pulse
[*pulse* beat]

pulse pressure (PRESH-ur)
[*pulse* beat]

pulse wave
[*pulse* beat]

continued on page 422

activity. For example, working muscles need a far greater blood supply than do resting muscles. This is because tissues and organs (such as working muscles) with a greater metabolic rate obviously need more oxygen. And that oxygen is delivered by the RBCs of flowing blood in the capillaries. Because the amount of blood in our bodies is relatively constant, this means that the flow of blood to specific areas must be managed according to their activities. We will begin with a discussion of the physiology of the heart and then examine blood flow throughout the cardiovascular system.

A&P CONNECT

Whether blood flows straight through vessels or in a swirling, turbulent pattern has great clinical significance. For example, turbulent flow may signal a partially blocked artery and may promote the formation of dangerous blood clots. Learn more about this concept in **Focus on Turbulent Blood Flow** online at **A&P Connect.**

evolve

THE HEART AS A PUMP

Recall from Chapter 17 that the left side of the heart services the *systemic circulatory route,* which supplies blood flow to the entire body, except to the lungs. The right side of the heart services the *pulmonary circulatory route,* which supplies blood flow to and from the lungs. Both routes require coordination so that they function as a single pumping structure. To do this, the impulses (action potentials) that trigger con-

traction must be coordinated. The heart must have a system to generate rhythmic impulses. It must then distribute these impulses to the different regions of the myocardium. This distribution is accomplished by the *impulse-conducting pathway.* Four structures make up the core of the **conduction system of the heart:**

1. Sinoatrial (SA) node
2. Atrioventricular (AV) node
3. AV bundle (bundle of His)
4. Subendocardial branches (Purkinje fibers)

You can see these structures represented in Figure 18-1. Please refer to this figure as you read through the following discussion.

The structures that make up the heart's conduction system are composed of cells that differ in function from those of ordinary cardiac muscle: *The heart conduction cells cannot contract strongly.* Instead, they permit the generation or rapid conduction of an action potential throughout the myocardium.

Normally, the cardiac impulses that control heart contraction begin in the **sinoatrial (SA) node.** This node, also called the "pacemaker," is located just below the atrial epicardium at its junction with the superior vena cava (Figure 18-1, *A*). The pacemaker cells in the SA node have an *intrinsic rhythm.* This means that they do not require nervous input from the brain or spinal cord. They themselves initiate impulses at regular intervals. In fact, if you remove pacemaker cells and put them in a nutrient solution, they will continue to beat! They do *not require* nervous or hormonal stimulation to con-

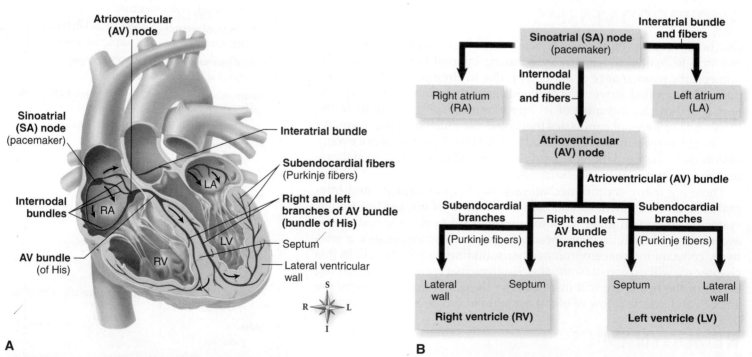

A

B

▲ **FIGURE 18-1** **Conduction system of the heart.** Specialized cardiac muscle cells *(boldface type)* in the wall of the heart rapidly initiate or conduct an electrical impulse throughout the myocardium. Both the sketch of the conduction system **(A)** and the flowchart **(B)** show the origin and path of conduction. The signal is initiated by the SA node (pacemaker) and spreads directly to the rest of the right atrial myocardium. From there, it travels to the left atrial myocardium by way of a bundle of interatrial conducting fibers, and then to the AV node by way of three internodal bundles. The AV node then initiates a signal that is conducted through the ventricular myocardium by way of the AV bundle (of His) and subendocardial branches (Purkinje fibers).

tract. However, in a living heart, there *are* nervous and hormonal components that influence the cells of the pacemaker.

Here's an overview of how the heart's conduction system works (Figure 18-1, *B*). Each impulse generated at the sinoatrial (SA) node travels swiftly throughout the muscle fibers of both atria. (An *interatrial bundle* of conducting fibers provides rapid conduction to the left atrium.) Thus, the atria begin to contract. The action potential next enters the **atrioventricular (AV) node** via three *internodal bundles.* These bundles are also composed of conducting fibers. Here the conduction of the *impulse slows* considerably. This allows the atria to contract completely before the conduction reaches the ventricles below.

After passing slowly through the AV node, the conduction velocity increases again as the impulse is relayed through the **atrioventricular (AV) bundle** (also called the *bundle of His*). At this point, the right and left *bundle branches* and the **subendocardial branches** (also called the *Purkinje fibers*) in which they terminate conduct the impulses throughout the muscle of both ventricles. This impulse stimulates the ventricular muscle fibers to contract almost simultaneously.

Thus the SA node initiates each heartbeat and sets the basic pace—it is the heart's own natural pacemaker. Under the influence of autonomic and endocrine controls, the SA node will normally "fire" at an intrinsic rhythmical rate of 70 to 75 beats per minute under resting conditions. However, if for any reason the SA node loses its ability to generate an impulse,

pacemaker activity will shift to another excitable component of the conduction system. These might include the AV node or the subendocardial branches. Pacemakers other than the SA node are abnormal. They are called **ectopic pacemakers.** Unfortunately, such ancillary pacemakers usually set a much slower rhythm—only 40 to 60 beats per minute. If the heart's own pacemaker fails to maintain a healthy heart rhythm, an artificial pacemaker can be implanted to restore normal function.

A&P CONNECT

Artificial pacemakers can help keep individuals with damaged hearts alive for many years. Check out **Artificial Cardiac Pacemakers** online at **A&P Connect** to see how these devices work and how they are implanted.

ⓔvolve

Electrocardiogram (ECG)

Electrocardiography

Impulse conduction in the heart generates tiny electrical currents that spread through surrounding tissues to the surface of the body. This fact has great clinical importance because these currents can be measured with an *electrocardiograph.* An electrocardiogram (ECG) is produced by attaching electrodes of a recording voltmeter (the electrocardiograph) to the chest and/or limbs of the subject (Figure 18-2, *A*). (The abbreviation for an

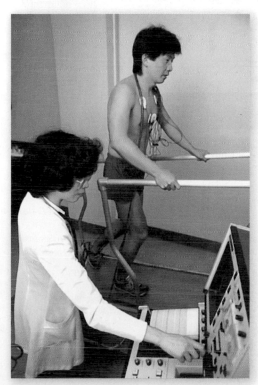

A

▲ **FIGURE 18-2** **Electrocardiogram. A,** A nurse monitors a patient's ECG as he exercises on a treadmill. **B,** Idealized ECG deflections represent depolarization and repolarization of cardiac muscle tissue. **C,** Principal ECG intervals between P, QRS, and T waves. Note that the P-R interval is measured from the start of the P wave to the start of the Q wave.

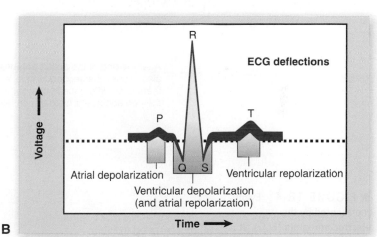

B

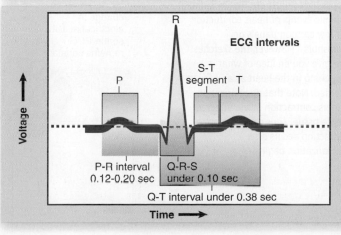

C

▶ **FIGURE 18-3** The basic theory of electrocardiography.

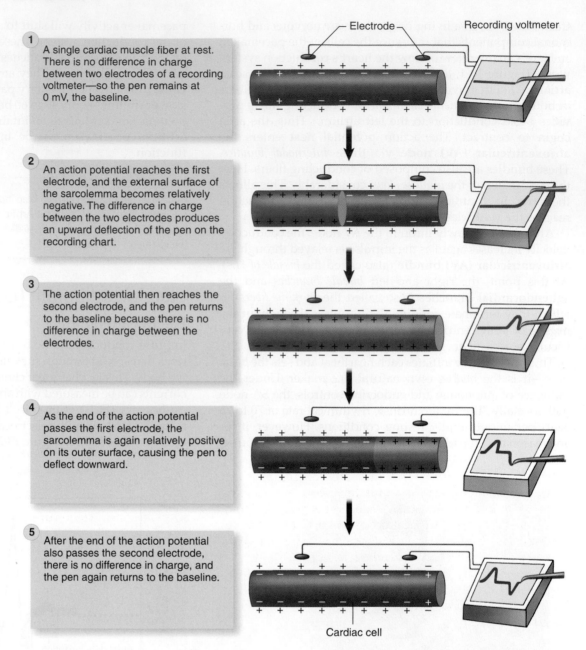

1 A single cardiac muscle fiber at rest. There is no difference in charge between two electrodes of a recording voltmeter—so the pen remains at 0 mV, the baseline.

2 An action potential reaches the first electrode, and the external surface of the sarcolemma becomes relatively negative. The difference in charge between the two electrodes produces an upward deflection of the pen on the recording chart.

3 The action potential then reaches the second electrode, and the pen returns to the baseline because there is no difference in charge between the electrodes.

4 As the end of the action potential passes the first electrode, the sarcolemma is again relatively positive on its outer surface, causing the pen to deflect downward.

5 After the end of the action potential also passes the second electrode, there is no difference in charge, and the pen again returns to the baseline.

Electrode

Recording voltmeter

Cardiac cell

▶ **FIGURE 18-4** Events represented by the electrocardiogram (ECG). It is impossible to illustrate the invisible, dynamic events of heart conduction in a few cartoon panels or "snapshots." However, the sketches here give you an idea of what is happening in the heart as an ECG is recorded. Note that depolarization triggers contraction in the affected muscle tissue. Thus cardiac muscle contraction occurs *after* depolarization begins.

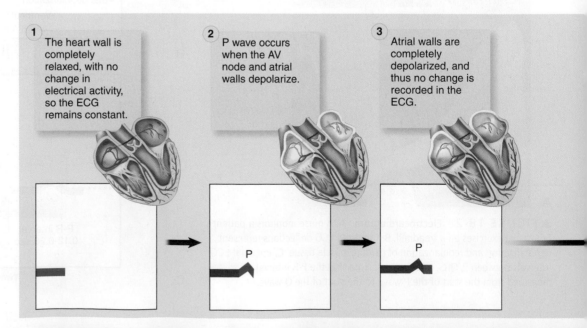

1 The heart wall is completely relaxed, with no change in electrical activity, so the ECG remains constant.

2 P wave occurs when the AV node and atrial walls depolarize.

3 Atrial walls are completely depolarized, and thus no change is recorded in the ECG.

P

P

electrocardiogram is **ECG** when it is written and **EKG** when it is spoken.) The ECG is not a record of the heart's contractions but of the *electrical events* that precede the contractions.

You can see changes in voltage (which represent changes in the heart's electrical activity) as deflections of the line in a recording voltmeter (Figure 18-3). We've simplified the situation somewhat by showing only a single cardiac muscle fiber with the two electrodes of a recording voltmeter nearby.

Before the action potential reaches either electrode, there is no difference in charge between them. Thus, no change in voltage is recorded on the voltmeter graph (Figure 18-3, step 1).

As an action potential reaches the first electrode, the external surface of the sarcolemma becomes relatively negative. This results in an *upward deflection* of the pen of the recording chart (Figure 18-3, step 2). (Note that the voltmeter records the *difference* in charge between the two electrodes.) When the action potential also reaches the second electrode, the pen returns to the zero baseline. This happens because there is no difference in charge once again between the two electrodes (Figure 18-3, step 3).

As the end of the action potential passes the first electrode, the sarcolemma is again relatively positive on its outer surface. This causes the pen to again deflect away from the baseline. This time, however, because the direction of the negative and positive electrodes is *reversed*, the pen deflects *downward* (Figure 18-3, step 4). After the end of the action potential also passes the second electrode, the pen again returns to the zero baseline (Figure 18-3, step 5).

In short, when ECG electrodes are set up this way, *depolarization* of the cardiac muscle causes a deflection *upward; repolarization* causes a deflection *downward*. However, depending on the location of electrodes relative to heart muscle, the direction of ECG waves can vary. It is this activity that is detected in an electrocardiogram. Many ECG setups in use today show these voltage fluctuations in real time on a video monitor and at the same time records them in a computer file and on a paper chart (like the one represented in Figure 18-3).

ECG Waves

We will limit our discussion of electrocardiography to normal ECG *deflection waves* and the *ECG intervals* between them. As we do so, please refer to Figure 18-2 and the sequence of events presented in Figure 18-4.

As you can see in Figures 18-2 and 18-4, the normal ECG is composed of deflection waves called the *P wave, QRS complex,* and *T wave.* (The letters do not represent any words; they were simply chosen as an arbitrary sequence of letters!)

The **P wave** represents depolarization of the atria. It measures the deflection caused by the passage of an electrical impulse from the SA node through the musculature of the atria.

The **QRS complex** represents the depolarization of the ventricles. This is a multi-step process. First there is the depolarization of the interventricular septum. This is followed by the spread of depolarization by the subendocardial branches (Purkinje fibers) through the lateral ventricular walls. All three deflections, then—Q, R, and S together—represent the entire process of depolarization of the ventricles.

There is a slight catch. At the same time that the ventricles are depolarizing, the atria are repolarizing. You might expect to see some indication of this in the ECG tracing. However, the massive ventricular depolarization that is occurring at the same time literally "drowns out" the relatively small voltage fluctuation produced by atrial repolarization. For this reason, the QRS complex represents the *net deflection due to both ventricular depolarization and atrial repolarization.*

The **T wave** reflects repolarization of the ventricles. Depending on exactly where the electrodes are placed relative to the direction of electrical activity in the ventricular myocardium, this repolarization wave may deflect the ECG trace in the same direction as in depolarization.

ECG Intervals

The principal ECG intervals between P, QRS, and T waves are illustrated for you in Figure 18-2, *C.* Measurement of these

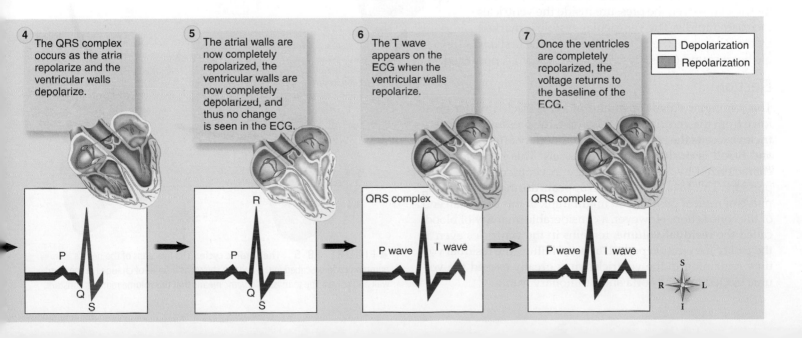

4 The QRS complex occurs as the atria repolarize and the ventricular walls depolarize.

5 The atrial walls are now completely repolarized, the ventricular walls are now completely depolarized, and thus no change is seen in the ECG.

6 The T wave appears on the ECG when the ventricular walls repolarize.

7 Once the ventricles are completely repolarized, the voltage returns to the baseline of the ECG.

☐ Depolarization
■ Repolarization

intervals can provide important information concerning the rate of conduction of an action potential through the heart.

Cardiac Cycle

The **cardiac cycle** describes a complete heartbeat or a single pumping cycle. It consists of contraction **(systole)** and relaxation **(diastole)** of both atria and both ventricles. First, the two atria contract simultaneously. Then, as the atria relax, the two ventricles contract and relax. As a result, all the chambers of the heart do *not* contract as a single unit. This alternation of contraction and relaxation imparts a pumping rhythm to the heart.

During the following discussion, please refer often to Figure 18-5, which shows the major phases of the cardiac cycle.

Atrial Systole

The contracting myocardium of the atria forces the blood into the ventricles below. The atrioventricular (cuspid) valves are opened during this phase to allow for the passage of blood into the relaxed ventricles. The semilunar valves are closed, preventing blood return from the aorta or the pulmonary trunk. This part of the cycle is represented by the P wave on an ECG. Passage of the electrical wave of depolarization is then followed almost immediately by actual contraction of the atrial musculature.

Ventricular Contraction

During the brief period of ventricular contraction (between the start of ventricular systole and the opening of the semilunar valves), the volume of blood in the ventricles remains constant (isovolumetric). However, the pressure inside the ventricles increases rapidly. The ventricular systole is marked by the R wave on the ECG. At this time, the first audible heart sound (often described as a "lubb") is produced.

Ejection

The semilunar valves open and blood is ejected under great force from the ventricles. At this point, the pressure in the ventricles exceeds the pressure in the pulmonary artery and aorta, and blood is pushed into these vessels. This *rapid ejection* is characterized by a marked increase in ventricular and aortic pressure. The T wave of the ECG appears during the later, long phase of *reduced ejection*. You might think of this as the tail end of the contraction. However, a considerable quantity of blood, called the **residual volume,** remains in the ventricles even at the end of the ejection period. In heart failure, the residual volume remaining in the ventricles may greatly exceed the volume ejected into the aorta and pulmonary trunk.

Ventricular Relaxation

Diastole—the relaxation of the ventricles—begins with this period between closing of the semilunar valves and the opening of the atrioventricular valves. At the end of ventricular contraction, the semilunar valves close so that blood cannot re-enter back into the ventricles from the great vessels. The second heart sound (described as a "dupp") is now heard.

The atrioventricular valves do not open until the pressure in the atrial chambers exceeds the pressure in the relaxed ventricles. The result is a dramatic fall in intraventricular pressure—*but no change in blood volume*. This is an isovolumetric phase—both sets of valves are closed.

Passive Ventricular Filling

The continuing return of venous blood from the venae cavae and the pulmonary veins increases pressure within both atria until the atrioventricular valves are forced open. When this happens, blood rushes into the relaxed ventricles. This rapid influx of blood lasts only about 0.1 second but results in a

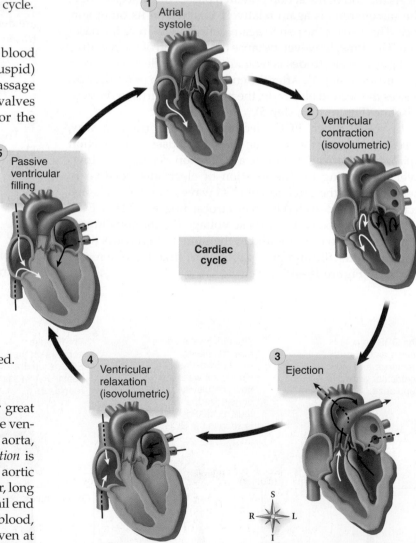

▲ **FIGURE 18-5** **The cardiac cycle.** The five steps of the heart's pumping cycle described in the text are shown as a series of changes in the heart wall and valves. The term *isovolumetric* means that the volume remains constant.

dramatic increase in the volume of blood in the ventricle. The abrupt inflow of blood that occurs immediately after opening of the AV valves is followed by a slow and continuous flow of venous blood into the atria. This blood then flows through the open AV valves and into the ventricles, slowly building up the blood pressure and volume within the ventricles.

Heart Sounds

The first "lubb" or systolic sound is caused largely by the contraction of the ventricles and by the closing atrioventricular valves. It is longer and lower than the second or diastolic sound, which is short and sharp. Vibrations of the closing semilunar valves cause this second "dupp" sound.

Heart sounds have clinical significance and can provide information about the valves. Any variation from normal "lubb-dupp" sounds may suggest imperfect valve function. A **heart murmur** is one commonly heard type of abnormal heart sound. Sometimes it is described as a "swishing" sound. This may indicate an incomplete closing of the valves or *stenosis* (constriction or narrowing) of the valves.

QUICK ✓ CHECK

6. Briefly outline the steps of the cardiac cycle.
7. When are heart sounds of medical importance?

PRIMARY PRINCIPLE OF CIRCULATION

In order for blood to flow within the circulatory system, there must be a *gradient* from high pressure to low pressure (Figure 18-6). This is sometimes called the **primary principle of circulation.** For example, blood enters an arteriole at 85 mm Hg and leaves at 35 mm Hg. The blood thus moves *down* a pressure gradient (from 85 mm Hg to 35 mm Hg) as it flows through the arteriole. The pressure difference *drives* the flow of blood.

ARTERIAL BLOOD PRESSURE

High pressure in the arteries must be maintained to keep blood flowing through the circulatory system. The volume of blood within the arteries largely determines arterial blood pressure. Thus, an increase in arterial blood volume tends to increase arterial pressure. Likewise, a decrease in arterial volume tends to decrease arterial pressure. However, many factors determine arterial pressure through their influence on arterial volume. Two of the most important, *cardiac output* and *peripheral resistance*, are directly proportional to blood volume, as we will see below.

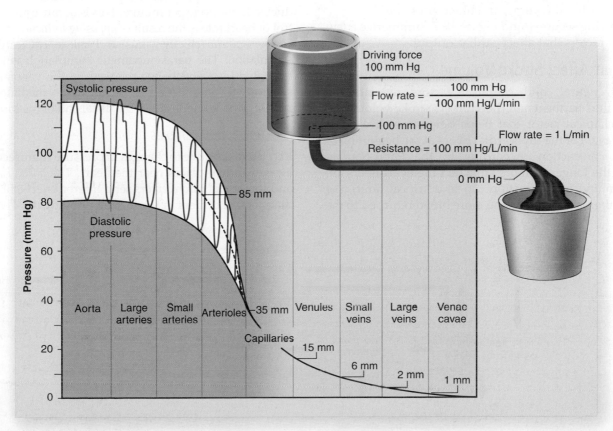

▲ **FIGURE 18-6** **The primary principle of circulation.** Fluid always travels from an area of high pressure to an area of low pressure. Water flows from an area of high pressure in the tank (100 mm Hg) toward the area of low pressure above the bucket (0 mm Hg). Blood tends to move from an area of high average pressure at the beginning of the aorta (100 mm Hg) toward the area of lowest pressure at the end of the venae cavae (0 mm Hg). Blood flow between any two points in the circulatory system can always be predicted by the pressure gradient.

Cardiac Output

Cardiac output (CO) is the amount of blood that flows out of a ventricle per unit of time. For example, the resting cardiac output from the left ventricle into the systemic arteries is about 5,000 ml/min. The cardiac output influences the flow rate to the various organs of the body. For the sake of our discussion, we will focus on the cardiac output from the left ventricle into the systemic loop.

Cardiac output is determined by the volume of blood pumped out of a ventricle by each beat (**stroke volume** or **SV**) and by **heart rate (HR).** Because contraction of the heart is called *systole,* sometimes the volume of blood pumped by one contraction is called the *systolic discharge.* Stroke volume, or volume pumped per heartbeat, is one of two major factors that determine cardiac output. CO can be determined by the following equation:

SV (volume/beat) × HR (beats/min) = CO (volume/min)

Thus the greater the stroke volume, the greater the CO (but only if the heart rate remains constant). Anything that changes the rate of the heartbeat or its stroke volume tends to change CO. This means that anything that makes the heart beat *faster* or *stronger* (increases its stroke volume) *tends* to increase CO and therefore arterial blood volume and pressure. Conversely, anything that causes the heart to beat more *slowly* or more *weakly* (decreases its stroke volume) *tends* to decrease CO, arterial volume, and blood pressure.

The following sections, and Figure 18-7, summarize a few of the major factors that affect cardiac output.

Factors that Affect Stroke Volume

Mechanical, neural, and chemical factors regulate the strength of the heartbeat and therefore its stroke volume.

One mechanical factor that helps determine stroke volume is the *length* of the myocardial fibers at the beginning of ventricular contraction. According to the *Frank-Starling mechanism,* the longer or more stretched the heart fibers are at the beginning of contraction (up to a critical limit), the stronger is their contraction. The more blood returned to the heart per minute, the more stretched will be their fibers. This will lead to stronger contractions and a larger volume of blood ejected with each contraction.

However, if too much blood stretches the heart beyond a certain critical limit, the myocardial muscle seems to lose its elasticity. As a result the heart contracts less vigorously.

We can now see that the heart pumps out what it receives. That is, within certain limits, the strength of myocardial contraction matches the pumping load. This is contrary to most mechanical pumps that do *not* adjust themselves to their input with every stroke. In the case of a human heart, under ordinary conditions, it automatically adjusts output (stroke volume) to input (venous return to the heart).

Other factors that influence stroke volume are neural and endocrine chemical factors. Norepinephrine (released by sympathetic fibers in the cardiac nerve) and epinephrine (released into the blood by the adrenal medulla) can both increase the strength of contraction, or **contractility,** of the myocardium. This increased contractility of the heart muscle forces more blood volume out of the heart per cardiac stroke, thus increasing the stroke volume.

Factors that Affect Heart Rate

Although the sinoatrial node normally initiates each heartbeat, the heart rate it sets can be altered. In fact, various factors can and do change the rate of the heartbeat. One major modifier is the ratio of sympathetic and parasympathetic impulses conducted to the node per minute. This is because autonomic control of heart rate is the result of *opposing influences* between the *parasympathetic* (chiefly vagus) and *sympathetic* (cardiac nerve) stimulation. The parasympathetic stimulation is *inhibitory*—mediated by acetylcholine released by the vagus nerve. The sympathetic stimulation is *stimulatory*—mediated by the release of norepinephrine at the distal end of the cardiac nerve.

Cardiac Pressoreflexes

Receptors sensitive to changes in pressure **(baroreceptors)** are located in two places near the heart. The *aortic baroreceptors* and the *carotid baroreceptors* send afferent nerve fibers to cardiac control centers in the medulla oblongata. These stretch

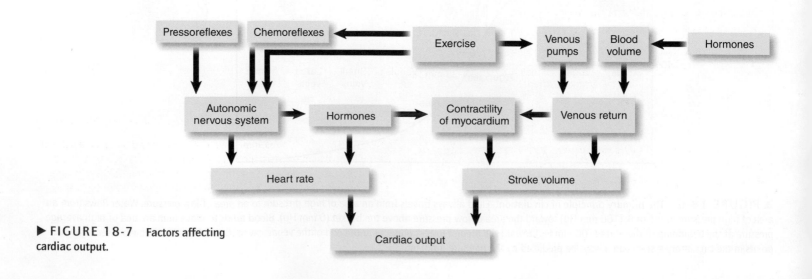

▶ FIGURE 18-7 **Factors affecting cardiac output.**

receptors, located in the aorta and carotid sinus, respectively, are vitally important to controlling heart rate. Baroreceptors operate with integrators in the cardiac control centers in negative feedback loops called *pressoreflexes.* These oppose changes in pressure by adjusting heart rate.

Other Reflexes that Influence Heart Rate

Reflexes involving important factors such as emotions, exercise, hormones, blood temperature, pain, and stimulation of various *exteroceptors* also influence heart rate. Anxiety, fear, and anger often make the heart beat faster to the point where it seems to pound within the chest. Interestingly, grief tends to slow heart rate. Emotions produce changes in the heart rate through the influence of impulses from the "higher centers" (in the cerebrum and hypothalamus) on the cardiac control centers in the brainstem.

As you know, heart rate accelerates during intense exercise. Amazing as it may sound, the mechanism for this acceleration is still not completely understood. However, it certainly involves mechanisms that influence the brainstem's cardiac control centers. The hormone epinephrine is also believed to be an important cardiac accelerator.

Increased blood temperature or stimulation of skin heat receptors also tends to increase the heart rate. Decreased blood temperature or stimulation of skin cold receptors tends to slow it. Sudden, intense stimulation of pain receptors in visceral structures such as the gallbladder, ureters, or intestines can result in slowing of the heart rate to such an extent that fainting can result.

Finally, reflexive increases in heart rate often result from an increase in sympathetic stimulation of the heart. Sympathetic impulses originate in the cardiac control center of the medulla and reach the heart by way of sympathetic fibers. Norepinephrine released as a result of sympathetic stimulation increases heart rate and the strength of cardiac muscle contraction.

QUICK CHECK

8. Discuss the role of pressure gradient and the flow of blood.

9. What factors determine cardiac output?

10. List some factors that affect stroke volume.

11. List some factors that affect heart rate.

12. How can emotions affect heart rate and stroke volume?

Peripheral Resistance

How Resistance Influences Blood Pressure

Peripheral resistance—the resistance to blood flow caused by the *friction* of blood striking the walls of the vessels— helps maintain arterial blood pressure. The friction that produces peripheral resistance develops partly because of the *viscosity* (resistance to flow) of blood and partly from the small diameter of arterioles and capillaries. The resistance created by arterioles, in particular, accounts for almost one

half of the total resistance in the systemic circulation. An increase in the number of blood cells can increase viscosity of blood and therefore resistance. Under certain abnormal conditions, such as severe anemia or hemorrhage, a decrease in blood viscosity may lower peripheral resistance and arterial pressure dangerously, even to the point of circulatory failure.

The muscular layer of the arterioles allows them to constrict or dilate and thus change the amount of resistance to blood flow. This muscular mechanism in the vessels is called the **vasomotor mechanism.** Reducing the vessel diameter by increasing the contraction of the muscular layer is called **vasoconstriction.** This process increases resistance to blood flow, and thus blood flow into tissues also decreases. **Vasodilation,** the relaxation of vascular muscles, decreases resistance to blood flow. As a result, blood flow increases to the tissues. As Figure 18-8 shows, even small changes in

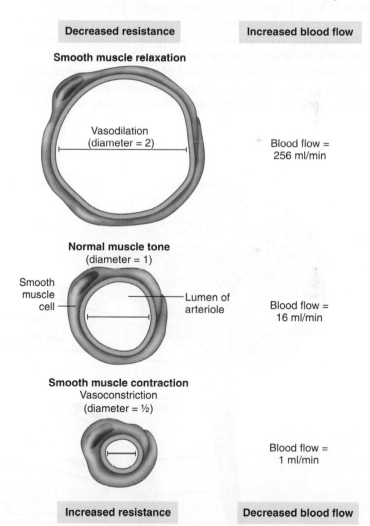

▲ FIGURE 18-8   Vessel diameter. The effect of changing diameter of arterioles on peripheral resistance and blood flow. The cross sections show vasodilation *(top),* normal diameter *(center),* and vasoconstriction *(bottom)* as tension in smooth muscle fibers changes. Note that reducing the diameter of a blood vessel to one half of normal does not reduce blood flow by half—it reduces blood flow to 1/16 of normal. Likewise, doubling the vessel diameter does not double the blood flow—it increases blood flow to 16 times over that of normal flow!

diameter can cause relatively large changes in resistance, and therefore large changes in local blood flow. As you can see, the vasomotor mechanism is well suited for quickly and dramatically changing blood flow under a variety of conditions.

Vasomotor Control Mechanism

Our blood pressure and even the amount of blood distributed to different organs can be influenced by the *diameter* of arterioles. The *control center* for this complex system lies in the **vasomotor center** of the medulla. When stimulated, the control system sends out impulses via sympathetic fibers, causing the constriction of smooth muscles surrounding some vessels. These include both *resistance vessels* (such as arterioles) and *capacitance vessels* (such as the venous networks of the spleen and liver and the veins surrounding blood reservoirs).

A relatively large volume of blood resides in the systemic veins and venules of a resting adult, compared to the volume of blood residing in other vessels of the body. This is why we call systemic veins and venules "reservoirs." Blood can be quickly moved out of blood reservoirs and shunted to arteries when skeletal muscles demand.

A sudden increase in arterial blood pressure stimulates aortic and carotid baroreceptors—the same ones that initiate cardiac reflexes. Not only does this stimulate the cardiac control center to *reduce* heart rate, but it also inhibits the vasomotor center. The parasympathetic fibers fire more impulses per second to the heart and blood vessels. As a result, the heartbeat slows and arterioles and venules of the blood reservoirs dilate. The nervous pathways involved in this mechanism are outlined for you in Figure 18-9.

You can readily see how interconnected these feedback systems are. A decrease in arterial pressure causes the aortic and carotid baroreceptors to send more impulses to the medulla's vasoconstrictor centers. The centers are stimulated and send out impulses via the sympathetic fibers to stimulate vascular smooth muscle and cause vasoconstriction. This squeezes more blood out of the blood reservoirs, increasing the amount of venous blood return to the heart. Ultimately, this "extra blood" from the reservoirs flows to active skeletal muscles and the heart because their arterioles become dilated (by another local mechanism). These mechanisms are especially important during high metabolic activity such as during strenuous exercise (Box 18-1).

There are also similar *chemoreceptor reflexes* originating in the aorta and carotid arteries that function to alter heart rate and blood flow when excess blood carbon dioxide and low oxygen content endanger the stability of the internal environment.

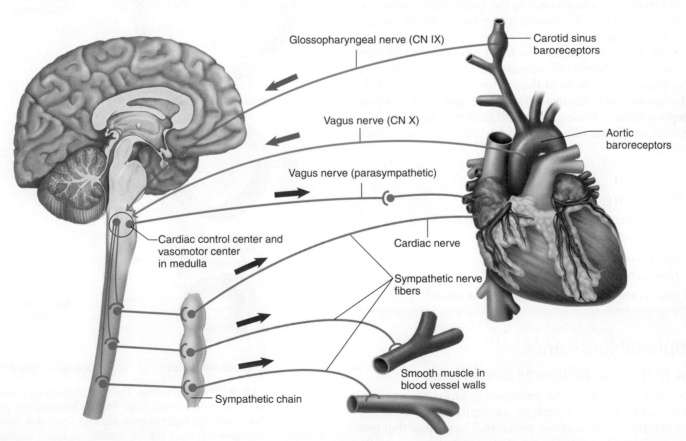

▲ **FIGURE 18-9   Vasomotor pressoreflexes.** Carotid sinus and aortic baroreceptors detect changes in blood pressure and feed the information back to the cardiac control center and the vasomotor center in the medulla. In response, these control centers alter the ratio between sympathetic and parasympathetic output. If the pressure is too high, a dominance of parasympathetic impulses will reduce it by slowing heart rate, reducing stroke volume, and dilating blood "reservoir" vessels. If the pressure is too low, a dominance of sympathetic impulses will increase it by increasing heart rate and stroke volume and constricting reservoir vessels.

| BOX 18-1 | Sports & Fitness |

The Cardiovascular System and Exercise

Exercise produces short-term and long-term changes in the cardiovascular system. Short-term changes involve negative feedback mechanisms that maintain set-point levels of blood oxygen, glucose, and other physiological variables. Because moderate to strenuous use of skeletal muscles greatly increases the body's overall rate of metabolism, oxygen and glucose are used up at a faster rate. This requires an increase in transport of oxygen and glucose by the cardiovascular system to maintain normal set-point levels of these substances. One response by the cardiovascular system is to increase the cardiac output (CO) from 5 to 6 L/min at rest to as much as 30 to 40 L/min during strenuous exercise. This represents a fivefold to eightfold increase in the blood output of the heart! Such an increase is accomplished by a reflexive increase in heart rate coupled with an increase in stroke volume. Exercise can also trigger a reflexive change in local distribution of blood flow to various tissues, as the figure shows. This results in a larger share of blood flow going to the skeletal muscles than to some other tissues. We've summarized a number of central and local regulatory effects that operate during exercise in the figure.

Long-term changes in the cardiovascular system come only when moderate to strenuous exercise occurs regularly over a long period.

Evidence indicates that 20 to 30 minutes of moderate aerobic exercise (such as cycling or running) three times per week produce profound, healthful changes in the cardiovascular system. Among these long-term cardiovascular changes are (1) an increase in the mass and contractility of the myocardial tissue, (2) an increase in the number of capillaries in the myocardium, (3) a lower resting heart rate, and (4) a decrease in peripheral resistance during rest. The lower resting heart rate coupled with an increase in stroke volume and increased myocardial mass and contractility produce a greater range of CO. Coupled with the other listed effects, a greater maximum CO is achieved. Perhaps you can see why exercise is also known to decrease the risk of various cardiovascular disorders, including arteriosclerosis, hypertension, and heart failure.

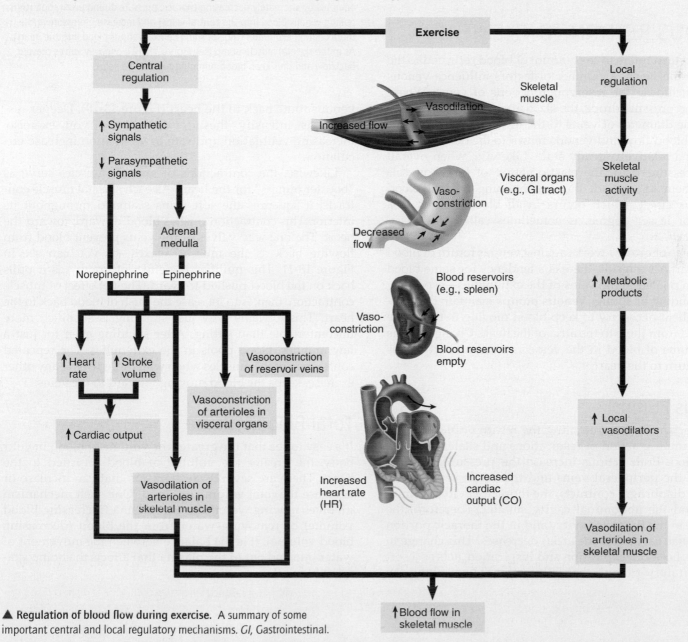

▲ **Regulation of blood flow during exercise.** A summary of some important central and local regulatory mechanisms. *GI, Gastrointestinal.*

Local Control of Arterioles

There are several kinds of *local mechanisms* that produce vasodilation in localized areas. They are not completely understood but we know that they function in times of increased tissue activity. They may account for increased blood flow into skeletal muscles during exercise, and also operate in oxygen-deprived tissues—serving as homeostatic mechanisms that restore normal blood flow.

VENOUS RETURN TO THE HEART

Venous return refers to the amount of blood returned to the heart via the veins. A number of factors influence venous return, including the reservoir functions of veins. Whenever blood pressure drops, the elasticity of the venous walls adapts the diameter of veins to the lower pressure. In this manner, blood flow and venous return to the heart is maintained at a relatively steady state. Likewise, when overall blood pressure rises, the elastic nature of blood vessels allows them to expand, thus maintaining normal blood flow. This effect, which occurs in all blood vessels to a greater or lesser degree, is sometimes called the *stress-relaxation effect*.

The force of gravity works against venous return of blood to the heart. As a result, there is a tendency for some blood to remain pooled in the veins of the extremities—especially when standing or sitting. **Venous pumps** maintain the pressure gradient necessary to keep blood moving to the venae cavae and from there to the atria of the heart. Changes in the total volume of blood in the vessels can also alter venous blood return to the heart.

Venous Pumps

An important factor promoting the return of blood is the blood-pumping action of respirations and skeletal muscle contraction. Both actions increase the pressure gradient between the peripheral veins and the venae cavae. Each time the diaphragm contracts, the thoracic cavity becomes larger and the abdominal cavity smaller. Therefore, the pressure in the thoracic cavity (and in the thoracic portion of the venae cavae and the atria) decreases. This change in pressure, between expiration and inspiration, acts as a sort of "respiratory pump" that helps move blood along the

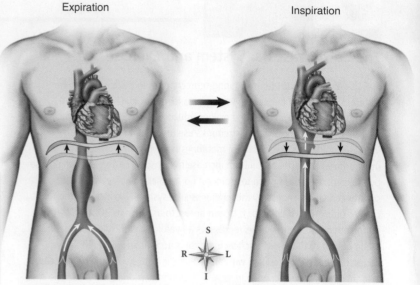

▲ FIGURE 18-10 Respiratory pump. This venous pumping mechanism operates by alternately decreasing thoracic pressure during inspiration (thus pulling venous blood into the central veins) and increasing pressure in the thorax during expiration (thus pushing central venous blood into the heart). As in the skeletal muscle pump (see Figure 18-11), one-way valves prevent backflow and thus keep blood pumping toward the heart.

venous route back to the heart (Figure 18-10). Deeper respirations intensify these effects. When you exercise, increasing ventilation and rate of ventilation increase circulation.

Likewise, the contractions of skeletal muscles serve as "booster pumps" for the heart. As each skeletal muscle contracts, it squeezes the soft veins scattered throughout its interior. This contraction pushes blood upward, toward the heart. The one-way valves of the veins prevent blood from flowing back as the muscle relaxes—as you can see in Figure 18-11. This nullifies the effect of gravity as it pulls back on the blood pushed forward. The net effect of muscle contraction, then, is to increase the push of blood back to the heart. This explains why just standing is so much more uncomfortable than sitting. After standing even for just a few minutes, blood pools in the extremities. The repeated contraction of the muscles when walking, or doing any other exercise, keeps the blood moving forward in the veins.

Total Blood Volume

It's easy to see that the greater the volume of blood in your body, the greater the volume of blood returned to the heart. There are several mechanisms that can increase or decrease the total volume of blood. One such mechanism involves moving water into the plasma (increasing blood volume) or removing water from the blood (decreasing blood volume). It is the balance between the movement of water into and out of the plasma that affects the homeostasis of blood flow.

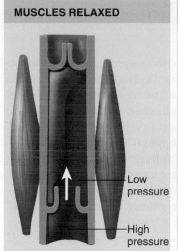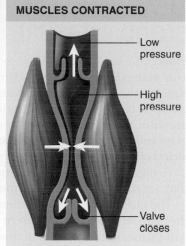

MUSCLES RELAXED

MUSCLES CONTRACTED

Low pressure

High pressure

Low pressure

High pressure

Valve closes

A   **B**

▲ **FIGURE 18-11** **Skeletal muscle pump.** This venous pumping mechanism operates by the alternate increase and decrease in peripheral venous pressure that normally occurs when the skeletal muscles are used for the activities of daily living. One-way valves aid pumping by preventing backflow of venous blood when pressure in a local area is low. **A,** Local high blood pressure pushes the flaps of the valve to the side of the vessel, allowing easy flow. **B,** When pressure below the valve drops, blood begins to flow backward but fills the "pockets" formed by the valve flaps, pushing the flaps together and thus blocking further backward flow.

Another endocrine reflex mechanism is the **renin-angiotensin-aldosterone system (RAAS).** The enzyme renin is released when blood pressure in the kidney is low. This triggers a series of events that leads to the secretion of aldosterone from the adrenal cortex. Aldosterone in turn promotes sodium retention by the kidneys, which stimulates the osmotic flow of water from kidney tubules back into the blood plasma. But this only occurs when ADH is present to allow the movement of water. Thus, low blood pressure increases the secretion of aldosterone. This in turn stimulates retention of water and an increase in blood volume back toward normal.

Yet another mechanism that can change blood plasma volume (and thus venous return of blood to the heart) is the **ANH mechanism.** ANH (atrial natriuretic hormone) is secreted by heart muscle cells located in the atrial wall. This takes place in response to overstretching of the atrial wall (such as when venous return to the heart is abnormally high). ANH adjusts venous return back down to its normal set point by promoting the loss of water from the plasma. As a result, there is a *decrease* in blood volume. ANH does this by increasing sodium loss from the blood into the urine, which causes water to follow by osmosis. The ANH mechanism opposes the effects of ADH and

Osmotic pressure tends to promote diffusion of fluid into the plasma. Increasing osmotic pressure of plasma, generated by increasing the number of large solute particles such as plasma proteins, can draw water into the plasma. Where arterioles join capillaries, there tends to be a net loss of blood volume to the surrounding interstitial fluid. However, this fluid is recaptured where the ends of capillaries join to create venules. As a result, the capillaries recover much of the fluid they have lost (Figure 18-12).

In actuality, there is a 10% loss of blood volume along the network of capillaries. However, this is recovered by the lymphatic system and returned to the venous blood before it returns to the heart. We will study this fluid recovery in greater detail in the following chapter.

Changes in Total Blood Volume

In Chapter 15, we studied the *endocrine reflexes* that help control water retention in the body. In particular, antidiuretic hormone (ADH) is released by the posterior pituitary and acts on the kidneys to reduce the amount of water loss. ADH does this by increasing the amount of water that the kidneys reabsorb from urine before it is excreted. The more ADH, the more water retention, and the greater the volume of the blood. Receptors in the body that detect the balance between water and solutes trigger the **ADH mechanism.**

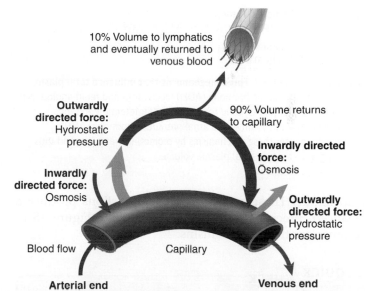

10% Volume to lymphatics and eventually returned to venous blood

90% Volume returns to capillary

Outwardly directed force: Hydrostatic pressure

Inwardly directed force: Osmosis

Inwardly directed force: Osmosis

Outwardly directed force: Hydrostatic pressure

Blood flow   Capillary

Arterial end   **Venous end**

▲ **FIGURE 18-12** **Starling's law of the capillaries.** At the arterial end of a capillary, the outward driving force of blood pressure (hydrostatic pressure of blood) is larger than the inwardly directed force of osmosis—thus fluid moves out of the vessel. At the venous end of a capillary, the inward driving force of osmosis is greater than the outwardly directed force of hydrostatic pressure—thus fluid enters the vessel. About 90% of the fluid leaving the capillary at the arterial end is recovered by the blood before it leaves the venous end. The remaining 10% is recovered by the venous blood eventually, by way of the lymphatic vessels (see Chapter 19).

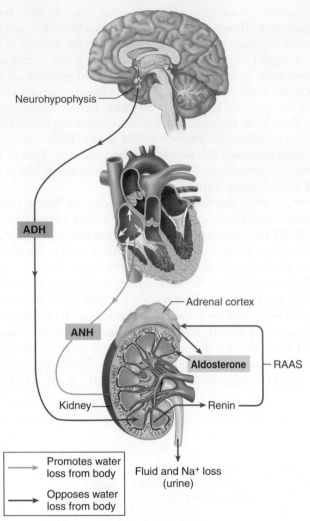

Neurohypophysis

ADH

Adrenal cortex

ANH

Aldosterone — RAAS

Kidney — Renin

→ Promotes water loss from body

➔ Opposes water loss from body

Fluid and Na⁺ loss (urine)

▲ **FIGURE 18-13   Three mechanisms that influence total plasma volume.** The antidiuretic hormone (ADH) mechanism and renin-angiotensin-aldosterone system (RAAS) tend to increase water retention and thus increase total plasma volume. The atrial natriuretic hormone (ANH) mechanism opposes these mechanisms by promoting water loss and thus promoting a decrease in total plasma volume.

aldosterone. These actions and counteractions produce a balanced, precise control of blood volume. Figure 18-13 summarizes these mechanisms.

QUICK ✓ CHECK

17. What is meant by venous return?

18. In general, how do venous pumps operate to promote the return of venous blood to the heart?

19. Discuss the several venous pumping mechanisms (the respiratory pump and the contraction/relaxation of skeletal muscles).

20. How does capillary exchange affect total blood volume?

21. Discuss several ways in which endocrine reflexes affect changes in total blood volume.

MEASURING ARTERIAL BLOOD PRESSURE

Blood pressure is measured with a **sphygmomanometer.** This makes it possible to measure the amount of air pressure equal to the blood pressure in an artery. The measurement is made in terms of how many millimeters (mm) high the air pressure raises a column of mercury (Hg) in an enclosed glass tube (Figure 18-14, *A*). The result is expressed as *millimeters of mercury (mm Hg).* Most sphygmomanometers today have mercury-free mechanical or electronic pressure sensors that are still calibrated to the mercury scale.

The cuff of the sphygmomanometer is wrapped around the arm over the brachial artery. Air is then pumped into the cuff by means of the bulb. In this way, air pressure in the cuff is exerted against the outside of the artery. Air is pumped in until the air pressure in the cuff *exceeds* the pressure within the artery. This compression of the artery by the air pressure in the cuff stops the flow of blood through the artery. This can be determined by a stethoscope placed over the brachial artery at the bend of the elbow, along the inner margin of the biceps muscle. No sounds can be heard.

Then the air is slowly released from the cuff as the examiner listens carefully for the first return of blood. This is the point at which blood flow in the artery *first* overcomes the pressure from the external cuff. The sharp "tapping" sound heard becomes increasingly louder as the pressure in the cuff is lowered even further. Finally the sounds become muffled and disappear altogether. Health professionals train themselves to hear these different sounds (Korotkoff sounds) and read the column of mercury at the same time (Figure 18-14, *B*).

The first tapping sounds when blood begins to return are a measure of the **systolic blood pressure.** This is the force with which the blood is pushing against the artery walls when the ventricles *are contracting.* The lowest point at which the sounds finally disappear is approximately equal to the **diastolic blood pressure.** This is the force of the blood when the ventricles *are relaxing.* Systolic pressure gives us valuable information about the *force* of the left ventricular contraction. Diastolic pressure gives valuable information about the *resistance* of the blood vessels.

Blood in the arteries of an adult with a blood pressure reading *at the high end* of the normal range exerts a pressure equal to that required to raise a column of mercury about 120 mm high in the glass tube during systole of the ventricles and 80 mm high in a glass tube during diastole. This is expressed as a *systolic/diastolic pressure* of "120 over 80" or 120/80. The first number indicates the systolic pressure and the second the diastolic pressure. The difference between systole and diastole is called **pulse pressure.** The pulse pressure typically increases in arteriosclerosis. This is because systolic pressure increases more than diastolic pressure.

As you've seen, blood exerts a comparatively high pressure in arteries but a very low pressure in veins. For this

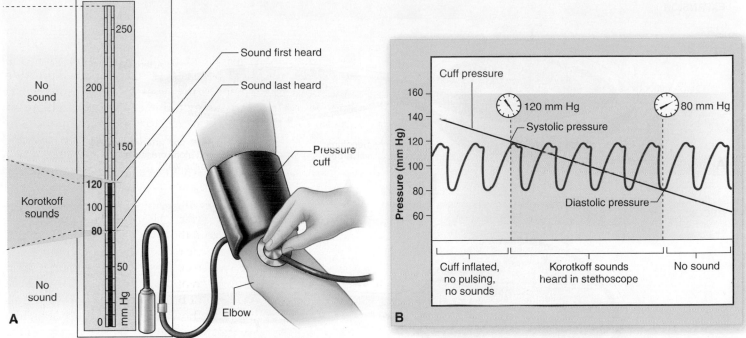

▲ FIGURE 18-14   Sphygmomanometer. This mercury-filled pressure sensor is used in clinical and research settings to quickly and accurately measure arterial blood pressure. **A,** The pressure cuff is pumped with air until the pressure inside the cuff exceeds the expected systolic pressure of the large arteries of the arm. No sound caused by pulsing of blood in the arteries can then be heard with a stethoscope. As the pressure inside the cuff is slowly released from a valve, the air pressure equals the maximum pressure of the pulse waves in the artery—thus the pulsing sounds can then be heard. **B,** The sounds of pulsing (Korotkoff sounds) continue as long as the cuff pressure is equal to pressures of the pulse wave. The sounds disappear at the point that cuff pressure drops below the minimum pulse pressure in the arteries.

reason, it gushes forth with considerable force from a cut artery, but seeps in a slow steady stream from a vein. As the ventricles contract, blood spurts forth forcefully, but when the ventricles relax, the flow ebbs to almost nothing because of the fall in pressure. In contrast, a steady pressure exists in the capillaries and veins.

PULSE AND PULSE WAVE

Pulse Wave

Pulse is defined as the alternating expansion and recoil of an artery. Two factors are responsible for a pulse that you can actually feel: (1) pulses of blood injected from the heart's left ventricle into the aorta (see Figure 18-6, p. 411), and (2) the elasticity of the arterial wall. If blood flowed steadily from the heart into the aorta, you would feel no pulse because the flow would be constant. Likewise, if our vessels were made from some inelastic material such as copper pipe, there might still be an alternating raising and lowering of pressure within the vessels, but the walls could not expand and recoil. Thus there would be no palpable pulse.

Each ventricular systole starts a new pulse that proceeds as a wave of expansion throughout the arteries. This is called the **pulse wave.** It gradually dissipates as it travels through the circulatory system, disappearing entirely in the capillaries. The pulse wave felt at the common carotid artery in the neck is large and powerful. Figure 18-15 shows that the carotid pulse wave begins during ventricular systole. Note that the closure of the aortic valve produces a detectable notch (a *dicrotic notch*) in the pulse wave. However, the pulse felt in the radial artery at the wrist does *not* coincide with the contraction of the ventricles. Instead, it follows the carotid pulse but with a detectable time lag because of its distance from the heart. The pulse reveals important information about the cardiovascular system, heart action, blood vessels, and circulation.

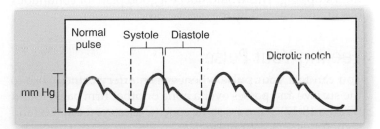

▲ FIGURE 18-15   Normal carotid pulse wave. This series of pulse waves shows rhythmic increases and decreases in pressure as measured at the common carotid artery in the neck. The dicrotic notch represents the pressure fluctuation generated by closure of the aortic valve.

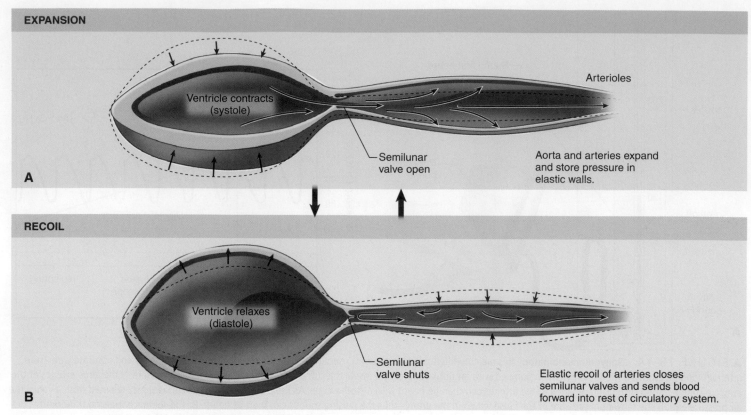

EXPANSION

Ventricle contracts (systole)

Arterioles

Semilunar valve open

Aorta and arteries expand and store pressure in elastic walls.

A

RECOIL

Ventricle relaxes (diastole)

Semilunar valve shuts

Elastic recoil of arteries closes semilunar valves and sends blood forward into rest of circulatory system.

B

▲ **FIGURE 18-16** **Functional role of the pulse wave.** The arterial pulse conserves energy by absorbing and storing force from the ventricular contraction. Force is stored by the elastic expansion of the arterial wall. The energy is used to maintain continued blood flow during ventricular relaxation by elastic recoil of the arterial wall—thus producing enough arterial pressure to keep blood flowing.

The pulse wave actually conserves energy produced by the pumping action of the heart (Figure 18-16). When the pressure of blood is ejected from the ventricle into the aorta, the walls of the aorta store part of this effort as potential energy, like a stretched rubber band. During ventricular diastole, the elastic nature of the aortic wall allows it to recoil. This recoil exerts pressure on the blood and thus keeps it moving. If the wall of the aorta were inelastic, it would *not* alternately expand and recoil. Thus, the blood would not move continuously. Instead, you'd see arterial blood as alternating as spurting then stopping, spurting then stopping. This would not be a good plan for continuous blood flow needed by the entire body!

Feeling Your Pulse

You can feel your pulse whenever an artery comes close to the surface and passes over a bone or other firm background (Figure 18-17). Here are some specific areas you can feel (and sometimes see) your pulse.

Radial artery: at the anterior, lateral surface of your wrist

Temporal artery: in front of your ear and above and to the outer side of your eye

Common carotid artery: along the anterior edge of the sternocleidomastoid muscle

Brachial artery: at the bend of your elbow, along the inner margin of the biceps

Posterior tibial artery: behind the medial malleolus (inner "ankle bone")

Femoral artery: in the middle of groin, where artery passes over pelvic bone; pulse can also be felt there

Dorsalis pedis artery: on the dorsal (upper) surface of your foot

QUICK CHECK

22. How is blood pressure measured?
23. What is the difference between systolic blood pressure and diastolic blood pressure?
24. What is the pulse wave? Where can it be measured?

Cycle of LIFE ⟳

As we've seen in the previous chapter, enormous changes occur in our circulatory system immediately after birth. Of course, there are physiological changes as well. For example, changes at the time of birth adapt the circulatory system from an aquatic environment to a terrestrial environment. As a result, changes in the blood pressure gradients take place that alter blood flow throughout much of our bodies.

ing heart rate for adults is about 72 beats per minute, but in older adults, resting heart rates range from 40 to 100 beats per minute. These ranges vary according to body weight and the degree to which a person is in good cardiovascular shape. A young cross-country runner may have a resting heart rate in the mid-40s. This rate may not change much as the person ages, depending on his or her aerobic fitness. ●

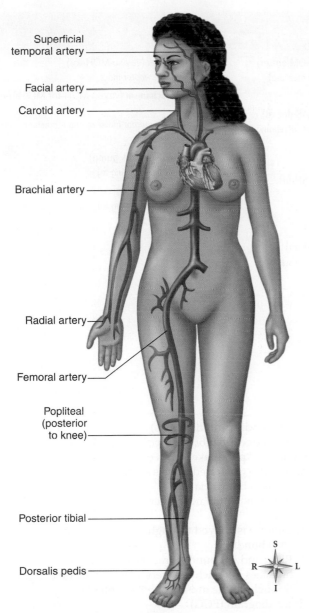

The BIG Picture

One of the basic and most important concepts of homeostasis is that renewable fluid comprises our internal environment. If we were not able to maintain the chemical nature and other physical characteristics of our internal fluid environment, we could not survive. To maintain this constancy, we must be able to shift and exchange nutrients, gases, hormones, waste products, agents of immunity, and other materials such as solutes around the body. As certain materials are depleted in one tissue and new materials enter the internal environment in another, redistribution must occur and *continue to take place*. As we have seen, the circulatory is the system by which the constancy of the internal environment is maintained.

Recall also from our earlier study of the integumentary and muscular systems that shifting the flow of blood to or away from warm tissues is essential to maintaining homeostasis of the body temperature. And, as we will see in Chapter 23, the ability of our blood to increase or decrease blood pressure in the kidney has a great impact on that organ's vital function of filtering the internal blood fluid. Understanding how blood flows and some of the basic physical parameters behind that flow is required for us to understand the fluid dynamics of our body.

▲ FIGURE 18-17 **Pulse points.** Each pulse point is named after the artery with which it is associated. (Some arteries in the figure have been enlarged to clarify the location of pulse points.)

Likewise, there are degenerative changes in our circulatory system that take place as we age. The heart has a reduced ability to maintain cardiac output and the arteries are less able to withstand high blood pressure.

Changes in arterial blood pressure are easily measured. In a newborn, normal arterial blood pressure is only about 90/55 mm Hg—much lower than values approaching 120/80 seen in many healthy adults. In older adults, arterial blood pressures commonly reach 150/90, so you can imagine the wear and tear on the elasticity of the vessels because of this dramatically increased blood pressure.

Another commonly observed change as we age is in the heart rate. A newborn may have heart rates ranging from 120 to 170 beats per minute. The resting heart rate of a preschooler can range from 80 to 160 beats per minute. The typical rest-

MECHANISMS OF DISEASE

As you might expect, cardiovascular disease is a major health issue in North America—affecting millions across the continent. Sometimes, the heart loses its normal rhythm—a type of condition called *dysrhythmia*. Dysrhythmia, or damage to the myocardium caused by blocked coronary blood flow, can lead to partial or full *heart failure*. Mechanisms that cause a sudden drop in blood pressure can produce *circulatory shock*. And factors that cause blood pressure to increase may result in chronic *hypertension (HTN)* (high blood pressure).

evolve To understand the heart better by exploring what can go wrong, check out *Mechanisms of Disease: Physiology of the Cardiovascular System* online.

LANGUAGE OF SCIENCE AND MEDICINE *continued from page 405*

P wave
[named for letter of Roman alphabet]

QRS complex (Q R S KOM-pleks)
[named for letters of Roman alphabet]

renin-angiotensin-aldosterone system (RAAS) (REE-nin-an-jee-oh-TEN-sin-al-DAH-stair-ohn SIS-tem)
[*ren-* kidney, *-in* substance, *angio-* vessel, *-tens-* pressure or stretch, *-in* substance, *aldo-* aldehyde, *-stero-* solid or steroid derivative, *-one* chemical, *system* organized whole]

residual volume (ree-ZID-yoo-al)
[*residu-* remainder, *-al* relating to]

sinoatrial (SA) node (sye-no-AY-tree-al)
[*sin-* hollow (sinus), *-atri-* entrance courtyard, *-al* relating to, *nod-* knot]

sphygmomanometer (sfig-moh-mah-NOM-eh-ter)
[*sphygmo-* pulse, *-mano-* thin, *-meter* measure]

stroke volume (SV)

subendocardial branch (sub-en-doh-KAR-dee-al)
[*sub-* under, *-endo-* within, *-cardi-* heart, *-al* relating to]

systole (SIS-toh-lee)
[*systole* contraction]

systolic blood pressure (sis-TOL-ik PRESH-ur)
[*systole-* contraction, *-ic* relating to]

T wave
[named for letter of Roman alphabet]

vasoconstriction (vay-soh-kon-STRIK-shun)
[*vaso-* vessel, *-constrict-* draw tight, *-tion* state]

vasodilation (vay-soh-dye-LAY-shun)
[*vaso-* vessel, *-dilat-* widen, *-tion* state]

vasomotor center (vay-so-MOH-tor)
[*vaso-* vessel, *-motor* move]

vasomotor mechanism (vay-so-MOH-tor MEK-ah-nih-zem)
[*vaso-* vessel, *-motor* move, *mechan-* machine, *-ism* state]

venous pump (VEE-nus pump)
[*ven-* vein, *-ous* relating to]

venous return (VEE-nus)
[*ven-* vein, *-ous* relating to]

CHAPTER SUMMARY

To download an MP3 version of the chapter summary for use with your iPod or other portable media player, access the Audio Chapter Summaries online at http://evolve.elsevier.com.

HINT *Scan this summary after reading the chapter to help you reinforce the key concepts. Later, use the summary as a quick review before your class or before a test.*

FUNCTION OF THE HEART AND BLOOD VESSELS

A. Homeostatic maintenance of our body is due to the continuous and controlled movement of blood throughout our circulatory system
B. Many control mechanisms help to regulate and integrate the diverse functions and component parts of our cardiovascular system

HEMODYNAMICS

A. Hemodynamics—refers to the various mechanisms that influence the movement of blood
B. Different organs may need vastly different amounts of blood flow, depending on their metabolic activity

THE HEART AS A PUMP

A. Heart must have a system to generate rhythmic impulses and distribute them to different regions of the myocardium
  1. Distribution is accomplished by the impulse-conducting pathway
  2. Four structures make up the core of the conduction system of the heart (Figure 18-1):
    a. Sinoatrial (SA) valve
    b. Atrioventricular (AV) node

c. AV bundle (bundle of His)
    d. Subendocardial branches (Purkinje fibers)
  3. SA node—"pacemaker"; has own intrinsic rhythm
    a. Impulse generated travels swiftly throughout the muscle fibers of both atria
    b. Action potential next enters the atrioventricular (AV) node
    c. Impulse is relayed through the atrioventricular (AV) bundle
    d. Right and left bundle branches and the subendocardial branches in which they terminate conduct the impulses throughout the muscle of both ventricles
B. Electrocardiogram (ECG)
  1. Electrocardiography—impulse currents can be measured with an electrocardiograph
    a. Electrocardiogram—record of the electrical events that precede the contractions
    b. Produced by attaching electrodes of a recording voltmeter (the electrocardiograph) to the chest and or limbs of the subject (Figure 18-2, *A*)
    c. Changes in the heart's electrical activity can be seen as deflections of the line from a video monitor (Figure 18-3)
  2. ECG waves—normal ECG is composed of deflection waves called the P wave, QRS complex, and T wave (Figures 18-2 and 18-4)
    a. P wave—represents depolarization of the atria
    b. QRS complex—represents depolarization of the ventricles
    c. T wave—reflects repolarization of the ventricles
  3. ECG intervals—provide important information concerning the rate of conduction of an action potential through the heart

C. Cardiac cycle—a complete heartbeat or a single pumping cycle (Figure 18-5)
1. Atrial systole—contracting myocardium of the atria forces the blood into the ventricles below; represented by the P wave on an ECG
2. Ventricular contraction—brief period between the start of ventricular systole and the opening of the semilunar valves; marked by the R wave on the ECG (isovolumetric)
3. Ejection—semilunar valves open and blood is ejected under great force from the ventricles
   a. Rapid ejection—characterized by a marked increase in ventricular and aortic pressure
   b. Reduced ejection—coincides with the T wave of the ECG
4. Ventricular relaxation (diastole)—period between closing of the semilunar valves and the opening of the atrioventricular valves (isovolumetric)
5. Passive ventricular filling—continuing return of venous blood from the venae cavae and the pulmonary veins increases pressure within both atria until the atrioventricular valves are forced open

D. Heart sounds
1. First "lubb" or systolic sound is caused largely by the contraction of the ventricles and by the closing atrioventricular valves
2. Vibrations of the closing semilunar valves cause the second "dupp" sound

PRIMARY PRINCIPLE OF CIRCULATION

A. In order for blood to flow within the circulatory system, there must be a gradient from high pressure to low pressure; primary principle of circulation (Figure 18-6)
B. Pressure difference drives the flow of blood

ARTERIAL BLOOD PRESSURE

A. The volume of blood within the arteries largely determines arterial blood pressure
1. Many factors determine arterial pressure through their influence on arterial volume; for example, cardiac output and peripheral resistance (Figure 18-7)
B. Cardiac output (CO)—amount of blood that flows out of a ventricle per unit of time
1. CO influences the flow rate to the various organs of the body
2. CO is determined by the volume of blood pumped out of a ventricle by each beat (stroke volume or SV) and by heart rate (HR)
3. CO (volume/min) = SV(volume/beat) × HR (beats/min)
C. Factors that affect stroke volume—mechanical, neural, and chemical factors regulate the strength of the heartbeat (stroke volume)
D. Factors that affect heart rate—sinoatrial (SA) node normally initiates each heartbeat
1. Various factors can change the rate of the heartbeat
   a. Ratio of sympathetic and parasympathetic impulses conducted to the node per minute
E. Cardiac pressoreflexes—receptors sensitive to changes in pressure are located in two places near the heart (Figure 18-9)
1. Aortic baroreceptors
2. Carotid baroreceptors

F. Other reflexes that influence heart rate—emotions, exercise, hormones, blood temperature, pain, and stimulation of various exteroceptors also influence heart rate
G. Peripheral resistance—the resistance to blood flow caused by the *friction* of blood striking the walls of the vessels
1. Vasomotor mechanism—muscular layer of the arterioles constricting or dilating and thus changing the amount of resistance to blood flow (Figure 18-8)
   a. Vasoconstriction—reducing the vessel diameter by increasing the contraction of the muscular layer
   b. Vasodilation—relaxation of vascular muscles, decreases resistance to blood flow
H. Vasomotor control mechanism
1. Control center for this complex system lies in the vasomotor center of the medulla
2. Upon stimulation, the control system sends out impulses causing the restriction of smooth muscles surrounding some vessels (Figure 18-9)
3. Sudden increase in arterial blood pressure stimulates aortic and carotid baroreceptors; results in arterioles and venules of the blood reservoirs dilating
4. Decrease in arterial pressure causes the aortic and carotid baroreceptors to send more impulses to the medulla's vasoconstrictor centers; causes the vascular smooth muscles to constrict
5. There are also chemoreceptor reflexes in the aorta and carotid arteries
   a. Function when excess blood carbon dioxide and low oxygen content endangers the stability of the internal environment
I. Local control of arterioles—several kinds of local mechanisms that produce vasodilation in localized areas
1. Function in times of increased tissue activity

VENOUS RETURN TO THE HEART

A. Venous return—amount of blood returned to the heart via the veins
1. Stress-relaxation effect—occurs when a change in blood pressure causes a change in vessel diameter
B. Venous pumps—blood-pumping action of respirations and skeletal muscle contractions facilitate venous return by increasing pressure gradient between peripheral veins and venae cavae
1. Change in pressure, between expiration and inspiration, helps move blood along the venous route back to the heart (Figure 18-10)
2. As each skeletal muscle contracts, it squeezes the soft veins scattered throughout its interior; contractions push blood upward, toward the heart (Figure 18-11)
   a. Repeated contraction of the muscles when walking or doing any other exercise keeps the blood moving forward in the veins aided by one-way valves (Figure 18-11)
C. Total blood volume—changes in total blood volume change the amount of blood returned to the heart
1. Balance between the movement of water into and out of the plasma that affects the homeostasis of blood flow
D. Changes in total blood volume
1. Receptors in the body that detect the balance between water and solutes trigger the ADH mechanism

a. ADH is released by the posterior pituitary and acts on the kidneys to reduce the amount of water loss

2. Renin-angiotensin-aldosterone system (RAAS)
   a. Renin is released when blood pressure in the kidney is low
   b. Release of renin triggers a series of events that leads to the secretion of aldosterone from the adrenal cortex
   c. Aldosterone promotes sodium retention by the kidneys; stimulates the osmotic flow of water from kidney tubules back into the blood plasma

3. ANH mechanism
   a. ANH is secreted by specialized cells in the atria
   b. ANH adjusts venous return back down to its normal set point by promoting the loss of water from the plasma
   c. ANH increases sodium loss from the urine; causes water to follow by osmosis

MEASURING ARTERIAL BLOOD PRESSURE

A. Blood pressure is measured with a sphygmomanometer (Figure 18-14, *A*)

B. Systolic blood pressure—force with which the blood is pushing against the artery walls when the ventricles are contracting

C. Diastolic blood pressure—force of the blood when the ventricles are relaxing

D. Pulse pressure—difference between systolic and diastolic blood pressure

PULSE AND PULSE WAVE

A. Pulse—alternating expansion and recoil of an artery
   1. Two factors are responsible for a pulse you can actually feel:
      a. Pulses of blood injected from the heart's left ventricle into the aorta
      b. Elasticity of the arterial wall
   2. Pulse wave (Figure 18-16)
      a. Each pulse starts with ventricular contraction and proceeds as a wave of expansion throughout the arteries
      b. Gradually dissipates as it travels through the circulatory system, disappearing entirely in the capillaries

B. Feeling your pulse—can feel your pulse whenever an artery comes close to the surface and passes over a bone or other firm background

REVIEW QUESTIONS

Write out the answers to these questions after reading the chapter and reviewing the Chapter Summary. If you simply think through the answer without writing it down, you won't retain much of your new learning.

HINT

1. Identify, locate, and describe the function of each of the following structures: SA node, AV node, AV bundle, and subendocardial branches (Purkinje fibers).

2. What does an electrocardiogram measure and record? List the normal ECG deflection waves and intervals. What do the various ECG waves represent?

3. What is meant by the term *cardiac cycle*?

4. List the "phases" of the cardiac cycle and briefly describe the events that occur in each.

5. What is meant by the term *residual volume* as it applies to the heart?

6. Describe and explain the origin of the heart sounds.

7. What is the primary determinant of arterial blood pressure?

8. List the two most important factors that indirectly determine arterial pressure by their influence on arterial volume.

9. How is cardiac output determined?

10. List and give the effect of several factors, such as grief or pain, on heart rate.

11. What mechanisms control peripheral resistance? Cite an example of the operation of one or more parts of this mechanism to increase resistance and to decrease it.

12. Explain how antidiuretic hormone can change the total blood volume.

13. What is the effect of low blood pressure in relation to aldosterone and antidiuretic hormone secretion?

14. Describe the measurement of arterial blood pressure.

15. Identify nine locations where the pulse point can be easily felt.

16. Describe the various types of cardiac dysrhythmias.

CRITICAL THINKING QUESTIONS

After finishing the Review Questions, write out the answers to these items to help you apply your new knowledge. Go back to sections of the chapter that relate to items that you find difficult.

HINT

1. What is an ectopic pacemaker? What would be the effect on the heart rate if an ectopic pacemaker took over for the SA node?

2. State in your own words the primary principle of circulation. How does it govern the various mechanisms involved in blood flow?

3. What is the Frank Starling mechanism (Starling's law) of the heart? What role does *venous return* play in this mechanism?

4. Explain how the heart rate is a good example of dual innervation in the autonomic nervous system. Include the nerves and neurotransmitters involved.

5. By the time blood gets to the veins, almost all the pressure from the contracting ventricles has been lost. What mechanisms does the body use to assist in returning blood to the heart?

6. Explain the forces that act on capillary exchange on both the arterial and venous ends of the capillary. Is the recovery of fluid at the venous end 100% effective?

continued from page 405

Now that you have read this chapter, see if you can answer these questions about the shock Bobby received in the Introductory Story.

When activated, the AED shocks the heart—the intended purpose being electrical stimulation of the cardiac muscle cells to elicit a response from the heart's pacemaker that will cause the cells to contract in unison to effectively pump blood.

1. What bundle of cells in Bobby's heart (and yours) is known as the pacemaker?
   a. SV node
   b. SA node
   c. AV node
   d. AV bundle

When the paramedics arrive, they rush Bobby to the hospital. "Let's get an EKG!" the attending physician calls out.

2. What is an EKG?
   a. Electrocardiogram
   b. Electrocirculogram
   c. Encephalocardiogram
   d. Enhancercardiogram

3. Ventricular depolarization is shown as which part of the EKG?
   a. V wave
   b. P wave
   c. T wave
   d. QRS complex

HINT To solve a case study, you may have to refer to the glossary or index, other chapters in this textbook, A&P Connect, Mechanisms of Disease, and other resources.

CHAPTER 19

Lymphatic and Immune Systems

STUDENT LEARNING OBJECTIVES

At the completion of this chapter, you should be able to do the following:

1. Briefly outline the components of the lymphatic and immune systems.
2. Contrast the composition of lymph with that of interstitial fluid.
3. Outline the general circulation of lymph through lymphatic vessels and nodes.
4. List several major groups of lymph nodes and their locations.
5. List the lymphatic functions of the following: tonsils, thymus, spleen.
6. Outline an overview of innate immunity.
7. List and briefly discuss the three lines of immune defense.
8. Discuss the significance of fever and inflammation.
9. Outline the roles of the following: macrophages, diapedesis, NK cells, interferon.
10. Give an overview of adaptive immunity.
11. Discuss the major types of immune system molecules and indicate how antibodies and complement proteins function.
12. Discuss the diversity of antibodies and their functions.
13. Discuss and contrast the development and functions of B and T cells.
14. Compare and contrast antibody-mediated and cell-mediated immunity.

KOSTAS remembered getting the flu (influenza) last winter: coughing, fever, achiness all over his body, watery eyes, and fatigue. He felt awful! But he argued, "What's the point of a flu shot, when all it does is give you the flu?" Kostas did not understand that the injected flu vaccine is a combination of several *inactivated* (killed) viruses injected into muscles in your body (usually in your arm). No active viruses are injected. So, as for the injected form of the vaccine "causing" the flu—people who claim that could already have been exposed to a flu virus before the vaccination or could have been exposed to one of the strains not included in that year's vaccine. Some people produce a mild immune reaction that can be mistaken for the flu. In the end, Kostas relented and got his flu shot!

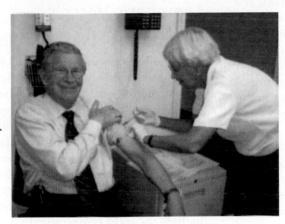

We're sure you'll enjoy reading about your own immune system, and how your body is programmed to protect you from disease. At the end of this chapter, you should be able to answer some questions about Kostas and his flu shot.

COMPONENTS OF THE LYMPHATIC AND IMMUNE SYSTEMS

We have combined two related systems in this chapter: the *lymphatic system* and the *immune system*.

The **lymphatic system** has at least three different functions. First, it serves to maintain the fluid balance of our internal body environment. Second, the lymphatic system serves to house the immune defenses of our body. Third, the lymphatic system also helps regulate the absorption of lipids from digested food in the small intestines and provides for their transport to the large systemic veins. As you will see, the vessels of the lymphatic system roughly parallel the vessels of the cardiovascular system.

The **immune system** serves to repel and destroy the hordes of microorganisms that threaten our lives every day. In addition, our immune system must defend us from our own abnormal cells that can cause cancerous tumors. Such tumors may damage surrounding tissues and spread cancer throughout the body. Without an internal "security force" to deal with such abnormal cells when they first appear, we would have very short lives indeed!

LANGUAGE OF SCIENCE AND MEDICINE

HINT *Before reading the chapter, say each of these terms out loud. This will help you avoid stumbling over them as you read.*

acquired immunity (ah-KWYERD ih-MYOO-nih-tee)
[*immun-* free, *-ity* state]

active immunity (AK-tiv ih-MYOO-nih-tee)
[*actus-* moving, *immun-* free, *-ity* state]

adaptive immunity (ah-DAP-tiv ih-MYOO-nih-tee)
[*adapt-* adjust, *-ive* relating to, *immun-* free, *-ity* state]

agglutination (ah-gloo-tin-AY-shun)
[*agglutin-* glue, *-ation* process]

aggregated lymphoid nodule (ag-rah-GAY I-ed LIM-foyd NOD-yool)
[*a(d)-* to, *-grega-* collect, *lymph-* water, *-oid* like, *nod-* knot, *-ule* small]

anastomosis (ah-nas-toh-MOH-sis)
[*ana-* anew, *-stomo-* mouth, *-osis* conditions of]
pl., anastomoses (ah-nas-toh-MOH-seez)

antibody (AN-tih-bod-ee)
[*anti-* against]

antibody-mediated immunity (AN-tih-bod-ee—MEE-dee-ayt-ed ih-MYOO-nih-tee)
[*anti-* against, *medi-* middle, *-ate* process, *immun-* free, *-ity* state]

antibody titer (AN-tih-bod-ee TYE-ter)
[*anti-* against, *titer* proportion (in a solution)]

antigen (AN-tih-jen)
[*anti-* against, *-gen* produce]

antigen-antibody complex (AN-tih-jen-AN-tih-bod-ee KOM-pleks)
[*anti-* against, *-gen* produce, *anti-* against, *complex* embrace]

antigen-presenting cell (APC) (AN-tih-jen sell)
[*anti-* against, *-gen* produce, *cell* storeroom]

autoimmunity (aw-toh-ih-MYOO-nih-tee)
[*auto-* self, *-immun-* free, *-ity* state]

axillary lymph node (AK-sih-lair-ee limf)
[*axilla-* wing, *-ary* relating to, *lymph* water, *nod-* knot]

B cell
[*B* bursa-equivalent tissue, *cell* storeroom]

booster shot

CD system (C D SIS-tem)
[*C* cluster, *D* differentiation, *system* organized whole]

cell-mediated immunity (sell-MEE-dee-ayt-ed ih-MYOO-nih-tee)
[*cell* storeroom, *medi-* middle, *-ate* process, *immun-* free, *-ity* state]

cellular immunity (SELL-yoo-lar ih-MYOO-nih-tee)
[*cell-* storeroom, *-ular* relating to, *immun-* free, *-ity* state]

chemotaxis (kee-moh-TAK-sis)
[*chemo-* chemical, *-taxis* movement]

continued on page 448

LYMPHATIC SYSTEM

Overview of the Lymphatic System

Figure 19-1 gives you an excellent start to your understanding of the lymphatic system.

As you can see, plasma filters into interstitial spaces from blood flowing through the capillaries. Most of this *interstitial fluid* is absorbed by tissue cells or is reabsorbed by the blood before it flows out of the tissue. However, a small amount of the interstitial fluid remains behind. It seems insignificant, but if even small amounts of extra fluid continued to accumulate in the tissues over time, the result would be tremendous **edema** (swelling). This would be followed by tissue destruction. The lymphatic system solves the problem of fluid retention in our tissues. In fact, the entire system acts as a *drainage system:* It continuously collects excess tissue fluid and returns it to the venous blood just before it reaches the heart.

The lymphatic system consists of a moving fluid (lymph) derived from the blood and tissue fluid as well as a group of vessels (lymphatics) that return the lymph to the blood. In addition to lymph and lymphatic vessels, the system includes various structures that contain **lymphoid tissue.** This tissue, as we will see, contains *lymphocytes* and other defensive cells of the immune system. For example, lymph nodes are located along the paths of the collecting lymphatic vessels. Additional lymphoid tissue is found in the intestinal wall, appendix, tonsils, thymus, spleen, and bone marrow (Figure 19-2, *A*).

The lymphatic system provides a unique transport function. It returns tissue fluid, proteins, fats, and other substances to the general circulatory system. However, unlike the circulatory system, the lymphatic vessels do *not* form a closed system of vessels. Instead they begin blindly in the intercellular spaces of the soft tissue of the body (see Figure 19-1).

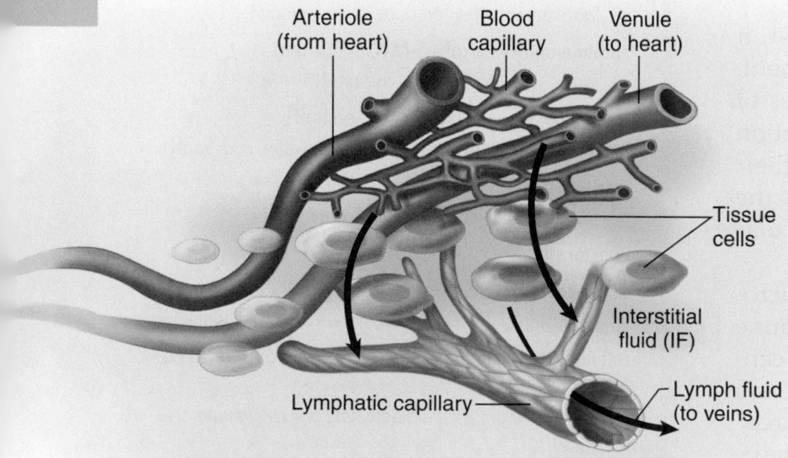

Lymph and Interstitial Fluid

Lymph is a clear fluid found in the lymphatic vessels, whereas **interstitial fluid (IF)** is the complex fluid that fills the spaces between the cells. Both lymph and interstitial fluid closely resemble blood plasma in composition. However, lymph cannot clot like blood. If the main lymphatic trunks in the thorax (see Figure 19-2) are damaged, the flow of lymph must be stopped surgically or death ensues.

Lymphatic Vessels

Distribution of Lymphatic Vessels

Lymphatic vessels begin as microscopic blind-ended **lymphatic capillaries.** If the lymphatic vessels originate in the villi of the small intestine, they are called **lacteals** (see Chapter 21). The wall of each lymphatic capillary consists of a single layer of flattened endothelial cells. We have extensive networks of lymphatic capillaries that branch and then rejoin repeatedly to form an elaborate network throughout the interstitial spaces of our bodies.

The lymphatic capillaries merge to form larger and larger vessels until main lymphatic trunks are formed. These include the **right lymphatic duct** and the **thoracic duct** seen in Figure 19-2. Lymph from the entire body, except from the upper right quadrant, eventually drains into the thoracic duct. This in turn drains into the left subclavian vein at the point where it joins the left internal jugular vein (see Figure 19-2, *B*). Lymph from the upper right quadrant empties into the right lymphatic duct and then into the right subclavian vein.

Note that the thoracic duct is considerably larger than the right lymphatic duct. This is because most of the body's lymph returns to the bloodstream via the thoracic duct.

Structure of Lymphatic Vessels

The walls of lymphatic capillaries have numerous openings or *clefts* between the cells. This makes them much more porous or permeable than blood capillaries. As lymph flows from the thin-walled lymphatic capillaries into vessels with larger diameters, the walls become thicker. Eventually these larger vessels have the three layers typical of arteries and veins.

One-way valves are abundant in lymphatic vessels of all sizes. These valves give the vessels a somewhat beaded appearance. Valves are present every few millimeters in large lymphatics and are even more numerous in the smaller vessels (Figure 19-3).

Function of Lymphatic Vessels

The lymphatics play a vital role in numerous homeostatic mechanisms. The great permeability of the lymphatic capillary wall permits very large molecules and even small particles to

▲ **FIGURE 19-1** **Role of the lymphatic system in fluid balance.** Fluid from plasma flowing through the capillaries moves into interstitial spaces. Although *most* of this interstitial fluid is either absorbed by tissue cells or resorbed by capillaries, *some* of the fluid tends to accumulate in the interstitial spaces. As this fluid builds up, it drains into lymphatic vessels that eventually return the fluid to the venous blood.

Labels in figure:
- Arteriole (from heart)
- Blood capillary
- Venule (to heart)
- Tissue cells
- Interstitial fluid (IF)
- Lymphatic capillary
- Lymph fluid (to veins)

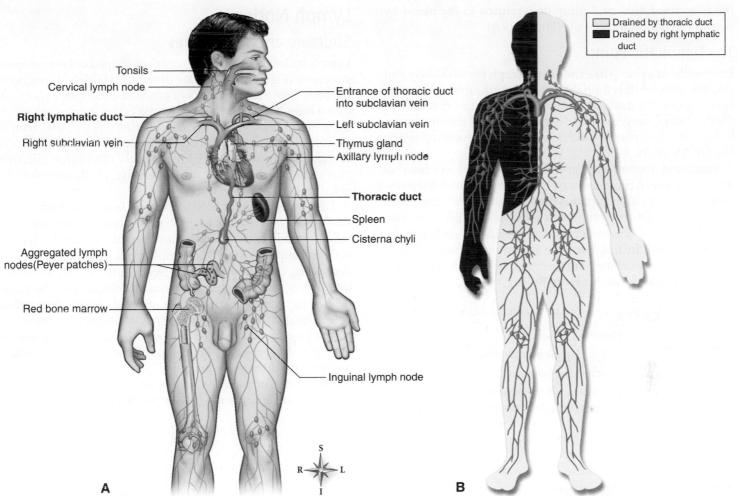

A

B

☐ Drained by thoracic duct
■ Drained by right lymphatic duct

Tonsils
Cervical lymph node
Right lymphatic duct
Right subclavian vein
Aggregated lymph nodes(Peyer patches)
Red bone marrow

Entrance of thoracic duct into subclavian vein
Left subclavian vein
Thymus gland
Axillary lymph node
Thoracic duct
Spleen
Cisterna chyli
Inguinal lymph node

▲ **FIGURE 19-2** **Lymphatic system. A,** Principal organs of the lymphatic system. **B,** The right lymphatic duct drains lymph from the upper right quadrant *(dark blue)* of the body into the right subclavian vein. The thoracic duct drains lymph from the rest of the body *(yellow)* into the left subclavian vein. The lymphatic fluid is thus returned to the systemic blood just before entering the heart.

be removed from the interstitial spaces. In fact, proteins that accumulate in the tissue spaces can return to blood *only* by the lymphatic system. This fact has great clinical importance. For example, if anything blocks the return of lymph for an extended period of time, blood protein concentration and blood osmotic pressure soon fall below normal. The result is fluid imbalance and death.

Lacteals from the villi of the small intestine are important in the absorption of fats and other nutrients. **Chyle**—the milky lymph found in lacteals after digestion—contains 1% to 2% fat.

Circulation of Lymph

Water and solutes continually filter out of capillary blood into the interstitial fluid (refer again to Figure 19-1). To balance this outflow from the system, fluid continually re-enters blood from the interstitial fluid. We now know that about 50% of the total blood protein leaks out of the capillaries into

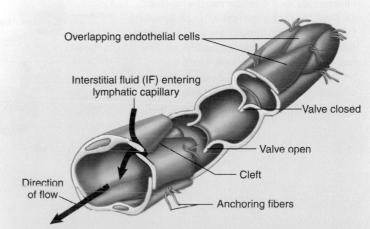

Overlapping endothelial cells
Interstitial fluid (IF) entering lymphatic capillary
Direction of flow
Valve closed
Valve open
Cleft
Anchoring fibers

▲ **FIGURE 19-3** **Structure of a typical lymphatic capillary.** Notice that interstitial fluid enters through clefts between overlapping endothelial cells that form the wall of the vessel. Valves ensure one-way flow of lymph out of the tissue.

the interstitial fluid and ultimately returns to the blood by way of the lymphatic vessels (Figure 19-4).

The Lymphatic Pump

Even without a pump like the heart, lymph moves slowly and steadily along in its lymphatic vessels into the general circulation at about 3 L/day. This occurs despite the fact that most of the flow is against gravity! It does so because of the large number of valves that permit fluid flow only in the general direction toward the heart.

Breathing movements and skeletal muscle contraction aid in this return movement of lymph to the circulatory system, just as they assist venous blood return, as we have seen (Chapter 10). During strenuous exercise, lymph flow may increase 10 to 15 times over normal because of skeletal muscle contraction. In this way, the continuous flow of lymph serves as an important homeostatic mechanism that maintains the constancy of our body fluids.

> ### QUICK CHECK
> 1. Where can you find lymphoid tissue?
> 2. Compare the composition of lymph and interstitial fluid.
> 3. Briefly describe the structure and function of lymphatic capillaries.
> 4. Briefly describe the factors that aid the movement of lymphatic fluid.

Lymph Nodes
Structure of Lymph Nodes

Lymph nodes (lymph glands) are oval-shaped or bean-shaped structures (Figure 19-5) that are distributed widely throughout the body. Some are as small as a pinhead; others are as large as a lima bean. The vast system of lymph nodes is linked together by the lymphatic vessels. Note in Figure 19-5, *C*, that lymph moves into a node by way of *several afferent* lymphatic vessels and emerges by *one or two efferent* vessels, creating an effective biological filter. One-way valves keep lymph flowing only in one direction.

Fibrous partitions or *trabeculae* extend from the covering capsule toward the center of a lymph node, creating compartments called *cortical nodules*. Each cortical nodule within the lymph node is composed of packed lymphocytes that surround a less dense area, the **germinal center** (see Figure 19-5, *C*). When an infection is present, germinal centers enlarge and the node begins to release lymphocytes. Special leukocytes called *B lymphocytes* (B cells) begin their final stages of maturation within the germinal center of the nodule. They are then pushed into the denser outer layers to mature before becoming antibody-producing **plasma cells.**

The center, or medulla, of a lymph node is composed of sinuses that separate medullary cords composed of plasma

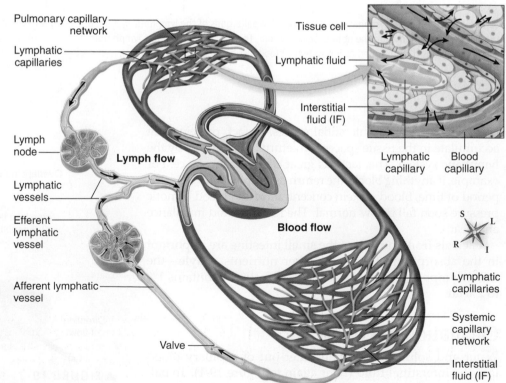

▶ **FIGURE 19-4** **Circulation plan of lymphatic fluid.** This diagram outlines the general scheme for lymphatic circulation. Fluids from the systemic and pulmonary capillaries leave the bloodstream and enter interstitial spaces, thus becoming part of the interstitial fluid (IF). The IF also exchanges materials with the surrounding tissues. Often, because less fluid is returned to the blood capillary than had left it, IF pressure increases—causing IF to flow into the lymphatic capillary. The fluid is then called lymph (lymphatic fluid) and is carried through one or more lymph nodes and finally to large lymphatic ducts. The lymph enters a subclavian vein, where it is returned to the systemic blood plasma. Thus fluid circulates through blood vessels, tissues, and lymphatic vessels in a sort of "open circulation."

cells and B cells. Both the cortical and medullary sinuses are lined with macrophages ready for phagocytosis.

Locations of Lymph Nodes

Most lymph nodes occur in groups, or clusters (see Figure 19-2), in certain areas, especially the head and neck. A total of approximately 500 to 600 lymph nodes are located throughout our bodies. Before you continue, take a moment to review the locations of the major lymph nodes in Figure 19-2.

Function of Lymph Nodes

Our lymph nodes defend our bodies from invading pathogens and also provide sites for the maturation of some types of lymphocytes.

Lymph flow slows as it passes through the sinus channels of the lymph nodes. This gives the special cells that line the channels time to remove microorganisms and other injurious particles. Here, the offending material is engulfed in the process of phagocytosis and destroyed. Thus, lymph nodes are the sites of both biological *and* mechanical filtration.

Sometimes, however, the lymph nodes are overwhelmed by massive numbers of infectious microorganisms. The nodes themselves then become sites of infection. Most people have experienced the pain of swollen lymph nodes. In addition, cancer cells breaking away from a malignant tumor may also enter the lymphatic system. They travel to the lymph nodes and may create cancerous growths that block the flow of lymph. This leaves too few channels for lymph to return to the blood and swelling results. For example, if tumors block **axillary lymph node** channels (located under our arms), fluid accumulates in the interstitial spaces of the arm, causing swelling and pain from the edema. Even viruses such as human immunodeficiency virus (HIV) and other types of pathogens can infect or infest lymph nodes.

Lymphoid tissues of lymph nodes also serve as the site for the final stages of maturation of some types of lymphocytes and monocytes.

Lymphatic Drainage of the Breast
Distribution of Lymphatics in the Breast

The mammary glands and surrounding tissues of the breast are drained by two sets of lymphatic vessels. There are lymphatics that originate in and drain the surface area and skin over the breast (excluding the areola and nipple areas). There are also lymphatics that originate in and drain the underlying tissue of the breast itself (including the skin of the areola and nipple).

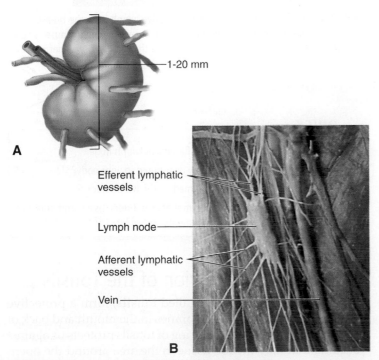

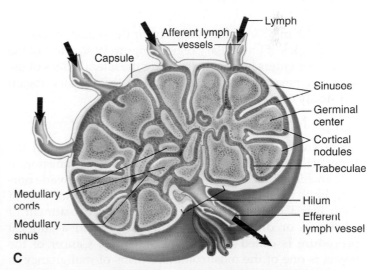

▲ FIGURE 19-5 Structure of a lymph node. A, A lymph node is typically a small structure into which afferent lymphatic ducts empty their lymph. Efferent lymphatic ducts drain the lymph from the node. An outer fibrous capsule maintains the structural integrity of the node. **B,** Photograph of a dissected cadaver shows a lymph node and its associated lymphatic vessels, along with nearby muscles, nerves, and blood vessels. **C,** Internal structure of a lymph node. Several afferent valved lymphatics bring lymph to the node. In this example, a single efferent lymphatic leaves the node at a concave area called the *hilum.* Note that the artery and vein also enter and leave at the hilum. *Arrows* show direction of lymph movement.

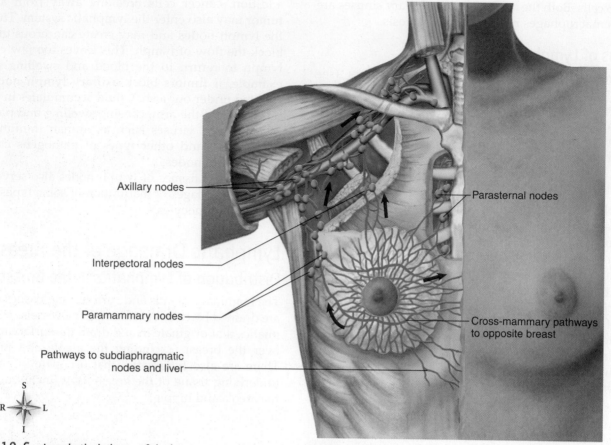

▲ **FIGURE 19-6** **Lymphatic drainage of the breast.** Note the extensive network of lymphatic vessels and nodes that receive lymph from the breast. Surgical procedures called mastectomies, in which some or all of the breast tissues are removed, are sometimes performed to treat breast cancer. Because cancer cells can spread so easily through the extensive network of lymphatic vessels associated with the breast, the lymphatic vessels and their nodes are sometimes also removed. Occasionally, such procedures cause swelling, or lymphedema.

More than 85% of the lymph from the breast enters the lymph nodes of the axillary region (Figure 19-6). Most of the remainder enters lymph nodes along the lateral edges of the sternum. Several very large nodes in the axillary region physically contact extensions of breast tissue.

Lymph Nodes Associated with the Breast

There are many **anastomoses** (connections) between the superficial lymphatics from both breasts. These anastomoses can allow cancerous cells from one breast to invade normal tissue from the other breast. Removal of a wide area of deep fascia is therefore required in surgical treatment of advanced or diffuse breast malignancy. Such a surgical procedure is called a **radical mastectomy.** Cancer of the breast is one of the most common forms of malignancy in women. However, it can also be found (although rarely) in men.

Breast infections are also a serious health concern, especially among women who nurse their infants. **Mastitis,** for example, is an inflammation of the mammary gland, usually caused by infectious agents. Breast infections, like cancer, can also spread easily through lymphatic pathways associated with the breast.

QUICK ✓ CHECK

5. Describe the overall structure of a typical lymph node.
6. How are lymph nodes generally distributed in your body?
7. What vital functions are performed by the lymph nodes?
8. How does the distribution of lymphatics in breast tissue and adjacent tissues relate to breast cancer and its spread?

Structure and Function of the Tonsils

Masses of lymphoid tissue, called **tonsils,** form a protective ring under the mucous membranes in the mouth and back of the throat (Figure 19-7). This ring of tonsils protects us against bacteria that may invade tissue in the area around the openings between the nasal and oral cavities. The **palatine tonsils** are located on each side of the throat. The **pharyngeal tonsils** (called adenoids when they become swollen) are near the posterior opening of the nasal cavity. A third type of tonsil, the **lingual tonsils,** lie near the base of the tongue. Other smaller tonsils are located near the opening of the auditory (eustachian) tube. Each tonsil has deep recesses that trap bacteria and expose them to the immune system.

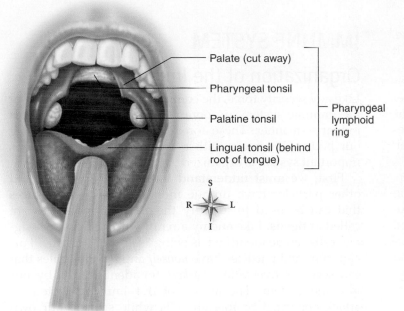

▲ **FIGURE 19-7** **Location of the tonsils.** Small segments of the roof and floor of the mouth have been removed to show the protective ring of tonsils (pharyngeal lymphoid ring) around the internal openings of the nose and throat.

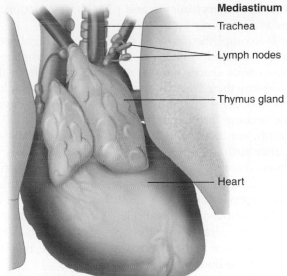

▲ **FIGURE 19-8** **Thymus.** Location of a child's thymus within the mediastinum.

The tonsils are part of our first line of defense from the external environment. As such, they are subject to chronic infection, or **tonsillitis.** In these cases, tonsils may be surgically removed **(tonsillectomy),** if non-surgical treatments prove ineffective. However, because of the critical immunological role played by the lymphatic tissue, the number of tonsillectomies performed annually continues to decrease.

Structure and Function of Aggregated Lymphoid Nodules

Aggregated lymphoid nodules, also called *Peyer patches,* are groups of small oval patches or groups of lymph nodes that form a single protective layer in the mucous membrane of the small intestine, especially the ileum. Because the entire gastrointestinal tract is potentially open to the external environment via the mouth, these patches are in a great location to provide immune surveillance in an area where massive numbers of potentially pathogenic bacteria can be found. The macrophages and other cells of the immune system prevent most of these bacteria from penetrating the gut wall. Aggregated lymphoid nodules and other lymphoid tissues are sometimes called *mucosa-associated lymphoid tissue* (MALT).

Structure and Function of the Thymus

The **thymus** is a primary organ of the lymphatic system. It consists of two pyramid-shaped lobes. The thymus is located in the mediastinum, extending up into the neck, close to the thyroid gland (Figure 19-8). It is largest (relative to body size) in a child about 2 years old. After puberty, however, the thymus gradually atrophies. In advanced old age, it may be largely replaced by fat, becoming yellow in color.

The thymus plays a critical part in the body's defenses against infection. Before birth, the thymus serves as the final site of lymphocyte development. Many lymphocytes leave the thymus and circulate to the spleen, lymph nodes, and other lymphoid tissue.

Soon after birth, the thymus assumes another function. It begins secreting a group of peptide hormones (collectively called **thymosin**) and other regulators that enable lymphocytes to develop into mature T cells. Only T lymphocytes that pass "immunological testing" are released into the bloodstream.

Structure and Function of the Spleen

The **spleen** is located below the diaphragm, just above most of the left kidney and behind the fundus of the stomach (see Figure 19-2). Roughly oval in shape (Figure 19-9), the spleen varies somewhat in size from individual to individual. For example, it enlarges *(hypertrophies)* during infectious disease and shrinks *(atrophies)* in old age.

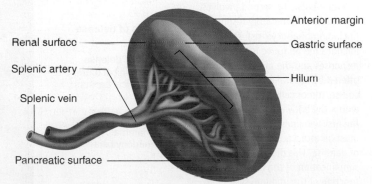

▲ **FIGURE 19-9** **Structure of the spleen.** Medial aspect of the spleen. Notice the concave surface that fits against the stomach within the abdominopelvic cavity.

The spleen has a variety of functions, including defense, hematopoiesis, and erythrocyte and platelet destruction. It also serves as a reservoir for blood.

Defense is accomplished as blood passes through highly permeable, enlarged blood vessels, called *sinusoids*. Macrophages lining these venous spaces remove microorganisms from the blood and destroy them by phagocytosis. Hematopoiesis takes place when monocytes and lymphocytes complete their development and become "activated" in the spleen.

Before birth, red blood cells are also formed in the spleen. However, after birth, the spleen forms red blood cells only in cases of severe anemia. Macrophages lining the spleen's sinusoids remove worn-out red blood cells and imperfectly formed platelets. Macrophages also break apart the hemoglobin molecules from the destroyed red blood cells. They salvage iron and globin content from destroyed erythrocytes and return these byproducts of destruction to the bloodstream. From here, they are sent to storage in the bone marrow and liver.

Finally, the spleen and its venous sinuses hold a considerable amount of blood. This blood reservoir can rapidly be added back into the circulatory system if it is needed. However, if the spleen is ruptured, as when the ribs are broken and pushed into the spleen, significant internal bleeding can occur, followed by death. Surgical repair or removal of the spleen is often required to stop the blood loss and save the patient's life.

Even though the spleen provides many useful functions, it is not a vital organ and can be removed without dire consequences.

QUICK CHECK

9. Where are the major tonsils located? What is their role?
10. What major roles does the thymus play in immunity?
11. What are the major functions of the spleen? Is it a "vital" organ?

► **FIGURE 19-10** **Lines of defense.** Immune function—that is, defense of the internal environment against foreign cells, proteins, and viruses—includes three layers of protection. The first line of defense is a set of barriers between the internal and external environments. The second involves the innate inflammatory response (including phagocytosis). The third includes the adaptive immune responses and the innate defense offered by natural killer cells. Of course, tumor cells that arise within the body are not affected by the first two lines of defense and must be attacked by the third line of defense. This diagram is a simplification of the complex function of the immune system. In reality, a great deal of crossover of mechanisms occurs between these "lines of defense."

IMMUNE SYSTEM

Organization of the Immune System

Like any security force, the components and mechanisms of the immune system are organized in an efficient—almost military—manner. These forces are continually patrolling our bodies. We begin with a brief overview of how this important system is organized.

First, we must understand that all cells, viruses, and other particles have unique molecules on their surfaces that can be used to identify them. These molecules are called **antigens.** Like enemy aircraft with distinctive insignia, cells can be identified as being "self" or "nonself." Foreign cells and particles have *nonself* antigen molecules that can serve as *recognition markers* for identification by our immune system. The ability of our immune system to attack abnormal or foreign cells while sparing our own cells is called **self-tolerance.**

Our bodies employ many different kinds of mechanisms to ensure the integrity and survival of our internal environment. All of these defense mechanisms are categorized either as *innate (nonspecific) immunity* or *adaptive (specific) immunity.*

Innate immunity is "in place" before you are exposed to a particular harmful particle or condition. It is naturally present at birth and is also called **nonspecific immunity** because it provides a *general defense* by acting against a wide variety of particles recognized as *nonself.*

Adaptive immunity, in contrast, involves mechanisms that program the body to recognize *specific* threatening agents. It "adapts" by targeting its response to these agents and to these agents alone. Because it targets only specific harmful particles, adaptive immunity is also called **specific immunity.** Adaptive immune mechanisms often take some

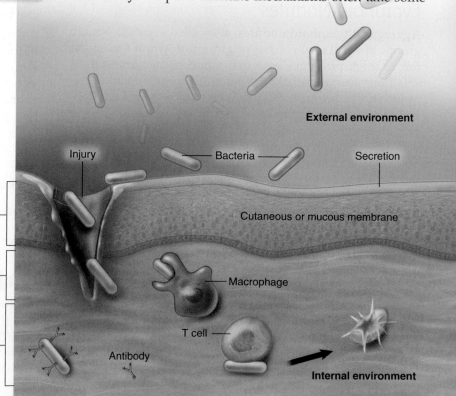

Lines of defense

First line of defense
• Mechanical barriers
• Chemical barriers

Second line of defense
• Inflammatory response
• Phagocytosis

Third line of defense
• Specific immune responses
• Natural killer cells

External environment

Injury — Bacteria — Secretion

Cutaneous or mucous membrane

Macrophage

T cell

Antibody

Internal environment

time to recognize their targets before they can react with sufficient force to overcome the threat.

As in any body system, cells or substances made by cells do the work of the immune system. The primary types of cells involved in innate immunity are these: *epithelial barrier cells, phagocytic cells (neutrophils, macrophages),* and aptly named *natural killer (NK) cells.* The primary types of cells involved in adaptive immunity are two types of lymphocytes called *T cells* and *B cells.*

Cytokines are chemicals released from cells to trigger or regulate innate and adaptive immune responses. Examples of cytokines include *interleukins (ILs), leukotrienes,* and *interferons (IFNs).* We will describe interferons more fully later in this chapter.

Human immune systems are such that there is a type of **species resistance** in which the genomes of specific organisms may be resistant to particular pathogens. This species resistance protects us from diseases such as canine distemper. As you might expect, however, our closer living primate relatives such as chimpanzees and bonobos have immune systems almost identical to our own, which means that some of their infections also affect us.

Our internal army of cells and molecules can be described in one word: *incredible.* Over one trillion lymphocytes and over 100 million trillion plasma protein molecules (antibodies) patrol our internal environment every minute. The basic components of innate immunity and adaptive immunity will be explained in more detail in the following pages.

Overview of Innate Immunity

There are numerous aspects of our innate (nonspecific) defensive mechanisms. The major players in this system are discussed below.

Mechanical and Chemical Barriers

Our internal environment is protected by a continuous mechanical barrier created by the cutaneous membrane (skin) and mucous membranes, as we have seen in Chapter 7. Often called the *first line of defense,* these membranes provide several layers of densely packed cells and other material. Together these cells and materials form a sort of "castle wall" against entry (Figure 19-10).

Besides forming a protective barrier, the skin and mucous membranes provide additional immune functions. For example, substances such as *skin surface film, sebum, mucus, enzymes,* and even *hydrochloric acid* (produced by the stomach lining) all serve to deter or destroy invading pathogens. The epithelial barriers of our bodies are essentially innate, nonspecific defenses.

Inflammatory Response and Fever

If bacteria or other invaders break through our mechanical and chemical barriers formed by the membranes and their secretions, the body has a *second line of defense* ready: the *inflammatory response* (see Figure 19-10). The **inflammatory** response elicits a number of actions that promote returning your body to a normal state. You can follow a flowchart illustrating how a local inflammatory response works in Figure 19-11. The diagram shows bacteria causing tissue damage. In turn, this abnormal condition triggers the release of various *inflammation mediators* from cells such as *mast cells* found in connective tissues. These inflammation mediators include *histamine, prostaglandins, leukotrienes, interleukins,* and other related compounds. They function to attract

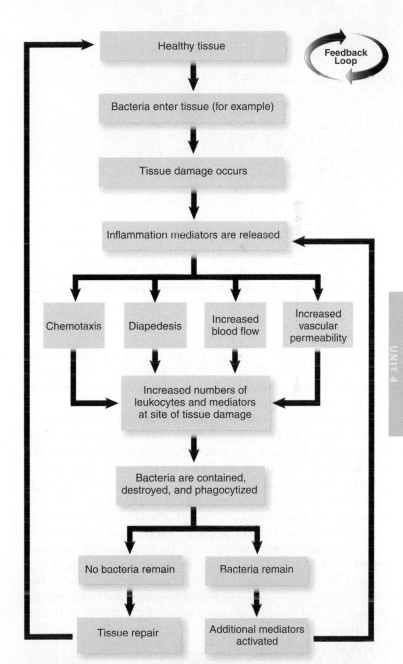

▲ **FIGURE 19-11** **Example of the inflammatory response.** Tissue damage caused by bacteria triggers a series of events that produce the inflammatory response. This promotes phagocytosis at the site of injury. These responses tend to inhibit or destroy the bacteria, eventually bringing the tissue back to its healthy state. Similar reactions will occur in the presence of other abnormal or injurious particles or conditions.

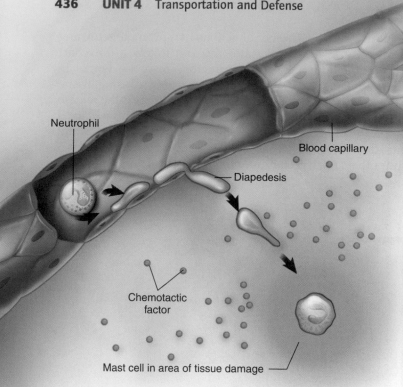

Neutrophil

Blood capillary

Diapedesis

Chemotactic factor

Mast cell in area of tissue damage

▲ **FIGURE 19-12** **Chemotaxis and diapedesis.** In this example, a neutrophil is attracted by agents released by a mast cell in a damaged or infected tissue. After adhering to the inside of the blood capillary, the neutrophil exits the capillary by diapedesis. Through chemotaxis (movement directed by chemical attraction), the neutrophil migrates toward the highest concentration of chemotactic factor—the site of the injury—where it can then begin its immune functions.

leukocytes to the area in a process called **chemotaxis** (Figure 19-12), which directs these cells to the site of inflammation. The characteristic signs of inflammation are heat, redness, pain, and swelling.

Besides local inflammation, systemic inflammation may occur when the inflammation mediators trigger responses that occur on a body-wide basis. A body-wide inflammatory response may be a **fever**—a state of elevated body temperature. Often a fever is accompanied by high neutrophil counts. A fever really results from a "reset" of the body's thermostat in the hypothalamus. This temporarily increases the set point or target temperature to a higher-than-normal value. Our bodies may shiver and feel cold. An elevated body temperature may facilitate some immune reactions. High fevers may also inhibit the reproduction of some microbial pathogens. However, the truth is that we really don't know for sure what fevers actually do!

Phagocytosis and Phagocytic Cells

A major component of the body's *second line of defense* is **phagocytosis**—the ingestion and destruction of microorganisms and other small particles by cells called *phagocytes*.

There are many types of phagocytes, but the basic mechanism of engulfing foreign substances is the same. Phagocytes extend footlike projections called *pseudopods* toward the invading organism. Soon the pseudopods encircle the pathogen and create a **phagosome.** The phagosome then moves into the interior of the cell. Here a lysosome fuses with it, releases digestive enzymes and hydrogen peroxide, and destroys the microorganism.

Because phagocytosis defends our bodies against a number of pathological agents, it is classified as an innate defense. As we will soon see, phagocytes play an important role in adaptive immunity as well.

The most numerous type of phagocyte is the **neutrophil.** *Chemotactic factors* (chemicals released at the site of infection) cause neutrophils and other phagocytes to adhere to the endothelial lining of capillaries servicing the affected area. The phagocytes then pass between the endothelial cells making up the capillary wall, dissolve the underlying basement membrane, and enter into the inflamed area. The movement of phagocytes from blood vessels to the site of inflammation is called **diapedesis** (see Figure 19-12).

Phagocytes have a very short life span. Dead phagocytic cells tend to "pile up" at the inflammation site, creating much of the white substance we call **pus.**

Another common type of phagocyte is the **macrophage.** These large phagocytic monocytes grow to several times their original size after migrating out of the bloodstream.

Other types of phagocytic cells are present in many areas of the body. For example, highly branched phagocytes called *dendritic cells (DCs)* are found in the interstitial fluid of most tissues of the body. Some phagocytes are even found on the outside surface of some mucous membranes (for example, in the respiratory tract).

Natural Killer Cells

Natural killer (NK) cells are large granular lymphocytes that patrol our blood and lymph and provide important innate defensive functions for our bodies. In fact, these cells kill many types of tumor cells and cells infected by different kinds of viruses. Natural killer cells are produced in the red bone marrow and make up about 15% of the total lymphocyte number. Because they have a broad-based action and do not have to be activated by specific foreign antigens to become active, we usually include NK cells among the body's innate immune system. They are not phagocytic, however. They release chemicals called *perforins* that cause the targeted cell's membrane to rupture and disintegrate.

Natural killer cells can attack a large range of invading cells simply by recognizing markers on the surface membrane of invading cells or defective cells. Some of the receptors on NK cells are "killer-inhibiting" in nature. If the killer-inhibiting receptor of an NK cell happens to bind to a major histocompatibility complex (MHC) protein also, then the killing action is stopped. (Box 19-1 explains that MHCs are surface proteins on all normal cells and are unique to each individual.) Thus, only abnormal and foreign cells fail to bind to the killer-inhibiting centers—and therefore are killed by the NK cell.

Interferon

Several types of cells, if invaded by viruses, respond rapidly by synthesizing and releasing glycoproteins called **interferons (IFNs).** Interferon proteins *interfere* with the ability of viruses to replicate and cause disease in the host's cells. These proteins induce the activation of antiviral genes in neighboring cells. In this way, interferons allow virus-infected cells to send out an "alarm" to nearby cells that protects the uninfected cells. The presence of interferons may also account for symptoms such as sore muscles, body aches, and fever when a virus invades our body.

Interferons come in several varieties, each with somewhat different antiviral actions. Three major types of interferons have now been produced with gene-splicing techniques, and studies exploring antiviral and anticancer activities are still being conducted.

Complement

Complement is a name that applies to a group of about 20 inactive enzymes found in the plasma and on cell membranes. Individual complement proteins are often designated by C (complement) followed by a number, such as C1, C2, and so on. Complement molecules are activated in a cascade of chemical reactions triggered by either adaptive or innate mechanisms. Ultimately, the complement *lyses* (breaks apart) the foreign cell that triggered the response. Complement also marks microbes for destruction by phagocytic cells. This process, called **opsonization,** promotes the inflammatory response in the body's affected tissues.

 BOX 19-1 FYI

Major Histocompatibility Complex

The **major histocompatibility complex (MHC)** is a set of genes that code for antigen-presenting proteins and other immune system proteins. *Antigens are proteins that potentially trigger a specific immune response.* Their function is to present different protein fragments (peptides) at the surface of the cell for possible recognition as either self or nonself antigens by immune system cells.

The MHC class I proteins function to present protein fragments from within the cell at the surface as antigens. An immune cell will then recognize the presented antigen as a *self-antigen* or as a *nonself-antigen* (see figure). Self-antigens are normally ignored by the immune cell. Nonself-antigens are instead recognized as abnormal and attacked. If a normal cell becomes infected with a virus or becomes cancerous, it may present some abnormal antigens on the surface and thus be identified by the immune system. MHC class I proteins are also involved in the mechanism by which natural killer (NK) cells recognize abnormal cells.

MHC class II proteins are expressed in immune cells that specialize in presenting antigens. These "professional" **antigen-presenting cells (APCs)** include macrophages and dendritic cells (DCs), for example. The APCs use their MHC class II proteins to present fragments of proteins that they've brought in from outside the cell. Thus they alert the immune system to the presence of these invaders and trigger certain adaptive (specific) immune responses.

MHC class III proteins include a wide variety of different immune-related proteins such as complement components and a number of immune and nonimmune proteins.

The major histocompatibility complex (MHC) first came to the attention of researchers who were trying to find out why transplants and tissue grafts were often rejected by the recipient. They found that individuals with

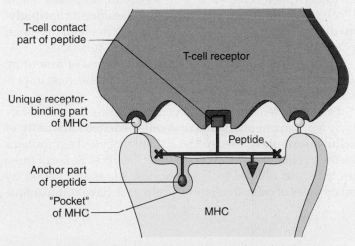

▲ **MHC function.** This simplified diagram shows that the MHC protein displays an antigen (protein fragment or peptide) on the surface of the cell. A receptor on the surface of a T cell may then bind to the unique receptor-binding part of the MHC and to a complementary part of the T cell. Antigens (peptides) presented this way can then be recognized by immune cells as being "self" or "nonself."

different MHC genes rejected tissues transplanted from one to the other. Thus they coined the term *histocompatibility* for this set of genes because the genes seemed to regulate the compatibility of transplants and grafts.

There are hundreds of different versions or *alleles* of the principal MHC genes—far more genetic variability than in any other group of genes in the human genome! Scientists are still trying to find a satisfactory explanation for this tremendous variation.

QUICK CHECK

12. Why are the skin and mucous membranes included as major players in the *first line of defense?*

13. Describe the basic inflammatory response. How does the inflammatory response protect the body?

14. What is phagocytosis? Name some phagocytic cells.

15. How do interferons and complement protect the body?

Overview of Adaptive Immunity

A variety of adaptive (specific) immune mechanisms are geared to attack specific agents that the body recognizes as abnormal or nonself. **Adaptive immunity**—part of the body's *third line of defense*—is provided by two different types of *lymphocytes* (Figure 19-13). Lymphocytes originally are produced in the red bone marrow of the fetus from **hematopoietic stem cells.** However, the cells that eventually become lymphocytes follow two different developmental paths (Figure 19-14). For this reason, there are two major classes of lymphocytes: *B lymphocytes* (**B cells**) and *T lymphocytes* (**T cells**).

B cells do *not* attack pathogens directly. Instead, they produce molecules called **antibodies** that attack the pathogens, or direct other cells, such as phagocytes, to attack them. B-cell mechanisms are therefore often classified as **antibody-mediated immunity.** Antibodies disperse freely in the blood plasma, where they perform their immune functions.

To help you understand the essential terms of immunity as you continue, we've created a short list of the most important ones in Box 19-2.

Because T cells attack pathogens more directly, we classify their mechanism of operation as **cell-mediated immunity** or **cellular immunity** (Figure 19-15). Lymphocytes bear proteins on their cellular surfaces called *surface markers.* Some of these proteins are unique to lymphocytes; some are shared by other types of cells. B cells and T cells each have some unique

BOX 19-2 FYI

The Language of Adaptive Immunity

Learning the mechanisms of adaptive (specific) immunity will be easier if you first become familiar with these terms:

Antigens—macromolecules (large molecules) that induce the immune system to respond in specific ways. Most antigens are foreign proteins. However, some are polysaccharides and some are nucleic acids. *Haptens*, sometimes called "incomplete antigens," are very small molecules that must first bind to a protein before they can induce an immune response. Many antigens that enter the body are macromolecules located in the walls or outer membranes of microorganisms or the outer coats of viruses. Of course, antigens on the surfaces of some tumor cells (tumor markers) are not really from outside the body but are "foreign" in the sense that they are recognized as "not belonging." The membrane molecules that identify all the normal cells of the body are called *major histocompatibility complex (MHC) antigens.*

Antibodies—plasma proteins of the class called *immunoglobulins.* Unlike most antigens, all antibodies are native molecules; that is, they are normally present in the body.

▲ **FIGURE 19-13** **Lymphocytes.** Color-enhanced scanning electron micrograph showing lymphocytes in *yellow*, red blood cells in *red*, and platelets in *green*.

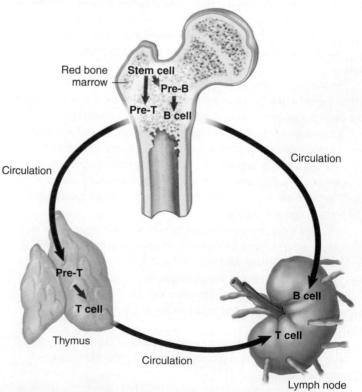

▲ **FIGURE 19-14** **Development of B cells and T cells.** Both types of lymphocytes originate from stem cells in the red bone marrow. Pre-B cells that are formed by dividing stem cells develop in the special tissues of the yolk sac, fetal liver, and bone marrow. Pre-T cells migrate to the thymus, where they continue developing. Once they are formed, B cells and T cells circulate to the lymph nodes and spleen.

Combining sites—two small concave regions on the surface of an antibody molecule. The unique shapes of the combining sites allow the antigen to bind to the antibody to form an **antigen-antibody complex.**

Clone—family of cells, all of which have descended from one cell.

Complement—a group of proteins that, when activated, work together to destroy foreign cells.

Effector cell—a B cell or T cell that is actively producing an immune response, such as secreting antibodies (effector B cells) or directly attacking other cells (effector T cells). Effector cells usually die during or just after their immune response. Effector B cells are also called *plasma cells.*

Memory cell—a B or T cell that has been activated (no longer naïve) but is *not* an effector cell producing an active response. A memory cell survives for a long period in the lymph nodes and, if later exposed to the same specific antigen, forms a clone of cells that rapidly produce a specific immune response.

Naïve—refers to a B or T cell that is inactive, because it has not yet been exposed to (or had an opportunity to react with) a specific antigen. The term is synonymous with "inactive."

surface markers that not only distinguish B cells from T cells, but also subdivide these categories even further into *subsets.*

There is actually an international system for naming the subset of surface markers on blood cells. It's called the **CD system** (CD stands for *cluster* of *differentiation*). The number after "CD" refers to a single, defined surface marker protein. For example, the number of cells in the CD4 and CD8 T-cell subsets are clinically important in *diagnosing* AIDS (Figure 19-16).

The densest populations of lymphocytes occur in the bone marrow, thymus gland, spleen, and lymph nodes. Lymphocytes pour into the bloodstream from these structures. In this way they are distributed throughout the body. After meandering through the tissue spaces, they eventually find their way into lymphatic capillaries. Lymph flow then transports the lymphocytes through a succession of lymph nodes and lymph vessels and then, as we have seen, they empty into the thoracic and right lymphatic ducts into the subclavian veins. In this manner they are returned to the blood. Now they embark on still another long journey—through blood, tissue spaces, and lymph and then back to the bloodstream. This recirculation and widespread distribution of lymphocytes allows these cells to search out, recognize, and destroy foreign invaders anywhere in the body.

Take a moment to re-read the information in Box 19-2 before you continue.

QUICK CHECK

16. What is adaptive immunity?

17. What cells are involved in antibody-mediated immunity?

18. Why is the CD system so important to diseases such as AIDS?

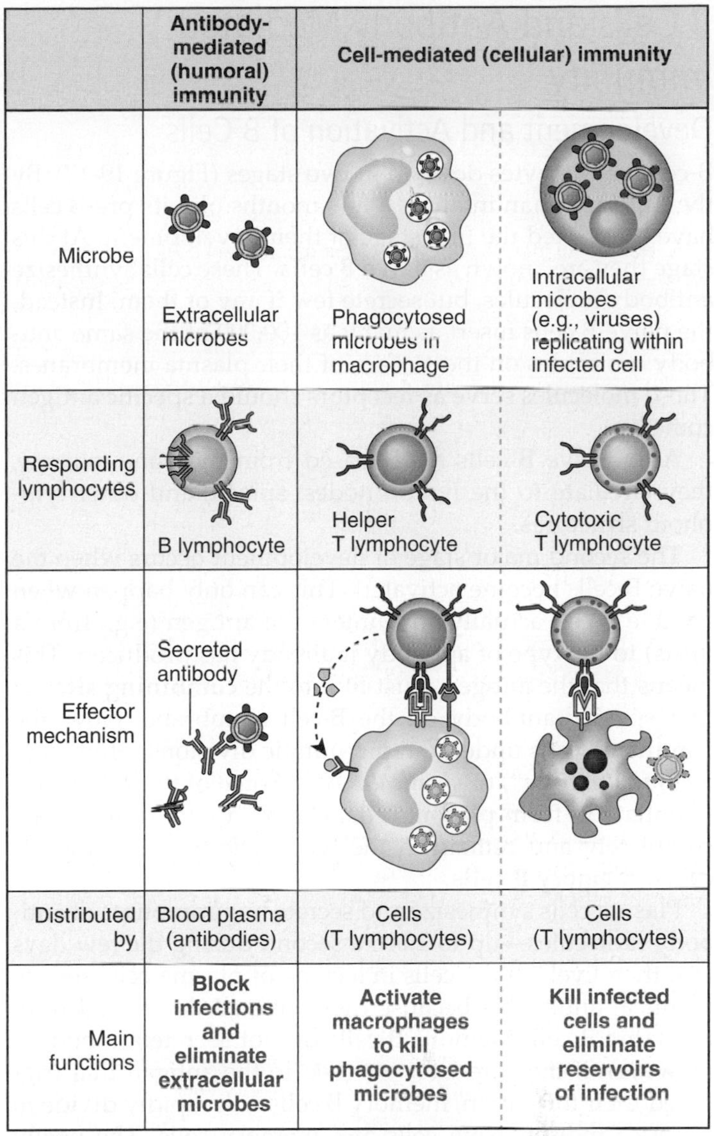

	Antibody-mediated (humoral) immunity	Cell-mediated (cellular) immunity	
Microbe	Extracellular microbes	Phagocytosed microbes in macrophage	Intracellular microbes (e.g., viruses) replicating within infected cell
Responding lymphocytes	B lymphocyte	Helper T lymphocyte	Cytotoxic T lymphocyte
Effector mechanism	Secreted antibody		
Distributed by	Blood plasma (antibodies)	Cells (T lymphocytes)	Cells (T lymphocytes)
Main functions	Block infections and eliminate extracellular microbes	Activate macrophages to kill phagocytosed microbes	Kill infected cells and eliminate reservoirs of infection

▲ **FIGURE 19-15** **Two strategies of adaptive immunity.** Simplified summary of antibody-mediated (humoral) immunity and cell-mediated (cellular) immunity.

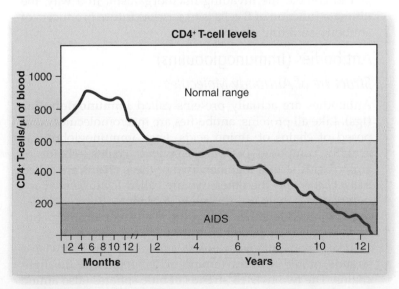

▲ **FIGURE 19-16** **Clinical progression of HIV/AIDS.** Changing numbers of CD4 T cells as an HIV infection progresses to AIDS.

B Cells and Antibody-Mediated Immunity

Development and Activation of B Cells

B-cell lymphocytes develop in two stages (Figure 19-17). By the time a human infant is a few months old, its pre-B cells have completed the first stage of their development. At this stage they are known as **naïve B cells.** These cells synthesize antibody molecules, but secrete few if any of them. Instead, the naïve B cells insert as many as 100,000 of the same antibody molecules on the surface of their plasma membranes. These molecules serve as receptors should a specific antigen come by.

After naïve B cells are released from the bone marrow, they circulate to the lymph nodes, spleen, and other lymphoid structures.

The second major stage of development occurs when the naïve B cells become activated. This can only happen when a naïve B cell actually encounters the antigen (e.g., from a virus) to the type of antibody it already has produced. This means that the antigen must fit into the **combining sites** of that specific antibody on the B-cell membrane. Now, the stimulated cells undergo rapid mitotic divisions, producing a **clone** ("family") of *identical B cells.* Some of these cells differentiate to form **plasma cells.** Others do not differentiate completely and remain in the lymphatic tissue as the so-called **memory B cells.**

Plasma cells synthesize and secrete huge amounts of antibody molecules—up to 2,000 a second during the few days that they live! All the cells in a clone of plasma cells secrete identical antibodies because they have all descended from the same B cell. Memory B cells do not secrete antibodies. However, if they are later exposed to the antigen that triggered their formation, memory B cells will rapidly divide to produce more plasma cells and memory cells. The newly formed plasma cells then quickly secrete antibodies. As before, these antibodies can combine with the initiating antigen to combat the invading microorganism. In a way, the ultimate function of B cells is to serve as future producers of antibody-secreting plasma cells.

Antibodies (Immunoglobulins)

Structure of Antibody Molecules

Antibodies are actually proteins called **immunoglobulins (Igs).** Like all proteins, antibodies are macromolecules composed of chains of amino acids. Each immunoglobulin is actually composed of four *polypeptide* chains—chains of amino acids strung together. Two of these chains are called *heavy chains* and the other two are called *light chains.* Each polypeptide chain is intricately folded to form globular regions joined together so that the resulting immunoglobulin molecule is shaped like a Y.

Take a moment now to look at Figure 19-18, *A.* The twisted strands of red spheres (amino acids) represent the light chains. The two twisted strands of blue spheres (also amino

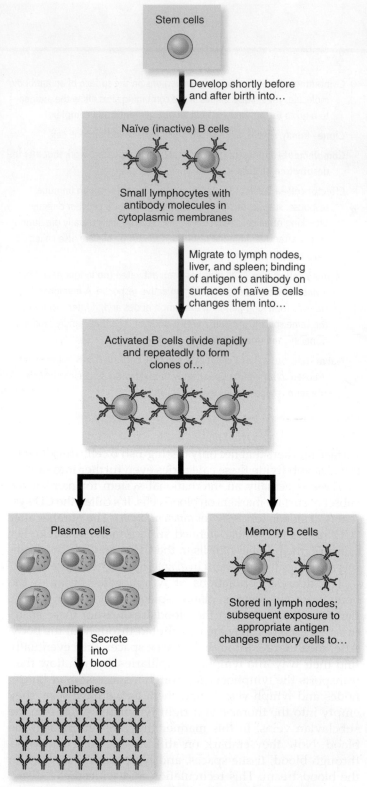

▲ FIGURE 19-17 **B cell development.** B cell development takes place in two stages. First stage: Shortly before and after birth, stem cells develop into naïve B cells. Second stage (occurs only if naïve B cell contacts its specific antigen): Naïve B cell develops into activated B cell, which divides rapidly and repeatedly to form a clone of plasma cells and a clone of memory cells. Plasma cells secrete antibodies capable of combining with specific antigens that cause naïve B cells to develop into active B cells. Stem cells maintain a constant population of newly differentiating cells.

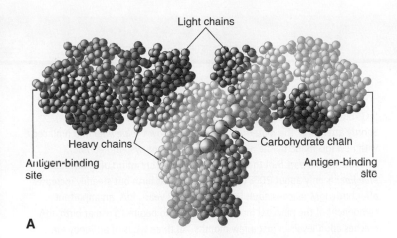

A

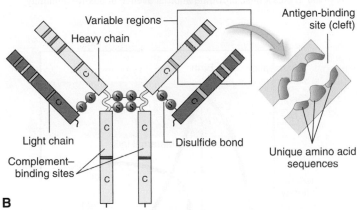

B

▲ **FIGURE 19-18**   **Structure of the antibody molecule. A,** In this molecular model of a typical antibody molecule, the light chains are represented by strands of red spheres (each represents an individual amino acid). Heavy chains are represented by strands of blue spheres. Notice that the heavy chains can join with a carbohydrate chain. **B,** This simplified diagram shows the variable regions, highlighted by colored bars, that represent amino acid sequences unique to that molecule. Constant regions of the heavy and light chains are marked with the letter C. The inset shows that the variable regions at the end of each arm of the molecule form a cleft that serves as an antigen-binding site.

acids) represent the heavy chains. Heavy chains are about twice as long and weigh about twice as much as light chains, hence their name.

The regions with colored bars seen in Figure 19-18, *B,* represent **variable regions.** This means that the amino acid sequence in these regions can *vary* between different antibody molecules. Because the sequence of amino acids varies in these regions, so does the final shape of the binding sites in these areas. At the end of each "arm" of the Y-shaped antibody molecule, the unique shapes of the variable regions form a molecular opening or *cleft.* This cleft serves as the antibody's binding site for antigens. It is this incredible structural feature that enables our antibodies to recognize and combine with specific antigens. This is the first crucial step in our body's defense against invading microorganisms and other foreign cells.

In addition to its variable region, each light chain of an antibody also has a **constant region.** The constant region

comprises an amino acid sequence that is identical in all antibody molecules. Each heavy chain of an antibody molecule consists of three constant regions in addition to its one variable region. Note in Figure 19-18, *B,* the location of two *complement-binding sites* on the antibody molecule (one on each heavy chain).

Diversity of Antibodies

Every normal baby is born with an enormous number of *different* clones of B cells. Populations of these cells are found in the bone marrow, lymph nodes, and spleen. All the cells of each clone are already committed to synthesizing a specific antibody. The sequence of amino acids in the variable regions of a specific antibody is *different* from the sequence in all the countless clones of B cells in the baby's body. This enormous variability may be due to the way the genes encode for the antibodies. In a way, the system acts like a genetic lottery, producing millions of unique genes by combining different gene segments to produce the unique polypeptides.

Classes of Antibodies

There are five classes of immunoglobulin antibodies, identified by letter names as immunoglobulins M, G, A, E, and D (Figure 19-19). Immunoglobulin M (written *IgM*) is the antibody that immature B cells synthesize and insert into their plasma membranes. It is also the predominant class of antibody produced after initial contact with an antigen. However, the most abundant circulating antibody, making up about 75% of all antibodies in your blood, is *IgG.* Production of IgG increases after later contacts with a given antigen. The IgG antibodies are also those that cross the placental barrier during pregnancy. This is how natural **passive immunity** is given to the baby (Box 19-3).

There are three other types of immunoglobulins: *IgA, IgE,* and *IgD.* IgA is the major class of antibody in the mucous membranes of the body. It is also found in saliva and in tears. IgE, although present in very small amounts, can produce the major symptoms of allergies. IgD is also present in our blood, but in very small amounts. We don't yet know its function. As Figure 19-19 shows, some immunoglobulin

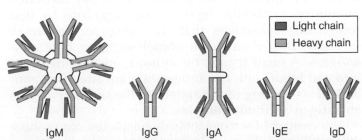

IgM   IgG   IgA   IgE   IgD

▲ **FIGURE 19-19**   **Classes of antibodies.** Antibodies are classified into five major groups: immunoglobulin M (IgM), immunoglobulin G (IgG), immunoglobulin A (IgA), immunoglobulin E (IgE), and immunoglobulin D (IgD). Notice that each IgM molecule has five Y-shaped basic antibody units, IgA has two basic antibody units, and the others have a single basic antibody unit.

BOX 19-3 **FYI**

Prenatal Immunity

Without direct access to external antigens, it is no wonder that the immune system is not very capable (on its own) of a vigorous defense during its maturation process before birth (prenatal development). However, as part *A* of the figure shows, certain antibodies from the mother (maternal antibodies) can be actively transported across the maternal-fetal blood barrier (tropho-blast). Only IgG antibodies in the mother's blood can bind to the receptors. This binding then triggers endocytosis and transports each IgG antibody across to the fetal bloodstream. This mechanism provides passive natural immunity before and shortly after birth.

Part *B* of the figure shows that, at birth, the newborn has adult levels of IgG—but nearly all of it came from the mother (maternal IgG, *blue line*). Shortly after birth, the maternal IgG is broken down *(blue line)* and replaced with new IgG made by the newborn's own immune system *(red line)*.

Note also in part *B* of the figure that the concentration of IgM *(broken blue line)* is only about 20% of the adult level at birth but steadily increases after birth. IgM reaches adult levels in about 2 years. IgA, an important component of the mucosal immune system, also begins to rise at birth. IgA reaches adult levels in just a few months. All three types of antibody are also found in breast milk, providing another avenue of passive immunity after birth.

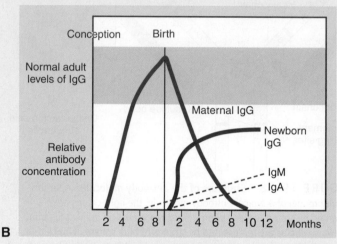

▲ A, Transport of antibodies across the placenta. B, Antibody concentrations before and after birth.

molecules are created by the joining of several basic anti-body units.

Functions of Antibodies

Antigen-Antibody Reactions

The function of antibody molecules—some 100 million trillion of them—is to produce **antibody-mediated immunity.** Antibodies fight disease organisms by first recognizing substances that are foreign or abnormal. In effect, they distinguish nonself-antigens from self-antigens. A small region, the **epitope,** on an antigen fits into and binds with the antigen-binding site of an anti-body. This binding of the antigen to the antibody creates an **antigen-antibody complex.**

This creation of antigen-antibody complexes may produce several effects. For example, it can transform antigens that are toxins into harmless substances. In other cases, antibodies bind to antigens on the surface of microorganisms. This makes the organisms stick together in clumps *(agglutinate).* The **agglutination** of the invaders (whether they are toxins or microorganisms) allows macrophages and other phagocytes to dispose of them rapidly as a sticky group, rather than one

at a time. The binding of antigens to antibodies often produces still another effect. It *alters the shape of the antibody molecule,* not very much, but enough to expose the mole-cule's previously hidden *complement-binding sites.* This may seem like a small effect, but it is not, as we will see below. In fact, this action initiates an incredible series of reactions that culminate in the destruction of microorganisms and other foreign cells (Figure 19-20).

Complement

Complement is a component of blood plasma that consists of about 20 *inactive* enzyme compounds. They are activated in a sequence and together catalyze a series of intricately linked reactions.

Here's how it works. The binding of an antibody to an antigen located on the surface of a cell alters the shape of the antibody molecule. By doing so, the complement-binding site is exposed. Through a complex series of reactions—a *complement cascade*—other series of events, such as inflam-mation, take place (see Figure 19-20). One of the more spec-tacular results of the complement cascade is the creation of **membrane attack complexes (MACs).** Molecules formed in

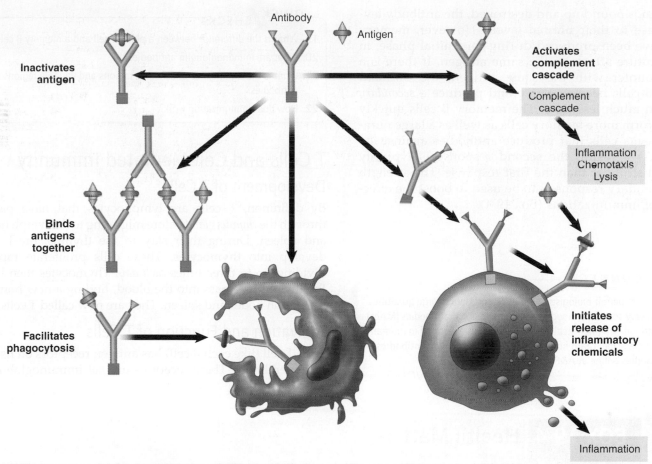

▲ FIGURE 19-20 **Actions of antibodies.** Antibodies act on antigens by inactivating and bending them together to facilitate phagocytosis, and by initiating inflammation and activating the complement cascade.

this complement cascade assemble themselves on the enemy cell's surface in the form of a doughnut with a hole in it. Ions and water then rush into the foreign cell through the hole created by the MAC. As a result, the foreign cell swells and bursts *(cytolysis)*.

Other complement cascades serve other functions. Some attract neutrophils to the site of infection and help with phagocytosis. The complement cascade can also be initiated

by innate immune mechanisms, causing *lysis* (rupture) of the foreign cells.

Primary and Secondary Responses

As you can see in Figure 19-21, the first encounter with a specific antigen produces a *primary response*. This response increases antibody production in a few days. As

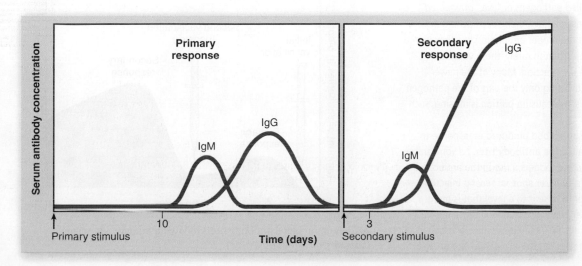

◀ FIGURE 19-21
Antibody response times. The initial encounter with a specific antigen (primary stimulus) produces a primary response (increased production of IgM and IgG) in a few days. A later encounter (secondary stimulus) produces a secondary response in much less time. Note that both IgM production and IgG production occur more quickly in the secondary response—and IgG production also increases in the total amount of antibody produced.

the antigen is bound up and destroyed, the antibody levels decrease to their normal levels. However, memory B cells have been produced during this initial phase, in wait for future attack by the same antigen. If there is a later encounter with the same antigen, the dormant memory B cells become active and produce a *secondary response* in much less time. The memory B cells quickly divide to form more memory cells as well as a large number of plasma cells that produce antibodies against the antigen. As a result, the second response is typically faster and stronger than the first response. The strength of the secondary response can be used to boost the effectiveness of immunizations (Box 19-4).

A&P CONNECT

Techniques that permit biologists to produce and isolate large quantities of pure and very specific antibodies called monoclonal antibodies (MAbs) and tiny antibody fragments called nanobodies have resulted in dramatic advances in medicine. Learn how this works in **Monoclonal Antibodies and Nanobodies** online at **A&P Connect.**

evolve

QUICK✓CHECK

19. What is the difference between a plasma cell and a memory B cell?
20. What are immunoglobulin antibodies?
21. What is the significance of constant regions and variable regions in antibodies?
22. How does immunization work?

T Cells and Cell-Mediated Immunity

Development of T Cells

By definition, T cells are lymphocytes that have passed through the *thymus gland* before migrating to the lymph nodes and spleen. During their stay in the thymus, pre-T cells develop into **thymocytes.** These cells proliferate rapidly, dividing up to three times *each day*! Thymocytes then leave the thymus and pass into the blood, finding a new home in the lymph nodes and spleen. They are now called T cells.

Activation and Function of T Cells

Each T cell (like each B cell) has antigen receptors on its surface membrane. These receptors are *not* immunoglobulins,

---

BOX 19-4 Health Matters

Immunization

Active immunity can be established artificially by using a technique called **vaccination.** The first vaccine was a live cowpox virus that was injected into healthy people to cause a mild cowpox infection. The term *vaccine* literally means "cow substance." Because the cowpox virus is similar to the deadly *smallpox virus,* vaccinated individuals developed antibodies that imparted immunity against both cowpox and smallpox viruses.

Modern vaccines work on a similar principle: Substances that trigger the formation of antibodies against specific pathogens are introduced orally or by injection. Some of these vaccines are killed pathogens or "live," *attenuated* (weakened) pathogens. Such pathogens still have their specific antigens intact, so they can trigger formation of the proper antibodies, but they are no longer *virulent* (able to cause disease). Although rarely, these vaccines sometimes backfire and actually cause an infection. Many of the newer vaccines avoid this potential problem by using only the part of the pathogen that contains antigens. Because the disease-causing portion is missing, such vaccines cannot cause infection.

The amount of antibodies in a person's blood produced in response to vaccination or an actual infection is called the **antibody titer.** As you can see in the figure, the initial injection of vaccine triggers a rise in the antibody titer that gradually diminishes. Often, a **booster shot,** or second injection, is given to keep the antibody titer high or to raise it to a level that is more likely to prevent infection. The secondary response is more intense than the primary response because memory B cells are ready to produce a large number of antibodies at a moment's notice. A later accidental exposure to

the pathogen will trigger an even more intense response—thus preventing infection.

Toxoids are similar to vaccines but use an altered form of a bacterial toxin to stimulate production of antibodies. Injection of toxoids imparts protection against toxins, whereas administration of vaccines imparts protection against pathogenic organisms and viruses.

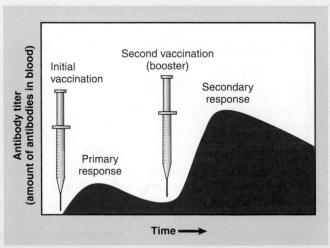

▲ Changes in blood antibody titers following primary and secondary (booster) vaccinations.

but proteins similar to them. When an antigen (*presented* by phagocytes) encounters a naïve T cell, the antigen binds to the T cell's receptors, but only if the surface receptors fit the antigen's epitope (see earlier discussion above).

Here is where we see one of several differences between antibody-mediated immunity and cell-mediated immunity. Antibodies can react to antigens dissolved in the plasma, *but T cells can only react to protein fragments presented on the surface of antigen-presenting cells (APCs)*. Thus T cells react to cells *that are already infected—or have engulfed the antigen*. B cells, in contrast, react primarily to antigens that are in the plasma.

The "presentation" of an antigen by an antigen-presenting cell activates or *sensitizes* the T cell. The T cell then divides repeatedly to form a clone of identical *sensitized T cells* that form **effector T cells** and **memory T cells.** Effector T cells include *cytotoxic T cells (killer T cells)* and *helper T cells*. Memory T cells ultimately produce additional active T cells. We've summarized the process of T-cell development and activation for you in Figure 19-22.

The effector T cells then travel to the site where the antigens originally entered the body. There, in the inflamed tissue, the sensitized T cells bind to antigens of the same kind that led to their formation. However, T cells bind to their specific antigen *only* if the antigen is presented by an APC such as a macrophage or *dendritic cell (DC)*.

As we saw earlier, the chemical messengers released by T cells are called *cytokines* (sometimes called *lymphokines* when secreted by lymphocytes). These messengers perform a variety of immune functions. For example, cytokines such as *lymphotoxins* quickly kill any cell they attack, including cancer cells. Cytokines also serve as signals that trigger additional immune responses. Cytokines released from helper T cells help activate both cytotoxic T cells and B cells when they are presented with antigens.

The general function of T cells is to produce cell-mediated immunity. These cells search out, recognize, and bind to appropriate antigens located on the surfaces of cancerous cells or cells that have been invaded by viruses. This action kills the cells—the ultimate function of killer T cells. Foreign cells—from transplanted tissue, for example—may also be attacked unless the killer T cells are suppressed. T cells also serve as overall regulators of adaptive immune mechanisms.

QUICK ✓ CHECK

23. From what structure do T cells derive their name?
24. What is the difference between effector T cells and memory T cells?

Types of Adaptive Immunity

Remember that *innate immunity* occurs when *nonspecific* immune mechanisms are formed *before birth*—during the early stages of human development in the uterus. However, *adaptive immunity* is a specific kind of resistance that develops *after birth*. Adaptive immunity is, therefore, **acquired immunity.** Acquired immunity may be further classified as either *natural immunity* or *artificial immunity,* depending on how the body is exposed to a specific antigen (Table 19-1).

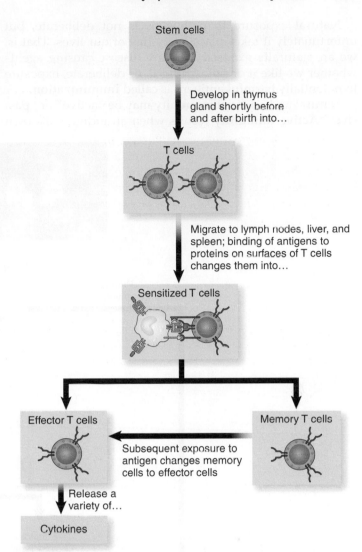

▲ FIGURE 19-22   T-cell development. The first stage occurs in the thymus gland shortly before and after birth. Stem cells maintain a constant population of newly differentiating cells as they are needed. The second stage occurs only if a T cell is presented an antigen, which combines with certain proteins on the T cell's surface.

TABLE 19-1   Types of Adaptive Immunity

TYPE	DESCRIPTION OR EXAMPLE
Natural Immunity	Exposure to the causative agent is not deliberate.
Active (exposure)	A child develops measles and acquires an immunity to a subsequent infection.
Passive (exposure)	A fetus receives protection from the mother through the placenta, or an infant receives protection through the mother's milk.
Artificial Immunity	Exposure to the causative agent is deliberate.
Active (exposure)	Injection of the causative agent, such as a vaccination against polio, confers immunity.
Passive (exposure)	Injection of protective material (antibodies) that was developed by another individual's immune system confers immunity.

Natural exposure to pathogens is not deliberate, but unfortunately it takes place every day of our lives. That is, we are naturally exposed to many disease-causing agents whether we like it or not. Artificial, or deliberate, exposure to potentially harmful antigens is called **immunization.**

Natural and artificial immunity may be "active" or "passive." **Active immunity** results when an individual's own immune system responds to a harmful agent—regardless of how it was encountered. **Passive immunity** results when immunity to a disease that has developed in another individual is transferred to an individual who was *not* previously immune. For example, antibodies in a mother's milk impart passive immunity to her nursing infant (see Box 19-3). Active immunity generally lasts longer than

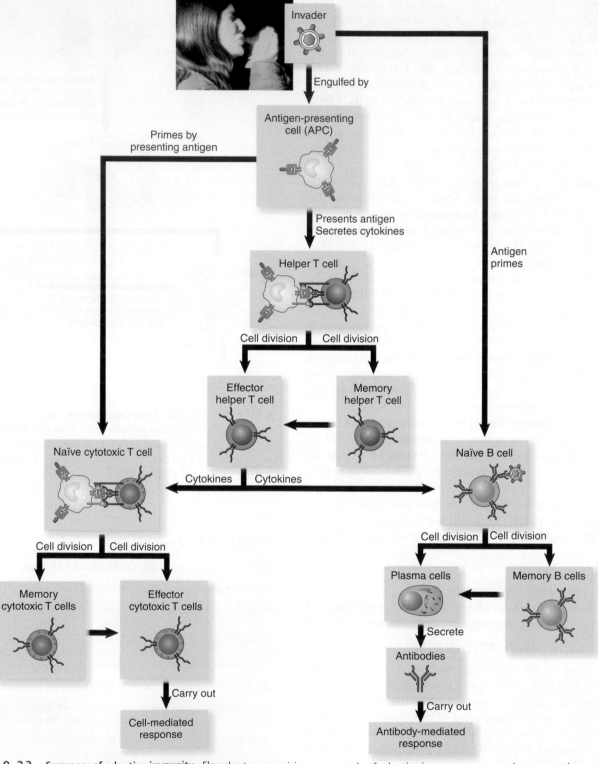

▲ **FIGURE 19-23** **Summary of adaptive immunity.** Flowchart summarizing an example of adaptive immune response when exposed to a microbial pathogen.

passive immunity. Passive immunity, although temporary, provides immediate protection.

Summary of Adaptive Immunity

Adaptive immunity is specific immunity—that is, it targets specific antigens. Two special types of lymphocytes play a major role in immunity: B cells and T cells. B cells recognize specific antigens and produce specific antibodies (immunoglobulins) to destroy the antigen. *This is antibody-mediated immunity.* T cells recognize antigens presented on cell surfaces to attack infected and abnormal cells in several ways. *This is cell-mediated immunity.*

Adaptive immunity progresses along a pathway of stages. First, B cells and T cells recognize a specific antigen. Next, the B and T cells are activated. They expand their population (a clone) and thus produce **effector cells** and **memory cells.** Then, the effector cells attack the source or sources of the antigen. When there are no longer enough antigens to continue stimulating these immune responses, the effector B and effector T cells die *(apoptosis)*. This represents a return to a homeostatic balance. However, a number of memory cells remain—ready to quickly engage the antigen should it reappear later (Figure 19-23).

QUICK CHECK

25. Describe the basic process of acquired immunity.
26. What is the difference between natural and artificial immunity?
27. What is the difference between active and passive immunity?

A&P CONNECT

The *mucosal immune system* is a set of innate and adaptive mechanisms that defend our body at the mucous membrane barriers to the outside world. Understanding this unique system has led to new strategies of immunization. Learn why this is so in **Mucosal Immunity** online at **A&P Connect.**

*e*volve

Cycle of LIFE ○

Most of the organs containing masses of developing lymphocytes appear before birth and continue growing through most of childhood. At puberty, this growth typically slows dramatically. After puberty, the lymphoid organs typically begin to slowly atrophy and are much reduced in size by adulthood. These organs—including the thymus, lymph nodes, tonsils, and other lymphoid structures—shrink in size and become fatty or fibrous. The notable exception is the spleen, which develops early in life and remains intact until very late adulthood.

As we've seen, the overall function of the immune system is maintained throughout maturity. However, deficiency of the immune system creates a greater risk of infections and cancer. Likewise, hypersensitivity of the immune system may make autoimmune conditions more likely to occur. ●

The BIG Picture

You can think of the lymphatic system as a vast, systemic filtration system that drains away excess lymph from the vascular system. The treatment sites in this filtration system are the lymph nodes. Contaminants are removed from lymph and the recycled fluid is eventually returned to the bloodstream. The lymphatic system not only prevents dangerous fluid buildup in our tissues, it also is vital in the production of lymphocytes to fight infections. During our later years, deficiency of the immune system permits a much greater risk from infection and cancer.

The agents of the immune system—antibodies, lymphocytes, and other substances and cells—are everywhere in the body. Without the constant defensive activity of the immune system, our internal homeostasis would be constantly challenged by cancer, infections, and even minor injuries.

We've seen some of the basic intricacies of the immune system, yet this is just the beginning. Recent research suggests that our immune system is regulated to some degree by the nervous and endocrine systems. A new field, *neuroimmunology*, is emerging. This is more evidence that all systems in our bodies are highly integrated. The agents of the immune system are not a separate, distinct group of cells and substances. They include blood cells, skin cells, mucosal cells, brain cells, liver cells, and many other types of cells and their secretions. Thus the immune system is more like a self-defense force made up of ordinary citizens who work shoulder-to-shoulder—with military specialists!

MECHANISMS OF DISEASE

We are afflicted with a variety of disorders and diseases of the lymphatic and immune systems, many of which can be life threatening. For example, there are disorders such as lymphedema associated with lymphatic vessels as well as diseases that cause acute inflammation of the lymphatic vessels. There are also a number of disorders associated with lymph nodes and the lymphatic organs, including the well-known ones such as tonsillitis and several types of lymphomas such as Hodgkin disease.

Disorders of the immune system include a variety of allergies and autoimmune conditions, such as systemic lupus erythematosus. Other conditions involve **autoimmunity,** an inappropriate and excessive response to self-antigens, and deficiencies of the immune system such as those seen in congenital immune deficiency and acquired immune deficiency.

*e*volve Find out more about these disorders and diseases of the lymphatic and immune systems online at *Mechanisms of Disease: Lymphatic and Immune Systems.*

chyle (kile)
[*chyl-* juice]

clone (klohn)
[*clon* a plant cutting]

combining site

complement
[*comple-* complete, *-ment* result of action]

constant region (KONS-tent REE-jun)

cytokine (SYE-toh-kyne)
[*cyto-* cell, *-kine* movement]

diapedesis (dye-ah-peh-DEE-sis)
[*dia-* through, *-pedesis* an oozing]

edema (eh-DEE-mah)
[*edema* swelling]

effector cell (ah-FEK-tor sell)
[*effect-* accomplish, *-or* agent, *cell* storeroom]

effector T cell
[*effect-* accomplish, *-or* agent, *T* thymus gland, *cell* storeroom]

epitope (EP-ih-tope)
[*epi-* on or upon, *-tope* place]

fever (FEE-ver)

germinal center (JER-mih-nal SEN-ter)
[*germ-* sprout, *-al* relating to]

hematopoietic stem cell (hee-mah-toh-poy-ET-ik)
[*hema-* blood, *-poie-* make, *-ic* relating to, *cell* storeroom]

immune system (ih-MYOON SIS-tem)
[*immun-* free, *system* organized whole]

immunization (ih-myoo-nih-ZAY-shun)
[*immun-* free (immunity), *-tion* process of]

immunoglobulin (Ig) (ih-myoo-noh-GLOB-yoo-lin)
[*immuno-* free (immunity), *-glob-* ball, *-ul-* small, *-in* substance]

inflammatory response (in-FLAM-ah-toh-ree)
[*inflamm-* set afire, *-ory* relating to]

innate immunity (IN-ayt ih-MYOO-nih-tee)
[*innat-* inborn, *immun-* free, *-ity* state]

interferon (IFN) (in-ter-FEER-on)
[*inter-* between, *-fer-* strike, *-on* substance]

interstitial fluid (IF) (in-ter-STISH-al FLOO-id)
[*inter-* between, *-stit-* stand, *-al* relating to]

lacteal (LAK-tee-al)
[*lact-* milk, *-al* relating to]

lingual tonsil (LING-gwal TAHN-sil)
[*lingua-* tongue, *-al* relating to]

lymph (limf)
[*lymph* water]

lymphatic capillary (lim-FAT-ik KAP-ih-lair-ee)
[*lymph-* water, *-atic* relating to, *capill-* hair, *-ary* relating to]

lymphatic system (lim-FAT-ik SIS-tem)
[*lymph-* water, *-atic* relating to, *system* organized whole]

lymphatic vessel (lim-FAT-ik)
[*lymph-* water, *-atic* relating to]

lymph node (limf)
[*lymph* water, *nod-* knot]

lymphoid tissue (LIM-foyd TISH-yoo)
[*lymph-* water, *-oid* like, *tissu* fabric]

macrophage (MAK-roh-fayj)
[*macro-* large, *-phage* eat]

major histocompatibility complex (MHC) (his-toh-kom-pat-ih-BIL-ih-tee KOM-pleks)
[*histo-* tissue, *-compatibil-* agreeable, *-ity* state, *complex* embrace]

mastitis (mass-TYE-tis)
[*mast-* breast, *-itis* inflammation]

membrane attack complex (MAC) (MEM-brayne KOM-pleks)
[*membran-* thin skin, *complex* embrace]

memory B cell
[*B* bursa-equivalent tissue, *cell* storeroom]

memory cell

memory T cell
[*T* thymus gland, *cell* storeroom]

naïve (nye-EVE)
[*naïve* natural]

naïve B cell (nye-EVE B sell)
[*naïve* natural, *B* bursa-equivalent tissue, *cell* storeroom]

natural killer (NK) cell

neutrophil (NOO-troh-fil)
[*neuter-* neither, *-phil* love]

nonspecific immunity (non-speh-SIF-ik ih-MYOO-nih-tee)
[*non-* not, *-specif-* form or kind, *-ic* relating to, *immun-* free, *-ity* state]

opsonization (op-son-ih-ZAY-shun)
[*opson-* condiment, *-ization* process]

palatine tonsil (PAL-ah-tyne TAHN-sil)
[*palat-* palate, *-ine* relating to]

passive immunity (PAS-iv ih-MYOO-nih-tee)
[*immun-* free, *-ity* state]

phagocytosis (fag-oh-sye-TOH-sis)
[*phago-* eating, *-cyt-* cell, *-osis* condition]

phagosome (FAG-oh-sohm)
[*phago-* eat, *-some* body]

pharyngeal tonsil (fah-RIN-jee-al TAHN-sil)
[*pharyng-* throat, *-al* relating to]

plasma cell (PLAZ-mah sell)
[*plasma* something molded (blood plasma), *cell* storeroom]

pus

radical mastectomy (RAD-ih-kal mas-TEK-toh-mee)
[*radic-* root, *-al* relating to, *mast-* breast, *-ec-* out, *-tom-* cut, *-y* action]

right lymphatic duct (lim-FAT-ik)
[*lymph-* water, *-atic* relating to]

self-tolerance

species resistance (SPEE-sheez ree-ZIS-tens)
[*species* form or kind]

specific immunity (speh-SIF-ik ih-MYOO-nih-tee)
[*specif-* form or kind, *-ic* relating to, *immun-* free, *-ity* state]

spleen

T cell
[*T* thymus gland, *cell* storeroom]

thoracic duct (thoh-RAS-ik)
[*thorac-* chest (thorax), *-ic* relating to]

thymocyte (THY-moh-syte)
[*thymo-* thyme flower (thymus gland), *-cyte* cell]

thymosin (THY-moh-sin)
[*thymos-* thyme flower (thymus gland), *-in* substance]

thymus (THY-mus)
[*thymus* thyme flower] *pl.,* thymuses

tonsil (TAHN-sil)

tonsillectomy (tahn-sih-LEK-toh-mee)
[*tonsil-* tonsil, *-ec-* out, *-tom-* cut, *-y* action]

tonsillitis (tahn-sih-LYE-tis)
[*tonsil-* tonsil, *-itis* inflammation]

toxoid (TOK-soyd)
[*tox-* poison, *-oid* like]

vaccination (vak-sih-NAY-shun)
[*vaccin-* cow (cowpox), *-ation* process]

variable region (VAIR-ee-ah-bel REE-jun)

CHAPTER SUMMARY

To download an MP3 version of the chapter summary for use with your iPod or other portable media player, access the Audio Chapter Summaries online at http://evolve.elsevier.com.

HINT Scan this summary after reading the chapter to help you reinforce the key concepts. Later, use the summary as a quick review before your class or before a test.

COMPONENTS OF THE LYMPHATIC AND IMMUNE SYSTEMS

A. The lymphatic system has at least three different functions:
　1. Serves to maintain the fluid balance of our internal body environment
　2. Serves as part of our immune system
　3. Helps regulate the absorption of lipids from digested food in the small intestines and their transport to the large systemic veins
B. The immune system serves as an internal "security force" to deal with abnormal cells
　1. Repels and destroys microorganisms
　2. Defends us from our own abnormal cells that can cause cancer

LYMPHATIC SYSTEM

A. Overview of the lymphatic system (Figure 19-1)
　1. The lymphatic system solves the problem of fluid retention in tissues
　　a. Acts as a drainage system
　　b. Collects excess tissue fluid and returns it to the venous blood just before it reaches the heart
　2. Lymphatic system is a part of the circulatory system
　　a. Consists of moving fluid derived from the blood and tissue fluid as well as a group of vessels that return the lymph to the blood
　3. In addition to lymph and lymphatic vessels, the system includes various structures that contain lymphoid tissue (Figure 19-2)
　4. Lymphatic system provides a unique transport function
　　a. Returns tissue fluid, proteins, fats, and other substances to the general circulatory system
　5. Lymphatic vessels do not form a closed system of vessels
　　a. They begin blindly in the intercellular spaces of the soft tissue of the body (Figure 19-1)
B. Lymph and interstitial fluid
　1. Lymph—clear fluid found in the lymphatic vessels
　2. Interstitial fluid—complex fluid that fills the spaces between the cells
　3. Both fluids closely resemble blood plasma in composition
　　a. Difference is that lymph cannot clot like blood
C. Lymphatic vessels
　1. Lymphatic vessels—microscopic blind-ended lymphatic capillaries; wall of each lymphatic capillary consists of a single layer of flattened endothelial cells
　　a. Networks of lymphatic capillaries branch and then rejoin repeatedly to form a network throughout the interstitial spaces of our bodies

　　b. Lymphatic capillaries merge to form larger and larger vessels until main lymphatic trunks are formed—right lymphatic duct and thoracic duct (Figure 19-2, B)
　2. Structure of lymphatic vessels (Figure 19-3)
　　a. Walls of lymphatic capillaries
　　　(1) Have numerous openings or clefts between the cells
　　　(2) As lymph flows from the thin-walled capillaries into vessels with larger diameters, the walls become thicker
　　b. Eventually these larger vessels have the three layers typical of arteries and veins
　　c. One-way valves are abundant in lymphatic vessels of all sizes
　3. Functions of lymphatic vessels
　　a. Permeability of the lymphatic capillary wall permits very large molecules and even small particles to be removed from the interstitial spaces
　　b. Lacteals in small intestines are important in the absorption of fats and other nutrients
　　c. Chyle is the milky lymph found in lacteals after digestion
D. Circulation of lymph
　1. About 50% of the total blood protein leaks out of the capillaries into the interstitial fluid; ultimately returns to the blood by way of the lymphatic vessels (Figure 19-4)
　2. The lymphatic pump
　　a. Lymph moves slowly and steadily along in its vessels into the general circulation at about 3 L/day
　　　(1) Lymph flow is possible because of the large number of valves that permit fluid flow only in the general direction toward the heart
　　　(2) Breathing movements and skeletal muscle contraction aid in this motion
E. Lymph nodes
　1. Structure of lymph nodes (Figure 19-5)
　　a. Oval-shaped or bean-shaped structures; widely distributed throughout the body
　　b. Lymph nodes are linked together by the lymphatic vessels
　　c. Fibrous partitions or trabeculae extend from the covering capsule toward the center of a lymph node, creating compartments called cortical nodules
　　d. Center, or medulla, of a lymph node is composed of sinuses; separate medullary cords composed of plasma cells and B cells
　2. Locations of lymph nodes
　　a. Most lymph nodes occur in groups, or clusters (Figure 19-2)
　　b. Approximately 500 to 600 lymph nodes are located in the body
　3. Functions of lymph nodes
　　a. Defend our bodies from invading pathogens; sites of both biological and mechanical filtration
　　b. Provide sites for the maturation of some types of lymphocytes
F. Lymphatic drainage of the breast
　1. Distribution of lymphatics in the breast
　　a. Mammary glands and surrounding tissues of the breast are drained by two sets of lymphatic vessels
　　　(1) Some originate in and drain the surface area and skin over the breast

(2) Others originate in and drain the underlying tissue of the breast itself

2. Lymph nodes associated with the breast
   a. Many anastomoses between the superficial lymphatics from both breasts
   b. Breast infections (e.g., mastitis) are serious health concern, and can spread easily because of the lymphatic pathways of the breast

G. Structure and function of the tonsils
   1. Tonsils—form a protective ring under the mucous membranes in the mouth and back of the throat (Figure 19-7)
      a. Protects against bacteria that may invade tissue in the area around the openings between the nasal and oral cavities; first line of defense from the external environment
      b. Tonsils:
         (1) Palatine—located on each side of the throat
         (2) Pharyngeal (adenoids)—lie near the posterior opening of the nasal cavity
         (3) Lingual—lie near the base of the tongue

H. Structure and function of aggregated lymphoid nodules
   1. Also called Peyer patches—small oval patches or groups of lymph nodes that form protective layer in mucous membrane of the small intestine
      a. Provide protection in a location that is potentially open to the external environment via the mouth
      b. Macrophages and other cells prevent most bacteria from penetrating the gut wall

I. Structure and function of the thymus
   1. Thymus—a primary organ of the lymphatic system
      a. Consists of two pyramid-shaped lobes
      b. Located in the mediastinum, extending up into the neck, close to the thyroid gland (Figure 19-8)
      c. Thymus plays a critical part in the body's defenses against infection
      d. Soon after birth, the thymus begins secreting a group of hormones that enable lymphocytes to develop into mature T cells

J. Structure and function of the spleen
   1. Spleen—located directly below the diaphragm, just above most of the left kidney and behind the fundus of the stomach (Figure 19-2)
      a. Roughly oval in shape (Figure 19-9)
      b. Spleen has variety of functions:
         (1) Defense
         (2) Hematopoiesis
         (3) Erythrocyte and platelet destruction
         (4) Reservoir for blood

IMMUNE SYSTEM

A. Organization of the immune system
   1. Identification of cells and other particles
      a. Antigens—cells, viruses, and other particles with unique molecules on their surfaces
         (1) Self
         (2) Nonself—molecules on the surface of foreign or abnormal cells or particles that serve as recognition markers for identification by our immune system
         (3) Self-tolerance—ability of our immune system to attack abnormal or foreign cells while sparing our own cells

   b. Two categories of defense mechanisms:
      (1) Innate immunity
      (2) Adaptive immunity

   2. Innate (nonspecific) and adaptive (specific) immunity
      a. Innate immunity—"in place" before you are exposed to a particular harmful particle or condition; nonspecific immunity
      b. Adaptive immunity—involves mechanisms that program the body to recognize specific threatening agents; specific immunity
         (1) Primary types of cells involved in innate immunity:
            (a) Epithelial barrier cells
            (b) Phagocytic cells
            (c) Natural killer (NK) cells
         (2) Primary types of cells involved in adaptive immunity:
            (a) T cells
            (b) B cells
         (3) Cytokines—chemicals released from cells to trigger or regulate innate and adaptive immune responses
            (a) Interleukins (ILs)
            (b) Leukotrienes
            (c) Interferons (IFNs)
         (4) Human immune systems are such that there is a type of "species resistance"; genomes of specific organisms may be resistant to particular pathogens

B. Overview of innate (nonspecific) immunity
   1. Mechanical and chemical barriers—first line of defense
      a. Internal environment is protected by a cutaneous membrane (skin) and mucous membranes (Figure 19-10)
      b. Skin and mucous membranes provide additional immune mechanisms; sebum, mucus, enzymes, and hydrochloric acid produced by the stomach
   2. Inflammatory response and fever—second line of defense
      a. Inflammatory response—elicits a number of actions that promote returning your body to a normal state (Figure 19-11)
         (1) Abnormal tissue damage triggers the release of various inflammation mediators
         (2) Inflammation mediators—function to attract leukocytes to the area in a process called chemotaxis (Figure 19-12)
         (3) Characteristic signs of inflammation are heat, redness, pain, and swelling
      b. Fever—a state of elevated body temperature
         (1) Results from a "reset" of the body's thermostat in the hypothalamus
         (2) Fever is thought to increase immune functions and inhibit the reproduction of some microbial pathogens
   3. Phagocytosis and phagocytic cells
      a. Phagocytosis—ingestion and destruction of microorganisms and other small particles by cells called phagocytes
      b. Phagocytosis is classified as an innate defense; also play an important role in adaptive immunity
      c. The most numerous type of phagocyte is the neutrophil; other types include macrophages.
      d. Chemotactic factors cause neutrophils and other phagocytes to adhere to the endothelial lining of capillaries servicing the affected area (Figure 19-12)
         (1) After this, phagocytic cells squeeze through the wall of a blood vessel to get to the site of the injury or infection; diapedesis

4. Natural killer cells—provide important innate defensive functions for our bodies
   a. Kill many types of tumor cells and cells infected by different kinds of viruses
   b. Produced in the red bone marrow and make up about 15% of the total lymphocyte number
   c. Recognize markers on surface membrane of invading or defective cells
      (1) Release chemicals called perforins that cause the cell's membrane to rupture and disintegrate
5. Interferon—glycoprotein that interferes with the ability of viruses to replicate and cause disease in the host's cells
   a. Induces the activation of antiviral genes in neighboring cells
      (1) Allows virus-infected cells to send out an "alarm" to nearby cells that protects the uninfected cells
   b. Three major types of interferons have now been produced with gene-splicing techniques
6. Complement—name that applies to a group of about 20 inactive enzymes in the plasma and on cell membranes
   a. Complement molecules are activated in a cascade of chemical reactions triggered by either adaptive or innate mechanisms
   b. Complement also marks microbes for destruction by phagocytic cells; opsonization
C. Overview of adaptive (specific) immunity
   1. Adaptive immunity—body's third line of defense; provided by two different types of lymphocytes (Figure 19-13)
      a. Two major classes of lymphocytes:
         (1) B lymphocytes (B cells)
         (2) T lymphocytes (T cells)
      b. B cells produce molecules called antibodies that attack the pathogens or direct other cells, such as phagocytes, to attack them; antibody-mediated immunity
      c. T cells attack pathogens more directly and thus operate as cell-mediated immunity (cellular immunity) (Figure 19-15)
      d. Lymphocytes bear proteins on their cellular surfaces called surface markers
         (1) International system for naming the surface markers on blood cells; CD system (CD stands for cluster of differentiation)
         (2) Densest populations of lymphocytes occur in the bone marrow, thymus gland, spleen, and lymph nodes
D. B cells and antibody-mediated immunity
   1. Development and activation of B cells
      a. B-cell lymphocytes develop in two stages (Figure 19-17)
         (1) Pre-B cells develop by a few months of age; naïve B cells
         (2) Second major stage of development then occurs when the naïve B cells become activated
   2. Antibodies (immunoglobulins)
   3. Structure of antibody molecules—consist of two heavy and two light polypeptide chains; each molecule has two antigen-binding sites and two complement-binding sites (Figure 19-18)
   4. Diversity of antibodies—every normal baby is born with an enormous number of different clones of B cells
      a. Cells of each clone are already committed to synthesizing a specific antibody

5. Classes of antibodies—five classes of immunoglobulin antibodies; identified by letter names as immunoglobulins M, G, A, E, and D (Figure 19-19)
   a. IgM—antibody that immature B cells synthesize and insert into their plasma membranes
   b. IgG—most abundant circulating antibody; crosses placental barrier during pregnancy to give passive immunity to baby (Table 19-1 and Box 19-3)
   c. IgA—major class of antibody in the mucous membranes of the body; also in saliva and tears
   d. IgE—produces the major symptoms of allergies
   e. IgD—small amount in blood; function unknown
6. Functions of antibodies
   a. Antigen-antibody reactions—antibody molecules produce antibody-mediated immunity
      (1) Antibodies first recognize substances that are foreign or abnormal
      (2) Distinguish nonself-antigens from self-antigens
      (3) Epitopes—bind to an antibody molecule's antigen-binding sites; form an antigen-antibody complex
   b. Complement—component of blood plasma that consists of about 20 inactive enzyme compounds
      (1) They are activated in a complex sequence; catalyze a series of intricately linked reactions
      (2) Complement cascade—series of reactions that take place as a result of the binding of an antibody to an antigen; membrane attack complexes (MACs) (Figure 19-20)
   c. Primary and secondary responses (Figure 19-21)
      (1) Primary response—initial encounter with a specific antigen triggers the formation and release of specific antibodies that reaches its peak in a few days
      (2) Secondary response—later encounter with the same antigen triggers a much quicker response; memory B cells rapidly divide, producing more plasma cells and thus more antibodies
E. T cells and cell-mediated immunity
   1. Development of T cells
      a. T cells—lymphocytes that have passed through the thymus gland before migrating to the lymph nodes and spleen
         (1) Pre-T cells develop into thymocytes while in thymus
         (2) Thymocytes leave the thymus and pass into the blood, to lymph nodes and spleen; now called T cells
   2. Activation and functions of T cells
      a. T cells have antigen receptors on their surface membrane
      b. The "presentation" of an antigen by an antigen-presenting cell activates or sensitizes the T cell
      c. T cell then divides repeatedly; forms a clone of identical sensitized T cells that form effector T cells and memory T cells (Figure 19-22)
         (1) Effector T cells—travel to the site where the antigens originally entered the body and begin their attack
            (a) Cytotoxic (killer) T cells—destroy infected or cancerous cells
            (b) Helper T cells—signal activation of cytotoxic T cells and B cells when presented with antigen
         (2) Memory T cells—ultimately produce additional effector T cells
      d. Cytokines—chemical messengers
         (1) Perform a variety of immune functions
         (2) Quickly kill any cell they attack, including cancer cells
         (3) Trigger activation of other immune responses

F. Types of adaptive immunity (Table 19-1)
1. Innate immunity—occurs when nonspecific immune mechanisms are formed before birth
2. Adaptive immunity—specific kind of resistance that develops after birth
3. Acquired immunity—classified as either natural immunity or artificial immunity
   a. Natural immunity—results from nondeliberate exposure to antigens
   b. Artificial immunity—results from deliberate exposure to antigens; immunizations
4. Natural and artificial immunity may be "active" or "passive
   a. Active immunity—results when an individual's own immune system responds to a harmful agent, regardless of how it was encountered
   b. Passive immunity—results when immunity to a disease that has developed in another individual is transferred to an individual who was not previously immune (Box 19-3)
G. Summary of adaptive immunity
1. Adaptive immunity is specific immunity; targets specific antigens
2. Two special types of lymphocytes play a major role in immunity:
   a. B cells—antibody-mediated immunity
   b. T cells—cell-mediated immunity
3. Adaptive immunity progresses along a pathway of stages
   a. Recognition of antigen
   b. Activation of lymphocytes
   c. Effector phase (immune attack)
   d. Decline of antigen causes lymphocyte death
   e. Memory cells remain for later response if needed

REVIEW QUESTIONS

HINT *Write out the answers to these questions after reading the chapter and reviewing the Chapter Summary. If you simply think through the answer without writing it down, you won't retain much of your new learning.*

1. List the anatomical components of the lymphatic system.
2. How do interstitial fluid and lymph differ from blood plasma?
3. How do lymphatic vessels originate?
4. Briefly describe the anatomy of the lymphatic capillary wall.
5. Lymph from what body areas enters the general circulation by way of the thoracic duct? By way of the right lymphatic duct?
6. Where does lymph enter the blood vascular system?
7. In general, lymphatics resemble veins in structure. List three exceptions to this general rule.
8. What are the unique lymphatic vessels that originate in the villi of the small intestine called?
9. Locate the thymus, and describe its appearance and size at birth, at maturity, and in old age.
10. Describe the location and functions of the spleen.
11. List several mechanisms of innate defense, and give a brief description of each one.
12. Identify the body's first line of defense.
13. Activated B cells develop into clones of what two kinds of cells?
14. What cells synthesize and secrete copious amounts of antibodies?
15. Explain the function of memory cells.
16. Differentiate between the classifications of natural and artificial immunity.
17. Describe the process behind the functioning of natural and artificial immunity.

CRITICAL THINKING QUESTIONS

HINT *After finishing the Review Questions, write out the answers to these items to help you apply your new knowledge. Go back to sections of the chapter that relate to items that you find difficult.*

1. Even though the lymphatic system is a component of the circulatory system, why is the term *circulation* not the most appropriate term to describe the flow of lymph?
2. Explain how lymph is formed.
3. Discuss the importance of valves in the lymphatic system.
4. Explain the role of the lymphatic system in the spread of breast cancer.
5. If a person had a mutation that prevented the formation of the complement proteins, what capabilities would be lessened in the immune system?
6. Why would you think the development of cancer can be seen as a failure of the immune system?

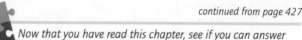
continued from page 427

Now that you have read this chapter, see if you can answer these questions about the flu shot Kostas received in the Introductory Story.

1. Which of Kostas' cells will respond to the flu antigens introduced by the vaccine?
   a. Erythrocytes
   b. Thrombocytes
   c. Lymphocytes
   d. Platelets

2. Which specific cell types will begin producing antibodies to the antigens?
   a. Z cells
   b. T cells
   c. A cells
   d. B cells

3. Which antibody is primarily involved in this response to vaccine?
   a. IgM
   b. IgE
   c. IgG
   d. IgD

4. Kostas' fever during the previous winter's flu was caused by the release of ___, molecules that help increase his body's "set point" to higher than normal.
   a. Vulcanogens
   b. Histamine
   c. Pyrogens
   d. Nanobodies

5. What would you call the specific type of immunity Kostas developed as a result of the vaccination?
   a. Natural active immunity
   b. Artificial active immunity
   c. Natural passive immunity
   d. Artificial passive immunity

HINT *To solve a case study, you may have to refer to the glossary or index, other chapters in this textbook, A&P Connect, Mechanisms of Disease, and other resources.*

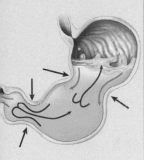

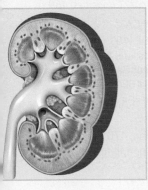

UNIT 5

Respiration, Nutrition, and Excretion

The chapters in Unit Five deal with respiration Chapter 20, digestion Chapter 21, processing of nutrients Chapter 22, and excretion of wastes by the urinary system Chapter 23. Throughout these chapters, you will see that our homeostatic mechanisms function to maintain a constant environment at the cellular level. Consider the oxygen (O_2) and carbon dioxide (CO_2) content of the blood, for example. Under normal conditions the O_2 and CO_2 content of blood does not change much over time. However, the quantity of O_2 and CO_2 that enters and exits the blood can vary widely, depending on a person's type and level of activity. Delivery of oxygen and elimination of carbon dioxide and other wastes resulting from the metabolism of nutrients must be regulated within narrow limits so that cellular function remains normal regardless of our level of activity. Maintaining a constant environment of fluid balance, as well as electrolyte and acid-base balance at the cellular level, is also required for survival. The anatomical structures and functional control mechanisms discussed in this unit all relate to homeostasis—maintaining the constant internal environment—of the entire body.

CHAPTER OUTLINE

 Scan this outline before you begin to read the chapter, as a preview of
HINT *how the concepts are organized.*

STUDENT LEARNING OBJECTIVES

At the completion of this chapter, you should be able to do the
following:

1. Outline the general flow of air through the respiratory system.
2. Describe the functions of the following: nose, pharynx, larynx.
3. Compare the organs of the upper respiratory tract with those
   of the lower respiratory tract.
4. Describe the structure of the bronchi and alveoli, and give
   their functions.
5. Describe the structure and function of the lungs.
6. Describe the process of pulmonary ventilation.
7. Outline the various measures of pulmonary volumes and
   capacities.
8. Discuss what is meant by partial pressure of gases and its
   significance in gas exchange.
9. Describe the process of the transport of blood gases.
10. Discuss how pulmonary function is regulated.

AFTER having surgery to remove a stomach tumor, Derrick woke up in the recovery room in extreme pain. It hurt to move; it hurt to blink; it hurt to take even a little breath. And here was this nurse demanding that he take a deep breath and cough. Was she crazy?

What Derrick didn't realize was that his shallow respirations were not getting rid of as much carbon dioxide as usual. As a result, concentration of carbon dioxide in his bloodstream was building to a level that would negatively affect the homeostasis of his entire body.

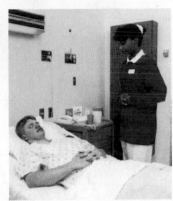

Before reading this chapter, write down what you think was going on in Derrick's respiratory system. Then compare your answer to that of the physician's as we continue Derrick's story at the end of this chapter.

ANATOMY AND PHYSIOLOGY OF THE RESPIRATORY SYSTEM

Your respiratory system has a number of vital roles. First, it functions as a gas distributor and exchanger. It does this by supplying oxygen to your cells and removing carbon dioxide from them. Because the majority of cells are too far from the lungs to absorb oxygen directly, the lungs first use tiny sacs, called *alveoli,* to exchange gases from the respiratory system with those from the blood of the circulatory system.

Second, the respiratory system filters, warms, and humidifies the air we breathe, in addition to providing us with vocal communication and *olfaction* (the sense of smell).

Third, the respiratory system plays a vital physiological role in our bodies by regulating homeostasis of metabolism, circulation, electrolyte and water balance, and acidity of the blood. These processes are critical to proper body functioning.

In this chapter, we will explore both the structural and physiological aspects of the respiratory system.

STRUCTURAL PLAN OF THE RESPIRATORY SYSTEM

The respiratory system can be divided into upper and lower tracts. The organs of the upper tract are located outside of the thorax (in the head and neck), while the lower tract is located almost entirely within the thorax (Figure 20-1).

The **upper respiratory tract** is composed of the nose, nasopharynx, oropharynx, laryngopharynx, and larynx. The **lower respiratory tract** consists of the trachea, the bronchial tree, and the lungs. There are also accessory structures such as the oral cavity, the rib cage, as well as respiratory musculature such as the intercostals and the diaphragm.

HINT *Before reading the chapter, say each of these terms out loud. This will help you avoid stumbling over them as you read.*

alveolus (al-VEE-oh-lus)
[*alve-* hollow, *-olus* little] *pl.,* alveoli (al-VEE-oh-lye)

apex (AY-peks)
[*apex* tip] *pl.,* apices (AY-pih-seez)

base (BAYS)
[*bas-* foundation]

bicarbonate (bye-KAR-boh-nayt)
[*bi-* two, *-carbon-* coal (carbon), *-ate* oxygen compound]

bronchiole (BRONG-kee-ohl)
[*bronch-* windpipe, *-ol-* little]

bronchopulmonary segment (brong-koh-PUL-moh-nair-ee)
[*bronch-* windpipe, *-pulmon-* lung, *-ary* relating to]

carbaminohemoglobin (kahr-bam-ih-no-hee-moh-GLOH-bin)
[*carb-* coal (carbon), *-amino-* ammonia compound (amino acid), *-hemo-* blood, *-glob-* ball, *-in* substance]

carbon monoxide (CO) (KAR-bon mon-OKS-ide)
[*mono-* single, *-ox-* sharp (oxygen), *-ide* chemical]

chronic obstructive pulmonary disease (COPD) (KRON-ik ob-STRUK-tiv PUL-moh-nair-ee)
[*chron-* time, *-ic* relating to, *pulmon-* lung, *-ary* relating to]

cleft palate (kleft PAL-ett)

concha (KONG-kah)
[*concha* sea shell] *pl.,* conchae (KONG-kee)

cough reflex (REE-fleks)
[*re-* back or again, *-flex* bend]

cribriform plate (KRIB-rih-form)
[*cribr-* sieve, *-form* shape]

diving reflex (REE-fleks)
[*re-* back or again, *-flex* bend]

elastic recoil (eh-LAS-tik REE-koyl)
[*elast-* drive or propel, *-ic* relating to]

endotracheal intubation (en-doh-TRAY-kee-al in-too-BAY-shun)
[*endo-* within, *-trache-* rough duct, *-al* relating to, *in-* within, *-tub-* tube, *-ation* process]

epiglottis (ep-ih-GLOT-is)
[*epi-* upon, *-glottis* tongue] *pl.,* epiglottides or epiglottises (ep-ih-GLOT-ih-deez, ep-ih-GLOT-ih-seez)

expiration (eks-pih-RAY-shun)
[*ex-* out, *-[s]pir-* breathe, *-ation* process]

expiratory reserve volume (ERV) (eks-PYE-rah-tor-ee)
[*ex-* out of, *-[s]pir-* breathe, *-tory* relating to]

functional residual capacity (FRC) (FUNK-shun-al ree-ZID-yoo-al kah-PASS-ih-tee)

glottis (GLOT-is)
[*glottis* tongue] *pl.,* glottides or glottises (GLOT-ih-deez, GLOT-ih-seez)

continued on page 477

▶ FIGURE 20-1 Structural plan of the respiratory system.

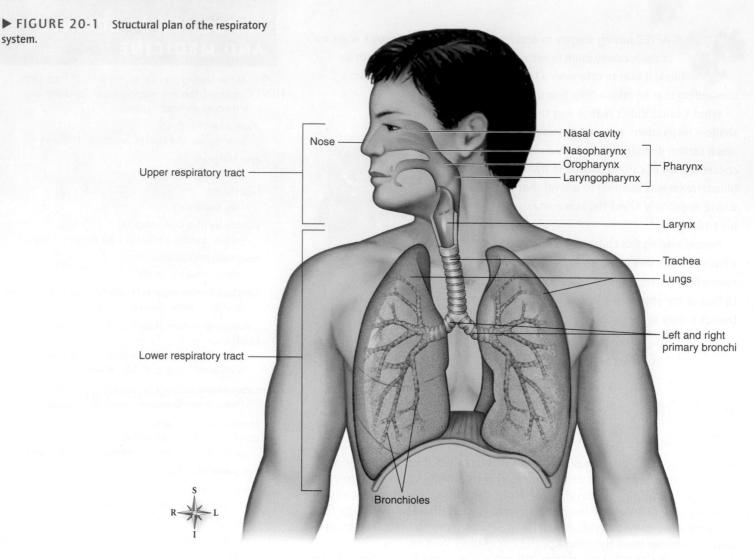

Nose

Upper respiratory tract

Nasal cavity

Nasopharynx
Oropharynx
Laryngopharynx

Pharynx

Larynx

Trachea

Lungs

Lower respiratory tract

Left and right primary bronchi

Bronchioles

UPPER RESPIRATORY TRACT

Nose

Structure of the Nose

The external portion of the nose protrudes from the face. It consists of a bony and cartilaginous framework overlaid by skin with many sebaceous glands (Figure 20-2). The two nasal bones meet the frontal bone of the skull at their superior end to form the root of the nose. They are surrounded laterally and inferiorly by the maxilla. The flaring cartilaginous expansion that forms and supports the outer side of each nostril opening is called the *ala*.

The palatine bones, which comprise the roof of the mouth, also form the base of the nasal cavity, separating it from the mouth. Sometimes the palatine bones fail to unite at their center, producing a condition called **cleft palate.** When this abnormality arises, the mouth is only partially separated from the nasal cavity. This can cause serious difficulty in swallowing and speaking unless surgically repaired.

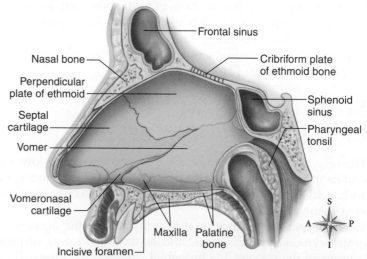

Frontal sinus

Nasal bone

Cribriform plate of ethmoid bone

Perpendicular plate of ethmoid

Sphenoid sinus

Septal cartilage

Pharyngeal tonsil

Vomer

Vomeronasal cartilage

Maxilla Palatine bone

Incisive foramen

▲ FIGURE 20-2 **Nasal septum.** The nasal septum consists of the perpendicular plate of the ethmoid bone, the vomer, and the septal and vomeronasal cartilages.

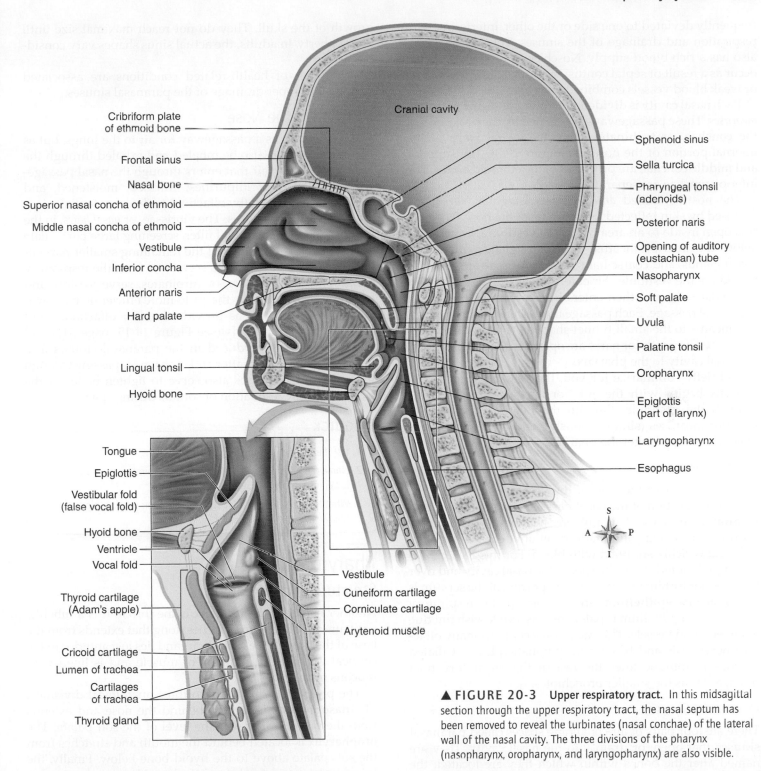

Cribriform plate of ethmoid bone

Frontal sinus

Nasal bone

Superior nasal concha of ethmoid

Middle nasal concha of ethmoid

Vestibule

Inferior concha

Anterior naris

Hard palate

Lingual tonsil

Hyoid bone

Cranial cavity

Sphenoid sinus

Sella turcica

Pharyngeal tonsil (adenoids)

Posterior naris

Opening of auditory (eustachian) tube

Nasopharynx

Soft palate

Uvula

Palatine tonsil

Oropharynx

Epiglottis (part of larynx)

Laryngopharynx

Esophagus

Tongue

Epiglottis

Vestibular fold (false vocal fold)

Hyoid bone

Ventricle

Vocal fold

Thyroid cartilage (Adam's apple)

Cricoid cartilage

Lumen of trachea

Cartilages of trachea

Thyroid gland

Vestibule

Cuneiform cartilage

Corniculate cartilage

Arytenoid muscle

S
A — P
I

▲ FIGURE 20-3 **Upper respiratory tract.** In this midsagittal section through the upper respiratory tract, the nasal septum has been removed to reveal the turbinates (nasal conchae) of the lateral wall of the nasal cavity. The three divisions of the pharynx (nasopharynx, oropharynx, and laryngopharynx) are also visible.

The roof of the nose is separated from the cranial cavity by a portion of the ethmoid bone called the **cribriform plate** (see Figures 20-2 and 20-3). The cribriform plate is perforated by many small openings that permit branches of the olfactory nerve to enter, allowing for our sense of smell. The cribriform plate is fragile and can be damaged from trauma to the nose. Such damage makes it possible for infectious material to pass directly from the nasal cavity into the cra-

nial fossa. Here it can infect the brain and its membranous coverings.

The hollow nasal cavity is separated by a midline partition, the **nasal septum,** to form left and right cavities (see Figure 20-2). The nasal septum is composed of four main structures: the perpendicular plate of the ethmoid bone above, along with the vomer bone and the septal nasal and vomeronasal cartilages below. The nasal septum of adults is frequently deviated to one side or the other, interfering with respiration and drainage of the sinuses. The nasal septum also has a rich blood supply. Nosebleeds, or *epistaxis,* often occur as a result of septal contusions from a blow to the nose or weak blood vessels combined with high blood pressure.

Each nasal cavity is divided into three passageways called *meatuses.* These passageways are created by the projection of the **conchae,** or **turbinates,** from the lateral walls of the internal portion of the nose (see Figure 20-3). The superior and middle conchae are processes of the ethmoid bones. The inferior conchae are separate bones.

The nostrils, called *anterior nares* (singular, *naris*), are enclosed by skin reflected from the ala of the nose. The nostrils open inside to an area called the **vestibule,** located just below the inferior meatus. Sebaceous glands, numerous sweat glands, and coarse hairs called **vibrissae** are found in the skin of the vestibule. Air enters the nose, passes over the vestibular skin, and then enters the **respiratory portion** of each nasal passage. Each passage then extends from the inferior meatus to the small funnel-shaped orifices of the *posterior nares.* The posterior nares are openings that pass air from the nasal cavity to the pharynx.

To briefly summarize for you, the passage of air to the pharynx begins with the anterior nares (nostrils), passes through the vestibule, through the inferior, middle, and superior meatuses (simultaneously), and then through the posterior nares to enter the pharynx.

Nasal Mucosa

After passing over the skin of the vestibule, air enters the respiratory portion of the nasal passage and passes over the **respiratory mucosa.** This mucous membrane has a pseudostratified ciliated columnar epithelium dense with goblet cells and is richly supplied with blood. For this reason, it is bright pink or red. Near the roof of the nasal cavity and over the superior turbinate and opposing portion of the septum is the **olfactory epithelium.** In contrast to the respiratory mucosa, this epithelium is paler and has a yellowish tint due to fewer blood vessels. This membrane contains many olfactory nerve cells and has a rich lymphatic plexus. Ciliated mucous membrane lines the rest of the respiratory tract down as far as the smaller bronchioles.

Paranasal Sinuses

There are four pairs of air-containing spaces called **paranasal sinuses** that open, or drain into the nasal cavity. They are named after the bones within which they are located: the frontal, maxillary, ethmoid, and sphenoid sinuses (see Figure 9-7, page 159). The sphenoid sinuses lie in the body of the sphenoid bone on either side of the midline close to the optic nerves and pituitary gland. Each sinus is lined by ciliated respiratory mucosa, which sweep their mucous secretions into the nose. The ethmoid sinus is actually a group of many small air cells, whereas the remaining sinuses are larger and interconnected.

The paired and often asymmetrical sinuses are small or poorly formed at birth. However, they increase in size with growth of the skull. They do not reach maximal size until after puberty. In adults, the actual sinus shapes vary considerably.

A number of health-related conditions are associated with the improper drainage of the paranasal sinuses.

Function of the Nose

The nose serves as a passageway for air to the lungs, but as you know, air can also be inhaled and exhaled through the mouth. However, air that enters through the nasal passageways is filtered of impurities, warmed, moistened, and chemically examined (by olfaction) to detect potentially irritating or toxic substances. The vibrissae, or *nasal hairs,* in the vestibule serve as an initial filter, screening large particulate matter from the air. Many of the remaining smaller particulates are filtered by mucous secretions from the respiratory membrane. The conchae, or turbinates, serve to slow and stir the air, which allows the air to more efficiently pass over the mucosae, as well as provide time for olfaction. Fluid from the lacrimal glands (see Figure 14-15, page 311) and additional mucus produced in the paranasal sinuses also help trap particulate matter and moisten air passing through the nose. These sinuses also serve to lighten bones of the skull and allow resonation of sounds during speech.

> **QUICK ✓ CHECK**
>
> 1. List the functions of the respiratory system.
> 2. Name the principal organs of the upper respiratory tract and the lower respiratory tract.
> 3. What are the paranasal sinuses? What is their relationship to the nose?

Pharynx

Structure of the Pharynx

Another name for the **pharynx** is the throat. It is a tubelike structure about 12.5 cm (5 inches) long that extends from the base of the skull to the esophagus and lies just anterior to the cervical vertebrae. It is made of muscle and is lined with mucous membrane.

The pharynx is divided into three anatomical divisions. The **nasopharynx** is located behind the nose and extends from the posterior nares to the level of the soft palate. The **oropharynx** is located behind the mouth and stretches from the soft palate above to the hyoid bone below. Finally, the **laryngopharynx** extends from the hyoid bone to the esophagus.

Seven openings are found in the pharynx (see Figure 20-3):

- Right and left auditory (eustachian) tubes that open into the nasopharynx
- Two posterior nares that open into the nasopharynx
- The opening from the mouth into the oropharynx
- The opening into the larynx from the laryngopharynx
- The opening into the esophagus from the laryngopharynx

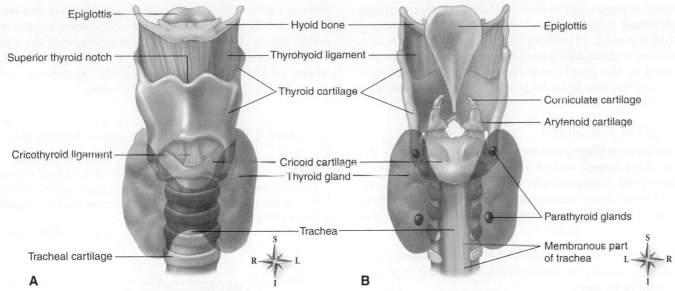

▲ **FIGURE 20-4** **Laryngeal cartilages.** Some softer tissues of the larynx and surrounding structures have been removed to make it possible to see the cartilages of the larynx. Note the position of the nearby thyroid gland. **A,** Anterior view. **B,** Posterior view.

The **pharyngeal tonsils** are located in the nasopharynx on its posterior wall opposite the posterior nares and are referred to as *adenoids* when they are enlarged. The oral and laryngeal divisions are able to collapse while the nasopharynx does not. However, the nasopharynx may become obstructed by enlarged adenoids, making it difficult or even impossible for air to travel from the nose into the throat.

Two pairs of tonsils are found in the oropharynx: the **palatine tonsils,** located back in the oropharynx, and the **lingual tonsils,** located at the base of the tongue. Palatine tonsils are generally the most commonly removed tonsils (see Figure 19-7, page 433).

Function of the Pharynx

The pharynx serves as a common pathway for the respiratory and digestive tracts, because both air and food must pass through this structure before reaching the appropriate tubes: the trachea for air and the esophagus for food. It also modifies speech production by changing shape, allowing humans to make different sounds, especially the variable sounds of vowels.

Larynx

Location of the Larynx

The **larynx,** or voice box, lies between the root of the tongue and the upper end of the trachea just below and in front of the lowest part of the pharynx (see Figure 20-1). It is like a vestibule opening into the trachea from the pharynx and normally extends between the third and sixth cervical vertebrae. However, the larynx is often higher in females and during childhood of both sexes. The lateral lobes of the thyroid gland and the carotid artery (within its sheath) touch the sides of the larynx.

Structure of the Larynx

The larynx is triangular and lined with ciliated mucous membrane. It consists predominantly of cartilages that are attached to one another and surrounding structures by muscles or fibrous and elastic tissue components (Figure 20-4). The cavity of the larynx extends from its triangular inlet at the epiglottis to the circular outlet at the lower border of the cricoid cartilage, where it is continuous with the lumen of the trachea (see Figure 20-3, *B*).

The mucous membrane is composed of two pairs of lateral folds called **vestibular folds** ("vocal cords"). The upper thick vestibular folds ("false vocal folds) have a minimal role in vocalization (Figure 20-5). In some cultures, they are used in a fascinating form of throat singing that produces many musical overtones. The false vocal folds protect the lower folds, the **vocal folds,** which do contribute to sound

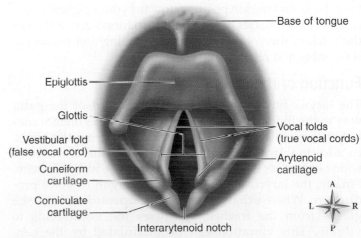

▲ **FIGURE 20-5** **Vocal folds.** Vocal folds ("cords") viewed from above.

production during speech. The vocal folds of most females are only about 70% as large as those of males, explaining much of the gender differences in human vocal pitch.

The vocal folds and the space between them are together designated as the **glottis** (see Figure 20-5). The laryngeal cavity above the vestibular folds is called the **vestibule** (see Figure 20-3, *B*).

Cartilages of the Larynx

There are nine cartilages that form the framework of the larynx. The three largest—the thyroid cartilage, the epiglottis, and the cricoid cartilage—are single (or fused) structures, while the other six are three pairs of smaller accessory cartilages named the arytenoid, corniculate, and cuneiform cartilages.

- The **thyroid cartilage** (Adam's apple) is the largest in the larynx and usually is larger, with less fat lying over it, in men than in women.

- The **epiglottis** is a small leaf-shaped cartilage that projects upward behind the tongue and hyoid bone. It is attached to the thyroid cartilage. The epiglottis is mobile and blocks the trachea when swallowing food or liquids.

- The smaller cartilages, along with the cricoid cartilage, articulate and form attachments for the vocal folds (see Figure 20-4).

Muscles of the Larynx

The larynx and its components are moved by *intrinsic* muscles (with origins and insertions on the larynx), as well as by *extrinsic* muscles (which attach to the larynx but have their origins from another structure, such as the hyoid bone). The intrinsic muscles are important for controlling vocal fold length and tension and in regulating the shape of the laryngeal inlet. Contractions of the extrinsic muscles physically move the larynx and its parts. However, both intrinsic and extrinsic muscles are important during speech, respiration, and swallowing. During swallowing, for example, the muscles that connect the arytenoid cartilage with the epiglottis *raise* the larynx and help prevent entry of food or fluid into the trachea. You can feel this for yourself: swallow and feel how the larynx rises in position, toward your epiglottis. Specific intrinsic muscles of the larynx also function to influence the pitch of the voice by either lengthening and tensing or shortening and relaxing the vocal folds.

Function of the Larynx

The larynx functions in respiration as part of the pathway to the lungs. It is ciliated, and its mucous membranes help filter, moisten, and warm the air entering the lungs. In addition, it protects the airway from the entrance of solids or liquids during swallowing, as we have just seen. Finally, the larynx also serves as the organ of voice production. When exhaled air passes upward toward the glottis from the trachea, it causes the vocal folds to vibrate. This vibration can be modulated by the controlled passage of air from below the larynx and by a number of muscles that increase or decrease the tension of the vocal cords. The sounds produced by the vocal folds can be further altered by the shape of the mouth, nose, and sinuses during vocal production. The size and shape of the nose, mouth, pharynx, and bony sinuses help determine the quality of the voice.

A&P CONNECT

Edema (swelling) of the mucosa covering the vocal folds and other laryngeal tissues can be a potentially lethal condition. Even a moderate amount of swelling can obstruct the glottis so much that air cannot get through and asphyxiation results. To find out more, check out **Swollen Larynx** online at **A&P Connect**.

ⓔvolve

QUICK ✓ CHECK

4. Describe the three main divisions of the pharynx.

5. Where are the tonsils located?

6. Distinguish between the true and false vocal folds and their functions.

LOWER RESPIRATORY TRACT

Trachea

Structure of the Trachea

The **trachea,** or windpipe, is a tube about 11 cm (4.5 inches) long and 2.5 cm (1 inch) wide that extends from the larynx in the neck to the primary bronchi in the thoracic cavity (see Figure 20-1). The wall of the trachea is composed of strong C-shaped cartilaginous rings (Figure 20-6, *A*) that are embedded in smooth muscle. The epithelium of the trachea, like most of the rest of the respiratory tract, is pseudostratified ciliated columnar epithelium (Figure 20-6, *B*).

Function of the Trachea

The trachea provides a sturdy open passageway from the upper respiratory tract into the lungs so that air can pass through unobstructed. The cilia on its epithelium also continue to filter particulates from the air as it passes to the lungs. *Endotracheal intubation,* a procedure that keeps the trachea open, is described for you in Box 20-1.

Bronchi and Alveoli

Structure of the Bronchi

The trachea divides into two **primary bronchi,** the right bronchus being slightly larger and more vertical than the left. The walls of the bronchi have incomplete cartilaginous rings for support outside of the lungs, and complete rings within the lungs. Ciliated mucosae line the bronchi. Upon

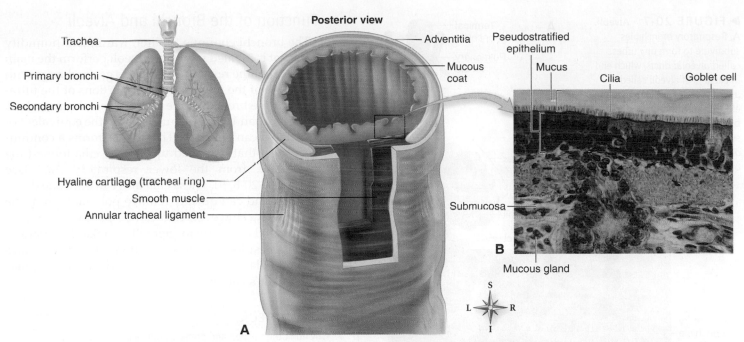

Posterior view

Trachea

Primary bronchi

Secondary bronchi

Hyaline cartilage (tracheal ring)

Smooth muscle

Annular tracheal ligament

Adventitia

Mucous coat

Pseudostratified epithelium

Mucus

Cilia

Goblet cell

Submucosa

B

Mucous gland

A

S
L — R
I

▲ **FIGURE 20-6 Cross section of the trachea. A,** The inset at the top shows from where the section was cut. **B,** The micrograph shows a transverse section of the trachea. Note the mucosa of ciliated epithelium. Hyaline cartilage occurs below the glandular submucosa and is not visible in this section (×70).

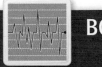

BOX 20-1 | Health Matters

Keeping the Trachea Open

Often, a tube is placed through the mouth, pharynx, and larynx into the trachea before surgery or before patients leave the operating room, especially if they have been given a muscle relaxant. This procedure is called **endotracheal intubation.** The purpose of the tube is to ensure an open airway (see parts *A* and *B* of the figure). To ensure that the tube enters the trachea rather than the nearby esophagus (which leads to the stomach),

anatomical landmarks such as the vocal folds are visualized. Likewise, the distinct feel of the V-shaped groove called the *interarytenoid notch* (see Figure 20-5) can help guide the proper insertion of the tube.

Another procedure done frequently in today's modern hospitals is a **tracheostomy,** that is, the cutting of an opening into the trachea (part *C* of the figure). A surgeon may perform this procedure so that a suction device can be inserted to remove secretions from the bronchial tree or so that mechanical ventilation can be used to improve ventilation of the lungs.

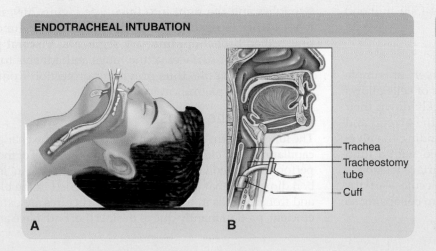

ENDOTRACHEAL INTUBATION

Trachea

Tracheostomy tube

Cuff

A

B

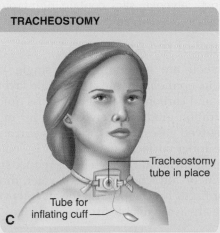

TRACHEOSTOMY

Tracheostomy tube in place

Tube for inflating cuff

C

▶ **FIGURE 20-7**   Alveoli. **A,** Respiratory bronchioles subdivide to form tiny tubes called alveolar ducts, which end in clusters of alveoli called alveolar sacs. **B,** Scanning electron micrograph of a bronchiole, alveolar ducts, and surrounding alveoli. The *arrowhead* indicates the opening of alveoli into the alveolar duct.

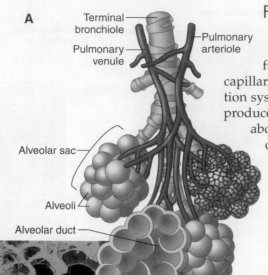

Function of the Bronchi and Alveoli

The bronchi continue to filter, warm, and humidify air that is inhaled while the alveoli perform the main function of the respiratory system: gas exchange with capillaries. One of the most important portions of the filtration system of the bronchi is the mucous membrane, which produces more than 125 ml of mucus daily (the equivalent of about half a can of soda!). This mucus forms a continuous sheet that is moved upward by cilia toward the pharynx from the lower respiratory tract (see Figure 20-6, *B*). These cilia move in only one direction and can quickly remove pollutants from the airway. However, cilia can be paralyzed by prolonged exposure to cigarette smoke. This paralysis results in accumulation of mucus and particulates and, of course, produces the familiar "smoker's cough."

QUICK ✓ CHECK

7. Why don't the trachea and primary bronchi collapse during inspiration?

8. What is meant by the term *bronchial tree*?

9. Describe the characteristics of the alveoli that enable them to exchange gases.

Lungs

Structure of the Lungs

The lungs are cone-shaped organs that fill the pleural portion of the thoracic cavity completely (Figure 20-8). The medial walls of the lungs are concave and provide room for the heart. The primary bronchi and pulmonary blood vessels are bound together by connective tissue and enter each lung through a slit on the medial surface called the **hilum.** The inferior surfaces of the lungs rest on the muscular diaphragm, which generates the "suction" of air by making the lungs larger for a time, thus creating a vacuum. The broad inferior surface of the lung (which rests on the diaphragm) constitutes the **base,** whereas the pointed upper margin is the **apex** (Figure 20-9).

Each lung is divided into lobes by fissures: the left into two lobes and the right into three (see Figure 20-9, *B*). The lobes of the lungs can be further divided into functional units called **bronchopulmonary segments. Visceral pleura** covers the outer surfaces of the lungs and adheres to it like the skin of an apple, thus providing protection from abrasion within the pleural cavity.

Function of the Lungs

The lungs perform both air distribution and gas exchange. The capillaries of the circulatory system and the alveoli of the lungs perform the essential function of exchanging oxygen and carbon dioxide so that these gases can be carried in the blood to and from the cells in the body. These structures are extremely efficient, predominantly due to the fact that both have an enor-

---

entering the lungs, each primary bronchus divides into smaller branches called **secondary bronchi.** These then branch to form tertiary bronchi, and these subdivide into **bronchioles** (see Figure 20-1). Bronchioles continue to branch into smaller and smaller tubes until they end in microscopic *terminal bronchioles,* which pass air into *respiratory bronchioles* and then to one or more alveolar sacs (Figure 20-7). In our two lungs, alveolar sacs contain numerous (300 million!) smaller sacs called **alveoli.**

Structure of the Alveoli

The alveoli are made up of a single layer of simple squamous epithelial tissue. This single layer of cells allows oxygen and carbon dioxide gas to pass quickly down their concentration gradients between the air in the alveoli and the dissolved gases in the capillaries surrounding the alveoli. The inner surface of the alveoli is coated with a fluid containing **surfactant,** which helps reduce surface tension—the force of attraction between water molecules—of the alveolar fluid. This helps keep the alveoli from collapsing so they are always available for gas exchange.

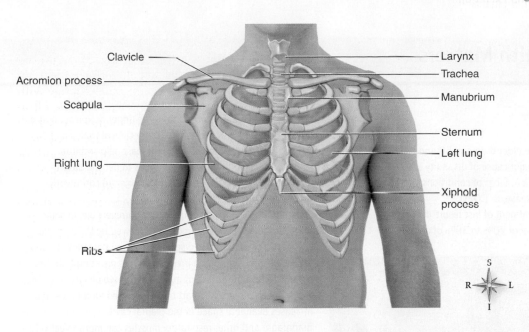

mous surface area relative to their volume. In fact, it has been estimated that if the lungs' 300 million alveoli were opened up and flattened out, they would cover the surface of a tennis court! It is this surface area that allows for rapid diffusion of gases between alveolar and capillary membranes.

Box 20-2 describes how portions of the lung can be surgically removed in extreme cases of lung damage.

A&P CONNECT

Some types of lung cancer may be curable if detected early. Learn how light can be used to treat lung cancer! Check out **Photodynamic Therapy** online at **A&P Connect.**

evolve

Thorax
Structure of the Thoracic Cavity

The thoracic cavity is divided by **pleura** to form three divisions. The *mediastinum* occupies the middle of the cavity, and the lateral spaces occupied by the lungs are called *pleural divisions.*

The **parietal pleura** lines the entire thoracic cavity by attaching to the inside of the ribs and superior surface of the diaphragm. Lying within the parietal pleura and continuous with it is the **visceral pleura,** which covers the outer surface of the lungs. An **intrapleural space** lies between the parietal and visceral pleura. It contains fluid for lubrication so that

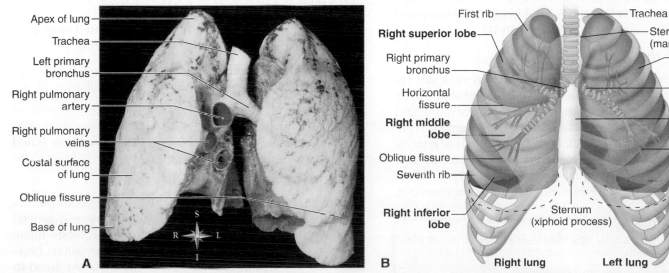

▲ **FIGURE 20-9** Anterior view of trachea, bronchi, and lungs. **A,** The lower respiratory tract has been dissected from a cadaver and its organs separated to show them clearly. **B,** Lobes of the lungs.

BOX 20-2 | Health Matters

Lung Volume Reduction Surgery

More than 2 million Americans, most of whom are older than age 50 and are current or former smokers, have emphysema—a major cause of disability and death in the United States. Emphysema is one of a number of conditions classified as a **chronic obstructive pulmonary disease**, or **COPD**.

Lung volume reduction surgery (LVRS) is a "treatment of last resort" for severe cases of emphysema. It involves the removal of 20% to 30% of each lung. Diseased tissue is generally removed from the upper or apical areas of the superior lobes. Evidence from a number of large clinical trials has now shown that the LVRS procedure may benefit or at least help stabilize emphysema patients whose lung function continues to decline despite aggressive pulmonary rehabilitation efforts and other more conservative forms of treatment.

Although lung damage caused by emphysema is irreversible, in some cases the disease may be halted or its progression slowed by LVRS. In the end stages of this chronic disease, breathing becomes labored as the lungs fill with large irregular spaces resulting from the enlargement and rupture of many alveoli (see illustration). The LVRS procedure removes part of the diseased lung tissue and increases available space in the pleural cavities. As a result, the diaphragm and other respiratory muscles can more effectively move air into and out of the remaining lung tissue, thereby improving pulmonary function and making breathing easier.

LVRS may reduce the need for lung transplantation procedures and enhance the effectiveness of supporting medical treatments such as nutritional supplementation and exercise training in the treatment of selected late-stage emphysema patients. Newer and less invasive techniques involving smaller incisions and tiny video equipment inserted into the thoracic cavity (video-assisted thoracic surgery) are now being used for many LVRS procedures. As a result, the relatively long hospital stays and home recovery periods previously required after more traditional open-chest surgery have been shortened.

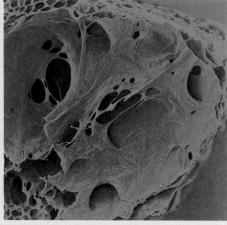

A   **B**

▲ **Emphysema. A,** Scanning electron micrograph (SEM) of normal lung with many small alveoli. **B,** SEM of lung tissue affected by emphysema. Alveoli have merged into large air spaces, thereby reducing the surface area available for gas exchange.

the lungs can move with ease in the thoracic cavity. You should note that each lung is encased in its own pleura and is separate from the other lung (Figure 20-10).

Function of the Thoracic Cavity

The thoracic cavity must enlarge greatly to create space for incoming air during respiration. This is accomplished as the ribs lift upward during inhalation so that both the depth and the width of the thorax are enlarged. The diaphragm also contracts during this action, which flattens it and makes it move downward. This creates an even larger change in the volume of the thoracic cavity. When the diaphragm relaxes, it moves upward, compressing the lungs and pushing waste gases out.

QUICK CHECK

10. What is a "lobe" of a lung? How many lobes are there in the left and right lungs?

11. How does the diaphragm enable ventilation?

RESPIRATORY PHYSIOLOGY

Functionally, the respiratory system is composed of an integrated set of processes that include the following:

- External respiration: pulmonary ventilation (breathing) and gas exchange in the pulmonary capillaries of the lungs

- Transport of gases by the blood

- Internal respiration: gas exchange in the systemic blood capillaries and cellular respiration

- Overall regulation of respiration

Figure 20-11 summarizes the essential processes of pulmonary function. This set of processes will serve as a general framework for understanding respiratory physiology. If you wish to review the biochemistry of cell respiration, go to Chapter 4. Chapter 22 will review metabolism in even greater detail.

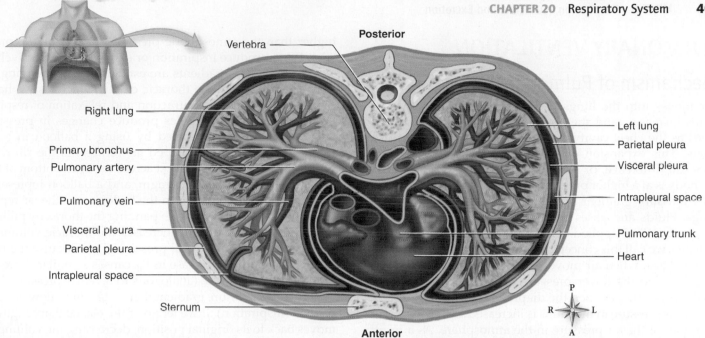

Posterior

Vertebra

Right lung

Primary bronchus

Pulmonary artery

Pulmonary vein

Visceral pleura

Parietal pleura

Intrapleural space

Sternum

Left lung

Parietal pleura

Visceral pleura

Intrapleural space

Pulmonary trunk

Heart

Anterior

P

R — L

A

▲ **FIGURE 20-10** **Lungs and pleura (transverse section).** Note the parietal pleura lining the right and left pleural divisions of the thoracic cavity before folding inward near the bronchi to cover the lungs as the visceral pleura. The intrapleural space separates the parietal and visceral pleura. The heart, esophagus, and aorta are shown in the central mediastinum.

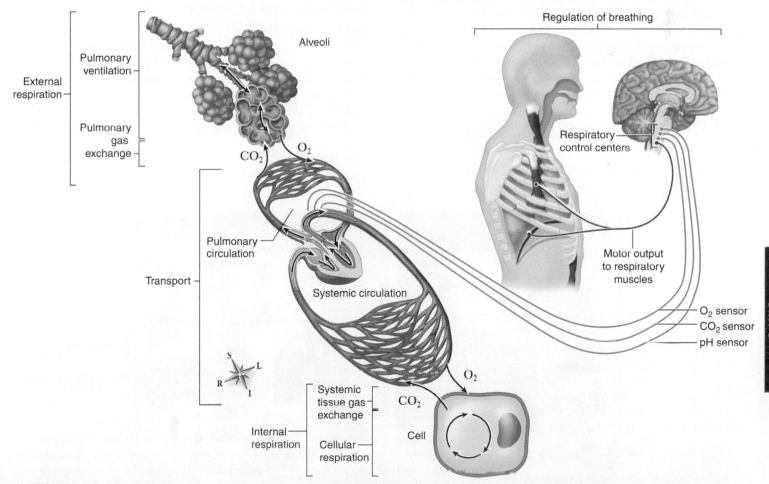

Regulation of breathing

Alveoli

Pulmonary ventilation

External respiration

Pulmonary gas exchange

CO_2 O_2

Pulmonary circulation

Transport

Systemic circulation

S   L

R   I

Respiratory control centers

Motor output to respiratory muscles

O_2 sensor

CO_2 sensor

pH sensor

O_2

CO_2

Systemic tissue gas exchange

Internal respiration

Cellular respiration

Cell

▲ **FIGURE 20-11** **Overview of respiratory physiology.** Respiratory function includes external respiration (ventilation and pulmonary gas exchange), transport of gases by blood, and internal respiration (systemic tissue gas exchange and cellular respiration). Cellular respiration is discussed separately in Chapter 4. Regulatory mechanisms centered in the brainstem use feedback from blood gas sensors to regulate ventilation.

PULMONARY VENTILATION

Mechanism of Pulmonary Ventilation

Air moves into the lungs because the volume within the lungs increases (and the air pressure within them is lowered) as the chest cavity expands and the diaphragm contracts. This expansion of the pulmonary cavity creates a pressure difference, or *pressure gradient:* The air outside of the body is at a higher pressure than the air within the lungs. This pressure gradient in turn causes air to rush into the lungs. Fluids and gases always move *down* their pressure gradients. When applied to the flow of air in the pulmonary airways, we call this concept the **primary principle of ventilation.** Thus, when air moves from the atmosphere of high pressure into the lower pressure of the lungs, **inspiration** occurs. In contrast, when the diaphragm pushes against the lungs, air pressure in the lungs is increased so that it is *higher* than that of the air pressure in the atmosphere. As a result, air again rushes *down* the pressure gradient and outside of the body, and **expiration** occurs.

Under standard conditions, air in the atmosphere exerts a pressure of 760 mm Hg. Air in the alveoli at the end of one expiration and before the beginning of another inspiration also exerts a pressure of 760 mm Hg. This explains why, at that moment, air is neither entering nor leaving the lungs. So, in effect, the mechanism that produces pulmonary ventilation is one that creates a gas pressure gradient between the atmosphere and the alveolar air. The pulmonary ventilation mechanism, therefore, modifies the alveolar pressure (P_A, pressure within the alveoli of the lungs) to be either lower or higher than the atmospheric pressure (or barometric pressure, P_B) to produce inspiration or exhalation, respectively.

These pressure gradients are established by enlarging or reducing the size of the thoracic cavity, which, as we have seen, is caused by the contraction and relaxation of respiratory muscles. These muscles produce changes in pressure that can be roughly modeled by using a balloon in a jar (Figure 20-12). The bell-shaped jar represents the rib cage (thoracic cavity), a rubber sheet across the open bottom of the bell jar represents the diaphragm, and a balloon represents the lungs. The space between the balloon and the jar represents the intrapleural space. Expanding the thorax by pulling the diaphragm downward increases the thoracic volume—thus decreasing intrapleural pressure (P_{IP}). Because the balloon is stretchy, the decrease in P_{IP} causes a similar decrease in the balloon pressure (analogous to alveolar pressure, P_A). This creates a pressure gradient that results in airflow *into* the balloon (inspiration). If we let go of the elastic diaphragm, it moves back to its original position, decreasing the volume of the chest cavity, thus *increasing* the air pressure in the lung so that it is higher than that of the atmosphere. This causes air to rush *out of* the balloon (expiration).

Inspiration and Expiration

Contraction of the diaphragm alone, or contraction of both the diaphragm and the external intercostal muscles together, produces quiet inspiration. Contraction of the diaphragm (thus lowering its position) increases the volume of the thoracic cavity, and at the same time, the contraction of the external intercostals pulls the anterior end of each rib up and out (Figure 20-13). This action *enlarges* the

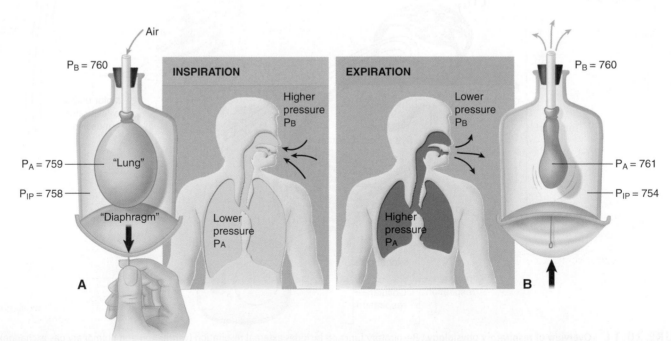

▲ **FIGURE 20-12**   **Balloon model of ventilation.** The cartoons show a classic model in which a jar represents the rib cage (thoracic cavity), a rubber sheet represents the diaphragm, and a balloon represents the alveoli of the lungs. The space between the jar and balloon represents the intrapleural space. **A,** Inspiration, caused by downward movement of the diaphragm. **B,** Expiration, caused by elastic recoil of the diaphragm upward (Pressure values are expressed in mm Hg.)

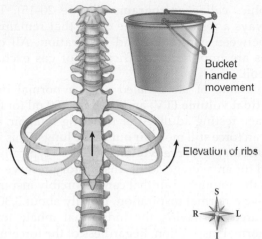

▲ **FIGURE 20-13** **Movement of the rib cage during breathing.** Inspiratory muscles pull the ribs upward and thus outward, as in a bucket handle.

thorax from front to back as well as from side to side. As the thorax enlarges, the lungs expand to fill this space, causing the pressure in the bronchioles and alveoli to lower. This creates the pressure gradient necessary to have high-pressure air in the atmosphere move *into* the lungs in

order to equalize the pressures inside and outside the lungs. Quiet expiration is a passive process that begins when the pressure gradients that resulted from inspiration are reversed by the relaxation of the inspiratory muscles. During forced expiration, abdominal and intercostal muscles can shrink the thoracic cavity such that pressure of air in the alveoli (alveolar pressure) is much larger than atmospheric pressure, thus causing a larger volume of air to move more quickly out of the lungs.

Regardless of the force exerted on the lungs and the thoracic cavity during breathing, a phenomenon called **elastic recoil** causes the lungs to return to their typical volume *before* the next inspiration.

Figure 20-14 gives you an overview of the mechanical aspects of inspiration and expiration. Please review this figure before continuing.

QUICK✓CHECK

12. What is meant by *pulmonary ventilation*?
13. How does expansion of the thoracic cavity affect air pressure inside the lungs?
14. What is meant by the "primary principle of ventilation"?

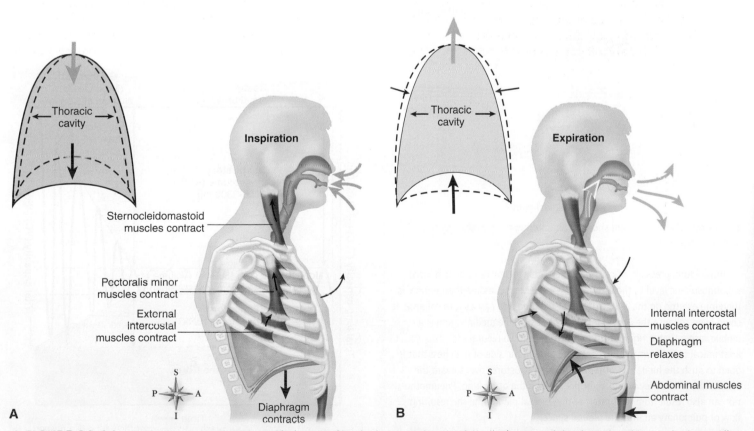

▲ **FIGURE 20-14** **Inspiration and expiration. A,** Mechanism of inspiration. Note the role of the diaphragm and the chest-elevating muscles (pectoralis minor and external intercostals) in increasing thoracic volume, which decreases pressure in the lungs and thus draws air inward. **B,** Mechanism of expiration. Note that relaxation of the diaphragm plus contraction of chest-depressing muscles (internal intercostals) reduces thoracic volume, which increases pressure in the lungs and thus pushes air outward.

UNIT 5

Pulmonary Volumes and Capacities

Pulmonary Volumes

Differing volumes of air move into and out of the lungs depending on the force with which one breathes. These volumes can be measured by a **spirometer** and recorded as graphics called **spirograms** (Figure 20-15). There is also always a residual volume of air that remains in the lungs between expiration and inspiration. All of these volumes are important to efficiency of gas exchange in the alveoli.

The volume of air exhaled after a normal breath is termed **tidal volume (TV)** and is about 500 ml (or 0.5 L) in an average resting adult. After releasing tidal air, an individual can force still more air out of the lungs. This is called the **expiratory reserve volume (ERV)**—typically about 1,000 ml for an average adult. **Inspiratory reserve volume (IRV)** is the amount of air that can be forcibly inspired over and above a normal inspiration, usually about 3,300 ml. It is measured by having the individual inhale forcefully after a normal inspiration. Regardless of the forcefulness of an exhalation, *some* air remains in the lungs, trapped in the alveoli. This air is called the **residual volume (RV)** and amounts to about 1,200 ml. Between breaths, an exchange of oxygen and carbon dioxide occurs between the trapped residual air in the alveoli and the blood. This process helps keep amounts of oxygen and carbon dioxide constant in the blood.

In a condition called *pneumothorax* (Box 20-3), the RV is eliminated when the lung collapses. Even after the RV is forced out, the collapsed lung has a porous, spongy texture and floats in water because of trapped air (equal to about 40% of the RV).

BOX 20-3 | Health Matters

Pneumothorax

Air in the pleural space may accumulate when the visceral pleura ruptures and air from the lung rushes out, or when atmospheric air rushes in through a wound in the chest wall and parietal pleura. In either case, the lung collapses and normal respiration is impaired. Air in the thoracic cavity is a condition known as **pneumothorax** (see figure). To apply some of the information you have learned about the respiratory mechanism, let us suppose that a surgeon makes an incision through the chest wall into the pleural space, as is done in one of the dramatic, modern open-chest operations. What change, if any, can you deduce takes place in respirations? Compare your deductions with those in the next paragraph.

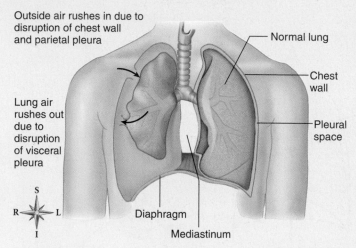

▲ **Pneumothorax.** Diagram showing air entering the thoracic cavity, causing lung collapse.

Intrapleural pressure, of course, immediately increases from its normal subatmospheric level to the atmospheric level. More pressure than normal is therefore exerted on the outer surface of the lung and causes it to collapse. It could even collapse the other lung. Why? Because the mediastinum is a mobile rather than a rigid partition between the two pleural sacs. This anatomical fact allows the increased pressure in the side of the chest that is open to push the heart and other mediastinal structures over toward the intact side, where they would exert pressure on the other lung. Pneumothorax can also result from disruption of the visceral pleura and the resulting flow of pulmonary air into the pleural space.

Pneumothorax results in many respiratory and circulatory changes. They are of great importance in determining medical and nursing care but lie beyond the scope of this book.

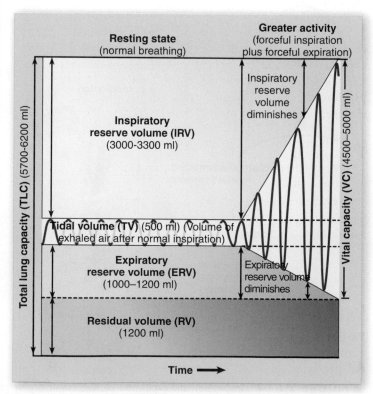

▲ **FIGURE 20-15** Pulmonary ventilation volumes and capacities. A spirogram.

Pulmonary Capacities

A *pulmonary capacity* is the sum of two or more pulmonary "volumes." The **vital capacity (VC),** for example, is the sum of the IRV, TV, and ERV (see Figure 20-15). The vital capacity represents the largest volume of air an individual can move in and out of the lungs. In general, a larger person has a larger vital capacity than a smaller person. VC can depend on the size of a person's thoracic cavity, his or her posture, and a number of other factors. For example, an individual has a larger VC standing up than lying down. The volume of blood in the lungs can also affect VC. The vital capacity may be lowered if the lungs contain more blood than normal and the alveolar air space is diminished. Excess fluid in the pleural or abdominal cavities also decreases vital capacity, as does the disease *emphysema*. In emphysema, alveoli become stretched and lose their elasticity and are unable to recoil during expiration, leaving an increased RV in the lungs, thus making each inspiration and expiration require more effort.

In diagnosing lung disorders, a physician may need to know the **inspiratory capacity (IC),** which is the *maximal amount* of air an individual can inspire after a normal expiration. As you can determine from Figure 20-15, IC is the sum of the TV and the IRV. **Functional residual capacity (FRC)** is the amount of air left in the lungs at the end of a normal expiration and is thus the sum of the ERV and the RV. The **total lung capacity (TLC)** is the sum of all four lung volumes (see Figure 20-15).

Table 20-1 summarizes typical pulmonary volumes and capacities for you.

Remember that only the volume of air that reaches the alveoli can be involved in gas exchange. This means that the rest of the air filling the pharynx, larynx, trachea, and bronchi is effectively "dead space." When alveoli cannot perform their function due to disorders such as **chronic obstructive pulmonary disease (COPD),** this dead space is increased, making gas exchange less efficient and breathing a difficulty.

QUICK CHECK

15. What is the difference between a pulmonary *volume* and a pulmonary *capacity*?
16. What is meant by the term *vital capacity*? What is it equal to?
17. What is a spirogram?

PULMONARY GAS EXCHANGE

Partial Pressure

Partial pressure means the pressure exerted by any one gas in a mixture of gases or in a liquid. According to the law of partial pressures, the partial pressure of a gas in a mixture of gases is directly related to the concentration of that gas in the mixture and to the total pressure of the mixture. About 78% of the atmosphere is made up of nitrogen, which means the par-

TABLE 20-1   Pulmonary Volumes and Capacities

NAME	DESCRIPTION	TYPICAL VALUE
Volumes		
Tidal volume (TV)	Volume moved into or out of the respiratory tract during a normal respiratory cycle	500 ml (0.5 L)
Inspiratory reserve volume (IRV)	Maximum volume that can be moved into the respiratory tract after a normal inspiration	3,000-3,300 ml (3.0-3.3 L)
Expiratory reserve volume (ERV)	Maximum volume that can be moved out of the respiratory tract after a normal expiration	1,000-1,200 ml (1.0-1.2 L)
Residual volume (RV)	Volume remaining in the respiratory tract after maximum expiration	1,200 ml (1.2 L)
Capacities		
Vital capacity (VC)	Largest volume of air that can be moved in and out of the lungs: TV + IRV + ERV	4,500-5,000 ml (4.5-5.0 L)
Inspiratory capacity (IC)	Maximal amount of air that can be inspired after a normal expiration: TV + IRV	3,500-3,800 ml (3.5-3.8 L)
Functional residual capacity (FRC)	Amount of air left in the lungs at the end of a normal expiration: ERV + RV	2,200-2,400 ml (2.2-2.4 L)
Total lung capacity (TLC)	Total volume of air that the lungs can hold: TV + IRV + ERV + RV	5,700-6,200 ml (5.7-6.2 L)

tial pressure of nitrogen (P_{N_2}) is 78% of the *total* atmospheric pressure. Thus, P_{N_2} is 592.8 mm (that is, 78% of 760 mm).

The partial pressure of a gas such as oxygen (O_2) in a liquid such as blood is directly proportional to the amount of that gas dissolved in the liquid, which in turn is determined by the partial pressure of the gas in the environment of the liquid. The partial pressure of oxygen (P_{O_2}) in both the alveoli and arterial blood is about 100 mm Hg, while in venous blood, it is about 37 mm Hg. Gas molecules will diffuse into alveolar blood from its gaseous environment in the alveoli and dissolve until the partial pressure of the gas in the blood becomes equal to its partial pressure in the alveoli. Therefore, arterial blood P_{O_2} and partial pressure of carbon dioxide (P_{CO_2}) are usually very nearly equal to alveolar P_{O_2} and P_{CO_2}.

Exchange of Gases in the Lungs

As we've seen, the exchange of gases in the lungs takes place between alveolar air and blood flowing through lung capillaries. It is important to note that the alveoli present not only a membrane that exchanges these gases efficiently but also a barrier between the internal world of the body and the outside environment. With this interpretation, the airways can be viewed simply as inward extensions of the external

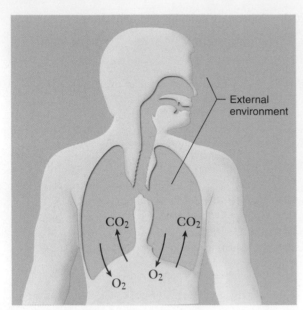

▲ **FIGURE 20-16** **External-internal barrier.** The respiratory membranes of the lung represent an interface or barrier that gases must cross to enter or exit the body's internal environment. The pulmonary airway is merely an extension of the external environment.

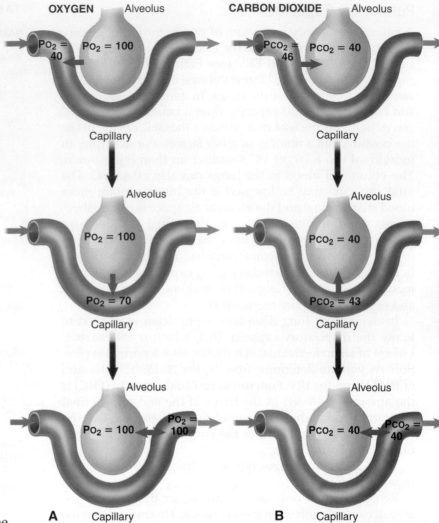

▲ **FIGURE 20-17** **Pulmonary gas exchange. A,** As blood enters a pulmonary capillary, oxygen diffuses down its pressure gradient (into the blood). Oxygen continues diffusing into the blood until equilibration has occurred (or until the blood leaves the capillary). **B,** As blood enters a pulmonary capillary, carbon dioxide diffuses down its pressure gradient (out of the blood). As with oxygen, carbon dioxide continues diffusing as long as there is a pressure gradient. P_{O_2} and P_{CO_2} remain relatively constant in a continually ventilated alveolus.

environment (Figure 20-16). This means that inspired air is *not* part of the internal environment.

Gases in the alveoli are exchanged in both directions across the respiratory membrane. This is because the blood returning from the body is low in oxygen but high in carbon dioxide, so its P_{O_2} is much lower than that of alveolar air while its P_{CO_2} is much higher (Figure 20-17). This causes oxygen to diffuse rapidly into the blood while carbon dioxide diffuses out of the blood. Gas exchange in the alveoli is facilitated by their very thin walls (only 0.004 mm thick) and the fact that alveolar and capillary surface areas are extremely large relative to their volume.

QUICK CHECK

18. How does the partial pressure of a gas relate to its concentration?
19. What determines the pressure gradient for oxygen and carbon dioxide?

HOW BLOOD TRANSPORTS GASES

Blood transports oxygen and carbon dioxide either as solutes or combined with other chemicals. Because fluids can dissolve only small amounts of gases, most of the oxygen and carbon dioxide transported in the blood is chemically united (soon after being dissolved) with other compounds such as hemoglobin, plasma protein, or water. Once gas molecules are bound to another molecule, their plasma concentration decreases so that more gas can diffuse from the alveoli into the plasma. This mechanism greatly increases the overall amount of gas that can be transported by the blood. The primary gas transport molecule of the blood is hemoglobin.

Hemoglobin

Hemoglobin (Hb) is a conjugated protein made of four polypeptide chains (two alpha chains and two beta chains). In Figure 20-18, you can see that each chain is associated with an iron-containing *heme* group, and each iron atom can bind an oxygen molecule (O_2). In a way, hemoglobin acts

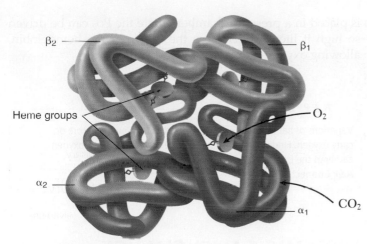

▲ FIGURE 20-18 **Hemoglobin.** Sketch showing that hemoglobin is a quaternary protein consisting of four different tertiary (folded) polypeptide chains—two alpha (α) chains and two beta (β) chains. Each chain has an associated iron-containing heme group. Oxygen (O_2) can bind to the iron of the heme group, or carbon dioxide (CO_2) can bind to amine groups of the amino acids in the polypeptide chains.

like an oxygen sponge. However, it can also act as a carbon dioxide sponge by absorbing carbon dioxide molecules from solution.

Gas Transport in the Blood

Oxygen

Oxygen travels in two forms: as dissolved oxygen in the plasma and as oxygen associated with hemoglobin (oxyhemoglobin).

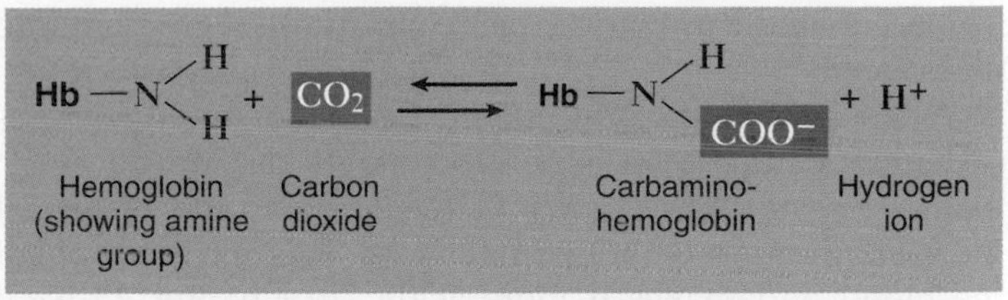

▲ FIGURE 20-19 **Carbon dioxide–hemoglobin reaction.** Carbon dioxide can bind to an amine group (NH_2) in an amino acid within a hemoglobin (Hb) molecule to form carbaminohemoglobin (HbNCOOH$^-$) and a hydrogen ion. The highlighted areas show where the original carbon dioxide molecule is in each part of the equation.

Of these two forms of transport, oxyhemoglobin carries the vast majority of the total oxygen transported by the blood.

Carbon Dioxide

Only about 10% of the carbon dioxide in blood travels as dissolved gas; the rest travels as **carbaminohemoglobin** (Figure 20-19) and as **bicarbonate**, according to the equation below:

$$CO_2 + H_2O \longleftrightarrow H_2CO_3 \longleftrightarrow H^+ + HCO_3^-$$

Carbon dioxide   Water   Carbonic acid   Hydrogen ion   Bicarbonate ion

Carbaminohemoglobin is an important carrier of blood carbon dioxide and is formed when CO_2 binds to an amine group on hemoglobin. This association is made more quickly and efficiently when there is more CO_2 in the blood and is also slowed when there is a lowering of P_{CO_2} in the blood.

Figure 20-20, which summarizes all three forms of carbon dioxide transport, shows that once bicarbonate ions are

▼ FIGURE 20-20 **Carbon dioxide transport in the blood.** As the illustration shows, CO_2 dissolves in the plasma. Some of the dissolved CO_2 enters red blood cells (RBCs) and combines with hemoglobin (Hb) to form carbaminohemoglobin (HbCO$_2$). Some of the CO_2 entering RBCs combines with H_2O to form carbonic acid (H_2CO_3), a process facilitated by the enzyme carbonic anhydrase (CA) present inside each cell. Carbonic acid then dissociates to form H^+ and bicarbonate (HCO$_3^-$). The H^+ combines with Hb, whereas the HCO$_3^-$ diffuses down its concentration gradient into the plasma. As HCO$_3^-$ leaves each RBC, Cl^- enters and prevents an imbalance in charge—a phenomenon called the *chloride shift.*

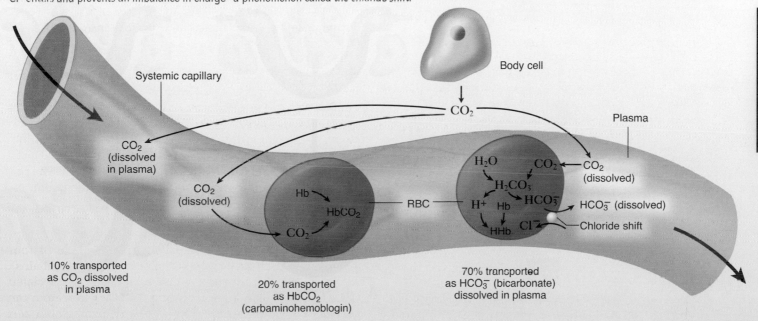

UNIT 5

formed, they diffuse down their concentration gradient into the plasma.

When carbon dioxide enters the blood and generates carbaminohemoglobin or bicarbonate, it also produces H^+ ions as a result. Thus, when CO_2 enters into the system, it increases the amount of hydrogen ions. This *lowers* the pH of the blood and makes it more acidic. In Chapter 23 we'll explore, briefly, why this is significant.

Carbon Monoxide

Gases other than oxygen and carbon dioxide can also bind to hemoglobin. For example, **carbon monoxide (CO)** binds more than 200 times more strongly to hemoglobin than oxygen does. (Carbon monoxide is a molecule produced by incomplete combustion in furnaces and engines.) In effect, carbon monoxide can replace oxygen at its binding sites with hemoglobin, greatly reducing the amount of oxygen carried in your blood. In addition, CO binds so strongly that it is hard to remove. This can cause a deadly situation. One treatment is to administer 100% oxygen (with a tight mask) to an afflicted person. Sometimes, under critical circumstances, the patient is placed in a pressure chamber where the Po_2 can be driven so high it literally displaces the CO from the hemoglobin, allowing oxygen once again to bind to its carrier.

A & P CONNECT

Variations of hemoglobin exist in the body to temporarily store or carry oxygen. Find out why we need more than one type of oxygen carrier in the body in **Oxygen-binding Proteins** online at **A&P Connect.**

evolve

SYSTEMIC GAS EXCHANGE

Exchange of gases in tissues takes place between arterial blood flowing through tissue capillaries (Figure 20-21), according to their pressure gradients. Thus, oxygen diffuses out of arterial ends of the capillaries and into cells, while carbon dioxide diffuses out of cells and into venous ends of the

▶ FIGURE 20-21 Systemic gas exchange. **A**, As blood enters a systemic capillary, O_2 diffuses down its pressure gradient (out of the blood). O_2 continues diffusing out of the blood until equilibration has occurred (or until the blood leaves the capillary). **B**, As blood enters a systemic capillary, CO_2 diffuses down its pressure gradient (into the blood). As with O_2, CO_2 continues diffusing as long as there is a pressure gradient.

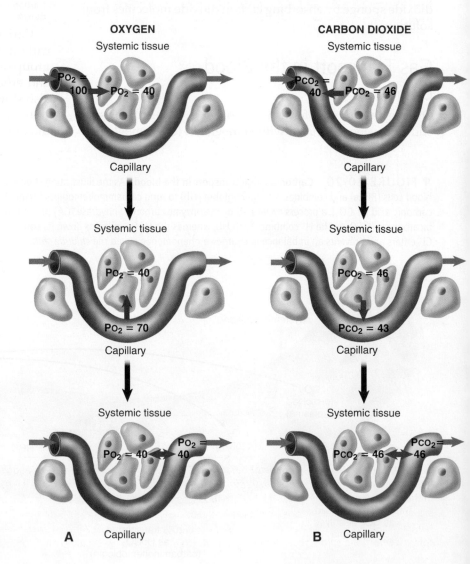

OXYGEN

CARBON DIOXIDE

Systemic tissue

$Po_2 = 100$ $Po_2 = 40$

Capillary

Systemic tissue

$Po_2 = 40$

$Po_2 = 70$

Capillary

Systemic tissue

$Po_2 = 40$ $Po_2 = 40$

A   Capillary

Systemic tissue

$Pco_2 = 40$ $Pco_2 = 46$

Capillary

Systemic tissue

$Pco_2 = 46$

$Pco_2 = 43$

Capillary

Systemic tissue

$Pco_2 = 46$ $Pco_2 = 46$

B   Capillary

capillaries. As cells use oxygen to metabolize sugars and other organic compounds, the P_{O_2} of the cells lowers compared to the P_{O_2} of arterial capillaries. This causes a movement of oxygen *into* the cells, to equalize the pressure. The arterial blood also has a low P_{CO_2} in comparison to metabolizing cells, which are producing more and more carbon dioxide. As a result, carbon dioxide travels down its pressure gradient *into* the venous capillaries, which then transport blood back to the lungs to rid the body of its waste CO_2.

As activity increases in any tissue, its cells necessarily use oxygen more rapidly. This decreases intracellular P_{O_2}, which means there is a larger pressure gradient between blood and tissues, which accelerates oxygen diffusion out of the tissue capillaries. Similarly, there is an increase in CO_2 production in the cells, which increases the difference in pressures between the cells and the capillaries, so more CO_2 moves out of the cells and into the venous capillaries. The increasing P_{CO_2} and decreasing P_{O_2} produce two effects—they favor oxygen dissociation from oxyhemoglobin and carbon dioxide associating with hemoglobin to generate carbaminohemoglobin. This means that, as activity increases, carbon dioxide is more efficiently removed from the system while oxygen is more efficiently unloaded from hemoglobin in order to supply the highly active cells.

REGULATION OF PULMONARY FUNCTION

Respiratory Control Centers

Various mechanisms maintain the relative constancy of the blood P_{O_2} and P_{CO_2}. This homeostasis of blood gases is maintained primarily by means of *changes in ventilation*—the rate and depth of breathing. The main integrators that control the nerves that affect the inspiratory and expiratory muscles are located within the brainstem. Together these integrators are simply called the **respiratory centers** (Figure 20-22) and they serve together to regulate our breathing patterns.

The basic rhythm of the respiratory cycle of inspiration and expiration is generated by the **medullary rhythmicity area.** This area contains two regions of interconnected control centers: the *dorsal respiratory group (DRG)* and the *ventral respiratory group (VRG)*. It is thought, based on research with animal models, that the VRG controls the basic rhythm of breathing while the DRG integrates information from chemoreceptors for P_{CO_2} and signals the VRG to alter breathing rhythm to restore homeostasis.

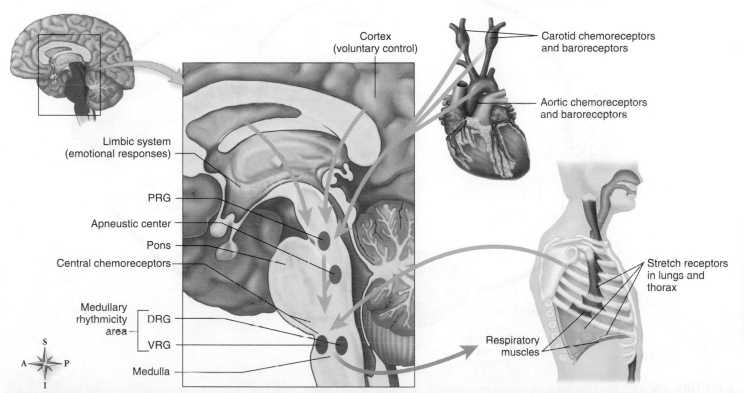

▲ **FIGURE 20-22   Regulation of breathing.** The dorsal respiratory group (DRG) and ventral respiratory group (VRG) of the medulla represent the medullary rhythmicity area. The pontine respiratory group (PRG) and apneustic center of the pons influence the basic respiratory rhythm by means of neural input to the medullary rhythmicity area. The brainstem also receives input from other parts of the body; information from chemoreceptors, baroreceptors, and stretch receptors can alter the basic breathing pattern, as can emotional (limbic) and sensory input. Despite these subconscious reflexes, the cerebral cortex can override the "automatic" control of breathing to some extent to do such activities as sing or blow up a balloon. *Green arrows* show flow of information to the respiratory control centers. The *purple arrow* shows the flow of information from the control centers to the respiratory muscles that drive breathing.

Additional regulatory centers in the pons, the *pontine respiratory group (PRG)* and the *apneustic center* can influence the activity of the medullary rhythaicity area.

Box 20-4 examines unusual breathing reflexes such as coughing and sneezing.

Factors that Influence Breathing

Feedback information to the medullary rhythmicity area comes from sensors throughout the nervous system, as well as from other control centers. For example, changes in pH and the partial pressures of carbon dioxide and oxygen within the systemic arterial blood all influence the medullary rhythmicity area.

The normal P_{CO_2} (about 38 to 40 mm Hg) in the blood acts on chemoreceptors in the medulla, which also monitor the pH of blood. When CO_2 increases or pH decreases (becoming more acidic) even slightly above the homeostatic values, it stimulates *central chemoreceptors,* which are present throughout the brainstem. Larger increases in arterial carbon dioxide also stimulate *peripheral chemoreceptors* in the carotid bodies and the aorta. Both of these will result in faster breathing relative to the increase of CO_2 in the blood. Figure 20-23 summarizes this negative feedback response (which acts much like a thermostat in your house).

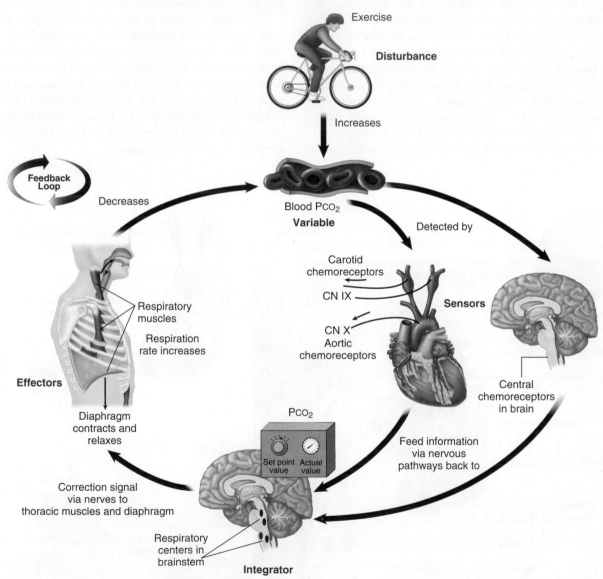

Exercise

Disturbance

Increases

Feedback Loop

Decreases

Blood P_{CO_2}
Variable

Detected by

Carotid chemoreceptors

CN IX

Sensors

CN X
Aortic chemoreceptors

Central chemoreceptors in brain

Respiratory muscles

Respiration rate increases

Effectors

Diaphragm contracts and relaxes

P_{CO_2}

Set point value   Actual value

Feed information via nervous pathways back to

Correction signal via nerves to thoracic muscles and diaphragm

Respiratory centers in brainstem

Integrator

▲ **FIGURE 20-23** **Negative feedback control of respiration.** This diagram summarizes the feedback loop that operates to increase the respiratory rate in response to high plasma P_{CO_2}. Increased cellular respiration during exercise causes a rise in plasma P_{CO_2}, which is detected by central chemoreceptors in the brain and perhaps peripheral chemoreceptors in the carotid sinus and aorta. Feedback information is relayed to integrators in the brainstem that respond to the increase in P_{CO_2} above the set-point value by sending nervous correction signals to the respiratory muscles, which act as effectors. The effector muscles increase their alternating contraction and relaxation, thus increasing the rate of respiration. As the respiration rate increases, the rate of CO_2 loss from the body increases and P_{CO_2} drops accordingly. This brings the plasma P_{CO_2} back to its set-point value.

BOX 20-4 FYI

Unusual Breathing Reflexes

The **cough reflex** is stimulated by foreign matter in the trachea or bronchi. The epiglottis and glottis reflexively close, and contraction of the expiratory muscles causes air pressure in the lungs to increase. The epiglottis and glottis then open suddenly, resulting in an upward burst of air that removes the offending contaminants—a cough.

The **sneeze reflex** is similar to the cough reflex, except that it is stimulated by contaminants in the nasal cavity. A burst of air is directed through the nose and mouth, forcing the contaminants (and mucus) out of the respiratory tract. Droplets from a sneeze can travel more than 161 km/hr (100 miles/hr) and travel 3 m (9.8 ft). Research suggests that many pathogenic microbes produce symptoms that trigger sneezing in order to spread themselves to other people. Scientists call this *altered host behavior* and identify it as a mechanism of microbes to efficiently spread themselves to additional human hosts.

The term **hiccup** is used to describe an involuntary, spasmodic contraction of the diaphragm. When such a contraction occurs, generally at the beginning of an inspiration, the glottis suddenly closes, producing the characteristic sound. Hiccups lasting for extended periods can be disabling. They may be produced by irritation of the phrenic nerve or the sensory nerves in the stomach or by direct injury or pressure on certain areas of the brain. Fortunately, most cases of hiccups last only a few minutes and are harmless.

A **yawn** is slow, deep inspiration through an unusually widened mouth. Yawns were once thought to be reflexes that increase ventilation when blood oxygen content is low, but newer evidence suggests that this is unlikely. A current theory states that we yawn for the same reason we occasionally stretch—to prepare our muscles and our circulatory system for action. Recent alternate hypotheses suggest that yawning cools the brain or otherwise regulates body temperature—or that yawning is triggered by neurotransmitters related to mood. The variety of hypotheses show one thing for certain: *we do not currently understand the physiology of yawning!*

A protective physiological response called the **diving reflex** is responsible for the astonishing recovery of apparent drowning victims—including some who may have been submerged for more than 40 minutes! Survivors are most often preadolescent children who have been immersed in water below 20° C (68° F). Apparently, the colder the water, the better the chance of survival. Victims initially appear dead when pulled from the water. Breathing has stopped; they have fixed, dilated pupils; they are cyanotic; and their pulse has stopped.

Studies have shown that, when the head and face are immersed in ice-cold water, there is immediate shunting of blood to the core body areas with peripheral vasoconstriction and slowing of the heart *(bradycardia)*. Metabolism is slowed, and tissue requirements for oxygen and nutrients decrease. The diving reflex is a protective response of the body to cold water immersion and is a function of such physiological and environmental parameters as water temperature, age, lung volume, and posture.

Control of Respirations During Exercise

Respirations increase abruptly at the beginning of exercise and decrease even more markedly as it ends. This much is known. The mechanism that accomplishes this increased ventilation rate, however, is not known. It is not identical to the one that produces moderate increases in breathing. Numerous studies have shown that arterial blood P_{CO_2}, P_{O_2}, and pH do not change enough during exercise to produce the degree of *hyperpnea* (faster, deeper respirations) observed. Presumably, many chemical and nervous factors and temperature changes operate as a complex, but still unknown, mechanism for regulating respirations during exercise (Figure 20-24).

QUICK CHECK

20. In what form is most oxygen transported in the blood?
21. In what form is most carbon dioxide transported in the blood?
22. How does carbon dioxide affect the pH of blood?

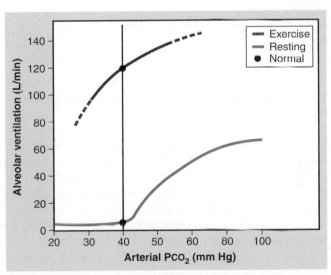

▲ **FIGURE 20-24** **Normal effects of maximum exercise in an athlete.** This graph shows that the breathing rate (vertical axis) is much higher in an athlete exercising maximally than would be expected for any given blood carbon dioxide pressure (P_{CO_2}) (horizontal axis). As you can see at the normal points of a P_{CO_2} of 40 mm Hg, the exercising athlete's breathing (ventilation) rate is 120 L/min. However, at rest the athlete's breathing rate is only about 5 or 6 L/min at the same P_{CO_2}—thus showing that P_{CO_2} is not the major factor causing an increased rate of breathing during exercise.

UNIT 5

MECHANISMS OF DISEASE

There are many common disorders and diseases of the respiratory system. These include disorders associated with the upper respiratory tract, such as rhinitis, laryngitis, and tonsillitis, as well as anatomical disorders that can cause sleep apnea and chronic nosebleeds. Disorders and diseases of the lower respiratory tract are also common, and include acute bronchitis, pneumonia, and less common diseases such as tuberculosis. Even more significant are various types of lung cancer, which are among the most common and deadly of all cancers.

Disorders associated with respiratory function include restrictive pulmonary disorders, obstructive pulmonary disorders, and chronic obstructive pulmonary diseases (COPD) such as bronchitis, emphysema, and asthma.

evolve Find out more about these diseases and disorders of the respiratory system online at *Mechanisms of Disease: Respiratory System.*

Cycle of LIFE ○

Respiratory exchange of oxygen and carbon dioxide must occur between air in the lungs and blood and, following this, between blood and every body cell. Hemoglobin plays a vital role in the respiration process, in conjunction with the structural components of the body through which the respiratory gases must pass. Each component of the system may be affected by developmental defects, by age-related structural changes, or by loss of function during the life cycle.

Premature birth can cause potentially fatal respiratory problems. A very low–birth weight baby may have inadequate blood flow to the lungs, an inability to ventilate properly, and inadequate quantities of surfactant. Other diseases that cause serious respiratory problems include cystic fibrosis and asthma and certain types of obstructive pulmonary disease and emphysema in older adults. Pneumothorax occurs more frequently in young adult females as a complication of endometriosis in the thorax (see Chapter 25).

The BIG Picture

Understanding the relationship of structure to function is critical to an understanding of homeostasis in all of the body organ systems. The anatomy of the respiratory system components permits the distribution of air and the exchange of respiratory gases. This dual function ultimately allows for both exchange of gases between environmental air and blood in the lungs and, finally, gas exchange between blood and individual body cells. In addition to delivery of air to the tiny terminal air passageways and alveoli for gas exchange with blood, components of the upper respiratory tract effectively filter, warm, and humidify the air we breathe.

Respiratory functions are dependent on the structural organization of the system parts and on the interrelationship of those components with other body systems, including the nervous, cardiovascular, muscular, and immune systems. For example, nerves regulate the thoracic and abdominal muscles that drive breathing, as well as the smooth muscles that regulate airflow through the bronchial tree. The immune system guards against airborne pathogens and irritants.

The homeostatic balance of the entire body, and thus the survival of each and every cell, depends on the proper functioning of the respiratory system. This is because the mitochondria in each cell require oxygen for their energy conversions. In addition, each cell produces toxic carbon dioxide as a waste product from these energy conversions, and the internal environment must continually acquire new oxygen and discard carbon dioxide. If each cell were immediately adjacent to the external environment—that is, atmospheric air—this would require no special system. However, because almost every one of the 100 trillion cells that make up the body are far removed from the outside air, another method of satisfying this condition must be employed; this is where the respiratory system comes in.

By the physical process of ventilation, fresh external air continually flows less than a hair's breadth away from the circulating fluid of the body—the blood. By means of diffusion, oxygen enters the internal environment and carbon dioxide leaves. The efficiency of this process is enhanced by hemoglobin, which immediately takes oxygen molecules out of solution in the plasma so that more oxygen can rapidly diffuse into the blood. The blood, the circulating fluid tissue of the cardiovascular system, carries the blood gases throughout the body—picking up gases where there is an excess and unloading them where there is a deficiency. In this manner, each cell of the body is continually bathed in a fluid environment that offers a constant supply of oxygen and an efficient system for removing carbon dioxide.

Specific mechanisms involved in respiratory function show the interdependence between body systems observed throughout our study of the human body. For example, without blood and the maintenance of blood flow by the cardiovascular system, blood gases could not be transported between the gas exchange tissues of the lungs and the various systemic tissues of the body. Without regulation by the nervous system, ventilation could not be adjusted to compensate for changes in the oxygen or carbon dioxide content of the internal environment. Without the skeletal muscles of the thorax, the airways could not maintain the flow of fresh air that is so vital to respiratory function. The skeleton itself provides a firm outer housing for the lungs and has an arrangement of bones that facilitates the expansion and recoil of the thorax, which is needed to accomplish inspiration and expiration. Without the immune system, pathogens from the external environment could easily colonize the respiratory tract and possibly cause a fatal infection.

Even more subtle interactions between the respiratory system and other body systems can be found. For example, the language function of the nervous system is limited without the speaking ability provided by the larynx and other structures of the respiratory tract. The homeostasis of pH, which is regulated by a variety of systems, is influenced by the respiratory system's ability to adjust the body's carbon dioxide levels (and thus the levels of carbonic acid).

Numerous age-related changes affect vital capacity, make ventilation difficult, or reduce the oxygen or carbon dioxide carrying capacity of blood. For example, in older adulthood the ribs and sternum tend to become more fixed and less able to expand during inspiration, the respiratory muscles are less effective, and hemoglobin levels are often reduced. The result is a general reduction in respiratory efficiency in old age. ●

LANGUAGE OF SCIENCE AND MEDICINE *continued from page 455*

hemoglobin (Hb) (hee-moh-GLOH-bin)
[*hemo-* blood, *-glob-* ball, *-in* substance]

hiccup (HIK-up)
[imitation of hiccup sound]

hilum (HYE-lum)
[*hilum* least bit] *pl.,* hila (HYE-lah)

inspiration (in-spih-RAY-shun)
[*in-* in, *-spir-* breathe, *-ation* process]

inspiratory capacity (IC) (in-SPY-rah-tor-ee kah-PASS-ih-tee)
[*in-* in, *-spir-* breathe, *-tory* relating to]

inspiratory reserve volume (IRV) (in-SPY-rah-tor-ee)
[*in-* in, *-spir-* breathe, *-tory* relating to]

intrapleural space (in-trah-PLOO-ral)
[*intra-* within, *-pleura-* rib, *-al* relating to]

laryngopharynx (lah-rin-goh-FAIR-inks)
[*laryng-* voicebox (larynx), *-pharynx* throat] *pl.,* laryngopharynges or laryngopharynxes (lah-rin-goh-FAIR-in-jeez, lah-rin-goh-FAIR-inks-ehz)

larynx (LAIR-inks)
[*larynx* voicebox] *pl.,* larynges or larynxes (lah-RIN-jeez, LAIR-inks-ehz)

lingual tonsil (LING-gwal TAHN-sil)
[*ling-* tongue, *-al* relating to]

lower respiratory tract (RES-pih-rah-tor-ee TRAKT)
[*re-* again, *-spir-* breathe, *-tory* relating to, *tractus* trail]

medullary rhythmicity area (MED-oo-lair-ee rith-MIH-sih-tee)
[*medulla-* middle, *-ary* relating to, *rhythm-* rhythm, *-ic-* relating to, *-ity* condition]

nasal septum (NAY-zal SEP-tum)
[*nas-* nose, *-al* relating to, *septum* partition]

nasopharynx (nay-zoh-FAIR-inks)
[*naso-* nose, *-pharynx* throat] *pl.,* nasopharynges or nasopharynxes (nay-zoh-FAIR-in-jeez, nay-zoh-FAIR-inks-ehz)

olfactory epithelium (ohl-FAK-tor-ee ep-ih THEE-lee-um)
[*olfact-* smell, *-ory* relating to, *epi-* upon, *-theli-* nipple, *-um* thing] *pl.,* epithelia (ep-ih-THEE lee-ah)

oropharynx (or-oh-FAIR-inks)
[*oro-* mouth, *-pharynx* throat] *pl.,* oropharynges or oropharynxes (or-oh-FAIR-in-jeez, or-oh-FAIR-inks-ehz)

palatine tonsil (PAL-ah-tyne TAHN-sil)
[*palat-* palate, *-ine* relating to]

paranasal sinus (pair-ah-NAY-zal SYE-nus)
[*para-* beside, *-nas-* nose, *-al* relating to, *sinus* hollow]

parietal pleura (pah-RYE-ih-tal PLOO-rah)
[*parie-* wall, *-al* relating to, *pleura* rib] *pl.,* pleurae (PLOOR-ee)

partial pressure (PAR-shal PRESH-ur)

pharyngeal tonsil (fah-RIN-jee-al TAHN-sil)
[*pharyng-* throat, *-al* relating to]

pharynx (FAIR-inks)
[*pharynx* throat] *pl.,* pharynges or pharynxes (FAH-rin-jeez, FAIR-inks-ehz)

pleura (PLOO-rah)
[*pleura* rib] *pl.,* pleurae (PLOOR-ee)

pneumothorax (noo-moh-THOH-raks)
[*pneumo-* air or wind, *-thorax* chest]

primary bronchus (BRONG-kus)
[*prim-* first, *-ary* relating to, *bronchus* windpipe] *pl.,* bronchi (BRONG-kye)

primary principle of ventilation (PRY-mair-ee PRIN-sip-al of ven-tih-LAY-shun)
[*prim-* first, *-ary* relating to, *princip-* foundation, *vent-* fan or create wind, *-tion* process]

residual volume (RV) (ree-ZID yoo-al)

respiratory center (RES-pih-rah-tor-ee SEN-ter)
[*re-* again, *-spir-* breathe, *-tory* relating to]

respiratory mucosa (RES-pih-rah-tor-ee myoo-KOH-sah)
[*re-* again, *-spir-* breathe, *-tory* relating to, *mucus* slime]

respiratory portion (RES pih rah-tor-ee POR-shun)
[*re-* again, *-spir-* breathe, *-tory* relating to]

secondary bronchus (SEK-on-dair-ee BRONG-kus)
[*second-* second, *-ary* relating to, *bronchus* windpipe] *pl.,* bronchi (BRONG-kye)

sneeze reflex (sneez REE-fleks)
[*re-* back or again, *-flex* bend]

spirogram (SPY-roh-gram)
[*spir-* breathe, *-gram* drawing]

spirometer (spih-ROM eh-ter)
[*spir-* breathe, *-meter* measurement]

surfactant (sur-FAK-tant)
[combination of surf(ace) act(ive) a(ge)nt]

thyroid cartilage (THY-royd KAR-tih-lij)
[*thyr-* shield, *-oid* like]

tidal volume (TV) (TYE-dal)
[*tid-* time, *-al* relating to]

total lung capacity (TLC)

trachea (TRAY-kee-ah)
[*trachea* rough duct] *pl.,* tracheae or tracheas (TRAY-kee-ee, TRAY-kee-ahz)

tracheostomy (tray-kee-OS-toh-mee)
[*trache-* rough duct, *-os-* mouth or opening, *-tom-* cut, *-y* action]

turbinate (TUR-bih-nayt)
[*turbin-* top (spinning toy), *-ate* of or like]

upper respiratory tract (RES-pih-rah-tor-ee TRAKT)
[*re-* again, *-spir-* breathe, *-tory* relating to, *tract-* trail]

vestibular fold (ves-TIB-yoo-lar)
[*vestibul-* entrance hall, *-al* relating to]

vestibule (VES-tih-byool)
[*vestibul-* entrance hall]

vibrissa (vye-BRISS-ah)
[*vibrissa* nostril hair] *pl.,* vibrissae (vye-BRISS-ee)

visceral pleura (VISS-er-al PLOO-rah)
[*viscer-* internal organ, *-al* relating to, *pleura* rib] *pl.,* pleurae (PLOOR-ee)

vital capacity (VC) (VYE-tal kah-PASS-ih-tee)
[*vita-* life, *-al* relating to]

vocal fold
[*voca-* voice, *-al* relating to]

yawn
[*yawn* gape]

CHAPTER SUMMARY

To download an MP3 version of the chapter summary for use with your iPod or other portable media player, access the Audio Chapter Summaries online at http://evolve.elsevier.com.

HINT   Scan this summary after reading the chapter to help you reinforce the key concepts. Later, use the summary as a quick review before your class or before a test.

ANATOMY AND PHYSIOLOGY OF THE RESPIRATORY SYSTEM

A. Respiratory system functions as a gas distributor and exchanger
B. Respiratory system filters, warms, and humidifies the air we breath, in addition to providing us with vocal communication and olfaction
C. Respiratory system also plays a vital physiological role in our bodies by regulating homeostasis of metabolism, circulation, electrolyte and water balance, and acidity of the blood

STRUCTURAL PLAN OF THE RESPIRATORY SYSTEM

A. Respiratory system can be divided into upper and lower tracts:
  1. Upper respiratory tract—composed of the nose, nasopharynx, oropharynx, laryngopharynx, and larynx
  2. Lower respiratory tract—consists of the trachea, the bronchial tree, and the lungs
  3. Accessory structures include the oral cavity, rib cage, intercostals, and diaphragm

UPPER RESPIRATORY TRACT

A. Nose
  1. Structure of the nose
    a. External portion consists of a bony and cartilaginous framework overlaid by skin with many sebaceous glands
    b. Two nasal bones meet the frontal bone of the skull at their superior end to form the root
    c. Palatine bones form the base of the nasal cavity
      (1) Cleft palate—condition when the palatine bones fail to unite at their center
    d. Cribriform plate—separates the roof of the nose from the cranial cavity (Figures 20-2 and 20-3)
    e. Nasal septum—separates the nasal cavity into right and left cavities; made up of four main structures (Figure 20-2):
      (1) Perpendicular plate of the ethmoid bone
      (2) Vomer bone
      (3) Septal nasal cartilage
      (4) Vomeronasal cartilage
    f. Nasal cavity is divided into three passageways (meatuses): created by the projection of the conchae (Figure 20-3)
    g. Nostrils (anterior nares)—enclosed by skin reflected from the ala of the nose; open inside to the vestibule
    h. Passage of air to the pharynx begins with the anterior nares, passes through the vestibule, through the inferior, middle, and superior meatuses (simultaneously), and then through the posterior nares to enter the pharynx
  2. Nasal mucosa—mucous membrane that air passes over; it contains a rich blood supply
    a. Olfactory epithelium—contains many olfactory nerve cells and has a rich lymphatic plexus

    b. Ciliated mucous membrane lines the rest of the respiratory tract down as far as the smaller bronchioles
  3. Paranasal sinuses—four pairs of air-containing spaces that open or drain into the nasal cavity; each sinus is lined by ciliated respiratory mucosa
  4. Functions of the nose
    a. Serves as a passageway for air to the lungs
    b. Air that enters through the nasal passageways is filtered of impurities, warmed, moistened, and chemically examined
    c. Sinuses serve to lighten bones of the skull and allow resonation of sounds during speech
B. Pharynx (throat)—tubelike structure that extends from the base of the skull to the esophagus; made of muscle and is lined with mucous membrane
  1. Pharynx is divided into the three anatomical divisions:
    a. Nasopharynx—located behind the nose and extends from the posterior nares to the level of the soft palate
    b. Oropharynx—located behind the mouth and stretches from the soft palate above to the hyoid bone below
    c. Laryngopharynx—extends from the hyoid bone to the esophagus
  2. Seven openings are found in the pharynx (Figure 20-3):
    a. Right and left auditory (eustachian) tubes that open into the nasopharynx
    b. Two posterior nares that open into the nasopharynx
    c. The opening from the mouth into the oropharynx
    d. The opening into the larynx from the laryngopharynx
    e. The opening into the esophagus from the laryngopharynx
  3. Pharyngeal tonsils—located in the nasopharynx; referred to as adenoids when they are enlarged
  4. Palatine tonsils—located back in the oropharynx; most commonly removed
  5. Lingual tonsils—located at the base of the tongue
  6. Pharynx serves as a common pathway for the respiratory and digestive tracts
C. Larynx (voice box)—lies between the root of the tongue and the upper end of the trachea; below and in front of the lowest part of the pharynx (Figure 20-1)
  1. Lined with ciliated mucous membrane
  2. Consists predominantly of cartilages attached to one another; surrounded by muscles or fibrous and elastic tissue components (Figure 20-4)
    a. Vestibular folds (false vocal cords)
    b. Vocal folds—contribute to sound production during speech
  3. Cartilages of the larynx—nine cartilages form the framework of the larynx (Figure 20-4):
    a. Thyroid cartilage
    b. Epiglottis
    c. Cricoid cartilage
    d. Arytenoid (two)
    e. Corniculate (two)
    f. Cuneiform (two)
  4. Muscles of the larynx—intrinsic and extrinsic muscles
    a. Intrinsic muscles—control vocal fold length and tension and are important in regulating the shape of the laryngeal inlet
    b. Extrinsic muscles—physically move the larynx and its parts
  5. Larynx forms part of the pathway to the lungs and produces voice

LOWER RESPIRATORY TRACT

A. Trachea (windpipe)—tube that extends from the larynx in the neck to the primary bronchi in the thoracic cavity (Figure 20-8)
 1. Wall of the trachea is composed of C-shaped cartilaginous rings (Figure 20-6)
 2. Provides a sturdy open passageway from the upper respiratory tract into the lungs
B. Bronchi and Alveoli
 1. Trachea divides into two primary bronchi; right bronchus slightly larger and more vertical than the left
 2. Primary bronchi divide into smaller branches called secondary bronchi
  a. Branch to form tertiary bronchi and then further on into bronchioles
  b. Bronchioles continue to branch into microscopic terminal bronchioles; pass air into respiratory bronchioles and then alveolar sacs
  c. Alveolar sacs contain numerous smaller sacs called alveoli
 3. Alveoli are made up of a single layer of simple squamous epithelial tissue
  a. Allows oxygen and carbon dioxide gas to pass quickly from alveoli to capillary
  b. Surface of the alveoli is coated with a fluid containing surfactant; keeps the alveoli from collapsing
 4. Functions of the bronchi and alveoli—distribute air to the lung's interior
C. Lungs—cone-shaped organs that fill the pleural portion of the thoracic cavity completely (Figures 20-8 and 20-9)
 1. Hilum—slit on the lung's medial surface where the primary bronchi and pulmonary blood vessels enter
 2. Base—broad inferior surface of the lung
 3. Apex—pointed upper margin of the lung
 4. Each lung is divided into lobes by fissures; the left into two lobes and the right into three
 5. Lobes of the lungs can be further divided into functional units; bronchopulmonary segments
 6. Visceral pleura—covers the outer surfaces of the lungs; provides protection from abrasion within the pleural cavity
 7. Lungs perform both air distribution and gas exchange
D. Thorax—thoracic cavity is divided by pleura to form three divisions
 1. Mediastinum—middle of thoracic cavity
 2. Pleural divisions—the part occupied by the lungs
  a. Parietal pleura—lines the entire thoracic cavity by attaching to the inside of the ribs and superior surface of the diaphragm
  b. Visceral pleura—covers the lungs entirely
 3. Functions of the thorax—brings about inspiration and expiration

RESPIRATORY PHYSIOLOGY

A. Respiratory system is composed of an integrated set of processes (Figure 20-11):
 1. External respiration
 2. Transport of gases by the blood
 3. Internal respiration
 4. Overall regulation of respiration

PULMONARY VENTILATION

A. Mechanism of pulmonary ventilation—air moves into the lungs because the volume within the lungs increases and the air pressure within them is lowered

 1. Pressure gradient—air outside of the body is at a higher pressure than the air within the lungs
  a. Pressure gradient causes air to rush into the lungs
 2. Primary principle of ventilation
  a. When air moves from the high pressure of the atmosphere into the lower pressure of the lungs, inspiration occurs
  b. When the diaphragm pushes against the lungs, air pressure in the lungs is increased so that it is higher than that of the air pressure in the atmosphere; air rushes down the pressure gradient and outside of the body, and expiration occurs
 3. Pressure gradients are established by enlarging or reducing the size of the thoracic cavity
 4. Inspiration and expiration (Figure 20-14)
  a. Inspiration—contraction of the diaphragm produces inspiration; as it contracts, it makes the thoracic cavity larger (Figure 20-13)
   (1) Expansion of the thorax results in lungs expanding to fill space, causing pressure in bronchioles and alveoli to lower
   (2) Air moves into the lungs in order to equalize pressure inside and outside
  b. Expiration
   (1) Quiet expiration—passive process that begins when the pressure gradients that resulted from inspiration are reversed by the relaxation of the inspiratory muscles
   (2) Forced expiration—abdominal and intercostals muscles shrink thoracic cavity so pressure in alveoli is much large than atmospheric pressure, causing large volume of air to move quickly out of lungs
   (3) Elastic recoil—lungs to return to their typical volume before the next inspiration
B. Pulmonary volumes—differing volumes of air move into and out of the lungs depending on the force with which one breathes
 1. Volumes can be measured by a spirometer; recorded as graphics called spirograms (Figure 20-15)
 2. Tidal volume (TV)—volume of air exhaled after a normal breath; about 500 ml
 3. Expiratory reserve volume (ERV)—the volume of air an individual can force out of the lungs after releasing tidal air; about 1,000 ml
 4. Inspiratory reserve volume (IRV)—amount of air that can be forcibly inspired over and above a normal inspiration; about 3,300 ml
 5. Residual volume (RV)—amount of air that cannot be forcibly exhaled (1,200 ml)
C. Pulmonary capacity—sum of two or more pulmonary "volumes" (Figure 20-15 and Table 20-1)
 1. Vital capacity (VC)—sum of the IRV, TV, and ERV
 2. Inspiratory capacity (IC)—sum of the TV and the IRV
 3. Functional residual capacity (FRC)—sum of the ERV and the RV
 4. Total lung capacity (TLC)—sum of all four lung volumes

PULMONARY GAS EXCHANGE

A. Partial pressure—pressure exerted by any one gas in a mixture of gases or in a liquid
 1. Law of partial pressure—the partial pressure of a gas in a mixture of gases is directly related to the concentration of that gas in the mixture and to the total pressure of the mixture

B. Exchange of gases in the lungs—exchange of gases in the lungs takes place between alveolar air and blood flowing through lung capillaries (Figure 20-16)

HOW BLOOD TRANSPORTS GASES

A. Blood transports oxygen and carbon dioxide either as solutes or combined with other chemicals
B. Hemoglobin (Hb)—quaternary protein made up of four polypeptide chains (two alpha chains and two beta chains)
  1. Each chain is associated with an iron-containing heme group, and each iron atom can bind an oxygen molecule (O_2) (Figure 20-18)
  2. Gases other than oxygen and carbon dioxide can also bind to hemoglobin
    a. Carbon monoxide
C. Gas transport in the blood
  1. Oxygen—travels as dissolved oxygen in plasma and as oxygen associated with hemoglobin—(oxyhemoglobin)
    a. Oxyhemoglobin carries the vast majority of the total oxygen transported by the blood
  2. Carbon dioxide (Figure 20-20)
    a. Only about 10% of the carbon dioxide in blood travels as dissolved gas
    b. Carbaminohemoglobin is an important carrier of blood carbon dioxide (20% of total); formed when CO_2 binds to an amine group on hemoglobin (Figure 20-19)
    c. About 70% of total blood CO_2 is transported as bicarbonate
    d. When CO_2 enters the blood, it also produces hydrogen ions
  3. Carbon monoxide
    a. CO binds more strongly to hemoglobin than oxygen does
    b. Hard to remove CO from hemoglobin, which can be deadly since it replaces oxygen at the binding sites

SYSTEMIC GAS EXCHANGE

A. Exchange of gases in tissues takes place between arterial blood flowing through tissue capillaries (Figure 20-21)
  1. Oxygen diffuses out of arterial capillaries and into cells according to their pressure gradients
  2. Carbon dioxide diffuses out of cells and into venous capillaries according to their pressure gradients
  3. Increasing P_{CO_2} and decreasing P_{O_2} produce two effects:
    a. Oxygen dissociation from oxyhemoglobin
    b. Carbon dioxide associating with hemoglobin to generate carbaminohemoglobin

REGULATION OF PULMONARY FUNCTION

A. Respiratory control centers—main integrators that control the nerves that affect the inspiratory and expiratory muscles; located within the brainstem (Figure 20-22)
B. Basic rhythm of the respiratory cycle of inspiration and expiration is generated by the medullary rhythmicity area
  1. Contains two regions of interconnected control centers:
    a. Dorsal respiratory group (DRG)—integrates information from chemoreceptors for P_{CO_2}; signals the VRG to alter breathing rhythm to restore homeostasis
    b. Ventral respiratory group (VRG)—controls the basic rhythm of breathing
C. Factors that influence breathing—sensors from the nervous system and other control centers provide feedback to the medullary rhythmicity area

  1. Changes in pH
    a. Chemoreceptors—monitor pH in blood
  2. Changes in partial pressures of carbon dioxide and oxygen
    a. Peripheral chemoreceptors—stimulated upon increased arterial carbon dioxide
D. Control of respirations during exercise
  1. Respirations increase abruptly at the beginning of exercise and decrease even more markedly as it ends
  2. Mechanism that accomplishes this increased ventilation rate is not known

REVIEW QUESTIONS

HINT Write out the answers to these questions after reading the chapter and reviewing the Chapter Summary. If you simply think through the answer without writing it down, you won't retain much of your new learning.

1. Identify the major anatomical structures of the nose.
2. How are the conchae arranged in the nose? What are they?
3. The pharynx is common to what two systems?
4. What is the voice box? Of what is it composed? What is the Adam's apple?
5. What is the epiglottis? What is its function?
6. Describe the structure and function of the trachea.
7. List the organs that are included in the upper respiratory tract. Do the same for the lower respiratory tract.
8. Identify the separate volumes that make up the total lung capacity.
9. Normally, about what percentage of the tidal volume fills the anatomical dead space?
10. What factors influence the amount of oxygen that diffuses into the blood from the alveoli?
11. Identify the major factors that influence breathing.
12. Describe the changes in respiration during a period of exercise.

CRITICAL THINKING QUESTIONS

HINT   After finishing the Review Questions, write out the answers to these items to help you apply your new knowledge. Go back to sections of the chapter that relate to items that you find difficult.

1. How would you describe the structure and function of the respiratory mucosa? Include the types of cells it contains and where these cells are located in the respiratory system.
2. Why do you think mucus production is especially important in the olfactory epithelium?
3. Make the distinction between air distribution and gas exchange in the respiratory system. Identify the organs that serve as air distributors and gas exchangers.
4. What is pulmonary ventilation? What evidence can you find to describe whether the lungs are active or passive during this process?
5. After strenuous exercise, inexperienced athletes will quite often attempt to recover and resume normal breathing by bending over or sitting down. Using the mechanics of ventilation, how would you modify the recovery practices of these athletes?

continued from page 455

Remember Derrick from the Introductory Story? See if you can answer the following questions about him now that you have read this chapter.

1. Which of these muscles would **not** contract when Derrick complied with his nurse's instructions?
   a. Diaphragm
   b. Serratus anterior
   c. Rectus abdominis
   d. External intercostals

2. Which statement best describes the "mechanics" of Derrick's inhalations?
   a. The thoracic cavity decreases in size, lowering the alveolar pressure, and air flows from high (atmosphere) pressure to low (alveolar) pressure.
   b. The thoracic cavity increases in size, lowering the alveolar pressure, and air flows from high (atmospheric) pressure to low (alveolar) pressure.
   c. Air flows from high (atmospheric) pressure to low (alveolar) pressure and expands the thoracic cavity.
   d. Air flows from high (intrapleural) pressure to low (alveolar) pressure and expands the thoracic cavity.

3. The increased carbon dioxide will make Derrick's blood _____.
   a. More acidic
   b. More basic
   c. More neutral
   d. None of the above

4. How is carbon dioxide transported in Derrick's blood?
   a. Dissolved in the plasma
   b. Bound to hemoglobin
   c. In the form of bicarbonate
   d. All of the above

HINT   *To solve a case study, you may have to refer to the glossary or index, other chapters in this textbook, A&P Connect, Mechanisms of Disease, and other resources.*

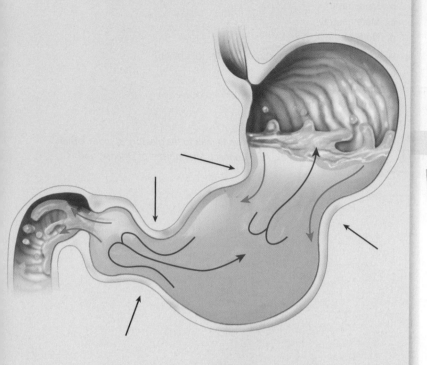

Digestive System

STUDENT LEARNING OBJECTIVES

At the completion of this chapter, you should be able to do the following:

1. Outline the organization of the digestive system, discussing its main components.
2. Describe the primary layers of the typical gastrointestinal wall.
3. Describe the functions of the following by making a table: oral cavity, tongue, salivary glands, teeth, pharynx, esophagus, stomach, small intestine, large intestine, and rectum.
4. Describe the functions of the peritoneum and its mesenteries.
5. Outline the functions of the following organs: liver, gallbladder, pancreas.
6. Describe the overall process of mechanical digestion.
7. Discuss the roles of peristalsis and segmentation throughout the gastrointestinal tract.
8. Explain how motility of food is regulated.
9. Explain the functions of digestive enzymes, discussing the role of proenzymes.
10. Making a table, outline the digestion of carbohydrates, fats, and proteins.
11. Explain the roles that saliva, gastric juice, pancreatic juice, bile, and intestinal juice play in digestion.
12. Explain how digestive gland secretions are controlled.
13. Outline the process of absorption and elimination.

IT took only 3 seconds. Sangetha turned away from her sewing basket to pick up her cell phone. When she looked back, she saw a handful of buttons disappear into her 24-month-old daughter's mouth. She quickly dropped down and removed the buttons from her daughter Jinder's mouth—but Sangetha had no way of knowing how many had gone in. Had any already gone down Jinder's throat? She was coughing slightly and drooling. Sangetha immediately took Jinder to the emergency department, thankfully a very short drive away, where they took radiographs of Jinder's neck and chest. On the x-ray film, they found one button in Jinder's lower esophagus.

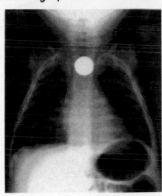

Before you read this chapter, try to hypothesize what will happen to the button in Jinder's digestive system. Will it pass through the entire digestive system and be safely expelled, or is there another, potentially more serious scenario? What would you do? See if you can answer the questions concerning Jinder's story at the end of this chapter.

This chapter deals with the anatomy and physiology of the digestive system. Working together, the organs of the digestive system modify food in both chemical composition and physical state so that nutrients can be absorbed and used by the body cells. This process is called *digestion*. The primary purpose of the digestive process is to bring essential nutrients into the internal environment so that they are available to each cell of the body. In this chapter, we will investigate the structures and functions of the digestive system.

ORGANIZATION OF THE DIGESTIVE SYSTEM

Organs of Digestion

The **digestive system** comprises the digestive tract and other organs that aid digestion. The main organs of the **digestive tract** (Figure 21-1) form a hollow tubelike system, open at both ends, which passes through the ventral cavities of the body. The digestive tract is often referred to as the **alimentary canal** or **gastrointestinal (GI) tract** (though this technically only refers to the stomach and intestines). Ingested food passing through the lumen of the GI tract is actually *outside* the internal environment of the body. In fact, you can think of the body as a tube within a tube: The outer skin "tube" folds inward through the mouth and again meets the outer skin at the anus, thus making the "inside" of the GI tract continuous with the *external environment!*

In Box 21-1 we list the main organs of the digestive system as well as the accessory organs located in the main digestive organs or opening into them. Organs such as the larynx, trachea, diaphragm, and spleen are

> **HINT** Before reading the chapter, say each of these terms out loud. This will help you avoid stumbling over them as you read.

absorption (ab-SORP-shun)
[*absorp-* swallow, *-tion* process]

alimentary canal (al-eh-MEN-tar-ee kah NAL)
[*aliment-* nourishment, *-ary* relating to, *canal* channel]

amylase (AM-eh-layz)
[*amyl-* starch, *-ase* enzyme]

anal canal (AY-nal kah-NAL)
[*an-* ring (anus), *-al* relating to, *canal* channel]

ascending colon (ah-SEND-ing KOH-lon)
[*ascend-* climb, *colon* large intestine]

bile (byle)

bile salt (byle)

body (BOD-ee)

brush border

cardia (KAR-dee-ah)
[*cardia* heart]

cementum (sih-MEN-tem)
[*cementum* rough stone]

cephalic phase (seh-FAL-ik fayz)
[*cephal-* head, *-ic* relating to]

cheek

chemical digestion (KEM-ih-kal dye-JES-chun)

chief cell

cholecystokinin (CCK) (koh-leh-sis-toh-KYE-nin)
[*chole-* bile, *-cyst-* bladder, *-kin-* move, *-in* substance]

chyme (kyme)
[*chym-* juice]

chymotrypsin (kye-moh-TRIP-sin)
[*chymo-* juice, *-tryps-* pound, *-in* substance]

colon (KOH-lon)
[*colon* large intestine]

constipation (kon-stih-PAY-shun)
[*constipa-* crowd together, *-tion* process]

crown

deciduous teeth (deh-SID-yoo-us)
[*decid-* fall off, *-ous* relating to]

defecation (def-eh-KAY-shun)
[*de-* remove, *-feca-* waste (feces), *-tion* process]

deglutition (deg-loo-TISH-un)
[*deglut-* swallow, *-tion* process]

dentin (DEN-tin)
[*dent-* tooth, *-in* substance]

dentition (den-TISH-en)
[*dent-* tooth, *-tion* state]

descending colon (dee-SEND-ing KOH-lon)
[*descend-* move downward, *colon* large intestine]

continued on page 510

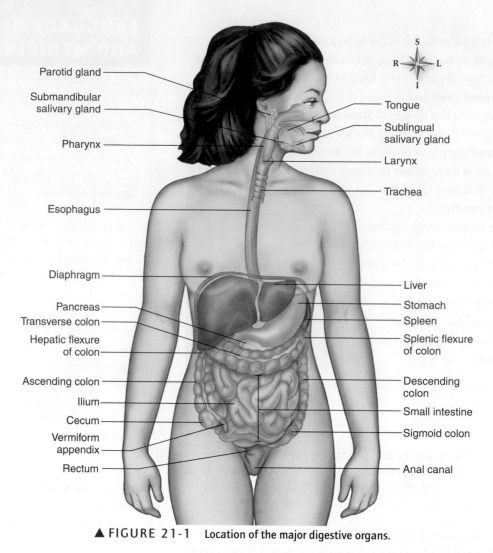

▲ FIGURE 21-1   Location of the major digestive organs.

also labeled in Figure 21-1, but they are *not* digestive organs. They are shown to assist your understanding of the orientation of the digestive organs with respect to other important body structures.

Wall of the GI Tract

Layers

The GI tract is fashioned from four layers of tissues: (1) an inner layer of mucosa, (2) a submucous coat of connective tissue that contains the main blood vessels of the tract, (3) a muscular layer, and (4) an outer fibroserous layer (Figure 21-2). Blood vessels and nerves travel through the mesentery to reach the digestive tube throughout most of its length.

Mucosa

The innermost layer of the GI wall—the layer facing the *lumen* (open space) of the tube—is called the **mucosa** or *mucous layer*. Note in Figure 21-2 that the mucosa is made up of three layers—an inner *mucous epithelium,* a layer of loose connective tissue called the *lamina propria,* and a thin layer of smooth muscle called the *muscularis mucosae.*

Submucosa

The **submucosa** layer of the digestive tube is composed of connective tissue that is thicker than the mucosal layer. It contains numerous small glands, blood vessels, and parasympathetic nerves that form the *submucosal plexus.*

Muscularis

The **muscularis**—or *muscular layer*—is a thick layer of muscle tissue that wraps around the submucosa. This portion of the wall is characterized by an inner layer of circular and an outer layer of longitudinal smooth muscle. Like the submucosa, the muscularis contains nerves organized into a plexus called the *myenteric plexus.*

Serosa

The **serosa**—or *serous layer*—is the outermost layer of the GI wall. The serosa is actually the *visceral layer* of the peritoneum—

BOX 21-1 | Organs of the Digestive System

Segments of the Gastrointestinal Tract

Mouth
Oropharynx
Esophagus
Stomach
Small intestine
- Duodenum
- Jejunum
- Ileum
Large intestine
- Cecum
- Colon
  - Ascending colon
  - Transverse colon
  - Descending colon
  - Sigmoid colon
- Rectum
Anal canal

Accessory Organs

Salivary glands
- Parotid
- Submandibular
- Sublingual
Tongue
Teeth
Liver
Gallbladder
Pancreas
Vermiform appendix

the serous membrane that lines the abdominopelvic cavity and covers its organs. The lining attached to and covering the walls of the abdominopelvic cavity is called the *parietal layer* of the peritoneum. The fold of serous membrane shown in Figure 21-2 that connects the parietal and visceral portions is called a *mesentery*. Notice that nerves and vessels servicing the wall of the GI tract are distributed within the mesentery.

Modifications of Layers

Although the same four tissue layers form the various organs of the digestive tract, their structures vary in different regions of the tube throughout its length. Variations in the epithelial layer of the mucosa, for example, range from stratified layers of squamous cells that provide protection from abrasion in the upper part of the esophagus to the simple columnar epithelium, designed for absorption and secretion, throughout the remainder of the tract.

> **QUICK CHECK**
> 1. What is the basic function of the digestive tract?
> 2. Name the four layers of the digestive tract wall.

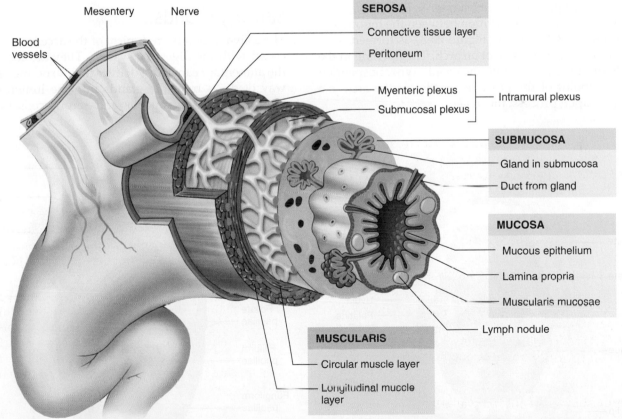

▲ **FIGURE 21-2   Wall of the GI tract.** The wall of the gastrointestinal (GI) tract is made up of four layers, shown here in a generalized diagram of a segment of the GI tract. Notice that the serosa is continuous with a fold of serous membrane called a *mesentery*. Notice also that digestive glands may empty their products into the lumen of the GI tract by way of ducts.

MOUTH

Structure of the Oral Cavity

The mouth or *oral cavity* (Figure 21-3) is created by the following structures: (1) the lips, which surround the orifice of the mouth and form the anterior boundary of the oral cavity, (2) the cheeks (side walls), and (3) the hard palate and soft palate (roof). The tongue and its muscle foundation also create part of the oral cavity.

The **lips** are covered externally by skin and internally by mucous membrane that continues into the oral cavity. Besides keeping food in the mouth while it is being chewed, the lips help sense the temperature and texture of food before it enters the mouth. They also participate in forming many sounds of speech.

The mucous membrane lining of the **cheeks** forms the lateral boundaries of the oral cavity and is continuous with the lips in front. Our cheeks are also lined by mucous membrane that is reflected onto the *gingiva,* or gums, and the soft palate. The bulk of the cheeks is formed by the buccinator muscle. This large muscle is sandwiched (along with a considerable amount of adipose tissue) between the outer skin and mucous membrane lining. Numerous small mucus-secreting glands are located between the mucous membrane and the buccinator muscle; their ducts open opposite the last molar teeth.

The **hard palate** consists of portions of four bones: two maxillae and two palatines (see Chapter 9, Figure 9-5). The **soft palate,** which forms a partition between the mouth and nasopharynx (see Chapter 20, Figure 20-3), is formed from muscle arranged in the shape of an arch. The opening in the arch leads from the mouth into the oropharynx. Suspended from the midpoint of the posterior border of the arch is a small cone-shaped process, the *uvula.*

Tongue

The **tongue,** which forms the base of the oral cavity and whose free end protrudes into it, is a solid mass of skeletal muscle (intrinsic muscles) covered by a mucous membrane. The tongue has a blunt *root*, a *tip*, and a *central body*. The upper, or dorsal, surface of the tongue is normally moist, pink, and covered by rough elevations, called *papillae*, which possess sensory organs called *taste buds* (Figure 21-4).

The *lingual frenulum* is a fold of mucous membrane in the midline of the ventral surface of the tongue that helps anchor the tongue to the floor of the mouth. The floor of the mouth and ventrum (lower surface) of the tongue are richly supplied with blood vessels. In this region, many vessels are superficial and covered only by a very thin layer of mucosa. This explains why some soluble drugs, such as aspirin or nitroglycerin (used during a heart attack), are absorbed into the circulation rapidly if placed under the tongue.

The tongue is also important for moving food material during *mastication* (chewing) and *deglutition* (swallowing), as well as for speech—especially during the formation of syllables. These activities are accomplished both by intrinsic muscles making up the tongue as well as by extrinsic muscles that insert into the tongue (but have their origins on the hyoid bone or one of the bones of the skull).

Salivary Glands

The salivary glands are typical of the accessory glands associated with the digestive system. They are located outside the alimentary canal and deliver their exocrine secretions by way of ducts from the glands into the lumen of the tract (Figure 21-5). The mucous and serous cells found in the

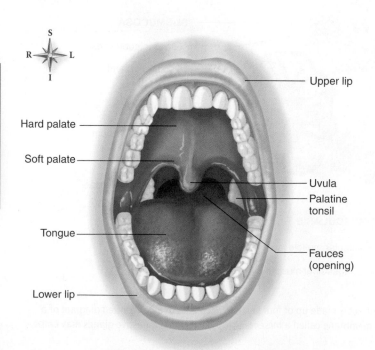

▲ **FIGURE 21-3** **The oral cavity.**

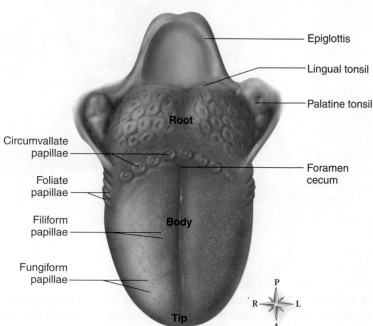

▲ **FIGURE 21-4** **Dorsal surface of tongue.** Sketch showing the three divisions of the tongue.

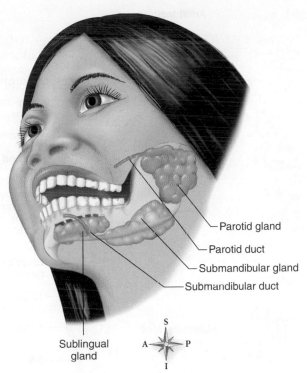

▲ FIGURE 21-5   Salivary glands. Location of the salivary glands.

glands together secrete a mixture of fluids that are then modified by the duct cells on their way out of the salivary gland.

Three pairs of major salivary glands (see Figure 21-5)—the *parotid, submandibular,* and *sublingual glands*—secrete a major portion (about 1 liter) of the saliva produced each day.

The **parotids** are the largest of the paired salivary glands. Shaped generally like pyramids, they are located between the skin and underlying masseter muscle in front of and below the external ear. The parotids produce a watery, or serous, type of saliva containing enzymes but not mucus. The parotid ducts penetrate the buccinator muscle on each side and open into the mouth opposite the upper second molars.

Submandibular glands contain both enzyme and mucus-producing elements. These glands are located just below the mandibular angle. The submandibular gland is irregular in form and about the size of a walnut. The ducts of the submandibular glands open into the mouth on either side of the lingual frenulum.

Sublingual glands are the smallest of the main salivary glands. They lie in front of the submandibular glands, under the mucous membrane covering the floor of the mouth. Each sublingual gland is drained by 8 to 20 ducts that open into the floor of the mouth. Unlike the other salivary glands, the sublingual glands produce *only* a mucous type of saliva.

Teeth

The teeth are designed to cut, tear, and grind ingested food so that it can be mixed with saliva before it is swallowed.

During the process of mastication, food is ground into small bits, which increases the surface area that can be acted on by the digestive enzymes.

Typical Tooth

A typical tooth (Figure 21-6) consists of three main parts: *crown, neck,* and *root.* The **crown** is the exposed portion of a tooth. It is covered by **enamel**—the hardest and chemically most stable tissue in the body. Enamel consists of approximately 97% calcified (inorganic) material and only 3% organic material and water. Its color varies from light yellow to grayish white. The thickness of enamel is typically greatest on the cusp and least at its border near the gum. The **neck** of a tooth is the section that connects the crown with the root below it.

The **root** of a tooth fits into the socket of the alveolar process of either the upper or lower jaw. The root of a tooth may be a single peglike structure or may comprise two or three separate conical projections. The root is not rigidly anchored to the alveolar process but is suspended in the socket by the fibrous *periodontal membrane.*

In addition to enamel, the outer shell of each tooth is composed of two additional dental tissues—*dentin* and *cementum* (see Figure 21-6). **Dentin** makes up the greatest proportion of the tooth shell. It is covered by enamel in the crown and by **cementum** in the root area. The dentin contains a **pulp cavity** that consists of connective tissue, blood and lymphatic vessels, as well as sensory nerves.

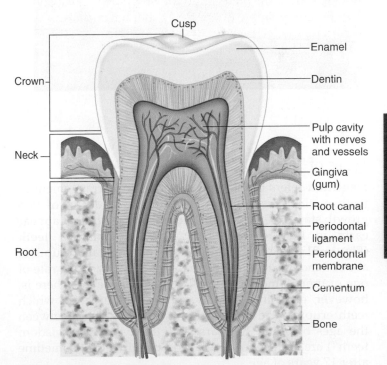

▲ FIGURE 21-6   Typical tooth. A molar tooth sectioned to show its bony socket and details of its three main parts: crown, neck, and root. Enamel (over the crown) and cementum (over the neck and root) surround the dentin layer. The pulp contains nerves and blood vessels.

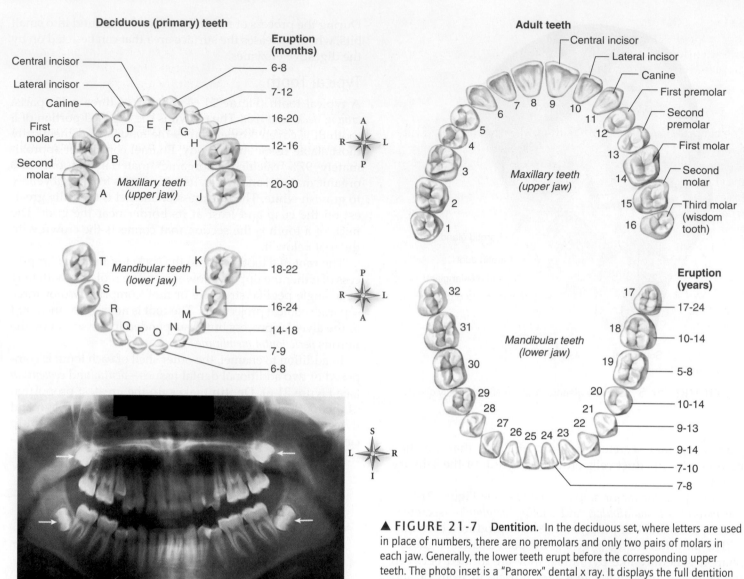

▲ **FIGURE 21-7** **Dentition.** In the deciduous set, where letters are used in place of numbers, there are no premolars and only two pairs of molars in each jaw. Generally, the lower teeth erupt before the corresponding upper teeth. The photo inset is a "Panorex" dental x ray. It displays the full dentition in a single "flattened-out" image. *Arrows* show the third molars or "wisdom teeth" that have not yet erupted.

Types of Teeth

Dentition refers to the type, number, and arrangement of teeth in the jaws. Human dentition is made up of two growth stages: 20 **deciduous teeth** (baby teeth) that appear early in life, replaced later by 32 **permanent teeth** (Figure 21-7). The first deciduous tooth usually erupts at about 6 months of age. The remainder follow at the rate of one or more a month until all 20 have appeared. There is, however, great individual variation in the age at which teeth erupt. Deciduous teeth are generally shed between the ages of 6 and 13 years. The third molars ("wisdom teeth") are the last to appear, and usually erupt sometime after 17 years of age.

PHARYNX

The act of swallowing moves a rounded mass of food, called a *bolus*, from the mouth to the stomach. As the food bolus passes from the mouth, it enters the oropharynx (the second division of the pharynx) and then passes down the digestive tube proper—the portion of the digestive tract that serves only the digestive system. The anatomy of the pharynx is discussed in more detail on page 458, Chapter 20.

ESOPHAGUS

The **esophagus**—a collapsible, muscular, mucosa-lined tube about 25 cm (10 inches) long—extends from the pharynx to the stomach and pierces the diaphragm in its descent from the thoracic cavity to the abdominal cavity (Figure 21-8, *A*). It lies posterior to the trachea and heart and serves as a muscular passageway for food, pushing the food toward the stomach.

The esophagus is the first segment of the digestive tube proper, and the four layers that form the wall of the GI tract tube can be identified there (Figure 21-8, *B*). The esophagus is normally flattened, and thus the lumen is practically nonexistent in the resting state. The inner circular and outer longitudinal layers of the muscular layer are striated (voluntary) in the upper third, mixed (striated and smooth) in the middle third, and smooth (involuntary) in the lower third of the tube. You have control over swallowing in the upper third of your esophagus, but once the bolus passes the middle third of your esophagus, it cannot be voluntarily regurgitated.

Each end of the esophagus is encircled by muscular sphincters that act as valves to regulate passage of material. The **upper esophageal sphincter (UES)** in the cervical part of the esophagus helps prevent air from entering the esophagus during respiration. Relaxation of the UES is what permits *belching* (or *burping*), which is the sudden escape of air trapped in the stomach and esophagus.

The **lower esophageal sphincter (LES)** is also called the *cardiac sphincter*. The *esophageal hiatus* is an opening in the diaphragm located near the junction between the terminal portion of the esophagus and the stomach. The esophageal hiatus may become enlarged to the point where part of the stomach pushes upward through the diaphragm and into the chest. This condition is called a **hiatal hernia** (Box 21-2).

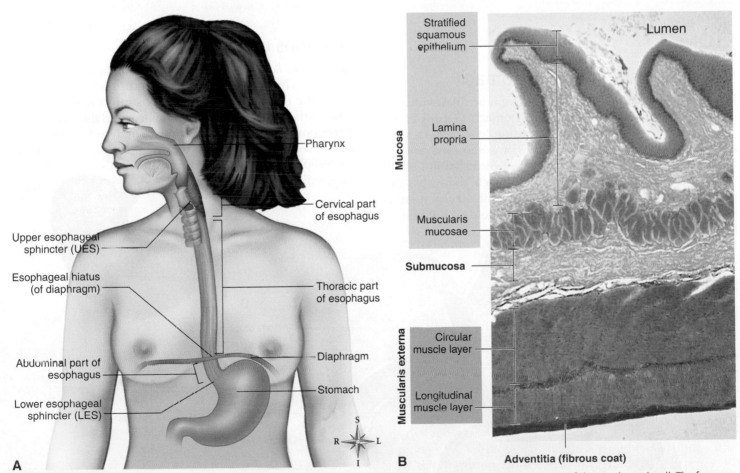

▲ **FIGURE 21-8** **Esophagus.** **A,** Diagram showing the major features of the esophagus. **B,** Microscopic appearance of the esophageal wall. The four layers of the gastrointestinal (GI) wall are easily identified.

BOX 21-2   Diagnostic Study

Upper Gastrointestinal X-Ray Study

An upper GI (UGI) study consists of a series of radiographs of the lower part of the esophagus, stomach, and duodenum. Barium sulfate is usually used as the contrast medium. The test is used to detect ulcerations, tumors, inflammation, or anatomical malpositions such as *hiatal hernia* (portion of stomach pushed through the hiatus [opening] of the diaphragm). Obstruction of the upper GI tract is also easily detected.

In this test the patient is asked to drink a flavored liquid containing barium sulfate. As the contrast medium travels through the system, the lower portion of the esophagus, gastric wall, pyloric channel, and duodenum are each evaluated for defects. Benign peptic ulcer is a common pathological condition affecting these areas. Tumors, cysts, or enlarged organs near the stomach can also be identified by an anatomical distortion of the outline of the upper GI tract. Can you identify the outline of the stomach in the figure?

▶ **X-ray film of the lower part of the esophagus, stomach, and duodenum.**

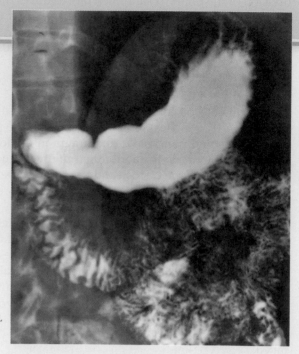

STOMACH

Size and Position of the Stomach

Just below the diaphragm, the digestive tube dilates into an elongated pouchlike structure, the stomach (Figure 21-9). The size of the stomach varies according to several factors, notably the amount of distention. For some time after a meal, the stomach is enlarged because of distention of its walls, but as food passes out of the stomach, the walls partially collapse, leaving the organ about the size of a large sausage. In adults, the stomach usually holds a volume of up to 1 to 1.5 liters.

Divisions of the Stomach

The *fundus, body,* and *pylorus* are the major divisions of the stomach. The **fundus** is the enlarged portion to the left and above the opening of the esophagus into the stomach. The **body** is the central part of the stomach, and the **pylorus** is its lower portion (see Figure 21-9). The small collar or margin of the stomach at its junction with the esophagus is often called the **cardia** or *cardiac part.*

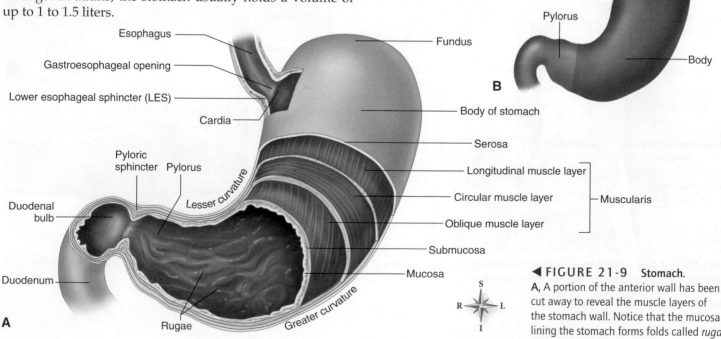

◀ **FIGURE 21-9   Stomach.**
A, A portion of the anterior wall has been cut away to reveal the muscle layers of the stomach wall. Notice that the mucosa lining the stomach forms folds called *rugae.*
B, Divisions of the stomach.

Sphincter Muscles

Sphincter muscles regulate passage of material at both stomach openings. Sphincter muscles consist of circular fibers arranged to form an opening in the center of them when they are relaxed and no opening when they are fully contracted. The *lower esophageal sphincter (LES)*, or *cardiac sphincter*, controls the opening of the esophagus into the stomach, and the **pyloric sphincter** controls the opening from the pyloric portion of the stomach into the first part of the small intestine (duodenum).

Stomach Wall

Each of the four layers of the stomach wall suits the function of this organ, as shown for you in Figures 21-9 and 21-10. Of particular interest are the modifications to the stomach mucosa and muscularis, both of which are briefly described below.

Gastric Mucosa

The epithelial lining of the stomach consists largely of folds, called *rugae*. Within these folds are many marked depressions called *gastric pits*. Numerous coiled tubular-type glands, or *gastric glands*, are found below the level of the pits,

particularly in the fundus and body of the stomach. Figure 21-10 illustrates the anatomical relationship of the gastric pits and gastric glands. The glands secrete most of the gastric juice, a mucous fluid containing digestive enzymes and hydrochloric acid (HCl).

In addition to the mucus-producing cells that cover the entire surface of the stomach and line the pits, the gastric glands contain three major secretory cells—**chief cells, parietal cells,** and **endocrine cells** (see Figure 21-10). Chief cells (zymogenic cells) secrete the enzymes of gastric juice. Parietal cells secrete hydrochloric acid and are also thought to produce the important substance known as *intrinsic factor*. Intrinsic factor binds to vitamin B_{12} molecules to protect them from digestive juices until they reach the small intestine—and then facilitates the absorption of B_{12}. Endocrine cells secrete *ghrelin (GHRL)*—a hormone that stimulates the hypothalamus to secrete growth hormone and increase appetite—and *gastrin*, which influences digestive functions.

Gastric Muscle

The thick layer of muscle in the stomach wall—the *muscularis*—is made of three distinct sublayers of smooth muscle tissue. As Figure 21-9, *A* illustrates, there is the usual layer of longitudinal muscles and circular muscles, as well as an underlying oblique layer. The crisscrossing pattern of smooth muscle fibers formed by this arrangement gives the stomach wall the ability to contract strongly at many angles—thus making mixing very efficient.

Function of the Stomach

The stomach performs the following functions:

- Serves as a food reservoir until the food can be partially digested and moved farther along the gastrointestinal tract.
- Secretes *gastric juice*, which aids in the digestion of food.
- Churns food, breaking it into small particles and mixing them well with the gastric juice.
- Secretes *intrinsic factor*, important for vitamin B_{12} absorption.
- Performs a limited amount of absorption.
- Produces the hormones *gastrin*, which helps regulate digestive functions, and *ghrelin (GHRL)*, which increases appetite.
- Helps protect the body by destroying pathogenic bacteria swallowed with food or with mucus from the respiratory tract.

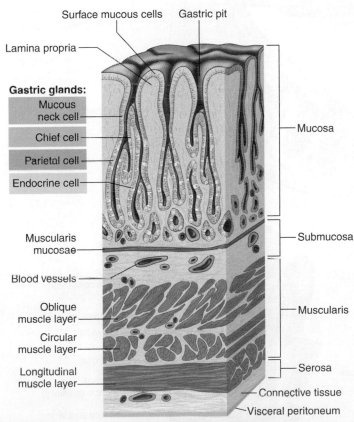

Surface mucous cells · Gastric pit

Lamina propria

Gastric glands:
Mucous neck cell
Chief cell
Parietal cell
Endocrine cell

Mucosa

Muscularis mucosae

Blood vessels

Submucosa

Oblique muscle layer
Circular muscle layer
Longitudinal muscle layer

Muscularis

Serosa

Connective tissue
Visceral peritoneum

▲ **FIGURE 21-10 Gastric pits and gastric glands.** Gastric pits are depressions in the epithelial lining of the stomach. At the bottom of each pit is one or more tubular *gastric glands.* Chief cells produce the enzymes of gastric juice, and parietal cells produce stomach acid. Endocrine cells secrete the appetite-boosting hormone ghrelin.

QUICK CHECK

8. Describe the passage of the esophagus from the oral cavity to the stomach.

9. What are the three main divisions of the stomach?

10. What is the function of gastric pits?

UNIT 5

SMALL INTESTINE

Size and Position of the Small Intestine

The small intestine is a tube measuring about 2.5 cm (1 inch) in diameter and 6 meters (20 feet) in length. Its coiled loops fill most of the abdominal cavity (see Figure 21-1). It is divided into the duodenum, the jejunum, and the ileum. The **duodenum** is the uppermost division and the part to which the pyloric end of the stomach attaches. It is about 25 cm (10 inches) long and is shaped roughly like the letter C. The duodenum *becomes* the **jejunum** at the point where the tube turns abruptly forward and downward. The jejunal portion continues for approximately the next 2.5 meters (8 feet), at the end of which it *becomes* the **ileum,** but without any clear line of demarcation between the two divisions. The ileum is about 3.5 meters (12 feet) long.

Wall of the Small Intestine

Notice in Figure 21-11 that the intestinal lining has circular *plicae* (folds) that have many tiny projections called **villi.** Millions of these projections, each about 1 mm in height, give the intestinal mucosa a velvety appearance. Each villus contains an arteriole, venule, and lymph vessel (lacteal).

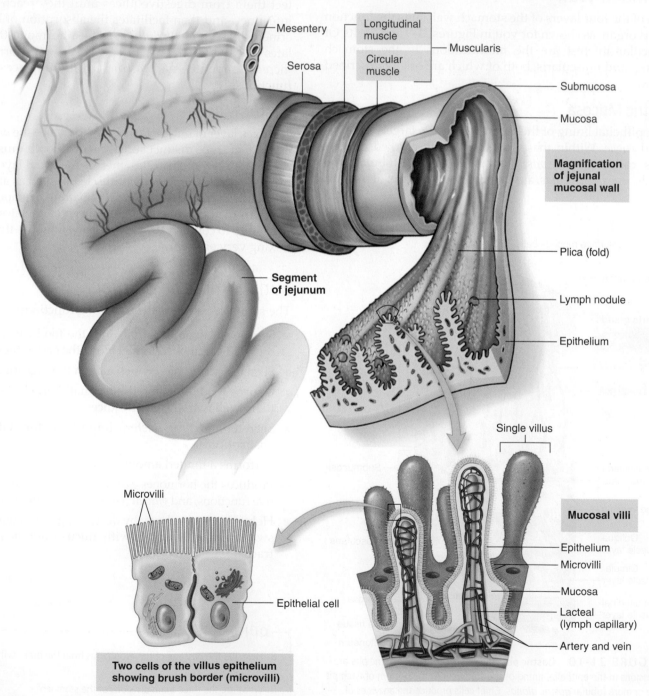

▲ **FIGURE 21-11**   **Wall of the small intestine.** Note folds of mucosa are covered with villi and each villus is covered with epithelium, which increases the surface area for absorption of food.

Mucus-secreting goblet cells are found in large numbers on villi. Endocrine cells that produce intestinal hormones are also found in the villi. The presence of villi and microvilli greatly increases the surface area of the small intestine hundreds of times. The increase in surface area allows this organ to efficiently digest and absorb most of the nutrients we ingest.

QUICK ✓ CHECK

11. What are the three main divisions of the small intestine?
12. What are intestinal villi? What is their function?

LARGE INTESTINE

Size of the Large Intestine

The lower part of the alimentary canal bears the name *large intestine* because its diameter is noticeably larger than that of the small intestine. Its length, however, is much less, being about 1.5 to 1.8 meters (5 or 6 feet). Its average diameter is approximately 6 cm (2.3 inches), but the diameter decreases toward the lower end of the tube. Box 21-3 describes a common test used for imaging the large intestine.

Divisions of the Large Intestine

The large intestine is divided into the cecum, colon, and rectum (Figure 21-12). The first 5 to 8 cm (2 or 3 inches) of the large intestine is the *cecum*. It is a blind pouch located in the lower right quadrant of the abdomen (see Figure 21-1).

The **colon** is divided into the following portions: *ascending, transverse, descending,* and *sigmoid* (see Figure 21-12).

■ The **ascending colon** lies in a vertical position, on the right side of the abdomen, and extends up to the lower border of the liver. The ileum joins the large intestine at the junction of the cecum and ascending colon (see Figure 21-12). The *ileocecal valve* permits material to pass from the ileum into the large intestine, but not usually in the reverse direction.

■ The **transverse colon** passes horizontally across the abdomen, below the liver, stomach, and spleen. Note that this part of the colon is *above* the small intestine (see Figure 21-1). The transverse colon extends from the *hepatic flexure* (near the liver) to the *splenic flexure* (near the spleen). At these two points the colon bends on itself to form roughly 90-degree angles.

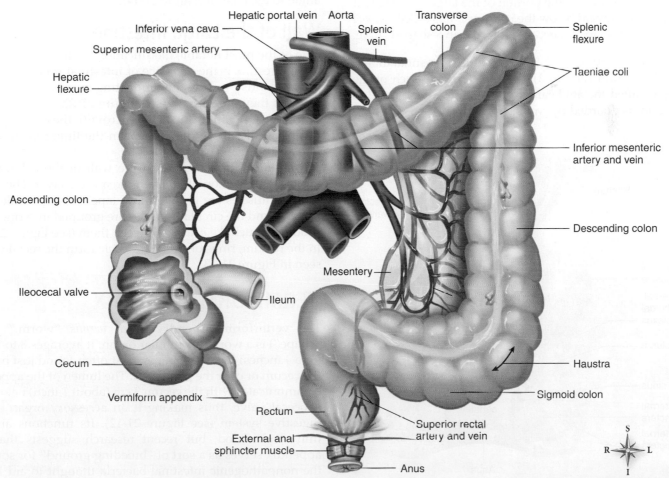

▲ **FIGURE 21-12** **Divisions of the large intestine.** Illustration showing divisions of the large intestine and adjacent vascular structures.

BOX 21-3 Diagnostic Study

Barium Enema Study

The barium enema (BE) study, or lower GI series, consists of a series of x-ray films of the colon that are used to detect and locate polyps, tumors, and *diverticula* (abnormal "pouches" in the lining of the intestine). Abnormalities in organ position can also be detected.

The test begins with an enema of approximately 500 to 1,500 ml of fluid containing barium sulfate. The patient is placed in various positions, and the progress of the barium's flow through the intestine is monitored on a fluoroscope. Small polyps and early changes in ulcerative colitis are more easily detected with an *air-contrast barium enema study.* In this study, after the bowel is outlined with a thin coat of barium, air is added to enhance the contrast and outline of small lesions. After the x-ray films are taken, the patient is allowed to expel the barium.

The figure shows a colorized radiograph of a barium enema produced by a special technique that produces a clear image of the large intestine and its position relative to the skeleton. Can you identify the divisions of the large intestine in the figure?

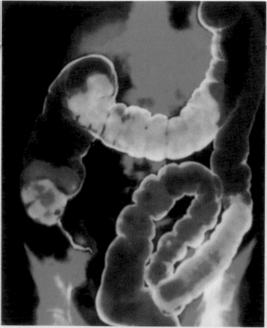

▲ Colorized radiograph of a barium enema.

■ The **descending colon** lies in the vertical position, on the left side of the abdomen, and extends from a point below the stomach and spleen to the level of the iliac crest.

■ The **sigmoid colon** is the portion of the large intestine that courses downward below the iliac crest. It is called *sigmoid* (meaning "S-shaped") because it forms an S-shaped curve.

The last 17 to 20 cm (7 or 8 inches) of the intestinal tube is called the **rectum** (Figure 21-13). The terminal inch of the rectum is called the **anal canal.** The opening of the canal to the exterior is guarded by two sphincter muscles—an inter-

nal one of smooth muscle and an external one of striated muscle. The opening itself is called the *anus.* The anus is directed slightly posteriorly and is therefore at almost a right angle to the rectum (Figure 21-14).

Wall of the Large Intestine

One of the most notable modifications of the GI wall in the large intestine is the presence of intestinal mucous glands. These glands produce the lubricating mucus that coats the feces as they are formed (see Figure 21-2). Although cells lining the large intestine have microvilli, these cells do not form villi like those that appear in the lining of the small intestine.

Another notable feature of the wall of the colon is the uneven distribution of fibers in the muscle layer. The longitudinal muscles are grouped into tapelike strips called *taeniae coli,* and the circular muscles are grouped into rings that produce pouchlike *haustra* between them (see Figure 21-12). In the rectum, rings of circular muscle form the rectal valves seen in Figure 21-13.

VERMIFORM APPENDIX

The **vermiform appendix** (from *vermis* "worm," *forma* "shape") is a wormlike tubular organ. It averages 8 to 10 cm (3 to 4 inches) in length and is most often found just behind the cecum or over the pelvic rim. The lumen of the appendix communicates with the cecum 3 cm (about 1 inch) below the ileocecal valve, thus making it an accessory organ of the digestive system (see Figure 21-12). Its functions are not fully understood, but recent research suggests that the appendix serves as a sort of "breeding ground" for some of the nonpathogenic intestinal bacteria thought to aid in the digestion or absorption of nutrients.

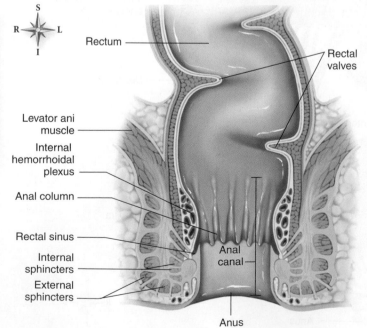

▲ **FIGURE 21-13** **The rectum and anus.** Lumen of GI tract.

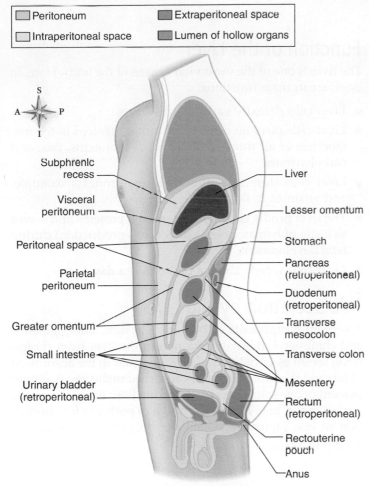

Peritoneum	Extraperitoneal space
Intraperitoneal space	Lumen of hollow organs

▲ **FIGURE 21-14** **Peritoneum.** Sagittal view of the abdomen showing the peritoneum and its reflections. Intraperitoneal spaces are shown in *yellow* and extraperitoneal spaces in *green*. The portion of the extraperitoneal space along the posterior wall of the abdomen is often called the *retroperitoneal space*.

PERITONEUM

The **peritoneum** is a large, continuous sheet of serous membrane that covers most of the organs just described and holds them loosely in place. It lines the walls of the entire abdominal cavity (parietal layer) and also forms the serous outer coat of the organs (visceral layer), as you can see in

Figure 21-14. All the space *outside* the parietal peritoneum is called *extraperitoneal space*.

In several places the peritoneum forms reflections, or extensions, that bind the abdominal organs together (see Figure 21-14). The **mesentery** is a fan-shaped projection of the parietal peritoneum that extends from the lumbar region of the posterior abdominal wall. The mesentery allows free movement of each coil of the intestine and helps prevent strangulation of the long tube.

The **greater omentum** is a continuation of the serosa of the greater curvature of the stomach and the first part of the duodenum to the transverse colon. It is a fold of two layers of peritoneal serous tissue that largely fuse together by adulthood. This ventral fold of the greater omentum hangs down as an "apron" in front of the abdominal organs (see Figure 21-14) and can become filled with fatty deposits. In cases of localized abdominal inflammation such as appendicitis, the greater omentum envelops the inflamed area, walling it off from the rest of the abdomen. The **lesser omentum** attaches from the liver to the lesser curvature of the stomach and the first part of the duodenum. Take a few moments to examine the relationships of peritoneal extensions in Figure 21-14.

> **QUICK CHECK**
>
> **13.** What are the four main divisions of the colon?
>
> **14.** What are haustra? How do they differ from taeniae coli?
>
> **15.** Where is the vermiform appendix located?
>
> **16.** What are some of the functions of the greater omentum?

LIVER

Structure of the Liver

The liver is the largest *gland* in the body. It weighs about 1.5 kg (3 to 4 pounds), lies immediately under the diaphragm, and occupies most of the right hypochondrium and part of the epigastrium (see Figure 21-1).

The liver consists of two lobes separated by the falciform ligament (Figure 21-15). The **left lobe** forms about one sixth of

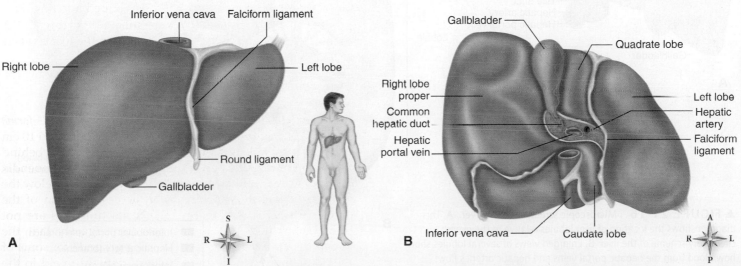

▲ **FIGURE 21-15** **Gross structure of the liver.** Diagrams of a normal liver. **A,** Anterior view. **B,** Inferior view.

the liver, and the **right lobe** makes up the remainder. The right lobe has three parts: the *right lobe proper*, the *caudate lobe*, and the *quadrate lobe*. Each lobe is divided into numerous lobules by small blood vessels and by fibrous strands that form a supporting framework called the *perivascular fibrous capsule*.

The **hepatic lobules** (Figure 21-16), the anatomical units of the liver, are tiny hexagonal or pentagonal cylinders about 2 mm high and 1 mm in diameter. Blood vessels—including portal veins carrying blood from the GI tract—arranged around a lobule feed blood into sinusoids that extend through the liver tissue, which processes the blood. The blood then drains into a central vein, which in turn drains into hepatic veins. Meanwhile, **bile** secreted by liver tissue collects in tiny bile ducts that surround each lobule.

Bile Ducts

The small bile ducts within the liver join to form two larger ducts that emerge from the undersurface of the organ as the *right* and *left hepatic ducts*. These ducts immediately join to form one common hepatic duct. The common hepatic duct merges with the *cystic duct* from the gallbladder to form the *common bile duct* (Figure 21-17), which opens into the duodenum in a small raised area called the *major duodenal papilla*. This papilla is located 7 to 10 cm (2 to 4 inches) below the pyloric opening from the stomach.

Function of the Liver

The liver is one of the most vital organs of the body. Here, in brief, are its main functions:

- Liver cells detoxify various substances.
- Liver cells carry on numerous important steps in the metabolism of all three kinds of foods—proteins, fats, and carbohydrates.
- Liver cells store several substances—iron, for example, and vitamins A, B_{12}, and D.
- The liver produces important plasma proteins and serves as a site of hematopoiesis (blood cell production) during fetal development.
- Liver cells secrete about a pint of bile a day.

Bile Secretion by the Liver

The main components of bile are bile salts, bile pigments, and cholesterol. Bile salts (formed in the liver from cholesterol) are an essential part of bile. They aid in the absorption of fats and then are themselves absorbed in the ileum. Eighty percent of bile salts are recycled in the liver to again become part of "new" bile. Bile also serves as a pathway for elimination of bile pigments, which are breakdown products of erythrocytes.

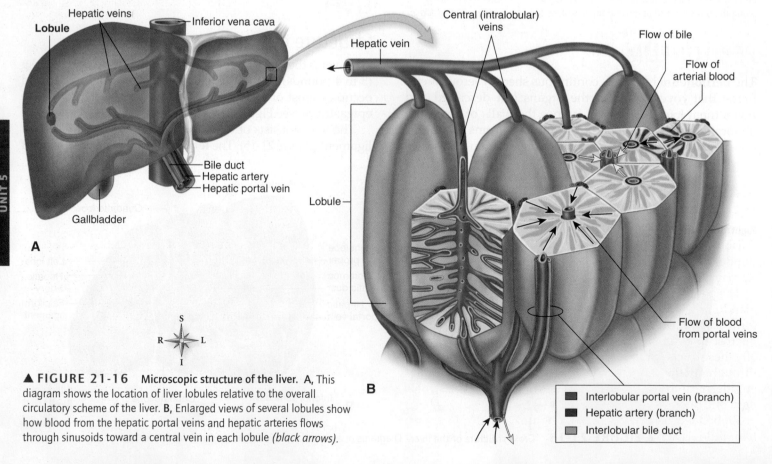

▲ **FIGURE 21-16** **Microscopic structure of the liver.** **A,** This diagram shows the location of liver lobules relative to the overall circulatory scheme of the liver. **B,** Enlarged views of several lobules show how blood from the hepatic portal veins and hepatic arteries flows through sinusoids toward a central vein in each lobule *(black arrows).*

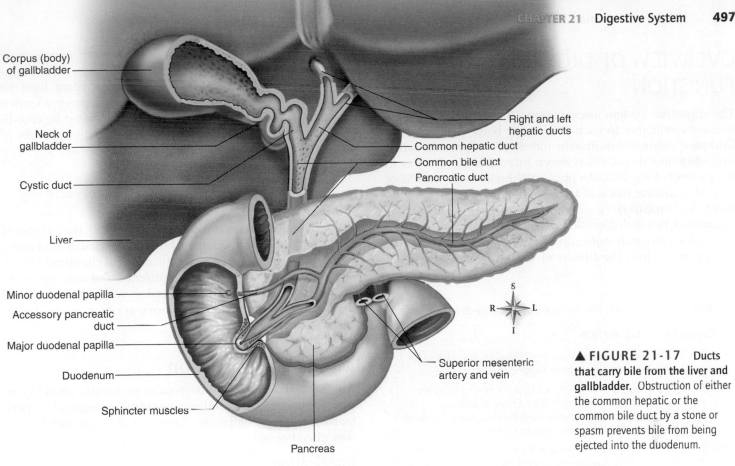

Corpus (body) of gallbladder

Neck of gallbladder

Cystic duct

Liver

Minor duodenal papilla

Accessory pancreatic duct

Major duodenal papilla

Duodenum

Sphincter muscles

Pancreas

Right and left hepatic ducts

Common hepatic duct

Common bile duct

Pancreatic duct

Superior mesenteric artery and vein

▲ **FIGURE 21-17** Ducts **that carry bile from the liver and gallbladder.** Obstruction of either the common hepatic or the common bile duct by a stone or spasm prevents bile from being ejected into the duodenum.

GALLBLADDER

The gallbladder is a pear-shaped sac 7 to 10 cm (3 to 4 inches) long and 3 cm broad at its widest point (see Figure 21-17). It can hold 30 to 50 ml of bile. The gallbladder lies on the undersurface of the liver and is attached there by areolar connective tissue.

The gallbladder stores bile that enters it by way of the hepatic and cystic ducts. During this time, the gallbladder concentrates bile 5-fold to 10-fold. Later, when digestion occurs in the stomach and intestines, the gallbladder contracts and ejects the concentrated bile into the duodenum.

PANCREAS

Structure of the Pancreas

The pancreas is a grayish pink–colored gland about 12 to 15 cm (6 to 9 inches) long, weighing about 60 g. It resembles a fish with its head and neck in the C-shaped curve of the duodenum, its body extending horizontally behind the stomach, and its tail touching the spleen (see Figure 21-17).

The pancreas is composed of both exocrine and endocrine glandular tissue (Figure 15-18, p. 342). The exocrine tissues form *acinar units* that release their collective secretions into microscopic ducts that unite to form larger ducts. Eventually these ducts join the main pancreatic duct, which extends throughout the length of the gland. The pancreatic duct empties into the duodenum at the same point as the common bile duct.

Embedded between the exocrine units of the pancreas, like little islands, lie clusters of endocrine cells called **pancreatic islets.** Although there are about a million islets, they constitute only about 2% of the total mass of the pancreas. Several kinds of cells make up the islets. They are secreting cells, but their secretion passes into blood capillaries rather than into ducts. Thus the pancreas is a dual gland—an exocrine, or duct, gland because of the acinar units and an endocrine, or ductless, gland because of the pancreatic islets.

Function of the Pancreas

The pancreas has a number of important functions that affect how nutrients are processed by the body.

■ The acinar units of the pancreas secrete the digestive enzymes found in pancreatic juice. Hence the pancreas plays an important part in digestion.

■ Beta cells of the pancreatic islets secrete **insulin,** a hormone that exerts a major control over carbohydrate and fat metabolism.

■ Alpha cells of the pancreatic islets secrete **glucagon,** a hormone that also helps control carbohydrate and fat metabolism.

The endocrine functions of the pancreas are also discussed for you in detail in Chapter 15, page 342.

> **QUICK ✓ CHECK**
>
> 17. Where is the liver located? How many lobes does it have?
> 18. Name three of the many functions of the liver.
> 19. Trace the route of bile from the gallbladder to the duodenum.
> 20. What is the function of the acinar units of the pancreas?

UNIT 5

OVERVIEW OF DIGESTIVE FUNCTION

The digestive system uses various mechanisms to make nutrients available to each cell of our bodies (Table 21-1). Complex foods must first be taken in—a process called **ingestion** and then broken down into simpler nutrients in the process of *digestion*. To physically break large chunks of food into smaller bits and to move it along the tract, movement (or **motility**) of the gastrointestinal (GI) wall is required. Chemical digestion—that is, breakdown of large molecules into small molecules—requires **secretion** of digestive enzymes into the lumen of the GI tract. After being

digested, nutrients are ready for the process of *absorption,* or movement through the GI mucosa into the internal environment. The material that is not absorbed must then be excreted to make room for more material—a process known as *elimination*. Of course, all these activities must be coordinated, which we have already learned is the process of *regulation*. Some of the major digestive processes are summarized in Table 21-1 and Figure 21-18.

DIGESTION

After food is ingested (taken into the mouth), the process of digestion begins immediately. **Digestion** is the overall name for all the processes that mechanically and chemically break complex foods into simpler nutrients that can be easily absorbed. Let's begin our discussion with an overview of *mechanical digestion* and then move on to a discussion of *chemical digestion*.

Mechanical Digestion

Mechanical digestion consists of all movement (motility) of the digestive tract, including mastication, deglutition, peristalsis, and segmentation. These mechanical actions are described below.

Mastication

Mechanical digestion begins in the mouth when the particle size of ingested food material is reduced by chewing movements, or **mastication.** The tongue, cheeks, and lips play an important role in keeping food material between the cutting or grinding surfaces of the teeth when a person is biting off or chewing food. In addition to reducing particle size, chewing movements serve to mix food with saliva in preparation for swallowing.

Deglutition

The process of swallowing, or **deglutition,** involves three main steps, or stages, that may be divided into the formation

TABLE 21-1   Primary Mechanisms of the Digestive System

MECHANISM	DESCRIPTION
Ingestion	Process of taking food into the mouth, starting it on its journey through the digestive tract
Digestion	A group of processes that break complex nutrients into simpler ones, thus facilitating their absorption; mechanical digestion physically breaks large chunks into small bits; chemical digestion breaks molecules apart
Motility	Movement by the muscular components of the digestive tube, including processes of mechanical digestion; examples include peristalsis and segmentation
Secretion	Release of digestive juices (containing enzymes, acids, bases, mucus, bile, or other products that facilitate digestion); some digestive organs also secrete endocrine hormones that regulate digestion or metabolism of nutrients
Absorption	Movement of digested nutrients through the gastrointestinal (GI) mucosa and into the internal environment
Elimination	Excretion of the residues of the digestive process (feces) from the rectum, through the anus; defecation
Regulation	Coordination of digestive activity (motility, secretion, etc.)

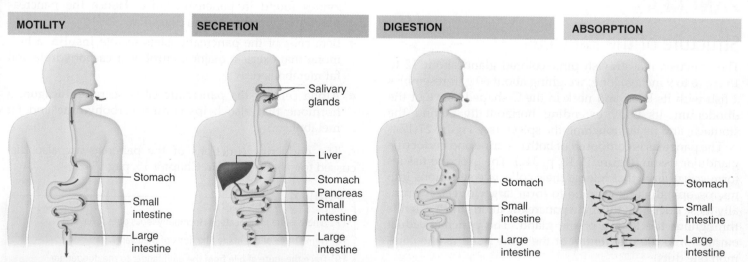

▲ **FIGURE 21-18   Overview of digestive functions.** Several important digestive functions are summarized in these diagrams. Note that the digestive tract is an extension of the external environment—extending like a tunnel through the body.

and then movement of a food bolus from the mouth to the stomach (Figure 21-19):

1. Oral stage (mouth to oropharynx)
2. Pharyngeal stage (oropharynx to esophagus)
3. Esophageal stage (esophagus to stomach)

First, the *oral stage,* which is voluntary and under control of the cerebral cortex, involves the formation of a food bolus that is to be swallowed. During the oral stage, the bolus is pressed against the palate by the tongue and then moved back into the oropharynx. The involuntary *pharyngeal* and *esophageal stages* that follow both consist of movement of food from the pharynx into the esophagus and, finally, into the stomach.

To propel food from the pharynx into the esophagus, three openings must be blocked: mouth, nasopharynx, and larynx. Continued elevation of the tongue seals off the mouth. The soft palate, including the uvula, is elevated and tensed, causing the nasopharynx to be closed off. Food is prevented from entering the larynx by muscle action that causes the epiglottis to block this opening. As a result, the bolus slips over the back of the epiglottis to enter the laryngopharynx. Contractions of the pharynx and esophagus compress the bolus into and through the esophageal tube.

Peristalsis and Segmentation

After food enters the lower portion of the esophagus, smooth muscle tissue in the wall of the GI tract is responsible for its movement forward. The motility produced by smooth muscle is of two main types: *peristalsis* and *segmentation.*

Peristalsis is often described as a wavelike ripple of the muscle layer of a hollow organ. The diagram in Figure 21-20 shows step by step how peristalsis occurs. A bolus stretches the GI wall, triggering a reflex contraction of circular muscle that pushes the bolus forward. This, in turn, triggers a reflex contraction in that location, pushing the bolus even farther. This process continues as long as the stretch reflex is activated by the presence of food.

Segmentation is most easily described as mixing movements. Segmentation occurs when digestive reflexes cause a *forward-and-backward* movement within a single region, or

ORAL STAGE

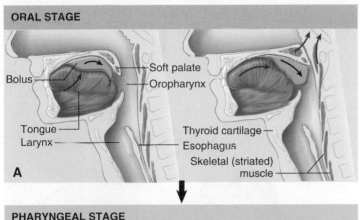

PHARYNGEAL STAGE

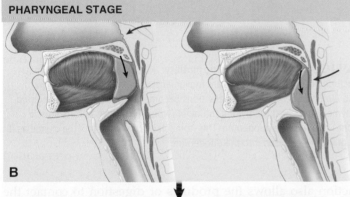

ESOPHAGEAL STAGE

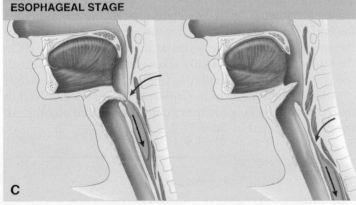

▲ **FIGURE 21-19** Deglutition. **A,** *Oral stage.* During this stage of deglutition (swallowing), a bolus of food is voluntarily formed on the tongue and pushed against the palate and then into the oropharynx. Notice that the soft palate acts as a valve that prevents food from entering the nasopharynx. **B,** *Pharyngeal stage.* After the bolus has entered the oropharynx, involuntary reflexes push the bolus down toward the esophagus. Notice that upward movement of the larynx and downward movement of the bolus close the epiglottis and thus prevent food from entering the lower respiratory tract. **C,** *Esophageal stage.* Involuntary reflexes of skeletal (striated) and smooth muscle in the wall of the esophagus move the bolus through the esophagus toward the stomach.

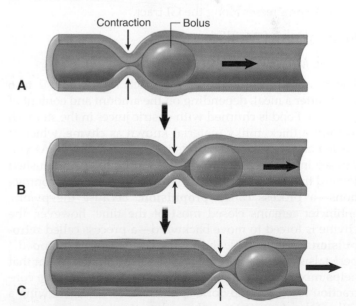

▲ **FIGURE 21-20** Peristalsis. Peristalsis is a progressive type of movement in which material is propelled from point to point along the gastrointestinal (GI) tract. **A,** A ring of contraction occurs where the GI wall is stretched, and the bolus is pushed forward. **B,** The moving bolus triggers a ring of contraction in the next region that pushes the bolus even farther along. **C,** The ring of contraction moves like a wave along the GI tract to push the bolus forward.

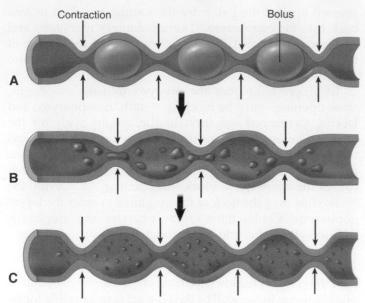

▲ **FIGURE 21-21** **Segmentation.** Segmentation is a back-and-forth action that breaks apart chunks of food and mixes in digestive juices. **A,** Ringlike regions of contraction occur at intervals along the gastrointestinal (GI) tract. **B,** Previously contracted regions relax and adjacent regions now contract, effectively "chopping" the contents of each segment into smaller chunks. **C,** The location of the contracted regions continues to alternate back and forth, chopping and mixing the contents of the GI lumen.

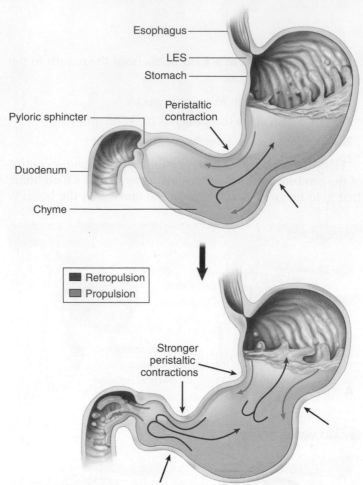

▲ **FIGURE 21-22** **Gastric motility.** Mixing actions in the stomach include both propulsion (forward movement) and retropulsion (backward movement). As peristaltic contractions become stronger, some of the liquid chyme squirts past the pyloric sphincter (which has decreased its muscle tone) and into the duodenum. The stomach continues to mix the chyme as it is gradually released into the duodenum.

segment, of the GI tract (Figure 21-21). These mixing movements help mechanically break down food particles, mix food and digestive juices thoroughly, and bring digested food in contact with intestinal mucosa to facilitate absorption.

Peristalsis and segmentation can occur in an alternating sequence. When this happens, food is churned *and* mixed as it slowly progresses along the GI tract.

Regulation of Motility

Gastric Motility

The process of emptying the stomach takes about 2 to 6 hours after a meal, depending on the amount and content of the meal. Food is churned with gastric juices in the stomach to form a thick, milky material known as **chyme,** which is ejected about every 20 seconds into the duodenum. As you can see in Figure 21-22, chyme is continually being pushed toward the pyloric sphincter by waves of peristaltic contractions—a process called **propulsion.** Because the pyloric sphincter remains closed most of the time, however, the chyme is forced to move backward—a process called **retropulsion.** Thus, because the chyme is temporarily "trapped," peristalsis creates a sort of "back-and-forth" movement that helps mix the chyme and gastric juice. Eventually, the contraction force of the pyloric sphincter decreases, allowing a little of the chyme to pass through to the duodenum.

Intestinal Motility

Intestinal motility includes both peristaltic contractions and segmentation. Segmentation in the duodenum and upper jejunum mixes the incoming chyme with digestive juices from the pancreas, liver, and intestinal mucosa. This mixing action also allows the products of digestion to contact the intestinal mucosa, where they can be absorbed into the internal environment. Peristalsis continues as the chyme nears the end of the jejunum—moving the food through the rest of the small intestine and into the large intestine. After leaving the stomach, chyme normally takes about 5 hours to pass all the way through the small intestine.

A list of definitions of the different processes involved in mechanical digestion, along with the organs that accomplish them, is presented in Table 21-2.

QUICK CHECK

21. What is meant by the term *motility?*
22. Is deglutition a voluntary or involuntary process? Explain.
23. What is the purpose of peristalsis and segmentation?

Chemical Digestion

Chemical digestion entails all the changes in chemical composition that transform foods during their travel through the digestive tract. These changes result largely from the hydrolysis of foods. **Hydrolysis** is a chemical process in which a compound unites with water and then splits into

TABLE 21-2   Processes of Mechanical Digestion

ORGAN	MECHANICAL PROCESS	NATURE OF PROCESS
Mouth (teeth and tongue)	Mastication	Chewing movements—reduce size of food particles and mix them with saliva
	Deglutition	Swallowing—movement of food from mouth to stomach
Pharynx	Deglutition	
Esophagus	Deglutition	
	Peristalsis	Rippling movements that squeeze food downward in digestive tract; a constricted ring forms first in one section, then the next, etc., causing waves of contraction to spread along entire canal
Stomach	Churning	Forward and backward movement of gastric contents, mixing food with gastric juices to form chyme
	Peristalsis	Wave starting in body of stomach about three times per minute and sweeping toward closed pyloric sphincter; at intervals, strong peristaltic waves press chyme past sphincter into duodenum
Small intestine	Segmentation (mixing contractions) Peristalsis	Forward and backward movement within segment of intestine; serves to mix food and digestive juices thoroughly and to bring all digested food into contact with intestinal mucosa to facilitate absorption; purpose of peristalsis, on the other hand, is to propel intestinal contents along digestive tract
Large intestine		
Colon	Segmentation Peristalsis	Churning movements within haustral sacs
Descending colon	Mass peristalsis	Entire contents moved into sigmoid colon and rectum; occurs three or four times a day, usually after a meal
Rectum	Defecation	Emptying of rectum, so-called bowel movement

simpler compounds (see Chapter 2). Numerous enzymes in the various digestive juices catalyze the hydrolysis of foods.

Digestive Enzymes

Although our interest until now has primarily concerned *intracellular enzymes* (see Chapter 2), our current discussion focuses on extracellular digestive enzymes as well. In the following paragraphs we briefly review enzymes in general and outline some characteristics of *digestive enzymes* in particular.

Recall that enzymes can be defined simply as "organic catalysts"—that is, they are organic compounds (proteins) that accelerate chemical reactions without being "consumed" by the reactions themselves.

As with any type of enzyme, digestive enzymes are *specific in their action;* that is, they act only on a specific substrate. This is attributed to a "key-in-a-lock" kind of action—the configuration of the enzyme molecule fitting the configuration of some part of the substrate molecule.

Digestive enzymes are continually being destroyed or eliminated from the body and therefore have to be continually synthesized, even though they are not used up in the reactions they catalyze. However, it is important to note that most digestive enzymes are synthesized and secreted as inactive **proenzymes.** Enzymes that break apart proenzymes and thus convert them to active enzymes are often called *kinases.* For example, enterokinase is a kinase that changes inactive trypsinogen into active trypsin.

Although we eat six main types of chemical substances (carbohydrates, proteins, fats, vitamins, mineral salts, and water), only the first three have to be chemically digested to be absorbed.

Carbohydrate Digestion

Carbohydrates are compounds composed of carbon, hydrogen and oxygen that ordinarily occur in the ratio 1:2:1 and exist both as *monosaccharides* and more complex forms. Monosaccharides are called "simple sugars" because they are each a single saccharide group. Simple sugars with 6 carbon atoms, such as glucose ($C_6H_{12}O_6$), are called *hexoses.* Other important hexoses include fructose and galactose. *Disaccharides* or "double sugars" include sucrose (glucose + fructose) and lactose (glucose + galactose). *Polysaccharides,* notably starches and glycogen, are long chains or polymers of linked simple sugars.

Polysaccharides are hydrolyzed to disaccharides by enzymes known as **amylases,** found in saliva and pancreatic juice. The enzymes that catalyze the final steps in carbohydrate digestion are *sucrase, lactase,* and *maltase* (Figure 21-23).

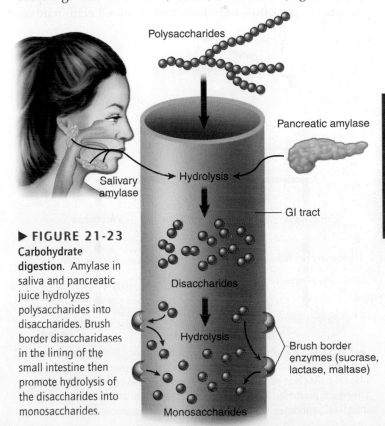

▶ **FIGURE 21-23**
Carbohydrate digestion. Amylase in saliva and pancreatic juice hydrolyzes polysaccharides into disaccharides. Brush border disaccharidases in the lining of the small intestine then promote hydrolysis of the disaccharides into monosaccharides.

These enzymes are located in the cell membrane of epithelial cells covering the villi and, therefore, lining the intestinal lumen. The resulting end products of digestion, mainly glucose, are thus conveniently made at the site of absorption (and are not floating around somewhere in the lumen).

Protein Digestion

Protein compounds have very large molecules made up of folded or twisted chains of amino acids, often hundreds in number. Enzymes called **proteases** catalyze the hydrolysis of proteins first into a variety of intermediate compounds called proteoses and peptides, which are simply shorter strands of amino acids. Then finally, proteases break these shorter molecules into individual amino acids (Figure 21-24).

The main proteases are **pepsin** in gastric juice, **trypsin** and **chymotrypsin** in pancreatic juice, and **peptidases** of the intestinal brush border (Box 21-4). Peptidases are also present within each intestinal cell, where they break apart dipeptides and tripeptides absorbed into these cells. Each kind of protease catalyzes the breaking apart of a specific kind of peptide bond. Because different amino acid combinations within a protein or polypeptide can have slightly different kinds of peptide bonds holding them together, a whole arsenal of different proteases is needed for efficient protein digestion.

Fat Digestion

Because fats are insoluble in water, they must be **emulsified**—that is, dispersed into very small droplets—before they can be digested. Two substances found in bile, **lecithin** and **bile salts,** emulsify dietary oils and fats in the lumen of the small intestine.

Lecithin is a phospholipid similar to other phospholipids that make up the bulk of cellular membranes. Lecithin mixes with lipids and water, forming tiny spheres called **micelles.** The tiny micelles permit fat components of foods to be soluble in water. Bile salts, which are derived from the lipid cholesterol, emulsify fats by forming micelles in the same manner.

The mechanical process of emulsification facilitates chemical digestion of fats by breaking large fat drops into small droplets. This process provides a greater contact area between fat molecules and pancreatic **lipases,** the main fat-digesting enzymes (Figure 21-25). The final products of fat digestion are fatty acids and glycerol (see Chapter 2).

You can find a summary of chemical digestion in Table 21-3.

Residues of Digestion

Certain components of food cannot be digested and are eliminated from the intestines in the **feces.** Among these *residues of digestion* are cellulose (a carbohydrate, also known as "dietary fiber") and undigested connective tissue from meat (mostly collagen). These substances remain undigested because humans lack the enzymes required to hydrolyze them. Residues of digestion also include undigested fats. Some fat molecules also have minerals such as calcium and

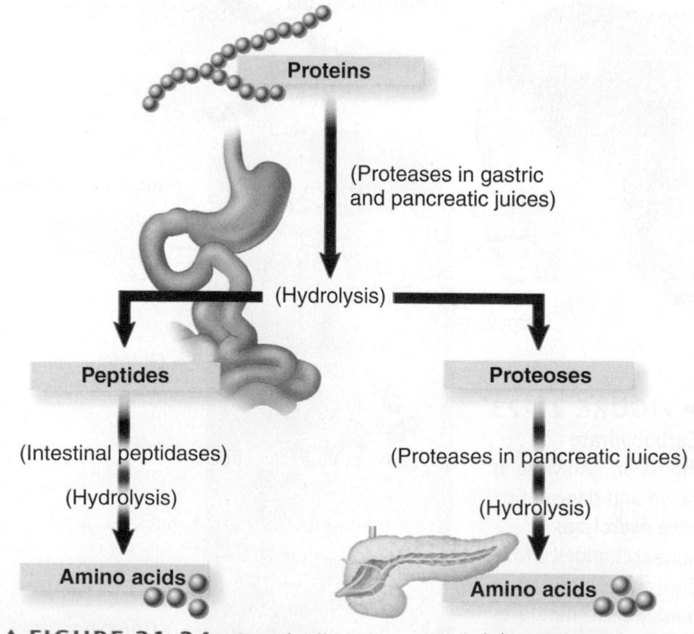

▲ **FIGURE 21-24** **Protein digestion.** Gastric juice protease (pepsin) and pancreatic juice proteases (trypsin and chymotrypsin) hydrolyze proteins into proteoses and peptides. Protein digestion is then completed by pancreatic proteases, which hydrolyze proteoses into amino acids, and by intestinal peptidases, which hydrolyze peptides into amino acids.

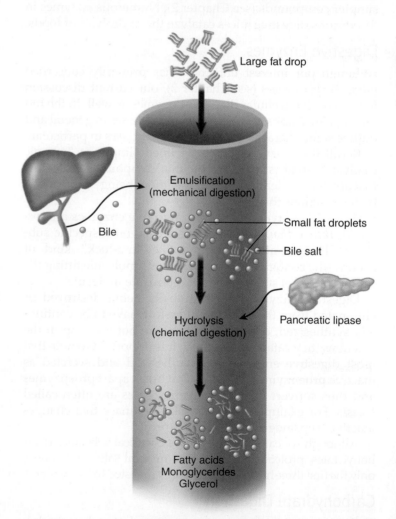

▲ **FIGURE 21-25** **Fat digestion.** Hydrolysis by the enzyme lipase is facilitated by prior emulsion of fats by bile (lecithin and bile salts). Though not pictured, colipase is needed to anchor lipase molecules to the inner face of each micelle.

BOX 21-4 | The Brush Border

The term **brush border** refers to the microvilli on the epithelial mucous cells that line the small intestine, visible in the figure. These microvilli are on the apical surfaces of the epithelial cells—the surfaces that face the interior of the intestinal lumen. Because, when viewed under high magnification, the microvilli look like the bristles of a brush, the surface of the intestinal mucosa was nicknamed the "brush border."

The brush border represents the boundary between the external environment (the lumen of the alimentary canal) and the internal environment of the body. It is across this border that molecules must pass if they are going to be absorbed into the body.

The brush border possesses such an incredibly large surface area because the microvilli increase the apical surface area. There are usually 2,000 to 3,000 microvilli on each cell! Of course, the presence of intestinal villi, circular folds (plicae circulares), and numerous loops also adds to the intestinal surface area (see Figure 21-11).

The large surface area of the brush border provides sites for digestive enzymes on the plasma membranes of intestinal cells—the *brush border enzymes*. The efficiency of the last stages of digestion is thus enhanced by having more surface area for more digestive enzymes.

The large surface area also provides more opportunities for the absorption of digested nutrients. More surface area allows for more phospholipid bilayer to absorb fats and more carrier molecules to carry amino acids, peptides, monosaccharides, and other nutrients.

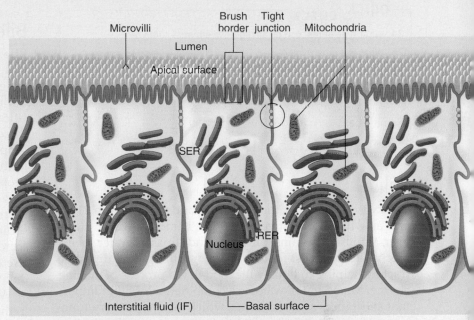

▲ **Intestinal epithelium.** The diagram shows intestinal epithelial cells joined by tight junctions and covered with microvilli on their apical (lumen-facing) surfaces. *RER,* Rough endoplasmic reticulum; *SER,* smooth endoplasmic reticulum.

TABLE 21-3   Chemical Digestion

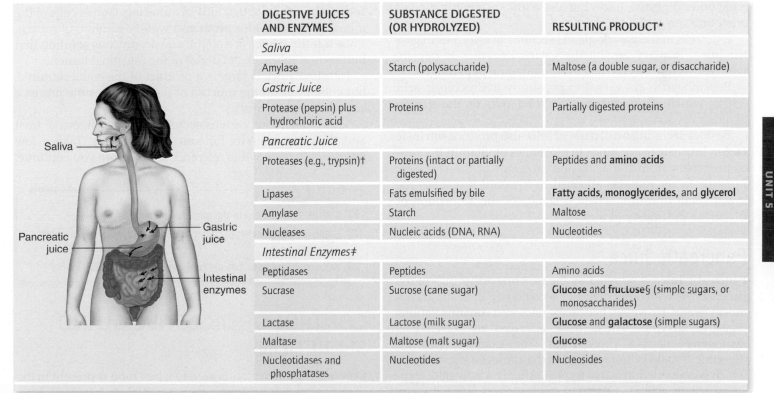

DIGESTIVE JUICES AND ENZYMES	SUBSTANCE DIGESTED (OR HYDROLYZED)	RESULTING PRODUCT*
Saliva		
Amylase	Starch (polysaccharide)	Maltose (a double sugar, or disaccharide)
Gastric Juice		
Protease (pepsin) plus hydrochloric acid	Proteins	Partially digested proteins
Pancreatic Juice		
Proteases (e.g., trypsin)†	Proteins (intact or partially digested)	Peptides and **amino acids**
Lipases	Fats emulsified by bile	**Fatty acids, monoglycerides, and glycerol**
Amylase	Starch	Maltose
Nucleases	Nucleic acids (DNA, RNA)	Nucleotides
Intestinal Enzymes‡		
Peptidases	Peptides	Amino acids
Sucrase	Sucrose (cane sugar)	**Glucose** and **fructose**§ (simple sugars, or monosaccharides)
Lactase	Lactose (milk sugar)	**Glucose** and **galactose** (simple sugars)
Maltase	Maltose (malt sugar)	**Glucose**
Nucleotidases and phosphatases	Nucleotides	Nucleosides

*Substances in **boldface type** are end products of digestion (that is, completely digested nutrients ready for absorption).
†Secreted in inactive form (trypsinogen); activated by enterokinase, an enzyme in the intestinal brush border.
‡Brush border enzymes.
§Glucose is also called *dextrose;* fructose is also called *levulose.*

magnesium, which render the fats indigestible. In addition to these wastes, feces consist of bacteria, pigments, water, and mucus.

SECRETION

Saliva

Saliva is secreted by the salivary glands (see Figure 21-5), and as with all digestive secretions, is mostly water. However, it also contains amylase and lipase. Water helps mechanically digest food as it moves through the digestive tract by helping to liquefy the food in the process of making chyme. Mixed in the water is a combination of other important substances such as **mucus,** which has the primary functions of protecting the digestive mucosa and lubricating food as it passes through the alimentary canal.

Gastric Juice

Gastric juice is secreted by exocrine *gastric glands*. Gastric juice contains not only the basic water and mucus mixture of most other digestive juices but also a unique combination of other substances.

Chief cells in the gastric glands secrete the enzymes in gastric juice. Primary among the gastric enzymes is pepsin, which is secreted as the inactive proenzyme **pepsinogen.**

Pepsinogen is converted to pepsin by hydrochloric acid (HCl), which is produced by *parietal cells* of the gastric glands.

Besides secreting acid, parietal cells also produce **intrinsic factor.** Intrinsic factor binds to molecules of vitamin B_{12}, protecting them from the acids and enzymes of the stomach. Intrinsic factor remains attached to B_{12} until it reaches the lower small intestine, where it facilitates the absorption of B_{12} across the intestinal wall.

Pancreatic Juice

Pancreatic juice is secreted by the exocrine *acinar cells* of the pancreas. As with other digestive secretions, pancreatic juice is mostly water. However, pancreatic juice also contains various digestive enzymes. All of these enzymes are secreted as **zymogens**—inactive proenzymes. For example, protein-digesting trypsin is first released as the zymogen trypsinogen, which is converted to active trypsin. After it is activated, trypsin can then activate other enzymes such as chymotrypsin (and other protein-digesting enzymes), amylase, various

lipases, and **nucleases** (RNA- and DNA-digesting enzymes). This process changes each molecule's shape to the active enzyme form. The advantage of this system is that the enzymes do not digest the cells that make them.

Bile

Bile is an interesting mixture of many different substances that is secreted by the liver and stored and concentrated by the gallbladder.

Bile contains several substances that aid in digestion, specifically lecithin and bile salts. As we saw earlier, both of these substances break down large drops of fat into smaller droplets, thus making the fats more easily digestible. Both lecithin and bile salts wrap a hydrophilic shell around the droplets, making them water soluble and therefore able to move freely through the watery chyme in the lumen of the gastrointestinal tract. Bile also contains a small amount of sodium bicarbonate, which as with the sodium bicarbonate secreted by pancreatic duct cells, helps neutralize chyme. Excreted substances in bile include cholesterol, products of detoxification, and bile pigments.

We will discuss the diverse functions of the liver further in the next chapter.

Intestinal Juice

The term **intestinal juice** refers to the sum of intestinal secretions, rather than to a premixed combination of substances, that enters the gastrointestinal lumen by way of a duct. Most intestinal cells produce a water-based solution of sodium bicarbonate that aids in buffering. Goblet cells of the intestinal mucosa also produce a watery solution of mucus. Thus intestinal juice is a slightly basic, mucous solution that buffers and lubricates material in the intestinal lumen.

Intestinal juice is largely a product of the small intestine, but goblet cells in the mucosa of the large intestine produce some lubricating mucus.

We've listed the various secretions of the digestive tract and their components for you in Table 21-4. Take a few moments to review this reference table before you continue.

CONTROL OF DIGESTIVE GLAND SECRETION

Exocrine digestive glands secrete when food is present in the digestive tract or when it is seen, smelled, or imagined. Integrated nervous and hormonal reflex mechanisms control

TABLE 21-4 **Digestive Secretions**

DIGESTIVE JUICE	SOURCE	SUBSTANCE	FUNCTIONAL ROLE*
Saliva	Salivary glands	Mucus	*Lubricates bolus of food; facilitates mixing of food*
		Amylase	**Enzyme; begins digestion of starches**
		Sodium bicarbonate	Increases pH (for optimum amylase function)
		Water	*Dilutes food and other substances; facilitates mixing*
Gastric juice	Gastric glands	Pepsin	**Enzyme; digests proteins**
		Hydrochloric acid	Denatures proteins; decreases pH (for optimum pepsin function)
		Intrinsic factor	**Protects and allows later absorption of vitamin B$_{12}$**
		Mucus	*Lubricates chyme; protects stomach lining*
		Water	*Dilutes food and other substances; facilitates mixing*
Pancreatic juice	Pancreas (exocrine portion)	Proteases (trypsin, chymotrypsin, collagenase, elastase, etc.)	**Enzymes; digest proteins and polypeptides**
		Lipases (lipase, phospholipase, etc.)	**Enzymes; digest lipids**
		Colipase	**Coenzyme; helps lipase digest fats**
		Nucleases	**Enzymes; digest nucleic acids (RNA and DNA)**
		Amylase	**Enzyme; digests starches**
		Water	*Dilutes food and other substances; facilitates mixing*
		Mucus	*Lubricates*
		Sodium bicarbonate	**Increases pH** (for optimum enzyme function)
Bile	Liver (stored and concentrated in gallbladder)	Lecithin and bile salts	*Emulsify lipids*
		Sodium bicarbonate	**Increases pH** (for optimum enzyme function)
		Cholesterol	Excess cholesterol from body cells, to be excreted with feces
		Products of detoxification	From detoxification of harmful substances by hepatic cells, to be excreted with feces
		Bile pigments (mainly bilirubin)	Products of breakdown of heme groups during hemolysis, to be excreted with feces
		Mucus	*Lubrication*
		Water	Dilutes food and other substances; facilitates mixing
Intestinal juice	Mucosa of small and large intestine	Mucus	*Lubrication*
		Sodium bicarbonate	**Increases pH** (for optimum enzyme function)
		Water	Small amount to carry mucus and sodium bicarbonate

***Boldface type** indicates a chemical digestive process; *italic type* indicates a mechanical digestive process.

the flow of digestive juices in such a way that they appear in proper amounts when and for as long as needed.

Control of Salivary Secretion

As far as is known, only reflex mechanisms control the secretion of saliva. Chemical, mechanical, olfactory, and visual stimuli initiate afferent impulses to centers in the brainstem that send out efferent impulses to the salivary glands, stimulating them. Chemical and mechanical stimuli come from the presence of food in the mouth. Olfactory and visual stimuli come, of course, from the smell and sight of food.

Control of Gastric Secretion

Stimulation of gastric juice secretion occurs in three phases that are controlled by reflex and chemical mechanisms. Because stimuli that activate these mechanisms arise in the head, stomach, and intestines, the three phases are known as the *cephalic, gastric,* and *intestinal phases,* respectively. As you read the description of each phase, glance at the diagrams shown in Figure 21-26.

The **cephalic phase** is also spoken of as the "psychic phase" because psychic (mental) factors activate the mechanism. For example, the sight, smell, taste, or even thought of food that is pleasing to an individual activates control centers in the medulla oblongata from which parasympathetic

1 **Cephalic Phase** Sensations of thoughts about food are relayed to the brainstem, where parasympathetic signals to the gastric mucosa are initiated. This directly stimulates gastric juice secretion and also stimulates the release of gastrin, which prolongs and enhances the effect.

☐ Hormonal mechanism
■ Nervous mechanism

2 **Gastric Phase** The presence of food in the stomach, specifically the distention it causes, triggers local and parasympathetic nervous reflexes that increase secretion of gastric juice and gastrin (which further amplifies gastric juice secretion).

3 **Intestinal Phase** As food moves into the duodenum, the presence of fats, carbohydrates, and acid stimulates hormonal and nervous reflexes that inhibit stomach activity.

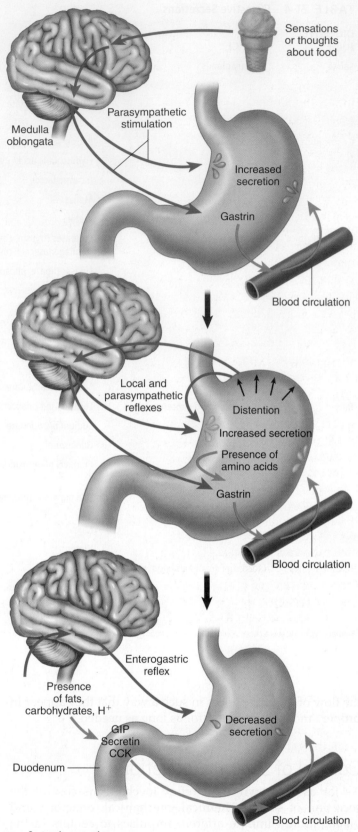

▲ FIGURE 21-26   Phases of gastric secretion.

fibers of the vagus nerve conduct efferent impulses to the gastric glands. Vagal nerve impulses also stimulate the production of **gastrin,** which stimulates gastric secretion, thus prolonging and enhancing the response.

During the **gastric phase** of gastric secretion, products of protein digestion in foods that have reached the pyloric portion of the stomach stimulate its mucosa to release *gastrin* into the blood within stomach capillaries. When it circulates to the gastric glands, gastrin greatly accelerates their secretion of gastric juice, which has a high pepsinogen and hydrochloric acid content. Gastrin release is also stimulated by distention of the stomach (caused by the presence of food).

The **intestinal phase** of gastric juice secretion is less clearly understood than the other two phases. Various different mechanisms seem to adjust gastric juice secretion as chyme passes to and through the intestinal tract. Experiments show that gastric secretions are inhibited when chyme containing fats, carbohydrates, and acid (low pH) is present in the duodenum.

Gastric secretion—and thus chemical digestion in the stomach—can be slowed when the duodenum becomes full. This prevents the stomach from finishing its task before the small intestine is ready to receive the chyme.

Control of Pancreatic Secretion

Several hormones released by intestinal mucosa are known to stimulate pancreatic secretion. *Secretin* induces the production of pancreatic fluid low in enzyme content but high in bicarbonate (HCO_3^-).This alkaline fluid acts to neutralize the acid (chyme) entering the duodenum.

The other intestinal hormone, known as **cholecystokinin (CCK),** was originally thought to be two separate substances. It has now been identified as one chemical with several important functions: (1) it causes the pancreas to increase exocrine secretions high in enzyme content; (2) it opposes the influence of gastrin on gastric parietal cells, thus inhibiting hydrochloric acid secretion by the stomach; and (3) it stimulates contraction of the gallbladder so that bile can pass into the duodenum.

Control of Bile Secretion

Bile is secreted continually by the liver and is stored in the gallbladder until needed by the duodenum. The hormones secretin and CCK, as described, stimulate ejection of bile from the gallbladder.

Control of Intestinal Secretion

Our knowledge about the regulation of intestinal exocrine secretions continues to grow. We know that intestinal secretions contain bicarbonate, which along with pancreatic bicarbonate, neutralizes acid from the stomach. Bicarbonate secretion is regulated by a reflex sensitive to changes in pH of the chyme. Neural mechanisms also help control the secretion of intestinal juice.

QUICK CHECK
30. How is salivary secretion controlled?
31. Name the three phases of gastric secretion.

ABSORPTION

Process of Absorption

Absorption is the passage of substances (notably digested foods, water, salts, and vitamins) through the intestinal mucosa into the blood or lymph. As stated earlier, most absorption occurs in the small intestine, where the large surface area provided by the intestinal villi and microvilli (see Figure 21-11) facilitates this process.

Mechanisms of Absorption

Absorption of water is simple and straightforward: osmosis. Some substances depend on other active or passive transport mechanisms (or both) for absorption. Movement of sodium and glucose out of the GI tract is a good example. Primary active transport of sodium by an ATP-driven sodium pump in intestinal cells creates a concentration gradient that drives the passive co-transport of glucose along with sodium in a process called *secondary active transport*. Amino acids and several other compounds are thought to also be absorbed by such a secondary active transport mechanism.

Fatty acids, monoglycerides (products of fat digestion), and cholesterol are transported with the aid of lecithin and bile salts from fat droplets in the intestinal lumen to absorbing cells on villi. Lecithin and bile salts form microscopic spheres called *micelles,* which contain simple lipids (see Figure 21-25).

As Figure 21-27 illustrates for you, water-soluble micelles formed in the lumen of the intestine approach the brush border of absorbing cells. There, simple lipid molecules are released to pass through the plasma membrane by simple diffusion. After they are inside the cell, fatty acids are rapidly reunited with monoglycerides to form triglycerides (neutral fats).

Vitamins A, D, E, and K, known as the "fat-soluble vitamins," also depend on bile salts for their absorption. Some water-soluble vitamins, such as certain of the B group, are small enough to be absorbed by simple diffusion; however, most require carrier-mediated transport. Many drugs (sedatives, analgesics, antibiotics) appear to be absorbed by simple diffusion because they are lipid soluble.

Note that, after absorption, most nutrients do not pass directly into the general circulation. **Lacteals** conduct fats along a series of lymphatic vessels and through many lymph nodes before releasing them into the venous blood flowing through the subclavian veins. Nutrients that are absorbed into the blood, such as amino acids and monosaccharides, first travel by way of the hepatic portal system to the liver.

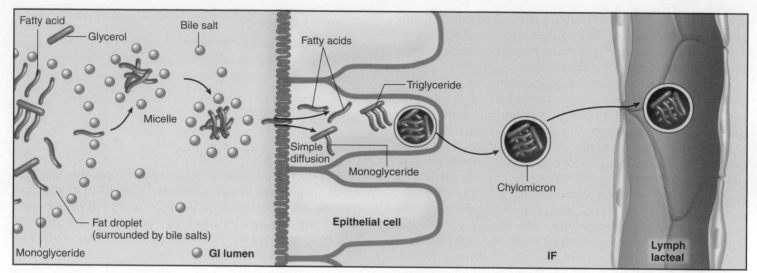

▲ **FIGURE 21-27** **Absorption of fats.** Fats such as triglycerides are chemically digested within emulsified fat droplets, yielding fatty acids, monoglycerides, and glycerol *(left).* Fatty acids and other lipid-soluble compounds (such as cholesterol) leave the fat droplets in small spheres coated with bile salts (micelles). When a micelle reaches the plasma membrane of an absorptive cell, individual fat-soluble molecules diffuse directly into the cytoplasm. The endoplasmic reticulum of the cell resynthesizes fatty acids and monoglycerides into triglycerides. A Golgi body within the cell packages the fats into water-soluble micelles called chylomicrons, which then exit the absorptive cell by exocytosis and enter a lymphatic lacteal. *IF,* Interstitial fluid.

ELIMINATION

The process of **elimination** is simply the expulsion of the residues of digestion—*feces*—from the digestive tract. Formation of feces is the primary function of the colon. The act of expelling feces is called **defecation.**

Defecation is a reflex brought about by stimulation of receptors in the rectal mucosa. Normally, the rectum is empty until mass peristalsis moves fecal matter out of the colon into the rectum. This distends the rectum and produces the desire to defecate. Also, it stimulates colonic peristalsis and initiates reflex relaxation of the internal sphincters of the anus. Voluntary straining efforts and relaxation of the external anal sphincter may then follow as a result of the desire to defecate. Note that this is a reflex partly under voluntary control. If one voluntarily inhibits defecation, the rectal receptors soon become depressed and the urge to defecate does not usually recur until hours later, when mass peristalsis again takes place.

Constipation occurs when the contents of the lower part of the colon and rectum move at a rate that is slower than normal. Extra water is absorbed from the fecal mass, producing a hardened, or constipated, stool.

Diarrhea may occur as a result of increased motility of the small intestine. Chyme moves through the small intestine too quickly, reducing the amount of absorption of water and electrolytes. Diarrhea may also result from bacterial toxins that damage the water reabsorption mechanisms of the intestinal mucosa. The large volume of material arriving in the large intestine exceeds the limited capacity of the colon for absorption, so a watery stool results. Prolonged diarrhea can be particularly serious, even fatal, in infants because they have a minimal reserve of water and electrolytes (Box 21-5).

QUICK CHECK

32. Explain the term secondary active transport.
33. Describe how fatty acids are absorbed by cells of the GI mucosa.
34. What triggers the defecation reflex?

BOX 21-5 **Health Matters**

Infant Diarrhea

Severe diarrhea caused by a *rotavirus*, an intestinal infection, kills more than 600,000 infants and young children worldwide each year. Death results from severe dehydration caused by 20 or more episodes of diarrhea in a single day. More than three million U.S. children suffer symptoms of rotavirus intestinal infection annually, and 65,000 require hospitalization. Good medical care in this country has limited the number of U.S. infant deaths caused by the disease each year to about 50.

Unfortunately, in developing countries, rotavirus-induced diarrhea remains one of the leading causes of infant mortality. The first major attempt at a rotavirus vaccine provided good protection against the virus but was withdrawn from the world market at the close of the twentieth century because of side effects. New rotavirus vaccines are now available and others are in the final stages of testing. Some have already been licensed outside the United States.

Until safe and effective vaccines become widely available and administered, one of the best treatment options available in many areas of the world involves oral administration of liberal doses of a simple, easily prepared solution containing sugar and salt. Called **oral rehydration therapy (ORT),** the salt-sugar solution replaces nutrients and electrolytes lost in diarrheal fluid. Because the replacement fluid can be prepared from readily available and inexpensive ingredients, it is particularly valuable in the treatment of infant diarrhea in developing countries.

The BIG Picture

The digestive system's primary contribution to overall homeostasis is its ability to maintain a constancy of nutrient concentration in the internal environment. It accomplishes this by breaking large, complex nutrients into smaller, simpler nutrients so they can be absorbed (see figure). The digestive system also provides the means of absorption—the cellular mechanisms that operate in the absorptive cells of the intestinal mucosa. The digestive system also provides some secondary, less vital functions. For example, the teeth and tongue aid the nervous system and respiratory system in producing spoken language. Also, acid in the stomach assists the immune system by destroying potentially harmful bacteria. Some of the various vital and nonvital roles played by the different organs that make up the digestive system are summarized in the figure.

The digestive system requires functional contributions by other systems of the body. Regulation of digestive motility and secretion requires the active participation of both the nervous system and the endocrine system. The oxygen needed for digestive activity requires the proper functioning of both the respiratory system and the circulatory system. The body's framework (integumentary and skeletal systems) is required to support and protect the digestive organs. The skeletal muscles must function if ingestion, mastication, deglutition, and defecation are to occur normally. As you've seen, the digestive system cannot operate alone—nor can any other system or organ, for that matter. The body is truly an integrated system, not a collection of independent components.

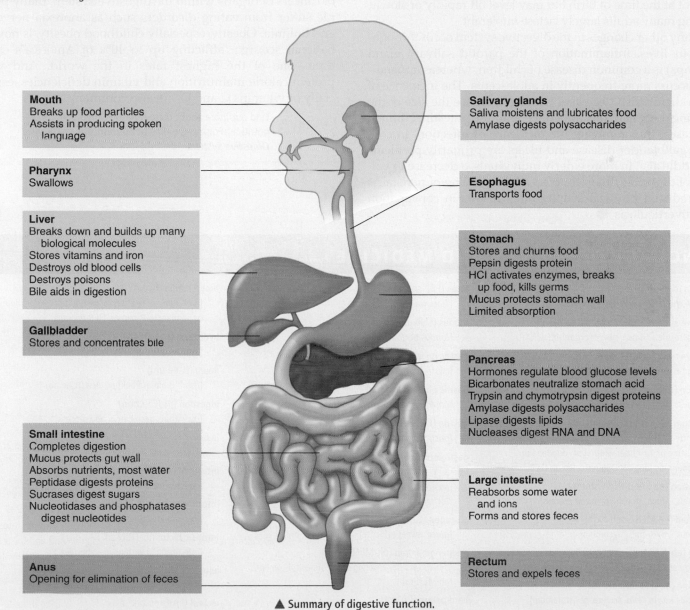

Mouth
Breaks up food particles
Assists in producing spoken
    language

Pharynx
Swallows

Liver
Breaks down and builds up many
    biological molecules
Stores vitamins and iron
Destroys old blood cells
Destroys poisons
Bile aids in digestion

Gallbladder
Stores and concentrates bile

Small intestine
Completes digestion
Mucus protects gut wall
Absorbs nutrients, most water
Peptidase digests proteins
Sucrases digest sugars
Nucleotidases and phosphatases
    digest nucleotides

Anus
Opening for elimination of feces

Salivary glands
Saliva moistens and lubricates food
Amylase digests polysaccharides

Esophagus
Transports food

Stomach
Stores and churns food
Pepsin digests protein
HCl activates enzymes, breaks
    up food, kills germs
Mucus protects stomach wall
Limited absorption

Pancreas
Hormones regulate blood glucose levels
Bicarbonates neutralize stomach acid
Trypsin and chymotrypsin digest proteins
Amylase digests polysaccharides
Lipase digests lipids
Nucleases digest RNA and DNA

Large intestine
Reabsorbs some water
    and ions
Forms and stores feces

Rectum
Stores and expels feces

▲ Summary of digestive function.

Cycle of LIFE ⭕

Significant changes in both the structure and function of the digestive system can occur at different times in the human life cycle. Such changes result in numerous diseases or pathological conditions and may occur in any segment of the intestinal tract from the mouth to the anus. In addition, life cycle changes also involve the accessory organs of digestion, such as the teeth, salivary glands, liver, gallbladder, and pancreas.

Because of the immaturity of intestinal mucosa in young infants, some types of intact proteins can pass through the epithelial cells that line the tract. The result may be an early allergic response caused by the protein triggering the baby's immune system. Lactose intolerance is another age-related example of a common digestive system problem. Intestinal lactase, needed for the digestion of lactose, or milk sugar, is almost always present at the time of birth but may level off rapidly or slowly, leaving many adults largely lactose intolerant.

Many other changes in the digestive system occur throughout our lives. Inflammation of the parotid salivary gland (mumps) is a common disease of children, whereas appendicitis occurs more frequently in adolescents. The incidence of appendicitis then decreases with age because the size of the opening between the appendix and the intestinal lumen decreases, thus reducing the incidence of infection. In contrast, gallbladder disease and ulcers are primarily problems of middle age. In more elderly individuals, a decrease in volume of digestive fluids coupled with a slowing of peristalsis and reduced physical activity often results in constipation and diverticulosis. ⬤

MECHANISMS OF DISEASE

A number of physiological, metabolic, and nutritional disorders can affect human digestion and nutrition. These range from genetic conditions and abnormalities of digestive tract development to metabolic disorders that undermine your body's ability to maintain homeostasis. Disorders of the digestive system abound, including disorders of the mouth and esophagus, tooth decay, gingivitis, and periodontitis. Gastroesophageal reflux disease also is a common and serious problem for many Americans.

Beyond common afflictions such as ulcers, diarrhea, constipation, and other more serious diseases, such as a variety of cancers of organs within the digestive system, many people suffer from eating disorders such as anorexia nervosa and bulimia. Obesity, especially childhood obesity, is now a national scourge, afflicting up to 40% of America's children—one of the highest rates in the world. And yet, protein-calorie malnutrition and vitamin deficiencies, especially of vitamins C and D, are also commonplace.

evolve   Find out more about these digestive, metabolic, and nutritional diseases online at *Mechanisms of Disease: Digestive System*.

LANGUAGE OF SCIENCE AND MEDICINE *continued from page 483*

diarrhea (dye-ah-REE-ah)
  [*dia-* through, *-rrhea* flow]

digestive system (dye-JES-tiv SIS-tem)
  [*digest-* break down, *system* organized whole]

digestive tract (dye-JES-tiv TRAKT)
  [*digest-* break down, *tract* trail]

digestion (dye-JES-chun)
  [*digest-* break down, *-tion* process]

duodenum (doo-oh-DEE-num)
  [shortened from *intestinum duodenum digitorum* intestine of 12 finger-widths] *pl.,* duodena or duodenums (doo-oh-DEE-nah, doo-oh-DEE-numz)

elimination (ee-lim-ih-NAY-shun)
  [*e-* out, *-limen-* threshold, *-ation* process]

emulsified (ee-MULL-seh-fyde)
  [*e-* out, *-muls-* milk, *-i-* combining form, *-fy* process]

enamel (eh-NAM-el)
  [*enamel* hard glossy coating]

endocrine cell (EN-doh-krin sell)
  [*endo-* within, *-crin-* secrete, *cell* storeroom]

esophagus (eh-SOF-ah-gus)
  [*es-* will carry, *-phagus* food] *pl.,* esophagi (eh-SOF-ah-gye)

feces (FEE-seez)
  [*feces* waste]

fundus (FUN-duss)
  [*fundus* bottom] *pl.,* fundi (FUN-dye)

gastric juice (GAS-trik)
  [*gastr-* stomach, *-ic* relating to]

gastric phase (GAS-trik)
  [*gastr-* stomach, *-ic* relating to]

gastrin (GAS-trin)
  [*gastr-* stomach, *-in* substance]

gastrointestinal (GI) tract (gas-troh-in-TES-tin-ul TRAKT)
  [*gastr-* stomach, *-intestin-* intestine, *-al* relating to, *tractus* trail]

glucagon (GLOO-kah-gon)
  [*gluca-* sweet (glucose), *-agon* lead or bring]

greater omentum (oh-MEN-tum)
  [*omentum* fatty covering of intestines] *pl.,* omenta (oh-MEN-tah)

hard palate (PAL-et)

hepatic lobule (heh-PAT-ik LOB-yool)
  [*hepa-* liver, *-ic* relating to, *lobus-* pod, *-ule* small]

hiatal hernia (hye-AY-tal HER-nee-ah)
  [*hiat-* gap, *-al* relating to, *hernia* rupture] *pl.,* herniae or hernias (HER-nee-ee, HER-nee-ahz)

hydrolysis (hye-DROHL-ih-sis)
  [*hydro-* water, *-lysis* loosening]

ileum (IL-ee-um)
  [*ileum* groin or flank] *pl.,* ilea (IL-ee-ah)

ingestion (in-JES-chun)
  [*in-* within, *-gest-* carry, *-tion* process]

insulin (IN-suh-lin)
  [*insul-* island, *-in* substance]

intestinal juice (in-TES-tih-nal)
  [*intestin-* intestine, *-al* relating to]

intestinal phase (in-TES-tih-nal)
  [*intestin-* intestine, *-al* relating to]

intrinsic factor (in-TRIN-sik FAK-tor)
  [*intr-* inside or within, *-insic* beside]

jejunum (jeh-JOO-num)
  [*jejunus* empty]

lacteal (LAK-tee-al)
  [*lact-* milk, *-al* relating to]

LANGUAGE OF SCIENCE AND MEDICINE *continued from page 510*

continued from page 510

lecithin (LES-ih-thin)
[*lecith-* yolk, *-in* substance]

left lobe

lesser omentum (oh-MEN-tum)
[*omentum* fatty covering of intestines]

lip

lipase (LYE-pays)
[*lip-* fat, *-ase* enzyme]

lower esophageal sphincter (LES) (eh-SOF-eh-JEE-ul SFINGK-ter)
[*es-* will carry, *-phag-* food (eat), *-al* relating to, *sphinc-* bind tight, *-er* agent]

mastication (mas-tih-KAY-shun)
[*masticat-* chew, *-tion* process]

mechanical digestion (dye-JES-chun)
[*digest-* break down, *-tion* process]

mesentery (MEZ-en-tair-ee)
[*mes-* middle, *-enter-* intestine]

micelle (my-SELL)
[*mic-* grain, *-elle* small]

motility (moh-TIL-ih-tee)
[*mot-* move, *-il-* relating to, *-ity* state]

mucosa (myoo-KOH-sah)
[*muc-* slime, *-osa* relating to] *pl.,* mucosae (myoo-KOH-see)

mucus (MYOO-kus)
[*mucus* slime]

muscularis (mus-kyoo-LAIR-is)
[*mus-* mouse, *-cul-* little, *-aris* relating to] *pl.,* musculares (mus-kyoo-LAIR-eez)

neck

nuclease (NOO-klee-ayz)
[*nucle-* nut or kernel (nucleic acid), *-ase* enzyme]

oral rehydration therapy (ORT) (OR-al ree-hye-DRAY-shun THER-ah-pee)
[*or-* mouth, *-alis* relating to, *re-* back again, *-hydra-* water, *-ation* process]

pancreatic islet (pan-kree-AT-ik EYE-let)
[*pan-* all, *-creat-* flesh, *-ic* relating to, *islet* island]

pancreatic juice (pan-kree-AT-ik)
[*pan-* all, *-creat-* flesh, *-ic* relating to]

parietal cell (pah-RYE-ih-tal sell)
[*paires-* wall, *-al* relating to, *cell* storeroom]

parotid (peh-RAH-tid)
[*par-* beside, *-ot-* ear, *-id* relating to]

pepsin (PEP-sin)
[*peps-* digestion, *-in* substance]

pepsinogen (pep-SIN-oh-jen)
[*peps-* digestion, *-in-* substance, *-o-* combining form, *-gen* produce]

peptidase (PEP-tyd-ayz)
[*pept-* digestion, *-ide-* chemical, *-ase* enzyme]

peristalsis (pair-ih-STAL-sis)
[*peri-* around, *-stalsis* contraction]

peritoneum (pair-ih-toh-NEE-um)
[*peri-* around, *-tone-* stretched, *-um* thing] *pl.,* peritonea (pair-ih-toh-NEE-ah)

proenzyme (proh-EN-zyme)
[*pro-* first, *-en-* in, *-zyme* ferment]

propulsion (proh-PUL-shen)
[*pro-* in front, *-pul-* drive, *-sion* process]

protease (PROH-tee-ayz)
[*prote-* protein, *-ase* enzyme]

pulp cavity (KAV-ih-tee)
[*pulp* flesh, *cav-* hollow, *-ity* condition]

pyloric sphincter (pye-LOR-ik SFINGK-ter)
[*pyl-* gate, *-or-* guard, *-ic* relating to, *sphinc-* bind tight, *-er* agent]

pylorus (pye-LOR-us)
[*pyl-* gate, *-orus* guard]

rectum (REK-tum)
[*rect-* straight, *-um* thing]

retropulsion (ret-roh-PUL-shen)
[*retro-* backward, *-pul-* drive, *-sion* process]

right lobe

root

saliva (sah-LYE-vah)

secretion (seh-KREE-shun)
[*secret-* separate, *-tion* process]

segmentation (seg-men-TAY-shun)
[*segment-* cut section, *-ation* process]

serosa (see-ROH-sah)
[*ser-* watery fluid, *-osa* relating to] *pl.,* serosae (see-ROH-see)

sigmoid colon (SIG-moyd KOH-lon)
[*sigm-* sigma (Σ or σ) eighteenth letter of Greek alphabet (Roman S), *-oid* like, *colon* large intestine]

soft palate (PAL-et)
[*palat-* palate (roof of mouth)]

sublingual gland (sub-LING-gwall)
[*sub-* under, *-lingua-* tongue, *-al* relating to, *gland* acorn]

submandibular gland (sub-man-DIB-yoo-lar)
[*sub-* under, *-mandibul-* chew (mandible or jawbone), *-ar* relating to, *gland* acorn]

submucosa (sub-myoo-KOH-sah)
[*sub-* under, *-muc-* slime, *-osa* relating to] *pl.,* submucosae (sub-myoo-KOH-see)

tongue

transverse colon (trans-VERS KOH-lon)
[*trans-* across, *-vers-* turn, *colon* large intestine]

trypsin (TRIP-sin)
[*tryps-* pound, *-in* substance]

upper esophageal sphincter (UES) (eh-SOF-ah-JEE-ul SFINGK-ter)
[*es-* will carry, *-phag-* food (eat), *-al* relating to, *sphinc-* bind tight, *-er* agent]

vermiform appendix (VERM-ih-form ah-PEN-diks)
[*vermi-* worm, *-form* shape, *append-* hang upon, *-ix* thing] *pl.,* appendices (ah-PEN-dih-seez)

villus (VIL-us)
[*villus* shaggy hair], *pl.,* villi (VIL-eye)

zymogen

CHAPTER SUMMARY

To download an MP3 version of the chapter summary for use with your iPod or other portable media player, access the Audio Chapter Summaries *online at http://evolve.elsevier.com.*

HINT   *Scan this summary after reading the chapter to help you reinforce the key concepts. Later, use the summary as a quick review before your class or before a test.*

ORGANIZATION OF THE DIGESTIVE SYSTEM

A. The digestive system comprises the digestive tract and other organs that aid digestion

B. The main organs of the digestive tract form a hollow tubelike system, which passes through the ventral cavities of the body (Figure 21-1)

C. Walls of the GI tract—fashioned from four layers of tissues (Figure 21-2):
1. Mucosa—inner mucous lining, innermost layer of the GI wall
2. Submucosa—layer of connective tissue that contains the main blood vessels of the tract
3. Muscularis—muscular layer characterized by an inner layer of circular and an outer layer of longitudinal smooth muscle
4. Serosa—also called serous layer, outer fibrous layer

D. Modifications of layers—layers vary in different regions of the tube throughout its length

MOUTH

A. Structure of the oral cavity
   1. Lips—covered externally by skin and internally by mucous membrane
   2. Cheeks—form the lateral boundaries of the oral cavity and mucous membrane lining is continuous with the lips in front
      a. Bulk of the cheeks is formed largely by the buccinator muscle
      b. Contain mucus-secreting glands
   3. Hard palate—consists of portions of four bones; two maxillae and two palatines
   4. Soft palate—forms a partition between the mouth and nasopharynx; made of muscle arranged in an arch (Figure 23-3)
   5. Tongue—solid mass of skeletal muscle covered by a mucous membrane; forms the base of the oral cavity
      a. Tongue has a blunt root, a tip, and a central body
      b. Covered by rough elevations (papillae); possess sensory organs called taste buds (Figure 21-4)
      c. Lingual frenulum—fold of mucous membrane in the midline of the ventral surface of the tongue; helps anchor the tongue to the floor of the mouth
      d. Important for mastication (chewing) and deglutition (swallowing)
   6. Salivary glands—located outside the alimentary canal; deliver their exocrine secretions by way of ducts from the glands into the lumen of the tract (Figure 21-5)
      a. Three major pairs: parotid, submandibular, and sublingual glands
   7. Teeth—designed to cut, tear, and grind ingested food
      a. Typical tooth—consists of three main parts: crown, neck, and root (Figure 21-6)
         (1) Crown—exposed portion of a tooth; covered by enamel
         (2) Neck—section that connects the crown with the root below it
         (3) Root—fits into the socket of the alveolar process
      b. Types of teeth (Figure 21-7)
         (1) Deciduous teeth (20 baby teeth)
         (2) Permanent teeth (32)

PHARYNX

A. Tube through which a food bolus passes when moved from the mouth to the esophagus

ESOPHAGUS

A. Collapsible, muscular, mucosa-lined tube; extends from the pharynx to the stomach (Figure 21-8)
   1. First segment of the digestive tube proper
   2. Four layers form the wall (Figure 21-8, B)
   3. Each end of the esophagus is encircled by muscular sphincters
      a. Upper esophageal sphincter (UES)
      b. Lower esophageal sphincter (LES) (cardiac sphincter)

STOMACH

A. Size and position of the stomach (Figure 21-9)
   1. Just below the diaphragm
   2. Size of the stomach varies; notably the amount of distention
B. Divisions of the stomach
   1. Fundus—enlarged portion to the left and above the opening of the esophagus into the stomach
   2. Body—central part of the stomach
   3. Pylorus—lower portion
   4. Cardia—small collar or margin of the stomach at its junction with the esophagus
C. Sphincter muscles—regulate passage of material at both stomach openings
   1. Consist of circular fibers arranged to form an opening in the center of them when they are relaxed; no opening when they are fully contracted
      a. Lower esophageal sphincter (LES) (cardiac sphincter)—controls the opening of the esophagus into the stomach
      b. Pyloric sphincter—controls the opening from the pyloric portion of the stomach into the first part of the small intestine
D. Stomach wall
   1. Epithelial lining of the stomach consists largely of folds (rugae)
   2. Gastric pits—marked depressions within the rugae
   3. Gastric glands—secrete most of the gastric juice (Figure 21-10)
   4. Three major secretory cells:
      a. Chief cells—secrete the enzymes of gastric juice
      b. Parietal cells—secrete hydrochloric acid and intrinsic factor
      c. Endocrine cells—secrete ghrelin (GHRL); hormone that stimulates the hypothalamus to increase appetite and release gastrin
   5. Gastric muscle
      a. Muscularis is made of three sublayers of smooth muscle tissue (Figure 21-9)
         (1) Longitudinal muscles
         (2) Circular muscles
         (3) Oblique muscles
E. Functions of the stomach
   1. Serves as a food reservoir
   2. Secretes gastric juice
   3. Breaks up food into small particles and mixes it with the gastric juice
   4. Secretes intrinsic factor
   5. Performs a limited amount of absorption
   6. Produces the hormones gastrin and ghrelin

SMALL INTESTINE

A. Size and position of the small intestine
   1. Tube measuring about 2.5 cm (1 inch) in diameter and 6 meters (20 feet) in length
   2. Its coiled loops fill most of the abdominal cavity (Figure 21-1)
   3. Divisions of the small intestine:
      a. Duodenum—uppermost division; part to which the pyloric end of the stomach attaches; 25 cm (10 inches) long
      b. Jejunum—point where the tube turns abruptly forward and downward; 2.5 meters (8 feet)
      c. Ileum—3.5 meters (12 feet) long
B. Wall of the small intestine
   1. Intestinal lining has plicae with villi (Figure 21-11)
   2. Villi—tiny projections on the intestinal mucosa
      a. Each villus contains an arteriole, venule, and lymph vessel (lacteal)
      b. Contain mucus-secreting goblet cells
      c. Each villus is covered by microvilli
      d. Presence of villi and microvilli greatly increases the surface area of the small intestine

LARGE INTESTINE

A. Size of the large intestine
1. Diameter is noticeably larger than that of the small intestine
2. Length is about 1.5 to 1.8 meters (5 or 6 feet) (Box 21-3)

B. Divisions of the large intestine (Figure 21-12)
1. Cecum
2. Colon (Figure 21-12)
   a. Ascending
   b. Transverse
   c. Descending
   d. Sigmoid
3. Rectum—last 17 to 20 cm (7 or 8 inches) of the intestinal tube
   a. Anal canal—terminal inch of the rectum
   b. Anus—opening of the canal (Figure 21-14)

C. Wall of the large intestine
1. Intestinal glands—produce the lubricating mucus that coats the feces as they are formed (Figure 21-2)
2. Uneven distribution of fibers in the muscle layer

VERMIFORM APPENDIX

A. Wormlike tubular organ; 8 to 10 cm (3 to 4 inches) in length
B. Most often found just behind the cecum or over the pelvic rim
C. Communicates with the cecum 3 cm (about 1 inch) below the ileocecal valve

PERITONEUM

A. Large, continuous sheet of serous membrane that covers most of the digestive organs
B. Forms reflections, or extensions, that bind the abdominal organs together (Figure 21-14)
1. Mesentery—fan-shaped projection of the parietal peritoneum that extends from the lumbar region of the posterior abdominal wall
2. Greater omentum—continuation of the serosa of the greater curvature of the stomach and the first part of the duodenum to the transverse colon
3. Lesser omentum—attaches from the liver to the lesser curvature of the stomach and the first part of the duodenum

LIVER

A. Structure of the liver
1. Largest gland in the body
2. Lies immediately under the diaphragm; occupies most of the right hypochondrium and part of the epigastrium (Figure 21-1)
3. Consists of two lobes separated by the falciform ligament (Figure 21-15)
   a. Left lobe—forms about one sixth of the liver
   b. Right lobe—divides into right lobe proper, caudate lobe, and quadrate lobe
   c. Each lobe is divided into numerous lobules by small blood vessels and by fibrous strands (perivascular fibrous capsule)
   d. Hepatic lobules—anatomical units of the liver (Figure 21-16)

B. Bile ducts
1. Small bile ducts form right and left hepatic ducts
2. These ducts immediately join to form one common hepatic duct
   a. Common hepatic duct merges with the cystic duct from the gallbladder to form the common bile duct (Figure 21-17)

C. Functions of the liver
1. Detoxify various substances
2. Carry on numerous important steps in the metabolism of all three kinds of foods
3. Store several substances such as iron and vitamins
4. Produce important plasma proteins
5. Secrete about a pint of bile a day

D. Bile secretion by the liver
1. Main components of bile are bile salts, bile pigments, and cholesterol

GALLBLADDER

A. Pear-shaped sac 7 to 10 cm (3 to 4 inches) long and 3 cm broad at its widest point (Figure 21-17); lies on the undersurface of the liver
B. Serous, muscular, and mucous layers compose the wall of the gallbladder; mucosal lining has rugae
C. Stores bile that enters it by way of the hepatic and cystic ducts

PANCREAS

A. Structure of the pancreas
1. Grayish pink–colored gland about 12 to 15 cm (6 to 9 inches) long, weighing about 60 grams; its body extends horizontally behind the stomach with its tail touching the spleen
2. Composed of exocrine and endocrine glandular tissue
3. The pancreatic duct empties into the duodenum at the same point as the common bile duct; major duodenal papilla
4. Pancreatic islets—clusters of endocrine cells, making the pancreas a dual gland (exocrine, or duct, gland and endocrine, or ductless, gland)

B. Functions of the pancreas
1. Acinar units of the pancreas secrete the digestive enzymes
2. Beta cells of the pancreas secrete insulin
3. Alpha cells secrete glucagon

OVERVIEW OF DIGESTIVE FUNCTION

A. Digestive system uses various mechanisms to make nutrients available to each cell of our bodies (Table 21-1)
1. Ingestion—food is taken in
2. Digestion—breakdown of food into simpler nutrients
3. Motility of the GI wall—physically breaks large chunks of food into smaller bits and moves it along the tract
4. Secretion of digestive enzymes into the lumen of the GI tract
5. Absorption—movement through the GI mucosa into the internal environment
6. Elimination—excretion of material not absorbed
7. Regulation—coordination of various functions of the digestive system

DIGESTION

A. Processes that chemically and mechanically break complex foods into simpler nutrients that can be easily absorbed
B. Mechanical digestion—movement (motility) of the digestive tract (Table 21-2)
1. Mastication—chewing movements
   a. Reduces size of food particles
   b. Mixes food with saliva in preparation for swallowing
2. Deglutition—process of swallowing; three stages (Figure 21-19)
   a. Oral stage (mouth to oropharynx)
   b. Pharyngeal stage (oropharynx to esophagus)
   c. Esophageal stage (esophagus to stomach)

3. Peristalsis and segmentation—motility produced by smooth muscle in the GI tract
   a. Peristalsis—wavelike ripple of the muscle layer of a hollow organ (Figure 21-20)
   b. Segmentation—mixing movements (Figure 21-21)
4. Regulation of motility
   a. Gastric motility—emptying the stomach takes about 2 to 6 hours after a meal
      (1) Food is churned with gastric juices in the stomach to form a thick, milky material; chyme
      (2) Chyme is churned (propulsion and retropulsion) and mixed with gastric juices (Figure 21-22)
   b. Intestinal motility—includes both peristaltic contractions and segmentation
      (1) Segmentation in the duodenum and upper jejunum mixes chyme with digestive juices from the pancreas, liver, and intestinal mucosa
      (2) Peristalsis continues as the chyme nears the end of the jejunum
C. Chemical digestion—changes in chemical composition that transform foods during their travel through the digestive tract; changes are a result of hydrolysis
   1. Hydrolysis—chemical process in which a compound unites with water and then splits into simpler compounds
   2. Digestive enzymes
      a. Enzymes—"organic catalysts"; organic compounds that accelerate chemical reactions
      b. Specific in their action (Figure 21-5)
      c. Continually being destroyed or eliminated from the body
      d. Most digestive enzymes are synthesized and secreted as inactive proenzymes (Figure 21-5)
   3. Carbohydrate digestion
      a. Carbohydrates are compounds made of saccharides
      b. Polysaccharides are hydrolyzed to disaccharides by enzymes known as amylases
      c. Enzymes that catalyze the final steps in carbohydrate digestion are sucrase, lactase, and maltase (Figure 21-23)
   4. Protein digestion
      a. Protein compounds have very large molecules made up of folded or twisted chains of amino acids
      b. Proteases catalyze the hydrolysis of proteins into individual amino acids (Figure 21-24)
      c. Main proteases: pepsin, trypsin, chymotrypsin, and peptidases (Box 21-4)
   5. Fat digestion
      a. Fats must be emulsified by bile
         (1) Emulsification facilitates chemical digestion of fats by breaking large fat drops into small droplets
      b. Lipase is the main fat-digesting enzyme (Figure 21-25)
   6. Residues of digestion—certain components of food cannot be digested and are eliminated from the intestines in the feces

SECRETION (TABLE 21-4)

A. Saliva—secreted by the salivary glands (Figure 21-5)
   1. Mucus—lubricates food and protects the digestive mucosa
B. Gastric juice—secreted by exocrine gastric glands (Figure 21-10)
   1. Gastric juice contains the basic water and mucus mixture of most other digestive juices and combinations of other substances
   2. Pepsin—primary gastric enzyme; secreted by chief cells
      a. Secreted as the inactive proenzyme pepsinogen

   b. Pepsinogen is converted to pepsin by hydrochloric acid (HCl), which is produced by parietal cells
   3. Parietal cells also produce intrinsic factor
      a. Intrinsic factor—binds to molecules of vitamin B_{12}, protecting them from the acids and enzymes of the stomach; later facilitates its absorption
C. Pancreatic juice—secreted by the exocrine acinar cells of the pancreas
   1. Pancreatic juice contains various digestive enzymes; secreted as zymogens (proenzymes)
      a. Trypsin—protein-digesting enzyme
      b. Amylase
      c. Lipase
      d. Nuclease—RNA- and DNA-digesting enzyme
D. Bile—mixture of many different substances that is secreted by the liver; stored in gallbladder
   1. Lecithin and bile salts—break down large drops of fat into smaller droplets; makes the fats more easily digestible
   2. Contains a small amount of sodium bicarbonate; neutralizes chyme
E. Intestinal juice—sum of intestinal secretions that enters the gastrointestinal lumen by way of a duct
   1. Largely a product of the small intestine
   2. Goblet cells in the mucosa of the large intestine produce some lubricating mucus

CONTROL OF DIGESTIVE GLAND SECRETION

A. Control of salivary secretion
   1. Reflex mechanisms control the secretion of saliva
   2. Chemical and mechanical stimuli come from the presence of food in the mouth
   3. Olfactory and visual stimuli come, of course, from the smell and sight of food
B. Control of gastric secretion—three phases (Figure 21-26)
   1. Cephalic phase ("psychic phase")—mental factors activate the mechanism
   2. Gastric phase—when products of protein digestion reach the pyloric portion of the stomach, they stimulate release of gastrin
      a. Gastrin greatly accelerates their secretion of gastric juice
   3. Intestinal phase—various mechanisms seem to adjust gastric juice secretion as chyme passes to and through the intestinal tract
C. Control of pancreatic secretion—stimulated by several hormones; secretin, CCK
D. Control of bile secretion—secreted continually by liver and stored in the gallbladder; secretin and CCK stimulate ejection of bile
E. Control of intestinal secretion—little is known about the regulation of intestinal exocrine secretions

ABSORPTION

A. Process of absorption
   1. Absorption—passage of substances through the intestinal mucosa into the blood or lymph
   2. Most absorption occurs in the small intestine
B. Mechanisms of absorption
   1. Absorption of water is simple and straightforward: osmosis
   2. Some substances depend on more complex mechanisms to be absorbed
      a. Secondary active transport
   3. Fatty acids, monoglycerides, and cholesterol are transported with the aid of lecithin and bile salts (Figure 21-27)

ELIMINATION

A. Elimination—expulsion of the residues of digestion (feces) from the digestive tract

B. Defecation is a reflex brought about by stimulation of receptors in the rectal mucosa

C. Constipation—occurs when the contents of the lower part of the colon and rectum move at a rate that is slower than normal; extra water is absorbed

D. Diarrhea—occurs as a result of increased motility of the small intestine; may also result from bacterial toxins that damage the water reabsorption mechanisms of the intestinal mucosa

REVIEW QUESTIONS

 HINT *Write out the answers to these questions after reading the chapter and reviewing the Chapter Summary. If you simply think through the answer without writing it down, you won't retain much of your new learning.*

1. List the component parts or segments of the GI tract and the accessory organs of digestion.
2. Name and describe the four tissue layers that form the wall of GI tract organs.
3. Identify the structures that form the mouth.
4. Define the following terms associated with the mouth: *hard palate and soft palate, uvula, lingual frenulum*.
5. Define the term *papillae*.
6. List and give the location of the paired salivary glands.
7. What type of saliva is produced by the parotid glands?
8. Describe a typical tooth.
9. Compare and contrast deciduous and permanent teeth.
10. List the divisions of the stomach. What is the difference between gastric pits and gastric glands?
11. Identify the three major cell types of the gastric glands. What cell type produces hydrochloric acid? Gastric enzymes? Gastrin? Ghrelin?
12. Describe the seven functions of the stomach.
13. List the divisions of the small intestine from proximal to distal.
14. In what area of the GI tract do you find villi? Haustra? Taeniae coli?
15. List the divisions of the large intestine.
16. What is believed to be the function of the vermiform appendix?
17. Discuss the peritoneum and its reflections.
18. Discuss the anatomy of a typical liver lobule.
19. Identify the ducts of the liver.
20. Explain the functions of the gallbladder.
21. Differentiate between endocrine and exocrine functions of the pancreas.
22. Define the actions of mechanical digestion: *mastication, deglutition, peristalsis,* and *segmentation*.
23. Discuss the process of chemical digestion. Describe carbohydrate, protein, and fat digestion.
24. Differentiate between the different digestive secretions: *saliva, gastric juice, pancreatic juice, bile,* and *intestinal juice*.
25. Compare and contrast absorption and elimination.

CRITICAL THINKING QUESTIONS

 HINT *After finishing the Review Questions, write out the answers to these items to help you apply your new knowledge. Go back to sections of the chapter that relate to items that you find difficult.*

1. Explain the role of the tongue's intrinsic and extrinsic muscles.
2. Describe the unique muscular layer of the esophagus.
3. Increasing the interior surface area of the small intestine allows it to absorb nutrients more efficiently. What examples can you find that add to the interior surface area of the small intestine?
4. If an elderly patient had abdominal pain, why would it be unlikely that it is caused by appendicitis?
5. How would you explain the two types of processes within the digestive system?
6. Why do you think emulsification is an example of mechanical digestion rather than chemical digestion to emulsify fats?

continued from page 483

Remember Jinder's button swallowing from the Introductory Story? See if you can answer the following questions about her now that you have read this chapter.

1. With what layer of the esophageal lining would the button be in contact?
   a. Serosa
   b. Muscularis
   c. Submucosa
   d. Mucosa

Because the button was rounded, with no sharp points, the decision was made to let it "pass" through Jinder's system naturally.

2. As the button makes its way through Jinder's alimentary tract, it will go through which sequence of sphincters and valves?
   a. Lower esophageal, pyloric, ileocecal, anal
   b. Anal, pyloric, lower esophageal, ileocecal
   c. Lower esophageal, ileocecal, pyloric, anal
   d. Pyloric, ileocecal, anal, lower esophageal

3. Imagine a camera is attached to the button. As the button travels through Jinder's small intestines, the viewing monitor shows many tiny projections that look like gently moving, fuzzy fingers in the lining of the small intestine. What are these projections?
   a. Plicae
   b. Rugae
   c. Villi
   d. Crowns

 HINT *To solve these questions, you may have to refer to the glossary or index, other chapters in this textbook, A&P Connect, Mechanisms of Disease, and other resources.*

UNIT 5

CHAPTER

22

Nutrition and Metabolism

CHAPTER OUTLINE

Scan this outline before you begin to read the chapter, as a preview of how the concepts are organized.

STUDENT LEARNING OBJECTIVES

At the completion of this chapter, you should be able to do the following:

1. Define and outline the differences between nutrition and metabolism.
2. Define these terms: *assimilation, catabolism, anabolism.*
3. Outline the process of carbohydrate metabolism.
4. Discuss the roles of glycolysis, the citric acid cycle, and the electron transport chain in the production of cellular energy.
5. List the hormones involved in the control of glucose metabolism.
6. Outline the role of lipids, their transport, and their metabolism.
7. Outline the role of proteins and their metabolism.
8. Discuss the difference between vitamins and minerals and their roles in metabolism.
9. Discuss the factors that control and influence metabolic rate.

WALTER heard that pizza was a complete meal because it contained all the major macronutrients. "Maybe I could invent a pizza diet and make millions," he thought. He went over the macronutrients in his head, "carbohydrates, lipids . . . hmm, and then there are macronutrients such as sodium, carbon, potassium . . . ."

A friend pointed out that some of his assumptions concerning nutrients were incorrect. Do you know which ones should not be part of Walter's list?

Chapter 21 explained the processes of getting nutrients into our internal environment. This chapter takes the story further by discussing how our body manages the nutrients after they are absorbed—that is, how they are stored and how they are used by the cells of our bodies.

OVERVIEW OF NUTRITION AND METABOLISM

Nutrition refers to the foods that we eat and the types of nutrients they contain. As you undoubtedly know, healthful nutrition requires a balance of different nutrients in appropriate amounts. In contrast, *malnutrition* is a deficiency or imbalance in the consumption of food, vitamins, and minerals. As a matter of convenient communication, many nutrition experts divide the essential (required) nutrients into two major categories:

1. **Macronutrients**—usually include those nutrients that we need in *large amounts,* such as carbohydrates, fats, and proteins. Water and minerals that we need in large quantities to remain in good health are often included among the macronutrients. For example, sodium, chloride, potassium, calcium, magnesium, and phosphorus are sometimes considered to be macronutrients.

2. **Micronutrients**—usually include nutrients that we need in very *small amounts,* such as vitamins and some minerals. Minerals in this group include iron, iodine, zinc, manganese, cobalt, and a few others. Mineral micronutrients are also called *trace elements.*

There are many small differences in individual genetic makeup, as well as differences in individual lifestyles and environments, that influence how nutrients affect our bodies. Fortunately, we have some advice that we can rely on to help us make healthy choices. For example, the United States government makes use of an individually customized food

LANGUAGE OF SCIENCE AND MEDICINE

HINT *Before reading the chapter, say each of these terms out loud. This will help you avoid stumbling over them as you read.*

amino acid (ah-MEE-no ASS-id)
[*amino* NH_2, *acid* sour]

anabolism (ah-NAB-oh-liz-em)
[*anabol-* build up, *-ism* action]

antioxidant (an-tee-OK-seh-dent)
[*anti-* against, *-oxi-* sharp (oxygen), *-ant* agent]

appetite center

assimilation (ah-sim-ih-LAY-shun)
[*assimila-* make alike, *-tion* process]

ATP synthase (SIN-thays)
[*ATP* adenosine triphosphate, *syn-* together, *-ase* enzyme]

basal metabolic rate (BMR) (BAY-sal met-ah-BAHL-ik)
[*bas-* basis, *-al* relating to, *metabol-* change, *-ic* relating to]

calcitriol (kal-SIT-ree-ol)
[*calci-* lime (calcium), *-tri-* three, *-ol* alcohol (after *1,25-D₃* or *1,25-dihydroxycholecalciferol*)]

catabolism (kah-TAB-oh-liz-em)
[*catabol-* break down, *-ism* action]

cellulose (SFII-yoo-lohs)
[*cell-* storeroom (cell), *-ul-* small, *-ose* carbohydrate]

chylomicron (kye-loh-MYE-kron)
[*chylo-* juice (chyle), *-micro-* small, *-on* particle]

citric acid cycle (SIT-rik ASS-id SYE-kul)
[*citr-* citron tree, *-ic* relating to, *acidus* sour, *kyklos* circle]

coenzyme (koh-EN-zyme)
[*co-* together, *-en-* in, *-zyme* ferment]

deamination (dee-am-ih-NAY-shun)
[*de-* undo, *-amin-* ammonia compound, *-ation* process]

electron transport system (eh-LEK-tron TRANZ-port SIS-tem)
[*electr-* electric, *-on* unit, *trans-* across, *-port* carry, *system* organized whole]

essential fatty acid
[*acid* sour]

free fatty acid (FFA)
[*acid* sour]

glucose phosphorylation (GLOO-kohs fos-for-ih-LAY-shun)
[*gluco-* sweet, *-ose* carbohydrate (sugar), *phos-* light, *-phor-* carry, *-yl-* chemical, *-ation* process]

glycolysis (glye-KOHL-ih-sis)
[*glyco-* sweet (glucose), *-o-* combining form, *-lysis* loosening]

hyperglycemia (hye-per-gly-SEE-mee-ah)
[*hyper-* above, *-glyc-* sweet (glucose), *-emia* blood condition]

hypoglycemia (hye-poh-gly-SEE-mee-ah)
[*hypo-* below, *-glyc-* sweet (glucose), *-emia* blood condition]

lipid (LIP-id)
[*lip-* fat, *-id* form]

continued on page 530

pyramid as a general nutrition guide (Figure 22-1). The Canadian government uses a similar individualized food guide to advise eating a healthy, balanced diet.

Now let's see how nutrients and *metabolism* are related.

Metabolism refers to the complex interactions of chemical processes that make life possible. It is essentially how the body uses foods and their nutrients after they have been digested, absorbed, and transported to the cells of our bodies. Your body cells use nutrients from food in several ways: as fuel (energy), as material for growth and maintenance, and for regulation of body functions. Before they can be used in these different ways, nutrients have to be assimilated. **Assimilation** occurs when nutrient molecules enter cells and undergo many chemical changes.

Metabolism is a complex process made up of many other processes. Two of the major metabolic processes are termed **catabolism** and **anabolism.** Each of these processes, in turn, consists of a series of enzyme-driven chemical reactions known as *metabolic pathways.*

Catabolism breaks food molecules down into smaller molecular compounds and, in so doing, releases energy from them. Anabolism does the opposite. It builds nutrient molecules up into larger molecular compounds and, in so doing, uses energy. Thus catabolism is a decomposition process, whereas anabolism is a synthesis process. Both catabolism and anabolism take place inside cells. Both catabolic and anabolic processes go on continually and concurrently.

Catabolism releases energy in two forms: heat energy and chemical energy. The amount of heat generated is relatively large—so large, in fact, that it would effectively "cook" cells if it were released in one large burst! Fortunately, catabolism is regulated by enzymes so that heat is released in frequent small bursts. Most of this heat is used to maintain the

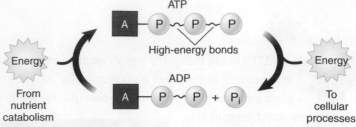

▲ **FIGURE 22-2**   **The role of ATP in metabolism.** ATP temporarily stores energy in its last high-energy phosphate bond. When water is added and phosphate breaks free, energy is released to do cellular work. The ADP and phosphate groups that result can be resynthesized into ATP, capturing additional energy from nutrient catabolism. This cycle is called the ATP/ADP system.

homeostasis of body temperature. In contrast, chemical energy released by catabolism is more obviously useful. It cannot, however, be used directly for biological reactions. First, it must be transferred to the high-energy molecule adenosine triphosphate (ATP). ATP supplies energy directly to the energy-using reactions of all cells in all living cellular organisms.

Look now at Figure 22-2. The structural formula at the top of the diagram shows three phosphate groups attached to the rest of the ATP molecule, two of them by high-energy bonds. Adding water to ATP yields an inorganic phosphate group (P_i), adenosine diphosphate (ADP), and energy, which is used for anabolism and other cell work. The diagram also shows that P_i and ADP then use energy released by catabolism to recombine to form ATP.

Metabolism is not identical in all cells. More active cells have a higher metabolic rate than do less active cells. In addition, anabolism in different kinds of cells produces different compounds. In liver cells, for example, anabolism synthesizes various blood protein compounds. But in beta cells of the pancreas, anabolism produces insulin.

Metabolism is a broad and complex mix of biological chemistry. This chapter discusses only the *essential* concepts related to the many and varied metabolic pathways of the human body.

CARBOHYDRATES

Dietary Sources of Carbohydrates

Carbohydrates are found in most of the foods that we eat. *Polysaccharides*—such as starches in vegetables, grains, and other plant tissues—are broken down into simpler carbohydrates before they are absorbed.

Cellulose, a major component of most plant tissues, is an important exception. Because we do not make enzymes that chemically digest this complex carbohydrate, it passes through our digestive system without being broken down. Also called *dietary fiber* or "roughage," cellulose and other indigestible polysaccharides mix with chyme and keep it thick enough to push easily through our digestive system. Most biologists now agree that a high-fiber diet has many health benefits.

Disaccharides such as those in refined sugar must also be chemically digested before they can be absorbed. *Monosaccharides* in fruits and some "diet foods" are already

▲ **FIGURE 22-1**   **United States Food Guide Pyramid.** Simple pyramid diagrams help educate the public on building a diet with a balance of foods from different categories illustrated in the diagram. This is an abbreviated version of the comprehensive food guide that can be found at www.mypyramid.gov. The full version includes recommended servings per day and other nutrition advice.

in an absorbable form, so they can move directly into the internal environment without initially being processed. As you will see, the monosaccharide *glucose* is the carbohydrate that is most useful to the typical human cell.

A&P CONNECT

Nutritionists often talk about the "energy value" of food—that is, how much energy the body can get from that food. Do you know what it means when a label states that food energy in calories? Do you know the difference between a *calorie* and a *Calorie?* Or a calorie and a *joule* or *kilojoule?* Find answers to these questions, and also learn the energy values of major nutrients and the amount of energy expended by different physical activities, in **Measuring Energy** online at **A&P Connect.**

evolve

Carbohydrate Metabolism

The body metabolizes carbohydrates by both catabolic and anabolic processes. Most of our cells use carbohydrates—mainly glucose—as their first or preferred energy fuel. When the amount of glucose entering cells is inadequate for their energy needs, they may make more glucose by using a pathway that catabolizes fats or proteins.

As you read through the following sections outlining the basic process of carbohydrate metabolism, remember the ultimate result of catabolism: *the transfer of energy from a nutrient molecule to ATP.* It is the continued production of ATP, the energy currency of the cell, which makes nutrient catabolism so incredibly vital to all of life's processes.

Glucose Transport and Phosphorylation

Carbohydrate metabolism begins with the movement of glucose through cell membranes. In the interior of a cell, glucose reacts with ATP to form glucose 6-phosphate, which cannot move back across the cell membrane. This step, named **glucose phosphorylation,** prepares glucose for further metabolic reactions. **Phosphorylation** is the process of adding a phosphate group to a molecule. Depending on their energy needs of the moment, cells either catabolize (break apart) or anabolize (bind together) glucose 6-phosphate.

Glycolysis

Glycolysis is the first step in the process of carbohydrate catabolism. This pathway consists of a series of anaerobic chemical reactions that take place in the cytoplasm (see Figure 4-13, page 71). In the end, glycolysis breaks apart one glucose molecule (made of six carbon atoms) to form two pyruvic acid molecules, each of which has three carbon atoms (Figure 22-3). As you can see, a specific enzyme catalyzes each of these reactions. Glycolysis is an essential process because it produces a small

amount of ATP (a net of two molecules for every sugar molecule) and also prepares glucose for the second step in catabolism, namely, the citric acid cycle. As we will see below, glucose itself cannot enter the cycle: It must first be converted to a compound called *acetyl coenzyme A (acetyl CoA).*

Citric Acid Cycle

Essentially, the **citric acid cycle** is a series of chemical reactions mediated by enzymes that converts the two acetyl molecules from each six-carbon glucose to four carbon dioxide and six water molecules (see Figure 4-14, page 72). The citric

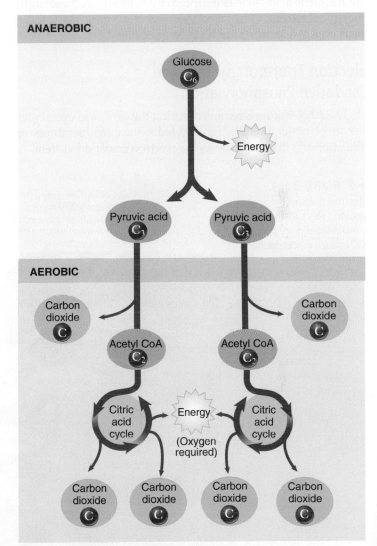

▲ **FIGURE 22-3 Catabolism of glucose.** Glycolysis splits one molecule of glucose (six carbon atoms) into two molecules of pyruvic acid (three carbon atoms each). The glycolytic pathway does not require oxygen, so it is termed *anaerobic.* The removal of a carbon dioxide molecule converts each pyruvic acid molecule into a two-carbon acetyl group that is "escorted" by coenzyme A (CoA) into the citric acid cycle, where it joins a four-carbon compound (oxaloacetic acid) to form a six-carbon compound (citric acid). Now, two more carbon dioxide molecules (one carbon atom each) are released from each citric acid molecule formed. The carbon and oxygen atoms in the original glucose molecule are thus released as waste products. However, the real metabolic prize is energy, which is released as the molecule is broken down. Because this part of the pathway requires oxygen, it is termed *aerobic.*

UNIT 5

acid cycle occurs in the mitochondria (recall that glycolysis takes place only in the cytoplasm).

Before it can enter the citric acid cycle, each pyruvic acid molecule combines with coenzyme A, thus forming acetyl CoA. Coenzyme A then detaches from acetyl CoA, leaving a two-carbon acetyl group, which enters the citric acid cycle by combining with oxaloacetic acid to form citric acid. This is what gives the citric acid cycle its name.

Each pyruvic acid molecule generates three CO_2 molecules, some ATP, and many high-energy electrons while going through the citric acid cycle. Most of the energy leaving the citric acid cycle is temporarily "stored" in these high-energy electrons. The next section describes how these high-energy electrons are used to generate ATP.

Electron Transport System and Oxidative Phosphorylation

High-energy electrons removed during the citric acid cycle enter a chain of carrier molecules embedded in the inner membrane of mitochondria that is known as the **electron transport system.**

Figure 22-4 shows that high-energy electrons—along with their accompanying protons (H^+)—are shuttled to the electron transport system during the citric acid cycle by carrier molecules called *nicotinamide adenine dinucleotide* (NAD) and *flavin adenine dinucleotide* (FAD). The electrons quickly move down the chain, from one membrane protein complex to the next, and eventually to their final acceptor, oxygen.

As the electrons are transported, some of their energy is used to pump their accompanying protons (H^+) to the intramembrane space between the inner and outer membranes of the mitochondrion. This creates a concentration gradient of protons, and the intermembrane space thus becomes a reservoir of protons. Like water behind a dam, the reservoir of protons temporarily stores energy. In much the same way as water flows through a dam and turns wheels to generate energy, the inner membrane has "proton wheels"—in the form of **ATP synthase.** ATP synthase is an enzyme that uses the proton movement down the concentration gradient to bind together ADP and a phosphate group to generate ATP (Figure 22-5).

▶ **FIGURE 22-4**

Electron transport system. This system of energy transfer takes place entirely within each mitochondrion.

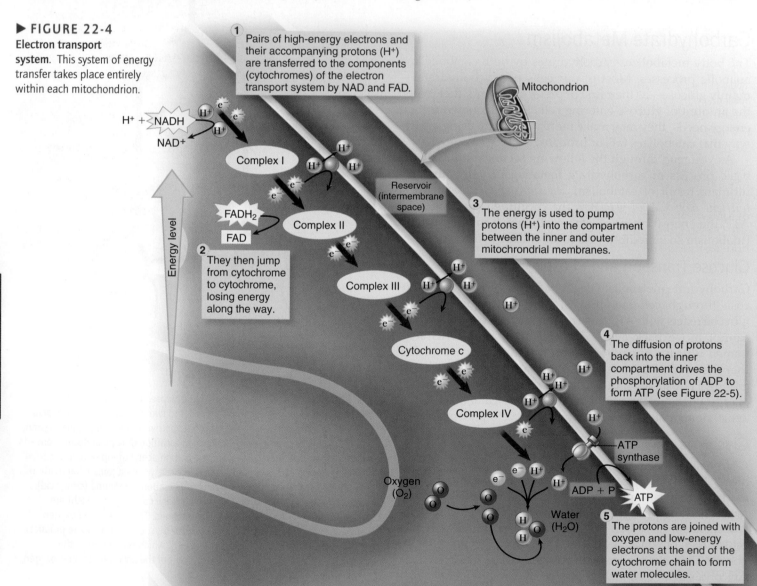

1 Pairs of high-energy electrons and their accompanying protons (H^+) are transferred to the components (cytochromes) of the electron transport system by NAD and FAD.

Mitochondrion

$H^+ + $ NADH

H^+ e^-

e^-

H^+

NAD$^+$

Complex I

H^+

H^+ H^+

Reservoir (intermembrane space)

e^- e^-

Energy level

FADH$_2$

FAD

Complex II

2 They then jump from cytochrome to cytochrome, losing energy along the way.

e^- e^-

3 The energy is used to pump protons (H^+) into the compartment between the inner and outer mitochrondrial membranes.

Complex III H^+ H^+

H^+

e^- e^-

Cytochrome c

4 The diffusion of protons back into the inner compartment drives the phosphorylation of ADP to form ATP (see Figure 22-5).

e^- e^-

H^+ H^+

H^+

Complex IV

H^+

e^-

ATP synthase

e^- e^- H^+

H^+

Oxygen (O_2)

O O

O O

e^-

ADP + P

ATP

O O

H O

H

Water (H_2O)

5 The protons are joined with oxygen and low-energy electrons at the end of the cytochrome chain to form water molecules.

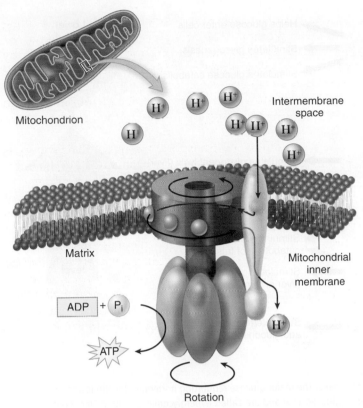

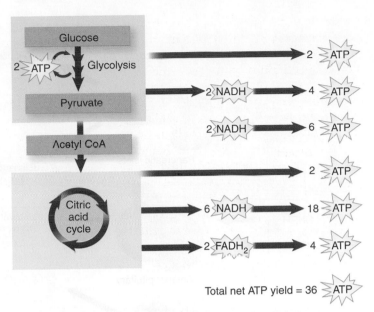

Total net ATP yield = 36 ATP

▲ **FIGURE 22-7** **Energy extracted from glucose.** Energy released from the breakdown of glucose is released mostly as heat, but some of it is transferred to a usable form—the high-energy bonds of ATP. In most human cells, one glucose molecule produces enough usable chemical energy to synthesize or "charge up" 36 ATP molecules. Some cells, such as heart and liver cells, shuttle electrons more efficiently and may be able to synthesize up to 38 ATP molecules. This represents an energy conversion efficiency of 38% to 44%, much better than the 20% to 25% typical of most machines.

▲ **FIGURE 22-5** **Generation of ATP by ATP synthase.** This simplified model of the proton "wheel" in the mitochondrial inner membrane shows how protons (H⁺) moving down their concentration gradient drive the rotation of a molecular machine. The energy of rotation then phosphorylates (adds phosphate to) ADP to become ATP.

At this time, the low-energy electrons (e^-) and their protons (H^+) join oxygen, forming water. This oxygen-requiring joining of a phosphate group to ADP to form ATP is called **oxidative phosphorylation.** As you can see, although oxygen is not needed until the very last step of aerobic respiration, its role is vital. Without oxygen to oxidize the hydrogen

into water, the energy generation pathway would stop. In effect, oxygen serves as an "electron dump," ridding the body of spent electrons derived from the breakdown of glucose.

The breakdown of ATP molecules, of course, provides virtually all the energy that does cellular work. Therefore, oxidative phosphorylation is the crucial part of glucose catabolism (Figures 22-6). The energy extracted during the various steps of the breakdown of glucose is given for you in Figure 22-7.

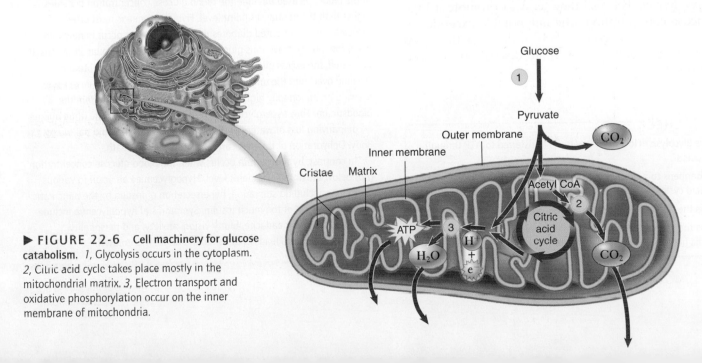

▶ **FIGURE 22-6** **Cell machinery for glucose catabolism.** *1,* Glycolysis occurs in the cytoplasm. *2,* Citric acid cycle takes place mostly in the mitochondrial matrix. *3,* Electron transport and oxidative phosphorylation occur on the inner membrane of mitochondria.

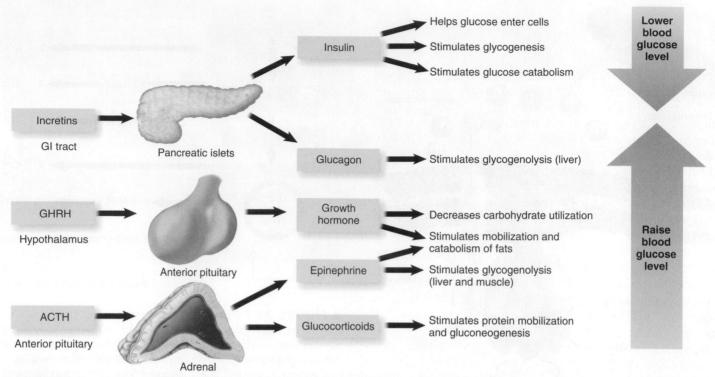

▲ **FIGURE 22-8   Hormonal control of blood glucose level.** Simplified view of some of the major glucose-regulating hormones. Insulin lowers the blood glucose level and is therefore hypoglycemic. Most hormones shown here raise the blood glucose level and are called hyperglycemic, or anti-insulin, hormones.

We can now summarize the long series of chemical reactions in glucose catabolism with one short equation:

$$C_6H_{12}O_6 + 6\,O_2 \rightarrow 6\,CO_2 + 6\,H_2O + 36 \text{ (or 38) ATP} + \text{Heat}$$

Control of Glucose Metabolism

Levels of sugar in the blood are under hormone control as shown in Figure 22-8, which shows that most hormones cause the glucose blood level to rise. These hormones are called *hyperglycemic* because they tend to promote a high blood glucose concentration. The one notable exception is insulin, which is *hypoglycemic* (tends to decrease the blood glucose level). See Box 22-1 for more discussion of blood glucose problems.

QUICK ✓ CHECK

4. What is glycolysis? How much energy is transferred to ATP through this process?

5. What happens to a nutrient molecule as it proceeds through the citric acid cycle?

6. What is the purpose of the electron transport system?

7. What is the difference between hyperglycemic hormones and hypoglycemic hormones?

BOX 22-1 FYI

Abnormal Blood Glucose Concentration

The term **hyperglycemia**, which literally means "condition of too much sugar in the blood," is used anytime the blood glucose concentration becomes higher than the normal set point level. Hyperglycemia is most often associated with untreated diabetes mellitus, but it can occur in newborns when too much intravenous glucose is given or in other similar situations. If untreated, the excess glucose leaves the blood in the kidney—literally "spilling over" into the urine. This increases the osmotic pressure of urine, drawing an abnormally high amount of water into the urine from the bloodstream. Thus hyperglycemia causes loss of glucose in the urine and its accompanying loss of water—potentially threatening the fluid balance of the body. Dehydration of this sort can ultimately lead to death.

In contrast, **hypoglycemia** occurs when the blood glucose concentration dips below the normal set point level. Hypoglycemia can occur in various conditions, including starvation, hypersecretion of insulin by the pancreatic islets, or injection of too much insulin. Symptoms of hypoglycemia include weakness, hunger, headache, blurry vision, anxiety, and personality changes—sometimes leading to coma and death if untreated.

LIPIDS

Dietary Sources of Lipids

Recall from Chapter 2 that **lipids** are a class of organic compounds that includes fats, oils, and related substances. The most common lipids in the diet are **triglycerides,** which are composed of a *glycerol* subunit attached to three *fatty acids.* Other important dietary lipids include *phospholipids* and *cholesterol.*

Dietary fats are often classified as either **saturated** or **unsaturated.** Saturated fats contain fatty acid chains in which there are no double bonds—that is, all available bonds of its hydrocarbon chain are filled (saturated) with hydrogen atoms (see Figure 2-16, p. 32). Saturated fats are usually solid at room temperature. Unsaturated fats contain fatty acid chains in which there are double bonds, meaning that not all sites for hydrogen are filled. Because the double bonds change the shape of unsaturated fats, the molecules usually do not "fit" together as well and so are usually liquid at room temperature.

Triglycerides are found in nearly every food that we eat. However, the amount of triglycerides in each type of food varies considerably, as does the proportion of saturated to unsaturated types. Phospholipids are also found in nearly all foods because they make up the cellular membranes of all living organisms. Cholesterol, however, is found only in foods of animal origin. Cholesterol concentration also varies. For example, it is particularly high in liver, shrimp, and the yolks of eggs.

Transport of Lipids

Lipids are transported in blood as chylomicrons, lipoproteins, and free fatty acids. **Chylomicrons** are small fat droplets found in blood soon after fat absorption takes place. Fatty acids and monoglyceride products of fat digestion combine during absorption to again form fats (triglycerides, or triacylglycerols). These triglycerides plus small amounts of cholesterol and phospholipids compose the chylomicrons.

Lipoproteins are produced mainly in the liver and, as their name suggests, consist of lipids and protein. Blood contains three types of lipoproteins: very-low-density lipoproteins, low-density lipoproteins, and high-density lipoproteins. Usually, they are designated by their abbreviations: VLDL, LDL, and HDL, respectively. Diets high in saturated fats and cholesterol tend to produce an increase in blood LDL concentration, which in turn is associated with a high incidence of coronary artery disease (CAD) and atherosclerosis (Figure 22-9 and Box 22-2). A high blood HDL concentration, in contrast, is associated with a low incidence of heart disease. You can remember this by thinking of the *LDLs* as the "*lethal* lipoproteins" and the *HDLs* as the "*healthy* lipoproteins." Considerable evidence indicates that exercise

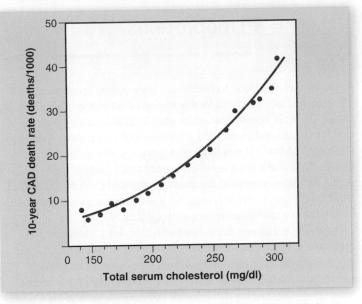

▲ **FIGURE 22-9** **Cholesterol and heart disease.** The graph shows a relationship between the total serum (blood plasma) cholesterol level and coronary artery disease (CAD).

tends to elevate HDL concentration and reduce the likelihood of coronary heart disease.

Fatty acids, on entering the blood from adipose tissue or other cells, combine with albumin to form **free fatty acids (FFAs).** Fatty acids are transported from cells of one tissue to those of another in the form of free fatty acids.

Lipid Metabolism

Lipid Catabolism

Lipid catabolism, like carbohydrate catabolism, consists of several processes. Each of these processes, in turn, consists of a series of chemical reactions. Triglycerides are first hydrolyzed to yield fatty acids and glycerol. Glycerol is then converted to *glyceraldehyde 3-phosphate,* which may be converted to glucose or it may enter the glycolysis pathway directly (see Figure 4-13, p. 71). Fatty acids are broken down into two-carbon pieces—the familiar acetyl CoA. These molecules are then catabolized via the citric acid cycle. The final process of lipid catabolism therefore consists of the same reactions as does carbohydrate catabolism. Catabolism of lipids, however, yields considerably more energy than does catabolism of carbohydrates. Whereas catabolism of 1 gram of carbohydrates yields only 4.1 kcal of heat, catabolism of 1 gram of fat yields 9 kcal.

Lipid Anabolism

Lipid anabolism, also called **lipogenesis,** consists of the synthesis of various types of lipids, notably *triglycerides, cholesterol, phospholipids,* and *prostaglandins.* Triglycerides

BOX 22-2 Lipoproteins

As you've seen, high blood concentrations of low-density lipoproteins (LDLs) are associated with a high risk for atherosclerosis. Atherosclerosis is a form of "hardening of the arteries" that occurs when lipids accumulate in cells lining the blood vessels and promote the development of a plaque that eventually impedes blood flow and may trigger clot formation. Atherosclerosis may also weaken the wall of a blood vessel to the point that it ruptures. In any case, a person with atherosclerosis of the coronary arteries risks a heart attack when blood flow to cardiac muscle is impaired. If vessels in the brain are affected, there is risk of a *cerebrovascular accident (CVA)*, or "stroke."

According to a current model (see part *A* of figure), LDL delivers cholesterol to cells for use in synthesizing steroid hormones and stabilizing the plasma membrane. Most, if not all, cells have many LDL receptors embedded in the outer surface of their plasma membranes. These receptors attract cholesterol-bearing LDL. Once the LDL molecule binds to the receptor, specific mechanisms operate to release the cholesterol it carries into the cell. Excess cholesterol is stored in droplets near the center of the cell. It seems that, in some individuals at least, cells have so few LDL receptors that they accumulate too much cholesterol in the blood. Some mechanism in endothelial cells moves this excess LDL into the wall of blood vessels. This has been proposed as a cause for the lipid accumulation characteristic of atherosclerosis.

High blood concentrations of high-density lipoproteins (HDLs) have been associated with a low risk of developing atherosclerosis and its many possible complications. Although the exact details of how this works have yet to be worked out, scientists have made progress toward that end. According to one model (see part *B* of figure), HDL molecules are attracted to HDL receptors embedded in the plasma membranes. Once they bind to their receptors, the cell is stimulated to release some of its cholesterol from storage. The released cholesterol migrates to the plasma membrane, where it may attach to the HDL molecule and be whisked away to the liver for excretion in bile.

High blood LDL levels (more than 180 mg LDL per 100 ml of blood) signify that a large amount of cholesterol is being delivered to cells. High blood HDL levels (more than 60 mg HDL per 100 ml of blood) indicate that a large amount of cholesterol is being removed from cells and delivered to the liver for excretion from the body. Currently, researchers are using this information to develop treatments that may prevent—or even cure—atherosclerosis and the disorders it causes.

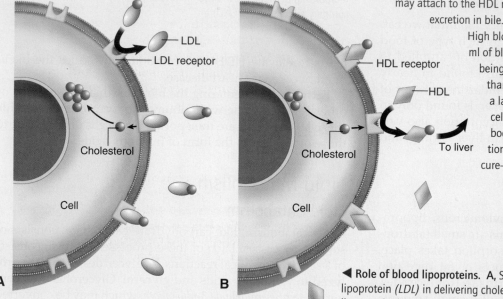

◀ **Role of blood lipoproteins. A,** Simplified diagram of the role of low-density lipoprotein *(LDL)* in delivering cholesterol to cells. **B,** Proposed role of high-density lipoprotein *(HDL)* in removing cholesterol from cells.

and structural lipids (e.g., phospholipids that make up our cell membranes) are synthesized from fatty acids and glycerol or from excess glucose or amino acids. This is why it is possible to "get fat" from foods other than fat! Triglycerides are stored mainly in adipose tissue cells. These fat deposits constitute the body's largest reserve energy source. Enormous amounts of fat can be stored in our bodies. In contrast, only a few hundred grams of carbohydrates can be stored as liver and muscle glycogen.

Most fatty acids can be synthesized by the body. A certain number of the unsaturated fatty acids must be provided by the diet and are thus called **essential fatty acids.** Some of the essential fatty acids serve as a source within the body for synthesis of an important group of lipids called *prostaglandins.* These hormone-like compounds, first discovered in the 1930s from prostate fluids forming semen, have in recent years gained increasing recognition for their occurrence in various tissues, where they support a wide spectrum of biological activity. Certain essential fatty acids are also necessary for manufacturing the phospholipids in cell membranes (see Chapter 4) and the myelin in nerve tissue (see Chapter 11).

Control of Lipid Metabolism

Lipid metabolism is controlled mainly by the following hormones:

- Insulin
- Adrenocorticotropic hormone (ACTH)
- Growth hormone
- Glucocorticoids

As you may recall, these help regulate fat metabolism such that the rate of fat catabolism is inversely related to the rate of carbohydrate catabolism. If some condition such as diabetes mellitus causes carbohydrate catabolism to decrease below energy needs, increased secretion of growth hormone, ACTH, and glucocorticoids soon follows. These hormones, in turn, bring about an increase in fat catabolism. But, when carbohydrate catabolism equals energy needs, fats are not mobilized out of storage and catabolized. Instead, they are spared and stored in adipose tissue. So, it can be said that "Excessive carbohydrates have a 'fat-storing' effect."

QUICK CHECK

8. In what forms are lipids transported to cells?
9. How can glycerol and fatty acids enter the citric acid cycle?
10. Which fatty acids cannot be made by the body?
11. Which hormones are involved in lipid metabolism?

PROTEINS

Sources of Proteins

Recall from Chapter 2 that proteins are very large molecules composed of chemical subunits called **amino acids.** Proteins are assembled from 20 different kinds of amino acids. If any one type of amino acid is deficient, vital proteins cannot be synthesized—a serious health threat. One way your body maintains a constant supply of amino acids is by synthesizing them from other compounds already present in the body. However, only about half of the required 20 types of amino acids can be made by the body. The remaining types of amino acids must be supplied in the diet. Nutritionists often refer to the amino acids that must be in the diet as essential amino acids. Table 22-1 lists amino acids according to whether they are considered essential in the diet or nonessential in the diet (synthesized by the body). Box 22-3 investigates the link between blood levels of amino acids and disease.

Proteins are obtained in the diet from various sources. Muscle meat and other animal tissues particularly high in proteins contain the essential amino acids. Food from a single plant or other *nonanimal* source does not usually contain

TABLE 22-1 Amino Acids

ESSENTIAL	NONESSENTIAL
Histidine*	Alanine
Isoleucine	Arginine
Leucine	Asparagine
Lysine	Aspartic acid
Methionine	Cysteine
Phenylalanine	Glutamic acid
Threonine	Glutamine
Tryptophan	Glycine
Valine	Proline
	Serine
	Tyrosine†

*Essential in infants and, perhaps, adult males.
†Can be synthesized from phenylalanine; therefore, nonessential as long as phenylalanine is in the diet.

an adequate amount of all the essential amino acids. Therefore, it is important to include meat (or other animal tissues), or a mixture of different vegetables that provide all the amino acids needed by the body, in the diet. Plant tissues that are particularly high in protein content include cereal grains, nuts, and legumes such as peas and beans.

Protein Metabolism

Protein Anabolism

Every cell synthesizes its own structural proteins and its own enzymes using ribosomes to read the DNA code and construct polypeptides. In addition, many cells, such as liver and glandular cells, synthesize special proteins for export. For example, liver cells manufacture the plasma proteins found in our blood. Any particular cell's genes, under the

BOX 22-3 Health Matters

Amino Acids and Disease

Recent research shows that the balance of amino acids circulating in the blood is associated with various diseases. High blood levels of homocysteine, one of several alternate forms of the amino acid cysteine (see Table 22-1), have been linked to heart disease, stroke, and dementias such as Alzheimer disease. Whether such abnormalities in homocysteine levels are the direct cause of these conditions is uncertain. Despite this uncertainty, many physicians recommend lowering abnormally high blood homocysteine levels to reduce the possible risk for these devastating conditions. Homocysteine can be reduced to normally low blood levels when there is adequate vitamin B_6, B_{12}, or B_9 (folic acid) in the diet.

influence of signaling mechanisms, determine the specific proteins to be synthesized for that cell or for other body cells. Protein anabolism is truly "big business" in our body. For example, red blood cell replacement alone requires the production of millions of cells per second and by itself creates huge demands for protein anabolism.

Protein Catabolism

The first step in protein catabolism takes place in liver cells. Called **deamination,** it consists of the splitting off of an amino (NH_2) group from an amino acid molecule to form a molecule of ammonia and one of keto acid. Most of the ammonia is converted by liver cells to *urea* and later excreted in the urine. The keto acid may be oxidized via the citric acid cycle or may be converted to glucose or to fat. Both protein catabolism and anabolism go on continually. Only their rates differ from time to time. With a protein-deficient diet, for example, protein catabolism exceeds protein anabolism. Various hormones, as we shall see below, also influence the rates of protein catabolism and anabolism.

Control of Protein Metabolism

Protein metabolism, like that of carbohydrates and fats, is controlled largely by hormones rather than by the nervous system. Growth hormone and the male hormone testosterone both have a stimulating effect on protein synthesis, or anabolism. For this reason, they are referred to as *anabolic* hormones. The protein *catabolic* hormones of greatest consequence are glucocorticoids. They speed up the hydrolysis of cell proteins to amino acids, their entry into the blood, and their subsequent catabolism. ACTH functions indirectly as a protein catabolic hormone because of its stimulating effect on glucocorticoid secretion.

Thyroid hormone is necessary for and promotes protein anabolism and therefore growth when plenty of carbohydrates and fats are available for energy production. Under different conditions—for example, when the amount of thyroid hormone is excessive or when the energy foods are deficient—this hormone may then promote protein mobilization and catabolism.

QUICK CHECK

12. What is meant by the term *essential amino acid?*
13. What happens when an amino acid is deaminated?
14. What is the purpose of the process of amino acid deamination?
15. How is protein metabolism controlled?

VITAMINS AND MINERALS

Vitamins

Vitamins are organic molecules needed in small quantities for normal metabolism throughout the body. Most vitamin molecules attach to enzymes or coenzymes and help them

work properly. **Coenzymes** are organic, nonprotein catalysts that often act as "molecule carriers." Many enzymes or coenzymes are not functional without the appropriate vitamins attached to them. This attachment gives coenzymes the proper functional shape. For example, coenzyme A (CoA)—an important carrier molecule associated with the citric acid cycle—has *pantothenic acid* (vitamin B_5) as one of its major components.

Not all vitamins are involved directly with enzymes and coenzymes. Vitamins A, D, and E play a variety of different, but no less important, roles in the chemistry of the body. The form of vitamin A called *retinal*, for example, plays an important role in detecting light in sensory cells of the retina. Vitamin D can be converted to the hormone **calcitriol,** which plays a role in the regulation of calcium homeostasis in the body. One role of vitamin E (and vitamin C) is to serve as an antioxidant that prevents free radicals (highly reactive oxygen atoms) from damaging electron-dense molecules in the cell membranes and DNA molecules.

All but one vitamin, vitamin D, cannot be made by the body itself. Recent research suggests that vitamin D supplements may reduce risks for a range of diseases, including cancers of the breast, colon, ovaries, and prostrate. Bacteria living in the colon make two more: vitamin K and biotin. We must eat vitamins, or molecules we can convert into vitamins, in our food to get the rest. The body can store fat-soluble vitamins—A, D, E, and K—in the liver for later use. Because the body cannot store significant amounts of water-soluble vitamins such as B vitamins and vitamin C, they must be continually supplied in the diet. Table 22-2 lists some common vitamins, their sources and functions, and symptoms of deficiency.

Minerals

Minerals are at least as important as vitamins in our diet. **Minerals** are inorganic elements or salts that are found naturally in the earth. Like vitamins, mineral ions can attach to enzymes or other organic molecules and help them work. Of course, minerals such as sodium, chloride, and potassium are essential in relatively large amounts for maintaining the fluid/ion composition of the internal fluid environment.

Minerals such as sodium and calcium also function in nerve conduction and in the contraction of muscle fibers. Without these minerals, the brain, heart, and respiratory tract would cease to function. Iron is needed to manufacture hemoglobin in red blood cells, and iodine is needed to make thyroid hormones T_3 and T_4. Calcium, phosphorus, and magnesium are required to build the strong structural components of the skeleton. Information about some of the more important minerals is summarized for you in Table 22-3. Like vitamins, minerals are beneficial only when taken in the proper amounts.

Recommended *adequate intakes (AIs)* of minerals can change over the life span. For example, calcium intake should increase throughout childhood and remain high throughout adulthood. However, the actual intake of

TABLE 22-2 Major Vitamins

VITAMIN	DIETARY SOURCE	FUNCTIONS	SYMPTOMS OF DEFICIENCY
Vitamin A	Green and yellow vegetables, dairy products, and liver	Maintains epithelial tissue and produces visual pigments	Night blindness and flaking skin
B-complex vitamins			
B$_1$ (thiamine)	Grains, meat, and legumes	Helps enzymes in the citric acid cycle	Nerve problems (beriberi), heart muscle weakness, and edema
B$_2$ (riboflavin)	Green vegetables, organ meats, eggs, and dairy products	Aids enzymes in the citric acid cycle	Inflammation of skin and eyes
B$_3$ (niacin)	Meat and grains	Helps enzymes in the citric acid cycle	Pellagra (scaly dermatitis and mental disturbances) and nervous disorders
B$_5$ (pantothenic acid)	Organ meat, eggs, and liver	Aids enzymes that connect fat and carbohydrate metabolism	Loss of coordination (rare), decreased gut motility
B$_6$ (pyridoxine)	Vegetables, meats, and grains	Helps enzymes that catabolize amino acids	Convulsions, irritability, and anemia
B$_9$ (folic acid)	Vegetables	Aids enzymes in amino acid catabolism and blood production	Digestive disorders and anemia
B$_{12}$ (cyanocobalamin)	Meat and dairy products	Involved in blood production and other processes	Pernicious anemia
Biotin (vitamin H)	Vegetables, meat, and eggs	Helps enzymes in amino acid catabolism and fat and glycogen synthesis	Mental and muscle problems (rare)
Vitamin C (ascorbic acid)	Fruits and green vegetables	Helps in manufacture of collagen fibers; antioxidant	Scurvy and degeneration of skin, bone, and blood vessels
Vitamin D (calciferol)	Dairy products and fish liver oil; also made in the body from cholesterol	Aids in calcium absorption	Rickets and skeletal deformity
Vitamin E (tocopherol)	Green vegetables and seeds	Protects cell membranes from being destroyed; antioxidant	Muscle and reproductive disorders (rare)

TABLE 22-3 Major Minerals

MINERAL	DIETARY SOURCE	FUNCTIONS	SYMPTOMS OF DEFICIENCY
Calcium (Ca)	Dairy products, legumes, and vegetables	Helps blood clotting, bone formation, and nerve and muscle function	Bone degeneration and nerve and muscle malfunction
Chlorine (Cl$^-$)	Salty foods	Aids in stomach acid production and acid-base balance	Acid-base imbalance
Cobalt (Co)	Meat	Helps vitamin B$_{12}$ in blood cell production	Pernicious anemia
Copper (Cu)	Seafood, organ meats, and legumes	Involved in extracting energy from the citric acid cycle and in blood production	Fatigue and anemia
Iodine (I)	Seafood and iodized salt	Required for thyroid hormone synthesis	Goiter (thyroid enlargement) and decrease in metabolic rate
Iron (Fe)	Meat, eggs, vegetables, and legumes	Involved in extracting energy from the citric acid cycle and in blood production	Fatigue and anemia
Magnesium (Mg)	Vegetables and grains	Helps many enzymes	Nerve disorders, blood vessel dilation, and heart rhythm problems
Manganese (Mn)	Vegetables, legumes, and grains	Helps many enzymes	Muscle and nerve disorders
Phosphorus (P)	Dairy products and meat	Aids in bone formation and is used to make ATP, DNA, RNA, and phospholipids	Bone degeneration and metabolic problems
Potassium (K)	Seafood, milk, fruit, and meats	Helps muscle and nerve function	Muscle weakness, heart problems, and nerve problems
Sodium (Na)	Salty foods	Aids in muscle and nerve function and fluid balance	Weakness and digestive upset
Zinc (Zn)	Many foods	Helps many enzymes	Inadequate growth

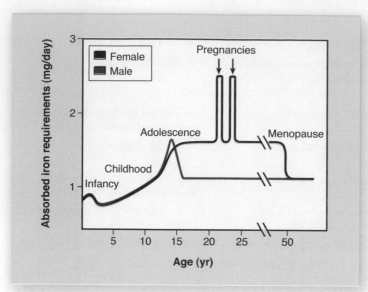

▲ **FIGURE 22-10** **Iron intake requirements.** The chart compares male and female absorbable iron requirements over the life span.

calcium among females in the United States tends to fall short during adulthood—thereby increasing the risk for osteoporosis and other disorders.

Figure 22-10 shows the requirement for iron over the life span for both men and women. Although both males and females require a large amount of iron during the spurt of growth in the teenage years, the iron requirement remains high only in women during the rest of adulthood. This difference is explained by the fact that adult women must continually replace the iron lost in the menstrual flow. Notice that female iron requirements drop to the level of males after menopause. Notice also that the iron requirement peaks during pregnancies—when fetal blood development requires large amounts of iron.

A&P CONNECT

One of the hottest areas in the field of nutrition today is that of *functional foods*. Find out what they are and why you may want to include them in your diet in **Functional Foods** online at **A&P Connect.**

evolve

QUICK ✓ CHECK

16. What is a vitamin? How do they function in the body?

17. List two functions of minerals in the body.

18. Why should the intake of iron be more important for women during the childbearing years?

METABOLIC RATES

Metabolic rate refers to the amount of energy released in the body in a given time by catabolism. Metabolic rates are

expressed in either of two ways: (1) in terms of the number of kilocalories of heat energy expended per hour or per day, or (2) as normal or as a definite percentage above or below normal.

Basal Metabolic Rate

The **basal metabolic rate (BMR)** is the body's rate of energy expenditure under "basal conditions," namely, when the individual is tested under the following conditions:

- Awake but resting, that is, lying down and, as far as possible, not moving a muscle.

- In the postabsorptive state (12 to 18 hours after the last meal).

- In a comfortably warm environment (the so-called *thermoneutral zone,* a temperature range at which metabolism is independent of ambient temperature).

Note that the BMR is not the minimum metabolic rate. It does not indicate the smallest amount of energy that must be expended to sustain life. It does, however, indicate the smallest amount of energy expenditure that can sustain life and also maintain the waking state and a normal body temperature in a comfortably warm environment.

Factors Influencing Basal Metabolic Rate

The BMR is not identical for all individuals because of the influence of various factors, some of which are described in the following paragraphs.

Most individuals have the same BMR *per square meter* of body surface, if other conditions are equal. However, because a larger individual has more square meters of surface area than does a smaller person, the BMR is greater than that of a smaller individual. Likewise, the higher the ratio of lean tissue to fat tissue in a person, the higher the BMR (Figure 22-11).

Other factors affecting BMR have to do with sex and age. Men oxidize their food approximately 5% to 7% faster than women do. This explains why their BMRs are about 5% to 7% higher for a given size and age. This gender difference in BMR probably results from the difference in the proportion

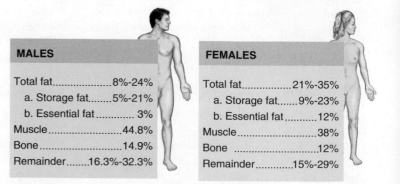

MALES		
Total fat		8%-24%
a. Storage fat		5%-21%
b. Essential fat		3%
Muscle		44.8%
Bone		14.9%
Remainder		16.3%-32.3%

FEMALES		
Total fat		21%-35%
a. Storage fat		9%-23%
b. Essential fat		12%
Muscle		38%
Bone		12%
Remainder		15%-29%

▲ **FIGURE 22-11** **Body composition.** Estimated values in healthy men and women.

of body fat, which is determined by sex hormones. In general, the younger the individual, the higher the BMR for a given size and sex.

Other factors affecting BMR are fever, drugs, and a person's physiological state. For example, fever increases the BMR. For every degree Celsius increase in body temperature, metabolism increases about 13%. A decrease in body temperature (hypothermia) has the opposite effect. In addition, certain drugs, such as caffeine, amphetamine, and levothyroxine, increase the BMR. Other factors, such as *emotions*, *pregnancy*, and *lactation* (milk production), also influence basal metabolism.

Total Metabolic Rate

Total metabolic rate is the amount of energy used or expended by the body in a given amount of time. It is often expressed in kilocalories per hour or per day. The main direct determinants of total metabolic rate are as follows:

Factor 1—the basal metabolic rate, which usually constitutes about 55% to 60% of the total metabolic rate.

Factor 2—the energy used to do skeletal muscle work.

Factor 3—the thermic effect of foods. The metabolic rate increases for several hours after a meal, apparently because of the energy needed for metabolizing foods.

Energy Balance and Body Weight

Our bodies maintain a state of energy balance, in which the body's energy input equals its energy output. Energy input per day equals the total calories (kilocalories) in the food ingested per day. Energy output equals the total metabolic rate expressed in kilocalories. If calorie intake and energy output are not equal, changes in body weight may occur:

- Body weight remains *constant* (except for possible variations in water content) when the total calories in the food ingested equal the total metabolic rate.

- Body weight *increases* when energy input exceeds energy output.

- Body weight *decreases* when energy input is less than energy output—when the total number of calories in the food eaten is less than the total metabolic rate.

Foods are stored primarily as glycogen and fats. Many cells (except for skeletal muscle) catabolize carbohydrates first, then fats. If there is no food intake, almost all of the glycogen is estimated to be used up in a matter of 1 or 2 days. Then, with no more carbohydrate to act as a fat sparer, fat is catabolized. The amount of fat available determines the length of time that an individual can catabolize fat as a reserve source of energy. Finally, with no more fat available, tissue proteins are catabolized. Because significant amounts of protein are not "stored" for use in catabolism, important structural and functional proteins are quickly depleted. For this reason, severe starvation will eventually lead to death.

MECHANISMS FOR REGULATING FOOD INTAKE

Mechanisms for regulating food intake are still not clearly established, though it is understood that the hypothalamus plays a major role in these mechanisms. A cluster of neurons in the lateral hypothalamus function as an **appetite center**—meaning that impulses from them bring about increased appetite.

It is likely that a group of neurons in the ventral medial nucleus of the hypothalamus functions as a **satiety center**—meaning that impulses from these neurons decrease appetite so that we feel satiated or, "full."

The temperature of the blood circulating to the hypothalamus is important in regulating the action of these centers. Another factor is blood glucose concentration and the rate of glucose use.

The hypothalamus also produces several hormones and neurotransmitters that affect the feeding centers. Some appetite-altering hormones and neurotransmitters are produced in many other organs, such as the liver, adipose tissue, pancreas, GI tract, and vagal nerve. Of course, factors such as daily eating habits or patterns, emotional responses, the sensations of food, and many others must also be involved in regulating or affecting appetite.

QUICK CHECK

19. Give one of the two ways in which metabolic rates can be expressed.
20. Name three of the factors that influence basal metabolic rate.
21. Distinguish between basal metabolic rate and total metabolic rate.
22. In which division of the brain would you find the control center for regulating food intake?

Cycle of LIFE ○

The importance of proper nutrition to an individual's well-being begins at the moment of conception and continues until death. In the womb, various nutrients must be obtained from the mother's blood in sufficient quantity to ensure normal growth and development.

One critical nutrient during fetal development, infancy, and childhood is protein. Sufficient proteins, containing all the essential amino acids, are required to permit normal development of the nervous system, muscle tissues, and other vital structures.

Another critical nutrient during the early years of life is the mineral calcium. Large quantities of calcium are needed by a growing body to maintain normal development of the skeleton and other tissues. In the womb, a steady supply of calcium in the mother's blood is maintained by increased levels of parathyroid hormone (PTH). PTH increases blood calcium levels by removing it from storage in the bones.

The BIG Picture

Every cell in the body must maintain the operation of its metabolic pathways to ensure its survival. Anabolic pathways are required to build the various structural and functional components of the cells. Catabolic pathways are required to convert energy to a usable form. Catabolic pathways are also needed to degrade large molecules into small subunits that can be used in anabolic pathways. These processes require the correct amounts of carbohydrates, fats, proteins, vitamins, and minerals in order to produce the structural and functional components necessary for cellular metabolism.

Various body systems operate to make sure that essential nutrients reach the cells as needed to maintain metabolism and homeostasis. For example, the nervous, skeletal, and muscular systems make it possible for us to take in complex foods from our external environment. The digestive system reduces these complex nutrients to simpler, more usable nutrients—then provides the

mechanisms that allow their absorption into the internal environment. The circulatory and lymphatic systems transport absorbed nutrients to individual cells for immediate use or to the liver or other organs for temporary storage. The endocrine system regulates the balance between immediate use and storage. The respiratory and circulatory systems provide the oxygen needed for oxidative phosphorylation to generate ATP. These two systems also provide a mechanism for removing waste CO_2 generated by the catabolism of nutrient molecules. Likewise, the urinary system provides a mechanism for removing waste urea generated by protein catabolism. Even the integumentary system becomes involved, by producing vitamin D in the presence of sunlight.

It should be easy for you to see now why metabolism is simply the sum total of all the biochemical processes required by a living organism!

Unless a pregnant woman consumes enough calcium to replace this calcium lost from bones, she may suffer from the bone-softening effects of calcium deficiency. If proteins, calcium, or other necessary nutrients are in short supply anytime before the beginning of adulthood, the consequences may be permanent. For example, bone deformities resulting from a lack of calcium during childhood could become permanent if not corrected or compensated for before the skeleton ossifies completely.

In late adulthood, the number of food calories needed declines because the metabolic rate declines due to changes in the balance of some metabolic hormones. Even though the number of required food calories declines, the overall balance of nutrients consumed must be maintained to preserve proper metabolic function. Some nutrients, such as calcium, may be needed in greater quantity in older adults to compensate for age-related bone loss. ●

MECHANISMS OF DISEASE

Nutritional and metabolic disorders are numerous and widespread in the United States. A number of inherited conditions are well known, but not common. Perhaps the most common disorder is obesity, followed by eating disorders such as anorexia nervosa and bulimia, which are well known because of media coverage. Beyond these conditions, however, there are serious nutritional disorders, including protein-calorie malnutrition and a number of vitamin deficiency disorders that were once common in this country and still can be seen in many developing countries around the world.

evolve   Find out more about these nutritional and metabolic disorders and diseases online at *Mechanisms of Disease: Nutrition and Metabolism.*

LANGUAGE OF SCIENCE AND MEDICINE   *continued from page 517*

lipogenesis (lip-oh-JEN-eh-sis)
  [*lipo-* fat, *-gen-* produce, *-esis* process]

lipoprotein (lip-oh-PROH-teen)
  [*lipo-* fat, *prote-* primary, *-in* substance]

macronutrient (MAK-roh-NOO-tree-ent)
  [*macro-* large, *-nutri-* nourish, *-ent* agent]

metabolic rate (met-ah-BOL-ik)
  [*metabol-* change, *-ic* relating to]

metabolism (meh-TAB-oh-liz-em)
  [*metabol-* change, *-ism* process]

micronutrient (MYE-kroh-NOO-tree-ent)
  [*micro-* small, *-nutri-* nourish, *-ent* agent]

mineral
  [*mineral-* mine]

nutrition (noo-TRIH-shun)
  [*nutri-* nourish, *-tion* process]

oxidative phosphorylation (ahk-sih-DAY-tiv fos-for-ih-LAY-shun)
  [*oxi-* sharp (oxygen), *-id-* chemical *(-ide)*, *-at-* action of *(-ate)*, *-ive* relating to, *phos-* light, *-phor-* carry, *-yl-* chemical, *-ation* process]

phosphorylation (fos-for-ih-LAY-shun)
  [*phos-* light, *-phor-* carry, *-yl-* chemical, *-ation* process]

satiety center (sah-TYE-eh-tee SEN-ter)
  [*sati-* enough, *-ety* state]

saturated (SATCH-yoo-ray-ted)

total metabolic rate (met-ah-BOL-ik)
  [*metabol-* change, *-ic* relating to]

triglyceride (try-GLISS-er-yde)
  [*tri-* three, *-glycer-* sweet, *-ide* chemical]

unsaturated (un-SATCH-yoo-ray-ted)

vitamin (VYE-tah-min)
  [*vita-* life, *-amin-* ammonia compound]

CHAPTER SUMMARY

To download an MP3 version of the chapter summary for use with your iPod or other portable media player, access the Audio Chapter Summaries online at http://evolve.elsevier.com.

HINT Scan this summary after reading the chapter to help you reinforce the key concepts. Later, use the summary as a quick review before your class or before a test.

OVERVIEW OF NUTRITION AND METABOLISM

A. Nutrition refers to the foods that we eat and the types of nutrients they contain
   1. Malnutrition—deficiency or imbalance in the consumption of food, vitamins, and minerals
B. Categories of nutrients
   1. Macronutrients—nutrients that we need in large amounts
      a. Carbohydrates, fats, and proteins
      b. Water
      c. Minerals
   2. Micronutrients—nutrients that we need in very small amounts
      a. Vitamins
      b. Minerals (trace elements)
C. Metabolism—complex, interactive set of chemical processes that make life possible; the use the body makes of foods and their nutrients after they have been digested, absorbed, and transported to the cells of our bodies
   1. Assimilation—occurs when nutrient molecules enter cells and undergo many chemical changes
   2. Two major metabolic processes:
      a. Catabolism—breaks food molecules down into smaller molecular compounds; releases energy from them (heat and chemical)
      b. Anabolism—a synthesis process
      c. Both catabolism and anabolism go on continually and concurrently and take place inside cells
   3. ATP supplies energy directly to the energy-using reactions of all cells in all living cellular organisms (Figure 22-2)

CARBOHYDRATES

A. Dietary sources of carbohydrates
   1. Polysaccharides—starches found in vegetables, grains, and other plant tissues
   2. Cellulose—major component of most plant tissues; passes through our digestive system without being broken down
   3. Disaccharides—found in refined sugar; must be chemically digested before they can be absorbed
   4. Monosaccharides—found in fruits; move directly into the internal environment without being processed; glucose
B. Carbohydrate metabolism—body metabolizes carbohydrates by both catabolic and anabolic processes
C. Glucose transport and phosphorylation—glucose reacts with ATP to form glucose 6-phosphate
   1. Glucose phosphorylation—prepares glucose for further metabolic reactions
   2. Phosphorylation—process of adding phosphate group to a molecule

D. Glycolysis—first step in the process of carbohydrate catabolism (Figure 22-3)
   1. Occurs in cytoplasm
   2. Produces small amount of ATP
   3. Prepares glucose for the citric acid cycle
E. Citric acid cycle—series of chemical reactions mediated by enzymes; converts two acetyl molecules to four carbon dioxide and six water molecules
   1. Occurs in the mitochondria
   2. Most of the energy leaving the citric acid cycle is "stored" in high-energy electrons
F. Electron transport and oxidative phosphorylation
   1. High-energy electrons (along with their protons) removed during the citric acid cycle enter a chain of carrier molecules (Figure 22-4)
   2. As electrons are transported, some of their energy is used to pump their accompanying protons (H^+) to the intramembrane space between the inner and outer membranes of the mitochondrion
   3. Protons temporarily store energy; move down their concentration gradient across the inner membrane, driving ATP synthase
   4. Low-energy electrons (e^-) and their protons (H^+) join oxygen, forming water; joining of a phosphate group to ADP to form ATP is called oxidative phosphorylation
G. Control of glucose metabolism—levels of sugar in the blood are under hormone control (Figure 22-8)
   1. Hyperglycemic hormones—promote a high blood glucose concentration
   2. Hypoglycemic hormones—decrease blood glucose level

LIPIDS

A. Dietary sources of lipids
   1. Triglycerides—most common lipid; composed of a glycerol subunit attached to three fatty acids
   2. Phospholipids—important lipids; found in nearly all foods
   3. Cholesterol—an important lipid; found only in foods of animal origin
   4. Dietary fats
      a. Saturated—contain fatty acid chains in which there are no double bonds
      b. Unsaturated—contain fatty acid chains in which there are double bonds
B. Transport of lipids—lipids are transported in blood as chylomicrons, lipoproteins, and free fatty acids
   1. Chylomicrons—small fat droplets found in blood soon after fat absorption
   2. Lipoproteins are produced by the liver and consist of lipids and proteins
   3. Fatty acids are transported from cells of one tissue to those of another in the form of free fatty acids
C. Lipid metabolism
   1. Lipid catabolism—consists of several processes
   2. Lipid anabolism (lipogenesis)—consists of the synthesis of various types of lipids; triglycerides, cholesterol, phospholipids, and prostaglandins
D. Control of lipid metabolism—controlled mainly by the following hormones:
   1. Insulin
   2. ACTH
   3. Growth hormone
   4. Glucocorticoids

PROTEINS

A. Sources of proteins
1. Proteins are assembled from 20 different kinds of amino acids
2. Body synthesizes amino acids from other compounds in the body
3. Only about half of the required 20 types of amino acids can be made by the body (nonessential amino acids); rest must be supplied through diet (essential amino acids)

B. Protein metabolism
1. Protein anabolism—process by which proteins are synthesized by the ribosomes of the cells
2. Protein catabolism—deamination takes place in liver cells
   a. Consists of the splitting off of an amino (NH_2) group from an amino acid molecule to form a molecule of ammonia and one of keto acid (e.g., alpha-ketoglutaric acid)
3. Control of protein metabolism—protein metabolism is controlled largely by hormones

VITAMINS AND MINERALS

A. Vitamins—organic molecules needed in small quantities for normal metabolism throughout the body; attach to enzymes or coenzymes (Table 22-2)
1. Coenzymes—organic, nonprotein catalysts that often act as "molecule carriers"
2. The body does not make most of the necessary vitamins; they must be obtained through diet

B. Minerals—inorganic elements or salts that are found naturally in the earth; attach to enzymes or other organic molecules and help them work (Table 22-3)
1. Recommended adequate intakes (AIs) of minerals can change over the life span (Figure 22-10)

METABOLIC RATES

A. Metabolic rate—refers to the amount of energy released in the body in a given time by catabolism
1. Metabolic rates are expressed in either of two ways:
   a. In terms of number of kilocalories of heat energy expended per hour or per day
   b. As normal or as a definite percentage above or below normal

B. Basal metabolic rate (BMR)—the body's rate of energy expenditure under these "basal conditions":
1. Awake but resting
2. In the postabsorptive state
3. In a comfortably warm environment

C. Factors influencing basal metabolic rate—BMR is not identical for all individuals because of the influence of various factors
1. Most people have the same BMR per square meter of body surface, but larger people have larger body surface, so BMR is greater
2. Gender differences based on proportions of body fat
3. Other factors are age, fever, drugs, physiological state

D. Total metabolic rate—the amount of energy used or expended by the body in a given amount of time
1. Factor 1—basal metabolic rate
2. Factor 2—energy used to do skeletal muscle work
3. Factor 3—thermic effect of foods

E. Energy balance and body weight—the body maintains a state of energy balance
1. Energy input equals its energy output
   a. Body weight remains constant when the total calories in the food ingested equals the total metabolic rate
   b. Body weight increases when energy input exceeds energy output
   c. Body weight decreases when energy input is less than energy output
   d. In starvation, the carbohydrates are used up first, then fats, then proteins

MECHANISMS FOR REGULATING FOOD INTAKE

A. Hypothalamus plays a major role in the mechanism of regulating food intake
1. Appetite center—cluster of neurons in the lateral hypothalamus function to bring about increased appetite
2. Satiety center—impulses from a group of neurons in the ventral medial nucleus of the hypothalamus decrease appetite; we feel satiated, or "full"

REVIEW QUESTIONS

Write out the answers to these questions after reading the chapter and reviewing the Chapter Summary. If you simply think through the answer without writing it down, you won't retain much of your new learning.

HINT

1. What is metabolism? Nutrition?
2. What two processes make up the process of metabolism?
3. Does the body digest dietary fiber? Why or why not?
4. Briefly describe glycolysis, the first process of carbohydrate catabolism.
5. Where does glycolysis occur?
6. How are dietary fats classified?
7. Explain how lipids are transported in blood.
8. List the hormones involved in the control of lipid metabolism.
9. What are the essential amino acids?
10. What does the term *metabolic rate* mean?
11. List the various factors that influence basal metabolic rate.
12. Describe various factors that influence the amount of food a person eats.

CRITICAL THINKING QUESTIONS

After finishing the Review Questions, write out the answers to these items to help you apply your new knowledge. Go back to sections of the chapter that relate to items that you find difficult.

HINT

1. How would you describe the mitochondria, and why do you think they are referred to as the "power plants" of the cells?
2. Can you identify and explain the processes and hormones involved in maintaining the homeostatic level of glucose in the blood?
3. State in your own words the process of lipid catabolism. How is it similar to the carbohydrate pathway?
4. Describe protein catabolism in your own words.
5. How would you compare and contrast the functions of proteins, carbohydrates, and fats?
6. What is the function of most vitamins in the body? What examples can you find that do not have this function? What functions do these vitamins have?
7. What is the difference between basal metabolic rate and total metabolic rate?

continued from page 517

Now that you have read this chapter, see if you can answer these questions about the new "pizza diet" Walter wanted to invent.

1. Which item doesn't belong in Walter's list?
   a. Protein
   b. Lipid
   c. Vitamin C
   d. Carbohydrate

2. Which of the minerals Walter listed is NOT considered a macronutrient?
   a. Sodium
   b. Carbon
   c. Potassium
   d. All are considered macronutrients

3. The olive oil on pizza crust is mostly triglyceride lipids containing monounsaturated fatty acids. What is the first step in catabolizing the triglycerides in the olive oil?
   a. Conversion to glycerol and three fatty acids
   b. Glycolysis
   c. Lipogenesis
   d. Deamination

HINT *To answer these questions, you may have to refer to the glossary or index, other chapters in this textbook, A&P Connect, Mechanisms of Disease, and other resources.*

UNIT 5

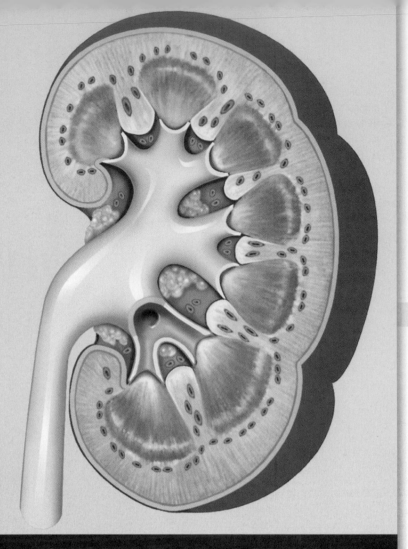

STUDENT LEARNING OBJECTIVES

At the completion of this chapter, you should be able to do the following:

1. Describe the overall anatomy of the urinary system, listing the primary structures.
2. Explain how the components of the urinary system *work together* to maintain fluid homeostasis.
3. Describe the basic structure of the kidney.
4. List the microscopic structures of the nephron and give their functions.
5. Give an overview description of the following terms: *filtration, tubular reabsorption, tubular secretion.*
6. Discuss the mechanism of filtration and the formation of urine.
7. Discuss the importance of fluid and electrolyte balance in the body.
8. List the ways in which water enters and leaves the body.
9. Discuss the significance of water and electrolyte regulation in the ICF.
10. Discuss the significance of the regulation of sodium and potassium levels in the body fluids.
11. List the mechanisms that control pH of body fluids.

CHAPTER

23

Urinary System and Fluid Balance

CHAPTER OUTLINE

 Scan this outline before you begin to read the chapter, as a preview of how the concepts are organized.

THE southern sun beat down on Whitney's head as she and her classmates cleared rubble around what used to be their college's courtyard before the hurricane. Her shirt was soaked with sweat, and she kept having to wipe her forehead as sweat dripped into her eyes. Whitney knew she should be drinking more water, but the cooler was on the bus—two blocks away. She just didn't have the energy to face that walk. "We'll be finished in just another hour," she thought. As she bent down to pick up a cement block, she suddenly felt dizzy and sat down quickly.

From what you know from your life experiences, what is your best diagnosis for Whitney's condition? Based on what you think you know, what would you do for Whitney? Try to answer the questions at the end of this chapter and see if you are right on target (or need to review!).

• • •

The urinary system not only produces urine, it also balances the composition of our blood plasma. The water content is adjusted, and important ions circulating in the blood (such as sodium and potassium) are maintained at appropriate levels. Even the pH of the blood can be controlled. In these ways, the urinary system regulates the content of blood plasma so that the homeostasis of the entire internal fluid environment is maintained continuously within normal limits.

In this chapter, we will explore the basic principles of urinary structure and function. We will also look at the role of the kidneys and other organs that maintain fluid and ion homeostasis. Finally, we will briefly tour the mechanisms of homeostasis of body fluid and electrolyte levels. You'll see that fluid balance is critical to our survival. The volume of fluid and the electrolyte concentrations inside our cells, in the interstitial spaces, and in our blood vessels must all remain relatively constant.

Last, we discuss the importance of acid-base balance and pH control of body fluids and the mechanisms that provide for homeostatic control of these processes.

ANATOMY OF THE URINARY SYSTEM

Gross Structure

The principal organs of the urinary system are the **kidneys,** which process blood and form urine as a waste to be *excreted*—that is, removed from the body (Box 23-1). The excreted urine travels from the kidneys to the outside of the body via accessory organs: the *ureters, urinary bladder,* and *urethra.*

Kidney

Our kidneys really do resemble kidney beans. As you can see from Figure 23-1, they are roughly oval and about the size of a mini-football. Usually the right kidney is slightly smaller and a little lower than the left, because the liver takes up some of the space above the right kidney.

LANGUAGE OF SCIENCE AND MEDICINE

HINT *Before reading the chapter, say each of these terms out loud. This will help you avoid stumbling over them as you read.*

acid-base balance (ASS-id bays BAL-ents)
[*acid-* sour, *-base* foundation]

acidity (ah-SID-ih-tee)
[*acid-* sour, *-ity* state]

acidosis (ass-ih-DOH-sis)
[*acid-* sour, *-osis* condition]

aldosterone (AL-doh-steh-rohn or al-DAH-stair-ohn)
[*aldo-* aldehyde, *-stero-* solid or steroid derivative, *-one* chemical]

alkaline (AL-kah-lin)
[*alkal-* ashes, *-ine* relating to]

alkalinity (al-kah-LIN-ih-tee)
[*alkal-* ashes, *-in-* relating to, *-ity* state]

alkalosis (al-kah-LOH-sis)
[*alkal-* ashes, *-osis* condition]

anion (AN-eye-on)
[*ana-* up, *-ion* go (ion)]

antidiuretic hormone (ADH) (an-tee-dye-yoo-RET-ik HOR-mohn)
[*anti-* against, *-dia-* through, *uret* urination, *-ic* relating to, *hormon-* excite]

atrial natriuretic hormone (ANH) (AY-tree-al nay-tree-yoo-RET-ik HOR-mohn)
[*atrium* entrance courtyard (atrium of heart), *natri-* sodium, *-ure-* urine, *-ic* relating to, *hormon-* excite]

Bowman capsule (BOH-men KAP-sul)
[*William Bowman* English anatomist]

calyx (KAY-liks)
[*calyx* cuplike] *pl.,* calyces (KAY-lih-seez)

cation (KAT-eye-on)
[*cat-* down, *-ion* go (ion)]

collecting duct (CD)

cortical nephron (KOHR-tih-kal NEF-ron)
[*cortic-* bark (cortex), *-al* relating to, *nephro-* kidney, *-on* unit]

countercurrent mechanism (kown-ter-KER-ent MEK-ah-nih-zem)
[*counter-* against, *-current* flow, *mechan-* machine, *-ism* state]

dehydration (dee-hye-DRAY-shun)
[*de-* remove, *-hydro-* water, *-ation* process]

detrusor (dee-TROO-sor)
[*detrus-* thrust, *-or* agent]

dissociate (dih-SOH-see-ayt)
[*dis-* apart, *-socia-* unite, *-ate* action]

distal convoluted tubule (DCT) (DIS-tall KON-vo-LOO-ted TOO-byool)
[*dist-* distance, *-al* relating to, *con-* together, *-volut-* roll, *tub-* tube, *-ul* little]

continued on page 556

BOX 23-1 FYI

Excretion

The urinary system's chief function is to regulate the volume and composition of body fluids and excrete unwanted material. However, it is not the only system in the body that is able to excrete unneeded substances.

The table below compares the excretory functions of several systems. Although all of these systems contribute to the body's effort to remove wastes, only the urinary system can finely adjust the water and electrolyte balance to the degree required for normal homeostasis of body fluids.

SYSTEM	ORGAN	EXCRETION
Urinary	Kidney	Nitrogen compounds; Toxins; Water; Electrolytes
Integumentary	Skin—sweat glands	Nitrogen compounds; Electrolytes; Water
Respiratory	Lung	Carbon dioxide; Water
Digestive	Intestine	Digestive wastes; Bile pigments; Salts of heavy metals

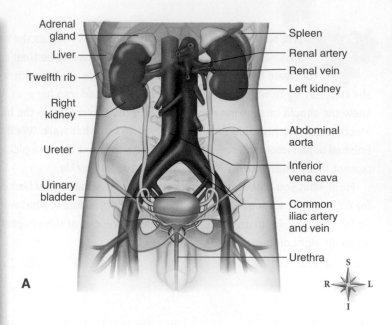

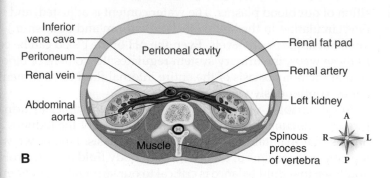

The medial surface of each kidney has a concave notch called the **hilum** (Figure 23-2, *A*). Vessels and other structures enter or leave the kidney through this notch.

The kidneys lie in a *retroperitoneal* position (meaning *behind* the parietal peritoneum) against the posterior wall of the abdomen (see Figure 23-1, *B*). Note in Figure 23-1, *A*, that the superior or upper portions (poles) of both kidneys extend above the level of the twelfth rib and the lower edge of the thoracic parietal pleura. In addition to being partly protected by the ribs, this anatomical relationship has important clinical implications (Box 23-2).

A heavy cushion of fat—the *renal fat pad*—encases each kidney and helps to hold it in position. Connective tissue, the *renal fascia*, anchors the kidneys to surrounding structures and also helps maintain their normal position. A tough, white fibrous capsule encases each kidney (see Figure 23-2, *A*).

The coronal section dividing the right kidney in Figure 23-2, *A*, into anterior and posterior sections shows you the major

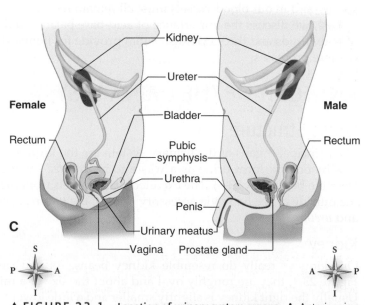

▲ **FIGURE 23-1** **Location of urinary system organs. A,** Anterior view of the urinary organs with the peritoneum and visceral organs removed. **B,** Horizontal (transverse) section of the abdomen showing the retroperitoneal position of the kidneys. **C,** Sagittal view of the female and male urinary tract. Each shows a distended (stretched) bladder.

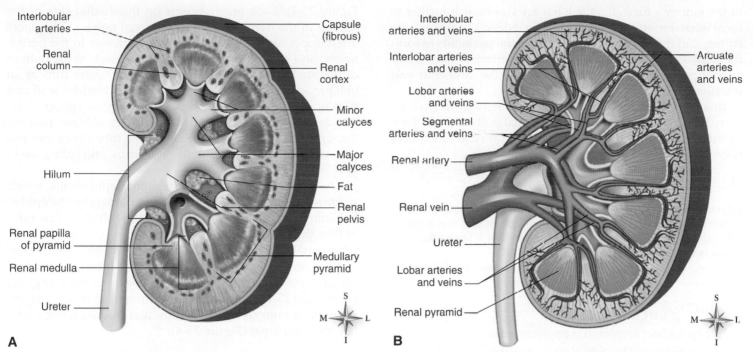

▲ **FIGURE 23-2** **Internal structure of the kidney. A,** Coronal section of the right kidney in an artist's rendering. **B,** Circulation of blood through the kidney. Diagram showing the major arteries and veins of the renal circulation.

internal structures of the kidney. Stop for a moment and identify the **renal cortex,** or outer region, and the **renal medulla,** or inner region. Note that a dozen or so distinct triangular wedges, the **renal pyramids,** make up much of the medullary tissue. The *base* of each pyramid faces outward, and the narrow *papilla* of each faces toward the hilum. Each renal papilla has multiple openings that release urine. Notice that the cortical tissue dips into the medulla between the pyramids, forming areas known as **renal columns.**

Each renal papilla (point of a pyramid) juts into a cuplike structure called a **calyx.** The calyces are considered the beginnings of the "plumbing system" of the urinary system. Urine (leaving the renal papilla) is collected by the calyces for transport out of the kidney. The cups that drain

the renal papillae directly are called *minor calyces.* These minor calyces are stemlike branches that join together to form larger branches called *major calyces.* The major calyces join together to form a large collection "basin" called the **renal pelvis.** The pelvis of the kidney narrows as it exits the hilum to become the tubelike ureter, which then transports urine to the bladder.

Blood Vessels of the Kidneys

Use Figure 23-2, *B,* to help you as you read the following paragraph. First, notice that a large branch of the abdominal aorta—the *renal artery*—brings blood into each kidney. As it nears the kidney, it divides into *segmental arteries,* which then divide to become *lobar arteries.* Between the pyramids

BOX 23-2 | Health Matters

Kidney Biopsy

Suspected disease of the kidney, such as renal cancer, often requires a *needle biopsy* to confirm the diagnosis. In these procedures, a hollow needle is inserted through the skin surface and then guided into the diseased organ to withdraw a tissue sample for analysis. In a renal biopsy, tissue is removed from the lower rather than the upper or superior end of the diseased kidney. This avoids the possibility of the biopsy needle damaging the pleura and thereby causing a *pneumothorax* (air or gas collects in the pleural cavity).

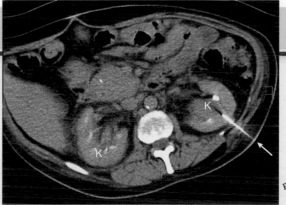

▲ **Needle biopsy of kidney.** Magnetic resonance image (MRI) of biopsy needle *(arrow)* inserted into the left kidney to remove tissue. *K,* Kidney.

of the kidney's medulla, the lobar arteries branch further to form *interlobar arteries*. These arteries extend out toward the cortex, and then arch over the bases of the pyramids to form the *arcuate arteries*. From the arcuate arteries, *interlobular arteries* penetrate the cortex. (Note that the interlobar and interlobular arteries are *different* arteries.)

Branches of the interlobular arteries called *afferent arterioles* carry blood directly to the tiny functional units of the kidney called *nephrons*. We will discuss the structure and function of nephrons later in this chapter.

A&P CONNECT

Knowing the pathway of blood flow in the kidney is important for understanding how the kidneys work. We continue the story later in the chapter. We put all the pieces together in an overview in **Tracing Blood Flow in the Kidney** online at **A&P Connect**.

evolve

QUICK CHECK

1. Name the basic components of the urinary system.
2. Name two general functions of the urinary system.
3. Distinguish between the renal cortex and the renal medulla.
4. Which branch of the aorta brings blood to the kidney?

Ureter

The two **ureters** (about 28 to 34 cm in length) are tubes that transport urine from the kidneys to the urinary bladder (see Figure 23-1). Each ureter begins on the medial side of the kidney, at the narrow outlet of the renal pelvis. The ureters are retroperitoneal (they lie behind or dorsal to the peritoneum) and pass downward until they attach to the bottom of the bladder (see Figure 23-1, *C*). Each ureter runs at an oblique angle for about 2 cm through the bladder wall and then opens at the outside angles of the *trigone* (floor) of the bladder (Figure 23-3, *A*). Because of its oblique passage through the bladder wall, the end of the tube closes and acts as a valve when the bladder is full, thus preventing backflow of urine (Figure 23-3, *B*).

The ureter is lined with transitional epithelium, which permits significant stretching without damage to the epithelial lining. This feature also permits either high or low rates of flow through the ureters.

In females, the ureters are in close proximity to the ovaries and cervix of the uterus. In males, the ureters are in close proximity to the seminal vesicles and near the prostate gland (see Figure 23-1, *C*). Each ureter is composed of three layers of tissue: a mucous lining, a muscular middle layer, and a fibrous outer layer (Figure 23-4).

Urinary Bladder

The urinary bladder is a muscular, collapsible bag located directly behind the pubic symphysis and in front of the rectum (see Figure 23-1, *C*). It lies below the parietal peritoneum, which covers only its superior surface (see Figure 23-3, *A*). In women, the bladder sits on the anterior surface of the vagina and in front of the uterus, whereas in men, it rests on the prostate.

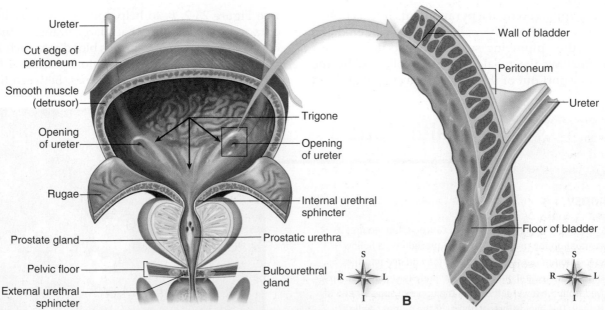

▲ **FIGURE 23-3** **Structure of the urinary bladder.** **A,** Frontal view of a dissected urinary bladder (male) in a fully distended state. **B,** The oblique path of the ureter through wall of the bladder, seen in frontal section, permits the junction to act as a valve—both reducing flow into the bladder as it becomes full and eliminating backflow into the kidney.

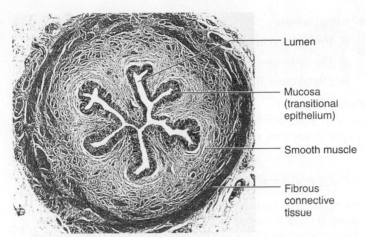

Lumen

Mucosa (transitional epithelium)

Smooth muscle

Fibrous connective tissue

▲ **FIGURE 23-4   Ureter (cross section).** Low-power micrograph. Note the convoluted folds, covered by a transitional mucous lining, that almost fill the lumen. The thick muscular layer is surrounded by a tough fibrous coat.

The wall of the bladder is made mostly of smooth muscle tissue collectively called the **detrusor** (see Figure 23-3, *A*). This smooth muscle layer is formed by a crisscrossing network of circular, oblique, and longitudinal bundles. The bladder is lined with mucous transitional epithelium that forms folds called *rugae* (see Figure 23-3, *A*). Because of the folds and the extensibility of transitional epithelium, the bladder can distend considerably. There are three openings in the floor of the bladder—two from the incoming ureters and one from the outgoing **urethra.** The ureter openings lie at the posterior corners of the triangle-shaped floor—the **trigone.** The urethral opening lies at the anterior, lower corner.

The bladder performs two major functions:

1. It serves as a reservoir for urine before it leaves the body.
2. Aided by the urethra, it expels urine from the body.

Urethra

The urethra is a small tube lined with mucous membrane. It leads from the floor of the bladder *(trigone)* to the exterior of the body. In females, the urethra extends downward and forward from the bladder for a distance of about 3 cm (1.2 inches), and ends at the external urinary meatus (see Figure 23-1, *C*). The female urethral tract is separate from the lower reproductive tract (vagina), which lies just behind the urethra.

The male urethra, on the other hand, extends along a winding path for about 20 cm (7.9 inches) (see Figure 23-1, *C*). It passes through the center of the *prostate gland* just after leaving the bladder. Within the prostate, it is joined by two *ejaculatory ducts*. After leaving the prostate, the urethra first extends downward, then forward, then upward to enter the base of the penis. The urethra then travels through the center of the penis and ends as a *urinary meatus* at the tip of the penis.

Because the male urethra is joined by the ejaculatory ducts, it also serves as a pathway for *semen* (fluid containing sperm) as it is ejaculated out of the body through the penis. In reality, then, the male urethra is a part of two *different* systems: the urinary system (when it is used to void urine) and the reproductive system (when it is used to ejaculate semen). Urine is prevented from mixing with semen during ejaculation by a reflex closure of sphincter muscles guarding the bladder's opening (see Figure 23-3, *A*).

Micturition

The mechanism for **micturition** (urination) begins with involuntary contractions of the detrusor muscle. As the pressure of urine against the inside of the bladder wall increases with urine volume, involuntary micturition contractions develop. This rapid succession of involuntary contractions gets stronger and stronger as the bladder fills and the urine volume and pressure increase. At the same time, the *internal urethral sphincter* muscles relax.

The internal urethral sphincters include a ringlike part of the detrusor muscle of the bladder wall, as you can see in Figure 23-3, *A*. The relaxation of these internal sphincters along with the micturition contractions of the bladder wall can force urine out of the bladder and through the urethra. Note that the skeletal muscles of the pelvic floor, including the *levator ani* muscle, act as *external urethral sphincters*.

A&P CONNECT

The urinary tract can be examined using medical imaging techniques. Check out **Visualizing the Urinary Tract** online at **A&P Connect** to find out how.

⊝volve

QUICK ✓ CHECK

5. Where does the ureter enter the bladder? Why is its angle through the wall significant?
6. What type of mucous epithelium lines most of the urinary tract? What is the functional advantage of this type of epithelium?
7. What are the two major functions of the bladder?

Voluntary control of micturition is not possible until the nervous system matures sufficiently. For this reason, voluntary control of urination is not possible during infancy and very early childhood.

Microscopic Structure

Over a million microscopic functional units called **nephrons** make up the bulk of each kidney. The shape of the nephron is uniquely suited to its function of blood plasma processing and

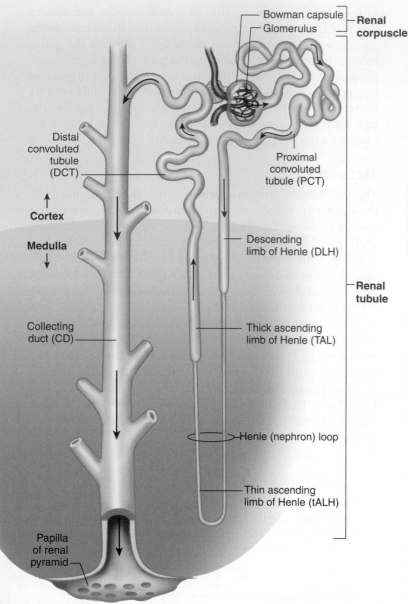

▲ **FIGURE 23-5** **Nephron.** The nephron is the basic functional unit of the kidney. *Arrows* show the direction of flow within the nephron.

urine formation (Figure 23-5). It resembles a tiny funnel with a long, winding stem about 3 cm (1.2 inches) long. Notice that each nephron is made up of two main regions: the *renal corpuscle* and the *renal tubule.* Fluid is filtered out of the blood plasma in the renal corpuscle. Then the **filtrate** (filtered fluid) flows through the renal tubule and collecting duct—where much of the filtrate is returned to the blood. The remaining filtrate leaves the collecting duct as urine. We will come back to the processing of filtrate later. For now, let's explore the structure of the nephron and collecting duct. Here are the main structures, listed in the order in which fluid flows through them:

Nephron

- Renal corpuscle
  - Glomerulus (raspillaries)
  - Bowman capsule

- Renal tubule
  - Proximal convoluted tubule
  - Henle loop
  - Distal convoluted tubule

Collecting duct

As you read the brief description of each of these microscopic structures, refer often to Figure 23-5.

> **A&P CONNECT**
>
> For our purposes, the simplified scheme of microscopic renal anatomy shown in Figure 23-5 works well. However, some renal biologists prefer the more elaborate scheme sketched out in a **Detailed Map of Nephron** online at **A&P Connect.**
>
> *evolve*

Nephron

Renal Corpuscle

The **renal corpuscle,** comprising the *Bowman capsule* and the *glomerulus,* is the first part of the nephron. Formation of a renal corpuscle is sometimes compared to pushing your fist into the end of an inflated balloon, as illustrated for you in Figure 23-6, *A.* Note that as the glomerular tuft of capillaries pushes into the balloon, it becomes surrounded by a double-walled cup with *parietal* (outer) and *visceral* (inner) walls—the Bowman capsule (Figure 23-6, *B*). Fluid from the blood plasma first filters out of the glomerulus and then into the Bowman capsule.

Bowman Capsule

The **Bowman capsule** *(glomerular capsule)* is the cup-shaped mouth of a nephron. The capsule is formed by two layers of epithelial cells with a space, called the *Bowman space (capsular space),* between them. Fluids, waste products, and electrolytes that pass through the porous glomerular capillaries and enter this space constitute the *filtrate.* These substances will be processed in the nephron to form urine.

The *parietal wall* is composed of simple squamous epithelium. It plays no role in the production of glomerular filtrate. However, the *visceral wall* is quite different. It is composed of special epithelial cells called **podocytes** (meaning "cells with feet"). Look at the scanning electron micrograph in Figure 23-6, *C.* Notice that the primary branches extending from the cell bodies divide into a network of smaller branches that end in little "feet" called *pedicels.* The pedicels are packed so closely together that only narrow slits of space lie between them. These spaces are called **filtration slits.** The slits are not merely open spaces, however. Within them is a mesh of fine connective tissue fibers collectively called the *slit diaphragm* that prevents the slits from enlarging under pressure and at the same time maintains permeability of the slit. The slit diaphragm is vital to filtration because it prevents many large macromolecules, such as proteins, from passing through.

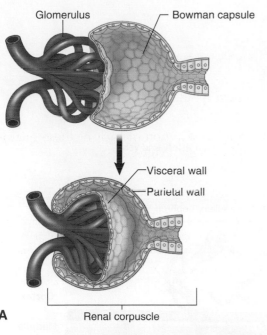

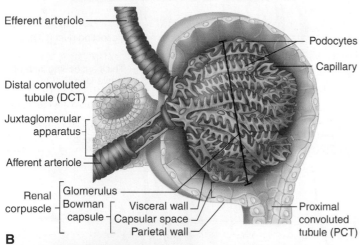

A FIGURE 23-6 Renal corpuscle. A, Model of the anatomical relationship of the glomerulus to the Bowman capsule. **B,** Structure of the renal corpuscle. **C,** Scanning electron micrograph (SEM) of the glomerular capillaries. A podocyte cell body *(CB)* and pedicels *(P)* are shown on the outer surface of a capillary endothelial cell.

Glomerulus

The **glomerulus** capillary network is vital to our survival. Its relationship to the Bowman capsule is clearly visible in Figure 23-6, *A* and *B*. Notice in both figures that an afferent arteriole leads into the glomerular network and an efferent arteriole leads out.

Like all capillaries, glomerular capillaries have thin, membranous walls composed of a single layer of endothelial cells. Many pores, or *fenestrations* ("windows"), are present in the glomerular capillary endothelium. These pores are not present in regular capillaries. This increased porosity is necessary for filtration to occur at the rate required for normal kidney function.

Renal Tubule

The renal tubule is a winding, hollow tube extending from the renal corpuscle to the end of the nephron. There it joins a collecting duct shared in common with other nearby nephrons. The renal tubule is divided into different regions: the *proximal convoluted tubule*, the *Henle loop*, and the *distal convoluted tubule*. Follow along in Figure 23-5 as we briefly explore these regions.

Proximal Convoluted Tubule

The **proximal convoluted tubule (PCT),** or more simply *proximal tubule*, is the second part of the nephron. It is also the first segment of the *renal tubule*. As its name suggests, the proximal convoluted tubule is proximal, or nearest, to the Bowman capsule. Because it follows a winding, convoluted course, it is called a *convoluted* tubule. Its wall consists of one layer of epithelial cells that have a brush border facing the lumen of the tubule. Thousands of microvilli form the brush border and greatly increase its luminal surface area—a structural fact of importance to its function, as we shall see.

Henle Loop

The **Henle loop,** or *nephron loop*, is the segment of renal tubule just beyond the proximal tubule. It consists of a thin *descending limb*, a sharp turn, and an *ascending limb*. Note in Figure 23-5 that the ascending limb has *two regions* of different wall thickness: the **thin ascending limb of Henle (tALH)** and the thick ascending limb (TAL). The length of the Henle loop is important in the production of highly concentrated *or* very dilute urine.

Distal Convoluted Tubule

The **distal convoluted tubule (DCT)** is a *convoluted* portion of the tubule beyond (distal to) the Henle loop. The distal convoluted tubule conducts filtrate out of the nephron and into a collecting duct.

The **juxtaglomerular apparatus** ("structure near the glomerulus") is found at the point where the afferent arteriole brushes past the distal convoluted tubule (see Figure 23-6, *B*). This structure is important in maintaining homeostasis of blood flow. This is because the juxtaglomerular complex quickly secretes **renin** when blood pressure in the afferent

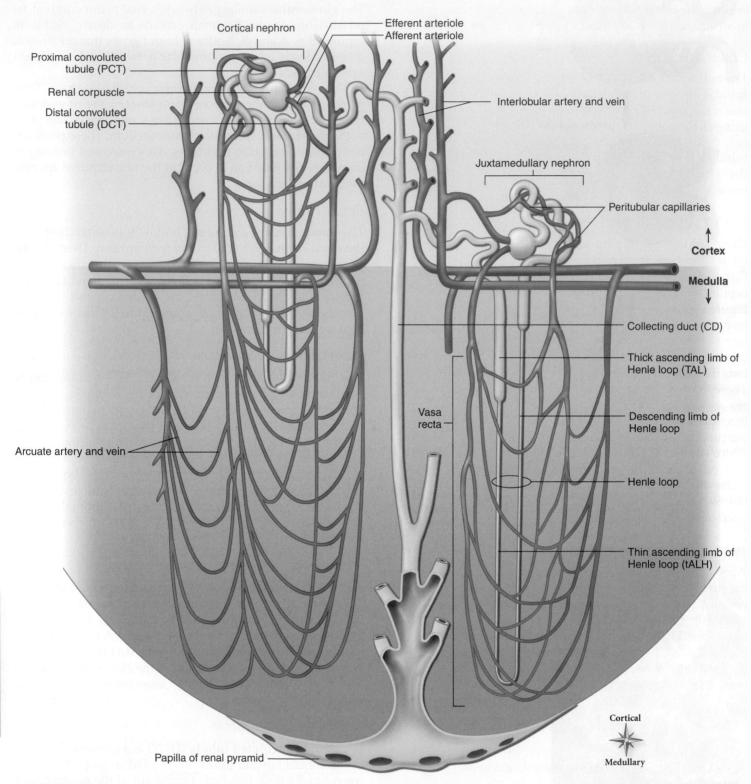

▲ **FIGURE 23-7** **Blood supply of nephrons.** Two types of nephrons (cortical and juxtaglomerular) are shown surrounded by the peritubular blood supply.

arteriole drops. Renin triggers a mechanism that produces *angiotensin*. This substance causes vasoconstriction and results in an increase in blood pressure (see Figure 23-10, p. 547).

Collecting Duct

The **collecting duct (CD)** is formed by the joining of renal tubules of several nephrons. All the collecting ducts of one renal pyramid converge at a renal papilla and release urine through their openings into one of the minor calyces. Note that Bowman capsules and both convoluted tubules lie entirely *within the cortex* of the kidney, whereas the Henle loops and collecting ducts extend *into the medulla* (Figure 23-7).

Blood Supply of the Nephron

Earlier in this chapter, we traced renal blood flow from the renal artery and its branches to the afferent arteriole. Blood then flows from the afferent arteriole into the glomerular capillaries and exits through an *efferent arteriole* (see Figure 23-7). The efferent arteriole then enters another capillary network that runs alongside the renal tubule. The capillaries of this network are called **peritubular capillaries.** Some of the blood from the efferent arteriole flows through long hairpin-shaped loops that follow the nephron loop. These long, looping arterioles are called **vasa recta** (singular, *vas rectum*) or *straight arteriole*. Blood flows very slowly through the vasa recta, a fact that plays an important role in the function of these vessels.

As you can see in Figure 23-7, blood flows through the efferent arteriole to the peritubular capillaries and vasa recta—the *peritubular blood* supply. It then flows back toward the heart through *interlobular veins* and *arcuate veins* that head toward the large *renal veins*.

Types of Nephrons

About 85% of all nephrons are located almost entirely in the renal cortex and are called **cortical nephrons.** The remainder, called **juxtamedullary nephrons,** are found adjoining (*juxta*) the medulla. Juxtamedullary nephrons have long Henle loops that dip far into the medulla (see Figure 23-7). The special role of these long Henle loops of juxtamedullary nephrons in concentrating urine is discussed later.

A&P CONNECT

Review the overall plan of renal blood flow in **Tracing Blood Flow in the Kidney** online at **A&P Connect.**

⊝volve
learning system

QUICK ✓ CHECK

8. Name the segments of the nephron in the order in which fluid flows through them.
9. What characteristics of the glomerular capsular membrane permit filtration?
10. What is the name of the blood supply that surrounds the nephron?
11. What is the other name for the straight arterioles that follow the nephron loop?

PHYSIOLOGY OF THE URINARY SYSTEM

Overview of Kidney Function

The chief functions of the kidney are to process blood plasma and excrete urine. These functions are vital because they maintain the homeostatic balance of our bodies. For example, the kidneys are the most important organs in the body for maintaining fluid-electrolyte and acid-base balance. The kidneys do this by varying the amount of water and electrolytes leaving the blood in the urine so that they equal the amounts of these substances entering the blood from various other avenues. Nitrogenous wastes from protein metabolism, notably *urea*, leave the blood by way of the kidneys.

Here are just a few of the blood constituents that cannot be held within their normal concentration ranges if the kidneys fail:

- Sodium
- Potassium
- Chloride
- Nitrogenous wastes (especially urea)

The kidneys also perform other important functions. They influence the rate of secretion of the hormones antidiuretic hormone (ADH) and aldosterone. They also synthesize the active form of vitamin D, the hormone *erythropoietin*, and certain prostaglandins.

As you now know, the basic functional unit of the kidney is the nephron. It has two main parts—the renal corpuscle and renal tubule—that form urine by means of three processes:

1. *Filtration*—movement of water and protein-free solutes from blood plasma in the glomerulus, across the glomerular capsular membrane, and into the capsular space of the Bowman capsule

2. *Tubular reabsorption*—movement of molecules out of the various segments of the tubule and into the peritubular blood

3. *Tubular secretion*—movement of molecules out of peritubular blood and into the tubule for excretion

These three mechanisms are used in concert to process blood plasma and form urine. First, a hydrostatic pressure gradient drives the filtration of much of the plasma into the nephron (Figure 23-8). Because the filtrate contains materials that the body must *conserve* (save), the walls of the tubules start reabsorbing these materials back into the blood. As the filtrate (urine) begins to leave the nephron, the kidney may secrete a few "last minute" items into the urine for excretion. In short, the kidney does not selectively filter out only harmful or excess material. It first filters out much of the plasma, then reabsorbs what should not be "thrown out" before the filtrate reaches the end of the tubule and becomes urine.

Filtration

Our first step, filtration, is a *physical process* that occurs in the kidneys' 2.5 million renal corpuscles (see Figures 23-5 and 23-6, *B*). As blood flows through the glomerular capillaries, water and small solutes filter out of the blood into Bowman capsules. The only blood constituents that do not move out are the blood cells and most plasma proteins. The result is about 180 liters of glomerular filtrate being formed each day.

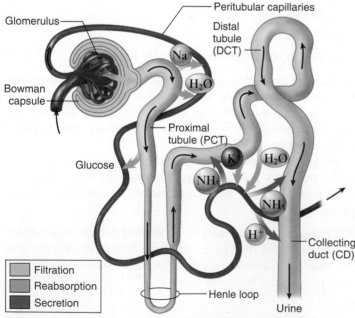

▲ **FIGURE 23-8** **Overview of urine formation.** The diagram shows the basic mechanisms of urine formation—filtration, reabsorption, and secretion—and where they occur in the nephron. Details of these steps are revealed later in this chapter.

This filtration takes place through the glomerular capsular membrane.

Filtration from glomeruli into Bowman capsules occurs because a *pressure gradient* exists between them. (Note: This is not the same as the concentration gradient of diffusion.) The main factor establishing the pressure gradient between the blood in the glomeruli and the filtrate in the Bowman capsule is the *hydrostatic pressure* of glomerular blood. This pressure gradient causes the filtration to occur from the glomerular blood plasma into Bowman capsules. The intensity of glomerular hydrostatic pressure is influenced by systemic blood pressure and the resistance to blood flow through the glomerular capillaries.

Tubular Reabsorption

Reabsorption, the second step in urine formation, requires passive and active transport mechanisms from all parts of the renal tubules. However, a major portion of water and electrolytes and (normally) all nutrients are reabsorbed from the proximal convoluted tubules. The rest of the renal tubule reabsorbs comparatively little of the filtrate. Researchers are still investigating the exact mechanisms of reabsorption in the various segments of the nephron. We have summarized only the essential principles of some of the current concepts in the following paragraphs.

Reabsorption in the Proximal Convoluted Tubule

More than two thirds of the filtrate is reabsorbed before it reaches the end of the proximal convoluted tubule. Proximal convoluted tubules are thought to reabsorb sodium ions (Na^+) by actively transporting them out of the lumen of the tubule and into peritubular blood.

Through the process of osmosis, water diffuses rapidly from the tubule fluid into peritubular blood, thus making the two fluids isotonic to each other. In short, transport of sodium ions out of the proximal convoluted tubules causes osmosis of water out of the tubules as well.

Proximal convoluted tubules reabsorb nutrients from the tubule fluid, notably glucose and amino acids, into peritubular blood by a special type of active transport mechanism called **sodium cotransport.** Normally, all the glucose that has filtered out of the glomeruli returns to the blood by this sodium cotransport mechanism. Therefore, glucose is not normally lost in urine.

Urea is a nitrogen-containing waste formed as a result of protein catabolism (see Chapter 22). Actually, toxic ammonia is formed first, but much of it is quickly transformed into the less toxic urea. Urea in the tubule fluid remains in the proximal convoluted tubule as sodium, chloride, and water are reabsorbed into blood. Once these are gone, a tubule fluid high in urea is left. Because the urea concentration in the tubule is then greater than its concentration in peritubular blood, urea passively diffuses into the blood. About half the urea present in the tubule fluid leaves the proximal convoluted tubule this way.

Reabsorption in the proximal convoluted tubules can be summarized in the following manner:

1. Sodium is actively transported out of the tubule fluid and into blood.

2. Glucose and amino acids "hitch a ride" with sodium and passively move out of the tubule fluid by means of the sodium cotransport mechanism.

3. About half the urea present in the tubule fluid passively moves out of the tubule, with half the urea thus left to move on to the Henle loop.

4. The total content of the filtrate has been reduced greatly by the time it is ready to leave the proximal convoluted tubule. Most of the water and solutes have been recovered by the blood. Only a small volume of fluid is left to continue to the next portion of the tubule, the Henle loop.

Reabsorption in the Henle Loop

In juxtamedullary nephrons (those lying low in the cortex, near the medulla), the Henle loop and its vasa recta participate in a very unique process called a **countercurrent mechanism.** A countercurrent structure is any set of parallel passages in which the contents flow in opposite directions (Figure 23-9). The Henle loop is a countercurrent structure because the contents of the ascending limb travel in a direction opposite to the direction of flow in the descending limb. The vasa recta also have a countercurrent structure because arterial blood flows down into the medulla and venous blood flows up toward the cortex. The kidney's countercurrent mechanism functions to keep the solute concentration of the medulla extremely high.

The primary functions of the Henle loop are summarized as follows:

- The Henle loop reabsorbs water from the tubule fluid (and picks up urea from the interstitial fluid) in its descending limb. It reabsorbs sodium and chloride from the tubule fluid in the thick ascending limb.

- By reabsorbing salt from its thick ascending limb, it makes the tubule fluid dilute (hypotonic).

- Reabsorption of salt in the thick ascending limb also creates and maintains a high osmotic pressure, or high solute concentration, of the medulla's interstitial fluid.

> ### QUICK CHECK
>
> 12. What are the three basic processes a nephron uses to form urine?
> 13. How is urea removed from the proximal convoluted tubule?
> 14. What is a countercurrent mechanism?
> 15. How does the function of the descending limb of the Henle loop differ from the function of the thick ascending limb?

Reabsorption in the Distal Tubules and Collecting Ducts

The distal convoluted tubule is similar to the proximal convoluted tubule in that it also reabsorbs some sodium by active transport, but in much smaller amounts.

Given no other circumstances, the kidney produces and excretes only very dilute (hypotonic) urine. However, if this indeed happened, it would be catastrophic. Why? Because the body would soon dehydrate. In fact, another regulatory mechanism centered outside the kidney normally prevents excessive loss of water. This mechanism involves **antidiuretic hormone (ADH),** a hormone released by the neurohypophysis (posterior pituitary). ADH targets cells of the distal tubules and collecting ducts and causes them to become more permeable to water. The solute concentration of the urine excreted depends in large part on the amount of ADH present.

Tubular Secretion

In addition to reabsorption, tubule cells also secrete certain substances. Tubular secretion is the movement of substances out of the blood and into tubular fluid. Recall that the descending limb of the Henle loop removes urea by means of diffusion. The distal tubules and collecting ducts secrete potassium, hydrogen, and ammonium ions. They actively transport potassium ions (K^+) or hydrogen ions (H^+) out of the blood into tubule fluid in exchange for sodium ions (Na^+), which diffuse back into the blood. Potassium secretion increases when the blood *aldosterone* concentration increases. **Aldosterone,** a hormone of the adrenal cortex, targets distal tubule and collecting duct cells and causes them to increase the activity of the sodium-potassium pumps that move sodium out of the tubule and potassium into the tubule.

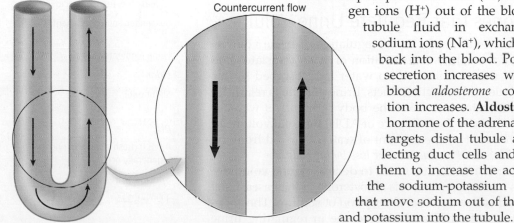

▶ **FIGURE 23-9 Concept of countercurrent flow.** Countercurrent flow simply refers to flow in opposite directions, as the *inset* shows. Tubule filtrate in the Henle loop flows in a countercurrent manner, as does blood flowing through vasa recta of the peritubular blood supply.

Countercurrent flow

TABLE 23-1 Summary of Nephron Function

PART OF NEPHRON	FUNCTION	SUBSTANCE MOVED
Renal corpuscle	Filtration (passive)	Water
		Smaller solute particles (ions, glucose, etc.)
Proximal convoluted tubule (PCT)	Reabsorption (active)	Active transport: Na$^+$
		Cotransport: glucose and amino acids
	Reabsorption (passive)	Diffusion: Cl$^-$, PO$_4^-$, urea, other solutes
		Osmosis: water
Henle loop		
Descending limb (DLH)	Reabsorption (passive)	Osmosis: water
Thin ascending limb (tALH)	Secretion (passive)	Diffusion: urea
	Reabsorption (active)	Active transport: Na$^+$
Thick ascending limb (TAL)	Reabsorption (passive)	Diffusion: Cl$^-$
Distal convoluted tubule (DCT)	Reabsorption (active)	Active transport: Na$^+$
	Reabsorption (passive)	Diffusion: Cl$^-$, other anions
		Osmosis: water (only in the presence of ADH)
	Secretion (passive)	Diffusion: ammonia
	Secretion (active)	Active transport: K$^+$, H$^+$, some drugs
Collecting duct (CD)	Reabsorption (active)	Active transport: Na$^+$
	Reabsorption (passive)	Diffusion: urea
		Osmosis: water (only in the presence of ADH)
	Secretion (passive)	Diffusion: ammonia
	Secretion (active)	Active transport: K$^+$, H$^+$, some drugs

Table 23-1 summarizes the functions of the different parts of the nephron in forming urine.

Hormones that Regulate Urine Volume

ADH has a central role in the regulation of urine volume. Control of the solute concentration of urine translates into control of urine volume. If no water is reabsorbed by the distal tubule and collecting ducts, urine volume is relatively high—and water loss from the body is high. As water is reabsorbed under the influence of ADH, the total volume of urine is reduced by the amount of water removed from the tubules. *Thus ADH reduces water loss by the body.*

Another hormone that tends to decrease urine volume—and thus conserves water—is aldosterone. It increases distal tubule and collecting duct absorption of sodium. The cooperative roles of ADH and aldosterone in regulating urine volume—and thus regulating fluid balance in the whole body—are explained in more detail in Figure 23-10.

Atrial natriuretic hormone (ANH) also influences water reabsorption in the kidney. ANH is secreted by specialized muscle fibers in the atrial wall of the heart. Its name implies its function: ANH promotes *natriuresis* (loss of Na$^+$ via urine). ANH indirectly acts as an antagonist of aldosterone, by promoting the secretion of sodium into the kidney tubules rather than sodium reabsorption. ANH also inhibits the secretion of aldosterone and opposes the aldosterone-ADH mechanism to reabsorb less water and therefore produce more urine. In short, ANH inhibits the ADH mechanism—thus inhibiting water conservation by the internal environment and increasing urine volume.

Urine volume also relates to the total amount of solutes other than sodium excreted in urine. Generally, the more solutes, the more urine. Probably the best-known example of this principle occurs in untreated diabetes mellitus. The symptom that often brings a person with undiagnosed diabetes to a physician is the voiding of abnormally large amounts of urine. Excess glucose "spills over" into urine, thereby increasing the solute concentration of urine (and decreasing the solute concentration of plasma), which in turn leads to *diuresis*, the excessive formation and excretion of urine.

Urine Composition

The physical characteristics of normal urine are listed in Table 23-2. Notice that normal and abnormal characteristics are listed.

TABLE 23-2 Characteristics of Urine

NORMAL CHARACTERISTICS	ABNORMAL CHARACTERISTICS
Color	
Transparent yellow, amber, or straw color	Abnormal colors or cloudiness, which may indicate the presence of blood, bile, bacteria, drugs, food pigments, or a high solute concentration
Compounds	
Mineral ions (for example, Na$^+$, Cl$^-$, K$^+$)	Acetone
Nitrogenous wastes: ammonia, creatinine, urea, uric acid	Albumin
Suspended solids (sediment)*	Bile, bacteria, blood cells, casts (solid matter)
Urine pigments	Glucose
Odor	
Slight odor	Acetone odor, which is common in diabetes mellitus
pH	
4.6-8.0 (freshly voided urine is generally acidic)	High in alkalosis; low in acidosis
Specific Gravity	
1.001-1.035	High specific gravity can cause precipitation of solutes and the formation of kidney stones

*Occasional trace amounts.

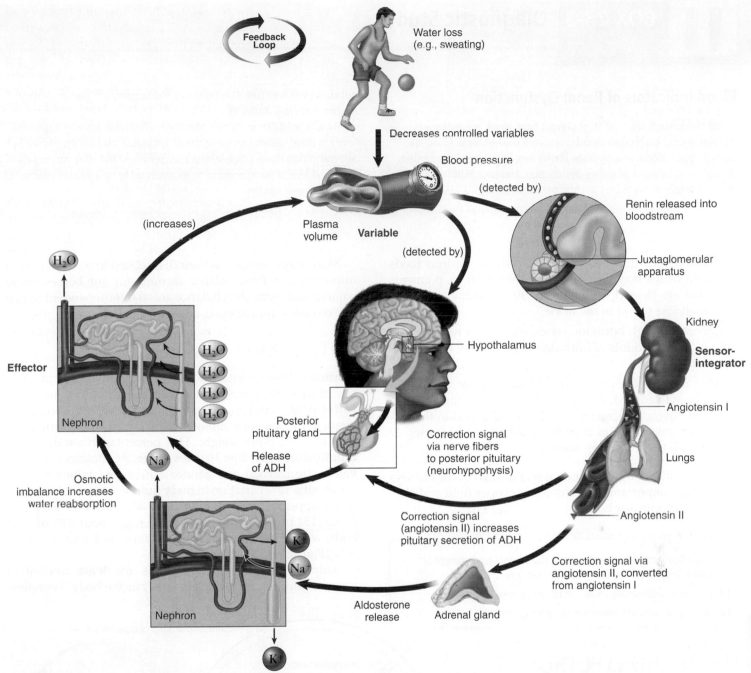

▲ **FIGURE 23-10   Cooperative roles of ADH and aldosterone in regulating urine and plasma volume.** The drop in blood pressure that accompanies loss of fluid from the internal environment triggers the hypothalamus to rapidly release ADH from the posterior pituitary gland. ADH increases water reabsorption by the kidney by increasing the water permeability of the distal tubules and collecting ducts. The drop in blood pressure is also detected by each nephron's juxtaglomerular apparatus, which responds by secreting renin. Renin triggers the formation of angiotensin II, which stimulates the release of aldosterone from the adrenal cortex. Aldosterone then slowly boosts water reabsorption by the kidneys by increasing reabsorption of Na^+. Because angiotensin II also stimulates the secretion of ADH, it serves as an additional link between the ADH and aldosterone mechanisms.

Urine is approximately 95% water, in which are dissolved several kinds of substances; the most important are discussed below:

Nitrogenous wastes—(resulting from protein catabolism) such as urea (the most abundant solute in urine), uric acid, ammonia, and creatinine (Box 23-3).

Electrolytes—mainly the following ions: sodium, potassium, ammonium, chloride, bicarbonate, phosphate,

and sulfate. The amounts and kinds of minerals vary with diet and other factors.

Toxins—during disease, bacterial poisons leave the body in urine. One reason for "forcing fluids" on patients suffering with infectious diseases is the need to dilute the toxins that might damage the kidney cells if eliminated in a concentrated form.

Pigments—especially *urochromes*, yellowish bile pigments derived from products of the breakdown of old red

BOX 23-3 **Diagnostic Study**

Blood Indicators of Renal Dysfunction

Renal clearance is the volume of plasma from which a substance is removed by the kidneys per minute. Elevated urea levels in blood, as measured in a blood urea nitrogen (BUN) test, were one of the earliest clinical measurements of kidney dysfunction. Elevated BUN levels indicate failure of the kidney to clear urea and, therefore, other

substances as well (see BUN online at *Mechanisms of Disease: Urinary System and Fluid Balance*).

Blood levels of creatinine are also used to test renal function. Creatinine levels in blood seldom change significantly because they are determined by skeletal muscle mass—which seldom changes much. Therefore, an increase in the blood level of plasma creatinine is considered to be a reliable indicator of depressed renal function.

---

blood cells in the liver and elsewhere. Various foods and drugs may contain, or be converted to, pigments that are cleared from plasma by the kidneys and are therefore found in the urine.

Hormones—high hormone levels sometimes result in significant amounts of hormone in the filtrate (and therefore in urine).

Abnormal constituents—such as blood, glucose, albumin (a plasma protein), casts (chunks of dead cellular material covered by secretions that harden inside the urinary passages and then are washed out in urine), or calculi (small kidney stones).

As we will see in the following sections, what the kidneys do is vitally important to maintaining proper fluid and electrolyte balance in our bodies.

QUICK CHECK

16. Does ADH promote water loss from the internal environment or water conservation by the internal environment?
17. How does aldosterone cause the body to conserve water?
18. What gives urine its characteristic yellowish color?

FLUID AND ELECTROLYTE BALANCE

Several of the physical properties of matter discussed in Chapter 2 explain the mechanisms of fluid and electrolyte balance. The concept of *chemical bonding* is a good example. The type of chemical bonds between molecules of certain chemical compounds, such as sodium chloride (NaCl), permits breakup, or dissociation, into separate particles (Na^+ and Cl^-). Recall that such compounds are known as **electrolytes.** The dissociated particles of an electrolyte are called **ions** and carry an electrical charge. Organic substances such as glucose, however, have a type of bond that does not permit the compound to break up, or **dissociate,** in solution. These compounds are called **nonelectrolytes.**

Many electrolytes and their dissociated ions are of critical importance in fluid balance throughout our bodies. Fluid balance and electrolyte balance are so interdependent that, if one deviates from normal, so does the other.

TOTAL BODY WATER

Normal values for total body water expressed as a percentage of total body weight will vary between 45% and 75%. Differences occur because of age, fat content of the body, and gender. In newborn infants, total body water represents about 75% of body weight. This percentage then decreases rapidly during the first 10 years of life. At adolescence, adult values are reached and gender differences, which account for about a 10% variation in body fluid volumes between the sexes, appear. In young, nonobese adults, males weighing 70 kg (154 pounds) will have on average about 60% of their body weight as water (nearly 40 liters) and females about 50% (Figure 23-11 and Table 23-3).

Adipose, or fat, tissue contains the least amount of water of any tissue (including bone) in the body. Therefore,

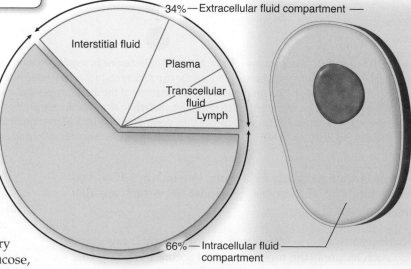

▲ **FIGURE 23-11** **Distribution of total body water.** Expressed as percentage of total fluid volume of body.

TABLE 23-3 Volumes of Body Fluid Compartments*

BODY FLUID	INFANT	ADULT MALE	ADULT FEMALE
Extracellular fluid			
Plasma	4	4	4
Interstitial fluid	26	16	11
Intracellular fluid	45	40	35
TOTAL	75	60	50

*Percentage of body weight.

regardless of age, obese individuals, with their high body fat content, have less body water per kilogram of weight than slender people do. In aged individuals of either sex, body water content may decrease to 45% of total body weight. One reason for this is that old age is often accompanied by a decrease in muscle mass (65% water) and an increase in fat (20% water). In addition, with advancing age the kidneys are less able to produce concentrated urine, and sodium-conserving responses become less effective. This affects our body fluid volume.

BODY FLUID COMPARTMENTS

On a functional level, we can divide our body water into two major fluid compartments: the *extracellular* and the *intracellular fluid compartments.* **Extracellular fluid (ECF)** consists mainly of the *plasma* found in the blood vessels and the *interstitial fluid* that surrounds the cells. In addition, the lymph and so-called *transcellular fluid*—such as cerebrospinal fluid, joint fluids, and humors of the eye—are also considered extracellular fluid. **Intracellular fluid (ICF)** refers to the water *inside* the cells. The distribution of body water by compartment is shown in Figure 23-11.

Extracellular fluid makes up the internal environment of the body. It serves the dual vital functions of providing a relatively constant environment for cells and transporting substances to and from them. Intracellular fluid, on the other hand (because it is a solvent) functions to facilitate intracellular chemical reactions that maintain life.

If we were to compare the relative volumes of these fluids, we'd see that intracellular fluid is the largest (25 liters), plasma the smallest (3 liters), and interstitial fluid in between (12 liters). Table 23-3 lists volumes of the body fluid compartments for infants and both sexes as a percentage of body weight.

ELECTROLYTES IN BODY FLUIDS

As we've seen, an electrolyte is a compound that will break up or dissociate into charged particles called ions when placed in solution. Sodium chloride, when dissolved in water, provides a positively charged sodium ion (Na^+) and a negatively charged chloride ion (Cl^-).

If two electrodes charged with a weak current are placed in an electrolyte solution, the ions will move, or migrate, in opposite directions according to their charge. Positive ions such as Na^+ will be attracted to the negative electrode (cathode) and are called **cations.** Negative ions such as Cl^- will migrate to the positive electrode (anode) and are called **anions.** Various anions and cations serve critical nutrient or regulatory roles in the body. Important cations include sodium (Na^+), calcium (Ca^{++}), potassium (K^+), and magnesium (Mg^{++}). Important anions include chloride (Cl^-), bicarbonate (HCO_3^-), phosphate (HPO_4^-), and many proteins.

Extracellular vs. Intracellular Fluids

Compared chemically, the two extracellular fluids (plasma and interstitial fluid) are almost identical. Intracellular fluid, on the other hand, shows striking differences when compared to either of the two extracellular fluids. Let us examine first the chemical structure of plasma and interstitial fluid, as shown in Figure 23-12.

In Figure 23-12 you can see that blood plasma contains a slightly larger total of electrolytes (ions) than interstitial fluid. Take a moment to compare the two fluids, ion for ion, to discover for yourself the most important difference between blood plasma and interstitial fluid. Look at the anions (negative ions) in these two extracellular fluids. Note that blood contains an appreciable amount of protein anions. *Interstitial fluid, in contrast, contains hardly any protein anions.* This is the only functionally important difference between blood and interstitial fluid. It exists because the normal capillary membrane is practically impermeable to proteins. For this reason, almost all protein anions remain behind in the blood instead of filtering out into the interstitial fluid. Because proteins remain in the blood, certain other differences also exist between blood and interstitial fluid—notably, blood contains more sodium ions and fewer chloride ions than interstitial fluid does.

Extracellular fluids and intracellular fluid are more unlike than alike chemically. Considerable chemical differences exist between the extracellular and intracellular fluids.

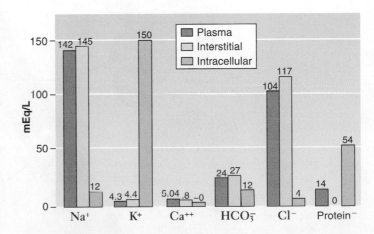

▲ **FIGURE 23-12 Electrolyte and protein concentrations in body fluid compartments.** This illustration compares individual electrolyte and protein concentration in the three fluid compartments.

Study Figure 23-12 and see if you can make some generalizations about the main chemical differences between the extracellular and intracellular fluids. For example: What is the most abundant cation in extracellular fluids? In intracellular fluid? What is the most abundant anion in extracellular fluids? In intracellular fluid? What about the relative concentrations of protein anions in extracellular fluids and intracellular fluid?

HOW WATER ENTERS AND LEAVES THE BODY

Water enters the body, as you undoubtedly know, by way of the digestive tract—in the liquids one drinks and in the foods one eats (Figure 23-13). But, in addition, water is added to the body's total fluid volume by metabolic activity of its billions of cells. Each cell produces water as it catabolizes foods, and this water enters the bloodstream. Water normally leaves the body in four basic ways: kidneys (urine), lungs (water in expired air), skin (by diffusion and by sweat), and intestines (feces). In accord with the cardinal principle of fluid balance, the total volume of water entering the body normally equals the total volume leaving. In short, fluid intake normally equals fluid output. Figure 23-13 illustrates the portals of water entry and exit, and their normal

volumes. These volumes, however, can vary considerably and still be considered normal.

QUICK ✓ CHECK

19. List three important cations and anions that serve critical nutrient or regulatory roles in the body.

20. Name the most abundant chemical constituent in blood plasma, interstitial fluid, and intracellular fluid.

21. Explain why extracellular fluids and intracellular fluids are not alike chemically.

22. List the major "portals" of water entry and exit from the body.

GENERAL PRINCIPLES OF FLUID BALANCE

Obviously, if more or less water leaves the body than enters it, an imbalance will result. Total fluid volume will increase or decrease but cannot remain constant under these conditions.

Mechanisms for varying output so that it equals intake are the most important means of maintaining fluid balance. However, mechanisms for adjusting intake to output also operate. Figure 23-14 summarizes for you the role that the *renin-angiotensin-aldosterone system (RAAS)* has in decreasing

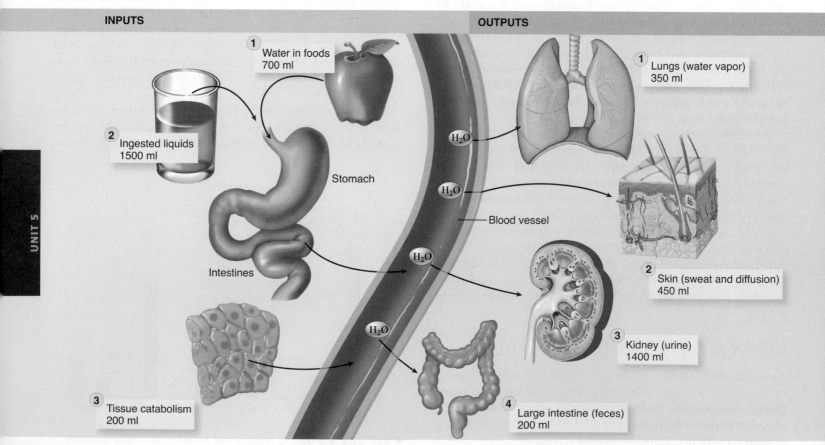

INPUTS

1. Water in foods 700 ml
2. Ingested liquids 1500 ml

Stomach

Intestines

3. Tissue catabolism 200 ml

H_2O

H_2O

Blood vessel

H_2O

H_2O

OUTPUTS

1. Lungs (water vapor) 350 ml
2. Skin (sweat and diffusion) 450 ml
3. Kidney (urine) 1400 ml
4. Large intestine (feces) 200 ml

▲ FIGURE 23-13 Sources of fluid intake and output.

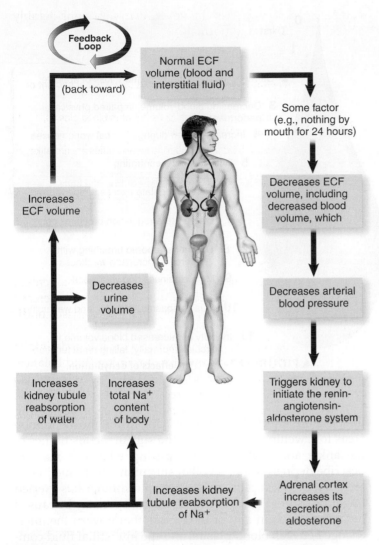

▲ **FIGURE 23-14** **Role of aldosterone in ECF homeostasis.** Aldosterone tends to restore normal extracellular fluid *(ECF)* volume when it decreases below normal. Excess aldosterone, however, leads to excess ECF volume—that is, excess blood volume (hypervolemia) and excess interstitial fluid volume (edema)—as well as an excess of the total Na^+ content of the body.

fluid output (urine volume) in order to compensate for decreased water intake.

Mechanisms for controlling water movement between the fluid compartments of the body are the most rapid-acting fluid balance processes. They serve first of all to maintain normal blood volume at the expense of interstitial fluid volume.

HOMEOSTASIS OF TOTAL FLUID VOLUME

Under normal conditions, homeostasis of the total volume of water in the body is maintained or restored primarily by mechanisms that adjust output (urine volume) to intake and secondarily by mechanisms that adjust fluid intake.

Regulation of Fluid Intake

A detailed explanation of the mechanism for controlling fluid intake and output is still beyond our understanding. However, research has shown that nerve cells located in the roof of the third ventricle of the brain act as critical regulators of fluid homeostasis. Cells in the hypothalamus are involved in antidiuretic hormone (ADH) production, which is important in conservation of body water when fluid intake is restricted.

Physiologists now identify these ventricles and hypothalamic cells as important **osmoreceptors,** which together make up the functional **thirst center** of the brain. Osmoreceptors are cells able to detect an increase in solute concentration in extracellular fluid caused by water loss. Signals generated by osmoreceptors in the ventricular roof and hypothalamus stimulate ADH secretion. They also affect a number of other body functions, including a decrease in the secretion of saliva.

Signals from the nuclei are sent to the cerebrum, where they trigger a conscious sense of dry mouth and thirst. This in turn initiates complex behaviors and thought processes. In most individuals these behaviors include a perceived need to increase the consumption of water in particular. Have you ever heard someone say, "Soda pop tastes good but nothing satisfies my thirst like a glass of water"? The end result is an overall increase in fluid intake to offset increased loss, regardless of cause, and this tends to restore fluid balance (Figure 23-15). If, however, an individual takes nothing by mouth for several days, fluid balance cannot be maintained despite every effort of homeostatic mechanisms to compensate for the zero intake. Obviously, under this

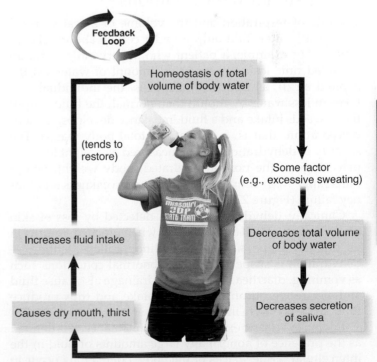

▲ **FIGURE 23-15** **Homeostasis of the total volume of body water.** A basic mechanism for adjusting intake to compensate for excess output of body fluid is diagrammed.

condition, the only way balance could be maintained would be for fluid output to also decrease to zero. But this cannot occur. Some output is obligatory. Why? Because as long as respirations continue, some water leaves the body by way of expired air. Also, as long as life continues, an irreducible minimum of water diffuses through the skin.

Regulation of Urine Volume

Urine volume is determined by the glomerular filtration rate and the rate of water reabsorption by the renal tubules. The glomerular filtration rate, except under abnormal conditions, remains fairly constant—hence it does not normally cause urine volume to fluctuate. The rate of tubular reabsorption of water, on the other hand, fluctuates considerably. The rate of tubular reabsorption, therefore, rather than the glomerular filtration rate, normally adjusts urine volume to fluid intake. The amount of antidiuretic hormone (ADH) and aldosterone secreted regulates the amount of water reabsorbed by the kidney tubules (see Figure 23-14). In other words, urine volume is regulated chiefly by hormones released by the posterior pituitary (ADH) and secreted by the adrenal cortex (aldosterone), and by atrial natriuretic hormone (ANH).

Although changes in the volume of fluid loss via the skin, the lungs, and the intestines also affect the fluid intake-output ratio, these volumes are not automatically adjusted to intake volume, as is the volume of urine.

Factors that Alter Fluid Loss Under Abnormal Conditions

The rate of respiration and the volume of sweat secreted may greatly alter fluid output under certain abnormal conditions. For example, a patient who hyperventilates for an extended time loses an excessive amount of water via the expired air. If, as frequently happens, the individual also takes in less water by mouth than normal, the fluid output then exceeds intake and a fluid imbalance develops, namely, **dehydration** (that is, a decrease in total body water). The severity of dehydration can be measured by weight loss as a percentage of the normal (hydrated) body weight. Symptoms range from simple thirst to muscle weakness and kidney failure (Figure 23-16).

Clinically, dehydration is often detected by loss of skin elasticity or *turgor*. If a fold of skin, when pinched, returns to its original shape *slowly*—a condition called "tenting"—dehydration is suspected. Other abnormal conditions such as vomiting, diarrhea, or intestinal drainage also cause fluid and electrolyte output to exceed intake and thus produce fluid and electrolyte imbalances.

Edema is a classic example of fluid imbalance. It is defined as the presence of abnormally large amounts of fluid in the intercellular tissue spaces of the body. Edema may occur in any organ or tissue of the body. However, the lungs, brain, and dependent body areas such as the legs and lower part of

0 Thirst

1

2 Stronger thirst, vague discomfort, loss of appetite

3 Decreasing blood volume, impaired physical performance

4 Increased effort during physical work; nausea

5 Difficulty in concentrating

6 Failure to regulate excess temperature

7 Temperature regulation problems continue

8 Dizziness, labored breathing with exercise, increased weakness

9 More dizziness and weakness

10 Muscle spasms, delirium, and wakefulness

11 Inability of decreased blood volume to circulate normally; failing renal function

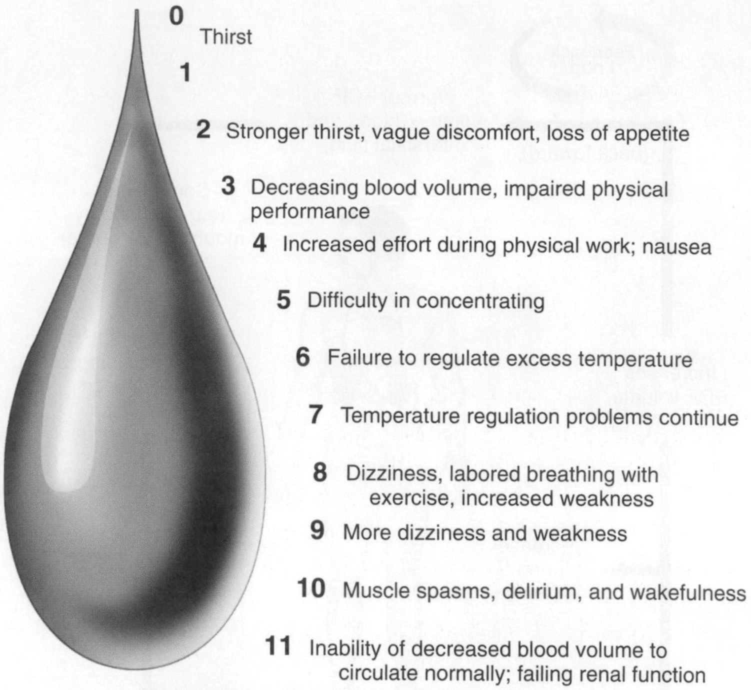

▲ FIGURE 23-16   The effects of dehydration.

the back are affected most often. One of the most common areas for swelling to occur is in the subcutaneous tissues of the ankle and foot. The term **pitting edema** is used to describe depressions in swollen subcutaneous tissue in this area that do not rapidly refill after an examiner has exerted finger pressure (Figure 23-17). The condition may be caused by disturbances in any of the factors that govern the interchange between blood plasma and the interstitial fluid compartments.

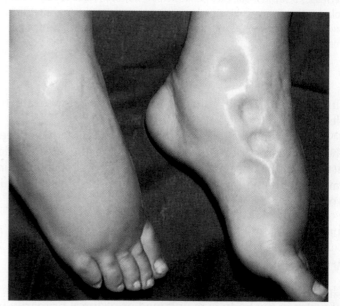

▲ FIGURE 23-17   **Pitting edema.** Note the finger-shaped depressions that do not rapidly refill after an examiner has exerted pressure.

REGULATION OF WATER AND ELECTROLYTE LEVELS IN ICF

The plasma membrane plays a critical role in the regulation of intracellular fluid composition. The mechanism that regulates water movement through cell membranes is similar to the one that regulates water movement through capillary membranes. As Figure 23-12 shows, most of the body sodium is outside the cells, and sodium is by far the chief electrolyte in *interstitial fluid*. In contrast, the main electrolyte of *intracellular fluid* is potassium. Therefore, a change in the sodium or the potassium concentration of either of these fluids causes the exchange of fluid between them to become unbalanced.

Any change in the solute concentration of extracellular fluid will have a direct effect on water movement across the cell membrane in one direction or another. If for any reason dehydration occurs, the concentration of solutes in the extracellular fluid will increase, and osmosis will cause water to move from the intracellular space into the extracellular space. In severe dehydration, the increasing concentration of intracellular fluid caused by water loss to the extracellular space results in abnormal metabolism or even cell death. Increased movement of water into the cell is caused by a decreased concentration of solutes in the extracellular fluids. Thus, fluid balance depends on electrolyte balance. Conversely, electrolyte balance depends on fluid balance. An imbalance in one produces an imbalance in the other.

REGULATION OF SODIUM AND POTASSIUM LEVELS IN BODY FLUIDS

A normal sodium concentration in interstitial fluid and potassium concentration in intracellular fluid depend on many factors, especially on the amount of ADH and aldosterone secreted. As shown in Figure 23-18, ADH regulates extracellular fluid electrolyte concentration and (colloid) osmotic pressure by regulating the amount of water reabsorbed into blood by renal tubules. Aldosterone, on the other hand, regulates extracellular fluid volume by regulating the amount of sodium reabsorbed into blood by renal tubules (see Figure 23-14).

If for any reason body sodium must be conserved, the normal kidney is capable of excreting essentially sodium-free urine. For this reason, the kidney is the chief regulator of sodium levels in body fluids. Sodium lost from sweat can

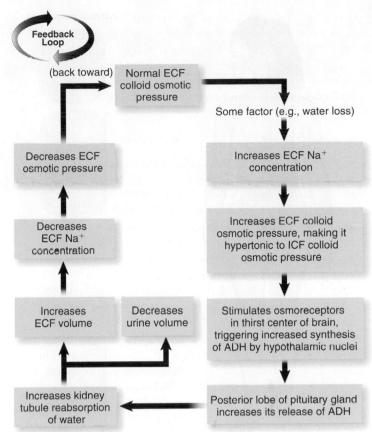

▲ **FIGURE 23-18** **Antidiuretic hormone (ADH) mechanism for ECF homeostasis.** The ADH mechanism helps maintain homeostasis of extracellular fluid *(ECF)* colloid osmotic pressure by regulating its volume and thereby its electrolyte concentration, that is, mainly ECF Na+ concentration. *ICF,* Intracellular fluid.

become significant when we experience elevated environmental temperatures or fever. However, the thirst that results may lead to replacement of water but not the lost sodium. In fact, because of the increased fluid intake, the remaining sodium pool may be diluted even more. Sodium loss in sweat is therefore not considered to be a normal means of regulation.

In addition to the well-regulated movement of sodium into and out of the body and between the three primary fluid compartments, there is a continuous movement or circulation of this important electrolyte between a number of internal secretions. More than 8 liters of various internal secretions such as saliva, gastric and intestinal secretions, bile, and pancreatic fluid are produced every day (Figure 23-19). You can see why the precise regulatory and conservation mechanisms for sodium are required for survival.

Chloride is the most important extracellular anion and is almost always linked to sodium. Generally ingested together, they provide in large part for the isotonicity of extracellular fluid. Chloride ions are usually excreted in the urine as a potassium salt, and therefore chloride deficiency—**hypochloremia**—is often found in cases of potassium loss. The body may lose a third to a half of its intracellular potassium reserves before the loss is reflected in lowered plasma potassium levels.

UNIT 5

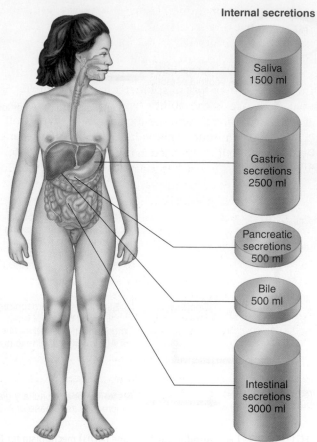

Internal secretions

Saliva
1500 ml

Gastric
secretions
2500 ml

Pancreatic
secretions
500 ml

Bile
500 ml

Intestinal
secretions
3000 ml

▲ **FIGURE 23-19** **Sodium-containing internal secretions.** Average volumes are shown. Depending on circumstances, the actual total volume of these secretions may reach 8,000 or more milliliters in a 24-hour period.

Potassium deficit, or **hypokalemia,** occurs whenever there is cell breakdown, as in starvation, burns, trauma, or dehydration. As individual cells disintegrate, potassium enters the extracellular fluid and is rapidly excreted because it is not reabsorbed efficiently by the kidney.

ACID-BASE BALANCE

Acid-base balance is one of the most important of the body's homeostatic mechanisms. The term refers to regulation of hydrogen ion concentration in the body fluids, and the study of acid-base physiology is, in a very real sense, the study of the hydrogen ion (H^+). Many of the body's most biologically important molecules contain chemical groups that can either "donate" or "accept" a hydrogen ion (H^+) and thus behave as a weak acid or base (for a review of acid, base, and buffer, see Chapter 2). As a molecule's pH changes, so does its shape—and thus also its biological activity. The functional ability of ion channels, membrane receptors, and a variety of enzymes and other important body proteins that influence protein conformation and structure, such as hemoglobin and chaperonins, closely depends on the maintenance of precise regulation of hydrogen ion concentration.

Thus, even slight deviations from the normal pH range in body cells and fluids result in pronounced, systemic, and potentially fatal changes in metabolic activity. For example, activity of the Na^+-K^+ pump, arguably the most important active transport mechanism in the cell membrane, falls by 50% when pH decreases by approximately 1 pH unit. An even more dramatic effect is seen in the activity of a key enzyme involved in the breakdown of glucose in the absence of O_2 during anaerobic catabolism (glycolysis). The biological activity of this key enzyme falls by approximately 90% when the pH decreases by only 0.1 unit! Maintaining acid-base balance within narrow and precise ranges is necessary for survival.

MECHANISMS THAT CONTROL pH OF BODY FLUIDS

Recall from Chapter 2 that water and all water solutions contain hydrogen ions (H^+) and hydroxide ions (OH^-). pH is a symbol used to represent the negative logarithm (exponent of 10) of the number of hydrogen ions (H^+) present in 1 liter of a solution. It is expressed as a number between 0 and 14. In Figure 23-20, the pH value is shown on the right side of the scale and the corresponding logarithmic value is on the left.

Take a moment to review the pH unit. **pH** indicates the degree of **acidity** or **alkalinity** of a solution. As the concentration of hydrogen ions increases, the pH goes down and the solution becomes more acid; a decrease in hydrogen ion concentration makes the solution more **alkaline** and the pH goes up. A pH of 7 indicates neutrality (equal amounts of H^+ and OH^-), a pH of less than 7 indicates acidity (more H^+ than OH^-), and a pH greater than 7 indicates alkalinity (more OH^- than H^+).

With a pH of about 1.5, gastric juice is the most acid substance in the body. With a pH of 7.0, intracellular fluid is essentially neutral.

Arterial and venous blood is slightly alkaline because both have a pH slightly higher than 7.0. The slight increase in acidity of venous blood (pH 7.36) compared with arterial blood (pH 7.40) results primarily from carbon dioxide entering venous blood as a waste product of cellular metabolism. Although any pH value above 7.0 is considered chemically basic, in clinical medicine the term **acidosis** is used to describe an arterial blood pH of less than 7.35 and **alkalosis** is used to describe an arterial blood pH greater than 7.45.

Buffers in the blood are chemicals that automatically neutralize small amounts of acid or base that enter the bloodstream. Although buffers act quickly, they are soon used up—so the body needs additional mechanisms to maintain stable blood pH.

The urinary system can alter its secretion of acids and bases into the urine. This results in a net gain or loss of these substances by the blood—thus correcting the blood's pH.

The lungs remove the equivalent of more than 30 *liters* of carbonic acid each day from the venous blood by elimination of carbon dioxide, and yet 1 liter of venous blood contains only about $1/100,000,000$ gram more hydrogen ions than does 1 liter of arterial blood. What incredible constancy! The pH homeostatic mechanism does indeed control effectively.

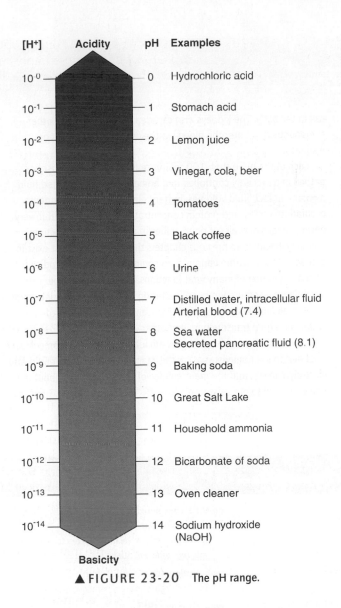

[H⁺]	Acidity	pH	Examples

10^0 — 0 — Hydrochloric acid

10^{-1} — 1 — Stomach acid

10^{-2} — 2 — Lemon juice

10^{-3} — 3 — Vinegar, cola, beer

10^{-4} — 4 — Tomatoes

10^{-5} — 5 — Black coffee

10^{-6} — 6 — Urine

10^{-7} — 7 — Distilled water, intracellular fluid
Arterial blood (7.4)

10^{-8} — 8 — Sea water
Secreted pancreatic fluid (8.1)

10^{-9} — 9 — Baking soda

10^{-10} — 10 — Great Salt Lake

10^{-11} — 11 — Household ammonia

10^{-12} — 12 — Bicarbonate of soda

10^{-13} — 13 — Oven cleaner

10^{-14} — 14 — Sodium hydroxide
(NaOH)

Basicity

▲ FIGURE 23-20   The pH range.

Cycle of LIFE 〇

The kidney plays a critical role in homeostasis by regulating the levels of many substances in the blood. Primary renal functions include filtration, reabsorption, and secretion. All are interrelated by homeostatic control systems involving central nervous system activity and hormonal secretions. More than 1 million nephron units in each kidney serve as the structural framework permitting normal function to occur.

Normally, life cycle changes in kidney structure and function occur only within rather narrow limits. Significant structural changes, such as dramatic decreases in the number of nephron units, almost always indicate serious disease or result from trauma such as crush injuries. Functionally, the kidney is able to operate normally throughout life under a wide array of conditions. If, however, the kidneys cannot cope with extreme conditions, such as water deprivation or disease, death will occur from the buildup of toxins in the blood.

Initially, kidney function in a newborn is less efficient than in an older child or adult. As a result, the urine is less concentrated because the regulatory mechanisms required to retain water are not fully operative. *Urinary incontinence*, or an inability to control urination, is normal in very young children. Reflex emptying occurs when the bladder fills, but normal sphincter activity keeps urine in the bladder until filling occurs. In contrast, many older adults have problems with incontinence because of loss of sphincter tone, or control.

Renal clearance is the ability of the kidneys to *clear*, or cleanse, the blood of a certain substance in a given unit of time, generally 1 minute. This value for certain substances tends to decrease with advanced age, thus indicating deterioration of kidney function. Changes in the porosity of the filtration membrane also occur in the elderly. Loss of functional nephron units is yet another consequence of aging. It contributes to the gradual decline in renal function in this age-group.

Although the amount of total water in the body does not vary much from day to day, the proportions of water to fat and dry solids in the body change considerably over our life spans. We start with over two thirds of our body mass as water and progress to about half of our body mass as water in adulthood. Because muscles and other metabolically active cells are high in water content, active adults have more water content than nonactive adults. Often we become less active during late adulthood. This also contributes to less water in our bodies as we age. Advanced age can also bring on some of the kidney problems mentioned earlier. ●

MECHANISMS OF DISEASE

Many people are afflicted by disorders and diseases of the urinary system, sometimes chronically, throughout their lives. These range from the discomfort and pain of common bladder infections to much more serious chronic and acute diseases.

Renal hypertension, for example is often caused by atherosclerotic plaque, which interferes with blood flow to the kidneys. There are also a number of renal obstructive disorders that interfere with normal urine flow at different locations within the urinary tract. Renal calculi, or kidney stones, are a common and very painful condition (sometimes likened by mothers to the closest pain to that of childbirth men will ever experience!). Neurogenic disorders and overactive bladders can also be chronic and embarrassing conditions. In addition to these, urinary tract infections (UTIs) such as urethritis, cystitis, and nephritis are caused by microorganisms that infect the urethra, bladder, ureter, and/or kidneys. Finally, there are a number of glomerular disorders that can cause progressive kidney damage, and of course, there are a number of cancers that afflict the kidneys and bladder. Kidney failure can result from a number of these disorders and diseases.

evolve   Find out more about these diseases and disorders of the urinary system online at *Mechanisms of Disease: Urinary System and Fluid Balance.*

The BIG Picture

As our study of the urinary system and fluid balance has shown us, homeostasis of water and electrolytes in body fluids depends largely on proper functioning of the kidneys. Each nephron within the kidney processes blood plasma in a way that adjusts its content to maintain a dynamic constancy of the internal environment of the body. Without renal processing, blood plasma characteristics would soon move out of their set point range. On the other hand, without the blood pressure generated by cardiovascular mechanisms, the kidney could not filter blood plasma and therefore could not process blood plasma. Thus the urinary system and the cardiovascular system are interdependent.

Regulation of urinary function and fluid balance, we have seen, is often centered outside the kidney—mainly in the form of endocrine hormone action. Urinary function is also regulated to some extent by nerve reflexes. Thus both the endocrine system and the nervous system must operate properly to ensure efficient kidney function. The urinary system also interacts with many other body systems and tissues. For example, the kidneys clear the blood plasma of nitrogenous wastes and excess metabolic acids produced by the chemical activity of nearly every cell in the body. The kidneys also can clear some toxins and other compounds that enter the blood via the digestive tract, skin, or respiratory tract.

Each of the more than 100 trillion cells of our bodies must be bathed in a precisely controlled and homeostatically regulated fluid medium. In fact, fluid volumes, buffers, electrolyte levels, nutrients, circulating wastes, and protein concentrations must exist within very narrow ranges of their set point values in order for us to function normally. Recall from previous chapters that many ions must exist in precise balances within various fluid compartments of our bodies for normal operation of many vital functions. For example, proper calcium ion balance is required in bone formation or reabsorption, contraction of all three muscle types, synaptic transmission, some types of endocrine signal transduction, and other vital functions.

Also recall that acids and bases are ions, and thus pH homeostasis is related to ion homeostasis. For this reason, homeostasis of fluid and electrolyte levels, and the maintenance of proper pH throughout our bodies, is vital to our existence.

LANGUAGE OF SCIENCE AND MEDICINE *continued from page 535*

edema (eh-DEE-mah)
[*edema* swelling]

electrolyte (eh-LEK-troh-lyte)
[*electro-* electricity, *-lyt* loosening]

extracellular fluid (ECF) (eks-trah-SELL-yoo-lar FLOO-id)
[*extra-* outside, *-cell-* storeroom, *-ular* relating to, *fluid* flow]

filtrate (FIL-trayt)
[*filtr-* filter, *-ate* characterized by]

filtration slit (fil-TRAY-shen slit)
[*filtr-* filter, *-at-* characterized by, *-tion* process, *slit* split]

glomerulus (gloh-MAIR-yoo-lus)
[*glomer-* ball, *-ulus* little] *pl.,* glomeruli (gloh-MAIR-yoo-leye)

Henle loop (HEN-lee)
[*Friedrich Gustave Henle* German anatomist]

hilum (HYE-lum)
[*hilum* least bit] *pl.,* hila (HYE-lah)

hypochloremia (hye-poh-kloh-REE-mee-ah)
[*hypo-* deficient, *-chlor-* green (chlorine), *-emia* blood condition]

hypokalemia (hye-poh-kah-LEE-mee-ah)
[*hypo-* deficient, *-kal-* potassium, *-emia* blood condition]

intracellular fluid (ICF) (in-trah-SELL-yoo-lar FLOO-id)
[*intra-* occurring within, *-cell-* storeroom, *-ular* relating to, *fluid* flow]

ion (EYE-on)
[*ion* go]

juxtaglomerular apparatus (juks-tah-gloh-MER-yoo-lar app-ah-RAT-us)
[*juxta-* near or adjoining, *-glomer-* ball, *-ul-* little, *-ar* relating to]

juxtamedullary nephron (juks-tah-MED-oo-lair-ee NEF-ron)
[*juxta-* near or adjoining, *-medulla-* middle, *-ary* relating to, *nephro-* kidney, *-on* unit]

kidney

micturition (mik-too-RISH-un)
[*mictur-* urinate, *-tion* process]

nephron (NEF-ron)
[*nephro-* kidney, *-on* unit]

nonelectrolyte (non-eh-LEK-troh-lyte)
[*non-* not, *-electro-* electricity, *-lyt* loosening]

osmoreceptor (os-moh-ree-SEP-tor)
[*osmos-* impulse, *-recept-* receive, *-or* agent]

peritubular capillary (pair-ee-TOOB-yoo-lar KAP-ih-lair-ee)
[*peri-* around, *-tub-* tube, *-ul-* little, *-ar* relating to, *capill-* hair, *-ary* relating to]

pH (pee-AYCH)
[abbreviation for *potenz* power, *hydrogen* hydrogen]

pitting edema (eh-DEE-mah)

podocyte (POD-oh-syte)
[*pod-* foot, *-cyte* cell]

proximal convoluted tubule (PCT) (PROK-sih-mal KON-voh-LOO-ted TOO-byool)
[*proxima-* near, *-al* relating to, *con-* together, *-volut-* roll, *tub-* tube, *-ul* little]

reabsorption (ree-ab-SORP-shun)
[*re-* back again, *-absorp-* swallow, *-tion* process]

renal clearance (REE-nal)
[*ren-* kidney, *-al* relating to]

renal column (REE-nal KOH-lum)
[*ren-* kidney, *-al* relating to]

renal corpuscle (REE-nal KOR-pus-ul)
[*ren-* kidney, *-al* relating to, *corpus-* body, *-cle* little]

renal cortex (REE-nal KOR-teks)
[*ren-* kidney, *-al* relating to, *cortex* bark] *pl.,* cortices (KOR-tih-sees)

renal medulla (REE-nal meh-DUL-ah)
[*ren-* kidney, *-al* relating to, *medulla* middle] *pl.,* medullae or medullas (meh-DUL-ee, meh-DUL-ahz)

renal pelvis (REE-nal PEL-vis)
[*ren-* kidney, *-al* relating to, *pelvis* basin]

renal pyramid (REE-nal PIR-ah-mid)
[*ren-* kidney, *-al* relating to]

renin (REE-nin)
[*ren-* kidney, *-in* substance]

sodium cotransport (SOH-dee-um koh-TRANZ-port)
[*sod-* soda, *-ium* chemical ending, *co-* with, *-trans-* across, *-port* carry]

LANGUAGE OF SCIENCE AND MEDICINE *continued from page 556*

thick ascending limb (TAL) (thik ah-SEND-ing lim)
  [*a[d]*- toward, *-scend*- climb]

thin ascending limb of Henle (LALH) (thin ah-SEND-ing lim ov HEN-lee)
  [*a[d]*- toward, *-scend*- climb, *Friedrich Gustave Henle* German anatomist]

thirst center

trigone (TRY-gohn)
  [*tri* three, *-gon* corner]

ureter (YOOR-eh-ter)
  [*ure*- urine, *-ter* agent or channel]

urethra (yoo-REE-thrah)
  [*ure*- urine, *-thr* agent or channel]

vasa recta (VAH-sah REK-tah)
  [*vasa* vessels, *recta* straight] *sing.,* vas rectum

CHAPTER SUMMARY

To download an MP3 version of the chapter summary for use with your iPod or other portable media player, access the Audio Chapter Summaries *online at http://evolve.elsevier.com.*

HINT *Scan this summary after reading the chapter to help you reinforce the key concepts. Later, use the summary as a quick review before your class or before a test.*

INTRODUCTION

A. The urinary system not only produces urine, it also balances the composition of our blood plasma
B. The urinary system regulates the content of blood plasma so that the homeostasis, or "dynamic constancy," of the entire internal fluid environment can be maintained within normal limits

ANATOMY OF THE URINARY SYSTEM

A. Gross structure—principal organs of the urinary system are the kidneys
  1. Kidney
    a. Kidneys resemble kidney beans; roughly oval, about the size of a mini-football (Figure 23-1)
    b. Right kidney is often slightly smaller and lower than the left
    c. Kidneys lie in a retroperitoneal position
    d. Lie on either side of the vertebral column between T12 and L3
    e. Renal fat pad—encases and cushions each kidney and helps to hold it in position
    f. Hilum—concave notch on medial surface where vessels and tubes enter kidney
    g. Internal structures
      (1) Renal cortex—outer region
      (2) Renal medulla—inner region
      (3) Renal pyramids—make up much of the medullary tissue
      (4) Renal columns—cortical tissue between the pyramids
      (5) Calyx—cuplike structure that marks the beginning of the "plumbing system"
      (6) Renal pelvis—region where the calyces join together; large collection "basin"
  2. Blood vessels of the kidneys—highly vascular organs (Figure 23-2, *B*)
    a. Renal artery—brings blood into each kidney
    b. Segmental arteries—division of renal arteries; divide to become lobar arteries

    c. Interlobar arteries—branches of the lobar arteries that extend out toward the cortex
    d. Arcuate arteries—formed from the interlobar arteries
    e. Interlobular arteries—branches of the arcuate arteries; penetrate the cortex
    f. Afferent arterioles—carry blood directly to the nephrons
  3. Ureter—tube running from each kidney to the urinary bladder (Figure 23-1)
    a. Lined with transitional epithelium; permits significant stretching
    b. Composed of three layers of tissue (Figure 23-4):
      (1) Mucous lining
      (2) Muscular middle layer
      (3) Fibrous outer layer
  4. Urinary bladder—muscular, collapsible bag located directly behind the pubic symphysis and in front of the rectum (Figures 23-1, *C*, and 23-3)
    a. Lies below the parietal peritoneum
    b. Wall of the bladder is made mostly of smooth muscle tissue
    c. Three openings in the floor of the bladder:
      (1) Two from the incoming ureters
      (2) One from the outgoing urethra
    d. Bladder performs two major functions:
      (1) It serves as a reservoir for urine before it leaves the body
      (2) Aided by the urethra, it expels urine from the body
  5. Urethra—small tube lined with mucous membrane; leads from the floor of the bladder to the exterior of the body
    a. In females it extends downward and forward from the bladder for a distance of about 3 cm
    b. Male urethra extends along a winding path for about 20 cm; passes through the center of the prostate gland just after leaving the bladder; extends downward, then forward, then upward to enter the base of the penis and ends as a urinary meatus at the tip of the penis
      (1) Also serves as a pathway for semen
  6. Micturition (urination)
    a. As bladder volume increases, micturition contractions (of detrusor muscle) increase and the internal urethral sphincter relaxes
    b. Relaxation of the internal sphincters along with the micturition contractions of the bladder wall can force urine out of the bladder and through the urethra
B. Microscopic structure
  1. Nephron—microscopic functional unit; nephrons make up bulk of each kidney; each nephron is made up of two regions and connects to a shared collecting duct (Figure 23-5)

UNIT 5

a. Renal corpuscle—first part of the nephron; made up of the Bowman capsule and the glomerulus (Figure 23-6, *A*)
b. Bowman capsule—cup-shaped mouth of a nephron
  (1) Formed by two layers of epithelial cells with a capsular space (Bowman space)
  (2) Pedicels in the visceral layer are packed closely together to form filtration slits; a slit diaphragm prevents filtration slits from enlarging under pressure
c. Glomerulus—network of fine capillaries surrounded by Bowman capsule (Figure 23-6, *A* and *B*)
  (1) Glomerular capillaries have thin, membranous walls composed of a single layer of endothelial cells; fenestrations (pores) are present in the glomerular endothelium
d. Renal tubule—winding, hollow tube extending from the renal corpuscle to the end of the nephron; divided into different regions (Figure 23-5)
  (1) Proximal convoluted tubule
  (2) Henle loop
  (3) Distal convoluted tubule
e. Proximal convoluted tubule (proximal tubule)—first part of the renal tubule nearest the Bowman capsule; follows a winding, convoluted course
f. Henle loop—segment of renal tubule just beyond the proximal tubule; two regions (Figure 23-5):
  (1) Descending limb
  (2) Ascending limb
g. Distal convoluted tubule—convoluted portion of the tubule beyond (distal to) the Henle loop; conducts filtrate out of the nephron and into a collecting duct
  (1) Juxtaglomerular apparatus—found at the point where the afferent arteriole brushes past the distal convoluted tubule; important in maintaining homeostasis of blood flow (Figure 23-6, *B*)
h. Collecting duct—formed by the joining of renal tubules of several nephrons
  (1) Release urine through their openings into one of the minor calyces
i. Blood supply of nephron (Figure 23-7)
  (1) Afferent arteriole enters glomerular capillary network
  (2) Efferent arteriole leaves glomerulus and extends to the peritubular blood supply
j. Types of nephrons
  (1) Cortical nephrons—located almost entirely in the renal cortex
  (2) Juxtamedullary nephrons—adjoin the medulla

PHYSIOLOGY OF THE URINARY SYSTEM

A. Overview of kidney function—chief functions of the kidneys is to process blood plasma and excrete urine; maintain homeostasis
  1. Kidneys are the most important organs in the body for maintaining fluid-electrolyte and acid-base balance
  2. Nitrogenous wastes from protein metabolism leave the blood by way of the kidneys
  3. Kidneys influence the rate of secretion of the hormones antidiuretic hormone (ADH) and aldosterone
  4. Kidneys synthesize the active form of vitamin D, the hormone erythropoietin, and certain prostaglandins
  5. Basic functional unit of the kidney is the nephron; forms urine by means of three processes (Table 23-1):
    a. Filtration—movement of water and protein-free solutes from blood plasma in the glomerulus into the capsular space of the Bowman capsule
    b. Tubular absorption—movement of molecules out of the various segments of the tubule and into the peritubular blood
    c. Tubular secretion—movement of molecules out of peritubular blood into the tubule for excretion
B. Filtration—first step in blood processing; occurs in renal corpuscles
  1. Blood flows through the glomerular capillaries; water and small solutes filter out of the blood into Bowman capsules
C. Mechanism of filtration
  1. Filtration from glomeruli into Bowman capsules occurs because a pressure gradient exists between them
  2. Intensity of glomerular hydrostatic pressure is influenced by systemic blood pressure and the resistance to blood flow through the glomerular capillaries
D. Tubular reabsorption—second step in urine formation; requires passive and active transport mechanisms from all parts of the renal tubules
  1. Major portion of water and electrolytes and (normally) all nutrients are reabsorbed from the proximal convoluted tubules
E. Reabsorption in the proximal convoluted tubule
  1. Sodium is actively transported out of the tubule fluid and into blood
  2. Glucose and amino acids "hitch a ride" with sodium and passively move out of the tubule fluid by means of the sodium cotransport mechanism
  3. About half the urea present in the tubule fluid passively moves out of the tubule, with half the urea thus left to move on to the Henle loop
  4. The total content of the filtrate has been reduced greatly by the time it is ready to leave the proximal convoluted tubule
F. Reabsorption in the Henle loop
  1. Two countercurrent mechanisms
    a. Henle loop is a countercurrent structure because the contents of the ascending limb travel in a direction opposite to the flow of urine in the descending limb
    b. Vasa recta also have a countercurrent structure because arterial blood flows down into the medulla and venous blood flows up toward the cortex
  2. The Henle loop reabsorbs water from the tubule fluid (and picks up urea from the interstitial fluid) in its descending limb; it reabsorbs sodium and chloride from the tubule fluid in the ascending limb
    a. By reabsorbing salt from its ascending limb, it makes the tubule fluid dilute (hypotonic)
    b. Reabsorption of salt in the ascending limb also creates and maintains a high osmotic pressure, or high solute concentration, of the medulla's interstitial fluid
G. Reabsorption in the distal tubules and collecting ducts
  1. Distal convoluted tubule is similar to the proximal convoluted tubule; it also reabsorbs some sodium by active transport, but in smaller amounts
  2. ADH is released by the posterior pituitary and targets the cells of the distal tubules and collecting ducts to make them more permeable to water
H. Tubular secretion—movement of substances out of the blood and into tubular fluid
  1. Descending limb of the Henle loop removes urea by means of diffusion

2. Distal tubules and collecting ducts secrete potassium, hydrogen, and ammonium ions
3. Aldosterone—hormone that targets the cells of the distal tubule and collecting duct cells; causes increased activity of the sodium-potassium pump

I. Regulation of urine volume
   1. ADH has a central role in the regulation of urine volume; reduces water loss by the body
   2. Aldosterone increases distal tubule and collecting duct absorption of sodium, thus promoting the reabsorption of water from the tubule
   3. Atrial natriuretic hormone (ANH)—influences water reabsorption in the kidney
      a. ANH indirectly acts as an antagonist of aldosterone; promotes the secretion of sodium into the kidney tubules rather than sodium reabsorption
   4. Urine volume also relates to the total amount of solutes other than sodium excreted in urine

J. Urine composition (Table 23-2)
   1. 95% water
   2. Nitrogenous wastes
   3. Electrolytes
   4. Toxins
   5. Pigments
   6. Hormones
   7. Abnormal constituents

FLUID AND ELECTROLYTE BALANCE

A. Electrolytes—have chemical bonds that allow dissociation into ions, which carry an electrical charge; of critical importance in fluid balance
B. Fluid balance and electrolyte balance are interdependent

TOTAL BODY WATER

A. Normal values for total body water will vary between 45% and 75% of total body weight
B. Differences occur because of age, fat content of the body, and gender

BODY FLUID COMPARTMENTS

A. Two major fluid compartments (Figure 23-11)
B. Extracellular fluid (ECF)—makes up the internal environment of the body
   1. Consists mainly of the plasma found in the blood vessels and the interstitial fluid that surrounds the cells; also lymph, cerebrospinal fluid, joint fluids, and humors of the eye
   2. ECF serves the dual vital functions of providing a relatively constant environment for cells and transporting substances to and from them
C. Intracellular fluid (ICF)—water inside the cells
   1. Functions to facilitate intracellular chemical reactions that maintain life
   2. By volume, ICF is the largest body fluid compartment

ELECTROLYTES IN BODY FLUIDS

A. Extracellular vs. intracellular fluids
   1. Two extracellular fluids (plasma and interstitial fluid) are almost identical; intracellular fluid shows striking differences
   2. Extracellular fluids
      a. Difference between blood and interstitial fluid—blood contains a slightly larger total of ions than interstitial fluid does
      b. Functionally important difference between blood and interstitial fluid is the number of protein anions
   3. Extracellular fluids and intracellular fluid are more unlike than alike chemically
   4. Chemical structure of the three fluids helps control water and electrolyte movement between them

HOW WATER ENTERS AND LEAVES THE BODY

A. Water enters the body by way of the digestive tract (Figure 23-13)
B. Water is added to the body's total fluid volume by metabolic activity of its billions of cells
C. Water normally leaves the body in four basic ways:
   1. As urine through the kidney
   2. As water in expired air through the lungs
   3. As sweat through its skin
   4. In feces from the intestine

GENERAL PRINCIPLES OF FLUID BALANCE

A. Mechanisms for varying output so that it equals intake are the most important means of maintaining fluid balance; e.g. renin-angiotensin-aldosterone system (RAAS) (Figure 23-14)
B. Mechanisms for controlling water movement between the fluid compartments of the body are the most rapid-acting fluid balance processes

HOMEOSTASIS OF TOTAL FLUID VOLUME

A. Under normal conditions, homeostasis of the total volume of water in the body is maintained or restored primarily by adjusting urine volume and secondarily by fluid intake
B. Regulation of fluid intake—decrease in fluid intake causes osmoreceptors in "thirst center" to increase ADH secretion
C. Regulation of urine volume—determined by the glomerular filtration rate and the rate of water reabsorption by the renal tubules (Figure 23-14)
D. Factors that alter fluid loss under normal conditions—rate of respiration and the volume of sweat secreted may greatly alter fluid output
   1. Vomiting, diarrhea, or intestinal drainage also produce fluid and electrolyte imbalances

REGULATION OF WATER AND ELECTROLYTE LEVELS IN ICF

A. Plasma membrane plays a critical role in the regulation of intracellular fluid composition
B. Mechanism that regulates water movement through cell membranes is similar to the one that regulates water movement through capillary membranes
C. Change in the sodium or the potassium concentration of either the interstitial or intracellular fluids causes the exchange of fluid between them to become unbalanced
D. Fluid balance depends on electrolyte balance

REGULATION OF SODIUM AND POTASSIUM LEVELS IN BODY FLUIDS

A. Normal sodium concentration in interstitial fluid and potassium concentration in intracellular fluid depend on many factors, especially on the amount of ADH and aldosterone secreted
   1. ADH regulates extracellular fluid electrolyte concentration and (colloid) osmotic pressure by regulating the amount of water reabsorbed into blood by renal tubules (Figure 23-18)

2. Aldosterone regulates extracellular fluid volume by regulating the amount of sodium reabsorbed into blood by renal tubules (Figure 23-14)

B. When conservation of body sodium is required, the kidneys excrete an essentially sodium-free urine; kidneys are considered the chief regulator of sodium levels

C. Chloride—most important extracellular anion and is almost always linked to sodium
   1. Chloride ions are usually excreted in the urine as a potassium salt
      a. Hypochloremia—chloride deficiency
      b. Hypokalemia—potassium deficiency; occurs whenever there is cell breakdown, as in starvation, burns, trauma, or dehydration; potassium enters the extracellular fluid and is rapidly excreted because it is not reabsorbed efficiently by the kidney

ACID-BASE BALANCE

A. Acid-base balance is one of the most important of the body's homeostatic mechanisms

B. Acid-base balance refers to the regulation of hydrogen ion concentration in the body fluids

C. Precise regulation of pH at the cellular level is necessary for survival

D. Slight pH changes have dramatic effects on cellular metabolism

MECHANISMS THAT CONTROL pH OF BODY FLUIDS

A. Review of pH concept
   1. pH is a symbol used to represent the negative logarithm (exponent of 10) of the number of hydrogen ions (H^+) present in 1 liter of a solution (Figure 23-20)
   2. Acidity—increase in concentration of hydrogen ions
   3. Alkalinity—decrease in concentration of hydrogen ions
   4. pH of 7.0—intracellular fluid is essentially neutral
      a. Acidosis—arterial blood pH less than 7.35
      b. Alkalosis—arterial blood pH greater than 7.45

B. Mechanisms that regulate pH
   1. Buffers—blood chemicals that quickly absorb excess acids or bases
   2. Urinary system—stabilizes blood pH by secreting acids or bases into urine
   3. Respiratory system—removes excess acid by releasing CO_2 from the lungs

REVIEW QUESTIONS

 HINT *Write out the answers to these questions after reading the chapter and reviewing the Chapter Summary. If you simply think through the answer without writing it down, you won't retain much of your new learning.*

1. List the principal and accessory organs of the urinary system.
2. Name, locate, and give the main function(s) of each organ of the urinary system.
3. Describe the microscopic structure of the kidney.
4. Diagram the flow of blood through the kidney.
5. Define the terms *filtration, tubular reabsorption,* and *tubular secretion.*
6. What happens to sodium and chloride in the ascending limb of the Henle loop?

7. Identify three body systems in addition to the urinary system that also excrete unneeded substances.
8. Discuss the changes in total body water content from infancy to adulthood.
9. How does total body water content differ in men and women?
10. List the compartments of extracellular fluid.
11. What is the cardinal principle regarding fluid balance?
12. How is urine volume regulated?

CRITICAL THINKING QUESTIONS

 HINT *After finishing the Review Questions, write out the answers to these items to help you apply your new knowledge. Go back to sections of the chapter that relate to items that you find difficult.*

1. What would result if the nerves supplying the bladder and urethra were damaged?
2. Can you describe the mechanism of urine formation? How is each step related to the part of the nephron that performs it?
3. If the proximal convoluted tubules were unable to transport sodium ions into blood, why would you expect to find high concentrations of both sodium and chloride ions in urine?
4. Why do you think ADH prevents rapid dehydration of the body?
5. How is the function of ANH related to the increase in urine volume?
6. How would you summarize the role of aldosterone in thirst?
7. What information would you use to support the view that the homeostasis of fluid and electrolyte levels is one of the most crucial requirements for the maintenance of life itself?

continued from page 535

Now that you have read this chapter, see if you can answer these questions about Whitney from the Introductory Story.

1. What is your best diagnosis for Whitney's condition?
   a. Allergic reaction to the sun
   b. Dehydration
   c. Kidney malfunction
   d. Too much extracellular water

Whitney's teacher rushed over, felt her skin and pulse and said, "You haven't been drinking any water today, have you?"

2. What would be the best fluid to give Whitney now?
   a. Distilled water
   b. Cold beer
   c. Cold soda
   d. A sports drink

3. What two hormones will be released to compensate for Whitney's condition?
   a. ADH and aldosterone
   b. Estrogen and renin
   c. Aldosterone and TSH
   d. Renin and angina

HINT *To solve these questions, you may have to refer to the glossary or index, other chapters in this textbook, A&P Connect, Mechanisms of Disease, and other resources.*

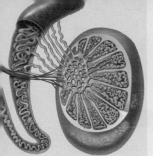

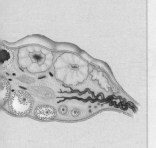

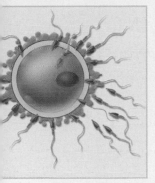

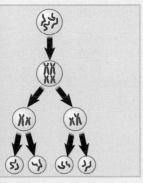

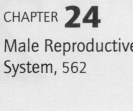

UNIT **6**

Reproduction and Development

U nit Six deals with human reproduction, growth, development, genetics, and heredity. You will see how the anatomical structures and complex control mechanisms of the male and female reproductive systems ensure survival of our genes. Our reproductive systems permit development of sperm or ova, followed by fertilization, normal embryonic and fetal development, and ultimately, birth. In this series of chapters you will also be introduced to human development and heredity.

Chapters 24 and 25 cover the male and female reproductive systems, respectively; Chapter 26 is devoted to human growth and development, and Chapter 27 outlines human genetics and heredity.

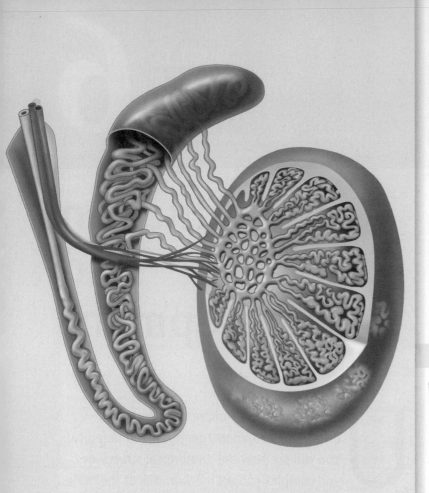

CHAPTER 24

Male Reproductive System

STUDENT LEARNING OBJECTIVES

At the completion of this chapter, you should be able to do the following:

1. Explain how sexual reproduction and asexual reproduction differ.
2. Briefly outline the male reproductive organs.
3. Discuss the structure and function of the testes.
4. Describe how testosterone works in the male body.
5. Outline the roles of FSH and LH in the male reproductive system.
6. Discuss the functions of the various reproductive ducts (epididymis, vas deferens, ejaculatory duct, and urethra)
7. Outline the role of the accessory reproductive glands.
8. List some factors that affect male fertility.

CHAPTER OUTLINE

Scan this outline before you begin to read the chapter, as a preview of how the concepts are organized.

CARLOS and his wife had been trying for years to have a baby with no success. Carlos had always assumed they just had bad timing. But recently they had started tracking Maria's cycle and found everything seemed to be on schedule. Finally, at Maria's request, they made an appointment with an infertility specialist. Carlos was expecting them to order expensive tests. But after the introductions, one of the first things the doctor asked about was what kind of underwear and pants Carlos typically wore. "What business is that of yours?" Carlos thought. Then the specialist added, " . . . because that may affect the average temperature of the testes."

It may seem odd to you, but Carlos wearing tight underwear and tight pants may really affect Maria's chance of getting pregnant. You may already know something about testes and temperature, but in this chapter you'll get "the rest of the story."

The importance of reproductive system function is notably different from that of any other organ system of the body. Ordinarily, body systems function to maintain the relative stability and survival of the individual organism. The reproductive system, on the other hand, ensures survival not of the individual but of the genes that characterize the human species. In both sexes, organs of the reproductive system are adapted for the specific sequence of functions that are concerned primarily with transferring genes to a new generation of offspring. A male reproductive system in one parent and a female reproductive system in another parent are needed to reproduce.

This chapter begins with a brief description of the male reproductive system. Chapter 26 then follows with the story of the female reproductive system.

SEXUAL REPRODUCTION

During **sexual reproduction,** a male and female each contribute *half the number of chromosomes* required to create the next generation of children. (**Asexual reproduction** requires just one parent who produces an offspring identical to it—a **clone.**) One advantage of sexual reproduction is that the process allows for the exchange and mixing of genes as sex cells are made and then recombined. Mixing the genetic deck of cards, so to speak, allows us tremendous, almost infinite variability in our children. This is vitally important to the survival and success of our species. Why is this important? Because such natural variation makes it more likely that at least some individuals will be able to survive new and evolving pathogens or other life-threatening changes that may occur over time in our internal or external environments.

Our reproductive systems also produce hormones that regulate the development of secondary sex characteristics that promote successful reproduction. For example, a variety of hormones creates structural and behavioral differences in the sexes. These differences permit adults to form sexual attractions with the opposite sex. In fact, reproductive hormones and other regulatory mechanisms provide us with the urge to have sex. Our sex drives are thus essential to successful reproduction.

HINT · Before reading the chapter, say each of these terms out loud. This will help you avoid stumbling over them as you read.

accessory organ (ak-SES-oh-ree OR-gan)
[*access-* extra, *-ory* relating to, *organ* instrument]

acrosome (AK-roh-sohm)
[*acro-* top or tip, *-some* body]

anal triangle (AY-nal)
[*an-* ring (anus), *-al* relating to]

androgen (AN-droh-jen)
[*andro-* male, *-gen* produce]

androgen-binding protein (ABP) (AN-droh-jen-BYND-ing PRO-teen)
[*andro-* male, *-gen* produce, *prote-* first rank, *-in* substance]

asexual reproduction (ay-SEK-shoo-al re-proh-DUK-shun)
[*a-* without, *sexu-* sex, *-al* relating to, *re-* again, *-produc-* bring forth, *-tion* process]

benign prostatic hypertrophy (BPH) (be-NYNE proh-STAT-ik hye-PER-troh-fee)
[*benign* kind, *pro-* before, *-stat-* set or place, *-ic* relating to, *hyper-* excessive or above, *-troph-* nourishment, *-y* state]

bulbourethral gland (BUL-boh-yoo-REE-thral)
[*bulb-* swollen root, *-ure-* urine, *thr* agent or channel (urethra), *-al* relating to]

capacitation (kah-pass-ih-TAY-shun)

clone (klohn)
[*clon* a plant cutting]

corpus cavernosum (KOHR-pus kav-er-NO-sum)
[*corpus* body, *cavern-* large hollow, *-os-* relating to, *-um* thing] *pl.,* corpora cavernosa (KOHR-poh-rah kav-er-NO-sah)

corpus spongiosum (KOHR-pus spun-jee-OH-sum)
[*corpus* body, *spong-* sponge, *-os-* relating to, *-um* thing] *pl.,* corpora spongiosa (KOHR-poh-rah spun-jee-OH-sah)

cremaster muscle (kreh-MASS-ter MUSS-el)
[*cremastos-* hanging, *mus-* mouse, *-cle* little]

ejaculation (ee-jak-yoo-LAY-shun)
[*e-* out or away, *-jacula-* throw, *-ation* process]

ejaculatory duct (ee-JAK-yoo-lah-toh-ree)
[*e-* out or away, *-jacula* throw, *-ory* relating to, *ducere-* lead]

emission (ee-MISH-un)
[*e-* out or away, *-mis-* send, *-sion* process]

epididymis (ep-ih-DID-ih-mis)
[*epi-* upon, *-didymis* pair] *pl.,* epididymides (ep-ih-DID-ih-mih-deez)

erection (ee-REK-shun)

essential organ (OR-gan)
[*organ* instrument]

external genitalia (eks-TER-nal jen-ih-TAIL-yah)
[*extern-* outside, *-al* relating to, *gen-* produce, *-al* relating to]

continued on page 574

563

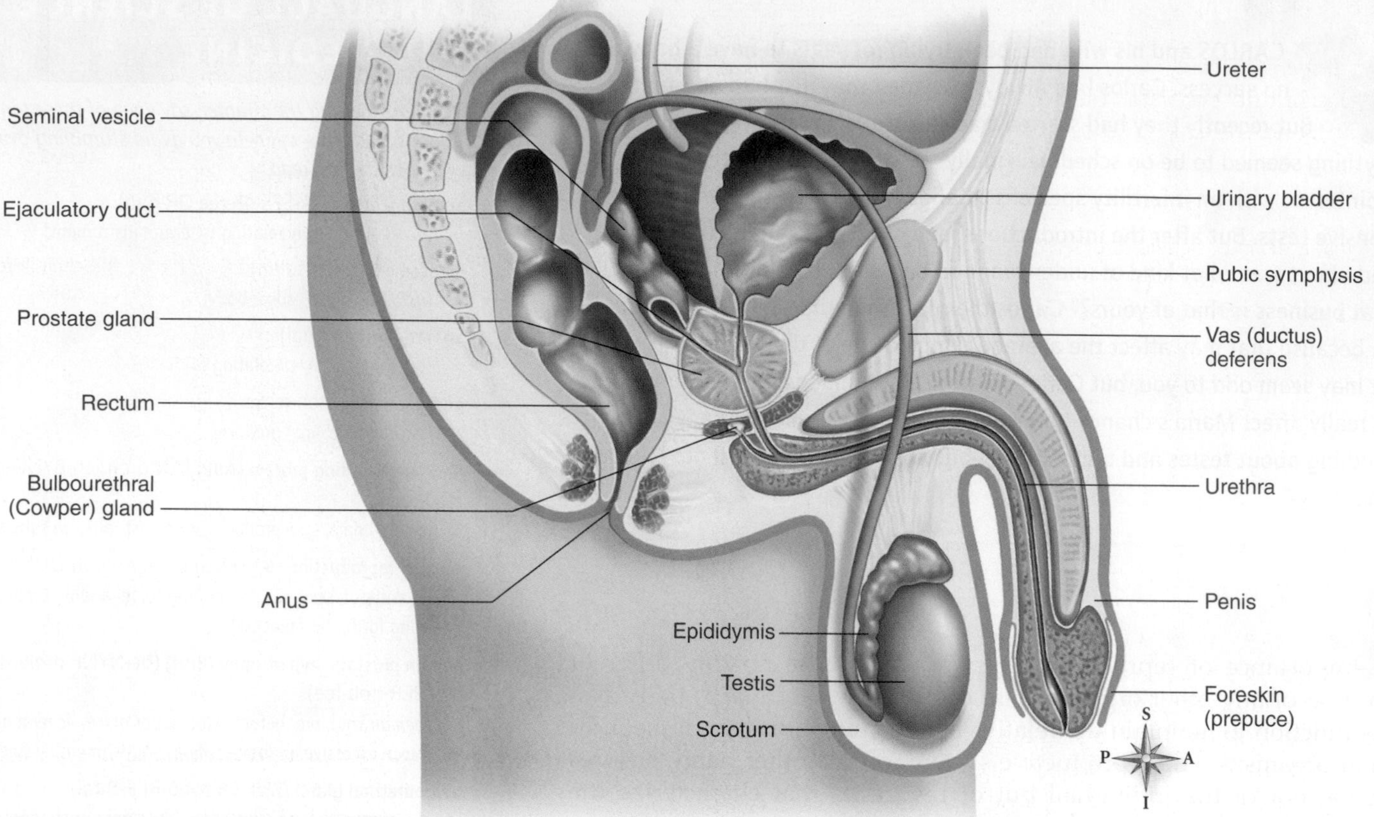

▲ **FIGURE 24-1** **Male reproductive organs.** Sagittal section of inferior abdominopelvic cavity showing placement of male reproductive organs.

MALE REPRODUCTIVE ORGANS

The male reproductive system consists of organs whose functions are to produce, transfer, and introduce mature sperm into the female reproductive tract. Here, the genes from each parent join to form a new individual.

Organs of the male reproductive system (Figure 24-1) are classified as (1) **essential organs** (primary organs) for the production of **gametes** (sex cells or sperm) and (2) **accessory organs** (secondary organs) that support gamete formation and viability.

The essential organs or **gonads** of a male are the *testes*. The accessory organs of male reproduction include the genital ducts, glands, and other supportive structures. *Reproductive ducts* (genital ducts) together are responsible for delivering sperm outside the body. The ducts include a pair of *epididymides* (singular, *epididymis*), the paired *vasa deferentia* (singular, *vas deferens*), a pair of *ejaculatory ducts*, and the *urethra*. *Accessory glands* in the reproductive system produce secretions that serve to nourish, transport, and mature sperm. The glands include a pair of *seminal vesicles*, a *prostate*, and a pair of *bulbourethral glands*. Supporting structures include the *scrotum*, the *penis*, and a pair of *spermatic cords*. You may be familiar with a number of these structures, at least in name, but we will go over each in some detail.

Perineum

The **perineum** in the male is an area between the thighs, shaped roughly like a diamond (Figure 24-2). It extends from the pubic symphysis anteriorly to the coccyx posteriorly. Its most lateral boundary on either side is the *ischial*

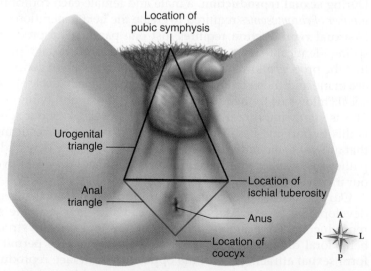

▲ **FIGURE 24-2** **Male perineum.** Sketch showing outline of the urogenital triangle *(red)* and anal triangle *(blue)*.

tuberosity (see Chapter 9, page 168). A line drawn between the two ischial tuberosities divides the perineal area into a larger *urogenital triangle* and a smaller *anal triangle*. The **urogenital triangle** contains the external genitals (penis and scrotum), and the **anal triangle** surrounds the anus.

> **QUICK CHECK**
>
> 1. What is the most significant difference between the reproductive system and other systems of your body?
> 2. Identify the essential and accessory organs of the male reproductive system.
> 3. Describe the perineum and its triangles.

TESTES

Structure and Location

The **testes** (singular, *testis*) are small, egg-shaped glands. They are about 4 to 5 cm in length and weigh 10 to 15 grams each. In a normal male, both testes are enclosed in a supporting sac, the *scrotum.* Both testes are suspended in the scrotum by attachments to the scrotal wall and by the *spermatic cords* (Figure 24-3). In addition to the vas deferens, note that the nerves, blood vessels, and lymphatics to the testis pass and are contained within the spermatic cord.

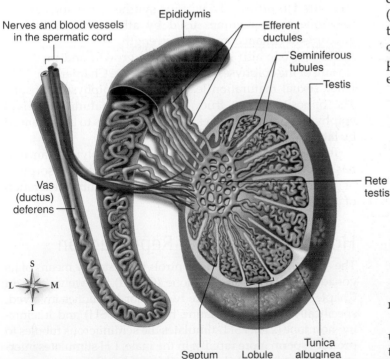

▲ **FIGURE 24-3 Tubules of the testis and epididymis.** Illustration showing epididymis lifted free of testis. The ducts and tubules are exaggerated in size.

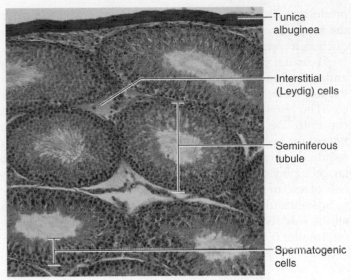

▲ **FIGURE 24-4 Testis.** Low-power view showing several seminiferous tubules surrounded by septa containing interstitial (Leydig) cells.

A dense, white, fibrous capsule called the **tunica albuginea** encases each testis and then enters each gland. It sends dividing walls called *septa* that extend into the interior of the testis, dividing the gland into 200 or more cone-shaped *lobules.* Each lobule contains scattered *interstitial cells* and one to three tiny, coiled *seminiferous tubules.* Unraveled, each of these minute tubules would stretch more than 75 cm (2 feet) in length! The tubules from each lobule come together to form a network called the *rete testis.* Sperm ducts called *efferent ductules* drain the rete testis. The tubes then pass through the tunica albuginea to enter the head of the epididymis.

Microscopic Anatomy of the Testis

Figure 24-4 shows a low-power view of testicular tissue. Note that a number of seminiferous tubules have been cut. This reveals numerous **interstitial cells** *(Leydig cells)* in the surrounding connective septa. Maturing sperm appear as dense nuclei; their *flagella* or "tails" project into the lumen of the tubule. The wall of each seminiferous tubule may contain five or more layers of these cells.

At puberty, when sexual maturity begins, sperm-forming cells in different stages of development appear. At this time, the hormone-producing interstitial cells become much more prominent in the surrounding septa.

The **sustentacular cells** (Sertoli or nurse cells) are long, irregular cells. They provide mechanical support and protection for the developing sperm attached to their surface. Sustentacular cells also secrete the hormone **inhibin,** which *inhibits* follicle-stimulating hormone (FSH) production in the anterior pituitary (see Chapter 15, p. 333). A drop in FSH lowers the rate of sperm production. This starts a negative feedback mechanism in which the supportive sustentacular cells can slow down sperm production, if conditions require.

At sexual maturity, sustentacular cells begin to secrete **androgen-binding protein (ABP).** This protein adheres to the steroid hormone *testosterone,* making it more water soluble. The *testosterone-ABP complex* increases the testosterone concentration within the seminiferous tubules. This is important because high concentrations of testosterone are required for normal germ cell maturation. Thus sustentacular cells play an important role in *spermatogenesis* (the process of sperm formation, discussed later).

Sustentacular cells extend from the basement membrane all the way to the surface facing the lumen of the seminiferous tubules (Figure 24-5). *Tight junctions* (see Chapter 3, p. 56)

exist between the sustentacular cells. These junctions divide the wall of the tubule into two compartments. The compartment near the basement membrane houses sperm-producing cells called *spermatogonia.* The compartment near the surface facing the lumen houses meiotically active cells.

Function of Testes and Testosterone

The testes perform two primary functions: *spermatogenesis* and *secretion of hormones.*

Spermatogenesis is the production of **spermatozoa** (sperm)—the male reproductive cell. The sperm are produced in the seminiferous tubules. The cross section of a seminiferous tubule in Figure 24-5 shows two cell divisions that result in a *reduction of chromosomes* from 46 in a normal body cell to 23 in a normal sperm. You'll find a complete discussion of this special type of division—called *meiosis*—in Chapter 26.

As you probably know, **testosterone** is the major **androgen** (masculinizing hormone) of males. This steroid hormone is produced by interstitial cells. Actually, testosterone has a number of important functions. First, it promotes "maleness." By this we mean the development and maintenance of male secondary sexual characteristics and accessory organs such as the prostate and seminal vesicles. Testosterone also develops and maintains adult male sexual behavior.

Testosterone also helps regulate metabolism. In fact, it stimulates protein *anabolism* (see Chapter 2, page 26), which in turn promotes growth of skeletal muscle. This, of course, is responsible for greater male muscular development and strength. Unfortunately, various synthetic versions of testosterone are sometimes used by athletes in ill-advised attempts to enhance muscular strength.

Testosterone also stimulates bone growth and promotes closure of the epiphyses in long bones (see Chapter 8, p. 144). Early sexual maturation leads to early epiphyseal closure. The opposite is also true: Late sexual maturation delays epiphyseal closure. As a result, tallness tends to be enhanced by late epiphyseal closure.

Testosterone also affects fluid and electrolyte balance. It has a mild stimulating effect on kidney tubule reabsorption of sodium and water, and promotes kidney tubule excretion of potassium.

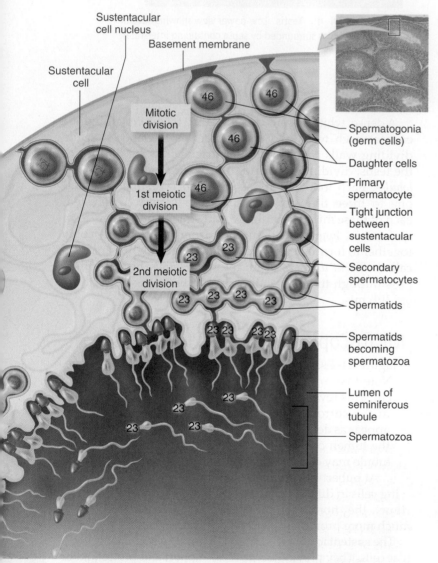

▲ FIGURE 24-5   Seminiferous tubule. Wedge from a cross section of the tubule, showing spermatogenesis and the relationship of the developing spermatozoa (sperm cells) to the sustentacular (Sertoli) cells. Mitotic cell division was explained in Chapter 5. Meiotic cell division, which reduces the number of chromosomes by half, will be explained further in Chapter 26.

Hormonal Control of Reproduction

The anterior pituitary gland controls the testes by means of its gonadotropin-releasing hormone (GnRH). As we've seen in Chapter 15 (p. 341), there are two major hormones involved, specifically follicle-stimulating hormone (FSH) and luteinizing hormone (LH). FSH stimulates the seminiferous tubules to produce sperm more rapidly. In the male, LH stimulates interstitial cells to increase their secretions of testosterone.

Note the negative feedback mechanism in Figure 24-6. If the blood concentration level of testosterone reaches a high level, it will *inhibit* secretion of GnRH from the hypothalamus. As a

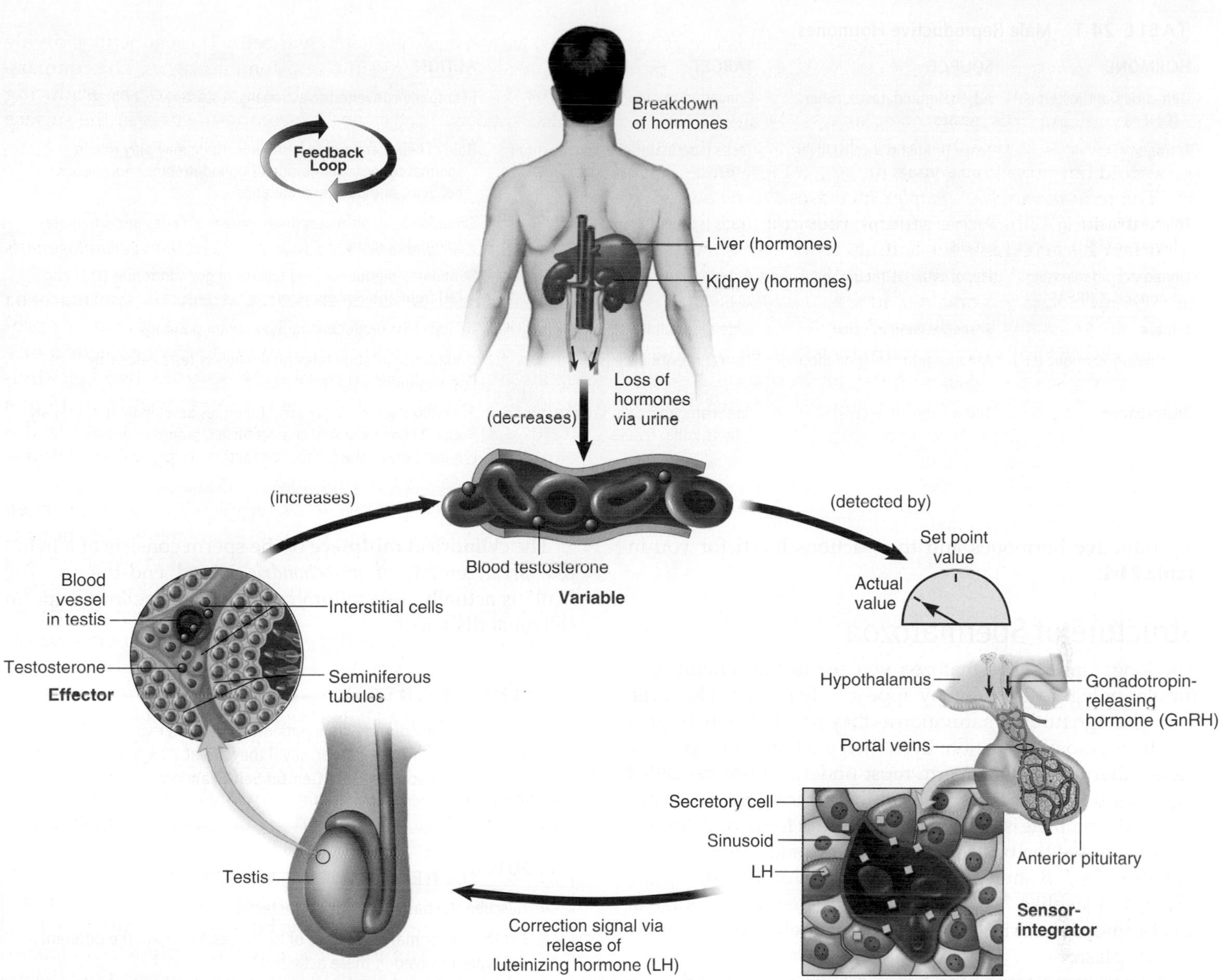

▲ **FIGURE 24-6   Negative feedback loop controlling testosterone secretion.** Diagram shows the negative feedback mechanism that controls anterior pituitary gland secretion of LH and interstitial cell secretion of testosterone. A similar negative feedback loop exists between inhibin-secreting sustentacular cells in the testis and FSH-secreting cells in the anterior pituitary gland.

result, the anterior pituitary secretion of LH will *decrease* and testosterone levels will return to the normal set point value.

Increasing blood levels of inhibin, produced by the sustentacular cells, will selectively *decrease* FSH secretion by the anterior pituitary and *decrease* the rate of sperm production. However, if sperm counts decrease below the normal set point, inhibin secretion will *decrease*, FSH secretion will *increase*, and sperm numbers will *increase* back to normal levels.

The negative feedback loops regulating testosterone secretion involve the hypothalamus (GnRH), the anterior pituitary gland (FSH and LH), and the hormone-producing cells of the testes (testosterone and inhibin).

Small but measurable amounts of *estrogen* are present in healthy adult males. In fact, much of the estrogen, a steroid hormone derived from testosterone, is made in the interstitial cells. However, estrogen in males is also made in the liver and other tissues. Possible roles for estrogen in men include (1) regulation of spermatogenesis, (2) feedback inhibition of FSH and LH, and (3) promotion of normal male sexual behavior. We are sure to learn much more about the role of estrogen in males. Recent research suggests that, in addition to gonadotrophins and testosterone, estrogens are likely playing a relevant role in spermatogenesis and human male gamete maturation. Take a moment to review the male

TABLE 24-1 Male Reproductive Hormones

HORMONE	SOURCE	TARGET	ACTION
Dehydroepiandrosterone (DHEA)	Adrenal gland, testis, other tissues	Converted to other hormones	Eventually converted to estrogens, testosterone, or both
Estrogen	Testis (interstitial cells), liver, other tissues	Testis (spermatogenic tissue), other tissues	Role of estrogen in men is still uncertain; may play role in spermatogenesis, inhibition of gonadotropins, male sexual behavior and partner preference
Follicle-stimulating hormone (FSH)	Anterior pituitary (gonadotroph cells)	Testis (spermatogenic tissue)	Gonadotropin; promotes development of testes and stimulates spermatogenesis
Gonadotropin-releasing hormone (GnRH)	Hypothalamus (neuroendocrine cells)	Anterior pituitary (gonadotroph cells)	Stimulates production and release of gonadotropins (FSH and LH) from anterior pituitary
Inhibin	Testis (interstitial cells)	Anterior pituitary (gonadotroph cells)	Inhibits FSH production in the anterior pituitary
Luteinizing hormone (LH)	Anterior pituitary (gonadotroph cells)	Testis (interstitial cells)	Gonadotropin; stimulates production of testosterone by interstitial cells of testis
Testosterone	Testis (interstitial cells)	Spermatogenic cells, skeletal muscle, bone, other tissues	Stimulates spermatogenesis, stimulates development of primary and secondary sexual characteristics, promotes growth of muscle and bone (anabolic effect)

reproductive hormones and their actions listed for you in Table 24-1.

Structure of Spermatozoa

The long, "tailed" spermatozoa you see in the seminiferous tubules of Figure 24-7, *A*, may appear fully formed. However, they undergo further maturation as they pass through the genital ducts before ejaculation. Even then the process is not complete. After ejaculation, sperm must undergo a process called **capacitation,** which takes place in the vagina. Only after this process is complete is a sperm cell capable of fertizing an ovum.

You can see the basic features of a normal spermatozoon in Figure 24-7, *B* and *C*. Each is composed of a head, middle piece, and lashlike "tail." The **head** is a compact package of 23 chromosomes. The head has no organelles and virtually no cytoplasm.

An **acrosome** containing *hydrolytic enzymes* forms the cap over the head of the sperm. These hydrolytic enzymes first break down the cervical mucus, allowing sperm to pass into the uterus and uterine tubes.

The cylindrical **midpiece** of the sperm consists of a helixlike arrangement of *mitochondria* joined end-to-end. The "tail" is actually a flagellum capable of propelling a sperm cell great distances.

A&P CONNECT

Many physicians encourage male patients to perform regular self-examination of their testes, especially if they are at a high risk for a getting a disorder. Check out **Male Genital Self-Examination** online at **A&P Connect.**

evolve

QUICK CHECK

4. Describe the basic features of the testis.
5. List the two primary functions of the testes. What are the different cell types involved in these activities?
6. List several important functions of testosterone outside those of reproduction.
7. Identify the structural components of a mature sperm.

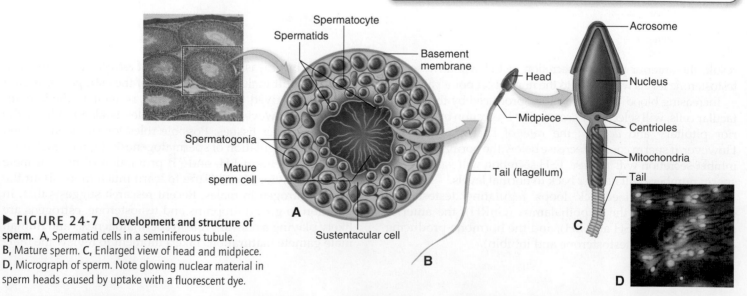

► **FIGURE 24-7** **Development and structure of sperm. A,** Spermatid cells in a seminiferous tubule. **B,** Mature sperm. **C,** Enlarged view of head and midpiece. **D,** Micrograph of sperm. Note glowing nuclear material in sperm heads caused by uptake with a fluorescent dye.

REPRODUCTIVE DUCTS

Epididymis

Each **epididymis** is a single, tightly coiled tube enclosed in a fibrous casing. Although its diameter is just barely visible with the naked eye, the tube measures 6 meters (20 feet) when uncoiled! It lies along the top of and behind the testis (see Figure 24-3). Shaped roughly like a comma, the epididymis is divided into several sections. The *head* is connected to the testis by the efferent ductules from the testis. A *central body* separates the head from the *tail*—a tapered portion that is continuous with the vas deferens.

Sperm must pass through the epididymis from the testis to the vas deferens. Each epididymis stores sperm, nourishing them with nutrients, from 1 to 3 weeks. The epididymal secretions also eventually become a small portion of the seminal fluid (semen) that is ejaculated during intercourse. After about 3 weeks, any unused sperm break down and are reabsorbed by the body.

Vas Deferens

The **vas deferens** (plural, *vasa deferentia*) is also a tube but, unlike the epididymis, it has a thick, muscular wall (see Figure 24-3). It can be felt (palpated) in the scrotal sac as a smooth, movable cord.

The vas deferens has a layered, muscular wall. Contractions of the muscles in the wall of the vasa deferentia help propel sperm through the duct system. The vas deferens from each testis ascends from the scrotum and passes *through* the inguinal canal as part of the *spermatic cord*. This cord, enclosed by fibrous connective tissue, contains muscle, blood vessels, nerves, and lymphatics, as we've seen (see Figure 24-3). The vas deferens continues into the abdominal cavity, where it extends over the top and down the posterior surface of the bladder. Here an enlarged and winding portion called the *ampulla* joins the duct from the seminal vesicle to form the *ejaculatory duct* (Figures 24-1 and 24-8, *A*).

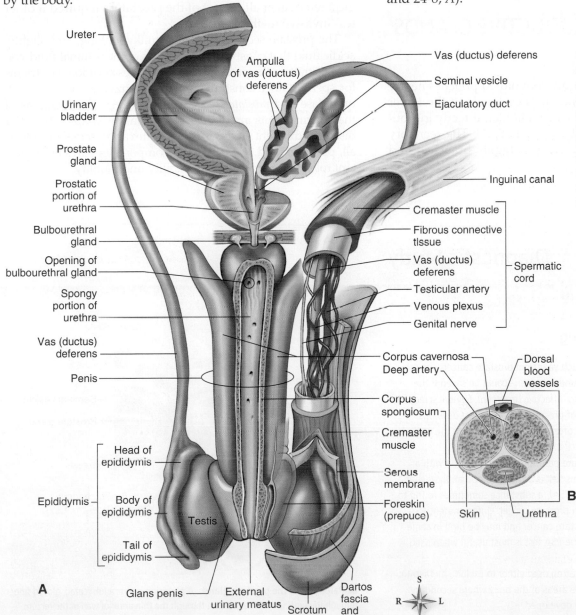

◀ **FIGURE 24-8 The male reproductive system.**
A, Illustration shows the testes, epididymis, vas (ductus) deferens, and glands of the male reproductive system in an isolation/dissection format.
B, Cross section of the shaft of the penis. Note the urethra within the substance of the corpus spongiosum.

Functionally, the vas deferens connects the epididymis with the ejaculatory duct. Sperm may remain in the vas deferens for varying periods of time, depending on the degree of sexual activity and the frequency of ejaculations. Storage time may exceed 1 month with no loss of fertility. A **vasectomy** (severing or clamping off of the vas deferens) makes a man sterile because it effectively stops the flow of sperm to the urethra.

Ejaculatory Duct and Urethra

The two **ejaculatory ducts** are short tubes about 1 cm long that pass through the prostate gland and terminate in the urethra. As you can see in Figure 24-8, these ducts are formed by the union of the vas deferens with the ducts from the seminal vesicle.

The male **urethra** serves a double function in males. It transfers both urine from the bladder and semen with sperm from the reproductive ducts.

ACCESSORY REPRODUCTIVE GLANDS

Seminal Vesicles

The **seminal vesicles** are highly convoluted pouches nearly 15 cm in length when extended. They lie along the lower part of the posterior surface of the bladder, directly in front of the rectum (see Figures 24-1 and 24-8, *A*). The secretory epithelium of the seminal vesicles is highly branched and convoluted.

The seminal vesicles secrete an alkaline, viscous, creamy-yellow liquid that makes up about 60% of the semen volume. The alkalinity helps neutralize the acid pH environment of the terminal male urethra and of the vagina. Fructose in the semen serves as an energy source for sperm motility after ejaculation. Other important components include prostaglandins.

Prostate Gland

The **prostate** lies just below the bladder and is shaped roughly like a doughnut. The fact that the urethra passes through the small hole in the center of the prostate is clinically important. This is because many older men suffer from a *noncancerous* enlargement of this gland known as **benign prostatic hypertrophy (BPH).** As the prostate enlarges, it squeezes the urethra, frequently closing it so completely that urination becomes nearly impossible. Urinary retention results, which can be uncomfortable and even painful. Surgical removal of all or part of the prostate is required if other less invasive methods fail.

The prostate secretes a watery, milky-looking, and slightly acidic fluid that constitutes about 30% of the seminal fluid volume. Citrate in the prostatic fluid provides additional nutrients for sperm. Other constituents include enzymes such as hyaluronidase and *prostate-specific antigen (PSA)*. The functions of these components are discussed later in this chapter. Box 24-1 discusses different methods for prostate cancer screening. Overall, prostatic fluid with its many components plays an important role in sperm activation, viability, and motility.

BOX 24-1 | Diagnostic Study

Prostate Cancer Screening

Many of the 32,000 men who die each year from prostate cancer—the most common nonskin type of cancer in American men—could be saved if the cancer were detected early enough for effective treatment. Several screening tests are available for the detection of prostate cancer once it develops. Cancerous growths in the gland can often be palpated through the wall of the rectum (see figure).

Sometimes, rectal examinations are performed in conjunction with a screening test called the *PSA test.* This test is a type of blood analysis that screens for *prostate-specific antigen (PSA)*, a substance sometimes found to be elevated in the blood of men with prostate cancer. Unfortunately, PSA levels may not be elevated with prostate cancer and may be high in some men without prostate cancer. Thus the PSA test is most useful when used with other screening methods.

A nuclear medicine *bone scan* is often used either to exclude metastatic spread of prostate cancer or to locate areas of the body where secondary prostate cancer tumors have already developed.

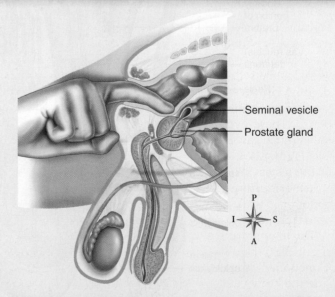

▲ **Palpation of the prostate gland.** A physician inserts a lubricated, gloved finger through the anus to feel the prostate through the thin anterior wall of the rectum.

Bulbourethral Glands

The two **bulbourethral glands** *(Cowper's glands)* resemble peas in size and shape. You can see the location of these compound glands in Figure 24-8, *A.* A duct approximately 2.5 cm (1 inch) long connects each gland with the penile portion of the urethra.

Like the seminal vesicles, the bulbourethral glands secrete an alkaline fluid. This fluid is important for counteracting the acid environment of the male urethra and the female vagina. Mucus produced in these glands serves to lubricate the urethra and helps protect sperm from damage due to friction during ejaculation.

A&P CONNECT

Infections of the reproductive tract, often acquired through sexual contact with infected individuals, can progress into conditions that may cause sterility—or even death. These *sexually transmitted diseases (STDs)* are discussed in **Sexually Transmitted Diseases** online at **A&P Connect.**

evolve

QUICK CHECK

8. List, in order, the reproductive ducts that sperm must pass through from their formation to ejaculation.

9. Describe the problems associated with the relationship of the prostate gland and the urethra.

10. Briefly compare the pH and composition of the secretions produced by the accessory reproductive glands.

SUPPORTING STRUCTURES

Scrotum

The **scrotum** is a skin-covered pouch suspended from the perineal region. Internally, it is divided into two sacs by a septum. Each sac contains a testis, epididymis, and lower part of a spermatic cord. Just below the skin lie the *dartos fascia* and *dartos muscles.* Contraction of the dartos muscle wrinkles the scrotal skin and can elevate or move the testes slightly. However, it is the **cremaster muscles** that are primarily responsible for testicular movement within the scrotal pouch. These two bands of skeletal muscle extend through the inguinal canal on either side as part of the spermatic cord (see Figure 24-8, *A*), and then attach to the posterior aspect of the testes. When contracted these "suspender" muscles, which arise from the internal oblique muscles of the lower abdominal wall, can dramatically elevate the testes during sexual arousal, exposure to cold, or threat of injury.

The temperature required for optimal sperm formation is about 3° C below normal body temperature. This is the "functional" reason that justifies placement of the testes outside the body cavity (where they are constantly exposed to potential environmental shock and traumatic injury). In a warm environment, the scrotum becomes elongated and its skin appears loose. This permits the testes to descend in the sac away from the body, thereby keeping them cool. However, in the cold, the scrotum elevates and becomes heavily wrinkled. Contraction of the cremaster muscles effectively pulls the testes upward toward the body wall, keeping them warmer. Both actions help maintain the temperature of the testes at a more constant level. Of course, factors other than temperature, including blood flow dynamics and even sexual selection, are also cited as reasons for the scrotal placement of the testes.

Penis and Spermatic Cords

The **penis** (see Figure 24-8) is composed of three cylindrical masses of *erectile tissues.* These cavernous tissues are enclosed in a separate fibrous covering and held together by a covering of skin. The two larger and uppermost of these cylinders are the **corpora cavernosa.** The smaller, lower cylinder, which contains the urethra, is called the **corpus spongiosum** (see Figure 24-8, *B*). The distal part of the corpus spongiosum *overlaps* the terminal end of the two corpora cavernosa. Here it forms a slightly bulging structure, the **glans penis.** A loose-fitting, retractable **prepuce** (foreskin) encloses most of the glans penis but leaves the urethral opening unobstructed for urination.

The penis contains the *urethra*—the terminal duct for both urinary and reproductive tracts. During sexual arousal, the erectile tissue of the penis fills with blood. This causes the organ to become rigid and enlarged in both diameter and length. The end result is called an *erection,* which allows the penis to penetrate the vagina during intercourse. The scrotum and penis together constitute the **external genitalia** of males.

The **spermatic cords** are cylindrical casings of white, fibrous tissue located in the *inguinal canals* between the scrotum and the abdominal cavity. They enclose the vasa deferentia, blood vessels, lymphatics, and nerves (see Figure 24-8, *A*).

COMPOSITION OF SEMINAL FLUID

Let's summarize the components of the **semen** *(seminal fluid)* we discussed earlier:

1. The testes and epididymis secretions comprise less than 5% of the seminal fluid volume.

2. Seminal vesicles secrete approximately 60% of the seminal fluid volume.

3. The prostate gland secretes about 30% of the seminal fluid volume.

4. The bulbourethral glands secrete less than 5% of the seminal fluid volume.

The seminal fluid serves to lubricate, protect, provide nourishment, and aid in the process of maturing sperm for ejaculation and survival. Note that sperm originate in the testes (glands located outside the body), travel *inside* the abdominal cavity, and then are expelled *outside.* In Box 24-2, we outline the basic neural controls of the male sexual response.

BOX 24-2 FYI

Neural Control of the Male Sexual Response

Recall that all body functions but one have for their ultimate goal survival of the individual. Only the function of reproduction serves a different, longer range, and (in nature's scheme) more important purpose—survival of the human species.

Male functions in reproduction consist of the production of male sex cells (spermatogenesis) and introduction of these cells into the female body *(coitus or sexual intercourse)*. For coitus to take place, erection of the penis must first occur, and for sperm to enter the female body, both the sex cells and secretions from the accessory glands must be introduced into the urethra (emission) and semen must be ejaculated from the penis.

Erection is a parasympathetic reflex initiated mainly by certain tactile, visual, and mental stimuli. It consists of dilation of the arteries and arterioles of the penis, which in turn floods and distends spaces in its erectile tissue and compresses its veins. Therefore, more blood enters the penis through the dilated arteries than leaves it through the constricted veins. As a result, the penis becomes larger and rigid: erection occurs.

Emission is the reflex movement of sex cells, or spermatozoa, and secretions from the genital ducts and accessory glands into the prostatic urethra. Once emission has occurred, ejaculation will follow.

Ejaculation of semen is also a reflex response. It is the usual outcome of the same stimuli that initiate erection. Ejaculation and various other responses—notably accelerated heart rate, increased blood pressure, hyperventilation, dilated skin blood vessels, and intense sexual excitement—characterize the male **orgasm**, or sexual climax.

MALE FERTILITY

Male fertility depends on many factors, but primarily on the number of sperm ejaculated as well as their size, shape, and *motility* (activity). In fact, fertile sperm typically have a uniform size and shape and are highly motile. Although it takes just one sperm (and only one sperm) to fertilize an egg, it appears now that millions of sperm must be ejaculated for this to occur. According to one recent estimate, when the sperm count falls below about 25 million/ml of semen, *functional sterility* can result.

One hypothesis that may explain why so many sperm must be ejaculated is that enough sperm must be present to secrete sufficient **hyaluronidase** and other hydrolytic enzymes. These enzymes liquefy the intercellular substance between the cells that encase each ovum. Without this, a single sperm cannot penetrate the layer and thus cannot fertilize the egg. Apparently it takes a large number of sperm to ensure fertilization. In effect, fertilization is a community effort!

If an *ovum* (egg) is present in the female reproductive tract when *semen* is introduced, then the release of additional capacitation enzymes from the multitude of sperm come into play. This mass release of hydrolytic enzymes is vital to fertilization because it allows the *first sperm* contacting the plasma membrane of the egg to actually *enter* the egg. Once the plasma membrane of the egg is penetrated, a series of events take place that eventually culminates in fertilization.

Infertility can also be caused by the production of antibodies some men make against their own sperm. This is called *immune infertility* and is caused by an antigen-antibody reaction.

Figure 24-9 shows that, as average plasma testosterone levels increase during puberty, sperm production begins.

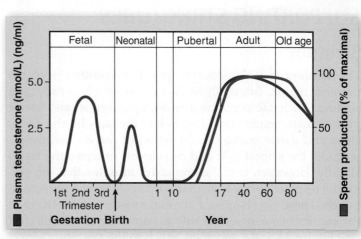

▲ **FIGURE 24-9** **Testosterone levels and sperm production.** Plasma testosterone levels *(red line)* rise during fetal development, when testosterone stimulates early development of male sexual organs. Testosterone rises again briefly around the time of birth, which facilitates descent of the testes into the scrotum. Then at puberty, testosterone rises enough to support sperm production *(blue line)* and later tapers off in advanced old age.

Testosterone levels—and thus sperm production—reach a peak in early adulthood and remain high into old age. In advanced old age, testosterone production tapers off, causing a drop in sperm count and fertility.

Cycle of LIFE ⟳

Our reproductive systems are unlike other systems in our bodies with regard to normal changes that occur throughout our life spans. All other systems perform their functions from the time they develop in utero until advanced old age. However, both male and female reproductive systems are "delayed" in that they cannot function until puberty.

Initial development of the male reproductive organs begins before birth. At about the seventh week of embryonic development, genes in the Y chromosome trigger the production of enough testosterone to stimulate the development of male reproductive organs from the undifferentiated reproductive tissues. Without the early secretion of testosterone, the organs would instead develop into their female counterparts.

Several months before birth, the immature testes descend from behind the parietal peritoneum and down into the scrotum (Figure 24-10). At this point, each testis is guided in its descent by the threadlike, fibrous *gubernaculum*. It is not uncommon for the testes to be late in their descent. And sometimes, they fail to descend until several weeks after birth. There is a spurt of testosterone levels around the time of birth—this stimulates the descent of the testes.

The testes and other reproductive organs remain in an immature state until puberty, when high levels of reproductive hormones stimulate the final stages of their development. From puberty until advanced old age, the male reproductive system continues to operate successfully. In fact, men can sire children until the time of death! ●

MECHANISMS OF DISEASE

Disorders of the male reproductive system include a variety of conditions that cause infertility and even sterility. In addition, there are occasional disorders resulting from reproductive tract infections that cause decreased sperm production. Older males may suffer from benign prostatic hypertrophy, an enlargement of the prostate, and many older males exhibit various stages of prostate cancers. Beyond these conditions and diseases are disorders of the penis and scrotum, including erectile dysfunction, hydrocele, and hernias.

ᴇvolve Find out more about these diseases and disorders of the male reproductive system online at *Mechanisms of Disease: Male Reproductive System.*

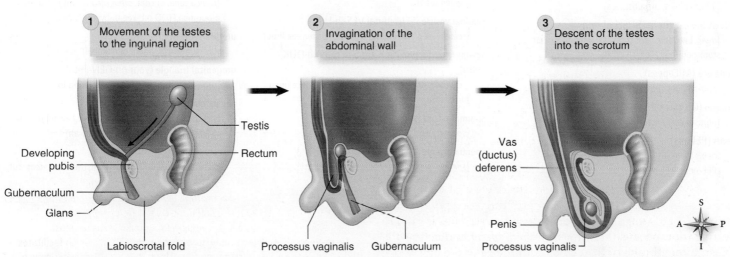

| 1 Movement of the testes to the inguinal region | 2 Invagination of the abdominal wall | 3 Descent of the testes into the scrotum |

Testis

Developing pubis — Rectum

Vas (ductus) deferens

Gubernaculum

Glans

Penis

Labioscrotal fold — Processus vaginalis — Gubernaculum — Processus vaginalis

▲ **FIGURE 24-10** **Descent of the testes.** Prior to birth, the testes move from their location near the kidneys and through the inguinal canal to the scrotum.

The BIG Picture

Reproduction of genes by individual humans provides the potential contribution of genes to the gene pool of the next generation of humans—truly a "big picture!" In males, the reproductive and urinary tracts converge terminally so that sometimes they are referred to as the *genitourinary tract* (or urogenital tract). This "sharing" also means *functional* sharing as well. For example, the urethra conducts urine during *micturition* but conducts semen during ejaculation. Nervous regulation of the muscles controlling the bladder, urethra, and ejaculatory duct prevents the flow of urine *from* the bladder and backflow of semen *into* the bladder during sexual activity.

As we've seen, both the primary and secondary sexual functions in males depend on complex interrelationships involving nervous, endocrine, muscular, urinary, and circulatory system structures. Even the skin can be perceived as a sexual organ—it receives many of the stimuli needed to produce the sexual response.

LANGUAGE OF SCIENCE AND MEDICINE *continued from page 563*

continued from page 563

gamete (GAM-eet)
[*gamete* marriage partner]

glans penis (glans PEE-nis)
[*glans* acorn, *penis* male sex organ] *pl.,* glandes penes (GLAN-deez PEE-neez)

gonad (GO-nad)
[*gon-* offspring, *-ad* relating to]

head

hyaluronidase (hye-al-yoo-RAHN-id-ayz)
[*hyal-* glass, *-uron-* urine, *-id-* relating to, *-ase* enzyme]

inhibin (in-HIB-in)
[*inhib-* inhibit, *-in* substance]

interstitial cell (in-ter-STISH-al sell)
[*inter-* between, *-stit-* stand, *-al* relating to, *cell* storeroom]

midpiece (MID-pees)
[*mid-* middle, *-piece* portion]

orgasm (OR-gaz-um)
[*orgasm* excitement]

penis (PEE-nis)
[*penis* male sex organ] *pl.,* penes or penises (PEE-neez, PEE-nis-ez)

perineum (pair-ih-NEE-um)
[*peri-* around, *-ine-* excrete, *-um* thing] *pl.,* perinea (pair-ih-NEE-ah)

prepuce (PREE-pus)
[*pre-* before, *-puc-* penis]

prostate (PROSS-tayt)
[*pro-* before, *-stat-* set or place]

scrotum (SKROH-tum)
[*scrotum* bag] *pl.,* scrota or scrotums (SKROH-tah, SKROH-tumz)

semen (SEE-men)
[*semen* seed]

seminal vesicle (SEM-ih-nal VES-ih-kul)
[*semen-* seed, *-al* relating to, *vesic-* blister, *-cle* little]

sexual reproduction (SEK-shoo-al re-proh-DUK-shun)
[*sexu-* sex, *-al* relating to, *re-* again, *-produce* bring forth, *-tion* process]

spermatic cord (sper-MAT-ik kord)
[*sperma-* seed, *-ic* relating to]

spermatogenesis (sper-mah-toh-JEN-eh-sis)
[*sperma-* seed, *-gen-* produce, *-esis* process]

spermatozoon (sper-mah-tah-ZOH-on)
[*sperma-* seed, *-zoon* animal] *pl.,* spermatozoa (sper-mah-tah-ZOH-ah)

sustentacular cell (sus-ten-TAK-yoo-lar sell)
[*sustent-* support, *-acular* relating to, *cell* storeroom]

testis (TES-tis)
[*testis* witness (male gonad)] *pl.,* testes (TES-teez)

testosterone (tes-TOS-teh-rohn)
[*test-* witness (testis), *-stero-* solid or steroid derivative, *-one* chemical]

tunica albuginea (TOO-nih-kah al-byoo-JIN-ee-ah)
[*tunica* tunic or coat, *albuginea* white] *pl.,* tunicae albuginea (TOO-nih-kee al-byoo-JIN-ee-ah)

urethra (yoo-REE-thrah)
[*ure-* urine, *-thr-* agent or channel]

urogenital triangle (yoor-oh-GEN-ih-tal)
[*uro-* urine, *-gen-* produce, *-al* relating to]

vas deferens (vas DEF-er-enz)
[*vas* duct or vessel, *deferens* carrying away] *pl.,* vasa deferentia (VAS-ah def-er-EN-shee-ah)

vasectomy (vah-SEK-toh-mee)
[*vas-* duct or vessel (vas deferens), *-ec-* out, *-tom-* cut, *-y* action]

CHAPTER SUMMARY

SEXUAL REPRODUCTION

A. The reproductive system is an important part of our individual homeostasis
   1. Vital part of our continuing survival and evolution as humans
   2. Organs of the reproductive system are adapted to transferring genes from parents to their offspring
   3. Reproductive systems produce hormones that regulate the development of secondary sex characteristics that promote successful reproduction
B. Sexual reproduction—male and female each contribute half the number of chromosomes required to create the next generation of children
   1. Advantage of sexual reproduction is that the process allows for the exchange and mixing of genes as sex cells are made and then recombined

MALE REPRODUCTIVE ORGANS

A. Functions are to produce, transfer, and introduce mature sperm into the female reproductive tract (Figure 24-1)
   1. Classified as essential organs (primary organs) and accessory organs (secondary organs)
      a. Essential organs or gonads of a male are the testes
      b. Accessory organs of male reproduction include the genital ducts, glands, and other supportive structures
B. Perineum—in males, it is an area between the thighs, shaped roughly like a diamond; extends from the pubic symphysis anteriorly to the coccyx posteriorly (Figure 24-2)
   1. Urogenital triangle—contains the external genitals (penis and scrotum)
   2. Anal triangle—surrounds the anus

TESTES

A. Structure and location
   1. Small, egg-shaped glands enclosed in a supporting sac called the scrotum
   2. Suspended in the scrotum by attachments to the scrotal wall and by the spermatic cords (Figure 24-3)
   3. Dense, white, fibrous capsule called the tunica albuginea encases each testis and then enters each gland
   4. Seminiferous tubules in testis open into a plexus called the rete testis, which is drained by a series of efferent ductules that emerge from the top of the organ and enter the head of epididymis
B. Microscopic anatomy of the testis
   1. Interstitial (Leydig) cells—hormone-producing cells between the seminiferous tubules

   2. Sustentacular cells (Sertoli or nurse cells)—provide mechanical support and protection for the developing sperm attached to their surface
      a. Secrete inhibin—inhibits follicle-stimulating hormone (FSH) production in the anterior pituitary
      b. Produce androgen-binding protein that adheres to the steroid hormone testosterone; makes it more water soluble
      c. Sustentacular cells play an important role in spermatogenesis
      d. Tight junctions exist between sustentacular cells to divide the wall of the tubule into two compartments
C. Functions of testis and testosterone (Figure 24-5)
   1. Spermatogenesis
      a. Production of spermatozoa (sperm)
      b. Involves meiosis—a special type of cell division that halves the number of chromosomes (see Chapter 26)
   2. Secretion of hormones by interstitial cells
      a. Testosterone—major androgen (masculinizing hormone)
      b. Functions of testosterone include: develops and maintains male secondary sexual characteristics and accessory organs; develops and maintains adult male sexual behavior; stimulates protein anabolism; affects fluid and electrolyte balance
D. Hormonal control of reproduction
   1. Anterior pituitary gland controls the testes by means of its gonadotropin-releasing hormone (GnRH)
   2. Two major hormones
      a. Follicle-stimulating hormone (FSH)—stimulates the seminiferous tubules to produce sperm more rapidly
      b. Luteinizing hormone (LH)—stimulates interstitial cells to increase their secretions of testosterone
E. Structure of spermatozoa (Figure 24-7)
   1. Consists of a head (covered by acrosome), neck, midpiece, and tail (Figure 24-7, *B* and *C*)

REPRODUCTIVE DUCTS

A. Epididymis—single, tightly coiled tube enclosed in a fibrous casing
   1. Lies along the top of and behind the testis (Figure 24-3)
   2. Anatomical divisions include head, central body, and tail
   3. Each epididymis stores sperm, nourishing them with nutrients from 1 to 3 weeks
   4. Epididymal secretions also eventually become a small portion of the seminal fluid (semen)
B. Vas deferens—tube but, unlike the epididymis, it has a thick, muscular wall
   1. Contractions of the muscles in the wall of the vasa deferentia help propel sperm through the duct system
   2. Functionally, the vas deferens connects the epididymis with the ejaculatory duct
C. Ejaculatory duct and urethra
   1. Formed by the union of the vas deferens with the ducts from the seminal vesicle (Figure 24-8)
   2. Urethra serves a double function in males; transfers both urine from the bladder and semen with sperm from the reproductive ducts

ACCESSORY REPRODUCTIVE GLANDS

A. Seminal vesicles
1. Convoluted pouches nearly 15 cm in length when extended
2. Lie along the lower part of the posterior surface of the bladder, directly in front of the rectum (Figures 24-1 and 24-8)
3. Secrete an alkaline, viscous, creamy-yellow liquid that makes up about 60% of the semen volume

B. Prostate gland
1. Lies just below the bladder; shaped roughly like a doughnut
2. Secretes a watery, milky-looking, and slightly acidic fluid that constitutes about 30% of the seminal fluid volume

C. Bulbourethral glands (Cowper's glands) (Figure 24-8)
1. Resemble peas in size and shape
2. A duct approximately 2.5 cm (1 inch) long connects each gland with the penile portion of the urethra
3. Secrete an alkaline fluid; important for counteracting the acid environment of the male urethra and the female vagina

SUPPORTING STRUCTURES

A. Scrotum
1. Skin-covered pouch suspended from the perineal region (Figure 24-8)
2. Divided internally into two sacs by a septum
3. Each sac contains a testis, epididymis, and lower part of a spermatic cord
4. Dartos wrinkles the scrotal skin and cremaster muscles elevate the scrotal pouch

B. Penis and spermatic cords
1. Penis is composed of three cylindrical masses of erectile tissues (Figure 24-8)
2. Functions—contains the urethra, the terminal duct for both urinary and reproductive tracts; during sexual arousal, penis becomes erect, serving as a penetrating copulatory organ during sexual intercourse
3. Spermatic cords—cylindrical casings of white, fibrous tissue located in the inguinal canals between the scrotum and the abdominal cavity
   a. Enclose the vasa deferentia, blood vessels, lymphatics, and nerves (Figure 24-8)

COMPOSITION OF SEMINAL FLUID

A. Seminal fluid serves to lubricate, protect, provide nourishment, and aid in the process of maturing sperm for ejaculation and survival

MALE FERTILITY

A. Depends on many factors, but primarily on the number of sperm ejaculated, size, shape, and motility
B. Functional sterility—when the sperm count falls below about 25 million/ml of semen
C. Sufficient numbers of sperm must be present to secrete enough hyaluronidase (enzymes that liquefy the substance that encases an ovum) so that one sperm can penetrate the ovum
D. Infertility can also be caused by the production of antibodies some men make against their own sperm—immune infertility
E. Male fertility begins at puberty and extends into old age (Figure 24-9)

REVIEW QUESTIONS

 HINT *Write out the answers to these questions after reading the chapter and reviewing the Chapter Summary. If you simply think through the answer without writing it down, you won't retain much of your new learning.*

1. Name the accessory glands of the male reproductive system.
2. List the genital ducts in the male.
3. List the supporting structures of the male reproductive system.
4. What is the tunica albuginea? How does it aid in dividing the testis into lobules?
5. What are the two primary functions of the testes?
6. What are the general functions of testosterone?
7. Discuss the structure of a mature spermatozoon.
8. What is meant by the term *capacitation?*
9. List the three functions of the epididymis.
10. List the anatomical divisions of the epididymis.
11. Discuss the formation of the ejaculatory ducts.
12. Discuss the type of secretion typical of the prostate gland and seminal vesicles.
13. What and where are the bulbourethral glands?
14. Describe the structure, location, and function or functions of the scrotum.
15. Name the three cylindrical masses of erectile, or cavernous, tissue in the penis.
16. What and where is the glans penis? The prepuce, or foreskin?
17. What is the spermatic cord? From what does it extend, and what does it contain?

CRITICAL THINKING QUESTIONS

HINT *After finishing the Review Questions, write out the answers to these items to help you apply your new knowledge. Go back to sections of the chapter that relate to items that you find difficult.*

1. How does the function of reproduction differ from all other body functions?
2. Can you identify the functions of the male reproductive system?
3. What is the relationship between the rete testis, seminiferous tubules, and efferent ductules?
4. How is the prostate gland related to the urethra? What problems can result from this relationship?
5. Can you list the structures in the reproductive system that contribute to the formation of seminal fluid?
6. Trace the course of seminal fluid from its formation to ejaculation.
7. What is the chemical in seminal fluid that is important to fertility? What is its function?
8. How is the structure of the spermatozoon related to its function?

continued from page 563

Remember Carlos and Maria from the Introductory Story? See if you can answer the following questions about Carlos' fertility now that you have read this chapter.

1. Sperm production occurs optimally at what temperature?
   a. 3° C above body temperature
   b. At body temperature
   c. 3° C below body temperature
   d. Optimal temperature changes with the seasons

Next, Carlos was asked to provide a sperm sample. "We're going to analyze the sperm count and morphology," said the doctor.

2. What number should Carlos' sperm count be above for that factor to be ruled out as a cause of the couple's infertility?
   a. 250 million/ml
   b. 25 million/ml
   c. 2500/ml
   d. 250/ml

3. Which is the correct pathway the sperm would take during ejaculation?
   a. Seminiferous tubules, rete testis, efferent ductules, epididymis, vas deferens, ejaculatory duct, urethra
   b. Rete testis, seminiferous tubules, efferent ductules, epididymis, vas deferens, urethra, ejaculatory duct
   c. Epididymis, vas deferens, seminiferous tubules, rete testis, ejaculatory duct, urethra
   d. Seminiferous tubules, rete testis, epididymis, vas deferens, efferent ductules, urethra, ejaculatory duct

4. What hormone directly stimulates sperm production?
   a. Estrogen
   b. Progesterone
   c. LH
   d. Testosterone

HINT *To solve these questions, you may have to refer to the glossary or index, other chapters in this textbook, A&P Connect, Mechanisms of Disease, and other resources.*

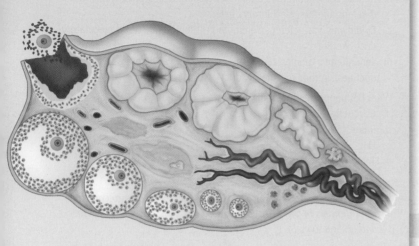

Female Reproductive System

CHAPTER OUTLINE

Scan this outline before you begin to read the chapter, as a preview of how the concepts are organized.

STUDENT LEARNING OBJECTIVES

At the completion of this chapter, you should be able to do the following:

1. Briefly describe the functions of the female reproductive system.
2. Differentiate between essential organs and accessory organs of the female reproductive system.
3. Describe the structure of the ovaries and list their functions.
4. Make an outline of oogenesis, listing the major structures involved.
5. Discuss the layers comprising the walls of the uterus, and the functions of these layers.
6. Describe the basic functions of the following: uterus, uterine tubes, vagina.
7. Outline the major components of the external genitalia and describe their basic functions.
8. Outline in general the recurring cycles of the female reproductive system.
9. Discuss the roles of hormones in the recurring cycles of female reproduction.
10. Identify the factors that affect female fertility.
11. Describe the structures involved in breast milk production and identify the hormones that affect its production.

CARLOS and his wife had been trying for years to have a baby with no success. After a visit to an infertility specialist, they found that Carlos had a low sperm count. Before suggesting a solution, the physicians will also check Maria's reproductive system to confirm that there is an open pathway for the egg.

Finding no blockage in Maria's reproductive tract, the physicians recommended an intrauterine insemination (IUI). To increase the chances of a sperm encountering an egg, a medication called Clomid *(clomiphene)* was prescribed for Maria. Clomid works as an ovulatory stimulant and acts as an antiestrogen agent, causing the body to perceive low estrogen levels. It is given on about days 5 to 10 of the menstrual cycle.

As you read the rest of this chapter, keep Carlos and Maria in mind, and see if you can answer questions about their situation at the end of the chapter.

OVERVIEW OF THE FEMALE REPRODUCTIVE SYSTEM

Function of the Female Reproductive System

The female reproductive system produces gametes called **ova** (eggs). The haploid nucleus of the egg must combine with the haploid nucleus of the sperm if successful fertilization is to occur. The process and function of sex cell formation emphasizes the similarity between the male and female reproductive systems. However, this is where the similarity ends. Unlike the male system, the female reproductive system also provides protection and nutrition to the developing offspring for up to several years after conception, as we shall see.

Structural Plan of the Female Reproductive System

A number of organs make up the female reproductive system, making it somewhat complex. For this reason, we need to look first at the structural plan of the system as a whole (Figure 25-1). As we stated in the previous chapter, reproductive organs can be classified as **essential organs** or **accessory organs,** depending on how directly they are involved in producing offspring. The gonads of women are the paired **ovaries** (the essential organs), which produce the ova. The accessory organs of reproduction in women consist of the following structures:

- A series of ducts or modified duct structures that includes the *uterine tubes, uterus,* and *vagina*
- The *vulva,* or external reproductive organs
- Additional glands, including the *mammary glands* (highly modified sebaceous glands), which secrete milk for the nourishment of newborn children

LANGUAGE OF SCIENCE AND MEDICINE

HINT *Before reading the chapter, say each of these terms out loud. This will help you avoid stumbling over them as you read.*

accessory organ (ak-SES-oh-ree OR-gan)
[*access-* extra, *-ory* relating to, *organ* instrument]

alveolus (al-VEE-oh-lus)
[*alve-* hollow, *-olus* little] *pl.,* alveoli (al-VEE-oh-lye)

ampulla (am-PUL-ah)
[*ampu-* flask, *-ulla* little] *pl.,* ampullae (am-PUL-ee)

anal triangle (AY-nal)
[*an-* ring (anus), *-al* relating to]

anterior fornix (an-TEER-ee-or FOR-niks)
[*ante-* front, *-er-* more, *-or* quality, *fornix* arch] *pl.,* fornices (FOR-nih-seez)

areola (ah-REE-oh-lah)
[*are-* area or space, *-ola* little] *pl.,* areolae, areoles, or areolas (ah-REE-oh-lee, ah-REE-ohlz, ah-REE-oh-lahz)

body [of the uterus] (BOD-ee)

cervix (SER-viks)
[*cervix* neck] *pl.,* cervices or cervixes (SER-veh-seez, SER-viks-ehz)

clitoris (KLIT-oh-ris) *pl.,* clitorides (klit-OH-rih-deez)

corpus albicans (KOHR-pus AL-bih-kanz)
[*corpus* body, *albicans* whitening] *pl.,* corpora albicantia (KOHR-pohr-ah al-bih-KAN-shee-ah)

corpus luteum (KOHR-pus LOO-tee-um)
[*corpus* body, *lute-* yellow, *-um* thing] *pl.,* corpora lutea (KOHR-pohr-ah LOO-tee-ah)

cortex (KOHR-teks)
[*cortex* bark] *pl.,* cortices (KOR-tih-sees)

ectopic pregnancy (ek-TOP-ik)
[*ec-* out of, *-top-* place, *-ic* relating to]

endometrium (en-doh-MEE-tree-um)
[*endo-* within, *-metr-* womb, *-um* thing] *pl.,* endometria (en-doh-MEE-tree-ah)

episiotomy (eh-piz-ee-OT-oh-mee)
[*episi-* vulva, *-tom-* cut, *-y* action]

essential organ (OR-gan)
[*organ* instrument]

estrogen (ES-troh-jen)
[*estr-* frenzy, *-gen* produce]

fimbria (FIM-bree-ah)
[*fimbria* fringe] *pl.,* fimbriae (FIM-bree-ee)

follicle-stimulating hormone (FSH) (FOL-ih-kul-STIM yoo lay-ting HOR-mohn)
[*foll-* bag, *-icle* little, *hormon-* excite]

follicular phase (foh-LIK-yoo-lar fayz)
[*foll-* bag, *-icul-* little, *-ar* relating to]

fornix (FOR-niks)
[*fornix* arch] *pl.,* fornices (FOR-nih-seez)

continued on page 596

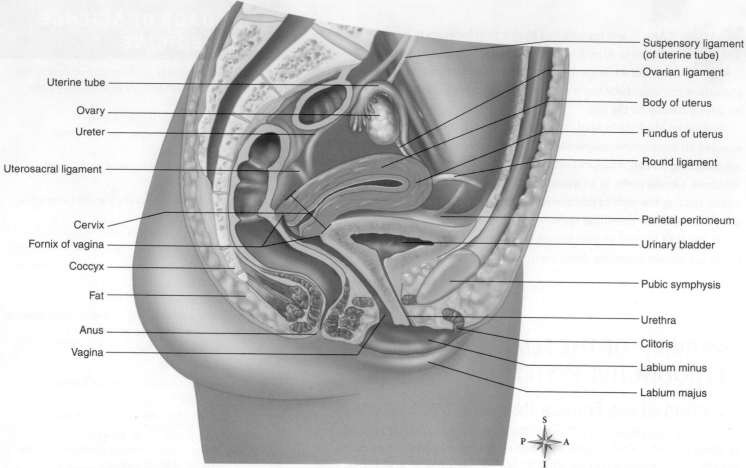

▲ **FIGURE 25-1   Female reproductive organs.** Diagram (sagittal section) of pelvis showing location of female reproductive organs.

You can see most of the essential and accessory organs of the female reproductive system in Figures 25-1, 25-2, and 25-3. Refer to these figures often as you read through the following pages.

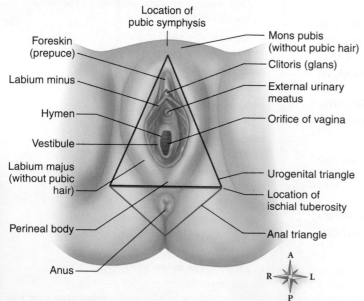

▲ **FIGURE 25-2   Female perineum.** Sketch showing outline of the urogenital triangle *(red)* and anal triangle *(blue)*.

Perineum

The female **perineum** is a muscular region within a diamond-shaped area between the thighs and the vaginal orifice and the anus (see Figure 25-2). It extends from the pubic symphysis anteriorly to the coccyx posteriorly. Its lateral boundary on either side is the ischial tuberosity. A line drawn between the ischial tuberosities divides the area into two triangles. The larger **urogenital triangle** contains the external genitals (labia, vaginal orifice, clitoris) and urinary opening; the smaller **anal triangle** surrounds the anus.

The perineum has great clinical importance because it may be torn during childbirth. Such tears are often deep, have irregular edges, and may extend all the way through the perineum, through the muscular *perineal body,* and even through the anal sphincter. Such damage may result in seepage from the rectum until the laceration is repaired. To avoid these possibilities in a woman prone to such injuries, a surgical incision known as an **episiotomy** may be made in the perineum, particularly at the birth of a first baby. In current medical practice, episiotomy procedures are decreasing in frequency and are no longer performed on a routine basis preceding vaginal delivery of a baby.

QUICK ✓ CHECK

1. What are the *essential organs* of the female reproductive system?

2. List the *major accessory organs* of the female reproductive system.

3. What is the purpose of an episiotomy?

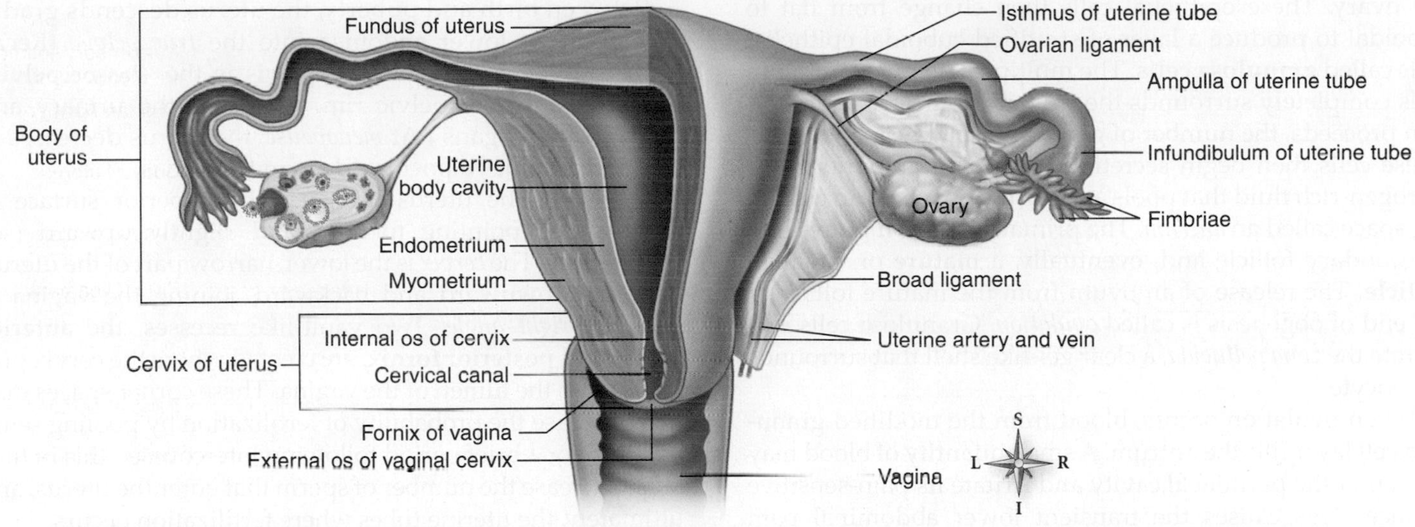

Fundus of uterus

Isthmus of uterine tube

Ovarian ligament

Ampulla of uterine tube

Body of uterus

Uterine body cavity

Infundibulum of uterine tube

Ovary

Fimbriae

Endometrium

Myometrium

Broad ligament

Internal os of cervix

Cervix of uterus

Cervical canal

Uterine artery and vein

Fornix of vagina

External os of vaginal cervix

Vagina

▲ **FIGURE 25-3** **Internal female reproductive organs.** Posterior view. Diagram shows left side of uterus and upper portion of the vagina and the left uterine tube and ovary in a frontal section. The broad ligament has been removed from the posterior surface of the uterus and adjacent structures.

OVARIES

Location of the Ovaries

The ovaries are *homologous* (share the same embryonic origin) to the testes of the male. They are nodular oval glands with a puckered, uneven surface. After puberty, they resemble large almonds in size and shape. One ovary lies on each side of the uterus, below and behind the uterine tubes.

Each ovary weighs about 3 grams and is attached to the posterior surface of the broad ligament by the *mesovarian ligament (mesovarium).* This structure carries blood vessels, nerves, and lymphatics. The *ovarian ligament* anchors the ovary to the uterus. The distal portion of the uterine tube has **fimbriae** that form a cup of fingerlike extensions around the ovary. Note, however, that most of the fimbriae do not actually attach to it (see Figure 25-3). Only one of these, the *ovarian fimbria,* actually attaches directly with the ovary. Unfortunately, this configuration makes it possible for a pregnancy to begin in the pelvic cavity instead of in the uterus, as is normal. Development of the fetus in a location other than the uterus is referred to as an **ectopic pregnancy** (from the Greek *ektopos,* "displaced").

Structure and Function of the Ovaries

The ovary, like many organs in the body, consists of two major layers of tissue: an outer **cortex** and an inner **medulla.** Covering the outer cortex is a layer of flattened epithelial cells called the *germinal epithelium.* Deep to the surface layer of germinal epithelium is a tough layer of connective tissue called the *tunica albuginea.* This tough layer covers the ovarian cortex. Hundreds of thousands of microscopic **ovarian follicles** are embedded in the connective tissue matrix of the cortex. Each follicle contains an immature female sex cell, or *oocyte,* as well as its surrounding cells. After puberty, the oocytes and the specialized cells that surround them are present in varying stages of development. The **ovarian medulla** contains

supportive connective tissue cells, blood vessels, nerves, and lymphatics.

Overview of Oogenesis

Now look at Figure 25-4 for a moment and follow the development of a female sex cell from its origin through its release (ovulation).

Throughout the process of ovarian development the oocyte grows in size. So, too, does the number of cell layers surrounding it. Initially, there is a single layer of flat epithelial cells that originate from the surface epithelium covering

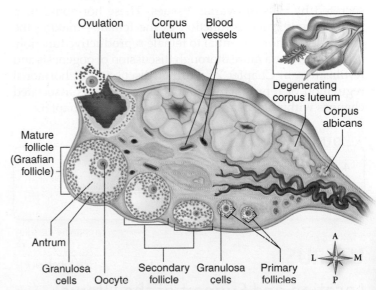

Ovulation

Corpus luteum

Blood vessels

Degenerating corpus luteum

Corpus albicans

Mature follicle (Graafian follicle)

Antrum

Granulosa cells

Oocyte

Secondary follicle

Granulosa cells

Primary follicles

▲ **FIGURE 25-4** **Stages of ovarian follicle development.** Artist's rendition shows the successive stages of ovarian follicle and oocyte development. Begin with the first stage *(primary follicle)* and follow clockwise to the final stage *(degenerating corpus luteum).* Note that all the stages shown occur over time to a *single* follicle. The presence of all these stages at a single point in time is an artificial arrangement for learning purposes only.

UNIT 6

the ovary. These epithelial cells then change from flat to cuboidal to produce a layer of stratified cuboidal epithelial cells called **granulosa cells.** The multiple layers of granulosa cells completely surrounds the primary follicle. As maturation proceeds, the number of granulosa cell layers increases. These cells then begin secreting increasing amounts of an estrogen-rich fluid that pools around the oocyte in an enlarging space called an *antrum.* The primary follicle matures into a secondary follicle and, eventually, a **mature** or Graafian **follicle.** The release of an ovum from the mature follicle at the end of oogenesis is called *ovulation.* Granulosa cells also secrete the *zona pellucida,* a clear gel-like shell that surrounds the oocyte.

When ovulation occurs, blood from the modified granulosa cell layer fills the antrum. A small quantity of blood may also enter the peritoneal cavity and irritate its pain-sensitive surface. This causes the transient lower abdominal pain many women experience at the time of ovulation. Proliferating granulosa cells soon replace the blood filling the antrum, forming a yellow body called the **corpus luteum.** In turn, the corpus luteum secretes the hormones progesterone, inhibin, relaxin, and limited amounts of estrogen. Progesterone and inhibin suppress follicle-stimulating hormone (FSH) secretion. They also prevent the continued development of new follicles during the functional life of the corpus luteum. The small amounts of relaxin secreted by the corpus luteum each month help "quiet" or "calm" uterine contractions. This action improves the chances for successful implantation if fertilization should occur. If pregnancy does occur, larger amounts of these hormones continue to be produced by the *placenta,* as we shall see.

In addition to oogenesis, the ovaries are also endocrine organs, secreting the female sex hormones. **Estrogens** (chiefly estradiol and estrone) and **progesterone** are secreted by cells of ovarian tissues. These hormones help regulate reproductive function in the female—making the ovaries even more essential to female reproductive function.

We will have a more thorough discussion of oogenesis and fertilization in Chapter 26. Further discussion of hormonal regulation of reproductive functions, as well as associated changes within the ovaries, appears later in this chapter.

QUICK ✓ CHECK

4. Briefly describe the location and shape of the ovaries.
5. What is the function of the ovarian follicles?
6. List the major functions of the ovaries.

UTERUS

Location and Support of the Uterus

The uterus is located in the pelvic cavity between the urinary bladder in front and the rectum behind. However, age, pregnancy, and distention of related pelvic viscera such as the bladder may alter the position of the uterus.

Between birth and puberty, the uterus descends gradually from the lower abdomen into the *true pelvis.* (Recall from Chapter 9 that the true pelvis is the "lesser pelvis" located below the pelvic rim. It houses the urinary and reproductive organs.) At *menopause,* the uterus decreases in size and assumes a position deep in the pelvis.

Normally the uterus lies over the superior surface of the bladder, pointing forward and slightly upward (see Figure 25-1). The *cervix* is the lower, narrow part of the uterus: It points downward and backward, joining the vagina at nearly a *right angle.* Two vault-like recesses, the **anterior fornix** and **posterior fornix,** are created where the cervix protrudes into the lumen of the vagina. These corner spaces may help increase the probability of fertilization by pooling seminal fluid for a brief period following intercourse. This in turn helps increase the number of sperm that enter the uterus, and ultimately, the uterine tubes where fertilization occurs.

Several ligaments hold the uterus in place but allow its body considerable movement. In addition, fibers from several muscles that form the pelvic floor converge to form a node called the **perineal body** (see Figure 25-2). This structure also serves an important role in support of the uterus.

Eight *uterine ligaments* (three pairs, two single ones) hold the uterus in its normal position by anchoring it in the pelvic cavity. These ligaments include the broad (paired), uterosacral (paired), posterior (single), anterior (single), and round (paired) ligaments. Six of these so-called ligaments are actually extensions of the parietal peritoneum running in different directions. However, the round ligaments are fibromuscular cords. You can see most of these structures in Figures 25-1 and 25-3.

The uterus may lie in any one of several abnormal positions, largely because the ligaments hold it so loosely. A common abnormal position is **retroflexion,** in which the entire organ is tilted backward. Retroflexion may allow the uterus to *prolapse,* or descend, into the vaginal canal, which can cause chronic discomfort and pain.

Shape and Structure of the Uterus

In a woman who has never been pregnant, the **uterus** is pear shaped and measures approximately 7 cm (3 inches) in length, 5 cm (2 inches) in width at its widest part, and 3 cm (1 inch) in thickness. Note in Figure 25-3 that the uterus has two main parts: a wide, upper portion (the **body**), and a lower, narrow "neck" (the **cervix**). The body of the uterus rounds into a bulging prominence, the **fundus.** The dome-shaped fundus is superior to the points of entry of the uterine tubes on both sides.

Three layers comprise the walls of the uterus: (1) the inner endometrium, (2) a middle myometrium, and (3) an outer incomplete layer of parietal peritoneum.

Endometrium

A ciliated mucous membrane called the **endometrium** lines the uterus. During menstruation and after delivery of a baby, the outer layers of the endometrium slough off. The endometrium varies in thickness from 0.5 mm just after the menstrual flow to about 5 mm near the end of the endometrial cycle.

The endometrium has a rich supply of blood capillaries. It also has numerous exocrine *uterine glands* that secrete mucus and other substances onto the endometrial surface. The mucous glands in the lining of the cervix produce mucus that changes in consistency during the female reproductive cycle. Most of the time, cervical mucus acts as a barrier to sperm. Around the time of ovulation, however, cervical mucus becomes more slippery and actually facilitates the movement of sperm through the cervix and into the body of the uterus.

Myometrium

The **myometrium** is the thick, middle layer of the uterine wall. It consists of three layers of smooth muscle fibers. These muscle fibers extend in all directions: longitudinally, transversely, and obliquely—and thus give the uterus great strength. The bundles of smooth muscle fibers are interlaced with elastic and connective tissue components. The result is a blending into the endometrial lining with no sharp boundary between the two layers. The myometrium is thickest in the fundus and thinnest in the cervix—a good example of the principle of structural adaptation to function. The fundus must contract more forcibly than the lower part of the uterine wall to expel the fetus; the cervix must stretch or dilate to accommodate the fetus.

Perimetrium

The outermost serous layer of the uterus (the visceral peritoneal covering), is called the **perimetrium.** This layer does not completely cover the surface of the uterus. It is absent over the entire cervix and the lower one-fourth of the anterior surface of the uterine body. Look carefully at Figure 25-1 and note that the parietal peritoneum of the anterior pelvic wall folds back on itself and becomes the visceral peritoneum covering the top of the bladder. It then turns upward to cover the upper three fourths of the anterior surface of the uterine body and continues up over the fundus and down the posterior surface of the uterus to the top of the cervix where it is then reflected back to cover the rectum. The perimetrium, although continuous with the peritoneal lining, is incomplete in that it does not cover the entire surface of the uterus. The fact that the entire uterus is *not* covered with peritoneum may seem silly to point out, but it has clinical significance. It makes it possible to perform operations on this organ without the same risk of infection that occurs in procedures that cut through the peritoneum.

Function of the Uterus

The uterus has many functions important to successful reproduction. It serves as part of the female reproductive tract, permitting sperm from the male to ascend toward the uterine tubes. If fusion of gametes (*fertilization*, or *conception*) occurs, the developing offspring implants in the endometrial lining of the uterus and continues its development during the term of pregnancy (*gestation*). The tiny endometrial glands produce nutrient secretions—sometimes called "uterine milk"—to sustain the developing offspring until a *placenta* can be produced. The **placenta** is a unique organ that permits the exchange of materials between the offspring's blood and the maternal blood. A rich network of endometrial capillaries promotes efficiency of this exchange function. Regular contractions of the myometrium, or **labor,** are inhibited during gestation but become rhythmic and intense as the time of delivery approaches.

If conception or the successful implantation of the embryo fails, then the outer layers of the endometrium are shed during *menstruation*. **Menstruation** is a regular event of the female reproductive cycle. It permits the endometrium to renew itself in anticipation of conception and implantation during the next cycle. The myometrial contractions seem to aid menstruation by promoting the complete sloughing of the outer endometrial layers (the "period"). Fatigue of the myometrial muscle tissues may contribute to the abdominal cramping sometimes associated with menstruation.

UTERINE TUBES

Position and Structure of the Uterine Tubes

The uterine tubes are also sometimes called *fallopian tubes,* or *oviducts.* They are about 10 cm (4 inches) long and are attached to the uterus at its upper outer angles (see Figures 25-1 and 25-3). You can see that the uterine tubes lie in the upper *free margin* of the broad ligaments. From here they extend upward and outward toward the sides of the pelvis before curving downward and backward toward the uterus.

The same three layers (mucous, smooth muscle, and serous) of the uterus also comprise the uterine tubes. In fact, the mucosal lining of the tubes is continuous with the peritoneum lining the pelvic cavity. This has great clinical significance because the tubal mucosa is also continuous with that of the uterus and vagina. As a result, the continuous reproductive lining can become infected by gonococci or other organisms introduced into the vagina. Inflammation of the tubes **(salpingitis)** may readily spread to become inflammation of the peritoneum **(peritonitis)**—a very serious condition. Inflammation of the uterine tubes may also lead to scarring and partial or complete closure of the lumen. This can happen even if the original infection is cured with antibiotics. (Note: In the male, there is no such direct route by which microorganisms can reach the peritoneum from the exterior.)

Divisions and Tissues of the Uterine Tubes

Each uterine tube consists of three divisions (see Figure 25-3): (1) a connecting part called the **isthmus;** (2) a dilated portion called the **ampulla;** and (3) a funnel-shaped end called the **infundibulum.** The infundibulum lies just above and extends laterally over the ovary. It opens directly into the peritoneal cavity, dividing into fringelike projections called *fimbriae.*

BOX 25-1 FYI

Tubal Ligation

Tubal ligation literally means "tying a tube." For this reason, this surgical procedure is often referred to as "having one's tubes tied." Tubal ligation involves tying a piece of suture material around each uterine tube in two places, then cutting each tube between these two points (see figure). Because sperm and eggs are thus blocked from meeting, fertilization and subsequent pregnancy are prevented. Tubal ligation is also called *surgical sterilization* and is comparable to vasectomy in the male.

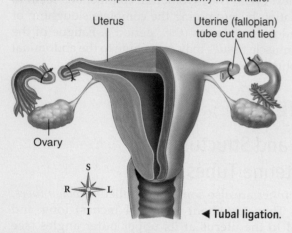

◄ Tubal ligation.

Function of the Uterine Tubes

The uterine tubes are really extensions of the uterus that communicate loosely with the ovaries. This arrangement allows an ovum released from the surface of the ovary to be collected by the fimbriae. From here the ovum is swept along the uterine tube by ciliary action toward the body of the uterus.

However, the uterine tubes serve as more than mere transport channels. The uterine tube is also the *site of fertilization.* Sperm and ova most often meet, and fertilization occurs, in the *ampulla* of the uterine tube. A relatively small number of the sperm deposited in the vagina during sexual intercourse move up the uterine tube, where they meet the ovum being swept down toward them. Here is where fertilization usually takes place. Totally blocking the openings into either the distal (abdominal) or proximal (uterine) ends of both uterine tubes, for any reason, results in sterility (Box 25-1).

QUICK CHECK

7. Describe the three principal layers of the uterine wall.
8. Describe the anatomical position of the uterus. How is it held in place?
9. List the major functions of the uterus.
10. What are the functions of the uterine (fallopian) tubes?

VAGINA

Structure of the Vagina

The **vagina** is a collapsible tube about 8 cm (3 inches) long, situated between the rectum, and the urethra, and the bladder. It is capable of enormous distention during delivery of a baby. It is composed mainly of smooth muscle and is lined with mucous membrane arranged in rugged folds called *rugae.* The vaginal mucosa contains numerous tiny exocrine mucous glands that secrete lubricating fluid during the female sexual response.

Note that the anterior wall of the vagina is shorter than the posterior wall because of the way the cervix protrudes into the uppermost portion of the tube (see Figure 25-1). In some cases—especially in young girls—a fold of mucous membrane, the **hymen,** forms a border around the external opening of the vagina, partially closing the orifice. Occasionally, this structure completely covers the vaginal outlet, a condition referred to as **imperforate hymen.** Perforation must be performed at puberty before the menstrual flow can escape.

Function of the Vagina

The *vagina* has several important functions. During sexual intercourse, the lining of the vagina lubricates and stimulates the glans penis, which in turn triggers the ejaculation of semen. Thus the vagina also serves as a receptacle for semen, which often pools in the anterior or posterior **fornix** of the vagina. Here the semen meets the cervix of the uterus. Sperm within the semen may move further into the female reproductive tract by "climbing" along fibrous strands of mucus in the cervical canal.

The vagina also serves as the lower portion of the birth canal. At the time of delivery, the baby is pushed from the body of the uterus, through the cervical canal, and finally through the vagina and out of the mother's body. The placenta, or "afterbirth," is also expelled through the vagina.

Another important function of the vagina is transport of blood and tissue shed from the lining of the uterus during menstruation.

VULVA

Structure of the Vulva

Figure 25-5 shows you the structures that, together, constitute the female *external genitalia.* Collectively, these structures are called the **vulva.** We've summarized the various components and their functions for you in the following paragraphs.

The **mons pubis** is a skin-covered pad of fat over the pubic symphysis. Coarse pubic hairs appear on this structure at puberty and persist throughout life.

The **labia majora** (Latin, "large lips") are covered with pigmented skin and hair on the outer surface and are smooth

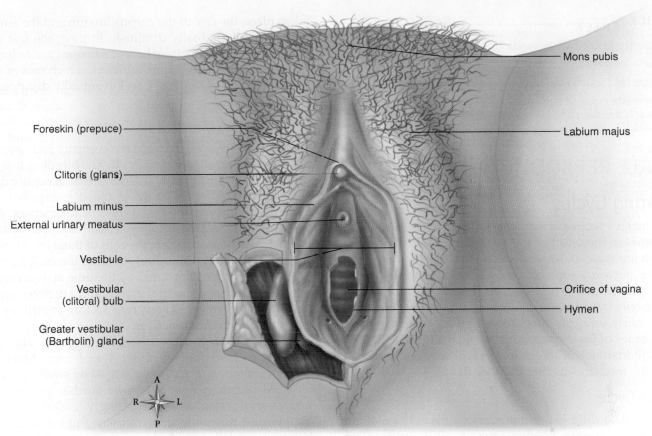

Mons pubis

Foreskin (prepuce)

Labium majus

Clitoris (glans)

Labium minus

External urinary meatus

Vestibule

Vestibular
(clitoral) bulb

Orifice of vagina

Hymen

Greater vestibular
(Bartholin) gland

A
R — L
P

▲ FIGURE 25-5 **Vulva (pudendum).** Sketch showing major features of the external female genitals (genitalia).

and free from hair on the inner surface. Each labium majus is composed mainly of fat and connective tissue with numerous sweat and sebaceous glands on the inner surface. Together the labia majora are homologous to the scrotum in the male.

The **labia minora** (Latin, "small lips") are located medially to the labia majora. Each labium minus is covered with hairless skin.

The **clitoris** is composed of erectile tissue. A small portion of it is visible just behind the junction of the labia minora. Most of the erectile tissue lies buried beneath the skin of the vulva. The structure of this organ is homologous to the penile structure of the male. Like the erectile tissue of the male, the clitoris becomes engorged with blood during the sexual response.

The **glans clitoris** is the only visible part of the erectile structures of the clitoris. It is equivalent to the *glans penis* in the male. The glans clitoris is covered with highly sensitive skin that, during sexual stimulation, produces most of the female sexual response.

A clitoral *foreskin* or **prepuce** forms a hood over the superior surface of the glans clitoris.

The *external urinary meatus* (urethral orifice) is the small opening of the urethra, situated between the clitoris and the vaginal orifice. The **vaginal orifice** has a much larger opening than the urinary meatus. It is located posterior to the meatus.

The **greater vestibular glands** are two bean-shaped glands, one on each side of the vaginal orifice. Each gland opens by means of a single, long duct into the space between the hymen and the labium minus. These glands, which are also called *Bartholin glands*, are of clinical importance because they can be infected (bartholinitis or Bartholin abscess), particularly by gonococci. They are homologous to the bulbourethral glands in the male.

Function of the Vulva

The various components of the external genitals of the female operate alone or separately to accomplish several functions important to successful reproduction. For example, the protective features of the mons pubis and labia help prevent injury to the delicate tissues of the clitoris and vestibule. The clitoris becomes erect during sexual stimulation. Like the male glans, it possesses a large number of sensory receptors that feed information back to the sexual response areas of the brain.

A&P CONNECT

There are many sexually transmitted diseases (STDs) that can affect the female reproductive tract. Review examples of important STDs in **Sexually Transmitted Diseases** online at **A&P Connect.**

evolve
learning system

QUICK ✓ CHECK

11. List several functions of the vagina.
12. What is another name for the external genitals of the female?
13. List the basic features and functions of the external female genitalia.
14. How are the clitoris of the female and the glans penis of the male similar in structure and function? Can you explain this?

FEMALE REPRODUCTIVE CYCLES

Recurring Cycles

Many changes recur periodically in the female during the years between the onset of the menses (*menarche*) and their cessation (*menopause*). Most obvious, of course, is menstruation—the outward sign of changes in the endometrium. Most women also note periodic changes in their breasts. But these are only two of many changes that occur over and over again at fairly uniform intervals during the approximately three decades of female reproductive maturity.

We will first look at the major cyclical changes, and then discuss the mechanisms that produce them.

Ovarian Cycle

Before a female child is born, precursor cells in her ovarian tissue, called **oogonia,** begin a type of cell division called *meiosis,* which *reduces the number of chromosomes* in the daughter cells by half (review Chapter 5, p. 87). By the time the child is born, her ovaries contain about 250,000 **primary follicles,** each containing an oocyte that has temporarily suspended the meiotic process before it is complete.

Once each month, on about the first day of menstruation, the oocytes within several primary follicles resume meiosis. At the same time, the follicular cells surrounding them increase in number and start to secrete estrogens (and tiny amounts of progesterone). Usually, only one of these developing follicles *matures and migrates* to the surface of the ovary. Just before *ovulation,* the meiosis within the oocyte of the mature follicle *stops again.* It is this cell (which has not quite completed meiosis) that is expelled from the ruptured wall of the mature follicle during ovulation. Meiosis is completed only when, and if, the head of a sperm cell is later drawn into the ovum during the process of fertilization.

When does ovulation occur? This is a question of great practical importance and one that in the past was given many answers. Today it is known that ovulation usually occurs 14 days before the next menstrual period begins. (Only in a 28-day menstrual cycle is this also 14 days *after* the beginning of the preceding menstrual cycle, as explained later in this chapter.)

Immediately after ovulation, cells of the ruptured follicle enlarge. Because of the appearance of lipid-like substances in them, they are transformed into a golden-colored body, the *corpus luteum.* The corpus luteum grows for 7 or 8 days. During this time, it secretes progesterone in increasing amounts. Then, provided fertilization of the ovum has not taken place, the size of the corpus luteum and the amount of its secretions gradually diminish. In time, the last components of each nonfunctional corpus luteum are reduced to a white scar called the **corpus albicans,** which moves into the central portion of the ovary and eventually disappears (see Figure 25-4).

Endometrial (Menstrual) Cycle

During menstruation, parts of the compact and spongy layers of the endometrium slough off. The bleeding that ensues produces a dark menstrual discharge that generally does not clot. Between 30 and 100 ml of blood is expelled, with a majority lost during the first 3 days of the menses. As with the length of the menstrual cycle, considerable variation is normal. After menstruation, the cells of these layers *proliferate* (increase in size and number), causing the endometrium to reach a thickness of 2 or 3 mm by the time of ovulation. During this period, endometrial glands and arterioles grow longer and more coiled. These two factors contribute to the thickening of the endometrium.

After ovulation, the endometrium grows still thicker, reaching a maximum of about 4 to 6 mm. Most of this increase, however, is probably caused by swelling produced by fluid retention rather than by further proliferation of endometrial cells. The increasingly coiled endometrial glands start to secrete their nutrient fluid during the time between ovulation and the next menses. Then, the day before menstruation starts again, a drop in progesterone causes muscle in the walls of the tightly coiled arterioles to constrict, producing *endometrial ischemia.* This leads to death of the tissue, sloughing, and once again, menstrual bleeding.

The menstrual cycle is customarily divided into phases, named for major events occurring in each: *menses, postmenstrual phase, ovulation,* and *premenstrual phase.*

1. The **menses,** or **menstrual period,** occur on days 1 to 5 of a new cycle. There is some individual variation, however.

2. The **postmenstrual phase** occurs between the end of the menses and ovulation. It is also called the *preovulatory phase* as well as the **proliferative phase.** In a 28-day cycle, it usually includes cycle days 6 to 13 or 14. However, the length of this phase varies more than the others. It lasts longer in long cycles and ends sooner in short cycles. This phase is also called the **follicular phase,** because of the high blood estrogen level resulting from secretion by the developing follicle. Increases in estrogen levels cause predictable changes in the appearance, amount, and consistency of cervical mucus. Collectively, these changes can be used as a fertility sign to predict ovulation (Box 25-2).

3. **Ovulation** is the rupture of the mature follicle with expulsion of its ovum into the pelvic cavity (Figure 25-6). It occurs most often on cycle day 14 in a 28-day cycle. However, ovulation can occur on different days in cycles of different length, depending on the length of the preovulatory phase. For example, in a 32-day cycle the

Fertility Signs Used in Predicting the Time of Ovulation

Many rhythmic and recurring events that a woman may recognize on almost a monthly schedule during her reproductive years are called "fertility signs." These "signs" represent the body changes required to permit successful reproduction. They include cyclical changes in (1) the ovaries, (2) the amount and consistency of the cervical mucus produced during each cycle, (3) the myometrium, (4) the vagina, (5) gonadotropin secretion, (6) body temperature, and (7) mood or "emotional tone." Accurately predicting the time of ovulation in any given menstrual cycle by recognizing one or more of these recurring fertility signs would obviously be of help in either *avoiding* or *achieving* conception. However, knowing the length of a previous cycle or even a series of cycles cannot ensure with any degree of accuracy the time of appearance of other fertility signs in a current cycle. Nor can it predict how many days the preovulatory phase will last in the next or some future cycle.

Unfortunately, this means that prior cycle length is not an accurate fertility sign. This fact accounts for most of the unreliability of the *calendar rhythm method* of fertility planning. Other more sophisticated *natural family planning (NFP)* methods are available that are not based on a knowledge of previous cycle lengths to predict the day of ovulation. Instead, such natural methods base their judgments about fertility at any point in a woman's cycle on other changes. For example, women hopeful of becoming pregnant can predict their general state of receptiveness through the measurement of *basal body temperature*—body temperature taken after awakening at the same time each day. They can also monitor the cyclical changes in the amount and consistency of cervical mucus during their cycle. Changes in basal body temperature and the amount and consistency of cervical mucus occur in response to changes in circulating hormones that control ovulation. Typically, use of NFP for 1 year to avoid pregnancy will result in approximately 25 of every 100 women becoming pregnant.

The time of ovulation also can be approximated by over-the-counter urine tests that detect the high levels of luteinizing hormone (LH) associated with ovulation ("LH surge").

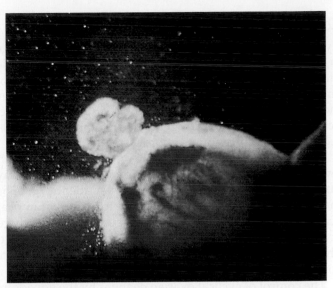

▲ **FIGURE 25-6** **Ovulation.** The rupture of a mature follicle on the surface of an ovary results in the release of an ovum into the pelvic cavity. This process of *ovulation* often occurs on day 14 in a 28-day menstrual cycle, but its exact timing depends on the length of the postmenstrual (preovulatory) phase. Notice in this photograph that the ovum released during ovulation is surrounded by a mass of cells.

14 days—or cycle days 15 to 28 in a 28-day cycle. Differences in length of the total menstrual cycle therefore exist mainly because of differences in duration of the postmenstrual rather than of the premenstrual phase.

Gonadotropic Cycle

As we saw in Chapter 15, the anterior pituitary gland secretes two hormones called *gonadotropins* that influence female reproductive cycles. Their names are **follicle-stimulating hormone (FSH)** and **luteinizing hormone (LH).** The amount of each gonadotropin secreted varies with a rhythmic regularity that can be related, as we shall see, to the rhythmic ovarian and uterine changes just described.

preovulatory phase probably lasts until cycle day 18. Ovulation would then occur on cycle day 19 instead of 14. Because the majority of women show some month-to-month variation in the length of their cycles, the day of ovulation in a current or future cycle cannot be predicted with accuracy based on the length of previous cycles (see again Box 25-2). *However, there is typically a decrease in basal body temperature just before ovulation and a rise in temperature at the time of ovulation.* This constitutes yet another "fertility sign" (see Figure 25-9).

4. The **premenstrual phase** occurs between ovulation and the onset of the menses. This phase is also called the **luteal phase** or **secretory phase,** because the corpus luteum secretes progesterone only during this time. The length of the premenstrual phase is fairly constant, lasting usually

---

QUICK ✓ CHECK

15. Define *menarche* and *menopause.*
16. What is the function of the corpus luteum?
17. What is the difference between the proliferative phase and the luteal or secretory phase?
18. Briefly describe the four phases of the menstrual cycle.

Control of Female Reproductive Cycles

Hormones play a major role in producing the cyclical changes characteristic of women during their reproductive years. The following paragraphs provide a brief description of the mechanisms that produce cyclical changes in the ovaries and uterus and in the amounts of gonadotropins secreted.

Control of Cyclical Changes in the Ovaries

Cyclical changes in the ovaries result from cyclical changes in the amounts of gonadotropins secreted by the anterior pituitary gland. An increasing FSH blood level has two effects: (1) it stimulates one or more primary follicles and their oocytes to start growing, and (2) it stimulates the follicular cells to secrete **estrogens.** (Developing follicles also secrete very small amounts of progesterone.)

Because of the influence of FSH on follicle secretion, the level of estrogens in blood increases gradually for a few days during the postmenstrual phase. Then suddenly, on about the twelfth cycle day, it leaps upward to a maximum peak. Scarcely 12 hours after this "estrogen surge," an "LH surge" occurs and presumably triggers ovulation a day or two later. This hormone surge is the basis of the over-the-counter "ovulation test" (see Box 25-2).

The control of cyclical ovarian changes by the gonadotropins FSH and LH is summarized for you in Figure 25-7. Refer to this diagram as you read the following description of cyclical changes in the ovary.

1. Completion of growth of the follicle and oocyte maturation with increasing secretion of estrogens before ovulation. LH and FSH act as synergists to produce these effects.

2. Rupturing of the mature follicle with expulsion of its ovum (ovulation). Because of this function, LH is sometimes also called "the ovulating hormone."

3. Formation of a yellowish body, the corpus luteum, in the ruptured follicle (process called **luteinization**). The name *luteinizing hormone* refers, obviously, to this LH function—a function to which, experiments have shown, FSH also contributes.

The corpus luteum functions as a *temporary* endocrine gland. It secretes only during the luteal (postovulatory, or premenstrual) phase of the menstrual cycle. It secretes progesterone and estrogen. The blood level of progesterone rises rapidly after the "LH surge" described earlier. It remains at a high level for about a week, and then decreases to a very low level

approximately 3 days before menstruation begins again. This low blood level of progesterone persists during both the menstrual and the postmenstrual phases. What are its sources? Not the corpus luteum, which secretes only during the luteal phase, but the developing follicles and the adrenal cortex. Blood's estrogen content increases during the luteal phase but to a lower level than develops before ovulation.

If pregnancy does not occur, lack of sufficient LH and FSH causes the corpus luteum to regress in about 14 days. The corpus luteum is then replaced by the *corpus albicans.* To make sure you've understood this process, review again Figure 25-4, which shows the cyclical changes in the ovarian follicles.

Control of Cyclical Changes in the Uterus

Changing blood concentrations of estrogens and progesterone also bring about cyclical changes in the uterus. As blood estrogens increase during the postmenstrual phase of the menstrual cycle, they produce the following changes in the uterus:

- Thickening of the endometrium
- Growth of glands and spiral arteries within the endometrium
- Increase in the water content of the endometrium
- Increase of myometrial contractions

Increasing blood progesterone concentration during the premenstrual phase of the menstrual cycle produces changes in the uterus due to the actions of progesterones. These changes are favorable for pregnancy—specifically the following:

- Preparation of the endometrium for the implantation of a fertilized ovum
- Increase in the water content of the endometrium
- Decrease of myometrial contractions

As we mentioned earlier, low levels of FSH and LH cause *atrophy* of the corpus luteum if pregnancy does not occur. This in turn causes a drop in estrogen and progesterone levels. As a result, the maintenance of a thick, vascular endometrium

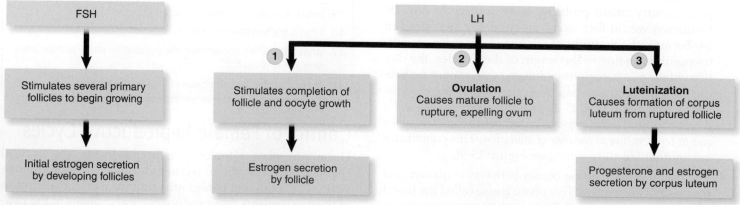

▲ **FIGURE 25-7** **The primary effects of gonadotropins on the ovaries.** Follicle-stimulating hormone *(FSH)* gets its name from the fact that it triggers development of primary ovarian follicles and stimulates follicular cells to secrete estrogens. Luteinizing hormone *(LH)* has several effects on ovaries: *(1)* LH acts with FSH to enhance its effects on follicular development and secretion; *(2)* LH presumably triggers ovulation—hence it is called "the ovulating hormone"; and *(3)* LH has a luteinizing effect (for which the hormone was named); recent evidence shows that FSH is also necessary for luteinization.

ceases. Finally, a drop in estrogen and progesterone levels at the end of the premenstrual phase triggers the endometrial sloughing that characterizes the menstrual phase (the "period").

Control of Cyclical Changes in Gonadotropin Secretion

Both negative and positive feedback mechanisms help control anterior pituitary secretion of the gonadotropins FSH and LH. These mechanisms involve the ovaries' secretion of inhibin, estrogens, and progesterone. They also involve the secretion of releasing hormones by the hypothalamus. Figure 25-8 shows you the negative feedback mechanism that controls gonadotropin secretion. Examine it carefully. Notice the effects of a sustained high blood concentration of estrogens and progesterone on anterior pituitary gland secretion. Note also the effect of a low blood concentration of FSH on follicular development. *Essentially, follicles do not mature and ovulation does not occur under these conditions.*

Several observations and animal experiments suggest that sustained high blood levels of estrogens, progesterone, and inhibin decrease pituitary secretion of FSH and LH.

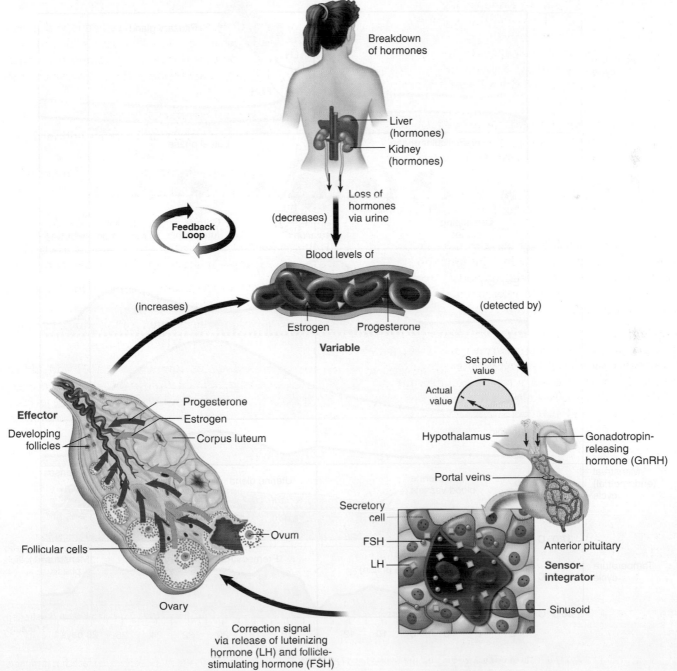

▲ **FIGURE 25-8**   **Control of FSH and estrogen secretion.**   A negative feedback mechanism controls anterior pituitary secretion of follicle-stimulating hormone (FSH) and ovarian secretion of estrogens. A high blood level of FSH stimulates estrogen secretion, whereas the resulting high estrogen level inhibits FSH secretion.

These ovarian hormones appear to inhibit the hypothalamus from secreting gonadotropin-releasing hormone (GnRH). Without the stimulating effects of these releasing hormones, the pituitary's secretion of FSH and LH decreases.

A positive feedback mechanism may also control LH secretion. There is a sudden and marked increase in blood estrogen level that occurs late in the follicular phase of the menstrual cycle. This is thought to stimulate the hypothalamus to secrete GnRH. In turn, GnRH stimulates the release of LH by the anterior pituitary. This would in turn account for the "LH surge" that triggers ovulation.

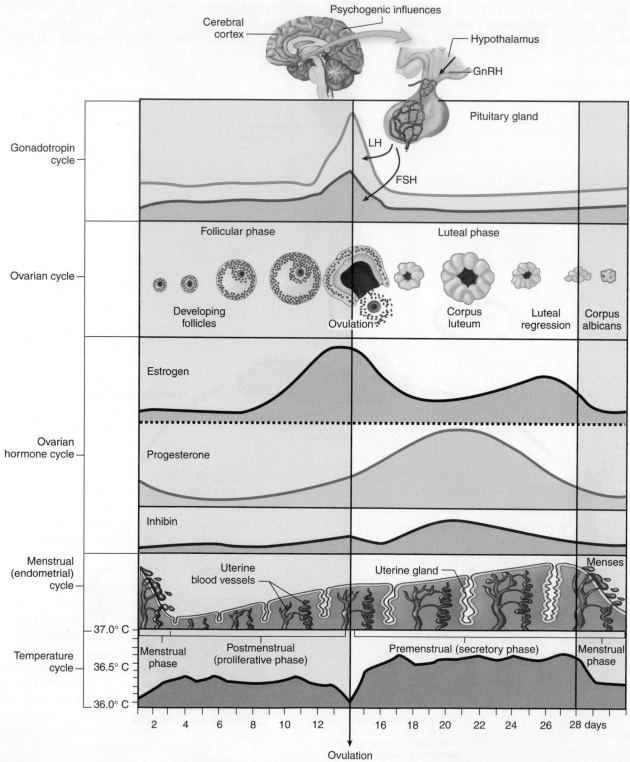

▲ **FIGURE 25-9**   **Female reproductive cycles.** This diagram illustrates the interrelationships among the cerebral, hypothalamic, pituitary, ovarian, and uterine functions throughout a standard 28-day menstrual cycle. The variations in basal body temperature are also illustrated. The labels at the bottom of the diagram are the phases of the menstrual cycle.

The fact that a part of the brain—the *hypothalamus*—secretes gonadotropin-releasing hormones has interesting implications. This may be part of the pathway by which changes in a woman's environment or in her emotional state can alter her menstrual cycle. That this occurs is a matter of common observation. Stress, for example—such as intense fear of either becoming or not becoming pregnant—often delays menstruation.

Importance of Female Reproductive Cycles

The female reproductive cycles play several important roles. As Figure 25-9 illustrates for you, the changes associated with the different cycles are all closely interrelated. The primary role of the ovarian cycle, for example, is to produce an ovum at regular enough intervals to make reproductive success likely. The ovarian cycle's secondary role is to regulate the endometrial (menstrual) cycle by means of the sex hormones estrogen and progesterone. The role of the endometrial cycle, in turn, is to ensure that the lining of the uterus is suitable for the implantation of an embryo if fertilization of the ovum occurs. The constant renewal of the endometrium makes successful implantation more likely.

Human fertility is further limited by the fact that sperm usually cannot survive in the female reproductive tract for more than a few days. Such limited fertility increases the likelihood that conception will occur only when the woman's body is at its reproductive peak. Box 25-3 discusses some common methods for managing fertility.

Infertility and Use of Fertility Drugs

Infertility is often defined as failure to conceive after 1 year of regular unprotected intercourse. Infertility may be caused by a wide variety of medical, environmental, and even lifestyle factors, such as smoking or alcohol abuse. Causal factors may be traced to various problems in either the male or female partner, each accounting for about 40% of cases. Of

---

BOX 25-3 Health Matters

Methods of Contraception

Hormonal methods of contraception began with establishment of the relationship between sex hormone levels and ovulation. Continuing research in this area led to the development of **oral contraceptives**—often collectively called "the Pill." Numerous oral contraceptive products are now available that contain different types, combinations, and dosages of estrogen and/or progesterone. The so-called minipill contains only synthetic progesterone.

Most hormonal contraceptives were developed to prevent pregnancy by initiating negative feedback inhibition of FSH and LH secretion. As a result, mature follicles do not develop, and LH levels required to initiate ovulation do not occur. The next menses, however, does take place if the progesterone and estrogen dosage is stopped in time to allow their blood levels to decrease as they normally do near the end of the cycle to bring on menstruation. For this reason, the Pill can be used to regulate the menstrual cycle, as well as prevent pregnancy. If used correctly and consistently, the pill is an extremely effective contraceptive with an unintended pregnancy rate estimated at between 0.1% and 3%. The higher percentage reflects "typical" rather than "ideal" use, and underscores the impact of human error in using any form of birth control.

In addition to oral contraceptives taken in pill form, other types of hormonal birth control "delivery mechanisms" are available. They include hormone-impregnated vaginal inserts, hormone injections, transcutaneous administration using skin "patches," and surgical insertion of hormone-containing implants under the skin (see figure).

The effects of hormonal contraceptives—indeed, of estrogens and progesterone—are much more complex than our explanation here indicates. They have widespread effects on the body quite independent of their action on the reproductive and endocrine systems and

are still not completely understood. Possible side effects—some extremely serious, including stroke and heart attack—may limit or prohibit use of these birth control methods by some women. Side effects and health risks of hormonal contraceptives are especially troublesome if these products are used for extended periods, by older women, by women who smoke, and by women with blood clotting problems or cardiovascular disease. On the other hand, long-term use has also been shown to have some beneficial health effects such as protection against uterine and ovarian cancer in some groups of women.

In addition to hormonal methods of contraception, many other methods, each with differing rates of effectiveness and unique advantages and disadvantages, are available. For example, *spermicidal methods* involve use of preparations (foams, jellies, and creams) that act to kill sperm, and *mechanical barrier methods* use devices such as condoms, diaphragms, and cervical caps to block sperm from entering the uterus. So-called *surgical methods* such as tubal ligation and male vasectomy result in permanent sterility.

The use of contraceptive methods to regulate reproductive function often involves many personal decisions. For example, the decision to use or avoid contraception—or employ any particular contraceptive method—at any point in time is often influenced by differing medical, social, cultural, ethical, and religious factors as well as by the cost, reliability, safety, or ease of use of a particular method. Informed and thoughtful decision making regarding this human behavior is critically important. It will often be necessary for some individuals to seek out a variety of information—from different but credible and knowledgeable sources—in order to make an informed decision that is "right" for those individuals. Regardless, seeking counsel and advice from a trusted health care provider early in the process is always recommended.

Hormone-containing contraceptive implants

TABLE 25-1 Some Female Reproductive Hormones*

HORMONE	SOURCE	TARGET	ACTION
Dehydroepiandrosterone (DHEA)	Adrenal gland, ovary, other tissues	Converted to other hormones	Eventually converted to estrogens, testosterone, or both
Estrogens (including estradiol [E_2] and estrone)	Ovary and placenta (small amounts in other tissues)	Uterus, breast, other tissues	Stimulates development of female sexual characteristics, breast development, bone and nervous system maintenance
Follicle-stimulating hormone (FSH)	Anterior pituitary (gonadotroph cells)	Ovary	Gonadotropin; promotes development of ovarian follicle; stimulates estrogen secretion
Gonadotropin-releasing hormone (GnRH)	Hypothalamus (neuroendocrine cells)	Anterior pituitary (gonadotroph cells)	Stimulates production and release of gonadotropins (FSH and LH) from anterior pituitary
Human chorionic gonadotropin (hCG)	Placenta	Ovary	Stimulates secretion of estrogen and progesterone during pregnancy
Inhibin	Ovary	Anterior pituitary (gonadotroph cells)	Inhibits FSH production in the anterior pituitary (perhaps by limiting GnRH)
Luteinizing hormone (LH)	Anterior pituitary (gonadotroph cells)	Ovary	Gonadotropin; triggers ovulation; promotes development of corpus luteum
Progesterone	Ovary and placenta	Uterus, mammary glands, other tissues	Helps maintain proper conditions for pregnancy
Relaxin	Placenta	Uterus and joints	Inhibits uterine contractions during pregnancy and softens pelvic joints to facilitate childbirth
Testosterone	Adrenal glands, ovaries	Nervous tissue, bone tissue, other tissues	May affect mood, sex drive, learning, sleep, protein anabolism, other functions

*The role of some hormones related to pregnancy, labor, and delivery are discussed in more detail in the next chapter (Chapter 26, Growth and Development).

the remaining 20% of affected couples, infertility in about 10% is due to problems shared by both partners and in about 10% the reason is never determined.

If testing identifies the female member of the couple as infertile, she joins a subset of about 25% of women in the overall population who will experience some period of infertility during their reproductive years. In many cases, infertility results from a *failure to ovulate*. This is often caused by a medical condition such as *polycystic ovary syndrome (PCOS)* (see *Mechanisms of Disease*, available at your Evolve site online). Significant numbers of infertile women who experience ovulatory dysfunction desire to become pregnant. After a sometimes long and complex medical workup and selection process, they may become candidates to receive so-called *fertility drugs*. These can be used alone or in combination with other "assisted reproductive procedures" such as artificial insemination.

Menarche and Menopause

The menstrual flow first occurs (**menarche**) at puberty, at about the age of 13 years, although there is individual variation according to race, nutrition, health, and heredity. Normally, it recurs about every 28 days for about three decades, except during pregnancy, and then ceases (**menopause**). The average age at which menstruation ceases is reported to have increased markedly—from about age 40 years a few decades ago to between ages 45 and 50 years more recently.

The changes just described relate to changes in hormone levels over the life span. Relatively low concentrations of

gonadotropins (FSH and LH) sustain a peak of estrogen secretion from menarche to menopause. After menopause, estrogen concentration decreases dramatically—which causes a negative feedback response that increases the gonadotropin levels. Because the follicular cells are no longer sensitive to gonadotropins after menopause, the increased gonadotropin level has no effect on estrogen secretion.

Table 25-1 summarizes some of the hormones important in female reproductive function.

QUICK CHECK

19. A surge in FSH and LH is associated with what major event of the ovarian cycle?
20. How does an increase in estrogen level affect the uterine lining?
21. What is infertility?

BREASTS

Location and Size of the Breasts

Two breasts lie over the pectoral muscles and are attached to them by a layer of connective tissue (Figure 25-10). Breasts are made up of milk-producing **mammary glands,** which are present in all mammals, along with extensive supporting tissues. They are present in both males and females—but only infrequently develop or produce milk in males.

Estrogens and progesterone both control breast development during puberty. Estrogens stimulate growth of the

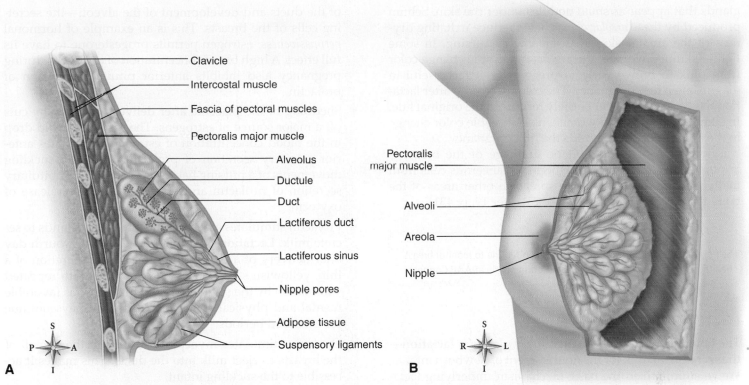

▲ FIGURE 25-10   The female breast. A, Sagittal section of a lactating breast. Notice how the glandular structures are anchored to the overlying skin and to the pectoral muscles by suspensory ligaments. Each lobule of glandular tissue is drained by a lactiferous duct that eventually opens through the nipple. **B,** Anterior view of a lactating breast. Overlying skin and connective tissue have been removed from the medial side to show the internal structure of the breast and underlying skeletal muscle. In nonlactating breasts, the glandular tissue is much less prominent, with adipose tissue making up most of each breast.

ducts of the mammary glands, whereas progesterone stimulates development of the actual *milk-secreting cells*. Breast size is determined more by the amount of fat around the glandular tissue than by the amount of glandular tissue itself. Hence the size of the breast is *not* related to its ability to produce milk.

Structure of the Breasts

Each breast consists of several lobes separated by septa (walls) of connective tissue. A lobe consists of several *lobules*, which, in turn, are composed of connective tissues. **Alveoli** or "pouches" of milk-secreting cells are imbedded in these tissues. The alveoli are arranged in grapelike clusters around a tiny ductule. Figure 25-11 shows one of the mammary alveoli and the milk-producing cells that form its walls. Modified epithelial cells called *myoepithelium* surround the outside of the alveolus. This type of cell contracts slightly, as if it were a muscle cell, thus squeezing milk out into the secretory duct.

The ductules from the various lobules unite, forming a single **lactiferous** "milk-carrying" **duct** for each lobe. There are between 15 and 20 lobes in each breast. These main lactiferous ducts converge toward the nipple, like the spokes of a wheel. They enlarge slightly into small *lactiferous sinuses* before reaching the **nipple** (see Figure 25-10, *A*). The lactiferous sinuses are positioned so that they will be squeezed by the suckling motion of a baby's jaws during breastfeeding. This allows the sinuses to act as little pumping chambers that

help milk flow out of the breast. Each of the main ducts terminates in a tiny opening on the surface of the nipple.

Typically a large amount of adipose tissue is deposited around the surface of the gland, just under the skin, as well as between the lobes. *Suspensory ligaments*, positioned throughout the connective tissue of the breast, help support the glandular and connective tissues of the entire structure, anchoring them to the underlying pectoral muscles.

The nipples are bordered by a circular pigmented area, the **areola** (see Figure 25-10, *B*). It contains numerous sebaceous

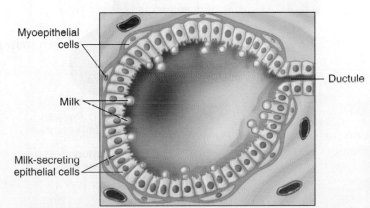

▲ FIGURE 25-11   Alveolus of the mammary gland. Notice the contractile myoepithelial cells that surround the milk-producing cells. Milk is released as vesicles of fluid pinched off from the cell.

glands that appear as small nodules under the skin. Sebum produced by these *areolar glands* helps reduce irritating dryness of the areolar skin associated with nursing. In some lighter-skinned women, the areola and nipple change color from pink to brown early in pregnancy—a fact useful in diagnosing a first pregnancy. The color decreases after lactation has ceased but never entirely returns to the original hue. In some darker-skinned women, no noticeable color change in the areola or nipple heralds the first pregnancy.

Knowledge of the lymphatic drainage of the breast is important in clinical medicine because cancerous cells from malignant breast tumors often spread to other areas of the body through the lymphatics (see Chapter 19, p. 431).

A&P CONNECT

Self-examination of breasts is a recommended routine to monitor breast health. Learn more in **Breast Self-Examination** online at **A&P Connect.**

evolve

Lactation

The obvious function of the mammary glands is **lactation**— the secretion of milk for the nourishment of newborn infants. We've summarized the basic mechanism underlying lactation below and we've highlighted the major structures and hormones for you in Figure 25-12. As you read the following, please refer to this figure.

- The ovarian hormones, estrogens and progesterone, act on the breasts to make them *structurally* ready to secrete milk. As we've discussed, estrogens promote development of the ducts of the breasts. Progesterone acts on the estrogen-primed breasts to promote completion of the development

of the ducts and development of the alveoli—the secreting cells of the breasts. This is an example of hormonal *permissiveness*; estrogen permits progesterone to have its full effect. A high blood concentration of estrogens during pregnancy also inhibits anterior pituitary secretion of prolactin.

- Shedding of the placenta after delivery of the baby cuts off a major source of estrogens. The resulting rapid drop in the blood concentration of estrogens stimulates anterior pituitary secretion of prolactin. Also, the suckling movements of a nursing baby stimulate anterior pituitary secretion of prolactin and posterior pituitary release of oxytocin.

- Prolactin stimulates alveoli of the mammary glands to secrete milk. Lactation starts about the third or fourth day after delivery of a baby. Milk replaces the secretion of a thin, yellowish secretion called *colostrum*. With repeated stimulation by the suckling infant, plus various favorable mental and physical conditions, lactation may continue for extended periods.

- Oxytocin stimulates myoepithelial cells in the alveoli of the breasts to eject milk into the ducts. This makes it accessible to the suckling infant.

This summary highlights only the major hormonal mechanisms that regulate lactation. However, there are *many* hormones that support the processes needed for successful lactation.

The Importance of Lactation and Breast Milk

The process of *lactation* plays an important role in the ultimate success of the reproductive system. The biological goal of

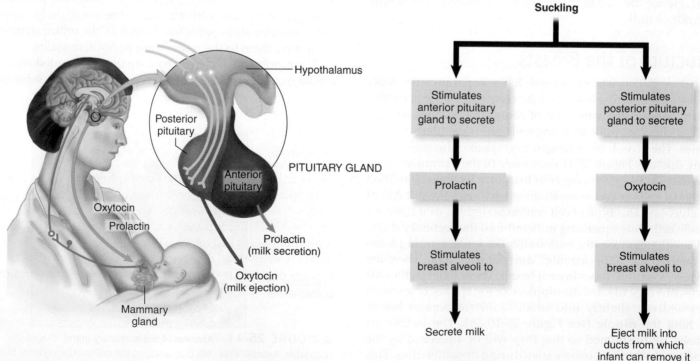

▲ **FIGURE 25-12** **Lactation.** The illustration and accompanying flowchart summarize the mechanisms that control the secretion and ejection of milk.

human reproduction does not lie solely in delivering a healthy infant—the infant must also survive until reproductive age. If a child does not survive to reproduce, then the genomes of the mother and father cannot be passed on to successive generations. Thus, the ultimate goal of reproduction will not have been met. Humans and other mammals help ensure the survival of offspring for up to several years by producing nutrient-rich milk. Nursing from the mother's breast provides several advantages for human offspring:

- Human milk is a rich source of proteins, fat, calcium, vitamins, and other nutrients in proportions needed by a young, developing body.

- Human milk provides passive immunity to the offspring in the form of maternal antibodies present in both colostrum and the milk.

- Nursing appears to enhance the emotional bond between mother and child. Such bonding may foster healthy psychological development in the child and strengthen family relationships that contribute to successful human development.

QUICK CHECK

22. Briefly describe the network of ducts and secreting cells that form the mammary glands.

23. List the hormones that prepare the breast structurally for lactation.

24. Which hormone causes milk to be ejected into the lactiferous ducts?

Cycle of LIFE

As we have seen, the reproductive system is unlike any other body system with regard to the normal changes that occur during the life span. For example, the female reproductive system does not begin to perform its functions until the teenage years (puberty). Furthermore, unlike the male reproductive system, the female reproductive system ceases its principal functions in middle adulthood.

The female organs begin their initial stages of development in the womb. As a matter of fact, the first stage of meiotic development of all the ova that will ever be produced by a woman is completed by the time she is born. However, full development of the reproductive organs—and the gametes within the ovaries—does not resume until puberty. At puberty, reproductive hormones stimulate the organs of the reproductive tract to become functional and produce one mature ovum one at a time. Reproductive function then continues in a cyclical fashion until menopause. Menopause is an event that is usually marked by the passage of at least one full year without menstruation. After that time, a woman may continue to enjoy normal sexual activity, but she cannot produce more offspring. ●

The BIG Picture

Perhaps you can see now how the significance of reproduction lies in the fact that it imparts virtual immortality to our genes. This is important not only to the survival of the human species but also to the survival of life itself. After all, life as we know it could not exist without a genetic code. The "big picture" of human procreation requires two reproductive systems—one reproductive system in each parent. The combined roles of the male and female reproductive systems will be explored as a single topic in the early part of the next chapter.

For now, let's briefly take a closer look at the female reproductive system and its relationships with other systems within a woman's body. As with any system, the female reproductive system cannot function without integrated support from the circulatory, immune, respiratory, digestive, and urinary systems. The female reproductive system shares a special anatomical relationship with the urinary system. These two systems develop in close proximity to each other and thus share a common structure: the vulva. A special anatomical relationship with the skeletal muscular system is evident in the structure known as the perineum. Of course the skeletal and muscular systems both support and protect the internal organs of the female reproductive system.

An even more special relationship with the integumentary system should be noted. The breasts, containing the milk-producing mammary glands, are actually modifications of the skin and their sweat glands. Structurally the breasts can be thought of as belonging to the integumentary system. However, functionally they are best considered as a part of the reproductive system. We have outlined the nervous and endocrine regulation of female reproductive function in this chapter. We will explore these and other connections further in the next chapter.

MECHANISMS OF DISEASE

The intricacies of the female reproductive system make it susceptible to a large number of conditions, disorders, and diseases. Hormonal and menstrual disorders such as dysmenorrhea, amenorrhea, and dysfunctional uterine bleeding are very common. Fibroid cysts and a collection of symptoms called premenstrual syndrome afflict millions of women. There are also numerous infectious diseases that are particularly debilitating, including pelvic inflammatory disease, salpingitis, and vaginitis. The list of afflictions becomes longer when tumors and related conditions of the female reproductive system are considered. These include polycystic ovary syndrome, ovarian cysts, and endometriosis. Sadly, cancers of the breast, cervix, and ovaries are also common and deadly.

evolve   Find out more about these diseases and disorders of the female reproductive system online at *Mechanisms of Disease: Female Reproductive System.*

UNIT 6

LANGUAGE OF SCIENCE AND MEDICINE *continued from page 579*

fundus (FUN-duss)
[*fundus* bottom] *pl.,* fundi (FUN-dye)

glans clitoris (glans KLIT-oh-ris)
[*glans* acorn]

granulosa cell (gran-yoo-LOH-sah sell)
[*gran-* grain, *-ul-* little, *-osa* relating to, *cell* storeroom]

greater vestibular gland (ves-TIB-yoo-lar)
[*vestibul-* entrance hall, *-ar* relating to, *gland* acorn]

hymen (HYE-men)
[*hymen* Greek god of marriage]

imperforate hymen (im-PER-fah-rayt HYE-men)
[*im-* not, *-perfor-* pierce, *-ate* state, *hymen* Greek god of marriage]

infertility (in-fer-TIL-ih-tee)
[*in-* not, *-fertil-* fruitful, *-ity* state]

infundibulum (in-fun-DIB-yoo-lum)
[*infundibulum* funnel]

isthmus (iSS-muss)
[*ithmus* narrow connection or passage]

labia majora (LAY-bee-ah mah-JOH-rah)
[*labia* lips, *majora* large] *sing.,* labium majus (LAY-bee-um MAY-jus)

labia minora (LAY-bee-ah mih-NO-rah)
[*labia* lips, *minora* small] *sing.,* labium minor (LAY-bee-um MYE-nor)

labor (LAY-bor)

lactation (lak-TAY-shun)
[*lact-* milk, *-ation* process]

lactiferous duct (lak-TIF-er-us)
[*lact-* milk, *-fer-* bear or carry, *-ous* relating to, *duct* lead]

luteal phase (LOO-tee-al fayz)
[*lute-* yellow, *-al* relating to]

luteinization (loo-tee-in-ih-ZAY-shun)
[*lute-* yellow, *-ization* process]

luteinizing hormone (LH) (loo-tee-in-EYE-zing HOR-mohn)
[*lute-* yellow, *-izing* process, *hormon-* excite]

mammary gland (MAM-er-ee)
[*mamma-* breast, *-ry* relating to, *gland* acorn]

mature follicle

medulla (meh-DUL-ah)
[*medulla* middle] *pl.,* medullae or medullas (meh-DUL-ee, meh-DUL-ahz)

menarche (meh-NAR-kee)
[*men-* month, *-arche* beginning]

menopause (MEN-oh-pawz)
[*men-* month, *-paus-* cease]

menses (MEN-seez)
[*menses* months] *sing.,* mensis (MEN-sis)

menstrual period (MEN-stroo-al)
[*mens-* month, *-al* relating to]

menstruation (men-stroo-AY-shun)
[*mens-* month, *-ation* process]

mons pubis (monz PYOO-bis)
[*mons* mountain, *pubis* groin] *pl.,* montes pubis (MON-teez PYOO-bis)

myometrium (my-oh-MEE-tree-um)
[*myo-* muscle, *-metr-* womb, *um* thing]

nipple (NIP-el)
[*nip-* beak, *-le* small]

oogonium (oh-oh-GO-nee-um)
[*oo-* egg, *-gon-* offspring, *-um* thing] *pl.,* oogonia (oh-oh-GO-nee-ah)

oral contraceptive (OR-al kon-tra-SEP-tiv)
[*contra-* against, *-cept-* take or receive (conception), *-ive* agent]

ovarian follicle (oh-VAIR-ee-an FOL-ih-kul)
[*ov-* egg, *-arian* relating to, *foll-* bag, *-icle* little]

ovarian medulla (oh-VAIR-ee-an meh-DUL-ah)
[*ov-* egg, *-arian* relating to, *medulla* middle] *pl.,* medullae or medullas (meh-DUL-ee, meh-DUL-ahz)

ovary (OH-var-ee)
[*ov-* egg, *-ar-* relating to, *-y* location of process]

ovulation (ov-yoo-LAY-shun)
[*ov-* egg, *-ation* process]

ovum (OH-vum)
[*ovum* egg] *pl.,* ova (OH-vah)

perimetrium (pair-ih-MEE-tree-um)
[*peri-* around, *-metr-* womb, *-um* thing]

perineal body (pair-ih-NEE-al BOD-ee)
[*peri-* around, *-ine-* excrete (perineum), *-al* relating to]

perineum (pair-ih-NEE-um)
[*peri-* around, *-ine-* excrete, *-um* thing] *pl.,* perinea (pair-ih-NEE-ah)

peritonitis (pair-ih-toh-NYE-tis)
[*peri-* around, *-ton-* stretch (peritoneum), *-itis* inflammation]

placenta (plah-SEN-tah)
[*placenta* flat cake] *pl.,* placentae or placentas (plah-SEN-tee, plah-SEN-tahz)

posterior fornix (pohs-teer-ee-or FOR-niks)
[*poster-* behind, *-or* quality, *fornix* arch]

prepuce (PREE-pus)
[*pre-* before, *-puc-* penis]

primary follicle (PRY-mair-ee FOL-ih-kul)
[*prim-* first, *-ary* state, *folli-* bag, *-cle* small]

progesterone (pro-JES-ter-ohn)
[*pro-* provide for, *-gester-* bearing (pregnancy), *-stero-* solid or steroid derivative, *-one* chemical]

proliferative phase (PROH-lif-er-eh-tiv fayz)
[*proli-* offspring, *-fer-* bear or carry, *-at-* process, *-ive* relating to]

retroflexion (ret-roh-FLEK-shen)
[*retro-* backward, *-flex-* bend, *-ion* process]

salpingitis (sal-pin-JYE-tis)
[*salping-* tube, *-itis* inflammation]

secretory phase (SEEK-reh-toh-ree fayz)
[*secret-* separate, *-ory* relating to]

urogenital triangle (yoor-oh-GEN-ih-tal)
[*uro-* urine, *-gen-* produce, *-al* relating to]

uterus (YOO-ter-us)
[*uterus* womb]

vagina (vah-JYE-nah)
[*vagina* sheath]

vaginal orifice (VAH-jih-nal OR-ih-fis)
[*vagina-* sheath, *-al* relating to, *ori-* mouth, *-fice-* something made]

vulva (VUL-vah)
[*vulva* wrapper]

CHAPTER SUMMARY

To download an MP3 version of the chapter summary for use with your iPod or other portable media player, access the Audio Chapter Summaries online at http://evolve.elsevier.com.

HINT *Scan this summary after reading the chapter to help you reinforce the key concepts. Later, use the summary as a quick review before your class or before a test.*

OVERVIEW OF THE FEMALE REPRODUCTIVE SYSTEM

A. Function of the female reproductive system
   1. Female reproductive system produces gametes called ova (eggs)
   2. Female reproductive system also provides protection and nutrition to the developing offspring for up to several years after conception

B. Structural plan of the female reproductive system
1. Reproductive organs are classified as essential or accessory (Figure 25-1)
    a. Essential organs—gonads are the paired ovaries
    b. Accessory organs—uterine tubes, uterus, vagina, vulva, and mammary glands (Figures 25-1, 25-2, and 25-3)
C. Perineum—female perineum is a muscular region within a diamond-shaped area between the thighs and the vaginal orifice and the anus (Figure 25-2)
1. Extends from the pubic symphysis anteriorly to the coccyx posteriorly
2. A line drawn between the ischial tuberosities divides the area into two triangles
    a. Urogenital triangle—contains the external genitals (labia, vaginal orifice, clitoris) and urinary opening
    b. Anal triangle—surrounds the anus
3. Area that may be torn during childbirth

OVARIES

A. Location of the ovaries
1. Nodular glands located on each side of the uterus, below and behind the uterine tubes (Figure 25-3)
2. Weigh about 3 grams; attached to the posterior surface of the broad ligament by the mesovarian ligament (mesovarium)
3. Ovarian ligament anchors the ovary to the uterus
B. Structure and function of the ovaries
1. Covering the outer cortex is a layer of flattened epithelial cells; germinal epithelium
2. Ovarian follicles contain the developing female sex cell
C. Overview of oogenesis
1. Oogenesis—process that results in formation of a mature egg (Figure 25-4)
    a. Throughout the process of ovarian development, the oocyte grows in size
    b. Ovulation—release of an ovum from the mature follicle at the end of oogenesis
    c. Ovaries are also endocrine organs, secreting the female sex hormones

UTERUS

A. Location and support of the uterus
1. Uterus is located in the pelvic cavity between the urinary bladder in front and the rectum behind
2. Normally the uterus lies over the superior surface of the bladder, pointing forward and slightly upward (Figure 25-1)
3. Uterus may lie in any one of several abnormal positions, largely because the ligaments hold it so loosely
    a. Retroflexion—entire organ is tilted backward; may allow the uterus to prolapse, or descend, into the vaginal canal
B. Shape and structure of the uterus
1. Pear shaped and measures approximately 7 cm (3 inches) in length, 5 cm (2 inches) in width at its widest part, and 3 cm (1 inch) in thickness
2. Uterus has two main parts (Figure 25-3):
    a. Body—wide, upper portion,
    b. Cervix—lower, narrow "neck"
3. Three layers comprise the walls of the uterus:
    a. Inner endometrium
    b. Middle myometrium
    c. Outer perimetrium, incomplete layer of visceral peritoneum
C. Endometrium—A ciliated mucous membrane that lines the uterus

1. Varies in thickness from 0.5 mm just after the menstrual flow to about 5 mm near the end of the endometrial cycle
2. Endometrium has a rich supply of blood capillaries and numerous uterine glands that secrete mucus and other substances onto the endometrial surface
D. Myometrium—thick, middle layer of the uterine wall; consists of three layers of smooth muscle fibers
1. Thickest in the fundus and thinnest in the cervix
E. Perimetrium—incomplete external layer of visceral peritoneum
1. Location is partially retroperitoneal
F. Functions of the uterus
1. Uterus serves as part of the female reproductive tract, permitting sperm from the male to ascend toward the uterine tubes
2. If conception occurs, an offspring develops in the uterus
3. If conception or the successful implantation of the embryo fails, then the outer layers of the endometrium are shed during menstruation

UTERINE TUBES

A. Uterine tubes are also sometimes called fallopian tubes, or oviducts
B. Uterine tubes are about 10 cm (4 inches) long and are attached to the uterus at its upper outer angles (Figures 25-1 and 25-3)
C. Structure of the uterine tubes
1. Same three layers (mucous, smooth muscle, and serous) of the uterus also comprise the uterine tubes
2. Mucosal lining of the tubes is continuous with the peritoneum lining the pelvic cavity
3. Tubal mucosa is also continuous with that of the uterus and vagina
D. Divisions and tissues of the uterine tubes
1. Each uterine tube consists of three divisions (Figure 25-3)
    a. Ampulla
    b. Isthmus
    c. Infundibulum
E. Functions of the uterine tubes—serve as more than mere transport channels; also site of fertilization

VAGINA

A. Structure of the vagina
1. Collapsible tube about 8 cm (3 inches) long, capable of enormous distention during delivery of a baby
2. Composed mainly of smooth muscle and is lined with mucous membrane arranged in rugged folds called rugae
3. Anterior wall of the vagina is shorter than the posterior wall because the cervix protrudes into the uppermost portion of the tube (Figure 25-1)
4. Hymen—fold of mucous membrane that forms a border around the external opening of the vagina
B. Functions of the vagina
1. Lining of the vagina lubricates and stimulates the glans penis, which in turn triggers the ejaculation of semen
2. Serves as a receptacle for semen
3. Serves as the lower portion of the birth canal
4. Transports blood and tissue shed from the lining of the uterus during menstruation

VULVA

A. Structure of the vulva
1. Mons pubis—skin-covered pad of fat over the pubic symphysis

2. Labia majora—composed mainly of fat and connective tissue with numerous sweat and sebaceous glands on the inner surface
3. Labia minora—located medially to the labia majora; covered with hairless skin
4. Clitoris—composed of erectile tissue
5. Glans clitoris—only visible part of the erectile structures of the clitoris; covered with highly sensitive skin that during sexual stimulation produces most of the female sexual response
6. External urinary meatus—situated between the clitoris and the vaginal orifice
7. Vaginal orifice—much larger opening than the urinary meatus; located posterior to the meatus
8. Greater vestibular glands (Bartholin glands)—two bean-shaped glands, one on each side of the vaginal orifice

B. Functions of the vulva
1. Protective features of the mons pubis and labia help prevent injury to the delicate tissues of the clitoris and vestibule
2. Possesses a large number of sensory receptors that feed information back to the sexual response areas of the brain

FEMALE REPRODUCTIVE CYCLES

A. Recurring cycles—many changes recur periodically in the female during the years between the onset of the menses (menarche) and their cessation (menopause)
1. Menstruation—the outward sign of changes in the endometrium

B. Ovarian cycle—ovaries from birth contain oocytes in primary follicles in which the meiotic process has been suspended; at the beginning of menstruation each month, several of the oocytes resume meiosis; meiosis will stop again just before the cell is released during ovulation

C. Endometrial (menstrual) cycle—divided into four phases
1. Menses (menstrual period)—occur on days 1 to 5 of a new cycle
2. Proliferative phase—occurs between the end of the menses and ovulation
3. Ovulation—rupture of the mature follicle with expulsion of its ovum into the pelvic cavity (Figure 25-6)
4. Luteal phase (secretory phase)—occurs between ovulation and the onset of the menses

D. Gonadotropic cycle—two hormones called gonadotropins that influence female reproductive cycles
1. Follicle-stimulating hormone (FSH)
2. Luteinizing hormone (LH)

E. Control of female reproductive cycles—hormones control cyclical changes

F. Control of cyclical changes in the ovaries
1. Cyclical changes in the ovaries result from cyclical changes in the amounts of gonadotropins secreted by the anterior pituitary gland
2. Increasing FSH blood level has two effects (Figure 25-7):
   a. Stimulates one or more primary follicles and their oocytes to start growing
   b. Stimulates the follicular cells to secrete estrogens

G. Control of cyclical changes in the uterus
1. Cyclical changes in the uterus are caused by changes in estrogens and progesterone

H. Control of cyclical changes in gonadotropin secretion
1. Both negative and positive feedback mechanisms help control anterior pituitary secretion of the gonadotropins FSH and LH

I. Importance of female reproductive cycles
1. Ovarian cycle's primary function is to produce ova at regular intervals
   a. Secondary role is to regulate the endometrial cycle by means of the sex hormones estrogen and progesterone
2. Role of the endometrial cycle is to ensure that the lining of the uterus is suitable for the implantation of an embryo

J. Infertility and use of fertility drugs
1. Infertility—failure to conceive after 1 year of regular unprotected intercourse
   a. May be caused by a wide variety of medical, environmental, and even lifestyle factors
   b. Fertility drugs and other assisted reproductive procedures are available

K. Menarche and menopause
1. Menstrual flow first occurs (menarche) at puberty, at about the age of 13 years; individual variation according to race, nutrition, health, and heredity
   a. Recurs about every 28 days for about three decades, except during pregnancy, and then ceases (menopause)

BREASTS

A. Location and size of the breasts
1. Breasts lie over the pectoral muscles and are attached to them by a layer of connective tissue (Figure 25-10)
2. Made up of milk-producing mammary glands
3. Estrogens and progesterone both control breast development during puberty
4. Breast size is determined more by the amount of fat around the glandular tissue than by the amount of glandular tissue

B. Structure of the breasts
1. Each breast consists of several lobes separated by septa (walls) of connective tissue
   a. Each lobe consists of several lobules, which are composed of connective tissues
   b. Alveoli or "pouches" of milk-secreting cells are imbedded in these tissues (Figure 25-11)
   c. Alveoli are arranged in grapelike clusters around a tiny ductule
   d. Ductules from the various lobules unite, forming a single lactiferous duct for each lobe
2. Nipples are bordered by a circular pigmented area, the areola (Figure 25-10, B)
   a. Contains numerous sebaceous glands that appear as small nodules under the skin; sebum helps reduce irritating dryness of the areolar skin associated with nursing

C. Lactation—secretion of milk for the nourishment of newborn infants
1. Basic mechanism underlying lactation:
   a. The ovarian hormones, estrogens and progesterone, act on the breasts to make them structurally ready to secrete milk
   b. Shedding of the placenta after delivery of the baby cuts off a major source of estrogens; stimulates anterior pituitary secretion of prolactin
   c. Prolactin stimulates alveoli of the mammary glands to secrete milk
   d. Oxytocin stimulates myoepithelial cells in the alveoli of the breasts to eject milk into the ducts

D. The importance of lactation and breast milk
1. The process of lactation plays an important role in the ultimate success of the reproductive system

2. Nursing from the mother's breast provides several advantages for human offspring
   a. Human milk is a rich source of proteins, fat, calcium, vitamins, and other nutrients in proportions needed by a young, developing body
   b. Human milk provides passive immunity to the offspring in the form of maternal antibodies present in both colostrum and the milk
   c. Nursing appears to enhance the emotional bond between mother and child

REVIEW QUESTIONS

HINT *Write out the answers to these questions after reading the chapter and reviewing the Chapter Summary. If you simply think through the answer without writing it down, you won't retain much of your new learning.*

1. Identify the essential and accessory organs in the female reproductive system.
2. Describe the three layers that compose the wall of the uterus.
3. Identify the vessels that supply blood to the uterus.
4. List the eight ligaments that hold the uterus in a normal position.
5. How does the uterus serve as part of the female reproductive tract?
6. What and where are the uterine tubes? Approximately how long are they? What lines the uterine tubes? Their lining is continuous on their distal ends with what? With what on their proximal ends?
7. What hormones are secreted by the cells in ovarian tissue?
8. Identify all vaginal functions.
9. List all the structures that make up the female external genitals.
10. Define the term *episiotomy*.
11. Identify the advantages that nursing from the mother's breast provides offspring.
12. Describe the hormonal changes during menopause.

CRITICAL THINKING QUESTIONS

HINT *After finishing the Review Questions, write out the answers to these items to help you apply your new knowledge. Go back to sections of the chapter that relate to items that you find difficult.*

1. Name and explain the function of the various hormones that regulate lactation. Where are they produced, and how would you summarize their function and their influence on lactation?

2. List the phases of the menstrual cycle. Which of these phases shows the most variance in length of time? How do the events in each phase contribute to the overall function of the reproductive system?
3. How would you explain the interaction of the hormones that result in ovulation? From what is the name "luteinizing" hormone derived?
4. State in your own words the control of cyclical ovarian changes brought on by FSH and LH.
5. Explain, in your own words, the control of cyclical uterine changes brought on by the ovarian hormones. The drop in the level of these hormones triggers what event?
6. How would you correlate the events of the ovarian cycle with the events of the uterine cycle?

continued from page 579

Now that you have read this chapter, see if you can answer these questions about Carlos and Maria from the Introductory Story of this chapter and Chapter 24.

1. What effect will Clomid have on FSH production?
   a. Increase FSH production
   b. Decrease FSH production
   c. No change in FSH production
   d. Slight decrease in FSH production followed by a sharp increase

2. After ovulation, the follicular cells first transform into what?
   a. Corpus lucidum
   b. Corpus luteum
   c. Corpus rubrum
   d. Corpus albicans

3. Where in the female reproductive tract should the sperm and the oocyte meet (hopefully completing the process of fertilization)?
   a. In the cervix
   b. In the uterus
   c. In the uterine tubes
   d. In the ovaries

HINT *To solve these questions, you may have to refer to the glossary or index, other chapters in this textbook, A&P Connect, Mechanisms of Disease, and other resources.*

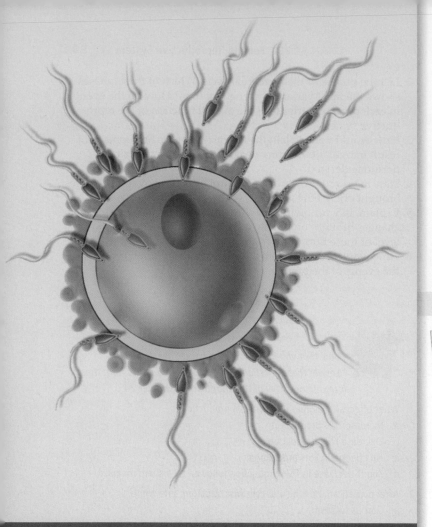

Growth and Development

STUDENT LEARNING OBJECTIVES

At the completion of this chapter, you should be able to do the following:

1. Outline the process of meiosis and the production of haploid sex cells.
2. Explain how the process of spermatogenesis differs from that of oogenesis.
3. Discuss ovulation and insemination.
4. Outline the actual process of fertilization.
5. Describe the basic aspects of the prenatal period.
6. Identify the following: placenta, hCG, blastocyst.
7. Describe the basic processes of embryonic development.
8. Explain how stem cells and primary germ layers function in embryonic development.
9. Describe the process of parturition and the hormones involved.
10. Define what is meant by the following terms: postnatal period, infancy, childhood, adolescence, adulthood, senescence.

CARLOS and his wife Maria had been trying for years to have a baby with no success. After a visit to an infertility specialist, they found that Carlos had a low sperm count. After looking at all of their options, they decided to try an intrauterine insemination (IUI). A day or so after the procedure (although they didn't know it yet), Carlos' sperm had indeed found Maria's ovum. The nucleus of the sperm and the nucleus of the egg had fused together, and fertilization had taken place. Maria was pregnant!

At 16 weeks, an ultrasound was performed to make sure everything was going smoothly. Maria was looking at the screen and could clearly see what appeared to be two separate images. "Is that what I think it is?" she asked the technician. "Officially, the doctor has to tell you—but I'll just let you watch the screen as I measure *two* heartbeats and make *two* sets of measurements," said the technician. "Looks like one boy and one girl," he added with a smile.

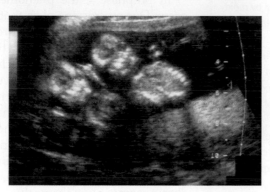

What type of twins is Maria expecting? How could this have happened? The answer lies within this chapter!

The day of your birth marks the end of one phase of life, called the *prenatal period*, and the beginning of a second phase called the *postnatal period*. The **prenatal period** begins at conception and ends at birth. The **postnatal period** begins at birth and continues until death. In reality, however, they are part of an ongoing and continuous process. The field of human **developmental biology**—the study of the many changes that occur during the cycle of life from conception to death—isolates periods such as infancy or old age for study (Box 26-1). However, life is not a series of stop-and-start events. Instead, it is a biological process that is characterized by continuous change.

This chapter discusses some of the important events and changes that occur in the continual development of the individual from conception to death. We will study development during the prenatal period and the process of labor and birth, and then follow these fascinating topics with a review of changes that occur during infancy through adulthood.

A NEW HUMAN LIFE

Meiosis and Production of Sex Cells

A new human offspring begins when the nucleus of a single *spermatozoon* (sperm) unites with the nucleus of an *ovum* (egg). As we will see, spermatozoa are produced in a process called *spermatogenesis*. In a similar fashion, ova are produced in a process called *oogenesis*. Although spermatogenesis and oogenesis may seem very different, both processes are

LANGUAGE OF SCIENCE AND MEDICINE

HINT Before reading the chapter, say each of these terms out loud. This will help you avoid stumbling over them as you read.

adolescence (ad-oh-LESS-ens)
[*adolesc-* grow up, *-ence* state]

adulthood

amniotic cavity (am-nee-OT-ik KAV-ih-tee)
[*amnio-* fetal membrane, *-ic* relating to, *cav-* hollow, *-ity* state]

Apgar score (AP-gar)
[*Virginia Apgar* American physician]

blastocyst (BLASS-toh-sist)
[*blasto-* bud, *-cyst* pouch]

cesarean section (seh-SAIR-ee-an SEK-shun)
[*Julius Caesar* Roman emperor, *-ean* of]

childhood

chorion (KOH-ree-on)
[*chorion* skin]

conception (kon-SEP-shun)
[*con-* together, *-cept-* take, *-ion* process]

developmental biology
[*bio-* life, *-log-* words (study of), *-y* activity]

diploid (DIP-loyd)
[*diplo-* twofold, *-oid* of or like]

ectoderm (EK-toh-derm)
[*ecto-* outside, *-derm* skin]

embryology (em-bree-OL-oh-gee)
[*em-* in, *-bryo-* fill to bursting, *-log-* words (study of), *-y* activity]

endoderm (EN-doh-derm)
[*endo-* within, *-derm* skin]

fertilization (FER-tih-lih-ZAY-shun)
[*fertil-* fruitful, *-ation* process]

first polar body (POH-lar BOD-ee)
[*pol-* pole, *-ar* relating to]

gerontology (jair-on-TAHL-oh-jee)
[*geronto-* old age, *-log-* words (study of), *-y* activity]

gestation period (jes-TAY-shun)
[*gesta-* bear, *-tion* process]

granulosa cell (gran-yoo-LOH-sah sell)
[*gran-* grain, *-ul-* little, *-osa* relating to, *cell* storeroom]

haploid (HAP-loyd)
[*haplo-* single, *-oid* of or like]

histogenesis (hiss-toh-JEN-eh-sis)
[*histo-* tissue, *-gen-* produce, *-esis* process]

human chorionic gonadotropin (hCG) (koh-ree-ON-ik go-nah-doh-TROH-pin)
[*chorion-* skin, *-ic* relating to, *gon-* offspring, *-ad-* relating to, *-trop-* nourishment, *-in* substance]

continued on page 619

BOX 3-1 FYI

Developmental Biology

Developmental biology is the name given to the branch of life science that studies the process of change over the life cycle. This process of change is called development. It is important to realize that, in developmental biology, the terms growth and development do not mean the same thing. Growth is simply an increase in body mass. Development, on the other hand, refers to the complex series of changes that occur at various times of life. Early stages of development—particularly the prenatal stages—are characterized by rapid growth, whereas later stages of development are characterized by little, if any, growth of body tissues.

In this chapter, we briefly discuss various subtopics within the field of human developmental biology. For example, prenatal development is studied by a branch of developmental biology called *embryology*. The biological changes observed during late adulthood are studied by a branch of developmental biology called *gerontology*.

Basic concepts of embryology, gerontology, and other subdisciplines of developmental biology seem to have taken on a greater practical importance during the past few decades than ever before. One reason is the explosion in knowledge of developmental processes and our ability to treat the abnormalities that we can now find. Procedures such as fetal surgery, electrocardiography, and ultrasound permit physicians to diagnose and treat the fetus much like any other patient. Recent discoveries about the processes of aging—as well as the rapidly growing population of aged individuals—have spawned new methods of recognizing and treating physical and psychological problems in the elderly. Another reason developmental biology has taken on great practical importance is that it is a field that serves to unify human biology into a framework that integrates anatomy, physiology, cell biology, molecular biology, medicine, and other disciplines. Thus developmental biology gives us an excellent view of the "big picture" of the human body.

forms of **meiosis**—the orderly reduction and distribution of chromosomes to make sperm and eggs. In effect, meiosis reduces the number of chromosomes in each daughter cell to *half the number present in the parent cell.*

Except for mature erythrocytes, each human body (somatic) cell contains 23 pairs of similar or *homologous* chromosomes, for a total of 46 chromosomes. This is known as the **diploid** number. If the male and female cells united without first *halving* their respective chromosome numbers, the resulting fertilized egg would contain *twice* as many chromosomes as normal. Mature ova and sperm therefore contain only 23 chromosomes, or half as many, as other human cells. This total of 23 chromosomes per sex cell is known as the **haploid** number of chromosomes.

Meiotic division consists of two nuclear divisions that take place one after the other in succession. In order, these nuclear divisions are referred to as meiosis I and meiosis II. In both divisions, prophase, metaphase, anaphase, and telophase occur (Figure 26-1). To help you understand the process of meiosis, only 4 of the 46 chromosomes are shown in Figure 26-1.

Interphase (a part of the cell cycle during which chromosomes are replicated) and the events preceding interphase are similar in both mitosis and meiosis. Specifically, each chromosome *replicates* (is copied) and thereby becomes a pair of *chromatids* that remain attached to each other at a common centromere. For this reason, we refer to these special chromosomes as single chromosomes, but *doubled.* We call them single chromosomes, because the two sister chromatids are still connected by a single centromere. (Note: The term *chromosome* applies to any condensed chromatin with its own centromere.) Early in prophase I of meiosis I, homologous pairs of chromosomes join together to form four-armed structures called *tetrads.*

In prophase I, the phenomenon of *"crossing over"* may occur. During crossing over, matched segments from *different* chromatids of a homologous chromosome pair cross over and become permanent parts of the other chromosome (see Figure 27-4, p. 627). Because each chromatid segment consists of specific genes, the crossing over of chromatids *reshuffles* the genes—that is, it exchanges some of them between homologous chromosomes. This exchange and recombination of genetic material can add an almost infinite variety to the ultimate genetic makeup of offspring. We will discuss the effects of crossing over in more detail in the next chapter.

Metaphase I follows prophase I, and as in mitosis, the chromosomes align themselves roughly along the equator of the cell, as you can see in Figure 26-1. However, in anaphase I the two chromatids that make up each chromosome *do not separate* from each other to form two new *single* chromosomes, as they do in mitosis. In anaphase I, only one of each pair of chromosomes moves to each pole of the parent cell.

Look again at Figure 26-1. Note that, when the parent cell divides to form two cells in meiotic division I, each daughter cell contains only two chromosomes, or half as many as the parent cell had. *However, remember that each chromosome still consists of two sister chromatids joined at the centromere. These chromosomes may have had their genes shuffled and recombined during crossing over, and so they may be quite different from the original pair.* Thus the daughter cells formed by meiotic division I contain a *haploid number* of (still doubled) chromosomes, or half as many as the diploid number in the parent cell.

As you can see in Figure 26-1, meiotic division II is essentially the same as mitotic division. In both spermatogenesis and oogenesis, the second meiotic division produces two cells each from the two cells formed during meiosis I. Thus, a total of four cells is produced, each with the haploid number of chromosomes. Note, however, that each cell produced has "single" chromosomes.

Spermatogenesis

Spermatogenesis is the process by which sperm-forming cells, or **spermatogonia,** (already present in the seminiferous tubules of a newborn baby boy) later become transformed into mature sperm. Spermatogenesis begins at about the time of puberty and usually continues throughout a man's life. Refer to Figure 24-5 (p. 566) to review the major steps of spermatogenesis. Carefully study each step in this diagram as you read the following paragraph.

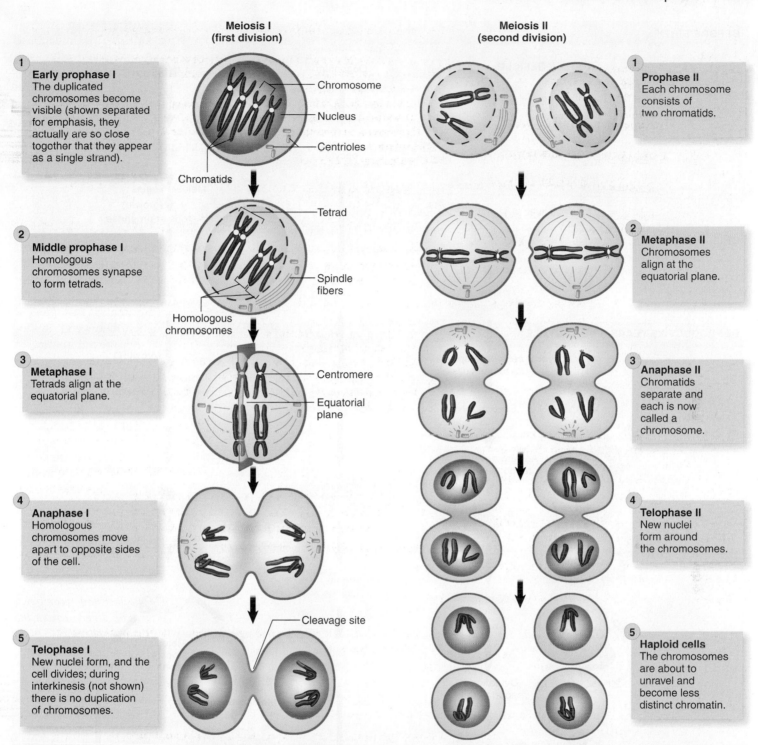

**Meiosis I
(first division)**

1 Early prophase I
The duplicated chromosomes become visible (shown separated for emphasis, they actually are so close together that they appear as a single strand).

Chromosome

Nucleus

Centrioles

Chromatids

2 Middle prophase I
Homologous chromosomes synapse to form tetrads.

Tetrad

Spindle fibers

Homologous chromosomes

3 Metaphase I
Tetrads align at the equatorial plane.

Centromere

Equatorial plane

4 Anaphase I
Homologous chromosomes move apart to opposite sides of the cell.

5 Telophase I
New nuclei form, and the cell divides; during interkinesis (not shown) there is no duplication of chromosomes.

Cleavage site

**Meiosis II
(second division)**

1 Prophase II
Each chromosome consists of two chromatids.

2 Metaphase II
Chromosomes align at the equatorial plane.

3 Anaphase II
Chromatids separate and each is now called a chromosome.

4 Telophase II
New nuclei form around the chromosomes.

5 Haploid cells
The chromosomes are about to unravel and become less distinct chromatin.

▲ **FIGURE 26-1   Meiotic cell division.** Meiosis is a series of events that involves two separate division processes called meiosis I and meiosis II. Notice that four daughter cells, each with the haploid number of chromosomes, are produced from each parent cell that enters meiotic cell division. For simplicity's sake, only four chromosomes (two homologous pairs) are shown in the parent cell instead of the usual 46.

Each **primary spermatocyte** undergoes meiosis I to form two **secondary spermatocytes,** each with a haploid number of chromosomes (23). These secondary spermatocytes still have doubled (recombined) chromosomes. Each secondary spermatocyte then undergoes meiosis II to form a total of four spermatids. Spermatids then differentiate to form heads and tails, eventually becoming mature sperm. Thus spermatogenesis forms four haploid sperm (each with only 23

single chromosomes) from one primary spermatocyte that had 23 pairs, or 46 chromosomes.

Oogenesis

Oogenesis is the process by which egg-forming cells, or **oogonia,** become mature eggs. As you read through the following paragraphs, trace the steps of oogenesis by referring back to Figure 25-4 (p. 581) and looking at Figure 26-2.

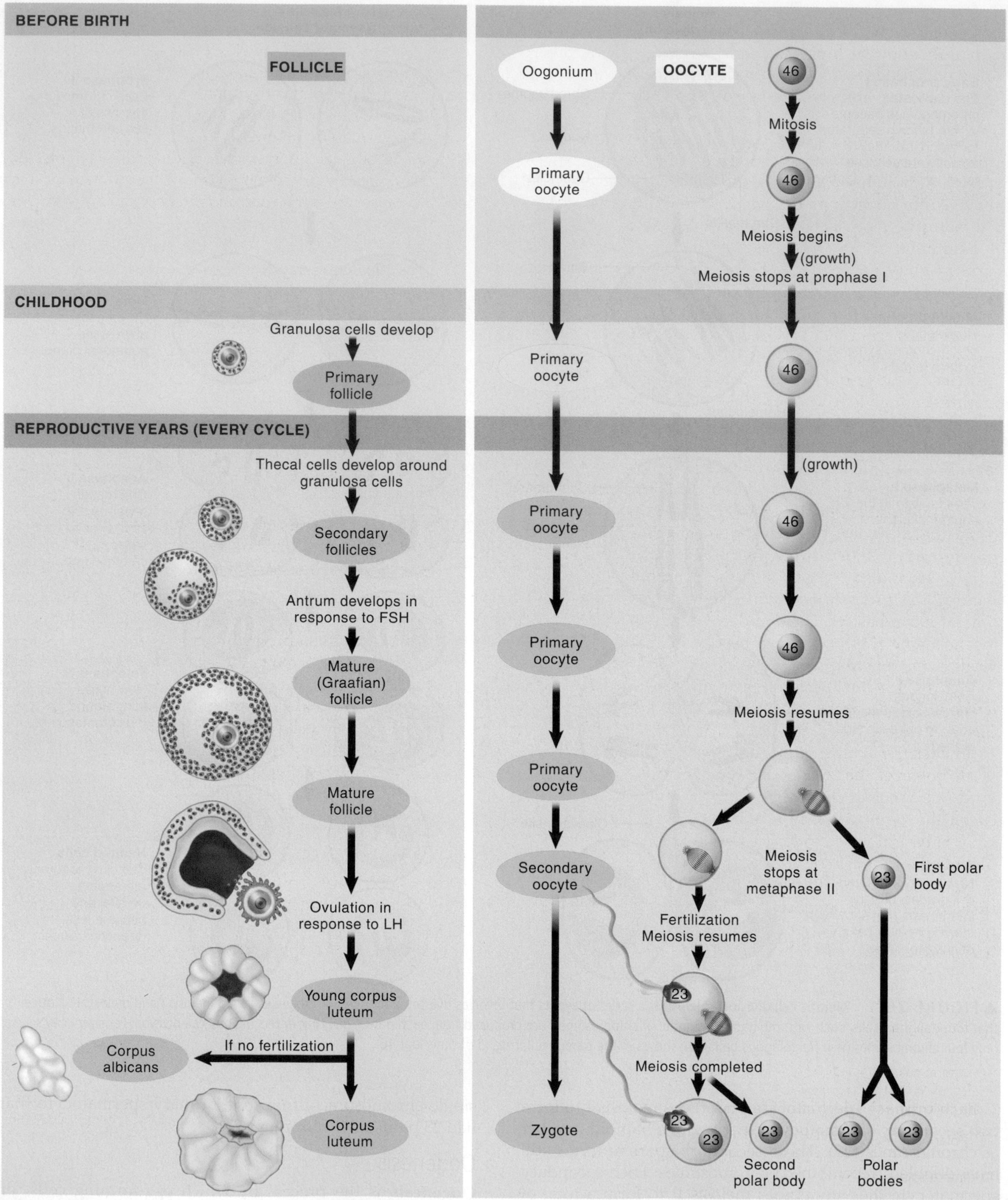

▲ **FIGURE 26-2** **Oogenesis.** Production of a mature ovum (oocyte) and subsequent fertilization are shown as a series of cell divisions. This process can be seen as a series of changes in the ovarian follicle in Figure 25-4 (p. 581).

During the fetal period, oogonia in the ovaries undergo mitotic division to form **primary oocytes**—about a half million of them by the time a baby girl is born! Most of the primary oocytes develop to prophase I of meiosis *before* birth. There they stay literally in a state of "suspended animation" until puberty.

During childhood, **granulosa cells** develop around each primary oocyte, forming a **primary follicle** (see Figure 25-4, p. 581). Although several thousand primary oocytes will not survive into puberty, by the time a female child reaches sexual maturity, about 400,000 primary oocytes still remain.

Beginning at puberty, during each cycle, about a thousand primary oocytes resume meiosis. Their surrounding follicles begin to mature and some of the outer granulosa cells differentiate to become **theca cells.** Theca cells produce *androgen,* a steroid that is converted by granulosa cells into estrogen. At this point, the follicles are known as **secondary follicles.** Usually only one follicle per cycle survives and matures enough to reach the surface of the ovary. This developing follicle, or **tertiary follicle,** can be seen as a fluid-filled bump, surrounded with granulosa cells, on the ovarian surface. The fluid-filled space *within* each follicle is called the *antrum.* A mature follicle ready to burst open from the ovary's surface is also called a *graafian follicle.*

By that time, meiosis has resumed inside the primary oocyte within the mature follicle. Meiosis I produces a *haploid* **secondary oocyte** and the **first polar body** (see Figure 26-2). Just before ovulation, meiosis again halts—this time at metaphase II. Then, under the influence of luteinizing hormone (LH), ovulation occurs. Ovulation, you may recall, is the release of an oocyte from a burst follicle.

Meiosis II in the released oocyte resumes only when, and if, the head of a sperm cell enters the secondary oocyte (at this point we'll call it the ovum). If fertilization does not occur, then the ovum simply degenerates. If fertilization does occur, however, then meiotic division of the secondary oocyte produces a second polar body and a mature, fertilized ovum called the **zygote.** The first polar body may also undergo what is basically a mitotic division to produce two haploid polar bodies with single chromosomes.

Note that, during oogenesis, the cytoplasm is *not* equally divided among the daughter cells. Of the four haploid daughter cells produced, only one is large enough to survive to become the actual egg. Thus each primary oocyte produces only one mature ovum, plus three tiny *polar bodies.* Both the ovum and the polar bodies are haploid with *single* chromosomes. The polar bodies quickly break down and are reabsorbed into nearby cells. A total of four mature sperm cells are formed from each primary spermatocyte in spermatogenesis. This difference may be accounted for by the fact that, for reproductive success, an ovum must have a huge store of cytoplasm with all of its organelles, nutrients, and regulatory molecules. As a result, nearly all the cytoplasm of the original primary oocyte is conserved by the one daughter oocyte that survives.

A&P CONNECT

What physiological advantage is gained by postponing the completion of meiotic division as an ovum matures? The surprising answer is found in **Arrest of Oocyte Development** online at **A&P Connect.**

evolve

Ovulation and Insemination

As we've just seen, the second step necessary for the conception of a new individual requires that a sperm and egg be brought together so that they can unite. Two processes are involved in this step:

1. *Ovulation,* or expulsion of the mature ovum from the mature ovarian follicle into the abdominopelvic cavity, from where it enters one of the uterine (fallopian) tubes.

2. *Insemination,* or expulsion of the seminal fluid from the male urethra into the female vagina. Millions of sperm enter the female reproductive tract with each ejaculation of semen. Sperm move by lashing movements of their tail-like flagella. Various processes in the female reproductive tract assist their swimming motions. In this manner, the sperm make their way into the *external os* of the cervix, through the cervical canal and uterine cavity, and into the uterine (fallopian) tubes.

Fertilization

After ovulation, the discharged ovum first enters the abdominopelvic cavity. It soon finds its way into the uterine (fallopian) tubes, where conception, or **fertilization,** may take place (Figure 26-3). Sperm cells "swim" up the uterus and into the uterine tubes toward the ovum. Now, look at the relationship

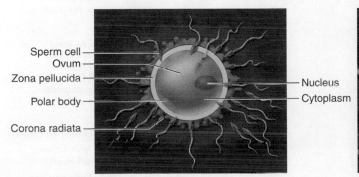

Sperm cell —
Ovum —
Zona pellucida —
Polar body —
Corona radiata —
— Nucleus
— Cytoplasm

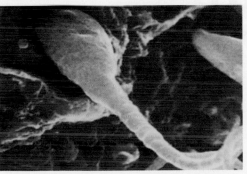

◀ **FIGURE 26-3**
Fertilization. Fertilization is a specific biological event. It occurs when the male and female sex cells fuse. After union between a sperm cell and the ovum has occurred, the cycle of life begins. The scanning electron micrograph shows spermatozoa attached to the surface of an ovum. Only one will enter the ovum.

UNIT 6

of the ovary, the uterine tube, and the uterus in Figure 26-4. Recall from Chapter 25 that each uterine tube extends outward from the uterus and ends in the abdominal cavity near the ovary (see Figure 26-4) in an opening surrounded by fringelike processes, the fimbriae. Sperm cells deposited in the vagina must enter and "swim" through the uterus and then move out of the uterine cavity and through the uterine tube to meet the ovum. Fertilization most often occurs in the outer one third of the oviduct, as shown in Figure 26-4.

The process of sperm movement is assisted by mechanisms within the female reproductive tract. For example, mucous strands in the cervical canal guide the sperm on their way into the uterus. Rhythmic contractions of the female reproductive tract and ciliary movement along the lining of the uterine tubes also assist the movement of sperm. In addition, sperm are attracted to the warmer temperatures of the uterine tubes. Despite all this, however, only a small fraction of the sperm deposited in the vagina ever reach the ovum. Only 50 to 100 sperm out of 250 million to 500 million ejaculated sperm actually reach their target.

The ovum also takes an active role in the process of fertilization. Experiments show that the ovum and its surrounding layers actually attract nearby sperm with special peptides. Once bound to receptor peptides, the acrosome in the "head" of the sperm releases enzymes that break down the outer layers surrounding the ovum. After the sperm reaches the surface of the ovum, the two plasma membranes fuse and the nucleus of the sperm moves inside the ovum. As soon as the head and neck of one sperm fuse with the egg, the flagellum degenerates. Immediately following this, complex mechanisms in the egg are activated by sperm proteins to ensure that no more sperm enter. This thick film becomes an impenetrable barrier called the *fertilization membrane*. The 23 chromosomes from the sperm nucleus then combine with the 23 chromosomes of the nucleus of the ovum to restore the diploid number of 46 chromosomes.

Because the ovum lives only a short time (about a day or so) after leaving the ruptured follicle, the fertilization "window" occurs only around the time of ovulation. However, sperm may live up to 6 days after entering the female tract. Thus, sexual intercourse any time from about 3 days before ovulation to a day or so after ovulation may result in fertilization.

The fertilized ovum, or *zygote,* is genetically complete; it represents the first cell of a genetically new individual. Time, nourishment, and a proper prenatal environment are all that are needed for the expression of characteristics such as sex, hair, and skin color that were determined at the time of fertilization.

A & P CONNECT

Artificial fertilization outside the body is sometimes used to enhance reproductive success. Read a brief description of this strategy and other **assisted reproductive technnologies (ARTs)** in **In Vitro Fertilization** online at **A&P Connect.**

ⓔvolve

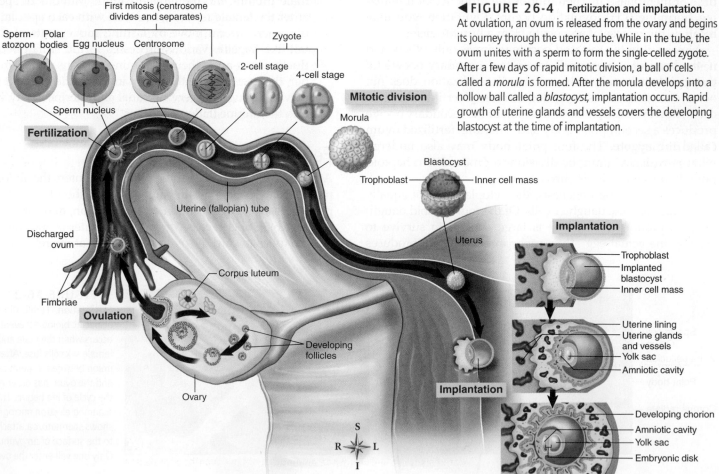

◀**FIGURE 26-4 Fertilization and implantation.** At ovulation, an ovum is released from the ovary and begins its journey through the uterine tube. While in the tube, the ovum unites with a sperm to form the single-celled zygote. After a few days of rapid mitotic division, a ball of cells called a *morula* is formed. After the morula develops into a hollow ball called a *blastocyst,* implantation occurs. Rapid growth of uterine glands and vessels covers the developing blastocyst at the time of implantation.

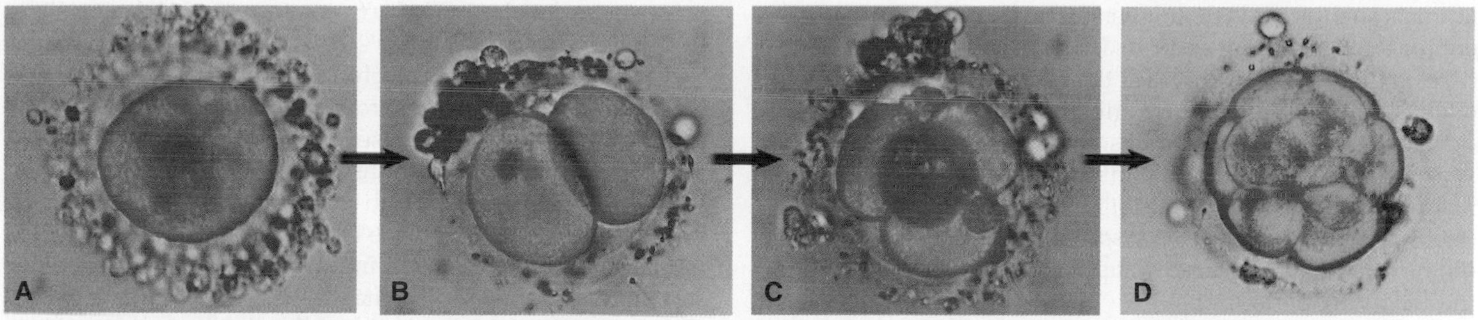

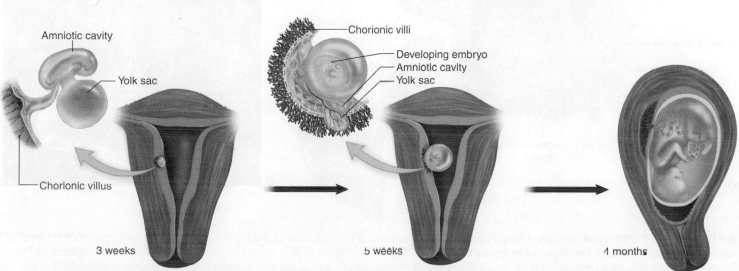

▲ **FIGURE 26-5** **Initial rapid cell division in human development.** **A,** Fertilized ovum, or zygote. **B** to **D,** Early cell divisions produce more and more cells. The solid mass of cells shown in **D** forms the morula—an early stage in embryonic development. NOTE: Although the number of cells is rapidly increasing, the size of the developing embryo does not increase until after it implants into the wall of the uterus as a blastocyst.

PRENATAL PERIOD

The *prenatal period* of development begins at the time of **conception,** or fertilization (i.e., at the moment the nuclei of the female ovum and a male sperm unite). The period of prenatal development continues until the birth of the child about 39 weeks later. The science of the development of the individual before birth is called **embryology.** It is an incredibly complex story of biological marvels, describing the means by which a new human life is begun and the steps by which a single microscopic cell is transformed into a complex human being.

Cleavage and Implantation

Once formed, the zygote immediately begins to *cleave,* or divide. In about 3 days, a solid mass of cells called a **morula**

forms (see Figure 26-4). The cells of the morula begin to form an inner cavity as they continue to divide, and by the time the developing embryo reaches the uterus, it is a hollow ball of cells with a fluid-filled center called a **blastocyst.** At about 1 week after fertilization, the process of **implantation** begins. In about 10 days from the time of fertilization, the blastocyst is completely implanted in the uterine lining, which secretes "uterine milk," or *histotrophe,* to nourish it. Of course, problems in development or implantation may occur at any stage—resulting in loss of the offspring and termination of the developmental process **(spontaneous abortion).**

The rapid cell division taking place up to the blastocyst stage occurs with no significant increase in total mass compared with the zygote (Figure 26-5). One of the specializations of the ovum (and its surrounding layers) is its incredible store of nutrients that support this embryonic development until implantation has occurred.

Note in Figure 26-4 that the blastocyst consists of an outer layer of cells and an **inner cell mass.** The outer wall of the blastocyst is called the **trophoblast.** As the blastocyst develops further, the inner cell mass forms a structure with two cavities called the **yolk sac** and **amniotic cavity** (see Figures 26-4 and 26-6). The yolk sac is most important in

Amniotic cavity

Yolk sac

Chorionic villus

Chorionic villi

Developing embryo
Amniotic cavity
Yolk sac

3 weeks

5 weeks

4 months

▲ **FIGURE 26-6** Development of the chorion and amniotic cavity to 4 months of gestation.

animals such as birds that depend heavily on yolk as a nutrient for the developing embryo. In these animals, the yolk sac digests the yolk and provides the resulting nutrients to the embryo. Because the uterine lining provides nutrients to the developing embryo in humans, the function of the yolk sac is not a nutritive one. Instead, it has other functions, including production of blood cells. *It is the inner cell mass that eventually forms the tissues of the offspring's body.* The trophoblast, on the other hand, forms the support structures described in the following paragraphs.

The amniotic cavity becomes a fluid-filled, shock-absorbing sac, sometimes called the "bag of waters," in which the embryo floats during development. The **chorion,** shown in Figures 26-4, 26-6, and 26-7, develops from the trophoblast to become an important fetal membrane in the *placenta.* The *chorionic villi* shown in Figures 26-6 and 26-7 are extensions of the blood vessels of the chorion that bring the embryonic circulation to the placenta. The placenta (see Figure 26-7) anchors the developing offspring to the uterus. It also provides a "bridge" for the exchange of nutrients, waste products, and carbon dioxide and oxygen between mother and baby. Because the lungs are not functional, the placenta serves that purpose, as we will see.

Placenta

The **placenta** is a unique structure that has a temporary but very important set of functions during pregnancy. It is composed of tissues from *both* mother and offspring and functions not only as a structural "anchor" and nutritional bridge but also as an excretory, respiratory, and endocrine organ.

Placental tissue normally separates the maternal and fetal blood supplies so that no intermixing occurs. The thin layer of placental tissue also serves as an effective "barrier" that protects the developing baby from many harmful substances that may enter the mother's bloodstream. Unfortunately, toxic substances such as alcohol and some infectious organisms may penetrate this protective placental barrier and injure the developing baby (see Box 17-2, Chapter 17, p. 398). For example, the virus responsible for rubella (German measles) can easily pass through the placenta and cause tragic developmental defects in the fetus.

Placental tissue also has important endocrine functions. As Figure 26-8 shows, placental tissue secretes large amounts of **human chorionic gonadotropin (hCG)** early in pregnancy. Secretion of hCG peaks about 8 or 9 weeks after fertilization, then drops to a continuous low level by about week 16. Human chorionic gonadotropin acts as a gonadotropin and thus stimulates the corpus luteum to continue its secretion of estrogen and progesterone. Recall from Figure 25-9 (p. 590) that reduced levels of the anterior pituitary gonadotropins (FSH and LH) after ovulation normally cause a corresponding reduction in luteal secretion of the estrogen and progesterone needed to sustain the uterine lining. The drop in estrogen and progesterone secretion results from the fact that the FSH and LH needed to maintain the corpus luteum are now in short supply.

Here's what happens to prevent menstruation and to allow successful implantation and development of the offspring: The cells of the trophoblast and, later, the placenta secrete enough hCG to maintain the corpus luteum and thus keep luteal estrogen and progesterone levels high.

As the placenta develops, it begins to secrete its own estrogen and progesterone. As Figure 26-8 shows you, as more estrogen and progesterone are secreted from the placenta, a corresponding decrease in hCG secretion produces a drop in luteal secretion of these hormones. After about 3 months, the corpus luteum has degenerated and the

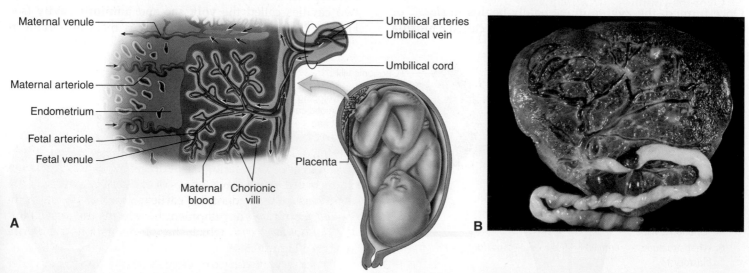

A

B

▲ **FIGURE 26-7**   **Structural features of the placenta.** The close proximity of the fetal blood supply and the maternal blood supply permits diffusion of nutrients and other substances. The placenta also forms a thin barrier that prevents diffusion of most harmful substances. No mixing of fetal and maternal blood occurs. **A,** Diagram showing a cross section of the placental structure. **B,** Photograph of a normal, full-term placenta (fetal side) showing the branching of the placental blood vessels.

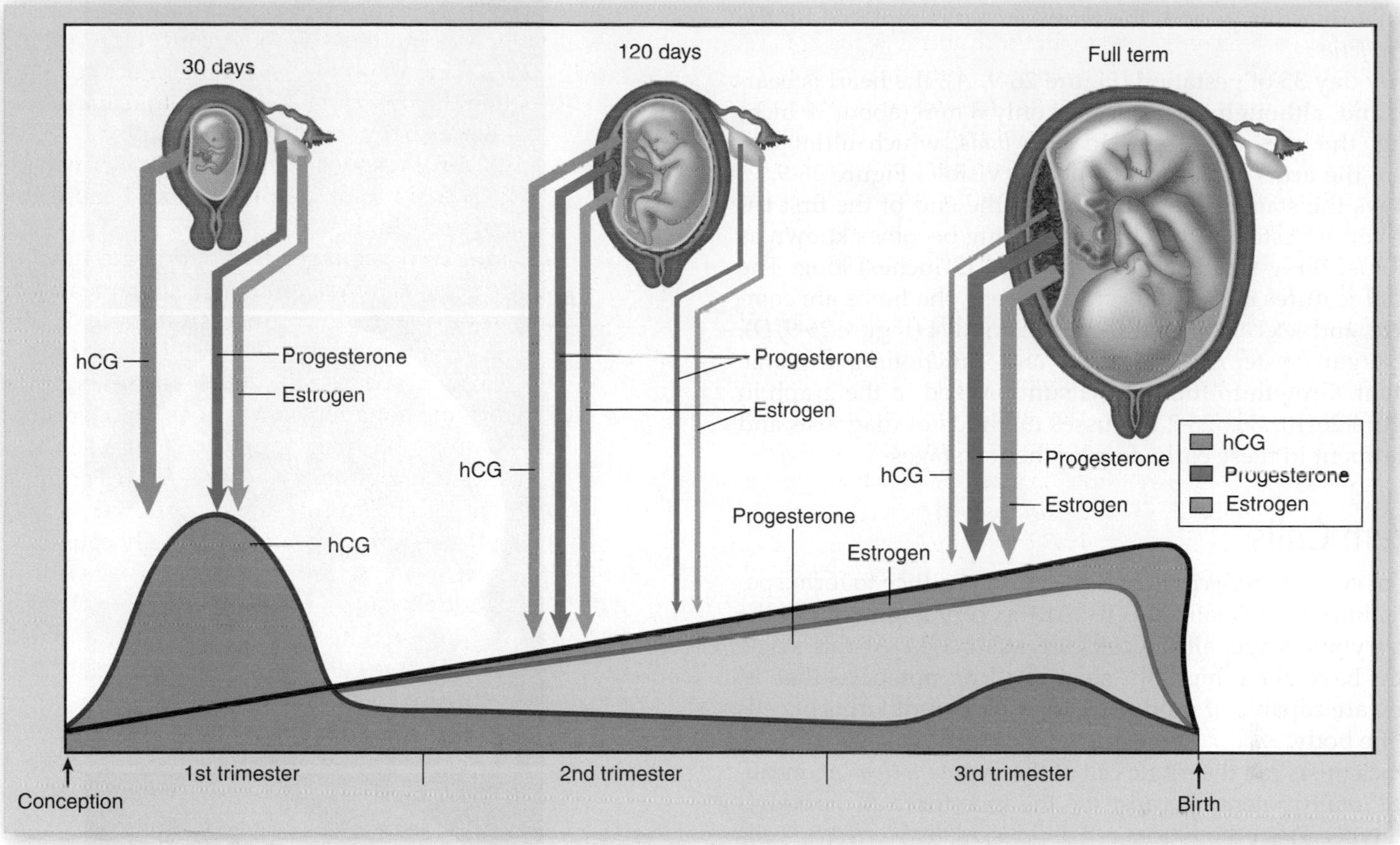

▲ **FIGURE 26-8** **Hormone levels during pregnancy.** Diagram showing the changes that occur in the blood concentration of human chorionic gonadotropin (hCG), estrogen, and progesterone during gestation. Note that high hCG levels produced by placental tissue early in pregnancy maintain estrogen and progesterone secretion by the corpus luteum. This prevents menstruation and promotes maintenance of the uterine lining. As the placenta takes over the job of secreting estrogen and progesterone, hCG levels drop, and the corpus luteum subsequently ceases secreting these hormones.

placenta has completely taken over the job of secreting the estrogen and progesterone needed to sustain the pregnancy.

Over-the-counter "early pregnancy" tests detect the presence of the hCG that is excreted in the urine during the first couple of months of a pregnancy. Such tests can detect hCG in the urine as early as 1 or 2 days after implantation occurs.

There are other hormones that support pregnancy, and their functions were discussed in previous chapters. During pregnancy some hormones, such as growth hormone (GH), remain stable, but others *increase*. All of these hormones directly or indirectly promote the physiological processes of fetal development and lactation. Cortisol, for example, also helps trigger important events during pregnancy—including the onset of labor.

QUICK CHECK

5. What is a morula? What is a blastocyst?

6. What structures are derived from the trophoblast (outer wall) of the blastocyst?

7. What is the placenta? What is its function?

8. What placental hormone maintains the corpus luteum during the early weeks of pregnancy?

Periods of Development

The length of pregnancy (about 39 weeks)—called the **gestation period**—is divided into three approximately 3-month segments called *trimesters*. The first trimester extends from the first day of the last menstrual period to the end of the twelfth week. The second trimester extends from the end of the twelfth to the twenty-eighth week of pregnancy. The third trimester extends from the twenty-eighth week of pregnancy until the baby is delivered.

Several terms are used to describe stages of development during the three trimesters of pregnancy. During the first trimester, or 3 months, of pregnancy, numerous terms are used. *Zygote* is used to describe the ovum just after fertilization by a sperm cell. After about 3 days of constant cell division, the solid mass of cells, identified earlier as the *morula*, enters the uterus. Continued development transforms the morula into the hollow *blastocyst*, which then implants into the uterine wall (see Figure 26-4).

The embryonic phase of development extends from fertilization until the end of week 8 of gestation. During this period in the first trimester, the term *embryo* is used to describe the developing individual. The fetal phase is used to indicate the development extending from week 8 to week

39. During this period, the term *embryo* is replaced by the term *fetus*.

By day 35 of gestation (Figure 26-9, *A*), the heart is beating and, although the embryo is only 8 mm (about ⅜ inch) long, the eyes and so-called *limb buds,* which ultimately form the arms and legs, are clearly visible. Figure 26-9, *C*, shows the stage of development at the end of the first trimester of gestation, when the offspring becomes known as a fetus. Body size is about 7 to 8 cm (3 inches) long. The facial features of the fetus are apparent, the limbs are complete, and sex can be identified. By month 4 (Figure 26-9, *D*), all organ systems are formed and functioning to some extent. Growth to full term is summarized in the graph in Figure 26-10. Box 26-2 discusses methods of diagnosis and treatment in these early developmental stages.

Stem Cells

Stem cells are *unspecialized* cells that reproduce to form specific lines of specialized cells. At the very beginning of the embryonic stage, all the cells are stem cells. At this stage, they have their highest "stemness" or potency—that is, they are capable of producing many different kinds of cells in the body.

Scientists call the single cell of the zygote *totipotent*, meaning "totally potent," because it is the ancestor to all the body's cell types. After the zygote cell divides, many *pluripotent* cells are formed—cells that can produce many (but not all) kinds of cells. It is these early pluripotent cells that are commonly referred to as *embryonic stem cells.* It is the embryonic stem cells that form the germ layers described in the next section.

Some stem cells remain throughout development and maturity. These *multipotent* stem cells, such as the hematopoietic stem cells found in adult bone marrow, can only produce a few types of cells. *Adult stem cells,* as these multipotent cells are usually called, are found in many tissues of the body. For example, adult stem cells are found in the skin, many glands, muscles, nerve tissue, bone, and the GI tract. Adult stem cells replace the specialized cells in a tissue and thus ensure stable, functional populations of the cell types needed for survival.

> **A&P CONNECT**
>
> What is all the fuss about stem cell research? Find out in **Stem Cell Research** online at **A&P Connect.**
>
> *evolve*

Formation of the Primary Germ Layers

Early in the first trimester of pregnancy, three layers of unique cells develop that embryologists call the **primary germ layers.** Cells of the *embryonic disk* seen in Figure 26-4 differentiate into distinct types that form each of these three primary germ layers. Pluripotent stem cells in each layer continue to differentiate and thus give rise to the various specific organs and systems of the body, such as the skin,

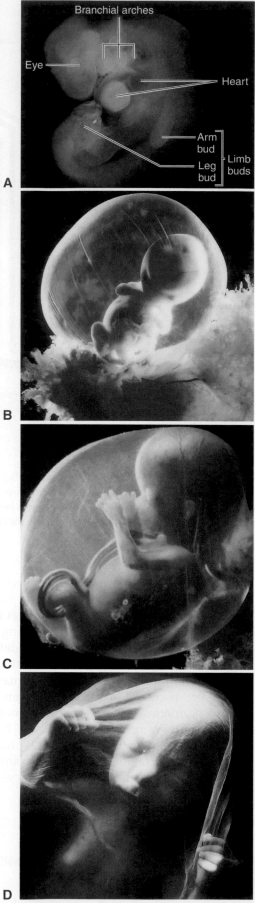

▲ **FIGURE 26-9**   **Human embryos and fetuses. A,** At 35 days. **B,** At 49 days. **C,** At the end of the first trimester. **D,** At 4 months.

BOX 26-2 Diagnostic Study

Antenatal Diagnosis and Treatment

Advances in **antenatal** (from the Latin *ante-*, "before" and *-natus*, "birth") **medicine** now permit extensive diagnosis and treatment of disease in the fetus much like any other patient. This new dimension in medicine began with techniques by which Rh-positive babies could be given transfusions before birth.

Current procedures using images provided by **ultrasonography** equipment (see the figure) allow physicians to prepare for and perform, before the birth of a baby, corrective surgical procedures such as bladder repair. These procedures also allow physicians to monitor the progress of other types of treatment on a developing fetus. Part *A* of the figure shows placement of the ultrasound transducer on the abdominal wall. The resulting image (see part *B* of the figure), called an *ultrasonogram*, shows a midsagittal view of a 20-week-old fetus.

▶ Ultrasonography.

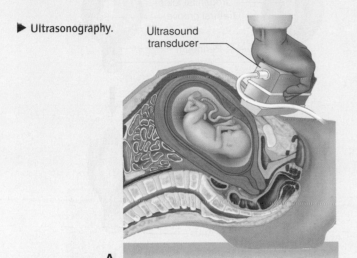

A

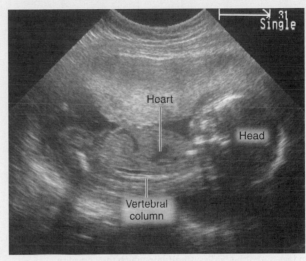

B

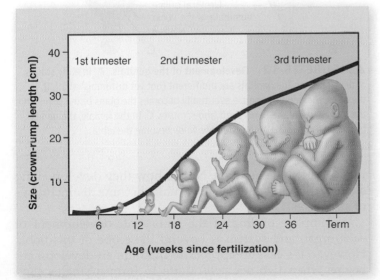

▲ **FIGURE 26-10 Increase in size during prenatal development.** This graph shows the usual progression in size and body shape during the three trimesters of fetal development.

nervous tissue, muscles, or digestive organs (Figure 26-11). As new tissues and organs develop, older cells often die through the process of apoptosis and thus make room for newer, more specialized cells. Each primary germ layer is called, respectively, *endoderm*, or inside layer; *ectoderm*, or outside layer; and *mesoderm*, or middle layer.

Ectoderm

The outer germ layer, or **ectoderm,** forms many of the structures around the periphery of the body. For example, the epidermis of the skin, enamel of the teeth, and cornea and lens of the eye are derived from the ectoderm. Besides these peripheral structures, various components of the nervous system—including the brain and the spinal cord—also have an ectodermal origin.

Endoderm

The inner germ layer, or **endoderm,** forms the linings of various tracts, as well as several glands. For example, the lining of the respiratory tract and GI tract, including some of the accessory structures such as tonsils, is derived from the

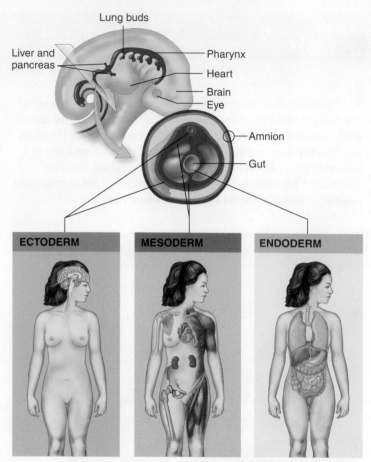

▲ **FIGURE 26-11** **The primary germ layers.** Illustration shows the primary germ layers and the body systems into which they develop.

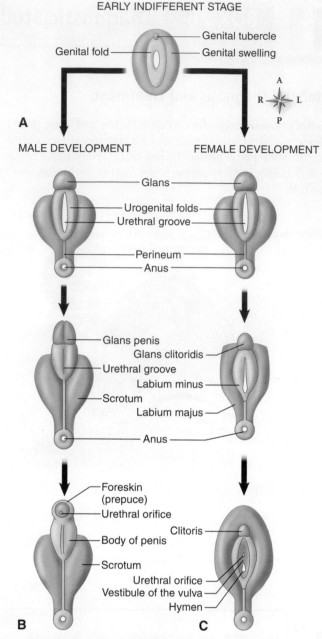

▲ **FIGURE 26-12** **Development of the genitals. A,** In early stages of development, the genitals are indifferent (not yet distinguishable). **B,** In the male, the genital tubercle eventually becomes the glans of the penis and the folds become the penis shaft and scrotum. **C,** In the female, the genital tubercle becomes the clitoris and the folds become the labia.

endoderm. The linings of the pancreatic ducts, hepatic ducts, and urinary tract also have an endodermal origin. The glandular epithelium of the thymus, thyroid, and parathyroid glands is also derived from the endoderm.

Mesoderm

The middle germ layer, or **mesoderm,** forms most of the organs and other structures between those formed by the endoderm and ectoderm. For example, the dermis of the skin, the skeletal muscles and bones, many of the glands of the body, kidneys, gonads, and components of the circulatory system are derived from the mesoderm. Look carefully at Figure 26-11 to discern the logical pattern exhibited by germ layer development and differentiation.

Embryonic Development

The process by which the primary germ layers develop into many different kinds of tissues is called **histogenesis.** The way these tissues arrange themselves into organs is called **organogenesis.**

A brief example of organogenesis that is particularly useful in our current discussion of reproduction and development is the differentiation and development of the genitals. Figure 26-12 outlines the development of the male and female genitals. Notice how they develop along slightly different paths to eventually become distinct types of structures.

A complete outline of the embryonic development of each organ and system is beyond the scope of this book. However, it's easy to appreciate that human development begins when two sex cells unite to form a single-celled zygote and that the new offspring's body grows and changes by a series of processes consisting of cell differentiation, multiplication, growth, apoptosis, and rearrangement. All of this must take place in a definite, orderly sequence! Development of structure and function go hand in hand, and from 4 months of gestation, when every organ system is in place and functioning to some extent, until term (about 280 days), development of the fetus is mainly a matter of growth.

Figure 26-13 shows the normal intrauterine position of a fetus just before birth in a full-term pregnancy. The large size of the pregnant uterus toward the end of pregnancy affects the normal function of the mother's body greatly. For example, you might be able to tell from Figure 26-13 that a woman's center of gravity is shifted forward. This can make walking and other movements of the body difficult—or even hazardous—because the sensory and motor control systems often do not compensate completely for this shift. The pregnant uterus presses on the bladder and the rectum, sometimes adversely affecting intestinal

motility, and thus may cause constipation and/or hemorrhoids. Pressure on the bladder reduces its urine-storing capacity, which results in frequent urination. Upward pressure pushes the abdominal organs against the diaphragm, making deep breathing difficult and sometimes causing the stomach to protrude into the thoracic cavity—a condition called *hiatal hernia*.

> **QUICK ✓ CHECK**
>
> 9. What is a *trimester*?
> 10. What is the difference between an *embryo* and a *fetus*?
> 11. Name the three primary germ layers.
> 12. What are some of the urinary and digestive problems faced by women late in pregnancy? Why?

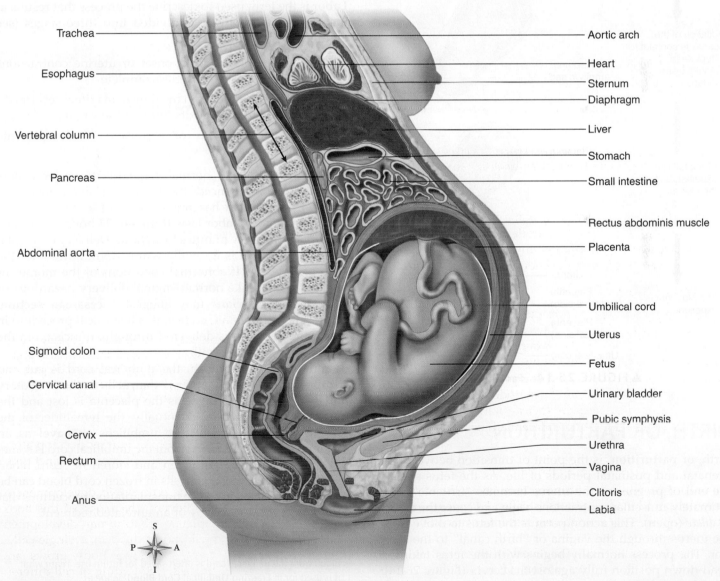

▲ **FIGURE 26-13** **Full-term pregnancy.** Notice that the mother's organs are being pushed by the developing fetus, placenta, and uterus and that the woman's center of gravity is now shifted forward.

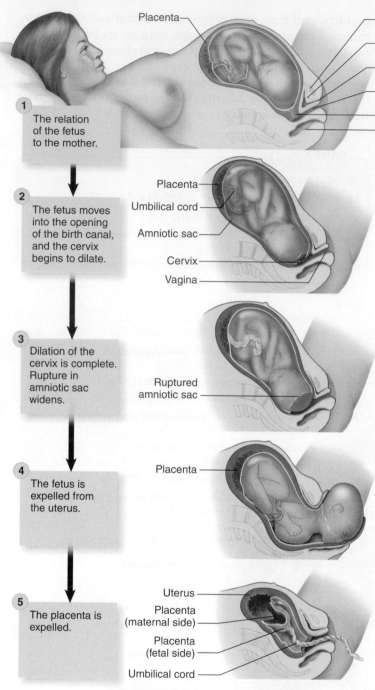

① The relation of the fetus to the mother.

② The fetus moves into the opening of the birth canal, and the cervix begins to dilate.

③ Dilation of the cervix is complete. Rupture in amniotic sac widens.

④ The fetus is expelled from the uterus.

⑤ The placenta is expelled.

▲ **FIGURE 26-14** Parturition.

BIRTH, OR PARTURITION

Birth, or **parturition,** is the point of transition between the prenatal and postnatal periods of life. As the fetus signals the end of pregnancy, the uterus becomes "irritable" and, ultimately, muscular contractions begin and cause the cervix to *dilate* (open). This action permits the fetus to move from the uterus through the vagina or "birth canal" to the exterior. The process normally begins with the fetus taking a head-down position fully against the cervix (Figure 26-14). When contractions occur, the amniotic sac, or "bag of waters," usually ruptures, and labor begins.

Several hormones help signal the time of labor and promote the processes needed for successful delivery. High levels of cortisol at the end of pregnancy trigger a drop in hCG, which in turn causes a drop in progesterone levels (see Figure 26-8). Progesterone inhibits the release of oxytocin (OT) earlier in the pregnancy—but at this point, the "brake" on the uterine muscle is released. Oxytocin is released in a positive feedback mechanism that amplifies the rate and strength of labor contractions (see illustration in Box 1-2, page 16). An injection of a synthetic OT preparation (Pitocin) can stimulate labor contractions in a difficult or delayed delivery. Prostaglandins E_2 and F_2 released by the placenta also play a role in the onset of labor by further sensitizing the myometrium of the uterus to OT.

Stages of Labor

Labor is the term used to describe the process that results in the birth of the baby. It is divided into three stages (see Figure 26-14):

1. Stage one—period from onset of uterine contractions until dilation of the cervix is complete
2. Stage two—period from the time of maximal cervical dilation until the baby exits through the vagina
3. Stage three—process of expulsion of the placenta through the vagina

The time required for normal vaginal birth varies widely and may be influenced by many variables, including whether the woman has previously had a child. In most cases, stage one of labor lasts from 6 to 24 hours, and stage two lasts from a few minutes to an hour. Delivery of the placenta (stage three) is normally within 15 minutes after the birth of the baby. If abnormal conditions of the mother or fetus (or both) make normal vaginal delivery hazardous or impossible, physicians may suggest a **cesarean section.** Often called simply a *C-section,* it is a surgical procedure in which the newborn is delivered through an incision in the abdomen and uterine wall.

Immediately after birth, the umbilical cord is cut and clamped. Recall from Chapter 17 that, at birth, the circulatory route of the infant changes as the placenta is lost and the lungs begin to function. Eventually, the remainder of the cord sloughs off—leaving the umbilicus or **navel** as an abdominal landmark. Blood from the umbilical cord is sometimes frozen in liquid nitrogen and stored in "cord blood banks" for future use. Stem cells in frozen cord blood can be harvested and used in future transplantation procedures that may benefit the donor baby or an unrelated recipient.

A&P CONNECT

Blood from the umbilical cord is often saved for future use. Learn what it is used for in **Freezing Umbilical Cord Blood** online at **A&P Connect.**

Multiple Births

The term *multiple birth* refers to the birth of two or more infants from the same pregnancy. The birth of twins is more common than the birth of triplets, quadruplets, or quintuplets. Multiple-birth babies are often born prematurely, so they are at a greater than normal risk of complications in infancy. However, premature infants who have modern medical care available have a much lower risk of complications than those without such care.

Twinning, or double births, can result from either of two different processes:

1. *Identical twins* result from the splitting of embryonic tissue from the same zygote early in development. One way this happens is that, during the early blastocyst stage of development, the inner cell mass divides into two masses. Each inner cell mass thus formed develops into a separate individual. As Figure 26-15, *A*, shows, identical twins usually share the same placenta but have separate umbilical cords. This is not surprising because in this type of twinning there is a single, shared trophoblast. Because they develop from the same fertilized egg, identical twins have the same genetic code. Despite this, identical twins are not absolutely identical in terms of structure and function. Different environmental factors and personal experiences lead to individuality even in genetically identical twins.

2. *Fraternal twins* result from the fertilization of two different ova by two different spermatozoa (Figure 26-15, *B*). Fraternal twinning requires the production of more than one mature ovum during a single menstrual cycle, a trait that is often inherited. Multiple ovulation may also occur in response to certain fertility drugs, especially the gonadotropin preparations. Fraternal twins are no more closely related genetically than any other sibling relationship. Because two separate fertilizations must occur, it is even possible for fraternal twins to have different biological fathers. Triplets, quadruplets, and other multiple births may be identical, fraternal, or any combination.

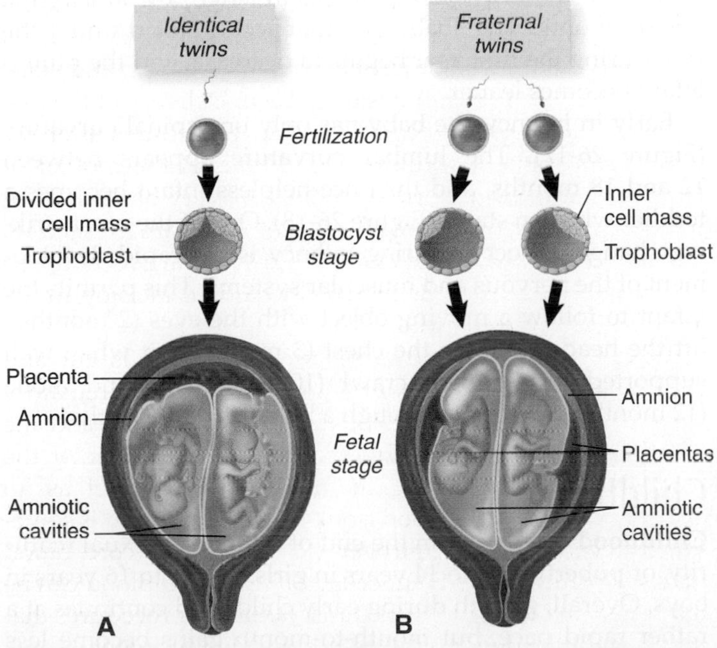

QUICK ✓ CHECK

13. Briefly, what happens during parturition?
14. Briefly describe the three stages of labor.
15. How does identical twinning differ from fraternal twinning?

POSTNATAL PERIOD

The *postnatal period* begins at birth and lasts until death. It is often divided into major periods (such as infancy and adulthood) for study. However, you must understand that growth and development are continuous processes that occur throughout the life cycle. Gradual changes in the physical appearance of the body as a whole and in the relative proportions of the head, trunk, and limbs are quite noticeable between birth and adolescence. Note in Figure 26-16 the obvious changes in the size of bones and in the proportionate sizes between different bones and body areas. The head, for example, becomes proportionately smaller. Whereas the infant head is approximately one fourth of the total height of the body, the adult head is only about one eighth of the total height. The facial bones also show several changes between

▲ **FIGURE 26-15** **Multiple births. A,** Identical twins develop when embryonic tissue from a single zygote splits to form two individuals. Notice that, because the trophoblast is shared, the placenta and the part of the amnion separating the amniotic cavities are shared by the twins. **B,** Fraternal twins develop when two ova are fertilized at the same time, producing two separate zygotes. Notice that each fraternal twin has its own placenta and amnion.

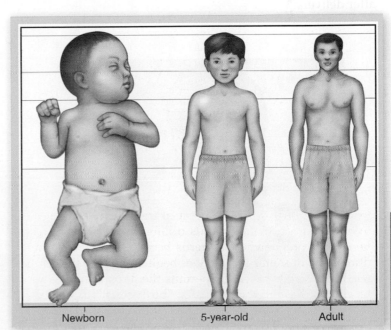

▲ **FIGURE 26-16** **Changes in the proportions of body parts from birth to maturity.** Note the dramatic differences in relative head size.

infancy and adulthood. In an infant, the face is one eighth of the skull surface, but in an adult the face is half of the skull surface. Another change in proportion involves the trunk and lower extremities. The legs become proportionately longer and the trunk proportionately shorter. In addition, the thoracic and abdominal contours change from round to elliptical.

Such changes are good examples of the ever-changing and ongoing nature of growth and development. It is unfortunate that many of the changes that occur in the later years of life are degenerative changes. We will explore some of these changes in the *Mechanisms of Disease* feature online.

The following are the most common postnatal periods: (1) *infancy,* (2) *childhood,* (3) *adolescence* and *adulthood,* and (4) *older adulthood.*

Infancy

The period of **infancy** begins abruptly at birth and lasts about 18 months. The first 4 weeks of infancy are often referred to as the **neonatal period.** Dramatic changes occur at a rapid rate during this short but critical period. **Neonatology** is the medical and nursing specialty concerned with the diagnosis and treatment of disorders of the newborn. Advances in this area have resulted in dramatically reduced infant mortality.

Many of the changes that occur in the cardiovascular and respiratory systems at birth are necessary for survival. Whereas the fetus totally depended on the mother for life support, the newborn infant, to survive, must become totally self-supporting in terms of blood circulation and respiration immediately after birth. A baby's first breath is deep and forceful. The stimulus to breathe results primarily from the increasing amounts of carbon dioxide (CO_2) that accumulate in the blood after the umbilical cord is cut shortly after delivery.

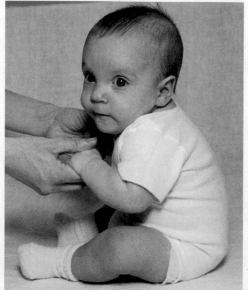

▲ **FIGURE 26-17 The infant spine.** Photograph showing the normal rounded curvature of the vertebral column in an infant.

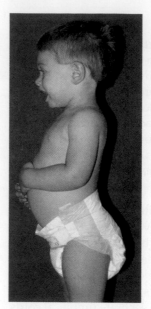

◀ **FIGURE 26-18 The toddler spine.** Photograph showing the normal curvature of the vertebral column in a toddler. The dark shadow emphasizes the distinct lumbar curvature that develops with the ability to walk (compare with Figure 26-17).

To assess the general condition of a newborn, a system that scores five health criteria is often used. The criteria are heart rate (HR), respiration, muscle tone, skin color, and response to stimuli. Each aspect is scored as 0, 1, or 2—depending on the condition of the infant. The resulting total score is called the **Apgar score.** The Apgar score in a completely healthy newborn is 10.

Many developmental changes occur between the end of the neonatal period and 18 months of age. Birth weight generally doubles during the first 4 to 6 months and then triples by 1 year. The baby also increases in length by 50% by the twelfth month. The "baby fat" that accumulated under the skin during the first year begins to decrease, and the plump infant becomes leaner.

Early in infancy, the baby has only one spinal curvature (Figure 26-17). The lumbar curvature appears between 12 and 18 months, and the once-helpless infant becomes a toddler who can stand (Figure 26-18). One of the most striking changes to occur during infancy is the rapid development of the nervous and muscular systems. This permits the infant to follow a moving object with the eyes (2 months); lift the head and raise the chest (3 months); sit when well supported (4 months); crawl (10 months); stand alone (12 months); and run, although a bit stiffly (18 months).

Childhood

Childhood extends from the end of infancy to sexual maturity, or puberty—12 to 14 years in girls and 14 to 16 years in boys. Overall, growth during early childhood continues at a rather rapid pace, but month-to-month gains become less consistent. By the age of 6 years, the child appears more like a preadolescent than an infant or toddler. The child becomes less chubby, the potbelly becomes flatter, and the face loses its babyish look. The nervous and muscular systems continue to develop rapidly during the middle years of childhood; by 10 years of age, the child has developed numerous motor and coordination skills.

The *deciduous teeth*, which began to appear at about 6 months of age, are lost during childhood, beginning at about 6 years of age. The *permanent teeth*, with the general exception of the third molars (wisdom teeth), have all erupted by age 14 years.

Adolescence and Adulthood

The average age range of **adolescence** varies, but generally the teenage years (13 to 19) are used as the standard age range. The period is marked by rapid and intense physical growth, which ultimately results in sexual maturity.

The stage of adolescence in which a person becomes sexually mature is called **puberty.** Many of the developmental changes that occur during this period are controlled by the secretion of gonadotropins (FSH and LH) and sex hormones such as testosterone and estrogen. Some of these changes involve development of the gonads themselves and are called **primary sex characteristics.** However, most of the more visible changes involve development of the **secondary sex characteristics** such as genital and skeletal changes, fat distribution patterns, growth of pubic and body hair, and growth of the larynx.

Breast development is often the first sign of approaching puberty in girls, beginning about age 9 or 10 years. Most girls begin to menstruate at 12 to 14 years of age. In boys, the first sign of puberty is often enlargement of the testes, which begins between 10 and 14 years of age. Both sexes show a spurt in height during adolescence (Figure 26-19). In girls, the spurt in height begins between the ages of 10 and 12 years and is nearly complete by 14 or 15 years. In boys, the period of rapid growth begins between 12 and 13 years and is generally complete by 16 or 17 years.

Many developmental changes that begin early in childhood are not completed until the early or middle years of adulthood. Examples include the maturation of bone, resulting in the full closure of the growth plates, and changes in the size and placement of other body components such as the sinuses. Many body traits do not become apparent for years after birth. Normal male balding patterns, for example, are determined at the time of fertilization by heredity but do not appear until maturity. As a rule, adulthood is characterized by the maintenance of existing body tissues. With the passage of years, the ongoing effort of maintenance and repair of body tissues becomes more and more difficult. As a result, degeneration begins. This is part of the process of aging, and it culminates in death.

Older Adulthood and Senescence

Most body systems are in peak condition and function at a high level of efficiency during early adulthood. As a person grows into **older adulthood,** or **senescence,** a gradual but certain decline takes place in the functioning of every major organ system in the body. The study of aging is called **gerontology.** Unfortunately, the mechanisms and causes of aging are not well understood.

Although the causes and basic mechanisms of aging are yet to be understood, at least many of the signs of aging are obvious. Figure 26-20 shows some of the biological changes associated with senescence.

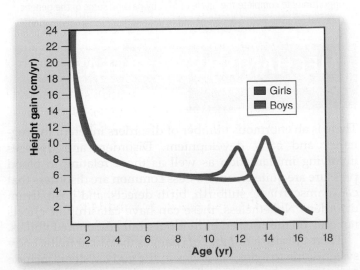

▲ FIGURE 26-19 **Growth in height.** The graph shows typical patterns of gain in height from birth to adulthood for girls and boys. Notice the rapid gain in height during the first few years, a period of slower growth, then another burst of growth during adolescence—finally ending at the beginning of adulthood.

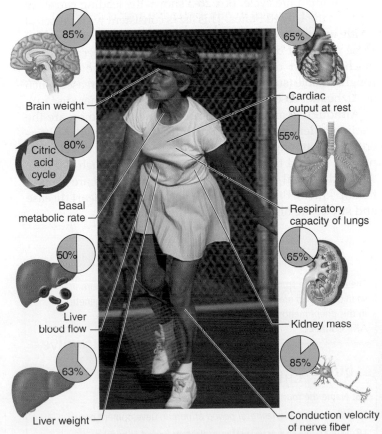

▲ FIGURE 26-20 **Some biological changes associated with maturity and aging.** Insets show proportion of remaining function in the organs of a person in late adulthood compared with a 20-year-old person.

BOX 26-3 | Health Matters

15 Leading Causes of Death in the United States

1. Diseases of the heart
2. Malignant neoplasms (cancer)
3. Cerebrovascular diseases (including stroke)
4. Chronic lower respiratory disease (including COPD)
5. Accidents (unintentional injuries)
6. Diabetes mellitus
7. Alzheimer disease
8. Influenza and pneumonia
9. Nephritis, nephritic syndrome, and nephrosis
10. Septicemia
11. Intentional self-harm (suicide)
12. Chronic liver disease and cirrhosis
13. Essential (primary) hypertension and hypertensive renal disease
14. Parkinson disease
15. Assault (homicide)

CAUSES OF DEATH

No matter what your age, death of the individual is also part of the human life cycle. Box 26-3 shows the leading causes of death in the United States. This list is consistent with causes of death in other economically developed countries. Some of these causes of death have been diminishing over the years (such as heart diseases and stroke), some have remained roughly stable (such as cancer), and some have increased (such as Alzheimer disease, Parkinson disease, and kidney disease).

Although heart disease, cancer, stroke, and so on are the leading killers in developed nations, it is a somewhat different story among the developing nations. In developing nations around the world, infectious diseases such as HIV/AIDS, diarrheal diseases, malaria, and measles are among the top killers. But even so, in developing areas heart disease and stroke are also near the very top of the list.

A&P CONNECT

The human life span has been increasing over time—even more so in developed areas of the world. Read about the role of genetics in this phenomenon in **Genes and Longevity** online at **A&P Connect.**

evolve
learning system

QUICK ✓ CHECK

16. Name the four major phases of the postnatal period.
17. When does the *neonatal period* of human development occur?
18. What signs characterize the *adolescent period* of human development?
19. What are some of the leading causes of death in the United States?

The BIG Picture

Understanding basic concepts of human growth and development is essential for understanding the dynamic, ever-changing nature of the human body. It is important to remember that life is a biological process characterized by continuous change in the structure and function of our body.

As we discussed in this and earlier chapters, production of offspring is vital to the survival of the collective human genome. Many biologists believe that this gives an important biological meaning to life in general—continued flow of genetic information from one generation to the next. Biologically speaking, production of offspring in humans is not successful unless and until a person grows and develops for many years—more than a decade—to the point where viable gametes can be produced. Gametogenesis, then, is the first process necessary to ready the genetic material of the parent to be passed along to the offspring.

Once gametes are available, the sexual union of a male and female may result in fusion of the gametes of the two adults to form the first cell of the offspring. Thus genetic information from two different individuals is now united in a new and unique way in the zygote.

The first cell of the offspring—the zygote—quickly divides again and again to eventually produce a mass of cells that will hopefully implant in the wall of the mother's uterus. Then a process unlike any other we have discussed takes over. The offspring and mother biologically connect to each other so that the mother's physiological mechanisms can sustain the offspring until its own systems are developed sufficiently. All the mother's systems alter their function to some degree to maintain the homeostatic balance of both the mother and the offspring.

After many weeks of growth and development—including histogenesis and organogenesis of a full complement of human structures—delivery of the offspring occurs. This event marks the beginning of a continuing series of changes: infancy, childhood, puberty, adolescence, adulthood, senescence, and death.

At some point in adolescence or adulthood, many of us have the opportunity to complete the "cycle of life" by passing some of the genetic code we received from our parents on to yet another human generation.

MECHANISMS OF DISEASE

There is an enormous number of disorders inherent to pregnancy and early development. Disorders and diseases involving implantation as well as those relating to blood pressure are quite common. Less common are disorders that cause miscarriage, stillbirth, birth defects, and postpartum disorders. Nonetheless, these can have catastrophic effects not only on the affected individuals, but also entire families.

There are also a number of changes that occur in the body systems due to aging. Changes in the body systems include reduction in bone mineral density, loss of muscle mass, wrinkles, cataracts, menopause, and much more.

evolve
learning system

Find out more about these diseases and disorders of human growth and development online at *Mechanisms of Disease: Growth and Development.*

Cycle of LIFE

Aging affects each individual in different ways. Environment, genetics, and perhaps even emotional and mental attitudes may affect the degree to which the structure and function of a person's body change through older adulthood. Despite advances in understanding how some of the effects of aging can be minimized or even avoided, one cannot completely avoid the fact that body structures degenerate and functions decrease as we get older. Figure 26-20 summarizes a few of the many biological changes that occur by the time we reach late adulthood.

You can read about the specific changes due to age in each body system online at *Mechanisms of Disease: Growth and Development.* ●

LANGUAGE OF SCIENCE AND MEDICINE *continued from page 601*

implantation (im-plan-TAY-shun)
  [*im*- in, *-planta*- set or place, *-ation* process]

infancy
  [*infan*- unable to speak, *-y* state]

inner cell mass

interphase (IN-ter-fayz)
  [*inter*- between, *-phase* stage]

labor (LAY-bor)

meiosis (my-OH-sis)
  [*meiosis* becoming smaller]

mesoderm (MEZ-oh-derm)
  [*meso*- middle, *-derm* skin]

morula (MOR-yoo-lah)
  [*mor*- mulberry, *-ula* little] *pl.*, morulae (MOR-yoo-lee)

navel (NAY-vel)

neonatal period (nee-oh-NAY-tal)
  [*neo*- new, *-nat*- birth, *-al* relating to]

neonatology (nee-oh-nay-TOL-oh-jee)
  [*neo*- new, *-nat*- born, *-log*- words (study of), *-y* activity]

older adulthood

oogenesis (oh-oh-JEN-eh-sis)
  [*oo*- egg, *-gen*- produce, *-esis* process]

oogonium (oh-oh-GOH-nee-um)
  [*oo*- egg, *-gonium* offspring] *pl.*, oogonia (oh-oh-GOH-nee-ah)

organogenesis (or-gah-noh-JEN-eh-sis)
  [*organ*- instrument (organ), *-gen*- produce, *-esis* process]

parturition (pahr-too-RIH-shun)
  [*parturi*- give birth, *-tion* process]

placenta (plah-SEN-tah)
  [*placenta* flat cake] *pl.*, placentae or placentas (plah-SEN-tee, plah-SEN-tahz)

postnatal period (post-NAY-tal)
  [*post*- after, *-nat*- birth, *-al* relating to]

prenatal period (pree-NAY-tal)
  [*pre*- before, *-nat*- birth, *-al* relating to]

primary follicle (PRY-mair-ee FOL-ih-kul)
  [*prim*- first, *-ary* state, *folli*- bag, *-cle* small]

primary germ layer
  [*prim*- first, *-ary* state, *germ* sprout]

primary oocyte (PRY-mair-ee OH-oh-syte)
  [*prim*- first, *-ary* state, *oo*- egg, *-cyte* cell]

primary sex characteristic
  [*prim*- first, *-ary* state, *charassein* engrave]

primary spermatocyte (PRY-mair-ee sper-MAH-toh-syte)
  [*prim*- first, *-ary* relating to, *sperm*- seed, *-cyte* cell]

puberty (PYOO-ber-tee)
  [*pubert*- age of maturity, *-y* state]

secondary follicle (SEK-on-dair-ee FOL-ih-kul)
  [*second*- second, *-ary* relating to, *folli*- bag, *-cle* small]

secondary oocyte (SEK-on-dair-ee OH-oh-site)
  [*second*- second, *-ary* relating to, *oo*- egg, *-cyte* cell]

secondary sex characteristic
  [*second*- second, *-ary* state, *charassein* engrave]

secondary spermatocyte (SEK-on-dair-ee sper-MAH-toh-syte)
  [*second*- second, *-ary* relating to, *sperm*- seed, *-cyte* cell]

senescence (seh-NES-enz)
  [*senesc*- grow old, *-ence* state]

spermatogenesis (sper-mah-toh-JEN-eh-sis)
  [*sperm*- seed, *-gen*- produce, *-esis* process]

spermatogonium (sper-mah-toh-GOH-nee-um)
  [*sperm*- seed, *-gonia* offspring] *pl.*, spermatogonia (sper-mah-toh-GOH-nee-ah)

spontaneous abortion (spon-TAY-nee-us ah-BOR-shun)
  [*ab*- away from, *-or*- be born, *-tion* process]

tertiary follicle (TER-she-air-ee FOL-ih-kul)
  [*terti*- third, *-ary* relating to, *folli*- bag, *-cle* small]

theca cell (THEE-kah sell)
  [*theca* sheath, *cell* storeroom]

trophoblast (TROH-foh-blast)
  [*tropho*- nourishment, *-blast* sprout]

ultrasonography (ul-trah-soh-NOG-rah-fee)
  [*ultra*- beyond, *-sono*- sound, *-graph*- draw, *-y* process]

yolk sac

zygote (ZYE-goht)
  [*zygot*- union or yoke]

CHAPTER SUMMARY

To download an MP3 version of the chapter summary for use with your iPod or other portable media player, access the Audio Chapter Summaries *online at http://evolve.elsevier.com.*

HINT *Scan this summary after reading the chapter to help you reinforce the key concepts. Later, use the summary as a quick review before your class or before a test.*

INTRODUCTION

A. Prenatal period—begins at conception and ends at birth
B. Postnatal period—begins at birth and continues until death
C. Developmental biology—study of the many changes that occur during the cycle of life from conception to death

A NEW HUMAN LIFE

A. Meiosis and production of sex cells
  1. A new human offspring begins when the nucleus of a single spermatozoon (sperm) unites with the nucleus of an ovum (egg)
  2. Meiosis—orderly reduction and distribution of chromosomes to make sperm and eggs
    a. Mature ova and sperm contain only 23 chromosomes, half as many as other human cells
    b. Meiotic division consists of two nuclear divisions that take place one after the other in succession; meiosis I and meiosis II (Figure 26-1)
B. Spermatogenesis—process by which sperm-forming cells later become transformed into mature sperm
  1. Meiotic division I—one primary spermatocyte forms two secondary spermatocytes, each with 23 chromosomes

2. Meiotic division II—each of the two secondary spermato-cytes forms a total of four spermatids

C. Oogenesis—process by which egg-forming cells become mature eggs (Figure 26-2)
 1. Mitosis—oogonia reproduce to form primary oocytes
 2. Once during each menstrual cycle, a few primary oocytes resume meiosis and migrate toward the surface of the ovary; usually only one oocyte matures enough for ovula-tion, and meiosis again halts at metaphase I
 3. Meiosis resumes only if the head of a sperm cell enters the ovum

D. Ovulation and insemination
 1. Ovulation—expulsion of the mature ovum from the mature ovarian follicle into the abdominopelvic cavity, from where it enters one of the uterine tubes
 2. Insemination—expulsion of the seminal fluid from the male urethra into the female vagina

E. Fertilization (conception) (Figure 26-3)
 1. Sperm cells "swim" up the uterine tubes toward the ovum
 2. Fertilization most often occurs in the outer one third of the oviduct (Figure 26-4)
 3. The process of sperm movement is assisted by mechanisms within the female reproductive tract
  a. Mucous strands in the cervical canal guide the sperm on their way into the uterus
  b. Rhythmic contractions of the female reproductive tract and ciliary movement along the lining of the uterine tubes also assist the movement of sperm
  c. Sperm are attracted to the warmer temperatures of the uterine tubes
 4. Ovum also takes an active role in the process of fertilization
  a. Ovum and its surrounding layers actually attract nearby sperm with special peptides
 5. Acrosome in the "head" of the sperm releases enzymes that break down the outer layers surrounding the ovum
 6. 23 chromosomes from the sperm head and 23 chromosomes in the ovum make up a total of 46 chromosomes
 7. Zygote—fertilized ovum; genetically complete

PRENATAL PERIOD

A. Begins with conception and continues until the birth of a child
B. Cleavage and implantation—once formed, the zygote immedi-ately begins to divide (Figure 26-4)
 1. Morula—solid mass of cells formed from the zygote (Figure 26-5)
 2. Blastocyst—hollow ball of cells of the developing embryo that implants into the uterine lining
 3. Blastocyst consists of an outer layer of cells and an inner cell mass
  a. Trophoblast—outer wall of blastocyst
  b. Inner cell mass—as blastocyst develops, yolk sac and amniotic cavity are formed (Figures 26-4 and 26-6)
 4. Chorion develops from the trophoblast to become an important fetal membrane in the placenta
C. Placenta—anchors the developing offspring to the uterus (Figure 26-7)
 1. Also provides a "bridge" for the exchange of nutrients, waste products, and carbon dioxide and oxygen between mother and baby
 2. Placental tissue normally separates the maternal and fetal blood supplies so that no intermixing occurs
 3. Has important endocrine functions—secretes large amounts of hCG, which stimulate the corpus luteum to continue to secrete estrogen and progesterone (Figure 26-8)

D. Periods of development (Figures 26-9 and 26-10)
 1. Gestation period—approximately 39 weeks; divided into three 3-month segments called trimesters
 2. Embryonic phase extends from fertilization until the end of week 8 of gestation
 3. Fetal phase—weeks 8 to 39
E. Stem cells
 1. Stem cells—unspecialized cells that reproduce to form specific lines of specialized cells
 2. Totipotent stem cell—can produce any type of cell; found in zygote
 3. Pluripotent stem cell—embryonic stem cell that can produce a broad range of cell types; found in embryonic germ layers
 4. Multipotent stem cell—adult stem cell found in some tissues that can produce a few cell types and thus maintain functional populations of specialized cells
F. Formation of the primary germ layers
 1. Three layers of developmental cells arise early in the first trimester of pregnancy
 2. Cells of the embryonic disk differentiate and form each of the three primary germ layers
 3. Each primary germ layer gives rise to specific organs and systems of the body (Figure 26-11)
  a. Endoderm—inside layer; forms many of the structures around the periphery of the body
  b. Ectoderm—outside layer; forms the linings of various tracts, as well as several glands
  c. Mesoderm—middle layer; forms most of the organs and other structures between those formed by the endoderm and ectoderm
G. Embryonic development
 1. Histogenesis—process by which primary germ layers develop into different kinds of tissues
 2. Organogenesis—how tissues arrange themselves into organs

BIRTH, OR PARTURITION

A. Birth (parturition)—transition between the prenatal and postnatal periods of life
B. Several hormones help signal the time of labor and promote the processes needed for successful delivery
 1. High levels of cortisol at the end of pregnancy trigger a drop in hCG, which in turn causes a drop in progesterone levels (Figure 26-8)
 2. Progesterone inhibits the release of oxytocin (OT) earlier in the pregnancy, but at this point, the "brake" on the uterine muscle is released
C. Stages of labor (Figure 26-14)
 1. Stage one—period from onset of uterine contractions until dilation of the cervix is complete
 2. Stage two—period from the time of maximal cervical dilation until the baby exits through the vagina
 3. Stage three—process of expulsion of the placenta through the vagina
D. Multiple births—birth of two or more infants from the same pregnancy
 1. Identical twins result from the splitting of embryonic tissue from the same zygote early in development (Figure 26-15, A)
 2. Fraternal twins result from the fertilization of two different ova by two different spermatozoa (Figure 26-15, B)

POSTNATAL PERIOD

A. Postnatal period—begins at birth and lasts until death (Figure 26-16)

B. Common postnatal periods:
   1. Infancy
   2. Childhood
   3. Adolescence and adulthood
   4. Older adulthood
C. Infancy—begins abruptly at birth and lasts about 18 months
   1. Neonatal period—first 4 months of infancy; dramatic changes occur at a rapid rate during this period
   2. Changes that occur in the cardiovascular and respiratory systems at birth are necessary for survival
   3. Apgar score assesses general condition of newborn infant
D. Childhood—extends from the end of infancy to sexual maturity, or puberty
   1. Growth during early childhood continues at a rather rapid pace, but month-to-month gains become less consistent
   2. By the age of 6 years, the child appears more like a preadolescent than an infant or toddler
   3. Nervous and muscular systems continue to develop rapidly during the middle years of childhood
E. Adolescence and adulthood
   1. Adolescence—generally the teenage years (13 to 19) are used as the standard age range
      a. This period is marked by rapid and intense physical growth, which ultimately results in sexual maturity
      b. Puberty—stage of adolescence in which a person becomes sexually mature
      c. Many of the developmental changes that occur during this period are controlled by the secretion of hormones
      d. Primary sex characteristics—maturation of gonads and reproductive tract
      e. Secondary sex characteristics—fat distribution patterns, growth of pubic and body hair, and growth of the larynx
   2. Adulthood—characterized by maintenance of existing body tissues
F. Older adulthood and senescence
   1. As a person grows older, a gradual but certain decline takes place in the functioning of every major organ system in the body

CAUSES OF DEATH

A. In developed countries such as the United States, heart disease, cancer, and stroke are among the leading causes of death (Box 26-3)
B. In developing nations around the world, infectious diseases such as HIV/AIDS, diarrheal diseases, malaria, and measles are among the top killers; heart disease and stroke are also leading causes of death

REVIEW QUESTIONS

HINT *Write out the answers to these questions after reading the chapter and reviewing the Chapter Summary. If you simply think through the answer without writing it down, you won't retain much of your new learning.*

1. Define the terms *developmental biology, growth,* and *development.*
2. Outline the major steps in spermatogenesis. Do the same for oogenesis.
3. Identify the two processes necessary to bring the sperm and ovum into proximity with each other.
4. During fertilization, how does the ovum attract sperm?
5. At what developmental stage does implantation occur?
6. Describe the structural differences between a morula and a blastocyst.

7. How does the placenta develop?
8. What functions does the placenta provide?
9. Outline the hormonal levels of human chorionic gonadotropin (hCG), estrogen, and progesterone at various stages during gestation.
10. During what period of growth is the term *embryo* replaced by the term *fetus?*
11. List the various structures derived from each of the three primary germ layers.
12. Explain the three stages of labor.
13. What is the difference between identical and fraternal twins?
14. What are the time spans of the following postnatal periods: infancy, childhood, adolescence?
15. During what postnatal period do the secondary sex characteristics develop? What initiates this development?

CRITICAL THINKING QUESTIONS

HINT *After finishing the Review Questions, write out the answers to these items to help you apply your new knowledge. Go back to sections of the chapter that relate to items that you find difficult.*

1. How is the process of meiosis different from mitosis?
2. If a diploid cell rather than a haploid cell were used for human reproduction, what would the number of chromosomes per cell be after three generations?
3. How do histogenesis and organogenesis differ? Which of these occurs first in development?
4. Explain the procedure a physician might use if a normal vaginal delivery would be dangerous for the mother or baby.

continued from page 601

With the knowledge you have gained from reading this chapter, try to answer these questions about Maria's pregnancy from the Introductory Story.

1. What is a newly fertilized ovum called?
   a. Zygote   c. Fetus
   b. Embryo   d. Blastocyst

2. How much time will elapse from fertilization to full implantation?
   a. 2 days   c. 24 hours
   b. 4 days   d. 10 days

3. What type of twins is Maria expecting?
   a. Maternal   c. Identical
   b. Fraternal   d. Sibling

At week 37, Maria starts feeling some slight contractions, but she passes them off as indigestion. Later that night, though, she notices a thin, watery fluid leaking from her vagina, and the contractions are getting stronger. She wakes up Carlos, and they head to the birthing center.

4. What was the watery fluid Maria noted?
   a. Placental fluid   c. Amniotic fluid
   b. Chorionic fluid   d. Urine

HINT *To solve these questions, you may have to refer to the glossary or index, other chapters in this textbook, A & P Connect, Mechanisms of Disease, and other resources.*

UNIT 6

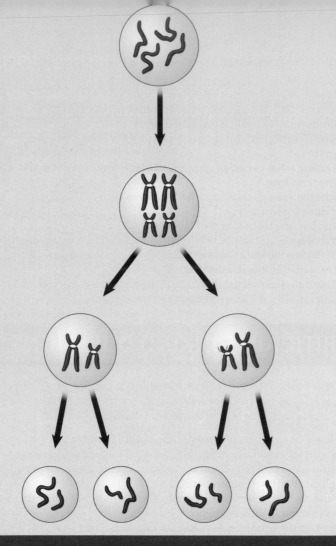

CHAPTER 27

Human Genetics and Heredity

CHAPTER OUTLINE

HINT *Scan this outline before you begin to read the chapter, as a preview of how the concepts are organized.*

STUDENT LEARNING OBJECTIVES

At the completion of this chapter, you should be able to do the following:

1. Describe the structure and function of genes and chromosomes.
2. Discuss the significance of decoding the human genome.
3. Discuss the significance of genomics and proteomics.
4. Outline the process of meiosis and the distribution of chromosomes to gametes.
5. Explain the principle of independent assortment.
6. Contrast dominant and recessive traits.
7. Identify the following terms: genotype, phenotype, homozygous, heterozygous.
8. Identify the significance of the following terms: deletion, insertion, mutagen.
9. Discuss the use of pedigrees and karyotyping in genetic counseling.

CARLOS and Maria were ecstatic to have two healthy babies. Within the first hour, the babies were cleaned and measured, and their heels were pricked for blood samples. These initial blood tests were to assess for phenylketonuria (PKU), thyroid hormone levels, cystic fibrosis, and several other metabolic disorders. To their surprise, their new daughter tested positive for PKU. Neither Carlos nor Maria had PKU. But what were the chances that their son would have PKU as well?

Thankfully, their son tested negative for PKU. But, as they were discussing genetic testing, Maria remembered that her father had been color blind. She wondered whether that meant her son would also be color blind.

What would you, with your knowledge of anatomy and physiology at this point, advise the couple? Try the questions at the end of this chapter to help you with your answers.

✕ ✕ ✕

Genetics, the scientific study of inheritance, is vital to all fields of human biology—especially anatomy, physiology, and medicine. Several years ago, scientists mapped virtually the entire human genome. Currently, scientists are working to understand the functions of all the proteins that it encodes as well as the ways in which our genes control our lives, from embryonic development to old age. In fact, health and science columns in newspapers and academic resources on the Internet keep us informed of the latest discoveries of genes involved with disease, human behavior, and even longevity. In this chapter, we briefly review the essential concepts of genetics and explain how heredity affects every structure and function in the body.

THE SCIENCE OF GENETICS

History shows that humans have been aware of patterns of inheritance—or *heredity*—for thousands of years. However, it was not until the 1860s that the scientific study of these patterns—*genetics*—was born. At that time, a monk named Gregor Mendel living in Brno, South Moravia (now located in the southern Czech republic) became the first to discover the basic mechanism by which traits are transmitted from parents to offspring. He showed that independent units (which we now call *genes*) are responsible for the inheritance of biological traits.

The science of genetics developed from Mendel's quest to explain how biological characteristics are inherited. As time went by and more genetic studies were done, it became clear that certain diseases also have a genetic basis. For example, the group of blood-clotting disorders called *hemophilia* is directly inherited by children from their parents. In this case, parents pass on a defective hemoglobin gene (or more specifically, *allele*) for hemophilia. Directly inherited diseases are often called "hereditary diseases." Other diseases involve environmental factors in addition to genetic inheritance. For example, certain forms of skin cancer are thought to have a genetic basis. However, a person who inherits the genes associated with skin cancer typically will develop the disease only if his or her skin is also heavily exposed to the ultraviolet radiation in sunlight.

LANGUAGE OF SCIENCE AND MEDICINE

HINT · Before reading the chapter, say each of these terms out loud. This will help you avoid stumbling over them as you read.

albinism (AL-bih-niz-em)
[*alb-* white, *-in-* characterized by, *-ism* state]

amniocentesis (AM-nee-oh-sen-TEE-sis)
[*amnio-* birth membrane, *-centesis* a pricking]

autosome (AW-toh-sohm)
[*auto-* self, *-som-* body]

carrier (KARE-ee-er)

chorionic villus sampling (CVS) (koh-ree-ON-ik VIL-us)
[*chorion-* skin, *-ic* relating to, *villus* shaggy hair]

chromatin (KROH-mah-tin)
[*chrom-* color, *-in* substance]

chromosome (KROH-meh-sohm)
[*chrom-* color, *-som-* body]

codominance (koh-DOM-ih-nans)
[*co-* together, *-domina-* rule, *-ance* state]

crossing over

deletion

diploid (DIP-loyd)
[*diplo-* twofold, *-oid* form of]

dominant gene
[*gene* produce or generate]

gamete (GAM-eet)
[*gamete* marriage partner]

gene (jeen)
[*gen-* produce or generate]

gene linkage
[*gen-* produce or generate]

genetic mutation (jeh-NET-ik myoo-TAY-shun)
[*gene-* produce or generate, *-ic* relating to, *muta-* change, *-ation* process]

genetics (jeh-NET-iks)
[*gen-* produce or generate, *-ic* relating to]

genome (JEE-nohm)
[*gen-* produce or generate, *-ome* entire collection]

genomics (jeh-NOH-miks)
[*gen-* produce or generate, *-om-* entire collection, *-ic* relating to]

genotype (JEN-oh-type)
[*gen-* produce or generate, *-type* kind]

haploid (HAP-loyd)
[*haplo-* single, *-oid* of or like]

heterozygous (het-er-oh-ZYE-gus)
[*hetero-* different, *-zygo-* union or yoke, *-ous* characterized by]

histone (HISS-tohn)
[*histo-* tissue, *-one* unit]

continued on page 633

CHROMOSOMES AND GENES

Mechanism of Gene Function

The genetic code is transmitted from parent to offspring in discrete, independent units called **genes.** Each gene is a sequence of nucleotide bases in the deoxyribonucleic acid (DNA) molecule. Figure 27-1 shows you a detailed artist's view of human DNA. Beginning at the top of the diagram, you can see a fully *condensed* **chromosome** unfold toward the right and downward, where eventually a single double-helix strand of DNA is visible. As the genetic codes of a DNA molecule's genes are being actively transcribed in a cell's nucleus, the DNA exists in the threadlike form called **chromatin.** Chromatin, as you can see, is actually a thread of DNA wound around little spools made of proteins called **histones.** The chromatin is thus organized into little "spooled" subunits called **nucleosomes.**

During cell division, each replicated strand of chromatin coils on itself to form a compact chromosome. Each DNA molecule can be called either a chromatin strand or a chromosome, depending on what form it is in. Throughout this chapter, we will use the term *chromosome* for DNA, regardless of its actual form, and the term *gene* for each distinct encoding segment within a DNA molecule.

Each gene in a chromosome contains a genetic code that the cell transcribes to a messenger ribonucleic acid (mRNA) molecule, which codes for a specific polypeptide. A transcribed mRNA molecule associates with a ribosome and the code is translated to form a specific polypeptide molecule. Through slight differences in editing of mRNA, one mRNA may perhaps actually produce several specific polypeptides. And the polypeptides may be complete tertiary-level proteins by themselves—or they may combine with any of several other polypeptides to form several different specific large quaternary-level proteins (see Figure 2-22, p. 36).

Many of the protein molecules formed from the polypeptides encoded by genes are *enzymes,* which catalyze specific chemical reactions. Because enzymes and other functional proteins such as hemoglobin regulate the biochemistry of the body, they also regulate the entire structure and function of the body. Other proteins, such as collagen and keratin, are important structural components of the body—and thus determine important structural characteristics of various body parts.

In effect, genes determine the structure and function of the human body by producing a set of specific structural proteins, along with many functional proteins and RNA molecules.

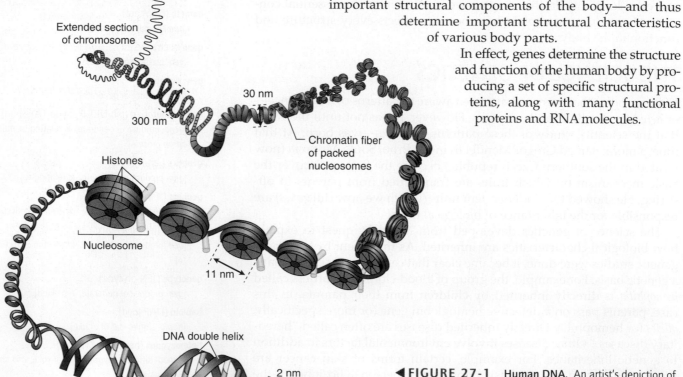

Condensed section
of chromosome

700 nm

1400 nm

Metaphase
chromosome

Extended section
of chromosome

300 nm

30 nm

Chromatin fiber
of packed
nucleosomes

Histones

Nucleosome

11 nm

DNA double helix

2 nm

Nucleotide
bases
(A, C, G, T)

◀ **FIGURE 27-1** **Human DNA.** An artist's depiction of human DNA shows a fully condensed chromosome on the left and the unfolding of it on the right—unfolded to the point of a single DNA strand (double helix). Note that the scale changes dramatically as you progress from one loop of the diagram to the next. (One nanometer [nm] equals a millionth of a centimeter—almost four ten-millionths of an inch.)

The Human Genome

The entire collection of genetic material in each typical cell of the human body is called the **genome.** The structure of the human genome is summarized in Figure 27-2. The typical human genome includes 46 individual nuclear chromosomes and one mitochondrial chromosome.

The *Human Genome Project* finished decoding most of the human genome in 2003. We now know that the human genome contains only about 23,000 genes. This is about one fifth to one quarter of the number originally estimated. Amazingly, it is roughly the same number of genes as in a rat or mouse—and only about one and one-half times as many genes as in a fruit fly!

Amazingly, less than 2% of human DNA carries protein-coding genes. Until recently, most of the rest of the genome was thought to be "junk DNA" that either is not used or is edited out of mRNA before it is used to make proteins. Some of this "junk DNA" codes for RNA now known to contain "micro-RNA" sequences that regulate protein synthesis without being translated into proteins. In addition, some of this noncoding DNA may actually be made up of broken bits of genes that are no longer functional—remnants of our evolutionary past. Termed **pseudogenes,** these bits of formerly functional genes are like genetic fossils: They have begun to reveal an interesting history of our genetic past.

The current "draft" of the human genome shows us that most coding genes tend to lie in clusters rich in cytosine (C) and guanine (G), separated by long stretches of noncoding DNA rich in adenine (A) and thymine (T). Chromosome 1 has the most genes, with nearly 3,000 genes, and the Y chromosome of males has the fewest, with just over 200 genes. Hundreds of the newly discovered genes in the human genome seem to be viral in origin, perhaps inserted there by bacteria or viruses that infected our distant ancestors.

Although we now have the essential picture of the details of the human genome, much work still lies ahead in the field of **genomics,** the analysis of the genome's code. In particular, a great deal of work remains in discovering all possible mutations (see the discussion later in this chapter) and all the proteins encoded by the genes that make up the human genome (Box 27-1).

Another investigative field called **proteomics** involves the analysis of the proteins encoded by the genome. The entire group of proteins encoded by the human genome is called the human **proteome.** The ultimate goal of proteomics is to understand the role of each protein (and how each protein interacts with other proteins) in the body. This knowledge

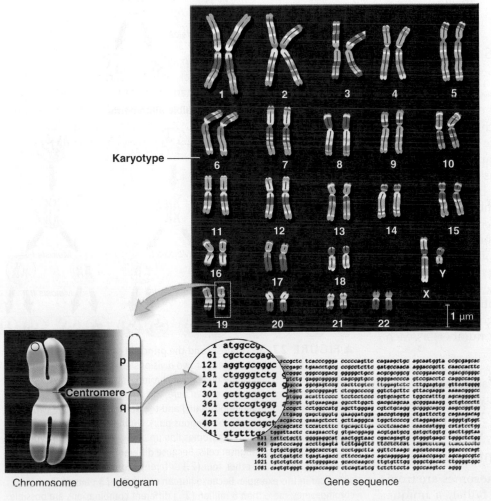

Karyotype

Chromosome   Ideogram   Gene sequence

◀ **FIGURE 27-2** **Human genome.** A cell taken from the body is stained and photographed. A photograph of nuclear chromosomes is then cut and pasted, arranging each of the 46 chromosomes into numbered pairs of decreasing size to form a chart called the karyotype. Each chromosome is a coiled mass of chromatin (DNA). In this figure, differentially stained bands in each chromosome appear as different, bright colors. Such bands are useful as reference points when identifying the locations of specific genes within a chromosome. The staining bands are also represented on an **ideogram,** or simple graph, of the chromosome as reference points to locate specific genes. The genes themselves are usually represented as the actual sequence of nucleotide bases, abbreviated as *a, c, g,* and *t.* In this figure, the sequence of one *exon* (segment) of a gene called GPI from chromosome 19 is shown. Each of the different images in this figure can be thought of as a different type of "genetic map."

will greatly enhance our understanding of normal body functions as well as the mechanisms of many diseases.

The analysis of the human genome and proteome has surged forward recently with the widespread use of *RNA interference (RNAi)* technology. These techniques "silence" particular genes in the laboratory setting in order to find out what proteins are transcribed from them.

Another technique that "knocks out" individual genes has been used for some time to demonstrate the effects of specific genes. Using embryonic stem cells from laboratory mice in which specific genes are targeted and disabled, a generation of genetically altered "knockout mice" can be produced. The mice are then studied to find out the effects of the gene(s) missing from the mouse genome.

Information obtained about the human genome can be expressed in a variety of ways. In Figure 27-2 you can see the relative position of the chromosome's centromere, which divides the chromosome into two segments. The shorter segment of the chromosome is called the **p-arm** and the longer segment is called the **q-arm.** Genes are mapped on either the p- or q-arm. The bands in a map of a chromosome show staining landmarks and help identify the regions of the chromosome. Sometimes physical maps of genes will show exact positions of individual genes on the p-arm and q-arm of a chromosome. A more detailed representation of a gene would show the actual sequence of nucleotide bases, abbreviated *A, T, C,* and *G* for *adenine, thymine, cytosine,* and *guanine,* and as shown in Figure 27-2.

Distribution of Chromosomes to Offspring

Meiosis

Each cell of the human body contains 46 chromosomes. The only exceptions to this principle are the **gametes**—male *spermatozoa* and female *ova.* Recall from Chapter 26 that a special form of nuclear division called **meiosis** (see Figure 26-1, p. 603) produces **haploid** gametes with only 23 chromosomes—exactly one half of the usual number. This process follows a basic principle of genetics first discovered by Gregor Mendel called the **principle of segregation.** This principle simply states that the two members of a pair of homologous chromosomes *segregate,* or separate, during meiosis.

When a sperm (with its 23 chromosomes) unites with an ovum (with its 23 chromosomes) at conception, a *zygote* with 46 chromosomes is formed. Thus the zygote has the same number of chromosomes (46, the **diploid** number) as each typical body cell in the parents.

As the karyotype in Figure 27-2 shows, the 46 human chromosomes can be arranged in 23 pairs according to size. One pair called the **sex chromosomes** may not match, but the remaining 22 pairs of **autosomes** always appear to be nearly identical to each other.

Principle of Independent Assortment

Because one half of an offspring's chromosomes are from the mother and one half are from the father, a unique composition of inherited traits is formed. According to another of Mendel's principles, each chromosome assorts itself independently during meiosis. This **principle of independent assortment** states that, as sperm or eggs are formed, and chromosome pairs separate (the principle of segregation), the maternal and paternal chromosomes get mixed up and redistribute themselves independently of the other chromosome pairs (Figure 27-3). Thus each sperm or egg is likely to receive a *different* set of 23 chromosomes. Independent assortment of chromosomes ensures that each offspring from a single set of parents is very likely to be genetically unique.

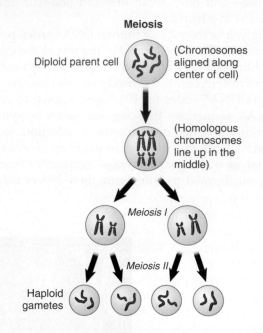

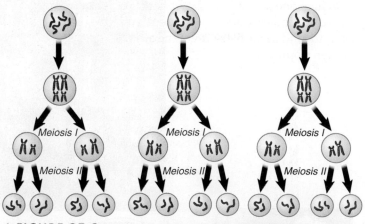

▲ **FIGURE 27-3   Meiosis and the principle of independent assortment.** In meiosis, a series of two divisions results in the production of gametes with one half of the number of chromosomes of the original parent cell. In both meiotic divisions shown here, the original cell has four chromosomes (two homologous pairs) and the gametes each have two chromosomes (one of each homologous pair). During the first division of meiosis, pairs of similar chromosomes line up along the cell's equator for even distribution to daughter cells. Because different pairs assort independently of each other, four (2^2) different alignments of chromosomes can occur in this example. Because human cells have 23 pairs of chromosomes, more than 8 million (2^{23}) different combinations are possible.

BOX 27-1 FYI

1000 Genomes Project

An international research partnership sponsored in part by the United States government is currently building a catalog of at least 1,000 individual human genomes. This effort, called the *1000 Genomes Project,* promises to reveal the many variant forms of genes present in the whole human genome. Although a number of complete individual genomes have been catalogued—most notably the genomes of gene pioneers Craig Venter and James Watson—much more information is needed. The project is cataloging the many small DNA-coding variations called *single nucleotide polymorphisms (SNPs)* as well as the larger structural variations in the human genome. Such information could lead to better understanding of the genetic basis of disease—and ultimately to effective treatments or cures.

According to the principle of **gene linkage,** genes on an individual chromosome tend to stay together. An important application of this principle occurs during one phase of meiosis, when pairs of matching chromosomes (chromosomes that have the same genes that code for the same traits) line up along the middle of the cell and exchange genes or groups of linked genes with one another. This process is called **crossing over** because genes from a particular location cross over to the same location on the matching chromosome (Figure 27-4). Crossing over introduces additional opportunities for genetic variation among the offspring of a single set of parents.

When one considers the genetic variation that is produced by independent assortment and crossing over, it is easy to understand the tremendous variation seen in the human population.

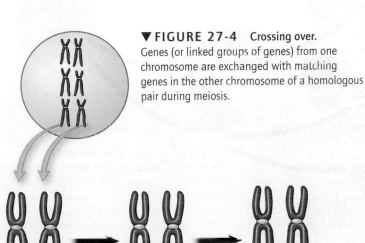

▼ **FIGURE 27-4** Crossing over. Genes (or linked groups of genes) from one chromosome are exchanged with matching genes in the other chromosome of a homologous pair during meiosis.

GENE EXPRESSION

Hereditary Traits

Dominant and Recessive Traits

Mendel discovered that the genetic units we now call genes may be expressed differently among individual offspring. After rigorous experimentation with pea plants, he discovered that each inherited trait is controlled by two sets of similar genes, one from each parent. Each autosome in a pair matches its partner in the type of genes it contains. In other words, if one autosome has a gene for hair color, its partner or *homologous* autosome will also have a gene for hair color—in the same location. Although both genes specify hair color, they may not specify the *same* hair color. Mendel discovered that some genes are *dominant* and some are *recessive.* A **dominant gene** is one whose effects are seen and whose effects are capable of masking the effects of a **recessive gene** for the same trait.

Consider the example of **albinism,** a total lack of melanin pigment in the skin and eyes. Because people with this condition lack dark pigmentation, they have difficulty seeing in bright light and must avoid direct sunlight to protect themselves from burns. The genes that cause albinism are recessive; the genes that cause normal melanin production are dominant. By convention, dominant genes are represented by uppercase letters and recessive genes by lowercase letters. One can represent the gene for albinism as *a* and one of the genes for normal skin pigmentation as *A.* An individual with the gene combination *AA* has two dominant genes—and so will exhibit a normal skin color. The code *AA* is called a **genotype.** A person with a genotype of two identical forms of a trait (*AA* or *aa* in our example) is said to be **homozygous** (from *homo-,* "same," and *-zygo,* "joined") for that trait.

The manner in which a genotype is expressed is called the **phenotype.** Thus a person who is homozygous dominant (*AA*) for skin color will have a normal phenotype (i.e., normal skin pigmentation). Someone with the gene combination *Aa* will also have normal skin color because the normal gene *A* is dominant over the recessive albinism gene *a.* A person with genotype *Aa* is said to be **heterozygous** (from *hetero-,* "different," and *-zygo,* "joined") and will express the normal phenotype. Only a person with the homozygous recessive genotype of *aa* will have the abnormal phenotype, albinism, because there is no dominant gene to mask the effects of the two recessive genes.

In the example of albinism, a person with the heterozygous genotype of *Aa* is said to be a genetic **carrier** of albinism. This means that the person can transmit the albinism gene, *a*, to offspring. Thus two normally pigmented parents, each having the heterozygous genotype *Aa*, can produce both normally pigmented children (*AA, Aa*) and children who have albinism (*aa*) (Figure 27-5).

Polygenic Traits

You should note that melanin pigmentation in human skin is actually governed by several different pairs of genes. The gene for the form of albinism discussed here involves just one of the gene pairs that regulate skin color. Inherited characteristics, such as skin color and height, which are determined by the combined effect of many different gene pairs, are often called **polygenic** ("many genes") traits to distinguish them from **monogenic,** or single-gene, traits.

Polygenic traits are often hard to study in the phenotype because they are so variable. You can think of a polygenic trait as a "combined trait" because it results from the combined activity of several different genes. Each gene may be dominant or recessive in character. Because each gene is only one of several that govern the combined trait, however, the phenotype may be any of a large number of different variations of the trait. Using skin color as an example, the form of albinism described above involved only one of several genes that govern pigmentation of the skin and eyes. But that one gene is critical; it negates the effects of all the other genes that govern skin color. However, if the dominant form of that critical gene is in place, then it is possible for variations in any of the other genes that regulate skin color to exert influence on skin pigmentation.

Codominant Traits

What happens if two different dominant genes occur together? Suppose there is a gene A^1 for light skin and a gene A^2 for dark skin. In a form of dominance called codominance, they will simply have roughly equal effects, and a person with the heterozygous genotype A^1A^2 will exhibit a phenotype of skin color that is something between light and dark. Recall from Chapter 16 (see Box 16-1, p. 360) that the genes for sickle cell anemia behave this way. A person with two sickle cell genes will have *sickle cell anemia*, whereas a person with one normal gene and one sickle cell gene will have a milder form of the disease called *sickle cell trait*.

The case of sickle cell inheritance is a good example of how the mechanism of codominance works. The hemoglobin molecules within all red blood cells (RBCs) are quaternary-level proteins that each comprise four polypeptide chains—two alpha chains and two beta chains (see Figure 20-18, p. 471). The sickle cell gene is actually an abnormal version of the gene that contains the code for the beta chains of the hemoglobin molecule. Any beta chain that is produced by this code has one amino acid (of 146 in total) replaced by the wrong amino acid—making it different enough to drastically alter its function. The RBCs of a person with one sickle cell gene and one normal beta-chain gene contain hemoglobin in which about one half of the total number of beta chains are abnormal and about one half are normal. The RBCs of a person with two sickle cell genes contain hemoglobin in which all the beta chains are abnormal. Thus in sickle cell trait only some hemoglobin molecules are defective, but in sickle cell anemia *all* of the hemoglobin molecules are defective.

The frequency of occurrence of the abnormal sickle cell gene is an example of an interesting epidemiological

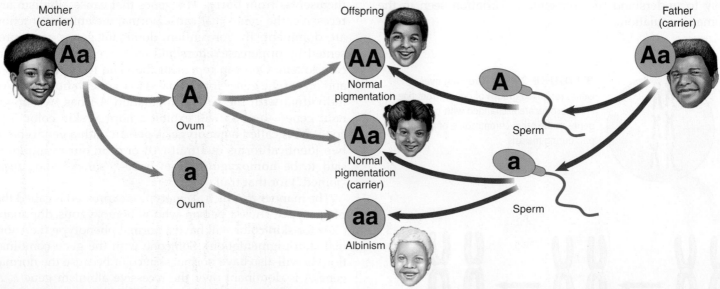

▲ **FIGURE 27-5** **Inheritance of albinism.** Albinism is a recessive trait, producing abnormalities only in those with two recessive genes *(a)*. Presence of the dominant gene *(A)* prevents albinism.

phenomenon. Because sickle cell trait provides resistance to the parasite that causes the most deadly form of **malaria** (*Plasmodium falciparum* malaria), sickle cell disorders persist in areas of the world in which malaria is still prevalent. Malaria is a sometimes fatal condition caused by blood cell parasites (*Plasmodium* species) and is characterized by fever, anemia, swollen spleen, and possible relapse months or years later. The unique distribution of *P. falciparum* malaria results from the fact that people without sickle cell trait more often die of this condition before producing offspring than those with the malaria-resistant sickle cell trait. Thus the "bad" sickle cell gene is more likely to be transmitted to the next generation than the "good" genes for normal hemoglobin.

The sickle cell–malaria relationship points to an important concept in medical genetics: "Disease" genes often provide some biological advantage for a human population in certain circumstances. It is only when circumstances change that the gene does more harm than good and may be eventually "weeded out" of the population by *natural selection*. Genes for many other hereditary diseases (e.g., *thalassemia* and *Tay-Sachs disease*) are now known to impart protection against pathogenic conditions in heterozygous individuals.

Sex-Linked Traits

Recall from our earlier discussion that, besides the 22 pairs of autosomes, there is one pair of sex chromosomes. Notice in the lower right portion of the karyotype in Figure 27-2 that the chromosomes of this pair do not have matching structures. The larger sex chromosome is called the *X chromosome*, and the smaller one is called the *Y chromosome*. The X chromosome is sometimes called the "female chromosome" because it includes genes that determine female sexual characteristics. If a person has two X chromosomes, she is genetically a female. The Y chromosome is often called the "male chromosome" because anyone possessing a Y chromosome is genetically a male. Because men produce both X-bearing and Y-bearing sperm, any two parents can produce male or female children (Figure 27-6).

The large X chromosome contains many genes besides those needed for female sexual traits. Genes for producing certain clotting factors, photopigments in the retina of the eye, and many other proteins are also found on the X chromosome. The tiny Y chromosome, on the other hand, contains few genes other than those that determine male sexual characteristics. Thus both males and females need at least one normal X chromosome—otherwise genes for clotting factors and other essential proteins would be missing. Traits carried on sex chromosomes are called **sex-linked traits.** Some sex-linked traits are called *X-linked traits* because they are determined by genes in the large X chromosome. Other sex-linked traits are called *Y-linked traits* because they are determined by genes within the Y chromosome.

Dominant X-linked traits appear in each person, as one would expect for any dominant trait. In females, recessive X-linked genes are masked if there are dominant genes on the other X chromosome. Only females with two recessive X-linked genes can exhibit the recessive trait. Because the Y chromosome has no genes that match the X chromosome and males inherit only one X chromosome (from the mother), the presence of only one recessive X-linked gene is enough to produce the recessive trait. For this reason, X-linked recessive traits appear much more frequently in males than in females.

An example of a recessive X-linked condition is *red-green color blindness*, which involves a deficiency of normal photopigments in the retina (see Box 14-3, p. 316). In this condition, male children of a parent who carries the recessive abnormal gene on an X chromosome may be color blind (Figure 27-7).

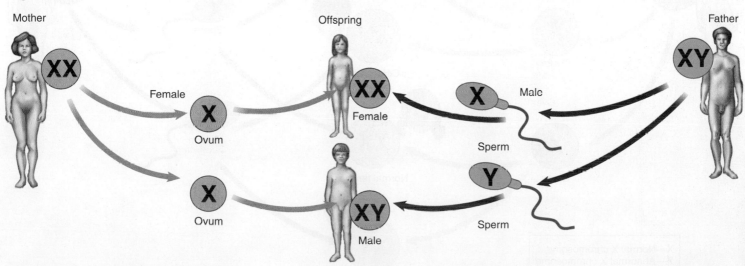

▲ **FIGURE 27-6**   **Sex determination.** The presence of the Y chromosome specifies maleness. In the absence of a Y chromosome, an individual develops into a female.

A female can inherit this form of color blindness only if her father is color blind *and* her mother is either color blind (homozygous recessive) or a color-blindness carrier (heterozygous).

Only one clinically significant Y-linked condition has been identified by geneticists. A missing part of the q-arm of the Y chromosome may cause inheritable problems with spermatogenesis—and possible infertility. This Y-linked condition is only passed from an afflicted father to his son.

Genetic Mutations

The term *mutation* simply means "change." A **genetic mutation** is a change in an individual's genetic code. Some mutations involve a change in the genetic code within a single gene, perhaps a slight rearrangement of the nucleotide sequence. A mutation called a **deletion** occurs when one or more nucleotide bases in a sequence are missing. An **insertion** mutation occurs when one or more additional nucleotides appear within the usual sequence of nucleotide bases in a gene. With either type of mutation, the cell cannot read the genetic code normally, and thus the encoded protein cannot be made in its usual form. Other mutations involve damage to a portion of a chromosome; a portion of a chromosome may completely break away.

Mutations may occur spontaneously without the influence of factors outside the DNA itself. However, most genetic mutations are thought to be caused by **mutagens**—agents that cause mutations in or damage to DNA molecules.

Genetic mutagens include chemicals, some forms of radiation, and even viruses. If mutations occur in reproductive cells or their precursors, they may be inherited by offspring. Beneficial mutations allow organisms to adapt to their environments. Because such mutant genes benefit survival, they tend to spread throughout a population over generations.

Harmful mutations, however, inhibit survival and therefore are not likely to spread widely through the population. Most harmful mutations kill the organism in which they occur or at least prevent successful reproduction—and so are never passed to offspring. Harmful mutations that are recessive, however, may persist at low frequencies in a population indefinitely because they do not cause problems for individuals who inherit just one of these genes. If a harmful dominant mutation is only mildly harmful, it may persist in a population over many generations.

The "weeding out" of harmful mutations and the promotion of beneficial mutations is the fundamental action of natural selection in evolution.

QUICK ✓ CHECK

5. What is a dominant genetic trait? A recessive trait?
6. What is codominance? Give an example.
7. What is X-linked inheritance?
8. How can a mutant gene benefit a human population?
9. What is a mutagen?

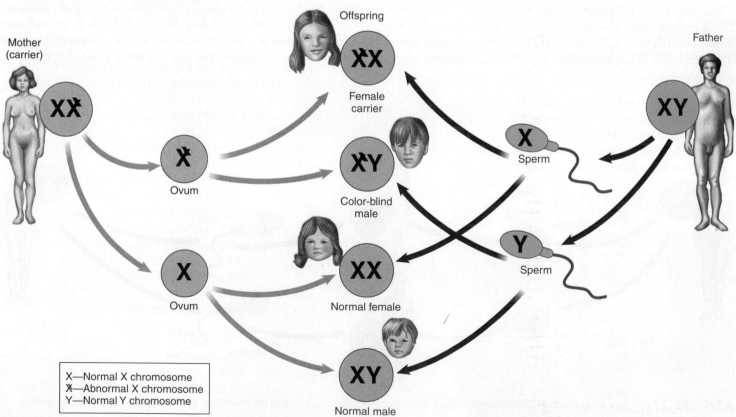

▲ **FIGURE 27-7** **Sex-linked inheritance.** Some forms of color blindness involve recessive X-linked genes. In this case, a female carrier of the abnormal gene can produce male children who are color blind.

PREVENTION OF GENETIC DISEASES

Genetic Counseling

The term *genetic counseling* refers to professional consultations with families regarding genetic diseases. Trained genetic counselors may help a family determine the risk of producing children with genetic diseases. Parents with a high risk of producing children with genetic disorders may decide to avoid having children of their own.

Genetic counselors may also help evaluate whether any offspring already have a genetic disorder and offer advice on treatment or care. A growing list of tools is available to genetic counselors, some of which are described in the following section.

Pedigree

A **pedigree** is a chart that illustrates genetic relationships in a family over several generations (Figure 27-8). Using medical records and family histories, genetic counselors assemble the chart beginning with the client and moving backward through as many generations as are known. Squares represent males; circles represent females. Fully shaded symbols represent affected individuals, and unshaded symbols represent normal individuals. Partially shaded symbols represent carriers of a recessive trait. A horizontal line between symbols designates a sexual relationship that produced offspring.

Pedigrees are useful in determining the possibility of producing offspring with certain genetic disorders. They also may tell a person whether he or she might have a genetic disorder that appears late in life, such as Huntington disease. In either case, a family can prepare emotionally, financially, and medically before a crisis occurs.

Punnett Square

The **Punnett square** is a grid used to determine the mathematical *probability* of inheriting genetic traits. As Figure 27-9, *A*, shows you, genes in the mother's gametes are represented along the horizontal axis of the grid and genes in the father's gametes along the vertical axis. The ratio of different gene

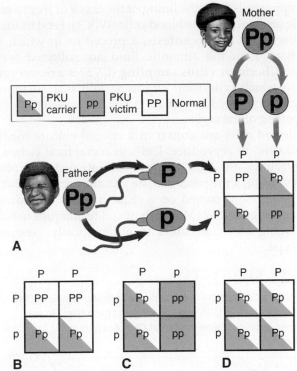

▲ **FIGURE 27-9   Punnett square.** The Punnett square is a grid used to determine relative probabilities of producing offspring with specific gene combinations. Phenylketonuria (PKU) is a recessive disorder caused by the gene *p; P* is the normal gene. **A,** Possible results of a cross between two PKU carriers. Because one in four of the offspring represented in the grid have PKU, a genetic counselor would predict a 25% chance that this couple will produce a PKU baby at each birth. **B,** Cross between a PKU carrier and a normal noncarrier. **C,** Cross between a person with PKU and a PKU carrier. **D,** Cross between a person with PKU and a normal noncarrier.

combinations in the offspring predicts their probability of occurrence in the next generation. Thus offspring produced by two carriers of *phenylketonuria (PKU)*—a recessive disorder—have a one in four (25%) chance of inheriting this recessive condition (see Figure 27-9, *A*).

The same grid shows that there is a two in four (50%) chance that a child produced will be a PKU carrier.

Figure 27-9, *B*, however, shows that offspring of a carrier and a noncarrier cannot inherit PKU. The grid in Figure 27-9, *C*, shows the probability of producing an affected offspring when a person with PKU and a PKU carrier have children. Figure 27-9, *D*, shows the genetic probability when a person with PKU and a noncarrier produce children.

Karyotype

Disorders that involve trisomy, monosomy, and broken chromosomes can be detected after a **karyotype** is produced.

The first step in producing a karyotype is getting a sample of cells from the individual to be tested. This can be done

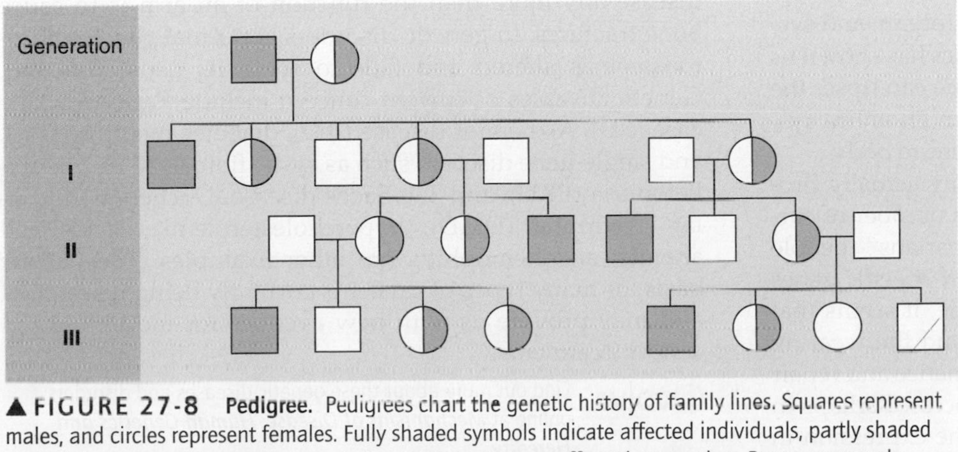

▲ **FIGURE 27-8   Pedigree.** Pedigrees chart the genetic history of family lines. Squares represent males, and circles represent females. Fully shaded symbols indicate affected individuals, partly shaded symbols indicate carriers, and unshaded symbols indicate unaffected noncarriers. Roman numerals indicate the order of generations. This pedigree reveals the presence of an X-linked recessive trait.

by scraping cells from the lining of the cheek or from a blood sample containing white blood cells (WBCs). Fetal tissue can be collected by **amniocentesis,** a procedure in which fetal cells floating in the amniotic fluid are collected with a syringe. **Chorionic villus sampling (CVS)** is a newer procedure in which cells from chorionic villi that surround a young embryo (see Chapter 26, p. 607) are collected through the opening of the cervix.

Collected cells are grown in a special culture medium and allowed to reproduce. Cells in metaphase (when the chromosomes are most distinct) are stained and photographed using a microscope. The chromosomes are cut out of the photo and pasted on a chart in pairs according to size, as in Figure 27-2. More advanced techniques use digital imaging and computers to automatically generate a karyotype.

A&P CONNECT

Many tools for discovering the secrets of a particular sample of DNA are available, such as DNA fingerprinting and gene chips. To see how these techniques work, check out **DNA Analysis** online at **A&P Connect.**

evolve

QUICK CHECK

10. What is genetic counseling?

11. How are pedigrees used by genetic counselors?

12. How is a Punnett square used to predict mathematical probabilities of inheriting specific genes?

13. What is chorionic villus sampling?

14. How is a karyotype prepared? What is its purpose?

Cycle of LIFE

As we've seen, our genes control every aspect of our lives, from meiosis to embryonic development and growth through development across the life span. For example, we know that DNA directs metabolism—the total chemical activity of the body. We also know that cellular metabolism is the foundation of the function of each tissue, organ, and system in the body. Our study of medical genetics has shown us that even one mistake in one codon of a gene can upset the chemistry of the body enough to shut down an entire system—and thus threaten the survival of the entire body.

It is possible that there are genes that may actually prolong human life in some individuals. And, a number of current genetic studies involving "super centenarians" (people over age 100) are attempting to determine if specific genes are related to longevity. At this point in time, it seems that aging is related to both genetic influences and the wearing down of our body over time and that maintenance and repair systems also wear down with the aging process. Perhaps in the future we'll have some control over gene expression in our bodies, but that day is still a long way off. ●

The BIG Picture

The importance of genetics extends from the molecular level all the way to the context of the whole human species and the environment. At the molecular level, the genetic perspective reveals the importance of nucleic acids—especially DNA. DNA molecules serve as tiny data storage devices that contain *all* the information needed to manufacture the molecules needed to build and maintain the human body. Each cell, tissue, organ, and system in the body is built of molecules that have been synthesized with "recipes" contained in the genes of each DNA molecule. DNA in concert with various forms of RNA directs the synthesis of structural and functional proteins. Some of the functional proteins, in turn, trigger and direct the synthesis of all other biomolecules that form and regulate the body—lipids, carbohydrates, and so on.

This vast collection of different kinds of molecules also includes the enzymes and other molecules needed to gather and assimilate chemicals from the external environment outside the body. These external chemicals include water, oxygen, vitamins, minerals, sugars, starches, amino acids, fats, and other nutrients. Some of the molecules that ultimately owe their existence in the body to DNA also include those that help us rid the body of wastes, such as urea, carbon dioxide, and bile.

MECHANISMS OF DISEASE

A large and growing number of genetic diseases and disorders has now been identified. But although we often hear of new "disease genes" being discovered—and the pace is rapidly increasing—the function of these genes is not to cause disease any more than the function of an arm is to cause bone fractures. In genetic disorders, a normal gene or chromosome is altered and fails to serve its usual function. Genetic diseases of current concern include Parkinson disease (PD), Alzheimer disease (AD), diabetes mellitus (DM), and single-gene diseases such as cystic fibrosis (CF), phenylketonuria (PKU), and Tay-Sachs disease. Duchenne muscular dystrophy (DMD), hypercholesterolemia, sickle cell anemia, and hemophilia are other examples. The genetic basis for many types of cancer is currently being researched and may provide us with new avenues for individualized cancer treatments.

evolve Find out more about these genetic diseases and disorders online at *Mechanisms of Disease: Human Genetics and Heredity.*

LANGUAGE OF SCIENCE AND MEDICINE *continued from page 623*

homozygous (hoh-moh-ZYE-gus)
[*homo-* same, *-zygo-* union or yoke, *-ous* characterized by]

ideogram (ID-ee-oh-gram)
[*ide-* idea, *-gram* drawing]

insertion

karyotype (KAIR-ee-oh-type)
[*karyo-* nucleus, *-type* kind]

malaria (mah-LAIR-ee-ah)
[*mal-* bad, *-ar-* air, *-ia* condition]

meiosis (my-OH-sis)
[*meiosis* becoming smaller]

monogenic (mon-oh-JEN-ik)
[*mono-* single, *-gen-* produce (gene), *-ic* relating to]

mutagen (MYOO-tah-jen)
[*muta-* change, *-gen* produce]

nucleosome (NOO-klee-oh-sohm)
[*nucleo-* kernel (nucleus), *-som-* body]

p-arm
[*p* petite (small)]

pedigree (PED-ih-gree)
[*pied de grue* crane's foot pattern]

phenotype (FEE-noh-type)
[*pheno-* appear, *-type* kind]

polygenic (pahl-ee-JEN-ik)
[*poly-* many, *-gen-* produce, *-ic* relating to]

principle of independent assortment

principle of segregation

proteome (PRO-tee-ohm)
[*prote-* protein, *-ome* entire collection]

proteomics (proh-tee-OH-miks)
[*prote-* first rank (protein), *-om-* entire collection, *-ic* relating to]

pseudogene (SOOD-oh-jeen)
[*pseudo-* false, *gene* produce or generate]

Punnett square (PUN-it)
[*Reginald C. Punnett* English geneticist]

q-arm
[*q* follows p in Roman alphabet]

recessive gene
[*recess-* retreat, *-ive* relating to, *gene* produce or generate]

sex chromosome (KROH-moh-sohm)
[*chrom-* color, *-som-* body]

sex-linked trait

CHAPTER SUMMARY

To download an MP3 version of the chapter summary for use with your iPod or other portable media player, access the Audio Chapter Summaries *online at http://evolve.elsevier.com.*

HINT *Scan this summary after reading the chapter to help you reinforce the key concepts. Later, use the summary as a quick review before your class or before a test.*

INTRODUCTION

A. Genetics—scientific study of inheritance; vital to all fields of human biology—especially anatomy, physiology, and medicine

THE SCIENCE OF GENETICS

A. The science of genetics developed from Gregor Mendel's quest to explain how biological characteristics are inherited

CHROMOSOMES AND GENES

A. Mechanism of gene function
1. Genetic code is transmitted from parent to offspring by way of genes; contain segments of DNA
2. DNA (Figure 27-1)
   a. Compact form of DNA is a chromosome
   b. Strand form of DNA is chromatin; a thread of DNA wound around little spools made of proteins called histones forms a nucleosome
3. Each gene in a chromosome contains a genetic code that the cell transcribes to a messenger ribonucleic acid (mRNA) molecule
4. Each mRNA molecule associates with a ribosome and the code is translated to form a specific polypeptide molecule
5. Genes determine the structure and function of the human body by producing a set of specific structural proteins, along with many functional proteins and RNA molecules

B. The human genome
1. Genome—entire collection of genetic material in each typical cell of the human body
   a. Map of the entire human genome was completed in 2003
   b. Contains about 23,000 genes and large amounts of noncoding DNA
   c. Less than 2% of human DNA carries protein-coding genes
   d. Most of the rest of the genome is thought to be "junk DNA" code that either is not used or is edited out of mRNA; pseudogenes
2. Genomics—analysis of the genome's code
3. Proteomics—analysis of the proteins encoded by the genome
   a. Proteome—entire group of proteins encoded by the human genome
4. Genomic information can be expressed in various ways
   a. Genes are often represented as their actual sequence of nucleotide bases expressed by the letters A, C, G, and T (Figure 27-2)

C. Distribution of chromosomes to offspring
1. Meiosis—produces haploid gametes with only 23 chromosomes (Figure 26-1)
2. Principle of independent assortment
   a. As sperm and ovum are formed during meiosis, two members of a pair of homologous chromosomes separate (the principle of segregation) and the maternal and paternal chromosomes get mixed up and redistributed independently in each gamete, with each having a different set of 23 chromosomes (Figure 27-3)
   b. Gene linkage—genes on an individual chromosome tend to stay together
   c. Crossing over—genes from a particular location cross over to the same location on the matching chromosome; introduces additional opportunities for genetic variation among the offspring (Figure 27-4)

GENE EXPRESSION

A. Hereditary traits
1. Dominant and recessive traits
    a. Dominant gene is one whose effects are seen and whose effects are capable of masking the effects of a recessive gene for the same trait
2. Genotype—combination of genes within the cells of an individual
    a. Homozygous—genotype with two identical forms of a gene
    b. Heterozygous—genotype with two different forms of a gene
3. Phenotype—manner in which a genotype is expressed; how an individual looks because of a genotype
4. Carrier—person who possesses the gene for a recessive trait but does not exhibit the trait
B. Polygenic traits—when more than one gene is involved in producing a particular trait
C. Codominant traits—when two different dominant genes occur together, each will have an equal effect
D. Sex-linked traits (Figures 27-6 and 27-7)
1. X chromosome—"female chromosome"; includes genes that determine female sexual characteristics, as well as nonsexual characteristics
2. Y chromosome—"male chromosome"; contains few genes other than male sexual characteristics
E. Genetic mutations
1. Mutation—change in an individual's genetic code
    a. Deletion—occurs when one or more nucleotide bases in a sequence are missing
    b. Insertion—occurs when one or more additional nucleotides appear within the usual sequence of nucleotide bases in a gene
    c. When deletion and insertion occur the cell cannot read the genetic code normally; encoded protein cannot be made in its usual form
2. Mutations may occur spontaneously without the influence of factors outside the DNA itself
3. Mutagens—agents that cause mutations in or damage to DNA molecules

PREVENTION OF GENETIC DISEASES

A. Genetic counseling—professional consultations with families regarding genetic diseases
B. Pedigree—chart that illustrates genetic relationships in a family over several generations (Figure 27-8)
C. Punnett square—grid used to determine the mathematical *probability* of inheriting genetic traits (Figure 27-9)
D. Karyotype—ordered arrangement of photographs of chromosomes from a single cell; used to detect genetic disorders

REVIEW QUESTIONS

 Write out the answers to these questions after reading the chapter and reviewing the Chapter Summary. If you simply think through the answer without writing it down, you won't retain much of your new learning.

1. Who was the first person to discover the basic mechanism by which traits are transmitted from parents to offspring?

2. Describe albinism in relation to dominance, recessiveness, and genotype.
3. Explain the difference between a genotype and a phenotype.
4. Define codominance, and give an example of a condition that demonstrates codominance.
5. If a certain trait is identified as X-linked recessive, describe the genotype of a female expressing the given trait.
6. What role do environmental factors play in relation to certain genetic diseases?
7. Define the term *genome*.

CRITICAL THINKING QUESTIONS

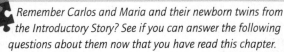

 After finishing the Review Questions, write out the answers to these items to help you apply your new knowledge. Go back to sections of the chapter that relate to items that you find difficult.

1. Differentiate among the text usage of chromatin, chromosomes, and genes.
2. Explain the processes that increase the variability of the genetic code in the offspring.
3. Explain why the structure of the X and Y chromosomes would predict a greater occurrence of sex-linked disorders in males.
4. Explain what is meant by a pedigree. Explain what is meant by a karyotype.

continued from page 623

Remember Carlos and Maria and their newborn twins from the Introductory Story? See if you can answer the following questions about them now that you have read this chapter.

1. What are the chances that their son will have PKU as well?
    a. 25%
    b. 50%
    c. 75%
    d. 0%

2. They could not tell yet, but what were the chances that Maria and Carlos' son would be color blind?
    a. 25%
    b. 50%
    c. 75%
    d. 0%

Carlos and Maria were so happy with their new twins. They were thinking about trying to have more children . . . maybe in a few years.

3. Having had one boy and one girl, what are Carlos and Maria's chances of having another boy?
    a. Depends on the timing of the fertilization
    b. 25%
    c. 50%
    d. 75%

To solve these questions, you may have to refer to the glossary or index, other chapters in this textbook, A&P Connect, Mechanisms of Disease, and other resources.

Glossary

A

A band region of striated muscle fiber with both thick (myosin) and thin (actin) filaments
abdominal aorta (ab-DOM-i-nal ay-OR-tah) portion of the descending aorta that passes through the abdominal cavity
abdominal cavity (ab-DOM-i-nal KAV-i-tee) the cavity containing the abdominal organs
abdominal reflex (ab-DOM-i-nal REE-fleks) drawing in of the abdominal wall in response to stroking the side of the abdomen
abducens nerve (ab-DOO-sens nerv) cranial nerve VI; motor nerve; controls movement of the eye and proprioception
absolute refractory period (AB-so-loot ree-FRAK-toh-ree PEER-ee-od) time during which the local area of the membrane has surpassed the threshold potential and will not respond to any stimulus
absorption (ab-SORP-shun) passage of a substance through a membrane such as skin or mucosa; often refers to passage of nutrients into blood
accessory nerve (ak-SES-oh-ree nerv) cranial nerve XI (motor nerve)
accessory organ (ak-SES-oh-ree OR-gan) an organ that assists other organs in accomplishing their functions
accommodation (ah-kom-oh-DAY-shun) mechanism that allows the normal eye to focus on objects closer than 20 feet
acetylcholine (ACh) (ass-ee-til-KOH-leen) type of neurotransmitter used by motor neurons at neuromuscular junctions to stimulate muscle contraction at or in some autonomic synapses (in ganglionic synapses, at all parasympathetic effectors, and at some sympathetic effectors)
acid (ASS-id) substance that ionizes in water to release hydrogen ions; substance with a pH of less than 7.0
acidity (ah-SID-i-tee) the amount of acid in a solution or the state of having a low pH
acidosis (ass-i-DOH-sis) condition in which there is an excessive proportion of acid in the blood and thus an abnormally low blood pH; opposite of **alkalosis**
acini (ASS-i-nee) any cell of a compound gland that secretes a watery fluid; also called *acinar cells*
acquired immunity (ah-KWYERD i-MYOO-ni-tee) immunity that is obtained after birth through exposure to a specific harmful agent; *see* **adaptive immunity**
acrosome (AK-roh-sohm) structure on the tip of the sperm head containing enzymes that break down the covering of the ovum to facilitate conception
action (AK-shun) activity or active function, as in *muscle action, hormone action, action potential*
action potential (AK-shun poh-TEN-shal) nerve impulse; membrane potential fluctuation of an actively conducting axon
active immunity (AK-tiv i-MYOO-ni-tee) a type of acquired immunity in which the body produces its own antibodies against disease-causing antigens
active site any location on the surface of a molecule that reacts with another molecule
active transport (AK-tiv TRANZ port) movement of a substance into or out of a living cell requiring the use of cellular energy
acute pain fiber (ah-KYOOT FYE-ber) sensory nerve fiber for pain often found in superficial areas of the body and associated with rapid-onset, intense pain; also known as *fast pain fiber* or *A fiber*
adaptation (ad-ap-TAY-shun) condition of many sensory receptors in which the magnitude of a receptor potential decreases over a period of time in response to a continuous stimulus
adaptive immunity (ah-DAP-tiv i-MYOO-ni-tee) type of immunity in which specific antigens or particles are recognized, then targeted for destruction; also called *acquired immunity* or *specific immunity*
adenosine triphosphate (ATP) (ah-DEN-oh-seen try-FOS-fayt) chemical compound that provides energy for use by body cells
ADH mechanism (A-D-H MEK-ah-niz-em) mechanism of fluid homeostasis in which antidiuretic hormone (ADH) triggers the conservation of water in the body; *see* **antidiuretic hormone**
adipocyte (AD-i-poh-syte) fat-storing cell
adipose tissue (AD-i-pohs TISH-yoo) fat tissue
adolescence (ad-oh-LESS-ens) period between puberty (the onset of reproductive maturity) and adulthood
adrenal cortex (ah-DREE-nal KOHR-teks) outer portion of adrenal gland that secretes hormones called *corticoids*
adrenal gland (ah-DREE-nal) endocrine gland that rests on the top of each kidney; made up of cortex and medulla regions
adrenaline (ah-DREN-ah-len) *see* **epinephrine**
adrenal medulla (ah-DREE-nal meh-DUL-ah) inner portion of adrenal gland that secretes epinephrine and norepinephrine
adrenocorticotropic hormone (ACTH) (ah-dree-no-kor-teh-koh-TROH-pic HOR-mohn) hormone that stimulates the adrenal cortex to secrete larger amounts of hormones
adulthood (ah-DULT-hood) developmental period after adolescence
aerobic (air-OH-bik) relating to the use of oxygen, as in aerobic respiration
aerobic respiration pathway (air-OH-bik res-pi-RAY-shun PATH-way) catabolic process; the stage of cellular respiration requiring oxygen
aerobic training (air-OH-bik TRAIN-ing) continuous vigorous exercise requiring the body to increase its consumption of oxygen and develop the muscles' ability to sustain activity over a long time; also known as *endurance training*
afferent (AF-fer-ent) carrying or conveying toward the center (e.g., an afferent neuron carries nerve impulses toward the central nervous system); opposite of **efferent**
agglutinate (ah-GLOO-tin-ayt) antibodies causing antigens to clump or stick together
agonist (AG-ah-nist) agent that works like or with (rather than against) another agent
agranulocyte (ah-GRAN-yoo-loh-syte) white blood cell without cytoplasmic granules
albinism (AL-bi-niz-em) recessive, inherited condition characterized by a lack of the dark brown pigment melanin in the skin, eyes, and hair, resulting in vision problems and susceptibility to sunburn and skin cancer; ocular albinism is a lack of pigment in the layers of the eyeball
aldosterone (AL-doh-steh-rohn or al-DAH-stair-ohn) hormone that stimulates the kidney to retain sodium, ions, and water; only physiologically important mineralocorticoid
alimentary canal (al-eh-MEN-tar-ee kah-NAL) the digestive tract as a whole
alkaline (AL-kah-lin) refers to a substance that in solution has a pH of greater than 7.0
alkalinity (al-kah-LIN-ih-tee) state of having a high pH; amount of base or alkali in a solution
alkalosis (al-kah-LOH-sis) condition in which there is an excessive proportion of alkali (base) in the blood; opposite of **acidosis**
allosteric effector (al-oh-STEER-ik eh-FEK-tor) an agent that alters the function of an enzyme by changing the shape of the enzyme's active site
alpha cell (AL-fah sell) pancreatic islet cell that secretes glucagon; also called *A cell*
alveolus (al-VEE-oh-lus) literally, a small cavity; alveoli of lungs are microscopic saclike dilations of terminal bronchioles; gas exchange in the lungs occurs across the membranes of the alveoli
amino acid (ah-MEE-no ASS-id) type of chemical unit from which proteins are built
amino acid derivative hormone (ah-MEE-no ASS-id deh-RIV-ah-tiv HOR-mohn) category of nonsteroid hormones; each hormone is derived from a single amino acid molecule
amniocentesis (am-nee-oh-sen-TEE-sis) procedure in which a sample of amniotic fluid is removed with a syringe for use in genetic testing, often to produce a karyotype of the fetus
amniotic cavity (am-nee-OT-ik KAV-i-tee) cavity within the blastocyst that will become a fluid-filled sac in which the embryo will float during development
amphiarthrosis (am-fee-ar-THROH-sis) slightly movable joint such as the one joining the two pubic bones
ampulla (am-PUL-ah) saclike dilation of a tube or duct; found at end of each semicircular duct, contains crista ampullaris
amylase (AM-eh-lays) enzyme that digests carbohydrates
anabolic hormone (an-ah-BOL-ik HOR-mohn) hormone that stimulates anabolism in the target organ
anabolic steroid (an-ah-BOL-ik STAYR-oyd) any steroid (lipid) hormone that promotes anabolism, such as testosterone
anabolism (ah-NAB-oh-liz-em) cells making complex molecules (for example, hormones) from simpler compounds (for example, amino acids); opposite of **catabolism**
anaerobic (an-air-OH-bik) relating to a process that requires no oxygen
anaerobic respiration pathway (an-air-OH-bik res-pi-RAY-shun PATH-way) catabolic process; stage of cellular respiration not requiring oxygen
anal canal (AY-nal kah-NAL) passage from the colon to the outside of the body
anal triangle (AY-nal TRY-ang-gul) area surrounding anus
anaphase (AN-ah-fayz) stage of mitosis; duplicate chromosomes move to poles of dividing cell
anaplasia (an-ah-PLAY-zha) growth of abnormal (undifferentiated) cells, as in a tumor or neoplasm
anastomosis (ah-nas-toh-MOH-sis) connection between vessels that allows collateral circulation
anatomical head (an-ah-TOM-i-kal hed) the part of a bone that articulates with another bone
anatomical position (an-ah-TOM-i-kal po-ZISH-un) the standard reference position for the body; standing with arms hanging at sides with palms forward, gives meaning to directional terms
anatomy (ah-NAT-oh-mee) the study of the structure of an organism and the relationships of its parts
androgen (AN-droh-jen) hormone that promotes development and maintenance of male characteristics
androgen-binding protein (ABP) (AN-droh-jen-BYND-ing PRO-teen) specialized protein that binds to testosterone and increases concentration within the seminiferous tubules
anemia (ah-NEE-me-ah) deficient number of red blood cells, or deficient hemoglobin
anesthesia (an-es-THEE-zhah) state in which a person lacks the feeling of pain
angiography (an-jee-AH-graf-ee) radiography in which radiopaque contrast medium is injected into a vessel to make it more visible in a medical image (angiogram); in arteries the image is called *arteriogram;* in veins, a *venogram* or *phlebogram;* in lymphatic vessels, a *lymphangiogram*
angular movement body movement that increases or decreases the angle of a joint
ANH mechanism (A-N-H MEK-ah-niz-em)
anion (AN-eye-on) negatively charged molecule (negative ion)
ankle jerk reflex extension of the foot in response to tapping of the Achilles tendon
antagonist (an-TAG-oh-nist) agent that has an opposing effect or works against another agent
anterior (an-TEER-ee-or) front or ventral; opposite of **posterior** or *dorsal*
anterior cavity (an-TEER-ee-or KAV-i-tee) any cavity in an anterior (forward) region of the body or an organ; for example, the anterior cavity of the eye is a fluid-filled space in front of the lens
anterior fornix (an-TEER-ee-or FOR-niks) space created by the protrusion of the cervix into the lumen of the vagina; increases probability of reproductive success by retaining seminal fluid
antibody (AN-ti-bod-ee) plasma protein produced by B lymphocytes that destroys or inactivates a specific substance (antigen) that has entered the body
antibody-mediated immunity (AN-ti-bod-ee-MEE-dee ayt-ed i-MYOO-ni-tee) immunity that is produced when antibodies render antigens unable to harm the body; also called *humoral immunity*
antibody titer (AN-ti-bod-ee TYE-ter) measurement of the amount of antibody in the plasma
anticoagulant drug (an-tee-koh-AG-yoo-lant) a substance that prevents blood from clotting
antidepressant (an-tee-deh-PRESS-ant) drug that inhibits feelings of depression or sadness
antidiuretic hormone (an-tee-dye-yoo-RET-ik HOR-mohn) hormone produced in the posterior pituitary gland to regulate the balance of water in the body by accelerating reabsorption of water in the kidney tubules; also called *arginine vasopressin (AVP)*
antigen (AN-ti-jen) substance, usually a protein fragment, that causes an immune response
antigen A (AN-ti-jen) cellular immune marker on the plasma membrane of red blood cells of blood types A and AB
antigen-antibody complex (AN-ti-jen AN-ti-bod-ee KOM-pleks) combination of an antigen with its complementary antibody
antigen B (AN-ti-jen B) cellular immune marker on the plasma membrane of red blood cells of blood types B and AB
antigen-presenting (APC) (AN-ti-jen) any of a variety of immune cells that present protein fragments (antigens) on their surface and thus allow recognition and reaction by other immune system cells; include macrophages and dendritic cells (DCs) and B cells
antioxidant (an-tee-OK-seh-dent) agent that inhibits chemical oxidation
antiplatelet drug (an-tee-PLAYT-let) any drug that inhibits the action of platelets (thrombocytes) and thereby inhibits blood clotting
aorta (ay-OR-tah) largest systemic artery; extends directly from the left ventricle
aortic aneurysm (ay-OR-tik AN-yoo-riz-em) a local widening or ballooning of the aorta
aortic arch (ay-OR-tik) 180-degree curve of the aorta, shortly after it leaves the left ventricle
apex (AY-peks) the point or tip of a structure; *see also* **apical**
Apgar score (AP-gar) system of assessing general health of newborn infant, in which heart rate, respiration, muscle tone, skin color, and response to stimuli are scored (a perfect total score is 10)
aphasia (ah-FAY-zhah) language deficit caused by damage in speech centers of the brain
apical (AY-pik-al) relating to the apex (tip) of an organ, cell, or other structure; in a cell, often refers to the surface facing the lumen of the organ
apocrine sweat glands (AP-oh-krin) sweat glands located in the axillae and genital regions; these glands enlarge and begin to function at puberty
aponeurosis (ap-oh-nyoo-ROH-sis) broad, flat sheet of connective tissue
apoptosis (ap-op-TOH-sis) nonpathological, programmed cell death in which specific biochemical steps within the cell lead to the fragmentation of the cell and removal of the pieces by phagocytic cells; occurs when cells are no longer needed and is the normal process by which our tissues remodel themselves
appendicular (ah-pen-DIK-yoo-lar) refers to the upper and lower extremities of the body
appendicular skeleton (ah-pen-DIK-yoo-lar SKEL-eh-ton) the bones of the upper and lower extremities of the body
appetite center cluster of neurons in the lateral hypothalamus whose impulses cause an increase in appetite
aqueous humor (AY-kwee-us HYOO-mohr) watery fluid that fills the anterior chamber of the eye, in front of the lens
arachnoid mater (ah-RAK-noyd MAH-ter) delicate, weblike middle membrane covering the brain, the meninges
arbor vitae (AR-bor VI-tay) internal white matter of the cerebellum
areola (ah-REE-oh-lah) region of skin of the breast that surrounds the nipple
areolar tissue (ah-REE-oh-lar TISH-yoo) a type of connective tissue consisting of fibers and a variety of cells embedded in a loose matrix of soft, sticky gel
arginine vasopressin (AVP) (AHR-jih-neen vas-oh-PRES-in) *see* **antidiuretic hormone**
arterial anastomosis (ar-TEER-ee-al ah-nas-toh-MOH-sis) junction between two arteries
arteriole (ar-TEER-ee-ohl) small branch of an artery
artery (AR-ter-ee) vessel carrying blood away from the heart
arthroplasty (AR-throh-plas-tee) the total or partial replacement of a diseased joint with an artificial device (prosthesis)

articular cartilage (ar-TIK-yoo-lar KAR-ti-lij) layer of hyaline cartilage covering the joint surfaces of epiphyses

articulation (ar-tik-yoo-LAY-shun) joint

ascending aorta (ah-SEND-ing ay-OR-tah) portion of the aorta extending from the aortic valve to the aortic arch

ascending colon (ah-SEND-ing KOH-lon) portion of the colon extending from the cecum to the hepatic flexure

ascending tract (ah-SEND-ing trakt) spinal cord tract that conducts impulses up the cord to the brain

ascites (ah-SYE-tees) effusion (inflow of watery fluid) in the abdominal cavity; abdominal bloating

assimilation (ah-sim-i-LAY-shun) the process of incorporation of a substance into the body

astrocyte (ASS-troh-syte) star-shaped neuroglial cell

atherosclerosis (ath-er-oh-skleh-ROH-sis) type of "hardening of the arteries" in which lipids and other substances build up on the inside wall of blood vessels

atom (AT-om) smallest particle of a chemical element that retains the properties of that element; particles that combine to form molecules (chemical building blocks)

atomic number (ah-TOH-mik NUM-ber) number of protons in an atom of an element

atomic weight (ah-TOM-ik) number of protons plus the number of neutrons in an atom of an element

ATP synthase (A-T-P SIN-thays) enzyme that produces ATP molecules

atrial natriuretic hormone (ANH) (AY-tree-all nay-tree-yoo-RET-ik HOR-mohn) hormone produced by the atrium that promotes secretion of sodium into the urine and therefore loss of water from the body

atrioventricular (AV) bundle (ay-tree-oh-ven-TRIK-yoo-lar BUN-del) bundle of specialized cardiac muscle fibers that extend from the AV node to the subendocardial branches (Purkinje fibers); involved in coordination of heart muscle contraction

atrioventricular node (ay-tree-oh-ven-TRIK-yoo-lar nohd) a small mass of special cardiac muscle tissue; part of the conduction system of the heart

atrioventricular (AV) valves (ay-tree-oh-ven-TRIK-yoo-lar valvs) either of two valves that separate the atrial chambers from the ventricles; also called *cuspid valve*

atrium (AY-tree-um) (*pl.,* atria [AY-tree-ah]) one of the upper chambers of the heart; receives blood from either the systemic or pulmonary circulation

atrophy (AT-roh-fee) wasting away of tissue; decrease in size of a part; sometimes referred to as *disuse atrophy*

auditory tube (AW-di-toh-ree toob) tube that connects the throat with the middle ear and equalizes pressure between the middle ear and the exterior; also known as *eustachian tube*

auricle (AW-ri-kul) part of the ear attached to the side of the head; earlike appendage of each atrium of the heart

autocrine hormone (AW-toh-krin HOR-mohn) a type of hormone that regulates activity in the secreting cell itself

autoimmunity (aw-toh-i-MYOO-ni-tee) immune system reaction against normal body structures (self-antigens)

autonomic nervous system (ANS) (aw-toh-NAHM-ik NER-vus SIS-tem) division of the nervous system that monitors and regulates subconscious (involuntary) functions

autonomic (visceral) reflex (aw-toh-NOM-ik REE-fleks) feedback regulatory mechanism of the autonomic nervous system in which autonomic (visceral) sensory receptors trigger a regulatory response

autorhythmic (aw-toh-RITH-mic) relating to the characteristic of certain cells (such as involuntary muscle fibers) to display a self-regulated repeating pattern of action, as when heart cells beat in a rhythm without external stimulation

autosome (AW-toh-sohm) one of the 44 chromosomes (22 pairs) in the human genome besides the two sex chromosomes; means "same body," in reference to each member of a pair of autosomes matching the other in size and other structural features

avascular (ah-VAS-kyoo-lar) free of blood vessels

axial (AK-see-all) refers to the head, neck, and torso or trunk of the body

axial skeleton (AK-see-all SKEL-eh-ton) the bones of the head, neck, and torso

axillary lymph node (AK-sil-lair-ee limf nohd) lymph node in the region at or near the armpit

axillary vein (AK-si-lair-ee vayn) vein of the armpit region

axon (AK-son) in a neuron, the single process that extends from the axon hillock and transmits impulses away from the cell body

axon hillock (AK-son HILL-ok) portion of the cell body from which the axon extends

B

Babinski reflex (bah-BIN-skee REE-fleks) *see* **plantar reflex**

ball-and-socket joint spheroid joint, such as shoulder or hip joint; most movable type of joint

baroreceptor (bar-oh-ree-SEP-tor) sensory neuron sensitive to changes in blood pressure

basal (BAY-sal) relating to the base or widest part of an organ or other structure; in a cell, pertains to the surface facing away from the lumen of an organ

basal cell carcinoma (BAY-sal sell car-sin-OH-mah) one of the most common forms of skin cancer; usually occurs on the upper face

basal metabolic rate (BMR) (BAY-sal met-ah-BOL-ik rayt) number of calories of heat that must be produced per hour by catabolism to keep the body alive, awake, and comfortably warm

basal nuclei (BAY-sal NOO-klee-eye) islands of gray matter located in the cerebrum of the brain that are responsible for automatic movements and postures; formerly known as *basal ganglia*; also known as *cerebral nuclei*

base (bayse) substance that ionizes in water to decrease the number of hydrogen ions; also known as *alkali*

basement membrane (BAYSE-ment MEM-brayne) the connective tissue layer of the serous membrane that holds and supports the epithelial cells

basilar membrane (BAYS-i-lar MEM-brayne) *see* **spiral membrane**

basophil (BAY-so-fil) white blood cell that stains readily with basic (alkaline) dyes

B cell (B sell) immune system cell that produces antibodies against specific antigens; same as *B lymphocyte*

benign (be-NYNE) refers to a tumor, neoplasm, or other condition that does not metastasize (spread to different tissues) or otherwise cause serious harm

benign prostatic hypertrophy (BPH) (be-NYNE pros-TAT-ik hye-PER-troh-fee) a nonmalignant enlargement of the prostate gland

beta cell (BAY-tah sell) pancreatic islet cell that secretes insulin; also called *B cell*

biaxial joint (bye-AK-see-al joynt) skeletal articulation that has two axes of movement

bicarbonate (bye-KAR-boh-nayte) *see* **bicarbonate ion**

bicarbonate ion (bye-KAR-boh-nayte EYE-on) HCO_3^-; an ion that serves an important role in maintaining normal blood pH

bicuspid valve (bye-KUSS-pid valv) one of the two AV valves; it is located between the left atrium and ventricle and is sometimes called the *mitral valve*

bilateral symmetry (bye-LAT-er-al SIM-eh-tree) concept of the right and left sides of the body being approximate mirror images of each other

bile (byle) mixture of excretions and secretions produced by the liver and released into the digestive tract

bile salt (byle) an emulsifying agent found in bile

biological clock (bye-oh-LOJ-i-kal) internal timing mechanism that governs hunger, sleeping, reproduction, and behavior

bipolar cell (bye-POH-lar sell) any cell that has exactly two primary extensions from a central cell body

bipolar neuron (bye-POH-lar NOO-ron) neuron with only one dendrite and only one axon

blackhead sebum that accumulates, darkens, and enlarges a duct of the sebaceous glands, as in the case of acne; also known as a *comedo*

blastocyst (BLASS-toh-sist) stage of developing embryo that implants in uterine wall; consists of hollow ball of cells plus an inner cell mass

blind spot point on retina where blood vessels and nerves exit the eyeball; "blind" portion of visual field resulting from no photoreceptors present in that portion of retina; also known as *optic disk*

blood boosting *see* **blood doping**

blood-brain barrier (BBB) (blud brayn BAYR-ee-er) structural and functional barrier formed by astrocytes and blood vessel walls in the brain; it prevents some substances from diffusing from the blood into brain tissue

blood doping (blud DOH-ping) practice of increasing the hematocrit by transfusing additional blood or by the use of erythropoietin to boost hematocrit

blood serum (blud SEER-um) in a sample of blood, the pale yellowish liquid left after a clot forms

blood types (blud) the different types of blood that are identified by certain antigens in red blood cells (A, B, AB, O, and Rh-negative or Rh-positive)

B lymphocyte (B LIM-foh-syte) immune system cell that produces antibodies against specific antigens

body (BOD-ee) the structure of the entire organism; also, the main part of an organ, cell, or other structure

body composition (BOD-ee kahm-poh-ZISH-un) percentages of the body made of lean tissue and fat tissue

body membrane (BOD-ee MEM-brayne) any of several types of membrane in the body that cover or line organs or cavities and are made up of more than one kind of tissue

body of uterus (BOD-ee of YOO-ter-us) part of the uterus above the isthmus, comprising about 70% of the organ

body plane (BOD-ee playn) imagined flat surface that cuts through the body at any of various angles; *see* **coronal plane, sagittal plane, transverse plane**

bone (bohn) type of connective tissue whose matrix is hard and calcified

bone marrow transplant (bohn MAIR-oh TRANZ-plant) medical procedure in which bone marrow tissue from a donor is placed in a recipient in hopes that it will produce healthy blood cells

booster shot additional vaccination that boosts the immune response against a particular antigen

Bowman capsule (BOH-men KAP-sul) in the kidney, the cup-shaped top of a nephron that surrounds the glomerulus; also called *glomerular capsule*

brachial plexus (BRAY-kee-all PLEK-sus) nerve plexus located deep in the shoulder that innervates the lower part of the shoulder and the entire arm

brachial vein (BRAY-kee-al vayn) vein of the arm

brachiocephalic artery (brayk-ee-oh-seh-FAL-ik AR-ter-ee) artery of the upper thorax

brachiocephalic vein (brayk-ee-oh-seh-FAL-ik vayn) vein of the upper thorax

brainstem part of brain containing the midbrain, pons, and medulla oblongata

bronchiole (BRONG-kee-ohl) small branch of a bronchus

bronchopulmonary segment (brong-koh-PUL-moh-nair-ee SEG-ment) region of the lung supplied by a tertiary bronchus

brush border lining of small intestine that resembles bristles of a brush; formed by microvilli

buffer (BUF-er) compound that combines with an acid or with a base to form a weaker acid or base, thereby lessening the change in hydrogen ion concentration that would occur without the buffer; often operates as buffer pairs

bulbourethral gland (bul-boh-yoo-REE-thral) either of two small glands at the base of the penis that contribute a small amount of fluid to semen; also called *Cowper gland*

bulbous corpuscle (BUL-bus KOR-pus-ul) sensory receptor that senses deep pressure and continuous touch; located in dermis of the skin; also known as *Ruffini corpuscle*

bursa (BER-sah) (*pl.,* bursae [BER-see]) small, cushion-like sac found between moving body parts, making movement easier

C

calcaneal tendon (kal-KAH-nee-al TEN-don) connects gastrocnemius to calcaneus; Achilles tendon

calcification (kal-si-fi-KAY-shun) process of depositing calcium crystals in a tissue

calcitonin (CT) (kal-si-TOH-nin) hormone secreted by the thyroid that decreases calcium levels in the blood

calcitriol (kal-SIT-ree-ol) hormone that regulates calcium homeostasis in the body

callus (KAL-us) bony tissue that forms a sort of collar around the broken ends of fractured bone during the healing process; in the skin, abnormally thick stratum corneum found at points of friction

calyx (KAY-liks) cup-shaped division of the renal pelvis

canaliculi (kan-ah-LIK-yoo-lye) an extremely narrow tubular passage or channel in compact bone; radiates from lacunae and connects with other lacunae and the central (haversian) canal of the osteon

cancellous bone (KAN-seh-lus bohn) bone containing tiny, branchlike trabeculae; also known as *spongy bone* or *trabecular bone*

cancellous bone tissue (KAN-seh-lus bohn TISH-yoo) *see* **cancellous bone**

canthus (KAN-thus) a corner of the eye where the eyelids meet

capacitance (kah-PASS-i-tens) the ease of stretch of a blood vessel wall; referring especially to veins where most of the blood volume is found

capacitation (kah-pass-i-TAY-shun) process occurring after ejaculation of semen and needed for a mature sperm to become capable of fertilizing an ovum

capillary (KAP-i-lair-ee) tiny vessel that connects arterioles and venules; gas exchange from blood to tissues occurs in capillaries

capitulum (kah-PITCH-uh-lum) rounded knob on the humerus bone below the lateral epicondyle

carbaminohemoglobin (karb-am-ee-no-hee-moh-GLOH-bin) carbonated form of hemoglobin

carbohydrates (kar-boh-HYE-drayts) organic compounds containing carbon, hydrogen, and oxygen in certain specific proportions; for example, sugars, starches, and cellulose

carbon monoxide (CO) (KAR-bon mon-OKS-ide) molecule made up of one carbon atom and one oxygen atom

cardia (KAR-dee-ah) small collarlike section of the stomach near the junction with the esophagus, so-called because of its nearness to the heart; also called *cardiac part* or *cardial part*

cardiac cycle (KAR-dee-ak SYE-kul) complete heartbeat consisting of diastole and systole of both atria and both ventricles

cardiac muscle tissue (KAR-dee-ak MUSS-el TISH-yoo) muscle tissue type that makes up the heart

cardiac output (KAR-dee-ak OUT-put) volume of blood pumped by one ventricle per minute (example, L/min or ml/min)

cardiovascular system (kar-dee-oh-VAS-kyoo-lar SIS-tem) the system that transports cells throughout the body by way of blood vessels; sometimes also called *circulatory system*

carpal bone (KAR-pul bohn) any of the bones of the wrist

carrier (KAIR-ee-er) in genetics, a person who possesses the gene for a recessive trait but who does not actually exhibit the trait; in cell biology, a protein structure in the membrane that facilitates transport of ions and other molecules

carrier-mediated passive transport (KAIR-ee-er MEE-dee-ayt-ed PASS-iv TRANZ-port) type of facilitated diffusion in which solutes move down their concentration gradient through carrier mechanisms in the membrane wall

cartilaginous joint (kar-ti-LAJ-in-us joynt) articulation between bones that is primarily composed of cartilage

catabolism (kah-TAB-oh-liz-em) breakdown of food compounds or cytoplasmic constituents into simpler compounds; opposite of **anabolism**

catalyst (KAT-ah-list) chemical that speeds up reactions without being changed itself

cation (KAT-eye-on) positively charged particle

cauda equina (KAW-da eh-KWINE-ah) lower end of spinal cord with its attached spinal nerve roots

CD system (C-D SIS-tem) international system for naming surface markers on red blood cells

cell (sell) basic unit of a living organism, comprising a cytoplasm surrounded by a plasma membrane and containing a nucleus and various organelles

cell body (sell BOD-ee) the main part of a cell, ordinarily containing the nucleus; in the neuron, the cell body is also called the *soma* or *perikaryon*

cell-mediated immunity (sell MEE-dee-ayt-ed i-MYOO-ni-tee) type of adaptive immunity in which T lymphocytes act directly on antigens to protect the body

cell theory (sell THEE-oh-ree) concept proposed more than 100 years ago that all living organisms are made up of biological units called *cells*; *see* **cell**

cellular immunity (SELL-yoo-lar i-MYOO-ni-tee) type of adaptive immunity in which T lymphocytes act directly on antigens to protect the body

cellular respiration (SELL-yoo-lar res-pi-RAY-shun) set of biochemical reactions of a cell that transfer energy from nutrient molecules to ATP molecules

cellulose (SEL-yoo-lohs) dietary fiber; major component of most plant tissue; nondigestible by humans

central (SEN-tral) relating to the center of the body, as in *central nervous system*
central canal (SEN-tral kah-NAL) canal in the osteon (haversian system) of bone that runs parallel to the long axis; contains blood vessels and nerves; also called *osteonal canal* or *haversian canal*
central nervous system (CNS) (SEN-tral NER-vus SIS-tem) the brain and spinal cord
centriole (SEN-tree-ohl) one of a pair of tiny cylinders in the centrosome of a cell; believed to be involved with the spindle fibers formed during mitosis
centrosome (SEN-troh-sohm) area of the cytoplasm near the nucleus that coordinates the building and breaking up of microtubules in the cell
cephalic phase (seh-FAL-ik fayz) stage in regulation of stomach secretion in which mental factors stimulate gastric juice secretion
cerebral cortex (seh-REE-bral KOHR-teks) thin layer of gray matter made up of neuron dendrites and cell bodies that compose the surface of the cerebrum
cerebral hemisphere (seh-REE-bral HEM-i-sfeer) either of the two right and left halves of the cerebrum
cerebral localization (seh-REE-bral) principle of brain function that states that specific cerebral functions are likely to be located in the cerebral cortex at specific, often predictable, locations
cerebral peduncle (seh-REE-bral peh-DUNG-kul) white matter tracts in the brain connecting the cerebellum to the cerebrum
cerebral plasticity (seh-REE-bral plas-TIS-i-tee) characteristic of the cerebrum that permits the relocation of functions from one area of the cerebral cortex to another area
cerebrospinal fluid (CSF) (sair-eh-broh-SPY-nal FLOO-id) plasmalike fluid that fills the subarachnoid space in the brain and spinal cord and in the cerebral ventricles
cerebrovascular accident (CVA) (SAIR-eh-broh-VAS-kyoo-lar) event in which hemorrhage or cessation of blood flow caused by an embolism or ruptured aneurysm in a brain blood vessel results in ischemia of brain tissue and destruction of neurons; commonly called a *stroke*
cerebrum (SAIR-eh-brum) largest and uppermost part of the human brain that controls consciousness, memory, sensations, emotions, and voluntary movements
cerumen (seh-ROO-men) earwax
ceruminous gland (seh-ROO-mi-nus) gland that produces a waxy substance called *cerumen* (earwax)
cervical plexus (SER-vi-kal PLEK-sus) plexus located deep within the neck; innervates muscles and skin of the neck, upper shoulder, and part of the head
cervical vertebra (SER-vi-kal VER-teh-bra) one of the seven spinal (vertebral) bones of the neck
cervix (SER-viks) neck; particularly the inferior necklike portion of the uterus
cesarean section (seh-SAIR-ee-an SEK-shun) surgical removal of a fetus through an incision of the skin and uterine wall; also called *C-section*
channel-mediated passive transport (CHAN-al MEE-dee-ayt-ed PASS-iv TRANZ-port) any membrane transport mechanism in which molecules moved down a concentration gradient through membrane channels
cheek facial prominence shaped by the underlying zygomatic bone
chemical bond (KEM-i-kal) energy relationship joining two or more atoms; involves sharing or exchange of electrons
chemical digestion (KEM-i-kal dye-JES-chun) changes in chemical composition of food as it passes through the alimentary canal
chemical reaction (KEM-i-kal ree-AK-shun) any event in which molecules are formed or modified
chemical synapse (KEM-i-kal SIN-aps) junction between a neuron and another cell that uses chemical messengers to transmit a signal between the cells
chemoreceptor (kee-moh-ree-SEP-tor) receptor that responds to chemicals; responsible for taste and smell and monitoring concentration of specific chemicals in the blood
chemotaxis (kee-moh-TAK-sis) process by which a substance attracts cells or organisms into (or away from) its vicinity; for example, when inflammation mediators attract white blood cells
chief cell cells lining the gastric glands of the stomach that secrete pepsinogen and intrinsic factor; also called *zymogenic cell*
childhood period of human development from infancy to puberty
cholecystokinin (CCK) (koh-leh-sis-toh-KYE-nin) hormone secreted from the intestinal mucosa of the duodenum that stimulates contraction of the gallbladder, resulting in bile flowing into the duodenum
chondrocyte (KON-droh-syte) cartilage cell
chordae tendinea (KOR-dee ten-DIN-ee) stringlike structures that attach the AV valves to the wall of the heart
chorion (KOH-ree-on) outermost fetal membrane; contributes to tissues in the placenta
chorionic villus sampling (koh-ree-ON-ik VIL-lus SAM-pling) procedure in which a tube is inserted through the (uterine) cervical opening and a sample of the chorionic tissue surrounding a developing embryo is removed for karyotyping
choroid plexus (KOH-royd PLEK-sus) tuft of capillaries in ventricles of the brain that secrete cerebrospinal fluid
chromatids (KROH-mah-tids) either of the two DNA strands joined by a centromere existing after DNA has replicated (before cell division) but before the centromere has divided
chromatin (KROH-mah-tin) threadlike form of DNA, making up the genetic material in the nucleus
chromosome (KROH-moh-sohm) compact, barlike bodies of chromatin (DNA) that have coiled to form a compact mass during mitosis or meiosis; each chromosome is composed of regions called *genes*, each of which transmits hereditary information
chronic obstructive pulmonary disease (COPD) (KRON-ik ob-STRUK-tiv PUL-moh-nair-ee di-ZEEZ) general term referring to a group of disorders characterized by progressive, irreversible obstruction of airflow in the lungs
chronic pain fiber (KRON-ik FYE-ber) sensory nerve fiber for pain often found in deep, internal areas of the body and associated with slow-onset, dull pain; also known as *slow pain fiber* or *B fiber*
chyle (kyle) milky fluid; the fat-containing lymph in the lymphatics of the intestine
chylomicron (kye-loh-MYE-kron) small fat droplet
chyme (kyme) partially digested food mixture leaving the stomach
chymotrypsin (kye-moh-TRIP-sin) pancreatic enzyme that digests proteins in the digestive tract
cilia (SIL-ee-ah) hairlike projections of cells
ciliary body (SIL-ee-air-ee BOD-ee) thickening of the choroid that is located between the anterior margin of the retina and the posterior margin of the iris
ciliate (SIL-ee-at) type of protozoan having cilia
circular movement arclike rotation of structure around an axis
citric acid cycle (SIT-rik ASS-id SYE-kul) second series of chemical reactions in the process of glucose metabolism in which carbon dioxide is formed and energy is released; it is an aerobic process; also known as *Krebs cycle*
clavicle (KLAV-i-kul) collar bone
cleft palate (kleft PAL-et) facial deformity that is an X-linked inherited condition; when the palatine bones fail to unite completely
clitoris (KLIT-oh-ris) small, erectile body located within the vestibule of the vagina; also called *glans clitoris*
clone (klone) any of a family of many identical cells descended from a single "parent" cell
coccygeus muscle (kohk-SIJ-ee-us MUSS-el) one of the muscles that form the pelvic floor
coccyx (KOK-siks) last bone of the vertebral column, made up of 4 or 5 vertebrae that have fused together
cochlea (KOHK-lee-ah) snail shell–like structure in the inner ear that houses the spiral organ (organ of Corti), which is responsible for sense of hearing
cochlear nerve (KOHK-lee-ar nerv) part of vestibulocochlear nerve (cranial nerve VIII); sensory nerve responsible for hearing
codominance (koh-DOM-i-nance) in genetics, a form of dominance in which two dominant versions of a trait are both expressed in the same individual
codon (KOH-don) in RNA, a triplet of three base pairs that codes for a particular amino acid
coenzyme (koh-EN-zyme) organic, nonprotein catalyst that acts as molecule carrier
cofactor (KOH-fak-ter) a nonprotein unit attached to an enzyme molecule than enables the enzyme to function properly
collagen (KAH-lah-jen) principal organic constituent of connective tissue
collagenous fiber (kah-LAJ-eh-nus FYE-ber) fiber made of the protein collagen; *see* **collagen**
collateral circulation (koh-LAT-er-al ser-kyoo-LAY-shun) circulation through surrounding nearby vessels when a primary vessel is blocked
collecting duct (CD) (koh-LEK-ting dukt) in the kidney, straight tubule joined by the distal tubules of several nephrons
colon (KOH-lon) division of the large intestine; divided into ascending, transverse, descending, and sigmoid portions
color blindness (KUL-or BLIND-ness) X-linked inherited condition in which one or more photopigments in the cones of the retina are abnormal or missing
columnar (kah-LUM-nar) cell classification by shape in which cells are higher than they are wide
coma (KOH-mah) altered state of consciousness from which an individual cannot be aroused
combining sites (kom-BINE-ing) two small concave regions on the surface of an antibody molecule
common carotid artery (kah-ROT-id AR-ter-ee) blood vessel that supplies the region of the head and neck
compact bone tissue (kom-PAKT bohn TISH-yoo) dense bone; contains structural units called *osteons* or *haversian systems*
complement (KOM-pleh-ment) any of several proteins normally present in blood plasma that, when activated, kill foreign cells by puncturing them
complementary base pairing (kom-pleh-MEN-tah-ree bayse PAIR-ing) bonding purines and pyrimidines in DNA; adenine always binds with thymine, and cytosine always binds with guanine
complete blood cell count (CBC) clinical blood test that usually includes standard red blood cell, white blood cell, and thrombocyte counts, the differential white blood cell count, hematocrit, and hemoglobin content
compound (KOM-pownd) a chemical combination of two or more elements
concentration gradient (kahn-sen-TRAY-shun GRAY-dee-ent) measurable difference in concentration from one area to another
concha (KONG-kah) shell-shaped structure; for example, bony projection into the nasal cavity
conduction system of the heart (kon-DUK-shun SIS-tem) the electrical conduction system that regulates the heart rate
condyloid joints (KON-di-loyd joynts) ellipsoidal joints
cone receptor cell located in the retina that is stimulated by bright light
connective tissue (koh-NEK-tiv TISH-yoo) most abundant and widely distributed tissue in the body
connective tissue membrane (koh-NEK-tiv TISH-yoo MEM-brayne) body membrane that lines movable joint cavities; for example, synovial
consciousness state of awareness of one's self and environment and other beings
constipation (kon-sti-PAY-shun) condition that results when extra water is absorbed from the fecal mass, producing a hardened stool
constriction (kon-STRIK-shun) narrowing of a passageway, as in *constriction of the pupil of the eye*
contraction (kon-TRAK-shun) tension of muscle fibers produced by sliding of cytoskeletal filaments and producing either shortening of the muscle fiber, pull on a load, or both
convergence (kon-VER-jens) a coming together, as in movement of the two eyeballs inward so that their visual axes come together at the same point on the object viewed; also, when more than one presynaptic axon synapses with a single postsynaptic neuron
convolution (kon-voh-LOO-shun) *see* **gyrus**
cornea (KOHR-nee-ah) transparent, anterior portion of the sclera
corneal reflex (KOHR-nee-al REE-fleks) blinking in response to the cornea being touched
coronal plane (ko-ROH-nal playn) frontal plane; divides the body into front and back portions
coronary artery (KOHR-oh-nair-ee AR-ter-ee) blood vessel that provides blood to the myocardial cells
coronary sinus (KOHR-oh-nair-ee SYE-nus) area that receives deoxygenated blood from the coronary veins and empties into the right atrium
coronoid fossa (KOHR-oh-noyd FOSS-ah) (*pl.*, fossae [FOSS-ee]) depression at the distal end of the humerus bone, into which the coronoid process of the ulna fits (last part of the humerus-ulna articulation)
corpora cavernosa (KOHR-pohr-ah kav-er-NOH-sah) *see* **corpus cavernosum**
corpora quadrigemina (KOHR-pohr-ah kwod-ri-JEM-i-nah) midbrain landmark composed of inferior and superior colliculi
corpus albicans (KOHR-pus AL-bi-kanz) white scar on ovary that replaces the degenerated corpus luteum
corpus callosum (KOHR-pus kah-LOH-sum) nerve tissue connecting the right and left cerebral hemispheres; also called *commissural tract*
corpus cavernosum (KOHR-pus kav-er-NOH-sum) either of two columns of erectile tissue in the shaft of the penis
corpus luteum (KOHR-pus LOO-tee-um) a hormone-secreting glandular structure that is transformed after ovulation from a ruptured follicle; it secretes chiefly progesterone, with some estrogen secreted as well
corpus spongiosum (KOHR-pus spun jee-OH-sum) a column of erectile tissue surrounding the urethra in the penis
cortex (KOHR-teks) outer part of an internal organ; for example, the outer part of the cerebrum and of the kidneys
cortical (KOHR-tik-al) relating to the cortex, or outer area of an organ or structure
cortical nephron (KOHR-tik-al NEF-ron) nephron located in the renal cortex
corticosteroid (kohr-ti-koh-STAYR-oyd) glucocorticoid secreted by zona fasciculata of the adrenal cortex
cortisol (KOHR-ti-sol) glucocorticoid secreted by zona fasciculata of the adrenal cortex; also known as *hydrocortisone*
cough reflex epiglottis and glottis close and contract in response to foreign matter in the trachea or bronchi
coumarin (KOO-mar-in) compound that retards blood coagulation; often given as medication to prevent heart attacks
countercurrent mechanism (MEK-ah-niz-em) system in which renal tubule filtrate flows in opposite directions; facilitates urine concentration
covalent bond (koh-VAYL-ent) chemical bond formed by two atoms sharing one or more pairs of electrons
coxal bone (KOK-sal bohn) hip bone
cranial nerve (KRAY-nee-al nerv) any of twelve pairs of nerves that attach to the undersurface of the brain and conduct impulses between the brain and structures in the head, neck, and thorax
cranium (KRAY-nee-um) bony vault, made up of eight bones, that encases the brain
cremaster muscle (kreh-MASS-ter MUSS-el) muscle responsible for elevating the testes
cretinism (KREE-tin-iz-em) dwarfism caused by hyposecretion of the thyroid gland
cribriform plate (KRIB-ri-form playt) perforated portion of ethmoid bone that separates the nasal and cranial cavities
crista ampullaris (KRIS-tah am-pyoo-LAIR-is) fold that serves as a sensory receptor organ located within the ampulla of the semicircular ducts; detects head movements
crossing over phenomenon that occurs during meiosis when pairs of homologous chromosomes synapse and exchange genes
crown (krown) topmost part of an organ or other structure
cuboidal (kyoo-BOYD-al) cell classification by shape in which cells resemble a cube
cupula (KYOO-pyoo-lah) structure found within the crista ampullaris of the semicircular duct; the gelatinous ridge in which the hair cells are embedded
Cushing syndrome (KOOSH-ing SIN-drohm) condition caused by the hypersecretion of glucocorticoids from the adrenal cortex
cusp (kusp) a projection, pointed or rounded, on the biting surface of a tooth
cutaneous membrane (kyoo-TAYN-ee-us MEM-brayne) primary organ of the integumentary system; the skin
cytokinesis (sye-toe-kin-EE-sis) process by which a dividing cell splits its cytoplasm and plasma membrane into two distinct daughter cells; cytokinesis happens along with mitosis (or meiosis) during the cell division process
cytoplasm (SYE-toh-plaz-em) gel-like substance of a cell exclusive of the nucleus and plasma membrane; includes organelles (except nucleus) and cytosol (intracellular fluid)
cytoskeleton (sye-toh-SKEL-eh-ton) cell's internal supporting framework

D

deamination (dee-am-i-NAY-shun) removal of an amino group from an amino acid to form a molecule of ammonia and one of keto acid; occurs in the liver as first step in protein catabolism
deciduous teeth (deh-SID-yoo-us) commonly referred to as "baby teeth"; 20 teeth that are

shed at a certain age before development of permanent teeth

decomposition reaction (dee-kahm-poh-ZISH-un ree-AK-shun) chemical reaction that breaks down a substance into two or more simpler substances

decussate (de-KUS-sayt) cross over from one side to another

deep farther away from the body's surface; opposite of **superficial**

defecation (def-eh-KAY-shun) expelling feces from the digestive tract

deglutition (deg-loo-TISH-un) swallowing

dehydration (dee-hye-DRAY-shun) an abnormal loss of fluid from the body's internal environment

deletion genetic mutation that occurs within a DNA molecule when one or more nucleotide bases in a sequence are missing, causing a misreading of the genetic code and a failure to produce a normal protein needed for body function

delta cell (DEL-tah sell) pancreatic islet cell that secretes somatostatin; also called *D cell*

denaturation (de-nayt-shur-AY-shen) any process that alters the shape of a protein by a change in pH, heat, or some other manner to change its chemical properties

dendrite (DEN-dryte) branching or treelike nerve cell process that receives input from other neurons and transmits impulses toward the cell body (or toward the axon in a unipolar neuron)

dendritic cell (DC) (DEN-dri-tik sell) phagocytic cell in the skin

dens (denz) upward projection from the body of the second cervical vertebra that furnishes an axis for rotating the head; also called *odontoid process*

dense fibrous tissue (dense FYE-brus TISH-yoo) tissue consisting of fibers packed densely in the matrix

dentate nucleus (DEN-tayt NOO-klee-us) either of paired cerebellar gray matter structures that are connected by tracts with the thalamus, as well as motor areas of the cerebral cortex

deoxyribonucleic acid (DNA) (dee-ok-see-rye-boh-noo-KLAY-ik ASS-id) genetic material of the cell that carries the chemical "blueprint" of the body

depolarization (dee-poh-lar-i-ZAY-shun) the electrical activity that triggers a contraction of the heart muscle

dermis (DER-mis) the deeper of the two major layers of the skin, composed of dense fibrous connective tissue interspersed with glands, nerve endings, and blood vessels

dermoepidermal junction (DEJ) (der-mo-ep-i-DER-mal JUNK-shen) thin, gluelike layer that binds the epidermis of the skin to the underlying dermis; also called *dermal-epidermal junction*

descending aorta (di-SEND-ing ay-OR-tah) conducts blood downward from the arch of the aorta to the abdominal cavity

descending colon (di-SEND-ing KOH-lon) portion of the colon that lies in the vertical position, on the left side of the abdomen; extends from below the stomach to the iliac crest

descending tract (di-SEND-ing trakt) bundle of axons in the spinal cord that conducts impulses down the cord from the brain

desmosome (DES-moh-sohm) category of cell junction that holds adjacent cells together; consists of dense plate or band of connecting structures at point of adhesion

detrusor (dee-TROO-sor) a smooth muscle tissue that makes up the wall of the bladder

developmental biology branch of biology that studies the process of change over the life cycle

diabetes insipidus (dye-ah-BEE-teez in-SIP-i-dus) condition resulting from hyposecretion of ADH in which large volumes of urine are formed

dialysis (dye-AL-i-sis) separation of smaller (diffusible) particles from larger (nondiffusable) particles through a semipermeable membrane

diapedesis (dye-ah-peh-DEE-sis) passage of any formed elements within blood through the vessel wall, as in movement of white cells into an area of injury and infection

diaphysis (dye-AF-i-sis) (*pl.*, diaphyses [dye-AF-i-seez]) shaft of a long bone

diarrhea (dye-ah-REE-ah) abnormally frequent defecation of liquid or semiliquid feces

diarthrosis (dye-ar-THROH-sis) freely movable joint

diastole (dye-ASS-toh-lee) relaxation of the heart (especially the ventricles), during which it fills with blood; opposite of **systole**

diastolic blood pressure (dye-ah-STOL-ik blud PRESH-ur) blood pressure in arteries during diastole (relaxation) of the heart; clinically more important than systolic pressure

diencephalon (dye-en-SEF-ah-lon) "between" brain; part of the brain between the cerebral hemispheres and the mesencephalon, or midbrain

differential WBC count (dif-er-EN-shawl) percentage enumeration of the different types of leukocytes in a stained blood smear

diffusion (di-FYOO-shun) spreading; natural tendency of small particles to spread out evenly within any given space; for example, scattering of dissolved particles

digestion (dye-JES-chun) breakdown of food materials either mechanically (through chewing) or chemically (via digestive enzymes)

digestive system (dye-JES-tiv SIS-tem) system composed of mouth, pharynx, esophagus, stomach, small intestine, large intestine, rectum, and anal canal

digestive tract (dye-JES-tiv trakt) the entire alimentary canal from mouth to anus

diploe (DIP-lo-ee) region of cancellous (spongy) bone within the wall of a flat bone of the cranium

diploid (DIP-loyd) normal number of chromosomes per somatic cell (46 in humans)

disaccharide (dye-SAK-ah-ryde) a double sugar molecule formed from the covalent bonding of two monosaccharides

dissection (dye-SEK-shun) cutting technique used to separate body parts for study

dissociate (di-SOH-see-ayt) when a compound breaks apart in solution, forming ions that are surrounded by solvent molecules

distal (DIS-tall) toward the end of a structure; opposite of **proximal**

distal convoluted tubule (DCT) (DIS-tall kon-voh-LOO-ted TOOB-yool) in the kidney, the part of the tubule distal to the ascending limb of the Henle loop that terminates in a collecting duct; main portion of *distal tubule*

disuse atrophy (DIS-yoos AT-roh-fee) loss of muscle tissue mass after a period of few or no contractions, resulting in weakness

divergence (dye-VER-jens) when a single presynaptic axon synapses with more than one different postsynaptic neuron

diving reflex protective physiological response of the body to cold water immersion

dominant gene gene whose effects are seen; capable of masking the effects of a recessive gene for the same trait

dorsal (DOR-sal) pertaining to the back; in a direction toward the back of the body; see posterior[*dors-* back, *-al* relating to]

dorsal cavity (DOR-sal KAV-i-tee) body cavity that includes the cranial cavity and the spinal cavity; not a standard anatomical term, but used here to help organize the body for the beginning student

dorsal (posterior) nerve root (DOR-sal nerv) bundle of nerve fibers that carry sensory information into the spinal cord

dorsal ramus (DOR-sal RAY-mus) branch of spinal nerve that supplies somatic motor and sensory fibers to several smaller nerves

dorsal root (DOR-sal) posterior branch of the attachment of a spinal nerve to the spinal cord

dorsal root ganglion (DOR-sal GANG-glee-on) small region of gray matter in dorsal nerve root made up of cell bodies of unipolar sensory neurons

double helix (HEE-lix) shape of DNA molecules; a double spiral

ductus arteriosus (DUK-tus ar-teer-ee-OH-sus) in the developing fetus, this arterial duct connects the aorta and the pulmonary artery, allowing most blood to bypass the fetus' developing lungs

ductus venosus (DUK-tus veh-NOH-sus) continuation of the umbilical vein that shunts blood returning from the placenta past the fetus' developing liver directly into the inferior vena cava

duodenum (doo-oh-DEE-num) first subdivision of the small intestine; where most chemical digestion occurs

dura mater (DOO-rah MAH-ter) literally "strong mother" or "hard mother"; outermost layer of the meninges

dynamic equilibrium (dye-NAM-ik ee-kwi-LIB-ree-um) maintaining balance when the head or body is rotated or suddenly moved

dysplasia (diss-PLAY-zha) abnormal changes in size, shape, and organization of cells in a tissue associated with neoplasms (tumors)

E

eccrine sweat gland (EK-rin) water-producing exocrine sweat glands widely dispersed throughout the skin

ectoderm (EK-toh-derm) outermost of the primary germ layers that develops early in the first trimester of pregnancy; gives rise to the skin and the nervous system

ectopic pacemaker (ek-TOP-ik PAYS-may-ker) pacemaker other than the SA node

ectopic pregnancy (ek-TOP-ik PREG-nan-see) a pregnancy in which the fertilized ovum implants someplace other than in the uterus

edema (eh-DEE-mah) accumulation of fluid in a tissue, as in inflammation; swelling

effector cell (eh-FEK-tor sell) type of lymphocyte that attacks antigens

effector T cell (eh-FEK-tor T sell) cell that differentiates from a T cell; causes contact killing of a target

efferent (EF-fer-ent) carrying from, as neurons that transmit impulses from the central nervous system to the periphery; opposite of **afferent**

efferent division (EF-fer-ent di-VI-zhun) the motor division (outgoing pathways) of the nervous system

efferent (motor) neuron (EF-fer-ent NOO-ron) nerve cell that transmits impulses away from the central nervous system to or toward muscles or glands

ejaculation (ee-jak-yoo-LAY-shun) sudden discharging of semen from the body

ejaculatory duct (ee-JAK-yoo-lah-toh-ree dukt) duct formed by the joining of the ductus deferens and the duct from the seminal vesicle that allows sperm to enter the urethra

elastic artery (eh-LAS-tik AR-ter-ee) largest artery; includes aorta and some of its branches

elastic cartilage (eh-LAS-tik KAR-ti-lij) cartilage with elastic, as well as collagenous, fibers; provides elasticity and firmness, as in, for example, the cartilage of the external ear

elastic filament (eh-LAS-tik FIL-ah-ment) in muscle fibers, microscopic protein filaments composed of *titin (connectin)* that anchor the ends of the thick filaments to the Z disk and give myofibrils their characteristic elasticity

elastic fiber (eh-LAS-tik FYE-ber) fiber common in connective tissue and made up of elastin protein, which is able to recoil when stretched

elastic recoil (eh-LAS-tik REE-koyl) tendency of the thorax and lungs to return to their preinspiration volume

elastin (eh-LAS-tin) protein found in elastic fiber

electrical synapse (eh-LEK-tri-kal SIN-aps) junction between a neuron and another cell characterized by gap junctions, which permit electrical potentials to travel directly from one cell to the next

electrocardiogram (ECG) (eh-lek-troh-KAR-dee-oh-gram) graphic record of the heart's action potentials

electroencephalogram (eh-lek-troh-en-SEF-ah-loh-gram) graphic representation of voltage changes in brain tissue used to evaluate nerve tissue function

electrolyte (eh-LEK-troh-lyte) substance that dissociates into ions in solution, rendering the solution capable of conducting an electrical current

electron (eh-LEK-tron) small, negatively charged subatomic particle found in the outer regions of an atom

electron transport system (eh-LEK-tron TRANZ-port SIS-tem) carrier molecules embedded in the inner membrane of the mitochondria that take high-energy electrons from the citric acid cycle and form water and energy for oxidative phosphorylation; also called *electron transport chain (ETC)*

element (EL-eh-ment) substance composed of only one type of atom that cannot be broken into simpler constituents by chemical means

elimination (ee-lim-i-NAY-shun) defecation

embryology (em-bree-OHL-oh-gee) study of the development of an individual from conception to birth

emission (ee-MISH-un) reflex movement of spermatozoa and secretions from genital ducts and accessory glands into prostatic urethra; precedes ejaculation

emulsified (ee-MULL-seh-fyde) dispersed in a liquid; for example, fat molecules formed into tiny droplets before they can be digested

endocardium (en-doh-KAR-dee-um) thin layer of very smooth tissue lining each chamber of the heart

endochondral ossification (en-doh-KON-drall os-i-fi-KAY-shun) process by which bones are formed by replacement of cartilage models

endocrine cell (EN-doh-krin sell) glandular secretory cells located in the pancreas; found in pancreatic islets

endocrine glands (EN-doh-krin) secretory structure that discharges hormones directly into the blood

endocrine hormone (EN-doh-krin HOR-mohn) substance secreted by an endocrine gland into the bloodstream that acts on a specific target tissue to produce a given response

endocrine system (EN-doh-krin SIS-tem) system composed of glands that secrete chemicals known as *hormones* directly into the blood

endocytosis (en-doh-sye-TOH-sis) process that allows extracellular material to enter the cell without actually passing through the plasma membrane

endoderm (EN-doh-derm) innermost layer of the primary germ layers that develops early in the first trimester of pregnancy; gives rise to digestive and urinary structures, as well as many other glands and organ parts

endometrium (en-doh-MEE-tree-um) mucous membrane lining the uterus

endomysium (en-doh-MEE-see-um) delicate connective tissue membrane covering the skeletal muscle fibers in a skeletal muscle organ

endoneurium (en-doh-NOO-ree-um) thin wrapping of fibrous connective tissue that surrounds each axon in a nerve

endoplasmic reticulum (ER) (en-doh-PLAZ-mik reh-TIK-yoo-lum) network of tubules and vesicles in cytoplasm that contributes to cellular protein manufacture (via attached ribosomes) and distribution

endorphin (en-DOR-fin) chemical in central nervous system that influences pain perception; a natural painkiller

endosteum (en-DOS-tee-um) fibrous membrane that lines the medullary cavity of long bones

endotracheal intubation (en-doh-TRAY-kee-al in-too-BAY-shun) placing a tube in the trachea to ensure an open airway

end-product inhibition (end-PROD-ukt in-hib-ISH-un) process in a biochemical pathway in which the chemical product at the end of the pathway (the end product) becomes an allosteric effector, inhibiting the function of one or more enzymes in the pathway and thus inhibiting further production of the end product

endurance training continuous vigorous exercise requiring the body to increase its consumption of oxygen and develop the muscles' ability to sustain activity over a prolonged period

energy level limited region surrounding the nucleus of an atom at a certain distance containing electrons; also called a *shell*

enzyme (EN-zyme) biochemical catalyst that allows chemical reactions to take place; functional proteins that regulate various metabolic pathways of the body

eosinophil (ee-oh-SIN-oh-fil) white blood cell, readily stained by eosin (a pinkish acid dye)

ependymal cells (eh-PEN-di-mal sells) cells that line the ventricles of the brain and the central canal of the spinal cord

epicardium (ep-i-KAR-dee-um) inner layer of the pericardium that covers the surface of the heart; it is also called the *visceral pericardium*

epidermis (ep-i-DER-miss) outermost layer of the skin; sometimes called the "false" skin

epididymis (ep-i-DID-i-miss) one of a pair of tightly coiled male reproductive tubes that carry sperm to the vas deferens

epidural space (ep-i-DOO-ral) in the brain, the space above the dura mater

epiglottis (ep-i-GLOT-iss) lidlike cartilage overhanging the entrance to the larynx

epimysium (ep-i-MIS-ee-um) coarse sheet of connective tissue that covers a muscle as a whole

epinephrine (Epi) (ep-i-NEF-rin) adrenaline; hormone secreted by the adrenal medulla

epineurium (ep-i-NOO-ree-um) a tough fibrous sheath that covers the whole nerve

epiphyseal line (ep-i-FEEZ-ee-all lyne) point of fusion seen in a mature bone that replaces the epiphyseal cartilage or growth plate that once separated the epiphysis and diaphysis of a growing bone

epiphyseal plate (ep-i-FEEZ-ee-all playt) cartilage plate that is between the epiphysis and the diaphysis and allows growth to occur; sometimes referred to as a *growth plate*

epiphysis (eh-PIF-i-sis) end of a long bone; also, the pineal body of the brain

episiotomy (eh-piz-ee-OT-oh-mee) surgical procedure used during birth to prevent a laceration of the mother's perineum or the vagina

epithelial membrane (ep-i-THEE-lee-al MEM-brayne) membrane composed of epithelial tissue with an underlying layer of connective tissue

epitope (EP-i-tope) specific portion of an antigen that elicits an immune response

equilibrium (ee-kwi-LIB-ree-um) a balance between opposing elements

erection (ee-REK-shun) condition of erectile tissue when filled with blood; often refers to enlargement of the penis during sexual arousal
erythroblastosis fetalis (ee-rith-roh-blas-TOH-sis fee-TAL-iss) condition of a fetus or infant caused by the mother's Rh antibodies reacting with the baby's Rh-positive red blood cells, characterized by massive agglutination of the blood and resulting in life-threatening circulatory problems for the infant
erythrocytes (eh-RITH-roh-sytes) red blood cells
erythropoiesis (eh-rith-roh-poy-EE-sis) process of red blood cell formation
erythropoietin (eh-rith-roh-POY-eh-tin) glycoprotein secreted to increase red blood cell production in response to oxygen deficiency
esophagus (eh SOF-ah-gus) muscular, mucus-lined tube that connects the pharynx with the stomach; also known as the *food pipe*
essential fatty acid (ee-SEN-shal ASS-id) unsaturated fatty acid that must be provided by the diet; serves as a source within the body for prostaglandin synthesis
essential organs (ee-SEN-shal OR-gans) primary organs; organs needed for the essential functions of a system
estrogen (ES-troh-jen) sex hormone secreted by the ovary that causes development and maintenance of female secondary sex characteristics and stimulates growth of the epithelial cells lining the uterus
ethmoid (ETH-moyd) irregular cranial bone that lies anterior to the sphenoid and posterior to the nasal bones
eustachian tube (yoo-STAY-shun toob) auditory tube
exchange reaction chemical reaction that breaks down a compound and then synthesizes a new compound by switching portions of the molecules; for example, AB + CD → AD + BC
excitability (ek-syte-eh-BIL-i-tee) ability of a muscle to be stimulated; also known as *irritability*
excitation (ek-sye-TAY-shun) electrical fluctuation (increase in voltage) occurring when a neuron or muscle fiber is stimulated and additional Na^+ channels open
exocrine gland (EK-soh-krin) secretory structure that discharges products into ducts
exocytosis (eks-o-sye-TOH-sis) process that allows large molecules to leave the cell without actually passing through the plasma membrane
exon (EKS-ahn) segment of a gene in DNA that is used directly for protein synthesis; in the mRNA transcript of a gene, the intervening intron (noncoding) segments are removed and the remaining exon (coding) segments are spliced together to form the final, edited version of the mRNA transcript; *see* **ribonucleic acid (RNA), transcription**
exophthalmos (ek-soff-THAL-mus) protrusion of the eyeballs resulting, in part, from edema of tissue at the back of the eye socket
expiration (eks-pi-RAY-shun) exhalation
expiratory reserve volume (ERV) (eks-PYE-rah-tor-ee ree-ZERV VOL-yoom) amount of air that can be forcibly exhaled after expiring the tidal volume (TV)
external anal sphincter muscle (eks-TER-nal AY-nal SFINGK-ter MUSS-el) circular muscle located around the anus
external ear (eks-TER-nal eer) outer part of the ear: auricle and external auditory canal
external genitalia (eks-TER-nal jen-i-TAIL-yah) any of the reproductive organs (usually the external organs); penis, scrotum, and related structures in males; vagina, vulva, and related structures in females
external iliac vein (eks TER-nal IL-ee-ak vayn) vein of the lower extremity
external jugular vein (eks-TER-nal JUG-yoo-lar vayn) vein of the neck
exteroceptor (ek-ster-oh-SEP-tor) somatic sense receptor located on the body surface
extracellular bone matrix (eks-trah-SEL-yoo-lar bohn MAY triks) material between cells in bone tissue, made up of water, a variety of proteins such as collagen, and inorganic hydroxyapatite
extracellular fluid (ECF) (eks-trah-SELL-yoo-lar FLOO-id) liquid found outside of cells, located in two compartments: between cells (interstitial fluid) and in blood (plasma); lymph, cerebrospinal fluid, and joint fluids are also *extracellular fluids*
extracellular matrix (ECM) (eks-trah-SEL-yoo lar MAY-triks) material between cells in a tissue, made up of water and a variety of proteins
extrinsic eye muscle (eks-TRIN-sik MUSS-el) voluntary muscle that attaches the eyeball to the socket and produces movement of the eyeball
extrinsic foot muscle (eks-TRIN-sik MUSS-el) leg muscle responsible for movement of the ankle and foot
extrinsic muscle (eks-TRIN-sik MUSS-el) muscle originating from outside of the part of the skeleton moved
extrinsic pathway (eks-TRIN-sik PATH-way) in the blood, the metabolic pathway for blood clot formation that is triggered by tissue clotting factors outside the blood itself
eyebrow (EYE-brow) patch of hair just above the opening of the eye, along a ridge of the frontal bone
eyelash (EYE-lash) hair protruding from the edge of an eyelid
eyelid (EYE-lid) skin flap that partly covers the anterior, exposed region of the eyeball

F

facet (fah-SET or FASS-et) flat, rounded face on a bone projection (as in vertebrae)
facial nerve (FAY-shal nerv) cranial nerve VII, mixed nerve
facilitated diffusion (fah-SIL-i-tay-ted di-FYOO-shun) special type of diffusion; when movement of a molecule is made more efficient by action of carrier or channel mechanisms in the plasma membrane
false pelvis (PEL-vis) structure above the pelvic inlet
falx cerebelli (falks ser-eh-BEL ee) small fold in the dura mater in the posterior cranial fossa, between the cerebellum and cerebrum
falx cerebri (falks SER-eh-bree) fold in the dura mater that separates the two cerebral hemispheres
fascia (FAY-shah) general name for the fibrous connective tissue masses located throughout the body
fascicle (FAS-i-kul) small bundle or cluster, especially of groups of skeletal muscle fibers bound together by perimysium
fat one of the three basic food types; primarily a source of energy
feces (FEE-seez) waste material discharged from the intestines
feedback control loop (FEED-bak kon-TROL loop) highly complex and integrated communication control network, classified as negative or positive; negative feedback loops are the most important and numerous homeostatic control mechanisms
femoral vein (FEM-or-al vayn) vein of the thigh
femur (FEE-mur) the thigh bone
fertilization (FER-ti-li-ZAY-shun) union of an ovum and a sperm; conception
fetal alcohol syndrome (FAS) (FEE-tal AL-koh-hol SIN-drohm) a condition that may cause congenital abnormalities in a baby; it results from a woman consuming alcohol during pregnancy
fever (FEE-ver) unusually high body temperature
fiber (FYE-ber) threadlike structure, as in a muscle fiber (a threadlike cell) or a collagen fiber (a threadlike protein strand)
fibrinolysis (fye-brin-OL-i-sis) physiological mechanism that dissolves clots
fibrocartilage (fye-broh-KAR-ti-lij) cartilage with the greatest number of collagenous fibers; strongest and most durable type of cartilage
fibrous joint (FYE-brus joynt) connection between bones made primarily by bands of fibrous connective tissue
fibrous pericardium (FYE-brus pair-i-KAR-dee-um) tough, loose-fitting, and inelastic sac around the heart
fibula (FIB-yoo-lah) lower leg bone; lateral to tibia
fight-or-flight response (fyte-or-flyte ree-SPONS) changes produced by increased sympathetic impulses allowing the body to deal with any type of stress
filtration (fil-TRAY-shun) movement of water and solutes through a membrane because of a higher hydrostatic pressure on one side
filum terminale (FYE-lum ter-mi NAL-ee) slender filament formed by the pia mater that blends with the dura mater and then the periosteum of the coccyx
fimbriae (FIM-bree-ee) (*sing.*, fimbria [FIM-bree-ah]) fringe of tiny fingerlike projections around the opening of each fallopian (uterine) tube; the projections help move an ovum into the fallopian tube
first-degree burn partial-thickness burn; actual tissue destruction is minimal
first polar body (PO lar BOD-ee) small, nonfunctional cell produced during meiotic divisions in the formation of the female gamete
fixator muscle (fik-SAY-tor MUSS-el) muscle that functions as a joint stabilizer
flagellum (flah-JEL-um) (*pl.*, flagella [flah-JEL-ah]) single projection extending from the cell surface similar to a cilium; only example in human is the "tail" of the male sperm
flat bone one of the four types of bone; the frontal bone is an example of a flat bone
flat feet condition in which the tendons and ligaments of the foot are weak, allowing the normally curved arch to flatten out
fluid mosaic model (FLOO-id mo-ZAY-ik MAHD-el) theory of plasma membrane composition in which molecules of the membrane are bound tightly enough to form a continuous layer but loosely enough so molecules can slip past one another
folia (FOH-lee-ah) thin, delicate gyri (raised ridges) of the surface of the cerebellum
follicles (FOL-li-kulz) general name for a pocket or bubble structure; for example, ovarian structure consisting of oocyte surrounded by numerous supporting cells (follicle cells); also, pocketlike structures from which a hair grows; also, a small hollow sphere with a wall of glandular epithelium, found in thyroid tissue
follicle-stimulating hormone (FSH) (FOL-li-kul STIM-yoo-lay-ting HOR-mohn) hormone present in males and females; in males, FSH stimulates the production of sperm; in females, FSH stimulates the ovarian follicles to mature and follicle cells to secrete estrogen
follicular phase (foh-LIK-yoo-lar fayz) postmenstrual phase of the female reproductive cycle; also called *preovulatory* or *estrogenic phase*
fontanels (FON-tah-nelz) "soft spots" where ossification in the cranium is incomplete at birth
foramen ovale (foh-RAY-men oh-VAL-ee) in the developing fetus, opening that shunts blood from the right atrium directly into the left atrium, allowing most blood to bypass the baby's developing lungs
formed elements any of the cells of the blood tissue: red blood cells, white blood cells, and platelets in blood
fornix (FOR-niks) corner of the vagina where it meets the cervix of the uterus
fourth degree burn burn involving underlying muscles, fasciae, or bone
fovea centralis (FOH-vee-ah sen-TRAL-is) small depression in the macula lutea where cones are most densely packed; vision is sharpest where light rays focus on the fovea
fracture (FRAK-chur) a break in a bone and/or cartilage
free fatty acid (FFA) fatty acid combined with albumin to be transported by the blood to other cells
friction ridge raised underlying dermal papillae; form fingerprints
frontal bone (FRON-tal bohn) forehead bone
frontal lobe (FRON-tal) the part of the brain located behind the forehead
functional group (FUNK-shun-al groop) small cluster of atoms in an organic molecule that gives the molecule particular functional characteristics such as certain chemical binding properties; often represented generically by the letter R
functional protein (FUNK-shun-al PRO teen) category of proteins that affect the functional operations of a cell; contrast with **structural protein**
functional residual capacity (FRC) (FUNK-shun-al reh-ZID-yoo-al kah-PASS-i-tee) amount of air left in the lungs at the end of a normal expiration
fundus (of stomach) (FUN-duss) one of three divisions of the stomach; enlarged portion to the left and above the opening of the esophagus into the stomach
fundus (of uterus) (FUN-duss ov YOO-ter-us) bulging prominence above level where uterine tubes attach to the body of the uterus
funiculus (fuh-NIK-yoo-lus) large bundle of nerve fibers divided into smaller bundles called *spinal tracts*

G

gametes (GAM-eets) sex cells; spermatozoa and ova
ganglion (GANG-lee-on) (*pl.*, ganglia [GANG-lee-ah]) in peripheral nerves, a region of gray matter made up of unmyelinated fibers
ganglion neuron (GANG-glee-on NOO-ron) sensory neurons in the retina of the eye that collect information from rods and cones and also act as photoreceptors themselves
gap junction (gap JUNK-shen) cell connection formed when membrane channels of adjacent plasma membranes adhere to each other
gastric juice (GAS-trik) stomach secretion containing acid and enzymes; aids in the digestion of food
gastric phase (GAS-trik fayz) phase during which the stomach releases gastrin, which accelerates secretion of gastric juice
gastrin (GAS-trin) gastrointestinal (GI) hormone that plays an important regulatory role in the digestive process by stimulating gastric secretion
gastrointestinal (GI) tract (gas troh in TESS-ti-nul trakt) alimentary canal; tube formed by the major organs of digestion
gene (jean) one of many segments of a chromosome (DNA molecule); each gene contains the genetic code for synthesizing a protein molecule such as an enzyme or hormone or to make a functional RNA molecule such as tRNA or rRNA
gene augmentation (jeen awg-men-TAY-shun) therapeutic technique that introduces genes with the hope that they will add to the production of a needed protein
gene linkage when a whole group of genes stay together during the crossing-over process
general sense organ structure that consists of microscopic receptors widely distributed throughout the body
gene replacement therapeutic technique that replaces genes that specify production of abnormal proteins with normal genes
gene therapy (jeen THER-ah-pee) manipulation of genes to cure genetic problems; most forms of gene therapy have not yet proven to be effective in humans
genetic mutation (jeh-NET-ik myoo-TAY-shun) change in the genetic material within a genome; may occur spontaneously or as a result of mutagens
genetics (jeh-NET-iks) scientific study of heredity and the genetic code
geniculate body (jeh-NIK-yoo-lit BOD-ee) either of two groups of cerebral nuclei comprising the thalamus; located in posterior region of each lateral mass; play role in processing auditory and visual input
genome (JEE-nome) entire set of chromosomes in a cell; the *human genome* refers to the entire set of human chromosomes
genomics (jeh-NOM-iks) field of endeavor involving the analysis of the genetic code contained in the human or another species' genome
genotype (JEN-oh-type) alleles present at one or more specific loci on a chromosome of a given individual; *see* **phenotype**
gerontology (jair-on-TAHL-oh-jee) study of the aging process
gestation period (jes-TAY-shun PEER-ee-od) the length of pregnancy, approximately 9 months in humans
ghrelin (GRAY-lin) hormone secreted by epithelial cells lining the stomach; ghrelin boosts appetite, slows metabolism, and reduces fat burning; may be involved in the development of obesity
glandular (GLAN-dyoo-lar) resembling a gland
glans clitoris (glans KLIT-oh-ris) *see* **clitoris**
glans penis (glans PEE-nis) slightly bulging structure formed by the distal end of the corpus spongiosum; head of the penis; covered by the foreskin in uncircumcised males
glaucoma (glaw-KOH-mah) disorder characterized by elevated pressure in the eye; can lead to permanent blindness
glia (GLEE-ah) nonexcitable supporting cells of nervous tissue; formerly called *neuroglia*
gliding joint a junction that allows the articular surface of one bone to move over the articular surface of another without any angular or circular movement
gliding movement movement that results when the articular surface of one bone moves over the articular surface of another without any angular or circular movement
globin (GLOH-bin) chain of four proteins; binds to a red pigment (heme) to form hemoglobin
glomerulus (gloh-MAIR-yoo-lus) compact cluster, particularly when referring to the tuft of capillaries forming part of the nephron
glossopharyngeal nerve (glos-oh-fah-RIN-jee-al nerv) cranial nerve IX; mixed nerve
glottis (GLOT-iss) composed of the true vocal cords and the space between them
glucose phosphorylation (GLOO-kohs fos-for-i-LAY-shun) process of converting glucose to glucose 6-phosphate; prepares glucose for further metabolic reactions
gluteal (GLOO tee al) pertaining to buttock muscle, which moves the thigh
glycogen (GLYE-koh-jen) polysaccharide (complex carbohydrate); main carbohydrate stored in animal cells

glycolysis (glye-KOHL-i-sis) first series of chemical reactions in carbohydrate catabolism; changes glucose to pyruvic acid in a series of anaerobic reactions
glycoprotein (glye-koh-PRO-teen) substance made of molecules that are a combined form of carbohydrate and protein
glycoprotein hormone (glye-koh-PRO-teen HOR-mohn) hormone made of molecules that are a combined form of carbohydrate and protein
Golgi apparatus (GOL-jee app-ah-RA-tus) organelle consisting of small sacs stacked on one another near the nucleus that makes carbohydrate compounds, combines them with protein molecules, and packages the product for distribution from the cell
Golgi tendon organ (GOL-jee TEN-don OR-gan) sensory organ embedded in muscle tendons made up of sensory neurons (Golgi tendon receptors) that are encapsulated with collagen bundles; also called *tendon organ*
gomphosis (gom-FOH-sis) fibrous joint where a process is inserted into a socket; for example, the joint between the tooth and mandible
gonadocorticoid (go-nad-oh-KOR-tih-koyd) a hormone of the adrenal cortex that influences the activity of gonads
gonadotropin (go-nah-doh-TROH-pin) any of the hormones (FSH and LSH) produced by the anterior pituitary or embryonic tissue (hCG) that stimulate growth and maintenance of the testes or ovaries
gonads (GO-nads) sex gland in which reproductive cells are formed; ovary in women, testis in men
graafian follicle (GRAH-fee-en FOL-i-kul) a secondary follicle; consists of a mature ovum surrounded by granulosa cells at boundary of fluid-filled antrum
granulocyte (GRAN-yoo-loh-syte) leukocyte with granules in cytoplasm
granulosa cell (gran-yoo-LOH-sah sell) develops around each primary oocyte, forming a primary follicle
Graves disease (gravz di-ZEEZ) inherited, possibly autoimmune endocrine disorder characterized by hyperthyroidism accompanied by exophthalmos (protruding eyes)
gray column any one of the three longitudinal areas (anterior, lateral, posterior) of the spinal cord that contain gray matter
gray matter (MAT-er) type of nerve tissue characterized by a relative lack of myelin and often includes many neuron cell bodies and synapses that process information; contrast with **white matter**
greater omentum (GRAYT-er oh-MEN-tum) pouch-like extension of the visceral peritoneum extending from greater curvature of the stomach to the transverse colon; often called "the lace apron"
greater vestibular glands (ves-TIB-yoo-lar) glands on each side of the vaginal orifice that secrete a lubricating fluid; also called *Bartholin's glands*
ground substance organic matrix of bone and cartilage
growth hormone (GH) (HOR-mohn) hormone secreted by the anterior pituitary gland that controls the rate of skeletal and visceral growth
gustatory (GUS-tah-tor-ee) refers to taste
gustatory cell (GUS-tah-tor-ee sell) chemoreceptors in tongue that sense taste
gyrus (JYE-rus) convoluted ridge, usually refers to rounded elevations of the brain surface; also called *convolution*

H

hair follicle (FOL-i-kul) small blind-ended tube extending from the dermis through the epidermis that contains the hair root and where hair growth occurs; sebaceous and apocrine skin glands have ducts leading into the follicle
haploid (HAP-loid) halved number of chromosomes in gametes resulting from meiosis; in humans, 23 chromosomes per sex cell
hard palate (PAL-et) portions of four bones (two maxillae and two palantines) that make up the roof of the mouth
heart (hart) organ of circulatory system that pumps the blood; composed of cardiac muscle tissue
heart murmur (hart MUR-mur) abnormal heart sound that may indicate valvular insufficiency or stenosing of the valve
heart rate (HR) (hart rayt) the number of heart beats, usually per minute
hematocrit (hee-MAT-oh-krit) volume percent of blood cells in whole blood; packed cell volume
hematopoietic stem cell (hee-mah-toh-poy-ET-ik sell) nucleated cell in the red bone marrow that develops into a red blood cell
hemocytoblast (hee-moh-SYE-toh-blast) bone marrow stem cell from which all formed elements of blood arise
hemodialysis (hee-moh-dye-AL-i-sis) therapy involving separation of smaller (diffusible) particles from larger (nondiffusible) particles in blood through a semipermeable membrane, usually employed when a patient's kidneys fail to remove these particles from the blood
hemodynamics (hee-moh-dye-NAM-iks) study of the mechanisms that influence dynamic circulation of blood
hemoglobin (HEE-moh-gloh-bin) iron-containing protein in red blood cells responsible for their oxygen-carrying capacity
hemolysis (hee-MAHL-i-sis) breakdown of blood cells
hemostasis (hee-moh-STAY-sis) stoppage of blood flow
Henle loop (HEN-lee loop) extension of the proximal tubule of the nephron; also called *loop of Henle* or *nephron loop*
heparin (HEP-ah-rin) substance obtained from the liver; inhibits blood clotting
hepatic (heh-PAT-ik) relating to the liver
hepatic portal vein (heh-PAT-ik POR-tal vayn) delivers blood directly from the gastrointestinal tract to the liver
heterozygous (het-er-oh-ZYE-gus) genotype with two different forms of a trait
hiatal hernia (hye-AYT-al HER-nee-ah) condition in which a portion of the stomach is pushed through the hiatus (opening) of the diaphragm, often weakening or expanding the cardiac sphincter at the inferior end of the esophagus
hiccup (HIK-up) involuntary spasmodic contraction of the diaphragm
high-energy bond chemical bond that requires an input of energy to form and, when broken, can result in the transfer of useful energy to cellular processes, as in ATP
hilum (HYE-lum) slit on medial surface of each lung where primary bronchi and pulmonary vessels enter; slit on medial surface of each kidney where blood vessels and other structures enter the kidney
hinge joint (hinj joynt) type of diarthrotic synovial joint that allows movement around a single axis in the manner of a hinge
histamine inflammatory chemical
histogenesis (hiss-toh-JEN-eh-sis) formation of tissues from primary germ layers of embryo
histology (hiss-TOL-oh-jee) branch of microscopic anatomy that studies tissues; biology of tissues
histone (HISS-tohn) protein that organizes chromatin into nucleosomes
holocrine gland (HOH-loh-krin) gland that collects secretory product inside its cells, which then rupture completely to release it; for example, sebaceous glands of the skin
homeostasis (hoh-mee-oh-STAY-sis) relative constancy of the normal body's internal (fluid) environment
homozygous (hoh-moh-ZYE-gus) genotype with two identical forms of a trait
hormone (HOR-mohn) substance secreted by an endocrine gland into the bloodstream that acts on a specific target tissue to produce a given response
human chorionic gonadotropin (hCG) (koh-ree-ON-ik go-nah-doh-TROH-pin) hormone secreted early in pregnancy by the placenta that serves to maintain the uterine lining
human engineered chromosome (HEC) gene augmentation procedure that inserts therapeutic genes into a separate strand of DNA that is inserted into cell's nucleus
Human Genome Project a worldwide collaborative effort of scientists and others to map out the entire human genome and study the biological, medical, and ethical aspects of their discoveries; the HGP is largely funded by U.S. government sources such as the DOE (Department of Energy) and the NIH (National Institutes of Health); *see* **genome, genomics**
human placental lactogen (hPL) (plah-SEN-tal lak-TOH-jen) hormone released by the placenta; promotes development of mammary glands during pregnancy and regulates energy balance in fetus
humerus (HYOO-mer-us) upper arm bone
hyaline cartilage (HYE-ah-lin KAR-ti-lij) most common type of cartilage; appears gelatinous and glossy
hydrogen bond (HYE-droh-jen) weak chemical bond that occurs between the partial positive charge on a hydrogen atom covalently bound to a nitrogen or oxygen atom and the partial negative charge of another polar molecule
hydrolysis (hye-DROHL-i-sis) chemical process in which a compound is split by addition of H^+ and OH^- portions of a water molecule
hydrostatic pressure (hye-droh-STAT-ik PRESH-ur) force of a fluid pushing against some surface
hymen (HYE-men) Greek for "membrane"; mucous membrane that may partially or entirely occlude the vaginal outlet
hyoid bone (HYE-oyd bohn) U-shaped bone of the neck between the mandible and the larynx
hyperglycemia (hye-per-glye-SEE-mee-ah) higher than normal blood glucose concentration
hypersecretion (hye-per-seh-KREE-shun) too much secretion of a substance
hypertonic (hye-per-TON-ik) solution containing a higher level of salt (NaCl) than is found in a living red blood cell (above 0.9% NaCl); causes cells to shrink
hypertrophy (hye-PER-troh-fee) increased size of a part caused by an increase in the size of its cells
hypochloremia (hye-poh-kloh-REE-mee-ah) abnormally low levels of chloride in the blood associated with potassium loss
hypodermis (hye-poh-DER-mis) loose layer of fascia rich in fat and areolar tissue located beneath the dermis; also called *subcutaneous layer* or *superficial fascia*
hypoglossal nerve (hye-poh-GLOS-al nerv) cranial nerve XII; motor nerve; responsible for tongue movement
hypoglycemia (hye-poh-gly-SEE-mee-ah) lower-than-normal blood glucose concentration
hypokalemia (hye-poh-kal-EE-mee-ah) abnormally low serum potassium level
hyposecretion (hye-poh-seh-KREE-shun) too little secretion of a substance
hypophyseal portal system (hye-poh-FIZ-ee-al POR-tal SIS-tem) complex of small blood vessels through which releasing hormones travel from the hypothalamus to the pituitary
hypothalamic releasing hormone (hye-poh-THAL-ah-mik re-LEE-sing HOR-mohn) hormone produced by the hypothalamus that causes the pituitary gland to release its hormones
hypothalamus (hye-poh-THAL-ah-muss) important autonomic and neuroendocrine control center located inferior to the thalamus in the brain
hypothesis (hye-POTH-eh-sis) idea or scientific concept, usually based on previous ideas or observations, that is proposed as a possible explanation of nature or a natural process; hypotheses undergo intense testing before being accepted widely in the scientific community; *see* **theory**
hypotonic (hye-poh-TON-ik) adjective used to describe a solution that has a lower potential osmotic pressure than a solution to which it is being compared; a hypotonic solution tends to have low osmolality and thus tends to gain water (by way of osmosis) from the solution to which it is compared
H zone central part of the A band containing myosin filaments

I

ileum (IL-ee-um) distal portion of the small intestine
immune system (i-MYOON SIS-tem) body's defense system against disease
immunization (i-myoo-ni-ZAY-shun) deliberate artificial exposure to disease for the purpose of producing acquired immunity
immunoglobulin (Ig) (i-myoo-noh-GLOB-yoo-lin) antibody
imperforate hymen (im-PER-fah-rayt HYE-men) condition in which the hymen completely covers the vaginal outlet
implantation (im-plan-TAY-shun) process in which developing offspring tissue connects to the uterine wall of the mother
infancy (IN-fan-see) period of human development from birth to about 18 months of age
inferior (in-FEER-ee-or) lower; opposite of **superior**
inferior nasal concha (in-FEER-ee-or NAY-zal KONG-kah) *see* **concha**
inferior vena cava (in-FEER-ee-or VEE-nah KAY-vah) vein of the thorax; drains blood from the lower trunk and extremity
infertility (in-fer-TIL-i-tee) failure to conceive after 1 year of regular unprotected intercourse
inflammatory response (in-FLAM-ah-toh-ree ree-SPONS) specific process involving tissues and blood vessels in response to injury
infundibulum (in-fun-DIB-yoo-lum) stalk that connects the pituitary gland to the hypothalamus
ingestion (in-JES-chun) taking in of complex foods, usually by mouth
inhibin (in-HIB-in) glycoprotein hormone produced by the ovary to regulate FSH secretion by the anterior pituitary (adenohypophysis)
inhibition (in-hib-ISH-un) any process that slows down or stops another process
innate immunity (IN-ayt i-MYOO-ni-tee) type of immunity that exists prior to exposure to a specific antigen and that can recognize and destroy a variety of harmful agents or conditions; also called *nonspecific immunity*
inner cell mass layer of cells in the blastocyst
inner ear region of the ear consisting of a bony labyrinth and a membranous labyrinth; also called *labyrinth*
inorganic compound (in-or-GAN-ik KOM-pownd) chemical constituents that do not contain both carbon and hydrogen; for example, water, carbon dioxide, and oxygen
insertion (in-SER-shun) (muscle insertion) attachment of a muscle to the bone that it moves when contraction occurs (as distinguished from its origin); (insertion mutation) a type of genetic mutation that occurs when one or more nucleotide bases appear within the usual sequence of nucleotide bases in a gene; in insertion mutations, the cell cannot read the genetic code normally and thus the encoded protein cannot be made in its usual form; *see* **deletion**
inspiration (in-spih-RAY-shun) process of bringing air into the lungs
inspiratory capacity (IC) (in-SPY-rah-tor-ee kah-PASS-i-tee) maximum amount of air an individual can inspire after a normal expiration
inspiratory reserve volume (IRV) (in-SPY-rah-tor-ee ree-SERV VOL-yoom) the amount of air that can be forcibly inspired over and above a normal respiration
insula (IN-soo-lah) lobe of the cerebral cortex; lies hidden from view in the lateral fissure
insulin (IN-suh-lin) hormone secreted by beta cells of the pancreatic islets that increases the uptake of glucose and amino acids by most body cells
integumentary system (in-teg-yoo-MEN-tar-ee SIS-tem) skin and its related structures
intercalated disk (in-TER-kah-lay-ted) any of the disk-like connections between ends of adjacent cardiac muscle fibers characterized by gap junctions and often appearing as tiny, darkly stained lines in micrographs
interferon (IFN) (in-ter-FEER-on) small protein produced by the immune system that inhibits virus multiplication
internal jugular vein (JUG-yoo-lar vayn) deep vein of the neck
interneuron (in-ter-NOO-ron) in a three-neuron reflex arc, nerve cell that conducts impulses from a sensory neuron to a motor neuron
interoceptors (in-ter-oh-SEP-tors) somatic sense receptors located in the internal visceral organs
interphase (IN-ter-fayz) mitotic phase immediately before visible condensation of the chromosomes during which the DNA of each chromosome replicates itself
interstitial cell (in-ter-STISH-al sell) any of the small cells in the testes found between seminiferous tubules and that secrete the male sex hormone, testosterone; also called *Leydig cell*
interstitial fluid (IF) (in-ter-STISH-al FLOO-id) fluid located in microscopic spaces between cells
intestinal juice (in-TES-ti-nal) refers to the sum total of intestinal secretions
intestinal phase (in-TES-ti-nal fayz) stage of gastric secretion triggered when material leaves the stomach and enters the small intestine
intracellular fluid (ICF) (in-trah-SELL-yoo-lar FLOO-id) water inside the cell
intramembranous ossification (in-trah-MEM-brah-nus os-i-fi-KAY-shun) process by which most flat bones are formed within connective tissue membranes
intravenous injection (in-trah-VEE-nus in-JEK-shun) administration of medication into the muscle
intrinsic eye muscle (in-TRIN-sik MUSS-el) involuntary muscle located within the eye; responsible for size of the iris and shape of the lens
intrinsic factor (in-TRIN-sik FAK-tor) binds to molecules of B_{12}, protecting them from the acids and enzymes of the stomach; secreted by parietal cells
intrinsic foot muscle (in-TRIN-sik MUSS-el) muscle located within the foot
intrinsic muscle (in-TRIN-sik MUSS-el) muscle that is actually within the part moved
intrinsic pathway (in-TRIN-sik PATH-way) in the blood, the metabolic pathway for blood

clot formation that is triggered by clotting factors from within the blood itself
ion (EYE-on) electrically charged atom or group of atoms
ionic bond (eye-ON-ik) electrocovalent bond; bond formed by transferring of electrons from one atom to another
iris (EYE-riss) colored portion of the eye
ischemic (is-KEE-mik) relating to ischemia
ischium (IS-kee-um) lowermost coxal bone
islets of Langerhans (EYE-lets ov lahn-GER-hans) pancreatic islets
isometric contraction (eye-soh-MET-rik kon-TRAK-shun) type of muscle contraction in which muscle does not alter the distance between two bones; *see* **isotonic contraction**
isotonic contraction (eye-soh-TON-ik kon-TRAK-shun) type of muscle contraction in which the muscle sustains the same tension or pressure and a change in the distance between two bones occurs
isotonic fluids (eye-soh-TON-ik FLOO-ids) two fluids that have the same potential osmotic pressure
isotope (EYE-soh-tope) atoms with the same atomic number but different atomic weights

J

jejunum (jeh-JOO-num) middle third of the small intestine
joint capsule (joynt CAP-sool) sleevelike extension of the periosteum of each of the bones at an articulation
joint cavity (joynt KAV-i-tee) small space between articulating surfaces of the two bones of the joint
juxtaglomerular (JG) apparatus (jux-tah-gloh-MAIR-yoo-lar app-ah-RAT-us) in the nephron, the complex of cells from the distal tubule and the afferent arteriole, which helps regulate blood pressure by secreting renin in response to blood pressure changes in the kidney; located near the glomerulus; also called *juxtaglomerular complex*
juxtamedullary nephron (jux-tah-MED-eh-lair-ee NEF-ron) region of the nephron that lies near the junction of the cortical and medullary layers of the kidney

K

Kaposi sarcoma (KS) (KAH-poh-see sar-KOH-mah) rare malignant neoplasm of the skin that often spreads to lymph nodes and internal organs; Kaposi sarcoma is often found in AIDS patients
karyotype (KAIR-ee-oh-type) ordered arrangement of photographs of chromosomes from a single cell used in genetic counseling to identify chromosomal disorders such as trisomy or monosomy
keratinization (kehr-ah-tin-i-ZAY-shun) process by which cells of the stratum corneum become fitted with keratin and move to the surface
keratinize (KEHR-ah-tin-eyes) process of converting cytoplasm to the protein keratin in epidermal cells
keratinocyte (keh-RAT-i-no-syte) epidermal cell responsible for synthesizing keratin
kidney (KID-nee) one of the two organs that cleanses the blood of waste products continually produced by metabolism; the kidneys produce urine
kinase (KYE-nayz) substance that converts proenzymes to active enzymes
knee jerk reflex extension of the lower leg in response to tapping of the patellar tendon; also called *patellar reflex*

L

labia majora (LAY-bee-ah mah-JO-rah) large lateral folds of the vulva
labia minora (LAY-bee-ah mi-NO-rah) small medial folds of the vulva
labor (LAY-bor) process of expulsion of the fetus and the placenta; childbirth
labyrinth (LAB-i-rinth) bony cavities and membranes of the inner ear
lacrimal apparatus (LAK-ri-mal app-ah-RAT-us) in the eye, the tear (lacrimal) gland plus associated ducts that form tears
lacrimal bone (LAK-ri-mal bohn) facial bone that joins the maxilla, frontal bone, and ethmoid bone
lactate (LAK-tayt) an energy-containing substrate produced by anaerobic respiration
lactation (lak-TAY-shun) milk production
lacteal (LAK-tee-al) lymphatic vessel located in each villus of the intestine; serves to absorb fat materials from chyme passing through the small intestine
lacuna (lah-KOO-nah) space or cavity; for example, lacunae in bone contain bone cells
lamella (lah-MEL-ah) (*pl.,* lamellae [lah-MEL-ee]) thin layer, as of bone
lamellar corpuscle (lah-MEL-ar KOR-pus-ul) sensory receptor with a layered encapsulation found deep in the dermis that detects pressure on the skin surface, also known as *Pacini corpuscle*
lanugo (lah-NOO-go) extremely fine and soft hair coat on developing fetus
laryngopharynx (lah-ring-go-FAIR-inks) lowest part of the pharynx
larynx (LAIR-inks) voice box located just below the pharynx; the largest piece of cartilage making up the larynx is the thyroid cartilage, commonly known as the *Adam's apple*
lateral (LAT-er-all) of or toward the side; opposite of **medial**
lecithin (LES-i-thin) substance in bile that emulsifies dietary oils and fats in the lumen of the small intestine
left lobe left segment of an organ
leptin (LEHP-tin) a protein hormone produced by fat-storing cells in adipose tissue that plays a role in inhibiting food intake by regulating the satiety center in the hypothalamus; leptin also plays a role in reducing fat storage in nonadipose cells in the liver and skeletal muscles; because of its role in fat storage, it may play a role in future treatments for diabetes, obesity, and other conditions related to fat metabolism; *see* **satiety center;** leptin also helps regulate some immune and neuroendocrine functions and plays a role in development
lesser omentum (oh-MEN-tum) extension of the peritoneum that is attached from the liver to the lesser curvature of the stomach and the first part of the duodenum
leukocyte (LOO-koh-syte) white blood cell
leukocytosis (loo-koh-sye-TOH-sis) abnormally high white blood cell numbers in the blood
leukopenia (loo-koh-PEE-nee-ah) abnormally low white blood cell numbers in the blood
lever any rigid bar free to turn about a fixed point
levodopa (LEV-oh-doh-pah) a molecule derived from tyrosine in neurons that is used to produce the neurotransmitter dopamine
ligament (LIG-ah-ment) band of white fibrous tissue connecting bones to other bones
limbic system (LIM-bik SIS-tem) parts of the brain involved in emotions and sense of smell; plays key role in coupling sensory inputs to short- and long-term memory; consists of the hippocampus, the hypothalamus, and several other structures
linea alba (LIN-ee-ah AL-bah) tough band of connective tissue that covers the rectus abdominus muscle
lingual tonsil (LING-gwal TAHN-sil) tonsil located at the base of the tongue
linked trait a trait inherited together with other traits on the same chromosome
lip region that lines the mouth and continues into the oral cavity
lipase (LYE-payse) fat-digesting enzymes
lipid (LIP-id) class of organic compounds that includes fats, oils, and related substances
lipogenesis (lip-oh-JEN-eh-sis) formation of body fat from food sources
lipoprotein (lip-oh-PRO-teen) substance that is part lipid and part protein; produced mainly in the liver
local potential (poh-TEN-shal) slight shift from resting membrane potential in a specific region of the plasma membrane
lock-and-key model (lok-and-kee MAHD-el) description of how a specific enzyme will fit into only a specific substrate, like a key fits into a lock
long bone bone that is characterized by its extended longitudinal axis and unique articular ends
long-term memory storage of information that can be retrieved days, or even years, later
loose connective tissue (LOOS koh-NEK-tiv TISH-yoo) *see* **areolar tissue**
lower esophageal sphincter (LES) (ee-SOFF-ah-JEE-ull SFINGK-ter) muscle located at the junction between the terminal portion of the esophagus and the stomach; also called *cardiac sphincter*
lower respiratory tract (RES-pi-rah-tor-ee trakt) region of the respiratory tract that consists of the trachea, all segments of the bronchial tree, and the lungs
lumbar plexus (LUM-bar PLEK-sus) spinal nerve plexus located in the low back
lumbar puncture (LUM-bar PUNK-chur) clinical procedure in which some cerebrospinal fluid is withdrawn from the subarachnoid space in the lumbar region of the spinal cord for analysis
lumbar vertebra (LUM-bar VER-teh-bra) lower five bones of the vertebral column; support the small of the back
lumen (LOO-men) the hollow area of a hollow organ such as the stomach, blood vessel, or urinary bladder; *see also* **luminal**
luminal (LOO-min-al) relating to the hollow part, or lumen, of a hollow organ such as the urinary bladder or small intestine; *see also* **lumen** [*lumin-* light, *-al* relating to]
luteal phase (LOO-tee-al fayz) phase of the menstrual cycle that occurs between ovulation and the onset of menses; also called *premenstrual phase, postovulatory phase,* or *secretory phase*
luteinization (loo-tee-in-i-ZAY-shun) formation of a golden body (corpus luteum) in the ruptured follicle
luteinizing hormone (LH) (loo-tee-in-EYE-zing HOR-mohn) in females, acts in conjunction with follicle-stimulating hormone (FSH) to stimulate follicle and ovum maturation, release of estrogen, and ovulation; known as the *ovulating hormone;* in males, causes testes to develop and secrete testosterone
lymph (limf) specialized fluid formed in the tissue spaces that returns excess fluid and protein molecules to the blood via lymphatic vessels
lymphatic capillaries (lim-FAT-ik KAP-i-lair-ees) microscopic blind-ended vessels that transport lymph
lymphatic system (lim-FAT-ik SIS-tem) a system that plays a critical role in the functioning of the immune system, moves fluids and large molecules from the tissue spaces and fat-related nutrients from the digestive system to the blood
lymphatic vessels (lim-FAT-ik VES-els) any vessel of a system of blind-ended vessels that collect lymph and deliver it to the circulatory system via the thoracic duct and the right lymphatic duct
lymphedema (lim-feh-DEE-mah) swelling caused by lymphatic vessel blockage
lymph node (limf nohd) small structure that performs biological filtration of lymph on its way to the circulatory system
lymphocyte (LIM-foh-syte) one type of white blood cell; *see* **B lymphocyte, T lymphocytes**
lymphoid tissue (LIM-foyd TISH-yoo) type of reticular tissue that contains lymphocytes and other specialized cells
lysosome (LYE-so-sohm) membranous organelle containing various enzymes that can dissolve most cellular compounds; called *digestive bags* or *suicide bags of cell*

M

macromolecule (mak-roh-MOL-eh-kyool) large, complex chemical made of combinations of molecules
macronutrients (mak-roh-NOO-tree-ents) nutrients needed in large amounts; carbohydrates, fats, and proteins
macrophage (MAK-roh-fayje) phagocytic cell in the immune system
macula (MAK-yoo-lah) strip of sensory epithelium in the utricle and saccule; provides information related to head position or acceleration
macula lutea (MAK-yoo-lah LOO-tee-ah) yellowish area near center of the retina where cones are densely distributed; also called simply "macula"
major histocompatibility complex (MHC) (his-toh-kom-pat-i-BIL-i-tee KOM-pleks) set of genes in chromosome 6 that all code for antigen-presenting proteins and other immune system proteins; proteins produced by MHC genes in class I and class II also are called *human leukocyte antigens (HLAs);* MHC proteins present different protein fragments (peptides) at the surface of the cell for possible recognition as either self or nonself antigens by immune system cells
malaria (mah-LAIR-ee-ah) sometimes fatal condition caused by blood parasites
malignant melanoma (mah-LIG-nant mel-ah-NOH-mah) most deadly type of skin cancer
malignant tumor (mah-LIG-nant TOO-mer) cancer that is not encapsulated and tends to spread to other regions of the body
mammary gland (MAM-ar-ee) milk-producing glands of the breasts; classified as external accessory sex organs in females
mandible (MAN-di-bal) jaw bone; largest and strongest bone of the face
mast cells immune system cell to which antibodies become attached in early stages of inflammation
mastication (mass-ti-KAY-shun) chewing
mastitis (mass-TYE-tis) breast inflammation
matrix (MAY-triks) extracellular substance of a tissue; for example, the matrix of bone is calcified, whereas that of blood is liquid; *see also* **extracellular matrix (ECM)**
matter anything that has mass and occupies space
maxilla (mak-SIL-ah) upper jaw bone
mechanical digestion process through which food is broken into smaller portions through chewing and movements of the alimentary canal; enables enzymes to act on a larger surface area to accomplish chemical digestion
mechanoreceptor (mek-an-oh-ree-SEP-tor) receptors that respond to physical movement in the environment, such as sound waves; for example, equilibrium and balance sensors in the ears
medial (MEE-dee-al) of or toward the middle; opposite of **lateral**
mediastinum (mee-dee-ass-TI-num) a portion of the thoracic cavity in the middle of the thorax
medulla (meh-DUL-ah) Latin for "marrow"; the inner portion of an organ, in contrast to the outer portion, or cortex
medulla oblongata (meh-DUL-ah ob-long-GAH-tah) lowest part of the brainstem; an enlarged extension of the spinal cord; the vital centers are located within this area
medullary (MED-eh-lair-ee) relating to the middle or center of an organ or structure
medullary cavity (MED-eh-lair-ee KAV-i-tee) hollow area inside the diaphysis of the bone that contains yellow bone marrow
medullary rhythmicity area (MED-eh-lair-ee rith-MI-si-tee) area in the brainstem that generates the basic rhythm of the respiratory cycle of inspiration and expiration
meiosis (my-OH-sis) nuclear division in which the number of chromosomes is reduced to half their original number through separation of homologous pairs; produces gametes
melanin (MEL-ah-nin) brown pigment primarily in skin and hair
melanocyte (MEL-ah-noh-syte) cell type in the stratum basale of the skin that produces melanin pigment granules, releasing them to other nearby skin cells
melanosome (MEL-ah-noh-sohm) pigment granule released by melanocytes
melatonin (mel-ah-TOH-nin) important hormone produced by the pineal gland; it is believed to regulate onset of puberty and the menstrual cycle; also referred to as the *third eye* because it responds to levels of light and is thought to be involved with the body's internal clock
membrane attack complex (MAC) (MEM-brayne KOM-pleks) molecules formed by the complement cascade; leads to cell lysis
membrane bone (MEM-brayne bohn) bone formed within membranous tissues, such as the flat bones of the skull, instead of indirectly through endochondral ossification
membrane potential (MEM-brayne poh-TEN-shal) difference in electrical charge between inside and outside of the plasma membrane
membrane pump (MEM-brayne pump) active transport pump embedded within the cell membrane
membranous (MEM-brah-nus) resembling a membrane
memory B cell (MEM-oh-ree B sell) cell that has been activated but is not an effector cell producing an active response
memory cell (MEM-oh-ree sell) cell that remains in reserve in the lymph nodes until their ability to secrete antibodies is needed
memory T cell (MEM-oh-ree T sell) T cell that has been activated but is not an effector cell producing an active response
menarche (meh-NAR-kee) first menses occurring at the onset of puberty
meninges (meh-NIN-jeez) fluid-containing membranes surrounding the brain and spinal cord
meningitis (men-in-JYE-tis) inflammation of the meninges caused by various factors, including bacterial infection, mycosis, viral infection, and tumors
meniscus (meh-NIS-kus) articular cartilage disk
menopause (MEN-oh-pawz) termination of menstrual cycles; also called *climacteric*
menses (MEN-seez) periodic shedding of endometrial lining in uterus; occurs in cycles of about 28 days
menstrual period (MEN-stroo-al PEER-ee-od) *see* **menses**
menstruation (men-stroo-AY-shun) menses; regular event of the female reproductive cycle that allows the endometrium to renew itself
merocrine gland (MER-oh-krin) gland that discharges secretions directly through the cell or plasma membrane

mesentery (MEZ-en-tair-ee) large double fold of peritoneal tissue that anchors the loops of the digestive tract to the posterior wall of the abdominal cavity
mesoderm (MEZ-oh-derm) middle layer of the primary germ layers; gives rise to such structures as muscle, bones, and blood vessels
messenger RNA (mRNA) duplicate copy of a gene sequence on the DNA that passes from the nucleus to the cytoplasm; used by ribosomes to create specific proteins
metabolic pathway (met-ah-BOL-ik PATH-way) sequence of chemical reactions
metabolism (meh-TAB-oh-liz-em) complex, intertwining set of chemical processes by which life is made possible for a living organism; *see* **anabolism, catabolism**
metacarpal bone (met-ah-KAR-pal bohn) bone of the hand
metaphase (MET-ah-fayz) second stage of mitosis, during which the nuclear membrane and nucleolus disappear and the chromosomes align on the equatorial plane
metarteriole (met-ar-TEER-ee-ohl) short connecting blood vessel that connects a true arteriole with the proximal end of dozens of capillaries
micelle (mye-SELL) droplet of lipid surrounded by bile salts, which makes the lipid temporarily water-soluble
microcirculation (mye-kroh-ser-kyoo-LAY-shun) flow of blood through the capillary bed
microglia (mye-KROG-lee-ah) type of small neuroglial cell of nerve tissue that serves an immune system function by becoming an active phagocyte when stimulated
micronutrients (mye-kroh-NOO-tree-ents) nutrients needed by the body in very small quantity
microvilli (mye-kroh-VIL-ee) brushlike border made up of epithelial cells on each villus in the small intestine; increases the surface area for absorption of nutrients
micturition (mik-too-RISH-un) urination, voiding
midbrain (MID-brayn) region of the brainstem between the pons and the diencephalon
middle ear (MID-ul eer) tiny and very thin epithelium-lined cavity in the temporal bone that houses the ossicles; in the middle ear, sound waves are amplified
minerals (MIN-er-als) inorganic element or salts occurring naturally in the earth, many of which are vital to proper functioning of the body and must be obtained in the diet
mineralocorticoid (MC) (min-er-al-oh-KOR-ti-koyd) hormone that influences mineral salt metabolism; secreted by adrenal cortex; aldosterone is the chief mineralocorticoid
miscarriage (mis-KARE-ij) loss of an embryo or fetus before the twentieth week of pregnancy; after 20 weeks, the event is termed a *stillbirth;* also called *spontaneous abortion*
mitochondria (mye-toh-KON-dree-ah) organelle in which ATP generation occurs; often termed "powerhouse of cell"
mitosis (mye-TOH-sis) complex process in which a cell's DNA is replicated and divided equally between two daughter cells
mitral valve (MYE-tral valv) located between the left atrium and ventricle, this valve prevents backflow of blood into the left atrium; named for its resemblance to a bishop's hat (miter); also known as the *bicuspid valve*
mixed cranial nerve (KRAY-nee-al nerv) bundle of axons that contain both sensory and motor neurons
mixed nerve nerve with axons of both sensory and motor neurons
M line (EM lyne) region of the sarcomere where myosin filaments are held together and stabilized by protein molecules
mobile-receptor model (MO-bil-ree-SEP-tor MAHD-el) theory that a hormone enters a target cell where it binds to a free (nonstationary) receptor, thereby activating a gene and thus the beginning of mRNA transcription
molecule (MOL-eh-kyool) formed when two or more atoms join
monocyte (MON-oh-syte) large white blood cell; an agranular leukocyte
monogenic (mon-oh-JEN-ik) single-gene
monosaccharide (mon-oh-SAK-ah-ryde) simple sugar, such as glucose or fructose; building block of carbohydrates
mons pubis (monz PYOO-bis) skin-covered pad of fat over the pubic symphysis in the female
morula (MOR-yoo-lah) solid mass of cells formed by the divisions of a fertilized egg
motility (moh-TIL-i-tee) ability to move spontaneously
motor cranial nerve (KRAY-nee-al nerv) nerve that consists mainly of motor neurons
motor end plate (MOH-ter end playt) point at which motor neurons connect to the sarcolemma of a muscle cell to form the neuromuscular junction
motor nerve (MOH-ter nerv) nerve containing motor neurons and thus transmitting nerve impulses from the brain and spinal cord to muscles and glandular epithelial tissues
motor neuron (MOH-ter NOO-ron) transmits nerve impulses from the brain and spinal cord to muscles and glandular epithelial tissues
motor program set of coordinated commands that control the programmed muscle activity mediated by extrapyramidal pathways
motor unit (MOH-ter YOO-nit) functional unit composed of a single motor neuron with the muscle cells it innervates
mucosa (myoo-KOH-sah) innermost layer of the gastrointestinal wall
mucous membrane (MYOO-kus MEM-brayne) epithelial membrane that lines body surfaces opening directly to the exterior and secretes mucus
mucus (MYOO-kus) thick, slippery material secreted by mucous membranes that keeps the membrane moist and protected
multiaxial joint (mul-tee-AK-see-al joynt) joint that permits movement around three or more axes and in three or more planes
multiple sclerosis (MS) (MUL-tih-pul skleh-ROH-sis) most common primary disease of the central nervous system, MS leads to demyelination of nerves, which commonly causes problems with vision, muscle control, and incontinence
multipolar neuron (mul-ti-POL-ar NOO-ron) neuron with only one axon but several dendrites
multiunit smooth muscle type of smooth muscle tissue composed of many independent single-fiber units that does not usually generate its own impulse but rather responds only to nervous input and is often found in bundles
muscle (MUSS-el) type of tissue characterized by long, fiberlike cells capable of contracting with force
muscle fatigue (MUSS-el fah-TEEG) state of exhaustion produced by strenuous muscular activity
muscle spindle stretch receptor in muscle cells involved in maintaining muscle tone
muscle tissue (MUSS-el TISH-yoo) tissue type that produces movement
muscle tone tonic contraction; characteristic of muscle of a normal individual who is awake
muscular artery (MUSS-kyoo-lar AR-ter-ee) artery that carries blood farther away from the heart to specific organs and areas of the body; also called *distributing artery*
muscularis (muss-kyoo-LAIR-is) two layers of muscle surrounding the digestive tube that produce wavelike, rhythmic contractions, called *peristalsis,* that move food material along the digestive tract
muscular system (MUSS-kyoo-lar SIS-tem) the muscles of the body
mutagen (MYOO-tah-jen) agent capable of causing mutation (alteration) of DNA
myelin (MYE-eh-lin) lipoprotein substance in the myelin sheath around many nerve fibers that contributes to high-speed conductivity of impulses
myelinated fiber (MYE-eh-li-nay-ted FYE-ber) axon surrounded by a sheath of myelin formed by Schwann cells (PNS) or oligodendrocytes (CNS)
myelination (mye-eh-li-NAY-shun) the process of wrapping an axon with a sheath of myelin produced by Schwann cells
myelin disorder (MYE-eh-lin) any of several disorders characterized by loss or improper development of the myelin sheath that surrounds many axons of the nervous system
myelin sheath (MYE-eh-lin sheeth) whitish, fatty sheath surrounding nerve fibers; substance produced by Schwann cells
myeloid tissue (MYE-eh-loyd TISH-yoo) red bone marrow; type of soft, diffuse connective tissue; the site of hematopoiesis
myocardial infarction (MI) (mye-oh-KAR-dee-al in-FARK-shun) death of cardiac muscle cells resulting from inadequate blood supply, as in coronary thrombosis; also called "heart attack"
myocardium (mye-oh-KAR-dee-um) (*pl.,* myocardia [mye-oh-KAR-dee-ah]) muscle of the heart
myofibril (mye-oh-FYE-bril) very fine longitudinal fibers found in skeletal muscle cells; composed of thick and thin filaments
myofilaments (mye-oh-FIL-ah-ments) ultramicroscopic, threadlike structures found in myofibrils; composed of myosin (thick) and actin (thin)
myoglobin (MYE-oh-glo-bin) large protein molecule in the sarcoplasm of muscle cells that attracts oxygen and holds it temporarily
myometrium (mye-oh-MEE-tree-um) muscle layer in the uterus
myxedema (mik-seh-DEE-mah) firm swelling (edema) of the skin caused by deficiency of thyroid hormone in adults

N

naïve (nye-EVE) refers to a B or T cell that is inactive
naïve B cell (nye-EVE B sell) *see* **naïve**
nasal bone (NAY-zal bohn) bone that gives shape to the nose; forms upper bridge of the nose
nasal septum (NAY-zal SEP-tum) a partition that separates the right and left nasal cavities
nasopharynx (nay-zoh-FAIR-inks) uppermost portion of the tube just behind the nasal cavities
natural killer (NK) cell type of lymphocyte that kills many types of tumor cells
necrosis (neh-KROH-sis) death of cells in a tissue, often resulting from ischemia (reduced blood flow)
negative feedback (NEG-ah-tiv FEED-bak) feedback control system in which the level of a variable is changed in the direction opposite to that of the initial stimulus
neonatal period (nee-oh-NAY-tal PEER-ee-od) period of development immediately after birth
neonatology (nee-oh-nay-TOL-oh-jee) diagnosis and treatment of disorders of the newborn
neoplasm (NEE-oh-plaz-em) tumor or abnormal growth; may be benign or malignant
nephron (NEF-ron) anatomical and functional unit of the kidney, consisting of the renal corpuscle and the renal tubule
nerve (nerv) bundle of nerve fibers, plus surrounding connective tissue, located outside the brain or spinal cord
nerve impulse (nerv IM-puls) self-propagating wave of electrical depolarization that carries information along nerves; also called *action potential*
nervous system (NER-vus SIS-tem) brain, spinal cord, and nerves
nervous tissue (NER-vus TISH-yoo) tissue type consisting of neurons and glia that provides rapid communication and control of body function
neural network (NOOR-al) a network of interconnected neurons in nervous tissue, forming a web of pathways to process information
neurilemma (noo-ri-LEM-mah) sheath of Schwann; formed by the nonmyelinated out layer of the Schwann cell (neurolemmocyte) around the axon of a neuron
neuroendocrine system (noo-roh-EN-doh-krin SIS-tem) endocrine and nervous systems working in concert to perform communication, integration, and control within the body
neuroglia (noo-ROG-lee-ah) nonexcitable supporting cells of nervous tissue; more properly called *glia*
neuromuscular junction (NMJ) (noo-roh-MUSS-kyoo-lar JUNK-shen) point of contact between nerve endings and muscle fibers; *see* **motor end plate**
neuron (NOO-ron) nerve cell, including its processes (axons and dendrites)
neurosecretory tissue (noo-roh-SEK-reh-tor-ee TISH-yoo) modified neurons that secrete chemical messengers that diffuse into the bloodstream rather than across a synapse; for example, in the hypothalamus
neurotransmitter (noo-roh-trans-MIT-ter) chemical by which neurons communicate; the substance is released by a neuron, diffuses across the synapse, and binds to the postsynaptic neuron
neutron (NOO-tron) neutral subatomic particle located in the nucleus of an atom
neutrophil (NOO-troh-fil) white blood cell that stains readily with neutral dyes
nipple (NIP-el) small projection at the center of the areola on the anterior surface of the breast where the lactiferous ducts release milk
nociceptor (noh-see-SEP-tor) receptor activated by intense stimuli of any type that results in tissue damage; pain receptor
nodes of Ranvier (nohds ov rahn-vee-AY) short space in the myelin sheath between adjacent Schwann cells; also called *myelin sheath gap*
nonelectrolyte (nahn-ee-LEK-troh-lyte) compound that does not dissociate in solution; for example, glucose
nonkeratinized (nahn-KEHR-ah-tin-eyezd) lacking the protein keratin
nonspecific immunity (nahn-speh-SIF-ik i-MYOO-ni-tee) mechanisms that resist various threatening agents or conditions, not just certain specific agents; *see* **innate immunity**
nonsteroid hormone (nahn-STAYR-oyd HOR-mohn) hormone synthesized primarily from amino acids rather than from cholesterol
norepinephrine (NE, NR) (nor-ep-i-NEF-rin) hormone secreted by adrenal medulla that increases cardiac output; neurotransmitter released by sympathetic nervous system cells
nuclear envelope (NOO-klee-ar AHN-vel-ohp) the boundary of a cell's nucleus, made up of a double layer of cellular membrane
nuclease (NOO-klee-ayz) RNA and DNA digesting enzyme
nucleolus (noo-KLEE-oh-lus) dense, well-defined but membraneless body within the nucleus; critical to protein formation because it "programs" the formation of ribosomes in the nucleus
nucleosome (NOO-klee-oh-sohm) tightly wound subunit of chromatin and proteins (histones)
nucleotide (NOO-klee-oh-tide) monomer made up of three types of chemical groups (sugar, phosphate, nitrogen base) that can act alone or to make up a polymer (nucleic acid)
nucleus (NOO-klee-us) membranous organelle that contains most of the genetic material of the cell; also, group of neuron cell bodies in the brain or spinal cord
nutrition (noo-TRI-shun) foods we eat and the nutrients they contain

O

obligatory base pairing (oh-BLIG-ah-tor-ee bayse PAIR-ing) same nitrogen bases in the DNA structure pairing off with each other; adenine to thymine and cytosine to guanine
occipital bone (awk-SIP-it-al bohn) posterior and inferior bone of the skull
occipital lobe (awk-SIP-it-al) posterior and inferior lobe of the cerebrum
octet rule (ok-TET rool) general principle in chemistry whereby atoms usually form bonds in ways that will provide each atom with an outer shell of eight electrons
oculomotor nerve (awk-yoo-loh-MOH-tor nerv) cranial nerve III; motor nerve; controls eye movements
odorant molecule (OH-doh-rent MOL-eh-kyool) any molecule capable of being interpreted as an odor
older adulthood postnatal period after adulthood
olecranon (oh-LEK-rah-non) scoop-shaped process at distal end of ulna that articulates with the humerus at the elbow joint
olecranon fossa (oh-LEK-rah-non FOSS-ah) depression in the posterior surface of the humerus bone to allow room for the olecranon of the ulna when the elbow joint extends
olfactory cilia (ohl-FAK-tor-ee SIL-ee-ah) odor-sensitive cilia on the sensory neurons that detect smell
olfactory epithelium (ohl-FAK-tor-ee ep-i-THEE-lee-um) lining of the upper surface of the nasal cavity
olfactory nerve (ol-FAK-tor-ee nerv) cranial nerve I; sensory nerve; responsible for the sense of smell
oligodendrocyte (ohl-i-goh-DEN-droh-syte) small astrocyte with few cell processes; helps to form myelin sheaths around axons within the central nervous system
oogenesis (oh-oh-JEN-eh-sis) production of female gametes
oogonium (oh-oh-GOH-nee-um) primitive cell from which oocytes derive meiosis
ophthalmoscope (off-THAL-mah-skohp) instrument used to examine the retinal surface and internal eye structures
opsin (OP-sin) protein produced by the breakdown of rhodopsin in rods and cones of the retina; involved in a chemical reaction that initiates an impulse that results in interpretation of light energy as vision
optic chiasm (OP-tik kye-AS-mah) region where right and left optic nerves enter the brain and cross each other, exchanging fibers
optic disc (OP-tik disk) area in the retina where the optic nerve fibers exit the eye and where therefore no rods or cones are present; also known as a *blind spot*
optic nerve (OP-tik nerv) cranial nerve II; sensory nerve; nerves that conduct visual information to the brain
oral contraceptive (OR-al kon-tra-SEP-tiv) medication that controls sex hormone levels and ovulation
oral rehydration therapy (ORT) (OR-al ree-hye-DRAY-shun THER-ah-pee) treatment of infant

diarrhea by the administration of a liberal dose of sugar and salt solution
organ (OR-gan) group of several tissue types that together perform a special function
organelle (or-gah-NELL) any of many cell "organs" or organized structures; for example, a ribosome or mitochondrion
organic (or-GAN-ik) referring to chemicals that contain covalently bound carbon and hydrogen atoms and are involved in metabolic reactions
organism (OR-gah-niz-em) any living entity considered as a whole; may be unicellular (one-celled) or composed of many different cells and body systems working together to maintain life
organ of Corti (OR-gan of KOR-tee) the organ of hearing located in the cochlea and filled with endolymph; also called *spiral organ*
organogenesis (or-gah-noh-JEN-eh-sis) formation of organs from the primary germ layers of the embryo
orgasm (OR-gaz-um) sexual climax
origin (OR-i-jin) attachment of a muscle to the bone, which does not move when contraction occurs; compare with **insertion**
oropharynx (oh-roh-FAIR-inks) the portion of the pharynx that is located behind the mouth
osmoreceptor (os-moh-ree-SEP-tor) special receptor near the supraoptic nucleus that detects decreased osmotic pressure of blood when the body dehydrates
osmosis (os-MOH-sis) movement of a fluid (usually water) through a semipermeable membrane from an area of lesser solute concentration to an area of greater concentration
osmotic pressure (os-MOT-ik PRESH-ur) water pressure that develops in a solution across a semipermeable membrane as a result of osmosis
ossification (os-i-fi-KAY-shun) bone formation
osteoblast (OS-tee-oh-blast) bone-forming cell
osteoclast (OS-tee-oh-klast) bone-absorbing cell
osteocyte (OS-tee-oh-syte) bone cell
osteogenesis (os-tee-oh-JEN-eh-sis) combined action of osteoblasts and osteoclasts to mold bones into adult shape
osteon (OS-tee-on) unit of compact bone tissue made up of a tapered cylinder with layered, concentric arrangements of calcified matrix and cells around a central canal for nerves and blood vessels; also called *haversian system*
otolith (OH-toh-lith) tiny "ear stone" composed of protein and calcium carbonate in the maculae of the ear, which, by responding to gravity and changes in body position, triggers hair cells that initiate impulses resulting in sense of balance
oval window (OH-val WIN-doh) small, membrane-covered opening that separates the middle and inner ear
ovarian follicle (oh-VAIR-ee-an FOL-i-kul) spherical configuration of cells in the ovary that contains a single oocyte
ovarian medulla (oh-VAIR-ee-an meh-DUL-ah) inner region of the ovary; contains supportive connective tissue cells, blood vessels, nerves, and lymphatics
ovaries (OH-var-ees) female gonads that produce ova (sex cells)
ovulation (ov-yoo-LAY-shun) release of an ovum from the ovary at the end of oogenesis
ovum (OH-vum) female sex cell (gamete)
oxidative phosphorylation (ahk-si-DAY-tiv fos-for-i-LAY-shun) reaction that joins a phosphate group to ADP to form ATP
oxygen debt (AHK-si-jen det) additional oxygen required for ATP synthesis to remove excess lactic acid following anaerobic exercise; also called *excess postexercise oxygen consumption (EPOC)*
oxytocin (OT) (ahk-see-TOH-sin) hormone secreted by the posterior pituitary gland before and after delivering a baby; thought to initiate and maintain labor, as well as cause the release of breast milk into ducts of the mammary glands

P

palatine bone (PAL-ah-tine bohn) bone that forms the posterior part of the hard palate and the lateral wall of the posterior part of each nasal cavity
palatine tonsils (PAL-ah-tine TAHN-sils) tonsils located behind and below the pillars of the fauces
pancreatic islets (pan-kree-AT-ik eye-lets) endocrine portion of the pancreas; made up of alpha and beta cells, among others; also called *islets of Langerhans*
pancreatic juice (pan-kree-AT-ik) digestive secretion; secreted by the exocrine acinar cells of the pancreas
pancreatic polypeptide (pan-kree-AT-ik pol-ee-PEP-tyde) hormone produced in the periphery of the pancreatic islets; influences digestion and distribution of food molecules
papillae (pah-PIL-ee) (*sing.*, papilla [pah-PIL-ah]) small, nipple-shaped elevations
papillary (PAP-i-lair-ee) relating to papillae, as in the papillary layer of the dermis located at the dermal papillae
papillary muscle (PAP-i-lair-ee MUSS-el) finger-like projections of the heart's myocardial muscle that help anchor the atrioventricular valves during ventricular contraction
paracrine hormone (PAIR-ah-krin HOR-mohn) hormone that regulates activity in nearby cells within the same tissue as their source
paralysis (pah-RAL-i-sis) loss of the power of motion, especially voluntary motion
paranasal sinuses (pair-ah-NAY-sal SYE-nus-ez) air-containing spaces that are connected by channels to the nasal cavity
parasympathetic division (pair-ah-sim-pah-THET-ik di-VI-zhun) part of the autonomic nervous system; ganglia are connected to the brainstem and the sacral segments of the spinal cord; controls many autonomic effectors under normal ("rest and repair") conditions
parasympathetic nervous system (PNS) (pair-ah-sim-pah-THET-ik NER-vus SIS-tem) part of the autonomic nervous system; ganglia are connected to the brainstem and the sacral segments of the spinal cord; controls many visceral effectors under normal conditions
parathyroid glands (pair-ah-THYE-royd) endocrine glands located in the neck on the posterior aspect of the thyroid gland; secrete parathyroid hormone
parathyroid hormone (PTH) (pair-ah-THYE-royd HOR-mohn) hormone released by the parathyroid gland that increases concentration of calcium in the blood
parenteral therapy (pah-REN-ter-al THER-ah-pee) administration of nutrients, special fluids, and/or electrolytes by injection or means other than through the digestive tract
parietal (pah-RYE-i-tal) refers to the walls of an organ or cavity; opposite of **visceral**
parietal bone (pah-RYE-i-tal bohn) a cranial bone between the frontal and occipital bones and forming the superior, lateral wall of the cranium
parietal cell (pah-RYE-i-tal sell) cell located on the basement membrane of gastric glands of the stomach that secretes hydrochloric acid
parietal lobe (pah-RYE-i-tal) lobe that occupies the lateral and medial surface of the cerebrum
parietal pleura (pah-RYE-i-tal PLOOR-ah) serous membrane that lines the entire thoracic cavity
Parkinson disease (PARK-in-son di-ZEEZ) nervous disorder characterized by abnormally low levels of the neurotransmitter dopamine in parts of the brain that control voluntary movement—victims usually exhibit involuntary trembling and muscle rigidity
p-arm (PEE-arm) the short segment of a chromosome that is divided into two segments by a centromere
parotid (peh-RAH-tid) largest of the paired salivary glands
partial pressure (PAR-shal PRESH-ur) pressure exerted by any one gas in a mixture of gases or in a liquid
parturition (pahr-too-RI-shun) act of giving birth
passive transport (PASS iv TRANZ-port) cellular process in which substances move through a cellular membrane with their own energy supplied directly by the cell or its membrane
patella (pah-TEL-ah) knee cap
pectoral girdle (PEK-toh-ral GIR-dul) incomplete ring of bones formed by the scapulae and clavicles and providing structural support for the upper limbs; the shoulder girdle
pedigree (PED-i-gree) chart used in genetic counseling to illustrate genetic relationships over several generations
pelvic girdle (PEL-vik GIR-dul) bony structure formed by the coxal bones; supports the trunk and attaches the lower extremities to it
pelvis (PEL-vis) basin- or funnel-shaped structure
penis (PEE-nis) mass of erectile tissue; forms part of the male genitalia, when sexually aroused, becomes stiff to enable it to enter and deposit sperm in the vagina
pepsin (PEP-sin) protein-digesting enzyme
pepsinogen (pep-SIN-oh-jen) proenzyme that is converted to pepsin by hydrochloric acid
peptidase (PEP-tyd-ayz) protease found in the intestinal brush border; hydrolyzes peptides to amino acids
peptide bond (PEP-tyde) bond that forms between the amino group of one amino acid and the carboxyl group of another
perception interpreting a sensation
pericardial fluid (pair-i-KAR-dee-al FLOO-id) lubricating fluid secreted by the serous membrane; in pericardial space
pericardial space (pair-i-KAR-dee-al) space between the visceral layer and the parietal layer surrounding the heart, filled with pericardial fluid
pericardium (pair-i-KAR-dee-um) membrane that surrounds the heart
perichondrium (pair-i-KON-dree-um) fibrous covering of cartilage structures
perimetrium (pair-i-MEE-tree-um) serous covering that partly covers the uterus; a portion of the parietal peritoneum
perimysium (pair-i-MEE-see um) tough, connective tissue surrounding fascicles
perineum (pair-i-NEE-um) area between anus and genitals
periosteum (pair-ee-OS-tee-um) tough, connective tissue covering the bone
peripheral (peh-RIF-er-all) relating to the periphery, or outer boundaries of the body, as in peripheral nervous system
peripheral nervous system (PNS) (peh-RIF-er-all NER-vus SIS-tem) nerves connecting brain and spinal cord to other parts of the body
peripheral resistance (peh-RIF-er-all ree-SIS-tens) resistance to blood flow caused by friction of blood passing through blood vessels
peristalsis (pair-i-STAWL-sis) wavelike, rhythmic contractions of the stomach and intestines that move food material along the digestive tract
peritoneum (pair-i-toh-NEE-um) large, moist, slippery sheet of serous membrane that lines the abdominopelvic cavity (parietal layer) and its organs (visceral layer)
peritonitis (pair-i-toh-NYE-tis) inflammation of the serous membranes in the abdominopelvic cavity; sometimes a serious complication of an infected appendix
peritubular capillary (pair-ee-TOO-byoo-lar KAP-i-lair-ee) capillary around the tubules in the kidney
permanent teeth set of 32 teeth that replaces deciduous teeth
permeate (PERM-ee-ayt) to move through or soak through, as water moving through the pores of a fabric or molecules diffusing through the pores of a cellular membrane
peroxisome (peh-ROKS-i-sohm) organelles that detoxify harmful substances that have entered cells
pH (p H) units by which acid and base concentrations (relative H^+ ion concentrations) are measured; scale ranges from 0 (extremely acidic; high H^+ concentration) to 14 (extremely basic, or alkaline; low H^+ concentration)
phagocytosis (fag-oh-sye-TOH-sis) ingestion and digestion of particles by a cell
phalanx (FAH-lanks) (*pl.*, phalanges [fah-LAN-jeez]) any of the bones of the fingers or toes
pharyngeal tonsil (fah-RIN-jee al TAHN-sil) tonsil located in the nasopharynx on its posterior wall; when enlarged, referred to as adenoids
pharynx (FAIR-inks) tubelike structure that extends from the base of the skull to the esophagus; also called *throat*
phenotype (FEE-noh-type) overt, observable expression of a genotype
phospholipid (fos-foh-LIP-id) phosphate-containing fat molecule; an important constituent of cell membranes
phosphorylation (fos-for-i-LAY-shun) process of adding a phosphate group to a molecule
photopigments (foh-toh-PIG-ments) chemicals in retinal cells that are sensitive to light
photoreceptor (FOH-toh-ree-sep-tor) receptor only in the eye; responds to light stimuli if the intensity is great enough to generate a receptor potential
phrenic nerve (FREN-ik nerv) nerve that stimulates the diaphragm to contract
physiological polycythemia (fiz-ee-oh-LOJ-i-kal pol-ee-sye-THEE-mee-ah) elevated red blood cell numbers and hematocrit values in healthy individuals who live and work in high altitudes
physiology (fiz-ee-OL-oh-jee) scientific study of an organism's body function
pia mater (PEE-ah MAH-ter) vascular innermost covering (meninx) of the brain and spinal cord
pineal body (PIN-ee-al BOD-ee) endocrine gland located in the diencephalon and thought to be involved with regulating the body's biological clock; produces melatonin; also called *pineal gland*
pineal gland (PIN-ee-al) *see* **pineal body**
pinocytosis (pin-oh-sye-TOH-sis) active transport mechanism used to transfer fluids or dissolved substances into cells
pitting edema (pitt-ing eh-DEE-mah) depressions in swollen subcutaneous tissue
pivot joint (PIV-ot joynt) a freely moving joint in which a bone rotates around another, permitting only rotational movement
placenta (plah-SEN-tah) structure that anchors the developing fetus to the uterus and provides a "bridge" for the exchange of nutrients and waste products between the mother and developing baby
plantar reflex (PLAN-tar REE-fleks) reflex in which the toes curl in flexion in response to stimulation of the outer margin of the foot; the Babinski sign is extension of the great toe (with our without fanning of the other toes) instead (normal in infants)
plasma (PLAZ mah) liquid part of the blood
plasma cell (PLAZ-mah sell) *see* **B lymphocyte**
plasma membrane (PLAZ-mah MEM-brayne) membrane that separates the contents of a cell from the tissue fluid, encloses the cytoplasm, and forms the outer boundary of the cell
plasmid (PLAZ-mid) small circular ring of bacterial DNA
plasminogen (plaz-MIN-oh-jen) an inactive form of plasmin
platelet (PLAYT-let) flattened cell fragment found in the blood that functions in hemostasis; thrombocyte
platelet plug (PLAYT-let) results when platelets undergo a change caused by an encounter with a damaged capillary wall, or with underlying connective tissue fibers; helps to stop the flow of blood into the tissues
pleura (PLOOR-ah) serous membrane in the thoracic cavity
plexus (PLEK-sus) complex network formed by converging and diverging nerves, blood vessels, or lymphatics
pneumothorax (noo-moh-THOH-raks) abnormal condition in which air is present in the pleural space surrounding the lung, possibly causing collapse of the lung
podocytes (POD-oh-sytes) special epithelial cells making up the visceral layer of the Bowman capsule
polar molecule (PO-lar MOL-eh-kyool) molecule in which the electrical charge is not evenly distributed, causing one side of the molecule to be more positive or negative than the other
polygenic (pahl-ee-JEN-ik) refers to traits that are determined by the combined effect of many different pair of genes
polyribosome (pahl-ee-RYE-boh-sohm) temporary structure formed within a cell by many ribosomes following one another along a strand of mRNA during the process of translation
polysaccharide (pahl-ee-SAK-ah-ryde) any large carbohydrate compound composed of many linked monosaccharide units
pons (ponz) part of the brainstem between the medulla oblongata and the midbrain
popliteal vein (pop-li-TEE-al vayn) vein that runs behind the knee joint
portal system (POR-tal SIS-tem) arrangement of blood vessels in which blood exiting one tissue is immediately carried to a second tissue before being returned to the heart and lungs for oxygenation and redistribution
positive feedback (POZ-it-iv FEED-bak) feedback control system that is stimulatory; tends to amplify or reinforce a change in the internal environment
posterior (pohs-TEER-ee-or) located behind; opposite of **anterior**
posterior cavity (pohs-TEER-ee-or KAV-i-tee) any cavity in a posterior (rear) region of the body or an organ; for example, the posterior cavity of the eye is a gel-filled space behind the lens
posterior fornix (pohs-TEER-ee-or FOR-niks) space created by the protrusion of the cervix into the lumen of the vagina; increases probability of reproductive success by retaining seminal fluid
posterior pituitary gland (pohs-TEER-ee-or pi-TOO-i-tair-ee) neurohypophysis; produces hormones ADH and oxytocin
postganglionic neuron (post-gang-glee-ON-ik NOO-ron) efferent autonomic neuron that conducts nerve impulses from a ganglion to effectors such as cardiac or smooth muscle or glandular epithelial tissue
postnatal period (post-NAY-tal PEER-ee-od) period after birth, ending at death

postsynaptic neuron (post-si-NAP-tik NOO-ron) in neuron-to-neuron communication, the neuron that receives a stimulus via an adjacent neuron's transmission of neurotransmitters across the synapse

posture (POS-chur) position of the body; often refers to the erect position of the body maintained unconsciously

PP cell cell in the periphery of the pancreatic islet that produces pancreatic polypeptide (a hormone)

precapillary sphincter (pree-KAP-pi-lair-ee SF-INGK-ter) smooth muscle cells that guard the entrance to the capillary

preganglionic neuron (pree-gang-glee-ON-ik NOO-ron) efferent autonomic neuron that conducts nerve impulses between the spinal cord and a ganglion

prenatal period (PREE-nay-tall PEER-ee-od) developmental period after conception until birth

prepuce (PREE-pus) foreskin, especially the covering fold of skin over the glans penis

primary bronchi (PRYE-mayr-ee BRONG-kye) first branches of the trachea (right and left primary bronchi)

primary follicles (PRYE-mayr-ee FOL-i-kuls) follicles present at puberty; formed from granulosa cells

primary germ layer (PRYE-mayr-ee jerm LAY-er) any of the three layers of developmental cells that give rise to definite structures as the embryo develops; *see* **ectoderm, endoderm, mesoderm**

primary oocyte (PRYE-mayr-ee OH-oh-syte) developmental stage of the ova; formed from oogonia during the fetal period

primary ossification center (PRYE-mayr-ee os-i-fi-KAY-shun) initial bone formation center of a developing bone; *see* **secondary ossification center**

primary principle of circulation (PRYE-mair-ee PRIN-sip-al ov ser-kyoo-LAY-shun) the concept that blood flows from an area of high pressure to an area of low pressure

primary principle of ventilation (PRYE-mayr-ee PRIN-sip-al ov ven-ti-LAY-shun) movement of air in the pulmonary airways from the area where the pressure is higher to the area where the pressure is lower

primary sexual characteristics (PRYE-mayr-ee SEK-shoo-al kair-ak-ter-ISS-tiks) changes that involve the development of the gonads

primary structure (PRYE-mayr-ee STRUK-cher) the amino acid sequence of the peptide chains

principle of independent assortment genetic principle that states that, as chromosomes pairs separate, the maternal and paternal chromosomes redistribute themselves independently of the other chromosome pairs

principle of segregation genetic principle that states that the two members of a pair of chromosomes separate during meiosis

proenzyme (proh-EN-zyme) inactive form in which many enzymes are synthesized

progeria (proh-JEE-ree-ah) rare, inherited condition in which a person appears to age rapidly as a result of abnormal, widespread degeneration of tissues; adult and childhood forms exist, with the childhood form resulting in death by age 20 or so

progesterone (proh-JES-ter-ohn) steroid hormone produced by the ovaries (particularly the corpus luteum) that helps prepare the uterus for implantation; along with estrogen, helps maintain normal uterine and mammary gland function

prolactin (PRL) (proh-LAK-tin) hormone secreted by the anterior pituitary gland during pregnancy to stimulate the breast development needed for lactation

proliferative phase (PROH-lif-eh-rah-tiv fayz) phase of menstrual cycle that begins after the menstrual flow ends and lasts until ovulation

prophase (PROH-fayz) first stage of mitosis during which chromosomes become visible

proprioceptors (proh-pree-oh-SEP-tors) perception of movement and position of the body

propulsion (proh-PUL-shen) continual pushing of chyme in the stomach toward the pyloric sphincter by peristaltic contractions

prostaglandins (PGs) (pross-tah-GLAN-dins) any of a group of naturally occurring lipid-based substances that act in a hormone-like way to affect many body functions, including vasodilation, uterine smooth muscle contraction, and the inflammatory response

prostate gland (PROSS-tayt) exocrine gland that lies just below the bladder in the male; secretes a fluid that constitutes about 30% of the seminal fluid volume; helps activate sperm and helps them maintain motility

prosthesis (pross-THEE-sis) artificial device that is used in the partial or total replacement of a diseased joint

protease (PROH-tee-ayz) enzyme that catalyzes the hydrolysis of proteins into intermediate compounds (proteoses and peptides)

proteasome (PROH-tee-ah-sohm) cell structure that breaks down individual proteins

protein (PROH-teen) large molecules formed by linkage of amino acids by peptide bonds; one of the basic building blocks of the body

protein hormone (PROH-teen HOR-mohn) long, folded chain of amino acids; for example, insulin and parathyroid hormone

protein synthesis (PROH-teen SIN-the-sis) the process by which proteins are manufactured; includes *transcription* and *translation* of genetic information

proteome (PROH-tee-ohm) the entire group of proteins produced by a cell or by the entire body

proteomics (proh-tee-OH-miks) the endeavor that involves the analysis of the proteins encoded by the genome, with the ultimate goal of understanding the role of each protein in the body

prothrombin (proh-THROM-bin) a protein present in normal blood that is required for blood clotting

proton (PROH-ton) positively charged subatomic particle

proximal (PROK-si-mal) next or nearest; located nearest the center of the body or the point of attachment of a structure; opposite of **distal**

proximal convoluted tubule (PCT) (PROK-si-mal kon-voh-LOO-ted TOOB-yool) second part of the nephron and the first segment of a renal tubule; major portion of the *proximal tubule*

pseudogene (SOOD-oh-jeen) nonfunctional "broken" genetic code found in "junk DNA" located between the functioning, coding genes of a DNA molecule; pseudogenes are thought to be genetic "fossils" remaining from our evolutionary past

pseudostratified columnar epithelium (SOOD-oh-STRAT-i-fyed KAHL-um-nar ep-i-THEE-lee-um) type of tissue similar to simple columnar epithelium; forms a membrane made up of a single layer of cells that are tall and narrow but have been squeezed together in a way that pushes the nuclei into two layers and thus gives the appearance that it is stratified

puberty (PYOO-ber-tee) stage of adolescence in which a person becomes sexually mature

pubis (PYOO-biss) most anterior coxal bone

pulmonary circulation (PUL-moh-nair-ee ser-kyoo-LAY-shun) blood flow from the right ventricle to the lung and returning to the left atrium

pulp cavity (KAV-i-tee) cavity in the dentin of a tooth that contains connective tissue, blood and lymphatic vessels, and sensory nerves

pulse rhythm of alternating expansion and contraction of the arteries created by the contractions of the heart

pulse pressure difference between systolic and diastolic blood pressure

pulse wave wave of alternating expansion and recoil of the arterial wall resulting from the alternate contraction and relaxation of the ventricles

punctum (PUNK-tum) opening into the lacrimal canals located at the inner canthus of the eye

Punnett square (PUN-it skwair) grid used in genetic counseling to determine the probability of inheriting genetic traits

pupil (PYOO-pill) opening in the center of the iris that regulates the amount of light entering the eye

pus accumulation of white blood cells, dead bacterial cells, and damaged tissue cells at site of an infection

P wave electrocardiogram deflection that represents depolarization of the atria

pyloric sphincter (pye-LOR-ik SFINGK-ter) sphincter (muscular valve) that prevents food from leaving the stomach and entering the duodenum

pylorus (pye-LOR-us) lower part of the stomach

pyramid (PEER-ah-mid) one of two bulges of white matter located on the ventral surface of the medulla

Q

q-arm (KYU-arm) the longer segment of the chromosome, which is divided into two "arms" by the centromere (the shorter segment is called the p-arm)

QRS complex (Q-R-S KOM-pleks) electrocardiogram deflection that represents depolarization of the ventricles

quaternary structure (KWAH-ter-nair-ee STRUK-cher) the quaternary structure of a protein, which is the fourth level of complexity in the structure of a protein molecule; a quaternary protein is a protein possessing a fourth level of complexity in its molecular structure

R

radical mastectomy (RAD-i-kal mas-TEK-toh-mee) procedure that removes the entire breast, nearby muscles, and lymph nodes

radius (RAY-dee-us) forearm bone; located on thumb side

ramus (RAY-mus) large branch of a spinal nerve as it emerges from the spinal cavity

reabsorption (ree-ab-SORP-shun) movement of substances back into the bloodstream, as when bone calcium is dissolved by osteoclasts and diffuses into the bloodstream, or when solutes in renal filtrate (in the nephron) move back into the bloodstream

receptor-mediated endocytosis (ree-SEP-tor-MEE-dee-ayt-ed en-doh-sye-TOH-sis) process in which cells use chemical receptor molecules embedded in the plasma membrane to trigger the engulfment of particular substances

receptor potential (ree-SEP-tor poh-TEN-shal) potential that develops in the receptor's membrane when an adequate stimulus has been received

recessive gene (ree-SES-iv) in genetics, refers to genes that have effects that do not appear in the offspring when they are masked by a dominant gene (recessive forms of a gene are represented by lowercase letters); *see* **dominant gene**

rectum (REK-tum) distal region of the intestinal tube

referred pain pain felt on or near the surface of the body that results from stimulation of nociceptors in deep structures; for example, experiencing pain in the left arm when heart muscle receptors signal pain, as during a heart attack

reflex (REE-fleks) automatic involuntary reaction to a stimulus resulting from a nerve impulse passing over a reflex arc

reflex arc (REE-fleks ark) impulse conduction route to and from the central nervous system; smallest portion of nervous system that can receive a stimulus and generate a response

refraction (ree-FRAK-shun) bending of a ray of light as it passes from a medium of one density to one of a different density; occurs as light rays pass through the eye

refractory period (ree-FRAK-tor-ee PEER-ee-od) the length of time it takes for an excitable membrane to react to a second stimulus

regeneration (ree-jen-er-AY-shun) process of replacing missing tissue with new tissue by means of cell division

relaxin (reh-LAK-sin) hormone that inhibits contractions during pregnancy and softens pelvic joints to facilitate childbirth

renal clearance (REE-nal) amount of a substance removed from the blood by the kidneys per minute

renal columns (REE-nal KAH-lums) within the kidneys, the cortical tissue in the medulla between the pyramids

renal corpuscle (REE-nal KOR-pus-ul) within the nephron, the glomerulus plus the Bowman capsule surrounding it

renal cortex (REE-nal KOHR-teks) outer portion of the kidney

renal medulla (REE-nal meh-DUL-ah) inner portion of the kidney

renal pelvis (REE-nal PEL-vis) basinlike upper end of the ureter that is located inside the kidney into which the calyces drain

renal pyramid (REE-nal PEER-ah-mid) any of the distinct triangular wedges that make up most of the medullary tissue in the kidney

renin (REE-nin) enzyme produced by the juxtaglomerular apparatus of the kidney nephrons that catalyzes the formation of angiotensin, a substance that increases blood pressure; *see* **renin-angiotensin-aldosterone system (RAAS)**

renin-angiotensin-aldosterone system (RAAS) (REE-nin-an-jee-oh-TEN-sin-al-DAH-stair-ohn SIS-tem) causes changes in blood plasma volume mainly by controlling aldosterone secretion

repolarization (ree-poh-lah-ri-ZAY-shun) phase of the action potential in which the membrane potential changes from its maximum degree of depolarization toward the resting state potential

residual volume (RV) (reh-ZID-yoo-al VOL-yoom) amount of air that remains in the lungs after the most forceful expiration

respiratory center (RES-pi-rah-tor-ee) control center located in the medulla and pons that stimulates muscles of respiration

respiratory mucosa (RES-pi-rah-tor-ee myoo-KOH-sah) mucus-covered membrane that lines tubes of the respiratory tree

respiratory portion (RES-pi-rah-tor-ee POR-shun) area within the nasal passage that extends from the inferior meatus to the small funnel-shaped orifices of the *posterior nares*

rest-and-repair the name of the parasympathetic division of the autonomic nervous system, referring to normal, nonstressful situations

resting membrane potential (RMP) (MEM-brayne poh-TEN-shal) membrane potential maintained by a nonconducting neuron's plasma membrane; approximately 70 mV

reticular (reh-TIK-yoo-lar) resembling a netlike pattern

reticular activating system (RAS) (reh-TIK-yoo-lar SIS-tem) complex processing system in the neural network (reticular formation) in the brain responsible for maintaining consciousness

reticular fiber (reh-TIK-yoo-lar FYE-ber) delicate fiber of the extracellular matrix made of a type of collagen called reticulin

reticular tissue (reh-TIK-yoo-lar TISH-yoo) meshwork of netlike tissue that forms the framework of the spleen, lymph nodes, and bone marrow

retina (RET-i-nah) innermost layer of the eyeball; contains rods and cones and continues posteriorly with the optic nerve

retinal (RET-i-nal) light-absorbing portion of all photopigments

retropulsion (ret-roh-PUL-shun) process of chyme being forced to move backward behind a closed pyloric sphincter

reversible reaction (ree-VER-si-buhl ree-AK-shun) when the products of a chemical reaction change back to the original reactants; generally, an equilibrium of products and reactants exists

Rh antigen (R-H AN-ti-jen) category of membrane antigens that may occur on the surfaces of red blood cells in some individuals; named for Rhesus monkeys, in which they were first discovered; also called D antigen

rhodopsin (roh-DOP-sin) photopigment in rods

ribonucleic acid (RNA) (rye-boh-noo-KLAY-ik ASS-id) nucleic acid found in both nucleus and cytoplasm of cells; involved in transmission of genetic information from nucleus to cytoplasm and in cytoplasmic assembly of proteins

ribosome (RYE-boh-sohm) organelle in the cytoplasm of cells that synthesizes proteins; sometimes called "protein factory"

right lobe the right portion of an organ, such as the liver

right lymphatic duct (lim-FAT-ik dukt) main lymphatic duct

RNA interference (RNAi) a regulatory process of the cell in which a small molecule of dsRNA (double-stranded RNA) called *siRNA* (short interfering RNA) joins with a RISC (RNA-induced silencing complex) protein structure to break down a specific mRNA (messenger RNA) transcript and thus effectively silence the gene encoded by the mRNA; RNAi is a natural regulatory process thought to be involved with regulating gene expression, as in inhibiting viral infections, but is also used as a research technique to study the human genome

RNAi therapy any medical procedure in which RNAi techniques are used to silence (disable) the effects of a disease-causing gene; *see also* **RNA interference (RNAi)**

rods photoreceptor cells responsible for night vision

root blunt tip of the tongue; portion of the tooth that fits into the socket of the alveolar process of either the upper or lower jaw

rotator cuff (roh-TAY-tor) musculotendinous cuff resulting from fusion of the tendons of the supraspinatus, infraspinatus, teres minor, and subscapularis (SITS muscles); adds to the stability of the glenohumeral (shoulder) joint

rough endoplasmic reticulum (ruf en-do-PLAZ-mik reh-TIK-yoo-lum) an organelle that forms a network of tubules and vesicles within a cell

round window opening into inner ear; covered by a membrane

rule of nines frequently used method to estimate extent of a burn injury in an adult; the body is divided into areas that are multiples and fractions of 9%

S

sacral plexus (SAY-kral PLEK-sus) plexus formed by fibers from the fourth and fifth lumbar nerves and the first four sacral nerves

sacrum (SAY-krum) bone of the lower vertebral column between the last lumbar vertebra and the coccyx, formed by the fusion of five sacral vertebrae

saddle joint (SAD-el joynt) a biaxial joint that can exhibit a variety of movements except axial rotation

sagittal plane (SAJ-i-tal playn) longitudinal plane that divides the body or a part into left and right sides

saliva (sah-LYE-vah) secretion of the salivary glands that is made up of water, mucus, amylase, sodium bicarbonate, and lipase

salpingitis (sal-pin-JYE-tis) inflammation of the uterine tubes

saltatory conduction (SAL-tah-tor-ee kon-DUK-shun) process in which a nerve impulse travels along a myelinated fiber by jumping from one node of Ranvier to the next

sarcolemma (sar-koh-LEM-ah) plasma membrane of a striated muscle fiber

sarcomere (SAR-koh-meer) contractile unit of muscle cells; length of a myofibril between two Z disks

sarcoplasm (SAR-koh-plaz-em) cytoplasm of muscle fibers

sarcoplasmic reticulum (SR) (sar-koh-PLAZ-mik reh-TIK-yoo-lum) network of tubules and sacs in muscle cells; similar to endoplasmic reticulum of other cells

satiety center (sah-TYE-eh-tee) cells in the hypothalamus that send impulses to decrease appetite so that an individual feels satisfied; *see also* **leptin**

saturated (SACH-yoo-ray-ted) condition when all available bonds of a hydrocarbon chain are filled with hydrogen atoms

scapula (SKAP-yoo-lah) upper extremity bone; shoulder blade

scar (skahr) dense fibrous mass of tissue that remains after a damaged tissue has been repaired

Schwann cells (shwon sells) any of the large nucleated cells that form myelin around the axons of neurons; also called *neurilemmocytes*

scientific method (sye-en-TIF-ik METH-od) any logical and systematic approach to discovering principles of nature, often involving testing of tentative explanations called *hypotheses*

scrotum (SKROH-tum) pouchlike sac that contains the testes

seasonal affective disorder (SAD) mental disorder in which a patient suffers severe depression only in winter; linked to the pineal gland

sebaceous gland (seh-BAY-shus) any of the oil-producing glands in the skin

sebum (SEE-bum) secretion of sebaceous glands

secondary bronchus (SEK-on-dair-ee BRON-kus) branch of the pulmonary airway formed when the primary bronchus enters the lung on its respective side and immediately splits into smaller bronchioles

secondary oocyte (SEK-on-dair-ee OH-oh-site) oocyte that arises form a primary oocyte after it completes meiosis I

secondary ossification center (SEK-on-dair-ee os-i-fi-KAY-shun) bone formation center of developing bone that is formed after an initial bone formation; *see* **primary ossification center**

secondary structure (SEK-on-dair-ee STRUK-cher) folded, three-dimensional structure of biological polymers such as nucleic acids and proteins from which a tertiary structure is formed

secondary sexual characteristics (SEK-on-dair-ee SEK-shoo-al kair-ak-ter-ISS-tiks) external physical characteristics of sexual maturity resulting from action of sex hormones; include male and female patterns of body hair and fat distribution, as well as development of external genitals

second-degree burn burn involving deep epidermal layers of the skin but not causing irreparable damage

second messenger model model that explains nonsteroid hormone mechanism of action; nonsteroid hormone is "first messenger" acting on cell membrane; intracellular "second messenger"—often cyclic AMP—triggers specific cellular action; also called *fixed-membrane-receptor model*

secretion (seh-KREE-shun) process by which a substance is released outside the cell

secretory phase (SEK-reh-tor-ee fayz) phase of menstrual cycle that begins at ovulation and lasts until the next menses begins

segmentation (seg-men-TAY-shun) occurs when digestive reflexes cause a forward-and-backward movement within a single region of the GI tract

selectively permeable (sel-EK-tiv-lee PERM-ee-ah-bil) adjective used to describe a living membrane that allows only certain substances to move through (permeate) it and only at certain times

self-tolerance ability of our immune system to attack abnormal or foreign cells but spare our own normal cells

semen (SEE-men) ejaculate from the penis that contains spermatozoa plus fluids from the testes, seminal vesicles, bulbourethral glands, and prostate

semicircular canals (sem-i-SIR-kyoo-lar kah-NALS) three bony, tubelike structures located in the temporal bone, making up part of the inner ear; each structure contains a membranous semicircular duct that functions in the sense of equilibrium

semilunar (SL) valve (sem-i-LOO-nar valv) valve located between each ventricle and the large artery that carries blood away from it; valves in the veins are sometimes referred to as *semilunar valves*

seminal vesicle (SEM-i-nal VES-i-kul) highly convoluted pouch that secretes an alkaline, viscous, creamy-yellow liquid that constitutes about 60% of semen volume

senescence (seh-NES-enz) older adulthood; aging

sensation interpretation of sensory nerve impulses by the brain as an awareness of an internal or external event; for example, feeling pain

sensory cranial nerve (SEN-sor-ee KRAY-nee-al nerv) cranial nerve that consists of only sensory axons

sensory nerve (SEN-soh-ree nerv) nerve that contains primarily sensory neurons

sensory pathway (SEN-soh-ree PATH-way) part of the nervous system that transmits impulses to the spinal cord and brain from all parts of the body

septum (SEP-tum) a wall that divides two areas; for example, nasal septum

serosa (seh-ROH-sah) outermost covering of the digestive tract; composed of the parietal pleura in the abdominal cavity

serous membrane (SEER-us MEM-brayne) two-layer epithelial membrane that lines body cavities and covers surfaces of organs

serous pericardium (SEER-us pair-i-KAR-dee-um) part of pericardial coverings of the heart; made up of a parietal layer and a visceral layer

sesamoid bones (SES-ah-moyd bohns) small seed-shaped bones embedded in tendons; the patella is the largest and most consistently found in the human skeleton

set point normal reading or range of normal

sex chromosomes (seks KRO-moh-sohms) pair of chromosomes in the human genome that determine gender; normal males have one X chromosome and one Y chromosome (XY); normal females have two X chromosomes (XX)

sex hormone (seks HOR-mohn) hormone that targets reproductive tissue; examples include estrogen, progesterone, and testosterone

short bone cube- or box-shaped bone that is about as broad as it is long; for example, carpals and tarsals

short-term memory storage of information over a few seconds or minutes; compare with **long-term memory**

sigmoid colon (SIG-moyd KOH-lon) portion of the large intestine that courses downward below the iliac crest

signal transduction (SIG-nal tranz-DUK-shen) process of changing a signal such as a hormone or neurotransmitter into another form such as enzymatic reaction within the cell receiving the signal (thus the extracellular hormone signal is transduced, or changed, to an intracellular enzymatic signal)

simple diffusion (SIM-ple di-FYOO-shun) movement of molecules through a membrane by means of the natural tendency of the molecules to spread and the ability of the spreading molecules to move through, or permeate, the membrane

simple epithelium (SIM-ple ep-i-THEE-lee-um) arrangement of epithelial cells in a single layer

simple goiter (SIM-ple GOY-ter) condition in which the thyroid enlarges because iodine is lacking in the diet

single-unit smooth muscle most common type of smooth muscle tissue; smooth muscle tissue in which gap junctions join individual smooth muscle fibers into large, continuous sheets that contract together in an autorhythmic fashion; also called *visceral muscle*

sinoatrial (SA) node (sye-noh-AY-tree-al nohd) the heart's pacemaker; where the impulse conduction of the heart normally starts; located in the wall of the right atrium near the opening of the superior vena cava

sinus (SYE-nus) space or cavity

skeletal muscle (SKEL-eh-tal MUSS-el) muscle under willed or voluntary control; *see* **skeletal muscle tissue**

skeletal muscle tissue (SKEL-eh-tal MUSS-el TISH-yoo) also known as *voluntary* or *striated voluntary muscle*; includes muscle fibers under willed or voluntary control

skin *see* cutaneous membrane

sliding filament model (SLY-ding FIL-ah-ment MAHD-el) theory of muscle contraction in which sliding of thin filaments toward the center of each sarcomere quickly shortens the muscle fiber and thereby the entire muscle

smooth endoplasmic reticulum (smooth en-doh-PLAZ-mik reh-TIK-yoo-lum) region of endoplasmic reticulum system that lacks associated ribosomes

smooth muscle (smooth MUSS-el) muscle fibers that are not under conscious control; also known as *involuntary* or *visceral muscle*; forms the walls of blood vessels and hollow organs

sneeze reflex (sneez REE-fleks) a burst of air directed through the nose and mouth; stimulated by contaminants in the nasal cavity

sodium cotransport (SO-dee-um koh-TRANZ-port) complex transport process in which carriers that bind both sodium ions and glucose molecules passively transport the molecules together out of the GI lumen

sodium-potassium pump (SO-dee-um-poh-TAS-ee-um) active transport pump that operates in the plasma membrane of all human cells; transports both sodium ions and potassium ions but in opposite directions and in a 3:2 ratio, thereby maintaining a gradient across the plasma membrane

soft palate (PAL-et) partition between the mouth and nasopharynx

solute (SOL-yoot) dissolved particles in solution

solution liquid made up of a mixture of molecule types, usually made of solutes (solids) scattered in a solvent (liquid), such as salt in water

solvent (SOL-vent) liquid portion of a solution in which a solute is dissolved

soma (SOH-mah) body; for example, the body of a neuron; also called *perikaryon*

somatic motor division (soh-MAH-tik MOH-tor di-VI-zhun) part of the somatic nervous system with pathways that travel from the central nervous system to skeletal muscles

somatic nervous system (SNS) (soh-MAH-tik NER-vus SIS-tem) motor neurons that control voluntary actions of skeletal muscles

somatic reflex (soh-MAH-tik REE-fleks) reflexive contraction of skeletal muscles

somatic sensory division (soh-MAH-tik SEN-sor-ee di-VI-zhun) division of the nervous system made up of afferent (incoming) pathways from somatic sensory receptors (receptors involved in conscious perception)

somatostatin (soh-mah-toh-STAT-in) hormone produced by delta cells of the pancreas that inhibits secretion of glucagon, insulin, and pancreatic polypeptide

somatotroph (soh-mah-toh-TROHF) cell type of the adenohypophysis (anterior pituitary) that secretes growth hormone (somatotropin)

somatotropin (STH) (soh-mah-toh-TROH-pin) growth hormone

special movement unique or unusual movement that occurs only in a very limited number of joints

special sense characterized by receptors grouped closely together or grouped in a complex sensory organ; for example, sense of smell, taste, hearing, equilibrium, or vision

special sense organ a complex sensory organ that detects any of the special senses; *see* **special sense**

species resistance (SPEE-sheez ree-SIS-tens) genetic characteristics common to all organisms of a particular species that provide natural inborn immunity to a certain disease

specific immunity (spi-SI-fik i-MYOO-ni-tee) protective mechanisms by which the immune system is able to recognize, remember, and destroy specific types of bacteria or toxins; *see* **adaptive immunity**

spermatic cord (sper-MAT-ik kord) cylindrical casings of white fibrous tissue formed by the ductus deferens and located in the inguinal canal between the scrotum and the abdominal cavity

spermatogenesis (sper-mah-toh-JEN-eh-sis) production of sperm cells

spermatogonia (sper-mah-toh-GOH-nee-ah) stem cells of a population that give rise to sperm cells

spermatozoa (sper-mah-tah-ZOH-ah) (*sing.*, spermatozoon [sper-mah-tah-ZOH-on]) mature male gametes; sperm cells

sphenoid bone (SFEE-noyd bohn) keystone bone of the cranium; resembles a bat

sphygmomanometer (sfig-moh-mah-NOM-eh-ter) device for measuring blood pressure in the arteries of a limb

spinal nerve (SPY-nal nerv) nerve that connects the spinal cord to peripheral structures such as the skin and skeletal muscles

spinal tracts (SPY-nal trakts) white columns of the spinal cord that provide conduction paths to and from the brain; ascending tracts carry information to the brain, whereas descending tracts conduct impulses from the brain

spinous process (SPY-nus PRAH-ses) (*pl.*, processes [PRAH-ses-eez]) sharp projection from the laminae of vertebral bones, pointing posteriorly and inferiorly

spiral membrane (SPY-rahl MEM-brayne) floor of the cochlear duct; this structure's vibrating response to sound frequencies leads to the transduction of sound waves into nerve impulses; also called *basilar membrane*

spirogram (SPY-roh-gram) graphic recording of the changing pulmonary volumes observed during breathing

spirometer (spih-ROM-eh-ter) instrument used to measure the amount of air exchanged in breathing

splanchnic nerve (SPLANK-nik nerv) nerve that innervates the viscera

spleen largest lymphoid organ; filters blood, destroys worn-out red blood cells, salvages iron from hemoglobin, and serves as a blood reservoir

spongy bone (SPUN-jee bohn) netlike arrangement of bone tissue found inside bones and which is often filled with red marrow; also called *cancellous bone* or *trabecular bone*

spontaneous abortion (spon-TAY-nee-us ah-BOR-shun) *see* **miscarriage**

squamous (SKWAY-muss) scalelike

squamous cell carcinoma (SKWAY-muss sell car-sih-NOH-mah) slow-growing skin cancer that arises in the epidermis

stem cells cells that have the ability to maintain a constant population of newly differentiating cells

sternum (STER-num) breastbone

steroid (STAYR-oyd) any of a class of lipids related to sterols and forming numerous reproductive and adrenal hormones

steroid hormone (STAYR-oyd HOR-mohn) lipid-soluble hormone that passes intact through the cell membrane of the target cell and influences cell activity by acting on specific genes

stratified epithelium (STRAT-i-fyde ep-i-THEE-lee-um) epithelial cells layered one on another

stratum (STRAH-tum) layer

stratum basale (STRAH-tum bay-SAH-lee) "base layer"; deepest layer of the epidermis; cells in this layer are able to reproduce themselves

stratum corneum (STRAH-tum KOR-nee-um) tough outer layer of the epidermis; cells are filled with keratin

stratum germinativum (STRAH-tum JER-mih-nah-tiv-um) synonym for stratum basale of the skin epidermis; sometimes used to denote stratum basale and stratum spinosum together

stratum granulosum (STRAH-tum gran-yoo-LOH-sum) "granular layer," layer in which the process of keratinization begins

stratum lucidum (STRAH-tum LOO-see-dum) "clear" layer of the epidermis, in thick skin between the stratum granulosum and the stratum corneum

stratum spinosum (STRAH-tum spi-NO-sum) "spiny layer"; layer of epidermis that is rich in RNA to aid in protein synthesis required for keratin production

strength training (STREN-th TRAIN-ing) contracting muscles against resistance to enhance muscle hypertrophy; *see* **aerobic training**, **endurance training**

streptokinase (SK) (strep-toh-KIN-ayz) a plasminogen-activating factor made by certain *Streptococcus* bacteria

stroke volume (SV) amount of blood that is ejected from the ventricles of the heart with each beat

structural protein (STRUK-shur-al PRO-teen) any of a category of proteins with the primary function of forming structures of the cell or tissue; contrast with **functional protein**

subarachnoid space (sub-ah-RAK-noyd) within the meninges, space under the arachnoid and outside the pia mater

subclavian artery (sub-KLAY-vee-an AR-ter-ee) artery of the upper extremity

subclavian vein (sub-KLAY-vee-an vayn) deep vein of the upper extremity

subcutaneous injection (sub-kyoo-TAY-nee-us in-JEK-shun) administration of nutrients, special fluids, and/or electrolytes into the spongy and porous subcutaneous layer beneath the skin

subcutaneous layer (sub-kyoo-TAY-nee-us LAY-er) *see* **hypodermis**

subdural space (sub-DOO-ral) within the meninges; the space between the dura mater and the arachnoid membrane

subendocardial branch (sub-en-doh-KAR-dee-al) conductive cardiac muscle fibers located in the walls of the ventricles; relay impulses from the AV node to the ventricles, causing them to contract

sublingual gland (sub-LING-gwall) smallest of the three major salivary glands; produces a mucous type of saliva

submandibular gland (sub-man-DIB-yoo-lar) salivary gland located just below the mandibular angle; contains enzyme and mucus-producing elements

submucosa (sub-myoo-KOH-sah) connective tissue layer containing blood vessels and nerves in the wall of the digestive tract

subtalar joint (sub-TAY-ler joynt) diarthrotic synovial joint formed by the talus bone of the ankle overlying the calcaneus bone that permits side-to-side motion of the foot

sudoriferous gland (soo-doh-RIF-er-us) epidermal sweat gland

superficial (soo-per-FISH-all) near the body surface; opposite of **deep**

superficial fascia (soo-per-FISH-all FAH-shah) hypodermis; subcutaneous layer beneath the dermis

superior (soo-PEER-ee-or) higher; opposite of **inferior**

superior vena cava (soo-PEER-ee-or VEE-nah KAY-vah) large vein of the upper extremity; drains blood into the right atrium of the heart

surface film layer on the outer surface of skin made up of a mixture of secretions from sweat and sebaceous glands along with epithelial cells being shed from the epidermis; works as a protective barrier

surfactant (sur-FAK-tant) substance covering the surface of the respiratory membrane inside the alveolus; it reduces surface tension and prevents the alveoli from collapsing

surgical neck (SER-jik-el nek) region of the humerus bone of the arm just below the tubercles, so named because of its liability to fracture

sustentacular cells (sus-ten-TAK-yoo-lar sells) cells of seminiferous tubule wall (of testis) that support sperm development

suture (SOO-chur) immovable joint, such as those between the bones of the skull

sweat gland (swet) gland in the skin that produces a transparent, watery liquid that eliminates ammonia and uric acid and helps maintain body temperature

sympathetic division (sim-pah-THET-ik di-VI-zhun) part of the autonomic nervous system; ganglia are connected to the thoracic and lumbar regions of the spinal cord; functions in "fight-or-flight" response

sympathetic postganglionic neurons (sim-pah-THET-ik post-gang-glee-ON-ik NOO-rons) dendrites and cell bodies that are in sympathetic ganglia and axons; travel to a variety of visceral effectors

sympathetic preganglionic neurons (sim-pah-THET-ik pree-gang-glee-ON-ik NOO-rons) dendrites and cell bodies that are located in the gray matter of the thoracic and lumbar segments of the spinal cord; leave the cord through an anterior root of a spinal nerve and terminate in a collateral ganglion

sympathetic trunk (sim-pah-THET-ik) arrangement of sympathetic axon collaterals that bridge the gap between adjacent ganglia that lie on the same side of the vertebral column

symphysis (SIM-fi-sis) joint characterized by the presence of a pad or disk of fibrocartilage connecting the two bones

synapse (SIN-aps) membrane-to-membrane junction between a neuron and another neuron, effector cell, or sensory cell; functions to propagate nerve impulses (via neurotransmitters); two types: electrical and chemical

synaptic cleft (si-NAP-tik kleft) space between a synaptic knob and the plasma membrane of a postsynaptic neuron

synaptic knob (si-NAP-tik nob) tiny bulge at the end of a terminal branch of a presynaptic neuron's axon that contains vesicles with neurotransmitters

synarthrosis (sin-ar-THROH-sis) joint in which fibrous connective tissue joins bones and holds them together tightly; commonly called *sutures*

synchondrosis (sin-kon-DROH-sis) joint characterized by the presence of hyaline cartilage between articulating bones

syncytium (sin-SISH-ee-em) continuous, electrically coupled mass of cardiac fibers; allows an efficient, coordinated pumping action

syndesmosis (sin-dez-MOH-sis) fibrous joint

synergist (SIN-er-jist) muscle that assists a prime mover

synovial fluid (sih-NOH-vee-all FLOO-id) thick, colorless lubricating fluid secreted by the synovial membrane in synovial joints

synovial joint (si-NOH-vee-all joynt) category of joint characterized by the presence of a synovial cavity that allows a significant range of motion

synovial membrane (sih-NOH-vee-all MEM-brayne) connective tissue membrane lining spaces between bones and joints that secretes synovial fluid

synthesis reaction (SIN-the-sis ree-AK-shun) reaction that combines two or more reactants to form a more complex structure

system (SIS-tem) a group of organs that functions as a coordinated team; also called a *body system*

systemic circulation (sis-TEM-ik ser-kyoo-LAY-shun) blood flow from the left ventricle to all parts of the body and back to the right atrium

systole (SIS-toh-lee) contraction of heart muscle

systolic blood pressure (sis-TOL-ik blud PRESH-ur) force with which blood pushes against artery walls when ventricles contract

T

tactile corpuscle (TAK-tyle KOR-pus-ul) large, encapsulated sensory neuron of the skin for light or discriminative touch; also known as *Meissner corpuscle*

tactile disk (TAK-tyle) flat-ended, unencapsulated sensory neuron of the skin for light or discriminative touch; also known as *Merkel disk* or *tactile meniscus*

talus (TAY-lus) (*pl.,* tali [TAY-lye]) bone overlying the calcaneus bone and articulating with the tibia and fibula of the leg to form the ankle joint

target cell cell that, when acted on by a particular hormone, responds because it has receptors to which the hormone can bind

tarsal bone (TAR-sal bohn) any of the bones of the posterior region of the foot

T cells *see* **T lymphocytes**

telophase (TEL-oh-fayz) last stage of mitosis

temporal bone (TEM-poh-ral bohn) cranial bone located on lower side of cranium and part of its floor

temporal lobe (TEM-poh-ral) curved, flat bones located on the sides and base of the skull

tendons (TEN-dons) bands or cords of fibrous connective tissue that attach a muscle to a bone or other structure

tentorium cerebelli (ten-TOR-ee-um sair-eh-BEL-lee) inward extension of the dura mater that separates the cerebellum from the cerebrum

terminal ganglion (TER-mi-nal GANG-glee-on) ganglia that consists of parasympathetic fibers near or in the effectors in the chest and abdomen

terminal hair (TUR-mi-nal) coarse pubic and axillary hair that develops at puberty

tertiary structure (TERSH-ee-air-ee STRUK-cher) the three-dimensional structure of a protein

testes (TES-teez) male gonad responsible for production of sex cells or gametes (sperm) and testosterone

testosterone (tes-TOS-teh-rohn) male sex hormone produced by interstitial cells in the testes; the "masculinizing hormone"

tetanus (TET-ah-nus) smooth, sustained muscular contraction caused by high-frequency stimulation

tetraiodothyronine (T_4) (tet-rah-eye-oh-doh-THY-roh-neen) one of two hormones that compose the thyroid hormone; also called *thyroxine*; *see also* **triiodothyronine**

thalamus (THAL-ah-muss) mass of gray matter located in diencephalon just above the hypothalamus; helps produce sensations, associates sensations with emotions, and plays a part in the arousal mechanism

theca cell (THEE-kah sell) specialized cell formed from the outer layer of granulosa cells; transforms into a fibrous capsule that secretes a hormone that ultimately converts into additional estrogen

theory (THEE-oh-ree) a scientific idea or explanation that has a reasonably high degree of confidence after rigorous testing or observation; compare with **hypothesis thermoreceptor**

(ther-moh-ree-SEP-tor) sensory receptor activated by heat or cold

thick ascending limb (TAL) (thik ah-SEND-ing lim) thick-walled region of the limbs of the hairpin nephron loop (Henle loop) that conducts filtrate toward the distal tubule

thin ascending limb of Henle (tALH) (thin ah-SEND-ing lim ov HEN-lee) thin-walled region of the limbs of the hairpin nephron loop (Henle loop) that conducts filtrate toward the distal tubule

third-degree burn full-thickness burn; most serious of the three degrees of burns

thirst center region of the hypothalamus containing osmoreceptors that can detect an increase in solute concentration in extracellular fluid caused by water loss

thoracic aorta (thoh-RASS-ik ay-OR-tah) downward arch of the aorta through the thorax and ending at the diaphragm

thoracic cavity (thoh-RASS-ik KAV-i-tee) hollow space within the larger ventral body cavity that contains the lungs (in pleural cavities) and heart (in the mediastinum)

thoracic duct (thoh-RASS-ik dukt) largest lymphatic vessel in the body

thoracic vertebra (thoh-RASS-ik VER-teh-bra) vertebral bone located in the posterior part of the chest

thorax (THOR-aks) chest

thoroughfare channel (THUR-oh-fair CHAN-al) the distal end of a metarteriole that is devoid of precapillary sphincters, allowing blood to bypass the capillary bed

threshold potential (THRESH-hold poh-TEN-shal) magnitude of voltage across a membrane at which an action potential, or nerve impulse, is produced

threshold stimulus (THRESH-hold STIM-yoo-lus) minimal level of stimulation required to cause a muscle fiber to contract

thrombocytes (THROM-boh-sytes) cell fragments that play a role in blood clotting; also called *platelets*

thrombopoiesis (throm-boh-poy-EE-sis) formation of platelets

thrombosis (throm-BOH-sis) condition resulting from a clot (thrombus) that stays in one place

thymocyte (THY-moh-syte) cell in the thymus that develops into a T lymphocyte

thymopoietin (thy-moh-POY-eh-tin) peptide that has a critical role in the development of the immune system; thought to stimulate the production of specialized lymphocytes involved in the immune response called *T cells*

thymosin (THY-moh-sin) hormone produced by the thymus that is vital to the development and functioning of the body's immune system

thymus gland (THY-muss) endocrine gland located in the mediastinum; vital part of the body's immune system

thyroid cartilage (THY-royd KAR-ti-lij) largest cartilage of the larynx; Adam's apple

thyroid gland (THY-royd) endocrine gland located in the neck that stores its hormones until needed; thyroid hormones regulate cellular metabolism

thyroid-stimulating hormone (TSH) (THY-royd-STIM-yoo-lay-ting HOR-mohn) a tropic hormone secreted by the anterior pituitary gland that stimulates the thyroid gland to increase its secretion of thyroid hormone

thyroxine (T_4) (thy-ROK-sin) thyroid hormone that stimulates cellular metabolism

tibia (TIB-ee-ah) larger, stronger, and more medially and superficially located of the two leg bones

tidal volume (TV) (TYE-dal VOL-yoom) amount of air breathed in and out with each breath

tight junction (tite JUNK-shen) connection between cells in which they are joined by "collars" of tightly fused membrane

tissue (TISH-yoo) group of similar cells that performs a common function

T lymphocytes (T LIM-foh-sytes) cells of the immune system that have undergone maturation in the thymus; produce cell-mediated immunity

tongue solid mass of skeletal muscle components covered by a mucous membrane; manipulates food in the mouth and contains taste buds

tonsillectomy (tahn-si-LEK-toh-mee) surgical removal of the tonsils

tonsillitis (tahn-si-LYE-tis) inflammation of the tonsils, usually due to infection

tonsils (TAHN-sils) masses of lymphoid tissue; protect against bacteria; three types: palatine tonsils, located on each side of the throat; pharyngeal tonsils (adenoids), near the posterior opening of the nasal cavity; and lingual tonsils, near the base of the tongue

total lung capacity (TLC) total volume of air a lung can hold

total metabolic rate (TMR) (TOH-tal met-ah-BOL-ik rayt) total amount of energy used by the body per day

toxoid (TOK-soyd) form of bacterial toxin that stimulates production of antibodies

trabeculae (trah-BEK-yoo-lee) tiny branchlike threads in a tissue, such as the beams of spongy (cancellous) bone, that surround a network of spaces

trachea (TRAY-kee-ah) windpipe; the tube extending from the larynx to the bronchi

tracheostomy (tray-kee-OS-toh-mee) surgical procedure in which an opening is cut into the trachea

transcription (trans-KRIP-shun) process in which DNA molecule is used as template to form mRNA

transfusion reaction fatal event resulting from a mixture of agglutinogens (antigens) and agglutinins (antibodies) resulting in the agglutination of the donor and recipient blood

transitional epithelium (tranz-I-shen-al ep-i-THEE-lee-um) stratified tissue typically found in body areas that are subjected to stress and tension changes

translation (trans-LAY-shun) process in which mRNA is used by ribosomes in the synthesis of a protein

transverse colon (trans-VERS KOH-lon) division of the colon that passes horizontally across the abdomen

transverse plane (tranz-VERS playn) horizontal plane that divides the body or any of its parts into upper and lower parts

transverse process (tranz-VERS PRAH-ses) (*pl.,* processes [PRAH-ses-eez]) any of the lateral projections of a vertebral bone

tricuspid valve (try-KUS-pid valv) heart valve located between right atrium and ventricle

trigeminal nerve (try-JEM-i-nal nerv) cranial nerve V; responsible for chewing movements and sensations of the head and face

triglyceride (try-GLISS-er-eyed) lipid that is synthesized from fatty acids and glycerol or from excess glucose or amino acids; stored mainly in adipose tissue cells

trigone (TRY-gon) triangular structure, as in the 3-cornered floor of the urinary bladder

triiodothyronine (T_3) (try-eye-oh-doh-THY-roh-neen) one of two components of thyroid hormone that stimulates cellular metabolism; *see also* **tetraiodothyronine**

trochlea (TROK-lee-ah) (*pl.,* trochleae [TROK-lee-ee]) spool-shaped projection on the distal end of the humerus bone of the arm that articulates with the olecranon of the ulna

trochlear nerve (TROK-lee-ar nerv) cranial nerve IV; motor nerve; responsible for eye movements

trophoblast (TROH-foh-blast) outer wall of the blastocyst; contributes in formation of the placenta

tropic hormone (TROH-pik HOR-mohn) hormone that stimulates another endocrine gland to secrete its hormones

true ankle joint (tru ANG-kel joynt) diarthrotic synovial joint formed at the distal ends of the medial malleolus of the tibia and the lateral malleolus of the fibula embracing the underlying talus, allowing up-and-down movement of the foot

true capillary (KAP-i-lair-ee) blood vessel that receives blood flowing out of metarterioles or other small arterioles

true pelvis (tru PEL-vis) (*pl.,* pelves or pelvises [PEL-veez, PEL-vis-ez]) structure forming a bony ring between the pelvic inlet and the pelvic outlet of the skeleton

trypsin (TRIP-sin) protein-digesting enzyme

trypsinogen (trip-SIN-oh-gen) inactive proenzyme that is subsequently converted to active trypsin by enterokinase in the intestinal lumen

T tubule (TEE TOOB-yool) transverse tubule unique to muscle cells; formed by inward extensions of the sarcolemma that allow electrical impulses to move deeper into the cell

tumor marker (TOO-mer) abnormal antigen on cancer cells; also called *tumor-specific antigen*

tumor-specific antigen (TOO-mer speh-SIF-ik AN-ti-jen) *see* **tumor marker**

tunica albuginea (TOO-ni-kah al-byoo-JIN-ee-ah) tough, whitish membrane that surrounds each testis and enters the gland to divide it into lobules

tunica externa (TOO-ni-kah eks-TER-nah) outermost layer of arteries, veins, lymphatic vessels, and other organs; made of strong, flexible fibrous connective tissue; sometimes called *tunica adventitia*

tunica intima (TOO-ni-kah IN-ti-mah) layer made up of endothelium that lines blood vessels; also called *tunica interna*

tunica media (TOO-ni-kah MEE-dee-ah) muscular middle layer in blood vessels; the tunica media of arteries is more muscular than that of veins

turbinate (TUR-bi-nayt) *see* **concha**

T wave electrocardiogram deflection that reflects the repolarization of the ventricles

tympanic membrane (tim-PAN-ik MEM-brayn) eardrum

U

ulna (UL-nah) forearm bone

ultrasonography (ul-trah-son-OG-rah-fee) imaging technique in which high-frequency sound waves are reflected off tissue to form an image

umbilical arteries (um-BIL-i-kul AR-ter-eez) two small arteries that carry oxygen-poor blood from the developing fetus to the placenta

umbilical cord (um-BIL-i-kul kord) flexible structure connecting the fetus with the placenta; contains umbilical arteries and vein

umbilical vein (um-BIL-i-kul vayn) large vein carrying oxygen-rich blood from the placenta to the developing fetus

uniaxial joint (yoo-nee-AK-see-al joynt) synovial joint that permits movement around only one axis and in only one plane

unipolar (pseudounipolar) neuron (yoo-nee-POH-lar [SOO-doh-yoo-nee-POH-lar] NOO-ron) structural category of neurons made up of cells that appear to have only one extension from the cell body

unmyelinated fiber (un-MYE-eh-lin-ay-ted FYE-ber) nerve fiber that does not have a myelin sheath; also referred to as *gray fiber*

unsaturated (un-SATCH-yoo-ray-ted) referring to an organic compound having at least one carbon-to-carbon double bond

upper esophageal sphincter (UES) (eh-SOF-ah-JEE-ul SFINGK-ter) ring of muscular tissue at proximal end of esophagus; helps prevent air from entering the esophagus during respiration

upper respiratory tract (RES-pi-rah-tor-ee trakt) respiratory organs that are not contained within the thorax; includes nasal cavity, pharynx, and associated structures

ureter (YOO-ree-ter) long tube that carries urine from kidney to bladder

urethra (yoo-REE-thrah) passageway from bladder to exterior; functions in elimination of urine; in males, also acts as a genital duct that carries sperm to the exterior

urethral sphincter (yoo-REE-thral SFINGK-ter) circular muscle of the pelvic floor that constricts around the urethra, thus regulating urine flow from the bladder and out of the body

urogenital triangle (yoor-oh-GEN-i-tal TRY-ang-gul) region of the perineum that contains the external genitals (labia, vaginal orifice, clitoris) and urinary opening, and the anal triangle, which surrounds the anus

uterus (YOO-ter-us) hollow, muscular organ that holds and sustains developing offspring until birth

V

vaccination (vak-si-NAY-shun) method used to achieve active immunity by triggering the body to form antibodies against specific pathogens

vagina (vah-JYE-nah) internal tube from uterus to vulva

vaginal orifice (VAH-ji-nal OR-i-fis) opening of the vagina to the outside of the body

vagus nerve (VAY-gus nerv) cranial nerve X; mixed nerve; sensations and movements of organs

valence electrons (VAY-lentz eh-LEK-tron) the outermost electrons that determine how an atom will react chemically

vasa recta (VAH-sah REK-tah) a long, hairpin-shaped arteriole of the kidney leading from the efferent arteriole and following the nephron loop; also called *straight arteriole (of kidney)*

vascular anastomosis (VAS-kyoo-lar ah-nas-toh-MOH-sis) condition when blood moves from veins to other veins or from arteries to other arteries without passing through an intervening capillary network

vas deferens (vas DEF-er-enz) reproductive duct that extends from the epididymis to the ejaculatory duct; also called *ductus deferens*

vasectomy (va-SEK-toh-mee) surgical severing of the vas deferens to render a male sterile

vasoconstriction (vay-soh-kon-STRIK-shun) reduction in vessel diameter caused by increased contraction of the muscular coat

vasodilation (vay-soh-dye-LAY-shun) increase in vessel diameter caused by relaxation of vascular muscles

vasomotor center (vay-soh-MOH-tor) nerve center in the brainstem that exerts control over smooth muscles in the blood vessels, permitting vasoconstriction and vasodilation

vasomotor mechanism (vay-soh-MOH-tor MEK-ah-niz-em) feedback regulation of the diameter of arterioles

vasopressin (vas-oh-PRES-in) *see* **arginine vasopressin**

vein (vayn) vessel carrying blood from capillaries toward the heart

vellus (VEL-us) strong, fine, and less pigmented hair

venous anastomosis (VEE-nus ah-nas-toh-MOH-sis) connection between two veins

venous pump (VEE-nus) blood-pumping action of respirations and skeletal muscle contractions facilitating venous return by increasing pressure gradient between peripheral veins and venae cavae

venous return (VEE-nus) amount of blood returned to the heart by the veins

ventral (anterior) nerve root (VEN-tral nerv) bundle of nerve fibers that carry motor information out of the spinal cord

ventral cavity (VEN-tral KAV-i-tee) body cavity that includes the thoracic cavity and abdominopelvic cavity; not a standard anatomical term, but used here to help organize the body for the beginning student

ventral ramus (VEN-tral RAY-mus) large, complex branch of each spinal nerve

ventral root (VEN-tral) motor branch of a spinal nerve, by which it is attached to the spinal cord

ventricle (VEN-tri-kul) a cavity, such as the large, fluid-filled spaces within the brain or the chambers of the heart

venule (VEN-yool) small blood vessel that collects blood from capillaries and joins to form veins

vermiform appendix (VERM-i-form ah-PEN-diks) hollow, tubular structure attached to the cecum (of the colon) and thought to be a breeding ground for beneficial intestinal bacteria

vermis (VER-mis) central section of the cerebellum

vertebrae (VER-teh-bree) any of the bones that make up the spinal column

vertebral foramen (ver-TEE-bral foh-RAY-men) the central opening in the vertebral column that contains the spinal cord

vertigo (VER-ti-goh) abnormal sensation of spinning; dizziness

vestibular fold (ves-TIB-yoo-lar) either of the lower of two pairs of lateral folds of the mucosa in the larynx, just above the vocal folds; also called *false vocal fold* or *false vocal cord*; compare with **vocal fold**

vestibular membrane (ves-TIB-yoo-lar MEM-brayne) roof of the cochlear duct; also called *Reissner's membrane*

vestibule (VES-ti-byool) located in the bony labyrinth of the inner ear; portion adjacent to the oval window between the semicircular canals and the cochlea

vestibulocochlear nerve (ves-TIB-yoo-loh-kok-lee-ar nerv) cranial nerve VIII; sensory nerve; responsible for hearing and equilibrium

vibrissa (vih-BRISS-ah) coarse hair found in the skin of the vestibule of the nose

villi (VIL-eye) fingerlike folds covering the plicae of the small intestines

visceral (VISS-er-al) relating to the viscera (internal organs); toward or on the internal organs; opposite of **parietal**

visceral pleura (VISS-er-al PLOOR-ah) the serous membrane that adheres to and covers the lung

visceral sensory division (VISS-er-al SEN-sor-ee di-VI-zhun) division of the nervous system made up of afferent (incoming) pathways from autonomic sensory receptors (receptors involved in subconscious perception) of the internal organs (viscera)

vital capacity (VC) (VYE-tal kah-PASS-i-tee) largest amount of air that can be moved in and out of the lungs in one inspiration and expiration

vitamins (VYE-tah-mins) organic molecules needed in small quantities to help enzymes operate effectively

vitreous humor (VIT-ree-us HYOO-mohr) jelly-like fluid in the eye, posterior to the lens

vocal fold lower pair of lateral folds of mucosa in the larynx, responsible for vocalization; also called *true vocal fold* or *true vocal cord* or *vocal cord*

vulva (VUL-vah) external genitals of the female

W

white matter (MAT-ter) nerves covered with white myelin sheath; contrast with **gray matter**

whole blood volume blood with all its components intact

Y

yawn slow, deep inspiration through an unusually widened mouth

yolk sac (yohk sak) in humans, involved with production of blood cells in the developing embryo

Z

Z disk microscopic structure within the myofibril of a muscle fiber where the thin filaments unite with each other and form a netlike disk; serves as boundary of sarcomere unit; also called *Z line*

zygomatic bone (zye-goh-MAT-ik bohn) cheek bone; also called *malar bone*

zygote (ZYE-goht) original cell of an offspring, formed by the union of an ovum and a sperm

Index

Note: Page numbers followed by b, f, and t indicate boxes, figures, and tables, respectively.

Illustration and Photograph Credits

UNIT 1

Seeing the Big Picture, p. xxii (photo): Copyright Kevin Patton, Lion Den Inc., Weldon Spring, MO.

Chapter 1

1-2: Copyright Kevin Patton, Lion Den Inc, Weldon Spring, MO. **A&P Connect Box:** From Goldman L, Ausiello D: *Cecil textbook of medicine,* ed 22, Philadelphia, 2004, Saunders.

Chapter 2

Box 2-3: From National Institute of General Medical Sciences, *The structures of life,* July 2007, retrieved November 2008 from http://publications.nigms.nih.gov/structlife/. **Introductory Story Box:** From Potter P, Perry A: *Basic nursing: essentials for practice,* ed 6, St Louis, 2006, Mosby.

Chapter 3

3-1, *B*: Courtesy A. Arlan Hinchee. **3-8, *B*:** Courtesy Charles Flickinger, University of Virginia.

Chapter 4

4-8: Adapted from McCance K, Huether S: *Pathophysiology,* ed 4, St Louis, 2002, Mosby. **4-10 (Electron micrographs):** Courtesy M.M. Perry and A.B. Gilbert, Edinburgh Research Center. **Box 4-1, *B*:** From Goldman L, Ausiello D, *Cecil textbook of medicine,* ed 22, Philadelphia, 2004, Saunders.

Chapter 5

5-1 (Photo): Cold Spring Harbor Laboratory.

Chapter 6

Table 6-2, *A, B, D, E, F, H*: Dennis Strete. **Table 6-2, *C*:** Adapted from Gartner L, Hiatt J: *Color textbook of histology,* ed 3, Philadelphia, 2006, Saunders. **Table 6-2, *G*:** Adapted from Young B, Lowe J, Stevens A, et al: *Wheater's functional histology,* ed 5, Philadelphia, 2006, Churchill Livingstone. **Table 6-4, *A, F, L*:** Adapted from Gartner L, Hiatt J: *Color textbook of histology,* ed 3, Philadelphia, 2006, Saunders. **Table 6-4, *B-E, G, I, J, K*:** Dennis Strete. **Table 6-4, *H*:** From Kerr J: *Atlas of functional histology,* London, 1999, Mosby. **Table 6-5, *A, B, D*:** Dennis Strete. **Table 6-5, *C*:** From Gartner L, Hiatt J: *Color Textbook of Histology,* ed 3, Philadelphia, 2006, Saunders. **Box 6-2:** From Zitelli B, Davis H: *Atlas of pediatric physical diagnosis,* ed 3, Philadelphia, 1997, Mosby. **Box 6-3:** From Linsley D: *Wardlaw's perspectives in nutrition,* ed 2, St Louis, 1993, Mosby–Year Book.

Chapter 7

7-1 (Photo): Ed Reschke. **7-5 (Gradient):** From McCance K, Huether S: *Pathophysiology,* ed 5, St Louis, 2006, Mosby. **7-6:** From Epstein O, Perkin GD, Cookson J, de Bono D: *Clinical examination,* ed 3, St Louis, 2003, Mosby. **7-7, *C*:** Copyright © by David Scharf, 1986, 1993. **7-8:** Copyright Kevin Patton, Lion Den Inc, Weldon Spring, MO. **7-9:** Courtesy Christine Olekyk. **Box 7-4, *A*:** From Goldman L, Ausiello D, *Cecil textbook of medicine,* ed 23, Philadelphia, 2003, Saunders. **Box 7-4, *B*:** From Noble J: *Textbook of primary care medicine,* ed 3, Philadelphia, 2001, Mosby. **Box 7-4, *C*:** From Townsend C, Beauchamp RD, Evers BM, Mattox K: *Sabiston textbook of surgery,* ed 18, Philadelphia, 2008, Saunders. **Box 7-4, *D*:** From Rakel R: *Textbook of family medicine,* ed 7, Philadelphia, 2007, Saunders. **Box 7-5:** From Emond R: *Color atlas of infectious diseases,* ed 4, Philadelphia, 2003, Mosby.

Chapter 8

8-2, *B*: From White T, *Human osteology,* ed 2, Philadelphia, 2000, Academic Press. **8-3, *B*:** From Moses K, Nava P, Banks J, Petersen D: *Moses atlas of clinical gross anatomy,* Philadelphia, 2005, Mosby. **8-3, *D*:** Dennis Strete.

8-9: From Booher JM, Thibodeau Ga: *Athletic injury assessment,* St Louis, 1985, Mosby.

Chapter 9

9-11, *A, B, D, E, F,* 9-12, 9-14, *B,* 9-15, *D,* 9-18, *C,* 9-27, *A, B,* 9-29, *C,* 9-30, *A, B*: Adapted from Drake R, Vogl AW, Mitchell AWM, et al: *Gray's atlas of anatomy,* ed 1, Philadelphia, 2007, Churchill Livingstone. **9-11, *C,* 9-14, *A*:** Adapted from Drake R, Vogl W, Mitchell AWM: *Gray's anatomy for students,* ed 1, Philadelphia, 2004, Churchill Livingstone. **9-17, *B,* 9-21, *B*:** Courtesy Vidic B, Suarez FR: *Photographic atlas of the human body,* St Louis, 1984, Mosby. **9-22 (Drawings):** From Yvonne Wylie Walston. **9-22 (Photo inset):** From Seidel HM, Ball JW, Dains JE, Benedict GW: *Mosby's guide to physical examination,* ed 5, St Louis, 2003, Mosby. **Box 9-2:** From Canale ST: *Campbell's operative orthopaedics,* ed 9, St Louis, 1998, Mosby. **Box 9-3 (Arthrograms):** From Abrahams P, Marks S, Hutchings R: *McMinn's color atlas of human anatomy,* ed 5, Philadelphia, 2003, Mosby. **Box 9-3 (Photo):** From Cummings N, Stanley-Green S, Higgs P: *Perspectives in athletic training,* St Louis, 2009, Mosby. **Introductory Story Box:** From Browner B, Jupiter J, Trafton P: *Skeletal trauma: basic science, management, and reconstruction,* ed 3, Philadelphia, 2003, Saunders.

Chapter 10

10-3, Courtesy Dr. H.E. Huxley. **10-12, *B*:** Courtesy Dr. Paul C. Letourneau, Department of Anatomy, Medical School, University of Minnesota, MN. **10-20:** Adapted from Muscolino J: *Kinesiology,* St Louis, 2006, Mosby.

Chapter 11

Box 11-1, *A*: Courtesy Marie Simar Couldwell, MD, and Maiken Nedergaard.

Chapter 12

12-7 (Photo): From Gosling J, Harris P, Whitmore I, Willan P: *Human anatomy,* ed 4, Philadelphia, 2002, Mosby. **12-9:** From Vidic B, Suarez FR: *Photographic atlas of the human body,* St Louis, 1984, Mosby. **12-11, *B*:** From Abrahams P, Marks S, Hutchings R: *McMinn's color atlas of human anatomy,* ed 5, Philadelphia, 2003, Mosby. **Introductory Story Questions:** From Forbes CD, Jackson WD: *Color atlas of text and clinical medicine,* ed 3, London, 2003, Mosby.

Chapter 13

13-4: From Beare P, Myers J: *Adult health nursing,* ed 3, St Louis, 1998, Mosby. **13-6, *A-D*:** From Seidel HM, Ball JW, Dains JE, Benedict GW: *Mosby's guide to physical examination,* ed 5, St Louis, 2003, Mosby.

Chapter 14

14-5, *D*: Photo Researchers Inc. **14-12:** Copyright Kevin Patton, Lion Den Inc, Weldon Spring, MO. **14-14, 14-20:** From Newell FW: *Ophthalmology: principles and concepts,* ed 7, St Louis, 1992, Mosby. **14-19, *C*:** Courtesy Dr. Scott Mittman, Johns Hopkins Hospital, Baltimore, MD. **14-22:** Adapted from Boron W, Boulpaep E: *Medical physiology,* updated version, ed 1, Philadelphia, 2005, Saunders. **Box 14-2:** From Swartz MH: *Textbook of physical diagnosis,* ed 4, Philadelphia, 2002, Saunders. **Box 14-3:** From *Ishihara's tests for colour deficiency,* Tokyo, Japan, 1973, Kanehara Trading Co, Copyright Isshinkai Foundation.

Chapter 15

Box 15-2, *A*: From Swartz MH: *Textbook of physical diagnosis,* ed 4, Philadelphia, 2002, Saunders. **Box 15-2, *B*:** From Stein Ha, Slatt BJ, Stein RM: *The ophthalmic assistant: fundamentals and clinical practice,* ed 7, Philadelphia, 2000, Mosby. **Box 15-2, *C*:** From Swartz MH: *Textbook of physical diagnosis history and examination,* ed 5, Philadelphia, 2006, Saunders. **Box 15-3:** Courtesy Gower Medical Publishers.

Chapter 16

16-3, 16-15, B: Copyright Dennis Kunkel Microscopy Inc. **16-6, 16-7, 16-9, 16-10:** Dennis Strete. **16-8:** Adapted from Young B, Lowe J, Stevens A, et al: *Wheater's functional histology*, ed 5, Philadelphia, 2006, Churchill Livingstone. **16-13 (Inset):** From Belcher AE: *Blood disorders*, St Louis 1993, Mosby. **Box 16-1:** Courtesy Bevelander G, Ramalay Ja: *Essentials of histology*, ed 8, St Louis, 1979, Mosby. **Introductory Story Box:** From Stevens ML: *Fundamentals of clinical hematology*, Philadelphia, 1997, Saunders.

Chapter 17

17-11: Adapted from McCance K, Huether S: *Pathophysiology*, ed 5, St Louis, 2006, Mosby. **Box 17-1:** From Goldman L, Ausiello D: *Cecil textbook of medicine*, ed 23, Philadelphia, 2008, Saunders. **Box 17-2:** Courtesy Simon C, Janner M: *Color atlas of pediatric diseases with differential diagnosis*, ed 2, Hamilton, Ontario, 1990, BC Decker. **Introductory Story Box:** Courtesy Dr. Daniel Simon and Mr. Paul Zambino.

Chapter 18

18-11: Adapted from McCance K, Huether S: *Pathophysiology*, ed 4, St Louis, 2002, Mosby. **18-15:** Adapted from Canobbio MM: *Cardiovascular disorders*, St Louis, 1990, Mosby. **Introductory Story Box:** From Hicks GH: *Cardiopulmonary anatomy and physiology*, Philadelphia, 2000, Saunders.

Chapter 19

19-5, A: Adapted from Mathers L, Chase R, Dolph J, Glasgow E: *CLASS clinical anatomy principles*, Philadelphia, 1996, Mosby. **19-5, B:** Courtesy Nielsen M: *Human anatomy lab manual and workbook*, ed 4, Dubuque, IA, 2002, Kendall/Hunt Publishing Company. **19-6:** From Seidel HM, Ball JW, Dains JE, Benedict GW: *Mosby's guide to physical examination*, ed 6, St Louis, 2006, Mosby. **19-13:** Copyright Dennis Kunkel Microscopy Inc. **19-15:** From Abbas A, Lichtman A: *Cellular and molecular immunology*, ed 5, Philadelphia, 2003, Saunders. **19-16, 19-21:** From Copstead-Kirkhorn L, Banasik J: *Pathophysiology*, ed 2, St Louis, 1999, Saunders. **Box 19-3:** Adapted from McCance K, Huether S: *Pathophysiology*, ed 4, St Louis, 2002, Elsevier. **Introductory Story Box:** From Mason DJ, Leavitt J, Chaffee M: *Policy and politics in nursing and health care*, ed 5, St Louis, 2007, Saunders.

Chapter 20

20-7, B: From Erlandsen SL, Magney J: *Color atlas of histology*, St Louis, 1992, Mosby. **20-8:** Adapted from Thompson JM, Wilson SF: *Health assessment for nursing practice*, St Louis, 1996, Mosby. **20-9, A:** From Vidic B, Suarez RF: *Photographic atlas of the human body*, St Louis, 1984, Mosby. **20-13:** From Drake R, Vogl AW, Mitchell A: *Gray's anatomy for students*, Philadelphia, 2005, Churchill Livingstone. **20-24:** Adapted from Guyton A, Hall J: *Textbook of medical physiology*, ed 11, Philadelphia, 2006, Saunders. **Box 20-4:** Copyright Kevin Patton, Lion Den Inc, Weldon Spring, MO. **Introductory Story Box:** From Mosby's Medical, Nursing, and Allied Health Dictionary, ed 6, St. Louis, 2002, Mosby.

Chapter 21

21-7 (Inset): From Weir J, Abrahams P: *Imaging atlas of the human anatomy*, ed 2, Philadelphia, 1997, Mosby. **Box 21-2:** From Daffner DH: *Clinical radiology: the essentials*, ed 3, Baltimore, 1992, Lippincott, Williams & Wilkins. **Box 21-3:** Photo Researchers Inc.

Chapter 22

22-1: US Department of Agriculture. **22-9:** Adapted from Report of the Expert Panel for Population Strategies for Blood Cholesterol Reduction: Bethesda, MD, November 1990, The National Cholesterol Education Program, National Heart Lung and Blood Institute, Public Health Service, US Department of Health and Human Services, NIH Publication No. 90-3046. **22-10, 22-11:** Adapted from Mahan LK, Escott-Stump S: *Krause's food, nutrition and diet therapy*, ed 11, St Louis, 2004, Saunders.

Chapter 23

23-2, A, 23-5: Adapted from Brundage DJ: Renal disorders. *Mosby's clinical nursing series*, St Louis, 1992, Mosby. **23-3, B, 23-6, A:** From Heylings D, Spence R, Kelly B: *Integrated anatomy*, Edinburgh, 2007, Churchill Livingstone. **23-4:** From Stevens A, Lowe J: *Human histology*, ed 3, Philadelphia, 2005, Mosby. **23-6, C:** From Boron W, Boulpaep E: *Medical physiology*, updated version, ed 1, Philadelphia, 2005, Saunders. **23-15:** Copyright Kevin Patton, Lion Den Inc, Weldon Spring, MO. **23-16:** Adapted from Mahan LK, Escott-Stump S: *Krause's food, nutrition and diet therapy*, ed 12, St Louis, 2007, Saunders. **23-17:** From Bloom A, Ireland J: *Color atlas of diabetes*, ed 2, St Louis, 1992, Mosby. **Box 23-2:** From Schmidbauer J, Remzi M, Memarsadeghi M et al: Diagnostic accuracy of computed tomography-guided percutaneous biopsy of renal masses, *Eur Urol* 53(5):869-1100, 2007.

Chapter 24

24-4: From Stevens A, Lowe J: *Human histology*, ed 3, Philadelphia, 2005, Mosby. **24-7, D:** Lennart Nilsson. **24-9:** Adapted from Guyton A, Hall J: *Textbook of medical physiology*, ed 11, Philadelphia, 2006, Saunders. **24-10:** Adapted from Boron W, Boulpaep E: *Medical physiology*, updated version, ed 1, Philadelphia, 2005, Saunders. **Box 24-1:** From Seidel HM, Ball JW, Dains JE, Benedict GW: *Mosby's guide to physical examination*, ed 6, St Louis, 2006, Mosby.

Chapter 25

25-6: Courtesy Dr. Richard Blandau, Department of Biological Structure, University of Washington. **25-11:** Adapted from Boron W, Boulpaep E: *Medical physiology*, updated version, ed 1, Philadelphia, 2005, Saunders. **Box 25-3:** Wyeth-ayerst Laboratories.

Chapter 26

26-3 (Photo), 26-9: Lennart Nilsson. **26-5:** Courtesy Lucinda L. Veeck, Jones Institute for Reproductive Medicine, Norfolk, Va. **26-7, B:** From Cotran R, Kumar V, Collins T: *Robbins pathologic basis of disease*, ed 6, Philadelphia, 1999, Saunders. **26-12:** Adapted from Boron W, Boulpaep E: *Medical physiology*, updated version, ed 1, Philadelphia, 2005, Saunders. **26-17:** From Hockenberry MJ, Wilson D: *Wong's essentials of pediatric nursing*, ed 8, St Louis, 2009, Mosby. **26-18:** Copyright Kevin Patton, Lion Den Inc, Weldon Spring, MO. **26-19:** Adapted from Mahan LK, Escott-Stump S: *Krause's food, nutrition and diet therapy*, ed 12, St Louis, 2007, Saunders. **Box 26-2, B:** Courtesy Kevin Patton, Lion Den Inc, Weldon Spring, MO. **Introductory Story Box:** From Hagen-ansert SL: *Textbook of diagnostic ultrasonography*, Vol 2, ed 6, St Louis, 2007, Mosby.

Chapter 27

27-1: Adapted from Boron W, Boulpaep E: *Medical physiology*, updated version, ed 1, Philadelphia, 2005, Saunders.